INTERNATIONAL ENERGY AGENCY
AGENCE INTERNATION

ENERGY BALANCES OF NON-OECD COUNTRIES

1996 - 1997

BILANS ENERGETIQUES DES PAYS NON-MEMBRES

OECD
OCDE

1999 Edition

INTERNATIONAL ENERGY AGENCY
9, RUE DE LA FÉDÉRATION, 75739 PARIS CEDEX 15, FRANCE

The International Energy Agency (IEA) is an autonomous body which was established in November 1974 within the framework of the Organisation for Economic Co-operation and Development (OECD) to implement an international energy programme.

It carries out a comprehensive programme of energy co-operation among twenty four* of the OECD's twenty nine Member countries. The basic aims of the IEA are:

- To maintain and improve systems for coping with oil supply disruptions;
- To promote rational energy policies in a global context through co-operative relations with non-member countries, industry and international organisations;
- To operate a permanent information system on the international oil market;
- To improve the world's energy supply and demand structure by developing alternative energy sources and increasing the efficiency of energy use;
- To assist in the integration of environmental and energy policies.

IEA Member countries: Australia, Austria, Belgium, Canada, Denmark, Finland, France, Germany, Greece, Hungary, Ireland, Italy, Japan, Luxembourg, the Netherlands, New Zealand, Norway, Portugal, Spain, Sweden, Switzerland, Turkey, the United Kingdom, the United States. The European Commission also takes part in the work of the IEA.

ORGANISATION FOR ECONOMIC CO-OPERATION
AND DEVELOPMENT

Pursuant to Article 1 of the Convention signed in Paris on 14th December 1960, and which came into force on 30th September 1961, the Organisation for Economic Co-operation and Development (OECD) shall promote policies designed:

- To achieve the highest sustainable economic growth and employment and a rising standard of living in Member countries, while maintaining financial stability, and thus to contribute to the development of the world economy;
- To contribute to sound economic expansion in Member as well as non-member countries in the process of economic development; and
- To contribute to the expansion of world trade on a multilateral, non-discriminatory basis in accordance with international obligations.

The original Member countries of the OECD are Austria, Belgium, Canada, Denmark, France, Germany, Greece, Iceland, Ireland, Italy, Luxembourg, the Netherlands, Norway, Portugal, Spain, Sweden, Switzerland, Turkey, the United Kingdom and the United States. The following countries became Members subsequently through accession at the dates indicated hereafter: Japan (28th April 1964), Finland (28th January 1969), Australia (7th June 1971), New Zealand (29th May 1973), Mexico (18th May 1994), the Czech Republic (21st December 1995), Hungary (7th May 1996), Poland (22nd November 1996) and the Republic of Korea (12th December 1996). The Commission of the European Communities takes part in the work of the OECD (Article 13 of the OECD Convention).

AGENCE INTERNATIONALE DE L'ÉNERGIE

9, RUE DE LA FEDERATION, 75739 PARIS CEDEX 15, FRANCE

L'agence internationale de l'énergie (AIE) est un organe autonome institué en novembre 1974 dans le cadre de l'Organisation de coopération et de développement économiques (OCDE) afin de mettre en œuvre un programme international de l'énergie.

Elle applique un programme général de coopération dans le domaine de l'énergie entre vingt-quatre* des vingt-neuf pays Membres de l'OCDE. Les objectifs fondamentaux de l'AIE sont les suivants :

- tenir à jour et améliorer des systèmes permettant de faire face à des perturbations des approvisionnements pétroliers ;
- œuvrer en faveur de politiques énergétiques rationnelles dans un contexte mondial grâce à des relations de coopération avec les pays non membres, l'industrie et les organisations internationales ;
- gérer un système d'information continue sur le marché international du pétrole ;
- améliorer la structure de l'offre et de la demande mondiales d'énergie en favorisant la mise en valeur de sources d'énergie de substitution et une utilisation plus rationnelle de l'énergie ;
- contribuer à l'intégration des politiques d'énergie et d'environnement.

* Pays Membres de l'AIE : Allemagne, Australie, Autriche, Belgique, Canada, Danemark, Espagne, États-Unis, Finlande, France, Grèce, Hongrie, Irlande, Italie, Japon, Luxembourg, Norvège, Nouvelle-Zélande, Pays-Bas, Portugal, Royaume-Uni, Suède, Suisse et Turquie. La Commission des Communautés européennes participe également aux travaux de l'AIE.

ORGANISATION DE COOPÉRATION
ET DE DÉVELOPPEMENT ÉCONOMIQUES

En vertu de l'article 1er de la Convention signée le 14 décembre 1960, à Paris, et entrée en vigueur le 30 septembre 1961, l'Organisation de Coopération et de Développement Économiques (OCDE) a pour objectif de promouvoir des politiques visant :

- à réaliser la plus forte expansion de l'économie et de l'emploi et une progression du niveau de vie dans les pays Membres, tout en maintenant la stabilité financière, et à contribuer ainsi au développement de l'économie mondiale ;
- à contribuer à une saine expansion économique dans les pays Membres, ainsi que les pays non membres, en voie de développement économique ;
- à contribuer à l'expansion du commerce mondial sur une base multilatérale et non discriminatoire conformément aux obligations internationales.

Les pays Membres originaires de l'OCDE sont : l'Allemagne, l'Autriche, la Belgique, le Canada, le Danemark, l'Espagne, les États-Unis, la France, la Grèce, l'Irlande, l'Islande, l'Italie, le Luxembourg, la Norvège, les Pays-Bas, le Portugal, le Royaume-Uni, la Suède, la Suisse et la Turquie. Les pays suivants sont ultérieurement devenus Membres par adhésion aux dates indiquées ci-après : le Japon (28 avril 1964), la Finlande (28 janvier 1969), l'Australie (7 juin 1971), la Nouvelle-Zélande (29 mai 1973), le Mexique (18 mai 1994), la République tchèque (21 décembre 1995), la Hongrie (7 mai 1996), la Pologne (22 novembre 1996) et la République de Corée (12 décembre 1996). La Commission des Communautés européennes participe aux travaux de l'OCDE (article 13 de la Convention de l'OCDE).

TABLE OF CONTENTS

SUMMARY TABLES AND ENERGY INDICATORS

TABLE DES MATIERES

PARTIE I: METHODOLOGIE

PARTIE II: DONNEES STATISTIQUES

TABLEAUX RECAPITULATIFS ET INDICATEURS ENERGETIQUES

ABBREVIATIONS

Btu:	British thermal unit
GWh:	gigawatt hour
kcal:	kilocalorie
kg:	kilogramme
kJ:	kilojoule
Mt:	million tonnes
m^3:	cubic metre
t:	metric ton = tonne = 1000 kg
TJ:	terajoule
toe:	tonne of oil equivalent = 10^7 kcal
CHP:	combined heat and power
GCV:	gross calorific value
HHV:	higher heating value = GCV
LHV:	lower heating value = NCV
NCV:	net calorific value
PPP:	purchasing power parity
EU:	European Union
IEA:	International Energy Agency
OECD:	Organisation for Economic Co-Operation and Development
OLADE:	Organización Latino Americana de Energía
UN:	United Nations
IPCC:	Intergovernmental Panel on Climate Change
ISIC:	International Standard Industrial Classification
UNIPEDE:	International Union of Producers and Distributers of Electrical Energy
-	not applicable, nil or not available

Note	**See multilingual pullout, including Chinese, at the end of the publication.**
Attention	**Voir le dépliant en plusieurs langues à la fin du présent recueil.**
Achtung	**Aufklappbarer Text auf der letzten Umschlagseite.**
Attenzione	**Riferirsi al glossario poliglotta alla fine del libro.**
注意	**巻末 の日本語の折り込みページを参照**
Nota	**Véase el glosario plurilingüe al final del libro.**
Примеч.	**Смотрите многоязычный словарь в конце книги.**

ABREVIATIONS

Btu :   British thermal unit
GWh :   gigawatt-heure
kcal :   kilocalorie
kg :   kilogramme
kJ :   kilojoule
Mt :   million de tonnes
m^3 :   mètre cube
t :   tonne métrique = 1000 kg
tep :   tonne d' équivalent pétrole = 10^7 kcal
TJ :   terajoule

PCI :   pouvoir calorifique inférieur
PCS :   pouvoir calorifique supérieur
PPA :   parité de pouvoir d'achat

AIE :   Agence internationale de l' énergie
OCDE :   Organisation de coopération et de développement économiques
OLADE :   Organización Latino Americana de Energía
ONU :   Organisation des nations unies
UE :   Union européenne

CITI :   Classification internationale type par industrie
GIEC :   Groupe d'experts intergouvernemental sur l' évolution du climat
UNIPEDE :   Union Internationale des Producteurs et Distributeurs d'Energie Electrique

-                   sans objet, néant ou non disponible

Attention	**Voir le dépliant en plusieurs langues, y compris en chinois, à la fin du présent recueil.**
Note	**See multilingual pullout at the end of the publication.**
Achtung	**Aufklappbarer Text auf der letzten Umschlagseite.**
Attenzione	**Riferirsi al glossario poliglotta alla fine del libro.**
注意	**巻末の日本語の折り込みページを参照**
Nota	**Véase el glosario plurilingüe al final del libro.**
Примеч.	**Смотрите многоязычный словарь в конце книги.**

INTRODUCTION

This new publication offers the same coverage on energy balances, trends and indicators as the homonymous publication for OECD countries. It provides statistics on production, trade and consumption in a common unit for each source of energy in more than 100 non-OECD countries and main regions including world. These countries cover developing countries, Central and Eastern European countries and the former USSR. The consistency and complementarity of OECD and non-OECD countries' statistics ensure an accurate picture of the global energy situation.

An analysis of energy problems requires a comprehensive presentation of basic statistics in original units such as tonnes of coal and kilowatt hours of electricity. For the first time, this type of presentation is published in *Energy Statistics of non-OECD Countries 1996-1997,* the sister volume to this publication. The usefulness of such basic data can be considerably improved by expressing them in a common unit suitable for uses such as estimation of total energy supply, forecasting and the study of substitution and conservation. The energy balance is a presentation of the basic supply and demand data for all fuels in a manner which shows the main fuels together but separately distinguished and expressed in a common energy unit. Both of these characteristics will allow the easy comparison of the contribution each fuel makes to the economy and their interrelationships through the conversion of one fuel into another.

This volume has been prepared in close collaboration with other international organisations including the Economic Commission for Europe of the United Nations (UN-ECE), the Organizacíon Latino Americana De Energía (OLADE), the Asia Pacific Energy Research Centre, the Statistical Office of the United Nations and the Forestry Department of the Food and Agriculture Organisation of the United Nations. It draws upon and complements the extensive work of the United Nations in the field of world energy statistics.

While every effort is made to ensure the accuracy of the data, quality is not homogeneous throughout the publication. In some countries data are based on secondary sources, and where incomplete or unavailable, on estimates. In general, data are likely to be more accurate for production, trade and total consumption than for individual sectors in transformation or final consumption. Consequently, energy balances are presented in two formats reflecting the degree of detail available. Moreover, the statistics showing contribution from renewable energies and energy from wastes are less detailed in this publication as in the *Energy Statistics of OECD Countries.* General issues of data quality, country notes and individual country data should be consulted when using regional aggregates.

The IEA published *World Energy Statistics and Balances* in 1989 and 1990 and *Energy Statistics and Balances of Non-OECD Countries* from 1991 to 1998. This book, for the first time published as a separate volume, revises and updates those publications.

Energy data on OECD and non-OECD countries are collected by the team in the Energy Statistics

Division (ESD) of the IEA Secretariat, headed by Mr Jean-Yves Garnier. Non-OECD countries statistics are currently the responsibility of Mr Coleman Nee, Mr Bruno Castellano and Mr Yannis Yaxas. Ms Nina Kousnetzoff has overall editorial responsibility. Secretarial support was supplied by Ms Sharon Michel and Ms Susan Stolarow.

Data from 1971 to 1997 are available on diskettes suitable for use on IBM-compatible personal computers. An order form has been provided in the back of this publication.

Enquiries about data or methodology should be addressed to the head of the non-OECD Countries Section, Energy Statistics Division, at:

Telephone: (+33-1) 40-57-66-34
Fax: (+33-1) 40-57-66-49
E-mail: wed@iea.org.

INTRODUCTION

Cette nouvelle publication offre la même couverture statistique que la publication homonyme pour les pays de l'OCDE. Elle fournit, dans une unité commune, des statistiques sur la production, les échanges et la consommation pour chaque source d'énergie, pour plus de 100 pays ne faisant pas partie de l'OCDE et les principales régions y compris le monde. Ces pays incluent des pays en développement, des pays d'Europe centrale et orientale et l'ex-URSS. La compatibilité et la complémentarité des statistiques des pays membres de l'OCDE et des pays ne faisant pas partie de l'OCDE garantissent la qualité de l'image de la situation énergétique mondiale.

L'analyse des questions énergétiques suppose une présentation détaillée de statistiques de base exprimées dans leurs différentes unités d'origine : tonnes de charbon et kilowatts-heure d'électricité, par exemple. Pour la première fois, ces informations sont présentées dans les *Statistiques de l'énergie des pays non-membres*, recueil publié parallèlement au présent document. On peut accroître considérablement l'utilité de ces données de base en adoptant, pour les exprimer, une unité commune qui permette de les exploiter, par exemple, pour estimer les approvisionnements totaux en énergie, établir des prévisions, ou étudier les possibilités de substitution ou d'économies d'énergie. Le bilan énergétique présente les données de base concernant l'offre et la demande pour toutes les formes d'énergie, selon une méthode permettant de les présenter groupées par grandes catégories, mais aussi de les indiquer séparément selon une unité commune

d'énergie. Ces deux caractéristiques permettent une comparaison aisée de la contribution des différentes formes d'énergie à l'économie et faciliteront l'étude de leurs relations réciproques grâce à l'utilisation de coefficients de conversion.

Ce document a été préparé en étroite collaboration avec d'autres organisations internationales dont la Commission Economique pour l'Europe des Nations Unies (CEE-ONU), l'Organización Latino Americana De Energía (OLADE), le Asia Pacific Energy Research Centre, le Bureau de Statistiques des Nations Unies, et le Département des forêts de l'Organisation des Nations Unies pour l'Alimentation et l'Agriculture. Il utilise et complète le travail considérable déjà accompli par les Nations Unies dans le domaine des statistiques énergétiques.

Bien que tout ait été mis en oeuvre pour assurer l'exactitude de ces données, la qualité des chiffres de cette publication n'est pas toujours homogène. Pour certains pays les données proviennent de sources secondaires et, lorsque les données manquent, sont basées sur des estimations. D'une façon générale, les chiffres sont sans doute plus exacts en ce qui concerne la production, les échanges et la consommation totale que pour les secteurs désagrégés de la transformation ou de la consommation finale. Les bilans énergétiques sont présentés en deux formats différents selon les détails disponibles pour chaque pays. De plus, les statistiques montrant la contribution des énergies renouvelables et des déchets sont moins détaillées dans cette publication que dans les

Statistiques de l'énergie des pays de l'OCDE. Il convient, lors de l'utilisation des agrégats régionaux, de consulter également la note générale sur la qualité des données, les notes relatives aux différents pays et les données par pays.

L'AIE a publié *Statistiques et bilans énergétiques mondiaux* en 1989 et 1990 et *Statistiques et bilans énergétiques des pays non-membres* de 1991 à 1998. Le présent recueil, publié pour la première fois en volume séparé, met à jour et revise les précédentes éditions.

Les données énergétiques sur les pays membres et non-membres de l'OCDE sont collectées par la Division des statistiques énergétiques (ESD) du Secrétariat de l'AIE, dirigée par M. Jean-Yves Garnier. M. Coleman Nee, M. Bruno Castellano et M. Yannis Yaxas sont actuellement

responsables des statistiques des pays non-membres de l'OCDE. Mme Nina Kousnetzoff est responsable de la publication. Mme Sharon Michel et Mme Susan Stolarow ont assuré le secrétariat d'édition et la saisie des textes.

Les données relatives aux années 1971 à 1997 sont disponibles sur disquettes exploitables sur ordinateurs personnels compatibles. Un formulaire de commande est fourni à la fin de cet ouvrage.

Les demandes de renseignements sur les données ou la méthodologie doivent être adressées au chef de la section des pays non-membres, division des statistiques de l'énergie :

Téléphone :	(+33-1) 40-57-66-34
Fax :	(+33-1) 40-57-66-49
E-mail :	wed@iea.org.

PART I: METHODOLOGY

PARTIE I : METHODOLOGIE

PART II
METHODOLOGY

PARTIE II
MÉTHODOLOGIE

1. ISSUES OF DATA QUALITY

A. Methodology

Considerable effort has been made to ensure that data presented in this publication adhere to the IEA definitions contained in the Explanatory Notes (Part I.2). These definitions are used by most of the international organisations that collect energy statistics. Nevertheless, the national energy statistics which are reported to international organisations are often collected using criteria and definitions which differ, sometimes considerably, from those employed by the international organisations. The extent to which the IEA Secretariat has identified these differences and, where possible, adjusted the data to meet international definitions, is outlined below. Recognized anomalies occurring in specific countries are presented in Part I.5, Country Notes and Sources. The Country Notes simply identify some of the more important and obvious deviations from IEA methodology in certain countries and are by no means a comprehensive list of anomalies by country.

B. Estimation

In addition to any adjustments undertaken to compensate for differences in definitions, estimations are sometimes required to complete major aggregates from which key statistics are missing. Examples may be found in the more detailed accounts given below. Except as concerns combustible renewables and waste (see E.6 below), it has been the Secretariat's aim to provide all of the elements of energy balances down to the level of Total Final Consumption (TFC) for all countries and years. This entails providing the elements of supply as well as inputs of primary fuels and outputs of secondary fuels to and from the main transformation activities such as oil refining and electricity generation. This has often required estimations prepared after consultation with national statistical offices, oil companies, electricity utilities and national energy experts.

For all countries and all years, the Secretariat also provides energy indicators computed with GDP and population data. When these economic data were not available from official international sources, the Secretariat used the estimates provided by the CHELEM-CEPII database (see I.5 Country Notes and Sources, General References and I.2.E Indicators).

C. Time Series and Political Changes

Energy balances for the republics of the former USSR have been constructed since 1992. These balances have been constructed from official data and, where necessary, estimates have been made based on information obtained from industry sources and other international organisations. Time series and energy balances for the country 'former USSR' continue to be published. While it remains difficult to collect data for all of the former USSR states on a consistent and common basis the regional aggregate forms a link with the past showing the overall effects of adaptation to the new conditions.

Energy balances for the individual countries of former Yugoslavia also appear in this publication. These have been compiled from official data. Again, in the interest of maintaining time series, figures for the country 'former Yugoslavia' continue to be published.

Energy balances for the Slovak Republic have been constructed since 1971. For the years 1989 to 1997, the balances for the Slovak Republic have been constructed from official data. For the years 1971 to 1988, estimates have been made where necessary (see Country Notes and Sources).

Energy statistics for some countries undergo countinuous changes in their coverage or methodology so that, for example, it is possible to present more detailed energy accounts for China beginning in 1980. Consequently, "breaks in series" are considered to be unavoidable.

The IEA Secretariat reviews its data bases each year. In the light of new assessments, important revisions are made to the time series of individual countries during the course of this review. Therefore, data in this publication have been substantially revised with respect to previous editions.

D. Classification of Fuel Uses

National statistics sources often lack adequate information on the consumption of fuels in different categories of end use. Many countries do not conduct annual surveys of the fuel consumption in the main sectors of economic activity and consequently published data are based on out-of-date surveys. Sectoral disaggregation of consumption for individual countries should therefore be interpreted with caution.

Previous to the reforms undertaken in the 1990's the sectoral classification of fuel consumption in the transition economies (eastern Europe and the countries of the former USSR) and China differed greatly from that practiced in market economies. Sectoral consumption was defined according to the economic branch to which the user of the fuel belonged rather than according to the purpose or use of the fuel. Consumption of gasoline in the vehicle fleet of an enterprise attached to the economic branch 'iron and steel' was classified as industrial

consumption of gasoline in the iron and steel industry. Where possible, the data have been adjusted to fit international classifications, for example, all gasoline is assumed to be consumed in the 'transport' sector and is reported in this publication as such. However, it has not been possible to reclassify products other than gasoline and jet fuel with the same facility, and few adjustments have been made to the remaining products.

E. Specific Issues by Fuel

1. Coal

Data on sectoral coal consumption are usually quoted in metric tonnes and the heat content of the different coal grades used in the various sectors of end use are often not available. In these cases the Secretariat makes estimates based on known heat values of coal for coal production, imports and exports.

2. Oil

The IEA Secretariat collects comprehensive statistics for oil supply and use, including refineries' own use of oil, oil delivered to ships' bunkers and oil for use as petrochemical feedstock. National statistics often do not report these amounts. Reported production of refined products frequently refers to net rather than gross refinery output and consumption of oil products may be limited to sales to domestic markets and may not include deliveries to international shipping or aircraft. Oil consumed as petrochemical feedstock in integrated refinery/petrochemical complexes is often not included in available official statistics.

Where possible, the Secretariat, in consultation with the oil industry, makes estimates of these unreported data. In the absence of any other indication of refinery fuel use the consumption has been estimated to be about 5 per cent of refinery throughput and has been divided equally between refinery gas and heavy fuel oil.

3. Natural Gas

The IEA defines natural gas production as marketable production, i.e. that which is net of field losses, flaring, venting and reinjection.

Furthermore, natural gas should be comprised mainly of methane and other gases such as ethane and higher hydrocarbons should be reported under the heading of 'oil'.

Readily available data for natural gas, however, often do not identify the separate elements of field losses, flaring, venting and reinjection. Moreover, reported data are frequently unaccompanied by adequate definitions so that it is difficult or impossible to identify gas at the different stages of its separation into dry gas (methane) and the heavier fractions in gas separation facilities.

Natural gas supply and demand statistics are normally reported in volumetric units and it is difficult to obtain accurate data on the calorific value of the gas. Where the heat value is unknown, a general gross calorific value of 38 TJ/million m3 has been applied.

It should also be noted that reliable consumption data for natural gas at a disaggregated level are often difficult to locate. This is especially true of some of the larger natural gas consuming countries in the Middle East. Industrial use of natural gas for these countries is therefore frequently missing from data appearing in this publication.

4. Electricity

As defined in the Explanatory Notes (Part I.2), an autoproducer of electricity is an establishment which, in addition to its main activities, generates electricity wholly or partly for its own use. Data on the basis of this definition are frequently unknown in non-OECD countries. In such cases the primary fuels inputs for autoproduced electricity are reported in the appropriate end use sector.

When statistics of the production of electricity from inputs of Combustible renewables and waste are available, they are shown in the corresponding column. These data are not comprehensive; for example, much of the electricity generated from waste biomass in sugar refining remains unreported.

Inputs of fuels for electricity generation are estimated (when unreported) using information on electricity output, fuel efficiency and the type of generation capacity.

5. Heat

The transition economies (eastern Europe and countries of the former USSR) and China employed a methodology with respect to heat that set them apart from common practice in market economies. The approach taken was to allocate the transformation of primary fuels (coal, oil and gas) by industry into heat *for consumption on site* to the transformation activity *Heat Production*, not to industrial consumption as in the IEA methodology[1]. The transformation output of *Heat* was then allocated to the various end use sectors. The losses occurring in the transformation of fuels into heat in industry were not included in final consumption of industry. Although a number of countries have recently switched to the practice of the international organisations, this important distinction reduces the validity of cross-country comparisons of sectoral end use consumption between transition economies and market economies.

6. Combustible Renewables and Waste

The IEA publishes production and primary supply of combustible renewables and waste for all non-OECD countries and all regions for the years 1974 to 1997 (1971 to 1997 on diskettes).

Data shown are often from secondary sources, inconsistent and may be of questionable quality, which makes comparisons between countries difficult. The historical data of many countries derive from surveys which were often irregular, irreconcilable and conducted at the local rather than national level making them incomparable between regions and time. Where historical series are incomplete or unavailable, they have been estimated using the methodology consistent with the projection framework of the IEA's 1998 edition of *World Energy Outlook* (September 1998). First, nationwide biomass and wastes domestic supply per capita was compiled or estimated for 1995. Secondly, per capita supply for the years 1971 to 1994 was estimated, using a log/log equation with either GDP per capita or the percentage of urban population as the exogenous variable, depending on the region. Thirdly, total biomass and waste supply for the years 1996 and 1997 was estimated assuming a growth rate either constant, or equal to the population growth, or following the 1971-1994 estimation. However, these estimated time-series should be treated very cautiously.

---

1 The international methodology restricts the inclusion of heat in transformation sector to that sold to third parties. See definition in Part I.2.

The chart below attempts to provide a broad indication of the data quality and estimation methodology by region.

Region	Main source of data	Data Quality	Exogenous Variables
Africa	FAO database and AfDB	Low	% urban population
Latin America	National and OLADE	High	None
Asia	Surveys	High to Low	GDP per capita
Non-OECD Europe	Questionnaires and FAO	High to Medium	None
Former USSR	National, questionnaires and FAO	High to Medium	None
Middle East	FAO	Medium to Low	None

AfDB: African Development Bank

Although the methods used for estimations are consistent with those used for biomass in the *World Energy Outlook 1998*, minor numerical discrepancies occur for a number of reasons. The reasons include later data revisions and the level of country disaggregations employed.

For the years 1994 to 1997, consumption by end-use has been compiled for all the countries. Combustible renewables and waste is shown in the same energy balance format as for fossil and non combustible renewable fuels for the years 1996 and 1997. The figures confirm the importance of vegetal fuels in the energy sector of many developing countries.

The IEA hopes that the inclusion of these data will encourage national administrations and other agencies active in the field to enhance the level and quality of data collection and coverage. More details on the methodology used by country may be provided on request, and comments are welcome.

2. EXPLANATORY NOTES

A. Unit

The IEA energy balance methodology is based on the calorific content of the energy commodities and a common unit of account. The unit of account adopted by the IEA is the tonne of oil equivalent (toe) which is *defined* as 10^7 kilocalories (41.868 gigajoules). This quantity of energy is, within a few per cent, equal to the net heat content of 1 tonne of crude oil. Throughout this publication 1 tonne means 1 metric ton or 1000 kg.

B. Conversion (from original units to toe)

The change from using the original units to tonnes of oil equivalent implies choosing coefficients of equivalence between different forms and sources of energy. This problem can be approached in many different ways. For example one could adopt a single equivalence for each major primary energy source in all countries, e.g. 29 307 kJ/kg (7 000 kcal/kg) for hard coal, 41 868 kJ/kg (10 000 kcal/kg) for oil, etc. The main objection to this method is that it results in distortions since there is a wide spread in calorific values between types of coal and individual coal products, and between calorific values of these fuels in different countries. The Secretariat has therefore adopted specific factors supplied by the national administrations for the main categories of each quality of coal and for each flow or use (i.e. production, imports, exports, electricity generation, coke ovens, blast furnaces and industry). For crude oil, specific factors have been

used based on consultations with experts from the national administrations, while petroleum products have a single set of conversion factors for all countries (a few exceptions are listed in Section 3). Gas data in *Energy Statistics of Non-OECD Countries* are presented in terajoules on **a gross calorific basis**. Data on *combustible renewables & waste* are presented in terajoules on **a net calorific basis**. See Section 3, Units and Conversions.

The balances are expressed in terms of "net" calorific value. The difference between the "net" and the "gross" calorific value for each fuel is the latent heat of vaporisation of the water produced during combustion of the fuel. For coal and oil, net calorific value is 5 per cent less than gross, for most forms of natural and manufactured gas the difference is 9-10 per cent, while for electricity there is no difference as the concept has no meaning in this case. The use of net calorific value is consistent with the practice of the Statistical Offices of the European Communities and the United Nations.

Electricity data are converted from original units of gigawatt hours to million tonnes of oil equivalent using the relationship: 1 terawatt hour = 0.086 Mtoe.

C. Primary Energy Conventions

When constructing an energy balance, it is necessary to adopt conventions for primary energy from several sources such as nuclear, geothermal, solar, hydro, wind, etc. Two types of assumptions have to be made:

I.   Choice of the primary energy form

For each of these sources, there is a need to define the form of primary energy to be considered; for instance, in the case of hydro energy, a choice must be made between the kinetic energy of falling water and the electricity produced. For nuclear energy, the choice is between the energy content of the nuclear fuel, the heat generated in the reactors and the electricity produced. For photovoltaic electricity, the choice is between the solar radiation received and the electricity produced.

The principle adopted by the IEA is that the primary energy form should be the first energy form downstream in the production process for which multiple energy uses are practical. The application of this principle leads to the choice of the following primary energy forms:

- **Heat** for nuclear heat and electricity production, geothermal heat and electricity production and solar heat production,

- **Electricity** for hydro, wind, wave/ocean and photovoltaic solar electricity production.

II. Calculation of the primary energy equivalent

There are essentially two methods that can be used to calculate the primary energy equivalent of the above energy sources: the partial substitution method and the physical energy content method.

- *The partial substitution method:* In this method, the primary energy equivalent of the above sources of electricity generation represents the amount of energy that would be necessary to generate an identical amount of electricity in conventional thermal power plants. The primary energy equivalent is calculated using an average generating efficiency of these plants. This method has several shortcomings including the difficulty of choosing an appropriate generating efficiency and the fact that it is not relevant for countries with a high share of hydro electricity. For these reasons, the IEA, as most of the international organisations, has now stopped using this method and adopted the physical energy content method.

- *The physical energy content method:* This method uses the physical energy content of the primary energy source as its primary energy equivalent. As a consequence, there is an obvious link between the principles adopted in defining the primary energy forms of energy sources and the primary energy equivalent of these sources. For instance, in the case of nuclear electricity production, as heat is the primary energy form selected by the IEA, the primary energy equivalent is the quantity of heat generated in the reactors. However, as the amount of heat produced is not always known, the IEA estimates the primary energy equivalent from the electricity generation by assuming an efficiency of 33%, which is the average of nuclear power plants in Europe. In the case of hydro, as electricity is the primary energy form selected, the primary energy equivalent is the physical energy content of the electricity generated in the plant, which amounts to assuming an efficiency of 100%. A more detailed presentation of the assumptions used by the IEA in establishing its energy balances is given in Section 3.

Since these two types of energy balances differ significantly in the treatment of electricity from solar, hydro, wind, etc., the share of renewables in total energy supply will appear to be very different depending on the method used. As a result, when looking at the percentages of various energy sources in total supply, it is important to understand the underlying conventions that were used to calculate the primary energy balances.

D.   Energy Balances

The energy balances are presented in tabular format: columns for the various sources of energy and rows for the different origins and uses. The following description refers to the layout of the full balance type presentation. In the case of the regional and abbreviated balances, the definition of the products is the same as in the full balances. However, sectoral breakdown has been restricted to main end use totals (Total Industry, etc.) and to selected main categories (Iron and Steel, etc.).

I.   Columns

Across the top of the table from left to right, there are eleven columns with the following headings:

Column 1: *Coal* includes all coal, both primary (including hard coal and lignite) and derived fuels (including patent fuel, coke oven coke, gas coke, BKB, coke oven gas, and blast furnace gas). Peat is also included in this category.

Column 2: *Crude Oil* comprises crude oil, natural gas liquids, refinery feedstocks, and additives as well as other hydrocarbons such as synthetic oil, mineral oils extracted from bituminous minerals (in the row *indigenous production*) and oils from coal and natural gas liquefaction (in the row *liquefaction*).

Column 3: *Petroleum products* comprise refinery gas, ethane, LPG, aviation gasoline, motor gasoline, jet fuels, kerosene, gas/diesel oil, heavy fuel oil, naphtha, white spirit, lubricants, bitumen, paraffin waxes, petroleum coke and other petroleum products.

Column 4: *Gas* includes natural gas (excluding natural gas liquids) and gas works gas. The latter appears as a positive figure in the "gas works" row but is not part of indigenous production.

Column 5: *Nuclear* shows the primary heat equivalent of the electricity produced by a nuclear power plant with an average thermal efficiency of 33 per cent. See Sections 2.C and 3.

Column 6: *Hydro* shows the energy content of the electricity produced in hydro power plants. Hydro output *excludes* output from pumped storage plants. See Sections 2.C and 3.

Column 7: *Geothermal, solar, etc.;* indigenous production of geothermal, solar, wind, tide and wave energy and the use of these energy forms for electricity generation. Other uses shown in this column relate only to geothermal and solar heat. Unless the actual efficiency of the geothermal process is known, the quantity of geothermal energy entering electricity generation is inferred from the electricity production at geothermal plants assuming an average thermal efficiency of 10 per cent. For solar, wind, tide and wave energy, the quantities entering electricity generation are equal to the electrical energy generated. See Sections 2.C and 3.

Column 8: *Combustible renewables & waste* comprises solid biomass and animal products, gas/liquids from biomass, industrial waste and municipal waste. Biomass is defined as any plant matter used directly as fuel or converted into fuels

(e.g. charcoal) or electricity and/or heat. Included here are wood, vegetal waste (including wood waste and crops used for energy production), ethanol, animal materials/wastes and sulphite lyes (Sulphite lyes are also known as "black liquor" and are an alkaline spent liquor from the digesters in the production of sulphate or soda pulp during the manufacture of paper. The energy is derived from the lignin removed from the wood pulp.). Municipal waste comprises wastes produced by the residential, commercial and public service sectors that are collected by local authorities for disposal in a central location for the production of heat and/or power. Hospital waste is included in this category.

Data under this heading are often based on small sample surveys or other incomplete information. Thus the data give only a broad impression of developments, and are not strictly comparable between countries. In some cases complete categories of vegetal fuel are omitted due to lack of information. Please refer to individual country data when consulting regional aggregates.

Column 9: *Electricity* shows final consumption and trade in electricity, which is accounted at the same heat value as electricity in final consumption (i.e. 1 GWh = 0.000086 Mtoe).

Column 10: *Heat* shows the disposition of heat produced for sale. The large majority of the heat included in this column results from the combustion of fuels although some small amounts are produced from electrically powered heat pumps and boilers. Any heat extracted from ambient air by heat pumps is shown as indigenous production.

Column 11: *TOTAL* = the total of Columns 1 to 10.

II. Rows

The categories on the left hand side of the table have the following functions:

Row 1: *Indigenous production* is the production of primary energy, i.e. hard coal, lignite, peat, crude oil, NGLs, natural gas, combustible renewables and wastes, nuclear, hydro, geothermal, solar and the heat from heat pumps that is extracted from the ambient environment. Production is calculated after removal of impurities (e.g. sulphur from natural gas). Calculation of production of hydro, geothermal, etc. and nuclear electricity is explained in Section 3, Units and Conversions.

Row 2/3: *Imports and exports* comprise amounts having crossed the national territorial boundaries of the country, whether or not customs clearance has taken place.

a)   Coal

Imports and exports comprise the amount of fuels obtained from or supplied to other countries, whether or not there is an economic or customs union between the relevant countries. Coal in transit should not be included.

b)   Oil and gas

Quantities of crude oil and oil products imported or exported under processing agreements (i.e. refining on account) are included. Quantities of oil in transit are excluded. Crude oil, NGL and natural gas are reported as coming from the country of origin; refinery feedstocks and oil products are reported as coming from the country of last consignment.

Re-exports of oil imported for processing within bonded areas are shown as exports of product from the processing country to the final destination.

c)   Electricity

Amounts are considered as imported or exported when they have crossed the national territorial boundaries of the country.

Row 4: *International marine bunkers* cover those quantities delivered to sea-going ships of all flags, including warships. Consumption by ships engaged in transport in inland and coastal waters is not included. See *internal navigation* Row 40 and *agriculture* Row 43.

Row 5: *Stock changes* reflect the difference between opening stock levels on the first day of the year and closing levels on the last day of the year of stocks on national territory held by producers, importers, energy transformation industries and large consumers. A stock build is shown as a negative number, and a stock draw as a positive number.

Row 6: *Total primary energy supply (TPES)* is made up of indigenous production (Row 1) + imports (Row 2) - exports (Row 3) - international marine bunkers (Row 4) ± stock changes (Row 5).

Row 7: *Transfers* include interproduct transfers, products transferred and recycled products (e.g. used lubricants which are reprocessed).

Row 8: *Statistical differences* is a category which includes the sum of the unexplained statistical differences for individual fuels, as they appear in the basic energy statistics. It also includes the statistical differences that arise because of the variety of conversion factors in the coal column. See introduction to *Energy Statistics of Non-OECD Countries 1996-1997* for further details.

Row 9: *Electricity plants* refers to plants which are designed to produce electricity only. If one or more units of the plant is a CHP unit (and the inputs and outputs can not be distinguished on a unit basis) then the whole plant is designated as a CHP plant. Both public[1] and autoproducer[2] plants are included here. Columns 1 through 8 show the use of primary and secondary fuels for the production of electricity as negative entries. Gross electricity produced (including power stations' own consumption) appears as a positive quantity in the electricity column. Transformation losses appear in the total column as a negative number.

Row 10: *Combined heat and power plants (CHP),* refers to plants which are designed to produce both heat and electricity). UNIPEDE refers to these as co-generation power stations. If possible, fuel inputs and electricity/heat outputs are on a unit basis rather than on a plant basis. However, if data are not available on a unit basis, the convention for defining a CHP plant noted above is adopted. Both public and autoproducer plants are included here. *Note that for autoproducer's CHP plants, all fuel inputs to electricity production are taken into account, while only the part of fuel inputs to heat **sold** is shown. Fuel inputs for the production of heat consumed within the autoproducer's establishment are **not** included here but are included with figures for the final consumption of fuels in the appropriate consuming sector.*

Columns 1 through 8, show the use of primary and secondary fuels for the production of electricity and heat as negative entries. Total gross electricity produced appears as a positive quantity in the

---

1   Public supply undertakings generate electricity and/or heat for sale to third parties, *as their primary activity.* They may be privately or publicly owned. Note that the sale need not take place through the public grid.

2   Autoproducer undertakings generate electricity and/or heat, wholly or partly for their own use as an activity which supports their primary activity. They may be privately or publicly owned.

electricity column and heat produced appears as a positive number in the heat column. Transformation losses appear in the total column as a negative number.

Row 11: *Heat plants* refers to plants (including heat pumps and electric boilers) designed to produce heat only, which is sold to a third party under the provisions of a contract. Both public and autoproducer plants are included here.

Columns 1 through 8, show the use of primary and secondary fuels in a heating system that transmits and distributes heat from one or more energy source to, among others, residential, industrial, and commercial consumers for space heating, cooking, hot water and industrial processes.

Row 12: Where there is production of gas at *Gas works* the treatment is similar to that for electricity generation, with the quantity produced appearing as a positive figure in the gas column, inputs as negative entries in the coal, petroleum products and gas columns, and conversion losses appearing in the total column.

Row 13: The row *Petroleum refineries* shows the use of primary energy for the manufacture of finished petroleum products and the corresponding output. Thus, the total reflects transformation losses. In certain cases the data in the total column are positive numbers. This can be due either to problems in the primary refinery balance, or to the fact that the IEA uses standardised net calorific values for petroleum products.

Row 14: *Coal transformation* contains losses in transformation of coal from primary to secondary fuels and from secondary to tertiary fuels (hard coal to coke, coke to blast furnace gas, lignite to BKB, etc.). It is often difficult to correctly account for all inputs and outputs in energy transformation industries, and to separate energy that is transformed from energy that is combusted. As a result, in certain cases the data in the total column are positive numbers, indicating a problem in the underlying energy data.

Row 15: *Liquefaction* includes diverse liquefaction processes, such as coal and natural gas liquefaction in South Africa.

Row 16: *Other transformation* covers non-specified transformation not shown elsewhere including losses in charcoal burners. It also includes

backflows from the petrochemical sector. Note, backflows from oil products that are used for non-energy purposes (i.e. white spirit and lubricants) are not included here, but in non-energy use.

Row 17: *Own use* contains the primary and secondary energy consumed by transformation industries for heating, pumping, traction and lighting purposes [ISIC[1] Divisions 10, 11, 12, 23 and 40]. These are shown as negative figures. Included here are, for example, coal mines' own use of energy, power plants' own consumption (which includes net electricity consumed for pumped storage), and energy used for oil and gas extraction.

Row 18: *Distribution and transmission losses* includes losses in gas distribution, electricity transmission and coal transport.

Row 19: *Total final consumption* (TFC) is the sum of consumption by the different end-use sectors. In final consumption, petrochemical feedstocks are shown as an *"of which"* item under *chemical industry*, while non-energy use of such oil products as white spirit, lubricants, bitumen, paraffin waxes and other products are shown under *non-energy use*, and are included in Total final consumption only. See Rows 47-50 (*Non-energy use*). Backflows from the petrochemical industry are not included in final consumption (see Row 16, *other transformation*).

Rows 20-34: Consumption in the *Industry sector* is specified in the following sub-sectors (energy used for transport by industry is not included here but is reported under transport):

- *Iron and steel* industry [ISIC Group 271 and Class 2731];

- *Chemical industry* [ISIC Division 24];

  of which: petrochemical feedstocks. The petrochemical industry includes cracking and reforming processes for the purpose of producing ethylene, propylene, butylene, synthesis gas, aromatics, butadene and other hydrocarbon-based raw materials in processes such as steam cracking, aromatics plants and steam reforming. [Part of ISIC Group 241]; See feedstocks under Rows 47-50 (*Non-energy use*);

1 International Standard Industrial Classification of All Economic Activities, Series M, No. 4/Rev. 3, United Nations, New York, 1990.

- *Non-ferrous metals* basic industries [ISIC Group 272 and Class 2732];

- *Non-metallic mineral* products such as glass, ceramic, cement, etc. [ISIC Division 26];

- *Transport equipment* [ISIC Divisions 34 and 35];

- *Machinery* comprises fabricated metal products, machinery and equipment other than transport equipment [ISIC Divisions 28, 29, 30, 31 and 32];

- *Mining (excluding fuels) and quarrying* [ISIC Divisions 13 and 14];

- *Food and tobacco* [ISIC Divisions 15 and 16];

- *Paper, pulp and print* [ISIC Divisions 21 and 22];

- *Wood and wood products* (other than pulp and paper) [ISIC Division 20];

- *Construction* [ISIC Division 45];

- *Textile and leather* [ISIC Divisions 17, 18 and 19];

- *Non-specified* (any manufacturing industry not included above) [ISIC Divisions 25, 33, 36 and 37].

Note: Most countries have difficulties supplying an industrial breakdown for all fuels. In these cases, the *non-specified* industry row has been used. *Regional aggregates of industrial consumption should therefore be used with caution.*

Rows 35-41: The *Transport sector* includes all fuels for transport except international marine bunkers [ISIC Divisions 60, 61 and 62]. It includes transport in the industry sector and covers road, railway, air, internal navigation (including small craft and coastal shipping not included under marine bunkers), fuels used for transport of materials by pipeline and non-specified transport. Fuel used for ocean, coastal and inland fishing should be included in agriculture (Row 43).

Rows 42-46: *Other sectors* cover agriculture (including ocean, coastal and inland fishing) [ISIC Divisions 01, 02 and 05], residential, commercial and public services [ISIC Divisions 41, 50, 51, 52, 55, 63, 64, 65, 66, 67, 70, 71, 72, 73, 74, 75, 80, 85, 90, 91, 92, 93, 95 and 99], and non-specified consumption. In many cases administrations find it impossible to distinguish energy consumption in *commercial and public services* from *residential*

consumption. Some cannot distinguish consumption in *agriculture* from that in *residential*. In these cases, the *residential sector* will also include consumption in *agriculture* and/or *commercial/ public services*. The *non-specified* row may include military use that is not shown in the transport sector. The *other sectors* total is, therefore, more accurate than its components. Please refer to individual country data when consulting regional aggregates.

Rows 47-50: *Non-energy* use covers use of *other petroleum products* such as white spirit, paraffin waxes, lubricants, bitumen and other products. It also includes the non-energy use of coal (excluding peat). They are shown separately in final consumption under the heading *non-energy use*. It is assumed that the use of these products is exclusively *non-energy use*. An exception to this treatment is the petroleum coke, which is shown as *non-energy use* only when there is evidence of such use; otherwise it is shown under energy use in industry or in other sectors. Non-energy use of coal includes carbon blacks, graphite electrodes, etc., and is also shown separately by sector.

Please note that feedstocks for the petrochemical industry are accounted for in industry under chemical industry (Row 22) and shown separately under: *of which: feedstocks* (Row 23). This covers all oil including naphtha (except the *other petroleum products* listed above) and gas used as petrochemical feedstock.

Rows 51-53: *Electricity generated* shows the total number of GWh generated by thermal power plants separated into electricity plants and CHP plants, as well as production by nuclear and hydro (excluding pumped storage production), geothermal, etc. (see, however, the notes on Rows 9 and 10). Please note that electricity produced from crude oil and NGL is included in the *petroleum products* column.

Rows 54-56: *Heat generated* shows the total amount of TJ generated by power plants separated into CHP plants and heat plants. Please note that the heat produced by electric boilers is shown in the *electricity* column and heat produced by heat pumps is shown in the *heat* column.

E. Indicators

Energy Production: total primary energy production expressed in Mtoe. See Section 2.D.II.1.

Total Primary Energy Supply: expressed in Mtoe. See Section 2.D.II.6.

Net Oil Imports: imports minus exports of oil expressed in Mtoe. See Section 2.D.II.2/3.

Oil Supply: primary supply of oil expressed in Mtoe. See Section 2.D.II.6.

Electricity Consumption: domestic consumption, i.e. gross production + imports - exports - distribution losses expressed in TWh.

The main source of the population and GDP data is *World Development Indicators*, World Bank Washington D.C., 1998 and 1999.

Population data for **Gibraltar, Former Yugoslav Republic of Macedonia, Former USSR, Former Yugoslavia** and the three regions **Other Africa, Other Latin America** and **Other Asia** are based on the CHELEM-CEPII CD ROM. Population data for **Chinese Taipei** come from the Council for Economic Planning and Development, Republic of China, *Taiwan Statistical Databook* 1998.

GDP figures for **Bosnia-Herzegovina, Croatia, Cuba, Gibraltar, Kuwait, Libya, Former Yugoslav Republic of Macedonia, Moldova, Netherlands Antilles, Qatar, Slovenia, Ukraine,** the **United Arab Emirates, Former USSR** (may differ from the sum of the 15 republics), **Federal Republic of Yugoslavia, Former Yugoslavia** and the three regions **Other Africa, Other Latin America** and **Other Asia** are from the CHELEM-CEPII CD ROM, Paris, 1999. GDP figures for **Angola** (1971-1984), **Bahrain** (1971-1979 and 1996-1997), **Bangladesh** (1997), **Bolivia** (1971-1979), **Brunei** (1971-1973 and 1996-1997), **Bulgaria** (1971-1979), **Cyprus** (1971-1974 and 1995-1997), **Ethiopia** (1971-1982), **Iran** (1971-1973 and 1996-1997), **Iraq** (1991-1997), **Jordan** (1971-1974), **Lebanon** (1971-1987), **Mozambique** (1971-1979), **Nicaragua** (1997), **Oman** (1996-1997), **Panama** (1997), **Romania** (1971-1974), **Sudan** (1995-1997), **Tajikistan** (1997), **Tanzania**

(1971-1987), **Turkmenistan** (1997) and **Yemen** (1971-1989) have been estimated based on the growth rates of the CHELEM-CEPII CD ROM. GDP figures for **Chinese Taipei** come from the Council for Economic Planning and Development, Republic of China, *Taiwan Statistical Databook* 1997. The GDP growth rate of the former Czechoslovakia was used to estimate GDP for the **Slovak Republic** from 1971 to 1983.

The GDP data have been compiled for individual countries at market prices in local currency and annual rates. These data have been scaled up/down to the price levels of 1990 and then converted to US dollars using the yearly average 1990 exchange rates or purchasing power parities (PPPs).

In recent years, there have been wide fluctuations in exchange rates, consequently there has been a growing need and interest in developing energy indicators using a measure of GDP which would avoid these fluctuations and better reflect the relative purchasing power of different currencies. As a result, this publication is presenting GDP calculated using PPPs as well as with the traditional exchange rates.

Purchasing power parities are the rates of currency conversion that equalize the purchasing power of different currencies. A given sum of money, when converted into different currencies at the PPP rates, buys the same basket of goods and services in all countries. In other words, PPPs are the rates of currency conversion which eliminate the differences in price levels between different countries. The PPPs selected to convert the GDP of non-OECD countries from national currencies to US dollars come from the CHELEM-CEPII CD ROM, Paris 1999. They were aggregated using the Geary-Khamis (GK) method and rebased on the United States. For a more detailed description of the methodology please see: Equipe CHELEM, *La base CHELEM-PIB*, CEPII, Paris, April 1998.

3. UNITS AND CONVERSIONS

General Conversion Factors for Energy

To:	TJ	Gcal	Mtoe	MBtu	GWh
From:	multiply by:				
TJ	1	238.8	2.388×10^{-5}	947.8	0.2778
Gcal	4.1868×10^{-3}	1	10^{-7}	3.968	1.163×10^{-3}
Mtoe	4.1868×10^{4}	10^{7}	1	3.968×10^{7}	11630
MBtu	1.0551×10^{-3}	0.252	2.52×10^{-8}	1	2.931×10^{-4}
GWh	3.6	860	8.6×10^{-5}	3412	1

Conversion Factors for Mass

To:	kg	t	lt	st	lb
From:	multiply by:				
kilogramme (kg)	1	0.001	9.84×10^{-4}	1.102×10^{-3}	2.2046
tonne (t)	1000	1	0.984	1.1023	2204.6
long ton (lt)	1016	1.016	1	1.120	2240.0
short ton (st)	907.2	0.9072	0.893	1	2000.0
pound (lb)	0.454	4.54×10^{-4}	4.46×10^{-4}	5.0×10^{-4}	1

Conversion Factors for Volume

To:	gal U.S.	gal U.K.	bbl	ft³	l	m³
From:	multiply by:					
U.S. gallon (gal)	1	0.8327	0.02381	0.1337	3.785	0.0038
U.K. gallon (gal)	1.201	1	0.02859	0.1605	4.546	0.0045
Barrel (bbl)	42.0	34.97	1	5.615	159.0	0.159
Cubic foot (ft³)	7.48	6.229	0.1781	1	28.3	0.0283
Litre (l)	0.2642	0.220	0.0063	0.0353	1	0.001
Cubic metre (m³)	264.2	220.0	6.289	35.3147	1000.0	1

Decimal Prefixes

10^1	deca (da)		10^{-1}	deci (d)
10^2	hecto (h)		10^{-2}	centi (c)
10^3	kilo (k)		10^{-3}	milli (m)
10^6	mega (M)		10^{-6}	micro (μ)
10^9	giga (G)		10^{-9}	nano (n)
10^{12}	tera (T)		10^{-12}	pico (p)
10^{15}	peta (P)		10^{-15}	femto (f)
10^{18}	exa (E)		10^{-18}	atto (a)

A. Gas

Energy Statistics of Non-OECD Countries expresses the following gases in terajoules, using their gross calorific value.

1 terajoule = 0.00002388 Mtoe.

To calculate the net heat content of a gas from its gross heat content, multiply the gross heat content by the appropriate following factor.

Natural gas	0.9
Gas works gas	0.9
Coke oven gas	0.9
Blast furnace gas	1.0
Oxygen Steel Furnace Gas	1.0

B. Electricity

Figures for electricity production, trade, and final consumption are calculated using the energy content of the electricity (i.e. at a rate of 1 TWh = 0.086 Mtoe). Hydro-electricity production (excluding pumped storage) and electricity produced by other non-thermal means (wind, tide, photovoltaic, etc.) are accounted for similarly using 1 TWh = 0.086 Mtoe. However, the primary energy equivalent of nuclear electricity is calculated from the gross generation by assuming a 33% conversion efficiency, i.e. 1 TWh = (0.086 ÷ 0.33) Mtoe. In the case of electricity produced from geothermal heat, if the actual geothermal efficiency is not known, then the primary equivalent is calculated assuming an efficiency of 10%, so 1 TWh = (0.086 ÷ 0.1) Mtoe.

C. Crude Oil

The Country Specific Net Calorific Values are shown by country in tables in Part II. These are average values used to convert production, imports, exports and consumption to heat values.

D. Petroleum Products

The following conversion factors are used for all years (toe per tonne):

	Specific values for certain countries				All other countries
	Algeria	Brazil	Malaysia	South Africa	
Refinery Gas	-	0.880	-	-	1.150
Ethane	-	-	-	-	1.130
LPG	1.180	1.175	1.088	1.117	1.130
Naphtha	-	1.132	1.054	1.073	1.075
Aviation Gasoline	-	1.129	1.050	1.088	1.070
Motor Gasoline	-	1.122	1.050	1.052	1.070
Jet Gasoline	-	-	-	0.973	1.070
Jet Kerosene	-	1.109	1.032	0.981	1.065
Other Kerosene	-	1.109	1.032	1.033	1.045
Gas/Diesel Oil	-	1.075	1.015	1.025	1.035
Heavy Fuel Oil	-	1.090	0.991	0.999	0.960
White Spirit	-	1.124	1.032	1.015	0.960
Lubricants	-	1.077	1.006	-	0.960
Bitumen	-	1.005	0.998	-	0.960
Paraffin Waxes	-	-	1.035	-	0.960
Petroleum Coke	-	0.850	0.869	-	0.740
Other Products	-	1.089	1.015	-	0.960

E. Coal

Coal has separate net calorific values (NCV) for production, imports, exports, inputs to public power plants and coal used in coke ovens, blast furnaces and industry. All other flows are converted using the average NCV. Each country's individual conversion factors for 1996 and 1997 can be found in Part II, Country Specific Net Calorific Values.

F. Combustible Renewables and Waste

The heat content of combustible renewables and waste, expressed in terajoules on a net calorific value basis, is presented in *Energy Statistics of Non-OECD Countries.*

1 terajoule = 0.00002388 Mtoe.

G. Heat

Information on heat is supplied in terajoules.

1 terajoule = 0.00002388 Mtoe.

H. Examples

The following examples indicate how to calculate the net calorific content (in Mtoe) of the quantities expressed in original units in *Energy Statistics of Non-OECD Countries, 1996-1997.*

Coking coal production (Colombia) for 1997 in thousand tonnes;
multiply by 0.00065= Mtoe.

Natural gas in terajoules (gross); *0.00021496 1*
multiply by 0.00002149= Mtoe.

Motor gasoline (except Brazil, Malaysia and South Africa) in thousand tonnes;
multiply by 0.0010700= Mtoe.

Heat in terajoules (net);
multiply by 0.00002388= Mtoe.

4. GEOGRAPHICAL COVERAGE

- **Africa** includes Algeria, Angola, Benin, Cameroon, Congo, Democratic Republic of Congo, Egypt, Ethiopia, Gabon, Ghana, Ivory Coast, Kenya, Libya, Morocco, Mozambique, Nigeria, Senegal, South Africa, Sudan, Tanzania, Tunisia, Zambia, Zimbabwe and **Other Africa**.

- **Other Africa** includes Botswana, Burkina-Faso, Burundi, Cape Verde, Central African Republic, Chad, Djibouti, Equatorial Guinea, Gambia, Guinea, Guinea-Bissau, Lesotho, Liberia, Madagascar, Malawi, Mali, Mauritania, Mauritius, Niger, Réunion, Rwanda, Sao Tome-Principe, Seychelles, Sierra Leone, Somalia, Swaziland, Togo and Uganda.

- **Latin America** includes Argentina, Bolivia, Brazil, Chile, Colombia, Costa Rica, Cuba, Dominican Republic, Ecuador, El Salvador, Guatemala, Haiti, Honduras, Jamaica, Netherlands Antilles, Nicaragua, Panama, Paraguay, Peru, Trinidad and Tobago, Uruguay, Venezuela and **Other Latin America**.

- **Other Latin America** includes Antigua and Barbuda, Bahamas, Barbados, Belize, Bermuda, Dominica, French Guiana, Grenada, Guadeloupe, Guyana, Martinique, St. Kitts-Nevis-Anguilla, Saint Lucia, St. Vincent-Grenadines and Surinam.

- **Asia** includes Bangladesh, Brunei, Chinese Taipei, India, Indonesia, DPR of Korea, Malaysia, Myanmar, Nepal, Pakistan, Philippines, Singapore, Sri Lanka, Thailand, Vietnam and **Other Asia**.

- **Other Asia and Other Oceania** include Afghanistan, Bhutan, Fiji, French Polynesia, Kiribati, Maldives, New Caledonia, Papua New Guinea, Samoa, Solomon Islands and Vanuatu.

- **China** includes the People's Republic of China and Hong Kong (China).

- **Non-OECD Europe** includes Albania, Bosnia-Herzegovina, Bulgaria, Croatia, Cyprus, the Former Yugoslav Republic of Macedonia (FYROM), Gibraltar, Malta, Romania, Slovak Republic, Slovenia and the Federal Republic of Yugoslavia.

- **Former USSR** includes Armenia, Azerbaijan, Belarus, Estonia, Georgia, Kazakhstan, Kyrgyzstan, Latvia, Lithuania, Moldova, Russia, Tajikistan, Turkmenistan, Ukraine and Uzbekistan.

- **Middle East** includes Bahrain, Iran, Iraq, Israel, Jordan, Kuwait, Lebanon, Oman, Qatar, Saudi Arabia, Syria, the United Arab Emirates and Yemen.

- The **Organisation for Economic Co-Operation and Development (OECD)** includes Australia, Austria, Belgium, Canada, the Czech Republic, Denmark, Finland, France, Germany, Greece, Hungary, Iceland, Ireland, Italy, Japan, Korea, Luxembourg, Mexico, the Netherlands, New Zealand, Norway, Poland, Portugal, Spain, Sweden, Switzerland, Turkey, the United Kingdom and the United States.

Within OECD:

Denmark excludes Greenland and the Danish Faroes.

France includes Monaco, and excludes overseas departments (Martinique, Guadeloupe, French Polynesia and Réunion).

Germany includes the new federal states of Germany.

Italy includes San Marino and the Vatican.

Japan includes Okinawa.

The **Netherlands** excludes Surinam and the Netherlands Antilles.

Portugal includes the Azores and Madeira.

Spain includes the Canary Islands.

Switzerland includes Liechtenstein.

United States includes Puerto Rico, Guam, the Virgin Islands and the Hawaiian Free Trade Zone.

The Organisation of the Petroleum Exporting Countries (OPEC) includes Algeria, Indonesia, Iran, Iraq, Kuwait, Libya, Nigeria, Qatar, Saudi Arabia, the United Arab Emirates and Venezuela.

Please note that the following countries have not been considered due to lack of data:

Africa: Comoros, Namibia, Saint Helena and Western Sahara.

America: Aruba, British Virgin Islands, Caymen Islands, Falkland Islands, Montserrat, Saint Pierre-Miquelon and Turks and Caicos Islands;

Asia and Oceania: American Samoa, Cambodia, Christmas Island, Cook Islands, Laos, Macau, Mongolia, Nauru, Niue, Pacific Islands (US Trust), Tonga and Wake Island.

5. COUNTRY NOTES AND SOURCES

General References –
Références Générales

Annual Bulletin of Coal Statistics for Europe, Economic Commission for Europe (ECE), New York, 1994.

Annual Bulletin of Electric Energy Statistics for Europe, Economic Commission for Europe (ECE), New York, 1994.

Annual Bulletin of Gas Statistics for Europe, Economic Commission for Europe (ECE), New York, 1994.

Annual Bulletin of General Energy Statistics for Europe, Economic Commission for Europe (ECE), New York, 1994.

Annual Report July 1991- June 1992, South African Development Community (SADC), Gaborone, 1993.

Annual Statistical Bulletin 1997, Organization of Petroleum Exporting Countries (OPEC), Vienna, 1998.

APEC Energy Database, Mosaic/ World Wide Web site: http://www.ieej.or.jp/apec/database

Arab Oil and Gas Directory 1998, Arab Petroleum Research Centre, Paris, 1999.

ASEAN Energy Review 1995 Edition, ASEAN-EC Energy Management Training and Research Centre (AEEMTRC), Jakarta, 1996.

Base CHELEM-PIB, Centre d'Etudes Prospectives et d'Informations Internationles (CEPII) Paris, 1999.

Eastern Bloc Energy, Tadcaster, various issues to May 1999.

Energy Indicators of Developing Member Countries, Asian Development Bank (ADB), Manila, 1994.

Energy Information Administration (EIA) Mosaic/World Wide Web site: http://www.eia.doe.gov

Energy-Economic Information System (SIEE), Latin American Energy Organization (OLADE), Ecuador, 1999.

Energy Statistics Yearbook 1990, South African Development Community (SADC), Luanda, 1992.

Energy Statistics Yearbook 1996, United Nations, New York, 1998.

Food and Agriculture Organisation of the United Nations, 1999 Forestry Data (via Internet), Rome, 1999.

Forests and Biomass Sub-sector in Africa, African Energy Programme of the African Development Bank, Abidjan, 1996.

International Coal Report, various issues to May 1999.

International Energy Annual 1990, 1991, 1992, 1993, Energy Information Administration (EIA), Washington, D.C., 1991-1994.

International Energy Data Report 1992, World Energy Council, London, 1993.

Les Centrales Nucléaires dans le Monde Commissariat à l'Énergie Atomique, Paris, 1998.

Middle East Economic Survey (MEES), Nicosia, various issues to June 1999.

Natural Gas in the World, 1998 Survey, Cedigaz, Paris, 1998.

Notes d'Information et Statistiques, Banque Centrale des Etats de l'Afrique de l'Ouest, Dakar, 1995.

Pétrole 1994, Comité Professionnel du Pétrole (CPDP), Paris, 1995.

PIW's Global Oil Stocks & Balances, New York, various issues to June 1995.

PlanEcon Energy Outlook for Eastern Europe and the Former Soviet Republics, Washington, October 1998.

Prospects of Arab Petroleum Refining Industry, Organization of Arab Petroleum Exporting Countries (OAPEC), Kuwait, 1990.

Review of Wood Energy Data in RWEDP Member Countries, Regional Wood Energy Development Programme in Asia, Food and Agriculture Organisation of the United Nations, Bangkok, 1997.

Statistical Handbook 1993 States of the Former USSR, The World Bank, Washington, 1993.

Statistical Yearbook of the Member States of the CMEA, Council of Mutual Economic Assistance (CMEA), Moscow, 1985 and 1990.

The United Nations Energy Statistics Database 1996, United Nations Statistical Office (UNSO), New York, 1998.

World Development Indicators 1999 on CD-ROM, The World Bank, Washington, 1999.

Note:

\-   The OLADE database was used for most of the Latin American countries.

\-   For the period 1971 to 1996, the UN database was the only source of information for the eleven individual countries which are not listed below, and for the regions Other Africa, Other Latin America and Other Asia and Oceania. It was also used in a number of other countries as a complementary source.

Albania

Large quantities of oil widely reported to have moved through Albania into former Yugoslavia are not included in oil trade for 1993. Although estimated to represent up to 100 pour cent of underlying domestic consumption, no reliable figures for this trade were available. Series have been revised since 1990.

Sources 1971-1997:

Combustible Renewables and Waste:

The UN Energy Statistics Database, 1995, UN ECE Energy Questionnaires and Secretariat estimates.

Sources up to 1995:

Aide Memoire of World Bank Mission to Albania May/June 1991.

The UN Energy Statistics Database, 1994.

UN ECE Energy Questionnaires 1994 and 1995.

Algeria

Sources 1971-1997:

Combustible Renewables and Waste:

The UN Energy Statistics Database, 1996.

Sources 1992-1997:

Direct communication to the Secretariat from the Ministry of Industry and Energy, Information Systems Management Department, Algiers.

Sources up to 1991:

Bilan Energétique National, Gouvernement Algérien, Algiers, 1984.

Algérie Energie, No 6, Ministère de l'Energie et des Industries Chimiques et Pétrochimiques, Algiers, 1979 to 1983.

Annuaire Statistique de l'Algérie 1980-1984, Office National des Statistiques, Algiers, 1985.

Angola

Sources 1971-1997:

Combustible Renewables and Waste:

Secretariat estimates based on 1991 data from African Energy Programme of the African Development Bank, *Forests and Biomass Sub-sector in Africa*, Abidjan, 1996.

Sources 1992-1997:

Direct communications to the Secretariat from oil industry sources.

Eskom Annual Statistical Yearbook 1993, 1994, 1995 (Johannesburg, 1994, 1995, 1996) citing Empresa Nacional de Electricidade, Luanda as source.

The UN Energy Statistics Database, 1995.

Sources up to 1991:

Le Pétrole et l'Industrie Pétrolière en Angola en 1985, Poste d'Expansion Economique de Luanda, Luanda, 1985.

Argentina

Sources 1971-1997:

Combustible Renewables and Waste:

SIEE, (OLADE).

Sources 1992-1997:

Secretaría de Energía Wide Web site: http://www.mecon.ar/energia/

Direct communication to the Secretariat from the Ministry of Economy and Public Services, Secretariat of Energy, Buenos Aires.

Annuario Estadístico de La Republica Argentina, Instituto Nacional de Estadistica y Censos, Buenos Aires, September 1997.

Sources up to 1991:

Anuario de Combustibles, Ministerio de Obras y Servicios Públicos, Secretaria de Energía, Buenos Aires, 1980 to 1984, 1986, 1988 to 1992, 1995, 1997.

Combustibles Boletin Mensual, Ministerio de Obras y Servicios Públicos, Secretaria de Energía, Buenos Aires, various editions.

Natural Gas Projection up to 2000, Gas del Estado Argentina, Buenos Aires, 1970, 1984 to 1986.

Anuario Estadístico de la Republica Argentina 1970-1981, Instituto Nacional de Estadístico y Censos, Secretaria de Planificación, Buenos Aires, 1982.

Anuario Energía Eléctrica, Ministerio de Obras y Servicios Públicos, Secretaria de Energía, Buenos Aires, 1987 to 1990.

Balance Energetico Nacional 1970-1985, Ministerio de Obras y Servicios Públicos, Secretaria de Energía, Buenos Aires, 1986.

Plan Energetico Nacional 1986-2000, Ministerio de Obras y Servicios Públicos, Secretaria de Energía, Subsecretaria de Planificación Energetica, Buenos Aires, 1985.

Anuario Estadístico, Yacimientos Petrolíferos Fiscales, Buenos Aires, 1984 to 1987.

Memoria Y Balance General, Yacimientos Petrolíferos Fiscales, Buenos Aires, 1984 to 1986.

Bahrain

Sources 1992-1997:

Direct communication to the Secretariat from oil industry sources.

Statistical Abstract, 1994, Council of Ministers, Control Statistics Organisation, Bahrain, 1995.

The UN Energy Statistics Database, 1995.

Sources up to 1991:

1986 Annual Report, Bahrain Monetary Agency, Bahrain, 1987.

B.S.C. Annual Report, Bahrain Petroleum Company, Bahrain, 1982, 1983 and 1984.

Foreign Trade Statistics, Council of Ministers, Central Statistics Organisation, Bahrain, 1985.

Bahrain in Figures, Council of Ministers, Central Statistics Organisation, Bahrain, 1983, 1984 and 1985.

Statistical Abstract 1990, Council of Ministers, Central Statistics Organisation, Bahrain, 1991.

Bangladesh

Energy statistics are reported for a fiscal year.

Sources 1971-1997:

Combustible Renewables and Waste:

Secretariat estimates based on a per capita average consumption from various surveys and studies.

Sources 1996-1997:

Statistical Yearbook of Bangladesh 1996,1997, 7th Edition, Ministry of Planning, Bangladesh Bureau of Statistics, Dhaka, 1997,1998.

Direct communication to the Secretariat from oil and gas industry sources and electricity utility.

Sources 1992-1995:

Statistical Pocket Book of Bangladesh, Ministry of Planning, Bangladesh Bureau of Statistics, Dhaka, 1986 to 1996.

The UN Energy Statistics Database, 1995.

Sources up to 1991:

Bangladesh Energy Balances 1976-1981, Government of Bangladesh, Dhaka, 1982.

Statistical Yearbook of Bangladesh 1991, Government of Bangladesh, Dhaka, 1976 to 1991.

Monthly Statistical Bulletin of Bangladesh, Ministry of Planning, Bangladesh Bureau of Statistics, Statistics Division, Dhaka, June 1986 and October 1989.

Benin

Sources 1971-1997:

Combustible Renewables and Waste:

Secretariat estimates based on 1991 data from *Forests and Biomass Sub-sector in Africa,* African Energy Programme of the African Development Bank, Abidjan, 1996.

Sources up to 1997:

Direct communication to the Secretariat from the Electricity utility 1998, 1999.

Direct communication to the Secretariat from the Direction de l' Energie, Cotonou, 1999.

The UN Energy Statistics Database, 1995.

Rapport sur l'Etat de l'Economie Nationale, Ministère de l' Economie, Cotonou, septembre 1993.

Bolivia

Sources 1971-1997:

Combustible Renewables and Waste:

SIEE, (OLADE).

Sources 1992-1997:

Informe Estadístico 1992, 1993, 1994, 1995, 1996 and 1997, Yacimientos Petrolíferos Fiscales Bolivianos, La Paz, 1993, 1994, 1995, 1996, 1997, 1998.

Memoria Anual 1992, Yacimientos Petrolíferos Fiscales Bolivianos, La Paz, 1993.

Sources up to 1991:

Boletin Estadístico 1973-1985, Banco Central de Bolivia, Division de Estudios Económicos, La Paz, 1986.

Diez Anos de Estadística Petrolera en Bolivia 1976-1986, Dirección de Planeamiento, Division de Estadística, La Paz, 1987.

Empresa Nacional de Electricidad S.A. 1986 Ende Memoria, Empresa Nacional de Electricidad, La Paz, 1987.

Brazil

Sources 1971-1997:

Combustible Renewables and Waste:

Ministério de Minas e Energia.

Sources 1992-1997:

Balanço Energético Nacional, Ministério de Minas e Energia, Brasilia, 1993, 1994, 1995, 1996, 1997, 1998.

Direct communication to the Secretariat from Petrobrás, Commercial Department, Marine Fuels Division.

Sources up to 1991:

Balanço Energético Nacional, Ministério de Minas e Energia, Brasilia, 1983 to 1992.

Anuario Estatistico, Conselho Nacional do Petroleo, Diretoria de Planejamento, Coordenadoria de Estatistica, Brasilia, 1982, 1987, 1988.

Brunei

Historical series have been considerably revised using direct communication from the Office of the Prime Minister, Petroleum Unit.

Sources 1971-1997:

Combustible Renewables and Waste:

The UN Energy Statistics Database, 1995.

Sources 1990-1997:

Direct communication to the secretariat from the Office of the Prime Minister, Petroleum Unit 1999.

Direct communication to the secretariat from the Ministry of Development, Electrical Services Department 1999.

Brunei Statistical Yearbook, 1992 to 1994, Ministry of Finance, Statistics Section, Brunei, 1993, 1995.

Direct communication to the Secretariat from the UN Energy Statistics Unit.

Sources up to 1991:

Fifth National Development Plan 1986-1990, Ministry of Finance, Economic Planning Unit, Bandar Seri Bagawan, 1985.

Bulgaria

See Part I.1, Issues of Data Quality above.

Sources 1971-1997:

Combustible Renewables and Waste:

The UN Energy Statistics Database, 1995 and UN ECE Energy Questionnaires.

Sources 1992-1997:

UN ECE Energy Questionnaires.

Energy Balances, National Statistical Institute, Sofia, 1995.

Sources up to 1991:

Energy Development of Bulgaria, Government of Bulgaria, Sofia, 1980 and 1984.

Energy in Bulgaria, Government of Bulgaria, Sofia, 1980 to 1983.

General Statistics in the Republic of Bulgaria 1989/1990, Government of Bulgaria, Sofia, 1991.

Cameroon

Sources 1971-1997:

Combustible Renewables and Waste:

Secretariat estimates based on 1991 data from *Forests and Biomass Sub-sector in Africa,* African Energy Programme of the African Development Bank, Abidjan, 1996.

Sources up to 1997:

Direct communication to the Secretariat from oil industry sources and the electricity utility.

The UN Energy Statistics Database, 1995.

Chile

Sources 1971-1997:

Combustible Renewables and Waste:

Comisión Nacional de Energía.

Sources 1992-1997:

Balance Nacional de Energía 1995, 1997 Comisión Nacional de Energía, Santiago, 1996, 1998.

Balance Nacional de Energía 1977-1996, Comisión Nacional de Energía, Santiago, September 1997.

Balance de Energía Preliminar 1993, Comisión Nacional de Energía, Santiago, 1994.

Balance de Energía 1973 - 1992, Comisión Nacional de Energía, Santiago, 1993, *1975-1994,* Santiago, 1995.

Sources up to 1991

Compendio Estadístico Chile 1985, Ministerio de Economía, Fomento Y Reconstrucción, Instituto Nacional de Estadísticas, Santiago, 1986.

China

See Part I.1, Issues of Data Quality above.

Coal production statistics refer to unwashed and unscreened coal. IEA coal statistics normally refer to coal after washing and screening for the removal of inorganic matter.

It is known that much of the agricultural use of diesel in China is for transport purposes but the national data have not been adjusted by the IEA Secretariat to take account of this.

See Part I.1, Issues of Data Quality for further information.

Sources 1971-1997:

Combustible Renewables and Waste:

Secretariat estimates based on a per capita average consumption from various surveys and studies.

Sources 1992-1997:

Energy Balances of China 1997, obtained from the APEC Energy Database.

Energy Balances of China, provided to the Secretariat by the State Statistical Bureau for 1993, 1994, 1995 and 1996.

China Statistical Yearbook 1995, State Statistical Bureau of the People's Republic of China, Beijing, 1995.

China Energy Databook, Lawrence Berkeley National Laboratory, Berkeley, 1996.

1995 Energy Report of China, State Planning Commission of the People's Republic of China, Beijing, 1995.

China Petroleum Newsletter Monthly, China Petroleum Information Institute, Beijing, various issues to June 1995.

China OGP, China Oil Gas and Petrochemicals/Xinhua News Agency, Beijing, various issues to June 1995.

Petroleum Data Monthly, China Oil Gas and Petrochemicals/Xinhua News Agency, Beijing, various issues to June 1995.

Energy of China, China International Book Trading Co, Beijing, various issues to June 1995.

Energy Commodity Account of China, Asian Development Bank, Manila, 1994.

China's Customs Statistics, State Statistical Bureau, Economic Information & Agency, Beijing, various editions to 1995.

China's Customs Statistics, General Administration of Customs, PRC, Economic Information and Agency, Hong Kong, various editions from 1991 to 1995.

Statistical Yearbook of China, State Statistical Bureau of the People's Republic of China, Economic Information & Agency, Hong Kong, various editions from 1981 to 1994.

Statistical Communique, State Statistical Bureau of the People's Republic of China, Hong Kong, March 1995.

China's Downstream Oil Industry in Transition, Fesharaki Associates Consulting and Technical Services, Inc., Honolulu, 1993.

Sources up to 1991:

Outline of Rational Utilization and Conservation of Energy in China, Bureau of Energy Conservation State Planning Commission, Beijing, June 1987.

China Coal Industry Yearbook, Ministry of Coal Industry, People's Republic of China, Beijing, 1983, 1984 and 1985.

Energy in China 1989, Ministry of Energy, People's Republic of China, Beijing, 1990.

Electric Industry in China in 1987, Ministry of Water Resources and Electric Power, Department of Planning, Beijing, 1988.

Statistical Yearbook of China 1991, State Statistical Bureau of the People's Republic of China, Beijing, 1992.

China: A Statistics Survey 1975-1984, State Statistical Bureau, Beijing, 1985.

China Petro-Chemical Corporation (SINOPEC) Annual Report, SINOPEC, Beijing, 1987.

Almanac of China's Foreign Economic Relations and Trade, The Editorial Board of the Almanac, Beijing, 1986.

Chinese Taipei

Autoproducer electricity includes only the inputs and outputs of the iron and steel industry and waste disposal plants. Energy used by the other autoproducers (mostly industrial cogeneration) was counted as final consumption.

Sources 1971-1997:

Combustible Renewables and Waste:

The UN Energy Statistics Database, 1995.

Sources 1992-1997:

Energy Balances in Taiwan, Ministry of Economic Affairs, Taipei, 1992 to 1997.

Yearbook of Energy Statistics, Ministry of Trade Industry and Energy, Taipei, 1996.

Sources up to 1991:

Industry of Free China 1975-1985, Council for Economic Planning and Development, Taipei, 1986.

Taiwan Statistical Data Book 1954-1985, Council for Economic Planning and Development, Taipei, 1986.

Energy Policy for the Taiwan Area, Ministry of Economic Affairs, Energy Committee, Taipei, 1984.

The Energy Situation in Taiwan, Ministry of Economic Affairs, Energy Committee, Taipei, 1986, 1987, 1988 and 1992.

Energy Balances in Taiwan, Ministry of Economic Affairs, Taipei, 1984 to 1991.

Energy Indicators Quarterly, Taiwan Area, Ministry of Economic Affairs, Taipei, 1986.

Taipower 1987 Annual Report, Taipower, Taipei, 1988.

Energy Data Report, 1986, WEC National Committee, Taipei, 1986.

Colombia

Sources 1971-1997:

Combustible Renewables and Waste:

Ministry of Mines and Energy, Energy Information Department.

Sources 1992-1997:

Direct communication to the Secretariat from the Ministry of Mines and Energy, Energy Information Department, Bogotá.

Sources up to 1991:

Estadísticas Basicas del Sector Carbón, Carbocol, Oficina de Planeación, Bogotá, various editions from 1980 to 1988.

Colombia Estadística 1985, DANE, Bogotá, 1970 to 1983 and 1987.

Empresa Colombiana de Petróleos, Informe Anual, Empresa Colombiana de Petróleos, Bogotá, 1979, 1980, 1981 and 1985.

Estadísticas de la Industria Petrolera Colombiana Bogota 1979-1984, Empresa Colombiana de Petróleos, Bogotá, 1985.

Informe Estadístico Sector Eléctrico Colombiano, Government of Colombia, Bogotá, 1987 and 1988.

La Electrificacion en Colombia 1984-1985, Instituto Colombiano de Energía Electrica, Bogotá, 1986.

Balances Energéticos 1975-1986, Ministerio de Minas y Energía, Bogota, 1987.

Energía y Minas Para el Progreso Social 1982-1986, Ministerio de Minas y Energía, Bogota, 1987.

Estadísticas Minero-Energéticas 1940-1990, Ministerio de Minas y Energía, Bogotá, 1990.

Boletin Minero-Energético, Ministerio de Minas y Energía, Bogota, December 1991.

Congo

Sources 1971-1997:

Combustible Renewables and Waste:

Secretariat estimates based on 1991 data from *Forests and Biomass Sub-sector in Africa,* African Energy Programme of the African Development Bank, Abidjan, 1996.

Sources up to 1997:

Annual Statistical Yearbook 1993, 1994, 1995, Eskom, Johannesburg, 1994, 1995, 1996, citing Empresa Nacional de Electricidade, Luanda as source.

L'Energie en Afrique, IEPE/ENDA, Paris, 1995, in turn sourced from the Direction des Etudes et de la Planification, Ministère des Mines et de l'Energie, and the Société Congolaise de Raffinage, Brazzaville.

Direct communication to the Secretariat from the UN Energy Statistics Unit.

Democratic Republic of Congo

Sources 1971-1997:

Combustible Renewables and Waste:

Secretariat estimates based on 1991 data from *Forests and Biomass Sub-sector in Africa,* African Energy Programme of the African Development Bank, Abidjan, 1996.

Sources up to 1997:

L'Energie en Afrique, IEPE/ENDA, Paris, 1995, in turn sourced from the *Annuaire Statistique Energétique 1990,* Communauté Economique des Pays des Grands Lacs, Bujumbura, 1990.

The UN Energy Statistics Database, 1996.

Cuba

Historical series have been considerably revised using *Anuario Estadístico de Cuba 1996,* from Oficina Nacional de Estadísticas.

Sources 1971-1997:

Combustible Renewables and Waste:

SIEE, (OLADE).

Anuario Estadístico de Cuba 1996, Oficina Nacional de Estadísticas, Havana, 1998.

Sources up to 1991:

Compendio estadístico de energía de Cuba 1989, Comite Estatal de Estadísticas, Havana, 1989.

Anuario Estadístico de Cuba, Comite Estatal de Estadísticas, Havana, various editions from 1978 to 1987.

Cyprus

Sources 1971-1997:

Combustible Renewables and Waste:

UN ECE Energy Questionnaires and Secretariat estimates.

Sources 1992-1997:

UN ECE Energy Questionnaires.

Electricity Authority of Cyprus Annual Report 1988, 1992, 1996, Electricity Authority of Cyprus, Nicosia, 1989, 1993, 1997.

Industrial Statistics 1988, Ministry of Finance, Department of Statistics, Nicosia, 1989.

Ecuador

Sources 1971-1997:

Combustible Renewables and Waste:

Ministerio de Energia y Minas and *SIEE,* OLADE.

Sources 1996:

Sector Energético Ecuatoriano, Edición No 7, Direccion de Planificacion, Ministerio de Energia y Minas, Quito, November 1997.

Sources 1992-1995:

Balance Energético Nacional 1995, Ministerio de Energia y Minas, Quito, December 1996.

Balances Energéticos 1988-1994, Ministerio de Energia y Minas, Quito, 1996.

Sources up to 1991: Ministerio de Energia y Minas,

Cuentas Nacionales, Banco Central del Ecuador, Quito, various editions from 1982 to 1987.

Memoria 1980-1984, Banco Central del Ecuador, Quito, 1985.

Ecuadorian Energy Balances 1974-1986, Instituto Nacional de Energía, Quito, 1987.

Informacion Estadística Mensual, No. 1610, Instituto Nacional de Energía, Quito, 1988.

Plan Maestro de Electrificación de Ecuador, Ministerio de Energía y Minas, Quito,1989.

Egypt

Sources 1971-1996:

Combustible Renewables and Waste:

The UN Energy Statistics Database, 1995.

Sources 1992-1997:

1995, 1997 Annual Report, Ministry of Petroleum, Egyptian General Petroleum Corporation, Cairo, 1996,1998.

Annual Report of Electricity Statistics 1996/1997, Ministry of Electricity and Energy, Egyptian Electricity Authority, Cairo, 1998.

Arab Oil and Gas, The Arab Petroleum Research Center, Paris, October 1997.

Middle East Economic Survey, Middle East Petroleum and Economic Publications, Nicosia, February 1994, June 1996, March 1998.

Direct submisson to the Secretariat from the Ministry of Petroleum.

A Survey of the Egyptian Oil Industry 1993, Embassy of the United States of America in Cairo, Cairo, 1994.

Sources up to 1991:

Annual Report of Electricity Statistics 1990/1991, Ministry of Electricity and Energy, Egyptian Electricity Authority, Cairo, 1992.

Statistical Yearbook of the Arab Republic of Egypt, Central Agency for Public Mobilisation and Statistics, Cairo, 1977 to 1986.

L'Electricité, l'Energie, et le Pétrole, République Arabe d'Egypte, Organisme Général de l'Information, Cairo, 1990.

Annual Report, The Egyptian General Petroleum Corporation, Cairo, 1985.

Ethiopia

Ethiopia is including Eritrea.

Sources 1971-1997:

Combustible Renewables and Waste:

Secretariat estimates based on 1992 data from Eshetu, L. and Bogale, W., *Power Restructuring in Ethiopia*, AFREPREN, Nairobi, 1996.

Sources 1992-1997

Direct communication to the Secretariat from the Ministry of Economic Development and Co-Operation, Addis Ababa, 1998, 1999.

Direct communication to the Secretariat from the UN Energy Statistics Unit.

Sources up to 1991:

Ten Years of Petroleum Imports, Refinery Products, and Exports, Ministry of Mines & Energy, Addis Ababa, 1989.

Energy Balance for the Year 1984, Ministry of Mines & Energy, Addis Ababa, 1985.

1983 Annual Report, National Bank of Ethiopia, Addis Ababa, 1984.

Quarterly Bulletin, National Bank of Ethiopia, Addis Ababa, various editions from 1980 to 1985.

Gabon

Sources 1971-1997:

Combustible Renewables and Waste:

Secretariat estimates based on 1991 data from *Forests and Biomass Sub-sector in Africa,* African Energy Programme of the African Development Bank, Abidjan, 1996.

Direct communication from the oil industry.

Sources 1992-1997:

Tableau de Bord de l' Economie, Situation 1997, Perspectives 1998-1999, Direction Générale de l'Economie, Ministère des Finance, de l' Economie, du Budget et des participations, chargé de la privatisation, Mai 1998.

Direct communication to the Secretariat from the Société Gabonaise de Raffinage, Port Gentil, 1997.

Rapport d'Activité Banque Gabonaise de Développement, Libreville, 1985, 1990, 1992 and 1993.

The UN Energy Statistics Database, 1995.

Sources up to 1991:

Tableau de Bord de l'Economie, Situation 1983 Perspective 1984-85, Ministère de l'Economie et des Finances, Direction Générale de l'Economie, Libreville, 1984.

Ghana

Sources 1971-1997:

Combustible Renewables and Waste:

Ministry of Mines and Energy and the UN Energy Statistics Database, 1995.

Sources 1992-1996

Direct communication to the Secretariat from the UN Energy Statistics Unit.

National Energy Statistics, Ministry of Energy and Mines, Accra, 1997.

Quarterly Digest of Statistics, Government of Ghana, Statistical Services, Accra, March 1990, March 1991, March 1992, March 1995.

Sources up to 1991:

Energy Balances, Volta River Authority, Accra, various editions from 1970 to 1985.

Hong Kong, China

Sources 1971-1997:

Combustible Renewables and Waste:

The UN Energy Statistics Database, 1995 and Secretariat estimates.

Sources 1992-1996:

Hong Kong Energy Statistics - Annual Report, Census and Statistics Department, Hong Kong, various editions to 1997.

Hong Kong Energy Statistics - Quarterly Supplement, Census and Statistics Department, Hong Kong, various editions to 1997.

Hong Kong Monthly Digest of Statistics, Census and Statistics Department, Hong Kong, various editions to 1994.

India

Natural gas supply and demand data are reported by fiscal year. Oil supply data are reported by calendar year, and the oil consumption data have been pro-rated using available fiscal year end use detail. Autoproducer's electricity has been assumed to be consumed in the industrial sector.

Sources 1971-1997:

Combustible Renewables and Waste:

Secretariat estimates based on a per capita average consumption from various surveys and studies.

Sources 1992-1997:

Indian Oil and Gas, Ministry of Petroleum and Natural Gas, Economics & Statistics Division, New Delhi, January 1985 to March 1998.

Direct communication to the Secretariat from the Ministry of Petroleum and Natural Gas.

Coal Directory of India, 1992-1993, 1993-1994, 1995-1996, 1996-1997, 1997-1998, Ministry of Coal, Coal Controller's Organization, Calcutta, 1994, 1995, 1996, 1997, 1998.

Annual Review of Coal Statistics, 1993-1994, 1995-1996, 1996-1997, 1997-1998, Ministry of Coal, Coal Controller's Organization, Calcutta, 1995, 1997, 1998, 1999.

Annual Report 1994-1995, 1995-1996, 1998-1999, Ministry of Coal, New Delhi, 1995, 1996, 1999.

Indian Petroleum and Natural Gas Statistics, Ministry of Petroleum and Natural Gas, Economics & Statistics Division, New Delhi, 1985 to 1997.

Energy Data Directory, Yearbook "TEDDY", and *Annual Report,* Tata Energy Research Institute "TERI", New Delhi, 1986, 1987, 1988, 1990, 1994/1995 and 1995/1996.

General Review, Public Electricity Supply, India Statistics, Central Electricity Authority, New Delhi, 1982 to 1985, 1995, 1996.

Monthly Abstract of Statistics, Ministry of Planning, Central Statistics Organisation, Department of Statistics, New Delhi, various editions from 1984 to March 1998.

Annual Report 1993-1994, 1998-1999 Ministry of Petroleum and Natural Gas, New Delhi, 1995, 1999.

India Monthly Coal Statistics, Ministry of Energy, Department of Coal, Coal Controllers Organisation, New Delhi, various editions to June 1996.

Annual Report 1994-1996, 1998-1999 Ministry of Energy, Department of Non-Conventional Energy, New Delhi, 1996, 1999.

India's Energy Sector, July 1995, , Center for Monitoring Indian Economy PVT Ltd., Bombay, 1995.

Monthly Review of the Indian Economy, Center for Monitoring Indian Economy PVT Ltd., New Delhi, various issues from 1994 to June 1999.

Sources up to 1991:

Indian Oil Corporation Limited 1987-88 Annual Report, Indian Oil Corporation Limited, New Delhi, 1989-1992.

Report 1986-87, Ministry of Energy, Department of Coal, New Delhi, 1981 to 1987.

Annual Report 1986-1987, Ministry of Energy, Department of Non-Conventional Energy, New Delhi, 1987.

Economic Survey, Ministry of Finance, New Delhi, various editions from 1975 to 1986.

Statistical Outline of India, Ministry of Finance, New Delhi, 1983, 1984, 1986, and 1987.

Monthly Coal Bulletin, vol xxxvi no.2., Ministry of Labour, Directorate General of Mines Safety, New Delhi, February 1986.

Indonesia

Although it has been estimated that electricity output by autoproducers is approximately equal to that generated by the national electricity utility, since this amount is not known with any certainty, autoproducer output refers only to that amount sold to the public electricity grid.

Sources 1971-1997:

Combustible Renewables and Waste:

The UN Energy Statistics Database, 1996 and Secretariat estimates.

Sources 1992-1997:

Direct communication to the Secretariat from the Ministry of Mines and Energy, Directorate General of Electricity and Energy Development.

The Petroleum Report Indonesia, U.S. Embassy in Jakarta, Jakarta, 1986 to 1996.

Oil Statistics of Indonesia, Direktorat Jenderal Minyak Dan Gas Bumi, Jakarta, 1981 to December 1998.

Statistik Dan Informasi Ketenagalistrikan Dan Energi, Direktorat Jenderal Listrik Dan Pengembangan Energi, Jakarta, December 1998.

Mining and Energy in Indonesia, 1995, Ministry of Mines and Energy, Jakarta, 1995.

Sources up to 1991:

Indonesian Financial Statistics, Bank of Indonesia, Jakarta, 1982.

Indikator Ekonomi 1980-1985, Biro Pusat Statistik, Jakarta, 1986.

Statistical Yearbook of Indonesia, Biro Pusat Statistik, Jakarta, 1978 to 1984, 1992.

Statistik Pertambangan Umum, 1973 - 1985, Biro Pusat Statistik, Jakarta, 1986.

Energy Planning for Development in Indonesia, Directorate General for Power, Ministry of Mines and Energy, Jakarta, 1981.

Commercial Information, Electric Power Corporation, Perusahaan Umum Listrik Negara, Jakarta, 1984 and 1985.

Iran

Energy statistics are reported on a fiscal year basis.

Sources 1971-1997:

Combustible Renewables and Waste:

The UN Energy Statistics Database, 1996 and FAO, Forestry Statistics, 1999 (via Internet).

Sources 1992-1997:

Direct communication to the Secretariat from the Ministry of Petroleum, 1999.

Direct communication to the Secretariat from the Ministry of Energy, Office of Deputy Minister for Energy, Teheran 1998.

Electric Power in Iran, Ministry of Energy, Power Planning Bureau, Statistics Section, Teheran, 1992.

Sources up to 1991:

Electric Power in Iran, Ministry of Energy, Power Planning Bureau, Statistics Section, Teheran, 1967 to 1977, 1988, 1990, 1991.

Israel

For reasons of confidentiality the annual *"Energy in Israel"* only reports aggregated data for kerosene and jet kerosene.

Sources 1971-1996:

Combustible Renewables and Waste:

FAO, Forestry Statistics, 1999.

Sources 1992-1997:

Direct communication to the Secretariat from the Ministry of Energy and Infrastructure, Jerusalem.

Energy in Israel, Ministry of Energy and Infrastructure, Central Bureau of Statistics, Jerusalem, 1992, 1993, 1994, 1995, 1996, 1997.

Statistical Report 1993, 1994, 1995, The Israel Electric Corporation, Haifa, April 1994, May 1995, April 1996.

Statistical Results 1992, The Israel Electric Corporation, Haifa, June 1993.

Central Bureau of Statistics (CBS) World Wide Web site: http://www.cbs.gov.il

Sources up to 1991:

Energy in Israel, Ministry of Energy and Infrastructure, Central Bureau of Statistics, Jerusalem, 1975 to 1991.

Statistical Abstract of Israel, Ministry of Energy and Infrastructure, Central Bureau of Statistics, Jerusalem, 1985.

Supplement to Monthly Bulletin of Statistics, Ministry of Energy and Infrastructure, Central Bureau of Statistics, Jerusalem, various editions from 1984 to 1986.

Ivory Coast

Sources 1971-1996:

Combustible Renewables and Waste:

Secretariat estimates based on 1991 data from *Forests and Biomass Sub-sector in Africa,* African Energy Programme of the African Development Bank, Abidjan, 1996.

Sources 1996-1997:

La Côte d'Ivoire en chiffres, Ministère de l'Economie et des Finances, edition 1996-97.

Direct communication to the Secretariat from oil industry and the Ministry of Energy, Abidjan, July 1998.

Sources 1992-1995:

Direct communication to the Secretariat from the Bureau des Economies d'Energie.

L'Energie en Afrique, IEPE/ENDA, Paris, 1995, in turn sourced from the Ministère des Mines et de L'Energie, Abidjan.

The UN Energy Statistics database, 1995.

Sources up to 1991:

Etudes & Conjoncture 1982 - 1986, Ministère de l'Economie et des Finances, Direction de la Planification et de la Prévision, Abidjan, 1987.

Jamaica

Sources 1971-1997:

Combustible Renewables and Waste:

SIEE, (OLADE).

Sources 1992-1994:

Economic and Social Survey Jamaica 1992 to 1995, Planning Institute of Jamaica, Kingston, 1993, 1994, April 1995, April 1996.

Sources up to 1991:

National Energy Outlook 1985-1989, Petroleum Corporation of Jamaica, Economics and Planning Division, Kingston, 1985.

Energy and Economic Review, Petroleum Corporation of Jamaica, Energy Economics Department, Kingston, September 1986, December 1986 and March 1987.

Production Statistics 1988, Planning Institute of Jamaica, Kingston,1989.

Statistical Digest, Research and Development Division, Bank of Jamaica, Kingston, 1984, 1985, 1986, 1989, 1990 and 1991.

Jordan

Sources 1971-1997:

Combustible Renewables and Waste:

FAO, Forestry Statistics, 1999 (via Internet).

Sources 1992-1997:

Annual Report 1992, 1993, 1995, 1996, Jordan Electricity Authority, Amman, 1993, 1994, 1996, 1997.

Energy and Electricity in Jordan 1992, 1993, 1994, 1995, Jordan Electricity Authority, Amman, 1993, 1994, 1995, 1996.

Statistical Yearbook, 1994, Department of Statistics, Amman, 1995.

Sources up to 1991:

Monthly Statistical Bulletin, Central Bank of Jordan, Department of Research Studies, Amman, various issues.

Statistical Yearbook, Department of Statistics, Amman, 1985, 1986 and 1988.

1986 Annual Report, Ministry of Energy and Mineral Resources, Amman, 1987.

1989 Annual Report, Ministry of Energy and Mineral Resources, Amman, 1990.

Kenya

Sources 1971-1997:

Combustible Renewables and Waste:

Secretariat estimates based on 1991 data from *Forests and Biomass Sub-sector in Africa,* African Energy Programme of the African Development Bank, Abidjan, 1996.

Sources 1992-1997:

Economic Survey, 1995, 1996, 1997, 1998, Central Bureau of Statistics, Nairobi.

The UN Energy Statistics Database, 1996.

Sources up to 1991:

Economic Survey, Government of Kenya, Nairobi, 1989.

Economic Survey 1991, Ministry of Planning and National Development, Central Bureau of Statistics, Nairobi, 1992.

Kenya Statistical Digest, Ministry of Planning and National Development, Central Bureau of Statistics, Nairobi, 1988.

Kuwait

Data include 50 per cent of Neutral Zone output.

Sources 1971-1997:

Combustible Renewables and Waste:

FAO, Forestry Statistics, 1999 (via Internet).

Sources 1992-1997:

Direct communication to the Secretariat from the Ministry of Oil, Safat.

Monthly Digest of Statistics, Ministry of Planning, Central Statistical Office, Kuwait, 1998.

A Survey of the Kuwait Oil Industry, Embassy of the United States of America in Kuwait City, Kuwait,1993.

Twelfth Annual Report 1991-1992, Kuwait Petroleum Corporation, Kuwait, 1993.

Sources up to 1991:

Quarterly Statistical Bulletin, Central Bank of Kuwait, Kuwait, various editions from 1986 and 1987.

The Kuwaiti Economy, Central Bank of Kuwait, Kuwait, various editions from 1980 to 1985.

Annual Statistical Abstract, Ministry of Planning, Central Statistical Office, Kuwait, 1986 and 1989.

Monthly Digest of Statistics, Ministry of Planning, Central Statistical Office, Kuwait, various editions from 1986 to 1990.

Economic and Financial Bulletin Monthly, Central Bank of Kuwait, Kuwait, various editions from 1983 to 1986.

Kuwait in Figures, The National Bank of Kuwait, Kuwait, 1986 and 1987.

Lebanon

There was no refinery production in 1993, 1994, 1995, 1996 and 1997.

Sources 1971-1997:

Combustible Renewables and Waste:

FAO, Forestry Statistics, 1999 (via Internet) and Secretariat estimates.

Sources 1992-1997:

L'Energie au Liban, Les Bilans Energétiques en 1997, Association Libanaise pour la Maîtrise de l'Energie, Beirut,1998.

L'Energie au Liban, le Défi, Association Libanaise pour la Maîtrise de l'Energie, Beirut, décembre 1996.

Les Bilans Energétiques au Liban, Association Libanaise pour la Maîtrise de l'Energie, Beirut, décembre 1994.

Direct communication to the Secretariat from Association Libanaise pour la Maîtrise de l'Energie.

Libya

Sources 1971-1997:

Combustible Renewables and Waste:

The UN Energy Statistics Database, 1996.

Sources up to 1991:

Statistical Abstract of Libya, 19th vol, Government of Libya, Tripoli, 1983.

Malaysia

Sources 1971-1997:

Combustible Renewables and Waste:

The UN Energy Statistics Database, 1996 and FAO, Forestry Statistics, 1999 (via Internet).

Sources 1992-1997:

Direct communication to the Secretariat from the Ministry of Energy, Telecommunications and Posts.

Information Malaysia, Ministry Telekom Dan Pos Malaysia, Kuala Lumpur, 1995, 1996.

National Energy Balance, Ministry of Energy, Telecommunications and Posts, Kuala Lumpur, 1980-1995.

Sources up to 1991:

National Energy Balances Malaysia, Ministry Telekom Dan Pos Malaysia, Kuala Lumpur, 1978 to 1991.

Malta

UN ECE Questionnaire on Oil 1995 to 1997.

UN ECE Questionnaire on Coal 1994 and 1995.

UN ECE Questionnaire on Electricity and Heat, 1994 to 1997.

Morocco

Sources 1971-1997:

Combustible Renewables and Waste:

The UN Energy Statistics Database, 1996.

Sources 1992-1997:

Annuaire Statistique du Maroc, Ministère du Plan, Direction de la Statistique, Rabat, 1980, 1984, 1986 to 1998.

Electricity consumption by economic sector by direct communication from the Ministère du Plan, Rabat.

Sources up to 1991:

Rapport d'Activité du Secteur Pétrolier 1983, Ministère de l'Energie et des Mines, Direction de l'Energie, Rabat, 1984.

Rapport sur les Données Energétiques Nationales 1979-1981, Ministère de l'Energie et des Mines, Rabat, 1982.

Le Maroc en Chiffres 1986, Ministère du Plan, Direction de la Statistique, Rabat, 1987.

Rapport d'Activité 1992, Office National de l'Electricité, Casablanca, 1993.

Rapport Annuel, Office National de Recherches et d'Exploitations Pétrolieres, Maroc, 1984.

Mozambique

Sources 1971-1996:

Combustible Renewables and Waste:

Secretariat estimates based on 1991 data from *Forests and Biomass Sub-sector in Africa,* African Energy Programme of the African Development Bank, Abidjan, 1996.

Sources 1992-1997:

Direct communication to the Secretariat from the Ministry of Energy and Mineral Resources, (Maputo June 1998, April 1999) and the electricity utility.

Annual Statistical Yearbook 1993, 1994, 1995, Eskom, Johannesburg, 1994, 1995, 1996, citing Electricidade de Mozambique, Maputo, as source.

The UN Energy Statistics Database, 1995.

Myanmar

Sources 1971-1997:

Combustible Renewables and Waste:

Secretariat estimates based on 1990 data from *UNDP Sixth Country Programme Union of*

Myanmar, World Bank, Programme Sectoral Review of Energy, by Sousing John, et. al., Washington, D.C., 1991.

Sources 1992-1997:

Direct communications to the Secretariat from the Ministry of Energy, Planning Department, Rangoon, 1996, 1997 and 1998, 1999.

Review of the Financial Economic and Social Conditions, Ministry of National Planning and Economic Development, Central Statistical Organization, Rangoon, 1995, 1996.

Statistical Yearbook, Ministry of National Planning and Economic Development, Central Statistical Organization, Rangoon, 1995, 1996.

The UN Energy Statistics Database, 1995.

Sources up to 1991:

Sectoral Energy Demand in Myanmar, UNDP Economic and Social Commission for Asia and The Pacific, Bangkok, 1992.

Selected Monthly Economic Indicators, paper no. 3, Ministry of Planning and Finance, Central Statistical Organization, Rangoon, 1989.

Nepal

Energy statistics are reported for a fiscal year.

Sources 1971-1997:

Combustible Renewables and Waste:

Water and Energy Commission Secretariat (WECS), Ministry of Water Resources.

Sources 1992-1997:

Asian Energy News, Asian Institute of Technology, Pathumthani, November 1997.

Nepal and the World, a Statistical Profile, Federation of Nepalese Chambers of Commerce and Industrie, Kathmandou, 1997.

Energy Balance Sheet of Nepal 1981-1992, Ministry of Water Resources, Water and Energy Commission, Katmandou, 1993.

Energy Synopsis Report 1994/95, Ministry of Water Resources, Water and Energy Commission, Katmandou, 1996.

Netherlands Antilles

Sources 1992-1994:

Direct communication to the Secretariat from the Central Bureau of Statistics, Fort Amsterdam, Curaçao.

Nicaragua

Historical series have been considerably revised using direct communication from Instituto Nicaraguense de Energía, Dirección General de Hidrocarburos.

Sources 1971-1997:

Combustible Renewables and Waste:

SIEE, (OLADE).

Direct submission to the Secretariat from the Instituto Nicaraguense de Energía, Dirección General de Hidrocarburos, Managua, 1999.

Informe Annual 1996: Datos Estadisticos del Sector Electrico, INE, Managua, 1999.

Direct submission to the Secretariat from the Electricity Utility.

Nigeria

Direct submission to the Secretariat from the Energy Commission of Nigeria.

Sources 1971-1997:

Combustible Renewables and Waste:

Secretariat estimates based on 1991 data from *Forests and Biomass Sub-sector in Africa,* African Energy Programme of the African Development Bank, Abidjan, 1996.

Sources 1992-1997:

Annual Report and statement of Accounts 1995, Central Bank of Nigeria, Lagos, 1996.

Direct communication from the oil industry. Statistical difference probably includes oil products smuggled into neighbouring countries for consumption.

Nigerian Petroleum News, Energy Publications, monthly reports, various issues up to May 1998.

Sources up to 1991:

Annual Report and Statement of Accounts, Central Bank of Nigeria, Lagos, various editions from 1981 to 1987.

Basic Energy Statistics for Nigeria, Nigerian National Petroleum Corporation, Lagos, 1984.

NNPC Annual Statistical Bulletin, Nigerian National Petroleum Corporation, Lagos, 1983 to 1987.

The Economic and Financial Review, Central Bank of Nigeria, Lagos, various editions.

Oman

Sources 1992-1997:

Direct communication to the Secretariat from the Ministry of Petroleum and Minerals, Muscat, October 1997, October 1998.

Direct communication to the Secretariat from the Ministry of Electricity & Water, Office of the Under Secretary, Ruwi, September 1998.

Quarterly Bulletin December 1994, Central Bank of Oman, Muscat, February 1995.

Annual Report 1992, Central Bank of Oman, Muscat, 1993.

Statistical Yearbook, 1994, 1995, 1996, 1997 Ministry of Development, Muscat, August 1995, August 1996, August 1997 and August 1998.

Sources up to 1991:

Quarterly Bulletin, Central Bank of Oman, Muscat, 1986, 1987, 1989 and 1995.

Annual Report to His Majesty the Sultan of Oman, Department of Information and Public Affairs, Petroleum Development, Muscat, 1981, 1982, and 1984.

Oman Facts and Figures 1986, Directorate General of National Statistics, Development Council, Technical Secretariat, Muscat, 1987.

Quarterly Bulletin on Main Economic Indicators, Directorate General of National Statistics, Muscat, March 1989.

Statistical Yearbook, Directorate General of National Statistics, Development Council, Muscat, 1985, 1986, 1988 and 1992.

Pakistan

Energy statistics are reported on a fiscal year basis.

Sources 1971-1997:

Combustible Renewables and Waste:

Secretariat estimates based on 1991 data from "Household Energy Strategy Study (HESS)" of 1991.

Sources 1992-1997:

Energy Year Book, Ministry of Petroleum and Natural Resources, Directorate General of New and Renewable Energy Resources, Islamabad, various editions from 1979 to 1998.

Pakistan Economic Survey 1994-1995, 1996, 1997, Government of Pakistan, Finance Division, Islamabad, 1995, 1997, 1998.

Statistical Supplement 1993/1994, Finance Division, Economic Adviser's Wing, Government of Pakistan, Islamabad, 1995.

Sources up to 1991:

As above.

Monthly Statistical Bulletin, no. 12, Federal Bureau of Statistics, Islamabad, December 1989.

1986 Bulletin, The State Bank of Pakistan, Islamabad, 1987.

Panama

Sources 1971-1997:

Combustible Renewables and Waste:

SIEE, (OLADE) and "Balance Energetico Nacional."

Sources 1992-1994:

Direct communication to the Secretariat from the Electricity and Hydro Resources Institute, Panama.

Sources up to 1991:

Balance Energetico Nacional, Serie Historica 1970-80, CONADE Peica Programa Energético del Istmo Centro Americano, Comisión Nacional de Energía, Panama, 1981.

Paraguay

The Itaipu hydroelectric plant, operating since 1984 and located on the Paraná river (which forms the border of Brazil and Paraguay) was formed as a joint venture between Eletrobrás and the Paraguayan government. Production is equally shared between Brazil and Paraguay. Consumption in Paraguay accounts for less than 5 per cent of the total power produced, the remaining 45 per cent is exported to Brazil, and has been accounted as such. See Brazil country notes for supply historical data from Itaipu.

Sources 1971-1996:

Combustible Renewables and Waste:

SIEE, (OLADE) and Secretariat estimates.

Sources 1997:

Direct communication to the Secretariat from the Electricity utility.

Sources up to 1991:

Boletin Estadístico no. 316, Banco Central del Paraguay, Departamento de Estudios Económicos, Asuncion, 1984.

Importaciones Por Productos Principales 1964-1986, Banco Central del Paraguay, Asuncion, 1987.

Peru

Sources 1971-1996:

Combustible Renewables and Waste:

SIEE, (OLADE).

Sources 1992-1997:

Direct communication to the Secretariat from the Ministry of Energy and Mines, Lima.

Balance Nacional de Energía 1997, Ministerio de Energía y Minas, Oficina Técnica de Energía, República del Perú.

Sources up to 1991:

Resena Económica, Banco Central de Reserva del Peru, Lima, 1984 and 1985.

Balance Nacional de Energía, Ministerio de Energía y Minas, Oficina Sectorial de Planificación, Lima, various editions from 1978 to 1987.

Annual Report, Petróleos del Peru, Lima, 1983 and 1984.

Estadísticas de las Operaciones Exploración/Producción 1985, Petróleos del Peru, Lima, 1986.

Philippines

Sources 1971-1997:

Combustible Renewables and Waste:

Department of Energy, National Economic and Development Authority and the UN Energy Statistics Database, 1995.

Sources 1996-1997:

Philippine Energy Bulletin 1997, 1998, Department of Energy, Metro Manila, 1998, 1999.

Sources 1992-1995:

Direct communication to the Secretariat from the Office of Energy Affairs, Metro Manila.

Philippine Statistical Yearbook 1977-1983, and *1993*, National Economic and Development Authority, Manila, 1978-1984, and 1994.

The APEC Energy Statistics 1994, Tokyo, October 1996.

The UN Energy Statistics Database, 1995.

Sources up to 1991:

1990 Power Developmemt Program (1990 - 2005), National Power Corporation, Manila, 1990.

Philippine Medium-term Energy Plan 1988 - 1992, Office of Energy Affairs, Manila, 1989.

1985 and *1989 Annual Report*, National Power Corporation, Manila, 1986, 1990.

Philippine Economic Indicators, National Economic and Development Authority, Manila, various editions of 1985.

Accomplishment Report: Energy Self-Reliance 1973-1983, Ministry of Energy, Manila, 1984.

Industrial Energy Profiles 1972-1979, vol. 1-4, Ministry of Energy, Manila, 1980.

National Energy Program, Ministry of Energy, Manila, 1982-1987 and 1986-1990.

Philippine Statistics 1974-1981, Ministry of Energy, Manila, 1982.

Energy Statistics, National Economic and Development Authority, Manila, 1983.

Quarterly Review, Office of Energy Affairs, Manila, various editions.

Qatar

Sources 1971-1997:

Combustible Renewables and Waste:

FAO, Forestry Statistics, 1998 (via Internet).

Sources 1992-1997:

Annual Statistical Abstract, Presidency of the Council of Ministers, Central Statistical Office, Doha, July 1994, July 1995, July 1996, July 1997, July 1998.

The UN Energy Statistics Database, 1995.

Sources up to 1991:

Qatar General Petroleum Corporation 1981-1985, General Petroleum Corporation, Doha, 1986.

Economic Survey of Qatar 1990, Ministry of Economy and Commerce, Department of Economic Affairs, Doha, 1991.

Statistical Report 1987 Electricity & Water, Ministry of Electricity, Doha, 1988.

State of Qatar Seventh Annual Report 1983, Qatar Monetary Agency, Department of Research and Statistics, Doha, 1984.

Romania

See Part I.1, Issues of Data Quality, above.

Sources 1971-1997:

Combustible Renewables and Waste:

UN ECE Energy Questionnaires and Secretariat estimates.

Sources 1992-1997:

UN ECE Energy Questionnaires.

Buletin Statistic de Informare Publica, Comisia Nationala Pentru Statistica, Bucharest, various issues to June 1995.

Renel Information Bulletin, Romanian Electricity Authority, Bucharest, 1990, 1991, 1992, 1993, 1994.

Sources up to 1991:

Anuarul Statistic al Republicii Socialiste Romania, Comisia Nationala Pentru Statistica, Bucharest, 1984, 1985, 1986, 1990 and 1991.

Saudi Arabia

The data include 50 per cent of Neutral Zone output.

Sources 1971-1997:

Combustible Renewables and Waste:

FAO, Forestry Statistics, 1999 (via Internet).

Sources 1992-1997:

Direct submissions from oil industry sources.

A Survey of the Saudi Arabian Oil Industry 1993, Embassy of the United States of America in Riyadh, Riyadh, January 1994.

Electricity Growth and Development in the Kingdom of Saudi Arabia up to the year of 1416H. (1996G.), Ministry of Industry and Electricity, Riyadh, 1997.

Sources up to 1991:

Annual Reports, ARAMCO, various issues.

Petroleum Statistical Bulletin 1983, Ministry of Petroleum and Mineral Resources, Riyadh, 1984.

Achievement of the Development Plans 1970-1984, Ministry of Planning, Riyadh, 1985.

The 1st, 2nd, 3rd and 4th Development Plans, Ministry of Planning, Riyadh, 1970, 1975, 1980 and 1985.

Annual Report, Saudi Arabian Monetary Agency, Research and Statistics Department, Riyadh, 1984, 1985, 1986, 1988 and 1989.

Statistical Summary, Saudi Arabian Monetary Agency, Research and Statistics Department, Riyadh, 1986.

Senegal

Sources 1971-1997:

Combustible Renewables and Waste:

Secretariat estimates based on 1994 data from *Forests and Biomass Sub-sector in Africa,* African Energy Programme of the African Development Bank, Abidjan, 1996, and from direct communication with ENDA, Senegal.

Sources 1992-1997:

Direct communication to the Secretariat from the Ministère de l'Energie, des Mines et de l'Industrie, Direction de l'Energie, Dakar, 1998, 1998.

Report of Senegal on the Inventory of Greenhouse Gases Sources, Ministère de l'Environnement et de la Protection de la Nature, Dakar, 1994.

Direct communication to the Secretariat from ENDA - Energy Program, Dakar, 1997.

Direct communicatons from oil industry sources.

The UN Energy Statistics Database, 1995.

Sources up to 1991:

Situation Economique 1985, Ministère de l'Economie et des Finances, Direction de la Statistique, Senegal, 1986.

Singapore

Official Singapore trade statistics do not show oil trade between Indonesia and Singapore. The quantity of this trade for crude oil has been estimated.

Sources 1971-1997:

Combustible Renewables and Waste:

The UN Energy Statistics Database, 1996.

Sources 1992-1997:

Direct communication to the Secretariat from the Public Utilities Board.

Direct submissions from oil industry sources.

Yearbook of Statistics Singapore 1993, 1995, 1996, Department of Statistics, Singapore, 1994, 1996, 1997.

The Strategist Oil Report, Singapore, various issues up to March 1999.

Singapore Trade Statistics, Department of Statistics, Singapore, various editions from 1985 to 1995.

Petroleum in Singapore 1993/1994, Petroleum Intelligence Weekly, Singapore, 1994.

AEEMTRC, 1996.

Sources up to 1991:

Monthly Digest of Statistics, Department of Statistics, Singapore, various editions from 1987 and 1989.

Yearbook of Statistics Singapore 1975/1985, Department of Statistics, Singapore, 1986.

Asean Oil Movements and Factors Affecting Intra-Asean Oil Trade, Institute of Southeast Asian Studies, Singapore, 1988.

The Changing Structure of the Oil Market and Its Implications for Singapore's Oil Industry, Institute of Southeast Asian Studies, Singapore, 1988.

Public Utilities Board Annual Report (1986 and 1989), Public Utilities Board, Singapore, 1987 and 1990.

Slovak Republic

See Part I.1, Issues of Data Quality, above.

Sources 1971-1997:

Combustible Renewables and Waste:

UN ECE Energy Questionnaires, the UN Energy Statistics Database, 1995 and from the Slovak government as a result of IEA in-depth energy survey.

Sources 1980-1997:

Direct submission from the Power Research Institute (EGU), Bratislava.   Primary source: Statistical Office of the Slovak republic:

1989-1995: complete balances.

1980-1988: electricity and heat balances; primary balances for coal, oil, gas and electricity.

UN ECE Energy Questionnaires 1995 to 1997.

Source 1971-1988:

Statistical Yearbook of the Czechoslovak Socialist Republic, Federal Czech and Slovak Statistical Office, Prague, 1981 to 1991.

South Africa

In the second half of 1995, the IEA undertook an in-depth energy survey of South Africa and the results were published in *Energy Policies of South Africa,* IEA/OECD, Paris, 1996. As a result of the survey, much of the data have been revised and improved. Natural gas production began in 1993; all gas is liquefied and refined into petroleum products. Input of coal to coal liquefaction is shown in the coal column of the balance, and the crude produced appears as an output in the crude column; this crude is included in refinery input.

Sources 1971-1997:

Combustible Renewables and Waste:

South African Energy Statistics 1950-1989, No. 1, National Energy Council, Pretoria, 1989 and Secretariat estimates.

Sources 1993-1997:

Direct submission from the Institute for Energy Studies, Rand Afrikaans University, Pretoria, 1998, 1999.

Digest of South African Energy Statistics 1998, Department of Minerals and Energy, Pretoria, 1999

Direct submissions from the Department of Mineral and Energy Affairs, Pretoria.

Direct submissions from the Energy Research Institute, University of Cape Town.

Eskom Annual Report, Electricity Supply Commission (ESKOM), South Africa, 1989 to 1994.

Statistical Yearbook, Electricity Supply Commission (ESKOM), South Africa, 1983 to 1994.

South Africa's Mineral Industry, Department of Mineral and Energy Affairs, Braamfontein, 1995.

South African Energy Statistics, 1950-1993, Department of Mineral and Energy Affairs, Pretoria, 1995.

Wholesale Trade Sales of Petroleum Products, Central Statistical Service, Pretoria, 1995.

South African Coal Statistics 1994, South African Coal Report, Randburg, 1995.

Energy Balances in South Africa 1970-1993, Energy Research Institute, Plumstead, 1995.

Sources up to 1991:

Statistical News Release 1981-1985, Central Statistical Service, South Africa, various editions from 1986 to 1989.

Annual Report Energy Affairs 1985, Department of Mineral and Energy Affairs, Pretoria, 1986.

Energy Projections for South Africa (1985 Balance), Institute for Energy Studies, Rand Afrikaans University, South Africa, 1986.

Sri Lanka

Sources 1971-1997:

Combustible Renewables and Waste:

Energy Conservation Fund and Ceylon Electricity Board.

Sources 1992-1997:

Sri Lanka Energy Balance 1997, Energy Conservation Fund, Colombo, November 1997.

Annual Report 1993, Central Bank of Sri Lanka, Colombo, July 1994.

Direct communication to the Secretariat from the Ceylon Electricity Board, *Sri Lanka Energy Balances, 1994.*

Sources up to 1991:

Energy Balance Sheet 1991, 1992, Energy Unit, Ceylon Electricity Board, Colombo, 1992, 1993.

Bulletin 1989, Central Bank of Sri Lanka, Colombo, July 1989.

Bulletin (monthly), Central Bank of Sri Lanka, Colombo, May 1992.

Sectoral Energy Demand in Sri Lanka, UNDP Economic and Social Commission for Asia and The Pacific, Bangkok, 1992.

External Trade Statistics 1992, Government of Sri Lanka, Colombo, 1993.

Sudan

Sources 1971-1997:

Combustible Renewables and Waste:

Secretariat estimates based on 1990 data from Bhagavan, M.R., Editor, *Energy Utilities and Institutions in Africa,* AFREPREN, Nairobi, 1996.

Sources 1992-1997:

Direct communication to the Secretariat from the Ministry of Energy and Mines, Khartoum, June 1998, April 1999.

Sources up to 1991:

Foreign Trade Statistical Digest 1990, Government of Sudan, Khartoum, 1991.

Syria

Sources 1971-1997:

Combustible Renewables and Waste:

FAO, Forestry Statistics, 1998 (via Internet) and Secretariat estimates.

Sources 1992-1997:

Statistical Abstract 1992-1997, Office of the Prime Minister, Central Bureau of Statistics, Damascus, 1993, 1996, 1997, 1998.

The UN Energy Statistics Database, 1995.

Sources up to 1991:

Quarterly Bulletin, Central Bank of Syria, Research Department, Damascus, 1984.

Tanzania

Sources 1971-1997:

Combustible Renewables and Waste:

Secretariat estimates based on 1990 data from *Energy Statistics Yearbook 1990,* SADC, Luanda, 1992.

Sources up to 1997:

Direct communication to the Secretariat from the electricity utility.

Tanzanian Economic Trends, Economic Research Bureau, University of Dar-es-Salaam, 1991.

Thailand

Sources 1971-1997:

Combustible Renewables and Waste:

Thailand Energy Situation, Ministry of Science, Technology and Energy, National Energy Administration.

Sources 1992-1997:

Electric Power in Thailand, Ministry of Science, Technology and Energy, National Energy Administration, Bangkok, 1985, 1986, 1988 to 1998.

Oil and Thailand, Ministry of Science, Technology and Energy, National Energy Administration, Bangkok, 1979 to 1998.

Thailand Energy Situation, Ministry of Science, Technology and Energy, National Energy Administration, Bangkok, 1978 to 1998.

Trinidad-and-Tobago

Sources 1971-1997:

Combustible Renewables and Waste:

SIEE, (OLADE).

Sources 1992-1997:

Direct communication to the Secretariat from the Ministry of Energy and Natural Resources, Port of Spain.

Annual Economic Survey 1994, 1995, Central Bank of Trinidad and Tobago, Port of Spain 1995, 1996.

Petroleum Industry Monthly Bulletin, Ministry of Energy and Natural Resources, Port of Spain, various issues to 1998.

Sources up to 1991:

Annual Statistical Digest, Central Statistical Office, Port of Spain, 1983 and 1984.

History And Forecast, Electricity Commission, Port of Spain, 1987.

Annual Report, Ministry of Energy and Natural Resources, Port of Spain, 1985 and 1986.

The National Energy Balances 1979-1983, Ministry of Energy and Natural Resources, Port of Spain, 1984.

Trinidad and Tobago Electricity Commission Annual Report, Trinidad and Tobago Electricity Commission, Port of Spain, 1984 and 1985.

Tunisia

Sources 1971-1997:

Combustible Renewables and Waste:

Secretariat estimates based on 1991 data from *Analyse du Bilan de Bois d'Energie et Identification d'un Plan d'Action,* Ministry of Agriculture, Tunis, 1998.

Sources 1992-1997:

Energy balance and electricity consumption and production data provided by direct submission from the Observatoire National de l'Energie, Agence pour la Maîtrise de l'Energie, Tunis.

Sources up to 1991:

Bilan Energétique de l'Année 1991, Banque Centrale de Tunisie, Tunis, September 1992.

Rapport d'Activité 1990, Observatoire National de l'Energie, Agence pour la Maîtrise de l'Energie, Tunis, 1991.

Rapport Annuel 1990, Banque Centrale de Tunisie, Tunis, 1991.

Activités du Secteur Pétrolier en Tunisie, Banque Centrale de Tunisie, Tunis, 1987.

Statistiques Financières, Banque Centrale de Tunisie, Tunis, 1986.

Entreprise Tunisienne d'Activités Pétrolières (ETAP), Tunis, 1987.

Annuaire Statistique de la Tunisie, Institut National de la Statistique, Ministère du Plan, Tunis, 1985 and 1986.

L'Economie de la Tunisie en Chiffres, Institut National de la Statistique, Tunis, 1984 and 1985.

Activités et Comptes de Gestion, Société Tunisienne de l'Electricité et du Gaz, Tunis, 1987.

United Arab Emirates

Estimates of annual sales for marine bunkers in the facilities offshore of Fujairah in the UAE have been made in consultation with the oil industry.

Sources 1992-1997:

Combustible Renewables and Waste:

FAO, Forestry Statistics, 1999 (via Internet).

Sources 1993-1997:

Statistical Yearbook 1995, Department of Planning, Abu Dhabi, November 1996.

Sources up to 1992:

Abu Dhabi National Oil Company, 1985 Annual Report, Abu Dhabi National Oil Company, Abu Dhabi, 1986.

United Arab Emirates Statistical Review 1981, Ministry of Petroleum and Mineral Resources, Abu Dhabi, 1982.

Annual Statistical Abstract, Ministry of Planning, Central Statistical Department, Abu Dhabi, various editions from 1980 to 1993.

Uruguay

The power produced from the Salto Grande hydroelectric plant, operating since 1980 and located on the Uruguay river (natural border of Argentina and Uruguay), is equally shared between the two countries. Electricity exports include power produced in Salto Grande and exported to Argentina.

Sources 1971-1997:

Combustible Renewables and Waste:

Direccion Nacional de Energia and *SIEE* (OLADE).

Sources 1992-1997:

Balance Energetico Nacional, Ministerio de Industria, Energía y Mineria, Dirección Nacional de Energía, Montevideo, 1992, 1993, 1994, 1995 1996 and 1997.

Sources up to 1991:

UTE Memoria Anual, Administración Nacional de Usinas y Trasmiones Eléctricas, Montevideo, 1978.

Boletín Estadístico 1975-1985, Banco Central del Uruguay, Departemento de Investigaciones Económicas, Montevideo, 1986.

Informaciones y Estadísticas Nacionales e Internationales no. 48, Centro de Estadísticas Nacionales y Comercio Internacional, Montevideo, 1989.

Boletín Mensual Energético, Ministerio de Industria y Energía, Dirección Nacional de Energía, Montevideo, various editions from 1986 to 1987.

Balance Energetico Nacional, Ministerio de Industria, Energía y Mineria, Dirección Nacional de Energía, Montevideo, 1981, 1985, 1986, 1990, 1991.

Former USSR

Coal production statistics refer to unwashed and unscreened coal. IEA coal statistics normally refer to coal after washing and screening for the removal of inorganic matter. Also see notes under 'Classification of Fuel Uses' and 'Heat', above.

The energy balances presented for the former USSR include Secretariat estimates of fuel consumption in the main categories of transformation. These estimates are based on secondary sources and on isolated references in FSU literature.

Sources 1991 to 1997 for NIS:

External trade of the region Former USSR: *PlanEcon Energy Outlook for the Former Soviet Republics,* Washington, June 1995 and 1996 and *PlanEcon Energy Outlook for Eastern Europe and the Former Soviet Republics,* Washington, September 1997 and October 1998.

Foreign Scouting Service, Commonwealth of Independent States, IHS Energy Group – IEDS Petroconsultants, Geneva, April 1999.

Statistical Bulletin, various editions, The State Committee of Statistics of the CIS, Moscow, 1993 and 1994.

External Trade of the Independent Republics and the Baltic States in 1991, The State Committee of Statistics of the CIS, Moscow, 1992.

Statistical Yearbook, 1993, 1994 and 1995, The State Committee of Statistics of the CIS, Moscow, 1994, 1995 and 1996.

Sources up to 1990 for the Former USSR:

Statistical Yearbook, The State Committee for Statistics of the USSR, Moscow, various editions from 1980 to 1989.

External Trade of the Independent Republics and the Baltic States, 1990 and 1991, The State Committee of Statistics of the CIS, Moscow, 1992.

External Trade of the USSR, annual and quarterly, various editions, The State Committee of Statistics of the USSR, Moscow, 1986 to 1990.

External Trade of the USSR, annual and quarterly, various editions, The State Committee of Statistics of the USSR, Moscow, 1986 to 1990.

CIR Staff Paper no. 14, 28, 29, 30, 32 and 36, Center for International Research, U.S. Bureau of the Census, Washington, 1986, 1987 and 1988.

Yearbook on Foreign Trade, The Ministry of Foreign Trade, Moscow, 1986.

Armenia

Data for Combustible renewables and waste: FAO, Forestry Statistics, 1998 (via Internet) and Secretariat estimates.

Direct communications to the Secretariat from the Ministry of Energy and Fuels, 1992 and 1996.

UN ECE Questionnaire on Electricity, 1992 to 1997.

UN ECE Questionnaire on Coal, 1995 and 1996.

Azerbaijan

Data for Combustible renewables and waste: FAO, Forestry Statistics, 1998 (via Internet) and Secretariat estimates.

Direct communications to the Secretariat from the Ministry of Economics, June 1999.

Direct communications to the Secretariat from the State Committee of Statistics and Analysis, May 1993, June 1994, May 1996 and June 1998 .

UN ECE Energy Questionnaires, 1992 to 1994.

UN ECE Questionnaires on Oil, Gas and Electricity 1995 and 1996.

Belarus

Data for Combustible renewables and waste: UN ECE Energy Questionnaires and Secretariat estimates.

Direct communication to the Secretariat from the Ministry of Statistics and Analysis, May 1996.

UN ECE Energy Questionnaires, 1990 to 1997.

Estonia

Data for Combustible renewables and waste: UN ECE Energy Questionnaires and Secretariat estimates.

Up to 1994, process heat from final consumption is included in the transformation sector.

UN ECE Questionnaires (Summary), 1970 to 1990.

UN ECE Questionnaire on Coal, 1990.

Statistical Office of Estonia, *Energy Balances, 1994*.

UN ECE Energy Questionnaires, 1991 to 1997.

Georgia

Data for Combustible renewables and waste. The UN Energy Statistics Database, 1995.

Official energy balance of Georgia 1990-1997, Ministry of Economy and Ministry of Energy, Tbilissi, November 1998.

Energy outlook of Georgia 1998-2010, draft, Georgian Energy Research Institute, Tbilissi.

Energy outlook of Georgia 1993-2015, Georgian Public Institute Techinform, Tbilissi.

Direct communication to the Secretariat from the Socio-Economical Information Committee, 1992.

Kazakhstan

Data for Combustible renewables and waste: FAO, Forestry Statistics, 1998 (via Internet).

Direct communication to the Secretariat from the State Committee of Statistics and Analysis, August 1994.

UN ECE Energy Questionnaires, 1993 to 1995.

Kyrgyzstan

Data for Combustible renewables and waste: The UN Energy Statistics Database, 1995.

Direct communication to the Secretariat, 1992.

UN ECE Energy Questionnaires, 1993 to 1996.

Latvia

Data for Combustible renewables and waste: UN ECE Energy Questionnaires and Secretariat estimates.

Balance of Latvian Energy Sources 1991 and 1992, EC PHARE Project Implementation Unit, Riga, 1994.

UN ECE Energy Questionnaire on Coal, 1993 to 1997.

UN ECE Questionnaire on Natural Gas, 1992 to 1997.

UN ECE Questionnaire on Electricity and Heat, 1993 to 1997.

UN ECE Questionnaire on Oil 1995 to 1997.

Lithuania

Data for Combustible renewables and waste: UN ECE Energy Questionnaires and Secretariat estimates.

Direct communications to the Secretariat of the Energy Agency, February 1994 and May 1996.

UN ECE Energy Questionnaires, 1992 to 1997.

Balances of electricity, heat, fuel and energy in Lithuania 1991 – 1993, Lithuanian Ministry of Energy Agency, Vilnius.

Lithuania Power Demand and Supply Options, The World Bank, Washington, 1993.

Energy Conservation Potentials in Lithuania and Latvia, Riso National Laboratory, Roskilde, 1992.

Energy in Lithuania (Power, Heat and Fuel Balances 1980 - 1992), Lithuanian Energy Institute, Vilnius, 1993.

Moldova

Data for Combustible renewables and waste: The UN Energy Statistics Database, 1995.

Direct communication to the Secretariat from the Ministry of Industry and Energy, July 1992.

UN ECE Questionnaire on Electricity and Heat, 1991 to 1997.

UN ECE Questionnaire on Coal, 1992 to 1997.

UN ECE Questionnaire on Oil, 1993 to 1997.

UN ECE Questionnaire on Natural Gas, 1991 to 1997.

Russia

Data for Combustible renewables and waste: The State Committee of Statistics of Russian Federation and Secretariat estimates.

Process heat from final consumption is included in the transformation sector.

Oil and Natural Gas: Direct communication to the Secretariat from the State Committee of Statistics of Russia, May 1995.

Energy trade: Direct communication to the Secretariat from the State Committee of Statistics of Russia, July 1994.

UN ECE Questionnaire on Coal, 1992 to 1996.

UN ECE Questionnaire on Natural Gas, 1991 to 1996.

UN ECE Questionnaire on Electricity and Heat, 1991 to 1996.

UN ECE Questionnaires on Oil, 1991 to 1996.

Statistical Yearbook of Russia 1994, 1997 and 1998, The State Committee of Statistics, Moscow, 1994, 1997 and 1998.

The Russian Federation in 1992, Statistical Yearbook, The State Committee of Statistics of Russia, Moscow, 1993.

Russian Federation External Trade, annual and quarterly various editions, The State Committee of Statistics of Russia, Moscow.

Statistical Bulletin, various editions, The State Committee of Statistics of the CIS, Moscow, 1993 and 1994.

Statistical Bulletin n° 3, The State Committee of Statistics of Russia, Moscow, 1992.

Fuel and Energy Balance of Russia 1990, The State Committee of Statistics of Russia, Moscow, 1991.

Direct communication to the Secretariat, June 1998.

Energetika, Energo-Atomisdat, Moscow, 1981 and 1987.

Tajikistan

UN ECE Questionnaire on Coal, 1994 and 1995.

UN ECE Questionnaire on Natural Gas, 1994 and 1995.

UN ECE Questionnaire on Electricity and Heat, 1994 and 1995.

UN ECE Questionnaire on Oil, 1994.

Ukraine

Coal production statistics refer to unwashed and unscreened coal.

Data for Combustible renewables and waste: Statistical Office in Kiev, World Bank and Secretariat estimates.

Direct communication to the Secretariat from the Ministry of Statistics, the Coal Ministry, the National Dispatching Company, November 1995.

Coal: Direct communications to the Secretariat from the State Mining University of Ukraine, May 1995 and 1996.

Natural Gas: Direct communication to the Secretariat from Ukrgazprom, February 1995.

Direct communication to the Secretariat from the Ministry of Statistics of the Ukraine, July 1994.

UN ECE Questionnaire on Coal, 1991, 1992 and 1995 to 1997.

UN ECE Questionnaire on Oil, 1992 and 1995 to 1997.

UN ECE Questionnaire on Natural Gas, 1992 and 1995 to 1997.

UN ECE Questionnaire on Electricity and Heat, 1991, 1992, and 1994 to 1997.

Ukraine in 1992, Statistical Handbook, Ministry of Statistics of the Ukraine, Kiev, 1993.

Ukraine Power Demand and Supply Options, The World Bank, Washington, 1993.

Power Industry in Ukraine, Ministry of Power and Electrification, Kiev, 1994.

Energy Issues Paper, Ministry of Economy, March 1995.

Ukraine Energy Sector Statistical Review 1993, 1994, 1995, 1996 and 1997, World Bank Regional Office, Kiev, 1994, 1995, 1996, 1997 and 1998.

Global Energy Saving Strategy for Ukraine, Commission of the European Communities, TACIS, Madrid, July 1995.

Uzbekistan

Up to 1994, crude oil includes NGL.

Direct communications to the Secretariat from the Institute of Power Engineering and Automation, March 1994 and June 1996.

UN ECE Questionnaires 1995 to 1997.

Venezuela

Sources 1971-1997:

Combustible Renewables and Waste:

The UN Energy Statistics Database, 1996.

Sources 1992-1997:

Direct communication to the Secretariat from the Ministry of Energy and Mines.

Petróleo y Otros Datos Estadísticos, Dirección General Sectorial de Hidrocarburos, Dirección de Economía de Hidrocarburos, Caracas, 1993, 1994, 1995 and 1997.

Sources up to 1991:

Petróleo y Otros Datos Estadísticos, Dirección General Sectorial de Hidrocarburos, Dirección de Planificación y Economía de Hidrocarburos, Caracas, 1983 to 1991.

Balance Energetico Consolidado de Venezuela 1970-1984, Ministerio de Energía y Minas, Dirección General Sectorial de Energía, División de Programación Energetica, Caracas, 1986.

Compendio Estadístico del Sector Eléctrico, Ministerio de Energía y Minas, Dirección de Electricidad, Carbón y Otras Energías, Caracas, 1984, 1989, 1990 and 1991.

Memoria Y Cuenta, Ministerio de Energía y Minas, Caracas, 1991.

Petróleos de Venezuela S.A. 1985 Annual Report, Petróleos de Venezuela, Caracas, 1991.

Vietnam

Sources 1971-1997:

Combustible Renewables and Waste:

Secretariat estimates based on 1992 data from *Vietnam Rural and Household Energy Issues and Options: Report No. 161/94,* World Bank, ESMAP, Washington, D.C., 1994.

Direct communications to the Secretariat from the Center for Energy-Environment Research and Development, Pathumthami, 1997, 1998 and 1999.

Data were supplied by RWEDP in Asia, Bangkok, Thailand.

Direct communication from the Oil Industry.

Sectoral Energy Demand in Vietnam, UNDP Economic and Social Commission for Asia and The Pacific, Bangkok, 1992.

Energy Commodity Account of Vietnam 1992, Asian Development Bank, Manila, 1994.

World Economic Problems No 2 (20), National Centre for Social Sciences of the S.R. Vietnam, Institute of World Economy, Hanoi, 1993.

Vietnam Energy Review, Institute of Energy, Hanoï, 1995, 1997, 1998.

Yemen

Sources 1971-1996:

Combustible Renewables and Waste:

The UN Energy Statistics Database, 1995 and Secretariat estimates.

Sources 1992-1997:

Statistical Yearbook 1993, 1994,1995, 1996, 1997, Ministry of Planning and Development, Central Statistical Organization, Republic of Yemen, Yemen, 1994, 1995, 1996 and 1997.

Statistical Indicators in the Electricity Sector, Ministry of Planning and Development, Central Statistical Organization, Republic of Yemen, Yemen, 1993.

Sources up to 1991:

Statistical Yearbook, Government of Yemen Arab Republic, Yemen, 1988.

Former Yugoslavia

From 1992, the energy balance for former Yugoslavia has been prepared by adding together the basic statistics of it's component countries: Croatia, Slovenia, Former Yugoslav Republic of Macedonia, Bosnia-Herzegovina, and the Federal Republic of Yugoslavia. In the case of the latter two, energy data beyond basic production flows are extremely poor. Trade between the component countries has been discounted. Oil imports into Former Yugoslavia breaking the UN embargo of Serbia & Montenegro are theoretically included, although the absolute levels of this trade are not known with any certainty. As such the balance for former Yugoslavia should be seen as an estimation that is necessary for completing regional aggregates.

Sources up to 1991:

Statisticki Godisnjak Yugoslavije, Socijalisticka Federativna Rebublika Jugoslavija, Savezni Zavod Za Statistiku, Beograd, 1985 to 1991.

Indeks, Socijalisticka Federativna Rebublika Jugoslavija, Beograd, 1990, 1991, 1992.

Bosnia-Herzegovina

Combustible Renewables and Waste: The UN Energy Statistics Database, 1995.

UN ECE Energy Questionnaires 1993, 1994.

Croatia

UN ECE Energy Questionnaires 1993 to 1997.

Former Yugoslav Republic of Macedonia (FYROM)

Combustible Renewables and Waste: The UN Energy Statistics Database, 1995 and FAO, Forestry Statistics, 1998 (via Internet).

UN ECE Energy Questionnaires, 1993 to 1995.

Slovenia

UN ECE Energy Questionnaires 1993 to 1997.

Zambia

Sources 1971-1997:

Combustible Renewables and Waste:

Secretariat estimates based on 1991 data from *Forests and Biomass Sub-sector in Africa,* African Energy Programme of the African Development Bank, Abidjan, 1996.

Sources 1992-1996:

Direction Communication to the Secretariat from oil sources.

Annual Statistical Yearbook 1993,1994 and 1995 (*Consumption in Zambia 1978-1983),* Eskom, Lusaka, 1984.

Zimbabwe

Sources 1971-1997:

Combustible Renewables and Waste:

Secretariat estimates based on 1991 data from *Forests and Biomass Sub-sector in Africa,* African Energy Programme of the African Development Bank, Abidjan, 1996.

Sources 1996-1997:

Direct communication to the Secretariat from the Ministry of Environment and Tourism, Harare, 1999.

Direct communication to the Secretariat from the electricity utility.

Electricity Statistics Information, Central Statistical Office, Causeway, February 1998.

Sources 1992-1995:

Eskom Annual Statistical Yearbook 1993, 1995 and 1995, Johannesburg, 1994, 1995, 1996, citing Zimbabwe Electricity Supply Authority, Harare as source.

The UN Energy Statistics Database, 1995.

Sources up to 1991:

Zimbabwe Statistical Yearbook 1986, Central Statistical Office, Harare, 1990.

Quarterly Digest of Statistics, Central Statistical Office, Harare, 1990.

Zimbabwe Electricity Supply Authority Annual Report, Zimbabwe Electricity Supply Authority, Harare, 1986 to 1991.

Other Africa, Other Latin America, and Other Asia and Oceania

The series for these 'sum' countries are made up by simple addition of their component countries. Intra-component country trade is therefore included as part of total trade, although to truly represent trade to and from the grouping, intra-component country movements should be discounted. Trade is therefore likely to be overstated.

1. DE LA QUALITE DES DONNEES

A. Méthodologie

Des efforts considérables ont été déployés pour veiller à ce que les données présentées dans cette publication correspondent aux définitions de l'AIE figurant dans les Notes générales (parties I.2). Ces définitions sont utilisées par la plupart des organisations internationales qui recueillent des statistiques sur l'énergie. Toutefois, les statistiques nationales sur l'énergie communiquées à ces organisations sont souvent fondées sur des critères et définitions qui diffèrent, parfois sensiblement, de ceux qu'elles emploient. On trouvera ci-après des informations sur les différences repérées par le Secrétariat de l'AIE et sur les ajustements qu'il a pu, le cas échéant, opérer pour faire correspondre les données aux définitions internationales. Les différences notoires relevées dans certains pays sont signalées dans la partie I, 5 Notes et sources par pays. Ces notes indiquent seulement quelques-unes des déviations les plus importantes et les plus évidentes constatées dans certains pays par rapport à la méthodologie de l'AIE, et ne constituent en aucun cas une liste exhaustive des différences et anomalies par pays.

B. Estimation

Outre les ajustements opérés pour compenser les différences de définition, des estimations se révèlent parfois nécessaires pour compléter certains agrégats importants pour lesquels il manque des statistiques essentielles. On en trouvera des exemples dans les notes plus détaillées proposées ci-dessous. Excepté

en ce qui concerne les combustibles renouvenables et les déchets, le Secrétariat s'est attaché à fournir tous les éléments des bilans par produit jusqu'à la Consommation finale totale pour tous les pays et toutes les années. Il a fallu pour cela indiquer les éléments de l'offre ainsi que les entrées de combustibles primaires et les sorties de combustibles secondaires correspondant aux principales activités de transformation telles que le raffinage du pétrole et la production d'électricité. Des estimations ont souvent dû être établies, après consultation d'offices statistiques nationaux, de compagnies pétrolières, de compagnies d'électricité et d'experts nationaux en matière d'énergie.

Le Secrétariat fournit également pour tous les pays et toutes les années des indicateurs énergétiques calculés à l'aide de données de PIB et de population. Lorsque ces données économiques n'étaient pas disponibles auprès de sources internationales officielles, le Secrétariat a utilisé les estimations fournies par la base de données CHELEM-CEPII (voir I.5 Notes et sources par pays, Références générales et I.2.E Indicateurs).

C. Séries chronologiques et changements politiques

Les bilans énergétiques des républiques de l'ex-URSS ont été établis à partir de 1992. Ces bilans ont été établis à l'aide de données officielles et, lorsque cela a été nécessaire, des estimations ont été effectuées à partir de données obtenues auprès de diverses industries et d'autres organisations internationales. Il est à noter que les séries chronologiques et les bilans énergétiques de la région "ex-URSS" continuent à être publiés. En effet tant qu'il demeurera difficile de collecter de façon cohérente et homogène des données pour tous les pays de l'ex-URSS, l'agrégat régional servira de point de comparaison avec la situation antérieure et d'indicateur d'adaptation au nouvel environnement.

Les bilans énergétiques des pays de l'ex-République yougoslave apparaissent également dans cette publication. Ils ont été établis à partir de données officielles. Ici encore, pour ne pas interrompre les séries temporelles, les chiffres de "l'ex-Yougoslavie" continuent à être publiés.

Les bilans énergétiques de la République slovaque ont été établis à partir de 1971. Pour les années 1989 à 1997, les bilans de la République slovaque ont été établis à partir de données officielles. Pour les années 1971 à 1988, des estimations ont été faites lorsque c'était nécessaire (voir Notes et sources par pays).

Les statistiques énergétiques de certains pays ont subi diverses modifications soit au niveau des données prises en compte soit de la méthodologie utilisée, de sorte que des données énergétiques plus détaillées ont pu être présentées, par exemple, pour la Chine à partir de 1980. En conséquence, certaines "ruptures de séries" apparaissent inévitables.

Le Secrétariat de l'AIE réexamine ses bases de données chaque année. A la lumière des nouvelles évaluations, d'importantes révisions sont apportées aux séries chronologiques des pays au cours de cet examen. En conséquence, les données de la présente publication ont été sensiblement révisées par rapport aux précédentes éditions.

D. Classification des utilisations des combustibles

Les sources statistiques nationales manquent souvent d'informations adéquates concernant la consommation de combustibles suivant les différentes catégories d'utilisations finales. Beaucoup de pays ne font pas d'enquêtes annuelles de la consommation de combustibles dans les principaux secteurs de l'économie, c'est pourquoi les données publiées sont fondées sur des enquêtes anciennes. La ventilation par secteurs de la consommation dans les différents pays doit donc être interprétée avec prudence.

Avant les réformes entreprises dans les années 1990, la classification sectorielle de la consommation de combustibles dans les économies en transition (d'Europe de l'Est et de l'ex-URSS) et en Chine était très différente de celle pratiquée dans les économies de marché. La consommation sectorielle était définie d'après la branche de l'économie à laquelle appartenait l'utilisateur du combustible et non d'après la vocation ou l'usage du combustible. Par exemple, la consommation d'essence des véhicules d'une entreprise appartenant au secteur de la sidérurgie faisait partie de la consommation industrielle d'essence du secteur de la sidérurgie. Les données ont été ajustées dans la mesure du possible pour correspondre aux classifications internationales: par exemple, la totalité de la consommation d'essence est supposée relever du secteur des transports et est reportée comme telle dans la présente publication. Cependant, il n'a pas été possible de reclasser tous les produits aussi facilement que l'essence et le carburéacteur, et peu d'ajustements ont été opérés sur les autres produits.

E. Problèmes particuliers par combustible

1. Charbon

Les données relatives à la consommation de charbon par secteur sont généralement indiquées en tonnes et le contenu calorifique des différentes qualités de charbon utilisées dans les divers secteurs d'utilisation finale n'est souvent pas connu. Dans ce cas, le Secrétariat procède à une estimation à partir du pouvoir calorifique connu du charbon dans le cas de la production, des importations et des exportations de charbon.

2. Pétrole

Le Secrétariat de l'AIE recueille des statistiques très complètes concernant l'offre et la consommation de pétrole, y compris l'autoconsommation des raffineries, les soutages maritimes et le pétrole utilisé en pétrochimie. Les statistiques nationales font rarement état de ces chiffres. La production indiquée de produits raffinés correspond souvent à la production nette, et non brute, des raffineries et la consommation de produits pétroliers se limite parfois aux ventes sur les marchés intérieurs et ne comprend pas les livraisons aux transports internationaux maritimes ou aériens. Le pétrole utilisé comme produit de base en pétrochimie dans des complexes intégrés de raffinage/pétrochimie n'est souvent pas comptabilisé dans les statistiques officielles disponibles.

Lorsque cela est possible, le Secrétariat procède, en consultation avec l'industrie pétrolière, à une estimation des données non communiquées. En l'absence de toute autre indication sur l'utilisation de combustibles par les raffineries, la consommation des raffineries a été estimée à environ 5 pour cent des entrées en raffinerie et divisée à parts égales entre le gaz de raffinerie et le fioul lourd.

Pour une description des ajustements apportés à la consommation sectorielle de produits pétroliers, se reporter au paragraphe ci-devant intitulé "Classification des utilisations des combustibles".

3. Gaz naturel

L'AIE définit la production de gaz naturel comme la production commercialisable, c'est-à-dire nette des pertes à l'extraction et des quantités brûlées à la torchère, rejetées et réinjectées. De plus, le gaz naturel doit être constitué essentiellement de méthane, et les autres gaz tels que l'éthane et les hydrocarbures à chaîne plus longue doivent être comptabilisés dans la rubrique "pétrole".

Toutefois, les données directement disponibles sur le gaz naturel n'indiquent souvent pas le détail des pertes à l'extraction ainsi que des quantités brûlées à la torchère, rejetées et réinjectées. Les données communiquées sont rarement accompagnées de définitions adéquates, c'est pourquoi il est difficile, voire impossible, d'identifier le gaz aux différents stades de sa séparation en gaz sec (méthane) et en fractions plus lourdes dans les installations de séparation.

Les statistiques concernant l'offre et la demande de gaz naturel sont généralement exprimées en unités de volume et il est difficile d'obtenir des données précises sur le pouvoir calorifique du gaz. Lorsque ce chiffre n'est pas connu, un pouvoir calorifique supérieur de 38 TJ/million de m3 a été appliqué.

Signalons par ailleurs qu'il est souvent difficile d'obtenir des données fiables sur la ventilation de la consommation de gaz naturel. Cette constatation s'applique tout particulièrement à certains des plus grands pays consommateurs de gaz naturel du Moyen-Orient. Pour cette raison, la consommation industrielle de gaz naturel de ces pays n'est souvent pas indiquée dans les données figurant dans la présente publication.

4. Electricité

Comme l'indiquent les Notes explicatives (partie I, 3), on appelle autoproducteur d'électricité une entreprise qui, en plus de ses activités principales, produit de l'énergie électrique destinée en totalité ou en partie à couvrir ses besoins propres. Les pays non membres de l'OCDE disposent rarement de données répondant à cette définition. Pour ces pays, les entrées de combustibles correspondant à l'autoproduction d'électricité sont indiquées dans la rubrique utilisation finale industrielle.

Lorsqu'il existe des statistiques concernant la production d'électricité à partir d'énergies renouvelables combustibles et de déchets, elles apparaissent dans la colonne correspondante. Ces données ne sont pas complètes : par exemple, l'électricité produite à partir des déchets végétaux des sucreries n'est généralement pas indiquée.

Les entrées de combustibles utilisés pour produire de l'électricité sont estimées (lorsqu'elles ne sont pas communiquées) à l'aide de données sur la production d'électricité, le rendement énergétique et le type d'installation de production.

5. Chaleur

Les économies en transition (d'Europe de l'Est et de l'ex-URSS) et la Chine utilisaient, s'agissant de la chaleur, une méthodologie qui les démarquait des pratiques communément adoptées dans les économies de marché. Cette méthodologie consistait à comptabiliser la transformation par l'industrie de combustibles primaires (charbon, pétrole et gaz) en chaleur destinée à la *consommation sur place* dans l'activité de transformation Production de chaleur et non dans la consommation industrielle, comme le prévoit la méthodologie de l'AIE[1.] La production de

---

1   La méthodologie internationale ne comptabilise la chaleur dans le secteur de transformation que dans la mesure où elle est vendue à des tiers. Voir la définition dans la partie I,2.

Chaleur était ensuite répartie entre les différents secteurs d'utilisation finale. Les pertes survenant pendant la transformation des combustibles en chaleur par l'industrie n'étaient pas comptabilisées dans la consommation industrielle finale. Bien que plusieurs pays aient récemment adopté la pratique des organisations internationales, cette différence non négligeable limite la validité des comparaisons entre les pays en transition et les économies de marché pour ce qui est de la consommation finale par secteur.

6. Energies renouvelables combustibles et déchets

L'AIE publie les données de production et d'approvisionnement intérieur en énergies renouvelables combustibles et en déchets de tous les pays non-membres de l'OCDE et toutes les régions pour les années 1974 à 1997 (1971 à 1997 sur disquettes).

Ces données proviennent souvent de sources secondaires, elles ne sont pas harmonisées et leur qualité est douteuse, ce qui rend difficile les comparaisons entre les pays. Les données historiques de nombreux pays proviennent d'enquêtes souvent irrégulières, non harmonisées et menées à un niveau local plutôt que national. Elles ne sont donc comparables ni entre régions ni dans le temps. Lorsque des séries chronologiques étaient incomplètes ou non disponibles, le Secrétariat a procédé à l'estimation des données selon une méthodologie compatible avec le cadre prévisionnel de l'édition 1998 de l'ouvrage de l'AIE intitulé *World Energy Outlook* (septembre 1998). Premièrement, l'approvisionnement intérieur par habitant de biomasse et de déchets a été calculé ou estimé pour 1995. Deuxièmement, l'approvisionnement par habitant a été estimé pour les années 1971 à 1994 en utilisant une équation log/log avec comme variable exogène soit le PIB par habitant, soit le taux d'urbanisation, selon la région. Troisièmement, l'approvisionnement total de biomasse et de déchets pour les années 1996 et 1997 a été estimé en supposant un taux de croissance soit constant, soit égal à la croissance de la population, soit conforme à l'estimation sur la période 1971-1994. Cependant, ces séries historiques estimées doivent être traitées avec prudence.

Le tableau ci-dessous fournit des indications sommaires sur la qualité des données et la méthode d'estimation par région.

Région	Principales sources de données	Qualité des données	Variable exogène
Afrique	base de données FAO et BAfD	Faible	taux d'urbanisation
Amérique latine	nationales et OLADE	Élevée	Aucune
Asie	enquêtes	Élevée à faible	PIB per habitant
Europe non-OCDE	questionnaires et FAO	Élevée à moyenne	Aucune
ex-USSR	questionnaires nationales et FAO	Élevée à moyenne	Aucune
Moyen-Orient	FAO	Moyenne à faible	Aucune

BAfD: Banque Africaine de Développement.

Bien qu'un soin tout particulier ait été apporté à la compatibilité des estimations avec celles réalisées dans la partie biomasse du *World Energy Outlook 1998*, des différences mineures subsistent pour certaines raisons. Parmi ces raisons, il convient de souligner des révisions plus tardives des données utilisées ainsi que des niveaux différents de désagrégations par pays.

Pour les années 1994 à 1997, la consommation par usage a été compilée pour tous les pays. Les énergies renouvelables combustibles et les déchets sont présentés dans un format identique à celui du bilan énergétique des énergies fossiles et renouvelables non combustibles pour les années 1996 et 1997. Les valeurs confirment l'importance des énergies d'origine végétale dans le secteur énergétique de nombreux pays en voie de développement.

L'AIE espère que l'inclusion de ces données va encourager les administrations et les autres agences spécialisées à améliorer la qualité et étendre la couverture de la collecte des données. Plus de détails sur la méthode suivie par pays peuvent être obtenus sur demande, et les commentaires sont bienvenus.

2. NOTES EXPLICATIVES

A. Unité

L'unité adoptée par l'AIE pour présenter les approvisionnements totaux en énergie primaire (ATEP) est la tonne d'équivalent pétrole (tep) definie comme étant égale à 10^7 kilocalories (41,868 gigajoules). Cette quantité d'énergie est équivalente, à quelques points de pourcentage près, au pouvoir calorifique inférieur d'une tonne de pétrole brut. Tout au long de cette publication, une tonne signifie une tonne métrique, soit 1000 kg.

B. Conversion (des unités d'origine en tep)

La conversion de l'unité d'origine en tonnes d'équivalent pétrole suppose le choix de coefficients d'équivalence entre les différentes formes et sources d'énergie. Il existe de nombreuses solutions à ce problème. On pourrait notamment adopter une seule équivalence pour chaque grande source d'énergie primaire dans tous les pays, par exemple 29 307 kJ/kg (7 000 kcal/kg) pour la houille, 41 868 kJ/kg (10 000 kcal/kg) pour le pétrole, etc. La principale objection que l'on peut opposer à cette méthode est qu'elle aboutit à des distorsions, car il existe de grandes différences entre les pouvoirs calorifiques des diverses catégories de charbon et de produits dérivés du charbon, ainsi qu'entre les pouvoirs calorifiques de ces combustibles selon les pays. Le Secrétariat a donc adopté les coefficients spécifiques communiqués par les administrations nationales pour les principales catégories de chaque

qualité de charbon et pour chaque flux ou utilisation (c'est-à-dire la production, les importations, les exportations, la production d'électricité, les cokeries, les hauts-fourneaux et l'industrie). Dans le cas du pétrole brut, les coefficients spécifiques utilisés ont fait l'objet de consultations avec les experts des administrations nationales. Les coefficients de conversion utilisés pour les produits pétroliers sont les mêmes pour tous les pays (une liste des exceptions est fournie dans la section 3). Les données relatives au gaz figurant dans la publication *Statistiques de l'énergie des pays de l'OCDE* sont exprimées en térajoules et fondées sur **le pouvoir calorifique supérieur**. Les données concernant les *énergies renouvelables combustibles et les déchets* sont fournies en térajoules et fondées sur **le pouvoir calorifique inférieur**. Voir également la section 3, Unités et coefficients de conversion.

Les bilans sont exprimés en pouvoir calorifique inférieur (PCI). Pour chaque combustible, la différence entre le pouvoir calorifique inférieur et le pouvoir calorifique supérieur correspond à la chaleur latente de vaporisation de la vapeur d'eau produite pendant la combustion. Pour le charbon et le pétrole, le pouvoir calorifique inférieur représente environ 5 pour cent de moins que le pouvoir calorifique supérieur et, pour la plupart des types de gaz naturel ou manufacturé, la différence est de 9 - 10 pour cent, tandis que, pour l'électricité, il n'y a pas de différence, la notion correspondante n'ayant alors aucune signification. L'emploi du pouvoir calorifique inférieur est conforme à la pratique des Bureaux de statistiques des Communautés Européennes et des Nations Unies.

Les données relatives à l'électricité sont fournies initialement en gigawatts-heure et sont converties en millions de tonnes d'équivalent pétrole au moyen de la relation suivante : 1 térawatt-heure = 0,086 Mtep.

C. Conventions sur l'énergie primaire

La construction d'un bilan énergétique nécessite l'adoption de conventions sur l'énergie primaire relatives à plusieurs sources d'énergie, et notamment à l'énergie nucléaire, géothermique, solaire, hydraulique, éolienne, etc. Deux hypothèses doivent être posées :

I. Le choix de la forme d'énergie primaire

Pour chacune des sources d'énergie, il convient de définir la forme d'énergie primaire à prendre en compte; par exemple dans le cas de l'énergie hydraulique, le choix doit être fait entre l'énergie cinétique de la chute d'eau et l'électricité produite. Dans le cas de l'énergie nucléaire, le choix est entre le contenu énergétique du combustible nucléaire, la chaleur produite dans les réacteurs et l'électricité produite. Dans le cas de l'électricité photovoltaïque, le choix est entre le rayonnement solaire capté et l'électricité produite.

Le principe adopté par l'AIE est que la forme d'énergie primaire à prendre en compte doit être la première forme d'énergie rencontrée au cours du processus de production pour laquelle il existe plusieurs usages énergétiques possibles. L'application de ce principe conduit au choix des formes d'énergie primaire suivantes :

- **la chaleur** pour la chaleur et l'électricité d'origine nucléaire, la chaleur et l'électricité d'origine géothermique et la chaleur d'origine solaire,
- **l'électricité** pour l'électricité d'origine hydraulique, éolienne, marémotrice et des océans ainsi que solaire photovoltaïque.

II. Le calcul de l'équivalence de l'énergie primaire

Il existe essentiellement deux méthodes de calcul de l'équivalence de l'énergie primaire des sources d'énergie citées plus haut: la méthode de la substitution partielle et celle du contenu énergétique.

- *La méthode de la substitution partielle :* dans cette méthode, l'équivalent en énergie primaire des sources de production d'électricité citées plus haut est représentée par la quantité d'énergie qui serait nécessaire pour produire la même quantité d'électricité dans une centrale thermique conventionnelle. L'équivalent en énergie primaire est calculé sur la base du rendement de production moyen de ce type de centrale. Parmi les limitations inhérentes à cette méthode il convient de citer la difficulté de choisir un rendement approprié et l'inadaptation de cette méthode aux pays dont la production d'électricité provient pour une large part de l'hydraulique. Pour ces raisons, l'AIE, à l'instar de la plupart des organisations internationales, n'utilise plus cette méthode et a adopté la méthode du contenu énergétique.

- *La méthode du contenu énergétique :* cette méthode consiste à comptabiliser l'équivalence en énergie primaire d'une source d'énergie en utilisant le contenu énergétique de la forme d'énergie primaire retenue. Il existe par conséquent un lien évident entre les principes adoptés pour le choix de la forme d'énergie primaire d'une source d'énergie et son équivalence en énergie primaire. Par exemple, dans le cas de la production d'électricité d'origine nucléaire, la chaleur étant la forme d'énergie primaire retenue par l'AIE, l'équivalent en énergie primaire est la quantité de chaleur produite par les réacteurs. Cependant, comme la production de chaleur n'est pas toujours connue, l'AIE estime l'équivalent en énergie primaire à partir de la production d'électricité, en appliquant un rendement de 33% (rendement moyen des centrales nucléaires en Europe). Dans le cas de l'énergie hydraulique, l'électricité étant la forme d'énergie primaire choisie, l'équivalent en énergie primaire est le contenu énergétique de l'électricité produite dans la centrale, ce qui revient à supposer un rendement de 100%. Une présentation plus détaillée des hypothèses utilisées par l'AIE pour la construction de ces bilans énergétiques est donnée en section 3.

En raison des différences significatives de traitement de l'électricité d'origine solaire, hydraulique, éolienne, etc. dans ces deux types de bilans énergétiques, la part des énergies renouvelables dans le total des approvisionnements en énergie peut faire apparaître des différences sensibles en fonction des méthodes utilisées. Il est par conséquent essentiel, lors de l'examen des contributions des diverses sources d'énergie dans l'approvisionnement total de comprendre les conventions implicites qui ont été utilisées pour calculer les bilans énergétiques primaires.

D. Bilans énergetiques

Les bilans énergétiques sont présentés sous forme de tableaux avec, en colonnes, les diverses sources d'énergie, et en lignes, les différentes origines et utilisations.

La description ci-dessous concerne les bilans complets. Dans le cas des bilans régionaux et des bilans agrégés les définitions des produits sont les mêmes que pour les bilans complets. Cependant, la ventilation de la consommation finale est réduite aux principaux totaux sectoriels (par exemple Secteur de l'industrie) et aux principales catégories de consommation (par exemple Industrie sidérurgique).

I. Colonnes

En haut du tableau, et de gauche à droite, on trouve onze colonnes avec les titres suivants :

Colonne 1 : *Charbon* - Comprend tous les charbons (y compris la houille et le lignite) et les produits dérivés (y compris les agglomérés, le coke de four à coke, le coke de gaz, les briquettes de lignite, le gaz de cokerie et le gaz de haut fourneau). La tourbe entre également dans cette catégorie.

Colonne 2 : *Pétrole brut* - Comprend le pétrole brut, les LGN, les produits d'alimentation des raffineries et les additifs ainsi que les autres hydrocarbures tels que le pétrole brut synthétique, les huiles minérales extraites des roches bitumineuses et le pétrole provenant de la liquéfaction du charbon et du gaz.

Colonne 3 : *Produits pétroliers* - Comprennent les gaz de raffineries, l'éthane, les gaz de pétrole liquéfiés, l'essence aviation, l'essence moteur, les carburéacteurs, le kérosène, le gazole/carburant diesel, le fioul lourd, les naphtas, le white spirit, les lubrifiants, le bitume, les paraffines, le coke de pétrole et autres produits pétroliers.

Colonne 4 : *Gaz* - Comprend le gaz naturel (à l'exception des LGN) et le gaz d'usine à gaz. Ce dernier est comptabilisé dans le tableau, affecté d'un signe positif, à la ligne "usines à gaz", mais il n'entre pas dans la production nationale.

Colonne 5 : *Nucléaire* - Indique l'équivalent en énergie primaire de l'électricité produite par les centrales nucléaires, sur la base d'un rendement thermique moyen de 33 pour cent. Voir les sections 2.C et 3.

Colonne 6 : *Hydraulique* - Indique l'équivalent en énergie primaire de l'électricité produite par les centrales hydroélectriques. La production d'énergie hydraulique *ne comprend pas* la production des centrales à accumulation par pompage (également appelé centrale de pompage). Voir les sections 2.C et 3.

Colonne 7 : *Géothermique, solaire, etc.* - Indique la production intérieure d'énergie géothermique, solaire, éolienne, marémotrice et houlomotrice ainsi que l'utilisation de ces formes d'énergie pour produire de l'électricité. Les autres utilisations figurant dans cette colonne ne concernent que la géothermie et la chaleur produite au moyen d'énergie solaire. Sauf dans les cas où le rendement effectif du procédé géothermique est connu, la quantité d'énergie géothermique employée pour la production d'électricité est calculée par induction à partir de la production d'électricité des centrales géothermiques, dans l'hypothèse d'un rendement thermique moyen de 10 pour cent. Pour l'énergie solaire, éolienne, marémotrice et houlomotrice, les quantités utilisées pour la production d'électricité sont égales à celles d'énergie électrique produite. Voir les sections 2.C et 3.

Colonne 8 : *Energies renouvelables combustibles et déchets* - Comprend la biomasse solide et les produits d'origine animale, les gaz/liquides tirés de la biomasse, les déchets industriels et les déchets urbains. La biomasse est, par définition, toute matière végétale utilisée directement comme combustible, ou bien transformée en combustibles (par exemple charbon de bois) ou en électricité et/ou chaleur. Cette définition recouvre le bois, les résidus végétaux (y compris les déchets de bois et les cultures destinées à la production d'énergie), l'éthanol, les matières/déchets d'origine animale et les lessives sulfitiques (les lessives sulfitiques sont également désignées par le terme "liqueur noire", ce sont les eaux alcalines de rejet issues des digesteurs lors de la production de pâte au sulfate ou à la soude dans la fabrication de la pâte à papier. L'énergie provient de la lignine extraite de la pâte chimique). Les déchets urbains correspondent aux déchets des secteurs résidentiel et commerce et services publics qui sont recueillis par les autorités municipales pour élimination dans une installation centralisée et pour la production de chaleur et/ou d'électricité. Les déchets hospitaliers sont indiqués dans cette catégorie.

Les données figurant sous ce titre sont souvent fondées sur le résultat d'enquêtes portant sur des échantillons réduits, ou sur d'autres informations incomplètes. Ainsi, elles ne fournissent qu'une indication générale des évolutions et ne sont pas strictement comparables d'un pays à l'autre. Dans certains cas, des catégories entières de combustibles végétaux sont omises faute d'information. Il est donc conseillé de consulter les données par pays lors de l'utilisation des agrégats régionaux.

Colonne 9 : Electricité - Indique la consommation finale et les échanges d'électricité (calculés sur la base du même pouvoir calorifique que l'électricité à la consommation finale, à savoir 1 GWh = 0,000086 Mtep).

Colonne 10 : Chaleur - Indique les quantités de chaleur produites pour la vente. La majeure partie de la chaleur figurant dans cette colonne provient de la combustion de combustibles, encore que de faibles quantités soient produites par des pompes à chaleur et des chaudières électriques. La chaleur extraite de l'air ambiant par les pompes à chaleur entre dans la production nationale.

Colonne 11 : TOTAL = total des colonnes 1 à 10.

II. Lignes

Les catégories figurant sur la partie gauche du tableau sont utilisées de la manière suivante :

Ligne 1 : La ligne *Production nationale* concerne la production d'énergie primaire, autrement dit houille, lignite, tourbe, pétrole brut et LGN, gaz naturel, énergies renouvelables combustibles et déchets, énergies nucléaire, hydraulique, géothermique et solaire, ainsi que la chaleur extraite du milieu ambiant par les pompes à chaleur. La production est calculée après élimination des impuretés (par exemple, élimination du soufre contenu dans le gaz naturel). Le mode de calcul de la production

d'énergie hydraulique, géothermique, etc., et d'électricité d'origine nucléaire est expliqué dans la section 3, Unités et coefficients de conversion.

Lignes 2 et 3 : Importations et exportations représentent les quantités ayant franchi les limites territoriales du pays, que le dédouanement ait été effectué ou non.

a)   Charbon

Les *importations et exportations* comprennent les quantités de combustibles obtenues d'autres pays ou fournies à d'autres pays, qu'il existe ou non une union économique ou douanière entre les pays en question. Le charbon en transit n'est pas pris en compte.

b)   Pétrole et gaz

Cette rubrique comprend les quantités de pétrole brut et de produits pétroliers importées ou exportées au titre d'accords de traitement (à savoir, raffinage à façon). Les quantités de pétrole en transit ne sont pas prises en compte. Le pétrole brut, les LGN et le gaz naturel sont indiqués comme provenant de leur pays d'origine. Pour les produits d'alimentation des raffineries et les produits pétroliers, en revanche, c'est le dernier pays de provenance qui est pris en compte.

Les réexportations de pétrole importé pour raffinage en zone franche sont comptabilisées comme des exportations de produits raffinés du pays où le traitement est effectué vers leur destination finale.

c)   Electricité

Les quantités sont considérées comme importées ou exportées lorsqu'elles ont franchi les limites territoriales du pays.

Ligne 4 : Les *soutages maritimes internationaux* correspondent aux quantités fournies aux navires de haute mer, y compris les navires de guerre et les navires de pêche, quel que soit leur pavillon. La consommation des navires assurant des transports par cabotage ou navigation intérieure n'est pas comprise. Voir *navigation intérieure*, ligne 40, et *agriculture*, ligne 43.

Ligne 5 : Les *variations des stocks* expriment la différence enregistrée entre le premier jour et le dernier jour de l'année dans le niveau des stocks détenus sur le territoire national par les producteurs, les importateurs, les entreprises de transformation de l'énergie et les gros consommateurs. Une augmentation des stocks est indiquée par un chiffre affecté d'un signe négatif, tandis qu'une diminution apparaît sous la forme d'un chiffre positif.

Ligne 6 : Les approvisionnements totaux en énergie primaire (ATEP) correspondent à la production nationale (ligne 1) + importations (ligne 2) - exportations (ligne 3) - soutages maritimes internationaux (ligne 4) ± variations des stocks (ligne 5).

Ligne 7 : Les *transferts* couvrent aussi bien le passage d'un produit d'une catégorie à une autre, le transfert matériel d'un produit et les produits recyclés (par exemple, les lubrifiants usés qui sont régénérés).

Ligne 8 : Les *écarts statistiques* correspondent à la somme des écarts statistiques inexpliqués pour les différents combustibles, tels qu'ils apparaissent dans les statistiques de base de l'énergie. Cette rubrique comprend également les écarts statistiques qui proviennent de l'utilisation de coefficients de conversion différents dans la colonne du charbon. Pour plus de détails, se reporter à l'introduction du document *Statistiques de l'énergie des pays non-membres 1996-1997*.

Ligne 9 : *Centrales électriques* désigne les centrales conçues pour produire uniquement de l'électricité. Si une unité ou plus de la centrale est une installation de cogénération (et que l'on ne peut pas comptabiliser séparément, sur une base unitaire, les combustibles utilisés et la production), elle est considérée comme une centrale de cogénération. Tant les centrales publiques[1] que les installations des autoproducteurs[2] entrent dans cette rubrique. Les colonnes 1 à 8 indiquent les quantités de combustibles primaires et secondaires utilisés pour la production d'électricité, les chiffres correspondants étant affectés d'un signe négatif. La production brute d'électricité (qui comprend la consommation propre des centrales) figure dans la

colonne de l'électricité, sous forme d'un chiffre positif. Les pertes de transformation sont indiquées dans la colonne du total, et sont affectées d'un signe négatif.

Ligne 10 : *Centrales de cogénération chaleur/ électricité* désigne les centrales conçues pour produire de la chaleur et de l'électricité. Dans la mesure du possible, les consommations de combustibles et les productions de chaleur/électricité doivent être exprimées sur la base des unités plutôt que des centrales. Cependant, à défaut de données disponibles exprimées sur une base unitaire, il convient d'adopter la convention indiquée ci-dessus pour la définition d'une centrale de cogénération. Tant les centrales publiques que les installations des autoproducteurs entrent dans cette rubrique. *On notera que, dans le cas des installations de cogénération chaleur/électricité des auto-producteurs, sont comptabilisés tous les combustibles utilisés pour la production d'électricité, tandis que seule la partie des combustibles utilisés pour la production de chaleur **vendue** est indiquée. Les combustibles utilisés pour la production de la chaleur destinée à la consommation interne des autoproducteurs **ne sont pas** comptabilisés dans cette rubrique mais dans les données concernant la consommation finale de combustibles du secteur de consommation approprié.*

Les colonnes 1 à 8 indiquent les quantités de combustibles primaires et secondaires utilisés pour la production d'électricité et de chaleur ; ces chiffres sont affectés d'un signe négatif. La production brute d'électricité figure dans la colonne de l'électricité, affectée d'un signe positif, et la production de chaleur apparaît dans la colonne de la chaleur, sous forme d'un chiffre positif. Les pertes de transformation sont indiquées dans la colonne du total, et sont affectées d'un signe négatif.

Ligne 11 : *Centrales calogènes* désigne les installations (pompes à chaleur et chaudières électriques comprises) conçues pour produire uniquement de la chaleur et qui en vendent à des tiers selon les termes d'un contrat. Cette rubrique comprend aussi bien les centrales publiques que les installations des autoproducteurs.

Les colonnes 1 à 8 indiquent les quantités de combustibles primaires et secondaires utilisés par les systèmes de chauffage qui transportent la chaleur, produite à partir d'une ou de plusieurs sources

---

1   La production publique désigne les installations dont la *principale activité* est la production d'électricité et/ou de chaleur pour la vente à des tiers. Elles peuvent appartenir au secteur privé ou public. Il convient de noter que les ventes ne se font pas nécessairement par l'intermédiaire du réseau public.

2   L'autoproduction désigne les installations qui produisent de l'électricité et/ou de la chaleur, en totalité ou en partie pour leur consommation propre, en tant qu'activité qui contribue à leur activité principale. Elles peuvent appartenir au secteur privé ou public.

d'énergie, et qui la distribuent à des consommateurs résidentiels, industriels et commerciaux, entre autres, pour le chauffage des locaux, la cuisson des aliments, la production d'eau chaude et les procédés industriels.

Ligne 12 : Lorsqu'il y a production de gaz dans des *usines à gaz*, les données sont traitées de la même manière que dans le cas de la production d'électricité : les quantités produites apparaissent sous forme de chiffres positifs dans la colonne du gaz naturel, les quantités utilisées sont représentées par des chiffres négatifs dans les colonnes du charbon, des produits pétroliers et du gaz naturel, et les pertes de transformation apparaissent dans la colonne du total.

Ligne 13 : La ligne *raffineries de pétrole* indique les quantités d'énergie primaire utilisées dans les raffineries et la production de produits pétroliers. Le total tient compte des pertes de transformation. Dans certains cas les données dans la colonne total sont des nombres positifs. Cela peut être dû soit à des incohérences du bilan primaire de raffinage, soit au fait que l'AIE utilise des pouvoirs calorifiques inférieurs standard pour les produits pétroliers.

Ligne 14 : La *transformation du charbon* comprend les pertes liées à la transformation du charbon de combustible primaire en combustible secondaire et de combustible secondaire en combustible tertiaire (de la houille en coke, du coke en gaz de haut fourneau, du lignite en briquettes de lignite, etc). Il est souvent difficile de prendre en compte correctement l'ensemble des entrées et des sorties des industries de transformation de l'énergie, et de faire la distinction entre énergie transformée et énergie brûlée. Par conséquent, dans certains cas les données dans la colonne du total sont des chiffres positifs, indiquant un problème dans les données.

Ligne 15 : La *liquéfaction* comprend divers procédés de liquéfaction comme, par exemple, la liquéfaction du charbon et du gaz naturel en Afrique du Sud.

Ligne 16 : La ligne *autres transformations* comprend les transformations non spécifiées ailleurs. Elle comprend aussi les retours de l'industrie pétrochimique. Il convient de noter que les retours en raffinerie des produits pétroliers utilisés à des fins non énergétiques (i.e. white spirit et lubrifiants) ne sont pas inclus sous cette rubrique, mais sous utilisations non énergétiques.

Ligne 17 : La ligne *consommation propre* indique la consommation d'énergie primaire et secondaire des industries de transformation pour le chauffage, le pompage, la traction et l'éclairage [Divisions 10, 11, 12, 23 et 40 de la CITI[1]], ces chiffres étant affectés d'un signe négatif. Cette rubrique comprend, par exemple, la consommation propre d'énergie des mines de charbon, celle des centrales électriques (y compris la quantité nette d'électricité consommée par les installations hydroélectriques à accumulation par pompage) et l'énergie employée pour l'extraction du pétrole et du gaz.

Ligne 18 : Les *pertes de distribution et de transport* comprennent les pertes dans la distribution du gaz ainsi que les pertes dans le transport de l'électricité et du charbon.

Ligne 19 : La ligne *consommation finale totale (CFT)* donne la somme des consommations des différents secteurs d'utilisation finale. Dans la consommation finale, les produits d'alimentation de l'industrie pétrochimique figurent sous la rubrique *industrie chimique*, sous-rubrique *dont : produits d'alimentation*, tandis que les utilisations non énergétiques de produits pétroliers comme le white spirit, les lubrifiants, le bitume, les paraffines et autres produits figurent dans les *utilisations non énergétiques* et ne sont prises en compte que dans la consommation finale totale. Voir lignes 47 à 50 *(Utilisations non énergétiques)*. Les retours de l'industrie pétrochimique ne sont pas comptabilisés dans la consommation finale (voir ligne 16, *autres transformations*).

Lignes 20 à 34 : La consommation du *secteur industrie* est répartie entre les sous-secteurs suivants (l'énergie utilisée par l'industrie pour les transports n'est pas prise en compte ici mais figure dans la rubrique transports) :

- *Industrie sidérurgique* [Groupe 271 et Classe 2731 de la CITI] ;

- *Industrie chimique* [Division 24 de la CITI] ; *dont* : produits d'alimentation de l'industrie pétrochimique. L'industrie pétrochimique comprend les opérations de craquage et de reformage destinées à la production de l'éthylène,

---

1 Classification internationale type par industries de toutes les branches d'activité économique, Série M, Nº 4/Rév. 3, Nations Unies, New York, 1990.

du propylène, du butylène, des gaz de synthèse, des aromatiques, du butadiène et d'autres matières premières à base d'hydrocarbures dans les procédés mis en oeuvre, par exemple, pour le vapocraquage, dans les installations d'élaboration d'aromatiques et pour le reformage à la vapeur [Partie du Groupe 241 de la CITI] ; voir produits d'alimentation, lignes 47 à 50 (Utilisations non énergétiques) ;

- Industries de base des *métaux non ferreux* [Groupe 272 et Classe 2732 de la CITI] ;

- *Produits minéraux non métalliques* tels que verre, céramiques, ciment, etc. [Division 26 de la CITI] ;

- *Matériel de transport* [Divisions 34 et 35 de la CITI] ;

- *Construction mécanique*. Ouvrages en métaux, machines et matériels autres que le matériel de transport [Divisions 28, 29, 30, 31 et 32 de la CITI] ;

- *Industries extractives* (à l'exclusion de l'extraction de combustibles) [Divisions 13 et 14 de la CITI] ;

- *Produits alimentaires, boissons et tabacs* [Divisions 15 et 16 de la CITI] ;

- *Papier, pâte à papier et imprimerie* [Divisions 21 et 22 de la CITI] ;

- *Production de bois et d'articles en bois* (sauf pâtes et papiers) [Division 20 de la CITI] ;

- *Construction* [Division 45 de la CITI] ;

- *Textiles et cuir* [Division 17, 18 et 19 de la CITI] ;

- *Non spécifiés* (tout autre secteur industriel non spécifié précédemment) [Division 25, 33, 36 et 37 de la CITI].

Note : La plupart des pays éprouvent des difficultés pour fournir une ventilation par branche d'activité pour tous les combustibles. Dans ces cas, la rubrique *non spécifiés* a été utilisée. *Les agrégats régionaux de la consommation industrielle doivent donc être employés avec précaution.*

Lignes 35 à 41 : Le secteur *transports* regroupe tous les combustibles utilisés pour les transports, à l'exception des soutages maritimes internationaux [Divisions 60, 61 et 62 de la CITI]. Elle englobe les transports dans le secteur industriel et couvre les transports routiers, ferroviaires et aériens ainsi que la navigation intérieure (y compris les petites embarcations, les bateaux de pêche et les caboteurs dont la consommation n'est pas comptabilisée dans la rubrique "soutages maritimes internationaux"), les combustibles utilisés pour le transport par conduites et les transports non spécifiés. Les combustibles utilisés pour la pêche en haute mer, sur le littoral et dans les eaux intérieures doivent être comptabilisés dans le secteur de l'agriculture (ligne 43).

Lignes 42 à 46 : La rubrique *autres secteurs* couvre le secteur de l'agriculture (y compris la pêche en haute mer, sur le littoral et dans les eaux intérieures) [Divisions 01, 02 et 05 de la CITI], le secteur résidentiel ainsi que le commerce et les services publics [Divisions 41, 50, 51, 52, 55, 63, 64, 65, 66, 67, 70, 71, 72, 73, 74, 75, 80, 85, 90, 91, 92, 93, 95 et 99 de la CITI], ainsi que les consommations non spécifiées. Dans bien des cas, les administrations n'arrivent pas à faire la ventilation de la consommation d'énergie entre le secteur *commerce et services publics* et le secteur *résidentiel*. D'autres administrations ne peuvent pas ventiler les consommations des secteurs *agriculture* et *résidentiel*. Dans les cas de ce genre, le secteur *résidentiel* comprend également la consommation du secteur *agriculture* et/ou celle du secteur *commerce et services publics*. La rubrique *non spécifiés* peut comprendre les utilisations à des fins militaires qui ne figurent pas dans le secteur des transports. Le total de la ligne *autres secteurs* est donc plus exact que les éléments qui le composent. Se reporter aux données par pays lors de l'utilisation des agrégats régionaux.

Lignes 47 à 50 : La rubrique *utilisations non énergétiques* regroupe la consommation des *autres produits pétroliers* comme le white spirit, les paraffines, les lubrifiants, le bitume et divers autres produits. Ils incluent également les utilisations non énergétiques du charbon (excepté pour la tourbe). Ces produits sont indiqués à part, dans la consommation finale sous la rubrique *utilisations non énergétiques*. Il est supposé que l'usage de ces produits est strictement *non énergétique*. Le coke de pétrole fait exception à cette règle et ne figure sous la rubrique *utilisations non énergétiques* que si cette utilisation est prouvée ; dans le cas contraire, ce produit est comptabilisé à la rubrique des utilisations énergétiques dans l'industrie ou dans d'autres secteurs. Les utilisations non énergétiques du charbon comprennent notamment la préparation de noirs de carbone, d'électrodes en graphite, etc. et sont par ailleurs indiquées séparément par secteur.

On notera que les chiffres concernant les produits d'alimentation de l'industrie pétrochimique sont comptabilisés dans le secteur de l'industrie, au titre de l'industrie chimique (ligne 22) et figurent séparément à la rubrique *dont : produits d'alimentation* (ligne 23). Sont compris dans cette rubrique tous les produits pétroliers, y compris les naphtas, à l'exception des *autres produits pétroliers* énumérés ci-dessus, ainsi que le gaz utilisé comme produit d'alimentation de l'industrie pétrochimique.

Lignes 51 à 53 : La rubrique *électricité produite* indique le nombre total de GWh produits par les centrales thermiques, ventilées entre centrales électriques et installations de cogénération, ainsi que la production des centrales nucléaires, hydroélectriques (à l'exclusion des centrales à accumulation par pompage), géothermiques, etc. (voir cependant les notes relatives aux lignes 9 et 10). Il convient de noter que l'électricité produite à partir du pétrole brut ou des LGN est prise en compte dans la colonne *produits pétroliers*.

Lignes 54 à 56 : La rubrique *chaleur produite* indique le nombre total de TJ produits dans les centrales, avec une distinction faite entre centrales de cogénération et centrales calogènes. Il convient de noter que la chaleur produite au moyen de chaudières électriques est comptabilisée à la colonne *électricité* et la chaleur obtenue au moyen de pompes à chaleur figure à la colonne *chaleur*.

E. Indicateurs

Production énergétique : la production énergétique primaire totale exprimée en Mtep. Voir section 2.D.II.1.

Approvisionnements totaux en énergie primaire : exprimés en Mtep. Voir section 2.D.II.6.

Importations nettes de pétrole : les importations moins les exportations de pétrole exprimées en Mtep. Voir section 2.D.II.2/3.

Approvisionnement de pétrole : l'approvisionnement primaire de pétrole exprimé en Mtep. Voir section 2.D.II.6.

Consommation d'électricité : consommation nationale, c'est-à-dire la production brute + les importations – les exportations – les pertes de distribution exprimée en TWh.

World Development Indicators, La Banque mondiale Washington D.C., 1998 et 1999 est la principale source de données concernant la population et le PIB.

Les données de population pour **Gibraltar**, l'**ex-République yougoslave de Macédoine**, l'**ex-URSS** et l'**ex-Yougoslavie** ainsi que pour les trois régions **Autre Afrique**, **Autre Amérique latine** et **Autre Asie** proviennent de la base Chelem-PIB du CEPII. Les données de population pour le **Taipei chinois** proviennent du Conseil de planification et de développement économiques, République de Chine, *Taiwan Statistical Databook*, 1998.

Les données du PIB pour les **Antilles néerlandaises**, la **Bosnie-Herzégovine**, la **Croatie**, **Cuba**, les **Emirats arabes unis**, **Gibraltar**, le **Koweit**, la **Libye**, l'**ex-République yougoslave de Macédoine**, la **Moldova**, le **Qatar**, la **Slovénie**, l'**Ukraine**, l'**ex-URSS** (qui peut différer de la somme des 15 républiques), la **République fédérative de Yougoslavie** et l'**ex-Yougoslavie** ainsi que pour les trois régions **Autre Afrique**, **Autre Amérique latine** et **Autre Asie** proviennent de la base Chelem-PIB du CEPII Paris, 1999. Les données du PIB pour l'**Angola** (1971-1984), **Bahrein** (1971-1979 et 1997), le **Bangladesh** (1997), la **Bolivie** (1971-1979), **Brunei** (1971-1973 et 1997), la **Bulgarie** (1971-1979), **Chypre** (1971-1974 et 1995-1997), l'**Ethiopie** (1971-1982), l'**Iran** (1971-1973 et 1996-1997), l'**Irak** (1991-1997), la **Jordanie** (1971-1974), le **Liban** (1971-1987), le **Mozambique** (1971-1979 et 1997), le **Nicaragua** (1997), **Oman** (1996-1997), **Panama** (1997), la **Roumanie** (1971-1974), le **Soudan** (1995-1997), le **Tadjikistan** (1997), la **Tanzanie** (1971-1987), le **Turkmenistan** (1997) et le **Yemen** (1971-1989) ont été estimées à partir des taux de croissance de la banque de données Chelem CEPII. Les données du PIB pour le **Taipei chinois** proviennent du Conseil de planification et de développement économiques, République de Chine, *Taiwan Statistical Databook*, 1998. Le taux de croissance du PIB de l'ex-

Tchécoslovaquie a été utilisé pour estimer le PIB de la **République slovaque** de 1971 à 1983.

Les données relatives au PIB ont été calculées pour chaque pays aux prix du marché en monnaie locale et en taux annuels. Ces données ont ensuite été recalées par rapport au niveau des prix de 1990, puis converties en dollars US en utilisant les taux de change moyens annuels pour 1990 ou les parités de pouvoir d'achat (PPA).

Au cours des dernières années, de larges fluctuations dans les taux de change ont été observées, si bien qu'il devient de plus en plus nécessaire et pertinent de développer des indicateurs énergétiques basés sur une mesure du PIB qui permet d'éviter ces fluctuations et de mieux refléter les pouvoirs d'achat relatifs des différentes monnaies. En conséquence, cette publication présente, d'une part, une valeur du PIB calculée de manière traditionnelle en utilisant les taux de change habituels, et, d'autre part, une seconde valeur du PIB basée, cette fois, sur les parités de pouvoir d'achat.

Les parités de pouvoir d'achat représentent les taux de conversion monétaire qui égalisent les pouvoirs d'achat des différentes monnaies. Ainsi, une somme donnée d'argent, une fois convertie en différentes unités monétaires à partir des taux PPA, permet d'acheter le même panier de biens et de services dans tous les pays. En d'autres termes, les PPA sont les taux de conversion monétaires qui permettent d'éliminer les différences dans les niveaux de prix entre pays. Les PPA retenues pour convertir le PIB des pays non membres de l'OCDE d'unités monétaires locales en dollars US proviennent du CD ROM CEPII-CHELEM, Paris, 1999. Ils sont agrégés selon la méthode de Geary-Khamis (GK) et recalés sur les Etats-Unis. Pour une description plus détaillée de cette méthodologie, il convient de se référer au document suivant : Equipe CHELEM, *La base CHELEM-PIB,* CEPII, Paris, avril 1998.

3. UNITES ET COEFFICIENTS DE CONVERSION

Coefficients de conversion généraux pour l'énergie

Vers :	TJ	Gcal	Mtep	MBtu	GWh
De :	multiplier par :				
TJ	1	238,8	$2,388 \times 10^{-5}$	947,8	0,2778
Gcal	$4,1868 \times 10^{-3}$	1	10^{-7}	3,968	$1,163 \times 10^{-3}$
Mtep	$4,1868 \times 10^{4}$	10^{7}	1	$3,968 \times 10^{7}$	11630
MBtu	$1,0551 \times 10^{-3}$	0,252	$2,52 \times 10^{-8}$	1	$2,931 \times 10^{-4}$
GWh	3,6	860	$8,6 \times 10^{-5}$	3412	1

Coefficients de conversion pour les mesures de masse

Vers :	kg	t	lt	st	lb
De :	multiplier par :				
kilogramme (kg)	1	0,001	$9,84 \times 10^{-4}$	$1,102 \times 10^{-3}$	2,2046
tonne (t)	1000	1	0,984	1,1023	2204,6
tonne longue (lt)	1016	1,016	1	1,120	2240,0
tonne courte (st)	907,2	0,9072	0,893	1	2000,0
livre (lb)	0,454	$4,54 \times 10^{-4}$	$4,46 \times 10^{-4}$	$5,0 \times 10^{-4}$	1

Coefficients de conversion pour les mesures de volume

Vers:	gal U.S.	gal U.K.	bbl	ft³	l	m³
De:	multiplier par:					
Gallon U.S. (gal)	1	0,8327	0,02381	0,1337	3,785	0,0038
Gallon U.K. (gal)	1,201	1	0,02859	0,1605	4,546	0,0045
Baril (bbl)	42,0	34,97	1	5,615	159,0	0,159
Pied cube (ft³)	7,48	6,229	0,1781	1	28,3	0,0283
Litre (l)	0,2642	0,220	0,0063	0,0353	1	0,001
Mètre cube (m³)	264,2	220,0	6,289	35,3147	1000,0	1

Préfixes décimaux

10^1	déca (da)		10^{-1}	déci (d)
10^2	hecto (h)		10^{-2}	centi (c)
10^3	kilo (k)		10^{-3}	milli (m)
10^6	méga (M)		10^{-6}	micro ( )
10^9	giga (G)		10^{-9}	nano (n)
10^{12}	téra (T)		10^{-12}	pico (p)
10^{15}	péta (P)		10^{-15}	femto (f)
10^{18}	exa (E)		10^{-18}	atto (a)

A. Gaz

Dans les *Statistiques de l'énergie des pays non-membres*, les gaz indiqués ci-après sont toujours comptabilisés en térajoules et le pouvoir calorifique utilisé est le pouvoir calorifique supérieur (PCS).

1 térajoule = 0,00002388 Mtep.

Pour convertir le pouvoir calorifique supérieur (PCS) d'un gaz en pouvoir calorifique inférieur (PCI), il convient de multiplier le PCS par le coefficient indiqué dans le tableau ci-dessous.

Gaz naturel	0,9
Gaz d'usine à gaz	0,9
Gaz de cokerie	0,9
Gaz de haut-fourneau	1,0
Gaz de convertisseur à l'oxygène	1,0

B. Electricité

Les données relatives à la production, aux échanges et à la consommation finale d'électricité sont calculées en fonction du contenu énergétique de l'électricité, c'est-à-dire selon le coefficient suivant : 1 TWh = 0,086 Mtep. La production hydro-électrique (production des centrales à accumulation par pompage non comprise) et l'électricité produite par d'autres moyens non thermiques (énergies éolienne, marémotrice, photovoltaïque, etc.) sont affectées du même coefficient. Cependant, l'équivalent en énergie primaire de l'électricité d'origine nucléaire est calculé à partir de la production brute, compte tenu d'un coefficient hypothétique de rendement de conversion des installations de 33 pour cent. En d'autres termes,

1 TWh = (0,086 ÷ 0,33) Mtep. Dans le cas de l'électricité d'origine géothermique, si le rendement géothermique effectif n'est pas connu, le coefficient de rendement pris pour hypothèse est de 10 pour cent, soit 1 TWh = (0,086 ÷ 0,1) Mtep.

C.   Pétrole brut

Les pouvoirs calorifiques inférieurs spécifiques par pays sont présentés dans les tableaux de la deuxième partie. Il s'agit de valeurs moyennes utilisées pour convertir en pouvoir calorifique les données relatives à la production, aux importations, aux exportations et à la consommation.

D.   Produits pétroliers

Les coefficients de conversion suivants sont utilisés pour toutes les années (tep/tonne).

	Valeurs particulières pour certains pays				Tous autres pays
	Algérie	Brésil	Malasie	Afrique du Sud	
Gaz de raffinerie	-	0.880	-	-	1.150
Ethane	-	-	-	-	1.130
GPL	1.180	1.175	1.088	1.117	1.130
Naphta	-	1.132	1.054	1.073	1.075
Essence aviation	-	1.129	1.050	1.088	1.070
Essence moteur	-	1.122	1.050	1.052	1.070
Carburéacteur type essence	-	-	-	0.973	1.070
Carburéacteur type kérosène	-	1.109	1.032	0.981	1.065
Autre kérosène	-	1.109	1.032	1.033	1.045
Gazole/carburant diesel	-	1.075	1.015	1.025	1.035
Fioul lourd	-	1.090	0.991	0.999	0.960
White Spirit	-	1.124	1.032	1.015	0.960
Lubrifiants	-	1.077	1.006	-	0.960
Bitume	-	1.005	0.998	-	0.960
Paraffines	-	-	1.035	-	0.960
Coke de pétrole	-	0.850	0.869	-	0.740
Autres produits non spécifiés	-	1.089	1.015	-	0.960

E.   Charbon

Les pouvoirs calorifiques inférieurs (PCI) utilisés pour le charbon diffèrent selon qu'il s'agit de la production, des importations, des exportations, de l'alimentation des centrales électriques publiques, du charbon employé dans les fours à coke et du charbon utilisé dans l'industrie. Pour tous les autres flux, la conversion est effectuée en utilisant le pouvoir calorifique inférieur (PCI) moyen. Les coefficients de conversion pour chaque pays pour les années 1996 et 1997 sont indiqués dans la deuxième partie, Pouvoirs calorifiques inférieurs spécifiques par pays.

F.   Energies renouvelables combustibles et déchets

Le pouvoir calorifique des énergies renouvelables combustibles et des déchets exprimé en térajoules sur la base du pouvoir calorifique inférieur, est présenté dans les *Statistiques de l'énergie des pays non-membres*.

1 térajoule = 0,00002388 Mtep.

G.   Chaleur

Les données sur la chaleur sont exprimées en térajoules.

1 térajoule = 0,00002388 Mtep.

H.   Exemples

Les exemples ci-après montrent comment calculer le contenu calorifique inférieur (en Mtep) des quantités exprimées en unités d'origine dans les *Statistiques de l'énergie des pays non-membres, 1996-1997*.

Production de charbon à coke (Colombie) pour 1997 en milliers de tonnes ; multiplier par 0,00065=Mtep.

Gaz naturel en térajoules (PCS) ; multiplier par 0,00002149=Mtep.

Essence moteur (sauf Brésil, Malaisie et Afrique du Sud) en milliers de tonnes ; multiplier par 0,0010700=Mtep.

Chaleur en térajoules (PCI) ; multiplier par 0,00002388=Mtep.

4. COUVERTURE GEOGRAPHIQUE

- **L'Afrique** comprend l'Afrique du Sud, l'Algérie, l'Angola, le Bénin, le Cameroun, le Congo, la République démocratique du Congo, l'Egypte, l'Ethiopie, le Gabon, le Ghana, la Côte d'Ivoire, le Kenya, la Libye, le Maroc, le Mozambique, le Nigéria, le Sénégal, le Soudan, la Tanzanie, la Tunisie, la Zambie, le Zimbabwe et les **autres pays d'Afrique.**

- **Les autres pays d'Afrique** comprennent le Botswana, le Burkina Faso, le Burundi, le Cap-Vert, la République centrafricaine, Djibouti, la Gambie, la Guinée, la Guinée-Bissau, la Guinée équatoriale, le Lesotho, le Libéria, Madagascar, le Malawi, le Mali, la Mauritanie, Maurice, le Niger, l'Ouganda, la Réunion, le Rwanda, Sao Tomé-et-Principe, les Seychelles, la Sierra Leone, la Somalie, le Swaziland, le Tchad et le Togo.

- **L'Amérique latine** comprend les Antilles néerlandaises, l'Argentine, la Bolivie, le Brésil, le Chili, la Colombie, le Costa Rica, Cuba, la République dominicaine, El Salvador, l'Equateur, le Guatemala, Haïti, le Honduras, la Jamaïque, le Nicaragua, Panama, le Paraguay, le Pérou, Trinité-et-Tobago, l'Uruguay, le Venezuela et les **autres pays d'Amérique latine.**

- **Les autres pays d'Amérique latine** comprennent Antigua-et-Barbuda, les Bahamas, la Barbade, le Belize, les Bermudes, la Dominique, la Grenade, la Guadeloupe, le Guyana, la Guyane française, la Martinique, Saint-Kitts-et-Nevis et Anguilla, Sainte-Lucie, Saint-Vincent-et-les-Grenadines et le Surinam.

- **L'Asie** comprend le Bangladesh, Brunei, la République populaire démocratique de Corée, l'Inde, l'Indonésie, la Malaisie, Myanmar, le Népal, le Pakistan, les Philippines, Singapour, le Sri Lanka, le Taipei chinois, la Thaïlande, le Viêt-Nam et les **autres pays d'Asie.**

- **Les autres pays d'Asie et d'Océanie** comprennent l'Afghanistan, le Bhoutan, les Fidji, Kiribati, les Maldives, la Nouvelle-Calédonie, la Papouasie-Nouvelle-Guinée, la Polynésie française, le Samoa, les Iles Salomon et Vanuatu.

- **La Chine** comprend la République populaire de Chine et Hong Kong (Chine).

- **Le Moyen-Orient** comprend l'Arabie saoudite, Bahreïn, les Emirats arabes unis, l'Iran, l'Iraq, Israël, la Jordanie, le Koweït, le Liban, Oman, le Qatar, la Syrie et le Yémen.

- **La région Europe hors OCDE** comprend l'Albanie, la Bosnie-Herzégovine, la Bulgarie, Chypre, la Croatie, Gibraltar, l'ex-République yougoslave de Macédoine (FYROM), Malte, la Roumanie, la Slovaquie, la Slovénie et la République fédérative de Yougoslavie.

- **L'ex-URSS** comprend l'Arménie, l'Azerbaïdjan, le Bélarus, l'Estonie, la Géorgie, le Kazakhstan, le Kirghizistan, la Lettonie, la Lituanie, la Moldavie, l'Ouzbékistan, la Fédération de Russie, le Tadjikistan, le Turkménistan et l'Ukraine.

- L'**Organisation de coopération et de développement économiques (OCDE)** comprend l'Allemagne, l'Australie, l'Autriche, la Belgique, le Canada, la Corée, le Danemark, l'Espagne, les Etats-Unis, la Finlande, la France, la Grèce, la Hongrie, l'Irlande, l'Islande, l'Italie, le Japon, le Luxembourg, le Mexique, la Norvège, la Nouvelle-Zélande, les Pays-Bas, la Pologne, le Portugal, la République tchèque, le Royaume-Uni, la Suède, la Suisse et la Turquie.

- Dans la zone de l'OCDE :

L'**Allemagne** tient compte des nouveaux Länder à partir de 1970.

Le Groenland et les Iles Féroé danoises ne sont pas pris en compte dans les données relatives au **Danemark**.

L'**Espagne** englobe les Iles Canaries.

Les **Etats-Unis** englobent Porto-Rico, Guam et les Iles Vierges ainsi que la zone franche d'Hawaï.

Dans les données relatives à la **France**, Monaco est pris en compte, mais non les départements d'outre-mer (Martinique, Guadeloupe, Polynésie française et Ile de la Réunion).

L'**Italie** englobe Saint-Marin et le Vatican.

Le **Japon** englobe Okinawa.

Ni le Surinam ni les Antilles néerlandaises ne sont pris en compte dans les données relatives aux **Pays-Bas**.

Le **Portugal** englobe les Açores et l'Ile de Madère.

La **Suisse** englobe le Liechtenstein.

- **L'Organisation des pays exportateurs de pétrole (OPEP)** comprend l'Algérie, l'Indonésie, l'Iran, l'Irak, le Koweit, la Libye, le Nigéria, Qatar, l'Arabie saoudite, les Emirats arabes unis et le Vénézuela.

On notera que les pays suivants n'ont pas été pris en compte par suite d'un manque de données :

Afrique : Comores, Namibie, Sainte-Hélène et Sahara Occidental.

Amérique : Aruba, Iles Vierges Britanniques, Iles Caïmanes, Iles Falkland, Montserrat, Saint-Pierre et Miquelon et les Iles Turks et Caïcos.

Asie et Océanie : Samoa américaines, Cambodge, Ile Christmas, Iles Cook, Laos, Macao, Mongolie, Nauru, Nioué, Iles du Pacifique (total. amér.), Tonga et Ile de Wake.

PART II:
STATISTICAL DATA

PARTIE II :
DONNEES STATISTIQUES

COUNTRY SPECIFIC NET CALORIFIC VALUES

POUVOIRS CALORIFIQUES INFERIEURS SPECIFIQUES PAR PAYS

1997

INTERNATIONAL ENERGY AGENCY

Country Specific Net Calorific Values (tonne of oil equivalent per tonne)
Pouvoirs calorifiques inférieurs spécifiques par pays (tonnes d'équivalent pétrole par tonne)

1997

	Bituminous Coal and Anthracite						Sub-Bituminous Coal					
	Production	Imports	Exports	Electr. Gen.	Industry	Average	Production	Imports	Exports	Electr. Gen.	Industry	Average
Africa												
Algeria	0.6150	0.6150	0.6150	0.6150	0.6150	0.6150	0.2000	0.2000	0.2000	0.2000	0.2000	0.2000
Angola	0.6150	0.6150	0.6150	0.6150	0.6150	0.6150	-	-	-	-	-	-
Benin	-	-	-	-	-	-	-	-	-	-	-	-
Cameroon	0.6150	0.6150	0.6150	0.6150	0.6150	0.6150	-	-	-	-	-	-
Congo	-	-	-	-	-	-	-	-	-	-	-	-
DR of Congo	0.6026	0.6026	0.6026	0.6026	0.6026	0.6026	0.2000	0.2000	0.2000	0.2000	0.2000	0.2000
Egypt	0.6150	0.6150	0.6150	0.6150	0.6150	0.6150	0.2000	0.2000	0.2000	0.2000	0.2000	0.2000
Ethiopia	-	-	-	-	-	-	-	-	-	-	-	-
Gabon	0.6150	0.6150	0.6150	0.6150	0.6150	0.6150	0.2000	0.2000	0.2000	0.2000	0.2000	0.2000
Ghana	0.6150	0.6150	0.6150	0.6150	0.6150	0.6150	-	-	-	-	-	-
Ivory Coast	0.6150	0.6150	0.6150	0.6150	0.6150	0.6150	0.2000	0.2000	0.2000	0.2000	0.2000	0.2000
Kenya	0.6150	0.6150	0.6150	0.6150	0.6150	0.6150	0.2000	0.2000	0.2000	0.2000	0.2000	0.2000
Libya	0.6150	0.6150	0.6150	0.6150	0.6150	0.6150	0.2000	0.2000	0.2000	0.2000	0.2000	0.2000
Morocco	0.5600	0.6600	0.5600	0.5900	0.5900	0.5900	0.2000	0.2000	0.2000	0.2000	0.2000	0.2000
Mozambique	0.6150	0.6150	0.6150	0.6150	0.6150	0.6150	0.2000	0.2000	0.2000	0.2000	0.2000	0.2000
Nigeria	0.6150	0.6150	0.6150	0.6150	0.6150	0.6150	0.2000	0.2000	0.2000	0.2000	0.2000	0.2000
Senegal	0.6150	0.6150	0.6150	0.6150	0.6150	0.6150	0.2000	0.2000	0.2000	0.2000	0.2000	0.2000
South Africa	0.5636	0.6448	0.6686	0.4800	0.6448	0.6448	-	-	-	-	-	-
Sudan	-	-	-	-	-	-	-	-	-	-	-	-
Tanzania	0.6150	0.6150	0.6150	0.6150	0.6150	0.6150	-	-	-	-	-	-
Tunisia	0.6150	0.6150	0.6150	0.6150	0.6150	0.6150	0.2000	0.2000	0.2000	0.2000	0.2000	0.2000
Zambia	0.5901	0.5901	0.5901	0.5901	0.5901	0.5901	0.2000	0.2000	0.2000	0.2000	0.2000	0.2000
Zimbabwe	0.6150	0.6150	0.6150	0.6150	0.6150	0.6150	0.2000	0.2000	0.2000	0.2000	0.2000	0.2000
Other Africa	0.6150	0.6150	0.6150	0.6150	0.6150	0.6150	0.2000	0.2000	0.2000	0.2000	0.2000	0.2000
Latin America												
Argentina	0.5900	0.7200	0.5900	0.5900	0.7200	0.5900	0.2000	0.2000	0.2000	0.2000	0.2000	0.2000
Bolivia	0.6150	0.6150	0.6150	0.6150	0.6150	0.6150	0.2000	0.2000	0.2000	0.2000	0.2000	0.2000
Brazil	0.3790	0.5600	0.4400	0.3490	0.4460	0.3820	-	-	-	-	-	-
Chile	0.6790	0.6790	0.6790	0.6790	0.6790	0.6790	0.4102	0.2000	0.4102	0.4102	0.4102	0.4102
Colombia	0.6500	0.6500	0.6500	0.6500	0.6500	0.6500	0.2000	0.2000	0.2000	0.2000	0.2000	0.2000
Costa Rica	0.6150	0.6150	0.6150	0.6150	0.6150	0.6150	0.2000	0.2000	0.2000	0.2000	0.2000	0.2000
Cuba	0.6150	0.6150	0.6150	0.6150	0.6150	0.6150	0.2000	0.2000	0.2000	0.2000	0.2000	0.2000
Dominican Republic	0.6150	0.6150	0.6150	0.6150	0.6150	0.6150	0.2000	0.2000	0.2000	0.2000	0.2000	0.2000
Ecuador	0.6150	0.6150	0.6150	0.6150	0.6150	0.6150	0.2000	0.2000	0.2000	0.2000	0.2000	0.2000
El Salvador	0.6150	0.6150	0.6150	0.6150	0.6150	0.6150	0.2000	0.2000	0.2000	0.2000	0.2000	0.2000
Guatemala	-	-	-	-	-	-	-	-	-	-	-	-
Haiti	0.6150	0.6150	0.6150	0.6150	0.6150	0.6150	0.2000	0.2000	0.2000	0.2000	0.2000	0.2000
Honduras	0.6150	0.6150	0.6150	0.6150	0.6150	0.6150	0.2000	0.2000	0.2000	0.2000	0.2000	0.2000
Jamaica	0.6150	0.6150	0.6150	0.6150	0.6150	0.6150	0.2000	0.2000	0.2000	0.2000	0.2000	0.2000
Netherlands Antilles	0.6150	0.6150	0.6150	0.6150	0.6150	0.6150	0.2000	0.2000	0.2000	0.2000	0.2000	0.2000
Nicaragua	0.6150	0.6150	0.6150	0.6150	0.6150	0.6150	0.2000	0.2000	0.2000	0.2000	0.2000	0.2000
Panama	0.6150	0.6150	0.6150	0.6150	0.6150	0.6150	0.2000	0.2000	0.2000	0.2000	0.2000	0.2000
Paraguay	0.6150	0.6150	0.6150	0.6150	0.6150	0.6150	0.2000	0.2000	0.2000	0.2000	0.2000	0.2000
Peru	0.7000	0.7000	0.7000	0.7000	0.7000	0.7000	-	-	-	-	-	-
Trinidad-and-Tobago	0.6150	0.6150	0.6150	0.6150	0.6150	0.6150	0.2000	0.2000	0.2000	0.2000	0.2000	0.2000
Uruguay	0.6150	0.6150	0.6150	0.6150	0.6150	0.6150	0.2000	0.2000	0.2000	0.2000	0.2000	0.2000
Venezuela	0.6150	0.6150	0.6150	0.6150	0.6150	0.6150	0.2000	0.2000	0.2000	0.2000	0.2000	0.2000
Other Latin America	0.6150	0.6150	0.6150	0.6150	0.6150	0.6150	0.2000	0.2000	0.2000	0.2000	0.2000	0.2000
Asia												
Bangladesh	0.4998	0.4998	0.4998	0.4998	0.4998	0.4998	0.2000	0.2000	0.2000	0.2000	0.2000	0.2000
Brunei	0.6150	0.6150	0.6150	0.6150	0.6150	0.6150	0.2000	0.2000	0.2000	0.2000	0.2000	0.2000
Chinese Taipei	0.6200	0.6550	0.6200	0.6400	0.6300	0.6400	0.2000	0.2000	0.2000	0.2000	0.2000	0.2000
India	0.4773	0.6150	0.4773	0.4803	0.4773	0.4773	0.2340	0.2000	0.2340	0.2340	0.2340	0.2340
Indonesia	0.6150	0.6150	0.6150	0.6150	0.6150	0.6150	0.2000	0.2000	0.2000	0.2000	0.2000	0.2000
DPR of Korea	0.6150	0.6150	0.6150	0.6150	0.6150	0.6150	0.4200	0.2000	0.4200	0.4200	0.4200	0.4200
Malaysia	0.7000	0.7000	0.7000	0.7000	0.7000	0.7000	0.2690	0.2690	0.2690	0.2690	0.2690	0.2690
Myanmar	0.6150	0.6150	0.6150	0.6150	0.6150	0.6150	0.2000	0.2000	0.2000	0.2000	0.2000	0.2000
Nepal	0.6000	0.6000	0.6000	0.6000	0.6000	0.6000	0.2000	0.2000	0.2000	0.2000	0.2000	0.2000
Pakistan	0.4474	0.6579	0.4474	0.4474	0.4474	0.4474	0.2000	0.2000	0.2000	0.2000	0.2000	0.2000
Philippines	0.4800	0.4900	0.4800	0.4800	0.4800	0.4800	0.2000	0.2000	0.2000	0.2000	0.2000	0.2000
Singapore	-	-	-	-	-	-	0.2000	0.2310	0.2000	0.2000	0.2000	0.2000
Sri Lanka	0.6150	0.6150	0.6150	0.6150	0.6150	0.6150	0.2000	0.2000	0.2000	0.2000	0.2000	0.2000
Thailand	0.6300	0.6300	0.6300	0.6300	0.6300	0.6300	0.2900	0.2900	0.2900	0.2580	0.4359	0.2900
Vietnam	0.4994	0.4994	0.4994	0.4994	0.4994	0.4994	0.2000	0.2000	0.2000	0.2000	0.2000	0.2000
Other Asia	0.6150	0.6150	0.6150	0.6150	0.6150	0.6150	0.2000	0.2000	0.2000	0.2000	0.2000	0.2000

Country Specific Net Calorific Values (tonne of oil equivalent per tonne)
Pouvoirs calorifiques inférieurs spécifiques par pays (tonnes d'équivalent pétrole par tonne)

1997

	Coking Coal						Lignite					
	Production	Imports	Exports	Coke Ovens	Industry	Average	Production	Imports	Exports	Electr. Gen.	Industry	Average
Africa												
Algeria	0.6150	0.6150	0.6150	0.6150	0.6150	0.6150	0.2000	0.2000	0.2000	0.2000	0.2000	0.2000
Angola	0.6150	0.6150	0.6150	0.6150	0.6150	0.6150	-	-	-	-	-	-
Benin	-	-	-	-	-	-	-	-	-	-	-	-
Cameroon	0.6150	0.6150	0.6150	0.6150	0.6150	0.6150	-	-	-	-	-	-
Congo	-	-	-	-	-	-	-	-	-	-	-	-
DR of Congo	0.6026	0.6026	0.6026	0.6026	0.6026	0.6026	0.2000	0.2000	0.2000	0.2000	0.2000	0.2000
Egypt	0.6150	0.6150	0.6150	0.6150	0.6150	0.6150	0.2000	0.2000	0.2000	0.2000	0.2000	0.2000
Ethiopia	-	-	-	-	-	-	-	-	-	-	-	-
Gabon	0.6150	0.6150	0.6150	0.6150	0.6150	0.6150	0.2000	0.2000	0.2000	0.2000	0.2000	0.2000
Ghana	0.6150	0.6150	0.6150	0.6150	0.6150	0.6150	-	-	-	-	-	-
Ivory Coast	0.6150	0.6150	0.6150	0.6150	0.6150	0.6150	0.2000	0.2000	0.2000	0.2000	0.2000	0.2000
Kenya	0.6150	0.6150	0.6150	0.6150	0.6150	0.6150	0.2000	0.2000	0.2000	0.2000	0.2000	0.2000
Libya	0.6150	0.6150	0.6150	0.6150	0.6150	0.6150	0.2000	0.2000	0.2000	0.2000	0.2000	0.2000
Morocco	0.5600	0.6600	0.5600	0.5900	0.5900	0.5900	0.2000	0.2000	0.2000	0.2000	0.2000	0.2000
Mozambique	0.6150	0.6150	0.6150	0.6150	0.6150	0.6150	0.2000	0.2000	0.2000	0.2000	0.2000	0.2000
Nigeria	0.6150	0.6150	0.6150	0.6150	0.6150	0.6150	0.2000	0.2000	0.2000	0.2000	0.2000	0.2000
Senegal	0.6150	0.6150	0.6150	0.6150	0.6150	0.6150	0.2000	0.2000	0.2000	0.2000	0.2000	0.2000
South Africa	0.7403	0.7403	0.7403	0.7403	0.7403	0.7403	0.2000	0.2000	0.2000	0.2000	0.2000	0.2000
Sudan	-	-	-	-	-	-	-	-	-	-	-	-
Tanzania	0.6150	0.6150	0.6150	0.6150	0.6150	0.6150	-	-	-	-	-	-
Tunisia	0.6150	0.6150	0.6150	0.6150	0.6150	0.6150	0.2000	0.2000	0.2000	0.2000	0.2000	0.2000
Zambia	0.5901	0.5901	0.5901	0.5901	0.5901	0.5901	0.2000	0.2000	0.2000	0.2000	0.2000	0.2000
Zimbabwe	0.6150	0.6150	0.6150	0.6150	0.6150	0.6150	0.2000	0.2000	0.2000	0.2000	0.2000	0.2000
Other Africa	0.6150	0.6150	0.6150	0.6150	0.6150	0.6150	0.2000	0.2000	0.2000	0.2000	0.2000	0.2000
Latin America												
Argentina	0.5900	0.7200	0.5900	0.7200	0.7200	0.5900	0.2000	0.2000	0.2000	0.2000	0.2000	0.2000
Bolivia	0.6150	0.6150	0.6150	0.6150	0.6150	0.6150	0.2000	0.2000	0.2000	0.2000	0.2000	0.2000
Brazil	0.6250	0.7330	-	0.7320	0.7330	0.7410	-	-	-	-	-	-
Chile	0.6790	0.6790	0.6790	0.6790	0.6790	0.6790	0.4102	0.2000	0.4102	0.4102	0.4102	0.4102
Colombia	0.6500	0.6500	0.6500	0.6500	0.6500	0.6500	0.2000	0.2000	0.2000	0.2000	0.2000	0.2000
Costa Rica	0.6150	0.6150	0.6150	0.6150	0.6150	0.6150	0.2000	0.2000	0.2000	0.2000	0.2000	0.2000
Cuba	0.6150	0.6150	0.6150	0.6150	0.6150	0.6150	0.2000	0.2000	0.2000	0.2000	0.2000	0.2000
Dominican Republic	0.6150	0.6150	0.6150	0.6150	0.6150	0.6150	0.2000	0.2000	0.2000	0.2000	0.2000	0.2000
Ecuador	0.6150	0.6150	0.6150	0.6150	0.6150	0.6150	0.2000	0.2000	0.2000	0.2000	0.2000	0.2000
El Salvador	0.6150	0.6150	0.6150	0.6150	0.6150	0.6150	0.2000	0.2000	0.2000	0.2000	0.2000	0.2000
Guatemala	-	-	-	-	-	-	-	-	-	-	-	-
Haiti	0.6150	0.6150	0.6150	0.6150	0.6150	0.6150	0.2000	0.2000	0.2000	0.2000	0.2000	0.2000
Honduras	0.6150	0.6150	0.6150	0.6150	0.6150	0.6150	0.2000	0.2000	0.2000	0.2000	0.2000	0.2000
Jamaica	0.6150	0.6150	0.6150	0.6150	0.6150	0.6150	0.2000	0.2000	0.2000	0.2000	0.2000	0.2000
Netherlands Antilles	0.6150	0.6150	0.6150	0.6150	0.6150	0.6150	0.2000	0.2000	0.2000	0.2000	0.2000	0.2000
Nicaragua	0.6150	0.6150	0.6150	0.6150	0.6150	0.6150	0.2000	0.2000	0.2000	0.2000	0.2000	0.2000
Panama	0.6150	0.6150	0.6150	0.6150	0.6150	0.6150	0.2000	0.2000	0.2000	0.2000	0.2000	0.2000
Paraguay	0.6150	0.6150	0.6150	0.6150	0.6150	0.6150	0.2000	0.2000	0.2000	0.2000	0.2000	0.2000
Peru	0.7000	0.7000	0.7000	0.7000	0.7000	0.7000	-	-	-	-	-	-
Trinidad-and-Tobago	0.6150	0.6150	0.6150	0.6150	0.6150	0.6150	0.2000	0.2000	0.2000	0.2000	0.2000	0.2000
Uruguay	0.6150	0.6150	0.6150	0.6150	0.6150	0.6150	0.2000	0.2000	0.2000	0.2000	0.2000	0.2000
Venezuela	0.6150	0.6150	0.6150	0.6150	0.6150	0.6150	0.2000	0.2000	0.2000	0.2000	0.2000	0.2000
Other Latin America	0.6150	0.6150	0.6150	0.6150	0.6150	0.6150	0.2000	0.2000	0.2000	0.2000	0.2000	0.2000
Asia												
Bangladesh	0.4998	0.4998	0.4998	0.4998	0.4998	0.4998	0.2000	0.2000	0.2000	0.2000	0.2000	0.2000
Brunei	0.6150	0.6150	0.6150	0.6150	0.6150	0.6150	0.2000	0.2000	0.2000	0.2000	0.2000	0.2000
Chinese Taipei	0.6300	0.6300	0.6300	0.6300	0.6300	0.6300	0.2900	0.2900	0.2900	0.2580	0.4359	0.2900
India	0.4773	0.6150	0.4773	0.4773	0.4773	0.4773	0.2340	0.2000	0.2340	0.2340	0.2340	0.2340
Indonesia	0.6150	0.6150	0.6150	0.6150	0.6150	0.6150	0.2000	0.2000	0.2000	0.2000	0.2000	0.2000
DPR of Korea	0.6150	0.6150	0.6150	0.6150	0.6150	0.6150	0.4200	0.2000	0.4200	0.4200	0.4200	0.4200
Malaysia	0.7000	0.7000	0.7000	0.7000	0.7000	0.7000	0.2690	0.2690	0.2690	0.2690	0.2690	0.2690
Myanmar	0.6150	0.6150	0.6150	0.6150	0.6150	0.6150	0.2000	0.2000	0.2000	0.2000	0.2000	0.2000
Nepal	0.6000	0.6000	0.6000	0.6000	0.6000	0.6000	0.2000	0.2000	0.2000	0.2000	0.2000	0.2000
Pakistan	0.4474	0.6579	0.4474	0.6579	0.4474	0.4474	0.2000	0.2000	0.2000	0.2000	0.2000	0.2000
Philippines	0.4800	0.4900	0.4800	0.4800	0.4800	0.4800	0.2000	0.2000	0.2000	0.2000	0.2000	0.2000
Singapore	-	-	-	-	-	-	0.2000	0.2310	0.2000	0.2000	0.2000	0.2000
Sri Lanka	0.6150	0.6150	0.6150	0.6150	0.6150	0.6150	0.2000	0.2000	0.2000	0.2000	0.2000	0.2000
Thailand	0.6300	0.6300	0.6300	0.6300	0.6300	0.6300	0.2900	0.2900	0.2900	0.2580	0.4359	0.2900
Vietnam	0.4994	0.4994	0.4994	0.4994	0.4994	0.4994	0.2000	0.2000	0.2000	0.2000	0.2000	0.2000
Other Asia	0.6150	0.6150	0.6150	0.6150	0.6150	0.6150	0.2000	0.2000	0.2000	0.2000	0.2000	0.2000

Country Specific Net Calorific Values (tonne of oil equivalent per tonne)
Pouvoirs calorifiques inférieurs spécifiques par pays (tonnes d'équivalent pétrole par tonne)

1997

	Peat	Coal Products			Oil		
		Patent Fuel	Coke Oven Coke	BKB	Crude Oil	Bbls/ton[1]	NGL
Africa							
Algeria	0.2000	0.7000	0.6500	0.4800	1.0340	8.1300	1.0340
Angola	0.2000	-	-	-	1.0210	7.4100	-
Benin	0.2000	-	-	-	1.0170	7.3000	-
Cameroon	0.2000	-	-	-	1.0140	7.2100	-
Congo	0.2000	-	-	-	1.0250	7.1000	-
DR of Congo	0.2000	0.7000	0.6500	0.4800	1.0070	7.3200	1.0200
Egypt	0.2000	-	0.6500	0.4800	1.0160	7.2600	1.0160
Ethiopia	0.2000	-	-	-	1.0180	7.3000	-
Gabon	0.2000	-	0.6500	0.4800	1.0180	7.3100	1.0145
Ghana	0.2000	-	0.6500	-	1.0180	7.2900	-
Ivory Coast	0.2000	-	0.6500	0.4800	1.0180	7.2900	1.0180
Kenya	0.2000	-	0.6500	0.4800	1.0050	7.3000	1.0200
Libya	0.2000	-	0.6500	0.4800	1.0270	7.5800	1.0270
Morocco	0.2000	0.6500	0.6500	0.4800	0.9300	7.6000	1.0270
Mozambique	0.2000	-	0.6500	0.4800	1.0070	7.3000	1.0200
Nigeria	0.2000	-	0.6500	0.4800	1.0210	7.5000	1.0210
Senegal	0.2000	-	0.6500	0.4800	1.0180	7.3000	1.0180
South Africa	0.2000	-	0.6329	0.4800	0.9678	8.5300	1.0209
Sudan	0.2000	-	0.6500	-	1.0180	7.3000	-
Tanzania	0.2000	-	0.6500	-	1.0180	7.3000	-
Tunisia	0.2000	-	0.6500	0.4800	1.0300	7.6900	1.0300
Zambia	0.2000	-	0.6500	0.4800	1.0070	7.3000	1.0200
Zimbabwe	0.2000	-	0.6500	0.4800	1.0070	7.3000	1.0200
Other Africa	0.2000	0.7000	0.6500	0.4800	1.0070	7.3000	1.0200
Latin America							
Argentina	0.2000	-	0.6797	0.4800	1.0100	7.1200	1.0150
Bolivia	0.2000	-	0.6500	0.4800	1.0350	7.8800	1.0350
Brazil	-	-	0.7300	-	1.0900	7.2450	-
Chile	0.2000	-	0.6790	0.4800	1.0870	7.6200	1.0240
Colombia	0.2000	-	0.4800	0.4800	1.0090	7.0800	1.0000
Costa Rica	0.2000	0.7000	0.6500	0.4800	1.0070	7.3000	1.0200
Cuba	0.2000	-	0.6500	0.4800	0.9830	6.4500	0.9960
Dominican Republic	0.2000	0.7000	0.6500	0.4800	1.0070	7.1500	1.0200
Ecuador	0.2000	-	0.6500	0.4800	1.0000	7.0000	1.0140
El Salvador	0.2000	0.7000	0.6500	0.4800	1.0070	6.9540	1.0200
Guatemala	0.2000	-	-	-	1.0140	6.7000	-
Haiti	0.2000	0.7000	0.6500	0.4800	1.0070	7.3000	1.0200
Honduras	0.2000	0.7000	0.6500	0.4800	1.0070	7.3000	1.0200
Jamaica	0.2000	-	0.6500	0.4800	1.0070	7.3000	1.0200
Netherlands Antilles	0.2000	-	0.6500	0.4800	1.0070	7.3000	1.0200
Nicaragua	0.2000	0.7000	0.6500	0.4800	1.0070	7.3000	1.0200
Panama	0.2000	-	0.6500	0.4800	1.0070	7.3000	1.0200
Paraguay	0.2000	-	0.6500	0.4800	1.0160	7.3000	1.0160
Peru	0.2000	-	0.6500	-	1.0210	7.4100	1.0210
Trinidad-and-Tobago	0.2000	-	0.6500	0.4800	1.0090	7.0800	1.0000
Uruguay	0.2000	-	0.6500	0.4800	1.0200	5.5970	1.0200
Venezuela	0.2000	-	0.6500	0.4800	1.0045	6.9540	1.0030
Other Latin America	0.2000	0.7000	0.6500	0.4800	1.0070	7.3000	1.0200
Asia							
Bangladesh	0.2000	-	0.6500	0.4800	1.0070	7.4500	1.0200
Brunei	0.2000	-	0.6500	0.4800	1.0210	7.3400	1.0210
Chinese Taipei	0.2000	0.6300	0.6500	0.4800	1.0180	7.3100	1.1190
India	0.2000	-	0.6500	0.4800	1.0220	7.4400	1.0270
Indonesia	0.2000	-	0.6500	0.4800	1.0190	7.3600	1.0215
DPR of Korea	0.2000	-	0.6500	0.4800	1.0070	-	1.0200
Malaysia	0.2000	-	0.6300	0.4800	1.0340	7.6000	1.0580
Myanmar	0.2000	-	0.6500	-	1.0090	7.0800	1.0200
Nepal	0.2000	-	0.6500	0.4800	1.0160	-	1.0160
Pakistan	0.2000	-	0.6500	0.4800	1.0240	7.6000	1.0240
Philippines	0.2000	-	0.6500	0.4800	1.0170	7.2900	1.0000
Singapore	0.2000	-	0.6500	-	1.0200	-	1.0200
Sri Lanka	0.2000	-	0.6500	0.4800	1.0070	-	1.0200
Thailand	0.2000	0.6300	0.6500	0.4800	1.0180	7.3100	1.1190
Vietnam	0.2000	0.7000	0.6500	0.4800	1.0178	7.3000	1.0200
Other Asia	0.2000	0.7000	0.6500	0.4800	1.0070	7.3000	1.0200

(1) Average barrel per tonne conversion factor for indigenous *production* of crude oil.

Country Specific Net Calorific Values (tonne of oil equivalent per tonne)
Pouvoirs calorifiques inférieurs spécifiques par pays (tonnes d'équivalent pétrole par tonne)

1997

	Bituminous Coal and Anthracite						Sub-Bituminous Coal					
	Production	Imports	Exports	Electr. Gen.	Industry	Average	Production	Imports	Exports	Electr. Gen.	Industry	Average
China												
People's Rep. of China	0.5000	0.5000	0.5310	0.4650	0.4990	0.4990	0.2000	0.2000	0.2000	0.2000	0.2000	0.2000
Hong Kong, China	0.6150	0.6150	0.6150	0.6150	0.6150	0.6150	0.2000	0.2000	0.2000	0.2000	0.2000	0.2000
Non-OECD Europe												
Albania	0.6500	0.6500	0.6500	0.6500	0.6500	0.6500	0.2351	0.2351	0.2351	0.2351	0.2351	0.2351
Bulgaria	0.5657	0.6184	0.6000	0.6000	0.6000	0.6000	0.1634	0.1650	0.1650	0.1650	0.1650	0.1630
Cyprus	0.6150	0.6150	0.6150	0.6150	0.6150	0.6150	-	-	-	-	-	-
Gibraltar	0.6150	0.6150	0.6150	0.6150	0.6150	0.6150	0.2000	0.2000	0.2000	0.2000	0.2000	0.2000
Malta	0.6150	0.6150	0.6150	0.6150	0.6150	0.6150	0.2000	0.2000	0.2000	0.2000	0.2000	0.2000
Romania	0.3900	0.6000	0.3900	0.5100	0.3900	0.3900	0.1730	0.1730	0.1730	0.1600	0.1730	0.1730
Slovak Republic	0.5829	0.5713	0.6684	0.4792	0.4900	0.5995	0.2928	0.2914	0.3644	0.2914	0.2713	0.2914
Former Yugoslavia	0.5624	0.7329	0.7197	0.5524	0.6340	0.6197	0.2124	0.4039	0.4037	0.1966	0.4079	0.3117
Bosnia-Herzegovina	*0.5624*	*0.7329*	*0.7197*	*0.5524*	*0.634*	*0.6197*	*0.2124*	*0.4039*	*0.4037*	*0.1966*	*0.4079*	*0.3117*
Croatia	*0.6*	*0.7001*	*-*	*0.6306*	*0.6708*	*0.6441*	*-*	*0.3486*	*-*	*-*	*0.3561*	*0.3486*
FYROM	*0.5624*	*0.7329*	*0.7197*	*0.5524*	*0.634*	*0.6197*	*0.2124*	*0.4039*	*0.4037*	*0.1966*	*0.4079*	*0.3117*
Slovenia	*0.5624*	*0.7329*	*0.7197*	*0.5524*	*0.634*	*0.6197*	*0.2124*	*0.4039*	*0.4037*	*0.1966*	*0.4079*	*0.3117*
FR of Yugoslavia	*0.5624*	*0.7329*	*0.7197*	*0.5524*	*0.634*	*0.6197*	*0.2124*	*0.4039*	*0.4037*	*0.1966*	*0.4079*	*0.3117*
Former USSR Republics												
Armenia	-	0.4438	0.4438	0.4438	0.4438	0.4438	0.3500	0.3500	0.3500	0.3500	0.3500	0.3500
Azerbaijan	-	0.4438	0.4438	0.4438	0.4438	0.4438	0.3500	0.3500	0.3500	0.3500	0.3500	0.3500
Belarus	-	0.6100	0.6100	0.6100	0.6100	0.6100	0.3500	0.3500	0.3500	0.3500	0.3500	0.3500
Estonia	-	0.4438	0.4438	0.4438	0.4438	0.4438	0.2254	0.2254	0.2254	0.2254	0.2254	0.2254
Georgia	0.4438	0.4438	0.4438	0.4438	0.4438	0.4438	0.3500	0.3500	0.3500	0.3500	0.3500	0.3500
Kazakhstan	0.4438	0.4438	0.4438	0.4438	0.4438	0.4438	0.3500	0.3500	0.3500	0.3500	0.3500	0.3500
Kyrgyzstan	0.4438	0.4438	0.4438	0.4438	0.4438	0.4438	0.3500	0.3500	0.3500	0.3500	0.3500	0.3500
Latvia	-	0.4438	0.5999	0.5999	0.5999	0.5999	0.3500	0.3500	0.3500	0.3500	0.3500	0.3500
Lithuania	-	0.4440	0.4440	0.4440	0.4440	0.4440	0.3500	0.3500	0.3500	0.3500	0.3500	0.3500
Moldova	-	0.4438	0.4438	0.4438	0.4438	0.4438	0.3500	0.3500	0.3500	0.3500	0.3500	0.3500
Russia	0.4438	0.4438	0.4438	0.4438	0.4438	0.4438	0.3500	0.3500	0.3500	0.3500	0.3500	0.3500
Tajikistan	0.4438	0.4438	0.4438	0.4438	0.4438	0.4438	0.3500	0.3500	0.3500	0.3500	0.3500	0.3500
Turkmenistan	-	0.4438	0.4438	0.4438	0.4438	0.4438	0.3500	0.3500	0.3500	0.3500	0.3500	0.3500
Ukraine	0.5156	0.6100	0.5156	0.5156	0.5156	0.5156	0.3500	0.3500	0.3500	0.3500	0.3500	0.3500
Uzbekistan	0.4438	0.4438	0.4438	0.4438	0.4438	0.4438	0.3500	0.3500	0.3500	0.3500	0.3500	0.3500
Former USSR	0.4438	0.4438	0.4438	0.4438	0.4438	0.4438	0.3500	0.3500	0.3500	0.3500	0.3500	0.3500
Middle East												
Bahrain	0.6150	0.6150	0.6150	0.6150	0.6150	0.6150	0.2000	0.2000	0.2000	0.2000	0.2000	0.2000
Iran	0.6150	0.6150	0.6150	0.6150	0.6150	0.6150	0.2000	0.2000	0.2000	0.2000	0.2000	0.2000
Iraq	0.6150	0.6150	0.6150	0.6150	0.6150	0.6150	0.2000	0.2000	0.2000	0.2000	0.2000	0.2000
Israel	0.6262	0.6262	0.6262	0.6262	0.6262	0.6262	0.2000	-	0.2000	0.2000	0.2000	0.2000
Jordan	0.6150	0.6150	0.6150	0.6150	0.6150	0.6150	0.2000	0.2000	0.2000	0.2000	0.2000	0.2000
Kuwait	0.6150	0.6150	0.6150	0.6150	0.6150	0.6150	0.2000	0.2000	0.2000	0.2000	0.2000	0.2000
Lebanon	0.6610	0.6610	0.6610	0.6610	0.6610	0.6610	0.2000	0.2000	0.2000	0.2000	0.2000	0.2000
Oman	0.6150	0.6150	0.6150	0.6150	0.6150	0.6150	0.2000	0.2000	0.2000	0.2000	0.2000	0.2000
Qatar	0.6150	0.6150	0.6150	0.6150	0.6150	0.6150	0.2000	0.2000	0.2000	0.2000	0.2000	0.2000
Saudi Arabia	0.6150	0.6150	0.6150	0.6150	0.6150	0.6150	0.2000	0.2000	0.2000	0.2000	0.2000	0.2000
Syria	0.6150	0.6150	0.6150	0.6150	0.6150	0.6150	0.2000	0.2000	0.2000	0.2000	0.2000	0.2000
United Arab Emirates	0.6150	0.6150	0.6150	0.6150	0.6150	0.6150	0.2000	0.2000	0.2000	0.2000	0.2000	0.2000
Yemen	0.6150	0.6150	0.6150	0.6150	0.6150	0.6150	0.2000	0.2000	0.2000	0.2000	0.2000	0.2000

Country Specific Net Calorific Values (tonne of oil equivalent per tonne)
Pouvoirs calorifiques inférieurs spécifiques par pays (tonnes d'équivalent pétrole par tonne)

1997

	Coking Coal						Lignite					
	Production	Imports	Exports	Coke Ovens	Industry	Average	Production	Imports	Exports	Electr. Gen.	Industry	Average
China												
People's Rep. of China	0.5000	0.5000	0.5310	0.5980	0.5410	0.4990	0.2000	0.2000	0.2000	0.2000	0.2000	0.2000
Hong Kong, China	0.6150	0.6150	0.6150	0.6150	0.6150	0.6150	0.2000	0.2000	0.2000	0.2000	0.2000	0.2000
Non-OECD Europe												
Albania	0.6500	0.6500	0.6500	0.6500	0.6500	0.6500	0.2351	0.2351	0.2351	0.2351	0.2351	0.2351
Bulgaria	0.5657	0.6184	0.6000	0.6000	0.6000	0.6000	0.1634	0.1650	0.1650	0.1650	0.1650	0.1630
Cyprus	0.6150	0.6150	0.6150	0.6150	0.6150	0.6150	-	-	-	-	-	-
Gibraltar	0.6150	0.6150	0.6150	0.6150	0.6150	0.6150	0.2000	0.2000	0.2000	0.2000	0.2000	0.2000
Malta	0.6150	0.6150	0.6150	0.6150	0.6150	0.6150	0.2000	0.2000	0.2000	0.2000	0.2000	0.2000
Romania	0.3900	0.6000	0.3900	0.6000	0.3900	0.3900	0.1730	0.1730	0.1730	0.1600	0.1730	0.1730
Slovak Republic	0.5829	0.5713	0.6684	0.6675	0.4900	0.5995	0.2928	0.2914	0.3644	0.2914	0.2713	0.2914
Former Yugoslavia	0.5624	0.7329	0.7197	0.7239	0.6340	0.6197	0.2124	0.4039	0.4037	0.1966	0.4079	0.3117
Bosnia-Herzegovina	*0.5624*	*0.7329*	*0.7197*	*0.7239*	*0.634*	*0.6197*	*0.2124*	*0.4039*	*0.4037*	*0.1966*	*0.4079*	*0.3117*
Croatia	*-*	*0.7001*	*-*	*0.7001*	*-*	*0.7001*	*-*	*0.3486*	*-*	*-*	*0.3561*	*0.3486*
FYROM	*0.5624*	*0.7329*	*0.7197*	*0.7239*	*0.634*	*0.6197*	*0.2124*	*0.4039*	*0.4037*	*0.1966*	*0.4079*	*0.3117*
Slovenia	*0.5624*	*0.7329*	*0.7197*	*0.7239*	*0.634*	*0.6197*	*0.2124*	*0.4039*	*0.4037*	*0.1966*	*0.4079*	*0.3117*
FR of Yugoslavia	*0.5624*	*0.7329*	*0.7197*	*0.7239*	*0.634*	*0.6197*	*0.2124*	*0.4039*	*0.4037*	*0.1966*	*0.4079*	*0.3117*
Former USSR Republics												
Armenia	-	0.4438	0.4438	0.4438	0.4438	0.4438	0.3500	0.3500	0.3500	0.3500	0.3500	0.3500
Azerbaijan	-	0.4438	0.4438	0.4438	0.4438	0.4438	0.3500	0.3500	0.3500	0.3500	0.3500	0.3500
Belarus	-	0.6100	0.6100	0.6100	0.6100	0.6100	0.3500	0.3500	0.3500	0.3500	0.3500	0.3500
Estonia	-	0.4438	0.4438	0.4438	0.4438	0.4438	0.2254	0.2254	0.2254	0.2254	0.2254	0.2254
Georgia	0.4438	0.4438	0.4438	0.4438	0.4438	0.4438	0.3500	0.3500	0.3500	0.3500	0.3500	0.3500
Kazakhstan	0.4438	0.4438	0.4438	0.4438	0.4438	0.4438	0.3500	0.3500	0.3500	0.3500	0.3500	0.3500
Kyrgyzstan	0.4438	0.4438	0.4438	0.4438	0.4438	0.4438	0.3500	0.3500	0.3500	0.3500	0.3500	0.3500
Latvia	-	0.4438	0.5999	0.5999	0.5999	0.5999	0.3500	0.3500	0.3500	0.3500	0.3500	0.3500
Lithuania	-	0.4440	0.4440	0.4440	0.4440	0.4440	0.3500	0.3500	0.3500	0.3500	0.3500	0.3500
Moldova	-	0.4438	0.4438	0.4438	0.4438	0.4438	0.3500	0.3500	0.3500	0.3500	0.3500	0.3500
Russia	0.4438	0.4438	0.4438	0.4438	0.4438	0.4438	0.3500	0.3500	0.3500	0.3500	0.3500	0.3500
Tajikistan	0.4438	0.4438	0.4438	0.4438	0.4438	0.4438	0.3500	0.3500	0.3500	0.3500	0.3500	0.3500
Turkmenistan	-	0.4438	0.4438	0.4438	0.4438	0.4438	0.3500	0.3500	0.3500	0.3500	0.3500	0.3500
Ukraine	0.5156	0.6100	0.5156	0.5156	0.5156	0.5156	0.3500	0.3500	0.3500	0.3500	0.3500	0.3500
Uzbekistan	0.4438	0.4438	0.4438	0.4438	0.4438	0.4438	0.3500	0.3500	0.3500	0.3500	0.3500	0.3500
Former USSR	0.4438	0.4438	0.4438	0.4438	0.4438	0.4438	0.3500	0.3500	0.3500	0.3500	0.3500	0.3500
Middle East												
Bahrain	0.6150	0.6150	0.6150	0.6150	0.6150	0.6150	0.2000	0.2000	0.2000	0.2000	0.2000	0.2000
Iran	0.6150	0.6150	0.6150	0.6150	0.6150	0.6150	0.2000	0.2000	0.2000	0.2000	0.2000	0.2000
Iraq	0.6150	0.6150	0.6150	0.6150	0.6150	0.6150	0.2000	0.2000	0.2000	0.2000	0.2000	0.2000
Israel	0.6262	0.6262	0.6262	0.6262	0.6262	0.6262	0.1000	-	0.1000	0.1000	0.1000	0.1000
Jordan	0.6150	0.6150	0.6150	0.6150	0.6150	0.6150	0.2000	0.2000	0.2000	0.2000	0.2000	0.2000
Kuwait	0.6150	0.6150	0.6150	0.6150	0.6150	0.6150	0.2000	0.2000	0.2000	0.2000	0.2000	0.2000
Lebanon	0.6150	0.6150	0.6150	0.6150	0.6150	0.6150	0.2000	0.2000	0.2000	0.2000	0.2000	0.2000
Oman	0.6150	0.6150	0.6150	0.6150	0.6150	0.6150	0.2000	0.2000	0.2000	0.2000	0.2000	0.2000
Qatar	0.6150	0.6150	0.6150	0.6150	0.6150	0.6150	0.2000	0.2000	0.2000	0.2000	0.2000	0.2000
Saudi Arabia	0.6150	0.6150	0.6150	0.6150	0.6150	0.6150	0.2000	0.2000	0.2000	0.2000	0.2000	0.2000
Syria	0.6150	0.6150	0.6150	0.6150	0.6150	0.6150	0.2000	0.2000	0.2000	0.2000	0.2000	0.2000
United Arab Emirates	0.6150	0.6150	0.6150	0.6150	0.6150	0.6150	0.2000	0.2000	0.2000	0.2000	0.2000	0.2000
Yemen	0.6150	0.6150	0.6150	0.6150	0.6150	0.6150	0.2000	0.2000	0.2000	0.2000	0.2000	0.2000

Country Specific Net Calorific Values (tonne of oil equivalent per tonne)
Pouvoirs calorifiques inférieurs spécifiques par pays (tonnes d'équivalent pétrole par tonne)

1997

	Peat	Coal Products			Oil		
		Patent Fuel	Coke Oven Coke	BKB	Crude Oil	Bbls/ton[1]	NGL
China							
People's Rep. of China	0.2000	0.4170	0.6800	0.4310	1.0000	7.3200	-
Hong Kong, China	0.2000	-	0.6500	0.4800	1.0070	-	1.0200
Non-OECD Europe							
Albania	0.2000	-	0.6500	-	0.9900	6.6000	1.0790
Bulgaria	0.2000	0.7000	0.6300	0.4380	1.0140	7.3300	1.0790
Cyprus	0.2000	-	0.6500	-	1.0145	-	-
Gibraltar	0.2000	-	0.6500	0.4800	1.0070	-	1.0200
Malta	0.2000	-	0.6500	0.4800	1.0070	-	1.0200
Romania	0.2000	0.3500	0.4970	0.3500	0.9708	7.5050	1.0790
Slovak Republic	0.2000	0.7000	0.6451	0.5082	0.9980	6.7800	1.0790
Former Yugoslavia	0.2000	0.7000	0.6425	0.4800	1.0210	7.4180	1.0800
Bosnia-Herzegovina	*0.2*	*0.7*	*0.6425*	*0.48*	*1.021*	*7.418*	*1.08*
Croatia	*0.2*	-	*0.7001*	*0.48*	*1.021*	*7.418*	*1.08*
FYROM	*0.2*	*0.7*	*0.6425*	*0.48*	*1.021*	*7.418*	*1.08*
Slovenia	*0.2*	*0.7*	*0.6425*	*0.48*	*1.021*	*7.418*	*1.08*
FR of Yugoslavia	*0.2*	*0.7*	*0.6425*	*0.48*	*1.021*	*7.418*	*1.08*
Former USSR Republics							
Armenia	0.2000	0.7000	0.6000	0.4800	1.0050	7.3000	1.0010
Azerbaijan	0.2000	0.7000	0.6000	0.4800	1.0050	7.3000	1.0010
Belarus	0.2000	0.7000	0.6000	0.2000	1.0050	7.3000	1.0010
Estonia	0.2000	0.7000	0.6000	0.2000	1.0050	7.3000	1.0010
Georgia	0.2000	0.7000	0.6000	0.4800	1.0050	7.3000	1.0010
Kazakhstan	0.2000	0.7000	0.6000	0.4800	1.0050	7.3000	1.0010
Kyrgyzstan	0.2000	0.7000	0.6000	0.4800	1.0050	7.3000	1.0010
Latvia	0.2000	0.7000	0.6000	0.2000	1.0050	7.3000	1.0010
Lithuania	0.2000	0.7000	0.6000	0.2000	1.0050	7.3000	1.0010
Moldova	0.2000	0.7000	0.6000	0.4800	1.0050	7.3000	1.0010
Russia	0.2000	0.7000	0.6000	0.4800	1.0050	7.3000	1.0010
Tajikistan	0.2000	0.7000	0.6000	0.4800	1.0050	7.3000	1.0010
Turkmenistan	0.2000	0.7000	0.6000	0.4800	1.0050	7.3000	1.0010
Ukraine	0.2000	0.7000	0.6000	0.4800	1.0050	7.3000	1.0010
Uzbekistan	0.2000	0.7000	0.6000	0.4800	1.0050	7.3000	1.0010
Former USSR	0.2000	0.7000	0.6000	0.4800	1.0050	7.3000	1.0010
Middle East							
Bahrain	0.2000	-	0.6500	0.4800	1.0200	7.3200	1.0200
Iran	0.2000	-	0.6500	0.4800	1.0190	7.3500	1.0160
Iraq	0.2000	-	0.6500	0.4800	1.0230	7.4300	1.0230
Israel	0.2000	-	0.6500	0.4800	1.0160	7.2500	1.0150
Jordan	0.2000	-	0.6500	0.4800	1.0170	7.3000	1.0200
Kuwait	0.2000	-	0.6500	0.4800	1.0160	7.2500	1.0180
Lebanon	0.2000	-	0.6500	0.4800	1.0070	-	1.0200
Oman	0.2000	-	0.6500	0.4800	1.0200	7.3300	1.0200
Qatar	0.2000	-	0.6500	0.4800	1.0240	7.5000	1.0270
Saudi Arabia	0.2000	-	0.6500	0.4800	1.0160	7.3230	1.0180
Syria	0.2000	-	0.6500	0.4800	1.0040	7.2900	1.0040
United Arab Emirates	0.2000	-	0.6500	0.4800	1.0180	7.5960	1.0180
Yemen	0.2000	-	0.6500	0.4800	1.0270	7.6200	1.0160

(1) Average barrel per tonne conversion factor for indigenous *production* of crude oil.

GRAPHS AND ENERGY BALANCE SHEETS

GRAPHIQUES ET BILANS ENERGETIQUES

1996-1997

REGIONAL TOTALS

TOTAUX REGIONAUX

INTERNATIONAL ENERGY AGENCY

World / Monde

Total Production of Energy by Region

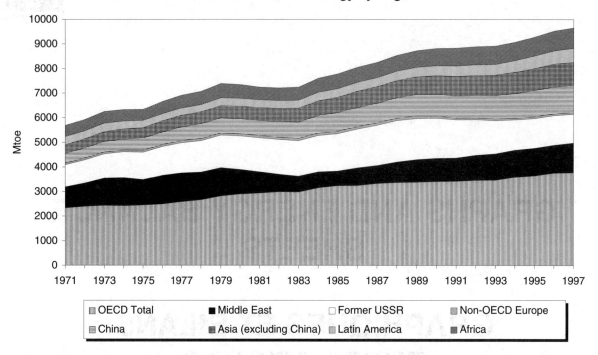

Total Primary Energy Supply (TPES) by Region

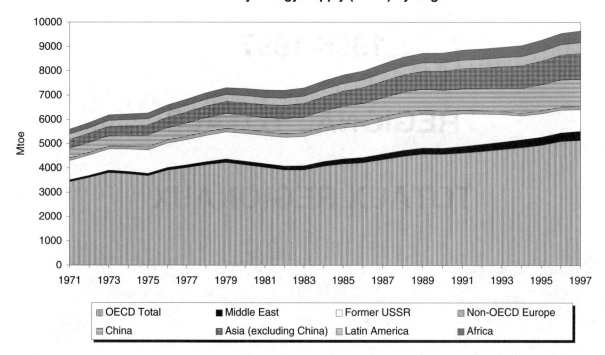

World / Monde

Production of Crude Oil and NGL by Region

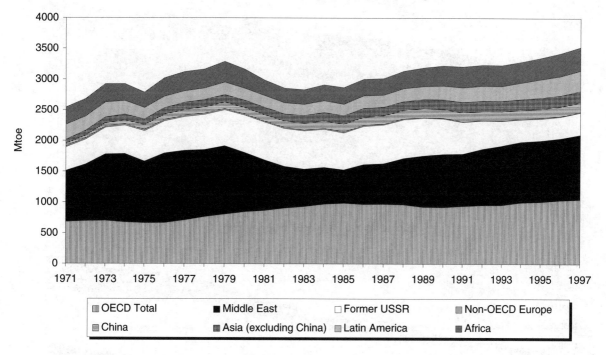

▦ OECD Total	■ Middle East	☐ Former USSR	▧ Non-OECD Europe
▤ China	▦ Asia (excluding China)	▧ Latin America	▨ Africa

Primary Supply of Oil by Region

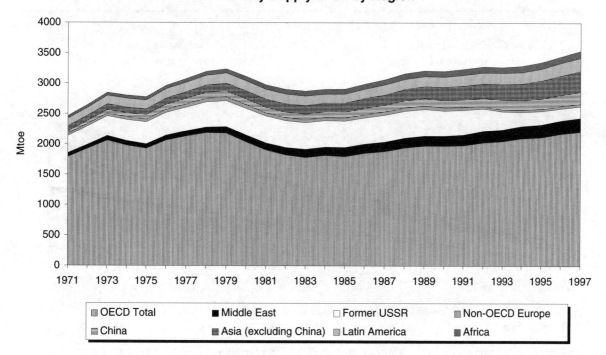

▦ OECD Total	■ Middle East	☐ Former USSR	▧ Non-OECD Europe
▤ China	▦ Asia (excluding China)	▧ Latin America	▨ Africa

World / Monde

Production of Coal by Region

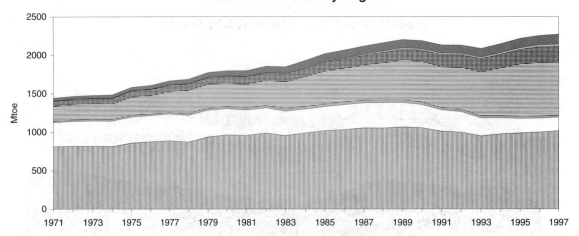

Production of Natural Gas by Region

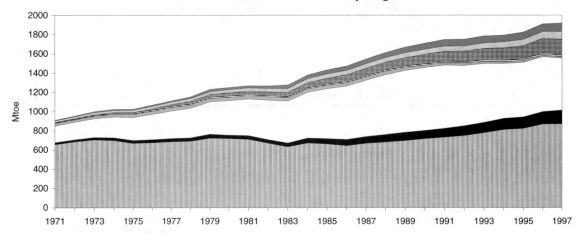

Electricity Generation* by Region

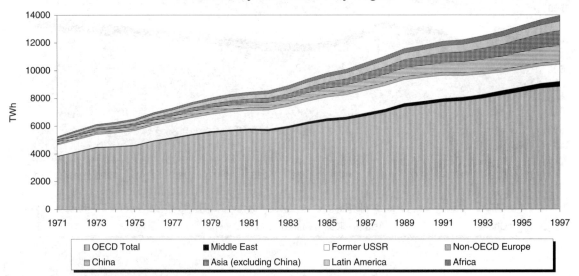

OECD Total	Middle East	Former USSR	Non-OECD Europe
China	Asia (excluding China)	Latin America	Africa

* Excludes production from pumped storage.

INTERNATIONAL ENERGY AGENCY

World / Monde

TPES/GDP in 1990 US$

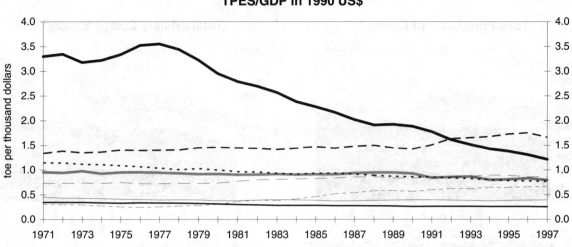

TPES/GDP in 1990 US$ PPP

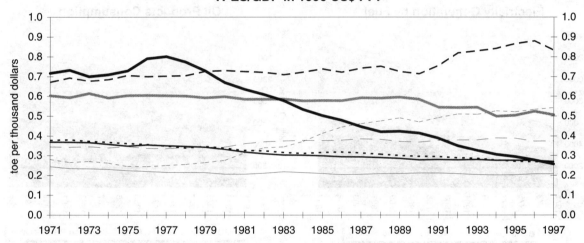

TPES/Population

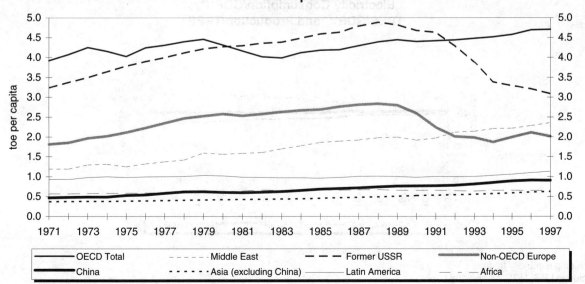

—— OECD Total	– – Middle East	– – – Former USSR	▬▬ Non-OECD Europe
▬ China	···· Asia (excluding China)	— Latin America	– – Africa

World / Monde

Total Production of Energy

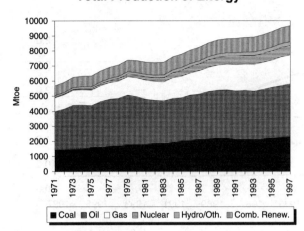

Total Primary Energy Supply*

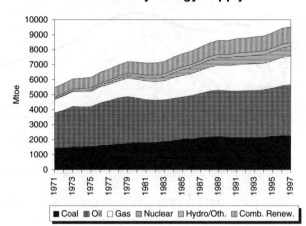

Electricity Generation by Fuel

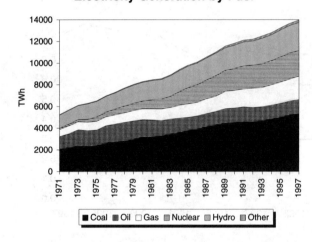

Oil Products Consumption

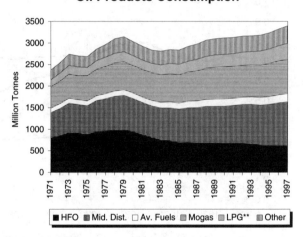

Electricity Consumption/GDP***, TPES/GDP*** and Production/TPES

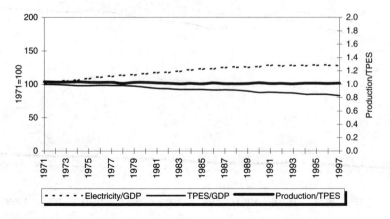

* Excluding electricity trade.
** Includes LPG, NGL, ethane and naphtha.
*** GDP in 1990 US dollars.

INTERNATIONAL ENERGY AGENCY

World / Monde : 1996

	Coal	Crude Oil	Petroleum Products	Gas	Nuclear	Hydro	Geotherm. Solar etc.	Combust. Renew. & Waste	Electricity	Heat	Total
SUPPLY AND CONSUMPTION											
APPROVISIONNEMENT ET DEMANDE	Charbon	Pétrole brut	Produits pétroliers	Gaz	Nucléaire	Hydro	Géotherm. solaire etc.	En. ren. combust. & déchets	Electricité	Chaleur	Total
Indigenous Production	2258.11	3428.39	-	1909.04	630.31	217.80	38.47	1053.07	-	0.31	9535.51
Imports	334.84	1858.35	651.87	445.29	-	-	-	1.16	36.42	-	3327.92
Exports	-339.67	-1784.99	-697.22	-444.94	-	-	-	-0.96	-36.50	-	-3304.29
Stock Changes	22.41	-13.66	3.52	-15.73	-	-	-	-0.22	-	-	-3.68
TPES	**2275.69**	**3488.10**	**-41.83**	**1893.66**	**630.31**	**217.80**	**38.47**	**1053.04**	**-0.09**	**0.31**	**9555.46**
Intl. Marine Bunkers	-	-	-129.12	-	-	-	-	-	-	-	-129.12
Transfers	-	-74.53	77.55	-	-	-	-	-0.16	-	-	2.86
Statistical Differences	21.06	15.73	-7.53	-0.13	-	-	-	0.11	0.64	0.65	30.52
Electricity Plants	-1132.64	-27.21	-220.30	-282.68	-625.78	-217.80	-35.75	-21.92	1028.83	-	-1535.25
CHP Plants	-227.40	-1.46	-26.87	-254.60	-4.53	-	-0.32	-44.37	145.84	158.99	-254.72
Heat Plants	-29.77	-0.12	-46.30	-71.41	-	-	-0.36	-8.97	-0.52	127.67	-29.79
Gas Works	-6.27	-0.31	-5.33	7.75	-	-	-	-	-	-	-4.15
Petroleum Refineries	-	-3398.94	3348.07	-0.07	-	-	-	-0.97	-	-	-51.91
Coal Transformation	-148.38	-	-	-	-	-	-	-	-	-	-148.38
Liquefaction	-14.17	8.00	-	-2.74	-	-	-	-	-	-	-8.91
Other Transformation	-	20.66	-24.59	-0.68	-	-	-	-44.04	-	-	-48.64
Own Use	-47.90	-7.51	-188.39	-213.55	-	-	-	-0.10	-108.63	-23.25	-589.32
Distribution Losses	-17.14	-6.21	-1.55	-21.51	-	-	-0.07	-0.18	-99.42	-16.85	-162.94
TFC	**673.09**	**16.20**	**2733.81**	**1054.04**	**-**	**-**	**1.96**	**932.45**	**966.65**	**247.52**	**6625.71**
INDUSTRY SECTOR	**488.53**	**13.23**	**519.22**	**466.72**	**-**	**-**	**0.42**	**130.67**	**413.14**	**102.12**	**2134.05**
Iron and Steel	-	-	-	-	-	-	-	-	-	-	-
Chemical and Petrochemical	-	-	-	-	-	-	-	-	-	-	-
of which: Feedstocks	-	-	-	-	-	-	-	-	-	-	-
Non-Ferrous Metals	-	-	-	-	-	-	-	-	-	-	-
Non-Metallic Minerals	-	-	-	-	-	-	-	-	-	-	-
Transport Equipment	-	-	-	-	-	-	-	-	-	-	-
Machinery	-	-	-	-	-	-	-	-	-	-	-
Mining and Quarrying	-	-	-	-	-	-	-	-	-	-	-
Food and Tobacco	-	-	-	-	-	-	-	-	-	-	-
Paper, Pulp and Printing	-	-	-	-	-	-	-	-	-	-	-
Wood and Wood Products	-	-	-	-	-	-	-	-	-	-	-
Construction	-	-	-	-	-	-	-	-	-	-	-
Textile and Leather	-	-	-	-	-	-	-	-	-	-	-
Non-specified	-	-	-	-	-	-	-	-	-	-	-
TRANSPORT SECTOR	**5.81**	**0.01**	**1536.76**	**40.27**	**-**	**-**	**-**	**8.10**	**18.81**	**-**	**1609.76**
Air	-	-	198.39	-	-	-	-	-	0.02	-	198.41
Road	-	-	1266.49	1.87	-	-	-	8.07	0.01	-	1276.44
Rail	-	-	-	-	-	-	-	-	-	-	-
Pipeline Transport	-	-	-	-	-	-	-	-	-	-	-
Internal Navigation	-	-	-	-	-	-	-	-	-	-	-
Non-specified	-	-	-	-	-	-	-	-	-	-	-
OTHER SECTORS	**155.81**	**1.41**	**500.27**	**547.05**	**-**	**-**	**1.54**	**793.69**	**534.70**	**145.39**	**2679.86**
Agriculture	-	-	-	-	-	-	-	-	-	-	-
Comm. and Publ. Services	-	-	-	-	-	-	-	-	-	-	-
Residential	-	-	-	-	-	-	-	-	-	-	-
Non-specified	-	-	-	-	-	-	-	-	-	-	-
NON-ENERGY USE	**22.94**	**1.54**	**177.56**	**-**	**-**	**-**	**-**	**-**	**-**	**-**	**202.04**
in Industry/Transf./Energy	-	-	-	-	-	-	-	-	-	-	-
in Transport	-	-	-	-	-	-	-	-	-	-	-
in Other Sectors	-	-	-	-	-	-	-	-	-	-	-
Electr. Generated - GWh	*5234615*	*-*	*1254920*	*2024486*	*2416589*	*2532516*	*54263*	*141568*	*-*	*-*	*13658957*
Electricity Plants	*-*	*-*	*-*	*-*	*-*	*-*	*-*	*-*	*-*	*-*	*11963128*
CHP Plants	*-*	*-*	*-*	*-*	*-*	*-*	*-*	*-*	*-*	*-*	*1695829*
Heat Generated - TJ	*-*	*-*	*-*	*-*	*-*	*-*	*-*	*-*	*-*	*-*	*12017245*
CHP Plants	*-*	*-*	*-*	*-*	*-*	*-*	*-*	*-*	*-*	*-*	*6663785*
Heat Plants	*-*	*-*	*-*	*-*	*-*	*-*	*-*	*-*	*-*	*-*	*5353460*

Million tonnes of oil equivalent / *Millions de tonnes d'équivalent pétrole*

World / Monde : 1997

Million tonnes of oil equivalent / *Millions de tonnes d'équivalent pétrole*

SUPPLY AND CONSUMPTION *APPROVISIONNEMENT ET DEMANDE*	Coal *Charbon*	Crude Oil *Pétrole brut*	Petroleum Products *Produits pétroliers*	Gas *Gaz*	Nuclear *Nucléaire*	Hydro *Hydro*	Geotherm. Solar etc. *Géotherm. solaire etc.*	Combust. Renew. & Waste *En. ren. combust. & déchets*	Electricity *Electricité*	Heat *Chaleur*	Total *Total*
Indigenous Production	2274.21	3530.47	-	1916.76	624.16	220.66	38.89	1063.68	-	0.32	9669.15
Imports	343.74	1962.14	669.71	440.73	-	-	-	1.00	36.91	-	3454.23
Exports	-347.88	-1870.77	-721.53	-440.68	-	-	-	-1.64	-36.60	-	-3419.10
Stock Changes	-15.10	-17.64	-10.89	-5.63	-	-	-	-1.26	-	-	-50.52
TPES	**2254.97**	**3604.20**	**-62.71**	**1911.17**	**624.16**	**220.66**	**38.89**	**1061.78**	**0.32**	**0.32**	**9653.75**
Intl. Marine Bunkers	-	-	-132.25	-	-	-	-	-	-	-	-132.25
Transfers	-	-74.99	79.32	-	-	-	-	-0.45	-	-	3.88
Statistical Differences	23.90	-2.86	-4.81	-1.28	-	-	-	-0.32	-0.37	0.47	14.72
Electricity Plants	-1158.74	-24.65	-227.78	-302.69	-619.12	-220.66	-36.04	-22.50	1055.79	-	-1556.38
CHP Plants	-214.82	-1.56	-24.75	-253.33	-5.04	-	-0.36	-45.06	143.80	151.81	-249.30
Heat Plants	-27.12	-0.11	-42.58	-68.73	-	-	-0.36	-9.70	-0.60	117.32	-31.88
Gas Works	-7.21	-0.22	-4.88	7.99	-	-	-	-	-	-	-4.33
Petroleum Refineries	-	-3505.48	3446.77	-0.07	-	-	-	-0.92	-	-	-59.70
Coal Transformation	-153.76	-	-	-	-	-	-	-	-	-	-153.76
Liquefaction	-13.79	7.83	-	-2.39	-	-	-	-	-	-	-8.36
Other Transformation	-0.03	25.43	-28.47	-0.63	-	-	-	-45.17	-	-	-48.88
Own Use	-51.93	-7.45	-191.40	-225.54	-	-	-	-0.12	-111.25	-21.95	-609.63
Distribution Losses	-15.96	-2.57	-1.42	-20.39	-	-	-0.07	-0.18	-101.01	-15.76	-157.37
TFC	**635.50**	**17.55**	**2805.05**	**1044.10**	**-**	**-**	**2.06**	**937.35**	**986.68**	**232.21**	**6660.50**
INDUSTRY SECTOR	**468.79**	**14.54**	**536.70**	**469.70**	**-**	**-**	**0.43**	**134.33**	**422.55**	**97.82**	**2144.87**
Iron and Steel	-	-	-	-	-	-	-	-	-	-	-
Chemical and Petrochemical	-	-	-	-	-	-	-	-	-	-	-
of which: Feedstocks	-	-	-	-	-	-	-	-	-	-	-
Non-Ferrous Metals	-	-	-	-	-	-	-	-	-	-	-
Non-Metallic Minerals	-	-	-	-	-	-	-	-	-	-	-
Transport Equipment	-	-	-	-	-	-	-	-	-	-	-
Machinery	-	-	-	-	-	-	-	-	-	-	-
Mining and Quarrying	-	-	-	-	-	-	-	-	-	-	-
Food and Tobacco	-	-	-	-	-	-	-	-	-	-	-
Paper, Pulp and Printing	-	-	-	-	-	-	-	-	-	-	-
Wood and Wood Products	-	-	-	-	-	-	-	-	-	-	-
Construction	-	-	-	-	-	-	-	-	-	-	-
Textile and Leather	-	-	-	-	-	-	-	-	-	-	-
Non-specified	-	-	-	-	-	-	-	-	-	-	-
TRANSPORT SECTOR	**6.81**	**0.01**	**1577.24**	**41.23**	**-**	**-**	**-**	**8.31**	**19.12**	**-**	**1652.71**
Air	-	-	204.64	-	-	-	-	-	0.02	-	204.66
Road	-	-	1296.24	1.84	-	-	-	8.28	0.08	-	1306.43
Rail	-	-	-	-	-	-	-	-	-	-	-
Pipeline Transport	-	-	-	-	-	-	-	-	-	-	-
Internal Navigation	-	-	-	-	-	-	-	-	-	-	-
Non-specified	-	-	-	-	-	-	-	-	-	-	-
OTHER SECTORS	**137.64**	**1.45**	**499.64**	**533.17**	**-**	**-**	**1.64**	**794.72**	**545.00**	**134.39**	**2647.65**
Agriculture	-	-	-	-	-	-	-	-	-	-	-
Comm. and Publ. Services	-	-	-	-	-	-	-	-	-	-	-
Residential	-	-	-	-	-	-	-	-	-	-	-
Non-specified	-	-	-	-	-	-	-	-	-	-	-
NON-ENERGY USE	**22.26**	**1.54**	**191.46**	**-**	**-**	**-**	**-**	**-**	**-**	**-**	**215.27**
in Industry/Transf./Energy	-	-	-	-	-	-	-	-	-	-	-
in Transport	-	-	-	-	-	-	-	-	-	-	-
in Other Sectors	-	-	-	-	-	-	-	-	-	-	-
Electr. Generated - GWh	*5337482*	*-*	*1282336*	*2157278*	*2393119*	*2565766*	*57588*	*155167*	*-*	*-*	*13948736*
Electricity Plants	*-*	*-*	*-*	*-*	*-*	*-*	*-*	*-*	*-*	*-*	*12276627*
CHP Plants	*-*	*-*	*-*	*-*	*-*	*-*	*-*	*-*	*-*	*-*	*1672109*
Heat Generated - TJ	*-*	*-*	*-*	*-*	*-*	*-*	*-*	*-*	*-*	*-*	*11283682*
CHP Plants	*-*	*-*	*-*	*-*	*-*	*-*	*-*	*-*	*-*	*-*	*6363550*
Heat Plants	*-*	*-*	*-*	*-*	*-*	*-*	*-*	*-*	*-*	*-*	*4920132*

OECD Total / Total OCDE

Total Production of Energy

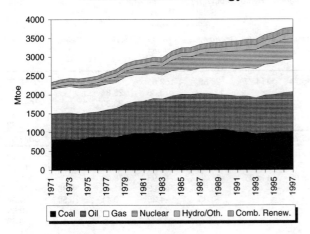

Coal ▦ Oil □ Gas ≣ Nuclear ▒ Hydro/Oth. ▥ Comb. Renew.

Total Primary Energy Supply*

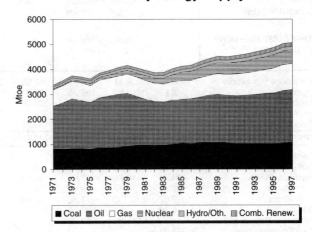

Coal □ Oil □ Gas ≣ Nuclear ▒ Hydro/Oth. ▥ Comb. Renew.

Electricity Generation by Fuel

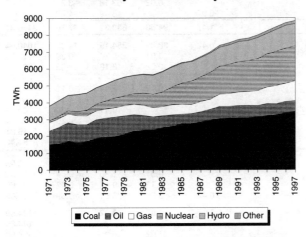

Coal ▦ Oil □ Gas ≣ Nuclear ▒ Hydro ▥ Other

Oil Products Consumption

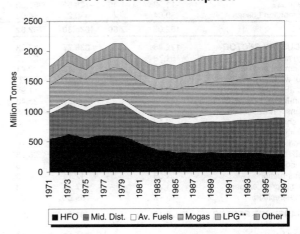

HFO ▦ Mid. Dist. □ Av. Fuels ≣ Mogas ▒ LPG** ▥ Other

Electricity Consumption/GDP***, TPES/GDP*** and Production/TPES

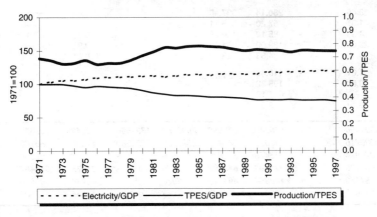

- - - - - Electricity/GDP ———— TPES/GDP ———— Production/TPES

* Excluding electricity trade.
** Includes LPG, NGL, ethane and naphtha.
*** GDP in 1990 US dollars.

INTERNATIONAL ENERGY AGENCY

OECD Total / Total OCDE : 1996

Million tonnes of oil equivalent / *Millions de tonnes d'équivalent pétrole*

SUPPLY AND CONSUMPTION / *APPROVISIONNEMENT ET DEMANDE*	Coal / *Charbon*	Crude Oil / *Pétrole brut*	Petroleum Products / *Produits pétroliers*	Gas / *Gaz*	Nuclear / *Nucléaire*	Hydro / *Hydro*	Geotherm. Solar etc. / *Géotherm. solaire etc.*	Combust. Renew. & Waste / *En. ren. combust. & déchets*	Electricity / *Electricité*	Heat / *Chaleur*	Total / *Total*
Indigenous Production	1000.14	1038.61	-	867.88	545.74	114.77	28.59	159.61	-	0.32	3755.66
Imports	240.80	1456.22	380.95	325.31	-	-	-	0.38	23.84	-	2427.51
Exports	-207.19	-406.11	-295.59	-153.36	-	-	-	-0.01	-23.33	-	-1085.60
Intl. Marine Bunkers	-	-	-77.31	-	-	-	-	-	-	-	-77.31
Stock Changes	9.22	-0.51	1.47	-3.85	-	-	-	-0.01	-	-	6.32
TPES	**1042.96**	**2088.20**	**9.52**	**1035.97**	**545.74**	**114.77**	**28.59**	**159.98**	**0.51**	**0.32**	**5026.57**
Transfers	-	-32.25	32.52	-	-	-	-	-0.16	-	-	0.12
Statistical Differences	4.95	8.95	1.04	-0.75	-	-	-	0.18	-	-	14.37
Electricity Plants	-694.44	-15.74	-107.88	-156.83	-545.65	-114.77	-26.38	-17.98	678.48	-	-1001.19
CHP Plants	-108.34	-0.87	-9.23	-84.44	-0.09	-	-0.32	-41.93	71.41	38.38	-135.44
Heat Plants	-8.67	-	-1.92	-5.43	-	-	-0.35	-4.36	-0.41	16.50	-4.64
Gas Works	-0.80	-	-3.74	3.55	-	-	-	-	-	-	-0.99
Petroleum Refineries	-	-2065.94	2063.38	-	-	-	-	-0.97	-	-	-3.53
Coal Transformation	-71.04	-	-	-	-	-	-	-	-	-	-71.04
Liquefaction	-	0.98	-	-1.20	-	-	-	-	-	-	-0.22
Other Transformation	-	20.66	-24.58	-0.65	-	-	-	-0.16	-	-	-4.74
Own Use	-11.90	-1.49	-128.53	-83.53	-	-	-	-0.02	-63.18	-3.22	-291.87
Distribution Losses	-0.64	-	-0.27	-4.08	-	-	-0.07	-	-48.49	-4.22	-57.77
TFC	**152.07**	**2.50**	**1830.30**	**702.62**	**-**	**-**	**1.47**	**94.58**	**638.32**	**47.76**	**3469.62**
INDUSTRY SECTOR	**122.48**	**2.50**	**333.77**	**282.63**	**-**	**-**	**0.42**	**38.03**	**254.15**	**14.65**	**1048.64**
Iron and Steel	51.78	-	9.73	27.67	-	-	-	0.01	28.77	0.39	118.35
Chemical and Petrochemical	12.10	2.50	193.10	109.41	-	-	-	0.82	48.18	4.83	370.93
of which: Feedstocks	-	2.48	173.51	20.76	-	-	-	-	-	-	196.74
Non-Ferrous Metals	2.78	-	4.41	11.98	-	-	-	0.12	22.44	0.11	41.84
Non-Metallic Minerals	27.88	-	19.52	25.43	-	-	-	0.58	13.32	0.13	86.87
Transport Equipment	0.54	-	2.92	5.96	-	-	-	-	10.41	0.60	20.43
Machinery	2.17	-	8.51	17.73	-	-	-	0.01	24.26	0.39	53.07
Mining and Quarrying	0.91	-	5.40	3.93	-	-	-	0.01	8.85	0.11	19.21
Food and Tobacco	6.08	-	13.51	25.75	-	-	-	4.02	17.20	0.83	67.40
Paper, Pulp and Printing	5.26	-	14.03	26.63	-	-	-	18.76	31.42	1.25	97.35
Wood and Wood Products	0.36	-	3.63	1.70	-	-	-	10.16	4.97	0.15	20.96
Construction	1.65	-	12.11	0.40	-	-	-	0.01	1.30	0.04	15.51
Textile and Leather	1.22	-	7.24	7.23	-	-	-	0.13	9.42	0.43	25.69
Non-specified	9.75	-	39.66	18.81	-	-	0.42	3.39	33.61	5.39	111.03
TRANSPORT SECTOR	**0.13**	**-**	**1091.93**	**22.59**	**-**	**-**	**-**	**1.02**	**8.92**	**-**	**1124.59**
Air	-	-	144.84	-	-	-	-	-	-	-	144.84
Road	-	-	905.23	0.55	-	-	-	1.02	-	-	906.80
Rail	0.05	-	18.20	-	-	-	-	-	7.60	-	25.85
Pipeline Transport	-	-	0.05	22.00	-	-	-	-	0.33	-	22.39
Internal Navigation	0.09	-	20.65	-	-	-	-	-	-	-	20.74
Non-specified	-	-	2.95	0.04	-	-	-	-	0.98	-	3.97
OTHER SECTORS	**28.46**	**-**	**286.14**	**397.40**	**-**	**-**	**1.05**	**55.52**	**375.25**	**33.11**	**1176.92**
Agriculture	1.87	-	57.20	5.11	-	-	0.06	1.46	6.54	0.28	72.52
Comm. and Publ. Services	4.38	-	85.10	114.55	-	-	0.05	2.44	169.67	7.30	383.49
Residential	22.03	-	139.00	264.42	-	-	0.93	49.98	197.61	22.48	696.47
Non-specified	0.16	-	4.83	13.32	-	-	0.01	1.64	1.43	3.05	24.44
NON-ENERGY USE	**1.01**	**-**	**118.46**	**-**	**-**	**-**	**-**	**-**	**-**	**-**	**119.47**
in Industry/Transf./Energy	0.83	-	108.32	-	-	-	-	-	-	-	109.15
in Transport	-	-	7.19	-	-	-	-	-	-	-	7.19
in Other Sectors	0.18	-	2.95	-	-	-	-	-	-	-	3.13
Electr. Generated - GWh	*3391590*	*-*	*656344*	*1073289*	*2093756*	*1334560*	*42349*	*127822*	*-*	*-*	*8719710*
Electricity Plants	*3020716*	*-*	*580866*	*749414*	*2093756*	*1334560*	*41876*	*68168*	*-*	*-*	*7889356*
CHP Plants	*370874*	*-*	*75478*	*323875*	*-*	*-*	*473*	*59654*	*-*	*-*	*830354*
Heat Generated - TJ	*1066126*	*-*	*231594*	*633793*	*3976*	*-*	*21275*	*316487*	*12801*	*25133*	*2311185*
CHP Plants	*784153*	*-*	*148384*	*494949*	*3976*	*-*	*6730*	*168860*	*1660*	*4144*	*1612856*
Heat Plants	*281973*	*-*	*83210*	*138844*	*-*	*-*	*14545*	*147627*	*11141*	*20989*	*698329*

OECD Total / Total OCDE : 1997

Million tonnes of oil equivalent / *Millions de tonnes d'équivalent pétrole*

SUPPLY AND CONSUMPTION / *APPROVISIONNEMENT ET DEMANDE*	Coal / *Charbon*	Crude Oil / *Pétrole brut*	Petroleum Products / *Produits pétroliers*	Gas / *Gaz*	Nuclear / *Nucléaire*	Hydro / *Hydro*	Geotherm. Solar etc. / *Géotherm. solaire etc.*	Combust. Renew. & Waste / *En. ren. combust. & déchets*	Electricity / *Electricité*	Heat / *Chaleur*	Total / *Total*
Indigenous Production	1016.85	1050.51	-	870.26	538.80	114.24	28.08	159.80	-	0.32	3778.85
Imports	251.72	1538.55	378.73	332.44	-	-	-	0.46	24.04	-	2525.94
Exports	-208.92	-427.94	-316.24	-155.71	-	-	-	-0.01	-23.18	-	-1132.00
Intl. Marine Bunkers	-	-	-78.54	-	-	-	-	-	-	-	-78.54
Stock Changes	-7.36	-4.55	-9.26	-5.55	-	-	-	-0.02	-	-	-26.73
TPES	**1052.28**	**2156.57**	**-25.31**	**1041.43**	**538.80**	**114.24**	**28.08**	**160.23**	**0.87**	**0.32**	**5067.52**
Transfers	-	-33.42	34.33	-	-	-	-	-0.45	-	-	0.47
Statistical Differences	1.05	-5.94	3.47	-3.85	-	-	-	-0.18	-	-	-5.46
Electricity Plants	-709.81	-12.54	-109.55	-169.89	-538.74	-114.24	-25.76	-18.18	687.98	-	-1010.71
CHP Plants	-103.85	-0.96	-8.50	-85.25	-0.06	-	-0.36	-42.63	72.14	38.07	-131.39
Heat Plants	-7.72	-	-1.63	-5.08	-	-	-0.34	-5.07	-0.44	16.23	-4.05
Gas Works	-0.73	-	-3.44	3.58	-	-	-	-	-	-	-0.60
Petroleum Refineries	-	-2126.01	2127.97	-	-	-	-	-	-	-	1.04
Coal Transformation	-72.81	-	-	-	-	-	-	-	-	-	-72.81
Liquefaction	-	0.81	-	-0.85	-	-	-	-	-	-	-0.04
Other Transformation	-	25.43	-28.44	-0.60	-	-	-	-0.18	-	-	-3.79
Own Use	-11.89	-1.41	-127.65	-85.73	-	-	-	-0.05	-62.47	-3.24	-292.43
Distribution Losses	-0.58	-	-0.24	-3.58	-	-	-0.07	-	-47.58	-4.29	-56.35
TFC	**145.96**	**2.52**	**1861.01**	**690.18**	**-**	**-**	**1.55**	**92.58**	**650.50**	**47.08**	**3491.40**
INDUSTRY SECTOR	**118.76**	**2.52**	**342.44**	**284.39**	**-**	**-**	**0.43**	**39.30**	**261.96**	**14.93**	**1064.73**
Iron and Steel	52.86	-	10.63	27.94	-	-	-	0.07	29.86	0.38	121.74
Chemical and Petrochemical	11.29	2.52	202.61	109.65	-	-	-	0.97	49.20	5.16	381.41
of which: Feedstocks	-	*2.50*	*184.02*	*20.96*	-	-	-	-	-	-	*207.48*
Non-Ferrous Metals	2.72	-	4.13	12.13	-	-	-	0.12	22.78	0.08	41.96
Non-Metallic Minerals	26.60	-	20.08	25.74	-	-	-	0.67	13.38	0.13	86.60
Transport Equipment	0.37	-	2.78	5.95	-	-	-	-	10.71	0.61	20.42
Machinery	1.74	-	8.19	18.02	-	-	-	0.01	25.09	0.36	53.41
Mining and Quarrying	0.92	-	5.43	3.82	-	-	-	0.01	8.83	0.12	19.13
Food and Tobacco	5.38	-	12.83	25.93	-	-	-	4.40	17.63	0.89	67.06
Paper, Pulp and Printing	5.44	-	13.26	27.00	-	-	-	19.59	32.29	1.24	98.81
Wood and Wood Products	0.29	-	3.65	1.74	-	-	-	10.27	5.11	0.17	21.22
Construction	1.94	-	11.25	0.39	-	-	-	0.01	1.17	0.05	14.80
Textile and Leather	0.88	-	6.97	7.31	-	-	-	0.12	9.86	0.43	25.57
Non-specified	8.33	-	40.64	18.77	-	-	0.43	3.07	36.05	5.31	112.60
TRANSPORT SECTOR	**0.14**	**-**	**1113.07**	**23.57**	**-**	**-**	**-**	**1.48**	**8.99**	**-**	**1147.26**
Air	-	-	148.62	-	-	-	-	-	-	-	148.62
Road	-	-	923.22	0.53	-	-	-	1.48	-	-	925.23
Rail	0.05	-	17.39	-	-	-	-	-	7.67	-	25.11
Pipeline Transport	-	-	0.01	23.01	-	-	-	-	0.35	-	23.37
Internal Navigation	0.09	-	20.86	-	-	-	-	-	-	-	20.95
Non-specified	-	-	2.97	0.04	-	-	-	-	0.97	-	3.98
OTHER SECTORS	**26.08**	**-**	**277.32**	**382.23**	**-**	**-**	**1.12**	**51.80**	**379.55**	**32.15**	**1150.25**
Agriculture	1.62	-	57.21	5.12	-	-	0.06	1.43	6.53	0.29	72.26
Comm. and Publ. Services	4.03	-	80.88	114.99	-	-	0.05	2.38	175.00	7.13	384.47
Residential	20.26	-	135.01	250.43	-	-	1.00	46.46	196.70	21.52	671.39
Non-specified	0.16	-	4.22	11.69	-	-	0.01	1.52	1.31	3.21	22.14
NON-ENERGY USE	**0.98**	**-**	**128.18**	**-**	**-**	**-**	**-**	**-**	**-**	**-**	**129.16**
in Industry/Transf./Energy	0.80	-	117.58	-	-	-	-	-	-	-	118.39
in Transport	-	-	7.49	-	-	-	-	-	-	-	7.49
in Other Sectors	0.18	-	3.10	-	-	-	-	-	-	-	3.28
Electr. Generated - GWh	*3437119*	*-*	*648676*	*1178488*	*2067254*	*1328316*	*44372*	*134465*	*-*	*-*	*8838690*
Electricity Plants	*3084723*	*-*	*571970*	*834441*	*2067254*	*1328316*	*43720*	*69384*	*-*	*-*	*7999808*
CHP Plants	*352396*	*-*	*76706*	*344047*	*-*	*-*	*652*	*65081*	*-*	*-*	*838882*
Heat Generated - TJ	*1042985*	*-*	*203786*	*634496*	*2685*	*-*	*21633*	*341493*	*14496*	*25573*	*2287147*
CHP Plants	*784333*	*-*	*130818*	*496434*	*2685*	*-*	*7248*	*172548*	*2326*	*3912*	*1600304*
Heat Plants	*258652*	*-*	*72968*	*138062*	*-*	*-*	*14385*	*168945*	*12170*	*21661*	*686843*

Non-OECD Total / Total non-OCDE

Total Production of Energy

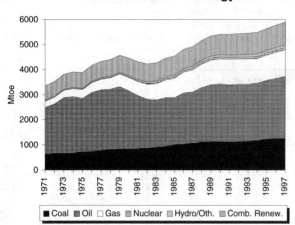

Total Primary Energy Supply*

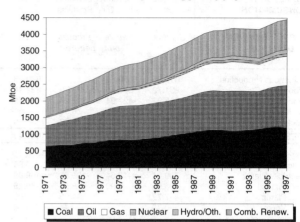

Electricity Generation by Fuel

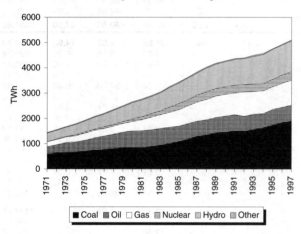

Oil Products Consumption

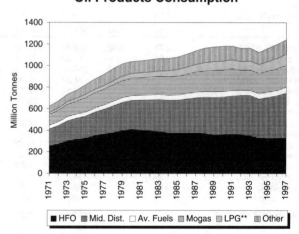

Electricity Consumption/GDP***, TPES/GDP*** and Production/TPES

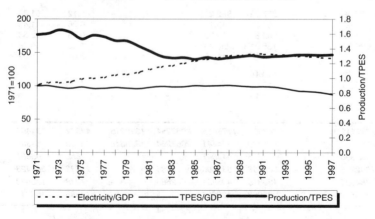

* Excluding electricity trade.
** Includes LPG, NGL, ethane and naphtha.
*** GDP in 1990 US dollars.

Non-OECD Total / Total non-OCDE : 1996

Million tonnes of oil equivalent / *Millions de tonnes d'équivalent pétrole*

SUPPLY AND CONSUMPTION / *APPROVISIONNEMENT ET DEMANDE*	Coal *Charbon*	Crude Oil *Pétrole brut*	Petroleum Products *Produits pétroliers*	Gas *Gaz*	Nuclear *Nucléaire*	Hydro *Hydro*	Geotherm. Solar etc. *Géotherm. solaire etc.*	Combust. Renew. & Waste *En. ren. combust. & déchets*	Electricity *Electricité*	Heat *Chaleur*	Total *Total*
Indigenous Production	1257.98	2389.78	-	1041.17	84.57	103.02	9.87	893.46	-	-	5779.85
Imports	94.04	402.14	270.92	119.97	-	-	-	0.77	12.58	-	900.42
Exports	-132.48	-1378.88	-401.63	-291.58	-	-	-	-0.95	-13.17	-	-2218.69
Intl. Marine Bunkers	-	-	-51.81	-	-	-	-	-	-	-	-51.81
Stock Changes	13.19	-13.14	2.04	-11.87	-	-	-	-0.21	-	-	-10.00
TPES	**1232.74**	**1399.90**	**-180.48**	**857.69**	**84.57**	**103.02**	**9.87**	**893.06**	**-0.59**	**-**	**4399.77**
Transfers	-	-42.28	45.03	-	-	-	-	-	-	-	2.75
Statistical Differences	16.11	6.78	-8.57	0.61	-	-	-	-0.06	0.64	0.65	16.15
Electricity Plants	-438.19	-11.47	-112.42	-125.85	-80.14	-103.02	-9.37	-3.94	350.34	-	-534.05
CHP Plants	-119.06	-0.60	-17.64	-170.16	-4.43	-	-	-2.44	74.43	120.62	-119.28
Heat Plants	-21.10	-0.12	-44.38	-65.98	-	-	-0.01	-4.61	-0.11	111.17	-25.15
Gas Works	-5.47	-0.31	-1.59	4.20	-	-	-	-	-	-	-3.17
Petroleum Refineries	-	-1333.00	1284.69	-0.07	-	-	-	-	-	-	-48.38
Coal Transformation	-77.34	-	-	-	-	-	-	-	-	-	-77.34
Liquefaction	-14.17	7.02	-	-1.54	-	-	-	-	-	-	-8.70
Other Transformation	-	-	-	-0.03	-	-	-	-43.88	-	-	-43.91
Own Use	-35.99	-6.02	-59.85	-130.01	-	-	-	-0.08	-45.45	-20.03	-297.45
Distribution Losses	-16.50	-6.21	-1.28	-17.43	-	-	-	-0.18	-50.93	-12.64	-105.18
TFC	**521.01**	**13.69**	**903.51**	**351.42**	**-**	**-**	**0.49**	**837.87**	**328.33**	**199.76**	**3156.08**
INDUSTRY SECTOR	**366.06**	**10.73**	**185.44**	**184.09**	**-**	**-**	**-**	**92.64**	**158.98**	**87.47**	**1085.40**
Iron and Steel	-	-	-	-	-	-	-	-	-	-	-
Chemical and Petrochemical	-	-	-	-	-	-	-	-	-	-	-
of which: Feedstocks	-	-	-	-	-	-	-	-	-	-	-
Non-Ferrous Metals	-	-	-	-	-	-	-	-	-	-	-
Non-Metallic Minerals	-	-	-	-	-	-	-	-	-	-	-
Transport Equipment	-	-	-	-	-	-	-	-	-	-	-
Machinery	-	-	-	-	-	-	-	-	-	-	-
Mining and Quarrying	-	-	-	-	-	-	-	-	-	-	-
Food and Tobacco	-	-	-	-	-	-	-	-	-	-	-
Paper, Pulp and Printing	-	-	-	-	-	-	-	-	-	-	-
Wood and Wood Products	-	-	-	-	-	-	-	-	-	-	-
Construction	-	-	-	-	-	-	-	-	-	-	-
Textile and Leather	-	-	-	-	-	-	-	-	-	-	-
Non-specified	-	-	-	-	-	-	-	-	-	-	-
TRANSPORT SECTOR	**5.67**	**0.01**	**444.83**	**17.68**	**-**	**-**	**-**	**7.07**	**9.89**	**-**	**485.17**
Air	-	-	53.55	-	-	-	-	-	0.02	-	53.56
Road	-	-	361.26	1.32	-	-	-	7.05	0.01	-	369.64
Rail	-	-	-	-	-	-	-	-	-	-	-
Pipeline Transport	-	-	-	-	-	-	-	-	-	-	-
Internal Navigation	-	-	-	-	-	-	-	-	-	-	-
Non-specified	-	-	-	-	-	-	-	-	-	-	-
OTHER SECTORS	**127.35**	**1.41**	**214.13**	**149.65**	**-**	**-**	**0.49**	**738.17**	**159.45**	**112.29**	**1502.94**
Agriculture	-	-	-	-	-	-	-	-	-	-	-
Comm. and Publ. Services	-	-	-	-	-	-	-	-	-	-	-
Residential	-	-	-	-	-	-	-	-	-	-	-
Non-specified	-	-	-	-	-	-	-	-	-	-	-
NON-ENERGY USE	**21.93**	**1.54**	**59.11**	**-**	**-**	**-**	**-**	**-**	**-**	**-**	**82.58**
in Industry/Transf./Energy	-	-	-	-	-	-	-	-	-	-	-
in Transport	-	-	-	-	-	-	-	-	-	-	-
in Other Sectors	-	-	-	-	-	-	-	-	-	-	-
Electr. Generated - GWh	*1843025*	*-*	*598576*	*951197*	*322833*	*1197956*	*11914*	*13746*	*-*	*-*	*4939247*
Electricity Plants	*-*	*-*	*-*	*-*	*-*	*-*	*-*	*-*	*-*	*-*	*4073772*
CHP Plants	*-*	*-*	*-*	*-*	*-*	*-*	*-*	*-*	*-*	*-*	*865475*
Heat Generated - TJ	*-*	*-*	*-*	*-*	*-*	*-*	*-*	*-*	*-*	*-*	*9706060*
CHP Plants	*-*	*-*	*-*	*-*	*-*	*-*	*-*	*-*	*-*	*-*	*5050929*
Heat Plants	*-*	*-*	*-*	*-*	*-*	*-*	*-*	*-*	*-*	*-*	*4655131*

Non-OECD Total / Total non-OCDE : 1997

Million tonnes of oil equivalent / *Millions de tonnes d'équivalent pétrole*

SUPPLY AND CONSUMPTION / *APPROVISIONNEMENT ET DEMANDE*	Coal / *Charbon*	Crude Oil / *Pétrole brut*	Petroleum Products / *Produits pétroliers*	Gas / *Gaz*	Nuclear / *Nucléaire*	Hydro / *Hydro*	Geotherm. Solar etc. / *Géotherm. solaire etc.*	Combust. Renew. & Waste / *En. ren. combust. & déchets*	Electricity / *Electricité*	Heat / *Chaleur*	Total / *Total*
Indigenous Production	1257.36	2479.96	-	1046.50	85.36	106.42	10.81	903.89	-	-	5890.29
Imports	92.02	423.59	290.98	108.29	-	-	-	0.54	12.87	-	928.30
Exports	-138.96	-1442.83	-405.29	-284.97	-	-	-	-1.64	-13.42	-	-2287.10
Intl. Marine Bunkers	-	-	-53.71	-	-	-	-	-	-	-	-53.71
Stock Changes	-7.74	-13.08	-1.64	-0.08	-	-	-	-1.25	-	-	-23.79
TPES	**1202.69**	**1447.63**	**-169.65**	**869.74**	**85.36**	**106.42**	**10.81**	**901.55**	**-0.55**	**-**	**4453.99**
Transfers	-	-41.58	44.99	-	-	-	-	-	-	-	3.41
Statistical Differences	22.85	3.08	-8.28	2.57	-	-	-	-0.14	-0.37	0.47	20.18
Electricity Plants	-448.93	-12.11	-118.23	-132.80	-80.38	-106.42	-10.28	-4.33	367.81	-	-545.67
CHP Plants	-110.97	-0.60	-16.25	-168.08	-4.98	-	-	-2.43	71.66	113.75	-117.91
Heat Plants	-19.41	-0.11	-40.95	-63.65	-	-	-0.01	-4.63	-0.16	101.09	-27.83
Gas Works	-6.48	-0.22	-1.44	4.41	-	-	-	-	-	-	-3.73
Petroleum Refineries	-	-1379.47	1318.80	-0.07	-	-	-	-	-	-	-60.75
Coal Transformation	-80.95	-	-	-	-	-	-	-	-	-	-80.95
Liquefaction	-13.79	7.02	-	-1.54	-	-	-	-	-	-	-8.32
Other Transformation	-0.03	-	-0.03	-0.03	-	-	-	-44.99	-	-	-45.08
Own Use	-40.05	-6.04	-63.75	-139.81	-	-	-	-0.08	-48.77	-18.71	-317.20
Distribution Losses	-15.37	-2.57	-1.18	-16.82	-	-	-	-0.18	-53.43	-11.47	-101.02
TFC	**489.54**	**15.03**	**944.04**	**353.92**	**-**	**-**	**0.52**	**844.77**	**336.18**	**185.13**	**3169.11**
INDUSTRY SECTOR	**350.04**	**12.02**	**194.26**	**185.31**	-	-	-	**95.02**	**160.60**	**82.89**	**1080.14**
Iron and Steel	-	-	-	-	-	-	-	-	-	-	-
Chemical and Petrochemical	-	-	-	-	-	-	-	-	-	-	-
of which: Feedstocks	-	-	-	-	-	-	-	-	-	-	-
Non-Ferrous Metals	-	-	-	-	-	-	-	-	-	-	-
Non-Metallic Minerals	-	-	-	-	-	-	-	-	-	-	-
Transport Equipment	-	-	-	-	-	-	-	-	-	-	-
Machinery	-	-	-	-	-	-	-	-	-	-	-
Mining and Quarrying	-	-	-	-	-	-	-	-	-	-	-
Food and Tobacco	-	-	-	-	-	-	-	-	-	-	-
Paper, Pulp and Printing	-	-	-	-	-	-	-	-	-	-	-
Wood and Wood Products	-	-	-	-	-	-	-	-	-	-	-
Construction	-	-	-	-	-	-	-	-	-	-	-
Textile and Leather	-	-	-	-	-	-	-	-	-	-	-
Non-specified	-	-	-	-	-	-	-	-	-	-	-
TRANSPORT SECTOR	**6.67**	**0.01**	**464.17**	**17.66**	-	-	-	**6.82**	**10.13**	**-**	**505.46**
Air	-	-	56.02	-	-	-	-	-	0.02	-	56.04
Road	-	-	373.01	1.32	-	-	-	6.79	0.08	-	381.20
Rail	-	-	-	-	-	-	-	-	-	-	-
Pipeline Transport	-	-	-	-	-	-	-	-	-	-	-
Internal Navigation	-	-	-	-	-	-	-	-	-	-	-
Non-specified	-	-	-	-	-	-	-	-	-	-	-
OTHER SECTORS	**111.56**	**1.45**	**222.32**	**150.94**	-	-	**0.52**	**742.92**	**165.45**	**102.23**	**1497.40**
Agriculture	-	-	-	-	-	-	-	-	-	-	-
Comm. and Publ. Services	-	-	-	-	-	-	-	-	-	-	-
Residential	-	-	-	-	-	-	-	-	-	-	-
Non-specified	-	-	-	-	-	-	-	-	-	-	-
NON-ENERGY USE	**21.28**	**1.54**	**63.29**	-	-	-	-	-	-	-	**86.11**
in Industry/Transf./Energy	-	-	-	-	-	-	-	-	-	-	-
in Transport	-	-	-	-	-	-	-	-	-	-	-
in Other Sectors	-	-	-	-	-	-	-	-	-	-	-
Electr. Generated - GWh	*1900363*	*-*	*633660*	*978790*	*325865*	*1237450*	*13216*	*20702*	*-*	*-*	*5110046*
Electricity Plants	*-*	*-*	*-*	*-*	*-*	*-*	*-*	*-*	*-*	*-*	*4276819*
CHP Plants	*-*	*-*	*-*	*-*	*-*	*-*	*-*	*-*	*-*	*-*	*833227*
Heat Generated - TJ	*-*	*-*	*-*	*-*	*-*	*-*	*-*	*-*	*-*	*-*	*8996535*
CHP Plants	*-*	*-*	*-*	*-*	*-*	*-*	*-*	*-*	*-*	*-*	*4763246*
Heat Plants	*-*	*-*	*-*	*-*	*-*	*-*	*-*	*-*	*-*	*-*	*4233289*

Africa / Afrique

Total Production of Energy

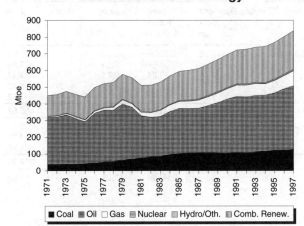

■ Coal ▦ Oil □ Gas ▤ Nuclear ▨ Hydro/Oth. ▥ Comb. Renew.

Total Primary Energy Supply*

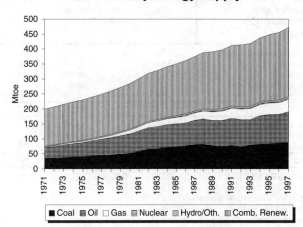

■ Coal ▦ Oil □ Gas ▤ Nuclear ▨ Hydro/Oth. ▥ Comb. Renew.

Electricity Generation by Fuel

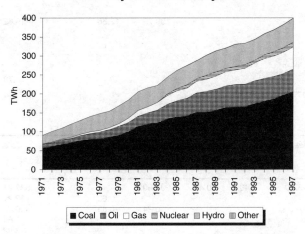

■ Coal ▦ Oil □ Gas ▤ Nuclear ▨ Hydro ▥ Other

Oil Products Consumption

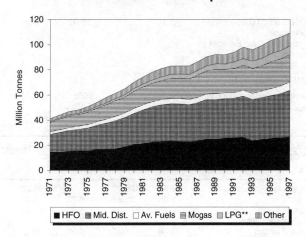

■ HFO ▦ Mid. Dist. □ Av. Fuels ▤ Mogas ▨ LPG** ▥ Other

Electricity Consumption/GDP***, TPES/GDP*** and Production/TPES

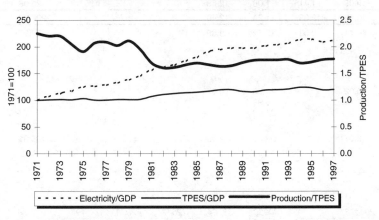

- - - - Electricity/GDP ——— TPES/GDP ——— Production/TPES

* Excluding electricity trade.
** Includes LPG, NGL, ethane and naphtha.
*** GDP in 1990 US dollars.

Africa / Afrique

Million tonnes of oil equivalent / *Millions de tonnes d'équivalent pétrole*

SUPPLY AND CONSUMPTION 1996	Coal	Crude Oil	Petroleum Products	Gas	Nuclear	Hydro	Geotherm. Solar etc.	Combust. Renew. & Waste	Electricity	Heat	Total
Indigenous Production	121.00	368.37	-	78.73	3.07	5.24	0.38	226.05	-	-	802.83
Imports	3.93	29.27	18.59	1.22	-	-	-	-	0.54	-	53.55
Exports	-40.69	-265.54	-36.86	-40.43	-	-	-	-0.24	-0.71	-	-384.47
Intl. Marine Bunkers	-	-	-8.42	-	-	-	-	-	-	-	-8.42
Stock changes	1.88	-9.14	-0.57	-	-	-	-	-	-	-	-7.82
TPES	86.12	122.96	-27.25	39.52	3.07	5.24	0.38	225.81	-0.18	-	455.67
Electricity and CHP Plants	-47.65	-	-13.44	-15.56	-3.07	-5.24	-0.38	-	32.92	-	-52.42
Petroleum Refineries	-	-126.94	123.18	-	-	-	-	-	-	-	-3.77
Other Transformation *	-20.87	4.01	-1.92	-11.99	-	-	-	-24.57	-6.44	-	-61.78
TFC	17.60	0.03	80.57	11.96	-	-	-	201.24	26.30	-	337.70
INDUSTRY SECTOR	9.61	0.03	12.88	8.39	-	-	-	20.44	13.21	-	64.57
Iron and Steel	-	-	-	-	-	-	-	-	-	-	-
Chemical and Petrochemical	-	-	-	-	-	-	-	-	-	-	-
Non-Metallic Minerals	-	-	-	-	-	-	-	-	-	-	-
Non-specified	-	-	-	-	-	-	-	-	-	-	-
TRANSPORT SECTOR	0.03	-	41.70	0.57	-	-	-	-	0.43	-	42.73
Air	-	-	6.58	-	-	-	-	-	-	-	6.58
Road	-	-	34.56	-	-	-	-	-	-	-	34.56
Non-specified	-	-	-	-	-	-	-	-	-	-	-
OTHER SECTORS	3.04	-	20.12	3.00	-	-	-	180.79	12.66	-	219.62
Agriculture	-	-	-	-	-	-	-	-	-	-	-
Comm. and Publ. Services	-	-	-	-	-	-	-	-	-	-	-
Residential	-	-	-	-	-	-	-	-	-	-	-
Non-specified	-	-	-	-	-	-	-	-	-	-	-
NON-ENERGY USE	4.92	-	5.87	-	-	-	-	-	-	-	10.78
Electricity Generated - GWh	196078	-	56667	56877	11775	60918	439	-	-	-	382754
Heat Generated - TJ	-	-	-	-	-	-	-	-	-	-	-

APPROVISIONNEMENT ET DEMANDE 1997	Charbon	Pétrole brut	Produits pétroliers	Gaz	Nucléaire	Hydro	Géotherm. solaire etc.	En. ren. combust. & déchets	Electricité	Chaleur	Total
Indigenous Production	128.51	379.76	-	88.22	3.30	5.40	0.43	231.60	-	-	837.21
Imports	3.86	28.39	18.72	0.62	-	-	-	-	0.62	-	52.21
Exports	-43.32	-273.38	-34.88	-47.67	-	-	-	-0.24	-0.83	-	-400.32
Intl. Marine Bunkers	-	-	-7.64	-	-	-	-	-	-	-	-7.64
Stock changes	-1.68	-7.07	-0.18	-	-	-	-	-	-	-	-8.92
TPES	87.37	127.70	-23.98	41.17	3.30	5.40	0.43	231.36	-0.22	-	472.53
Electricity and CHP Plants	-49.86	-	-14.38	-16.09	-3.30	-5.40	-0.43	-	34.33	-	-55.12
Petroleum Refineries	-	-129.08	123.68	-	-	-	-	-	-	-	-5.41
Other Transformation *	-19.09	1.41	-1.08	-12.68	-	-	-	-25.08	-6.74	-	-63.26
TFC	18.42	0.03	84.24	12.40	-	-	-	206.28	27.38	-	348.74
INDUSTRY SECTOR	10.41	0.03	13.72	8.80	-	-	-	20.97	13.65	-	67.59
Iron and Steel	-	-	-	-	-	-	-	-	-	-	-
Chemical and Petrochemical	-	-	-	-	-	-	-	-	-	-	-
Non-Metallic Minerals	-	-	-	-	-	-	-	-	-	-	-
Non-specified	-	-	-	-	-	-	-	-	-	-	-
TRANSPORT SECTOR	0.02	-	42.56	0.67	-	-	-	-	0.45	-	43.71
Air	-	-	6.99	-	-	-	-	-	-	-	6.99
Road	-	-	34.62	-	-	-	-	-	-	-	34.62
Non-specified	-	-	-	-	-	-	-	-	-	-	-
OTHER SECTORS	3.07	-	22.07	2.93	-	-	-	185.31	13.27	-	226.65
Agriculture	-	-	-	-	-	-	-	-	-	-	-
Comm. and Publ. Services	-	-	-	-	-	-	-	-	-	-	-
Residential	-	-	-	-	-	-	-	-	-	-	-
Non-specified	-	-	-	-	-	-	-	-	-	-	-
NON-ENERGY USE	4.91	-	5.89	-	-	-	-	-	-	-	10.80
Electricity Generated - GWh	204473	-	60030	58724	12647	62828	497	-	-	-	399199
Heat Generated - TJ	-	-	-	-	-	-	-	-	-	-	-

* Includes Transfers, Statistical Differences, Own Use and Distribution Losses.

INTERNATIONAL ENERGY AGENCY

Latin America / Amérique latine

Total Production of Energy

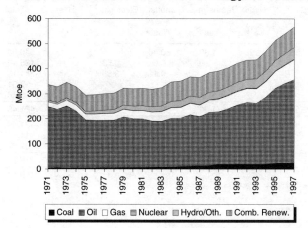

Coal ▦ Oil ▢ Gas ▥ Nuclear ▦ Hydro/Oth. ▥ Comb. Renew.

Total Primary Energy Supply*

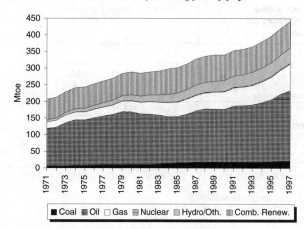

Coal ▦ Oil ▢ Gas ▥ Nuclear ▦ Hydro/Oth. ▥ Comb. Renew.

Electricity Generation by Fuel

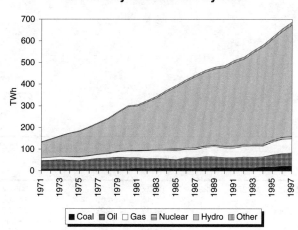

Coal ▦ Oil ▢ Gas ▥ Nuclear ▦ Hydro ▥ Other

Oil Products Consumption

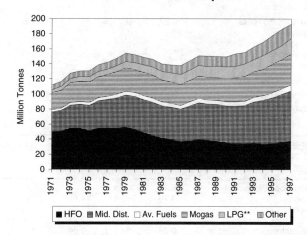

HFO ▦ Mid. Dist. ▢ Av. Fuels ▥ Mogas ▥ LPG** ▥ Other

Electricity Consumption/GDP***, TPES/GDP*** and Production/TPES

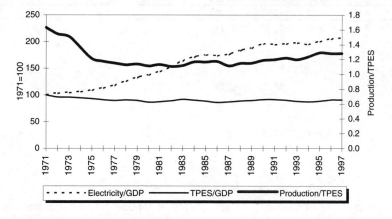

- - - - - Electricity/GDP ——— TPES/GDP ——— Production/TPES

* Excluding electricity trade.
** Includes LPG, NGL, ethane and naphtha.
*** GDP in 1990 US dollars.

Latin America / Amérique latine

Million tonnes of oil equivalent / *Millions de tonnes d'équivalent pétrole*

SUPPLY AND CONSUMPTION 1996	Coal	Crude Oil	Petroleum Products	Gas	Nuclear	Hydro	Geotherm. Solar etc.	Combust. Renew. & Waste	Electricity	Heat	Total
Indigenous Production	24.45	317.09	-	72.14	2.58	41.99	1.30	80.46	-	-	540.00
Imports	14.61	71.84	42.99	1.76	-	-	-	0.66	3.53	-	135.40
Exports	-18.24	-151.13	-66.03	-1.77	-	-	-	-0.11	-3.85	-	-241.13
Intl. Marine Bunkers	-	-	-7.01	-	-	-	-	-	-	-	-7.01
Stock changes	-0.66	-3.42	-1.47	-	-	-	-	-0.22	-	-	-5.78
TPES	20.16	234.37	-31.53	72.14	2.58	41.99	1.30	80.79	-0.32	-	421.48
Electricity and CHP Plants	-4.86	-0.09	-15.74	-17.73	-2.58	-41.99	-1.30	-3.72	56.52	-	-31.49
Petroleum Refineries	-	-220.36	208.85	-	-	-	-	-	-	-	-11.51
Other Transformation *	-7.12	-11.44	-3.09	-20.78	-	-	-	-5.98	-11.15	-	-59.57
TFC	8.17	2.48	158.49	33.62	-	-	-	71.09	45.05	-	318.91
INDUSTRY SECTOR	7.80	2.37	29.45	24.87	-	-	-	31.85	20.47	-	116.80
Iron and Steel	-	-	-	-	-	-	-	-	-	-	-
Chemical and Petrochemical	-	-	-	-	-	-	-	-	-	-	-
Non-Metallic Minerals	-	-	-	-	-	-	-	-	-	-	-
Non-specified	-	-	-	-	-	-	-	-	-	-	-
TRANSPORT SECTOR	-	-	90.67	0.99	-	-	-	7.06	0.17	-	98.89
Air	-	-	7.91	-	-	-	-	-	-	-	7.91
Road	-	-	78.89	0.94	-	-	-	7.05	0.01	-	86.88
Non-specified	-	-	-	-	-	-	-	-	-	-	-
OTHER SECTORS	0.23	0.11	29.23	7.76	-	-	-	32.19	24.42	-	93.93
Agriculture	-	-	-	-	-	-	-	-	-	-	-
Comm. and Publ. Services	-	-	-	-	-	-	-	-	-	-	-
Residential	-	-	-	-	-	-	-	-	-	-	-
Non-specified	-	-	-	-	-	-	-	-	-	-	-
NON-ENERGY USE	0.14	-	9.14	-	-	-	-	-	-	-	9.29
Electricity Generated - GWh	20460	-	59719	66094	9888	488215	1898	10938	-	-	657212
Heat Generated - TJ	-	-	-	-	-	-	-	-	-	-	-

APPROVISIONNEMENT ET DEMANDE 1997	Charbon	Pétrole brut	Produits pétroliers	Gaz	Nucléaire	Hydro	Géotherm. solaire etc.	En. ren. combust. & déchets	Electricité	Chaleur	Total
Indigenous Production	26.40	330.39	-	79.91	2.88	44.27	1.32	81.77	-	-	566.94
Imports	14.75	75.38	44.52	1.96	-	-	-	0.45	4.03	-	141.09
Exports	-20.59	-161.88	-69.79	-2.53	-	-	-	-0.69	-4.02	-	-259.50
Intl. Marine Bunkers	-	-	-7.41	-	-	-	-	-	-	-	-7.41
Stock changes	0.33	0.99	0.54	-	-	-	-	-1.28	-	-	0.59
TPES	20.89	244.89	-32.13	79.33	2.88	44.27	1.32	80.25	0.01	-	441.71
Electricity and CHP Plants	-5.48	-0.10	-16.18	-17.42	-2.88	-44.27	-1.32	-3.94	59.19	-	-32.39
Petroleum Refineries	-	-233.16	220.41	-	-	-	-	-	-	-	-12.76
Other Transformation *	-6.53	-9.09	-4.13	-24.92	-	-	-	-5.92	-11.63	-	-62.23
TFC	8.88	2.53	167.96	37.00	-	-	-	70.38	47.57	-	334.33
INDUSTRY SECTOR	8.44	2.42	32.45	27.39	-	-	-	32.99	21.65	-	125.34
Iron and Steel	-	-	-	-	-	-	-	-	-	-	-
Chemical and Petrochemical	-	-	-	-	-	-	-	-	-	-	-
Non-Metallic Minerals	-	-	-	-	-	-	-	-	-	-	-
Non-specified	-	-	-	-	-	-	-	-	-	-	-
TRANSPORT SECTOR	-	-	96.20	1.15	-	-	-	6.80	0.25	-	104.40
Air	-	-	8.52	-	-	-	-	-	-	-	8.52
Road	-	-	83.37	1.10	-	-	-	6.79	0.07	-	91.34
Non-specified	-	-	-	-	-	-	-	-	-	-	-
OTHER SECTORS	0.25	0.11	29.81	8.47	-	-	-	30.59	25.67	-	94.90
Agriculture	-	-	-	-	-	-	-	-	-	-	-
Comm. and Publ. Services	-	-	-	-	-	-	-	-	-	-	-
Residential	-	-	-	-	-	-	-	-	-	-	-
Non-specified	-	-	-	-	-	-	-	-	-	-	-
NON-ENERGY USE	0.19	-	9.50	-	-	-	-	-	-	-	9.69
Electricity Generated - GWh	21729	-	61409	65166	11061	514732	2106	12093	-	-	688296
Heat Generated - TJ	-	-	-	-	-	-	-	-	-	-	-

* Includes Transfers, Statistical Differences, Own Use and Distribution Losses.

Asia (excluding China) / Asie (Chine non incluse)

Total Production of Energy

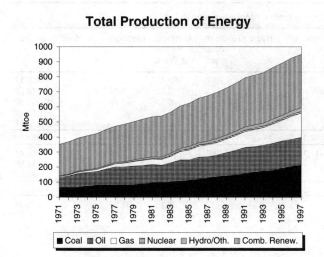

■ Coal ▦ Oil ☐ Gas ▤ Nuclear ▦ Hydro/Oth. ▥ Comb. Renew.

Total Primary Energy Supply*

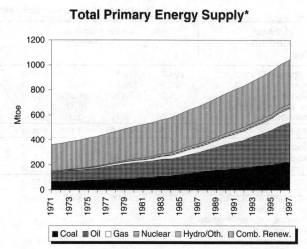

■ Coal ▦ Oil ☐ Gas ▤ Nuclear ▦ Hydro/Oth. ▥ Comb. Renew.

Electricity Generation by Fuel

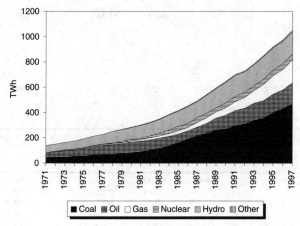

■ Coal ▦ Oil ☐ Gas ▤ Nuclear ▦ Hydro ▥ Other

Oil Products Consumption

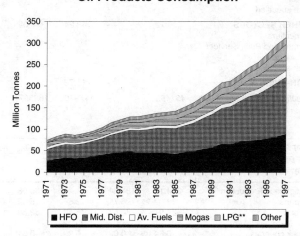

■ HFO ▦ Mid. Dist. ☐ Av. Fuels ▤ Mogas ▥ LPG** ▥ Other

Electricity Consumption/GDP***, TPES/GDP*** and Production/TPES

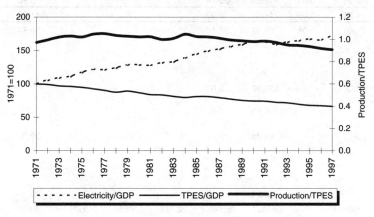

- - - - - Electricity/GDP ——— TPES/GDP ——— Production/TPES

* Excluding electricity trade.
** Includes LPG, NGL, ethane and naphtha.
*** GDP in 1990 US dollars.

Asia (excluding China) / Asie (Chine non incluse)

Million tonnes of oil equivalent / *Millions de tonnes d'équivalent pétrole*

SUPPLY AND CONSUMPTION 1996	Coal	Crude Oil	Petroleum Products	Gas	Nuclear	Hydro	Geotherm. Solar etc.	Combust. Renew. & Waste	Electricity	Heat	Total
Indigenous Production	204.74	180.12	-	154.08	12.16	15.26	7.67	352.44	-	-	926.47
Imports	35.14	192.82	108.38	4.35	-	-	-	0.02	0.24	-	340.97
Exports	-25.01	-79.10	-81.85	-53.75	-	-	-	-0.13	-0.17	-	-240.00
Intl. Marine Bunkers	-	-	-18.47	-	-	-	-	-	-	-	-18.47
Stock changes	0.88	-0.71	-0.90	-0.24	-	-	-	-	-	-	-0.97
TPES	215.76	293.13	7.16	104.45	12.16	15.26	7.67	352.34	0.07	-	1008.00
Electricity and CHP Plants	-129.52	-	-34.91	-38.62	-12.16	-15.26	-7.67	-0.21	83.94	-	-154.41
Petroleum Refineries	-	-284.48	272.62	-	-	-	-	-	-	-	-11.85
Other Transformation *	-12.42	-7.59	-9.50	-27.36	-	-	-	-13.22	-18.46	-	-88.54
TFC	73.83	1.06	235.38	38.46	-	-	-	338.90	65.56	-	753.19
INDUSTRY SECTOR	73.03	1.06	56.08	31.42	-	-	-	38.19	30.02	-	229.81
Iron and Steel	-	-	-	-	-	-	-	-	-	-	-
Chemical and Petrochemical	-	-	-	-	-	-	-	-	-	-	-
Non-Metallic Minerals	-	-	-	-	-	-	-	-	-	-	-
Non-specified	-	-	-	-	-	-	-	-	-	-	-
TRANSPORT SECTOR	0.08	-	124.33	0.02	-	-	-	-	0.64	-	125.06
Air	-	-	15.23	-	-	-	-	-	0.01	-	15.24
Road	-	-	103.30	-	-	-	-	-	-	-	103.31
Non-specified	-	-	-	-	-	-	-	-	-	-	-
OTHER SECTORS	0.72	-	43.24	7.02	-	-	-	300.71	34.89	-	386.59
Agriculture	-	-	-	-	-	-	-	-	-	-	-
Comm. and Publ. Services	-	-	-	-	-	-	-	-	-	-	-
Residential	-	-	-	-	-	-	-	-	-	-	-
Non-specified	-	-	-	-	-	-	-	-	-	-	-
NON-ENERGY USE	-	-	11.73	-	-	-	-	-	-	-	11.74
Electricity Generated - GWh	*431493*	-	*147970*	*162697*	*46671*	*177397*	*9547*	*265*	-	-	*976040*
Heat Generated - TJ	-	-	-	-	-	-	-	-	-	-	-

APPROVISIONNEMENT ET DEMANDE 1997	Charbon	Pétrole brut	Produits pétroliers	Gaz	Nucléaire	Hydro	Géotherm. solaire etc.	En. ren. combust. & déchets	Electricité	Chaleur	Total
Indigenous Production	214.11	184.64	-	160.08	12.17	15.18	8.51	355.44	-	-	950.13
Imports	38.07	196.71	113.58	5.02	-	-	-	0.01	0.23	-	353.63
Exports	-27.97	-80.87	-76.42	-52.68	-	-	-	-0.13	-0.17	-	-238.23
Intl. Marine Bunkers	-	-	-20.72	-	-	-	-	-	-	-	-20.72
Stock changes	-0.64	0.03	0.99	-0.31	-	-	-	-	-	-	0.07
TPES	223.57	300.51	17.43	112.12	12.17	15.18	8.51	355.33	0.06	-	1044.88
Electricity and CHP Plants	-136.54	-	-37.67	-43.35	-12.17	-15.18	-8.51	-0.38	90.60	-	-163.22
Petroleum Refineries	-	-294.40	279.52	-	-	-	-	-	-	-	-14.89
Other Transformation *	-12.21	-5.05	-12.25	-29.14	-	-	-	-13.86	-20.01	-	-92.52
TFC	74.81	1.06	247.03	39.63	-	-	-	341.09	70.64	-	774.25
INDUSTRY SECTOR	74.01	1.06	61.24	32.18	-	-	-	38.83	32.27	-	239.57
Iron and Steel	-	-	-	-	-	-	-	-	-	-	-
Chemical and Petrochemical	-	-	-	-	-	-	-	-	-	-	-
Non-Metallic Minerals	-	-	-	-	-	-	-	-	-	-	-
Non-specified	-	-	-	-	-	-	-	-	-	-	-
TRANSPORT SECTOR	0.04	-	130.97	0.02	-	-	-	-	0.69	-	131.71
Air	-	-	16.27	-	-	-	-	-	0.02	-	16.29
Road	-	-	108.46	-	-	-	-	-	-	-	108.46
Non-specified	-	-	-	-	-	-	-	-	-	-	-
OTHER SECTORS	0.77	-	44.28	7.44	-	-	-	302.26	37.69	-	392.43
Agriculture	-	-	-	-	-	-	-	-	-	-	-
Comm. and Publ. Services	-	-	-	-	-	-	-	-	-	-	-
Residential	-	-	-	-	-	-	-	-	-	-	-
Non-specified	-	-	-	-	-	-	-	-	-	-	-
NON-ENERGY USE	-	-	10.54	-	-	-	-	-	-	-	10.54
Electricity Generated - GWh	*468902*	-	*167947*	*180566*	*46715*	*176559*	*10583*	*2158*	-	-	*1053430*
Heat Generated - TJ	-	-	-	-	-	-	-	-	-	-	-

* Includes Transfers, Statistical Differences, Own Use and Distribution Losses.

China / Chine

Total Production of Energy

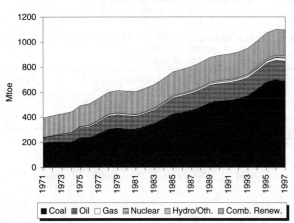

Total Primary Energy Supply*

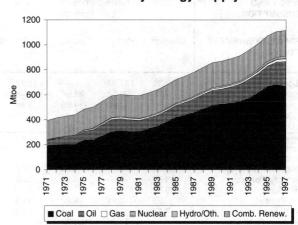

Electricity Generation by Fuel

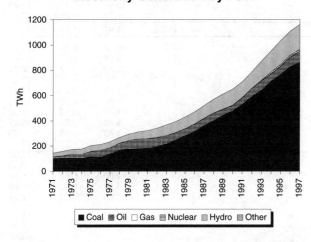

Oil Products Consumption

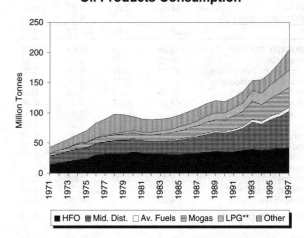

Electricity Consumption/GDP***, TPES/GDP*** and Production/TPES

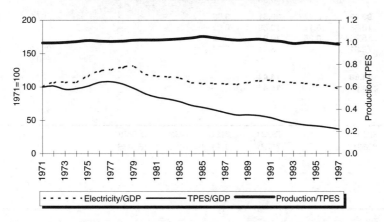

* Excluding electricity trade.
** Includes LPG, NGL, ethane and naphtha.
*** GDP in 1990 US dollars.

INTERNATIONAL ENERGY AGENCY

China / Chine : 1996

Million tonnes of oil equivalent / *Millions de tonnes d'équivalent pétrole*

SUPPLY AND CONSUMPTION / *APPROVISIONNEMENT ET DEMANDE*	Coal / *Charbon*	Crude Oil / *Pétrole brut*	Petroleum Products / *Produits pétroliers*	Gas / *Gaz*	Nuclear / *Nucléaire*	Hydro / *Hydro*	Geotherm. Solar etc. / *Géotherm. solaire etc.*	Combust. Renew. & Waste / *En. ren. combust. & déchets*	Electricity / *Electricité*	Heat / *Chaleur*	Total / *Total*
Indigenous Production	698.35	157.33	-	18.72	3.74	16.17	-	207.20	-	-	1101.51
Imports	5.95	22.62	38.08	1.48	-	-	-	0.01	0.68	-	68.81
Exports	-24.81	-20.40	-14.29	-1.48	-	-	-	-	-0.36	-	-61.34
Intl. Marine Bunkers	-	-	-3.32	-	-	-	-	-	-	-	-3.32
Stock Changes	3.65	-0.34	1.08	-	-	-	-	-	-	-	4.39
TPES	**683.14**	**159.21**	**21.55**	**18.72**	**3.74**	**16.17**	**-**	**207.21**	**0.31**	**-**	**1110.06**
Transfers	-	-	-	-	-	-	-	-	-	-	-
Statistical Differences	25.93	-0.56	4.34	-0.05	-	-	-	-	-	-0.05	29.61
Electricity Plants	-233.86	-0.68	-13.97	-0.69	-3.74	-16.17	-	-	95.33	-	-173.78
CHP Plants	-29.60	-	-	-	-	-	-	-	-	27.42	-2.18
Heat Plants	-0.65	-0.11	-4.23	-0.03	-	-	-	-	-	-	-5.01
Gas Works	-2.17	-	-1.11	2.96	-	-	-	-	-	-	-0.33
Petroleum Refineries	-	-151.34	153.82	-	-	-	-	-	-	-	2.49
Coal Transformation	-36.87	-	-	-	-	-	-	-	-	-	-36.87
Liquefaction	-	-	-	-	-	-	-	-	-	-	-
Other Transformation	-	-	-	-	-	-	-	-	-	-	-
Own Use	-32.47	-2.93	-13.02	-3.68	-	-	-	-	-13.84	-5.74	-71.67
Distribution Losses	-10.97	-1.60	-0.02	-0.53	-	-	-	-	-6.73	-0.36	-20.20
TFC	**362.48**	**1.99**	**147.37**	**16.71**	**-**	**-**	**-**	**207.21**	**75.07**	**21.28**	**832.11**
INDUSTRY SECTOR	**243.38**	**1.62**	**39.84**	**13.52**	**-**	**-**	**-**	**-**	**50.73**	**16.60**	**365.71**
Iron and Steel	48.85	0.05	3.80	1.17	-	-	-	-	8.51	2.67	65.05
Chemical and Petrochemical	47.97	1.46	25.18	8.04	-	-	-	-	12.72	6.67	102.04
of which: Feedstocks	-	*0.27*	*18.40*	*4.47*	-	-	-	-	-	-	*23.15*
Non-Ferrous Metals	6.35	0.01	0.79	0.24	-	-	-	-	4.00	0.76	12.15
Non-Metallic Minerals	69.66	0.06	4.21	0.46	-	-	-	-	5.64	0.18	80.21
Transport Equipment	3.79	-	0.42	0.08	-	-	-	-	1.45	0.56	6.30
Machinery	15.79	0.01	1.53	0.88	-	-	-	-	4.40	0.66	23.27
Mining and Quarrying	4.11	-	0.34	0.06	-	-	-	-	1.63	0.45	6.59
Food and Tobacco	17.31	0.01	0.66	0.12	-	-	-	-	3.04	1.27	22.40
Paper, Pulp and Printing	8.86	-	0.37	0.01	-	-	-	-	1.88	0.97	12.09
Wood and Wood Products	2.74	-	0.16	-	-	-	-	-	0.49	0.10	3.49
Construction	2.32	0.03	1.16	0.14	-	-	-	-	1.50	0.04	5.19
Textile and Leather	11.41	-	0.72	0.57	-	-	-	-	3.94	2.02	18.66
Non-specified	4.23	-	0.50	1.75	-	-	-	-	1.53	0.27	8.28
TRANSPORT SECTOR	**5.07**	**-**	**62.57**	**0.11**	**-**	**-**	**-**	**-**	**1.02**	**-**	**68.77**
Air	-	-	6.26	-	-	-	-	-	-	-	6.26
Road	-	-	46.77	0.11	-	-	-	-	-	-	46.88
Rail	5.04	-	3.58	-	-	-	-	-	0.93	-	9.55
Pipeline Transport	-	-	0.07	-	-	-	-	-	0.09	-	0.15
Internal Navigation	0.04	-	5.84	-	-	-	-	-	-	-	5.87
Non-specified	-	-	0.05	-	-	-	-	-	-	-	0.05
OTHER SECTORS	**99.43**	**0.37**	**30.51**	**3.08**	**-**	**-**	**-**	**207.21**	**23.32**	**4.67**	**368.59**
Agriculture	10.38	0.11	10.69	0.02	-	-	-	-	5.32	0.01	26.52
Comm. and Publ. Services	16.13	0.26	10.91	0.49	-	-	-	-	7.57	0.61	35.98
Residential	72.91	-	8.91	2.57	-	-	-	207.20	10.43	4.06	306.08
Non-specified	-	-	-	-	-	-	-	0.01	-	-	0.01
NON-ENERGY USE	**14.60**	**-**	**14.44**	**-**	**-**	**-**	**-**	**-**	**-**	**-**	**29.04**
in Industry/Transf./Energy	14.60	-	14.44	-	-	-	-	-	-	-	29.04
in Transport	-	-	-	-	-	-	-	-	-	-	-
in Other Sectors	-	-	-	-	-	-	-	-	-	-	-
Electr. Generated - GWh	***828012***	***-***	***75721***	***2421***	***14339***	***187966***	***-***	***-***	***-***	***-***	***1108459***
Electricity Plants	*828012*	*-*	*75721*	*2421*	*14339*	*187966*	*-*	*-*	*-*	*-*	*1108459*
CHP Plants	*-*	*-*	*-*	*-*	*-*	*-*	*-*	*-*	*-*	*-*	*-*
Heat Generated - TJ	***-***	***-***	***-***	***-***	***-***	***-***	***-***	***-***	***-***	***-***	***1148301***
CHP Plants	*-*	*-*	*-*	*-*	*-*	*-*	*-*	*-*	*-*	*-*	*1148301*
Heat Plants	*-*	*-*	*-*	*-*	*-*	*-*	*-*	*-*	*-*	*-*	*-*

China / Chine : 1997

	Million tonnes of oil equivalent / *Millions de tonnes d'équivalent pétrole*										
SUPPLY AND CONSUMPTION	Coal	Crude Oil	Petroleum Products	Gas	Nuclear	Hydro	Geotherm. Solar etc.	Combust. Renew. & Waste	Electricity	Heat	Total
APPROVISIONNEMENT ET DEMANDE	*Charbon*	*Pétrole brut*	*Produits pétroliers*	*Gaz*	*Nucléaire*	*Hydro*	*Géotherm. solaire etc.*	*En. ren. combust. & déchets*	*Electricité*	*Chaleur*	*Total*
Indigenous Production	686.41	160.74	-	21.13	3.76	16.85	-	208.31	-	-	1097.21
Imports	4.70	35.47	50.03	2.31	-	-	-	0.01	0.68	-	93.20
Exports	-23.51	-19.83	-17.90	-2.31	-	-	-	-	-0.67	-	-64.22
Intl. Marine Bunkers	-	-	-3.07	-	-	-	-	-	-	-	-3.07
Stock Changes	-6.00	-1.39	-2.69	-	-	-	-	-	-	-	-10.07
TPES	**661.59**	**175.00**	**26.38**	**21.13**	**3.76**	**16.85**	**-**	**208.32**	**0.02**	**-**	**1113.05**
Transfers	-	-	-	-			-	-	-	-	-
Statistical Differences	27.17	-1.06	7.08	0.12	-	-	-	-	-	0.29	33.60
Electricity Plants	-234.54	-0.68	-15.65	-1.91	-3.76	-16.85	-	-	100.05	-	-173.34
CHP Plants	-29.63	-	-	-	-	-	-	-	-	28.08	-1.55
Heat Plants	-0.79	-0.11	-3.62	-0.07	-	-	-	-	-	-	-4.59
Gas Works	-2.94	-	-1.08	3.10	-	-	-	-	-	-	-0.92
Petroleum Refineries	-	-166.07	166.43	-	-	-	-	-	-	-	0.35
Coal Transformation	-40.14	-	-	-	-	-	-	-	-	-	-40.14
Liquefaction	-	-	-	-	-	-	-	-	-	-	-
Other Transformation	-	-	-	-	-	-	-	-	-	-	-
Own Use	-36.53	-3.22	-15.95	-5.62	-	-	-	-	-16.71	-5.72	-83.74
Distribution Losses	-10.59	-1.76	-0.02	-0.79	-	-	-	-	-8.06	-0.35	-21.56
TFC	**333.61**	**2.09**	**163.57**	**15.96**	**-**	**-**	**-**	**208.32**	**75.31**	**22.30**	**821.15**
INDUSTRY SECTOR	**226.53**	**1.69**	**40.02**	**12.71**	**-**	**-**	**-**	**-**	**48.53**	**17.37**	**346.85**
Iron and Steel	50.11	0.05	3.59	1.09	-	-	-	-	8.14	2.79	65.78
Chemical and Petrochemical	43.63	1.52	26.24	6.79	-	-	-	-	12.17	6.98	97.33
of which: Feedstocks	-	*0.29*	*19.66*	*3.77*	-	-	-	-	-	-	*23.72*
Non-Ferrous Metals	5.88	0.01	0.75	0.23	-	-	-	-	3.83	0.79	11.49
Non-Metallic Minerals	62.85	0.06	3.97	0.41	-	-	-	-	5.39	0.19	72.88
Transport Equipment	3.43	-	0.39	0.07	-	-	-	-	1.39	0.59	5.88
Machinery	14.59	0.01	1.44	0.79	-	-	-	-	4.21	0.69	21.73
Mining and Quarrying	3.83	-	0.32	0.05	-	-	-	-	1.56	0.47	6.22
Food and Tobacco	15.59	0.01	0.62	0.10	-	-	-	-	2.90	1.33	20.55
Paper, Pulp and Printing	7.96	-	0.37	0.01	-	-	-	-	1.80	1.01	11.15
Wood and Wood Products	2.47	-	0.15	-	-	-	-	-	0.47	0.10	3.18
Construction	2.10	0.03	1.08	0.12	-	-	-	-	1.44	0.04	4.80
Textile and Leather	10.26	-	0.67	0.48	-	-	-	-	3.77	2.11	17.31
Non-specified	3.82	-	0.44	2.57	-	-	-	-	1.46	0.28	8.57
TRANSPORT SECTOR	**6.17**	**-**	**69.96**	**0.02**	**-**	**-**	**-**	**-**	**1.32**	**-**	**77.46**
Air	-	-	7.07	-	-	-	-	-	-	-	7.07
Road	-	-	49.37	0.02	-	-	-	-	-	-	49.39
Rail	6.13	-	3.93	-	-	-	-	-	1.20	-	11.26
Pipeline Transport	-	-	0.16	-	-	-	-	-	0.12	-	0.27
Internal Navigation	0.04	-	9.37	-	-	-	-	-	-	-	9.42
Non-specified	-	-	0.05	-	-	-	-	-	-	-	0.05
OTHER SECTORS	**87.07**	**0.41**	**34.09**	**3.23**	**-**	**-**	**-**	**208.32**	**25.46**	**4.93**	**363.50**
Agriculture	10.60	0.12	11.18	0.01	-	-	-	-	5.50	0.01	27.42
Comm. and Publ. Services	13.81	0.28	13.44	0.52	-	-	-	-	8.37	0.78	37.21
Residential	62.66	-	9.47	2.70	-	-	-	208.31	11.59	4.13	298.87
Non-specified	-	-	-	-	-	-	-	0.01	-	-	0.01
NON-ENERGY USE	**13.83**	**-**	**19.50**	**-**	**-**	**-**	**-**	**-**	**-**	**-**	**33.33**
in Industry/Transf./Energy	13.83	-	19.50	-	-	-	-	-	-	-	33.33
in Transport	-	-	-	-	-	-	-	-	-	-	-
in Other Sectors	-	-	-	-	-	-	-	-	-	-	-
Electr. Generated - GWh	*863134*	*-*	*83224*	*6657*	*14418*	*195983*	*-*	*-*	*-*	*-*	*1163416*
Electricity Plants	*863134*	*-*	*83224*	*6657*	*14418*	*195983*	*-*	*-*	*-*	*-*	*1163416*
CHP Plants	*-*	*-*	*-*	*-*	*-*	*-*	*-*	*-*	*-*	*-*	*-*
Heat Generated - TJ	*-*	*-*	*-*	*-*	*-*	*-*	*-*	*-*	*-*	*-*	*1175695*
CHP Plants	*-*	*-*	*-*	*-*	*-*	*-*	*-*	*-*	*-*	*-*	*1175695*
Heat Plants	*-*	*-*	*-*	*-*	*-*	*-*	*-*	*-*	*-*	*-*	*-*

Non-OECD Europe / Europe non-OCDE

Total Production of Energy

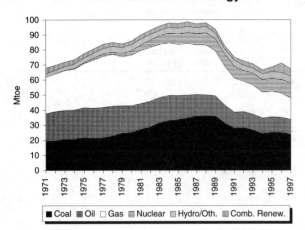

Coal ■ Oil ▦ Gas ☐ Nuclear ≣ Hydro/Oth. ▦ Comb. Renew. ▥

Total Primary Energy Supply*

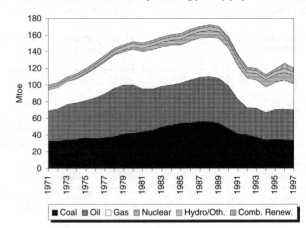

Coal ■ Oil ▦ Gas ☐ Nuclear ≣ Hydro/Oth. ▦ Comb. Renew. ▥

Electricity Generation by Fuel

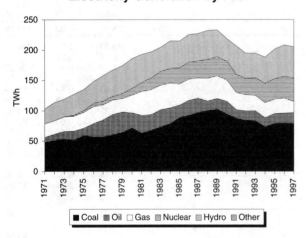

Coal ■ Oil ▦ Gas ☐ Nuclear ≣ Hydro ▦ Other ▥

Oil Products Consumption

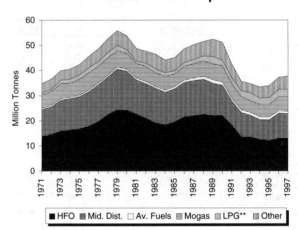

HFO ■ Mid. Dist. ▦ Av. Fuels ☐ Mogas ≣ LPG** ▦ Other ▥

Electricity Consumption/GDP***, TPES/GDP*** and Production/TPES

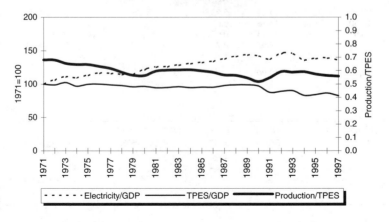

····· Electricity/GDP ——— TPES/GDP ——— Production/TPES

* Excluding electricity trade.
** Includes LPG, NGL, ethane and naphtha.
*** GDP in 1990 US dollars.

INTERNATIONAL ENERGY AGENCY

Non-OECD Europe / Europe non-OCDE

Million tonnes of oil equivalent / Millions de tonnes d'équivalent pétrole

SUPPLY AND CONSUMPTION 1996	Coal	Crude Oil	Petroleum Products	Gas	Nuclear	Hydro	Geotherm. Solar etc.	Combust. Renew. & Waste	Electricity	Heat	Total
Indigenous Production	24.96	10.06	-	16.07	9.20	4.48	0.01	6.44	-	-	71.22
Imports	9.66	26.45	11.39	18.77	-	-	-	0.03	1.32	-	67.62
Exports	-0.21	-0.11	-9.16	-	-	-	-	-0.02	-0.91	-	-10.40
Intl. Marine Bunkers	-	-	-1.26	-	-	-	-	-	-	-	-1.26
Stock changes	-0.64	-0.41	0.20	0.10	-	-	-	-0.01	-	-	-0.76
TPES	33.77	35.98	1.17	34.94	9.20	4.48	0.01	6.45	0.42	-	126.42
Electricity and CHP Plants	-23.92	-	-6.78	-9.19	-9.20	-4.48	-	-0.03	17.91	9.01	-26.68
Petroleum Refineries	-	-35.29	34.99	-	-	-	-	-	-	-	-0.30
Other Transformation *	-2.89	-0.70	-4.88	-5.16	-	-	-0.01	-0.10	-4.76	1.39	-17.10
TFC	6.95	-	24.50	20.59	-	-	-	6.32	13.57	10.41	82.34
INDUSTRY SECTOR	4.62	-	6.90	14.48	-	-	-	0.29	5.56	3.47	35.32
Iron and Steel	-	-	-	-	-	-	-	-	-	-	-
Chemical and Petrochemical	-	-	-	-	-	-	-	-	-	-	-
Non-Metallic Minerals	-	-	-	-	-	-	-	-	-	-	-
Non-specified	-	-	-	-	-	-	-	-	-	-	-
TRANSPORT SECTOR	0.09	-	12.49	0.09	-	-	-	0.01	0.42	-	13.09
Air	-	-	0.94	-	-	-	-	-	-	-	0.94
Road	-	-	10.93	0.08	-	-	-	-	-	-	11.02
Non-specified	-	-	-	-	-	-	-	-	-	-	-
OTHER SECTORS	1.95	-	3.60	6.02	-	-	-	6.02	7.60	6.94	32.13
Agriculture	-	-	-	-	-	-	-	-	-	-	-
Comm. and Publ. Services	-	-	-	-	-	-	-	-	-	-	-
Residential	-	-	-	-	-	-	-	-	-	-	-
Non-specified	-	-	-	-	-	-	-	-	-	-	-
NON-ENERGY USE	0.29	-	1.51	-	-	-	-	-	-	-	1.80
Electricity Generated - GWh	*79085*	-	*17008*	*24775*	*35291*	*52085*	-	*10*	-	-	*208254*
Heat Generated - TJ	-	-	-	-	-	-	-	*4445*	-	-	*528384*

APPROVISIONNEMENT ET DEMANDE 1997	Charbon	Pétrole brut	Produits pétroliers	Gaz	Nucléaire	Hydro	Géotherm. solaire etc.	En. ren. combust. & déchets	Electricité	Chaleur	Total
Indigenous Production	23.81	9.83	-	14.14	10.16	4.44	0.01	4.90	-	-	67.30
Imports	9.85	25.51	12.44	16.54	-	-	-	0.04	1.26	-	65.64
Exports	-0.09	-0.04	-8.62	-	-	-	-	-0.02	-0.97	-	-9.74
Intl. Marine Bunkers	-	-	-1.36	-	-	-	-	-	-	-	-1.36
Stock changes	-0.10	-	-1.10	0.01	-	-	-	-0.01	-	-	-1.21
TPES	33.46	35.30	1.36	30.68	10.16	4.44	0.01	4.92	0.29	-	120.63
Electricity and CHP Plants	-23.63	-	-6.65	-6.82	-10.16	-4.44	-	-0.02	17.65	8.55	-25.52
Petroleum Refineries	-	-34.65	34.38	-	-	-	-	-	-	-	-0.27
Other Transformation *	-3.26	-0.65	-3.58	-4.52	-	-	-0.01	-0.15	-4.54	0.71	-16.00
TFC	6.57	-	25.51	19.35	-	-	-	4.74	13.40	9.27	78.84
INDUSTRY SECTOR	4.55	-	6.87	12.93	-	-	-	0.36	5.59	2.99	33.29
Iron and Steel	-	-	-	-	-	-	-	-	-	-	-
Chemical and Petrochemical	-	-	-	-	-	-	-	-	-	-	-
Non-Metallic Minerals	-	-	-	-	-	-	-	-	-	-	-
Non-specified	-	-	-	-	-	-	-	-	-	-	-
TRANSPORT SECTOR	0.10	-	12.72	0.03	-	-	-	0.01	0.40	-	13.25
Air	-	-	1.05	-	-	-	-	-	-	-	1.05
Road	-	-	10.83	0.03	-	-	-	-	-	-	10.86
Non-specified	-	-	-	-	-	-	-	-	-	-	-
OTHER SECTORS	1.57	-	4.37	6.39	-	-	-	4.37	7.42	6.27	30.38
Agriculture	-	-	-	-	-	-	-	-	-	-	-
Comm. and Publ. Services	-	-	-	-	-	-	-	-	-	-	-
Residential	-	-	-	-	-	-	-	-	-	-	-
Non-specified	-	-	-	-	-	-	-	-	-	-	-
NON-ENERGY USE	0.35	-	1.56	-	-	-	-	-	-	-	1.91
Electricity Generated - GWh	*78983*	-	*18105*	*17533*	*38967*	*51608*	-	*13*	-	-	*205209*
Heat Generated - TJ	-	-	-	-	-	-	-	*2679*	-	-	*494551*

* Includes Transfers, Statistical Differences, Own Use and Distribution Losses.

INTERNATIONAL ENERGY AGENCY

Former USSR / Ex-URSS

Total Production of Energy

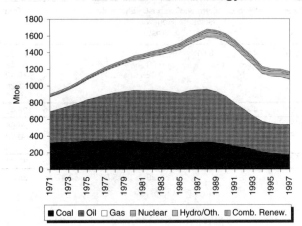

■ Coal ▦ Oil □ Gas ▥ Nuclear ▨ Hydro/Oth. ▨ Comb. Renew.

Total Primary Energy Supply*

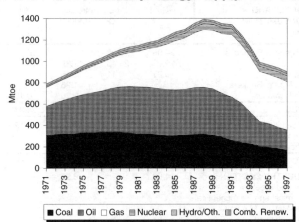

■ Coal ▦ Oil □ Gas ▥ Nuclear ▨ Hydro/Oth. ▨ Comb. Renew.

Electricity Generation by Fuel

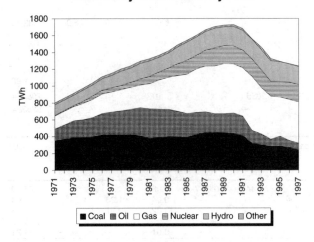

■ Coal ▦ Oil □ Gas ▥ Nuclear ▨ Hydro ▨ Other

Oil Products Consumption

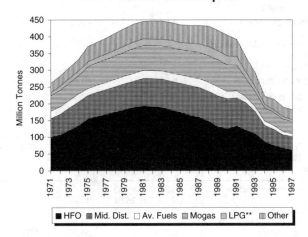

■ HFO ▦ Mid. Dist. □ Av. Fuels ▨ Mogas ▨ LPG** ▨ Other

Electricity Consumption/GDP***, TPES/GDP*** and Production/TPES

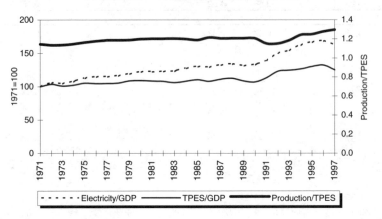

- - - Electricity/GDP —— TPES/GDP ══ Production/TPES

* Excluding electricity trade.
** Includes LPG, NGL, ethane and naphtha.
*** GDP in 1990 US dollars.

Former USSR / Ex-URSS : 1996

Million tonnes of oil equivalent / *Millions de tonnes d'équivalent pétrole*

SUPPLY AND CONSUMPTION	Coal	Crude Oil	Petroleum Products	Gas	Nuclear	Hydro	Geotherm. Solar etc.	Combust. Renew. & Waste	Electricity	Heat	Total
APPROVISIONNEMENT ET DEMANDE	*Charbon*	*Pétrole brut*	*Produits pétroliers*	*Gaz*	*Nucléaire*	*Hydro*	*Géotherm. solaire etc.*	*En. ren. combust. & déchets*	*Electricité*	*Chaleur*	*Total*
Indigenous Production	180.09	351.90	-	573.00	53.82	18.54	0.02	19.85	-	-	1197.22
Imports	0.53	-	5.05		-		-	-		-	5.58
Exports	-8.32	-112.04	-60.67	-100.06		-		-0.45	-0.88		-282.42
Intl. Marine Bunkers	-		-0.09								-0.09
Stock Changes	10.01	4.85	9.17	-7.54		-	-	0.02	-	-	16.51
TPES	**182.31**	**244.71**	**-46.55**	**465.41**	**53.82**	**18.54**	**0.02**	**19.41**	**-0.88**	**-**	**936.80**
Transfers	-	-0.03	0.10	-	-	-	-	-	-	-	0.08
Statistical Differences	-8.40	-1.58	-	-2.52	-	-	-	-0.09	0.08	-	-12.50
Electricity Plants	-1.09	-	-3.37	-19.46	-49.76	-18.54	-0.02	-	39.03	-	-53.21
CHP Plants	-81.48	-0.60	-13.16	-162.41	-4.07		-	-2.41	69.39	84.18	-110.54
Heat Plants	-20.03	-0.01	-38.55	-62.94	-		-	-4.52	-0.03	107.57	-18.51
Gas Works	-0.04	-0.31	-	0.36	-		-	-	-	-	0.01
Petroleum Refineries	-	-238.13	233.55	-0.07	-		-	-	-	-	-4.65
Coal Transformation	-14.79	-	-	-	-		-	-	-	-	-14.79
Liquefaction		-	-		-		-	-	-	-	
Other Transformation	-	-	-	-	-		-	-0.02	-	-	-0.02
Own Use	-0.79	-0.56	-6.43	-29.35	-		-	-0.04	-16.71	-12.76	-66.64
Distribution Losses	-5.06	-0.23	-0.06	-10.17	-		-	-0.14	-12.66	-10.91	-39.23
TFC	**50.63**	**3.28**	**125.55**	**178.84**	**-**	**-**	**-**	**12.20**	**78.23**	**168.07**	**616.81**
INDUSTRY SECTOR	**26.30**	**0.80**	**19.58**	**62.35**	**-**	**-**	**-**	**1.67**	**34.63**	**67.40**	**212.73**
Iron and Steel	-	-	-	-	-	-	-	-	-	-	-
Chemical and Petrochemical	-	-	-	-	-	-	-	-	-	-	-
of which: Feedstocks	-	-	-	-	-	-	-	-	-	-	-
Non-Ferrous Metals	-	-	-	-	-	-	-	-	-	-	-
Non-Metallic Minerals	-	-	-	-	-	-	-	-	-	-	-
Transport Equipment	-	-	-	-	-	-	-	-	-	-	-
Machinery	-	-	-	-	-	-	-	-	-	-	-
Mining and Quarrying	-	-	-	-	-	-	-	-	-	-	-
Food and Tobacco	-	-	-	-	-	-	-	-	-	-	-
Paper, Pulp and Printing	-	-	-	-	-	-	-	-	-	-	-
Wood and Wood Products	-	-	-	-	-	-	-	-	-	-	-
Construction	-	-	-	-	-	-	-	-	-	-	-
Textile and Leather	-	-	-	-	-	-	-	-	-	-	-
Non-specified	-	-	-	-	-	-	-	-	-	-	-
TRANSPORT SECTOR	**0.39**	**0.01**	**53.50**	**15.91**	**-**	**-**	**-**	**0.01**	**7.14**	**-**	**76.97**
Air	-	-	10.12	-	-	-	-	-	-	-	10.12
Road	-	-	33.74	0.18	-	-	-	-	-	-	33.92
Rail	-	-	-	-	-	-	-	-	-	-	-
Pipeline Transport	-	-	-	-	-	-	-	-	-	-	-
Internal Navigation	-	-	-	-	-	-	-	-	-	-	-
Non-specified	-	-	-	-	-	-	-	-	-	-	-
OTHER SECTORS	**21.94**	**0.94**	**42.02**	**100.58**	**-**	**-**	**-**	**10.51**	**36.46**	**100.67**	**313.13**
Agriculture	-	-	-	-	-	-	-	-	-	-	-
Comm. and Publ. Services	-	-	-	-	-	-	-	-	-	-	-
Residential	-	-	-	-	-	-	-	-	-	-	-
Non-specified	-	-	-	-	-	-	-	-	-	-	-
NON-ENERGY USE	**2.00**	**1.54**	**10.45**	**-**	**-**	**-**	**-**	**-**	**-**	**-**	**13.98**
in Industry/Transf./Energy	-	-	-	-	-	-	-	-	-	-	-
in Transport	-	-	-	-	-	-	-	-	-	-	-
in Other Sectors	-	-	-	-	-	-	-	-	-	-	-
Electr. Generated - GWh	*265500*	*-*	*84747*	*487478*	*204869*	*215603*	*29*	*2533*	*-*	*-*	*1260759*
Electricity Plants	*-*	*-*	*-*	*-*	*-*	*-*	*-*	*-*	*-*	*-*	*453866*
CHP Plants	*-*	*-*	*-*	*-*	*-*	*-*	*-*	*-*	*-*	*-*	*806893*
Heat Generated - TJ	*-*	*-*	*-*	*-*	*-*	*-*	*-*	*-*	*-*	*-*	*8029375*
CHP Plants	*-*	*-*	*-*	*-*	*-*	*-*	*-*	*-*	*-*	*-*	*3525126*
Heat Plants	*-*	*-*	*-*	*-*	*-*	*-*	*-*	*-*	*-*	*-*	*4504249*

Former USSR / Ex-URSS : 1997

Million tonnes of oil equivalent / Millions de tonnes d'équivalent pétrole											
SUPPLY AND CONSUMPTION / *APPROVISIONNEMENT ET DEMANDE*	Coal / *Charbon*	Crude Oil / *Pétrole brut*	Petroleum Products / *Produits pétroliers*	Gas / *Gaz*	Nuclear / *Nucléaire*	Hydro / *Hydro*	Geotherm. Solar etc. / *Géotherm. solaire etc.*	Combust. Renew. & Waste / *En. ren. combust. & déchets*	Electricity / *Electricité*	Heat / *Chaleur*	Total / *Total*
Indigenous Production	173.58	359.59	-	541.74	53.09	18.64	0.02	20.84	-	-	1167.51
Imports	1.42	-	4.84	-	-	-	-	-	-	-	6.26
Exports	-8.92	-115.71	-64.41	-94.55	-	-	-	-	-0.68	-	-284.26
Intl. Marine Bunkers	-	-	-0.16	-	-	-	-	-	-	-	-0.16
Stock Changes	-1.96	2.19	4.80	6.48	-	-	-	-0.52	-	-	11.00
TPES	**164.13**	**246.07**	**-54.93**	**453.68**	**53.09**	**18.64**	**0.02**	**20.32**	**-0.68**	**-**	**900.34**
Transfers	-	-0.03	0.08	-	-	-	-	-	-	-	0.05
Statistical Differences	-5.28	-2.46	-	-0.15	-	-	-	-0.09	-0.02	-	-8.01
Electricity Plants	-0.86	-	-3.05	-18.80	-49.52	-18.64	-0.02	-	39.53	-	-51.37
CHP Plants	-73.62	-0.60	-11.89	-162.20	-3.56	-	-	-2.41	66.62	77.12	-110.55
Heat Plants	-18.28	-	-35.94	-60.65	-	-	-	-4.58	-0.02	97.84	-21.63
Gas Works	-0.04	-0.22	-	0.29	-	-	-	-	-	-	0.03
Petroleum Refineries	-	-238.73	234.20	-0.07	-	-	-	-	-	-	-4.61
Coal Transformation	-14.92	-	-	-	-	-	-	-	-	-	-14.92
Liquefaction	-	-	-	-	-	-	-	-	-	-	-
Other Transformation	-0.03	-	-	-	-	-	-	-0.02	-	-	-0.05
Own Use	-0.73	-0.56	-6.65	-29.03	-	-	-	-0.04	-16.83	-11.52	-65.36
Distribution Losses	-4.43	-0.23	-0.06	-9.89	-	-	-	-0.14	-12.45	-9.87	-37.07
TFC	**45.93**	**3.24**	**121.76**	**173.17**	**-**	**-**	**-**	**13.05**	**76.16**	**153.56**	**586.87**
INDUSTRY SECTOR	**24.79**	**0.74**	**19.36**	**59.17**	**-**	**-**	**-**	**1.69**	**34.33**	**62.53**	**202.61**
Iron and Steel	-	-	-	-	-	-	-	-	-	-	-
Chemical and Petrochemical	-	-	-	-	-	-	-	-	-	-	-
of which: Feedstocks	-	-	-	-	-	-	-	-	-	-	-
Non-Ferrous Metals	-	-	-	-	-	-	-	-	-	-	-
Non-Metallic Minerals	-	-	-	-	-	-	-	-	-	-	-
Transport Equipment	-	-	-	-	-	-	-	-	-	-	-
Machinery	-	-	-	-	-	-	-	-	-	-	-
Mining and Quarrying	-	-	-	-	-	-	-	-	-	-	-
Food and Tobacco	-	-	-	-	-	-	-	-	-	-	-
Paper, Pulp and Printing	-	-	-	-	-	-	-	-	-	-	-
Wood and Wood Products	-	-	-	-	-	-	-	-	-	-	-
Construction	-	-	-	-	-	-	-	-	-	-	-
Textile and Leather	-	-	-	-	-	-	-	-	-	-	-
Non-specified	-	-	-	-	-	-	-	-	-	-	-
TRANSPORT SECTOR	**0.34**	**0.01**	**51.03**	**15.78**	**-**	**-**	**-**	**0.01**	**6.94**	**-**	**74.10**
Air	-	-	9.34	-	-	-	-	-	-	-	9.34
Road	-	-	32.40	0.17	-	-	-	-	-	-	32.57
Rail	-	-	-	-	-	-	-	-	-	-	-
Pipeline Transport	-	-	-	-	-	-	-	-	-	-	-
Internal Navigation	-	-	-	-	-	-	-	-	-	-	-
Non-specified	-	-	-	-	-	-	-	-	-	-	-
OTHER SECTORS	**18.80**	**0.94**	**40.99**	**98.23**	**-**	**-**	**-**	**11.35**	**34.88**	**91.04**	**296.22**
Agriculture	-	-	-	-	-	-	-	-	-	-	-
Comm. and Publ. Services	-	-	-	-	-	-	-	-	-	-	-
Residential	-	-	-	-	-	-	-	-	-	-	-
Non-specified	-	-	-	-	-	-	-	-	-	-	-
NON-ENERGY USE	**2.01**	**1.54**	**10.38**	**-**	**-**	**-**	**-**	**-**	**-**	**-**	**13.93**
in Industry/Transf./Energy	-	-	-	-	-	-	-	-	-	-	-
in Transport	-	-	-	-	-	-	-	-	-	-	-
in Other Sectors	-	-	-	-	-	-	-	-	-	-	-
Electr. Generated - GWh	*238327*	*-*	*79841*	*490925*	*202057*	*216757*	*29*	*6438*	*-*	*-*	*1234374*
Electricity Plants	*-*	*-*	*-*	*-*	*-*	*-*	*-*	*-*	*-*	*-*	*459709*
CHP Plants	*-*	*-*	*-*	*-*	*-*	*-*	*-*	*-*	*-*	*-*	*774665*
Heat Generated - TJ	*-*	*-*	*-*	*-*	*-*	*-*	*-*	*-*	*-*	*-*	*7326289*
CHP Plants	*-*	*-*	*-*	*-*	*-*	*-*	*-*	*-*	*-*	*-*	*3229305*
Heat Plants	*-*	*-*	*-*	*-*	*-*	*-*	*-*	*-*	*-*	*-*	*4096984*

Middle East / Moyen-Orient

Total Production of Energy

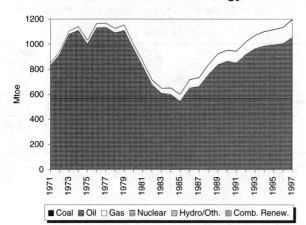

■ Coal ▨ Oil □ Gas ▤ Nuclear ▨ Hydro/Oth. ▥ Comb. Renew.

Total Primary Energy Supply*

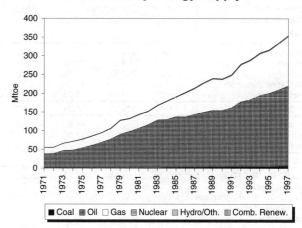

■ Coal ▨ Oil □ Gas ▤ Nuclear ▨ Hydro/Oth. ▥ Comb. Renew.

Electricity Generation by Fuel

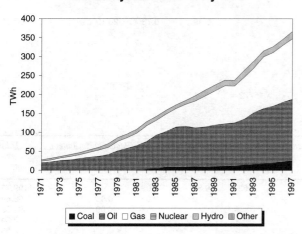

■ Coal ▨ Oil □ Gas ▤ Nuclear ▨ Hydro ▥ Other

Oil Products Consumption

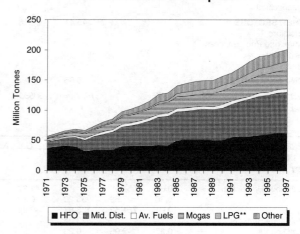

■ HFO ▨ Mid. Dist. □ Av. Fuels ▤ Mogas ▨ LPG** ▥ Other

Electricity Consumption/GDP***,
TPES/GDP*** and Production/TPES

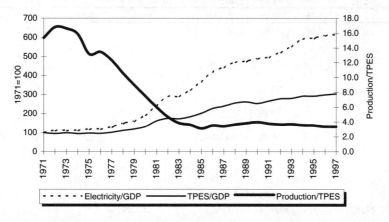

- - - - Electricity/GDP ——— TPES/GDP ━━━ Production/TPES

* Excluding electricity trade.
** Includes LPG, NGL, ethane and naphtha.
*** GDP in 1990 US dollars.

Middle East / Moyen-Orient

Million tonnes of oil equivalent / *Millions de tonnes d'équivalent pétrole*

SUPPLY AND CONSUMPTION 1996	Coal	Crude Oil	Petroleum Products	Gas	Nuclear	Hydro	Geotherm. Solar etc.	Combust. Renew. & Waste	Electricity	Heat	Total
Indigenous Production	0.61	1004.59	-	128.42	-	1.36	0.49	1.02	-	-	1136.48
Imports	4.98	24.85	29.69	0.20	-	-	-	0.03	0.06	-	59.81
Exports	-0.01	-719.50	-122.21	-6.10	-	-	-	-	-0.08	-	-847.91
Intl. Marine Bunkers	-	-	-13.25	-	-	-	-	-	-	-	-13.25
Stock changes	0.43	-0.76	0.73	-	-	-	-	-	-	-	0.39
TPES	6.01	309.17	-105.04	122.52	-	1.36	0.49	1.05	-0.03	-	335.53
Electricity and CHP Plants	-4.93	-10.70	-28.69	-32.34	-	-1.36	-	-	29.74	-	-48.28
Petroleum Refineries	-	-276.11	257.68	-	-	-	-	-	-	-	-18.43
Other Transformation *	-0.35	-17.52	7.70	-38.95	-	-	-	-0.14	-5.17	-	-54.42
TFC	0.73	4.85	131.65	51.23	-	-	0.49	0.92	24.54	-	214.40
INDUSTRY SECTOR	0.73	4.85	20.71	29.05	-	-	-	0.19	4.37	-	59.89
Iron and Steel	-	-	-	-	-	-	-	-	-	-	-
Chemical and Petrochemical	-	-	-	-	-	-	-	-	-	-	-
Non-Metallic Minerals	-	-	-	-	-	-	-	-	-	-	-
Non-specified	-	-	-	-	-	-	-	-	-	-	-
TRANSPORT SECTOR	-	-	59.58	-	-	-	-	-	0.08	-	59.65
Air	-	-	6.51	-	-	-	-	-	-	-	6.51
Road	-	-	53.07	-	-	-	-	-	-	-	53.07
Non-specified	-	-	-	-	-	-	-	-	-	-	-
OTHER SECTORS	-	-	45.41	22.18	-	-	0.49	0.73	20.10	-	88.91
Agriculture	-	-	-	-	-	-	-	-	-	-	-
Comm. and Publ. Services	-	-	-	-	-	-	-	-	-	-	-
Residential	-	-	-	-	-	-	-	-	-	-	-
Non-specified	-	-	-	-	-	-	-	-	-	-	-
NON-ENERGY USE	-	-	5.96	-	-	-	-	-	-	-	5.96
Electricity Generated - GWh	22397	-	156744	150855	-	15772	1	-	-	-	345769
Heat Generated - TJ	-	-	-	-	-	-	-	-	-	-	-

APPROVISIONNEMENT ET DEMANDE 1997	Charbon	Pétrole brut	Produits pétroliers	Gaz	Nucléaire	Hydro	Géotherm. solaire etc.	En. ren. combust. & déchets	Electricité	Chaleur	Total
Indigenous Production	0.61	1054.68	-	141.27	-	1.63	0.52	1.02	-	-	1199.73
Imports	5.56	27.59	33.20	0.20	-	-	-	0.03	0.05	-	66.64
Exports	-0.01	-761.53	-123.87	-9.85	-	-	-	-	-0.09	-	-895.35
Intl. Marine Bunkers	-	-	-13.34	-	-	-	-	-	-	-	-13.34
Stock changes	0.37	-2.95	0.23	-	-	-	-	-	-	-	-2.35
TPES	6.54	317.79	-103.78	131.62	-	1.63	0.52	1.05	-0.04	-	355.33
Electricity and CHP Plants	-5.46	-11.33	-28.99	-34.31	-	-1.63	-	-	31.49	-	-50.24
Petroleum Refineries	-	-282.99	260.19	-	-	-	-	-	-	-	-22.80
Other Transformation *	-0.35	-17.39	6.55	-40.90	-	-	-	-0.14	-5.73	-	-57.97
TFC	0.73	6.08	133.96	56.41	-	-	0.52	0.92	25.71	-	224.32
INDUSTRY SECTOR	0.73	6.08	20.60	32.15	-	-	-	0.19	4.57	-	64.31
Iron and Steel	-	-	-	-	-	-	-	-	-	-	-
Chemical and Petrochemical	-	-	-	-	-	-	-	-	-	-	-
Non-Metallic Minerals	-	-	-	-	-	-	-	-	-	-	-
Non-specified	-	-	-	-	-	-	-	-	-	-	-
TRANSPORT SECTOR	-	-	60.74	-	-	-	-	-	0.08	-	60.81
Air	-	-	6.78	-	-	-	-	-	-	-	6.78
Road	-	-	53.96	-	-	-	-	-	-	-	53.96
Non-specified	-	-	-	-	-	-	-	-	-	-	-
OTHER SECTORS	-	-	46.72	24.26	-	-	0.52	0.73	21.06	-	93.29
Agriculture	-	-	-	-	-	-	-	-	-	-	-
Comm. and Publ. Services	-	-	-	-	-	-	-	-	-	-	-
Residential	-	-	-	-	-	-	-	-	-	-	-
Non-specified	-	-	-	-	-	-	-	-	-	-	-
NON-ENERGY USE	-	-	5.91	-	-	-	-	-	-	-	5.91
Electricity Generated - GWh	24815	-	163104	159219	-	18983	1	-	-	-	366122
Heat Generated - TJ	-	-	-	-	-	-	-	-	-	-	-

* Includes Transfers, Statistical Differences, Own Use and Distribution Losses.

GRAPHS AND ENERGY BALANCE SHEETS

GRAPHIQUES ET BILANS ENERGETIQUES

1996-1997

COUNTRIES

PAYS

INTERNATIONAL ENERGY AGENCY

Albania / Albanie

Thousand tonnes of oil equivalent / *Milliers de tonnes d'équivalent pétrole*

SUPPLY AND CONSUMPTION 1996	Coal	Crude Oil	Petroleum Products	Gas	Nuclear	Hydro	Geotherm. Solar etc.	Combust. Renew. & Waste	Electricity	Heat	Total
Indigenous Production	24	484	-	19	-	489	-	60	-	-	1076
Imports	-	5	106	-	-	-	-	-	17	-	128
Exports	-	-	-	-	-	-	-	-	-	-	-
Intl. Marine Bunkers	-	-	-	-	-	-	-	-	-	-	-
Stock changes	-	-	-	-	-	-	-	-	-	-	-
TPES	24	489	106	19	-	489	-	60	17	-	1204
Electricity and CHP Plants	-	-	-44	-	-	-489	-	-	510	28	5
Petroleum Refineries	-	-489	433	-	-	-	-	-	-	-	-56
Other Transformation *	-	-	-46	-	-	-	-	-	-271	-25	-342
TFC	24	-	449	19	-	-	-	60	256	3	810
INDUSTRY SECTOR	-	-	45	16	-	-	-	-	69	-	129
Iron and Steel	-	-	-	-	-	-	-	-	-	-	-
Chemical and Petrochemical	-	-	-	7	-	-	-	-	-	-	7
Non-Metallic Minerals	-	-	-	1	-	-	-	-	-	-	1
Non-specified	-	-	45	8	-	-	-	-	69	-	122
TRANSPORT SECTOR	-	-	297	-	-	-	-	-	-	-	297
Air	-	-	-	-	-	-	-	-	-	-	-
Road	-	-	297	-	-	-	-	-	-	-	297
Non-specified	-	-	-	-	-	-	-	-	-	-	-
OTHER SECTORS	24	-	72	3	-	-	-	60	187	3	349
Agriculture	-	-	-	-	-	-	-	-	5	-	5
Comm. and Publ. Services	24	-	-	-	-	-	-	-	4	-	28
Residential	-	-	72	3	-	-	-	60	110	3	248
Non-specified	-	-	-	1	-	-	-	-	68	-	69
NON-ENERGY USE	-	-	35	-	-	-	-	-	-	-	35
Electricity Generated - GWh	-	-	242	-	-	5684	-	-	-	-	5926
Heat Generated - TJ	-	-	-	-	-	-	-	-	-	-	1189

APPROVISIONNEMENT ET DEMANDE 1997	Charbon	Pétrole brut	Produits pétroliers	Gaz	Nucléaire	Hydro	Géotherm. solaire etc.	En. ren. combust. & déchets	Electricité	Chaleur	Total
Indigenous Production	16	357	-	15	-	464	-	60	-	-	912
Imports	-	6	112	-	-	-	-	-	17	-	135
Exports	-	-	-	-	-	-	-	-	-	-	-
Intl. Marine Bunkers	-	-	-	-	-	-	-	-	-	-	-
Stock changes	-	-	-	-	-	-	-	-	-	-	-
TPES	16	363	112	15	-	464	-	60	17	-	1048
Electricity and CHP Plants	-	-	-37	-	-	-464	-	-	482	28	9
Petroleum Refineries	-	-363	356	-	-	-	-	-	-	-	-7
Other Transformation *	-	-	-34	-	-	-	-	-	-257	-25	-316
TFC	16	-	397	15	-	-	-	60	242	3	733
INDUSTRY SECTOR	-	-	21	13	-	-	-	-	65	-	99
Iron and Steel	-	-	-	-	-	-	-	-	-	-	-
Chemical and Petrochemical	-	-	-	5	-	-	-	-	-	-	5
Non-Metallic Minerals	-	-	-	1	-	-	-	-	-	-	1
Non-specified	-	-	21	7	-	-	-	-	65	-	93
TRANSPORT SECTOR	-	-	302	-	-	-	-	-	-	-	302
Air	-	-	-	-	-	-	-	-	-	-	-
Road	-	-	302	-	-	-	-	-	-	-	302
Non-specified	-	-	-	-	-	-	-	-	-	-	-
OTHER SECTORS	16	-	60	3	-	-	-	60	177	3	319
Agriculture	-	-	-	-	-	-	-	-	4	-	4
Comm. and Publ. Services	16	-	-	-	-	-	-	-	4	-	20
Residential	-	-	60	2	-	-	-	60	104	3	228
Non-specified	-	-	-	1	-	-	-	-	65	-	66
NON-ENERGY USE	-	-	13	-	-	-	-	-	-	-	13
Electricity Generated - GWh	-	-	209	-	-	5391	-	-	-	-	5600
Heat Generated - TJ	-	-	-	-	-	-	-	-	-	-	1189

* Includes Transfers, Statistical Differences, Own Use and Distribution Losses.

Algeria / Algérie

Total Production of Energy

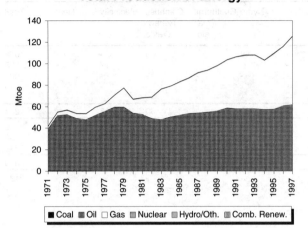

Total Primary Energy Supply*

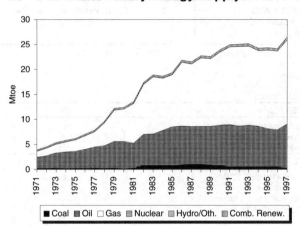

Electricity Generation by Fuel

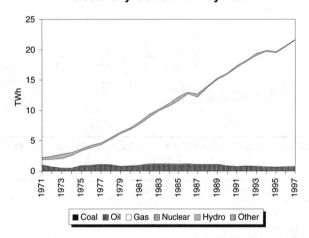

Oil Products Consumption

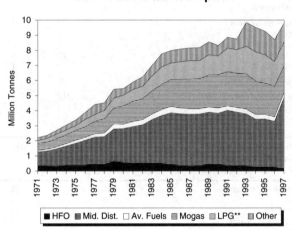

Electricity Consumption/GDP***, TPES/GDP*** and Production/TPES

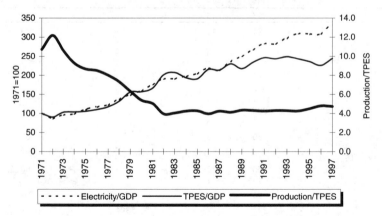

* Excluding electricity trade.
** Includes LPG, NGL, ethane and naphtha.
*** GDP in 1990 US dollars.

Algeria / Algérie : 1996

Thousand tonnes of oil equivalent / *Milliers de tonnes d'équivalent pétrole*

SUPPLY AND CONSUMPTION *APPROVISIONNEMENT ET DEMANDE*	Coal *Charbon*	Crude Oil *Pétrole brut*	Petroleum Products *Produits pétroliers*	Gas *Gaz*	Nuclear *Nucléaire*	Hydro *Hydro*	Geotherm. Solar etc. *Géotherm. solaire*	Combust. Renew. & Waste *En. ren. combust.*	Electricity *Electricité*	Heat *Chaleur*	Total *Total*
Indigenous Production	-	60719	-	54981	-	12	-	508	-	-	116219
Imports	468	364	132	-	-	-	-	-	24	-	989
Exports	-5	-35823	-17554	-39225	-	-	-	-	-36	-	-92643
Intl. Marine Bunkers	-	-	-329	-	-	-	-	-	-	-	-329
Stock Changes	54	30	-157	-	-	-	-	-	-	-	-74
TPES	**517**	**25290**	**-17908**	**15756**	**-**	**12**	**-**	**508**	**-12**	**-**	**24162**
Transfers	-	-5090	5809	-	-	-	-	-	-	-	719
Statistical Differences	-26	294	222	44	-	-	-	-	-	-	533
Electricity Plants	-	-	-235	-5596	-	-12	-	-	1776	-	-4067
CHP Plants	-	-	-	-	-	-	-	-	-	-	-
Heat Plants	-	-	-	-	-	-	-	-	-	-	-
Gas Works	-	-	-	-	-	-	-	-	-	-	-
Petroleum Refineries	-	-19530	19892	-	-	-	-	-	-	-	362
Coal Transformation	-149	-	-	-	-	-	-	-	-	-	-149
Liquefaction	-	-	-	-	-	-	-	-	-	-	-
Other Transformation	-	-	-	-	-	-	-	-	-	-	-
Own Use	-	-462	-278	-5189	-	-	-	-	-143	-	-6072
Distribution Losses	-197	-471	-	-160	-	-	-	-	-328	-	-1156
TFC	**145**	**29**	**7502**	**4855**	**-**	**-**	**-**	**508**	**1293**	**-**	**14331**
INDUSTRY SECTOR	**145**	**29**	**17**	**2299**	**-**	**-**	**-**	**-**	**538**	**-**	**3027**
Iron and Steel	140	-	-	245	-	-	-	-	54	-	438
Chemical and Petrochemical	-	-	-	844	-	-	-	-	62	-	906
of which: Feedstocks	-	-	-	*814*	-	-	-	-	-	-	*814*
Non-Ferrous Metals	-	-	-	-	-	-	-	-	-	-	-
Non-Metallic Minerals	-	-	-	-	-	-	-	-	-	-	-
Transport Equipment	-	-	-	-	-	-	-	-	-	-	-
Machinery	-	-	-	-	-	-	-	-	-	-	-
Mining and Quarrying	-	-	-	-	-	-	-	-	-	-	-
Food and Tobacco	-	-	-	-	-	-	-	-	-	-	-
Paper, Pulp and Printing	-	-	-	-	-	-	-	-	-	-	-
Wood and Wood Products	-	-	-	-	-	-	-	-	-	-	-
Construction	-	-	-	1008	-	-	-	-	123	-	1131
Textile and Leather	-	-	-	-	-	-	-	-	-	-	-
Non-specified	5	29	17	202	-	-	-	-	299	-	552
TRANSPORT SECTOR	**-**	**-**	**2569**	**567**	**-**	**-**	**-**	**-**	**29**	**-**	**3165**
Air	-	-	322	-	-	-	-	-	-	-	322
Road	-	-	2247	-	-	-	-	-	-	-	2247
Rail	-	-	-	-	-	-	-	-	20	-	20
Pipeline Transport	-	-	-	567	-	-	-	-	9	-	576
Internal Navigation	-	-	-	-	-	-	-	-	-	-	-
Non-specified	-	-	-	-	-	-	-	-	-	-	-
OTHER SECTORS	**-**	**-**	**4379**	**1989**	**-**	**-**	**-**	**508**	**726**	**-**	**7602**
Agriculture	-	-	-	-	-	-	-	-	-	-	-
Comm. and Publ. Services	-	-	-	-	-	-	-	-	-	-	-
Residential	-	-	1553	1989	-	-	-	508	726	-	4775
Non-specified	-	-	2827	-	-	-	-	-	-	-	2827
NON-ENERGY USE	**-**	**-**	**537**	**-**	**-**	**-**	**-**	**-**	**-**	**-**	**537**
in Industry/Transf./Energy	-	-	537	-	-	-	-	-	-	-	537
in Transport	-	-	-	-	-	-	-	-	-	-	-
in Other Sectors	-	-	-	-	-	-	-	-	-	-	-
Electr. Generated - GWh	**-**	**-**	**745**	**19770**	**-**	**135**	**-**	**-**	**-**	**-**	**20650**
Electricity Plants	-	-	745	19770	-	135	-	-	-	-	20650
CHP Plants	-	-	-	-	-	-	-	-	-	-	-
Heat Generated - TJ	**-**	**-**	**-**	**-**	**-**	**-**	**-**	**-**	**-**	**-**	**-**
CHP Plants	-	-	-	-	-	-	-	-	-	-	-
Heat Plants	-	-	-	-	-	-	-	-	-	-	-

Algeria / Algérie : 1997

Thousand tonnes of oil equivalent / *Milliers de tonnes d'équivalent pétrole*

SUPPLY AND CONSUMPTION *APPROVISIONNEMENT ET DEMANDE*	Coal *Charbon*	Crude Oil *Pétrole brut*	Petroleum Products *Produits pétroliers*	Gas *Gaz*	Nuclear *Nucléaire*	Hydro *Hydro*	Geotherm. Solar etc. *Géotherm. solaire*	Combust. Renew. & Waste *En. ren. combust.*	Electricity *Electricité*	Heat *Chaleur*	Total *Total*
Indigenous Production	-	61670	-	63381	-	6	-	519	-	-	125576
Imports	378	262	97	-	-	-	-	-	27	-	763
Exports	-	-34518	-18271	-46462	-	-	-	-	-27	-	-99278
Intl. Marine Bunkers	-	-	-229	-	-	-	-	-	-	-	-229
Stock Changes	-105	-	-230	-	-	-	-	-	-	-	-336
TPES	**272**	**27413**	**-18633**	**16919**	**-**	**6**	**-**	**519**	**-**	**-**	**26497**
Transfers	-	-6162	7032	-	-	-	-	-	-	-	870
Statistical Differences	-7	755	-187	-66	-	-	-	-	-	-	495
Electricity Plants	-	-	-245	-5898	-	-6	-	-	1865	-	-4285
CHP Plants	-	-	-	-	-	-	-	-	-	-	-
Heat Plants	-	-	-	-	-	-	-	-	-	-	-
Gas Works	-	-	-	-	-	-	-	-	-	-	-
Petroleum Refineries	-	-21282	21258	-	-	-	-	-	-	-	-24
Coal Transformation	-99	-	-	-	-	-	-	-	-	-	-99
Liquefaction	-	-	-	-	-	-	-	-	-	-	-
Other Transformation	-	-	-	-	-	-	-	-	-	-	-
Own Use	-	-368	-278	-5747	-	-	-	-	-166	-	-6560
Distribution Losses	-66	-328	-	-175	-	-	-	-	-273	-	-842
TFC	**101**	**29**	**8946**	**5033**	**-**	**-**	**-**	**519**	**1426**	**-**	**16053**
INDUSTRY SECTOR	**101**	**29**	**17**	**2502**	**-**	**-**	**-**	**-**	**661**	**-**	**3309**
Iron and Steel	101	-	-	229	-	-	-	-	45	-	375
Chemical and Petrochemical	-	-	-	1140	-	-	-	-	66	-	1206
of which: Feedstocks	-	-	-	*1112*	-	-	-	-	-	-	*1112*
Non-Ferrous Metals	-	-	-	-	-	-	-	-	-	-	-
Non-Metallic Minerals	-	-	-	-	-	-	-	-	-	-	-
Transport Equipment	-	-	-	-	-	-	-	-	-	-	-
Machinery	-	-	-	-	-	-	-	-	-	-	-
Mining and Quarrying	-	-	-	-	-	-	-	-	-	-	-
Food and Tobacco	-	-	-	-	-	-	-	-	-	-	-
Paper, Pulp and Printing	-	-	-	-	-	-	-	-	-	-	-
Wood and Wood Products	-	-	-	-	-	-	-	-	-	-	-
Construction	-	-	-	943	-	-	-	-	119	-	1062
Textile and Leather	-	-	-	-	-	-	-	-	-	-	-
Non-specified	-	29	17	190	-	-	-	-	430	-	666
TRANSPORT SECTOR	**-**	**-**	**2305**	**674**	**-**	**-**	**-**	**-**	**30**	**-**	**3009**
Air	-	-	351	-	-	-	-	-	-	-	351
Road	-	-	1953	-	-	-	-	-	-	-	1953
Rail	-	-	-	-	-	-	-	-	23	-	23
Pipeline Transport	-	-	-	674	-	-	-	-	7	-	681
Internal Navigation	-	-	-	-	-	-	-	-	-	-	-
Non-specified	-	-	-	-	-	-	-	-	-	-	-
OTHER SECTORS	**-**	**-**	**6124**	**1857**	**-**	**-**	**-**	**519**	**735**	**-**	**9234**
Agriculture	-	-	-	-	-	-	-	-	-	-	-
Comm. and Publ. Services	-	-	-	-	-	-	-	-	-	-	-
Residential	-	-	1576	1857	-	-	-	519	735	-	4687
Non-specified	-	-	4548	-	-	-	-	-	-	-	4548
NON-ENERGY USE	**-**	**-**	**501**	**-**	**-**	**-**	**-**	**-**	**-**	**-**	**501**
in Industry/Transf./Energy	-	-	501	-	-	-	-	-	-	-	501
in Transport	-	-	-	-	-	-	-	-	-	-	-
in Other Sectors	-	-	-	-	-	-	-	-	-	-	-
Electr. Generated - GWh	**-**	**-**	**773**	**20837**	**-**	**75**	**-**	**-**	**-**	**-**	**21685**
Electricity Plants	-	-	773	20837	-	75	-	-	-	-	21685
CHP Plants	-	-	-	-	-	-	-	-	-	-	-
Heat Generated - TJ	**-**	-	-	-	-	-	-	-	-	-	-
CHP Plants	-	-	-	-	-	-	-	-	-	-	-
Heat Plants	-	-	-	-	-	-	-	-	-	-	-

Angola

Thousand tonnes of oil equivalent / *Milliers de tonnes d'équivalent pétrole*

SUPPLY AND CONSUMPTION 1996	Coal	Crude Oil	Petroleum Products	Gas	Nuclear	Hydro	Geotherm. Solar etc.	Combust. Renew. & Waste	Electricity	Heat	Total
Indigenous Production	-	35417	-	457	-	80	-	5032	-	-	40986
Imports	-	-	29	-	-	-	-	-	-	-	29
Exports	-	-33717	-130	-	-	-	-	-	-	-	-33847
Intl. Marine Bunkers	-	-	-651	-	-	-	-	-	-	-	-651
Stock changes	-	-	-	-	-	-	-	-	-	-	-
TPES	-	1700	-751	457	-	80	-	5032	-	-	6518
Electricity and CHP Plants	-	-	-224	-	-	-80	-	-	88	-	-215
Petroleum Refineries	-	-1634	1539	-	-	-	-	-	-	-	-95
Other Transformation *	-	-66	-49	-	-	-	-	-1326	-29	-	-1471
TFC	-	-	515	457	-	-	-	3706	59	-	4737
INDUSTRY SECTOR	-	-	8	457	-	-	-	87	19	-	570
Iron and Steel	-	-	-	-	-	-	-	-	-	-	-
Chemical and Petrochemical	-	-	8	-	-	-	-	-	-	-	8
Non-Metallic Minerals	-	-	-	-	-	-	-	-	-	-	-
Non-specified	-	-	-	457	-	-	-	87	19	-	563
TRANSPORT SECTOR	-	-	387	-	-	-	-	-	-	-	387
Air	-	-	190	-	-	-	-	-	-	-	190
Road	-	-	198	-	-	-	-	-	-	-	198
Non-specified	-	-	-	-	-	-	-	-	-	-	-
OTHER SECTORS	-	-	89	-	-	-	-	3619	41	-	3749
Agriculture	-	-	-	-	-	-	-	-	-	-	-
Comm. and Publ. Services	-	-	-	-	-	-	-	-	-	-	-
Residential	-	-	89	-	-	-	-	3619	41	-	3749
Non-specified	-	-	-	-	-	-	-	-	-	-	-
NON-ENERGY USE	-	-	31	-	-	-	-	-	-	-	31
Electricity Generated - GWh	-	-	103	-	-	925	-	-	-	-	1028
Heat Generated - TJ	-	-	-	-	-	-	-	-	-	-	-

APPROVISIONNEMENT ET DEMANDE 1997	Charbon	Pétrole brut	Produits pétroliers	Gaz	Nucléaire	Hydro	Géotherm. solaire etc.	En. ren. combust. & déchets	Electricité	Chaleur	Total
Indigenous Production	-	35700	-	465	-	87	-	5178	-	-	41430
Imports	-	-	65	-	-	-	-	-	-	-	65
Exports	-	-33676	-321	-	-	-	-	-	-	-	-33997
Intl. Marine Bunkers	-	-	-651	-	-	-	-	-	-	-	-651
Stock changes	-	-	-	-	-	-	-	-	-	-	-
TPES	-	2025	-907	465	-	87	-	5178	-	-	6848
Electricity and CHP Plants	-	-	-268	-	-	-87	-	-	95	-	-259
Petroleum Refineries	-	-1958	1920	-	-	-	-	-	-	-	-38
Other Transformation *	-	-66	27	-	-	-	-	-1365	-31	-	-1435
TFC	-	-	772	465	-	-	-	3813	64	-	5115
INDUSTRY SECTOR	-	-	16	465	-	-	-	89	20	-	591
Iron and Steel	-	-	-	-	-	-	-	-	-	-	-
Chemical and Petrochemical	-	-	16	-	-	-	-	-	-	-	16
Non-Metallic Minerals	-	-	-	-	-	-	-	-	-	-	-
Non-specified	-	-	-	465	-	-	-	89	20	-	575
TRANSPORT SECTOR	-	-	627	-	-	-	-	-	-	-	627
Air	-	-	310	-	-	-	-	-	-	-	310
Road	-	-	317	-	-	-	-	-	-	-	317
Non-specified	-	-	-	-	-	-	-	-	-	-	-
OTHER SECTORS	-	-	98	-	-	-	-	3724	44	-	3866
Agriculture	-	-	-	-	-	-	-	-	-	-	-
Comm. and Publ. Services	-	-	-	-	-	-	-	-	-	-	-
Residential	-	-	98	-	-	-	-	3724	44	-	3866
Non-specified	-	-	-	-	-	-	-	-	-	-	-
NON-ENERGY USE	-	-	31	-	-	-	-	-	-	-	31
Electricity Generated - GWh	-	-	103	-	-	1006	-	-	-	-	1109
Heat Generated - TJ	-	-	-	-	-	-	-	-	-	-	-

* Includes Transfers, Statistical Differences, Own Use and Distribution Losses.

Argentina / Argentine

Total Production of Energy

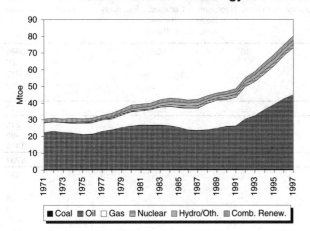

■ Coal ▦ Oil □ Gas ▤ Nuclear ▥ Hydro/Oth. ▩ Comb. Renew.

Total Primary Energy Supply*

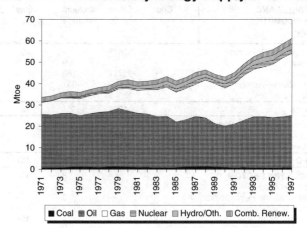

■ Coal ▦ Oil □ Gas ▤ Nuclear ▥ Hydro/Oth. ▩ Comb. Renew.

Electricity Generation by Fuel

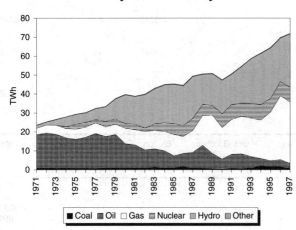

■ Coal ▦ Oil □ Gas ▤ Nuclear ▥ Hydro ▩ Other

Oil Products Consumption

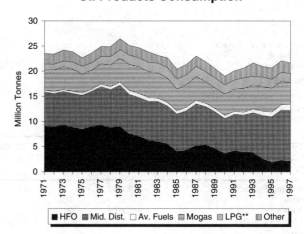

■ HFO ▦ Mid. Dist. □ Av. Fuels ▤ Mogas ▥ LPG** ▩ Other

Electricity Consumption/GDP***, TPES/GDP*** and Production/TPES

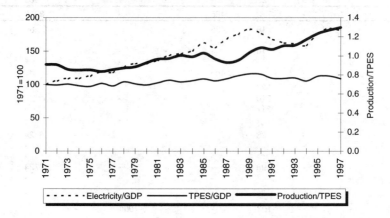

- - - - Electricity/GDP ——— TPES/GDP ——— Production/TPES

* Excluding electricity trade.
** Includes LPG, NGL, ethane and naphtha.
*** GDP in 1990 US dollars.

Argentina / Argentine : 1996

	Coal	Crude Oil	Petroleum Products	Gas	Nuclear	Hydro	Geotherm. Solar etc.	Combust. Renew. & Waste	Electricity	Heat	Total
SUPPLY AND CONSUMPTION											
APPROVISIONNEMENT ET DEMANDE	Charbon	Pétrole brut	Produits pétroliers	Gaz	Nucléaire	Hydro	Géotherm. solaire	En. ren. combust.	Electricité	Chaleur	Total
Indigenous Production	183	42379	-	25688	1944	1977	-	2690	-	-	74860
Imports	755	759	1711	1760	-	-	-	-	315	-	5299
Exports	-	-16858	-3662	-	-	-	-	-	-26	-	-20546
Intl. Marine Bunkers	-	-	-570	-	-	-	-	-	-	-	-570
Stock Changes	-50	-75	2	-	-	-	-	-	-	-	-123
TPES	**888**	**26205**	**-2520**	**27447**	**1944**	**1977**	**-**	**2690**	**289**	**-**	**58920**
Transfers	-	-1306	1440	-	-	-	-	-	-	-	134
Statistical Differences	175	215	-1148	-	-	-	-	-	-330	-	-1089
Electricity Plants	-548	-	-977	-8358	-1944	-1977	-	-135	5999	-	-7939
CHP Plants	-	-	-	-	-	-	-	-	-	-	-
Heat Plants	-	-	-	-	-	-	-	-	-	-	-
Gas Works	-	-	-	-	-	-	-	-	-	-	-
Petroleum Refineries	-	-24727	23985	-	-	-	-	-	-	-	-741
Coal Transformation	-165	-	-	-	-	-	-	-	-	-	-165
Liquefaction	-	-	-	-	-	-	-	-	-	-	-
Other Transformation	-	-	-	-	-	-	-	-314	-	-	-314
Own Use	-5	-28	-1059	-3724	-	-	-	-	-193	-	-5009
Distribution Losses	-8	-	-240	-2534	-	-	-	-	-1097	-	-3879
TFC	**337**	**358**	**19481**	**12832**	**-**	**-**	**-**	**2241**	**4668**	**-**	**39918**
INDUSTRY SECTOR	**337**	**358**	**1230**	**5897**	**-**	**-**	**-**	**1795**	**1916**	**-**	**11533**
Iron and Steel	224	-	-	-	-	-	-	-	-	-	224
Chemical and Petrochemical	-	358	657	198	-	-	-	-	-	-	1213
of which: Feedstocks	-	358	657	198	-	-	-	-	-	-	1213
Non-Ferrous Metals	-	-	-	-	-	-	-	-	-	-	-
Non-Metallic Minerals	-	-	-	-	-	-	-	-	-	-	-
Transport Equipment	-	-	-	-	-	-	-	-	-	-	-
Machinery	-	-	-	-	-	-	-	-	-	-	-
Mining and Quarrying	-	-	-	-	-	-	-	-	-	-	-
Food and Tobacco	-	-	-	-	-	-	-	-	-	-	-
Paper, Pulp and Printing	-	-	-	-	-	-	-	-	-	-	-
Wood and Wood Products	-	-	-	-	-	-	-	-	-	-	-
Construction	-	-	-	-	-	-	-	-	-	-	-
Textile and Leather	-	-	-	-	-	-	-	-	-	-	-
Non-specified	113	-	573	5699	-	-	-	1795	1916	-	10096
TRANSPORT SECTOR	**-**	**-**	**12637**	**906**	**-**	**-**	**-**	**-**	**36**	**-**	**13579**
Air	-	-	1148	-	-	-	-	-	-	-	1148
Road	-	-	11398	906	-	-	-	-	-	-	12304
Rail	-	-	-	-	-	-	-	-	36	-	36
Pipeline Transport	-	-	-	-	-	-	-	-	-	-	-
Internal Navigation	-	-	91	-	-	-	-	-	-	-	91
Non-specified	-	-	-	-	-	-	-	-	-	-	-
OTHER SECTORS	**-**	**-**	**4142**	**6028**	**-**	**-**	**-**	**447**	**2716**	**-**	**13333**
Agriculture	-	-	2693	-	-	-	-	-	40	-	2733
Comm. and Publ. Services	-	-	186	1098	-	-	-	-	1106	-	2389
Residential	-	-	1263	4931	-	-	-	386	1516	-	8095
Non-specified	-	-	-	-	-	-	-	61	54	-	115
NON-ENERGY USE	**-**	**-**	**1473**	**-**	**-**	**-**	**-**	**-**	**-**	**-**	**1473**
in Industry/Transf./Energy	-	-	1473	-	-	-	-	-	-	-	1473
in Transport	-	-	-	-	-	-	-	-	-	-	-
in Other Sectors	-	-	-	-	-	-	-	-	-	-	-
Electr. Generated - GWh	**1520**	**-**	**3800**	**33706**	**7459**	**22985**	**-**	**289**	**-**	**-**	**69759**
Electricity Plants	*1520*	*-*	*3800*	*33706*	*7459*	*22985*	*-*	*289*	*-*	*-*	*69759*
CHP Plants	*-*	*-*	*-*	*-*	*-*	*-*	*-*	*-*	*-*	*-*	*-*
Heat Generated - TJ	**-**	**-**	**-**	**-**	**-**	**-**	**-**	**-**	**-**	**-**	**-**
CHP Plants	*-*	*-*	*-*	*-*	*-*	*-*	*-*	*-*	*-*	*-*	*-*
Heat Plants	*-*	*-*	*-*	*-*	*-*	*-*	*-*	*-*	*-*	*-*	*-*

Thousand tonnes of oil equivalent / *Milliers de tonnes d'équivalent pétrole*

Argentina / Argentine : 1997

Thousand tonnes of oil equivalent / Milliers de tonnes d'équivalent pétrole

SUPPLY AND CONSUMPTION / APPROVISIONNEMENT ET DEMANDE	Coal / Charbon	Crude Oil / Pétrole brut	Petroleum Products / Produits pétroliers	Gas / Gaz	Nuclear / Nucléaire	Hydro / Hydro	Geotherm. Solar etc. / Géotherm. solaire	Combust. Renew. & Waste / En. ren. combust.	Electricity / Electricité	Heat / Chaleur	Total / Total
Indigenous Production	148	44801	-	28059	2057	2422	-	2647	-	-	80134
Imports	634	944	1307	1411	-	-	-	-	470	-	4766
Exports	-	-17275	-5039	-613	-	-	-	-	-24	-	-22951
Intl. Marine Bunkers	-	-	-570	-	-	-	-	-	-	-	-570
Stock Changes	74	22	235	-	-	-	-	-	-	-	331
TPES	**856**	**28493**	**-4067**	**28857**	**2057**	**2422**	**-**	**2647**	**446**	**-**	**61710**
Transfers	-	-1157	1525	-	-	-	-	-	-	-	368
Statistical Differences	37	137	-1016	-	-	-	-	-	-313	-	-1155
Electricity Plants	-368	-	-665	-8021	-2057	-2421	-	-100	6187	-	-7445
CHP Plants	-	-	-	-	-	-	-	-	-	-	-
Heat Plants	-	-	-	-	-	-	-	-	-	-	-
Gas Works	-	-	-	-	-	-	-	-	-	-	-
Petroleum Refineries	-	-27094	24847	-	-	-	-	-	-	-	-2247
Coal Transformation	-155	-	-	-	-	-	-	-	-	-	-155
Liquefaction	-	-	-	-	-	-	-	-	-	-	-
Other Transformation	-	-	-	-	-	-	-	-187	-	-	-187
Own Use	-	-	-1116	-4332	-	-	-	-	-227	-	-5675
Distribution Losses	-9	-	-	-2701	-	-	-	-	-1081	-	-3791
TFC	**361**	**379**	**19508**	**13803**	**-**	**-**	**-**	**2360**	**5012**	**-**	**41424**
INDUSTRY SECTOR	**361**	**379**	**1073**	**6236**	**-**	**-**	**-**	**1966**	**2073**	**-**	**12088**
Iron and Steel	227	-	-	-	-	-	-	-	-	-	227
Chemical and Petrochemical	-	379	655	209	-	-	-	-	-	-	1243
of which: Feedstocks	-	379	655	209	-	-	-	-	-	-	1243
Non-Ferrous Metals	-	-	-	-	-	-	-	-	-	-	-
Non-Metallic Minerals	-	-	-	-	-	-	-	-	-	-	-
Transport Equipment	-	-	-	-	-	-	-	-	-	-	-
Machinery	-	-	-	-	-	-	-	-	-	-	-
Mining and Quarrying	-	-	-	-	-	-	-	-	-	-	-
Food and Tobacco	-	-	-	-	-	-	-	-	-	-	-
Paper, Pulp and Printing	-	-	-	-	-	-	-	-	-	-	-
Wood and Wood Products	-	-	-	-	-	-	-	-	-	-	-
Construction	-	-	-	-	-	-	-	-	-	-	-
Textile and Leather	-	-	-	-	-	-	-	-	-	-	-
Non-specified	134	-	418	6026	-	-	-	1966	2073	-	10618
TRANSPORT SECTOR	**-**	**-**	**12905**	**1050**	**-**	**-**	**-**	**-**	**41**	**-**	**13996**
Air	-	-	1296	-	-	-	-	-	-	-	1296
Road	-	-	11537	1050	-	-	-	-	-	-	12587
Rail	-	-	-	-	-	-	-	-	41	-	41
Pipeline Transport	-	-	-	-	-	-	-	-	-	-	-
Internal Navigation	-	-	72	-	-	-	-	-	-	-	72
Non-specified	-	-	-	-	-	-	-	-	-	-	-
OTHER SECTORS	**-**	**-**	**4061**	**6517**	**-**	**-**	**-**	**394**	**2898**	**-**	**13870**
Agriculture	-	-	2737	-	-	-	-	-	41	-	2778
Comm. and Publ. Services	-	-	142	1435	-	-	-	-	1173	-	2750
Residential	-	-	1182	5082	-	-	-	333	1622	-	8220
Non-specified	-	-	-	-	-	-	-	61	62	-	123
NON-ENERGY USE	**-**	**-**	**1469**	**-**	**-**	**-**	**-**	**-**	**-**	**-**	**1469**
in Industry/Transf./Energy	-	-	1469	-	-	-	-	-	-	-	1469
in Transport	-	-	-	-	-	-	-	-	-	-	-
in Other Sectors	-	-	-	-	-	-	-	-	-	-	-
Electr. Generated - GWh	*754*	*-*	*2604*	*32325*	*7892*	*28157*	*-*	*215*	*-*	*-*	*71947*
Electricity Plants	*754*	*-*	*2604*	*32325*	*7892*	*28157*	*-*	*215*	*-*	*-*	*71947*
CHP Plants	*-*	*-*	*-*	*-*	*-*	*-*	*-*	*-*	*-*	*-*	*-*
Heat Generated - TJ	*-*	*-*	*-*	*-*	*-*	*-*	*-*	*-*	*-*	*-*	*-*
CHP Plants	*-*	*-*	*-*	*-*	*-*	*-*	*-*	*-*	*-*	*-*	*-*
Heat Plants	*-*	*-*	*-*	*-*	*-*	*-*	*-*	*-*	*-*	*-*	*-*

Armenia / Arménie

Thousand tonnes of oil equivalent / *Milliers de tonnes d'équivalent pétrole*

SUPPLY AND CONSUMPTION 1996	Coal	Crude Oil	Petroleum Products	Gas	Nuclear	Hydro	Geotherm. Solar etc.	Combust. Renew. & Waste	Electricity	Heat	Total
Indigenous Production	-	-	-	-	606	135	-	1	-	-	742
Imports	2	-	155	891	-	-	-	-	-	-	1048
Exports	-	-	-	-	-	-	-	-	-	-	-
Intl. Marine Bunkers	-	-	-	-	-	-	-	-	-	-	-
Stock changes	-	-	-	-	-	-	-	-	-	-	-
TPES	2	-	155	891	606	135	-	1	-	-	1790
Electricity and CHP Plants	-	-	-60	-685	-606	-135	-	-	534	53	-899
Petroleum Refineries	-	-	-	-	-	-	-	-	-	-	-
Other Transformation *	-	-	-	-	-	-	-	-	-241	19	-222
TFC	2	-	94	206	-	-	-	1	294	72	669
INDUSTRY SECTOR	-	-	-	106	-	-	-	-	70	30	207
Iron and Steel	-	-	-	-	-	-	-	-	-	-	-
Chemical and Petrochemical	-	-	-	-	-	-	-	-	14	-	14
Non-Metallic Minerals	-	-	-	-	-	-	-	-	7	-	7
Non-specified	-	-	-	106	-	-	-	-	49	30	185
TRANSPORT SECTOR	-	-	42	-	-	-	-	-	15	-	57
Air	-	-	20	-	-	-	-	-	-	-	20
Road	-	-	21	-	-	-	-	-	-	-	21
Non-specified	-	-	-	-	-	-	-	-	15	-	15
OTHER SECTORS	2	-	38	100	-	-	-	1	209	42	392
Agriculture	-	-	-	14	-	-	-	-	26	-	40
Comm. and Publ. Services	-	-	-	-	-	-	-	-	88	-	88
Residential	2	-	2	-	-	-	-	-	93	42	139
Non-specified	-	-	36	86	-	-	-	1	1	-	125
NON-ENERGY USE	-	-	14	-	-	-	-	-	-	-	14
Electricity Generated - GWh	-	-	127	2191	2324	1572	-	-	-	-	6214
Heat Generated - TJ	-	-	506	3256	-	-	-	-	-	-	3762

APPROVISIONNEMENT ET DEMANDE 1997	Charbon	Pétrole brut	Produits pétroliers	Gaz	Nucléaire	Hydro	Géotherm. solaire etc.	En. ren. combust. & déchets	Electricité	Chaleur	Total
Indigenous Production	-	-	-	-	417	119	-	1	-	-	537
Imports	2	-	155	1110	-	-	-	-	-	-	1267
Exports	-	-	-	-	-	-	-	-	-	-	-
Intl. Marine Bunkers	-	-	-	-	-	-	-	-	-	-	-
Stock changes	-	-	-	-	-	-	-	-	-	-	-
TPES	2	-	155	1110	417	119	-	1	-	-	1804
Electricity and CHP Plants	-	-	-60	-735	-417	-119	-	-	518	53	-761
Petroleum Refineries	-	-	-	-	-	-	-	-	-	-	-
Other Transformation *	-	-	-	-154	-	-	-	-	-146	19	-282
TFC	2	-	94	221	-	-	-	1	371	72	762
INDUSTRY SECTOR	-	-	-	114	-	-	-	-	62	30	206
Iron and Steel	-	-	-	-	-	-	-	-	-	-	-
Chemical and Petrochemical	-	-	-	-	-	-	-	-	17	-	17
Non-Metallic Minerals	-	-	-	-	-	-	-	-	8	-	8
Non-specified	-	-	-	114	-	-	-	-	37	30	181
TRANSPORT SECTOR	-	-	42	-	-	-	-	-	13	-	55
Air	-	-	20	-	-	-	-	-	-	-	20
Road	-	-	21	-	-	-	-	-	-	-	21
Non-specified	-	-	-	-	-	-	-	-	13	-	13
OTHER SECTORS	2	-	38	107	-	-	-	1	296	42	487
Agriculture	-	-	-	15	-	-	-	-	21	-	36
Comm. and Publ. Services	-	-	-	-	-	-	-	-	28	-	28
Residential	2	-	2	-	-	-	-	-	220	42	266
Non-specified	-	-	36	92	-	-	-	1	27	-	156
NON-ENERGY USE	-	-	14	-	-	-	-	-	-	-	14
Electricity Generated - GWh	-	-	127	2905	1600	1389	-	-	-	-	6021
Heat Generated - TJ	-	-	506	3256	-	-	-	-	-	-	3762

* Includes Transfers, Statistical Differences, Own Use and Distribution Losses.

Azerbaijan / Azerbaïdjan

Thousand tonnes of oil equivalent / *Milliers de tonnes d'équivalent pétrole*

SUPPLY AND CONSUMPTION 1996	Coal	Crude Oil	Petroleum Products	Gas	Nuclear	Hydro	Geotherm. Solar etc.	Combust. Renew. & Waste	Electricity	Heat	Total
Indigenous Production	-	9145	-	5108	-	132	-	2	-	-	14387
Imports	3	-	82	19	-	-	-	3	88	-	194
Exports	-	-	-2291	-	-	-	-	-	-50	-	-2341
Intl. Marine Bunkers	-	-	-	-	-	-	-	-	-	-	-
Stock changes	-	-	-	-	-	-	-	-	-	-	-
TPES	3	9145	-2209	5128	-	132	-	4	38	-	12240
Electricity and CHP Plants	-	-	-3549	-1539	-	-132	-	-	1470	499	-3251
Petroleum Refineries	-	-9145	7952	-	-	-	-	-	-	-	-1193
Other Transformation *	-	-	-115	-	-	-	-	-	-324	243	-196
TFC	3	-	2079	3589	-	-	-	4	1184	742	7601
INDUSTRY SECTOR	-	-	151	1503	-	-	-	-	412	742	2808
Iron and Steel	-	-	-	-	-	-	-	-	3	-	3
Chemical and Petrochemical	-	-	-	-	-	-	-	-	44	-	44
Non-Metallic Minerals	-	-	-	-	-	-	-	-	7	-	7
Non-specified	-	-	151	1503	-	-	-	-	359	742	2755
TRANSPORT SECTOR	3	-	1360	36	-	-	-	-	39	-	1437
Air	-	-	562	-	-	-	-	-	-	-	562
Road	-	-	789	-	-	-	-	-	-	-	789
Non-specified	3	-	9	36	-	-	-	-	39	-	86
OTHER SECTORS	-	-	500	2050	-	-	-	4	732	-	3287
Agriculture	-	-	38	479	-	-	-	-	367	-	885
Comm. and Publ. Services	-	-	12	-	-	-	-	-	74	-	87
Residential	-	-	11	1571	-	-	-	-	280	-	1861
Non-specified	-	-	438	-	-	-	-	4	11	-	454
NON-ENERGY USE	-	-	69	-	-	-	-	-	-	-	69
Electricity Generated - GWh	-	-	12450	3100	-	1538	-	-	-	-	17088
Heat Generated - TJ	-	-	31067	-	-	-	-	-	-	-	31067

APPROVISIONNEMENT ET DEMANDE 1997	Charbon	Pétrole brut	Produits pétroliers	Gaz	Nucléaire	Hydro	Géotherm. solaire etc.	En. ren. combust. & déchets	Electricité	Chaleur	Total
Indigenous Production	-	9066	-	4829	-	131	-	2	-	-	14027
Imports	3	261	74	-	-	-	-	3	120	-	461
Exports	-	-301	-2149	-	-	-	-	-	-52	-	-2502
Intl. Marine Bunkers	-	-	-	-	-	-	-	-	-	-	-
Stock changes	-	-	-	-	-	-	-	-	-	-	-
TPES	3	9026	-2075	4829	-	131	-	4	69	-	11987
Electricity and CHP Plants	-	-	-3536	-1353	-	-131	-	-	1445	499	-3076
Petroleum Refineries	-	-9026	7520	-	-	-	-	-	-	-	-1506
Other Transformation *	-	-	-115	-	-	-	-	-	-415	243	-287
TFC	3	-	1795	3476	-	-	-	4	1099	742	7118
INDUSTRY SECTOR	-	-	144	1510	-	-	-	-	383	742	2779
Iron and Steel	-	-	-	-	-	-	-	-	3	-	3
Chemical and Petrochemical	-	-	-	-	-	-	-	-	41	-	41
Non-Metallic Minerals	-	-	-	-	-	-	-	-	6	-	6
Non-specified	-	-	144	1510	-	-	-	-	333	742	2729
TRANSPORT SECTOR	3	-	1016	34	-	-	-	-	36	-	1089
Air	-	-	480	-	-	-	-	-	-	-	480
Road	-	-	526	-	-	-	-	-	-	-	526
Non-specified	3	-	10	34	-	-	-	-	36	-	83
OTHER SECTORS	-	-	566	1932	-	-	-	4	680	-	3182
Agriculture	-	-	34	447	-	-	-	-	341	-	821
Comm. and Publ. Services	-	-	12	-	-	-	-	-	69	-	81
Residential	-	-	10	1485	-	-	-	-	260	-	1754
Non-specified	-	-	511	-	-	-	-	4	10	-	526
NON-ENERGY USE	-	-	69	-	-	-	-	-	-	-	69
Electricity Generated - GWh	-	-	12235	3045	-	1520	-	-	-	-	16800
Heat Generated - TJ	-	-	31067	-	-	-	-	-	-	-	31067

* Includes Transfers, Statistical Differences, Own Use and Distribution Losses.

Bahrain / Bahrein

Thousand tonnes of oil equivalent / *Milliers de tonnes d'équivalent pétrole*

SUPPLY AND CONSUMPTION 1996	Coal	Crude Oil	Petroleum Products	Gas	Nuclear	Hydro	Geotherm. Solar etc.	Combust. Renew. & Waste	Electricity	Heat	Total
Indigenous Production	-	2376	-	5268	-	-	-	-	-	-	7644
Imports	-	11271	6	-	-	-	-	-	-	-	11277
Exports	-	-108	-11838	-	-	-	-	-	-	-	-11946
Intl. Marine Bunkers	-	-	-108	-	-	-	-	-	-	-	-108
Stock changes	-	-	831	-	-	-	-	-	-	-	831
TPES	-	13538	-11110	5268	-	-	-	-	-	-	7697
Electricity and CHP Plants	-	-	-	-1578	-	-	-	-	410	-	-1168
Petroleum Refineries	-	-13344	11452	-	-	-	-	-	-	-	-1892
Other Transformation *	-	-195	494	-937	-	-	-	-	-42	-	-680
TFC	-	-	836	2753	-	-	-	-	369	-	3958
INDUSTRY SECTOR	-	-	-	2753	-	-	-	-	60	-	2813
Iron and Steel	-	-	-	125	-	-	-	-	-	-	125
Chemical and Petrochemical	-	-	-	1165	-	-	-	-	-	-	1165
Non-Metallic Minerals	-	-	-	-	-	-	-	-	-	-	-
Non-specified	-	-	-	1462	-	-	-	-	60	-	1523
TRANSPORT SECTOR	-	-	757	-	-	-	-	-	-	-	757
Air	-	-	289	-	-	-	-	-	-	-	289
Road	-	-	468	-	-	-	-	-	-	-	468
Non-specified	-	-	-	-	-	-	-	-	-	-	-
OTHER SECTORS	-	-	51	-	-	-	-	-	308	-	360
Agriculture	-	-	-	-	-	-	-	-	2	-	2
Comm. and Publ. Services	-	-	-	-	-	-	-	-	96	-	96
Residential	-	-	51	-	-	-	-	-	210	-	261
Non-specified	-	-	-	-	-	-	-	-	-	-	-
NON-ENERGY USE	-	-	28	-	-	-	-	-	-	-	28
Electricity Generated - GWh	-	-	-	4771	-	-	-	-	-	-	4771
Heat Generated - TJ	-	-	-	-	-	-	-	-	-	-	-

APPROVISIONNEMENT ET DEMANDE 1997	Charbon	Pétrole brut	Produits pétroliers	Gaz	Nucléaire	Hydro	Géotherm. solaire etc.	En. ren. combust. & déchets	Electricité	Chaleur	Total
Indigenous Production	-	2375	-	5113	-	-	-	-	-	-	7487
Imports	-	12311	-	-	-	-	-	-	-	-	12311
Exports	-	-78	-11316	-	-	-	-	-	-	-	-11393
Intl. Marine Bunkers	-	-	-108	-	-	-	-	-	-	-	-108
Stock changes	-	-	190	-	-	-	-	-	-	-	190
TPES	-	14608	-11234	5113	-	-	-	-	-	-	8487
Electricity and CHP Plants	-	-	-	-1575	-	-	-	-	423	-	-1152
Petroleum Refineries	-	-14335	12018	-	-	-	-	-	-	-	-2317
Other Transformation *	-	-273	225	-899	-	-	-	-	-43	-	-990
TFC	-	-	1009	2639	-	-	-	-	381	-	4029
INDUSTRY SECTOR	-	-	-	2639	-	-	-	-	62	-	2701
Iron and Steel	-	-	-	120	-	-	-	-	-	-	120
Chemical and Petrochemical	-	-	-	1117	-	-	-	-	-	-	1117
Non-Metallic Minerals	-	-	-	-	-	-	-	-	-	-	-
Non-specified	-	-	-	1402	-	-	-	-	62	-	1464
TRANSPORT SECTOR	-	-	930	-	-	-	-	-	-	-	930
Air	-	-	268	-	-	-	-	-	-	-	268
Road	-	-	662	-	-	-	-	-	-	-	662
Non-specified	-	-	-	-	-	-	-	-	-	-	-
OTHER SECTORS	-	-	51	-	-	-	-	-	318	-	370
Agriculture	-	-	-	-	-	-	-	-	2	-	2
Comm. and Publ. Services	-	-	-	-	-	-	-	-	99	-	99
Residential	-	-	51	-	-	-	-	-	217	-	268
Non-specified	-	-	-	-	-	-	-	-	-	-	-
NON-ENERGY USE	-	-	28	-	-	-	-	-	-	-	28
Electricity Generated - GWh	-	-	-	4924	-	-	-	-	-	-	4924
Heat Generated - TJ	-	-	-	-	-	-	-	-	-	-	-

* Includes Transfers, Statistical Differences, Own Use and Distribution Losses.

Bangladesh

Total Production of Energy

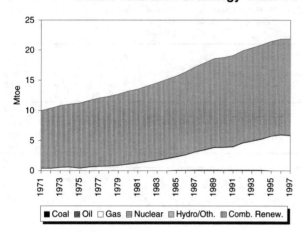

■ Coal ▥ Oil ☐ Gas ▤ Nuclear ▨ Hydro/Oth. ▥ Comb. Renew.

Total Primary Energy Supply*

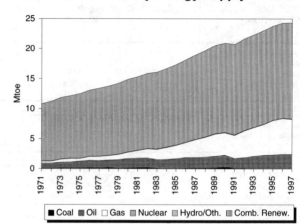

■ Coal ▥ Oil ☐ Gas ▤ Nuclear ▨ Hydro/Oth. ▥ Comb. Renew.

Electricity Generation by Fuel

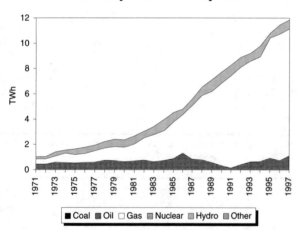

■ Coal ▥ Oil ☐ Gas ▤ Nuclear ▨ Hydro ▥ Other

Oil Products Consumption

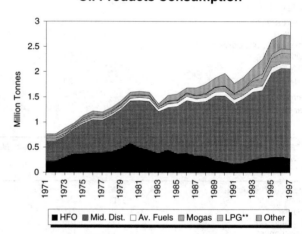

■ HFO ▥ Mid. Dist. ☐ Av. Fuels ▨ Mogas ▨ LPG** ▥ Other

Electricity Consumption/GDP***, TPES/GDP*** and Production/TPES

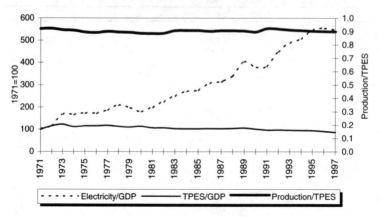

· · · · Electricity/GDP ——— TPES/GDP ——— Production/TPES

* Excluding electricity trade.
** Includes LPG, NGL, ethane and naphtha.
*** GDP in 1990 US dollars.

Bangladesh

Thousand tonnes of oil equivalent / *Milliers de tonnes d'équivalent pétrole*

SUPPLY AND CONSUMPTION 1996	Coal	Crude Oil	Petroleum Products	Gas	Nuclear	Hydro	Geotherm. Solar etc.	Combust. Renew. & Waste	Electricity	Heat	Total
Indigenous Production	-	10	-	5896	-	64	-	15902	-	-	21872
Imports	-	757	1676	-	-	-	-	-	-	-	2434
Exports	-	-	-	-	-	-	-	-	-	-	-
Intl. Marine Bunkers	-	-	-7	-	-	-	-	-	-	-	-7
Stock changes	-	-	-	-	-	-	-	-	-	-	-
TPES	-	767	1670	5896	-	64	-	15902	-	-	24299
Electricity and CHP Plants	-	-	-211	-2433	-	-64	-	-	987	-	-1721
Petroleum Refineries	-	-1151	1030	-	-	-	-	-	-	-	-121
Other Transformation *	-	384	28	-388	-	-	-	-	-213	-	-189
TFC	-	-	2517	3075	-	-	-	15902	774	-	22268
INDUSTRY SECTOR	-	-	217	2531	-	-	-	4819	600	-	8167
Iron and Steel	-	-	-	1	-	-	-	-	-	-	1
Chemical and Petrochemical	-	-	-	1913	-	-	-	-	-	-	1913
Non-Metallic Minerals	-	-	-	-	-	-	-	-	-	-	-
Non-specified	-	-	217	617	-	-	-	4819	600	-	6253
TRANSPORT SECTOR	-	-	1120	-	-	-	-	-	-	-	1120
Air	-	-	105	-	-	-	-	-	-	-	105
Road	-	-	725	-	-	-	-	-	-	-	725
Non-specified	-	-	290	-	-	-	-	-	-	-	290
OTHER SECTORS	-	-	934	544	-	-	-	11083	173	-	12735
Agriculture	-	-	444	-	-	-	-	-	15	-	459
Comm. and Publ. Services	-	-	-	68	-	-	-	-	36	-	105
Residential	-	-	490	476	-	-	-	-	115	-	1082
Non-specified	-	-	-	-	-	-	-	11083	7	-	11090
NON-ENERGY USE	-	-	246	-	-	-	-	-	-	-	246
Electricity Generated - GWh	-	-	741	9994	-	739	-	-	-	-	11474
Heat Generated - TJ	-	-	-	-	-	-	-	-	-	-	-

APPROVISIONNEMENT ET DEMANDE 1997	Charbon	Pétrole brut	Produits pétroliers	Gaz	Nucléaire	Hydro	Géotherm. solaire etc.	En. ren. combust. & déchets	Electricité	Chaleur	Total
Indigenous Production	-	7	-	5794	-	62	-	16031	-	-	21894
Imports	-	768	1672	-	-	-	-	-	-	-	2440
Exports	-	-	-	-	-	-	-	-	-	-	-
Intl. Marine Bunkers	-	-	-7	-	-	-	-	-	-	-	-7
Stock changes	-	-	-	-	-	-	-	-	-	-	-
TPES	-	775	1665	5794	-	62	-	16031	-	-	24327
Electricity and CHP Plants	-	-	-197	-2460	-	-62	-	-	1020	-	-1698
Petroleum Refineries	-	-1162	1041	-	-	-	-	-	-	-	-121
Other Transformation *	-	386	34	-366	-	-	-	-	-207	-	-153
TFC	-	-	2542	2968	-	-	-	16031	813	-	22355
INDUSTRY SECTOR	-	-	209	2389	-	-	-	4858	644	-	8100
Iron and Steel	-	-	-	1	-	-	-	-	-	-	1
Chemical and Petrochemical	-	-	-	1729	-	-	-	-	-	-	1729
Non-Metallic Minerals	-	-	-	-	-	-	-	-	-	-	-
Non-specified	-	-	209	659	-	-	-	4858	644	-	6370
TRANSPORT SECTOR	-	-	1132	-	-	-	-	-	-	-	1132
Air	-	-	109	-	-	-	-	-	-	-	109
Road	-	-	731	-	-	-	-	-	-	-	731
Non-specified	-	-	292	-	-	-	-	-	-	-	292
OTHER SECTORS	-	-	955	579	-	-	-	11173	169	-	12877
Agriculture	-	-	446	-	-	-	-	-	13	-	459
Comm. and Publ. Services	-	-	-	73	-	-	-	-	36	-	108
Residential	-	-	509	507	-	-	-	-	113	-	1129
Non-specified	-	-	-	-	-	-	-	11173	7	-	11180
NON-ENERGY USE	-	-	246	-	-	-	-	-	-	-	246
Electricity Generated - GWh	-	-	1118	10021	-	719	-	-	-	-	11858
Heat Generated - TJ	-	-	-	-	-	-	-	-	-	-	-

* Includes Transfers, Statistical Differences, Own Use and Distribution Losses.

INTERNATIONAL ENERGY AGENCY

Belarus / Bélarus : 1996

	Coal	Crude Oil	Petroleum Products	Gas	Nuclear	Hydro	Geotherm. Solar etc.	Combust. Renew. & Waste	Electricity	Heat	Total
SUPPLY AND CONSUMPTION											
APPROVISIONNEMENT ET DEMANDE	Charbon	Pétrole brut	Produits pétroliers	Gaz	Nucléaire	Hydro	Géotherm. solaire	En. ren. combust.	Electricité	Chaleur	Total
Indigenous Production	569	1869	-	207	-	1	-	516	-	-	3162
Imports	608	10698	262	11906	-	-	-	-	958	-	24433
Exports	-32	-301	-3330	-	-	-	-	-	-224	-	-3887
Intl. Marine Bunkers	-	-	-	-	-	-	-	-	-	-	-
Stock Changes	80	251	1055	-6	-	-	-	-	-	-	1380
TPES	**1225**	**12517**	**-2013**	**12107**	**-**	**1**	**-**	**516**	**735**	**-**	**25089**
Transfers	-	-	-	-	-	-	-	-	-	-	-
Statistical Differences	-	-	-	-	-	-	-	-	-	-	-
Electricity Plants	-	-	-491	-1955	-	-1	-	-	908	-	-1539
CHP Plants	-	-	-1845	-3800	-	-	-	-	1132	3099	-1413
Heat Plants	-293	-	-1454	-3166	-	-	-	-146	-	4699	-359
Gas Works	-	-	-	-	-	-	-	-	-	-	-
Petroleum Refineries	-	-11819	11320	-	-	-	-	-	-	-	-498
Coal Transformation	-147	-	-	-	-	-	-	-	-	-	-147
Liquefaction	-	-	-	-	-	-	-	-	-	-	-
Other Transformation	-	-	-	-	-	-	-	-	-	-	-
Own Use	-6	-	-575	-	-	-	-	-	-259	-	-840
Distribution Losses	-	-144	-32	-129	-	-	-	-	-323	-500	-1127
TFC	**780**	**555**	**4912**	**3058**	**-**	**-**	**-**	**369**	**2193**	**7299**	**19165**
INDUSTRY SECTOR	**19**	**555**	**877**	**1633**	**-**	**-**	**-**	**8**	**928**	**2999**	**7020**
Iron and Steel	-	-	1	49	-	-	-	-	79	-	129
Chemical and Petrochemical	-	555	259	1157	-	-	-	-	333	1000	3303
of which: Feedstocks	-	555	2	1035	-	-	-	-	-	-	1592
Non-Ferrous Metals	-	-	-	-	-	-	-	-	1	-	1
Non-Metallic Minerals	-	-	-	247	-	-	-	-	81	200	527
Transport Equipment	-	-	-	-	-	-	-	-	2	-	2
Machinery	5	-	16	89	-	-	-	-	198	400	709
Mining and Quarrying	-	-	-	-	-	-	-	-	-	-	-
Food and Tobacco	-	-	11	27	-	-	-	2	72	300	412
Paper, Pulp and Printing	-	-	10	-	-	-	-	-	18	200	228
Wood and Wood Products	-	-	-	7	-	-	-	6	26	200	239
Construction	13	-	327	-	-	-	-	1	30	100	470
Textile and Leather	1	-	2	7	-	-	-	-	44	200	254
Non-specified	-	-	250	51	-	-	-	-	45	400	746
TRANSPORT SECTOR	**16**	**-**	**1832**	**70**	**-**	**-**	**-**	**2**	**157**	**-**	**2076**
Air	-	-	-	-	-	-	-	-	-	-	-
Road	-	-	1454	32	-	-	-	-	-	-	1485
Rail	15	-	332	-	-	-	-	-	60	-	408
Pipeline Transport	-	-	1	38	-	-	-	-	68	-	107
Internal Navigation	-	-	19	-	-	-	-	-	-	-	19
Non-specified	1	-	26	-	-	-	-	2	29	-	58
OTHER SECTORS	**745**	**-**	**1732**	**1355**	**-**	**-**	**-**	**359**	**1109**	**4299**	**9598**
Agriculture	8	-	587	41	-	-	-	8	222	300	1166
Comm. and Publ. Services	-	-	-	-	-	-	-	-	310	-	310
Residential	292	-	957	910	-	-	-	281	437	3699	6576
Non-specified	445	-	188	403	-	-	-	70	139	300	1546
NON-ENERGY USE	**-**	**-**	**471**	**-**	**-**	**-**	**-**	**-**	**-**	**-**	**471**
in Industry/Transf./Energy	-	-	471	-	-	-	-	-	-	-	471
in Transport	-	-	-	-	-	-	-	-	-	-	-
in Other Sectors	-	-	-	-	-	-	-	-	-	-	-
Electr. Generated - GWh	*-*	*-*	*6910*	*16802*	*-*	*16*	*-*	*-*	*-*	*-*	***23728***
Electricity Plants	*-*	*-*	*2338*	*8207*	*-*	*16*	*-*	*-*	*-*	*-*	*10561*
CHP Plants	*-*	*-*	*4572*	*8595*	*-*	*-*	*-*	*-*	*-*	*-*	*13167*
Heat Generated - TJ	*-*	*-*	*-*	*-*	*-*	*-*	*-*	*-*	*-*	*-*	***326570***
CHP Plants	*-*	*-*	*-*	*-*	*-*	*-*	*-*	*-*	*-*	*-*	*129790*
Heat Plants	*-*	*-*	*-*	*-*	*-*	*-*	*-*	*-*	*-*	*-*	*196780*

Belarus / Bélarus : 1997

Thousand tonnes of oil equivalent / *Milliers de tonnes d'équivalent pétrole*

SUPPLY AND CONSUMPTION *APPROVISIONNEMENT ET DEMANDE*	Coal *Charbon*	Crude Oil *Pétrole brut*	Petroleum Products *Produits pétroliers*	Gas *Gaz*	Nuclear *Nucléaire*	Hydro *Hydro*	Geotherm. Solar etc. *Géotherm. solaire*	Combust. Renew. & Waste *En. ren. combust.*	Electricity *Electricité*	Heat *Chaleur*	Total *Total*
Indigenous Production	553	1831	-	204	-	2	-	685	-	-	3275
Imports	479	10513	346	13480	-	-	-	-	886	-	25705
Exports	-35	-402	-2703	-	-	-	-	-	-231	-	-3371
Intl. Marine Bunkers	-	-	-	-	-	-	-	-	-	-	-
Stock Changes	14	-95	-477	91	-	-	-	-	-	-	-468
TPES	**1011**	**11847**	**-2834**	**13775**	**-**	**2**	**-**	**685**	**655**	**-**	**25142**
Transfers	-	-	-	-	-	-	-	-	-	-	-
Statistical Differences	-	72	-	-	-	-	-	-	-	-	72
Electricity Plants	-	-	-301	-2579	-	-2	-	-	1098	-	-1784
CHP Plants	-	-	-1008	-4613	-	-	-	-	1143	3099	-1379
Heat Plants	-225	-	-1387	-3438	-	-	-	-139	-	4899	-290
Gas Works	-	-	-	-	-	-	-	-	-	-	-
Petroleum Refineries	-	-11255	11203	-	-	-	-	-	-	-	-52
Coal Transformation	-147	-	-	-	-	-	-	-	-	-	-147
Liquefaction	-	-	-	-	-	-	-	-	-	-	-
Other Transformation	-	-	-	-	-	-	-	-	-	-	-
Own Use	-6	-	-575	-	-	-	-	-	-268	-100	-948
Distribution Losses	-	-136	-26	-112	-	-	-	-	-327	-500	-1101
TFC	**634**	**529**	**5072**	**3034**	**-**	**-**	**-**	**546**	**2302**	**7399**	**19514**
INDUSTRY SECTOR	**17**	**529**	**993**	**1572**	**-**	**-**	**-**	**11**	**1081**	**2999**	**7202**
Iron and Steel	-	-	3	68	-	-	-	-	95	-	166
Chemical and Petrochemical	1	529	237	1003	-	-	-	-	399	1100	3268
of which: Feedstocks	-	*529*	*2*	*847*	-	-	-	-	-	-	*1377*
Non-Ferrous Metals	-	-	-	-	-	-	-	-	1	-	1
Non-Metallic Minerals	-	-	-	281	-	-	-	-	85	200	565
Transport Equipment	-	-	-	-	-	-	-	-	3	-	3
Machinery	1	-	23	94	-	-	-	-	215	500	832
Mining and Quarrying	-	-	-	-	-	-	-	-	-	-	-
Food and Tobacco	-	-	2	18	-	-	-	-	77	300	398
Paper, Pulp and Printing	2	-	10	-	-	-	-	1	21	200	233
Wood and Wood Products	-	-	-	32	-	-	-	-	31	200	262
Construction	12	-	356	8	-	-	-	6	31	100	514
Textile and Leather	1	-	1	9	-	-	-	3	50	200	264
Non-specified	1	-	361	59	-	-	-	1	75	200	697
TRANSPORT SECTOR	**7**	**-**	**1889**	**67**	**-**	**-**	**-**	**4**	**147**	**-**	**2115**
Air	-	-	-	-	-	-	-	-	-	-	-
Road	-	-	1501	30	-	-	-	-	-	-	1531
Rail	6	-	347	-	-	-	-	1	62	-	416
Pipeline Transport	-	-	1	37	-	-	-	-	55	-	94
Internal Navigation	-	-	20	-	-	-	-	-	-	-	20
Non-specified	1	-	21	-	-	-	-	3	30	-	55
OTHER SECTORS	**609**	**-**	**1650**	**1394**	**-**	**-**	**-**	**531**	**1073**	**4399**	**9658**
Agriculture	5	-	564	38	-	-	-	4	213	300	1124
Comm. and Publ. Services	1	-	-	-	-	-	-	-	256	-	257
Residential	276	-	913	922	-	-	-	441	460	3899	6912
Non-specified	327	-	173	434	-	-	-	86	145	200	1365
NON-ENERGY USE	**1**	**-**	**540**	**-**	**-**	**-**	**-**	**-**	**-**	**-**	**540**
in Industry/Transf./Energy	1	-	540	-	-	-	-	-	-	-	540
in Transport	-	-	-	-	-	-	-	-	-	-	-
in Other Sectors	-	-	-	-	-	-	-	-	-	-	-
Electr. Generated - GWh	*-*	*-*	*3865*	*22171*	*-*	*21*	*-*	*-*	*-*	*-*	*26057*
Electricity Plants	*-*	*-*	*1398*	*11350*	*-*	*21*	*-*	*-*	*-*	*-*	*12769*
CHP Plants	*-*	*-*	*2467*	*10821*	*-*	*-*	*-*	*-*	*-*	*-*	*13288*
Heat Generated - TJ	*-*	*-*	*-*	*-*	*-*	*-*	*-*	*-*	*-*	*-*	*334944*
CHP Plants	*-*	*-*	*-*	*-*	*-*	*-*	*-*	*-*	*-*	*-*	*129791*
Heat Plants	*-*	*-*	*-*	*-*	*-*	*-*	*-*	*-*	*-*	*-*	*205153*

Benin / Bénin

Thousand tonnes of oil equivalent / *Milliers de tonnes d'équivalent pétrole*

SUPPLY AND CONSUMPTION 1996	Coal	Crude Oil	Petroleum Products	Gas	Nuclear	Hydro	Geotherm. Solar etc.	Combust. Renew. & Waste	Electricity	Heat	Total
Indigenous Production	-	73	-	-	-	-	-	1792	-	-	1865
Imports	-	-	312	-	-	-	-	-	23	-	335
Exports	-	-76	-1	-	-	-	-	-	-	-	-77
Intl. Marine Bunkers	-	-	-5	-	-	-	-	-	-	-	-5
Stock changes	-	3	30	-	-	-	-	-	-	-	33
TPES	-	-	336	-	-	-	-	1792	23	-	2150
Electricity and CHP Plants	-	-	-6	-	-	-	-	-	4	-	-2
Petroleum Refineries	-	-	-	-	-	-	-	-	-	-	-
Other Transformation *	-	-	2	-	-	-	-	-37	-4	-	-38
TFC	-	-	331	-	-	-	-	1755	23	-	2110
INDUSTRY SECTOR	-	-	18	-	-	-	-	329	11	-	359
Iron and Steel	-	-	-	-	-	-	-	-	-	-	-
Chemical and Petrochemical	-	-	-	-	-	-	-	-	-	-	-
Non-Metallic Minerals	-	-	-	-	-	-	-	-	-	-	-
Non-specified	-	-	18	-	-	-	-	329	11	-	359
TRANSPORT SECTOR	-	-	271	-	-	-	-	-	-	-	271
Air	-	-	60	-	-	-	-	-	-	-	60
Road	-	-	212	-	-	-	-	-	-	-	212
Non-specified	-	-	-	-	-	-	-	-	-	-	-
OTHER SECTORS	-	-	42	-	-	-	-	1426	12	-	1480
Agriculture	-	-	-	-	-	-	-	-	-	-	-
Comm. and Publ. Services	-	-	-	-	-	-	-	-	1	-	1
Residential	-	-	42	-	-	-	-	1426	11	-	1480
Non-specified	-	-	-	-	-	-	-	-	-	-	-
NON-ENERGY USE	-	-	-	-	-	-	-	-	-	-	-
Electricity Generated - GWh	-	-	47	-	-	-	-	-	-	-	47
Heat Generated - TJ	-	-	-	-	-	-	-	-	-	-	-

APPROVISIONNEMENT ET DEMANDE 1997	Charbon	Pétrole brut	Produits pétroliers	Gaz	Nucléaire	Hydro	Géotherm. solaire etc.	En. ren. combust. & déchets	Electricité	Chaleur	Total
Indigenous Production	-	67	-	-	-	-	-	1830	-	-	1897
Imports	-	-	320	-	-	-	-	-	20	-	340
Exports	-	-65	-5	-	-	-	-	-	-	-	-70
Intl. Marine Bunkers	-	-	-5	-	-	-	-	-	-	-	-5
Stock changes	-	-2	22	-	-	-	-	-	-	-	20
TPES	-	-	331	-	-	-	-	1830	20	-	2182
Electricity and CHP Plants	-	-	-6	-	-	-	-	-	4	-	-2
Petroleum Refineries	-	-	-	-	-	-	-	-	-	-	-
Other Transformation *	-	-	-2	-	-	-	-	-37	-3	-	-42
TFC	-	-	324	-	-	-	-	1792	22	-	2137
INDUSTRY SECTOR	-	-	18	-	-	-	-	336	10	-	365
Iron and Steel	-	-	-	-	-	-	-	-	-	-	-
Chemical and Petrochemical	-	-	-	-	-	-	-	-	-	-	-
Non-Metallic Minerals	-	-	-	-	-	-	-	-	-	-	-
Non-specified	-	-	18	-	-	-	-	336	10	-	365
TRANSPORT SECTOR	-	-	269	-	-	-	-	-	-	-	269
Air	-	-	52	-	-	-	-	-	-	-	52
Road	-	-	217	-	-	-	-	-	-	-	217
Non-specified	-	-	-	-	-	-	-	-	-	-	-
OTHER SECTORS	-	-	37	-	-	-	-	1456	11	-	1504
Agriculture	-	-	-	-	-	-	-	-	-	-	-
Comm. and Publ. Services	-	-	-	-	-	-	-	-	1	-	1
Residential	-	-	37	-	-	-	-	1456	10	-	1503
Non-specified	-	-	-	-	-	-	-	-	-	-	-
NON-ENERGY USE	-	-	-	-	-	-	-	-	-	-	-
Electricity Generated - GWh	-	-	50	-	-	-	-	-	-	-	50
Heat Generated - TJ	-	-	-	-	-	-	-	-	-	-	-

* Includes Transfers, Statistical Differences, Own Use and Distribution Losses.

Bolivia / Bolivie

Thousand tonnes of oil equivalent / *Milliers de tonnes d'équivalent pétrole*

SUPPLY AND CONSUMPTION 1996	Coal	Crude Oil	Petroleum Products	Gas	Nuclear	Hydro	Geotherm. Solar etc.	Combust. Renew. & Waste	Electricity	Heat	Total
Indigenous Production	-	1797	-	2493	-	175	-	774	-	-	5239
Imports	-	-	138	-	-	-	-	-	-	-	138
Exports	-	-	-	-1768	-	-	-	-	-	-	-1769
Intl. Marine Bunkers	-	-	-	-	-	-	-	-	-	-	-
Stock changes	-	41	-17	-	-	-	-	-	-	-	25
TPES	-	1838	121	725	-	175	-	774	-	-	3633
Electricity and CHP Plants	-	-	-58	-331	-	-175	-	-24	278	-	-310
Petroleum Refineries	-	-1762	1607	-	-	-	-	-	-	-	-155
Other Transformation *	-	-77	-225	-138	-	-	-	-27	-36	-	-502
TFC	-	-	1445	256	-	-	-	723	242	-	2666
INDUSTRY SECTOR	-	-	86	254	-	-	-	170	98	-	608
Iron and Steel	-	-	-	-	-	-	-	-	-	-	-
Chemical and Petrochemical	-	-	-	-	-	-	-	-	-	-	-
Non-Metallic Minerals	-	-	-	-	-	-	-	-	-	-	-
Non-specified	-	-	86	254	-	-	-	170	98	-	608
TRANSPORT SECTOR	-	-	1006	-	-	-	-	-	-	-	1006
Air	-	-	136	-	-	-	-	-	-	-	136
Road	-	-	830	-	-	-	-	-	-	-	830
Non-specified	-	-	40	-	-	-	-	-	-	-	40
OTHER SECTORS	-	-	343	2	-	-	-	553	144	-	1042
Agriculture	-	-	51	-	-	-	-	-	-	-	51
Comm. and Publ. Services	-	-	-	-	-	-	-	-	42	-	42
Residential	-	-	292	2	-	-	-	553	101	-	949
Non-specified	-	-	-	-	-	-	-	-	-	-	-
NON-ENERGY USE	-	-	11	-	-	-	-	-	-	-	11
Electricity Generated - GWh	-	-	200	956	-	2037	-	39	-	-	3232
Heat Generated - TJ	-	-	-	-	-	-	-	-	-	-	-

APPROVISIONNEMENT ET DEMANDE 1997	Charbon	Pétrole brut	Produits pétroliers	Gaz	Nucléaire	Hydro	Géotherm. solaire etc.	En. ren. combust. & déchets	Electricité	Chaleur	Total
Indigenous Production	-	1876	-	3003	-	198	-	876	-	-	5953
Imports	-	-	243	-	-	-	-	-	1	-	244
Exports	-	-	-	-1918	-	-	-	-	-	-	-1918
Intl. Marine Bunkers	-	-	-	-	-	-	-	-	-	-	-
Stock changes	-	-	-24	-	-	-	-	-	-	-	-24
TPES	-	1876	219	1085	-	198	-	876	-	-	4254
Electricity and CHP Plants	-	-	-58	-306	-	-198	-	-27	295	-	-294
Petroleum Refineries	-	-1833	1662	-	-	-	-	-	-	-	-171
Other Transformation *	-	-43	-329	-433	-	-	-	-40	-34	-	-879
TFC	-	-	1494	346	-	-	-	809	261	-	2910
INDUSTRY SECTOR	-	-	80	344	-	-	-	305	106	-	835
Iron and Steel	-	-	-	-	-	-	-	-	-	-	-
Chemical and Petrochemical	-	-	-	-	-	-	-	-	-	-	-
Non-Metallic Minerals	-	-	-	-	-	-	-	-	-	-	-
Non-specified	-	-	80	344	-	-	-	305	106	-	835
TRANSPORT SECTOR	-	-	1016	-	-	-	-	-	-	-	1016
Air	-	-	145	-	-	-	-	-	-	-	145
Road	-	-	834	-	-	-	-	-	-	-	834
Non-specified	-	-	37	-	-	-	-	-	-	-	37
OTHER SECTORS	-	-	385	2	-	-	-	504	155	-	1046
Agriculture	-	-	48	-	-	-	-	-	-	-	48
Comm. and Publ. Services	-	-	-	-	-	-	-	-	46	-	46
Residential	-	-	338	2	-	-	-	504	109	-	953
Non-specified	-	-	-	-	-	-	-	-	-	-	-
NON-ENERGY USE	-	-	12	-	-	-	-	-	-	-	12
Electricity Generated - GWh	-	-	200	883	-	2306	-	44	-	-	3433
Heat Generated - TJ	-	-	-	-	-	-	-	-	-	-	-

* Includes Transfers, Statistical Differences, Own Use and Distribution Losses.

INTERNATIONAL ENERGY AGENCY

Bosnia-Herzegovina / Bosnie-Herzégovine

Thousand tonnes of oil equivalent / *Milliers de tonnes d'équivalent pétrole*

SUPPLY AND CONSUMPTION 1996	Coal	Crude Oil	Petroleum Products	Gas	Nuclear	Hydro	Geotherm. Solar etc.	Combust. Renew. & Waste	Electricity	Heat	Total
Indigenous Production	348	-	-	-	-	122	-	155	-	-	626
Imports	-	-	896	211	-	-	-	-	33	-	1140
Exports	-	-	-	-	-	-	-	-	-16	-	-16
Intl. Marine Bunkers	-	-	-	-	-	-	-	-	-	-	-
Stock changes	-	-	-	-	-	-	-	-	-	-	-
TPES	348	-	896	211	-	122	-	155	18	-	1750
Electricity and CHP Plants	-265	-	-34	-	-	-122	-	-	189	23	-209
Petroleum Refineries	-	-	-	-	-	-	-	-	-	-	-
Other Transformation *	7	-	-533	-211	-	-	-	-	-58	-	-795
TFC	90	-	328	-	-	-	-	155	150	23	747
INDUSTRY SECTOR	-	-	58	-	-	-	-	-	-	-	58
Iron and Steel	-	-	23	-	-	-	-	-	-	-	23
Chemical and Petrochemical	-	-	-	-	-	-	-	-	-	-	-
Non-Metallic Minerals	-	-	-	-	-	-	-	-	-	-	-
Non-specified	-	-	35	-	-	-	-	-	-	-	35
TRANSPORT SECTOR	90	-	270	-	-	-	-	-	-	-	361
Air	-	-	38	-	-	-	-	-	-	-	38
Road	-	-	228	-	-	-	-	-	-	-	228
Non-specified	90	-	4	-	-	-	-	-	-	-	95
OTHER SECTORS	-	-	-	-	-	-	-	155	150	23	328
Agriculture	-	-	-	-	-	-	-	-	-	-	-
Comm. and Publ. Services	-	-	-	-	-	-	-	-	-	-	-
Residential	-	-	-	-	-	-	-	155	-	-	155
Non-specified	-	-	-	-	-	-	-	-	150	23	173
NON-ENERGY USE	-	-	-	-	-	-	-	-	-	-	-
Electricity Generated - GWh	783	-	-	-	-	1420	-	-	-	-	2203
Heat Generated - TJ	966	-	-	-	-	-	-	-	-	-	966

APPROVISIONNEMENT ET DEMANDE 1997	Charbon	Pétrole brut	Produits pétroliers	Gaz	Nucléaire	Hydro	Géotherm. solaire etc.	En. ren. combust. & déchets	Electricité	Chaleur	Total
Indigenous Production	348	-	-	-	-	122	-	155	-	-	626
Imports	-	-	896	211	-	-	-	-	33	-	1140
Exports	-	-	-	-	-	-	-	-	-16	-	-16
Intl. Marine Bunkers	-	-	-	-	-	-	-	-	-	-	-
Stock changes	-	-	-	-	-	-	-	-	-	-	-
TPES	348	-	896	211	-	122	-	155	18	-	1750
Electricity and CHP Plants	-265	-	-34	-	-	-122	-	-	189	23	-209
Petroleum Refineries	-	-	-	-	-	-	-	-	-	-	-
Other Transformation *	7	-	-533	-211	-	-	-	-	-58	-	-795
TFC	90	-	328	-	-	-	-	155	150	23	747
INDUSTRY SECTOR	-	-	58	-	-	-	-	-	-	-	58
Iron and Steel	-	-	23	-	-	-	-	-	-	-	23
Chemical and Petrochemical	-	-	-	-	-	-	-	-	-	-	-
Non-Metallic Minerals	-	-	-	-	-	-	-	-	-	-	-
Non-specified	-	-	35	-	-	-	-	-	-	-	35
TRANSPORT SECTOR	90	-	270	-	-	-	-	-	-	-	361
Air	-	-	38	-	-	-	-	-	-	-	38
Road	-	-	228	-	-	-	-	-	-	-	228
Non-specified	90	-	4	-	-	-	-	-	-	-	95
OTHER SECTORS	-	-	-	-	-	-	-	155	150	23	328
Agriculture	-	-	-	-	-	-	-	-	-	-	-
Comm. and Publ. Services	-	-	-	-	-	-	-	-	-	-	-
Residential	-	-	-	-	-	-	-	155	-	-	155
Non-specified	-	-	-	-	-	-	-	-	150	23	173
NON-ENERGY USE	-	-	-	-	-	-	-	-	-	-	-
Electricity Generated - GWh	783	-	-	-	-	1420	-	-	-	-	2203
Heat Generated - TJ	966	-	-	-	-	-	-	-	-	-	966

* Includes Transfers, Statistical Differences, Own Use and Distribution Losses.

INTERNATIONAL ENERGY AGENCY

Brazil / Brésil

Total Production of Energy

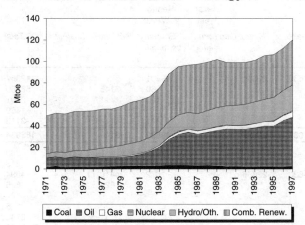

■ Coal ▦ Oil ☐ Gas ▤ Nuclear ▨ Hydro/Oth. ▥ Comb. Renew.

Total Primary Energy Supply*

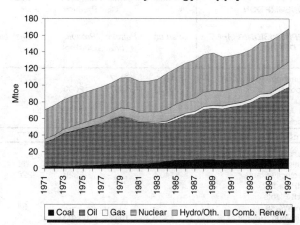

■ Coal ▦ Oil ☐ Gas ▤ Nuclear ▨ Hydro/Oth. ▥ Comb. Renew.

Electricity Generation by Fuel

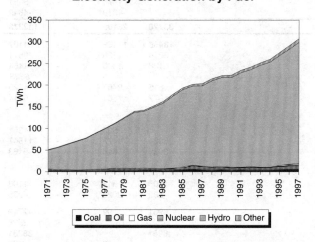

■ Coal ▦ Oil ☐ Gas ▤ Nuclear ▨ Hydro ▥ Other

Oil Products Consumption

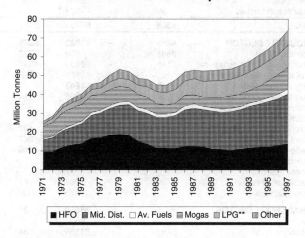

■ HFO ▦ Mid. Dist. ☐ Av. Fuels ▤ Mogas ▨ LPG** ▥ Other

Electricity Consumption/GDP***, TPES/GDP*** and Production/TPES

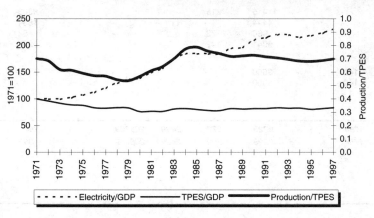

- - - - - Electricity/GDP ——— TPES/GDP ——— Production/TPES

* Excluding electricity trade.
** Includes LPG, NGL, ethane and naphtha.
*** GDP in 1990 US dollars.

Brazil / Brésil : 1996

Thousand tonnes of oil equivalent / *Milliers de tonnes d'équivalent pétrole*

SUPPLY AND CONSUMPTION *APPROVISIONNEMENT ET DEMANDE*	Coal *Charbon*	Crude Oil *Pétrole brut*	Petroleum Products *Produits pétroliers*	Gas *Gaz*	Nuclear *Nucléaire*	Hydro *Hydro*	Geotherm. Solar etc. *Géotherm. solaire*	Combust. Renew. & Waste *En. ren. combust.*	Electricity *Electricité*	Heat *Chaleur*	Total *Total*
Indigenous Production	1854	42947	-	4777	633	22856	-	39301	-	-	112367
Imports	10670	31167	12609	-	-	-	-	660	3145	-	58250
Exports	-	-113	-2479	-	-	-	-	-108	-1	-	-2701
Intl. Marine Bunkers	-	-	-1469	-	-	-	-	-	-	-	-1469
Stock Changes	-357	-2055	-481	-	-	-	-	-221	-	-	-3113
TPES	**12167**	**71947**	**8179**	**4777**	**633**	**22856**	**-**	**39632**	**3144**	**-**	**163334**
Transfers	-	-	360	-	-	-	-	-	-	-	360
Statistical Differences	-1	-	199	11	-	-	-	-	3	-	213
Electricity Plants	-1403	-	-2320	-228	-633	-22856	-	-1672	24418	-	-4694
CHP Plants	-	-	-	-	-	-	-	-	625	-	625
Heat Plants	-	-	-	-	-	-	-	-	-	-	-
Gas Works	-	-	-49	54	-	-	-	-	-	-	5
Petroleum Refineries	-	-71947	65203	-	-	-	-	-	-	-	-6743
Coal Transformation	-5853	-	-	-	-	-	-	-	-	-	-5853
Liquefaction	-	-	-	-	-	-	-	-	-	-	-
Other Transformation	-	-	-	-	-	-	-	-4240	-	-	-4240
Own Use	-363	-	-3857	-1085	-	-	-	-	-776	-	-6081
Distribution Losses	-39	-	-223	-77	-	-	-	-	-4310	-	-4649
TFC	**4508**	**-**	**67494**	**3451**	**-**	**-**	**-**	**33720**	**23105**	**-**	**132278**
INDUSTRY SECTOR	**4365**	**-**	**16922**	**3205**	**-**	**-**	**-**	**18456**	**11159**	**-**	**54107**
Iron and Steel	3595	-	459	722	-	-	-	3014	1820	-	9611
Chemical and Petrochemical	146	-	8539	1340	-	-	-	128	1300	-	11454
of which: Feedstocks	-	-	*5571*	*752*	-	-	-	-	-	-	*6323*
Non-Ferrous Metals	175	-	1243	25	-	-	-	615	2467	-	4525
Non-Metallic Minerals	257	-	2265	114	-	-	-	1977	501	-	5114
Transport Equipment	-	-	-	-	-	-	-	-	-	-	-
Machinery	-	-	-	-	-	-	-	-	-	-	-
Mining and Quarrying	11	-	730	107	-	-	-	-	503	-	1351
Food and Tobacco	85	-	1108	247	-	-	-	8934	1172	-	11547
Paper, Pulp and Printing	87	-	997	139	-	-	-	3084	856	-	5163
Wood and Wood Products	-	-	-	-	-	-	-	-	-	-	-
Construction	-	-	-	-	-	-	-	-	-	-	-
Textile and Leather	2	-	305	181	-	-	-	105	514	-	1108
Non-specified	7	-	1275	330	-	-	-	599	2025	-	4235
TRANSPORT SECTOR	**-**	**-**	**35658**	**31**	**-**	**-**	**-**	**6961**	**99**	**-**	**42749**
Air	-	-	2528	-	-	-	-	-	-	-	2528
Road	-	-	31389	31	-	-	-	6961	-	-	38381
Rail	-	-	395	-	-	-	-	-	99	-	493
Pipeline Transport	-	-	-	-	-	-	-	-	-	-	-
Internal Navigation	-	-	1346	-	-	-	-	-	-	-	1346
Non-specified	-	-	-	-	-	-	-	-	-	-	-
OTHER SECTORS	**-**	**-**	**11415**	**215**	**-**	**-**	**-**	**8303**	**11847**	**-**	**31780**
Agriculture	-	-	4444	-	-	-	-	1854	847	-	7145
Comm. and Publ. Services	-	-	930	73	-	-	-	152	5061	-	6216
Residential	-	-	6041	142	-	-	-	6297	5939	-	18419
Non-specified	-	-	-	-	-	-	-	-	-	-	-
NON-ENERGY USE	**142**	**-**	**3500**	**-**	**-**	**-**	**-**	**-**	**-**	**-**	**3642**
in Industry/Transf./Energy	142	-	3500	-	-	-	-	-	-	-	3642
in Transport	-	-	-	-	-	-	-	-	-	-	-
in Other Sectors	-	-	-	-	-	-	-	-	-	-	-
Electr. Generated - GWh	*4764*	*-*	*9328*	*973*	*2429*	*265769*	*-*	*7941*	*-*	*-*	*291204*
Electricity Plants	*4764*	*-*	*9328*	*973*	*2429*	*265769*	*-*	*669*	*-*	*-*	*283932*
CHP Plants	*-*	*-*	*-*	*-*	*-*	*-*	*-*	*7272*	*-*	*-*	*7272*
Heat Generated - TJ	*-*	*-*	*-*	*-*	*-*	*-*	*-*	*-*	*-*	*-*	*-*
CHP Plants	*-*	*-*	*-*	*-*	*-*	*-*	*-*	*-*	*-*	*-*	*-*
Heat Plants	*-*	*-*	*-*	*-*	*-*	*-*	*-*	*-*	*-*	*-*	*-*

Brazil / Brésil : 1997

	Coal	Crude Oil	Petroleum Products	Gas	Nuclear	Hydro	Geotherm. Solar etc.	Combust. Renew. & Waste	Electricity	Heat	Total
Thousand tonnes of oil equivalent / *Milliers de tonnes d'équivalent pétrole*											
SUPPLY AND CONSUMPTION / *APPROVISIONNEMENT ET DEMANDE*	*Charbon*	*Pétrole brut*	*Produits pétroliers*	*Gaz*	*Nucléaire*	*Hydro*	*Géotherm. solaire*	*En. ren. combust.*	*Electricité*	*Chaleur*	*Total*
Indigenous Production	2162	45988	-	5288	826	24000	-	41972	-	-	120236
Imports	10230	32187	13050	-	-	-	-	445	3481	-	59393
Exports	-	-140	-2454	-	-	-	-	-688	-1	-	-3282
Intl. Marine Bunkers	-	-	-1855	-	-	-	-	-	-	-	-1855
Stock Changes	-78	-1572	470	-	-	-	-	-1281	-	-	-2461
TPES	**12315**	**76464**	**9210**	**5288**	**826**	**24000**	**-**	**40448**	**3480**	**-**	**172030**
Transfers	-	-	327	-	-	-	-	-	-	-	327
Statistical Differences	-45	-	-103	-40	-	-	-	-	-	-	-188
Electricity Plants	-1606	-	-2544	-253	-826	-24000	-	-1787	25735	-	-5280
CHP Plants	-	-	-	-	-	-	-	-	693	-	693
Heat Plants	-	-	-	-	-	-	-	-	-	-	-
Gas Works	-	-	-49	43	-	-	-	-	-	-	-6
Petroleum Refineries	-	-76463	70297	-	-	-	-	-	-	-	-6166
Coal Transformation	-5686	-	-	-	-	-	-	-	-	-	-5686
Liquefaction	-	-	-	-	-	-	-	-	-	-	-
Other Transformation	-	-	-	-	-	-	-	-4319	-	-	-4319
Own Use	-337	-	-4378	-1171	-	-	-	-	-875	-	-6761
Distribution Losses	-27	-	-295	-32	-	-	-	-	-4493	-	-4848
TFC	**4614**	**-**	**72465**	**3833**	**-**	**-**	**-**	**34343**	**24540**	**-**	**139795**
INDUSTRY SECTOR	**4425**	**-**	**18890**	**3548**	**-**	**-**	**-**	**19281**	**11670**	**-**	**57814**
Iron and Steel	3807	-	485	691	-	-	-	3157	1852	-	9992
Chemical and Petrochemical	140	-	10152	1485	-	-	-	109	1417	-	13304
of which: Feedstocks	-	-	*6889*	*713*	-	-	-	-	-	-	*7602*
Non-Ferrous Metals	157	-	1262	35	-	-	-	580	2478	-	4512
Non-Metallic Minerals	167	-	2563	199	-	-	-	1975	551	-	5454
Transport Equipment	-	-	-	-	-	-	-	-	-	-	-
Machinery	-	-	-	-	-	-	-	-	-	-	-
Mining and Quarrying	-	-	743	151	-	-	-	-	542	-	1437
Food and Tobacco	69	-	1075	273	-	-	-	9624	1225	-	12266
Paper, Pulp and Printing	81	-	942	143	-	-	-	3141	884	-	5190
Wood and Wood Products	-	-	-	-	-	-	-	-	-	-	-
Construction	-	-	-	-	-	-	-	-	-	-	-
Textile and Leather	2	-	309	188	-	-	-	100	503	-	1101
Non-specified	2	-	1360	383	-	-	-	595	2217	-	4557
TRANSPORT SECTOR	**-**	**-**	**38331**	**40**	**-**	**-**	**-**	**6706**	**100**	**-**	**45177**
Air	-	-	2846	-	-	-	-	-	-	-	2846
Road	-	-	33437	40	-	-	-	6706	-	-	40183
Rail	-	-	368	-	-	-	-	-	100	-	467
Pipeline Transport	-	-	-	-	-	-	-	-	-	-	-
Internal Navigation	-	-	1681	-	-	-	-	-	-	-	1681
Non-specified	-	-	-	-	-	-	-	-	-	-	-
OTHER SECTORS	**-**	**-**	**11617**	**246**	**-**	**-**	**-**	**8356**	**12770**	**-**	**32989**
Agriculture	-	-	4576	-	-	-	-	1834	915	-	7324
Comm. and Publ. Services	-	-	994	103	-	-	-	150	5483	-	6731
Residential	-	-	6047	143	-	-	-	6372	6372	-	18934
Non-specified	-	-	-	-	-	-	-	-	-	-	-
NON-ENERGY USE	**189**	**-**	**3627**	**-**	**-**	**-**	**-**	**-**	**-**	**-**	**3816**
in Industry/Transf./Energy	189	-	3627	-	-	-	-	-	-	-	3816
in Transport	-	-	-	-	-	-	-	-	-	-	-
in Other Sectors	-	-	-	-	-	-	-	-	-	-	-
Electr. Generated - GWh	***5484***	**-**	***9723***	***1107***	***3169***	***279064***	**-**	***8755***	**-**	**-**	***307302***
Electricity Plants	*5484*	-	*9723*	*1107*	*3169*	*279064*	-	*692*	-	-	*299239*
CHP Plants	-	-	-	-	-	-	-	*8063*	-	-	*8063*
Heat Generated - TJ	**-**	**-**	**-**	**-**	**-**	**-**	**-**	**-**	**-**	**-**	**-**
CHP Plants	-	-	-	-	-	-	-	-	-	-	-
Heat Plants	-	-	-	-	-	-	-	-	-	-	-

Brunei

Thousand tonnes of oil equivalent / *Milliers de tonnes d'équivalent pétrole*

SUPPLY AND CONSUMPTION 1996	Coal	Crude Oil	Petroleum Products	Gas	Nuclear	Hydro	Geotherm. Solar etc.	Combust. Renew. & Waste	Electricity	Heat	Total
Indigenous Production	-	8768	-	8859	-	-	-	18	-	-	17646
Imports	-	-	91	-	-	-	-	-	-	-	91
Exports	-	-8550	-1	-7278	-	-	-	-	-	-	-15829
Intl. Marine Bunkers	-	-	-	-	-	-	-	-	-	-	-
Stock changes	-	87	-4	-	-	-	-	-	-	-	83
TPES	-	305	86	1581	-	-	-	18	-	-	1990
Electricity and CHP Plants	-	-	-4	-736	-	-	-	-	183	-	-558
Petroleum Refineries	-	-557	482	-	-	-	-	-	-	-	-76
Other Transformation *	-	252	-74	-845	-	-	-	-	-11	-	-678
TFC	-	-	489	-	-	-	-	18	171	-	679
INDUSTRY SECTOR	-	-	79	-	-	-	-	-	24	-	103
Iron and Steel	-	-	-	-	-	-	-	-	-	-	-
Chemical and Petrochemical	-	-	6	-	-	-	-	-	-	-	6
Non-Metallic Minerals	-	-	-	-	-	-	-	-	-	-	-
Non-specified	-	-	72	-	-	-	-	-	24	-	97
TRANSPORT SECTOR	-	-	362	-	-	-	-	-	-	-	362
Air	-	-	76	-	-	-	-	-	-	-	76
Road	-	-	283	-	-	-	-	-	-	-	283
Non-specified	-	-	3	-	-	-	-	-	-	-	3
OTHER SECTORS	-	-	22	-	-	-	-	18	147	-	188
Agriculture	-	-	-	-	-	-	-	-	-	-	-
Comm. and Publ. Services	-	-	-	-	-	-	-	-	106	-	106
Residential	-	-	22	-	-	-	-	-	42	-	64
Non-specified	-	-	-	-	-	-	-	18	-	-	18
NON-ENERGY USE	-	-	26	-	-	-	-	-	-	-	26
Electricity Generated - GWh	-	-	16	2107	-	-	-	-	-	-	2123
Heat Generated - TJ	-	-	-	-	-	-	-	-	-	-	-

APPROVISIONNEMENT ET DEMANDE 1997	Charbon	Pétrole brut	Produits pétroliers	Gaz	Nucléaire	Hydro	Géotherm. solaire etc.	En. ren. combust. & déchets	Electricité	Chaleur	Total
Indigenous Production	-	8719	-	8857	-	-	-	18	-	-	17595
Imports	-	-	106	-	-	-	-	-	-	-	106
Exports	-	-8167	-1	-7233	-	-	-	-	-	-	-15401
Intl. Marine Bunkers	-	-	-	-	-	-	-	-	-	-	-
Stock changes	-	-178	-15	-	-	-	-	-	-	-	-192
TPES	-	375	90	1624	-	-	-	18	-	-	2107
Electricity and CHP Plants	-	-	-5	-835	-	-	-	-	207	-	-633
Petroleum Refineries	-	-616	509	-	-	-	-	-	-	-	-107
Other Transformation *	-	241	-68	-789	-	-	-	-	-7	-	-623
TFC	-	-	526	-	-	-	-	18	200	-	744
INDUSTRY SECTOR	-	-	91	-	-	-	-	-	23	-	114
Iron and Steel	-	-	-	-	-	-	-	-	-	-	-
Chemical and Petrochemical	-	-	6	-	-	-	-	-	-	-	6
Non-Metallic Minerals	-	-	-	-	-	-	-	-	-	-	-
Non-specified	-	-	85	-	-	-	-	-	23	-	107
TRANSPORT SECTOR	-	-	387	-	-	-	-	-	-	-	387
Air	-	-	89	-	-	-	-	-	-	-	89
Road	-	-	296	-	-	-	-	-	-	-	296
Non-specified	-	-	2	-	-	-	-	-	-	-	2
OTHER SECTORS	-	-	24	-	-	-	-	18	177	-	220
Agriculture	-	-	1	-	-	-	-	-	-	-	1
Comm. and Publ. Services	-	-	-	-	-	-	-	-	131	-	131
Residential	-	-	23	-	-	-	-	-	46	-	70
Non-specified	-	-	-	-	-	-	-	18	-	-	18
NON-ENERGY USE	-	-	23	-	-	-	-	-	-	-	23
Electricity Generated - GWh	-	-	19	2388	-	-	-	-	-	-	2407
Heat Generated - TJ	-	-	-	-	-	-	-	-	-	-	-

* Includes Transfers, Statistical Differences, Own Use and Distribution Losses.

Bulgaria / Bulgarie

Total Production of Energy

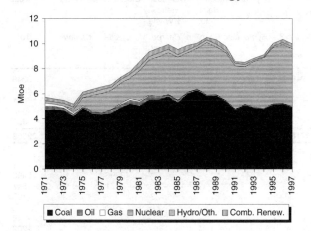

■ Coal ▦ Oil □ Gas ▤ Nuclear ▨ Hydro/Oth. ▥ Comb. Renew.

Total Primary Energy Supply*

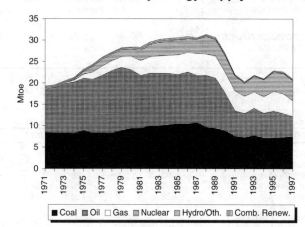

■ Coal ▦ Oil □ Gas ▤ Nuclear ▨ Hydro/Oth. ▥ Comb. Renew.

Electricity Generation by Fuel

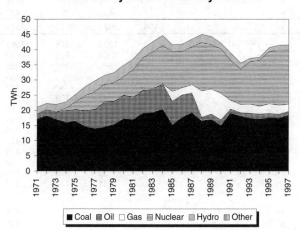

■ Coal ▦ Oil □ Gas ▤ Nuclear ▨ Hydro ▥ Other

Oil Products Consumption

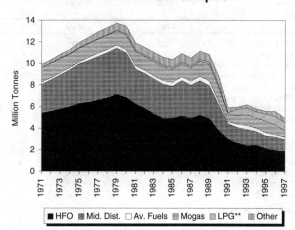

■ HFO ▦ Mid. Dist. □ Av. Fuels ▤ Mogas ▨ LPG** ▥ Other

Electricity Consumption/GDP***, TPES/GDP*** and Production/TPES

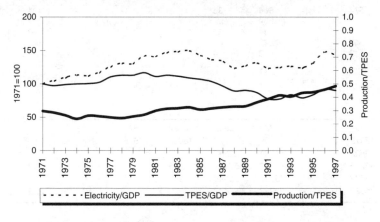

- - - - Electricity/GDP ——— TPES/GDP ━━━ Production/TPES

* Excluding electricity trade.
** Includes LPG, NGL, ethane and naphtha.
*** GDP in 1990 US dollars.

Bulgaria / Bulgarie : 1996

	Thousand tonnes of oil equivalent / Milliers de tonnes d'équivalent pétrole										
SUPPLY AND CONSUMPTION APPROVISIONNEMENT ET DEMANDE	Coal Charbon	Crude Oil Pétrole brut	Petroleum Products Produits pétroliers	Gas Gaz	Nuclear Nucléaire	Hydro Hydro	Geotherm. Solar etc. Géotherm. solaire	Combust. Renew. & Waste En. ren. combust.	Electricity Electricité	Heat Chaleur	Total Total
Indigenous Production	5170	32	-	33	4718	144	-	250	-	-	10348
Imports	2385	7094	468	4729	-	-	-	-	155	-	14830
Exports	-4	-	-1676	-	-	-	-	-7	-194	-	-1881
Intl. Marine Bunkers	-	-	-234	-	-	-	-	-	-	-	-234
Stock Changes	-268	-75	-34	-86	-	-	-	-3	-	-	-466
TPES	7283	7051	-1477	4676	4718	144	-	240	-39	-	22598
Transfers	-	-	-	-	-	-	-	-	-	-	-
Statistical Differences	5	1	-8	-27	-	-	-	-6	-26	189	129
Electricity Plants	-4005	-	-28	-	-4712	-144	-	-	2896	-	-5993
CHP Plants	-1859	-	-720	-1620	-6	-	-	-	671	1987	-1547
Heat Plants	-23	-1	-766	-715	-	-	-	-23	-	1359	-169
Gas Works	-	-	-	-	-	-	-	-	-	-	-
Petroleum Refineries	-	-7051	6904	-	-	-	-	-	-	-	-148
Coal Transformation	-202	-	-	-	-	-	-	-	-	-	-202
Liquefaction	-	-	-	-	-	-	-	-	-	-	-
Other Transformation	-	-	-	-23	-	-	-	-14	-	-	-37
Own Use	-98	-	-299	-80	-	-	-	-	-451	-383	-1310
Distribution Losses	-	-	-5	-93	-	-	-	-1	-480	-228	-807
TFC	1103	-	3602	2118	-	-	-	196	2571	2925	12514
INDUSTRY SECTOR	526	-	1422	2082	-	-	-	5	1054	1978	7067
Iron and Steel	338	-	26	278	-	-	-	-	169	125	936
Chemical and Petrochemical	69	-	896	1221	-	-	-	-	308	995	3489
of which: Feedstocks	-	-	729	632	-	-	-	-	-	-	1360
Non-Ferrous Metals	45	-	38	4	-	-	-	-	82	40	209
Non-Metallic Minerals	51	-	124	551	-	-	-	-	82	108	916
Transport Equipment	1	-	6	-	-	-	-	-	14	9	30
Machinery	7	-	34	24	-	-	-	1	120	104	291
Mining and Quarrying	9	-	47	-	-	-	-	-	75	13	145
Food and Tobacco	2	-	49	3	-	-	-	1	95	275	425
Paper, Pulp and Printing	-	-	6	1	-	-	-	-	29	117	153
Wood and Wood Products	-	-	7	-	-	-	-	1	11	49	68
Construction	3	-	119	-	-	-	-	1	22	3	148
Textile and Leather	1	-	12	-	-	-	-	-	41	120	175
Non-specified	-	-	56	-	-	-	-	-	7	19	83
TRANSPORT SECTOR	2	-	1501	-	-	-	-	-	70	-	1574
Air	-	-	196	-	-	-	-	-	-	-	196
Road	-	-	1245	-	-	-	-	-	-	-	1245
Rail	2	-	15	-	-	-	-	-	37	-	54
Pipeline Transport	-	-	-	-	-	-	-	-	-	-	-
Internal Navigation	-	-	6	-	-	-	-	-	-	-	6
Non-specified	-	-	38	-	-	-	-	-	33	-	72
OTHER SECTORS	574	-	668	35	-	-	-	191	1447	947	3863
Agriculture	2	-	309	-	-	-	-	2	52	77	442
Comm. and Publ. Services	4	-	44	32	-	-	-	3	141	21	244
Residential	556	-	242	-	-	-	-	171	988	665	2622
Non-specified	12	-	73	4	-	-	-	15	266	184	555
NON-ENERGY USE	-	-	11	-	-	-	-	-	-	-	11
in Industry/Transf./Energy	-	-	11	-	-	-	-	-	-	-	11
in Transport	-	-	-	-	-	-	-	-	-	-	-
in Other Sectors	-	-	-	-	-	-	-	-	-	-	-
Electr. Generated - GWh	17397	-	1335	2983	18082	1675	-	-	-	-	41472
Electricity Plants	13815	-	99	-	18082	1675	-	-	-	-	33671
CHP Plants	3582	-	1236	2983	-	-	-	-	-	-	7801
Heat Generated - TJ	36716	-	44957	57342	255	-	-	850	-	-	140120
CHP Plants	35346	-	16730	30896	255	-	-	-	-	-	83227
Heat Plants	1370	-	28227	26446	-	-	-	850	-	-	56893

Bulgaria / Bulgarie : 1997

	Thousand tonnes of oil equivalent / Milliers de tonnes d'équivalent pétrole										
SUPPLY AND CONSUMPTION	Coal	Crude Oil	Petroleum Products	Gas	Nuclear	Hydro	Geotherm. Solar etc.	Combust. Renew. & Waste	Electricity	Heat	Total
APPROVISIONNEMENT ET DEMANDE	*Charbon*	*Pétrole brut*	*Produits pétroliers*	*Gaz*	*Nucléaire*	*Hydro*	*Géotherm. solaire*	*En. ren. combust.*	*Electricité*	*Chaleur*	*Total*
Indigenous Production	4895	28	-	28	4633	146	-	251	-	-	9981
Imports	2392	5970	635	3851	-	-	-	-	68	-	12916
Exports	-3	-	-1586	-	-	-	-	-10	-373	-	-1972
Intl. Marine Bunkers	-	-	-333	-	-	-	-	-	-	-	-333
Stock Changes	208	50	-53	-180	-	-	-	-1	-	-	25
TPES	**7492**	**6049**	**-1336**	**3699**	**4633**	**146**	**-**	**240**	**-305**	**-**	**20616**
Transfers	-	-	-	-	-	-	-	-	-	-	-
Statistical Differences	51	1	18	18	-	-	-	-5	-21	217	280
Electricity Plants	-3821	-	-16	-	-4626	-146	-	-	2835	-	-5775
CHP Plants	-2324	-	-692	-1174	-7	-	-	-	740	1821	-1636
Heat Plants	-20	-1	-591	-657	-	-	-	-22	-	1137	-155
Gas Works	-	-	-	-	-	-	-	-	-	-	-
Petroleum Refineries	-	-6049	5996	-	-	-	-	-	-	-	-52
Coal Transformation	-236	-	-	-	-	-	-	-	-	-	-236
Liquefaction	-	-	-	-	-	-	-	-	-	-	-
Other Transformation	-	-	-26	-22	-	-	-	-15	-	-	-63
Own Use	-108	-	-231	-64	-	-	-	-	-441	-326	-1170
Distribution Losses	-	-	-3	-43	-	-	-	-2	-517	-210	-775
TFC	**1035**	**-**	**3119**	**1757**	**-**	**-**	**-**	**196**	**2290**	**2638**	**11035**
INDUSTRY SECTOR	**543**	**-**	**1393**	**1745**	**-**	**-**	**-**	**4**	**1009**	**1803**	**6498**
Iron and Steel	386	-	23	294	-	-	-	-	166	134	1002
Chemical and Petrochemical	63	-	872	991	-	-	-	-	294	937	3156
of which: Feedstocks	-	-	723	541	-	-	-	-	-	-	1265
Non-Ferrous Metals	44	-	39	4	-	-	-	-	68	44	198
Non-Metallic Minerals	34	-	176	427	-	-	-	-	74	89	801
Transport Equipment	-	-	6	-	-	-	-	-	14	5	25
Machinery	7	-	28	24	-	-	-	1	99	83	242
Mining and Quarrying	4	-	48	-	-	-	-	-	87	24	163
Food and Tobacco	2	-	51	4	-	-	-	1	63	230	350
Paper, Pulp and Printing	-	-	5	1	-	-	-	-	25	96	127
Wood and Wood Products	-	-	8	-	-	-	-	-	11	35	54
Construction	1	-	105	-	-	-	-	1	26	3	136
Textile and Leather	1	-	10	-	-	-	-	-	40	112	163
Non-specified	1	-	21	-	-	-	-	-	46	11	79
TRANSPORT SECTOR	**2**	**-**	**1087**	**-**	**-**	**-**	**-**	**-**	**57**	**-**	**1146**
Air	-	-	215	-	-	-	-	-	-	-	215
Road	-	-	823	-	-	-	-	-	-	-	823
Rail	2	-	2	-	-	-	-	-	2	-	6
Pipeline Transport	-	-	-	-	-	-	-	-	-	-	-
Internal Navigation	-	-	2	-	-	-	-	-	-	-	2
Non-specified	-	-	46	-	-	-	-	-	55	-	101
OTHER SECTORS	**490**	**-**	**630**	**12**	**-**	**-**	**-**	**191**	**1224**	**835**	**3382**
Agriculture	1	-	290	-	-	-	-	5	31	37	364
Comm. and Publ. Services	2	-	29	11	-	-	-	2	117	14	175
Residential	481	-	247	-	-	-	-	180	850	621	2379
Non-specified	5	-	64	1	-	-	-	4	227	163	463
NON-ENERGY USE	**-**	**-**	**9**	**-**	**-**	**-**	**-**	**-**	**-**	**-**	**9**
in Industry/Transf./Energy	-	-	9	-	-	-	-	-	-	-	9
in Transport	-	-	-	-	-	-	-	-	-	-	-
in Other Sectors	-	-	-	-	-	-	-	-	-	-	-
Electr. Generated - GWh	*18408*	*-*	*1269*	*2439*	*17751*	*1693*	*-*	*-*	*-*	*-*	*41560*
Electricity Plants	*13460*	*-*	*57*	*-*	*17751*	*1693*	*-*	*-*	*-*	*-*	*32961*
CHP Plants	*4948*	*-*	*1212*	*2439*	*-*	*-*	*-*	*-*	*-*	*-*	*8599*
Heat Generated - TJ	*41990*	*-*	*37618*	*43143*	*280*	*-*	*-*	*813*	*-*	*-*	*123844*
CHP Plants	*40721*	*-*	*16235*	*19008*	*280*	*-*	*-*	*-*	*-*	*-*	*76244*
Heat Plants	*1269*	*-*	*21383*	*24135*	*-*	*-*	*-*	*813*	*-*	*-*	*47600*

Cameroon / Cameroun

Thousand tonnes of oil equivalent / *Milliers de tonnes d'équivalent pétrole*

SUPPLY AND CONSUMPTION 1996	Coal	Crude Oil	Petroleum Products	Gas	Nuclear	Hydro	Geotherm. Solar etc.	Combust. Renew. & Waste	Electricity	Heat	Total
Indigenous Production	-	6068	-	-	-	247	-	4497	-	-	10811
Imports	-	-	9	-	-	-	-	-	-	-	9
Exports	-	-5175	-5	-	-	-	-	-	-	-	-5181
Intl. Marine Bunkers	-	-	-46	-	-	-	-	-	-	-	-46
Stock changes	-	-	-	-	-	-	-	-	-	-	-
TPES	-	892	-41	-	-	247	-	4497	-	-	5594
Electricity and CHP Plants	-	-	-9	-	-	-247	-	-	250	-	-6
Petroleum Refineries	-	-892	884	-	-	-	-	-	-	-	-8
Other Transformation *	-	-	-	-	-	-	-	-204	-50	-	-254
TFC	-	-	833	-	-	-	-	4293	199	-	5325
INDUSTRY SECTOR	-	-	49	-	-	-	-	737	113	-	899
Iron and Steel	-	-	-	-	-	-	-	-	-	-	-
Chemical and Petrochemical	-	-	-	-	-	-	-	-	-	-	-
Non-Metallic Minerals	-	-	-	-	-	-	-	-	-	-	-
Non-specified	-	-	49	-	-	-	-	737	113	-	899
TRANSPORT SECTOR	-	-	630	-	-	-	-	-	-	-	630
Air	-	-	60	-	-	-	-	-	-	-	60
Road	-	-	570	-	-	-	-	-	-	-	570
Non-specified	-	-	-	-	-	-	-	-	-	-	-
OTHER SECTORS	-	-	125	-	-	-	-	3556	86	-	3766
Agriculture	-	-	-	-	-	-	-	-	-	-	-
Comm. and Publ. Services	-	-	-	-	-	-	-	-	24	-	24
Residential	-	-	125	-	-	-	-	3556	28	-	3709
Non-specified	-	-	-	-	-	-	-	-	34	-	34
NON-ENERGY USE	-	-	30	-	-	-	-	-	-	-	30
Electricity Generated - GWh	-	-	*34*	-	-	*2868*	-	-	-	-	*2902*
Heat Generated - TJ	-	-	-	-	-	-	-	-	-	-	-

APPROVISIONNEMENT ET DEMANDE 1997	Charbon	Pétrole brut	Produits pétroliers	Gaz	Nucléaire	Hydro	Géotherm. solaire etc.	En. ren. combust. & déchets	Electricité	Chaleur	Total
Indigenous Production	-	6357	-	-	-	266	-	4627	-	-	11250
Imports	-	-	15	-	-	-	-	-	-	-	15
Exports	-	-5436	-18	-	-	-	-	-	-	-	-5454
Intl. Marine Bunkers	-	-	-55	-	-	-	-	-	-	-	-55
Stock changes	-	-	-	-	-	-	-	-	-	-	-
TPES	-	921	-58	-	-	266	-	4627	-	-	5756
Electricity and CHP Plants	-	-	-9	-	-	-266	-	-	269	-	-6
Petroleum Refineries	-	-921	911	-	-	-	-	-	-	-	-10
Other Transformation *	-	-	-1	-	-	-	-	-210	-53	-	-264
TFC	-	-	842	-	-	-	-	4417	216	-	5476
INDUSTRY SECTOR	-	-	47	-	-	-	-	759	123	-	928
Iron and Steel	-	-	-	-	-	-	-	-	-	-	-
Chemical and Petrochemical	-	-	-	-	-	-	-	-	-	-	-
Non-Metallic Minerals	-	-	-	-	-	-	-	-	-	-	-
Non-specified	-	-	47	-	-	-	-	759	123	-	928
TRANSPORT SECTOR	-	-	637	-	-	-	-	-	-	-	637
Air	-	-	54	-	-	-	-	-	-	-	54
Road	-	-	583	-	-	-	-	-	-	-	583
Non-specified	-	-	-	-	-	-	-	-	-	-	-
OTHER SECTORS	-	-	128	-	-	-	-	3659	94	-	3881
Agriculture	-	-	-	-	-	-	-	-	-	-	-
Comm. and Publ. Services	-	-	-	-	-	-	-	-	26	-	26
Residential	-	-	128	-	-	-	-	3659	31	-	3818
Non-specified	-	-	-	-	-	-	-	-	37	-	37
NON-ENERGY USE	-	-	30	-	-	-	-	-	-	-	30
Electricity Generated - GWh	-	-	*36*	-	-	*3092*	-	-	-	-	*3128*
Heat Generated - TJ	-	-	-	-	-	-	-	-	-	-	-

* Includes Transfers, Statistical Differences, Own Use and Distribution Losses.

INTERNATIONAL ENERGY AGENCY

Chile / Chili

Total Production of Energy

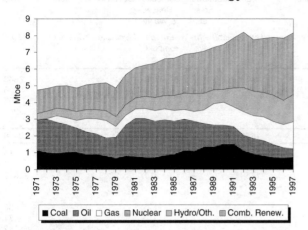

■ Coal ▦ Oil □ Gas ▤ Nuclear ▨ Hydro/Oth. ▥ Comb. Renew.

Total Primary Energy Supply*

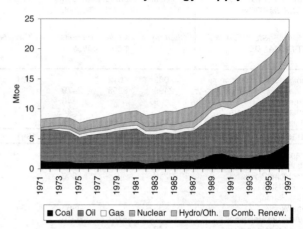

■ Coal ▦ Oil □ Gas ▤ Nuclear ▨ Hydro/Oth. ▥ Comb. Renew.

Electricity Generation by Fuel

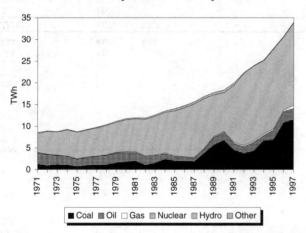

■ Coal ▦ Oil □ Gas ▤ Nuclear ▨ Hydro ▥ Other

Oil Products Consumption

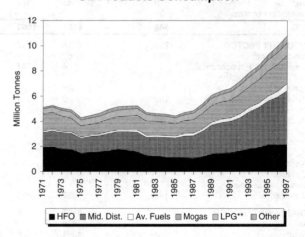

■ HFO ▦ Mid. Dist. □ Av. Fuels ▤ Mogas ▨ LPG** ▥ Other

Electricity Consumption/GDP***,
TPES/GDP*** and Production/TPES

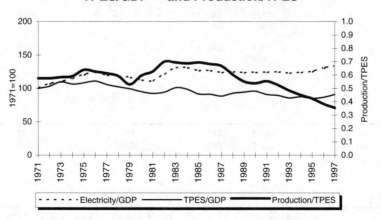

- - - - Electricity/GDP ——— TPES/GDP ——— Production/TPES

* Excluding electricity trade.
** Includes LPG, NGL, ethane and naphtha.
*** GDP in 1990 US dollars.

Chile / Chili : 1996

	Coal	Crude Oil	Petroleum Products	Gas	Nuclear	Hydro	Geotherm. Solar etc.	Combust. Renew. & Waste	Electricity	Heat	Total
SUPPLY AND CONSUMPTION											
APPROVISIONNEMENT ET DEMANDE	Charbon	Pétrole brut	Produits pétroliers	Gaz	Nucléaire	Hydro	Géotherm. solaire	En. ren. combust.	Electricité	Chaleur	Total
Indigenous Production	682	612	-	1391	-	1452	-	3708	-	-	7843
Imports	2499	8258	2479	-	-	-	-	-	-	-	13236
Exports	-9	-	-46	-	-	-	-	-	-	-	-55
Intl. Marine Bunkers	-	-	-244	-	-	-	-	-	-	-	-244
Stock Changes	116	64	-505	-	-	-	-	-	-	-	-324
TPES	3288	8934	1685	1391	-	1452	-	3708	-	-	20456
Transfers	-	-146	162	-	-	-	-	-	-	-	15
Statistical Differences	-2	-	10	-	-	-	-	-	-	-	8
Electricity Plants	-2212	-	-606	-75	-	-1452	-	-319	2648	-	-2016
CHP Plants	-	-	-	-	-	-	-	-	-	-	-
Heat Plants	-	-	-	-	-	-	-	-	-	-	-
Gas Works	-35	-	-65	64	-	-	-	-	-	-	-36
Petroleum Refineries	-	-8787	8316	-	-	-	-	-	-	-	-471
Coal Transformation	-142	-	-	-	-	-	-	-	-	-	-142
Liquefaction	-	-	-	-	-	-	-	-	-	-	-
Other Transformation	-	-	-	-	-	-	-	-	-	-	-
Own Use	-9	-	-745	-377	-	-	-	-31	-105	-	-1267
Distribution Losses	-39	-	-	-	-	-	-	-	-231	-	-270
TFC	848	-	8757	1003	-	-	-	3358	2311	-	16277
INDUSTRY SECTOR	761	-	2258	752	-	-	-	776	1578	-	6124
Iron and Steel	291	-	67	-	-	-	-	-	74	-	433
Chemical and Petrochemical	-	-	4	741	-	-	-	-	34	-	779
of which: Feedstocks	-	-	-	-	-	-	-	-	-	-	-
Non-Ferrous Metals	-	-	248	-	-	-	-	-	639	-	887
Non-Metallic Minerals	189	-	80	-	-	-	-	-	37	-	307
Transport Equipment	-	-	-	-	-	-	-	-	-	-	-
Machinery	-	-	-	-	-	-	-	-	-	-	-
Mining and Quarrying	18	-	308	-	-	-	-	-	18	-	344
Food and Tobacco	111	-	11	-	-	-	-	-	9	-	131
Paper, Pulp and Printing	1	-	153	-	-	-	-	437	215	-	806
Wood and Wood Products	-	-	-	-	-	-	-	-	-	-	-
Construction	-	-	-	-	-	-	-	-	-	-	-
Textile and Leather	-	-	-	-	-	-	-	-	-	-	-
Non-specified	149	-	1387	11	-	-	-	339	551	-	2437
TRANSPORT SECTOR	-	-	5172	6	-	-	-	-	17	-	5196
Air	-	-	470	-	-	-	-	-	-	-	470
Road	-	-	4261	6	-	-	-	-	8	-	4275
Rail	-	-	14	-	-	-	-	-	10	-	24
Pipeline Transport	-	-	-	-	-	-	-	-	-	-	-
Internal Navigation	-	-	427	-	-	-	-	-	-	-	427
Non-specified	-	-	-	-	-	-	-	-	-	-	-
OTHER SECTORS	87	-	1326	245	-	-	-	2582	716	-	4957
Agriculture	79	-	151	-	-	-	-	-	12	-	242
Comm. and Publ. Services	-	-	-	-	-	-	-	-	-	-	-
Residential	8	-	1071	245	-	-	-	2582	704	-	4610
Non-specified	-	-	104	-	-	-	-	-	-	-	104
NON-ENERGY USE	-	-	-	-	-	-	-	-	-	-	-
in Industry/Transf./Energy	-	-	-	-	-	-	-	-	-	-	-
in Transport	-	-	-	-	-	-	-	-	-	-	-
in Other Sectors	-	-	-	-	-	-	-	-	-	-	-
Electr. Generated - GWh	10800	-	2593	319	-	16878	-	200	-	-	30790
Electricity Plants	10800	-	2593	319	-	16878	-	200	-	-	30790
CHP Plants	-	-	-	-	-	-	-	-	-	-	-
Heat Generated - TJ	-	-	-	-	-	-	-	-	-	-	-
CHP Plants	-	-	-	-	-	-	-	-	-	-	-
Heat Plants	-	-	-	-	-	-	-	-	-	-	-

Thousand tonnes of oil equivalent / Milliers de tonnes d'équivalent pétrole

Chile / Chili : 1997

Thousand tonnes of oil equivalent / *Milliers de tonnes d'équivalent pétrole*

SUPPLY AND CONSUMPTION / *APPROVISIONNEMENT ET DEMANDE*	Coal / *Charbon*	Crude Oil / *Pétrole brut*	Petroleum Products / *Produits pétroliers*	Gas / *Gaz*	Nuclear / *Nucléaire*	Hydro / *Hydro*	Geotherm. Solar etc. / *Géotherm. solaire*	Combust. Renew. & Waste / *En. ren. combust.*	Electricity / *Electricité*	Heat / *Chaleur*	Total / *Total*
Indigenous Production	709	520	-	1619	-	1629	-	3691	-	-	8168
Imports	3184	8939	2362	545	-	-	-	-	-	-	15030
Exports	-12	-	-138	-	-	-	-	-	-	-	-150
Intl. Marine Bunkers	-	-	-296	-	-	-	-	-	-	-	-296
Stock Changes	359	-84	-16	-	-	-	-	-	-	-	260
TPES	**4240**	**9376**	**1913**	**2164**	**-**	**1629**	**-**	**3691**	**-**	**-**	**23012**
Transfers	-	-148	164	-	-	-	-	-	-	-	15
Statistical Differences	-	-	50	-	-	-	-	-	-	-	50
Electricity Plants	-2679	-	-597	-165	-	-1629	-	-328	2923	-	-2474
CHP Plants	-	-	-	-	-	-	-	-	-	-	-
Heat Plants	-	-	-	-	-	-	-	-	-	-	-
Gas Works	-35	-	-46	62	-	-	-	-	-	-	-19
Petroleum Refineries	-	-9227	8826	-	-	-	-	-	-	-	-401
Coal Transformation	-147	-	-	-	-	-	-	-	-	-	-147
Liquefaction	-	-	-	-	-	-	-	-	-	-	-
Other Transformation	-	-	-	-	-	-	-	-	-	-	-
Own Use	-9	-	-729	-379	-	-	-	-31	-117	-	-1265
Distribution Losses	-42	-	-	-	-	-	-	-	-277	-	-319
TFC	**1328**	**-**	**9580**	**1681**	**-**	**-**	**-**	**3332**	**2529**	**-**	**18451**
INDUSTRY SECTOR	**1221**	**-**	**2710**	**1419**	**-**	**-**	**-**	**621**	**1744**	**-**	**7715**
Iron and Steel	296	-	68	-	-	-	-	-	27	-	392
Chemical and Petrochemical	-	-	4	1159	-	-	-	-	49	-	1212
of which: Feedstocks	-	-	-	-	-	-	-	-	-	-	-
Non-Ferrous Metals	-	-	302	-	-	-	-	-	759	-	1061
Non-Metallic Minerals	199	-	57	-	-	-	-	-	41	-	297
Transport Equipment	-	-	-	-	-	-	-	-	-	-	-
Machinery	-	-	-	-	-	-	-	-	-	-	-
Mining and Quarrying	21	-	338	-	-	-	-	-	20	-	379
Food and Tobacco	62	-	12	-	-	-	-	-	9	-	82
Paper, Pulp and Printing	1	-	146	-	-	-	-	297	197	-	640
Wood and Wood Products	-	-	-	-	-	-	-	-	-	-	-
Construction	-	-	-	-	-	-	-	-	-	-	-
Textile and Leather	-	-	-	-	-	-	-	-	-	-	-
Non-specified	642	-	1783	260	-	-	-	325	642	-	3652
TRANSPORT SECTOR	**-**	**-**	**5507**	**5**	**-**	**-**	**-**	**-**	**18**	**-**	**5530**
Air	-	-	593	-	-	-	-	-	-	-	593
Road	-	-	4463	5	-	-	-	-	8	-	4476
Rail	-	-	13	-	-	-	-	-	10	-	23
Pipeline Transport	-	-	-	-	-	-	-	-	-	-	-
Internal Navigation	-	-	438	-	-	-	-	-	-	-	438
Non-specified	-	-	-	-	-	-	-	-	-	-	-
OTHER SECTORS	**107**	**-**	**1363**	**257**	**-**	**-**	**-**	**2711**	**767**	**-**	**5205**
Agriculture	94	-	140	-	-	-	-	-	15	-	249
Comm. and Publ. Services	-	-	120	36	-	-	-	-	-	-	155
Residential	13	-	985	222	-	-	-	2711	753	-	4683
Non-specified	-	-	118	-	-	-	-	-	-	-	118
NON-ENERGY USE	**-**	**-**	**-**	**-**	**-**	**-**	**-**	**-**	**-**	**-**	**-**
in Industry/Transf./Energy	-	-	-	-	-	-	-	-	-	-	-
in Transport	-	-	-	-	-	-	-	-	-	-	-
in Other Sectors	-	-	-	-	-	-	-	-	-	-	-
Electr. Generated - GWh	**11506**	**-**	**2608**	**730**	**-**	**18944**	**-**	**206**	**-**	**-**	**33994**
Electricity Plants	*11506*	*-*	*2608*	*730*	*-*	*18944*	*-*	*206*	*-*	*-*	*33994*
CHP Plants	*-*	*-*	*-*	*-*	*-*	*-*	*-*	*-*	*-*	*-*	*-*
Heat Generated - TJ	**-**	**-**	**-**	**-**	**-**	**-**	**-**	**-**	**-**	**-**	**-**
CHP Plants	*-*	*-*	*-*	*-*	*-*	*-*	*-*	*-*	*-*	*-*	*-*
Heat Plants	*-*	*-*	*-*	*-*	*-*	*-*	*-*	*-*	*-*	*-*	*-*

People's Republic of China / République populaire de Chine

Total Production of Energy

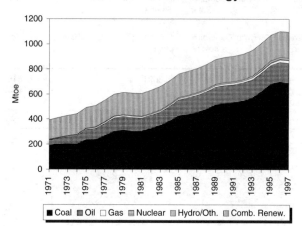

Total Primary Energy Supply*

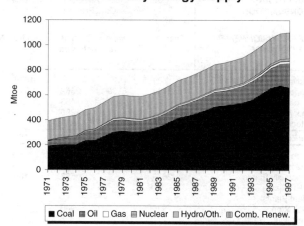

Electricity Generation by Fuel

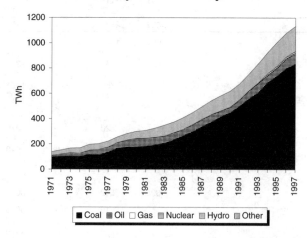

Oil Products Consumption

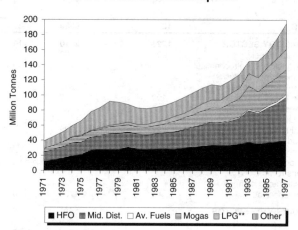

Electricity Consumption/GDP***, TPES/GDP*** and Production/TPES

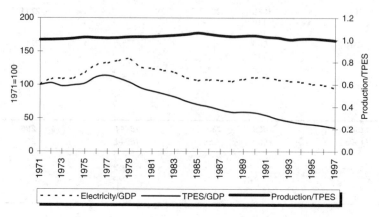

* Excluding electricity trade.
** Includes LPG, NGL, ethane and naphtha.
*** GDP in 1990 US dollars.

INTERNATIONAL ENERGY AGENCY

People's Republic of China / République populaire de Chine : 1996

Thousand tonnes of oil equivalent / *Milliers de tonnes d'équivalent pétrole*

SUPPLY AND CONSUMPTION	Coal	Crude Oil	Petroleum Products	Gas	Nuclear	Hydro	Geotherm. Solar etc.	Combust. Renew. & Waste	Electricity	Heat	Total
APPROVISIONNEMENT ET DEMANDE	*Charbon*	*Pétrole brut*	*Produits pétroliers*	*Gaz*	*Nucléaire*	*Hydro*	*Géotherm. solaire*	*En. ren. combust.*	*Electricité*	*Chaleur*	*Total*
Indigenous Production	698350	157334	-	18724	3737	16165	-	207153	-	-	1101463
Imports	1785	22617	19670	-	-	-	-	-	11	-	44082
Exports	-24809	-20402	-5666	-1479	-	-	-	-	-319	-	-52675
Intl. Marine Bunkers	-	-	-990	-	-	-	-	-	-	-	-990
Stock Changes	3653	-341	1201	-	-	-	-	-	-	-	4512
TPES	**678978**	**159208**	**14214**	**17245**	**3737**	**16165**	**-**	**207153**	**-309**	**-**	**1096392**
Transfers	-	-	-	-	-	-	-	-	-	-	-
Statistical Differences	25136	-557	4335	-46	-	-	-	-	-	-53	28814
Electricity Plants	-228900	-682	-13902	-694	-3737	-16165	-	-	92881	-	-171199
CHP Plants	-29600	-	-	-	-	-	-	-	-	27421	-2179
Heat Plants	-652	-108	-4227	-27	-	-	-	-	-	-	-5014
Gas Works	-2171	-	-558	2461	-	-	-	-	-	-	-268
Petroleum Refineries	-	-151339	153824	-	-	-	-	-	-	-	2485
Coal Transformation	-36872	-	-	-	-	-	-	-	-	-	-36872
Liquefaction	-	-	-	-	-	-	-	-	-	-	-
Other Transformation	-	-	-	-	-	-	-	-	-	-	-
Own Use	-32472	-2932	-13018	-3675	-	-	-	-	-13838	-5736	-71672
Distribution Losses	-10966	-1599	-16	-528	-	-	-	-	-6386	-355	-19850
TFC	**362480**	**1991**	**140652**	**14737**	**-**	**-**	**-**	**207153**	**72348**	**21277**	**820639**
INDUSTRY SECTOR	**243382**	**1623**	**39622**	**12025**	**-**	**-**	**-**	**-**	**50257**	**16604**	**363512**
Iron and Steel	48847	49	3805	1170	-	-	-	-	8511	2668	65050
Chemical and Petrochemical	47970	1461	25175	8042	-	-	-	-	12723	6668	102040
of which: Feedstocks	-	274	18401	4474	-	-	-	-	-	-	23149
Non-Ferrous Metals	6347	5	795	241	-	-	-	-	4001	758	12147
Non-Metallic Minerals	69658	60	4213	458	-	-	-	-	5637	181	80208
Transport Equipment	3789	1	420	78	-	-	-	-	1454	559	6301
Machinery	15789	8	1531	877	-	-	-	-	4401	663	23269
Mining and Quarrying	4112	2	339	58	-	-	-	-	1630	446	6586
Food and Tobacco	17309	5	665	119	-	-	-	-	3036	1268	22401
Paper, Pulp and Printing	8856	1	370	12	-	-	-	-	1882	968	12090
Wood and Wood Products	2744	-	158	-	-	-	-	-	489	96	3487
Construction	2325	26	1159	141	-	-	-	-	1501	38	5188
Textile and Leather	11409	3	717	573	-	-	-	-	3942	2020	18664
Non-specified	4226	2	276	256	-	-	-	-	1049	270	6080
TRANSPORT SECTOR	**5072**	**-**	**56309**	**106**	**-**	**-**	**-**	**-**	**1019**	**-**	**62506**
Air	-	-	3042	-	-	-	-	-	-	-	3042
Road	-	-	43732	106	-	-	-	-	-	-	43838
Rail	5035	-	3582	-	-	-	-	-	929	-	9547
Pipeline Transport	-	-	65	-	-	-	-	-	89	-	154
Internal Navigation	36	-	5838	-	-	-	-	-	-	-	5875
Non-specified	-	-	50	-	-	-	-	-	-	-	50
OTHER SECTORS	**99427**	**368**	**30454**	**2606**	**-**	**-**	**-**	**207153**	**21073**	**4674**	**365755**
Agriculture	10382	110	10686	21	-	-	-	-	5317	5	26522
Comm. and Publ. Services	16132	258	10914	277	-	-	-	-	6023	612	34216
Residential	72913	-	8854	2308	-	-	-	207153	9733	4057	305017
Non-specified	-	-	-	-	-	-	-	-	-	-	-
NON-ENERGY USE	**14600**	**-**	**14267**	**-**	**-**	**-**	**-**	**-**	**-**	**-**	**28866**
in Industry/Transf./Energy	14600	-	14267	-	-	-	-	-	-	-	28866
in Transport	-	-	-	-	-	-	-	-	-	-	-
in Other Sectors	-	-	-	-	-	-	-	-	-	-	-
Electr. Generated - GWh	*800132*	*-*	*75159*	*2421*	*14339*	*187966*	*-*	*-*	*-*	*-*	*1080017*
Electricity Plants	*800132*	*-*	*75159*	*2421*	*14339*	*187966*	*-*	*-*	*-*	*-*	*1080017*
CHP Plants	*-*	*-*	*-*	*-*	*-*	*-*	*-*	*-*	*-*	*-*	*-*
Heat Generated - TJ	*-*	*-*	*-*	*-*	*-*	*-*	*-*	*-*	*-*	*-*	*1148301*
CHP Plants	*-*	*-*	*-*	*-*	*-*	*-*	*-*	*-*	*-*	*-*	*1148301*
Heat Plants	*-*	*-*	*-*	*-*	*-*	*-*	*-*	*-*	*-*	*-*	*-*

People's Republic of China / République populaire de Chine : 1997

Thousand tonnes of oil equivalent / *Milliers de tonnes d'équivalent pétrole*

SUPPLY AND CONSUMPTION / *APPROVISIONNEMENT ET DEMANDE*	Coal / *Charbon*	Crude Oil / *Pétrole brut*	Petroleum Products / *Produits pétroliers*	Gas / *Gaz*	Nuclear / *Nucléaire*	Hydro / *Hydro*	Geotherm. Solar etc. / *Géotherm. solaire*	Combust. Renew. & Waste / *En. ren. combust.*	Electricity / *Electricité*	Heat / *Chaleur*	Total / *Total*
Indigenous Production	686410	160741	-	21134	3757	16855	-	208262	-	-	1097160
Imports	1183	35470	29870	-	-	-	-	-	8	-	66531
Exports	-23513	-19829	-7499	-2309	-	-	-	-	-620	-	-53769
Intl. Marine Bunkers	-	-	-970	-	-	-	-	-	-	-	-970
Stock Changes	-5998	-1386	-2637	-	-	-	-	-	-	-	-10020
TPES	**658083**	**174996**	**18765**	**18825**	**3757**	**16855**	**-**	**208262**	**-612**	**-**	**1098931**
Transfers	-	-	-	-	-	-	-	-	-	-	-
Statistical Differences	25595	-1058	7081	116	-	-	-	-	-	291	32026
Electricity Plants	-229454	-684	-15569	-1907	-3757	-16855	-	-	97565	-	-170661
CHP Plants	-29627	-	-	-	-	-	-	-	-	28076	-1551
Heat Plants	-788	-108	-3619	-74	-	-	-	-	-	-	-4589
Gas Works	-2943	-	-592	2586	-	-	-	-	-	-	-949
Petroleum Refineries	-	-166073	166427	-	-	-	-	-	-	-	354
Coal Transformation	-40143	-	-	-	-	-	-	-	-	-	-40143
Liquefaction	-	-	-	-	-	-	-	-	-	-	-
Other Transformation	-	-	-	-	-	-	-	-	-	-	-
Own Use	-36528	-3223	-15948	-5622	-	-	-	-	-16708	-5715	-83744
Distribution Losses	-10586	-1758	-20	-790	-	-	-	-	-7710	-354	-21219
TFC	**333610**	**2092**	**156524**	**13134**	**-**	**-**	**-**	**208262**	**72534**	**22298**	**808455**
INDUSTRY SECTOR	**226534**	**1687**	**39843**	**10379**	**-**	**-**	**-**	**-**	**48076**	**17372**	**343891**
Iron and Steel	50111	51	3589	1094	-	-	-	-	8142	2792	65778
Chemical and Petrochemical	43633	1520	26241	6787	-	-	-	-	12171	6976	97328
of which: Feedstocks	*-*	*285*	*19662*	*3771*	*-*	*-*	*-*	*-*	*-*	*-*	*23717*
Non-Ferrous Metals	5884	5	749	231	-	-	-	-	3827	793	11489
Non-Metallic Minerals	62847	62	3974	410	-	-	-	-	5392	190	72875
Transport Equipment	3433	1	393	72	-	-	-	-	1391	585	5875
Machinery	14594	8	1435	786	-	-	-	-	4210	694	21727
Mining and Quarrying	3827	2	316	49	-	-	-	-	1560	467	6220
Food and Tobacco	15589	5	623	100	-	-	-	-	2904	1326	20547
Paper, Pulp and Printing	7962	1	368	10	-	-	-	-	1801	1013	11155
Wood and Wood Products	2468	-	146	-	-	-	-	-	468	100	3183
Construction	2100	27	1078	118	-	-	-	-	1435	39	4798
Textile and Leather	10263	3	673	483	-	-	-	-	3771	2114	17307
Non-specified	3824	2	259	238	-	-	-	-	1004	283	5609
TRANSPORT SECTOR	**6173**	**-**	**63448**	**18**	**-**	**-**	**-**	**-**	**1317**	**-**	**70956**
Air	-	-	3659	-	-	-	-	-	-	-	3659
Road	-	-	46273	18	-	-	-	-	-	-	46292
Rail	6128	-	3933	-	-	-	-	-	1202	-	11263
Pipeline Transport	-	-	155	-	-	-	-	-	116	-	271
Internal Navigation	44	-	9372	-	-	-	-	-	-	-	9417
Non-specified	-	-	55	-	-	-	-	-	-	-	55
OTHER SECTORS	**87074**	**405**	**34030**	**2737**	**-**	**-**	**-**	**208262**	**23141**	**4926**	**360575**
Agriculture	10598	121	11176	10	-	-	-	-	5502	13	27420
Comm. and Publ. Services	13812	284	13439	292	-	-	-	-	6743	782	35353
Residential	62663	-	9415	2435	-	-	-	208262	10896	4131	297803
Non-specified	-	-	-	-	-	-	-	-	-	-	-
NON-ENERGY USE	**13830**	**-**	**19203**	**-**	**-**	**-**	**-**	**-**	**-**	**-**	**33033**
in Industry/Transf./Energy	13830	-	19203	-	-	-	-	-	-	-	33033
in Transport	-	-	-	-	-	-	-	-	-	-	-
in Other Sectors	-	-	-	-	-	-	-	-	-	-	-
Electr. Generated - GWh	***834519***	***-***	***82894***	***6657***	***14418***	***195983***	***-***	***-***	***-***	***-***	***1134471***
Electricity Plants	*834519*	*-*	*82894*	*6657*	*14418*	*195983*	*-*	*-*	*-*	*-*	*1134471*
CHP Plants	*-*	*-*	*-*	*-*	*-*	*-*	*-*	*-*	*-*	*-*	*-*
Heat Generated - TJ	***-***	***-***	***-***	***-***	***-***	***-***	***-***	***-***	***-***	***-***	***1175695***
CHP Plants	*-*	*-*	*-*	*-*	*-*	*-*	*-*	*-*	*-*	*-*	*1175695*
Heat Plants	*-*	*-*	*-*	*-*	*-*	*-*	*-*	*-*	*-*	*-*	*-*

Chinese Taipei / Taipei chinois

Total Production of Energy

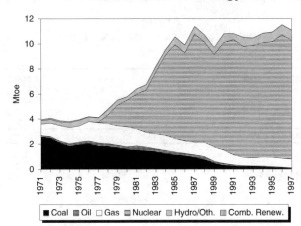

■ Coal ▦ Oil □ Gas ▤ Nuclear ▦ Hydro/Oth. ▥ Comb. Renew.

Total Primary Energy Supply*

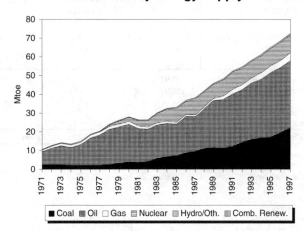

■ Coal ▦ Oil □ Gas ▤ Nuclear ▦ Hydro/Oth. ▥ Comb. Renew.

Electricity Generation by Fuel

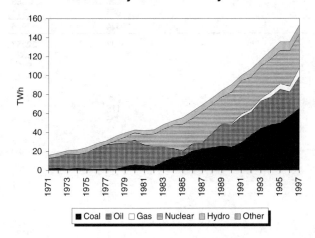

■ Coal ▦ Oil □ Gas ▤ Nuclear ▦ Hydro ▥ Other

Oil Products Consumption

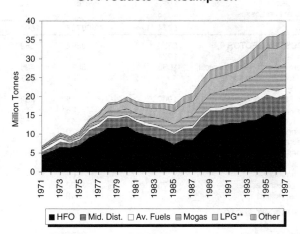

■ HFO ▦ Mid. Dist. □ Av. Fuels ▤ Mogas ▨ LPG** ▥ Other

Electricity Consumption/GDP***, TPES/GDP*** and Production/TPES

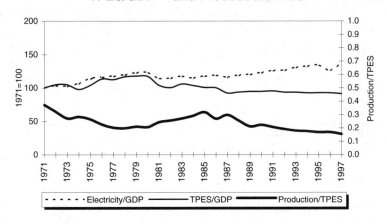

- - - - - Electricity/GDP ——— TPES/GDP ▬▬▬ Production/TPES

* Excluding electricity trade.
** Includes LPG, NGL, ethane and naphtha.
*** GDP in 1990 US dollars.

INTERNATIONAL ENERGY AGENCY

Chinese Taipei / Taipei chinois : 1996

	Thousand tonnes of oil equivalent / Milliers de tonnes d'équivalent pétrole										
SUPPLY AND CONSUMPTION	Coal	Crude Oil	Petroleum Products	Gas	Nuclear	Hydro	Geotherm. Solar etc.	Combust. Renew. & Waste	Electricity	Heat	Total
APPROVISIONNEMENT ET DEMANDE	*Charbon*	*Pétrole brut*	*Produits pétroliers*	*Gaz*	*Nucléaire*	*Hydro*	*Géotherm. solaire*	*En. ren. combust.*	*Electricité*	*Chaleur*	*Total*
Indigenous Production	91	56	-	722	9848	778	-	13	-	-	11507
Imports	20397	33107	7057	3060	-	-	-	-	-	-	63622
Exports	-	-	-2116	-	-	-	-	-	-	-	-2116
Intl. Marine Bunkers	-	-	-2290	-	-	-	-	-	-	-	-2290
Stock Changes	-934	-558	-416	-86	-	-	-	-	-	-	-1993
TPES	**19554**	**32606**	**2235**	**3696**	**9848**	**778**	**-**	**13**	**-**	**-**	**68729**
Transfers	-	1463	-1452	-	-	-	-	-	-	-	11
Statistical Differences	-360	-1	-5	-	-	-	-	-	575	-	208
Electricity Plants	-13005	-	-5950	-1357	-9848	-778	-	-	11634	-	-19303
CHP Plants	-	-	-	-	-	-	-	-	-	-	-
Heat Plants	-	-	-	-	-	-	-	-	-	-	-
Gas Works	-	-	-	-	-	-	-	-	-	-	-
Petroleum Refineries	-	-34067	33564	-	-	-	-	-	-	-	-504
Coal Transformation	-911	-	-	-	-	-	-	-	-	-	-911
Liquefaction	-	-	-	-	-	-	-	-	-	-	-
Other Transformation	-	-	-	-7	-	-	-	-	-	-	-7
Own Use	-	-	-1685	-672	-	-	-	-	-1226	-	-3583
Distribution Losses	-3	-	-	-53	-	-	-	-	-658	-	-714
TFC	**5274**	**-**	**26707**	**1608**	**-**	**-**	**-**	**13**	**10324**	**-**	**43926**
INDUSTRY SECTOR	**5274**	**-**	**10892**	**859**	**-**	**-**	**-**	**-**	**5332**	**-**	**22357**
Iron and Steel	1931	-	620	99	-	-	-	-	677	-	3327
Chemical and Petrochemical	1038	-	6823	424	-	-	-	-	1539	-	9823
of which: Feedstocks	-	-	*4726*	*393*	-	-	-	-	-	-	*5120*
Non-Ferrous Metals	-	-	109	-	-	-	-	-	30	-	138
Non-Metallic Minerals	1967	-	575	259	-	-	-	-	392	-	3193
Transport Equipment	-	-	26	-	-	-	-	-	115	-	141
Machinery	-	-	41	50	-	-	-	-	1023	-	1115
Mining and Quarrying	-	-	30	-	-	-	-	-	15	-	45
Food and Tobacco	-	-	419	9	-	-	-	-	265	-	693
Paper, Pulp and Printing	265	-	419	-	-	-	-	-	336	-	1020
Wood and Wood Products	-	-	-	-	-	-	-	-	49	-	49
Construction	-	-	167	-	-	-	-	-	42	-	210
Textile and Leather	73	-	1122	17	-	-	-	-	689	-	1901
Non-specified	-	-	540	-	-	-	-	-	161	-	702
TRANSPORT SECTOR	**-**	**-**	**11709**	**-**	**-**	**-**	**-**	**-**	**44**	**-**	**11753**
Air	-	-	1959	-	-	-	-	-	14	-	1973
Road	-	-	9436	-	-	-	-	-	1	-	9438
Rail	-	-	10	-	-	-	-	-	28	-	38
Pipeline Transport	-	-	-	-	-	-	-	-	-	-	-
Internal Navigation	-	-	303	-	-	-	-	-	-	-	304
Non-specified	-	-	-	-	-	-	-	-	-	-	-
OTHER SECTORS	**-**	**-**	**2871**	**749**	**-**	**-**	**-**	**13**	**4948**	**-**	**8581**
Agriculture	-	-	893	-	-	-	-	-	191	-	1084
Comm. and Publ. Services	-	-	666	137	-	-	-	-	1444	-	2247
Residential	-	-	1198	586	-	-	-	-	2372	-	4156
Non-specified	-	-	115	26	-	-	-	13	941	-	1095
NON-ENERGY USE	**-**	**-**	**1235**	**-**	**-**	**-**	**-**	**-**	**-**	**-**	**1235**
in Industry/Transf./Energy	-	-	1235	-	-	-	-	-	-	-	1235
in Transport	-	-	-	-	-	-	-	-	-	-	-
in Other Sectors	-	-	-	-	-	-	-	-	-	-	-
Electr. Generated - GWh	***57186***	**-**	***25265***	***5994***	***37788***	***9044***	**-**	**-**	**-**	**-**	***135277***
Electricity Plants	*57186*	-	*25265*	*5994*	*37788*	*9044*	-	-	-	-	*135277*
CHP Plants	-	-	-	-	-	-	-	-	-	-	-
Heat Generated - TJ	**-**	**-**	**-**	**-**	**-**	**-**	**-**	**-**	**-**	**-**	**-**
CHP Plants	-	-	-	-	-	-	-	-	-	-	-
Heat Plants	-	-	-	-	-	-	-	-	-	-	-

Chinese Taipei / Taipei chinois : 1997

Thousand tonnes of oil equivalent / *Milliers de tonnes d'équivalent pétrole*

SUPPLY AND CONSUMPTION / *APPROVISIONNEMENT ET DEMANDE*	Coal / *Charbon*	Crude Oil / *Pétrole brut*	Petroleum Products / *Produits pétroliers*	Gas / *Gaz*	Nuclear / *Nucléaire*	Hydro / *Hydro*	Geotherm. Solar etc. / *Géotherm. solaire*	Combust. Renew. & Waste / *En. ren. combust.*	Electricity / *Electricité*	Heat / *Chaleur*	Total / *Total*
Indigenous Production	61	49	-	688	9452	823	-	13	-	-	11086
Imports	23745	32949	7884	3747	-	-	-	-	-	-	68324
Exports	-	-	-2883	-	-	-	-	-	-	-	-2883
Intl. Marine Bunkers	-	-	-2719	-	-	-	-	-	-	-	-2719
Stock Changes	-1584	228	254	-164	-	-	-	-	-	-	-1265
TPES	**22223**	**33225**	**2536**	**4271**	**9452**	**823**	**-**	**13**	**-**	**-**	**72543**
Transfers	-	1950	-1969	-	-	-	-	-	-	-	-19
Statistical Differences	-351	-	2	-	-	-	-	-	-267	-	-616
Electricity Plants	-14600	-	-6205	-1908	-9452	-823	-	-	13183	-	-19804
CHP Plants	-	-	-	-	-	-	-	-	-	-	-
Heat Plants	-	-	-	-	-	-	-	-	-	-	-
Gas Works	-	-	-	-	-	-	-	-	-	-	-
Petroleum Refineries	-	-35175	34727	-	-	-	-	-	-	-	-449
Coal Transformation	-1548	-	-	-	-	-	-	-	-	-	-1548
Liquefaction	-	-	-	-	-	-	-	-	-	-	-
Other Transformation	-	-	-	-7	-	-	-	-	-	-	-7
Own Use	-	-	-1731	-622	-	-	-	-	-1261	-	-3615
Distribution Losses	-9	-	-	-108	-	-	-	-	-593	-	-711
TFC	**5715**	**-**	**27360**	**1626**	**-**	**-**	**-**	**13**	**11062**	**-**	**45775**
INDUSTRY SECTOR	**5715**	**-**	**11522**	**869**	**-**	**-**	**-**	**-**	**5841**	**-**	**23946**
Iron and Steel	2560	-	630	116	-	-	-	-	820	-	4126
Chemical and Petrochemical	1003	-	7344	420	-	-	-	-	1665	-	10432
of which: Feedstocks	-	-	*5079*	*380*	-	-	-	-	-	-	*5459*
Non-Ferrous Metals	-	-	127	-	-	-	-	-	33	-	160
Non-Metallic Minerals	1795	-	565	248	-	-	-	-	391	-	3000
Transport Equipment	-	-	29	-	-	-	-	-	120	-	150
Machinery	-	-	43	54	-	-	-	-	1198	-	1295
Mining and Quarrying	-	-	35	-	-	-	-	-	15	-	50
Food and Tobacco	-	-	417	10	-	-	-	-	265	-	692
Paper, Pulp and Printing	309	-	398	-	-	-	-	-	340	-	1047
Wood and Wood Products	-	-	-	-	-	-	-	-	50	-	50
Construction	-	-	159	-	-	-	-	-	42	-	201
Textile and Leather	49	-	1212	21	-	-	-	-	725	-	2007
Non-specified	-	-	563	-	-	-	-	-	174	-	737
TRANSPORT SECTOR	**-**	**-**	**11983**	**-**	**-**	**-**	**-**	**-**	**51**	**-**	**12034**
Air	-	-	1997	-	-	-	-	-	17	-	2014
Road	-	-	9662	-	-	-	-	-	2	-	9663
Rail	-	-	-	-	-	-	-	-	32	-	32
Pipeline Transport	-	-	-	-	-	-	-	-	-	-	-
Internal Navigation	-	-	324	-	-	-	-	-	-	-	325
Non-specified	-	-	-	-	-	-	-	-	-	-	-
OTHER SECTORS	**-**	**-**	**2602**	**757**	**-**	**-**	**-**	**13**	**5170**	**-**	**8542**
Agriculture	-	-	788	-	-	-	-	-	195	-	983
Comm. and Publ. Services	-	-	465	142	-	-	-	-	1551	-	2158
Residential	-	-	1229	590	-	-	-	-	2394	-	4214
Non-specified	-	-	120	25	-	-	-	13	1030	-	1188
NON-ENERGY USE	**-**	**-**	**1253**	**-**	**-**	**-**	**-**	**-**	**-**	**-**	**1253**
in Industry/Transf./Energy	-	-	1253	-	-	-	-	-	-	-	1253
in Transport	-	-	-	-	-	-	-	-	-	-	-
in Other Sectors	-	-	-	-	-	-	-	-	-	-	-
Electr. Generated - GWh	*65151*	*-*	*33505*	*8802*	*36269*	*9567*	*-*	*-*	*-*	*-*	*153294*
Electricity Plants	*65151*	*-*	*33505*	*8802*	*36269*	*9567*	*-*	*-*	*-*	*-*	*153294*
CHP Plants	*-*	*-*	*-*	*-*	*-*	*-*	*-*	*-*	*-*	*-*	*-*
Heat Generated - TJ	*-*	*-*	*-*	*-*	*-*	*-*	*-*	*-*	*-*	*-*	*-*
CHP Plants	*-*	*-*	*-*	*-*	*-*	*-*	*-*	*-*	*-*	*-*	*-*
Heat Plants	*-*	*-*	*-*	*-*	*-*	*-*	*-*	*-*	*-*	*-*	*-*

Colombia / Colombie

Total Production of Energy

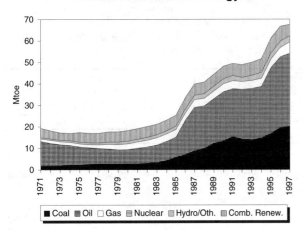

Total Primary Energy Supply*

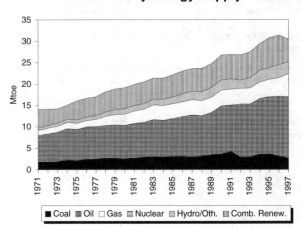

Electricity Generation by Fuel

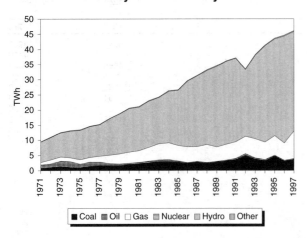

Oil Products Consumption

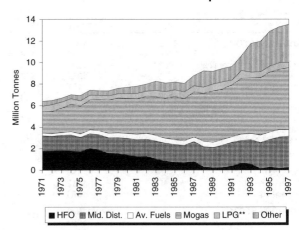

Electricity Consumption/GDP***, TPES/GDP*** and Production/TPES

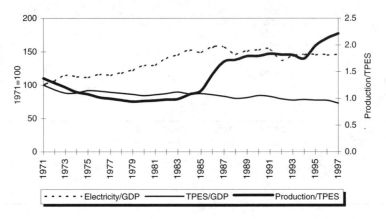

* Excluding electricity trade.
** Includes LPG, NGL, ethane and naphtha.
*** GDP in 1990 US dollars.

INTERNATIONAL ENERGY AGENCY

Colombia / Colombie : 1996

Thousand tonnes of oil equivalent / *Milliers de tonnes d'équivalent pétrole*

SUPPLY AND CONSUMPTION / *APPROVISIONNEMENT ET DEMANDE*	Coal / *Charbon*	Crude Oil / *Pétrole brut*	Petroleum Products / *Produits pétroliers*	Gas / *Gaz*	Nuclear / *Nucléaire*	Hydro / *Hydro*	Geotherm. Solar etc. / *Géotherm. solaire*	Combust. Renew. & Waste / *En. ren. combust.*	Electricity / *Electricité*	Heat / *Chaleur*	Total / *Total*
Indigenous Production	19542	32932	-	4343	-	3034	-	6888	-	-	66739
Imports	-	-	1202	-	-	-	-	-	14	-	1217
Exports	-16108	-16756	-2956	-	-	-	-	-	-	-	-35820
Intl. Marine Bunkers	-	-	-194	-	-	-	-	-	-	-	-194
Stock Changes	-303	-299	53	-	-	-	-	-	-	-	-548
TPES	3132	15877	-1895	4343	-	3034	-	6888	14	-	31393
Transfers	-	-329	363	-	-	-	-	-	-	-	34
Statistical Differences	207	2	-59	-31	-	-	-	1	162	-	282
Electricity Plants	-616	-14	-57	-1515	-	-3034	-	-128	3836	-	-1528
CHP Plants	-	-	-	-	-	-	-	-	-	-	-
Heat Plants	-	-	-	-	-	-	-	-	-	-	-
Gas Works	-	-	-	-	-	-	-	-	-	-	-
Petroleum Refineries	-	-14606	14382	-	-	-	-	-	-	-	-224
Coal Transformation	-608	-	-	-	-	-	-	-	-	-	-608
Liquefaction	-	-	-	-	-	-	-	-	-	-	-
Other Transformation	-	-	-	-	-	-	-	-108	-	-	-108
Own Use	-43	-63	-722	-1376	-	-	-	-	-60	-	-2264
Distribution Losses	-93	-	-	-	-	-	-	-	-838	-	-931
TFC	1979	867	12013	1420	-	-	-	6653	3114	-	26045
INDUSTRY SECTOR	1838	855	747	1046	-	-	-	1755	975	-	7215
Iron and Steel	63	52	47	29	-	-	-	-	164	-	355
Chemical and Petrochemical	101	83	68	441	-	-	-	27	141	-	861
of which: Feedstocks	-	-	-	-	-	-	-	-	-	-	-
Non-Ferrous Metals	322	-	-	-	-	-	-	-	-	-	322
Non-Metallic Minerals	699	163	132	401	-	-	-	18	122	-	1535
Transport Equipment	-	-	-	-	-	-	-	-	-	-	-
Machinery	-	73	74	26	-	-	-	-	40	-	213
Mining and Quarrying	-	-	-	-	-	-	-	-	79	-	79
Food and Tobacco	133	216	155	66	-	-	-	1481	166	-	2218
Paper, Pulp and Printing	300	1	21	50	-	-	-	227	75	-	674
Wood and Wood Products	1	107	4	18	-	-	-	-	13	-	142
Construction	-	13	80	-	-	-	-	-	5	-	98
Textile and Leather	219	94	42	11	-	-	-	1	136	-	503
Non-specified	-	52	124	5	-	-	-	-	34	-	215
TRANSPORT SECTOR	1	-	7942	47	-	-	-	-	-	-	7990
Air	-	-	739	-	-	-	-	-	-	-	739
Road	-	-	7041	-	-	-	-	-	-	-	7041
Rail	1	-	-	-	-	-	-	-	-	-	1
Pipeline Transport	-	-	-	-	-	-	-	-	-	-	-
Internal Navigation	-	-	162	-	-	-	-	-	-	-	162
Non-specified	-	-	-	47	-	-	-	-	-	-	47
OTHER SECTORS	140	12	1570	327	-	-	-	4898	2139	-	9086
Agriculture	-	5	453	-	-	-	-	1309	32	-	1799
Comm. and Publ. Services	-	7	369	49	-	-	-	-	568	-	993
Residential	140	-	748	278	-	-	-	3571	1268	-	6006
Non-specified	-	-	-	-	-	-	-	18	271	-	289
NON-ENERGY USE	-	-	1754	-	-	-	-	-	-	-	1754
in Industry/Transf./Energy	-	-	1754	-	-	-	-	-	-	-	1754
in Transport	-	-	-	-	-	-	-	-	-	-	-
in Other Sectors	-	-	-	-	-	-	-	-	-	-	-
Electr. Generated - GWh	*3080*	*-*	*241*	*5706*	*-*	*35281*	*-*	*297*	*-*	*-*	*44605*
Electricity Plants	*3080*	*-*	*241*	*5706*	*-*	*35281*	*-*	*297*	*-*	*-*	*44605*
CHP Plants	*-*	*-*	*-*	*-*	*-*	*-*	*-*	*-*	*-*	*-*	*-*
Heat Generated - TJ	*-*	*-*	*-*	*-*	*-*	*-*	*-*	*-*	*-*	*-*	*-*
CHP Plants	*-*	*-*	*-*	*-*	*-*	*-*	*-*	*-*	*-*	*-*	*-*
Heat Plants	*-*	*-*	*-*	*-*	*-*	*-*	*-*	*-*	*-*	*-*	*-*

Colombia / Colombie : 1997

Thousand tonnes of oil equivalent / *Milliers de tonnes d'équivalent pétrole*

SUPPLY AND CONSUMPTION / *APPROVISIONNEMENT ET DEMANDE*	Coal / *Charbon*	Crude Oil / *Pétrole brut*	Petroleum Products / *Produits pétroliers*	Gas / *Gaz*	Nuclear / *Nucléaire*	Hydro / *Hydro*	Geotherm. Solar etc. / *Géotherm. solaire*	Combust. Renew. & Waste / *En. ren. combust.*	Electricity / *Electricité*	Heat / *Chaleur*	Total / *Total*
Indigenous Production	19934	34164	-	5339	-	2821	-	5267	-	-	67524
Imports	-	-	1410	-	-	-	-	-	17	-	1427
Exports	-17222	-18028	-2750	-	-	-	-	-	-	-	-38000
Intl. Marine Bunkers	-	-	-175	-	-	-	-	-	-	-	-175
Stock Changes	-7	-159	-129	-	-	-	-	-	-	-	-295
TPES	**2704**	**15977**	**-1644**	**5339**	**-**	**2821**	**-**	**5267**	**17**	**-**	**30481**
Transfers	-	-225	143	-	-	-	-	-	-	-	-82
Statistical Differences	725	-302	524	-22	-	-	-	-13	6	-	919
Electricity Plants	-743	-18	-54	-2419	-	-2821	-	-129	3966	-	-2218
CHP Plants	-	-	-	-	-	-	-	-	-	-	-
Heat Plants	-	-	-	-	-	-	-	-	-	-	-
Gas Works	-	-	-	-	-	-	-	-	-	-	-
Petroleum Refineries	-	-14418	13903	-	-	-	-	-	-	-	-514
Coal Transformation	-522	-	-	-	-	-	-	-	-	-	-522
Liquefaction	-	-	-	-	-	-	-	-	-	-	-
Other Transformation	-	-	-	-	-	-	-	-55	-	-	-55
Own Use	-44	-72	-718	-1434	-	-	-	-	-66	-	-2333
Distribution Losses	-96	-	-	-	-	-	-	-	-876	-	-972
TFC	**2025**	**942**	**12154**	**1465**	**-**	**-**	**-**	**5070**	**3047**	**-**	**24704**
INDUSTRY SECTOR	**1888**	**929**	**773**	**1010**	**-**	**-**	**-**	**1731**	**970**	**-**	**7302**
Iron and Steel	61	90	40	31	-	-	-	-	163	-	386
Chemical and Petrochemical	104	57	65	436	-	-	-	35	141	-	838
of which: Feedstocks	-	-	-	-	-	-	-	-	-	-	-
Non-Ferrous Metals	332	-	-	-	-	-	-	-	-	-	332
Non-Metallic Minerals	720	180	120	397	-	-	-	16	121	-	1552
Transport Equipment	-	-	-	-	-	-	-	-	-	-	-
Machinery	-	80	120	-	-	-	-	-	40	-	239
Mining and Quarrying	-	-	-	-	-	-	-	-	78	-	78
Food and Tobacco	137	235	153	62	-	-	-	1451	166	-	2204
Paper, Pulp and Printing	309	1	28	51	-	-	-	227	75	-	691
Wood and Wood Products	-	117	4	20	-	-	-	-	13	-	154
Construction	-	12	81	-	-	-	-	-	5	-	98
Textile and Leather	226	102	41	10	-	-	-	1	136	-	516
Non-specified	-	57	121	2	-	-	-	-	33	-	214
TRANSPORT SECTOR	**1**	**-**	**8096**	**50**	**-**	**-**	**-**	**-**	**4**	**-**	**8150**
Air	-	-	667	-	-	-	-	-	-	-	667
Road	-	-	7261	-	-	-	-	-	-	-	7261
Rail	1	-	-	-	-	-	-	-	4	-	4
Pipeline Transport	-	-	-	-	-	-	-	-	-	-	-
Internal Navigation	-	-	168	-	-	-	-	-	-	-	168
Non-specified	-	-	-	50	-	-	-	-	-	-	50
OTHER SECTORS	**137**	**13**	**1436**	**405**	**-**	**-**	**-**	**3339**	**2073**	**-**	**7403**
Agriculture	-	5	448	-	-	-	-	1178	26	-	1657
Comm. and Publ. Services	-	8	372	61	-	-	-	-	584	-	1025
Residential	137	-	586	345	-	-	-	2141	1267	-	4474
Non-specified	-	-	30	-	-	-	-	20	196	-	246
NON-ENERGY USE	**-**	**-**	**1848**	**-**	**-**	**-**	**-**	**-**	**-**	**-**	**1848**
in Industry/Transf./Energy	-	-	1848	-	-	-	-	-	-	-	1848
in Transport	-	-	-	-	-	-	-	-	-	-	-
in Other Sectors	-	-	-	-	-	-	-	-	-	-	-
Electr. Generated - GWh	*3664*	*-*	*244*	*9108*	*-*	*32799*	*-*	*300*	*-*	*-*	*46115*
Electricity Plants	*3664*	*-*	*244*	*9108*	*-*	*32799*	*-*	*300*	*-*	*-*	*46115*
CHP Plants	*-*	*-*	*-*	*-*	*-*	*-*	*-*	*-*	*-*	*-*	*-*
Heat Generated - TJ	*-*	*-*	*-*	*-*	*-*	*-*	*-*	*-*	*-*	*-*	*-*
CHP Plants	*-*	*-*	*-*	*-*	*-*	*-*	*-*	*-*	*-*	*-*	*-*
Heat Plants	*-*	*-*	*-*	*-*	*-*	*-*	*-*	*-*	*-*	*-*	*-*

Congo

Thousand tonnes of oil equivalent / *Milliers de tonnes d'équivalent pétrole*

SUPPLY AND CONSUMPTION 1996	Coal	Crude Oil	Petroleum Products	Gas	Nuclear	Hydro	Geotherm. Solar etc.	Combust. Renew. & Waste	Electricity	Heat	Total
Indigenous Production	-	10629	-	3	-	37	-	843	-	-	11512
Imports	-	-	14	-	-	-	-	-	10	-	24
Exports	-	-9221	-299	-	-	-	-	-	-	-	-9520
Intl. Marine Bunkers	-	-	-10	-	-	-	-	-	-	-	-10
Stock changes	-	-783	-	-	-	-	-	-	-	-	-783
TPES	-	625	-294	3	-	37	-	843	10	-	1223
Electricity and CHP Plants	-	-	-1	-3	-	-37	-	-	37	-	-3
Petroleum Refineries	-	-625	626	-	-	-	-	-	-	-	1
Other Transformation *	-	-	-40	-	-	-	-	-33	-1	-	-74
TFC	-	-	291	-	-	-	-	810	46	-	1147
INDUSTRY SECTOR	-	-	4	-	-	-	-	134	25	-	163
Iron and Steel	-	-	-	-	-	-	-	-	-	-	-
Chemical and Petrochemical	-	-	-	-	-	-	-	-	-	-	-
Non-Metallic Minerals	-	-	-	-	-	-	-	-	-	-	-
Non-specified	-	-	4	-	-	-	-	134	25	-	163
TRANSPORT SECTOR	-	-	214	-	-	-	-	-	-	-	214
Air	-	-	74	-	-	-	-	-	-	-	74
Road	-	-	141	-	-	-	-	-	-	-	141
Non-specified	-	-	-	-	-	-	-	-	-	-	-
OTHER SECTORS	-	-	54	-	-	-	-	675	22	-	751
Agriculture	-	-	-	-	-	-	-	-	-	-	-
Comm. and Publ. Services	-	-	-	-	-	-	-	-	-	-	-
Residential	-	-	54	-	-	-	-	675	22	-	751
Non-specified	-	-	-	-	-	-	-	-	-	-	-
NON-ENERGY USE	-	-	19	-	-	-	-	-	-	-	19
Electricity Generated - GWh	-	-	3	3	-	429	-	-	-	-	435
Heat Generated - TJ	-	-	-	-	-	-	-	-	-	-	-

APPROVISIONNEMENT ET DEMANDE 1997	Charbon	Pétrole brut	Produits pétroliers	Gaz	Nucléaire	Hydro	Géotherm. solaire etc.	En. ren. combust. & déchets	Electricité	Chaleur	Total
Indigenous Production	-	12638	-	3	-	37	-	862	-	-	13540
Imports	-	-	14	-	-	-	-	-	10	-	24
Exports	-	-12013	-299	-	-	-	-	-	-	-	-12312
Intl. Marine Bunkers	-	-	-10	-	-	-	-	-	-	-	-10
Stock changes	-	-	-	-	-	-	-	-	-	-	-
TPES	-	625	-294	3	-	37	-	862	10	-	1242
Electricity and CHP Plants	-	-	-1	-3	-	-37	-	-	37	-	-3
Petroleum Refineries	-	-625	626	-	-	-	-	-	-	-	1
Other Transformation *	-	-	-40	-	-	-	-	-34	-1	-	-74
TFC	-	-	291	-	-	-	-	828	46	-	1165
INDUSTRY SECTOR	-	-	4	-	-	-	-	137	25	-	166
Iron and Steel	-	-	-	-	-	-	-	-	-	-	-
Chemical and Petrochemical	-	-	-	-	-	-	-	-	-	-	-
Non-Metallic Minerals	-	-	-	-	-	-	-	-	-	-	-
Non-specified	-	-	4	-	-	-	-	137	25	-	166
TRANSPORT SECTOR	-	-	214	-	-	-	-	-	-	-	214
Air	-	-	74	-	-	-	-	-	-	-	74
Road	-	-	141	-	-	-	-	-	-	-	141
Non-specified	-	-	-	-	-	-	-	-	-	-	-
OTHER SECTORS	-	-	54	-	-	-	-	691	21	-	766
Agriculture	-	-	-	-	-	-	-	-	-	-	-
Comm. and Publ. Services	-	-	-	-	-	-	-	-	-	-	-
Residential	-	-	54	-	-	-	-	691	21	-	766
Non-specified	-	-	-	-	-	-	-	-	-	-	-
NON-ENERGY USE	-	-	19	-	-	-	-	-	-	-	19
Electricity Generated - GWh	-	-	3	3	-	425	-	-	-	-	431
Heat Generated - TJ	-	-	-	-	-	-	-	-	-	-	-

* Includes Transfers, Statistical Differences, Own Use and Distribution Losses.

Democratic Republic of Congo / République démocratique du Congo

Thousand tonnes of oil equivalent / *Milliers de tonnes d'équivalent pétrole*

SUPPLY AND CONSUMPTION 1996	Coal	Crude Oil	Petroleum Products	Gas	Nuclear	Hydro	Geotherm. Solar etc.	Combust. Renew. & Waste	Electricity	Heat	Total
Indigenous Production	56	1381	-	-	-	527	-	12100	-	-	14064
Imports	185	120	961	-	-	-	-	-	5	-	1271
Exports	-	-1087	-25	-	-	-	-	-	-17	-	-1129
Intl. Marine Bunkers	-	-	-34	-	-	-	-	-	-	-	-34
Stock changes	-	-	1	-	-	-	-	-	-	-	1
TPES	241	414	903	-	-	527	-	12100	-12	-	14174
Electricity and CHP Plants	-	-	-46	-	-	-527	-	-	538	-	-35
Petroleum Refineries	-	-367	347	-	-	-	-	-	-	-	-20
Other Transformation *	-79	-47	-18	-	-	-	-	-554	-22	-	-721
TFC	162	-	1186	-	-	-	-	11546	504	-	13397
INDUSTRY SECTOR	102	-	56	-	-	-	-	2440	327	-	2925
Iron and Steel	20	-	-	-	-	-	-	-	-	-	20
Chemical and Petrochemical	-	-	-	-	-	-	-	-	-	-	-
Non-Metallic Minerals	-	-	-	-	-	-	-	-	-	-	-
Non-specified	82	-	56	-	-	-	-	2440	327	-	2905
TRANSPORT SECTOR	-	-	743	-	-	-	-	-	-	-	743
Air	-	-	169	-	-	-	-	-	-	-	169
Road	-	-	573	-	-	-	-	-	-	-	573
Non-specified	-	-	-	-	-	-	-	-	-	-	-
OTHER SECTORS	60	-	330	-	-	-	-	9106	177	-	9673
Agriculture	-	-	-	-	-	-	-	-	-	-	-
Comm. and Publ. Services	-	-	-	-	-	-	-	-	-	-	-
Residential	60	-	131	-	-	-	-	9106	168	-	9464
Non-specified	-	-	200	-	-	-	-	-	9	-	209
NON-ENERGY USE	-	-	57	-	-	-	-	-	-	-	57
Electricity Generated - GWh	-	-	130	-	-	6131	-	-	-	-	6261
Heat Generated - TJ	-	-	-	-	-	-	-	-	-	-	-

APPROVISIONNEMENT ET DEMANDE 1997	Charbon	Pétrole brut	Produits pétroliers	Gaz	Nucléaire	Hydro	Géotherm. solaire etc.	En. ren. combust. & déchets	Electricité	Chaleur	Total
Indigenous Production	56	1315	-	-	-	506	-	12488	-	-	14364
Imports	185	120	961	-	-	-	-	-	5	-	1271
Exports	-	-1021	-25	-	-	-	-	-	-17	-	-1063
Intl. Marine Bunkers	-	-	-34	-	-	-	-	-	-	-	-34
Stock changes	-	-	1	-	-	-	-	-	-	-	1
TPES	241	414	903	-	-	506	-	12488	-12	-	14539
Electricity and CHP Plants	-	-	-46	-	-	-506	-	-	517	-	-35
Petroleum Refineries	-	-367	347	-	-	-	-	-	-	-	-20
Other Transformation *	-79	-47	-18	-	-	-	-	-572	-21	-	-738
TFC	162	-	1186	-	-	-	-	11915	484	-	13746
INDUSTRY SECTOR	102	-	56	-	-	-	-	2518	314	-	2990
Iron and Steel	20	-	-	-	-	-	-	-	-	-	20
Chemical and Petrochemical	-	-	-	-	-	-	-	-	-	-	-
Non-Metallic Minerals	-	-	-	-	-	-	-	-	-	-	-
Non-specified	82	-	56	-	-	-	-	2518	314	-	2970
TRANSPORT SECTOR	-	-	743	-	-	-	-	-	-	-	743
Air	-	-	169	-	-	-	-	-	-	-	169
Road	-	-	573	-	-	-	-	-	-	-	573
Non-specified	-	-	-	-	-	-	-	-	-	-	-
OTHER SECTORS	60	-	330	-	-	-	-	9397	170	-	9957
Agriculture	-	-	-	-	-	-	-	-	-	-	-
Comm. and Publ. Services	-	-	-	-	-	-	-	-	-	-	-
Residential	60	-	131	-	-	-	-	9397	161	-	9749
Non-specified	-	-	200	-	-	-	-	-	9	-	208
NON-ENERGY USE	-	-	57	-	-	-	-	-	-	-	57
Electricity Generated - GWh	-	-	130	-	-	5880	-	-	-	-	6010
Heat Generated - TJ	-	-	-	-	-	-	-	-	-	-	-

* Includes Transfers, Statistical Differences, Own Use and Distribution Losses.

Costa Rica

Thousand tonnes of oil equivalent / *Milliers de tonnes d'équivalent pétrole*

SUPPLY AND CONSUMPTION 1996	Coal	Crude Oil	Petroleum Products	Gas	Nuclear	Hydro	Geotherm. Solar etc.	Combust. Renew. & Waste	Electricity	Heat	Total
Indigenous Production	-	-	-	-	-	328	441	374	-	-	1142
Imports	8	626	988	-	-	-	-	-	20	-	1642
Exports	-	-	-155	-	-	-	-	-	-9	-	-165
Intl. Marine Bunkers	-	-	-	-	-	-	-	-	-	-	-
Stock changes	-	-5	-14	-	-	-	-	-	-	-	-19
TPES	8	621	819	-	-	328	441	374	11	-	2601
Electricity and CHP Plants	-	-	-136	-	-	-328	-441	-	412	-	-492
Petroleum Refineries	-	-631	626	-	-	-	-	-	-	-	-6
Other Transformation *	-6	10	23	-	-	-	-	-7	-41	-	-21
TFC	1	-	1332	-	-	-	-	367	381	-	2082
INDUSTRY SECTOR	1	-	195	-	-	-	-	174	88	-	458
Iron and Steel	1	-	-	-	-	-	-	-	-	-	1
Chemical and Petrochemical	-	-	25	-	-	-	-	-	18	-	43
Non-Metallic Minerals	-	-	-	-	-	-	-	-	-	-	-
Non-specified	-	-	170	-	-	-	-	174	70	-	414
TRANSPORT SECTOR	-	-	985	-	-	-	-	-	-	-	985
Air	-	-	100	-	-	-	-	-	-	-	100
Road	-	-	862	-	-	-	-	-	-	-	862
Non-specified	-	-	23	-	-	-	-	-	-	-	23
OTHER SECTORS	-	-	135	-	-	-	-	193	294	-	621
Agriculture	-	-	73	-	-	-	-	-	1	-	73
Comm. and Publ. Services	-	-	28	-	-	-	-	1	117	-	147
Residential	-	-	34	-	-	-	-	192	176	-	402
Non-specified	-	-	-	-	-	-	-	-	-	-	-
NON-ENERGY USE	-	-	18	-	-	-	-	-	-	-	18
Electricity Generated - GWh	-	-	446	-	-	3812	533	-	-	-	4791
Heat Generated - TJ	-	-	-	-	-	-	-	-	-	-	-

APPROVISIONNEMENT ET DEMANDE 1997	Charbon	Pétrole brut	Produits pétroliers	Gaz	Nucléaire	Hydro	Géotherm. solaire etc.	En. ren. combust. & déchets	Electricité	Chaleur	Total
Indigenous Production	-	-	-	-	-	413	474	269	-	-	1157
Imports	-	630	992	-	-	-	-	-	9	-	1631
Exports	-	-	-112	-	-	-	-	-	-21	-	-133
Intl. Marine Bunkers	-	-	-	-	-	-	-	-	-	-	-
Stock changes	-	9	-2	-	-	-	-	-	-	-	7
TPES	-	639	878	-	-	413	474	269	-12	-	2663
Electricity and CHP Plants	-	-	-61	-	-	-413	-474	-	482	-	-467
Petroleum Refineries	-	-619	615	-	-	-	-	-	-	-	-4
Other Transformation *	-	-20	-27	-	-	-	-	-7	-67	-	-120
TFC	-	-	1405	-	-	-	-	263	403	-	2070
INDUSTRY SECTOR	-	-	220	-	-	-	-	183	94	-	498
Iron and Steel	-	-	-	-	-	-	-	-	-	-	-
Chemical and Petrochemical	-	-	29	-	-	-	-	-	19	-	48
Non-Metallic Minerals	-	-	-	-	-	-	-	-	-	-	-
Non-specified	-	-	191	-	-	-	-	183	75	-	450
TRANSPORT SECTOR	-	-	1016	-	-	-	-	-	-	-	1016
Air	-	-	110	-	-	-	-	-	-	-	110
Road	-	-	880	-	-	-	-	-	-	-	880
Non-specified	-	-	27	-	-	-	-	-	-	-	27
OTHER SECTORS	-	-	146	-	-	-	-	79	309	-	535
Agriculture	-	-	79	-	-	-	-	-	1	-	79
Comm. and Publ. Services	-	-	32	-	-	-	-	-	126	-	158
Residential	-	-	36	-	-	-	-	79	182	-	298
Non-specified	-	-	-	-	-	-	-	-	-	-	-
NON-ENERGY USE	-	-	22	-	-	-	-	-	-	-	22
Electricity Generated - GWh	-	-	171	-	-	4808	620	-	-	-	5599
Heat Generated - TJ	-	-	-	-	-	-	-	-	-	-	-

* Includes Transfers, Statistical Differences, Own Use and Distribution Losses.

Croatia / Croatie : 1996

Thousand tonnes of oil equivalent / *Milliers de tonnes d'équivalent pétrole*

SUPPLY AND CONSUMPTION / *APPROVISIONNEMENT ET DEMANDE*	Coal / *Charbon*	Crude Oil / *Pétrole brut*	Petroleum Products / *Produits pétroliers*	Gas / *Gaz*	Nuclear / *Nucléaire*	Hydro / *Hydro*	Geotherm. Solar etc. / *Géotherm. solaire*	Combust. Renew. & Waste / *En. ren. combust.*	Electricity / *Electricité*	Heat / *Chaleur*	Total / *Total*
Indigenous Production	38	1745	-	1458	-	622	-	329	-	-	4192
Imports	102	3700	249	718	-	-	-	-	341	-	5109
Exports	-	-113	-1614	-	-	-	-	-	-140	-	-1868
Intl. Marine Bunkers	-	-	-29	-	-	-	-	-	-	-	-29
Stock Changes	5	-54	-103	-9	-	-	-	-	-	-	-162
TPES	**145**	**5278**	**-1498**	**2167**	**-**	**622**	**-**	**329**	**200**	**-**	**7243**
Transfers	-	-245	254	-	-	-	-	-	-	-	9
Statistical Differences	2	-	-	-	-	-	-	-	-	-	2
Electricity Plants	-35	-	-391	-186	-	-622	-	-	794	-	-439
CHP Plants	-2	-	-351	-210	-	-	-	-2	113	263	-190
Heat Plants	-	-	-36	-33	-	-	-	-	-	65	-5
Gas Works	-	-	-11	9	-	-	-	-	-	-	-2
Petroleum Refineries	-	-4944	4928	-	-	-	-	-	-	-	-16
Coal Transformation	-	-	-	-	-	-	-	-	-	-	-
Liquefaction	-	-	-	-	-	-	-	-	-	-	-
Other Transformation	-	-	-	-	-	-	-	-	-	-	-
Own Use	-1	-89	-500	-233	-	-	-	-	-59	-22	-903
Distribution Losses	-	-	-	-83	-	-	-	-	-164	-46	-292
TFC	**110**	**-**	**2395**	**1432**	**-**	**-**	**-**	**327**	**884**	**261**	**5408**
INDUSTRY SECTOR	**97**	**-**	**554**	**924**	**-**	**-**	**-**	**-**	**228**	**88**	**1892**
Iron and Steel	6	-	9	41	-	-	-	-	20	-	76
Chemical and Petrochemical	5	-	301	539	-	-	-	-	47	17	909
of which: Feedstocks	-	-	*240*	*436*	-	-	-	-	-	-	*676*
Non-Ferrous Metals	-	-	4	-	-	-	-	-	6	-	10
Non-Metallic Minerals	50	-	70	158	-	-	-	-	37	3	317
Transport Equipment	-	-	3	1	-	-	-	-	10	1	15
Machinery	1	-	4	7	-	-	-	-	10	6	28
Mining and Quarrying	-	-	11	10	-	-	-	-	4	1	25
Food and Tobacco	32	-	60	90	-	-	-	-	31	26	240
Paper, Pulp and Printing	-	-	9	39	-	-	-	-	18	6	72
Wood and Wood Products	2	-	3	10	-	-	-	-	15	9	39
Construction	-	-	57	-	-	-	-	-	9	-	66
Textile and Leather	1	-	21	27	-	-	-	-	16	15	80
Non-specified	-	-	1	3	-	-	-	-	5	4	13
TRANSPORT SECTOR	**-**	**-**	**1254**	**-**	**-**	**-**	**-**	**-**	**21**	**-**	**1275**
Air	-	-	84	-	-	-	-	-	-	-	84
Road	-	-	1089	-	-	-	-	-	-	-	1089
Rail	-	-	30	-	-	-	-	-	12	-	42
Pipeline Transport	-	-	-	-	-	-	-	-	-	-	-
Internal Navigation	-	-	29	-	-	-	-	-	-	-	29
Non-specified	-	-	24	-	-	-	-	-	9	-	33
OTHER SECTORS	**13**	**-**	**501**	**507**	**-**	**-**	**-**	**327**	**636**	**172**	**2156**
Agriculture	-	-	159	21	-	-	-	-	6	1	186
Comm. and Publ. Services	5	-	95	102	-	-	-	-	208	20	430
Residential	8	-	248	385	-	-	-	327	421	152	1540
Non-specified	-	-	-	-	-	-	-	-	-	-	-
NON-ENERGY USE	**-**	**-**	**85**	**-**	**-**	**-**	**-**	**-**	**-**	**-**	**85**
in Industry/Transf./Energy	-	-	65	-	-	-	-	-	-	-	65
in Transport	-	-	17	-	-	-	-	-	-	-	17
in Other Sectors	-	-	3	-	-	-	-	-	-	-	3
Electr. Generated - GWh	*120*	*-*	*1872*	*1318*	*-*	*7228*	*-*	*10*	*-*	*-*	*10548*
Electricity Plants	*110*	*-*	*1113*	*780*	*-*	*7228*	*-*	*-*	*-*	*-*	*9231*
CHP Plants	*10*	*-*	*759*	*538*	*-*	*-*	*-*	*10*	*-*	*-*	*1317*
Heat Generated - TJ	*-*	*-*	*7864*	*5873*	*-*	*-*	*-*	*-*	*-*	*-*	*13737*
CHP Plants	*-*	*-*	*6374*	*4643*	*-*	*-*	*-*	*-*	*-*	*-*	*11017*
Heat Plants	*-*	*-*	*1490*	*1230*	*-*	*-*	*-*	*-*	*-*	*-*	*2720*

Croatia / Croatie : 1997

Thousand tonnes of oil equivalent / *Milliers de tonnes d'équivalent pétrole*

SUPPLY AND CONSUMPTION	Coal	Crude Oil	Petroleum Products	Gas	Nuclear	Hydro	Geotherm. Solar etc.	Combust. Renew. & Waste	Electricity	Heat	Total
APPROVISIONNEMENT ET DEMANDE	*Charbon*	*Pétrole brut*	*Produits pétroliers*	*Gaz*	*Nucléaire*	*Hydro*	*Géotherm. solaire*	*En. ren. combust.*	*Electricité*	*Chaleur*	*Total*
Indigenous Production	29	1800	-	1402	-	456	-	324	-	-	4011
Imports	157	3777	438	854	-	-	-	-	396	-	5621
Exports	-	-41	-1787	-	-	-	-	-	-57	-	-1884
Intl. Marine Bunkers	-	-	-24	-	-	-	-	-	-	-	-24
Stock Changes	72	-44	-92	-10	-	-	-	-	-	-	-74
TPES	258	5492	-1465	2246	-	456	-	324	340	-	7650
Transfers	-	-272	282	-	-	-	-	-	-	-	10
Statistical Differences	-8	-	-	-	-	-	-	-	-	-	-8
Electricity Plants	-145	-	-472	-104	-	-456	-	-	716	-	-461
CHP Plants	-6	-	-221	-310	-	-	-	-	117	256	-165
Heat Plants	-	-	-43	-32	-	-	-	-	-	63	-12
Gas Works	-	-	-15	10	-	-	-	-	-	-	-5
Petroleum Refineries	-	-5219	5178	-	-	-	-	-	-	-	-42
Coal Transformation	-	-	-	-	-	-	-	-	-	-	-
Liquefaction	-	-	-	-	-	-	-	-	-	-	-
Other Transformation	-	-	-	-	-	-	-	-	-	-	-
Own Use	-	-	-502	-224	-	-	-	-	-68	-30	-824
Distribution Losses	-	-	-	-60	-	-	-	-	-157	-44	-260
TFC	100	-	2744	1526	-	-	-	323	948	244	5885
INDUSTRY SECTOR	88	-	527	1012	-	-	-	-	259	76	1963
Iron and Steel	12	-	9	39	-	-	-	-	24	-	84
Chemical and Petrochemical	2	-	186	589	-	-	-	-	46	23	847
of which: Feedstocks	-	-	*106*	*479*	-	-	-	-	-	-	*585*
Non-Ferrous Metals	-	-	4	4	-	-	-	-	5	-	13
Non-Metallic Minerals	41	-	104	164	-	-	-	-	39	3	351
Transport Equipment	-	-	4	-	-	-	-	-	10	1	15
Machinery	-	-	9	13	-	-	-	-	15	9	47
Mining and Quarrying	-	-	13	8	-	-	-	-	4	1	26
Food and Tobacco	33	-	67	106	-	-	-	-	35	7	248
Paper, Pulp and Printing	-	-	4	48	-	-	-	-	20	16	88
Wood and Wood Products	-	-	4	7	-	-	-	-	12	-	23
Construction	-	-	94	-	-	-	-	-	20	-	114
Textile and Leather	-	-	19	26	-	-	-	-	16	17	78
Non-specified	-	-	8	8	-	-	-	-	14	-	30
TRANSPORT SECTOR	-	-	1409	-	-	-	-	-	21	-	1430
Air	-	-	88	-	-	-	-	-	-	-	88
Road	-	-	1225	-	-	-	-	-	-	-	1225
Rail	-	-	32	-	-	-	-	-	12	-	44
Pipeline Transport	-	-	-	-	-	-	-	-	1	-	1
Internal Navigation	-	-	37	-	-	-	-	-	-	-	37
Non-specified	-	-	27	-	-	-	-	-	8	-	35
OTHER SECTORS	11	-	615	514	-	-	-	323	667	168	2299
Agriculture	-	-	182	18	-	-	-	-	5	-	206
Comm. and Publ. Services	4	-	124	93	-	-	-	-	216	21	459
Residential	7	-	309	402	-	-	-	323	446	147	1634
Non-specified	-	-	-	-	-	-	-	-	-	-	-
NON-ENERGY USE	-	-	193	-	-	-	-	-	-	-	193
in Industry/Transf./Energy	-	-	164	-	-	-	-	-	-	-	164
in Transport	-	-	25	-	-	-	-	-	-	-	25
in Other Sectors	-	-	4	-	-	-	-	-	-	-	4
Electr. Generated - GWh	*511*	-	*2701*	*1172*	-	*5299*	-	*2*	-	-	*9685*
Electricity Plants	*498*	-	*2118*	*411*	-	*5299*	-	-	-	-	*8326*
CHP Plants	*13*	-	*583*	*761*	-	-	-	*2*	-	-	*1359*
Heat Generated - TJ	-	-	*5972*	*7355*	-	-	-	-	-	-	*13327*
CHP Plants	-	-	*4409*	*6292*	-	-	-	-	-	-	*10701*
Heat Plants	-	-	*1563*	*1063*	-	-	-	-	-	-	*2626*

Cuba

Thousand tonnes of oil equivalent / *Milliers de tonnes d'équivalent pétrole*

SUPPLY AND CONSUMPTION 1996	Coal	Crude Oil	Petroleum Products	Gas	Nuclear	Hydro	Geotherm. Solar etc.	Combust. Renew. & Waste	Electricity	Heat	Total
Indigenous Production	-	1451	-	16	-	8	-	5472	-	-	6947
Imports	11	1608	4644	-	-	-	-	-	-	-	6264
Exports	-	-	-	-	-	-	-	-	-	-	-
Intl. Marine Bunkers	-	-	-102	-	-	-	-	-	-	-	-102
Stock changes	-	-	143	-	-	-	-	-	-	-	143
TPES	11	3059	4685	16	-	8	-	5472	-	-	13251
Electricity and CHP Plants	-	-68	-3282	-2	-	-8	-	-686	1138	-	-2908
Petroleum Refineries	-	-2194	2132	-	-	-	-	-	-	-	-62
Other Transformation *	-	358	-352	133	-	-	-	-114	-296	-	-272
TFC	11	1155	3184	146	-	-	-	4672	842	-	10010
INDUSTRY SECTOR	11	1155	1069	24	-	-	-	4525	342	-	7126
Iron and Steel	11	-	-	-	-	-	-	-	9	-	19
Chemical and Petrochemical	-	-	-	-	-	-	-	-	27	-	27
Non-Metallic Minerals	-	-	-	-	-	-	-	-	-	-	-
Non-specified	-	1155	1069	24	-	-	-	4525	306	-	7079
TRANSPORT SECTOR	-	-	932	-	-	-	-	77	6	-	1015
Air	-	-	278	-	-	-	-	-	-	-	278
Road	-	-	654	-	-	-	-	77	-	-	731
Non-specified	-	-	-	-	-	-	-	-	6	-	6
OTHER SECTORS	-	-	870	122	-	-	-	70	494	-	1557
Agriculture	-	-	263	1	-	-	-	30	16	-	310
Comm. and Publ. Services	-	-	12	-	-	-	-	10	188	-	210
Residential	-	-	301	121	-	-	-	30	290	-	742
Non-specified	-	-	294	-	-	-	-	-	-	-	294
NON-ENERGY USE	-	-	312	-	-	-	-	-	-	-	312
Electricity Generated - GWh	-	-	11977	10	-	95	-	1154	-	-	13236
Heat Generated - TJ	-	-	-	-	-	-	-	-	-	-	-

APPROVISIONNEMENT ET DEMANDE 1997	Charbon	Pétrole brut	Produits pétroliers	Gaz	Nucléaire	Hydro	Géotherm. solaire etc.	En. ren. combust. & déchets	Electricité	Chaleur	Total
Indigenous Production	-	1685	-	18	-	8	-	5544	-	-	7255
Imports	47	1907	5023	-	-	-	-	-	-	-	6977
Exports	-	-	-	-	-	-	-	-	-	-	-
Intl. Marine Bunkers	-	-	-110	-	-	-	-	-	-	-	-110
Stock changes	-	-	151	-	-	-	-	-	-	-	151
TPES	47	3592	5064	18	-	8	-	5544	-	-	14273
Electricity and CHP Plants	-	-72	-3602	-2	-	-8	-	-702	1211	-	-3175
Petroleum Refineries	-	-2405	2213	-	-	-	-	-	-	-	-192
Other Transformation *	-18	-1	-384	133	-	-	-	-120	-218	-	-608
TFC	29	1114	3292	149	-	-	-	4721	993	-	10298
INDUSTRY SECTOR	29	1114	1109	26	-	-	-	4570	452	-	7300
Iron and Steel	29	-	-	-	-	-	-	-	12	-	41
Chemical and Petrochemical	-	-	-	-	-	-	-	-	36	-	36
Non-Metallic Minerals	-	-	-	-	-	-	-	-	-	-	-
Non-specified	-	1114	1109	26	-	-	-	4570	405	-	7224
TRANSPORT SECTOR	-	-	994	-	-	-	-	77	8	-	1079
Air	-	-	297	-	-	-	-	-	-	-	297
Road	-	-	696	-	-	-	-	77	-	-	773
Non-specified	-	-	-	-	-	-	-	-	8	-	8
OTHER SECTORS	-	-	865	122	-	-	-	74	533	-	1594
Agriculture	-	-	264	1	-	-	-	32	18	-	315
Comm. and Publ. Services	-	-	12	-	-	-	-	10	191	-	213
Residential	-	-	297	121	-	-	-	32	324	-	774
Non-specified	-	-	292	-	-	-	-	-	-	-	292
NON-ENERGY USE	-	-	324	-	-	-	-	-	-	-	324
Electricity Generated - GWh	-	-	12498	10	-	93	-	1486	-	-	14087
Heat Generated - TJ	-	-	-	-	-	-	-	-	-	-	-

* Includes Transfers, Statistical Differences, Own Use and Distribution Losses.

Cyprus / Chypre

Thousand tonnes of oil equivalent / *Milliers de tonnes d'équivalent pétrole*

SUPPLY AND CONSUMPTION 1996	Coal	Crude Oil	Petroleum Products	Gas	Nuclear	Hydro	Geotherm. Solar etc.	Combust. Renew. & Waste	Electricity	Heat	Total
Indigenous Production	-	-	-	-	-	-	-	14	-	-	14
Imports	10	816	1396	-	-	-	-	2	-	-	2224
Exports	-	-	-21	-	-	-	-	-	-	-	-21
Intl. Marine Bunkers	-	-	-89	-	-	-	-	-	-	-	-89
Stock changes	-	-45	39	-	-	-	-	-	-	-	-6
TPES	10	771	1325	-	-	-	-	16	-	-	2122
Electricity and CHP Plants	-	-	-681	-	-	-	-	-	223	-	-458
Petroleum Refineries	-	-771	766	-	-	-	-	-	-	-	-5
Other Transformation *	-	-	-112	-	-	-	-	-8	-25	-	-145
TFC	10	-	1298	-	-	-	-	8	198	-	1514
INDUSTRY SECTOR	10	-	382	-	-	-	-	-	35	-	427
Iron and Steel	-	-	-	-	-	-	-	-	-	-	-
Chemical and Petrochemical	-	-	-	-	-	-	-	-	4	-	4
Non-Metallic Minerals	10	-	215	-	-	-	-	-	14	-	240
Non-specified	-	-	167	-	-	-	-	-	17	-	183
TRANSPORT SECTOR	-	-	773	-	-	-	-	-	3	-	776
Air	-	-	265	-	-	-	-	-	-	-	265
Road	-	-	507	-	-	-	-	-	-	-	507
Non-specified	-	-	-	-	-	-	-	-	3	-	3
OTHER SECTORS	-	-	76	-	-	-	-	8	160	-	244
Agriculture	-	-	-	-	-	-	-	-	7	-	7
Comm. and Publ. Services	-	-	-	-	-	-	-	-	70	-	70
Residential	-	-	76	-	-	-	-	6	71	-	154
Non-specified	-	-	-	-	-	-	-	2	12	-	14
NON-ENERGY USE	-	-	67	-	-	-	-	-	-	-	67
Electricity Generated - GWh	-	-	*2592*	-	-	-	-	-	-	-	*2592*
Heat Generated - TJ	-	-	-	-	-	-	-	-	-	-	-

APPROVISIONNEMENT ET DEMANDE 1997	Charbon	Pétrole brut	Produits pétroliers	Gaz	Nucléaire	Hydro	Géotherm. solaire etc.	En. ren. combust. & déchets	Electricité	Chaleur	Total
Indigenous Production	-	-	-	-	-	-	-	14	-	-	14
Imports	18	1054	1087	-	-	-	-	3	-	-	2161
Exports	-	-	-	-	-	-	-	-	-	-	-
Intl. Marine Bunkers	-	-	-97	-	-	-	-	-	-	-	-97
Stock changes	-4	4	-4	-	-	-	-	-	-	-	-4
TPES	13	1058	986	-	-	-	-	17	-	-	2074
Electricity and CHP Plants	-	-	-719	-	-	-	-	-	233	-	-486
Petroleum Refineries	-	-1058	1050	-	-	-	-	-	-	-	-8
Other Transformation *	-	-	-20	-	-	-	-	-10	-28	-	-58
TFC	13	-	1297	-	-	-	-	7	205	-	1522
INDUSTRY SECTOR	13	-	355	-	-	-	-	-	34	-	402
Iron and Steel	-	-	-	-	-	-	-	-	-	-	-
Chemical and Petrochemical	-	-	-	-	-	-	-	-	1	-	1
Non-Metallic Minerals	12	-	180	-	-	-	-	-	14	-	206
Non-specified	2	-	175	-	-	-	-	-	19	-	195
TRANSPORT SECTOR	-	-	790	-	-	-	-	-	2	-	792
Air	-	-	261	-	-	-	-	-	-	-	261
Road	-	-	529	-	-	-	-	-	-	-	529
Non-specified	-	-	-	-	-	-	-	-	2	-	2
OTHER SECTORS	-	-	80	-	-	-	-	7	169	-	256
Agriculture	-	-	-	-	-	-	-	-	7	-	7
Comm. and Publ. Services	-	-	-	-	-	-	-	-	86	-	86
Residential	-	-	80	-	-	-	-	5	72	-	157
Non-specified	-	-	-	-	-	-	-	1	5	-	6
NON-ENERGY USE	-	-	72	-	-	-	-	-	-	-	72
Electricity Generated - GWh	-	-	*2711*	-	-	-	-	-	-	-	*2711*
Heat Generated - TJ	-	-	-	-	-	-	-	-	-	-	-

* Includes Transfers, Statistical Differences, Own Use and Distribution Losses.

INTERNATIONAL ENERGY AGENCY

Dominican Republic / République dominicaine

Thousand tonnes of oil equivalent / *Milliers de tonnes d'équivalent pétrole*

SUPPLY AND CONSUMPTION 1996	Coal	Crude Oil	Petroleum Products	Gas	Nuclear	Hydro	Geotherm. Solar etc.	Combust. Renew. & Waste	Electricity	Heat	Total
Indigenous Production	-	-	-	-	-	109	-	1303	-	-	1411
Imports	79	2125	1558	-	-	-	-	-	-	-	3762
Exports	-	-	-	-	-	-	-	-	-	-	-
Intl. Marine Bunkers	-	-	-	-	-	-	-	-	-	-	-
Stock changes	-	-	10	-	-	-	-	-	-	-	10
TPES	79	2125	1568	-	-	109	-	1303	-	-	5183
Electricity and CHP Plants	-79	-	-1540	-	-	-109	-	-20	589	-	-1158
Petroleum Refineries	-	-2125	2121	-	-	-	-	-	-	-	-3
Other Transformation *	-	-	-3	-	-	-	-	-202	-172	-	-377
TFC	-	-	2147	-	-	-	-	1081	417	-	3644
INDUSTRY SECTOR	-	-	352	-	-	-	-	288	107	-	746
Iron and Steel	-	-	-	-	-	-	-	-	-	-	-
Chemical and Petrochemical	-	-	-	-	-	-	-	-	-	-	-
Non-Metallic Minerals	-	-	-	-	-	-	-	-	-	-	-
Non-specified	-	-	352	-	-	-	-	288	107	-	746
TRANSPORT SECTOR	-	-	1097	-	-	-	-	-	-	-	1097
Air	-	-	63	-	-	-	-	-	-	-	63
Road	-	-	528	-	-	-	-	-	-	-	528
Non-specified	-	-	506	-	-	-	-	-	-	-	506
OTHER SECTORS	-	-	606	-	-	-	-	793	310	-	1708
Agriculture	-	-	36	-	-	-	-	-	-	-	36
Comm. and Publ. Services	-	-	-	-	-	-	-	-	-	-	-
Residential	-	-	569	-	-	-	-	793	310	-	1672
Non-specified	-	-	-	-	-	-	-	-	-	-	-
NON-ENERGY USE	-	-	93	-	-	-	-	-	-	-	93
Electricity Generated - GWh	296	-	5256	-	-	1265	-	30	-	-	6847
Heat Generated - TJ	-	-	-	-	-	-	-	-	-	-	-

APPROVISIONNEMENT ET DEMANDE 1997	Charbon	Pétrole brut	Produits pétroliers	Gaz	Nucléaire	Hydro	Géotherm. solaire etc.	En. ren. combust. & déchets	Electricité	Chaleur	Total
Indigenous Production	-	-	-	-	-	115	-	1308	-	-	1423
Imports	85	2230	1706	-	-	-	-	-	-	-	4021
Exports	-	-	-	-	-	-	-	-	-	-	-
Intl. Marine Bunkers	-	-	-	-	-	-	-	-	-	-	-
Stock changes	-	-1	10	-	-	-	-	-	-	-	9
TPES	85	2228	1716	-	-	115	-	1308	-	-	5453
Electricity and CHP Plants	-85	-	-1623	-	-	-115	-	-20	631	-	-1213
Petroleum Refineries	-	-2228	2215	-	-	-	-	-	-	-	-13
Other Transformation *	-	-	2	-	-	-	-	-190	-199	-	-387
TFC	-	-	2310	-	-	-	-	1098	432	-	3840
INDUSTRY SECTOR	-	-	376	-	-	-	-	289	120	-	786
Iron and Steel	-	-	-	-	-	-	-	-	-	-	-
Chemical and Petrochemical	-	-	-	-	-	-	-	-	-	-	-
Non-Metallic Minerals	-	-	-	-	-	-	-	-	-	-	-
Non-specified	-	-	376	-	-	-	-	289	120	-	786
TRANSPORT SECTOR	-	-	1146	-	-	-	-	-	-	-	1146
Air	-	-	66	-	-	-	-	-	-	-	66
Road	-	-	538	-	-	-	-	-	-	-	538
Non-specified	-	-	542	-	-	-	-	-	-	-	542
OTHER SECTORS	-	-	693	-	-	-	-	809	312	-	1814
Agriculture	-	-	37	-	-	-	-	-	-	-	37
Comm. and Publ. Services	-	-	-	-	-	-	-	-	-	-	-
Residential	-	-	656	-	-	-	-	809	312	-	1777
Non-specified	-	-	-	-	-	-	-	-	-	-	-
NON-ENERGY USE	-	-	94	-	-	-	-	-	-	-	94
Electricity Generated - GWh	321	-	5646	-	-	1338	-	30	-	-	7335
Heat Generated - TJ	-	-	-	-	-	-	-	-	-	-	-

* Includes Transfers, Statistical Differences, Own Use and Distribution Losses.
Note: Electricity generation from Petroleum products includes small amounts produced from Coal and Combustible Renewables and Waste

Ecuador / Equateur

Total Production of Energy

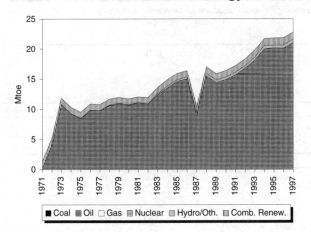

Total Primary Energy Supply*

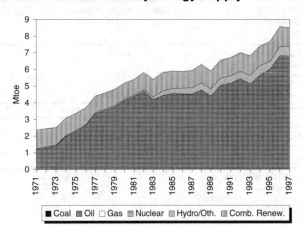

Electricity Generation by Fuel

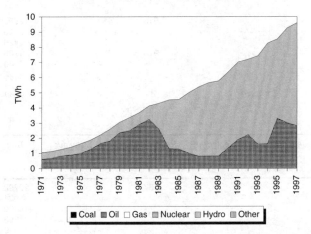

Oil Products Consumption

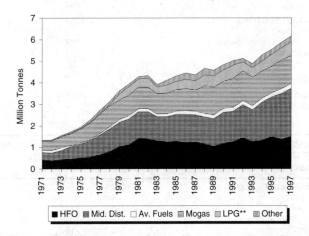

Electricity Consumption/GDP***, TPES/GDP*** and Production/TPES

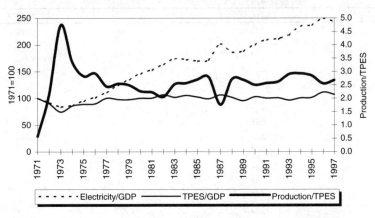

* Excluding electricity trade.
** Includes LPG, NGL, ethane and naphtha.
*** GDP in 1990 US dollars.

Ecuador / Equateur

Thousand tonnes of oil equivalent / *Milliers de tonnes d'équivalent pétrole*

SUPPLY AND CONSUMPTION 1996	Coal	Crude Oil	Petroleum Products	Gas	Nuclear	Hydro	Geotherm. Solar etc.	Combust. Renew. & Waste	Electricity	Heat	Total
Indigenous Production	-	20100	-	-	-	537	-	1216	-	-	21853
Imports	-	-	746	-	-	-	-	-	-	-	746
Exports	-	-12054	-1355	-	-	-	-	-	-	-	-13409
Intl. Marine Bunkers	-	-	-276	-	-	-	-	-	-	-	-276
Stock changes	-	-205	-145	-	-	-	-	-	-	-	-350
TPES	-	7841	-1029	-	-	537	-	1216	-	-	8566
Electricity and CHP Plants	-	-	-769	-	-	-537	-	-	796	-	-510
Petroleum Refineries	-	-7669	7132	-	-	-	-	-	-	-	-538
Other Transformation *	-	-172	-461	-	-	-	-	-	-192	-	-825
TFC	-	-	4873	-	-	-	-	1216	604	-	6693
INDUSTRY SECTOR	-	-	682	-	-	-	-	303	195	-	1179
Iron and Steel	-	-	-	-	-	-	-	-	-	-	-
Chemical and Petrochemical	-	-	-	-	-	-	-	-	-	-	-
Non-Metallic Minerals	-	-	-	-	-	-	-	-	-	-	-
Non-specified	-	-	682	-	-	-	-	303	195	-	1179
TRANSPORT SECTOR	-	-	2805	-	-	-	-	-	-	-	2805
Air	-	-	229	-	-	-	-	-	-	-	229
Road	-	-	2079	-	-	-	-	-	-	-	2079
Non-specified	-	-	497	-	-	-	-	-	-	-	497
OTHER SECTORS	-	-	1259	-	-	-	-	913	410	-	2582
Agriculture	-	-	348	-	-	-	-	-	-	-	348
Comm. and Publ. Services	-	-	249	-	-	-	-	-	177	-	426
Residential	-	-	626	-	-	-	-	913	228	-	1768
Non-specified	-	-	35	-	-	-	-	-	4	-	40
NON-ENERGY USE	-	-	127	-	-	-	-	-	-	-	127
Electricity Generated - GWh	-	-	3011	-	-	6249	-	-	-	-	9260
Heat Generated - TJ	-	-	-	-	-	-	-	-	-	-	-

APPROVISIONNEMENT ET DEMANDE 1997	Charbon	Pétrole brut	Produits pétroliers	Gaz	Nucléaire	Hydro	Géotherm. solaire etc.	En. ren. combust. & déchets	Electricité	Chaleur	Total
Indigenous Production	-	21072	-	-	-	583	-	1137	-	-	22792
Imports	-	-	924	-	-	-	-	-	-	-	924
Exports	-	-13376	-1514	-	-	-	-	-	-	-	-14890
Intl. Marine Bunkers	-	-	-303	-	-	-	-	-	-	-	-303
Stock changes	-	-25	16	-	-	-	-	-	-	-	-9
TPES	-	7671	-877	-	-	583	-	1137	-	-	8513
Electricity and CHP Plants	-	-	-673	-	-	-583	-	-	825	-	-431
Petroleum Refineries	-	-7766	7239	-	-	-	-	-	-	-	-526
Other Transformation *	-	94	-474	-	-	-	-	-	-198	-	-578
TFC	-	-	5215	-	-	-	-	1137	627	-	6979
INDUSTRY SECTOR	-	-	771	-	-	-	-	319	203	-	1292
Iron and Steel	-	-	-	-	-	-	-	-	-	-	-
Chemical and Petrochemical	-	-	-	-	-	-	-	-	-	-	-
Non-Metallic Minerals	-	-	-	-	-	-	-	-	-	-	-
Non-specified	-	-	771	-	-	-	-	319	203	-	1292
TRANSPORT SECTOR	-	-	2958	-	-	-	-	-	-	-	2958
Air	-	-	233	-	-	-	-	-	-	-	233
Road	-	-	2137	-	-	-	-	-	-	-	2137
Non-specified	-	-	588	-	-	-	-	-	-	-	588
OTHER SECTORS	-	-	1367	-	-	-	-	818	424	-	2610
Agriculture	-	-	376	-	-	-	-	-	-	-	376
Comm. and Publ. Services	-	-	273	-	-	-	-	-	185	-	458
Residential	-	-	681	-	-	-	-	818	235	-	1734
Non-specified	-	-	38	-	-	-	-	-	4	-	42
NON-ENERGY USE	-	-	118	-	-	-	-	-	-	-	118
Electricity Generated - GWh	-	-	2820	-	-	6775	-	-	-	-	9595
Heat Generated - TJ	-	-	-	-	-	-	-	-	-	-	-

* Includes Transfers, Statistical Differences, Own Use and Distribution Losses.

Egypt / Egypte

Total Production of Energy

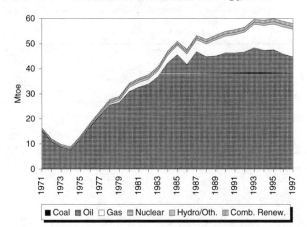

Mtoe

■ Coal ▦ Oil □ Gas ▤ Nuclear ▨ Hydro/Oth. ▥ Comb. Renew.

Total Primary Energy Supply*

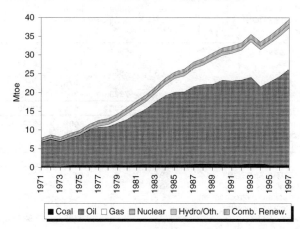

Mtoe

■ Coal ▦ Oil □ Gas ▤ Nuclear ▨ Hydro/Oth. ▥ Comb. Renew.

Electricity Generation by Fuel

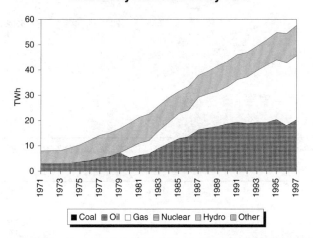

TWh

■ Coal ▦ Oil □ Gas ▤ Nuclear ▨ Hydro ▥ Other

Oil Products Consumption

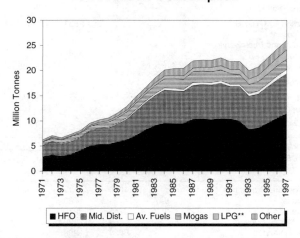

Million Tonnes

■ HFO ▦ Mid. Dist. □ Av. Fuels ▤ Mogas ▨ LPG** ▥ Other

Electricity Consumption/GDP***,
TPES/GDP*** and Production/TPES

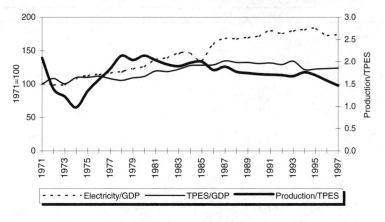

1971=100

Production/TPES

- - - - Electricity/GDP ——— TPES/GDP ▬▬ Production/TPES

* Excluding electricity trade.
** Includes LPG, NGL, ethane and naphtha.
*** GDP in 1990 US dollars.

INTERNATIONAL ENERGY AGENCY

Egypt / Egypte

Thousand tonnes of oil equivalent / *Milliers de tonnes d'équivalent pétrole*

SUPPLY AND CONSUMPTION 1996	Coal	Crude Oil	Petroleum Products	Gas	Nuclear	Hydro	Geotherm. Solar etc.	Combust. Renew. & Waste	Electricity	Heat	Total
Indigenous Production	-	45801	-	10820	-	994	-	1232	-	-	58847
Imports	947	-	1186	-	-	-	-	1	-	-	2134
Exports	-303	-6836	-5380	-	-	-	-	-18	-	-	-12537
Intl. Marine Bunkers	-	-	-2975	-	-	-	-	-	-	-	-2975
Stock changes	-	-7775	-345	-	-	-	-	-	-	-	-8121
TPES	644	31190	-7514	10820	-	994	-	1215	-	-	37349
Electricity and CHP Plants	-	-	-5005	-6790	-	-994	-	-	4682	-	-8107
Petroleum Refineries	-	-30200	27795	-	-	-	-	-	-	-	-2404
Other Transformation *	-514	-991	200	-875	-	-	-	-	-745	-	-2924
TFC	131	-	15477	3155	-	-	-	1215	3937	-	23914
INDUSTRY SECTOR	131	-	6438	2279	-	-	-	620	1856	-	11323
Iron and Steel	131	-	-	-	-	-	-	-	-	-	131
Chemical and Petrochemical	-	-	-	1454	-	-	-	-	-	-	1454
Non-Metallic Minerals	-	-	-	-	-	-	-	-	-	-	-
Non-specified	-	-	6438	825	-	-	-	620	1856	-	9738
TRANSPORT SECTOR	-	-	4956	-	-	-	-	-	-	-	4956
Air	-	-	866	-	-	-	-	-	-	-	866
Road	-	-	4090	-	-	-	-	-	-	-	4090
Non-specified	-	-	-	-	-	-	-	-	-	-	-
OTHER SECTORS	-	-	2831	876	-	-	-	595	2081	-	6382
Agriculture	-	-	-	-	-	-	-	-	141	-	141
Comm. and Publ. Services	-	-	-	-	-	-	-	-	-	-	-
Residential	-	-	2831	687	-	-	-	595	1437	-	5549
Non-specified	-	-	-	189	-	-	-	-	503	-	692
NON-ENERGY USE	-	-	1252	-	-	-	-	-	-	-	1252
Electricity Generated - GWh	-	-	*18009*	*24880*	-	*11555*	-	-	-	-	*54444*
Heat Generated - TJ	-	-	-	-	-	-	-	-	-	-	-

APPROVISIONNEMENT ET DEMANDE 1997	Charbon	Pétrole brut	Produits pétroliers	Gaz	Nucléaire	Hydro	Géotherm. solaire etc.	En. ren. combust. & déchets	Electricité	Chaleur	Total
Indigenous Production	-	44640	-	11070	-	1031	-	1256	-	-	57997
Imports	947	-	1729	-	-	-	-	1	-	-	2677
Exports	-303	-6241	-5276	-	-	-	-	-19	-	-	-11838
Intl. Marine Bunkers	-	-	-2929	-	-	-	-	-	-	-	-2929
Stock changes	-	-6624	298	-	-	-	-	-	-	-	-6326
TPES	644	31774	-6177	11070	-	1031	-	1238	-	-	39581
Electricity and CHP Plants	-	-	-5542	-6919	-	-1031	-	-	4958	-	-8534
Petroleum Refineries	-	-30777	28116	-	-	-	-	-	-	-	-2661
Other Transformation *	-514	-998	132	-831	-	-	-	-	-789	-	-3000
TFC	131	-	16529	3320	-	-	-	1238	4170	-	25386
INDUSTRY SECTOR	131	-	7014	2396	-	-	-	632	1965	-	12137
Iron and Steel	131	-	-	-	-	-	-	-	-	-	131
Chemical and Petrochemical	-	-	-	1365	-	-	-	-	-	-	1365
Non-Metallic Minerals	-	-	-	-	-	-	-	-	-	-	-
Non-specified	-	-	7014	1031	-	-	-	632	1965	-	10641
TRANSPORT SECTOR	-	-	5221	-	-	-	-	-	-	-	5221
Air	-	-	920	-	-	-	-	-	-	-	920
Road	-	-	4301	-	-	-	-	-	-	-	4301
Non-specified	-	-	-	-	-	-	-	-	-	-	-
OTHER SECTORS	-	-	2975	924	-	-	-	606	2204	-	6709
Agriculture	-	-	-	-	-	-	-	-	149	-	149
Comm. and Publ. Services	-	-	-	-	-	-	-	-	-	-	-
Residential	-	-	2975	704	-	-	-	606	1522	-	5806
Non-specified	-	-	-	220	-	-	-	-	533	-	754
NON-ENERGY USE	-	-	1319	-	-	-	-	-	-	-	1319
Electricity Generated - GWh	-	-	*20317*	*25352*	-	*11987*	-	-	-	-	*57656*
Heat Generated - TJ	-	-	-	-	-	-	-	-	-	-	-

* Includes Transfers, Statistical Differences, Own Use and Distribution Losses.

El Salvador

Thousand tonnes of oil equivalent / *Milliers de tonnes d'équivalent pétrole*

SUPPLY AND CONSUMPTION 1996	Coal	Crude Oil	Petroleum Products	Gas	Nuclear	Hydro	Geotherm. Solar etc.	Combust. Renew. & Waste	Electricity	Heat	Total
Indigenous Production	-	-	-	-	-	140	585	1862	-	-	2587
Imports	1	791	801	-	-	-	-	-	2	-	1594
Exports	-	-	-132	-	-	-	-	-	-	-	-132
Intl. Marine Bunkers	-	-	-	-	-	-	-	-	-	-	-
Stock changes	-	-	6	-	-	-	-	-	-	-	6
TPES	1	791	675	-	-	140	585	1862	2	-	4055
Electricity and CHP Plants	-	-	-268	-	-	-140	-585	-79	294	-	-778
Petroleum Refineries	-	-799	763	-	-	-	-	-	-	-	-36
Other Transformation *	-1	8	-15	-	-	-	-	-53	-44	-	-104
TFC	-	-	1155	-	-	-	-	1730	252	-	3137
INDUSTRY SECTOR	-	-	276	-	-	-	-	295	72	-	643
Iron and Steel	-	-	-	-	-	-	-	-	-	-	-
Chemical and Petrochemical	-	-	-	-	-	-	-	-	-	-	-
Non-Metallic Minerals	-	-	-	-	-	-	-	-	-	-	-
Non-specified	-	-	276	-	-	-	-	295	72	-	643
TRANSPORT SECTOR	-	-	734	-	-	-	-	-	-	-	734
Air	-	-	51	-	-	-	-	-	-	-	51
Road	-	-	683	-	-	-	-	-	-	-	683
Non-specified	-	-	-	-	-	-	-	-	-	-	-
OTHER SECTORS	-	-	103	-	-	-	-	1435	179	-	1717
Agriculture	-	-	-	-	-	-	-	-	-	-	-
Comm. and Publ. Services	-	-	-	-	-	-	-	-	88	-	88
Residential	-	-	103	-	-	-	-	1435	91	-	1629
Non-specified	-	-	-	-	-	-	-	-	-	-	-
NON-ENERGY USE	-	-	43	-	-	-	-	-	-	-	43
Electricity Generated - GWh	-	-	*1085*	-	-	*1633*	*680*	*20*	-	-	*3418*
Heat Generated - TJ	-	-	-	-	-	-	-	-	-	-	-

APPROVISIONNEMENT ET DEMANDE 1997	Charbon	Pétrole brut	Produits pétroliers	Gaz	Nucléaire	Hydro	Géotherm. solaire etc.	En. ren. combust. & déchets	Electricité	Chaleur	Total
Indigenous Production	-	-	-	-	-	103	620	1926	-	-	2649
Imports	1	810	632	-	-	-	-	-	8	-	1450
Exports	-	-	-65	-	-	-	-	-	-	-	-65
Intl. Marine Bunkers	-	-	-	-	-	-	-	-	-	-	-
Stock changes	-	13	48	-	-	-	-	-	-	-	61
TPES	1	823	615	-	-	103	620	1926	8	-	4095
Electricity and CHP Plants	-	-	-277	-	-	-103	-620	-84	313	-	-770
Petroleum Refineries	-	-935	910	-	-	-	-	-	-	-	-26
Other Transformation *	-1	113	-36	-	-	-	-	-56	-47	-	-27
TFC	-	-	1212	-	-	-	-	1786	274	-	3271
INDUSTRY SECTOR	-	-	286	-	-	-	-	308	78	-	672
Iron and Steel	-	-	-	-	-	-	-	-	-	-	-
Chemical and Petrochemical	-	-	-	-	-	-	-	-	-	-	-
Non-Metallic Minerals	-	-	-	-	-	-	-	-	-	-	-
Non-specified	-	-	286	-	-	-	-	308	78	-	672
TRANSPORT SECTOR	-	-	784	-	-	-	-	-	-	-	784
Air	-	-	51	-	-	-	-	-	-	-	51
Road	-	-	733	-	-	-	-	-	-	-	733
Non-specified	-	-	-	-	-	-	-	-	-	-	-
OTHER SECTORS	-	-	111	-	-	-	-	1478	196	-	1785
Agriculture	-	-	-	-	-	-	-	-	-	-	-
Comm. and Publ. Services	-	-	-	-	-	-	-	-	97	-	97
Residential	-	-	111	-	-	-	-	1478	99	-	1687
Non-specified	-	-	-	-	-	-	-	-	-	-	-
NON-ENERGY USE	-	-	31	-	-	-	-	-	-	-	31
Electricity Generated - GWh	-	-	*1706*	-	-	*1194*	*721*	*22*	-	-	*3643*
Heat Generated - TJ	-	-	-	-	-	-	-	-	-	-	-

* Includes Transfers, Statistical Differences, Own Use and Distribution Losses.
Note: Electricity generation from Petroleum products includes small amounts produced from Combustible Renewables and Waste

Estonia / Estonie : 1996

	Thousand tonnes of oil equivalent / Milliers de tonnes d'équivalent pétrole										
SUPPLY AND CONSUMPTION *APPROVISIONNEMENT ET DEMANDE*	Coal *Charbon*	Crude Oil *Pétrole brut*	Petroleum Products *Produits pétroliers*	Gas *Gaz*	Nuclear *Nucléaire*	Hydro *Hydro*	Geotherm. Solar etc. *Géotherm. solaire*	Combust. Renew. & Waste *En. ren. combust.*	Electricity *Electricité*	Heat *Chaleur*	Total *Total*
Indigenous Production	2912	349	-	-	-	-	-	591	-	-	3852
Imports	229	-	1384	636	-	-	-	-	21	-	2269
Exports	-36	-	-322	-	-	-	-	-9	-95	-	-462
Intl. Marine Bunkers	-	-	-92	-	-	-	-	-	-	-	-92
Stock Changes	104	-3	-23	-	-	-	-	1	-	-	79
TPES	**3209**	**346**	**946**	**636**	**-**	**-**	**-**	**583**	**-74**	**-**	**5646**
Transfers	-	-	1	-	-	-	-	-	-	-	1
Statistical Differences	-	-	-	-	-	-	-	-	-	-	-
Electricity Plants	-	-	-	-	-	-	-	-	-	-	-
CHP Plants	-2894	-	-84	-104	-	-	-	-1	783	308	-1993
Heat Plants	-36	-	-205	-254	-	-	-	-63	-12	483	-87
Gas Works	-	-	-	110	-	-	-	-	-	-	110
Petroleum Refineries	-	-346	337	-	-	-	-	-	-	-	-9
Coal Transformation	-43	-	-	-	-	-	-	-	-	-	-43
Liquefaction	-	-	-	-	-	-	-	-	-	-	-
Other Transformation	-	-	-	-	-	-	-	-	-	-	-
Own Use	-17	-	-11	-	-	-	-	-	-135	-	-163
Distribution Losses	-17	-	-4	-3	-	-	-	-	-147	-166	-338
TFC	**202**	**-**	**979**	**384**	**-**	**-**	**-**	**519**	**415**	**625**	**3125**
INDUSTRY SECTOR	**133**	**-**	**249**	**338**	**-**	**-**	**-**	**96**	**164**	**51**	**1031**
Iron and Steel	2	-	-	-	-	-	-	-	4	-	6
Chemical and Petrochemical	-	-	51	258	-	-	-	-	47	16	372
of which: Feedstocks	-	-	-	165	-	-	-	-	-	-	165
Non-Ferrous Metals	-	-	-	-	-	-	-	-	-	-	-
Non-Metallic Minerals	119	-	10	20	-	-	-	1	16	1	166
Transport Equipment	1	-	3	-	-	-	-	-	7	1	12
Machinery	1	-	7	2	-	-	-	1	13	1	26
Mining and Quarrying	-	-	-	-	-	-	-	-	2	-	2
Food and Tobacco	3	-	103	27	-	-	-	4	21	13	172
Paper, Pulp and Printing	-	-	9	18	-	-	-	23	5	4	59
Wood and Wood Products	-	-	25	-	-	-	-	19	12	4	60
Construction	4	-	17	-	-	-	-	1	10	1	34
Textile and Leather	2	-	9	4	-	-	-	1	17	6	37
Non-specified	1	-	15	7	-	-	-	45	11	5	85
TRANSPORT SECTOR	**1**	**-**	**538**	**1**	**-**	**-**	**-**	**-**	**9**	**-**	**549**
Air	-	-	17	-	-	-	-	-	-	-	17
Road	-	-	471	1	-	-	-	-	-	-	472
Rail	1	-	42	-	-	-	-	-	2	-	45
Pipeline Transport	-	-	-	-	-	-	-	-	-	-	-
Internal Navigation	-	-	7	-	-	-	-	-	-	-	7
Non-specified	-	-	-	-	-	-	-	-	7	-	7
OTHER SECTORS	**46**	**-**	**149**	**46**	**-**	**-**	**-**	**423**	**242**	**574**	**1479**
Agriculture	3	-	62	3	-	-	-	2	29	-	99
Comm. and Publ. Services	-	-	35	6	-	-	-	2	107	74	224
Residential	43	-	52	36	-	-	-	419	106	500	1156
Non-specified	-	-	-	-	-	-	-	-	-	-	-
NON-ENERGY USE	**22**	**-**	**44**	**-**	**-**	**-**	**-**	**-**	**-**	**-**	**66**
in Industry/Transf./Energy	22	-	37	-	-	-	-	-	-	-	59
in Transport	-	-	5	-	-	-	-	-	-	-	5
in Other Sectors	-	-	2	-	-	-	-	-	-	-	2
Electr. Generated - GWh	*8798*	*-*	*100*	*198*	*-*	*2*	*-*	*5*	*-*	*-*	*9103*
Electricity Plants	*-*	*-*	*1*	*-*	*-*	*2*	*-*	*-*	*-*	*-*	*3*
CHP Plants	*8798*	*-*	*99*	*198*	*-*	*-*	*-*	*5*	*-*	*-*	*9100*
Heat Generated - TJ	*9492*	*-*	*8910*	*12626*	*-*	*-*	*-*	*1867*	*237*	*-*	*33132*
CHP Plants	*8104*	*-*	*2094*	*2708*	*-*	*-*	*-*	*-*	*-*	*-*	*12906*
Heat Plants	*1388*	*-*	*6816*	*9918*	*-*	*-*	*-*	*1867*	*237*	*-*	*20226*

Estonia / Estonie : 1997

Thousand tonnes of oil equivalent / *Milliers de tonnes d'équivalent pétrole*											
SUPPLY AND CONSUMPTION ***APPROVISIONNEMENT ET DEMANDE***	Coal *Charbon*	Crude Oil *Pétrole brut*	Petroleum Products *Produits pétroliers*	Gas *Gaz*	Nuclear *Nucléaire*	Hydro *Hydro*	Geotherm. Solar etc. *Géotherm. solaire*	Combust. Renew. & Waste *En. ren. combust.*	Electricity *Electricité*	Heat *Chaleur*	Total *Total*
Indigenous Production	2795	367	-	-	-	-	-	625	-	-	3788
Imports	368	-	1688	618	-	-	-	-	18	-	2691
Exports	-35	-	-751	-	-	-	-	-37	-102	-	-925
Intl. Marine Bunkers	-	-	-100	-	-	-	-	-	-	-	-100
Stock Changes	40	5	53	-	-	-	-	4	-	-	102
TPES	**3168**	**372**	**890**	**618**	**-**	**-**	**-**	**592**	**-84**	**-**	**5556**
Transfers	-	-	1	-	-	-	-	-	-	-	1
Statistical Differences	-	-	-	-	-	-	-	-	-	-	-
Electricity Plants	-	-	-	-	-	-	-	-	-	-	-
CHP Plants	-2885	-	-94	-104	-	-	-	-2	792	300	-1992
Heat Plants	-53	-	-202	-246	-	-	-	-76	-9	479	-107
Gas Works	-	-	-	114	-	-	-	-	-	-	114
Petroleum Refineries	-	-372	365	-	-	-	-	-	-	-	-7
Coal Transformation	-36	-	-	-	-	-	-	-	-	-	-36
Liquefaction	-	-	-	-	-	-	-	-	-	-	-
Other Transformation	-	-	-	-	-	-	-	-	-	-	-
Own Use	-15	-	-14	-	-	-	-	-	-136	-	-165
Distribution Losses	-33	-	-1	-4	-	-	-	-	-130	-133	-301
TFC	**146**	**-**	**945**	**379**	**-**	**-**	**-**	**513**	**435**	**645**	**3063**
INDUSTRY SECTOR	**84**	**-**	**199**	**333**	**-**	**-**	**-**	**61**	**182**	**64**	**922**
Iron and Steel	-	-	-	-	-	-	-	-	1	-	1
Chemical and Petrochemical	2	-	40	255	-	-	-	-	45	17	359
of which: Feedstocks	-	-	-	162	-	-	-	-	-	-	162
Non-Ferrous Metals	-	-	-	-	-	-	-	-	-	-	-
Non-Metallic Minerals	77	-	4	20	-	-	-	1	22	1	124
Transport Equipment	-	-	1	-	-	-	-	-	5	1	8
Machinery	1	-	2	2	-	-	-	1	13	2	20
Mining and Quarrying	-	-	-	-	-	-	-	-	11	-	11
Food and Tobacco	2	-	91	27	-	-	-	3	21	16	160
Paper, Pulp and Printing	-	-	6	18	-	-	-	9	7	7	46
Wood and Wood Products	1	-	17	-	-	-	-	15	11	5	50
Construction	1	-	16	-	-	-	-	1	9	1	28
Textile and Leather	1	-	6	4	-	-	-	-	18	7	35
Non-specified	-	-	16	7	-	-	-	31	19	7	80
TRANSPORT SECTOR	**1**	**-**	**561**	**1**	**-**	**-**	**-**	**-**	**9**	**-**	**572**
Air	-	-	24	-	-	-	-	-	-	-	24
Road	-	-	494	1	-	-	-	-	-	-	495
Rail	1	-	37	-	-	-	-	-	1	-	39
Pipeline Transport	-	-	-	-	-	-	-	-	-	-	-
Internal Navigation	-	-	6	-	-	-	-	-	-	-	6
Non-specified	-	-	-	-	-	-	-	-	8	-	8
OTHER SECTORS	**41**	**-**	**146**	**45**	**-**	**-**	**-**	**453**	**243**	**581**	**1508**
Agriculture	2	-	49	3	-	-	-	6	20	-	80
Comm. and Publ. Services	5	-	31	6	-	-	-	10	120	76	248
Residential	33	-	66	36	-	-	-	437	104	505	1181
Non-specified	-	-	-	-	-	-	-	-	-	-	-
NON-ENERGY USE	**21**	**-**	**40**	**-**	**-**	**-**	**-**	**-**	**-**	**-**	**61**
in Industry/Transf./Energy	21	-	35	-	-	-	-	-	-	-	56
in Transport	-	-	3	-	-	-	-	-	-	-	3
in Other Sectors	-	-	2	-	-	-	-	-	-	-	2
Electr. Generated - GWh	***8788***	***-***	***178***	***241***	***-***	***3***	***-***	***8***	***-***	***-***	***9218***
Electricity Plants	-	-	*1*	-	-	*3*	-	-	-	-	*4*
CHP Plants	*8788*	-	*177*	*241*	-	-	-	*8*	-	-	*9214*
Heat Generated - TJ	***9858***	***-***	***8507***	***11673***	***-***	***-***	***-***	***2365***	***190***	***-***	***32593***
CHP Plants	*7875*	-	*2255*	*2413*	-	-	-	-	-	-	*12543*
Heat Plants	*1983*	-	*6252*	*9260*	-	-	-	*2365*	*190*	-	*20050*

Ethiopia / Ethiopie

Thousand tonnes of oil equivalent / *Milliers de tonnes d'équivalent pétrole*

SUPPLY AND CONSUMPTION 1996	Coal	Crude Oil	Petroleum Products	Gas	Nuclear	Hydro	Geotherm. Solar etc.	Combust. Renew. & Waste	Electricity	Heat	Total
Indigenous Production	-	-	-	-	-	96	42	15763	-	-	15902
Imports	-	772	709	-	-	-	-	-	-	-	1481
Exports	-	-	-67	-	-	-	-	-	-	-	-67
Intl. Marine Bunkers	-	-	-147	-	-	-	-	-	-	-	-147
Stock changes	-	-1	-	-	-	-	-	-	-	-	-1
TPES	-	771	495	-	-	96	42	15763	-	-	17167
Electricity and CHP Plants	-	-	-76	-	-	-96	-42	-	113	-	-101
Petroleum Refineries	-	-863	773	-	-	-	-	-	-	-	-90
Other Transformation *	-	93	-34	-	-	-	-	-529	-5	-	-475
TFC	-	-	1158	-	-	-	-	15234	108	-	16500
INDUSTRY SECTOR	-	-	207	-	-	-	-	-	70	-	277
Iron and Steel	-	-	-	-	-	-	-	-	-	-	-
Chemical and Petrochemical	-	-	-	-	-	-	-	-	-	-	-
Non-Metallic Minerals	-	-	-	-	-	-	-	-	-	-	-
Non-specified	-	-	207	-	-	-	-	-	70	-	277
TRANSPORT SECTOR	-	-	725	-	-	-	-	-	-	-	725
Air	-	-	283	-	-	-	-	-	-	-	283
Road	-	-	442	-	-	-	-	-	-	-	442
Non-specified	-	-	-	-	-	-	-	-	-	-	-
OTHER SECTORS	-	-	176	-	-	-	-	15234	38	-	15448
Agriculture	-	-	34	-	-	-	-	-	-	-	34
Comm. and Publ. Services	-	-	14	-	-	-	-	-	1	-	16
Residential	-	-	41	-	-	-	-	-	37	-	78
Non-specified	-	-	87	-	-	-	-	15234	-	-	15321
NON-ENERGY USE	-	-	50	-	-	-	-	-	-	-	50
Electricity Generated - GWh	-	-	148	-	-	1119	49	-	-	-	1316
Heat Generated - TJ	-	-	-	-	-	-	-	-	-	-	-

APPROVISIONNEMENT ET DEMANDE 1997	Charbon	Pétrole brut	Produits pétroliers	Gaz	Nucléaire	Hydro	Géotherm. solaire etc.	En. ren. combust. & déchets	Electricité	Chaleur	Total
Indigenous Production	-	-	-	-	-	101	89	16126	-	-	16316
Imports	-	471	490	-	-	-	-	-	-	-	962
Exports	-	-	-79	-	-	-	-	-	-	-	-79
Intl. Marine Bunkers	-	-	-67	-	-	-	-	-	-	-	-67
Stock changes	-	-	-	-	-	-	-	-	-	-	-
TPES	-	471	344	-	-	101	89	16126	-	-	17131
Electricity and CHP Plants	-	-	-37	-	-	-101	-89	-	115	-	-112
Petroleum Refineries	-	-471	431	-	-	-	-	-	-	-	-41
Other Transformation *	-	-	-29	-	-	-	-	-541	-5	-	-575
TFC	-	-	709	-	-	-	-	15585	110	-	16404
INDUSTRY SECTOR	-	-	101	-	-	-	-	-	71	-	173
Iron and Steel	-	-	-	-	-	-	-	-	-	-	-
Chemical and Petrochemical	-	-	-	-	-	-	-	-	-	-	-
Non-Metallic Minerals	-	-	-	-	-	-	-	-	-	-	-
Non-specified	-	-	101	-	-	-	-	-	71	-	173
TRANSPORT SECTOR	-	-	467	-	-	-	-	-	-	-	467
Air	-	-	290	-	-	-	-	-	-	-	290
Road	-	-	177	-	-	-	-	-	-	-	177
Non-specified	-	-	-	-	-	-	-	-	-	-	-
OTHER SECTORS	-	-	91	-	-	-	-	15585	38	-	15715
Agriculture	-	-	8	-	-	-	-	-	-	-	8
Comm. and Publ. Services	-	-	3	-	-	-	-	-	1	-	4
Residential	-	-	29	-	-	-	-	-	37	-	66
Non-specified	-	-	51	-	-	-	-	15585	-	-	15636
NON-ENERGY USE	-	-	50	-	-	-	-	-	-	-	50
Electricity Generated - GWh	-	-	66	-	-	1170	104	-	-	-	1340
Heat Generated - TJ	-	-	-	-	-	-	-	-	-	-	-

* Includes Transfers, Statistical Differences, Own Use and Distribution Losses.

INTERNATIONAL ENERGY AGENCY

Gabon

Thousand tonnes of oil equivalent / *Milliers de tonnes d'équivalent pétrole*

SUPPLY AND CONSUMPTION 1996	Coal	Crude Oil	Petroleum Products	Gas	Nuclear	Hydro	Geotherm. Solar etc.	Combust. Renew. & Waste	Electricity	Heat	Total
Indigenous Production	-	18606	-	71	-	63	-	848	-	-	19589
Imports	-	-	121	-	-	-	-	-	-	-	121
Exports	-	-17676	-249	-	-	-	-	-	-	-	-17925
Intl. Marine Bunkers	-	-	-67	-	-	-	-	-	-	-	-67
Stock changes	-	-191	-1	-	-	-	-	-	-	-	-192
TPES	-	739	-196	71	-	63	-	848	-	-	1526
Electricity and CHP Plants	-	-	-46	-43	-	-63	-	-	83	-	-68
Petroleum Refineries	-	-739	732	-	-	-	-	-	-	-	-7
Other Transformation *	-	-	-33	-28	-	-	-	-	-12	-	-73
TFC	-	-	457	1	-	-	-	848	72	-	1378
INDUSTRY SECTOR	-	-	116	1	-	-	-	170	37	-	323
Iron and Steel	-	-	-	-	-	-	-	-	-	-	-
Chemical and Petrochemical	-	-	11	-	-	-	-	-	-	-	11
Non-Metallic Minerals	-	-	-	-	-	-	-	-	-	-	-
Non-specified	-	-	105	1	-	-	-	170	37	-	312
TRANSPORT SECTOR	-	-	222	-	-	-	-	-	-	-	222
Air	-	-	71	-	-	-	-	-	-	-	71
Road	-	-	129	-	-	-	-	-	-	-	129
Non-specified	-	-	22	-	-	-	-	-	-	-	22
OTHER SECTORS	-	-	59	-	-	-	-	679	34	-	772
Agriculture	-	-	-	-	-	-	-	-	-	-	-
Comm. and Publ. Services	-	-	-	-	-	-	-	-	13	-	13
Residential	-	-	47	-	-	-	-	679	21	-	747
Non-specified	-	-	11	-	-	-	-	-	-	-	11
NON-ENERGY USE	-	-	61	-	-	-	-	-	-	-	61
Electricity Generated - GWh	-	-	*125*	*111*	-	*734*	-	-	-	-	*970*
Heat Generated - TJ	-	-	-	-	-	-	-	-	-	-	-

APPROVISIONNEMENT ET DEMANDE 1997	Charbon	Pétrole brut	Produits pétroliers	Gaz	Nucléaire	Hydro	Géotherm. solaire etc.	En. ren. combust. & déchets	Electricité	Chaleur	Total
Indigenous Production	-	18794	-	68	-	64	-	860	-	-	19786
Imports	-	-	218	-	-	-	-	-	-	-	218
Exports	-	-17855	-261	-	-	-	-	-	-	-	-18116
Intl. Marine Bunkers	-	-	-52	-	-	-	-	-	-	-	-52
Stock changes	-	-202	-	-	-	-	-	-	-	-	-202
TPES	-	738	-95	68	-	64	-	860	-	-	1635
Electricity and CHP Plants	-	-	-59	-41	-	-64	-	-	87	-	-77
Petroleum Refineries	-	-738	730	-	-	-	-	-	-	-	-8
Other Transformation *	-	-	-34	-27	-	-	-	-	-12	-	-73
TFC	-	-	541	1	-	-	-	860	75	-	1476
INDUSTRY SECTOR	-	-	130	1	-	-	-	172	39	-	342
Iron and Steel	-	-	-	-	-	-	-	-	-	-	-
Chemical and Petrochemical	-	-	11	-	-	-	-	-	-	-	11
Non-Metallic Minerals	-	-	-	-	-	-	-	-	-	-	-
Non-specified	-	-	120	1	-	-	-	172	39	-	331
TRANSPORT SECTOR	-	-	289	-	-	-	-	-	-	-	289
Air	-	-	99	-	-	-	-	-	-	-	99
Road	-	-	162	-	-	-	-	-	-	-	162
Non-specified	-	-	28	-	-	-	-	-	-	-	28
OTHER SECTORS	-	-	65	-	-	-	-	688	36	-	789
Agriculture	-	-	-	-	-	-	-	-	-	-	-
Comm. and Publ. Services	-	-	-	-	-	-	-	-	14	-	14
Residential	-	-	51	-	-	-	-	688	22	-	761
Non-specified	-	-	14	-	-	-	-	-	-	-	14
NON-ENERGY USE	-	-	57	-	-	-	-	-	-	-	57
Electricity Generated - GWh	-	-	*161*	*106*	-	*740*	-	-	-	-	*1007*
Heat Generated - TJ	-	-	-	-	-	-	-	-	-	-	-

* Includes Transfers, Statistical Differences, Own Use and Distribution Losses.

Georgia / Géorgie

Thousand tonnes of oil equivalent / *Milliers de tonnes d'équivalent pétrole*

SUPPLY AND CONSUMPTION 1996	Coal	Crude Oil	Petroleum Products	Gas	Nuclear	Hydro	Geotherm. Solar etc.	Combust. Renew. & Waste	Electricity	Heat	Total
Indigenous Production	10	129	-	2	-	520	-	35	-	-	696
Imports	22	-	924	631	-	-	-	-	19	-	1596
Exports	-	-114	-5	-	-	-	-	-	-12	-	-130
Intl. Marine Bunkers	-	-	-	-	-	-	-	-	-	-	-
Stock changes	-	-	-49	-	-	-	-	-	-	-	-49
TPES	32	15	870	634	-	520	-	35	7	-	2112
Electricity and CHP Plants	-	-	-85	-286	-	-520	-	-	621	476	207
Petroleum Refineries	-	-15	13	-	-	-	-	-	-	-	-2
Other Transformation *	-	-	-16	-	-	-	-	-	-92	-	-108
TFC	32	-	781	348	-	-	-	35	536	476	2209
INDUSTRY SECTOR	-	-	62	249	-	-	-	-	75	-	386
Iron and Steel	-	-	-	75	-	-	-	-	28	-	103
Chemical and Petrochemical	-	-	-	143	-	-	-	-	-	-	143
Non-Metallic Minerals	-	-	-	-	-	-	-	-	-	-	-
Non-specified	-	-	62	30	-	-	-	-	47	-	139
TRANSPORT SECTOR	-	-	536	-	-	-	-	-	22	-	558
Air	-	-	49	-	-	-	-	-	-	-	49
Road	-	-	477	-	-	-	-	-	-	-	477
Non-specified	-	-	11	-	-	-	-	-	22	-	32
OTHER SECTORS	32	-	183	99	-	-	-	35	440	476	1265
Agriculture	-	-	82	-	-	-	-	-	2	-	84
Comm. and Publ. Services	-	-	17	-	-	-	-	-	210	-	227
Residential	-	-	54	99	-	-	-	-	227	-	380
Non-specified	32	-	30	-	-	-	-	35	-	476	574
NON-ENERGY USE	-	-	-	-	-	-	-	-	-	-	-
Electricity Generated - GWh	-	-	80	1105	-	6041	-	-	-	-	7226
Heat Generated - TJ	-	-	-	-	-	-	-	-	-	-	19950

APPROVISIONNEMENT ET DEMANDE 1997	Charbon	Pétrole brut	Produits pétroliers	Gaz	Nucléaire	Hydro	Géotherm. solaire etc.	En. ren. combust. & déchets	Electricité	Chaleur	Total
Indigenous Production	2	135	-	2	-	520	-	35	-	-	694
Imports	-	27	993	765	-	-	-	-	56	-	1841
Exports	-	-132	-5	-	-	-	-	-	-40	-	-176
Intl. Marine Bunkers	-	-	-	-	-	-	-	-	-	-	-
Stock changes	-	-	-63	-	-	-	-	-	-	-	-63
TPES	2	30	924	767	-	520	-	35	16	-	2295
Electricity and CHP Plants	-	-	-82	-303	-	-520	-	-	617	476	189
Petroleum Refineries	-	-27	24	-	-	-	-	-	-	-	-3
Other Transformation *	-	-3	-14	-	-	-	-	-	-100	-	-117
TFC	2	-	853	464	-	-	-	35	533	476	2364
INDUSTRY SECTOR	-	-	65	351	-	-	-	-	71	-	487
Iron and Steel	-	-	-	59	-	-	-	-	25	-	84
Chemical and Petrochemical	-	-	-	167	-	-	-	-	-	-	167
Non-Metallic Minerals	-	-	-	-	-	-	-	-	-	-	-
Non-specified	-	-	65	125	-	-	-	-	47	-	237
TRANSPORT SECTOR	-	-	507	-	-	-	-	-	20	-	526
Air	-	-	51	-	-	-	-	-	-	-	51
Road	-	-	453	-	-	-	-	-	-	-	453
Non-specified	-	-	3	-	-	-	-	-	20	-	23
OTHER SECTORS	2	-	282	113	-	-	-	35	442	476	1351
Agriculture	-	-	80	-	-	-	-	-	1	-	81
Comm. and Publ. Services	-	-	79	-	-	-	-	-	218	-	297
Residential	-	-	79	113	-	-	-	-	223	-	416
Non-specified	2	-	44	-	-	-	-	35	-	476	557
NON-ENERGY USE	-	-	-	-	-	-	-	-	-	-	-
Electricity Generated - GWh	-	-	9	1119	-	6044	-	-	-	-	7172
Heat Generated - TJ	-	-	-	-	-	-	-	-	-	-	19950

* Includes Transfers, Statistical Differences, Own Use and Distribution Losses.

Ghana

Thousand tonnes of oil equivalent / *Milliers de tonnes d'équivalent pétrole*

SUPPLY AND CONSUMPTION 1996	Coal	Crude Oil	Petroleum Products	Gas	Nuclear	Hydro	Geotherm. Solar etc.	Combust. Renew. & Waste	Electricity	Heat	Total
Indigenous Production	-	361	-	-	-	514	-	4842	-	-	5717
Imports	2	1118	192	-	-	-	-	-	5	-	1316
Exports	-	-	-196	-	-	-	-	-	-39	-	-236
Intl. Marine Bunkers	-	-	-26	-	-	-	-	-	-	-	-26
Stock changes	-	-2	-	-	-	-	-	-	-	-	-2
TPES	2	1477	-31	-	-	514	-	4842	-34	-	6770
Electricity and CHP Plants	-	-	-21	-	-	-514	-	-	515	-	-20
Petroleum Refineries	-	-1054	1199	-	-	-	-	-	-	-	145
Other Transformation *	-	-423	-42	-	-	-	-	-1184	-66	-	-1715
TFC	2	-	1106	-	-	-	-	3658	414	-	5180
INDUSTRY SECTOR	-	-	104	-	-	-	-	333	291	-	727
Iron and Steel	-	-	-	-	-	-	-	-	-	-	-
Chemical and Petrochemical	-	-	-	-	-	-	-	-	-	-	-
Non-Metallic Minerals	-	-	-	-	-	-	-	-	-	-	-
Non-specified	-	-	104	-	-	-	-	333	291	-	727
TRANSPORT SECTOR	-	-	698	-	-	-	-	-	-	-	698
Air	-	-	61	-	-	-	-	-	-	-	61
Road	-	-	601	-	-	-	-	-	-	-	601
Non-specified	-	-	36	-	-	-	-	-	-	-	36
OTHER SECTORS	2	-	246	-	-	-	-	3326	123	-	3697
Agriculture	-	-	51	-	-	-	-	-	-	-	51
Comm. and Publ. Services	-	-	37	-	-	-	-	-	-	-	37
Residential	2	-	158	-	-	-	-	3326	123	-	3608
Non-specified	-	-	-	-	-	-	-	-	1	-	1
NON-ENERGY USE	-	-	58	-	-	-	-	-	-	-	58
Electricity Generated - GWh	-	-	7	-	-	5980	-	-	-	-	5987
Heat Generated - TJ	-	-	-	-	-	-	-	-	-	-	-

APPROVISIONNEMENT ET DEMANDE 1997	Charbon	Pétrole brut	Produits pétroliers	Gaz	Nucléaire	Hydro	Géotherm. solaire etc.	En. ren. combust. & déchets	Electricité	Chaleur	Total
Indigenous Production	-	361	-	-	-	529	-	4953	-	-	5843
Imports	2	1118	192	-	-	-	-	-	5	-	1316
Exports	-	-	-196	-	-	-	-	-	-39	-	-236
Intl. Marine Bunkers	-	-	-26	-	-	-	-	-	-	-	-26
Stock changes	-	-2	-	-	-	-	-	-	-	-	-2
TPES	2	1477	-31	-	-	529	-	4953	-34	-	6896
Electricity and CHP Plants	-	-	-21	-	-	-529	-	-	529	-	-20
Petroleum Refineries	-	-1054	1199	-	-	-	-	-	-	-	145
Other Transformation *	-	-423	-42	-	-	-	-	-1211	-69	-	-1744
TFC	2	-	1106	-	-	-	-	3742	427	-	5277
INDUSTRY SECTOR	-	-	104	-	-	-	-	340	300	-	744
Iron and Steel	-	-	-	-	-	-	-	-	-	-	-
Chemical and Petrochemical	-	-	-	-	-	-	-	-	-	-	-
Non-Metallic Minerals	-	-	-	-	-	-	-	-	-	-	-
Non-specified	-	-	104	-	-	-	-	340	300	-	744
TRANSPORT SECTOR	-	-	698	-	-	-	-	-	-	-	698
Air	-	-	61	-	-	-	-	-	-	-	61
Road	-	-	601	-	-	-	-	-	-	-	601
Non-specified	-	-	36	-	-	-	-	-	-	-	36
OTHER SECTORS	2	-	246	-	-	-	-	3402	127	-	3777
Agriculture	-	-	51	-	-	-	-	-	-	-	51
Comm. and Publ. Services	-	-	37	-	-	-	-	-	-	-	37
Residential	2	-	158	-	-	-	-	3402	127	-	3689
Non-specified	-	-	-	-	-	-	-	-	-	-	-
NON-ENERGY USE	-	-	58	-	-	-	-	-	-	-	58
Electricity Generated - GWh	-	-	7	-	-	6148	-	-	-	-	6155
Heat Generated - TJ	-	-	-	-	-	-	-	-	-	-	-

* Includes Transfers, Statistical Differences, Own Use and Distribution Losses.

INTERNATIONAL ENERGY AGENCY

Gibraltar

Thousand tonnes of oil equivalent / *Milliers de tonnes d'équivalent pétrole*

SUPPLY AND CONSUMPTION 1996	Coal	Crude Oil	Petroleum Products	Gas	Nuclear	Hydro	Geotherm. Solar etc.	Combust. Renew. & Waste	Electricity	Heat	Total
Indigenous Production	-	-	-	-	-	-	-	-	-	-	-
Imports	-	-	961	-	-	-	-	-	-	-	961
Exports	-	-	-	-	-	-	-	-	-	-	-
Intl. Marine Bunkers	-	-	-828	-	-	-	-	-	-	-	-828
Stock changes	-	-	-	-	-	-	-	-	-	-	-
TPES	-	-	133	-	-	-	-	-	-	-	133
Electricity and CHP Plants	-	-	-48	-	-	-	-	-	8	-	-40
Petroleum Refineries	-	-	-	-	-	-	-	-	-	-	-
Other Transformation *	-	-	-	-	-	-	-	-	-	-	-
TFC	-	-	85	-	-	-	-	-	8	-	93
INDUSTRY SECTOR	-	-	-	-	-	-	-	-	-	-	-
Iron and Steel	-	-	-	-	-	-	-	-	-	-	-
Chemical and Petrochemical	-	-	-	-	-	-	-	-	-	-	-
Non-Metallic Minerals	-	-	-	-	-	-	-	-	-	-	-
Non-specified	-	-	-	-	-	-	-	-	-	-	-
TRANSPORT SECTOR	-	-	71	-	-	-	-	-	-	-	71
Air	-	-	4	-	-	-	-	-	-	-	4
Road	-	-	67	-	-	-	-	-	-	-	67
Non-specified	-	-	-	-	-	-	-	-	-	-	-
OTHER SECTORS	-	-	-	-	-	-	-	-	8	-	8
Agriculture	-	-	-	-	-	-	-	-	-	-	-
Comm. and Publ. Services	-	-	-	-	-	-	-	-	-	-	-
Residential	-	-	-	-	-	-	-	-	-	-	-
Non-specified	-	-	-	-	-	-	-	-	8	-	8
NON-ENERGY USE	-	-	14	-	-	-	-	-	-	-	14
Electricity Generated - GWh	-	-	*93*	-	-	-	-	-	-	-	*93*
Heat Generated - TJ	-	-	-	-	-	-	-	-	-	-	-

APPROVISIONNEMENT ET DEMANDE 1997	Charbon	Pétrole brut	Produits pétroliers	Gaz	Nucléaire	Hydro	Géotherm. solaire etc.	En. ren. combust. & déchets	Electricité	Chaleur	Total
Indigenous Production	-	-	-	-	-	-	-	-	-	-	-
Imports	-	-	961	-	-	-	-	-	-	-	961
Exports	-	-	-	-	-	-	-	-	-	-	-
Intl. Marine Bunkers	-	-	-828	-	-	-	-	-	-	-	-828
Stock changes	-	-	-	-	-	-	-	-	-	-	-
TPES	-	-	133	-	-	-	-	-	-	-	133
Electricity and CHP Plants	-	-	-48	-	-	-	-	-	8	-	-40
Petroleum Refineries	-	-	-	-	-	-	-	-	-	-	-
Other Transformation *	-	-	-	-	-	-	-	-	-	-	-
TFC	-	-	85	-	-	-	-	-	8	-	93
INDUSTRY SECTOR	-	-	-	-	-	-	-	-	-	-	-
Iron and Steel	-	-	-	-	-	-	-	-	-	-	-
Chemical and Petrochemical	-	-	-	-	-	-	-	-	-	-	-
Non-Metallic Minerals	-	-	-	-	-	-	-	-	-	-	-
Non-specified	-	-	-	-	-	-	-	-	-	-	-
TRANSPORT SECTOR	-	-	71	-	-	-	-	-	-	-	71
Air	-	-	4	-	-	-	-	-	-	-	4
Road	-	-	67	-	-	-	-	-	-	-	67
Non-specified	-	-	-	-	-	-	-	-	-	-	-
OTHER SECTORS	-	-	-	-	-	-	-	-	8	-	8
Agriculture	-	-	-	-	-	-	-	-	-	-	-
Comm. and Publ. Services	-	-	-	-	-	-	-	-	-	-	-
Residential	-	-	-	-	-	-	-	-	-	-	-
Non-specified	-	-	-	-	-	-	-	-	8	-	8
NON-ENERGY USE	-	-	14	-	-	-	-	-	-	-	14
Electricity Generated - GWh	-	-	*93*	-	-	-	-	-	-	-	*93*
Heat Generated - TJ	-	-	-	-	-	-	-	-	-	-	-

* Includes Transfers, Statistical Differences, Own Use and Distribution Losses.

INTERNATIONAL ENERGY AGENCY

Guatemala

Thousand tonnes of oil equivalent / *Milliers de tonnes d'équivalent pétrole*

SUPPLY AND CONSUMPTION 1996	Coal	Crude Oil	Petroleum Products	Gas	Nuclear	Hydro	Geotherm. Solar etc.	Combust. Renew. & Waste	Electricity	Heat	Total
Indigenous Production	-	806	-	-	-	271	-	2982	-	-	4059
Imports	-	805	1404	-	-	-	-	-	-	-	2209
Exports	-	-728	-26	-	-	-	-	-	-	-	-754
Intl. Marine Bunkers	-	-	-124	-	-	-	-	-	-	-	-124
Stock changes	-	-	12	-	-	-	-	-	-	-	12
TPES	-	883	1266	-	-	271	-	2982	-	-	5402
Electricity and CHP Plants	-	-	-301	-	-	-271	-	-95	365	-	-301
Petroleum Refineries	-	-802	795	-	-	-	-	-3	-	-	-10
Other Transformation *	-	-81	-27	-	-	-	-	-54	-49	-	-210
TFC	-	-	1734	-	-	-	-	2831	317	-	4881
INDUSTRY SECTOR	-	-	383	-	-	-	-	181	107	-	671
Iron and Steel	-	-	-	-	-	-	-	-	-	-	-
Chemical and Petrochemical	-	-	-	-	-	-	-	-	-	-	-
Non-Metallic Minerals	-	-	-	-	-	-	-	-	-	-	-
Non-specified	-	-	383	-	-	-	-	181	107	-	671
TRANSPORT SECTOR	-	-	1010	-	-	-	-	-	-	-	1010
Air	-	-	50	-	-	-	-	-	-	-	50
Road	-	-	959	-	-	-	-	-	-	-	959
Non-specified	-	-	1	-	-	-	-	-	-	-	1
OTHER SECTORS	-	-	275	-	-	-	-	2650	209	-	3134
Agriculture	-	-	35	-	-	-	-	-	-	-	35
Comm. and Publ. Services	-	-	75	-	-	-	-	-	107	-	182
Residential	-	-	164	-	-	-	-	2650	103	-	2917
Non-specified	-	-	-	-	-	-	-	-	-	-	-
NON-ENERGY USE	-	-	65	-	-	-	-	-	-	-	65
Electricity Generated - GWh	-	-	*878*	-	-	*3153*	-	*216*	-	-	*4247*
Heat Generated - TJ	-	-	-	-	-	-	-	-	-	-	-

APPROVISIONNEMENT ET DEMANDE 1997	Charbon	Pétrole brut	Produits pétroliers	Gaz	Nucléaire	Hydro	Géotherm. solaire etc.	En. ren. combust. & déchets	Electricité	Chaleur	Total
Indigenous Production	-	1080	-	-	-	323	-	3030	-	-	4433
Imports	-	851	1477	-	-	-	-	-	-	-	2328
Exports	-	-988	-28	-	-	-	-	-	-	-	-1015
Intl. Marine Bunkers	-	-	-124	-	-	-	-	-	-	-	-124
Stock changes	-	-	11	-	-	-	-	-	-	-	11
TPES	-	943	1336	-	-	323	-	3030	-	-	5633
Electricity and CHP Plants	-	-	-313	-	-	-323	-	-96	421	-	-311
Petroleum Refineries	-	-859	833	-	-	-	-	-3	-	-	-29
Other Transformation *	-	-84	-24	-	-	-	-	-54	-56	-	-218
TFC	-	-	1832	-	-	-	-	2877	365	-	5075
INDUSTRY SECTOR	-	-	396	-	-	-	-	183	124	-	703
Iron and Steel	-	-	-	-	-	-	-	-	-	-	-
Chemical and Petrochemical	-	-	-	-	-	-	-	-	-	-	-
Non-Metallic Minerals	-	-	-	-	-	-	-	-	-	-	-
Non-specified	-	-	396	-	-	-	-	183	124	-	703
TRANSPORT SECTOR	-	-	1076	-	-	-	-	-	-	-	1076
Air	-	-	53	-	-	-	-	-	-	-	53
Road	-	-	1021	-	-	-	-	-	-	-	1021
Non-specified	-	-	1	-	-	-	-	-	-	-	1
OTHER SECTORS	-	-	291	-	-	-	-	2694	241	-	3226
Agriculture	-	-	37	-	-	-	-	-	-	-	37
Comm. and Publ. Services	-	-	82	-	-	-	-	-	123	-	205
Residential	-	-	172	-	-	-	-	2694	118	-	2983
Non-specified	-	-	-	-	-	-	-	-	-	-	-
NON-ENERGY USE	-	-	70	-	-	-	-	-	-	-	70
Electricity Generated - GWh	-	-	*917*	-	-	*3760*	-	*220*	-	-	*4897*
Heat Generated - TJ	-	-	-	-	-	-	-	-	-	-	-

* Includes Transfers, Statistical Differences, Own Use and Distribution Losses.

Haiti

Thousand tonnes of oil equivalent / *Milliers de tonnes d'équivalent pétrole*

SUPPLY AND CONSUMPTION 1996	Coal	Crude Oil	Petroleum Products	Gas	Nuclear	Hydro	Geotherm. Solar etc.	Combust. Renew. & Waste	Electricity	Heat	Total
Indigenous Production	-	-	-	-	-	23	-	1575	-	-	1598
Imports	-	-	368	-	-	-	-	-	-	-	368
Exports	-	-	-	-	-	-	-	-	-	-	-
Intl. Marine Bunkers	-	-	-	-	-	-	-	-	-	-	-
Stock changes	-	-	-	-	-	-	-	-	-	-	-
TPES	-	-	368	-	-	23	-	1575	-	-	1966
Electricity and CHP Plants	-	-	-78	-	-	-23	-	-10	54	-	-57
Petroleum Refineries	-	-	-	-	-	-	-	-	-	-	-
Other Transformation *	-	-	2	-	-	-	-	-289	-32	-	-319
TFC	-	-	292	-	-	-	-	1276	22	-	1590
INDUSTRY SECTOR	-	-	50	-	-	-	-	71	10	-	130
Iron and Steel	-	-	-	-	-	-	-	-	-	-	-
Chemical and Petrochemical	-	-	-	-	-	-	-	-	-	-	-
Non-Metallic Minerals	-	-	-	-	-	-	-	-	-	-	-
Non-specified	-	-	50	-	-	-	-	71	10	-	130
TRANSPORT SECTOR	-	-	207	-	-	-	-	-	-	-	207
Air	-	-	20	-	-	-	-	-	-	-	20
Road	-	-	83	-	-	-	-	-	-	-	83
Non-specified	-	-	104	-	-	-	-	-	-	-	104
OTHER SECTORS	-	-	33	-	-	-	-	1206	12	-	1251
Agriculture	-	-	-	-	-	-	-	-	-	-	-
Comm. and Publ. Services	-	-	-	-	-	-	-	48	2	-	50
Residential	-	-	33	-	-	-	-	1157	10	-	1200
Non-specified	-	-	-	-	-	-	-	-	-	-	-
NON-ENERGY USE	-	-	2	-	-	-	-	-	-	-	2
Electricity Generated - GWh	-	-	333	-	-	272	-	20	-	-	625
Heat Generated - TJ	-	-	-	-	-	-	-	-	-	-	-

APPROVISIONNEMENT ET DEMANDE 1997	Charbon	Pétrole brut	Produits pétroliers	Gaz	Nucléaire	Hydro	Géotherm. solaire etc.	En. ren. combust. & déchets	Electricité	Chaleur	Total
Indigenous Production	-	-	-	-	-	17	-	1281	-	-	1298
Imports	-	-	480	-	-	-	-	-	-	-	480
Exports	-	-	-	-	-	-	-	-	-	-	-
Intl. Marine Bunkers	-	-	-	-	-	-	-	-	-	-	-
Stock changes	-	-	-	-	-	-	-	-	-	-	-
TPES	-	-	480	-	-	17	-	1281	-	-	1779
Electricity and CHP Plants	-	-	-109	-	-	-17	-	-10	54	-	-82
Petroleum Refineries	-	-	-	-	-	-	-	-	-	-	-
Other Transformation *	-	-	-35	-	-	-	-	-299	-27	-	-361
TFC	-	-	335	-	-	-	-	973	27	-	1335
INDUSTRY SECTOR	-	-	79	-	-	-	-	66	11	-	156
Iron and Steel	-	-	-	-	-	-	-	-	-	-	-
Chemical and Petrochemical	-	-	-	-	-	-	-	-	-	-	-
Non-Metallic Minerals	-	-	-	-	-	-	-	-	-	-	-
Non-specified	-	-	79	-	-	-	-	66	11	-	156
TRANSPORT SECTOR	-	-	213	-	-	-	-	-	-	-	213
Air	-	-	20	-	-	-	-	-	-	-	20
Road	-	-	91	-	-	-	-	-	-	-	91
Non-specified	-	-	101	-	-	-	-	-	-	-	101
OTHER SECTORS	-	-	35	-	-	-	-	906	16	-	957
Agriculture	-	-	-	-	-	-	-	-	-	-	-
Comm. and Publ. Services	-	-	-	-	-	-	-	50	9	-	59
Residential	-	-	35	-	-	-	-	857	7	-	898
Non-specified	-	-	-	-	-	-	-	-	-	-	-
NON-ENERGY USE	-	-	9	-	-	-	-	-	-	-	9
Electricity Generated - GWh	-	-	405	-	-	200	-	20	-	-	625
Heat Generated - TJ	-	-	-	-	-	-	-	-	-	-	-

* Includes Transfers, Statistical Differences, Own Use and Distribution Losses.

INTERNATIONAL ENERGY AGENCY

Honduras

Thousand tonnes of oil equivalent / *Milliers de tonnes d'équivalent pétrole*

SUPPLY AND CONSUMPTION 1996	Coal	Crude Oil	Petroleum Products	Gas	Nuclear	Hydro	Geotherm. Solar etc.	Combust. Renew. & Waste	Electricity	Heat	Total
Indigenous Production	-	-	-	-	-	262	-	1522	-	-	1785
Imports	1	-	1193	-	-	-	-	-	-	-	1194
Exports	-	-	-	-	-	-	-	-	-	-	-
Intl. Marine Bunkers	-	-	-	-	-	-	-	-	-	-	-
Stock changes	-	-	-33	-	-	-	-	-	-	-	-33
TPES	1	-	1161	-	-	262	-	1522	-	-	2946
Electricity and CHP Plants	-	-	-4	-	-	-262	-	-	263	-	-3
Petroleum Refineries	-	-	-	-	-	-	-	-	-	-	-
Other Transformation *	-1	-	6	-	-	-	-	-12	-63	-	-70
TFC	-	-	1162	-	-	-	-	1510	200	-	2872
INDUSTRY SECTOR	-	-	478	-	-	-	-	279	67	-	824
Iron and Steel	-	-	-	-	-	-	-	-	-	-	-
Chemical and Petrochemical	-	-	-	-	-	-	-	-	-	-	-
Non-Metallic Minerals	-	-	-	-	-	-	-	-	-	-	-
Non-specified	-	-	478	-	-	-	-	279	67	-	824
TRANSPORT SECTOR	-	-	534	-	-	-	-	-	-	-	534
Air	-	-	26	-	-	-	-	-	-	-	26
Road	-	-	509	-	-	-	-	-	-	-	509
Non-specified	-	-	-	-	-	-	-	-	-	-	-
OTHER SECTORS	-	-	150	-	-	-	-	1231	133	-	1514
Agriculture	-	-	5	-	-	-	-	-	-	-	5
Comm. and Publ. Services	-	-	86	-	-	-	-	-	59	-	144
Residential	-	-	59	-	-	-	-	1231	75	-	1365
Non-specified	-	-	-	-	-	-	-	-	-	-	-
NON-ENERGY USE	-	-	-	-	-	-	-	-	-	-	-
Electricity Generated - GWh	-	-	9	-	-	3050	-	-	-	-	3059
Heat Generated - TJ	-	-	-	-	-	-	-	-	-	-	-

APPROVISIONNEMENT ET DEMANDE 1997	Charbon	Pétrole brut	Produits pétroliers	Gaz	Nucléaire	Hydro	Géotherm. solaire etc.	En. ren. combust. & déchets	Electricité	Chaleur	Total
Indigenous Production	-	-	-	-	-	279	-	1724	-	-	2003
Imports	1	-	1149	-	-	-	-	-	-	-	1150
Exports	-	-	-	-	-	-	-	-	-	-	-
Intl. Marine Bunkers	-	-	-	-	-	-	-	-	-	-	-
Stock changes	-	-	29	-	-	-	-	-	-	-	29
TPES	1	-	1178	-	-	279	-	1724	-	-	3182
Electricity and CHP Plants	-	-	-4	-	-	-279	-	-	283	-	-
Petroleum Refineries	-	-	-	-	-	-	-	-	-	-	-
Other Transformation *	-1	-	6	-	-	-	-	-16	-72	-	-82
TFC	-	-	1180	-	-	-	-	1709	212	-	3100
INDUSTRY SECTOR	-	-	485	-	-	-	-	299	63	-	847
Iron and Steel	-	-	-	-	-	-	-	-	-	-	-
Chemical and Petrochemical	-	-	-	-	-	-	-	-	-	-	-
Non-Metallic Minerals	-	-	-	-	-	-	-	-	-	-	-
Non-specified	-	-	485	-	-	-	-	299	63	-	847
TRANSPORT SECTOR	-	-	538	-	-	-	-	-	-	-	538
Air	-	-	26	-	-	-	-	-	-	-	26
Road	-	-	513	-	-	-	-	-	-	-	513
Non-specified	-	-	-	-	-	-	-	-	-	-	-
OTHER SECTORS	-	-	157	-	-	-	-	1409	149	-	1715
Agriculture	-	-	5	-	-	-	-	-	-	-	5
Comm. and Publ. Services	-	-	87	-	-	-	-	-	64	-	151
Residential	-	-	65	-	-	-	-	1409	85	-	1559
Non-specified	-	-	-	-	-	-	-	-	-	-	-
NON-ENERGY USE	-	-	-	-	-	-	-	-	-	-	-
Electricity Generated - GWh	-	-	48	-	-	3246	-	-	-	-	3294
Heat Generated - TJ	-	-	-	-	-	-	-	-	-	-	-

* Includes Transfers, Statistical Differences, Own Use and Distribution Losses.

Hong Kong, China / Hong Kong, Chine

Total Production of Energy

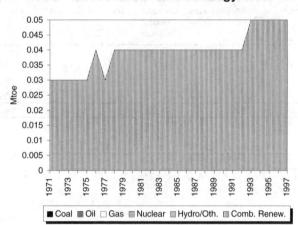

Total Primary Energy Supply*

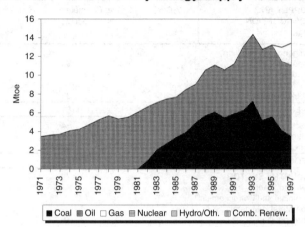

Electricity Generation by Fuel

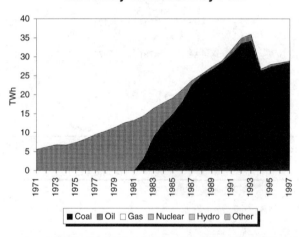

Oil Products Consumption

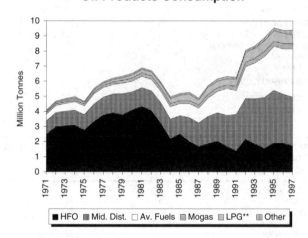

Electricity Consumption/GDP***,
TPES/GDP*** and Production/TPES

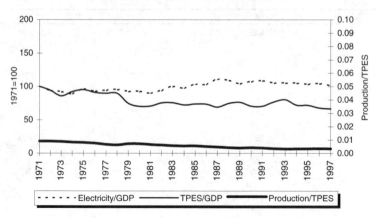

* Excluding electricity trade.
** Includes LPG, NGL, ethane and naphtha.
*** GDP in 1990 US dollars.

INTERNATIONAL ENERGY AGENCY

Hong Kong, China / Hong Kong, Chine : 1996

Thousand tonnes of oil equivalent / *Milliers de tonnes d'équivalent pétrole*

SUPPLY AND CONSUMPTION *APPROVISIONNEMENT ET DEMANDE*	Coal *Charbon*	Crude Oil *Pétrole brut*	Petroleum Products *Produits pétroliers*	Gas *Gaz*	Nuclear *Nucléaire*	Hydro *Hydro*	Geotherm. Solar etc. *Géotherm. solaire*	Combust. Renew. & Waste *En. ren. combust.*	Electricity *Electricité*	Heat *Chaleur*	Total *Total*
Indigenous Production	ι	-	-	-	-	-	-	48	-	-	48
Imports	4163	-	18407	1479	-	-	-	10	669	-	24728
Exports	-	-	-8621	-	-	-	-	-1	-46	-	-8668
Intl. Marine Bunkers	-	-	-2326	-	-	-	-	-	-	-	-2326
Stock Changes	-	-	-119	-	-	-	-	-	-	-	-119
TPES	**4163**	**-**	**7341**	**1479**	**-**	**-**	**-**	**57**	**623**	**-**	**13663**
Transfers	-	-	-	-	-	-	-	-	-	-	-
Statistical Differences	794	-	1	-	-	-	-	-	-	-	795
Electricity Plants	-4957	-	-67	-	-	-	-	-	2446	-	-2578
CHP Plants	-	-	-	-	-	-	-	-	-	-	-
Heat Plants	-	-	-	-	-	-	-	-	-	-	-
Gas Works	-	-	-556	494	-	-	-	-	-	-	-62
Petroleum Refineries	-	-	-	-	-	-	-	-	-	-	-
Coal Transformation	-	-	-	-	-	-	-	-	-	-	-
Liquefaction	-	-	-	-	-	-	-	-	-	-	-
Other Transformation	-	-	-	-	-	-	-	-	-	-	-
Own Use	-	-	-	-	-	-	-	-	-	-	-
Distribution Losses	-	-	-	-	-	-	-	-	-349	-	-349
TFC	**-**	**-**	**6719**	**1973**	**-**	**-**	**-**	**57**	**2721**	**-**	**11470**
INDUSTRY SECTOR	**-**	**-**	**220**	**1499**	**-**	**-**	**-**	**-**	**476**	**-**	**2195**
Iron and Steel	-	-	-	-	-	-	-	-	-	-	-
Chemical and Petrochemical	-	-	-	-	-	-	-	-	-	-	-
of which: Feedstocks	-	-	-	-	-	-	-	-	-	-	-
Non-Ferrous Metals	-	-	-	-	-	-	-	-	-	-	-
Non-Metallic Minerals	-	-	-	-	-	-	-	-	-	-	-
Transport Equipment	-	-	-	-	-	-	-	-	-	-	-
Machinery	-	-	-	-	-	-	-	-	-	-	-
Mining and Quarrying	-	-	-	-	-	-	-	-	-	-	-
Food and Tobacco	-	-	-	-	-	-	-	-	-	-	-
Paper, Pulp and Printing	-	-	-	-	-	-	-	-	-	-	-
Wood and Wood Products	-	-	-	-	-	-	-	-	-	-	-
Construction	-	-	-	-	-	-	-	-	-	-	-
Textile and Leather	-	-	-	-	-	-	-	-	-	-	-
Non-specified	-	-	220	1499	-	-	-	-	476	-	2195
TRANSPORT SECTOR	**-**	**-**	**6263**	**-**	**-**	**-**	**-**	**-**	**-**	**-**	**6263**
Air	-	-	3220	-	-	-	-	-	-	-	3220
Road	-	-	3043	-	-	-	-	-	-	-	3043
Rail	-	-	-	-	-	-	-	-	-	-	-
Pipeline Transport	-	-	-	-	-	-	-	-	-	-	-
Internal Navigation	-	-	-	-	-	-	-	-	-	-	-
Non-specified	-	-	-	-	-	-	-	-	-	-	-
OTHER SECTORS	**-**	**-**	**59**	**474**	**-**	**-**	**-**	**57**	**2244**	**-**	**2834**
Agriculture	-	-	-	-	-	-	-	-	-	-	-
Comm. and Publ. Services	-	-	1	217	-	-	-	-	1547	-	1765
Residential	-	-	57	258	-	-	-	48	697	-	1061
Non-specified	-	-	-	-	-	-	-	9	-	-	9
NON-ENERGY USE	**-**	**-**	**178**	**-**	**-**	**-**	**-**	**-**	**-**	**-**	**178**
in Industry/Transf./Energy	-	-	178	-	-	-	-	-	-	-	178
in Transport	-	-	-	-	-	-	-	-	-	-	-
in Other Sectors	-	-	-	-	-	-	-	-	-	-	-
Electr. Generated - GWh	***27880***	***-***	***562***	***-***	***-***	***-***	***-***	***-***	***-***	***-***	***28442***
Electricity Plants	*27880*	*-*	*562*	*-*	*-*	*-*	*-*	*-*	*-*	*-*	*28442*
CHP Plants	*-*	*-*	*-*	*-*	*-*	*-*	*-*	*-*	*-*	*-*	*-*
Heat Generated - TJ	***-***	***-***	***-***	***-***	***-***	***-***	***-***	***-***	***-***	***-***	***-***
CHP Plants	*-*	*-*	*-*	*-*	*-*	*-*	*-*	*-*	*-*	*-*	*-*
Heat Plants	*-*	*-*	*-*	*-*	*-*	*-*	*-*	*-*	*-*	*-*	*~*

Hong Kong, China / Hong Kong, Chine : 1997

				Thousand tonnes of oil equivalent / Milliers de tonnes d'équivalent pétrole							
SUPPLY AND CONSUMPTION	Coal	Crude Oil	Petroleum Products	Gas	Nuclear	Hydro	Geotherm. Solar etc.	Combust. Renew. & Waste	Electricity	Heat	Total
APPROVISIONNEMENT ET DEMANDE	*Charbon*	*Pétrole brut*	*Produits pétroliers*	*Gaz*	*Nucléaire*	*Hydro*	*Géotherm. solaire*	*En. ren. combust.*	*Electricité*	*Chaleur*	*Total*
Indigenous Production	-	-	-	-	-	-	-	48	-	-	48
Imports	3512	-	20165	2309	-	-	-	10	677	-	26674
Exports	-	-	-10397	-	-	-	-	-1	-48	-	-10447
Intl. Marine Bunkers	-	-	-2101	-	-	-	-	-	-	-	-2101
Stock Changes	-	-	-53	-	-	-	-	-	-	-	-53
TPES	**3512**	**-**	**7613**	**2309**	**-**	**-**	**-**	**57**	**629**	**-**	**14121**
Transfers	-	-	-	-	-	-	-	-	-	-	-
Statistical Differences	1575	-	2	-	-	-	-	-	-	-	1577
Electricity Plants	-5087	-	-82	-	-	-	-	-	2489	-	-2680
CHP Plants	-	-	-	-	-	-	-	-	-	-	-
Heat Plants	-	-	-	-	-	-	-	-	-	-	-
Gas Works	-	-	-487	514	-	-	-	-	-	-	27
Petroleum Refineries	-	-	-	-	-	-	-	-	-	-	-
Coal Transformation	-	-	-	-	-	-	-	-	-	-	-
Liquefaction	-	-	-	-	-	-	-	-	-	-	-
Other Transformation	-	-	-	-	-	-	-	-	-	-	-
Own Use	-	-	-	-	-	-	-	-	-	-	-
Distribution Losses	-	-	-	-	-	-	-	-	-345	-	-345
TFC	**-**	**-**	**7047**	**2823**	**-**	**-**	**-**	**57**	**2773**	**-**	**12700**
INDUSTRY SECTOR	**-**	**-**	**181**	**2329**	**-**	**-**	**-**	**-**	**453**	**-**	**2963**
Iron and Steel	-	-	-	-	-	-	-	-	-	-	-
Chemical and Petrochemical	-	-	-	-	-	-	-	-	-	-	-
of which: Feedstocks	-	-	-	-	-	-	-	-	-	-	-
Non-Ferrous Metals	-	-	-	-	-	-	-	-	-	-	-
Non-Metallic Minerals	-	-	-	-	-	-	-	-	-	-	-
Transport Equipment	-	-	-	-	-	-	-	-	-	-	-
Machinery	-	-	-	-	-	-	-	-	-	-	-
Mining and Quarrying	-	-	-	-	-	-	-	-	-	-	-
Food and Tobacco	-	-	-	-	-	-	-	-	-	-	-
Paper, Pulp and Printing	-	-	-	-	-	-	-	-	-	-	-
Wood and Wood Products	-	-	-	-	-	-	-	-	-	-	-
Construction	-	-	-	-	-	-	-	-	-	-	-
Textile and Leather	-	-	-	-	-	-	-	-	-	-	-
Non-specified	-	-	181	2329	-	-	-	-	453	-	2963
TRANSPORT SECTOR	**-**	**-**	**6509**	**-**	**-**	**-**	**-**	**-**	**-**	**-**	**6509**
Air	-	-	3410	-	-	-	-	-	-	-	3410
Road	-	-	3098	-	-	-	-	-	-	-	3098
Rail	-	-	-	-	-	-	-	-	-	-	-
Pipeline Transport	-	-	-	-	-	-	-	-	-	-	-
Internal Navigation	-	-	-	-	-	-	-	-	-	-	-
Non-specified	-	-	-	-	-	-	-	-	-	-	-
OTHER SECTORS	**-**	**-**	**57**	**494**	**-**	**-**	**-**	**57**	**2320**	**-**	**2928**
Agriculture	-	-	-	-	-	-	-	-	-	-	-
Comm. and Publ. Services	-	-	1	226	-	-	-	-	1629	-	1856
Residential	-	-	55	268	-	-	-	48	691	-	1063
Non-specified	-	-	-	-	-	-	-	9	-	-	9
NON-ENERGY USE	**-**	**-**	**300**	**-**	**-**	**-**	**-**	**-**	**-**	**-**	**300**
in Industry/Transf./Energy	-	-	300	-	-	-	-	-	-	-	300
in Transport	-	-	-	-	-	-	-	-	-	-	-
in Other Sectors	-	-	-	-	-	-	-	-	-	-	-
Electr. Generated - GWh	*28615*	*-*	*330*	*-*	*-*	*-*	*-*	*-*	*-*	*-*	*28945*
Electricity Plants	*28615*	*-*	*330*	*-*	*-*	*-*	*-*	*-*	*-*	*-*	*28945*
CHP Plants	*-*	*-*	*-*	*-*	*-*	*-*	*-*	*-*	*-*	*-*	*-*
Heat Generated - TJ	*-*	*-*	*-*	*-*	*-*	*-*	*-*	*-*	*-*	*-*	*-*
CHP Plants	*-*	*-*	*-*	*-*	*-*	*-*	*-*	*-*	*-*	*-*	*-*
Heat Plants	*-*	*-*	*-*	*-*	*-*	*-*	*-*	*-*	*-*	*-*	*-*

India / Inde

Total Production of Energy

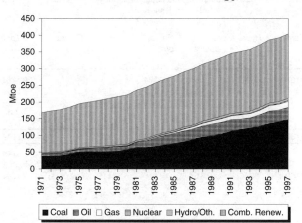

■ Coal ▥ Oil ☐ Gas ▤ Nuclear ▨ Hydro/Oth. ▦ Comb. Renew.

Total Primary Energy Supply*

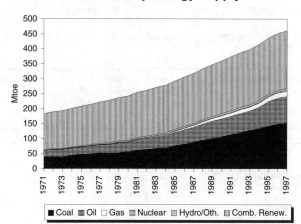

■ Coal ▥ Oil ☐ Gas ▤ Nuclear ▨ Hydro/Oth. ▦ Comb. Renew.

Electricity Generation by Fuel

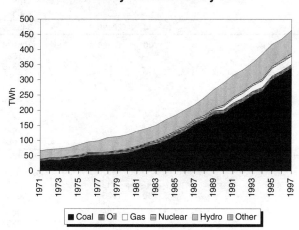

■ Coal ▥ Oil ☐ Gas ▤ Nuclear ▨ Hydro ▦ Other

Oil Products Consumption

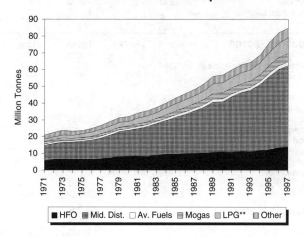

■ HFO ▥ Mid. Dist. ☐ Av. Fuels ▤ Mogas ▨ LPG** ▦ Other

Electricity Consumption/GDP***, TPES/GDP*** and Production/TPES

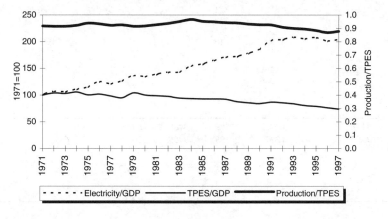

- - - - Electricity/GDP ——— TPES/GDP ——— Production/TPES

* Excluding electricity trade.
** Includes LPG, NGL, ethane and naphtha.
*** GDP in 1990 US dollars.

India / Inde : 1996

Thousand tonnes of oil equivalent / Milliers de tonnes d'équivalent pétrole										
SUPPLY AND CONSUMPTION *APPROVISIONNEMENT ET DEMANDE*	Coal *Charbon*	Crude Oil *Pétrole brut*	Petroleum Products *Produits pétroliers*	Gas *Gaz*	Nuclear *Nucléaire*	Hydro *Hydro*	Geotherm. Solar etc. *Géotherm. solaire*	Combust. Renew. & Waste *En. ren. combust.*	Electricity *Electricité*	Heat *Chaleur*	Total *Total*
Indigenous Production	141608	34906	-	17302	2189	5940	5	190474	-	-	392424
Imports	5812	34650	21098	-	-	-	-	-	144	-	61704
Exports	-62	-416	-2725	-	-	-	-	-	-11	-	-3214
Intl. Marine Bunkers	-	-	-533	-	-	-	-	-	-	-	-533
Stock Changes	1728	-	-	-	-	-	-	-	-	-	1728
TPES	**149085**	**69140**	**17840**	**17302**	**2189**	**5940**	**5**	**190474**	**133**	**-**	**452109**
Transfers	-	-1529	1683	-	-	-	-	-	-	-	153
Statistical Differences	-264	-2425	-627	-	-	-	-	-	-288	-	-3605
Electricity Plants	-99952	-	-2526	-5988	-2189	-5940	-5	-	37416	-	-79184
CHP Plants	-	-	-	-	-	-	-	-	-	-	-
Heat Plants	-	-	-	-	-	-	-	-	-	-	-
Gas Works	-	-	-	-	-	-	-	-	-	-	-
Petroleum Refineries	-	-64397	64365	-	-	-	-	-	-	-	-33
Coal Transformation	-5906	-	-	-	-	-	-	-	-	-	-5906
Liquefaction	-	-	-	-	-	-	-	-	-	-	-
Other Transformation	-	-	-	-	-	-	-	-	-	-	-
Own Use	-1637	-	-4295	-2207	-	-	-	-	-2387	-	-10526
Distribution Losses	-	-	-	-	-	-	-	-	-6685	-	-6685
TFC	**41326**	**789**	**76439**	**9107**	**-**	**-**	**-**	**190474**	**28190**	**-**	**346324**
INDUSTRY SECTOR	**41214**	**789**	**15689**	**8736**	**-**	**-**	**-**	**21700**	**12866**	**-**	**100994**
Iron and Steel	9807	-	835	-	-	-	-	-	1018	-	11660
Chemical and Petrochemical	1184	789	8514	8538	-	-	-	-	590	-	19615
of which: Feedstocks	-	789	3895	595	-	-	-	-	-	-	5279
Non-Ferrous Metals	-	-	173	-	-	-	-	-	-	-	173
Non-Metallic Minerals	4811	-	719	-	-	-	-	-	-	-	5530
Transport Equipment	-	-	-	-	-	-	-	-	-	-	-
Machinery	-	-	585	-	-	-	-	-	-	-	585
Mining and Quarrying	-	-	825	-	-	-	-	-	-	-	825
Food and Tobacco	-	-	1084	-	-	-	-	-	-	-	1084
Paper, Pulp and Printing	1675	-	-	-	-	-	-	-	-	-	1675
Wood and Wood Products	-	-	-	-	-	-	-	-	-	-	-
Construction	482	-	-	-	-	-	-	-	-	-	482
Textile and Leather	625	-	1107	-	-	-	-	-	892	-	2625
Non-specified	22629	-	1847	198	-	-	-	21700	10366	-	56739
TRANSPORT SECTOR	**67**	**-**	**39420**	**-**	**-**	**-**	**-**	**-**	**565**	**-**	**40052**
Air	-	-	2334	-	-	-	-	-	-	-	2334
Road	-	-	34589	-	-	-	-	-	-	-	34589
Rail	67	-	1829	-	-	-	-	-	565	-	2462
Pipeline Transport	-	-	-	-	-	-	-	-	-	-	-
Internal Navigation	-	-	658	-	-	-	-	-	-	-	658
Non-specified	-	-	9	-	-	-	-	-	-	-	9
OTHER SECTORS	**46**	**-**	**16434**	**371**	**-**	**-**	**-**	**168774**	**14758**	**-**	**200383**
Agriculture	-	-	331	153	-	-	-	-	7618	-	8103
Comm. and Publ. Services	-	-	-	-	-	-	-	-	2217	-	2217
Residential	46	-	15931	218	-	-	-	168774	4603	-	189572
Non-specified	-	-	172	-	-	-	-	-	319	-	491
NON-ENERGY USE	**-**	**-**	**4896**	**-**	**-**	**-**	**-**	**-**	**-**	**-**	**4896**
in Industry/Transf./Energy	-	-	4896	-	-	-	-	-	-	-	4896
in Transport	-	-	-	-	-	-	-	-	-	-	-
in Other Sectors	-	-	-	-	-	-	-	-	-	-	-
Electr. Generated - GWh	*318357*	*-*	*12000*	*27189*	*8400*	*69072*	*57*	*-*	*-*	*-*	*435075*
Electricity Plants	*318357*	*-*	*12000*	*27189*	*8400*	*69072*	*57*	*-*	*-*	*-*	*435075*
CHP Plants	*-*	*-*	*-*	*-*	*-*	*-*	*-*	*-*	*-*	*-*	*-*
Heat Generated - TJ	*-*	*-*	*-*	*-*	*-*	*-*	*-*	*-*	*-*	*-*	*-*
CHP Plants	*-*	*-*	*-*	*-*	*-*	*-*	*-*	*-*	*-*	*-*	*-*
Heat Plants	*-*	*-*	*-*	*-*	*-*	*-*	*-*	*-*	*-*	*-*	*-*

India / Inde : 1997

Thousand tonnes of oil equivalent / *Milliers de tonnes d'équivalent pétrole*

SUPPLY AND CONSUMPTION / *APPROVISIONNEMENT ET DEMANDE*	Coal / *Charbon*	Crude Oil / *Pétrole brut*	Petroleum Products / *Produits pétroliers*	Gas / *Gaz*	Nuclear / *Nucléaire*	Hydro / *Hydro*	Geotherm. Solar etc. / *Géotherm. solaire*	Combust. Renew. & Waste / *En. ren. combust.*	Electricity / *Electricité*	Heat / *Chaleur*	Total / *Total*
Indigenous Production	147233	37586	-	17815	2632	6431	5	192801	-	-	404503
Imports	5812	33079	20127	-	-	-	-	-	144	-	59162
Exports	-29	-610	-1726	-	-	-	-	-	-11	-	-2376
Intl. Marine Bunkers	-	-	-519	-	-	-	-	-	-	-	-519
Stock Changes	263	-	-	-	-	-	-	-	-	-	263
TPES	**153279**	**70056**	**17882**	**17815**	**2632**	**6431**	**5**	**192801**	**133**	**-**	**461032**
Transfers	-	-1732	1905	-	-	-	-	-	-	-	174
Statistical Differences	-725	1	-1143	-	-	-	-	-	-307	-	-2174
Electricity Plants	-102864	-	-2526	-6075	-2632	-6431	-5	-	39853	-	-80680
CHP Plants	-	-	-	-	-	-	-	-	-	-	-
Heat Plants	-	-	-	-	-	-	-	-	-	-	-
Gas Works	-	-	-	-	-	-	-	-	-	-	-
Petroleum Refineries	-	-67649	67558								-91
Coal Transformation	-5601	-	-	-	-	-	-	-	-	-	-5601
Liquefaction	-	-	-	-	-	-	-	-	-	-	-
Other Transformation	-	-	-	-	-	-	-	-	-	-	-
Own Use	-1475	-	-4477	-2315	-	-	-	-	-2542	-	-10809
Distribution Losses	-	-	-	-	-	-	-	-	-7118	-	-7118
TFC	**42614**	**676**	**79198**	**9425**	**-**	**-**	**-**	**192801**	**30018**	**-**	**354732**
INDUSTRY SECTOR	**42540**	**676**	**17328**	**9020**	**-**	**-**	**-**	**21965**	**13701**	**-**	**105229**
Iron and Steel	9936	-	864	-	-	-	-	-	1084	-	11885
Chemical and Petrochemical	1206	676	9884	8793	-	-	-	-	628	-	21188
of which: Feedstocks	-	*676*	*5111*	*639*	-	-	-	-	-	-	*6426*
Non-Ferrous Metals	-	-	179	-	-	-	-	-	-	-	179
Non-Metallic Minerals	4811	-	748	-	-	-	-	-	-	-	5560
Transport Equipment	-	-	-	-	-	-	-	-	-	-	-
Machinery	-	-	609	-	-	-	-	-	-	-	609
Mining and Quarrying	-	-	862	-	-	-	-	-	-	-	862
Food and Tobacco	-	-	1131	-	-	-	-	-	-	-	1131
Paper, Pulp and Printing	1637	-	-	-	-	-	-	-	-	-	1637
Wood and Wood Products	-	-	-	-	-	-	-	-	-	-	-
Construction	764	-	-	-	-	-	-	-	-	-	764
Textile and Leather	649	-	1148	-	-	-	-	-	950	-	2747
Non-specified	23537	-	1902	227	-	-	-	21965	11038	-	58668
TRANSPORT SECTOR	**29**	**-**	**40909**	**-**	**-**	**-**	**-**	**-**	**602**	**-**	**41540**
Air	-	-	2373	-	-	-	-	-	-	-	2373
Road	-	-	35931	-	-	-	-	-	-	-	35931
Rail	29	-	1912	-	-	-	-	-	602	-	2543
Pipeline Transport	-	-	-	-	-	-	-	-	-	-	-
Internal Navigation	-	-	684	-	-	-	-	-	-	-	684
Non-specified	-	-	9	-	-	-	-	-	-	-	9
OTHER SECTORS	**46**	**-**	**16909**	**405**	**-**	**-**	**-**	**170836**	**15715**	**-**	**203911**
Agriculture	-	-	342	180	-	-	-	-	8113	-	8634
Comm. and Publ. Services	-	-	-	-	-	-	-	-	2361	-	2361
Residential	46	-	16390	225	-	-	-	170836	4902	-	192398
Non-specified	-	-	178	-	-	-	-	-	340	-	517
NON-ENERGY USE	**-**	**-**	**4053**	**-**	**-**	**-**	**-**	**-**	**-**	**-**	**4053**
in Industry/Transf./Energy	-	-	4053	-	-	-	-	-	-	-	4053
in Transport	-	-	-	-	-	-	-	-	-	-	-
in Other Sectors	-	-	-	-	-	-	-	-	-	-	-
Electr. Generated - GWh	*338887*	*-*	*12000*	*27583*	*10100*	*74775*	*57*	*-*	*-*	*-*	*463402*
Electricity Plants	*338887*	*-*	*12000*	*27583*	*10100*	*74775*	*57*	*-*	*-*	*-*	*463402*
CHP Plants	*-*	*-*	*-*	*-*	*-*	*-*	*-*	*-*	*-*	*-*	*-*
Heat Generated - TJ	*-*	*-*	*-*	*-*	*-*	*-*	*-*	*-*	*-*	*-*	*-*
CHP Plants	*-*	*-*	*-*	*-*	*-*	*-*	*-*	*-*	*-*	*-*	*-*
Heat Plants	*-*	*-*	*-*	*-*	*-*	*-*	*-*	*-*	*-*	*-*	*-*

Indonesia / Indonésie

Total Production of Energy

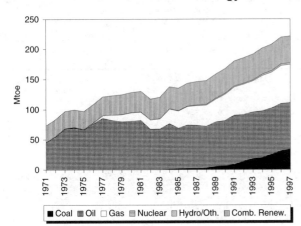

■ Coal ▦ Oil □ Gas ▤ Nuclear ▥ Hydro/Oth. ▨ Comb. Renew.

Total Primary Energy Supply*

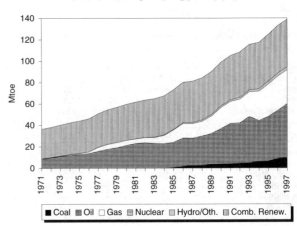

■ Coal ▦ Oil □ Gas ▤ Nuclear ▥ Hydro/Oth. ▨ Comb. Renew.

Electricity Generation by Fuel

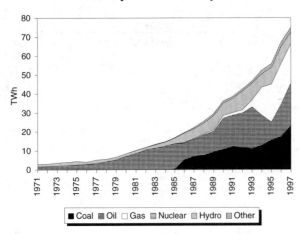

■ Coal ▦ Oil □ Gas ▤ Nuclear ▥ Hydro ▨ Other

Oil Products Consumption

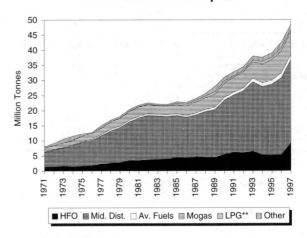

■ HFO ▦ Mid. Dist. □ Av. Fuels ▤ Mogas ▨ LPG** ▨ Other

Electricity Consumption/GDP***, TPES/GDP*** and Production/TPES

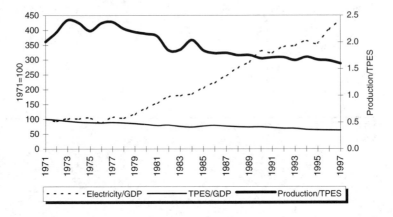

- - - - Electricity/GDP ——— TPES/GDP ——— Production/TPES

* Excluding electricity trade.
** Includes LPG, NGL, ethane and naphtha.
*** GDP in 1990 US dollars.

Indonesia / Indonésie : 1996

	Coal	Crude Oil	Petroleum Products	Gas	Nuclear	Hydro	Geotherm. Solar etc.	Combust. Renew. & Waste	Electricity	Heat	Total
	Charbon	Pétrole brut	Produits pétroliers	Gaz	Nucléaire	Hydro	Géotherm. solaire	En. ren. combust.	Electricité	Chaleur	Total

Thousand tonnes of oil equivalent / *Milliers de tonnes d'équivalent pétrole*

SUPPLY AND CONSUMPTION / *APPROVISIONNEMENT ET DEMANDE*

	Coal	Crude Oil	Petroleum Products	Gas	Nuclear	Hydro	Geotherm. Solar etc.	Combust. Renew. & Waste	Electricity	Heat	Total
Indigenous Production	30847	78514	-	63004	-	769	1988	44595	-	-	219717
Imports	253	9124	10342	-	-	-	-	-	-	-	19719
Exports	-22412	-38784	-13943	-31169	-	-	-	-102	-	-	-106410
Intl. Marine Bunkers	-	-	-332	-	-	-	-	-	-	-	-332
Stock Changes	254	-	4	-	-	-	-	-	-	-	258
TPES	**8941**	**48854**	**-3929**	**31835**	**-**	**769**	**1988**	**44493**	**-**	**-**	**132951**
Transfers	-	-2993	3311	-	-	-	-	-	-	-	318
Statistical Differences	-1205	325	105	-33	-	-	-	9	209	-	-591
Electricity Plants	-4585	-	-3297	-5773	-	-769	-1988	-	5767	-	-10645
CHP Plants	-	-	-	-	-	-	-	-	-	-	-
Heat Plants	-	-	-	-	-	-	-	-	-	-	-
Gas Works	-	-	-	473	-	-	-	-	-	-	473
Petroleum Refineries	-	-46186	44191	-	-	-	-	-	-	-	-1996
Coal Transformation	-	-	-	-	-	-	-	-	-	-	-
Liquefaction	-	-	-	-	-	-	-	-	-	-	-
Other Transformation	-	-	-	-	-	-	-	-122	-	-	-122
Own Use	-	-	-1995	-17177	-	-	-	-	-241	-	-19414
Distribution Losses	-	-	-	-	-	-	-	-	-712	-	-712
TFC	**3150**	**-**	**38385**	**9324**	**-**	**-**	**-**	**44381**	**5023**	**-**	**100263**
INDUSTRY SECTOR	**3150**	**-**	**7779**	**7566**	**-**	**-**	**-**	**-**	**2443**	**-**	**20938**
Iron and Steel	-	-	849	1062	-	-	-	-	-	-	1911
Chemical and Petrochemical	-	-	1151	5801	-	-	-	-	-	-	6951
of which: Feedstocks	-	-	-	*5801*	-	-	-	-	-	-	*5801*
Non-Ferrous Metals	-	-	-	-	-	-	-	-	-	-	-
Non-Metallic Minerals	3150	-	1129	130	-	-	-	-	-	-	4408
Transport Equipment	-	-	-	-	-	-	-	-	-	-	-
Machinery	-	-	69	-	-	-	-	-	-	-	69
Mining and Quarrying	-	-	1059	-	-	-	-	-	-	-	1059
Food and Tobacco	-	-	700	-	-	-	-	-	-	-	700
Paper, Pulp and Printing	-	-	-	-	-	-	-	-	-	-	-
Wood and Wood Products	-	-	-	-	-	-	-	-	-	-	-
Construction	-	-	337	-	-	-	-	-	-	-	337
Textile and Leather	-	-	1209	-	-	-	-	-	-	-	1209
Non-specified	-	-	1276	573	-	-	-	-	2443	-	4292
TRANSPORT SECTOR	**-**	**-**	**18875**	**-**	**-**	**-**	**-**	**-**	**-**	**-**	**18875**
Air	-	-	1708	-	-	-	-	-	-	-	1708
Road	-	-	15642	-	-	-	-	-	-	-	15642
Rail	-	-	-	-	-	-	-	-	-	-	-
Pipeline Transport	-	-	-	-	-	-	-	-	-	-	-
Internal Navigation	-	-	1525	-	-	-	-	-	-	-	1525
Non-specified	-	-	-	-	-	-	-	-	-	-	-
OTHER SECTORS	**-**	**-**	**10647**	**1759**	**-**	**-**	**-**	**44381**	**2579**	**-**	**59366**
Agriculture	-	-	1668	-	-	-	-	-	-	-	1668
Comm. and Publ. Services	-	-	309	236	-	-	-	-	529	-	1075
Residential	-	-	8670	1522	-	-	-	44381	1766	-	56339
Non-specified	-	-	-	-	-	-	-	-	284	-	284
NON-ENERGY USE	**-**	**-**	**1085**	**-**	**-**	**-**	**-**	**-**	**-**	**-**	**1085**
in Industry/Transf./Energy	-	-	1085	-	-	-	-	-	-	-	1085
in Transport	-	-	-	-	-	-	-	-	-	-	-
in Other Sectors	-	-	-	-	-	-	-	-	-	-	-
Electr. Generated - GWh	***17344***	***-***	***17018***	***21447***	***-***	***8941***	***2312***	***-***	***-***	***-***	***67062***
Electricity Plants	*17344*	*-*	*17018*	*21447*	*-*	*8941*	*2312*	*-*	*-*	*-*	*67062*
CHP Plants	*-*	*-*	*-*	*-*	*-*	*-*	*-*	*-*	*-*	*-*	*-*
Heat Generated - TJ	***-***	***-***	***-***	***-***	***-***	***-***	***-***	***-***	***-***	***-***	***-***
CHP Plants	*-*	*-*	*-*	*-*	*-*	*-*	*-*	*-*	*-*	*-*	*-*
Heat Plants	*-*	*-*	*-*	*-*	*-*	*-*	*-*	*-*	*-*	*-*	*-*

Indonesia / Indonésie : 1997

Thousand tonnes of oil equivalent / *Milliers de tonnes d'équivalent pétrole*

SUPPLY AND CONSUMPTION	Coal	Crude Oil	Petroleum Products	Gas	Nuclear	Hydro	Geotherm. Solar etc.	Combust. Renew. & Waste	Electricity	Heat	Total
APPROVISIONNEMENT ET DEMANDE	*Charbon*	*Pétrole brut*	*Produits pétroliers*	*Gaz*	*Nucléaire*	*Hydro*	*Géotherm. solaire*	*En. ren. combust.*	*Electricité*	*Chaleur*	*Total*
Indigenous Production	33888	77291	-	63043	-	515	2217	44595	-	-	221549
Imports	237	8706	14339	-	-	-	-	-	-	-	23282
Exports	-25507	-39383	-10119	-31537	-	-	-	-102	-	-	-106647
Intl. Marine Bunkers	-	-	-331	-	-	-	-	-	-	-	-331
Stock Changes	927	-	-	-	-	-	-	-	-	-	927
TPES	**9545**	**46614**	**3889**	**31506**	**-**	**515**	**2217**	**44493**	**-**	**-**	**138779**
Transfers	-	-2226	2462	-	-	-	-	-	-	-	236
Statistical Differences	-725	2496	-1907	76	-	-	-	9	231	-	180
Electricity Plants	-6081	-	-4862	-5603	-	-515	-2217	-	6436	-	-12843
CHP Plants	-	-	-	-	-	-	-	-	-	-	-
Heat Plants	-	-	-	-	-	-	-	-	-	-	-
Gas Works	-	-	-	473	-	-	-	-	-	-	473
Petroleum Refineries	-	-46884	44959	-	-	-	-	-	-	-	-1925
Coal Transformation	-	-	-	-	-	-	-	-	-	-	-
Liquefaction	-	-	-	-	-	-	-	-	-	-	-
Other Transformation	-	-	-	-	-	-	-	-122	-	-	-122
Own Use	-	-	-1984	-16580	-	-	-	-	-251	-	-18815
Distribution Losses	-	-	-	-	-	-	-	-	-741	-	-741
TFC	**2739**	**-**	**42557**	**9872**	**-**	**-**	**-**	**44381**	**5674**	**-**	**105222**
INDUSTRY SECTOR	**2739**	**-**	**10653**	**7869**	**-**	**-**	**-**	**-**	**2689**	**-**	**23950**
Iron and Steel	-	-	1616	985	-	-	-	-	-	-	2600
Chemical and Petrochemical	-	-	1530	5988	-	-	-	-	-	-	7518
of which: Feedstocks	-	-	-	*5988*	-	-	-	-	-	-	*5988*
Non-Ferrous Metals	-	-	-	-	-	-	-	-	-	-	-
Non-Metallic Minerals	2739	-	1756	112	-	-	-	-	-	-	4606
Transport Equipment	-	-	-	-	-	-	-	-	-	-	-
Machinery	-	-	75	-	-	-	-	-	-	-	75
Mining and Quarrying	-	-	1260	-	-	-	-	-	-	-	1260
Food and Tobacco	-	-	923	-	-	-	-	-	-	-	923
Paper, Pulp and Printing	-	-	-	-	-	-	-	-	-	-	-
Wood and Wood Products	-	-	-	-	-	-	-	-	-	-	-
Construction	-	-	362	-	-	-	-	-	-	-	362
Textile and Leather	-	-	1475	-	-	-	-	-	-	-	1475
Non-specified	-	-	1656	785	-	-	-	-	2689	-	5131
TRANSPORT SECTOR	**-**	**-**	**19866**	**-**	**-**	**-**	**-**	**-**	**-**	**-**	**19866**
Air	-	-	1787	-	-	-	-	-	-	-	1787
Road	-	-	16253	-	-	-	-	-	-	-	16253
Rail	-	-	-	-	-	-	-	-	-	-	-
Pipeline Transport	-	-	-	-	-	-	-	-	-	-	-
Internal Navigation	-	-	1826	-	-	-	-	-	-	-	1826
Non-specified	-	-	-	-	-	-	-	-	-	-	-
OTHER SECTORS	**-**	**-**	**10954**	**2003**	**-**	**-**	**-**	**44381**	**2985**	**-**	**60322**
Agriculture	-	-	1902	-	-	-	-	-	-	-	1902
Comm. and Publ. Services	-	-	332	236	-	-	-	-	616	-	1184
Residential	-	-	8720	1766	-	-	-	44381	2054	-	56921
Non-specified	-	-	-	-	-	-	-	-	315	-	315
NON-ENERGY USE	**-**	**-**	**1085**	**-**	**-**	**-**	**-**	**-**	**-**	**-**	**1085**
in Industry/Transf./Energy	-	-	1085	-	-	-	-	-	-	-	1085
in Transport	-	-	-	-	-	-	-	-	-	-	-
in Other Sectors	-	-	-	-	-	-	-	-	-	-	-
Electr. Generated - GWh	*23001*	*-*	*22447*	*20816*	*-*	*5990*	*2578*	*-*	*-*	*-*	*74832*
Electricity Plants	*23001*	*-*	*22447*	*20816*	*-*	*5990*	*2578*	*-*	*-*	*-*	*74832*
CHP Plants	*-*	*-*	*-*	*-*	*-*	*-*	*-*	*-*	*-*	*-*	*-*
Heat Generated - TJ	*-*	*-*	*-*	*-*	*-*	*-*	*-*	*-*	*-*	*-*	*-*
CHP Plants	*-*	*-*	*-*	*-*	*-*	*-*	*-*	*-*	*-*	*-*	*-*
Heat Plants	*-*	*-*	*-*	*-*	*-*	*-*	*-*	*-*	*-*	*-*	*-*

Iran

Thousand tonnes of oil equivalent / *Milliers de tonnes d'équivalent pétrole*

SUPPLY AND CONSUMPTION 1996	Coal	Crude Oil	Petroleum Products	Gas	Nuclear	Hydro	Geotherm. Solar etc.	Combust. Renew. & Waste	Electricity	Heat	Total
Indigenous Production	567	187299	-	34177	-	634	-	786	-	-	223464
Imports	368	-	9256	-	-	-	-	-	-	-	9624
Exports	-8	-121904	-4510	-82	-	-	-	-	-	-	-126504
Intl. Marine Bunkers	-	-	-540	-	-	-	-	-	-	-	-540
Stock changes	-	-3694	-	-	-	-	-	-	-	-	-3694
TPES	927	61702	4207	34096	-	634	-	786	-	-	102351
Electricity and CHP Plants	-	-	-8509	-10547	-	-634	-	-	7813	-	-11877
Petroleum Refineries	-	-61703	56836	-	-	-	-	-	-	-	-4866
Other Transformation *	-350	1	-2234	1968	-	-	-	-68	-1925	-	-2607
TFC	577	-	50301	25517	-	-	-	718	5888	-	83001
INDUSTRY SECTOR	577	-	7654	15185	-	-	-	186	1868	-	25470
Iron and Steel	116	-	-	-	-	-	-	-	-	-	116
Chemical and Petrochemical	-	-	374	4051	-	-	-	-	-	-	4425
Non-Metallic Minerals	-	-	-	-	-	-	-	-	-	-	-
Non-specified	461	-	7280	11135	-	-	-	186	1868	-	20930
TRANSPORT SECTOR	-	-	19445	-	-	-	-	-	-	-	19445
Air	-	-	730	-	-	-	-	-	-	-	730
Road	-	-	18715	-	-	-	-	-	-	-	18715
Non-specified	-	-	-	-	-	-	-	-	-	-	-
OTHER SECTORS	-	-	20265	10332	-	-	-	532	4020	-	35148
Agriculture	-	-	4085	-	-	-	-	-	493	-	4578
Comm. and Publ. Services	-	-	3789	1362	-	-	-	-	1373	-	6525
Residential	-	-	12391	8969	-	-	-	-	1913	-	23273
Non-specified	-	-	-	-	-	-	-	532	241	-	773
NON-ENERGY USE	-	-	2938	-	-	-	-	-	-	-	2938
Electricity Generated - GWh	-	-	33741	49734	-	7376	-	-	-	-	90851
Heat Generated - TJ	-	-	-	-	-	-	-	-	-	-	-

APPROVISIONNEMENT ET DEMANDE 1997	Charbon	Pétrole brut	Produits pétroliers	Gaz	Nucléaire	Hydro	Géotherm. solaire etc.	En. ren. combust. & déchets	Electricité	Chaleur	Total
Indigenous Production	567	184550	-	38398	-	634	-	786	-	-	224935
Imports	368	-	11711	-	-	-	-	-	-	-	12079
Exports	-8	-122106	-6193	-82	-	-	-	-	-	-	-128389
Intl. Marine Bunkers	-	-	-540	-	-	-	-	-	-	-	-540
Stock changes	-	203	-	-	-	-	-	-	-	-	203
TPES	927	62647	4979	38316	-	634	-	786	-	-	108289
Electricity and CHP Plants	-	-	-8172	-11865	-	-634	-	-	8238	-	-12433
Petroleum Refineries	-	-62646	56071	-	-	-	-	-	-	-	-6575
Other Transformation *	-350	-1	-2143	2256	-	-	-	-71	-2144	-	-2452
TFC	577	-	50735	28706	-	-	-	715	6094	-	86828
INDUSTRY SECTOR	577	-	7458	17083	-	-	-	186	1933	-	27238
Iron and Steel	116	-	-	-	-	-	-	-	-	-	116
Chemical and Petrochemical	-	-	377	4557	-	-	-	-	-	-	4934
Non-Metallic Minerals	-	-	-	-	-	-	-	-	-	-	-
Non-specified	461	-	7081	12526	-	-	-	186	1933	-	22188
TRANSPORT SECTOR	-	-	20142	-	-	-	-	-	-	-	20142
Air	-	-	732	-	-	-	-	-	-	-	732
Road	-	-	19410	-	-	-	-	-	-	-	19410
Non-specified	-	-	-	-	-	-	-	-	-	-	-
OTHER SECTORS	-	-	20190	11623	-	-	-	529	4161	-	36503
Agriculture	-	-	4222	-	-	-	-	-	510	-	4732
Comm. and Publ. Services	-	-	3685	1533	-	-	-	-	1421	-	6638
Residential	-	-	12283	10091	-	-	-	-	1980	-	24354
Non-specified	-	-	-	-	-	-	-	529	250	-	779
NON-ENERGY USE	-	-	2944	-	-	-	-	-	-	-	2944
Electricity Generated - GWh	-	-	32467	55951	-	7376	-	-	-	-	95794
Heat Generated - TJ	-	-	-	-	-	-	-	-	-	-	-

* Includes Transfers, Statistical Differences, Own Use and Distribution Losses.

Iraq / Irak

Thousand tonnes of oil equivalent / *Milliers de tonnes d'équivalent pétrole*

SUPPLY AND CONSUMPTION 1996	Coal	Crude Oil	Petroleum Products	Gas	Nuclear	Hydro	Geotherm. Solar etc.	Combust. Renew. & Waste	Electricity	Heat	Total
Indigenous Production	-	29196	-	3312	-	49	-	26	-	-	32584
Imports	-	-	-	-	-	-	-	-	-	-	-
Exports	-	-4427	-3131	-	-	-	-	-	-	-	-7557
Intl. Marine Bunkers	-	-	-	-	-	-	-	-	-	-	-
Stock changes	-	-	-	-	-	-	-	-	-	-	-
TPES	-	24770	-3131	3312	-	49	-	26	-	-	25027
Electricity and CHP Plants	-	-1841	-3264	-	-	-49	-	-	2494	-	-2660
Petroleum Refineries	-	-23035	21632	-	-	-	-	-	-	-	-1403
Other Transformation *	-	106	-1061	-	-	-	-	-11	-	-	-966
TFC	-	-	14177	3312	-	-	-	15	2494	-	19998
INDUSTRY SECTOR	-	-	2183	3312	-	-	-	-	-	-	5495
Iron and Steel	-	-	-	-	-	-	-	-	-	-	-
Chemical and Petrochemical	-	-	513	-	-	-	-	-	-	-	513
Non-Metallic Minerals	-	-	-	-	-	-	-	-	-	-	-
Non-specified	-	-	1670	3312	-	-	-	-	-	-	4983
TRANSPORT SECTOR	-	-	9005	-	-	-	-	-	-	-	9005
Air	-	-	439	-	-	-	-	-	-	-	439
Road	-	-	8566	-	-	-	-	-	-	-	8566
Non-specified	-	-	-	-	-	-	-	-	-	-	-
OTHER SECTORS	-	-	2269	-	-	-	-	15	2494	-	4778
Agriculture	-	-	-	-	-	-	-	-	-	-	-
Comm. and Publ. Services	-	-	-	-	-	-	-	-	-	-	-
Residential	-	-	2269	-	-	-	-	-	-	-	2269
Non-specified	-	-	-	-	-	-	-	15	2494	-	2509
NON-ENERGY USE	-	-	720	-	-	-	-	-	-	-	720
Electricity Generated - GWh	-	-	28430	-	-	570	-	-	-	-	29000
Heat Generated - TJ	-	-	-	-	-	-	-	-	-	-	-

APPROVISIONNEMENT ET DEMANDE 1997	Charbon	Pétrole brut	Produits pétroliers	Gaz	Nucléaire	Hydro	Géotherm. solaire etc.	En. ren. combust. & déchets	Electricité	Chaleur	Total
Indigenous Production	-	58311	-	3701	-	50	-	26	-	-	62088
Imports	-	-	-	-	-	-	-	-	-	-	-
Exports	-	-32320	-2677	-	-	-	-	-	-	-	-34997
Intl. Marine Bunkers	-	-	-	-	-	-	-	-	-	-	-
Stock changes	-	-	-	-	-	-	-	-	-	-	-
TPES	-	25991	-2677	3701	-	50	-	26	-	-	27091
Electricity and CHP Plants	-	-1886	-3357	-	-	-50	-	-	2542	-	-2751
Petroleum Refineries	-	-23595	22158	-	-	-	-	-	-	-	-1437
Other Transformation *	-	-509	-1566	-	-	-	-	-11	-	-	-2087
TFC	-	-	14558	3701	-	-	-	15	2542	-	20816
INDUSTRY SECTOR	-	-	2244	3701	-	-	-	-	-	-	5945
Iron and Steel	-	-	-	-	-	-	-	-	-	-	-
Chemical and Petrochemical	-	-	526	-	-	-	-	-	-	-	526
Non-Metallic Minerals	-	-	-	-	-	-	-	-	-	-	-
Non-specified	-	-	1718	3701	-	-	-	-	-	-	5419
TRANSPORT SECTOR	-	-	9250	-	-	-	-	-	-	-	9250
Air	-	-	451	-	-	-	-	-	-	-	451
Road	-	-	8800	-	-	-	-	-	-	-	8800
Non-specified	-	-	-	-	-	-	-	-	-	-	-
OTHER SECTORS	-	-	2326	-	-	-	-	15	2542	-	4884
Agriculture	-	-	-	-	-	-	-	-	-	-	-
Comm. and Publ. Services	-	-	-	-	-	-	-	-	-	-	-
Residential	-	-	2326	-	-	-	-	-	-	-	2326
Non-specified	-	-	-	-	-	-	-	15	2542	-	2557
NON-ENERGY USE	-	-	737	-	-	-	-	-	-	-	737
Electricity Generated - GWh	-	-	28980	-	-	581	-	-	-	-	29561
Heat Generated - TJ	-	-	-	-	-	-	-	-	-	-	-

* Includes Transfers, Statistical Differences, Own Use and Distribution Losses.

INTERNATIONAL ENERGY AGENCY

Israel / Israël

Total Production of Energy

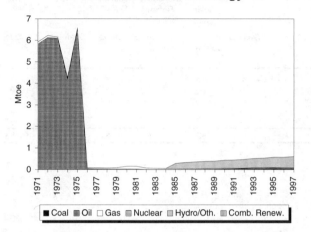

■ Coal ▦ Oil ☐ Gas ▤ Nuclear ▥ Hydro/Oth. ▨ Comb. Renew.

Total Primary Energy Supply*

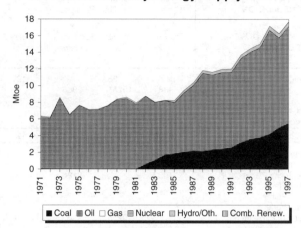

■ Coal ▦ Oil ☐ Gas ▤ Nuclear ▥ Hydro/Oth. ▨ Comb. Renew.

Electricity Generation by Fuel

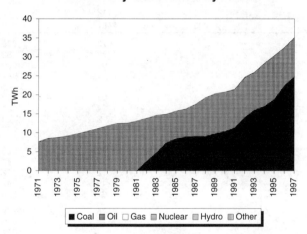

■ Coal ▦ Oil ☐ Gas ▤ Nuclear ▥ Hydro ▨ Other

Oil Products Consumption

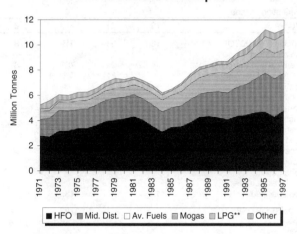

■ HFO ▦ Mid. Dist. ☐ Av. Fuels ▤ Mogas ▥ LPG** ▨ Other

Electricity Consumption/GDP***, TPES/GDP*** and Production/TPES

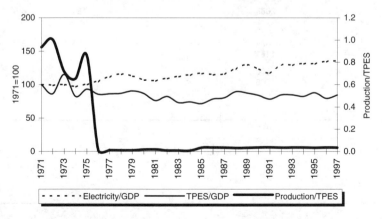

- - - Electricity/GDP —— TPES/GDP ▬▬ Production/TPES

* Excluding electricity trade.
** Includes LPG, NGL, ethane and naphtha.
*** GDP in 1990 US dollars.

Israel / Israël : 1996

Thousand tonnes of oil equivalent / *Milliers de tonnes d'équivalent pétrole*

SUPPLY AND CONSUMPTION *APPROVISIONNEMENT ET DEMANDE*	Coal *Charbon*	Crude Oil *Pétrole brut*	Petroleum Products *Produits pétroliers*	Gas *Gaz*	Nuclear *Nucléaire*	Hydro *Hydro*	Geotherm. Solar etc. *Géotherm. solaire*	Combust. Renew. & Waste *En. ren. combust.*	Electricity *Electricité*	Heat *Chaleur*	Total *Total*
Indigenous Production	42	25	-	12	-	5	485	3	-	-	573
Imports	4481	10251	2066	-	-	-	-	3	-	-	16801
Exports	-	-	-1399	-	-	-	-	-	-84	-	-1483
Intl. Marine Bunkers	-	-	-93	-	-	-	-	-	-	-	-93
Stock Changes	427	-	-40	-	-	-	-	-	-	-	387
TPES	**4950**	**10277**	**534**	**12**	**-**	**5**	**485**	**6**	**-84**	**-**	**16185**
Transfers	-	-	-	-	-	-	-	-	-	-	-
Statistical Differences	-1	-	574	-	-	-	-	-	10	-	583
Electricity Plants	-4933	-	-2196	-4	-	-5	-	-	2795	-	-4344
CHP Plants	-	-	-	-	-	-	-	-	-	-	-
Heat Plants	-	-	-	-	-	-	-	-	-	-	-
Gas Works	-	-	-	-	-	-	-	-	-	-	-
Petroleum Refineries	-	-10255	9862	-	-	-	-	-	-	-	-394
Coal Transformation	-	-	-	-	-	-	-	-	-	-	-
Liquefaction	-	-	-	-	-	-	-	-	-	-	-
Other Transformation	-	-	-	-	-	-	-	-	-	-	-
Own Use	-	-	-480	-	-	-	-	-	-121	-	-601
Distribution Losses	-	-	-	-	-	-	-	-	-112	-	-112
TFC	**16**	**21**	**8293**	**8**	**-**	**-**	**485**	**6**	**2487**	**-**	**11317**
INDUSTRY SECTOR	**16**	**21**	**1919**	**8**	**-**	**-**	**-**	**-**	**673**	**-**	**2638**
Iron and Steel	-	-	-	-	-	-	-	-	83	-	83
Chemical and Petrochemical	4	21	785	8	-	-	-	-	191	-	1009
of which: Feedstocks	-	*21*	*785*	*8*	-	-	-	-	-	-	*814*
Non-Ferrous Metals	-	-	-	-	-	-	-	-	-	-	-
Non-Metallic Minerals	-	-	-	-	-	-	-	-	62	-	62
Transport Equipment	-	-	-	-	-	-	-	-	24	-	24
Machinery	-	-	-	-	-	-	-	-	9	-	9
Mining and Quarrying	-	-	-	-	-	-	-	-	37	-	37
Food and Tobacco	-	-	-	-	-	-	-	-	86	-	86
Paper, Pulp and Printing	-	-	-	-	-	-	-	-	26	-	26
Wood and Wood Products	-	-	-	-	-	-	-	-	12	-	12
Construction	-	-	-	-	-	-	-	-	12	-	12
Textile and Leather	-	-	-	-	-	-	-	-	49	-	49
Non-specified	13	-	1135	-	-	-	-	-	81	-	1228
TRANSPORT SECTOR	**-**	**-**	**3783**	**-**	**-**	**-**	**-**	**-**	**75**	**-**	**3859**
Air	-	-	746	-	-	-	-	-	-	-	746
Road	-	-	3037	-	-	-	-	-	-	-	3037
Rail	-	-	-	-	-	-	-	-	-	-	-
Pipeline Transport	-	-	-	-	-	-	-	-	-	-	-
Internal Navigation	-	-	-	-	-	-	-	-	-	-	-
Non-specified	-	-	-	-	-	-	-	-	75	-	75
OTHER SECTORS	**-**	**-**	**2085**	**-**	**-**	**-**	**485**	**6**	**1739**	**-**	**4314**
Agriculture	-	-	-	-	-	-	-	-	123	-	123
Comm. and Publ. Services	-	-	-	-	-	-	-	-	528	-	528
Residential	-	-	624	-	-	-	485	3	743	-	1855
Non-specified	-	-	1461	-	-	-	-	3	345	-	1809
NON-ENERGY USE	**-**	**-**	**506**	**-**	**-**	**-**	**-**	**-**	**-**	**-**	**506**
in Industry/Transf./Energy	-	-	506	-	-	-	-	-	-	-	506
in Transport	-	-	-	-	-	-	-	-	-	-	-
in Other Sectors	-	-	-	-	-	-	-	-	-	-	-
Electr. Generated - GWh	***22397***	**-**	***10038***	**-**	**-**	***62***	**-**	**-**	**-**	**-**	***32497***
Electricity Plants	*22397*	-	*10038*	-	-	*62*	-	-	-	-	*32497*
CHP Plants	-	-	-	-	-	-	-	-	-	-	-
Heat Generated - TJ	**-**	**-**	**-**	**-**	**-**	**-**	**-**	**-**	**-**	**-**	**-**
CHP Plants	-	-	-	-	-	-	-	-	-	-	-
Heat Plants	-	-	-	-	-	-	-	-	-	-	-

Israel / Israël : 1997

Thousand tonnes of oil equivalent / Milliers de tonnes d'équivalent pétrole

SUPPLY AND CONSUMPTION APPROVISIONNEMENT ET DEMANDE	Coal Charbon	Crude Oil Pétrole brut	Petroleum Products Produits pétroliers	Gas Gaz	Nuclear Nucléaire	Hydro Hydro	Geotherm. Solar etc. Géotherm. solaire	Combust. Renew. & Waste En. ren. combust.	Electricity Electricité	Heat Chaleur	Total Total
Indigenous Production	42	26	-	13	-	6	511	3	-	-	601
Imports	5063	11728	2148	-	-	-	-	3	-	-	18942
Exports	-	-	-2054	-	-	-	-	-	-91	-	-2146
Intl. Marine Bunkers	-	-	-180	-	-	-	-	-	-	-	-180
Stock Changes	373	-	-	-	-	-	-	-	-	-	373
TPES	**5478**	**11754**	**-86**	**13**	**-**	**6**	**511**	**6**	**-91**	**-**	**17591**
Transfers	-	-	-	-	-	-	-	-	-	-	-
Statistical Differences	-	-	2	-	-	-	-	-	-	-	2
Electricity Plants	-5461	-	-2189	-4	-	-6	-	-	3018	-	-4641
CHP Plants	-	-	-	-	-	-	-	-	-	-	-
Heat Plants	-	-	-	-	-	-	-	-	-	-	-
Gas Works	-	-	-	-	-	-	-	-	-	-	-
Petroleum Refineries	-	-11733	11283	-	-	-	-	-	-	-	-450
Coal Transformation	-	-	-	-	-	-	-	-	-	-	-
Liquefaction	-	-	-	-	-	-	-	-	-	-	-
Other Transformation	-	-	-	-	-	-	-	-	-	-	-
Own Use	-	-	-480	-	-	-	-	-	-126	-	-606
Distribution Losses	-	-	-	-	-	-	-	-	-257	-	-257
TFC	**18**	**21**	**8531**	**8**	**-**	**-**	**511**	**6**	**2544**	**-**	**11639**
INDUSTRY SECTOR	**18**	**21**	**2343**	**8**	**-**	**-**	**-**	**-**	**688**	**-**	**3079**
Iron and Steel	-	-	-	-	-	-	-	-	85	-	85
Chemical and Petrochemical	4	21	941	8	-	-	-	-	196	-	1170
of which: Feedstocks	-	21	941	8	-	-	-	-	-	-	970
Non-Ferrous Metals	-	-	-	-	-	-	-	-	-	-	-
Non-Metallic Minerals	-	-	-	-	-	-	-	-	63	-	63
Transport Equipment	-	-	-	-	-	-	-	-	25	-	25
Machinery	-	-	-	-	-	-	-	-	10	-	10
Mining and Quarrying	-	-	-	-	-	-	-	-	38	-	38
Food and Tobacco	-	-	-	-	-	-	-	-	88	-	88
Paper, Pulp and Printing	-	-	-	-	-	-	-	-	27	-	27
Wood and Wood Products	-	-	-	-	-	-	-	-	12	-	12
Construction	-	-	-	-	-	-	-	-	13	-	13
Textile and Leather	-	-	-	-	-	-	-	-	50	-	50
Non-specified	14	-	1402	-	-	-	-	-	82	-	1498
TRANSPORT SECTOR	**-**	**-**	**3668**	**-**	**-**	**-**	**-**	**-**	**77**	**-**	**3745**
Air	-	-	862	-	-	-	-	-	-	-	862
Road	-	-	2806	-	-	-	-	-	-	-	2806
Rail	-	-	-	-	-	-	-	-	-	-	-
Pipeline Transport	-	-	-	-	-	-	-	-	-	-	-
Internal Navigation	-	-	-	-	-	-	-	-	-	-	-
Non-specified	-	-	-	-	-	-	-	-	77	-	77
OTHER SECTORS	**-**	**-**	**2112**	**-**	**-**	**-**	**511**	**6**	**1779**	**-**	**4408**
Agriculture	-	-	-	-	-	-	-	-	126	-	126
Comm. and Publ. Services	-	-	-	-	-	-	-	-	540	-	540
Residential	-	-	608	-	-	-	511	3	760	-	1882
Non-specified	-	-	1504	-	-	-	-	3	353	-	1860
NON-ENERGY USE	**-**	**-**	**407**	**-**	**-**	**-**	**-**	**-**	**-**	**-**	**407**
in Industry/Transf./Energy	-	-	407	-	-	-	-	-	-	-	407
in Transport	-	-	-	-	-	-	-	-	-	-	-
in Other Sectors	-	-	-	-	-	-	-	-	-	-	-
Electr. Generated - GWh	**24815**	**-**	**10218**	**-**	**-**	**65**	**-**	**-**	**-**	**-**	**35098**
Electricity Plants	*24815*	*-*	*10218*	*-*	*-*	*65*	*-*	*-*	*-*	*-*	*35098*
CHP Plants	*-*	*-*	*-*	*-*	*-*	*-*	*-*	*-*	*-*	*-*	*-*
Heat Generated - TJ	**-**	**-**	**-**	**-**	**-**	**-**	**-**	**-**	**-**	**-**	**-**
CHP Plants	*-*	*-*	*-*	*-*	*-*	*-*	*-*	*-*	*-*	*-*	*-*
Heat Plants	*-*	*-*	*-*	*-*	*-*	*-*	*-*	*-*	*-*	*-*	*-*

Ivory Coast / Côte d'Ivoire

Thousand tonnes of oil equivalent / *Milliers de tonnes d'équivalent pétrole*

SUPPLY AND CONSUMPTION 1996	Coal	Crude Oil	Petroleum Products	Gas	Nuclear	Hydro	Geotherm. Solar etc.	Combust. Renew. & Waste	Electricity	Heat	Total
Indigenous Production	-	900	-	-	-	161	-	3820	-	-	4881
Imports	-	2078	336	-	-	-	-	-	3	-	2417
Exports	-	-	-1018	-	-	-	-	-	-	-	-1018
Intl. Marine Bunkers	-	-	-166	-	-	-	-	-	-	-	-166
Stock changes	-	-619	-75	-	-	-	-	-	-	-	-694
TPES	-	2359	-923	-	-	161	-	3820	3	-	5419
Electricity and CHP Plants	-	-	-251	-	-	-161	-	-	259	-	-153
Petroleum Refineries	-	-2359	2186	-	-	-	-	-	-	-	-173
Other Transformation *	-	-	-115	-	-	-	-	-1423	-54	-	-1593
TFC	-	-	896	-	-	-	-	2397	208	-	3501
INDUSTRY SECTOR	-	-	137	-	-	-	-	-	104	-	241
Iron and Steel	-	-	-	-	-	-	-	-	-	-	-
Chemical and Petrochemical	-	-	-	-	-	-	-	-	19	-	19
Non-Metallic Minerals	-	-	-	-	-	-	-	-	-	-	-
Non-specified	-	-	137	-	-	-	-	-	85	-	222
TRANSPORT SECTOR	-	-	569	-	-	-	-	-	-	-	569
Air	-	-	107	-	-	-	-	-	-	-	107
Road	-	-	446	-	-	-	-	-	-	-	446
Non-specified	-	-	16	-	-	-	-	-	-	-	16
OTHER SECTORS	-	-	137	-	-	-	-	2397	105	-	2638
Agriculture	-	-	27	-	-	-	-	-	15	-	42
Comm. and Publ. Services	-	-	17	-	-	-	-	315	89	-	421
Residential	-	-	93	-	-	-	-	2081	-	-	2174
Non-specified	-	-	-	-	-	-	-	-	-	-	-
NON-ENERGY USE	-	-	53	-	-	-	-	-	-	-	53
Electricity Generated - GWh	-	-	1143	-	-	1872	-	-	-	-	3015
Heat Generated - TJ	-	-	-	-	-	-	-	-	-	-	-

APPROVISIONNEMENT ET DEMANDE 1997	Charbon	Pétrole brut	Produits pétroliers	Gaz	Nucléaire	Hydro	Géotherm. solaire etc.	En. ren. combust. & déchets	Electricité	Chaleur	Total
Indigenous Production	-	803	-	-	-	174	-	3931	-	-	4908
Imports	-	1750	320	-	-	-	-	-	3	-	2073
Exports	-	-	-1218	-	-	-	-	-	-	-	-1218
Intl. Marine Bunkers	-	-	-166	-	-	-	-	-	-	-	-166
Stock changes	-	-	-	-	-	-	-	-	-	-	-
TPES	-	2553	-1064	-	-	174	-	3931	3	-	5597
Electricity and CHP Plants	-	-	-261	-	-	-174	-	-	276	-	-159
Petroleum Refineries	-	-2553	2367	-	-	-	-	-	-	-	-186
Other Transformation *	-	-	-120	-	-	-	-	-1464	-58	-	-1642
TFC	-	-	922	-	-	-	-	2466	222	-	3610
INDUSTRY SECTOR	-	-	142	-	-	-	-	-	110	-	252
Iron and Steel	-	-	-	-	-	-	-	-	-	-	-
Chemical and Petrochemical	-	-	-	-	-	-	-	-	20	-	20
Non-Metallic Minerals	-	-	-	-	-	-	-	-	-	-	-
Non-specified	-	-	142	-	-	-	-	-	90	-	232
TRANSPORT SECTOR	-	-	578	-	-	-	-	-	-	-	578
Air	-	-	111	-	-	-	-	-	-	-	111
Road	-	-	450	-	-	-	-	-	-	-	450
Non-specified	-	-	17	-	-	-	-	-	-	-	17
OTHER SECTORS	-	-	140	-	-	-	-	2466	111	-	2718
Agriculture	-	-	28	-	-	-	-	-	17	-	44
Comm. and Publ. Services	-	-	17	-	-	-	-	325	95	-	436
Residential	-	-	95	-	-	-	-	2142	-	-	2237
Non-specified	-	-	-	-	-	-	-	-	-	-	-
NON-ENERGY USE	-	-	61	-	-	-	-	-	-	-	61
Electricity Generated - GWh	-	-	1191	-	-	2022	-	-	-	-	3213
Heat Generated - TJ	-	-	-	-	-	-	-	-	-	-	-

* Includes Transfers, Statistical Differences, Own Use and Distribution Losses.

Jamaica / Jamaïque

Thousand tonnes of oil equivalent / *Milliers de tonnes d'équivalent pétrole*

SUPPLY AND CONSUMPTION 1996	Coal	Crude Oil	Petroleum Products	Gas	Nuclear	Hydro	Geotherm. Solar etc.	Combust. Renew. & Waste	Electricity	Heat	Total
Indigenous Production	-	-	-	-	-	11	-	536	-	-	547
Imports	39	1037	2083	-	-	-	-	-	-	-	3159
Exports	-	-	-65	-	-	-	-	-	-	-	-65
Intl. Marine Bunkers	-	-	-38	-	-	-	-	-	-	-	-38
Stock changes	-	40	75	-	-	-	-	-	-	-	115
TPES	39	1077	2054	-	-	11	-	536	-	-	3718
Electricity and CHP Plants	-	-	-1713	-	-	-11	-	-276	519	-	-1481
Petroleum Refineries	-	-1077	1050	-	-	-	-	-	-	-	-27
Other Transformation *	-	-	-1	-	-	-	-	-192	-59	-	-252
TFC	39	-	1390	-	-	-	-	68	460	-	1958
INDUSTRY SECTOR	39	-	189	-	-	-	-	-	295	-	523
Iron and Steel	-	-	-	-	-	-	-	-	-	-	-
Chemical and Petrochemical	-	-	-	-	-	-	-	-	-	-	-
Non-Metallic Minerals	39	-	-	-	-	-	-	-	-	-	39
Non-specified	-	-	189	-	-	-	-	-	295	-	484
TRANSPORT SECTOR	-	-	796	-	-	-	-	-	-	-	796
Air	-	-	220	-	-	-	-	-	-	-	220
Road	-	-	401	-	-	-	-	-	-	-	401
Non-specified	-	-	174	-	-	-	-	-	-	-	174
OTHER SECTORS	-	-	394	-	-	-	-	68	166	-	627
Agriculture	-	-	243	-	-	-	-	-	-	-	243
Comm. and Publ. Services	-	-	30	-	-	-	-	-	98	-	128
Residential	-	-	121	-	-	-	-	68	67	-	256
Non-specified	-	-	-	-	-	-	-	-	2	-	2
NON-ENERGY USE	-	-	12	-	-	-	-	-	-	-	12
Electricity Generated - GWh	-	-	*5631*	-	-	*128*	-	*279*	-	-	*6038*
Heat Generated - TJ	-	-	-	-	-	-	-	-	-	-	-

APPROVISIONNEMENT ET DEMANDE 1997	Charbon	Pétrole brut	Produits pétroliers	Gaz	Nucléaire	Hydro	Géotherm. solaire etc.	En. ren. combust. & déchets	Electricité	Chaleur	Total
Indigenous Production	-	-	-	-	-	10	-	585	-	-	595
Imports	41	1133	2178	-	-	-	-	-	-	-	3352
Exports	-	-	-67	-	-	-	-	-	-	-	-67
Intl. Marine Bunkers	-	-	-38	-	-	-	-	-	-	-	-38
Stock changes	-	44	76	-	-	-	-	-	-	-	120
TPES	41	1177	2149	-	-	10	-	585	-	-	3963
Electricity and CHP Plants	-	-	-1757	-	-	-10	-	-317	538	-	-1546
Petroleum Refineries	-	-1177	1102	-	-	-	-	-	-	-	-75
Other Transformation *	-	-	-9	-	-	-	-	-198	-61	-	-268
TFC	41	-	1484	-	-	-	-	70	477	-	2073
INDUSTRY SECTOR	41	-	194	-	-	-	-	-	305	-	541
Iron and Steel	-	-	-	-	-	-	-	-	-	-	-
Chemical and Petrochemical	-	-	-	-	-	-	-	-	-	-	-
Non-Metallic Minerals	41	-	-	-	-	-	-	-	-	-	41
Non-specified	-	-	194	-	-	-	-	-	305	-	499
TRANSPORT SECTOR	-	-	865	-	-	-	-	-	-	-	865
Air	-	-	250	-	-	-	-	-	-	-	250
Road	-	-	435	-	-	-	-	-	-	-	435
Non-specified	-	-	179	-	-	-	-	-	-	-	179
OTHER SECTORS	-	-	414	-	-	-	-	70	172	-	656
Agriculture	-	-	246	-	-	-	-	-	-	-	246
Comm. and Publ. Services	-	-	32	-	-	-	-	-	101	-	133
Residential	-	-	135	-	-	-	-	70	69	-	274
Non-specified	-	-	-	-	-	-	-	-	2	-	2
NON-ENERGY USE	-	-	12	-	-	-	-	-	-	-	12
Electricity Generated - GWh	-	-	*5814*	-	-	*121*	-	*320*	-	-	*6255*
Heat Generated - TJ	-	-	-	-	-	-	-	-	-	-	-

* Includes Transfers, Statistical Differences, Own Use and Distribution Losses.

INTERNATIONAL ENERGY AGENCY

Jordan / Jordanie

Thousand tonnes of oil equivalent / *Milliers de tonnes d'équivalent pétrole*

SUPPLY AND CONSUMPTION 1996	Coal	Crude Oil	Petroleum Products	Gas	Nuclear	Hydro	Geotherm. Solar etc.	Combust. Renew. & Waste	Electricity	Heat	Total
Indigenous Production	-	2	-	176	-	2	-	2	-	-	181
Imports	-	3328	983	-	-	-	-	-	-	-	4311
Exports	-	-	-	-	-	-	-	-	-	-	-
Intl. Marine Bunkers	-	-	-	-	-	-	-	-	-	-	-
Stock changes	-	-	-5	-	-	-	-	-	-	-	-5
TPES	-	3330	977	176	-	2	-	2	-	-	4487
Electricity and CHP Plants	-	-	-1480	-176	-	-2	-	-	521	-	-1136
Petroleum Refineries	-	-3330	3325	-	-	-	-	-	-	-	-5
Other Transformation *	-	-	8	-	-	-	-	-	-80	-	-72
TFC	-	-	2831	-	-	-	-	2	441	-	3274
INDUSTRY SECTOR	-	-	485	-	-	-	-	-	152	-	637
Iron and Steel	-	-	-	-	-	-	-	-	1	-	1
Chemical and Petrochemical	-	-	-	-	-	-	-	-	16	-	16
Non-Metallic Minerals	-	-	-	-	-	-	-	-	33	-	33
Non-specified	-	-	485	-	-	-	-	-	103	-	587
TRANSPORT SECTOR	-	-	1354	-	-	-	-	-	-	-	1354
Air	-	-	325	-	-	-	-	-	-	-	325
Road	-	-	1029	-	-	-	-	-	-	-	1029
Non-specified	-	-	-	-	-	-	-	-	-	-	-
OTHER SECTORS	-	-	827	-	-	-	-	2	288	-	1118
Agriculture	-	-	17	-	-	-	-	-	79	-	96
Comm. and Publ. Services	-	-	-	-	-	-	-	-	75	-	75
Residential	-	-	444	-	-	-	-	2	134	-	580
Non-specified	-	-	366	-	-	-	-	-	-	-	366
NON-ENERGY USE	-	-	165	-	-	-	-	-	-	-	165
Electricity Generated - GWh	-	-	5306	729	-	22	1	-	-	-	6058
Heat Generated - TJ	-	-	-	-	-	-	-	-	-	-	-

APPROVISIONNEMENT ET DEMANDE 1997	Charbon	Pétrole brut	Produits pétroliers	Gaz	Nucléaire	Hydro	Géotherm. solaire etc.	En. ren. combust. & déchets	Electricité	Chaleur	Total
Indigenous Production	-	-	-	189	-	2	-	2	-	-	193
Imports	-	3548	1053	-	-	-	-	-	-	-	4602
Exports	-	-	-	-	-	-	-	-	-	-	-
Intl. Marine Bunkers	-	-	-	-	-	-	-	-	-	-	-
Stock changes	-	-	-	-	-	-	-	-	-	-	-
TPES	-	3548	1053	189	-	2	-	3	-	-	4795
Electricity and CHP Plants	-	-	-1573	-189	-	-2	-	-	539	-	-1224
Petroleum Refineries	-	-3548	3543	-	-	-	-	-	-	-	-5
Other Transformation *	-	-	9	-	-	-	-	-	-83	-	-74
TFC	-	-	3032	-	-	-	-	3	456	-	3491
INDUSTRY SECTOR	-	-	515	-	-	-	-	-	158	-	673
Iron and Steel	-	-	-	-	-	-	-	-	1	-	1
Chemical and Petrochemical	-	-	-	-	-	-	-	-	17	-	17
Non-Metallic Minerals	-	-	-	-	-	-	-	-	34	-	34
Non-specified	-	-	515	-	-	-	-	-	106	-	621
TRANSPORT SECTOR	-	-	1440	-	-	-	-	-	-	-	1440
Air	-	-	352	-	-	-	-	-	-	-	352
Road	-	-	1089	-	-	-	-	-	-	-	1089
Non-specified	-	-	-	-	-	-	-	-	-	-	-
OTHER SECTORS	-	-	902	-	-	-	-	3	298	-	1203
Agriculture	-	-	19	-	-	-	-	-	82	-	101
Comm. and Publ. Services	-	-	-	-	-	-	-	-	77	-	77
Residential	-	-	490	-	-	-	-	2	139	-	631
Non-specified	-	-	393	-	-	-	-	-	-	-	394
NON-ENERGY USE	-	-	175	-	-	-	-	-	-	-	175
Electricity Generated - GWh	-	-	5467	783	-	22	1	-	-	-	6273
Heat Generated - TJ	-	-	-	-	-	-	-	-	-	-	-

* Includes Transfers, Statistical Differences, Own Use and Distribution Losses.

Kazakhstan

Thousand tonnes of oil equivalent / *Milliers de tonnes d'équivalent pétrole*

SUPPLY AND CONSUMPTION 1996	Coal	Crude Oil	Petroleum Products	Gas	Nuclear	Hydro	Geotherm. Solar etc.	Combust. Renew. & Waste	Electricity	Heat	Total
Indigenous Production	33761	23082	-	5286	-	630	-	74	-	-	62833
Imports	1023	3174	884	4452	-	-	-	-	589	-	10123
Exports	-9719	-14460	-2200	-1897	-	-	-	-	-	-	-28276
Intl. Marine Bunkers	-	-	-	-	-	-	-	-	-	-	-
Stock changes	-	-	-	-	-	-	-	-	-	-	-
TPES	25065	11796	-1315	7840	-	630	-	74	589	-	44680
Electricity and CHP Plants	-17521	-	-1530	-2023	-	-630	-	-	5045	6	-16654
Petroleum Refineries	-	-11796	11109	-	-	-	-	-	-	-	-687
Other Transformation *	-397	-	-468	-1308	-	-	-	-	-1711	-	-3885
TFC	7146	-	7795	4509	-	-	-	74	3923	6	23453
INDUSTRY SECTOR	7146	-	-	-	-	-	-	-	1793	-	8940
Iron and Steel	100	-	-	-	-	-	-	-	688	-	788
Chemical and Petrochemical	-	-	-	-	-	-	-	-	270	-	270
Non-Metallic Minerals	-	-	-	-	-	-	-	-	-	-	-
Non-specified	7047	-	-	-	-	-	-	-	835	-	7882
TRANSPORT SECTOR	-	-	2773	-	-	-	-	-	295	-	3068
Air	-	-	404	-	-	-	-	-	-	-	404
Road	-	-	2369	-	-	-	-	-	-	-	2369
Non-specified	-	-	-	-	-	-	-	-	295	-	295
OTHER SECTORS	-	-	4921	4509	-	-	-	74	1834	6	11344
Agriculture	-	-	-	-	-	-	-	-	950	-	950
Comm. and Publ. Services	-	-	-	-	-	-	-	-	-	-	-
Residential	-	-	-	-	-	-	-	-	521	-	521
Non-specified	-	-	4921	4509	-	-	-	74	363	6	9872
NON-ENERGY USE	-	-	102	-	-	-	-	-	-	-	102
Electricity Generated - GWh	*42250*	-	*4277*	*4799*	-	*7331*	-	-	-	-	*58657*
Heat Generated - TJ	-	-	-	-	-	-	-	-	-	-	*256*

APPROVISIONNEMENT ET DEMANDE 1997	Charbon	Pétrole brut	Produits pétroliers	Gaz	Nucléaire	Hydro	Géotherm. solaire etc.	En. ren. combust. & déchets	Electricité	Chaleur	Total
Indigenous Production	32009	25568	-	6574	-	559	-	74	-	-	64784
Imports	970	621	723	1782	-	-	-	-	576	-	4673
Exports	-11091	-16476	-1316	-2155	-	-	-	-	-	-	-31038
Intl. Marine Bunkers	-	-	-	-	-	-	-	-	-	-	-
Stock changes	-	-	-	-	-	-	-	-	-	-	-
TPES	21888	9713	-594	6201	-	559	-	74	576	-	38418
Electricity and CHP Plants	-15402	-	-1206	-1600	-	-559	-	-	4472	6	-14289
Petroleum Refineries	-	-9713	9141	-	-	-	-	-	-	-	-572
Other Transformation *	-397	-	-368	-1035	-	-	-	-	-1533	-	-3333
TFC	6089	-	6973	3566	-	-	-	74	3515	6	20224
INDUSTRY SECTOR	6089	-	-	-	-	-	-	-	1607	-	7696
Iron and Steel	100	-	-	-	-	-	-	-	616	-	716
Chemical and Petrochemical	-	-	-	-	-	-	-	-	242	-	242
Non-Metallic Minerals	-	-	-	-	-	-	-	-	-	-	-
Non-specified	5989	-	-	-	-	-	-	-	748	-	6738
TRANSPORT SECTOR	-	-	2356	-	-	-	-	-	264	-	2620
Air	-	-	325	-	-	-	-	-	-	-	325
Road	-	-	2031	-	-	-	-	-	-	-	2031
Non-specified	-	-	-	-	-	-	-	-	264	-	264
OTHER SECTORS	-	-	4534	3566	-	-	-	74	1644	6	9824
Agriculture	-	-	-	-	-	-	-	-	851	-	851
Comm. and Publ. Services	-	-	-	-	-	-	-	-	-	-	-
Residential	-	-	-	-	-	-	-	-	467	-	467
Non-specified	-	-	4534	3566	-	-	-	74	325	6	8505
NON-ENERGY USE	-	-	84	-	-	-	-	-	-	-	84
Electricity Generated - GWh	*37455*	-	*3792*	*4254*	-	*6499*	-	-	-	-	*52000*
Heat Generated - TJ	-	-	-	-	-	-	-	-	-	-	*256*

* Includes Transfers, Statistical Differences, Own Use and Distribution Losses.

INTERNATIONAL ENERGY AGENCY

Kenya

Thousand tonnes of oil equivalent / *Milliers de tonnes d'équivalent pétrole*

SUPPLY AND CONSUMPTION 1996	Coal	Crude Oil	Petroleum Products	Gas	Nuclear	Hydro	Geotherm. Solar etc.	Combust. Renew. & Waste	Electricity	Heat	Total
Indigenous Production	-	-	-	-	-	283	335	10807	-	-	11426
Imports	55	1420	987	-	-	-	-	-	13	-	2475
Exports	-	-	-233	-	-	-	-	-	-	-	-233
Intl. Marine Bunkers	-	-	-213	-	-	-	-	-	-	-	-213
Stock changes	-	-	5	-	-	-	-	-	-	-	5
TPES	55	1420	545	-	-	283	335	10807	13	-	13459
Electricity and CHP Plants	-	-	-130	-	-	-283	-335	-	347	-	-401
Petroleum Refineries	-	-1770	1783	-	-	-	-	-	-	-	13
Other Transformation *	-	350	-99	-	-	-	-	-3060	-59	-	-2868
TFC	55	-	2099	-	-	-	-	7747	301	-	10203
INDUSTRY SECTOR	55	-	408	-	-	-	-	537	184	-	1184
Iron and Steel	-	-	-	-	-	-	-	-	-	-	-
Chemical and Petrochemical	-	-	-	-	-	-	-	-	-	-	-
Non-Metallic Minerals	55	-	-	-	-	-	-	-	-	-	55
Non-specified	-	-	408	-	-	-	-	537	184	-	1129
TRANSPORT SECTOR	-	-	1321	-	-	-	-	-	-	-	1321
Air	-	-	476	-	-	-	-	-	-	-	476
Road	-	-	804	-	-	-	-	-	-	-	804
Non-specified	-	-	41	-	-	-	-	-	-	-	41
OTHER SECTORS	1	-	323	-	-	-	-	7210	118	-	7652
Agriculture	-	-	78	-	-	-	-	583	-	-	661
Comm. and Publ. Services	-	-	-	-	-	-	-	46	26	-	72
Residential	1	-	170	-	-	-	-	6582	92	-	6844
Non-specified	-	-	75	-	-	-	-	-	-	-	75
NON-ENERGY USE	-	-	47	-	-	-	-	-	-	-	47
Electricity Generated - GWh	-	-	355	-	-	3295	390	-	-	-	4040
Heat Generated - TJ	-	-	-	-	-	-	-	-	-	-	-

APPROVISIONNEMENT ET DEMANDE 1997	Charbon	Pétrole brut	Produits pétroliers	Gaz	Nucléaire	Hydro	Géotherm. solaire etc.	En. ren. combust. & déchets	Electricité	Chaleur	Total
Indigenous Production	-	-	-	-	-	300	338	11013	-	-	11651
Imports	57	1843	951	-	-	-	-	-	12	-	2863
Exports	-	-	-170	-	-	-	-	-	-	-	-170
Intl. Marine Bunkers	-	-	-213	-	-	-	-	-	-	-	-213
Stock changes	-	-	8	-	-	-	-	-	-	-	8
TPES	57	1843	575	-	-	300	338	11013	12	-	14138
Electricity and CHP Plants	-	-	-130	-	-	-300	-338	-	364	-	-404
Petroleum Refineries	-	-1655	1668	-	-	-	-	-	-	-	13
Other Transformation *	-	-188	-102	-	-	-	-	-3118	-65	-	-3473
TFC	57	-	2010	-	-	-	-	7894	312	-	10274
INDUSTRY SECTOR	57	-	376	-	-	-	-	547	195	-	1174
Iron and Steel	-	-	-	-	-	-	-	-	-	-	-
Chemical and Petrochemical	-	-	-	-	-	-	-	-	-	-	-
Non-Metallic Minerals	57	-	-	-	-	-	-	-	-	-	57
Non-specified	-	-	376	-	-	-	-	547	195	-	1117
TRANSPORT SECTOR	-	-	1291	-	-	-	-	-	-	-	1291
Air	-	-	463	-	-	-	-	-	-	-	463
Road	-	-	788	-	-	-	-	-	-	-	788
Non-specified	-	-	40	-	-	-	-	-	-	-	40
OTHER SECTORS	1	-	303	-	-	-	-	7347	117	-	7768
Agriculture	-	-	72	-	-	-	-	594	-	-	666
Comm. and Publ. Services	-	-	-	-	-	-	-	47	21	-	68
Residential	1	-	157	-	-	-	-	6707	96	-	6960
Non-specified	-	-	73	-	-	-	-	-	-	-	73
NON-ENERGY USE	-	-	41	-	-	-	-	-	-	-	41
Electricity Generated - GWh	-	-	355	-	-	3490	393	-	-	-	4238
Heat Generated - TJ	-	-	-	-	-	-	-	-	-	-	-

* Includes Transfers, Statistical Differences, Own Use and Distribution Losses.

Dem. People's Rep. of Korea / Rép. pop. dém. de Corée

Thousand tonnes of oil equivalent / *Milliers de tonnes d'équivalent pétrole*

SUPPLY AND CONSUMPTION 1996	Coal	Crude Oil	Petroleum Products	Gas	Nuclear	Hydro	Geotherm. Solar etc.	Combust. Renew. & Waste	Electricity	Heat	Total
Indigenous Production	18107	-	-	-	-	1936	-	1019	-	-	21062
Imports	1750	1129	383	-	-	-	-	-	-	-	3261
Exports	-311	-	-	-	-	-	-	-	-	-	-311
Intl. Marine Bunkers	-	-	-	-	-	-	-	-	-	-	-
Stock changes	-	-	-	-	-	-	-	-	-	-	-
TPES	19546	1129	383	-	-	1936	-	1019	-	-	24013
Electricity and CHP Plants	-4969	-	-	-	-	-1936	-	-	3013	-	-3893
Petroleum Refineries	-	-1129	1149	-	-	-	-	-	-	-	20
Other Transformation *	-803	-	-199	-	-	-	-	-	-2531	-	-3534
TFC	13774	-	1332	-	-	-	-	1019	482	-	16606
INDUSTRY SECTOR	13774	-	190	-	-	-	-	-	-	-	13964
Iron and Steel	417	-	-	-	-	-	-	-	-	-	417
Chemical and Petrochemical	-	-	-	-	-	-	-	-	-	-	-
Non-Metallic Minerals	-	-	-	-	-	-	-	-	-	-	-
Non-specified	13357	-	190	-	-	-	-	-	-	-	13547
TRANSPORT SECTOR	-	-	1019	-	-	-	-	-	-	-	1019
Air	-	-	-	-	-	-	-	-	-	-	-
Road	-	-	1019	-	-	-	-	-	-	-	1019
Non-specified	-	-	-	-	-	-	-	-	-	-	-
OTHER SECTORS	-	-	122	-	-	-	-	1019	482	-	1623
Agriculture	-	-	-	-	-	-	-	-	-	-	-
Comm. and Publ. Services	-	-	-	-	-	-	-	-	-	-	-
Residential	-	-	122	-	-	-	-	-	-	-	122
Non-specified	-	-	-	-	-	-	-	1019	482	-	1501
NON-ENERGY USE	-	-	-	-	-	-	-	-	-	-	-
Electricity Generated - GWh	*12519*	-	-	-	-	*22517*	-	-	-	-	*35036*
Heat Generated - TJ	-	-	-	-	-	-	-	-	-	-	-

APPROVISIONNEMENT ET DEMANDE 1997	Charbon	Pétrole brut	Produits pétroliers	Gaz	Nucléaire	Hydro	Géotherm. solaire etc.	En. ren. combust. & déchets	Electricité	Chaleur	Total
Indigenous Production	17564	-	-	-	-	1878	-	1019	-	-	20461
Imports	1697	1095	371	-	-	-	-	-	-	-	3163
Exports	-301	-	-	-	-	-	-	-	-	-	-301
Intl. Marine Bunkers	-	-	-	-	-	-	-	-	-	-	-
Stock changes	-	-	-	-	-	-	-	-	-	-	-
TPES	18959	1095	371	-	-	1878	-	1019	-	-	23323
Electricity and CHP Plants	-4820	-	-	-	-	-1878	-	-	2923	-	-3776
Petroleum Refineries	-	-1095	1114	-	-	-	-	-	-	-	19
Other Transformation *	-778	-	-193	-	-	-	-	-	-2455	-	-3427
TFC	13361	-	1291	-	-	-	-	1019	468	-	16139
INDUSTRY SECTOR	13361	-	184	-	-	-	-	-	-	-	13545
Iron and Steel	404	-	-	-	-	-	-	-	-	-	404
Chemical and Petrochemical	-	-	-	-	-	-	-	-	-	-	-
Non-Metallic Minerals	-	-	-	-	-	-	-	-	-	-	-
Non-specified	12957	-	184	-	-	-	-	-	-	-	13141
TRANSPORT SECTOR	-	-	989	-	-	-	-	-	-	-	989
Air	-	-	-	-	-	-	-	-	-	-	-
Road	-	-	989	-	-	-	-	-	-	-	989
Non-specified	-	-	-	-	-	-	-	-	-	-	-
OTHER SECTORS	-	-	118	-	-	-	-	1019	468	-	1605
Agriculture	-	-	-	-	-	-	-	-	-	-	-
Comm. and Publ. Services	-	-	-	-	-	-	-	-	-	-	-
Residential	-	-	118	-	-	-	-	-	-	-	118
Non-specified	-	-	-	-	-	-	-	1019	468	-	1487
NON-ENERGY USE	-	-	-	-	-	-	-	-	-	-	-
Electricity Generated - GWh	*12143*	-	-	-	-	*21842*	-	-	-	-	*33985*
Heat Generated - TJ	-	-	-	-	-	-	-	-	-	-	-

* Includes Transfers, Statistical Differences, Own Use and Distribution Losses.

Kuwait / Koweit

Thousand tonnes of oil equivalent / *Milliers de tonnes d'équivalent pétrole*

SUPPLY AND CONSUMPTION 1996	Coal	Crude Oil	Petroleum Products	Gas	Nuclear	Hydro	Geotherm. Solar etc.	Combust. Renew. & Waste	Electricity	Heat	Total
Indigenous Production	-	108473	-	4431	-	-	-	-	-	-	112904
Imports	-	-	1	-	-	-	-	4	-	-	5
Exports	-	-64659	-33901	-	-	-	-	-	-	-	-98561
Intl. Marine Bunkers	-	-	-185	-	-	-	-	-	-	-	-185
Stock changes	-	-	-	-	-	-	-	-	-	-	-
TPES	-	43814	-34086	4431	-	-	-	4	-	-	14163
Electricity and CHP Plants	-	-	-1071	-4431	-	-	-	-	2191	-	-3311
Petroleum Refineries	-	-41074	36650	-	-	-	-	-	-	-	-4424
Other Transformation *	-	-2740	2424	-	-	-	-	-	-322	-	-638
TFC	-	-	3917	-	-	-	-	4	1869	-	5790
INDUSTRY SECTOR	-	-	1457	-	-	-	-	-	-	-	1457
Iron and Steel	-	-	-	-	-	-	-	-	-	-	-
Chemical and Petrochemical	-	-	-	-	-	-	-	-	-	-	-
Non-Metallic Minerals	-	-	-	-	-	-	-	-	-	-	-
Non-specified	-	-	1457	-	-	-	-	-	-	-	1457
TRANSPORT SECTOR	-	-	2255	-	-	-	-	-	-	-	2255
Air	-	-	414	-	-	-	-	-	-	-	414
Road	-	-	1841	-	-	-	-	-	-	-	1841
Non-specified	-	-	-	-	-	-	-	-	-	-	-
OTHER SECTORS	-	-	123	-	-	-	-	4	1869	-	1996
Agriculture	-	-	-	-	-	-	-	-	-	-	-
Comm. and Publ. Services	-	-	-	-	-	-	-	-	-	-	-
Residential	-	-	123	-	-	-	-	4	1869	-	1996
Non-specified	-	-	-	-	-	-	-	-	-	-	-
NON-ENERGY USE	-	-	82	-	-	-	-	-	-	-	82
Electricity Generated - GWh	-	-	5522	19953	-	-	-	-	-	-	25475
Heat Generated - TJ	-	-	-	-	-	-	-	-	-	-	-

APPROVISIONNEMENT ET DEMANDE 1997	Charbon	Pétrole brut	Produits pétroliers	Gaz	Nucléaire	Hydro	Géotherm. solaire etc.	En. ren. combust. & déchets	Electricité	Chaleur	Total
Indigenous Production	-	111633	-	4455	-	-	-	-	-	-	116087
Imports	-	-	1	-	-	-	-	4	-	-	5
Exports	-	-61870	-37872	-	-	-	-	-	-	-	-99742
Intl. Marine Bunkers	-	-	-185	-	-	-	-	-	-	-	-185
Stock changes	-	-	-	-	-	-	-	-	-	-	-
TPES	-	49762	-38056	4455	-	-	-	4	-	-	16165
Electricity and CHP Plants	-	-	-1282	-4455	-	-	-	-	2330	-	-3407
Petroleum Refineries	-	-45866	41179	-	-	-	-	-	-	-	-4687
Other Transformation *	-	-3896	2606	-	-	-	-	-	-325	-	-1615
TFC	-	-	4447	-	-	-	-	4	2005	-	6455
INDUSTRY SECTOR	-	-	1746	-	-	-	-	-	-	-	1746
Iron and Steel	-	-	-	-	-	-	-	-	-	-	-
Chemical and Petrochemical	-	-	-	-	-	-	-	-	-	-	-
Non-Metallic Minerals	-	-	-	-	-	-	-	-	-	-	-
Non-specified	-	-	1746	-	-	-	-	-	-	-	1746
TRANSPORT SECTOR	-	-	2469	-	-	-	-	-	-	-	2469
Air	-	-	453	-	-	-	-	-	-	-	453
Road	-	-	2017	-	-	-	-	-	-	-	2017
Non-specified	-	-	-	-	-	-	-	-	-	-	-
OTHER SECTORS	-	-	140	-	-	-	-	4	2005	-	2149
Agriculture	-	-	-	-	-	-	-	-	-	-	-
Comm. and Publ. Services	-	-	-	-	-	-	-	-	-	-	-
Residential	-	-	140	-	-	-	-	4	2005	-	2149
Non-specified	-	-	-	-	-	-	-	-	-	-	-
NON-ENERGY USE	-	-	91	-	-	-	-	-	-	-	91
Electricity Generated - GWh	-	-	7031	20060	-	-	-	-	-	-	27091
Heat Generated - TJ	-	-	-	-	-	-	-	-	-	-	-

* Includes Transfers, Statistical Differences, Own Use and Distribution Losses.

Kyrgyzstan / Kirghizistan

Thousand tonnes of oil equivalent / Milliers de tonnes d'équivalent pétrole

SUPPLY AND CONSUMPTION 1996	Coal	Crude Oil	Petroleum Products	Gas	Nuclear	Hydro	Geotherm. Solar etc.	Combust. Renew. & Waste	Electricity	Heat	Total
Indigenous Production	262	101	-	22	-	1054	-	4	-	-	1442
Imports	370	12	600	861	-	-	-	-	612	-	2455
Exports	-41	-21	-1	-	-	-	-	-	-791	-	-854
Intl. Marine Bunkers	-	-	-	-	-	-	-	-	-	-	-
Stock changes	-	-	-	-	-	-	-	-	-	-	-
TPES	592	91	599	883	-	1054	-	4	-179	-	3043
Electricity and CHP Plants	-217	-	-	-507	-	-1054	-	-	1183	458	-138
Petroleum Refineries	-	-91	87	-	-	-	-	-	-	-	-4
Other Transformation *	-	-	-	-	-	-	-	-	-422	-	-422
TFC	374	-	686	376	-	-	-	4	582	458	2479
INDUSTRY SECTOR	374	-	-	-	-	-	-	-	137	-	511
Iron and Steel	-	-	-	-	-	-	-	-	-	-	-
Chemical and Petrochemical	-	-	-	-	-	-	-	-	-	-	-
Non-Metallic Minerals	-	-	-	-	-	-	-	-	-	-	-
Non-specified	374	-	-	-	-	-	-	-	136	-	510
TRANSPORT SECTOR	-	-	341	-	-	-	-	-	11	-	352
Air	-	-	95	-	-	-	-	-	-	-	95
Road	-	-	246	-	-	-	-	-	-	-	246
Non-specified	-	-	-	-	-	-	-	-	11	-	11
OTHER SECTORS	-	-	332	376	-	-	-	4	435	458	1604
Agriculture	-	-	-	-	-	-	-	-	226	-	226
Comm. and Publ. Services	-	-	-	-	-	-	-	-	-	-	-
Residential	-	-	-	-	-	-	-	-	146	-	146
Non-specified	-	-	332	376	-	-	-	4	62	458	1231
NON-ENERGY USE	-	-	12	-	-	-	-	-	-	-	12
Electricity Generated - GWh	910	-	593	-	-	12255	-	-	-	-	13758
Heat Generated - TJ	-	-	-	-	-	-	-	-	-	-	19162

APPROVISIONNEMENT ET DEMANDE 1997	Charbon	Pétrole brut	Produits pétroliers	Gaz	Nucléaire	Hydro	Géotherm. solaire etc.	En. ren. combust. & déchets	Electricité	Chaleur	Total
Indigenous Production	320	85	-	34	-	965	-	4	-	-	1408
Imports	683	65	390	478	-	-	-	-	602	-	2218
Exports	-44	-40	-	-	-	-	-	-	-748	-	-833
Intl. Marine Bunkers	-	-	-	-	-	-	-	-	-	-	-
Stock changes	-	-	-	-	-	-	-	-	-	-	-
TPES	959	111	390	512	-	965	-	4	-146	-	2793
Electricity and CHP Plants	-226	-	-	-310	-	-965	-	-	1084	458	39
Petroleum Refineries	-	-111	106	-	-	-	-	-	-	-	-5
Other Transformation *	-	-	-	-	-	-	-	-	-390	-	-390
TFC	732	-	495	201	-	-	-	4	547	458	2437
INDUSTRY SECTOR	732	-	-	-	-	-	-	-	128	-	861
Iron and Steel	-	-	-	-	-	-	-	-	-	-	-
Chemical and Petrochemical	-	-	-	-	-	-	-	-	-	-	-
Non-Metallic Minerals	-	-	-	-	-	-	-	-	-	-	-
Non-specified	732	-	-	-	-	-	-	-	128	-	860
TRANSPORT SECTOR	-	-	190	-	-	-	-	-	10	-	200
Air	-	-	67	-	-	-	-	-	-	-	67
Road	-	-	123	-	-	-	-	-	-	-	123
Non-specified	-	-	-	-	-	-	-	-	10	-	10
OTHER SECTORS	-	-	294	201	-	-	-	4	408	458	1364
Agriculture	-	-	-	-	-	-	-	-	213	-	213
Comm. and Publ. Services	-	-	-	-	-	-	-	-	-	-	-
Residential	-	-	-	-	-	-	-	-	138	-	138
Non-specified	-	-	294	201	-	-	-	4	58	458	1014
NON-ENERGY USE	-	-	12	-	-	-	-	-	-	-	12
Electricity Generated - GWh	833	-	543	-	-	11224	-	-	-	-	12600
Heat Generated - TJ	-	-	-	-	-	-	-	-	-	-	19162

* Includes Transfers, Statistical Differences, Own Use and Distribution Losses.

Latvia / Lettonie : 1996

											Thousand tonnes of oil equivalent / *Milliers de tonnes d'équivalent pétrole*

SUPPLY AND CONSUMPTION	Coal	Crude Oil	Petroleum Products	Gas	Nuclear	Hydro	Geotherm. Solar etc.	Combust. Renew. & Waste	Electricity	Heat	Total
APPROVISIONNEMENT ET DEMANDE	*Charbon*	*Pétrole brut*	*Produits pétroliers*	*Gaz*	*Nucléaire*	*Hydro*	*Géotherm. solaire*	*En. ren. combust.*	*Electricité*	*Chaleur*	*Total*
Indigenous Production	78	-	-	-	-	160	-	760	-	-	998
Imports	99	-	2360	874	-	-	-	-	296	-	3629
Exports	-3	-	-37	-	-	-	-	-160	-18	-	-218
Intl. Marine Bunkers	-	-	-	-	-	-	-	-	-	-	-
Stock Changes	10	-	-259	-1	-	-	-	-12	-	-	-261
TPES	**185**	**-**	**2064**	**874**	**-**	**160**	**-**	**588**	**278**	**-**	**4147**
Transfers	-	-	25	-	-	-	-	-	-	-	25
Statistical Differences	33	-	-	-	-	-	-	-	-	-	33
Electricity Plants	-	-	-	-	-	-160	-	-	160	-	-
CHP Plants	-47	-	-293	-240	-	-	-	-	109	368	-103
Heat Plants	-60	-	-691	-120	-	-	-	-185	-	944	-113
Gas Works	-	-	-	-	-	-	-	-	-	-	-
Petroleum Refineries	-	-	-	-	-	-	-	-	-	-	-
Coal Transformation	-3	-	-	-	-	-	-	-	-	-	-3
Liquefaction	-	-	-	-	-	-	-	-	-	-	-
Other Transformation	-	-	-	-	-	-	-	-	-	-	-
Own Use	-	-	-	-25	-	-	-	-	-39	-46	-110
Distribution Losses	-	-	-	-24	-	-	-	-	-125	-165	-314
TFC	**108**	**-**	**1105**	**465**	**-**	**-**	**-**	**402**	**382**	**1101**	**3563**
INDUSTRY SECTOR	**7**	**-**	**65**	**217**	**-**	**-**	**-**	**49**	**146**	**331**	**815**
Iron and Steel	-	-	6	60	-	-	-	-	7	21	94
Chemical and Petrochemical	-	-	-	5	-	-	-	-	18	32	56
of which: Feedstocks	-	-	-	-	-	-	-	-	-	-	-
Non-Ferrous Metals	-	-	-	-	-	-	-	-	-	-	-
Non-Metallic Minerals	1	-	9	29	-	-	-	1	9	14	62
Transport Equipment	-	-	10	8	-	-	-	-	7	11	36
Machinery	-	-	-	2	-	-	-	4	14	14	33
Mining and Quarrying	1	-	1	-	-	-	-	-	1	-	3
Food and Tobacco	3	-	14	45	-	-	-	12	24	124	222
Paper, Pulp and Printing	-	-	2	3	-	-	-	1	3	6	15
Wood and Wood Products	-	-	-	12	-	-	-	27	18	39	95
Construction	1	-	9	-	-	-	-	1	-	29	40
Textile and Leather	1	-	1	7	-	-	-	1	17	29	55
Non-specified	1	-	13	45	-	-	-	3	30	10	103
TRANSPORT SECTOR	**-**	**-**	**824**	**1**	**-**	**-**	**-**	**-**	**15**	**-**	**840**
Air	-	-	34	-	-	-	-	-	-	-	34
Road	-	-	614	1	-	-	-	-	-	-	615
Rail	-	-	79	-	-	-	-	-	4	-	83
Pipeline Transport	-	-	-	-	-	-	-	-	3	-	3
Internal Navigation	-	-	97	-	-	-	-	-	-	-	97
Non-specified	-	-	-	-	-	-	-	-	8	-	8
OTHER SECTORS	**96**	**-**	**216**	**247**	**-**	**-**	**-**	**353**	**221**	**770**	**1903**
Agriculture	4	-	83	23	-	-	-	19	15	9	152
Comm. and Publ. Services	54	-	31	134	-	-	-	98	112	85	513
Residential	38	-	101	91	-	-	-	237	94	676	1237
Non-specified	-	-	1	-	-	-	-	-	-	-	1
NON-ENERGY USE	**5**	**-**	**-**	**-**	**-**	**-**	**-**	**-**	**-**	**-**	**5**
in Industry/Transf./Energy	5	-	-	-	-	-	-	-	-	-	5
in Transport	-	-	-	-	-	-	-	-	-	-	-
in Other Sectors	-	-	-	-	-	-	-	-	-	-	-
Electr. Generated - GWh	*96*	*-*	*643*	*524*	*-*	*1860*	*1*	*-*	*-*	*-*	*3124*
Electricity Plants	*-*	*-*	*-*	*-*	*-*	*1860*	*1*	*-*	*-*	*-*	*1861*
CHP Plants	*96*	*-*	*643*	*524*	*-*	*-*	*-*	*-*	*-*	*-*	*1263*
Heat Generated - TJ	*3530*	*-*	*34965*	*10206*	*-*	*-*	*-*	*6210*	*14*	*-*	*54925*
CHP Plants	*1372*	*-*	*8072*	*5964*	*-*	*-*	*-*	*-*	*-*	*-*	*15408*
Heat Plants	*2158*	*-*	*26893*	*4242*	*-*	*-*	*-*	*6210*	*14*	*-*	*39517*

Latvia / Lettonie : 1997

Thousand tonnes of oil equivalent / *Milliers de tonnes d'équivalent pétrole*

SUPPLY AND CONSUMPTION / *APPROVISIONNEMENT ET DEMANDE*	Coal / *Charbon*	Crude Oil / *Pétrole brut*	Petroleum Products / *Produits pétroliers*	Gas / *Gaz*	Nuclear / *Nucléaire*	Hydro / *Hydro*	Geotherm. Solar etc. / *Géotherm. solaire*	Combust. Renew. & Waste / *En. ren. combust.*	Electricity / *Electricité*	Heat / *Chaleur*	Total / *Total*
Indigenous Production	78	-	-	-	-	254	-	1303	-	-	1636
Imports	104	-	1564	1058	-	-	-	-	157	-	2883
Exports	-1	-	-5	-	-	-	-	-236	-	-	-242
Intl. Marine Bunkers	-	-	-	-	-	-	-	-	-	-	-
Stock Changes	-21	-	185	6	-	-	-	13	-	-	184
TPES	**160**	**-**	**1744**	**1065**	**-**	**254**	**-**	**1081**	**157**	**-**	**4460**
Transfers	-	-	-	-	-	-	-	-	-	-	-
Statistical Differences	34	-	-	-	-	-	-	-	-15	-	19
Electricity Plants	-	-	-	-	-	-254	-	-	254	-	-
CHP Plants	-46	-	-146	-493	-	-	-	-	133	441	-111
Heat Plants	-34	-	-376	-194	-	-	-	-237	-1	671	-171
Gas Works	-	-	-	-	-	-	-	-	-	-	-
Petroleum Refineries	-	-	-	-	-	-	-	-	-	-	-
Coal Transformation	-3	-	-	-	-	-	-	-	-	-	-3
Liquefaction	-	-	-	-	-	-	-	-	-	-	-
Other Transformation	-	-	-	-	-	-	-	-	-	-	-
Own Use	-6	-	-8	-14	-	-	-	-1	-41	-37	-107
Distribution Losses	-	-	-1	-25	-	-	-	-	-114	-161	-301
TFC	**105**	**-**	**1213**	**340**	**-**	**-**	**-**	**843**	**373**	**913**	**3787**
INDUSTRY SECTOR	**9**	**-**	**276**	**214**	**-**	**-**	**-**	**100**	**128**	**272**	**997**
Iron and Steel	-	-	26	94	-	-	-	-	9	17	146
Chemical and Petrochemical	-	-	45	5	-	-	-	3	17	27	96
of which: Feedstocks	-	-	-	-	-	-	-	-	-	-	-
Non-Ferrous Metals	-	-	-	-	-	-	-	-	-	-	-
Non-Metallic Minerals	2	-	54	15	-	-	-	-	8	12	92
Transport Equipment	-	-	5	3	-	-	-	-	6	9	23
Machinery	-	-	6	2	-	-	-	5	15	11	39
Mining and Quarrying	-	-	-	-	-	-	-	-	1	-	1
Food and Tobacco	4	-	83	67	-	-	-	15	28	102	299
Paper, Pulp and Printing	1	-	-	3	-	-	-	1	2	5	11
Wood and Wood Products	-	-	24	12	-	-	-	60	15	32	144
Construction	1	-	15	-	-	-	-	8	2	24	50
Textile and Leather	1	-	17	11	-	-	-	1	18	24	71
Non-specified	-	-	1	2	-	-	-	8	6	9	25
TRANSPORT SECTOR	**-**	**-**	**795**	**1**	**-**	**-**	**-**	**-**	**15**	**-**	**811**
Air	-	-	34	-	-	-	-	-	-	-	34
Road	-	-	608	1	-	-	-	-	-	-	608
Rail	-	-	83	-	-	-	-	-	5	-	87
Pipeline Transport	-	-	-	-	-	-	-	-	3	-	3
Internal Navigation	-	-	71	-	-	-	-	-	-	-	71
Non-specified	-	-	-	-	-	-	-	-	7	-	7
OTHER SECTORS	**90**	**-**	**142**	**125**	**-**	**-**	**-**	**744**	**230**	**641**	**1972**
Agriculture	3	-	66	14	-	-	-	13	16	-	112
Comm. and Publ. Services	48	-	41	37	-	-	-	88	121	54	391
Residential	39	-	35	74	-	-	-	643	93	587	1470
Non-specified	-	-	-	-	-	-	-	-	-	-	-
NON-ENERGY USE	**6**	**-**	**-**	**-**	**-**	**-**	**-**	**-**	**-**	**-**	**6**
in Industry/Transf./Energy	6	-	-	-	-	-	-	-	-	-	6
in Transport	-	-	-	-	-	-	-	-	-	-	-
in Other Sectors	-	-	-	-	-	-	-	-	-	-	-
Electr. Generated - GWh	*104*	*-*	*239*	*1205*	*-*	*2952*	*1*	*-*	*-*	*-*	*4501*
Electricity Plants	*-*	*-*	*-*	*-*	*-*	*2952*	*1*	*-*	*-*	*-*	*2953*
CHP Plants	*104*	*-*	*239*	*1205*	*-*	*-*	*-*	*-*	*-*	*-*	*1548*
Heat Generated - TJ	*2485*	*-*	*17195*	*19392*	*-*	*-*	*-*	*7449*	*19*	*-*	*46540*
CHP Plants	*1237*	*-*	*4676*	*12534*	*-*	*-*	*-*	*-*	*-*	*-*	*18447*
Heat Plants	*1248*	*-*	*12519*	*6858*	*-*	*-*	*-*	*7449*	*19*	*-*	*28093*

Lebanon / Liban

Thousand tonnes of oil equivalent / *Milliers de tonnes d'équivalent pétrole*

SUPPLY AND CONSUMPTION 1996	Coal	Crude Oil	Petroleum Products	Gas	Nuclear	Hydro	Geotherm. Solar etc.	Combust. Renew. & Waste	Electricity	Heat	Total
Indigenous Production	-	-	-	-	-	69	6	121	-	-	195
Imports	132	-	4359	-	-	-	-	1	59	-	4551
Exports	-	-	-	-	-	-	-	-	-	-	-
Intl. Marine Bunkers	-	-	-	-	-	-	-	-	-	-	-
Stock changes	-	-	-	-	-	-	-	-	-	-	-
TPES	132	-	4359	-	-	69	6	122	59	-	4747
Electricity and CHP Plants	-	-	-1490	-	-	-69	-	-	599	-	-960
Petroleum Refineries	-	-	-	-	-	-	-	-	-	-	-
Other Transformation *	-	-	-	-	-	-	-	-16	-79	-	-95
TFC	132	-	2869	-	-	-	6	107	579	-	3693
INDUSTRY SECTOR	132	-	628	-	-	-	-	-	151	-	911
Iron and Steel	-	-	-	-	-	-	-	-	-	-	-
Chemical and Petrochemical	-	-	-	-	-	-	-	-	-	-	-
Non-Metallic Minerals	132	-	-	-	-	-	-	-	-	-	132
Non-specified	-	-	628	-	-	-	-	-	151	-	779
TRANSPORT SECTOR	-	-	1605	-	-	-	-	-	-	-	1605
Air	-	-	114	-	-	-	-	-	-	-	114
Road	-	-	1491	-	-	-	-	-	-	-	1491
Non-specified	-	-	-	-	-	-	-	-	-	-	-
OTHER SECTORS	-	-	531	-	-	-	6	107	428	-	1072
Agriculture	-	-	-	-	-	-	-	-	-	-	-
Comm. and Publ. Services	-	-	-	-	-	-	-	-	79	-	79
Residential	-	-	531	-	-	-	6	-	316	-	853
Non-specified	-	-	-	-	-	-	-	107	33	-	140
NON-ENERGY USE	-	-	105	-	-	-	-	-	-	-	105
Electricity Generated - GWh	-	-	6167	-	-	798	-	-	-	-	6965
Heat Generated - TJ	-	-	-	-	-	-	-	-	-	-	-

APPROVISIONNEMENT ET DEMANDE 1997	Charbon	Pétrole brut	Produits pétroliers	Gaz	Nucléaire	Hydro	Géotherm. solaire etc.	En. ren. combust. & déchets	Electricité	Chaleur	Total
Indigenous Production	-	-	-	-	-	77	6	123	-	-	207
Imports	132	-	4852	-	-	-	-	2	52	-	5038
Exports	-	-	-	-	-	-	-	-	-	-	-
Intl. Marine Bunkers	-	-	-	-	-	-	-	-	-	-	-
Stock changes	-	-	-	-	-	-	-	-	-	-	-
TPES	132	-	4852	-	-	77	6	125	52	-	5244
Electricity and CHP Plants	-	-	-1687	-	-	-77	-	-	732	-	-1032
Petroleum Refineries	-	-	-	-	-	-	-	-	-	-	-
Other Transformation *	-	-	-	-	-	-	-	-15	-97	-	-112
TFC	132	-	3165	-	-	-	6	109	688	-	4100
INDUSTRY SECTOR	132	-	831	-	-	-	-	-	181	-	1143
Iron and Steel	-	-	-	-	-	-	-	-	-	-	-
Chemical and Petrochemical	-	-	-	-	-	-	-	-	-	-	-
Non-Metallic Minerals	132	-	-	-	-	-	-	-	-	-	132
Non-specified	-	-	831	-	-	-	-	-	181	-	1011
TRANSPORT SECTOR	-	-	1540	-	-	-	-	-	-	-	1540
Air	-	-	116	-	-	-	-	-	-	-	116
Road	-	-	1423	-	-	-	-	-	-	-	1423
Non-specified	-	-	-	-	-	-	-	-	-	-	-
OTHER SECTORS	-	-	710	-	-	-	6	109	507	-	1332
Agriculture	-	-	-	-	-	-	-	-	-	-	-
Comm. and Publ. Services	-	-	-	-	-	-	-	-	115	-	115
Residential	-	-	710	-	-	-	6	-	262	-	978
Non-specified	-	-	-	-	-	-	-	109	130	-	239
NON-ENERGY USE	-	-	84	-	-	-	-	-	-	-	84
Electricity Generated - GWh	-	-	7614	-	-	901	-	-	-	-	8515
Heat Generated - TJ	-	-	-	-	-	-	-	-	-	-	-

* Includes Transfers, Statistical Differences, Own Use and Distribution Losses.

Libya / Libye

Thousand tonnes of oil equivalent / *Milliers de tonnes d'équivalent pétrole*

SUPPLY AND CONSUMPTION 1996	Coal	Crude Oil	Petroleum Products	Gas	Nuclear	Hydro	Geotherm. Solar etc.	Combust. Renew. & Waste	Electricity	Heat	Total
Indigenous Production	-	71998	-	5619	-	-	-	125	-	-	77742
Imports	-	-	29	-	-	-	-	-	-	-	29
Exports	-	-55365	-6199	-1209	-	-	-	-	-	-	-62773
Intl. Marine Bunkers	-	-	-86	-	-	-	-	-	-	-	-86
Stock changes	-	-	-	-	-	-	-	-	-	-	-
TPES	-	16633	-6257	4410	-	-	-	125	-	-	14911
Electricity and CHP Plants	-	-	-3591	-	-	-	-	-	1563	-	-2028
Petroleum Refineries	-	-17295	15123	-	-	-	-	-	-	-	-2172
Other Transformation *	-	661	448	-2830	-	-	-	-	-	-	-1721
TFC	-	-	5722	1580	-	-	-	125	1563	-	8991
INDUSTRY SECTOR	-	-	1301	1580	-	-	-	-	-	-	2881
Iron and Steel	-	-	-	-	-	-	-	-	-	-	-
Chemical and Petrochemical	-	-	1003	1580	-	-	-	-	-	-	2583
Non-Metallic Minerals	-	-	-	-	-	-	-	-	-	-	-
Non-specified	-	-	298	-	-	-	-	-	-	-	298
TRANSPORT SECTOR	-	-	3522	-	-	-	-	-	-	-	3522
Air	-	-	335	-	-	-	-	-	-	-	335
Road	-	-	3187	-	-	-	-	-	-	-	3187
Non-specified	-	-	-	-	-	-	-	-	-	-	-
OTHER SECTORS	-	-	732	-	-	-	-	125	1563	-	2420
Agriculture	-	-	-	-	-	-	-	-	-	-	-
Comm. and Publ. Services	-	-	-	-	-	-	-	-	-	-	-
Residential	-	-	732	-	-	-	-	125	-	-	857
Non-specified	-	-	-	-	-	-	-	-	1563	-	1563
NON-ENERGY USE	-	-	167	-	-	-	-	-	-	-	167
Electricity Generated - GWh	-	-	*18180*	-	-	-	-	-	-	-	*18180*
Heat Generated - TJ	-	-	-	-	-	-	-	-	-	-	-

APPROVISIONNEMENT ET DEMANDE 1997	Charbon	Pétrole brut	Produits pétroliers	Gaz	Nucléaire	Hydro	Géotherm. solaire etc.	En. ren. combust. & déchets	Electricité	Chaleur	Total
Indigenous Production	-	73067	-	5750	-	-	-	125	-	-	78942
Imports	-	-	29	-	-	-	-	-	-	-	29
Exports	-	-56168	-6418	-1209	-	-	-	-	-	-	-63795
Intl. Marine Bunkers	-	-	-86	-	-	-	-	-	-	-	-86
Stock changes	-	-	-	-	-	-	-	-	-	-	-
TPES	-	16899	-6476	4541	-	-	-	125	-	-	15090
Electricity and CHP Plants	-	-	-3591	-	-	-	-	-	1563	-	-2028
Petroleum Refineries	-	-17595	15259	-	-	-	-	-	-	-	-2336
Other Transformation *	-	695	738	-2850	-	-	-	-	-	-	-1417
TFC	-	-	5929	1691	-	-	-	125	1563	-	9309
INDUSTRY SECTOR	-	-	1458	1691	-	-	-	-	-	-	3149
Iron and Steel	-	-	-	-	-	-	-	-	-	-	-
Chemical and Petrochemical	-	-	1031	1691	-	-	-	-	-	-	2722
Non-Metallic Minerals	-	-	-	-	-	-	-	-	-	-	-
Non-specified	-	-	427	-	-	-	-	-	-	-	427
TRANSPORT SECTOR	-	-	3539	-	-	-	-	-	-	-	3539
Air	-	-	315	-	-	-	-	-	-	-	315
Road	-	-	3224	-	-	-	-	-	-	-	3224
Non-specified	-	-	-	-	-	-	-	-	-	-	-
OTHER SECTORS	-	-	760	-	-	-	-	125	1563	-	2449
Agriculture	-	-	-	-	-	-	-	-	-	-	-
Comm. and Publ. Services	-	-	-	-	-	-	-	-	-	-	-
Residential	-	-	760	-	-	-	-	125	-	-	885
Non-specified	-	-	-	-	-	-	-	-	1563	-	1563
NON-ENERGY USE	-	-	172	-	-	-	-	-	-	-	172
Electricity Generated - GWh	-	-	*18180*	-	-	-	-	-	-	-	*18180*
Heat Generated - TJ	-	-	-	-	-	-	-	-	-	-	-

* Includes Transfers, Statistical Differences, Own Use and Distribution Losses.

INTERNATIONAL ENERGY AGENCY

Lithuania / Lituanie : 1996

Thousand tonnes of oil equivalent / *Milliers de tonnes d'équivalent pétrole*

SUPPLY AND CONSUMPTION *APPROVISIONNEMENT ET DEMANDE*	Coal *Charbon*	Crude Oil *Pétrole brut*	Petroleum Products *Produits pétroliers*	Gas *Gaz*	Nuclear *Nucléaire*	Hydro *Hydro*	Geotherm. Solar etc. *Géotherm. solaire*	Combust. Renew. & Waste *En. ren. combust.*	Electricity *Electricité*	Heat *Chaleur*	Total *Total*
Indigenous Production	15	156	-	-	3698	28	-	270	-	-	4168
Imports	184	4168	1367	2165	-	-	-	6	360	-	8249
Exports	-3	-124	-2432	-	-	-	-	-	-803	-	-3362
Intl. Marine Bunkers	-	-	-	-	-	-	-	-	-	-	-
Stock Changes	-31	-6	-79	3	-	-	-	-2	-	-	-116
TPES	**166**	**4194**	**-1145**	**2168**	**3698**	**28**	**-**	**273**	**-444**	**-**	**8939**
Transfers	-	-	47	-	-	-	-	-	-	-	47
Statistical Differences	-	-	-	-	-	-	-	-	-	-	-
Electricity Plants	-	-	-	-	-	-28	-	-	28	-	-
CHP Plants	-	-	-589	-678	-3698	-	-	-	1369	724	-2874
Heat Plants	-17	-6	-613	-524	-	-	-	-39	-9	1252	43
Gas Works	-	-	-	-	-	-	-	-	-	-	-
Petroleum Refineries	-	-4182	4249	-	-	-	-	-	-	-	68
Coal Transformation	-	-	-	-	-	-	-	-	-	-	-
Liquefaction	-	-	-	-	-	-	-	-	-	-	-
Other Transformation	-	-	-	-	-	-	-	-	-	-	-
Own Use	-1	-	-193	-	-	-	-	-	-231	-143	-568
Distribution Losses	-2	-	-15	-46	-	-	-	-	-153	-435	-651
TFC	**146**	**6**	**1741**	**919**	**-**	**-**	**-**	**234**	**560**	**1397**	**5004**
INDUSTRY SECTOR	**10**	**3**	**179**	**689**	**-**	**-**	**-**	**10**	**217**	**442**	**1550**
Iron and Steel	-	-	-	-	-	-	-	1	4	-	5
Chemical and Petrochemical	-	-	14	590	-	-	-	-	45	133	783
of which: Feedstocks	-	-	-	*548*	-	-	-	-	-	-	*548*
Non-Ferrous Metals	-	-	-	-	-	-	-	-	-	-	-
Non-Metallic Minerals	8	1	107	48	-	-	-	2	20	18	203
Transport Equipment	-	-	-	-	-	-	-	-	4	5	10
Machinery	-	-	-	21	-	-	-	1	33	24	80
Mining and Quarrying	-	-	-	4	-	-	-	-	2	-	6
Food and Tobacco	-	2	16	14	-	-	-	3	41	139	215
Paper, Pulp and Printing	1	-	-	1	-	-	-	-	8	32	42
Wood and Wood Products	-	-	2	7	-	-	-	2	7	5	24
Construction	-	-	27	1	-	-	-	-	6	2	36
Textile and Leather	-	-	-	-	-	-	-	-	31	53	85
Non-specified	1	-	12	3	-	-	-	1	16	28	62
TRANSPORT SECTOR	**1**	**-**	**1247**	**3**	**-**	**-**	**-**	**-**	**9**	**-**	**1261**
Air	-	-	32	-	-	-	-	-	-	-	32
Road	-	-	954	3	-	-	-	-	-	-	958
Rail	1	-	90	-	-	-	-	-	2	-	93
Pipeline Transport	-	-	-	-	-	-	-	-	-	-	-
Internal Navigation	-	-	1	-	-	-	-	-	-	-	1
Non-specified	-	-	170	-	-	-	-	-	6	-	177
OTHER SECTORS	**134**	**3**	**261**	**226**	**-**	**-**	**-**	**223**	**335**	**955**	**2138**
Agriculture	2	-	127	7	-	-	-	1	43	32	212
Comm. and Publ. Services	84	3	55	37	-	-	-	32	154	227	592
Residential	45	-	71	182	-	-	-	187	138	696	1319
Non-specified	4	-	9	-	-	-	-	3	-	-	16
NON-ENERGY USE	**-**	**-**	**54**	**-**	**-**	**-**	**-**	**-**	**-**	**-**	**54**
in Industry/Transf./Energy	-	-	54	-	-	-	-	-	-	-	54
in Transport	-	-	-	-	-	-	-	-	-	-	-
in Other Sectors	-	-	-	-	-	-	-	-	-	-	-
Electr. Generated - GWh	*-*	*-*	*1312*	*661*	*13942*	*326*	*-*	*-*	*-*	*-*	*16241*
Electricity Plants	*-*	*-*	*-*	*-*	*-*	*326*	*-*	*-*	*-*	*-*	*326*
CHP Plants	*-*	*-*	*1312*	*661*	*13942*	*-*	*-*	*-*	*-*	*-*	*15915*
Heat Generated - TJ	*1207*	*-*	*37428*	*39011*	*2724*	*-*	*-*	*1880*	*486*	*-*	*82736*
CHP Plants	*-*	*-*	*10497*	*17098*	*2724*	*-*	*-*	*-*	*-*	*-*	*30319*
Heat Plants	*1207*	*-*	*26931*	*21913*	*-*	*-*	*-*	*1880*	*486*	*-*	*52417*

Lithuania / Lituanie : 1997

Thousand tonnes of oil equivalent / *Milliers de tonnes d'équivalent pétrole*

SUPPLY AND CONSUMPTION / *APPROVISIONNEMENT ET DEMANDE*	Coal / *Charbon*	Crude Oil / *Pétrole brut*	Petroleum Products / *Produits pétroliers*	Gas / *Gaz*	Nuclear / *Nucléaire*	Hydro / *Hydro*	Geotherm. Solar etc. / *Géotherm. solaire*	Combust. Renew. & Waste / *En. ren. combust.*	Electricity / *Electricité*	Heat / *Chaleur*	Total / *Total*
Indigenous Production	18	213	-	-	3196	25	-	517	-	-	3970
Imports	122	5799	806	2001	-	-	-	-	389	-	9118
Exports	-21	-194	-3269	-	-	-	-	-1	-692	-	-4178
Intl. Marine Bunkers	-	-	-60	-	-	-	-	-	-	-	-60
Stock Changes	20	-105	40	-	-	-	-	2	-	-	-44
TPES	**139**	**5713**	**-2484**	**2001**	**3196**	**25**	**-**	**519**	**-303**	**-**	**8806**
Transfers	-	-	47	-	-	-	-	-	-	-	47
Statistical Differences	-	-	-	-	-	-	-	-	-	-	-
Electricity Plants	-	-	-	-	-	-25	-	-	25	-	-
CHP Plants	-	-	-451	-512	-3196	-	-	-	1212	669	-2278
Heat Plants	-14	-4	-561	-642	-	-	-	-40	-7	1162	-105
Gas Works	-	-	-	-	-	-	-	-	-	-	-
Petroleum Refineries	-	-5704	5681	-	-	-	-	-	-	-	-23
Coal Transformation	-2	-	-	-	-	-	-	-	-	-	-2
Liquefaction	-	-	-	-	-	-	-	-	-	-	-
Other Transformation	-	-	-	-	-	-	-	-	-	-	-
Own Use	-	-	-393	-	-	-	-	-	-212	-158	-763
Distribution Losses	-1	-1	-11	-41	-	-	-	-	-136	-347	-537
TFC	**122**	**4**	**1828**	**807**	**-**	**-**	**-**	**479**	**579**	**1326**	**5144**
INDUSTRY SECTOR	**12**	**2**	**183**	**616**	**-**	**-**	**-**	**8**	**239**	**409**	**1468**
Iron and Steel	1	-	-	-	-	-	-	-	5	-	6
Chemical and Petrochemical	-	-	2	525	-	-	-	-	43	126	696
of which: Feedstocks	-	-	-	471	-	-	-	-	-	-	471
Non-Ferrous Metals	-	-	-	-	-	-	-	-	-	-	-
Non-Metallic Minerals	5	-	121	39	-	-	-	3	19	14	202
Transport Equipment	-	-	-	1	-	-	-	-	4	3	7
Machinery	-	-	3	21	-	-	-	1	29	18	72
Mining and Quarrying	-	-	1	4	-	-	-	-	2	-	7
Food and Tobacco	3	2	18	14	-	-	-	2	42	138	219
Paper, Pulp and Printing	1	-	-	-	-	-	-	-	8	31	40
Wood and Wood Products	-	-	4	9	-	-	-	1	10	7	30
Construction	-	-	26	1	-	-	-	-	6	1	35
Textile and Leather	-	-	-	-	-	-	-	-	37	58	95
Non-specified	1	-	7	1	-	-	-	1	33	14	58
TRANSPORT SECTOR	**-**	**-**	**1253**	**1**	**-**	**-**	**-**	**-**	**9**	**-**	**1262**
Air	-	-	33	-	-	-	-	-	-	-	33
Road	-	-	1118	1	-	-	-	-	-	-	1119
Rail	-	-	83	-	-	-	-	-	2	-	85
Pipeline Transport	-	-	-	-	-	-	-	-	-	-	-
Internal Navigation	-	-	1	-	-	-	-	-	-	-	1
Non-specified	-	-	17	-	-	-	-	-	7	-	24
OTHER SECTORS	**110**	**2**	**224**	**191**	**-**	**-**	**-**	**471**	**332**	**916**	**2245**
Agriculture	1	-	94	7	-	-	-	1	37	37	178
Comm. and Publ. Services	63	2	42	35	-	-	-	34	147	241	565
Residential	46	-	87	148	-	-	-	436	148	637	1501
Non-specified	-	-	-	-	-	-	-	-	-	1	1
NON-ENERGY USE	**-**	**-**	**169**	**-**	**-**	**-**	**-**	**-**	**-**	**-**	**169**
in Industry/Transf./Energy	-	-	140	-	-	-	-	-	-	-	140
in Transport	-	-	29	-	-	-	-	-	-	-	29
in Other Sectors	-	-	-	-	-	-	-	-	-	-	-
Electr. Generated - GWh	*-*	*-*	*1564*	*504*	*12024*	*295*	*-*	*-*	*-*	*-*	*14387*
Electricity Plants	*-*	*-*	*-*	*-*	*-*	*295*	*-*	*-*	*-*	*-*	*295*
CHP Plants	*-*	*-*	*1564*	*504*	*12024*	*-*	*-*	*-*	*-*	*-*	*14092*
Heat Generated - TJ	*883*	*-*	*29575*	*41334*	*2619*	*-*	*-*	*2081*	*189*	*-*	*76681*
CHP Plants	*-*	*-*	*9682*	*15730*	*2619*	*-*	*-*	*-*	*-*	*-*	*28031*
Heat Plants	*883*	*-*	*19893*	*25604*	*-*	*-*	*-*	*2081*	*189*	*-*	*48650*

FYR of Macedonia / ex-République yougoslave de Macédoine

Thousand tonnes of oil equivalent / *Milliers de tonnes d'équivalent pétrole*

SUPPLY AND CONSUMPTION 1996	Coal	Crude Oil	Petroleum Products	Gas	Nuclear	Hydro	Geotherm. Solar etc.	Combust. Renew. & Waste	Electricity	Heat	Total
Indigenous Production	1518	-	-	-	-	73	13	187	-	-	1790
Imports	81	670	611	-	-	-	-	4	-	-	1366
Exports	-	-	-37	-	-	-	-	-	-	-	-37
Intl. Marine Bunkers	-	-	-	-	-	-	-	-	-	-	-
Stock changes	7	-	-	-	-	-	-	-	-	-	7
TPES	1606	670	574	-	-	73	13	190	-	-	3126
Electricity and CHP Plants	-1382	-	-29	-	-	-73	-	-	558	6	-920
Petroleum Refineries	-	-689	676	-	-	-	-	-	-	-	-13
Other Transformation *	-100	19	-201	-	-	-	-13	-	-120	118	-296
TFC	124	-	1021	-	-	-	-	190	438	124	1897
INDUSTRY SECTOR	109	-	260	-	-	-	-	-	173	67	610
Iron and Steel	64	-	33	-	-	-	-	-	94	3	194
Chemical and Petrochemical	-	-	57	-	-	-	-	-	12	15	84
Non-Metallic Minerals	-	-	28	-	-	-	-	-	2	-	31
Non-specified	46	-	142	-	-	-	-	-	65	49	302
TRANSPORT SECTOR	-	-	591	-	-	-	-	-	1	-	592
Air	-	-	28	-	-	-	-	-	-	-	28
Road	-	-	555	-	-	-	-	-	-	-	555
Non-specified	-	-	8	-	-	-	-	-	1	-	9
OTHER SECTORS	14	-	132	-	-	-	-	190	264	57	657
Agriculture	-	-	92	-	-	-	-	-	2	11	105
Comm. and Publ. Services	6	-	-	-	-	-	-	-	44	18	69
Residential	8	-	40	-	-	-	-	187	217	28	479
Non-specified	-	-	-	-	-	-	-	4	-	-	4
NON-ENERGY USE	-	-	38	-	-	-	-	-	-	-	38
Electricity Generated - GWh	5601	-	38	-	-	850	-	-	-	-	6489
Heat Generated - TJ	669	-	5011	-	-	-	528	-	-	-	6208

APPROVISIONNEMENT ET DEMANDE 1997	Charbon	Pétrole brut	Produits pétroliers	Gaz	Nucléaire	Hydro	Géotherm. solaire etc.	En. ren. combust. & déchets	Electricité	Chaleur	Total
Indigenous Production	1423	-	-	-	-	77	13	187	-	-	1700
Imports	73	408	661	-	-	-	-	4	-	-	1147
Exports	-	-	-5	-	-	-	-	-	-	-	-5
Intl. Marine Bunkers	-	-	-	-	-	-	-	-	-	-	-
Stock changes	-	-	-	-	-	-	-	-	-	-	-
TPES	1496	408	656	-	-	77	13	190	-	-	2841
Electricity and CHP Plants	-1288	-	-19	-	-	-77	-	-	578	6	-801
Petroleum Refineries	-	-408	387	-	-	-	-	-	-	-	-21
Other Transformation *	-88	-	-173	-	-	-	-13	-	-124	118	-279
TFC	120	-	851	-	-	-	-	190	454	124	1739
INDUSTRY SECTOR	106	-	186	-	-	-	-	-	179	67	539
Iron and Steel	61	-	24	-	-	-	-	-	97	3	185
Chemical and Petrochemical	-	-	37	-	-	-	-	-	12	15	65
Non-Metallic Minerals	-	-	22	-	-	-	-	-	2	-	25
Non-specified	45	-	103	-	-	-	-	-	67	49	264
TRANSPORT SECTOR	-	-	531	-	-	-	-	-	1	-	532
Air	-	-	32	-	-	-	-	-	-	-	32
Road	-	-	464	-	-	-	-	-	-	-	464
Non-specified	-	-	35	-	-	-	-	-	1	-	36
OTHER SECTORS	14	-	94	-	-	-	-	190	273	57	629
Agriculture	-	-	65	-	-	-	-	-	2	11	77
Comm. and Publ. Services	6	-	-	-	-	-	-	-	46	18	71
Residential	8	-	30	-	-	-	-	187	225	28	477
Non-specified	-	-	-	-	-	-	-	4	-	-	4
NON-ENERGY USE	-	-	39	-	-	-	-	-	-	-	39
Electricity Generated - GWh	5770	-	49	-	-	900	-	-	-	-	6719
Heat Generated - TJ	669	-	5011	-	-	-	528	-	-	-	6208

* Includes Transfers, Statistical Differences, Own Use and Distribution Losses.

Malaysia / Malaisie

Total Production of Energy

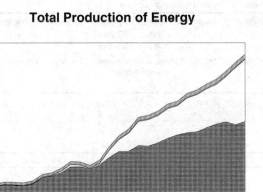

Coal ■ Oil ▦ Gas ☐ Nuclear ▤ Hydro/Oth. ▦ Comb. Renew. ▥

Total Primary Energy Supply*

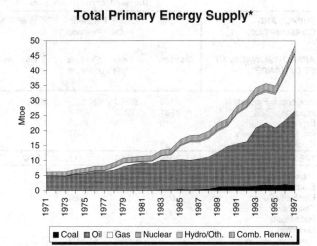

Coal ■ Oil ▦ Gas ☐ Nuclear ▤ Hydro/Oth. ▦ Comb. Renew. ▥

Electricity Generation by Fuel

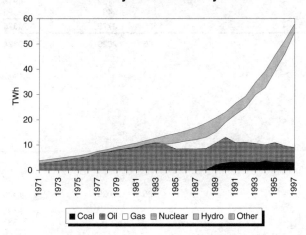

Coal ■ Oil ▦ Gas ☐ Nuclear ▤ Hydro ▦ Other ▥

Oil Products Consumption

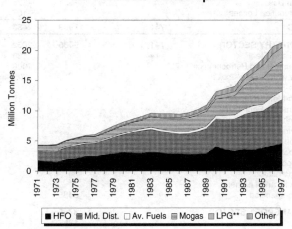

HFO ■ Mid. Dist. ▦ Av. Fuels ☐ Mogas ▤ LPG** ▦ Other ▥

Electricity Consumption/GDP***, TPES/GDP*** and Production/TPES

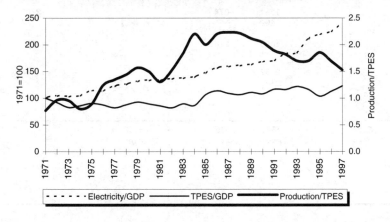

- - - - Electricity/GDP ——— TPES/GDP ━━━ Production/TPES

* Excluding electricity trade.
** Includes LPG, NGL, ethane and naphtha.
*** GDP in 1990 US dollars.

Malaysia / Malaisie : 1996

Thousand tonnes of oil equivalent / Milliers de tonnes d'équivalent pétrole

SUPPLY AND CONSUMPTION / APPROVISIONNEMENT ET DEMANDE	Coal Charbon	Crude Oil Pétrole brut	Petroleum Products Produits pétroliers	Gas Gaz	Nuclear Nucléaire	Hydro Hydro	Geotherm. Solar etc. Géotherm. solaire	Combust. Renew. & Waste En. ren. combust.	Electricity Electricité	Heat Chaleur	Total Total
Indigenous Production	153	36353	-	30251	-	446	-	2407	-	-	69610
Imports	1931	2210	8817	-	-	-	-	5	11	-	12973
Exports	-15	-17254	-8533	-15298	-	-	-	-21	-12	-	-41133
Intl. Marine Bunkers	-	-	-167	-	-	-	-	-	-	-	-167
Stock Changes	-23	-	-	-	-	-	-	-	-	-	-23
TPES	**2046**	**21308**	**117**	**14952**	**-**	**446**	**-**	**2390**	**-1**	**-**	**41259**
Transfers	-	-3629	3032	-	-	-	-	-	-	-	-597
Statistical Differences	-301	2184	66	120	-	-	-	-	-14	-	2055
Electricity Plants	-950	-	-2580	-8305	-	-446	-	-	4421	-	-7860
CHP Plants	-	-	-	-	-	-	-	-	-	-	-
Heat Plants	-	-	-	-	-	-	-	-	-	-	-
Gas Works	-	-	-	-	-	-	-	-	-	-	-
Petroleum Refineries	-	-19863	17903	-	-	-	-	-	-	-	-1960
Coal Transformation	-54	-	-	-	-	-	-	-	-	-	-54
Liquefaction	-	-	-	-	-	-	-	-	-	-	-
Other Transformation	-	-	-	-	-	-	-	-954	-	-	-954
Own Use	-	-	-551	-4226	-	-	-	-	-139	-	-4916
Distribution Losses	-	-	-	-107	-	-	-	-	-491	-	-598
TFC	**741**	**-**	**17989**	**2433**	**-**	**-**	**-**	**1436**	**3776**	**-**	**26375**
INDUSTRY SECTOR	**741**	**-**	**5435**	**2052**	**-**	**-**	**-**	**66**	**2029**	**-**	**10324**
Iron and Steel	14	-	-	136	-	-	-	-	-	-	150
Chemical and Petrochemical	-	-	-	1900	-	-	-	-	-	-	1900
of which: Feedstocks	-	-	-	1305	-	-	-	-	-	-	1305
Non-Ferrous Metals	-	-	-	-	-	-	-	-	-	-	-
Non-Metallic Minerals	727	-	-	16	-	-	-	-	-	-	744
Transport Equipment	-	-	-	-	-	-	-	-	-	-	-
Machinery	-	-	-	-	-	-	-	-	-	-	-
Mining and Quarrying	-	-	-	-	-	-	-	-	-	-	-
Food and Tobacco	-	-	-	-	-	-	-	-	-	-	-
Paper, Pulp and Printing	-	-	-	-	-	-	-	-	-	-	-
Wood and Wood Products	-	-	-	-	-	-	-	-	-	-	-
Construction	-	-	-	-	-	-	-	-	-	-	-
Textile and Leather	-	-	-	-	-	-	-	-	-	-	-
Non-specified	-	-	5435	-	-	-	-	66	2029	-	7530
TRANSPORT SECTOR	**-**	**-**	**8786**	**6**	**-**	**-**	**-**	**-**	**-**	**-**	**8792**
Air	-	-	1387	-	-	-	-	-	-	-	1387
Road	-	-	7380	-	-	-	-	-	-	-	7380
Rail	-	-	-	-	-	-	-	-	-	-	-
Pipeline Transport	-	-	-	-	-	-	-	-	-	-	-
Internal Navigation	-	-	19	-	-	-	-	-	-	-	19
Non-specified	-	-	-	6	-	-	-	-	-	-	6
OTHER SECTORS	**-**	**-**	**1581**	**374**	**-**	**-**	**-**	**1370**	**1747**	**-**	**5072**
Agriculture	-	-	492	-	-	-	-	-	-	-	492
Comm. and Publ. Services	-	-	-	-	-	-	-	-	1011	-	1011
Residential	-	-	1089	374	-	-	-	1370	736	-	3569
Non-specified	-	-	-	-	-	-	-	-	-	-	-
NON-ENERGY USE	**-**	**-**	**2187**	**-**	**-**	**-**	**-**	**-**	**-**	**-**	**2187**
in Industry/Transf./Energy	-	-	2187	-	-	-	-	-	-	-	2187
in Transport	-	-	-	-	-	-	-	-	-	-	-
in Other Sectors	-	-	-	-	-	-	-	-	-	-	-
Electr. Generated - GWh	*3285*	*-*	*6321*	*36615*	*-*	*5186*	*-*	*-*	*-*	*-*	*51407*
Electricity Plants	*3285*	*-*	*6321*	*36615*	*-*	*5186*	*-*	*-*	*-*	*-*	*51407*
CHP Plants	*-*	*-*	*-*	*-*	*-*	*-*	*-*	*-*	*-*	*-*	*-*
Heat Generated - TJ	*-*	*-*	*-*	*-*	*-*	*-*	*-*	*-*	*-*	*-*	*-*
CHP Plants	*-*	*-*	*-*	*-*	*-*	*-*	*-*	*-*	*-*	*-*	*-*
Heat Plants	*-*	*-*	*-*	*-*	*-*	*-*	*-*	*-*	*-*	*-*	*-*

Malaysia / Malaisie : 1997

	Thousand tonnes of oil equivalent / *Milliers de tonnes d'équivalent pétrole*										
SUPPLY AND CONSUMPTION *APPROVISIONNEMENT ET DEMANDE*	Coal *Charbon*	Crude Oil *Pétrole brut*	Petroleum Products *Produits pétroliers*	Gas *Gaz*	Nuclear *Nucléaire*	Hydro *Hydro*	Geotherm. Solar etc. *Géotherm. solaire*	Combust. Renew. & Waste *En. ren. combust.*	Electricity *Electricité*	Heat *Chaleur*	Total *Total*
Indigenous Production	153	38039	-	33096	-	283	-	2407	-	-	73979
Imports	1458	2611	9636	-	-	-	-	5	7	-	13718
Exports	-9	-17595	-7488	-13905	-	-	-	-21	-6	-	-39025
Intl. Marine Bunkers	-	-	-157	-	-	-	-	-	-	-	-157
Stock Changes	-21	-	-21	-	-	-	-	-	-	-	-42
TPES	**1581**	**23056**	**1970**	**19191**	**-**	**283**	**-**	**2390**	**1**	**-**	**48473**
Transfers	-	-3076	3138	-	-	-	-	-	-	-	62
Statistical Differences	109	-753	337	-1536	-	-	-	-	-6	-	-1849
Electricity Plants	-882	-	-2602	-10349	-	-283	-	-	4977	-	-9139
CHP Plants	-	-	-	-	-	-	-	-	-	-	-
Heat Plants	-	-	-	-	-	-	-	-	-	-	-
Gas Works	-	-	-	-	-	-	-	-	-	-	-
Petroleum Refineries	-	-19228	16350		-	-	-	-	-	-	-2877
Coal Transformation	-54	-	-		-	-	-	-	-	-	-54
Liquefaction	-	-	-		-	-	-	-	-	-	-
Other Transformation	-	-	-		-	-	-	-954	-	-	-954
Own Use	-	-	-549	-5063	-	-	-	-	-130	-	-5742
Distribution Losses	-	-	-	-24	-	-	-	-	-458	-	-483
TFC	**754**	**·**	**-**	**18644**	**2219**	**-**	**-**	**-**	**1436**	**4383**	**-**	**27437**
INDUSTRY SECTOR	**754**	**-**	**5958**	**1872**	**-**	**-**	**-**	**66**	**2422**	**-**	**11073**
Iron and Steel	14	-	-	124	-	-	-	-	-	-	138
Chemical and Petrochemical	-	-	-	1733	-	-	-	-	-	-	1733
of which: Feedstocks	-	-	-	*1190*	-	-	-	-	-	-	*1190*
Non-Ferrous Metals	-	-	-	-	-	-	-	-	-	-	-
Non-Metallic Minerals	741	-	-	15	-	-	-	-	-	-	756
Transport Equipment	-	-	-	-	-	-	-	-	-	-	-
Machinery	-	-	-	-	-	-	-	-	-	-	-
Mining and Quarrying	-	-	-	-	-	-	-	-	-	-	-
Food and Tobacco	-	-	-	-	-	-	-	-	-	-	-
Paper, Pulp and Printing	-	-	-	-	-	-	-	-	-	-	-
Wood and Wood Products	-	-	-	-	-	-	-	-	-	-	-
Construction	-	-	-	-	-	-	-	-	-	-	-
Textile and Leather	-	-	-	-	-	-	-	-	-	-	-
Non-specified	-	-	5958	-	-	-	-	66	2422	-	8447
TRANSPORT SECTOR	**-**	**-**	**9161**	**6**	**-**	**-**	**-**	**-**	**-**	**-**	**9167**
Air	-	-	1495	-	-	-	-	-	-	-	1495
Road	-	-	7645	-	-	-	-	-	-	-	7645
Rail	-	-	-	-	-	-	-	-	-	-	-
Pipeline Transport	-	-	-	-	-	-	-	-	-	-	-
Internal Navigation	-	-	21	-	-	-	-	-	-	-	21
Non-specified	-	-	-	6	-	-	-	-	-	-	6
OTHER SECTORS	**-**	**-**	**1743**	**341**	**-**	**-**	**-**	**1370**	**1961**	**-**	**5416**
Agriculture	-	-	535	-	-	-	-	-	-	-	535
Comm. and Publ. Services	-	-	-	-	-	-	-	-	1191	-	1191
Residential	-	-	1209	341	-	-	-	1370	770	-	3690
Non-specified	-	-	-	-	-	-	-	-	-	-	-
NON-ENERGY USE	**-**	**-**	**1781**	**-**	**-**	**-**	**-**	**-**	**-**	**-**	**1781**
in Industry/Transf./Energy	-	-	1781	-	-	-	-	-	-	-	1781
in Transport	-	-	-	-	-	-	-	-	-	-	-
in Other Sectors	-	-	-	-	-	-	-	-	-	-	-
Electr. Generated - GWh	***3050***	***-***	***5906***	***45625***	***-***	***3294***	***-***	***-***	***-***	***-***	***57875***
Electricity Plants	*3050*	*-*	*5906*	*45625*	*-*	*3294*	*-*	*-*	*-*	*-*	*57875*
CHP Plants	*-*	*-*	*-*	*-*	*-*	*-*	*-*	*-*	*-*	*-*	*-*
Heat Generated - TJ	***-***	***-***	***-***	***-***	***-***	***-***	***-***	***-***	***-***	***-***	***-***
CHP Plants	*-*	*-*	*-*	*-*	*-*	*-*	*-*	*-*	*-*	*-*	*-*
Heat Plants	*-*	*-*	*-*	*-*	*-*	*-*	*-*	*-*	*-*	*-*	*-*

Malta / Malte

Thousand tonnes of oil equivalent / *Milliers de tonnes d'équivalent pétrole*

SUPPLY AND CONSUMPTION 1996	Coal	Crude Oil	Petroleum Products	Gas	Nuclear	Hydro	Geotherm. Solar etc.	Combust. Renew. & Waste	Electricity	Heat	Total
Indigenous Production	-	-	-	-	-	-	-	-	-	-	-
Imports	-	-	971	-	-	-	-	-	-	-	971
Exports	-	-	-	-	-	-	-	-	-	-	-
Intl. Marine Bunkers	-	-	-77	-	-	-	-	-	-	-	-77
Stock changes	-	-	-	-	-	-	-	-	-	-	-
TPES	-	-	893	-	-	-	-	-	-	-	893
Electricity and CHP Plants	-	-	-490	-	-	-	-	-	143	-	-348
Petroleum Refineries	-	-	-	-	-	-	-	-	-	-	-
Other Transformation *	-	-	-	-	-	-	-	-	-27	-	-27
TFC	-	-	403	-	-	-	-	-	115	-	518
INDUSTRY SECTOR	-	-	-	-	-	-	-	-	44	-	44
Iron and Steel	-	-	-	-	-	-	-	-	-	-	-
Chemical and Petrochemical	-	-	-	-	-	-	-	-	-	-	-
Non-Metallic Minerals	-	-	-	-	-	-	-	-	-	-	-
Non-specified	-	-	-	-	-	-	-	-	44	-	44
TRANSPORT SECTOR	-	-	351	-	-	-	-	-	-	-	351
Air	-	-	114	-	-	-	-	-	-	-	114
Road	-	-	237	-	-	-	-	-	-	-	237
Non-specified	-	-	-	-	-	-	-	-	-	-	-
OTHER SECTORS	-	-	53	-	-	-	-	-	71	-	124
Agriculture	-	-	-	-	-	-	-	-	-	-	-
Comm. and Publ. Services	-	-	-	-	-	-	-	-	34	-	34
Residential	-	-	39	-	-	-	-	-	37	-	76
Non-specified	-	-	13	-	-	-	-	-	-	-	13
NON-ENERGY USE	-	-	-	-	-	-	-	-	-	-	-
Electricity Generated - GWh	-	-	*1658*	-	-	-	-	-	-	-	*1658*
Heat Generated - TJ	-	-	-	-	-	-	-	-	-	-	-

APPROVISIONNEMENT ET DEMANDE 1997	Charbon	Pétrole brut	Produits pétroliers	Gaz	Nucléaire	Hydro	Géotherm. solaire etc.	En. ren. combust. & déchets	Electricité	Chaleur	Total
Indigenous Production	-	-	-	-	-	-	-	-	-	-	-
Imports	-	-	1021	-	-	-	-	-	-	-	1021
Exports	-	-	-	-	-	-	-	-	-	-	-
Intl. Marine Bunkers	-	-	-78	-	-	-	-	-	-	-	-78
Stock changes	-	-	-	-	-	-	-	-	-	-	-
TPES	-	-	943	-	-	-	-	-	-	-	943
Electricity and CHP Plants	-	-	-499	-	-	-	-	-	145	-	-354
Petroleum Refineries	-	-	-	-	-	-	-	-	-	-	-
Other Transformation *	-	-	-	-	-	-	-	-	-28	-	-28
TFC	-	-	444	-	-	-	-	-	117	-	561
INDUSTRY SECTOR	-	-	-	-	-	-	-	-	39	-	39
Iron and Steel	-	-	-	-	-	-	-	-	-	-	-
Chemical and Petrochemical	-	-	-	-	-	-	-	-	-	-	-
Non-Metallic Minerals	-	-	-	-	-	-	-	-	-	-	-
Non-specified	-	-	-	-	-	-	-	-	39	-	39
TRANSPORT SECTOR	-	-	356	-	-	-	-	-	-	-	356
Air	-	-	119	-	-	-	-	-	-	-	119
Road	-	-	237	-	-	-	-	-	-	-	237
Non-specified	-	-	-	-	-	-	-	-	-	-	-
OTHER SECTORS	-	-	88	-	-	-	-	-	78	-	166
Agriculture	-	-	-	-	-	-	-	-	-	-	-
Comm. and Publ. Services	-	-	-	-	-	-	-	-	38	-	38
Residential	-	-	78	-	-	-	-	-	40	-	118
Non-specified	-	-	11	-	-	-	-	-	-	-	11
NON-ENERGY USE	-	-	-	-	-	-	-	-	-	-	-
Electricity Generated - GWh	-	-	*1686*	-	-	-	-	-	-	-	*1686*
Heat Generated - TJ	-	-	-	-	-	-	-	-	-	-	-

* Includes Transfers, Statistical Differences, Own Use and Distribution Losses.

Moldova : 1996

	Thousand tonnes of oil equivalent / *Milliers de tonnes d'équivalent pétrole*										
SUPPLY AND CONSUMPTION	Coal	Crude Oil	Petroleum Products	Gas	Nuclear	Hydro	Geotherm. Solar etc.	Combust. Renew. & Waste	Electricity	Heat	Total
APPROVISIONNEMENT ET DEMANDE	*Charbon*	*Pétrole brut*	*Produits pétroliers*	*Gaz*	*Nucléaire*	*Hydro*	*Géotherm. solaire*	*En. ren. combust.*	*Electricité*	*Chaleur*	*Total*
Indigenous Production	-	-	-	-	-	31	-	60	-	-4	87
Imports	487	-	836	2937	-	-	-	-	138	-	4399
Exports	-	-	-	-	-	-	-	-	-	-	-
Intl. Marine Bunkers	-	-	-	-	-	-	-	-	-	-	-
Stock Changes	24	-	133	-10	-	-	-	2	-	-	149
TPES	**511**	**-**	**970**	**2927**	**-**	**31**	**-**	**62**	**138**	**-4**	**4635**
Transfers	-	-	-	-	-	-	-	-	-	-	-
Statistical Differences	-	-	-	-	-	-	-	-	-	-	-
Electricity Plants	-328	-	-69	-774	-	-31	-	-	424	-	-778
CHP Plants	-16	-	-241	-507	-	-	-	-	103	264	-397
Heat Plants	-	-	-1	-53	-	-	-	-	-4	89	30
Gas Works	-	-	-	-	-	-	-	-	-	-	-
Petroleum Refineries	-	-	-	-	-	-	-	-	-	-	-
Coal Transformation	-3	-	-	-	-	-	-	-	-	-	-3
Liquefaction	-	-	-	-	-	-	-	-	-	-	-
Other Transformation	-	-	-	-	-	-	-	-	-	-	-
Own Use	-	-	-	-	-	-	-	-	-52	-	-52
Distribution Losses	-	-	-6	-121	-	-	-	-	-120	-25	-273
TFC	**163**	**-**	**653**	**1471**	**-**	**-**	**-**	**62**	**489**	**323**	**3162**
INDUSTRY SECTOR	**10**	**-**	**28**	**426**	**-**	**-**	**-**	**-**	**149**	**11**	**624**
Iron and Steel	-	-	-	-	-	-	-	-	49	-	49
Chemical and Petrochemical	-	-	-	-	-	-	-	-	-	-	-
of which: Feedstocks	-	-	-	-	-	-	-	-	-	-	-
Non-Ferrous Metals	-	-	-	-	-	-	-	-	-	-	-
Non-Metallic Minerals	2	-	1	128	-	-	-	-	18	-	148
Transport Equipment	-	-	-	-	-	-	-	-	-	-	-
Machinery	-	-	-	17	-	-	-	-	4	2	22
Mining and Quarrying	-	-	1	-	-	-	-	-	-	-	1
Food and Tobacco	8	-	18	8	-	-	-	-	35	6	75
Paper, Pulp and Printing	-	-	-	-	-	-	-	-	1	-	1
Wood and Wood Products	-	-	-	4	-	-	-	-	2	-	6
Construction	-	-	8	-	-	-	-	-	5	-	13
Textile and Leather	-	-	-	132	-	-	-	-	8	2	143
Non-specified	-	-	-	136	-	-	-	-	29	1	167
TRANSPORT SECTOR	**1**	**-**	**376**	**80**	**-**	**-**	**-**	**-**	**11**	**-**	**468**
Air	-	-	19	-	-	-	-	-	-	-	19
Road	-	-	328	11	-	-	-	-	-	-	339
Rail	1	-	27	-	-	-	-	-	4	-	32
Pipeline Transport	-	-	-	69	-	-	-	-	-	-	70
Internal Navigation	-	-	-	-	-	-	-	-	-	-	-
Non-specified	-	-	1	-	-	-	-	-	6	-	7
OTHER SECTORS	**152**	**-**	**233**	**966**	**-**	**-**	**-**	**62**	**329**	**312**	**2054**
Agriculture	2	-	171	80	-	-	-	-	68	1	322
Comm. and Publ. Services	60	-	8	474	-	-	-	-	93	6	641
Residential	87	-	32	326	-	-	-	-	168	306	918
Non-specified	4	-	21	86	-	-	-	62	-	-	173
NON-ENERGY USE	**-**	**-**	**16**	**-**	**-**	**-**	**-**	**-**	**-**	**-**	**16**
in Industry/Transf./Energy	-	-	-	-	-	-	-	-	-	-	-
in Transport	-	-	8	-	-	-	-	-	-	-	8
in Other Sectors	-	-	9	-	-	-	-	-	-	-	9
Electr. Generated - GWh	*1360*	*-*	*384*	*4012*	*-*	*366*	*-*	*-*	*-*	*-*	*6122*
Electricity Plants	*1339*	*-*	*211*	*3010*	*-*	*366*	*-*	*-*	*-*	*-*	*4926*
CHP Plants	*21*	*-*	*173*	*1002*	*-*	*-*	*-*	*-*	*-*	*-*	*1196*
Heat Generated - TJ	*-*	*-*	*2030*	*12568*	*-*	*-*	*-*	*-*	*-*	*-*	*14598*
CHP Plants	*-*	*-*	*1741*	*9315*	*-*	*-*	*-*	*-*	*-*	*-*	*11056*
Heat Plants	*-*	*-*	*289*	*3253*	*-*	*-*	*-*	*-*	*-*	*-*	*3542*

Moldova : 1997

Thousand tonnes of oil equivalent / *Milliers de tonnes d'équivalent pétrole*

SUPPLY AND CONSUMPTION / *APPROVISIONNEMENT ET DEMANDE*	Coal / *Charbon*	Crude Oil / *Pétrole brut*	Petroleum Products / *Produits pétroliers*	Gas / *Gaz*	Nuclear / *Nucléaire*	Hydro / *Hydro*	Geotherm. Solar etc. / *Géotherm. solaire*	Combust. Renew. & Waste / *En. ren. combust.*	Electricity / *Electricité*	Heat / *Chaleur*	Total / *Total*
Indigenous Production	-	-	-	-	-	33	-	67	-	-2	98
Imports	170	-	968	3132	-	-	-	-	170	-	4440
Exports	-	-	-	-	-	-	-	-	-2	-	-2
Intl. Marine Bunkers	-	-	-	-	-	-	-	-	-	-	-
Stock Changes	-19	-	-69	-5	-	-	-	-6	-	-	-99
TPES	**151**	**-**	**899**	**3127**	**-**	**33**	**-**	**61**	**168**	**-2**	**4436**
Transfers	-	-	-	-	-	-	-	-	-	-	-
Statistical Differences	125	-	-	-	-	-	-	-	-	-	125
Electricity Plants	-125	-	-48	-826	-	-33	-	-	345	-	-687
CHP Plants	-	-	-171	-542	-	-	-	-	109	242	-362
Heat Plants	-	-	-2	-57	-	-	-	-	-2	90	28
Gas Works	-	-	-	-	-	-	-	-	-	-	-
Petroleum Refineries	-	-	-	-	-	-	-	-	-	-	-
Coal Transformation	-1	-	-	-	-	-	-	-	-	-	-1
Liquefaction	-	-	-	-	-	-	-	-	-	-	-
Other Transformation	-35	-	-	-	-	-	-	-	-	-	-35
Own Use	-	-	-	-	-	-	-	-	-45	-	-45
Distribution Losses	-	-	-16	-130	-	-	-	-	-123	-33	-301
TFC	**115**	**-**	**662**	**1572**	**-**	**-**	**-**	**61**	**451**	**297**	**3159**
INDUSTRY SECTOR	**9**	**-**	**24**	**455**	**-**	**-**	**-**	**-**	**155**	**-**	**643**
Iron and Steel	-	-	-	-	-	-	-	-	-	-	-
Chemical and Petrochemical	-	-	-	-	-	-	-	-	-	-	-
of which: Feedstocks	-	-	-	-	-	-	-	-	-	-	-
Non-Ferrous Metals	-	-	-	-	-	-	-	-	-	-	-
Non-Metallic Minerals	2	-	1	137	-	-	-	-	-	-	139
Transport Equipment	-	-	-	-	-	-	-	-	-	-	-
Machinery	-	-	-	18	-	-	-	-	-	-	18
Mining and Quarrying	-	-	1	-	-	-	-	-	-	-	1
Food and Tobacco	7	-	14	9	-	-	-	-	-	-	30
Paper, Pulp and Printing	-	-	-	-	-	-	-	-	-	-	-
Wood and Wood Products	-	-	-	4	-	-	-	-	-	-	4
Construction	-	-	8	-	-	-	-	-	-	-	8
Textile and Leather	-	-	-	141	-	-	-	-	-	-	141
Non-specified	-	-	-	146	-	-	-	-	155	-	301
TRANSPORT SECTOR	**2**	**-**	**399**	**85**	**-**	**-**	**-**	**-**	**10**	**-**	**496**
Air	-	-	22	-	-	-	-	-	-	-	22
Road	-	-	351	11	-	-	-	-	-	-	363
Rail	2	-	23	-	-	-	-	-	4	-	29
Pipeline Transport	-	-	-	74	-	-	-	-	-	-	74
Internal Navigation	-	-	-	-	-	-	-	-	-	-	-
Non-specified	-	-	2	-	-	-	-	-	6	-	8
OTHER SECTORS	**105**	**-**	**224**	**1032**	**-**	**-**	**-**	**61**	**286**	**297**	**2005**
Agriculture	1	-	157	86	-	-	-	-	45	-	289
Comm. and Publ. Services	52	-	3	506	-	-	-	-	89	-	650
Residential	49	-	52	348	-	-	-	-	152	297	898
Non-specified	3	-	12	92	-	-	-	61	-	-	167
NON-ENERGY USE	**-**	**-**	**15**	**-**	**-**	**-**	**-**	**-**	**-**	**-**	**15**
in Industry/Transf./Energy	-	-	-	-	-	-	-	-	-	-	-
in Transport	-	-	7	-	-	-	-	-	-	-	7
in Other Sectors	-	-	9	-	-	-	-	-	-	-	9
Electr. Generated - GWh	***488***	***-***	***213***	***4190***	***-***	***382***	***-***	***-***	***-***	***-***	***5273***
Electricity Plants	*472*	*-*	*75*	*3082*	*-*	*382*	*-*	*-*	*-*	*-*	*4011*
CHP Plants	*16*	*-*	*138*	*1108*	*-*	*-*	*-*	*-*	*-*	*-*	*1262*
Heat Generated - TJ	***-***	***-***	***1142***	***12677***	***-***	***-***	***-***	***-***	***-***	***-***	***13819***
CHP Plants	*-*	*-*	*1001*	*9150*	*-*	*-*	*-*	*-*	*-*	*-*	*10151*
Heat Plants	*-*	*-*	*141*	*3527*	*-*	*-*	*-*	*-*	*-*	*-*	*3668*

Morocco / Maroc

Total Production of Energy

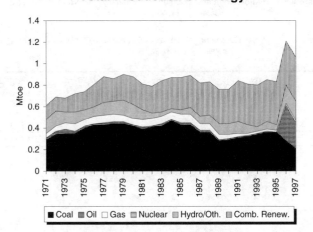

■ Coal ▦ Oil □ Gas ▤ Nuclear ▨ Hydro/Oth. ▥ Comb. Renew.

Total Primary Energy Supply*

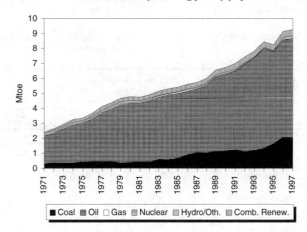

■ Coal ▦ Oil □ Gas ▤ Nuclear ▨ Hydro/Oth. ▥ Comb. Renew.

Electricity Generation by Fuel

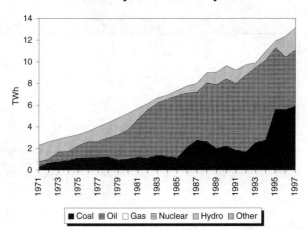

■ Coal ▦ Oil □ Gas ▤ Nuclear ▨ Hydro ▥ Other

Oil Products Consumption

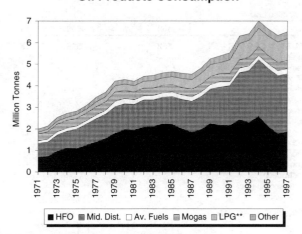

■ HFO ▦ Mid. Dist. □ Av. Fuels ▤ Mogas ▨ LPG** ▥ Other

Electricity Consumption/GDP***, TPES/GDP*** and Production/TPES

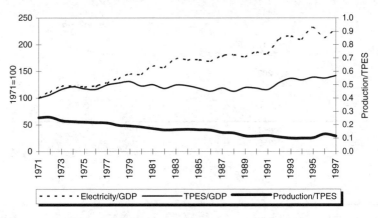

- - - - - Electricity/GDP ——— TPES/GDP ——— Production/TPES

* Excluding electricity trade.
** Includes LPG, NGL, ethane and naphtha.
*** GDP in 1990 US dollars.

Morocco / Maroc : 1996

	Coal	Crude Oil	Petroleum Products	Gas	Nuclear	Hydro	Geotherm. Solar etc.	Combust. Renew. & Waste	Electricity	Heat	Total
SUPPLY AND CONSUMPTION											
APPROVISIONNEMENT ET DEMANDE	*Charbon*	*Pétrole brut*	*Produits pétroliers*	*Gaz*	*Nucléaire*	*Hydro*	*Géotherm. solaire*	*En. ren. combust.*	*Electricité*	*Chaleur*	*Total*
Indigenous Production	283	329	-	15	-	167	-	407	-	-	1200
Imports	1795	4962	1297	-	-	-	-	-	11	-	8066
Exports	-	-	-368	-	-	-	-	-	-	-	-368
Intl. Marine Bunkers	-	-	-13	-	-	-	-	-	-	-	-13
Stock Changes	-	276	-7	-	-	-	-	-	-	-	269
TPES	**2079**	**5567**	**908**	**15**	**-**	**167**	**-**	**407**	**11**	**-**	**9154**
Transfers	-	-	-	-	-	-	-	-	-	-	-
Statistical Differences	-185	-	93	-	-	-	-	-	-	-	-91
Electricity Plants	-1332	-	-1265	-	-	-167	-	-	1063	-	-1701
CHP Plants	-	-	-	-	-	-	-	-	-	-	-
Heat Plants	-	-	-	-	-	-	-	-	-	-	-
Gas Works	-	-	-	-	-	-	-	-	-	-	-
Petroleum Refineries	-	-5567	5510	-	-	-	-	-	-	-	-58
Coal Transformation	-	-	-	-	-	-	-	-	-	-	-
Liquefaction	-	-	-	-	-	-	-	-	-	-	-
Other Transformation	-	-	-	-	-	-	-	-	-	-	-
Own Use	-	-	-257	-	-	-	-	-	-88	-	-345
Distribution Losses	-	-	-	-	-	-	-	-	-42	-	-42
TFC	**562**	**-**	**4990**	**15**	**-**	**-**	**-**	**407**	**943**	**-**	**6916**
INDUSTRY SECTOR	**562**	**-**	**536**	**15**	**-**	**-**	**-**	**62**	**427**	**-**	**1602**
Iron and Steel	-	-	-	-	-	-	-	-	-	-	-
Chemical and Petrochemical	-	-	-	-	-	-	-	-	32	-	32
of which: Feedstocks	-	-	-	-	-	-	-	-	-	-	-
Non-Ferrous Metals	-	-	-	-	-	-	-	-	-	-	-
Non-Metallic Minerals	562	-	-	-	-	-	-	-	56	-	617
Transport Equipment	-	-	-	-	-	-	-	-	-	-	-
Machinery	-	-	-	-	-	-	-	-	22	-	22
Mining and Quarrying	-	-	-	15	-	-	-	-	61	-	76
Food and Tobacco	-	-	-	-	-	-	-	-	43	-	43
Paper, Pulp and Printing	-	-	-	-	-	-	-	-	32	-	32
Wood and Wood Products	-	-	-	-	-	-	-	-	-	-	-
Construction	-	-	-	-	-	-	-	-	18	-	18
Textile and Leather	-	-	-	-	-	-	-	-	51	-	51
Non-specified	-	-	536	-	-	-	-	62	113	-	710
TRANSPORT SECTOR	**-**	**-**	**810**	**-**	**-**	**-**	**-**	**-**	**18**	**-**	**828**
Air	-	-	253	-	-	-	-	-	-	-	253
Road	-	-	401	-	-	-	-	-	-	-	401
Rail	-	-	-	-	-	-	-	-	18	-	18
Pipeline Transport	-	-	-	-	-	-	-	-	-	-	-
Internal Navigation	-	-	155	-	-	-	-	-	-	-	155
Non-specified	-	-	-	-	-	-	-	-	-	-	-
OTHER SECTORS	**-**	**-**	**3444**	**-**	**-**	**-**	**-**	**345**	**498**	**-**	**4288**
Agriculture	-	-	-	-	-	-	-	-	44	-	44
Comm. and Publ. Services	-	-	-	-	-	-	-	-	159	-	159
Residential	-	-	1017	-	-	-	-	345	295	-	1657
Non-specified	-	-	2427	-	-	-	-	-	-	-	2427
NON-ENERGY USE	**-**	**-**	**200**	**-**	**-**	**-**	**-**	**-**	**-**	**-**	**200**
in Industry/Transf./Energy	-	-	200	-	-	-	-	-	-	-	200
in Transport	-	-	-	-	-	-	-	-	-	-	-
in Other Sectors	-	-	-	-	-	-	-	-	-	-	-
Electr. Generated - GWh	*5550*	*-*	*4868*	*-*	*-*	*1938*	*-*	*-*	*-*	*-*	*12356*
Electricity Plants	*5550*	*-*	*4868*	*-*	*-*	*1938*	*-*	*-*	*-*	*-*	*12356*
CHP Plants	*-*	*-*	*-*	*-*	*-*	*-*	*-*	*-*	*-*	*-*	*-*
Heat Generated - TJ	*-*	*-*	*-*	*-*	*-*	*-*	*-*	*-*	*-*	*-*	*-*
CHP Plants	*-*	*-*	*-*	*-*	*-*	*-*	*-*	*-*	*-*	*-*	*-*
Heat Plants	*-*	*-*	*-*	*-*	*-*	*-*	*-*	*-*	*-*	*-*	*-*

Thousand tonnes of oil equivalent / *Milliers de tonnes d'équivalent pétrole*

Morocco / Maroc : 1997

Thousand tonnes of oil equivalent / *Milliers de tonnes d'équivalent pétrole*											
SUPPLY AND CONSUMPTION *APPROVISIONNEMENT ET DEMANDE*	Coal *Charbon*	Crude Oil *Pétrole brut*	Petroleum Products *Produits pétroliers*	Gas *Gaz*	Nuclear *Nucléaire*	Hydro *Hydro*	Geotherm. Solar etc. *Géotherm. solaire*	Combust. Renew. & Waste *En. ren. combust.*	Electricity *Electricité*	Heat *Chaleur*	Total *Total*
Indigenous Production	211	242	-	23	-	177	-	415	-	-	1067
Imports	1808	5594	1301	-	-	-	-	-	11	-	8714
Exports	-	-	-320	-	-	-	-	-	-	-	-320
Intl. Marine Bunkers	-	-	-13	-	-	-	-	-	-	-	-13
Stock Changes	-	-107	-66	-	-	-	-	-	-	-	-173
TPES	**2019**	**5729**	**901**	**23**	**-**	**177**	**-**	**415**	**11**	**-**	**9275**
Transfers	-	-	-	-	-	-	-	-	-	-	-
Statistical Differences	-9	-	-53	-	-	-	-	-	-	-	-63
Electricity Plants	-1418	-	-1340	-	-	-177	-	-	1129	-	-1807
CHP Plants	-	-	-	-	-	-	-	-	-	-	-
Heat Plants	-	-	-	-	-	-	-	-	-	-	-
Gas Works	-	-	-	-	-	-	-	-	-	-	-
Petroleum Refineries	-	-5729	5832	-	-	-	-	-	-	-	103
Coal Transformation	-	-	-	-	-	-	-	-	-	-	-
Liquefaction	-	-	-	-	-	-	-	-	-	-	-
Other Transformation	-	-	-	-	-	-	-	-	-	-	-
Own Use	-	-	-257	-	-	-	-	-	-98	-	-354
Distribution Losses	-	-	-	-	-	-	-	-	-48	-	-48
TFC	**591**	**-**	**5083**	**23**	**-**	**-**	**-**	**415**	**994**	**-**	**7105**
INDUSTRY SECTOR	**591**	**-**	**554**	**23**	**-**	**-**	**-**	**63**	**452**	**-**	**1682**
Iron and Steel	-	-	-	-	-	-	-	-	-	-	-
Chemical and Petrochemical	-	-	-	-	-	-	-	-	34	-	34
of which: Feedstocks	-	-	-	-	-	-	-	-	-	-	-
Non-Ferrous Metals	-	-	-	-	-	-	-	-	-	-	-
Non-Metallic Minerals	591	-	-	-	-	-	-	-	59	-	650
Transport Equipment	-	-	-	-	-	-	-	-	-	-	-
Machinery	-	-	-	-	-	-	-	-	23	-	23
Mining and Quarrying	-	-	-	23	-	-	-	-	66	-	89
Food and Tobacco	-	-	-	-	-	-	-	-	45	-	45
Paper, Pulp and Printing	-	-	-	-	-	-	-	-	34	-	34
Wood and Wood Products	-	-	-	-	-	-	-	-	-	-	-
Construction	-	-	-	-	-	-	-	-	19	-	19
Textile and Leather	-	-	-	-	-	-	-	-	54	-	54
Non-specified	-	-	554	-	-	-	-	63	119	-	736
TRANSPORT SECTOR	**-**	**-**	**823**	**-**	**-**	**-**	**-**	**-**	**18**	**-**	**841**
Air	-	-	263	-	-	-	-	-	-	-	263
Road	-	-	407	-	-	-	-	-	-	-	407
Rail	-	-	-	-	-	-	-	-	18	-	18
Pipeline Transport	-	-	-	-	-	-	-	-	-	-	-
Internal Navigation	-	-	153	-	-	-	-	-	-	-	153
Non-specified	-	-	-	-	-	-	-	-	-	-	-
OTHER SECTORS	**-**	**-**	**3492**	**-**	**-**	**-**	**-**	**352**	**524**	**-**	**4368**
Agriculture	-	-	-	-	-	-	-	-	47	-	47
Comm. and Publ. Services	-	-	-	-	-	-	-	-	167	-	167
Residential	-	-	1094	-	-	-	-	352	310	-	1756
Non-specified	-	-	2398	-	-	-	-	-	-	-	2398
NON-ENERGY USE	**-**	**-**	**214**	**-**	**-**	**-**	**-**	**-**	**-**	**-**	**214**
in Industry/Transf./Energy	-	-	214	-	-	-	-	-	-	-	214
in Transport	-	-	-	-	-	-	-	-	-	-	-
in Other Sectors	-	-	-	-	-	-	-	-	-	-	-
Electr. Generated - GWh	*5908*	*-*	*5158*	*-*	*-*	*2062*	*-*	*-*	*-*	*-*	*13128*
Electricity Plants	*5908*	*-*	*5158*	*-*	*-*	*2062*	*-*	*-*	*-*	*-*	*13128*
CHP Plants	*-*	*-*	*-*	*-*	*-*	*-*	*-*	*-*	*-*	*-*	*-*
Heat Generated - TJ	*-*	*-*	*-*	*-*	*-*	*-*	*-*	*-*	*-*	*-*	*-*
CHP Plants	*-*	*-*	*-*	*-*	*-*	*-*	*-*	*-*	*-*	*-*	*-*
Heat Plants	*-*	*-*	*-*	*-*	*-*	*-*	*-*	*-*	*-*	*-*	*-*

Mozambique

Thousand tonnes of oil equivalent / *Milliers de tonnes d'équivalent pétrole*

SUPPLY AND CONSUMPTION 1996	Coal	Crude Oil	Petroleum Products	Gas	Nuclear	Hydro	Geotherm. Solar etc.	Combust. Renew. & Waste	Electricity	Heat	Total
Indigenous Production	-	-	-	-	-	38	-	6925	-	-	6964
Imports	11	-	518	-	-	-	-	-	52	-	581
Exports	-	-	-2	-	-	-	-	-	-	-	-2
Intl. Marine Bunkers	-	-	-15	-	-	-	-	-	-	-	-15
Stock changes	-	-	-3	-	-	-	-	-	-	-	-3
TPES	11	-	498	-	-	38	-	6925	52	-	7524
Electricity and CHP Plants	-	-	-106	-	-	-38	-	-	41	-	-104
Petroleum Refineries	-	-	4	-	-	-	-	-	-	-	4
Other Transformation *	-	-	5	-	-	-	-	-734	-36	-	-765
TFC	11	-	401	-	-	-	-	6191	57	-	6659
INDUSTRY SECTOR	11	-	10	-	-	-	-	478	21	-	520
Iron and Steel	-	-	-	-	-	-	-	-	-	-	-
Chemical and Petrochemical	-	-	-	-	-	-	-	-	-	-	-
Non-Metallic Minerals	-	-	-	-	-	-	-	-	-	-	-
Non-specified	11	-	10	-	-	-	-	478	21	-	520
TRANSPORT SECTOR	-	-	91	-	-	-	-	-	-	-	91
Air	-	-	22	-	-	-	-	-	-	-	22
Road	-	-	69	-	-	-	-	-	-	-	69
Non-specified	-	-	-	-	-	-	-	-	-	-	-
OTHER SECTORS	-	-	292	-	-	-	-	5713	36	-	6041
Agriculture	-	-	32	-	-	-	-	-	-	-	32
Comm. and Publ. Services	-	-	164	-	-	-	-	-	13	-	177
Residential	-	-	20	-	-	-	-	5713	22	-	5756
Non-specified	-	-	77	-	-	-	-	-	-	-	77
NON-ENERGY USE	-	-	7	-	-	-	-	-	-	-	7
Electricity Generated - GWh	-	-	29	1	-	446	-	-	-	-	476
Heat Generated - TJ	-	-	-	-	-	-	-	-	-	-	-

APPROVISIONNEMENT ET DEMANDE 1997	Charbon	Pétrole brut	Produits pétroliers	Gaz	Nucléaire	Hydro	Géotherm. solaire etc.	En. ren. combust. & déchets	Electricité	Chaleur	Total
Indigenous Production	-	-	-	-	-	68	-	6925	-	-	6994
Imports	12	-	705	-	-	-	-	-	59	-	777
Exports	-	-	-1	-	-	-	-	-	-42	-	-43
Intl. Marine Bunkers	-	-	-62	-	-	-	-	-	-	-	-62
Stock changes	-	-	-2	-	-	-	-	-	-	-	-2
TPES	12	-	641	-	-	68	-	6925	17	-	7664
Electricity and CHP Plants	-	-	-156	-	-	-68	-	-	86	-	-138
Petroleum Refineries	-	-	5	-	-	-	-	-	-	-	5
Other Transformation *	-	-	32	-	-	-	-	-734	-37	-	-739
TFC	12	-	521	-	-	-	-	6191	67	-	6792
INDUSTRY SECTOR	12	-	11	-	-	-	-	478	25	-	526
Iron and Steel	-	-	-	-	-	-	-	-	-	-	-
Chemical and Petrochemical	-	-	-	-	-	-	-	-	-	-	-
Non-Metallic Minerals	-	-	-	-	-	-	-	-	-	-	-
Non-specified	12	-	11	-	-	-	-	478	25	-	526
TRANSPORT SECTOR	-	-	102	-	-	-	-	-	-	-	102
Air	-	-	27	-	-	-	-	-	-	-	27
Road	-	-	75	-	-	-	-	-	-	-	75
Non-specified	-	-	-	-	-	-	-	-	-	-	-
OTHER SECTORS	-	-	402	-	-	-	-	5713	42	-	6157
Agriculture	-	-	38	-	-	-	-	-	-	-	38
Comm. and Publ. Services	-	-	218	-	-	-	-	-	16	-	234
Residential	-	-	31	-	-	-	-	5713	26	-	5770
Non-specified	-	-	114	-	-	-	-	-	-	-	114
NON-ENERGY USE	-	-	7	-	-	-	-	-	-	-	7
Electricity Generated - GWh	-	-	212	2	-	791	-	-	-	-	1005
Heat Generated - TJ	-	-	-	-	-	-	-	-	-	-	-

* Includes Transfers, Statistical Differences, Own Use and Distribution Losses.

Myanmar

Thousand tonnes of oil equivalent / *Milliers de tonnes d'équivalent pétrole*

SUPPLY AND CONSUMPTION 1996	Coal	Crude Oil	Petroleum Products	Gas	Nuclear	Hydro	Geotherm. Solar etc.	Combust. Renew. & Waste	Electricity	Heat	Total
Indigenous Production	9	406	-	1322	-	142	-	10114	-	-	11993
Imports	-	368	329	-	-	-	-	-	-	-	697
Exports	-	-	-19	-	-	-	-	-	-	-	-19
Intl. Marine Bunkers	-	-	-5	-	-	-	-	-	-	-	-5
Stock changes	-	45	4	-157	-	-	-	-	-	-	-108
TPES	9	819	308	1164	-	142	-	10114	-	-	12557
Electricity and CHP Plants	-	-	-48	-909	-	-142	-	-	339	-	-760
Petroleum Refineries	-	-807	783	-	-	-	-	-	-	-	-24
Other Transformation *	-	-12	240	26	-	-	-	-1383	-135	-	-1264
TFC	9	-	1282	282	-	-	-	8732	204	-	10509
INDUSTRY SECTOR	9	-	425	277	-	-	-	291	77	-	1079
Iron and Steel	-	-	-	-	-	-	-	-	-	-	-
Chemical and Petrochemical	-	-	-	79	-	-	-	-	-	-	79
Non-Metallic Minerals	-	-	-	-	-	-	-	-	-	-	-
Non-specified	9	-	425	198	-	-	-	291	77	-	1000
TRANSPORT SECTOR	-	-	676	2	-	-	-	-	-	-	678
Air	-	-	58	-	-	-	-	-	-	-	58
Road	-	-	612	-	-	-	-	-	-	-	612
Non-specified	-	-	7	2	-	-	-	-	-	-	9
OTHER SECTORS	-	-	164	2	-	-	-	8441	127	-	8734
Agriculture	-	-	4	-	-	-	-	-	-	-	4
Comm. and Publ. Services	-	-	-	-	-	-	-	-	37	-	37
Residential	-	-	148	1	-	-	-	8441	91	-	8680
Non-specified	-	-	12	2	-	-	-	-	-	-	14
NON-ENERGY USE	-	-	17	-	-	-	-	-	-	-	17
Electricity Generated - GWh	-	-	*351*	*1943*	-	*1651*	-	-	-	-	*3945*
Heat Generated - TJ	-	-	-	-	-	-	-	-	-	-	-

APPROVISIONNEMENT ET DEMANDE 1997	Charbon	Pétrole brut	Produits pétroliers	Gaz	Nucléaire	Hydro	Géotherm. solaire etc.	En. ren. combust. & déchets	Electricité	Chaleur	Total
Indigenous Production	9	400	-	1428	-	142	-	10270	-	-	12249
Imports	-	653	300	-	-	-	-	-	-	-	953
Exports	-	-	-7	-	-	-	-	-	-	-	-7
Intl. Marine Bunkers	-	-	-4	-	-	-	-	-	-	-	-4
Stock changes	-	-1	-34	-147	-	-	-	-	-	-	-182
TPES	9	1051	256	1281	-	142	-	10270	-	-	13009
Electricity and CHP Plants	-	-	-56	-953	-	-142	-	-	362	-	-791
Petroleum Refineries	-	-1044	1016	-	-	-	-	-	-	-	-29
Other Transformation *	-	-7	248	3	-	-	-	-1404	-147	-	-1307
TFC	9	-	1463	330	-	-	-	8866	215	-	10883
INDUSTRY SECTOR	9	-	484	326	-	-	-	296	79	-	1194
Iron and Steel	-	-	-	-	-	-	-	-	-	-	-
Chemical and Petrochemical	-	-	-	65	-	-	-	-	-	-	65
Non-Metallic Minerals	-	-	-	-	-	-	-	-	-	-	-
Non-specified	9	-	484	261	-	-	-	296	79	-	1128
TRANSPORT SECTOR	-	-	779	2	-	-	-	-	-	-	781
Air	-	-	55	-	-	-	-	-	-	-	55
Road	-	-	717	-	-	-	-	-	-	-	717
Non-specified	-	-	7	2	-	-	-	-	-	-	8
OTHER SECTORS	-	-	183	2	-	-	-	8571	136	-	8892
Agriculture	-	-	9	-	-	-	-	-	-	-	9
Comm. and Publ. Services	-	-	-	-	-	-	-	-	38	-	38
Residential	-	-	161	-	-	-	-	8571	98	-	8829
Non-specified	-	-	14	2	-	-	-	-	-	-	15
NON-ENERGY USE	-	-	17	-	-	-	-	-	-	-	17
Electricity Generated - GWh	-	-	*511*	*2039*	-	*1655*	-	-	-	-	*4205*
Heat Generated - TJ	-	-	-	-	-	-	-	-	-	-	-

* Includes Transfers, Statistical Differences, Own Use and Distribution Losses.

Nepal / Népal

Thousand tonnes of oil equivalent / *Milliers de tonnes d'équivalent pétrole*

SUPPLY AND CONSUMPTION 1996	Coal	Crude Oil	Petroleum Products	Gas	Nuclear	Hydro	Geotherm. Solar etc.	Combust. Renew. & Waste	Electricity	Heat	Total
Indigenous Production	-	-	-	-	-	92	-	6464	-	-	6556
Imports	71	-	533	-	-	-	-	-	6	-	610
Exports	-	-	-15	-	-	-	-	-	-8	-	-22
Intl. Marine Bunkers	-	-	-	-	-	-	-	-	-	-	-
Stock changes	-	-	13	-	-	-	-	-	-	-	13
TPES	71	-	531	-	-	92	-	6464	-2	-	7157
Electricity and CHP Plants	-	-	-11	-	-	-92	-	-	102	-	-
Petroleum Refineries	-	-	-	-	-	-	-	-	-	-	-
Other Transformation *	1	-	1	-	-	-	-	-	-28	-	-26
TFC	73	-	521	-	-	-	-	6464	73	-	7130
INDUSTRY SECTOR	61	-	107	-	-	-	-	98	31	-	297
Iron and Steel	-	-	-	-	-	-	-	-	-	-	-
Chemical and Petrochemical	-	-	-	-	-	-	-	-	-	-	-
Non-Metallic Minerals	-	-	-	-	-	-	-	-	-	-	-
Non-specified	61	-	107	-	-	-	-	98	31	-	297
TRANSPORT SECTOR	3	-	200	-	-	-	-	-	-	-	203
Air	-	-	31	-	-	-	-	-	-	-	31
Road	-	-	169	-	-	-	-	-	-	-	169
Non-specified	3	-	-	-	-	-	-	-	-	-	3
OTHER SECTORS	8	-	207	-	-	-	-	6366	42	-	6624
Agriculture	-	-	13	-	-	-	-	-	2	-	15
Comm. and Publ. Services	8	-	32	-	-	-	-	36	7	-	83
Residential	1	-	161	-	-	-	-	6330	29	-	6521
Non-specified	-	-	-	-	-	-	-	-	5	-	5
NON-ENERGY USE	-	-	7	-	-	-	-	-	-	-	7
Electricity Generated - GWh	-	-	117	-	-	1074	-	-	-	-	1191
Heat Generated - TJ	-	-	-	-	-	-	-	-	-	-	-

APPROVISIONNEMENT ET DEMANDE 1997	Charbon	Pétrole brut	Produits pétroliers	Gaz	Nucléaire	Hydro	Géotherm. solaire etc.	En. ren. combust. & déchets	Electricité	Chaleur	Total
Indigenous Production	-	-	-	-	-	95	-	6464	-	-	6559
Imports	71	-	533	-	-	-	-	-	6	-	610
Exports	-	-	-15	-	-	-	-	-	-8	-	-22
Intl. Marine Bunkers	-	-	-	-	-	-	-	-	-	-	-
Stock changes	-	-	13	-	-	-	-	-	-	-	13
TPES	71	-	531	-	-	95	-	6464	-2	-	7160
Electricity and CHP Plants	-	-	-11	-	-	-95	-	-	105	-	-
Petroleum Refineries	-	-	-	-	-	-	-	-	-	-	-
Other Transformation *	1	-	1	-	-	-	-	-	-29	-	-27
TFC	73	-	521	-	-	-	-	6464	75	-	7133
INDUSTRY SECTOR	61	-	107	-	-	-	-	98	31	-	298
Iron and Steel	-	-	-	-	-	-	-	-	-	-	-
Chemical and Petrochemical	-	-	-	-	-	-	-	-	-	-	-
Non-Metallic Minerals	-	-	-	-	-	-	-	-	-	-	-
Non-specified	61	-	107	-	-	-	-	98	31	-	298
TRANSPORT SECTOR	3	-	200	-	-	-	-	-	-	-	203
Air	-	-	31	-	-	-	-	-	-	-	31
Road	-	-	169	-	-	-	-	-	-	-	169
Non-specified	3	-	-	-	-	-	-	-	-	-	3
OTHER SECTORS	8	-	207	-	-	-	-	6366	43	-	6625
Agriculture	-	-	13	-	-	-	-	-	2	-	15
Comm. and Publ. Services	8	-	32	-	-	-	-	36	7	-	84
Residential	1	-	161	-	-	-	-	6330	30	-	6522
Non-specified	-	-	-	-	-	-	-	-	5	-	5
NON-ENERGY USE	-	-	7	-	-	-	-	-	-	-	7
Electricity Generated - GWh	-	-	117	-	-	1109	-	-	-	-	1226
Heat Generated - TJ	-	-	-	-	-	-	-	-	-	-	-

* Includes Transfers, Statistical Differences, Own Use and Distribution Losses.

Netherlands Antilles / Antilles néerlandaises

Thousand tonnes of oil equivalent / *Milliers de tonnes d'équivalent pétrole*

SUPPLY AND CONSUMPTION 1996	Coal	Crude Oil	Petroleum Products	Gas	Nuclear	Hydro	Geotherm. Solar etc.	Combust. Renew. & Waste	Electricity	Heat	Total
Indigenous Production	-	-	-	-	-	-	-	-	-	-	-
Imports	-	14297	1667	-	-	-	-	-	-	-	15964
Exports	-	-524	-10996	-	-	-	-	-	-	-	-11520
Intl. Marine Bunkers	-	-	-1671	-	-	-	-	-	-	-	-1671
Stock changes	-	-113	-	-	-	-	-	-	-	-	-113
TPES	-	13661	-11000	-	-	-	-	-	-	-	2661
Electricity and CHP Plants	-	-	-75	-	-	-	-	-	87	-	13
Petroleum Refineries	-	-12683	12380	-	-	-	-	-	-	-	-304
Other Transformation *	-	-978	-489	-	-	-	-	-	-19	-	-1486
TFC	-	-	815	-	-	-	-	-	69	-	884
INDUSTRY SECTOR	-	-	132	-	-	-	-	-	38	-	169
Iron and Steel	-	-	-	-	-	-	-	-	-	-	-
Chemical and Petrochemical	-	-	-	-	-	-	-	-	-	-	-
Non-Metallic Minerals	-	-	-	-	-	-	-	-	-	-	-
Non-specified	-	-	132	-	-	-	-	-	38	-	169
TRANSPORT SECTOR	-	-	443	-	-	-	-	-	-	-	443
Air	-	-	66	-	-	-	-	-	-	-	66
Road	-	-	377	-	-	-	-	-	-	-	377
Non-specified	-	-	-	-	-	-	-	-	-	-	-
OTHER SECTORS	-	-	178	-	-	-	-	-	31	-	209
Agriculture	-	-	-	-	-	-	-	-	-	-	-
Comm. and Publ. Services	-	-	-	-	-	-	-	-	-	-	-
Residential	-	-	178	-	-	-	-	-	-	-	178
Non-specified	-	-	-	-	-	-	-	-	31	-	31
NON-ENERGY USE	-	-	62	-	-	-	-	-	-	-	62
Electricity Generated - GWh	-	-	*1017*	-	-	-	-	-	-	-	*1017*
Heat Generated - TJ	-	-	-	-	-	-	-	-	-	-	-

APPROVISIONNEMENT ET DEMANDE 1997	Charbon	Pétrole brut	Produits pétroliers	Gaz	Nucléaire	Hydro	Géotherm. solaire etc.	En. ren. combust. & déchets	Electricité	Chaleur	Total
Indigenous Production	-	-	-	-	-	-	-	-	-	-	-
Imports	-	14297	1667	-	-	-	-	-	-	-	15964
Exports	-	-524	-10996	-	-	-	-	-	-	-	-11520
Intl. Marine Bunkers	-	-	-1671	-	-	-	-	-	-	-	-1671
Stock changes	-	-113	-	-	-	-	-	-	-	-	-113
TPES	-	13661	-11000	-	-	-	-	-	-	-	2661
Electricity and CHP Plants	-	-	-75	-	-	-	-	-	90	-	16
Petroleum Refineries	-	-12683	12380	-	-	-	-	-	-	-	-304
Other Transformation *	-	-978	-489	-	-	-	-	-	-20	-	-1487
TFC	-	-	815	-	-	-	-	-	71	-	886
INDUSTRY SECTOR	-	-	132	-	-	-	-	-	39	-	171
Iron and Steel	-	-	-	-	-	-	-	-	-	-	-
Chemical and Petrochemical	-	-	-	-	-	-	-	-	-	-	-
Non-Metallic Minerals	-	-	-	-	-	-	-	-	-	-	-
Non-specified	-	-	132	-	-	-	-	-	39	-	171
TRANSPORT SECTOR	-	-	443	-	-	-	-	-	-	-	443
Air	-	-	66	-	-	-	-	-	-	-	66
Road	-	-	377	-	-	-	-	-	-	-	377
Non-specified	-	-	-	-	-	-	-	-	-	-	-
OTHER SECTORS	-	-	178	-	-	-	-	-	32	-	210
Agriculture	-	-	-	-	-	-	-	-	-	-	-
Comm. and Publ. Services	-	-	-	-	-	-	-	-	-	-	-
Residential	-	-	178	-	-	-	-	-	-	-	178
Non-specified	-	-	-	-	-	-	-	-	32	-	32
NON-ENERGY USE	-	-	62	-	-	-	-	-	-	-	62
Electricity Generated - GWh	-	-	*1052*	-	-	-	-	-	-	-	*1052*
Heat Generated - TJ	-	-	-	-	-	-	-	-	-	-	-

* Includes Transfers, Statistical Differences, Own Use and Distribution Losses.

Nicaragua

Thousand tonnes of oil equivalent / *Milliers de tonnes d'équivalent pétrole*

SUPPLY AND CONSUMPTION 1996	Coal	Crude Oil	Petroleum Products	Gas	Nuclear	Hydro	Geotherm. Solar etc.	Combust. Renew. & Waste	Electricity	Heat	Total
Indigenous Production	-	-	-	-	-	37	238	1220	-	-	1495
Imports	-	621	335	-	-	-	-	-	1	-	957
Exports	-	-	-27	-	-	-	-	-	-	-	-27
Intl. Marine Bunkers	-	-	-	-	-	-	-	-	-	-	-
Stock changes	-	41	-45	-	-	-	-	-	-	-	-4
TPES	-	663	262	-	-	37	238	1220	1	-	2421
Electricity and CHP Plants	-	-	-299	-	-	-37	-238	-89	165	-	-498
Petroleum Refineries	-	-656	617	-	-	-	-	-	-	-	-39
Other Transformation *	-	-7	-8	-	-	-	-	-43	-65	-	-123
TFC	-	-	572	-	-	-	-	1087	101	-	1760
INDUSTRY SECTOR	-	-	89	-	-	-	-	98	30	-	217
Iron and Steel	-	-	-	-	-	-	-	-	-	-	-
Chemical and Petrochemical	-	-	-	-	-	-	-	-	-	-	-
Non-Metallic Minerals	-	-	-	-	-	-	-	-	-	-	-
Non-specified	-	-	89	-	-	-	-	98	30	-	217
TRANSPORT SECTOR	-	-	392	-	-	-	-	-	-	-	392
Air	-	-	23	-	-	-	-	-	-	-	23
Road	-	-	338	-	-	-	-	-	-	-	338
Non-specified	-	-	31	-	-	-	-	-	-	-	31
OTHER SECTORS	-	-	67	-	-	-	-	989	71	-	1127
Agriculture	-	-	6	-	-	-	-	3	8	-	16
Comm. and Publ. Services	-	-	23	-	-	-	-	16	27	-	67
Residential	-	-	37	-	-	-	-	971	36	-	1044
Non-specified	-	-	-	-	-	-	-	-	-	-	-
NON-ENERGY USE	-	-	24	-	-	-	-	-	-	-	24
Electricity Generated - GWh	-	-	1169	-	-	431	277	43	-	-	1920
Heat Generated - TJ	-	-	-	-	-	-	-	-	-	-	-

APPROVISIONNEMENT ET DEMANDE 1997	Charbon	Pétrole brut	Produits pétroliers	Gaz	Nucléaire	Hydro	Géotherm. solaire etc.	En. ren. combust. & déchets	Electricité	Chaleur	Total
Indigenous Production	-	-	-	-	-	35	180	1314	-	-	1529
Imports	-	734	243	-	-	-	-	-	14	-	991
Exports	-	-	-16	-	-	-	-	-	-	-	-16
Intl. Marine Bunkers	-	-	-	-	-	-	-	-	-	-	-
Stock changes	-	102	-32	-	-	-	-	-	-	-	70
TPES	-	836	194	-	-	35	180	1314	14	-	2573
Electricity and CHP Plants	-	-	-308	-	-	-35	-180	-133	164	-	-493
Petroleum Refineries	-	-828	771	-	-	-	-	-	-	-	-57
Other Transformation *	-	-8	-10	-	-	-	-	-45	-63	-	-126
TFC	-	-	647	-	-	-	-	1135	115	-	1897
INDUSTRY SECTOR	-	-	65	-	-	-	-	118	33	-	217
Iron and Steel	-	-	-	-	-	-	-	-	-	-	-
Chemical and Petrochemical	-	-	-	-	-	-	-	-	-	-	-
Non-Metallic Minerals	-	-	-	-	-	-	-	-	-	-	-
Non-specified	-	-	65	-	-	-	-	118	33	-	217
TRANSPORT SECTOR	-	-	441	-	-	-	-	-	-	-	441
Air	-	-	30	-	-	-	-	-	-	-	30
Road	-	-	381	-	-	-	-	-	-	-	381
Non-specified	-	-	30	-	-	-	-	-	-	-	30
OTHER SECTORS	-	-	112	-	-	-	-	1018	81	-	1211
Agriculture	-	-	6	-	-	-	-	3	10	-	20
Comm. and Publ. Services	-	-	68	-	-	-	-	16	32	-	116
Residential	-	-	37	-	-	-	-	998	39	-	1075
Non-specified	-	-	-	-	-	-	-	-	-	-	-
NON-ENERGY USE	-	-	29	-	-	-	-	-	-	-	29
Electricity Generated - GWh	-	-	1248	-	-	407	209	43	-	-	1907
Heat Generated - TJ	-	-	-	-	-	-	-	-	-	-	-

* Includes Transfers, Statistical Differences, Own Use and Distribution Losses.
Note: Electricity generation from Petroleum products includes small amounts produced from Combustible Renewables and Waste

Nigeria / Nigéria

Thousand tonnes of oil equivalent / *Milliers de tonnes d'équivalent pétrole*

SUPPLY AND CONSUMPTION 1996	Coal	Crude Oil	Petroleum Products	Gas	Nuclear	Hydro	Geotherm. Solar etc.	Combust. Renew. & Waste	Electricity	Heat	Total
Indigenous Production	86	110301	-	4456	-	473	-	66721	-	-	182037
Imports	-	-	43	-	-	-	-	-	-	-	43
Exports	-	-96312	-1010	-	-	-	-	-	-	-	-97322
Intl. Marine Bunkers	-	-	-83	-	-	-	-	-	-	-	-83
Stock changes	-	-	-	-	-	-	-	-	-	-	-
TPES	86	13989	-1050	4456	-	473	-	66721	-	-	84675
Electricity and CHP Plants	-	-	-322	-1597	-	-473	-	-	1289	-	-1102
Petroleum Refineries	-	-17234	17219	-	-	-	-	-	-	-	-16
Other Transformation *	-80	3246	-7322	-2110	-	-	-	-1773	-449	-	-8488
TFC	6	-	8525	750	-	-	-	64948	840	-	75070
INDUSTRY SECTOR	6	-	701	750	-	-	-	6574	181	-	8212
Iron and Steel	-	-	-	56	-	-	-	-	-	-	56
Chemical and Petrochemical	-	-	-	303	-	-	-	-	-	-	303
Non-Metallic Minerals	6	-	-	-	-	-	-	-	-	-	6
Non-specified	-	-	701	390	-	-	-	6574	181	-	7846
TRANSPORT SECTOR	-	-	5067	-	-	-	-	-	-	-	5067
Air	-	-	519	-	-	-	-	-	-	-	519
Road	-	-	4507	-	-	-	-	-	-	-	4507
Non-specified	-	-	41	-	-	-	-	-	-	-	41
OTHER SECTORS	-	-	897	-	-	-	-	58375	659	-	59930
Agriculture	-	-	-	-	-	-	-	-	-	-	-
Comm. and Publ. Services	-	-	-	-	-	-	-	-	218	-	218
Residential	-	-	897	-	-	-	-	58375	441	-	59712
Non-specified	-	-	-	-	-	-	-	-	-	-	-
NON-ENERGY USE	-	-	1860	-	-	-	-	-	-	-	1860
Electricity Generated - GWh	-	-	3919	5572	-	5500	-	-	-	-	14991
Heat Generated - TJ	-	-	-	-	-	-	-	-	-	-	-

APPROVISIONNEMENT ET DEMANDE 1997	Charbon	Pétrole brut	Produits pétroliers	Gaz	Nucléaire	Hydro	Géotherm. solaire etc.	En. ren. combust. & déchets	Electricité	Chaleur	Total
Indigenous Production	86	117249	-	4429	-	481	-	68789	-	-	191034
Imports	-	-	43	-	-	-	-	-	-	-	43
Exports	-	-101332	-1010	-	-	-	-	-	-	-	-102342
Intl. Marine Bunkers	-	-	-83	-	-	-	-	-	-	-	-83
Stock changes	-	-	-	-	-	-	-	-	-	-	-
TPES	86	15916	-1050	4429	-	481	-	68789	-	-	88652
Electricity and CHP Plants	-	-	-322	-1624	-	-481	-	-	1305	-	-1121
Petroleum Refineries	-	-17346	17491	-	-	-	-	-	-	-	146
Other Transformation *	-80	1429	-7484	-2083	-	-	-	-1828	-455	-	-10499
TFC	6	-	8637	723	-	-	-	66962	850	-	77178
INDUSTRY SECTOR	6	-	723	723	-	-	-	6777	184	-	8413
Iron and Steel	-	-	-	54	-	-	-	-	-	-	54
Chemical and Petrochemical	-	-	-	292	-	-	-	-	-	-	292
Non-Metallic Minerals	6	-	-	-	-	-	-	-	-	-	6
Non-specified	-	-	723	376	-	-	-	6777	184	-	8060
TRANSPORT SECTOR	-	-	5147	-	-	-	-	-	-	-	5147
Air	-	-	524	-	-	-	-	-	-	-	524
Road	-	-	4582	-	-	-	-	-	-	-	4582
Non-specified	-	-	41	-	-	-	-	-	-	-	41
OTHER SECTORS	-	-	906	-	-	-	-	60184	667	-	61757
Agriculture	-	-	-	-	-	-	-	-	-	-	-
Comm. and Publ. Services	-	-	-	-	-	-	-	-	221	-	221
Residential	-	-	906	-	-	-	-	60184	446	-	61536
Non-specified	-	-	-	-	-	-	-	-	-	-	-
NON-ENERGY USE	-	-	1860	-	-	-	-	-	-	-	1860
Electricity Generated - GWh	-	-	3919	5667	-	5593	-	-	-	-	15179
Heat Generated - TJ	-	-	-	-	-	-	-	-	-	-	-

* Includes Transfers, Statistical Differences, Own Use and Distribution Losses.

Oman

Thousand tonnes of oil equivalent / Milliers de tonnes d'équivalent pétrole

SUPPLY AND CONSUMPTION 1996	Coal	Crude Oil	Petroleum Products	Gas	Nuclear	Hydro	Geotherm. Solar etc.	Combust. Renew. & Waste	Electricity	Heat	Total
Indigenous Production	-	46211	-	3997	-	-	-	-	-	-	50208
Imports	-	-	97	-	-	-	-	-	-	-	97
Exports	-	-42375	-1070	-405	-	-	-	-	-	-	-43850
Intl. Marine Bunkers	-	-	-26	-	-	-	-	-	-	-	-26
Stock changes	-	-43	-17	-	-	-	-	-	-	-	-60
TPES	-	3793	-1016	3592	-	-	-	-	-	-	6369
Electricity and CHP Plants	-	-	-409	-1750	-	-	-	-	585	-	-1574
Petroleum Refineries	-	-3794	3645	-	-	-	-	-	-	-	-149
Other Transformation *	-	1	-199	-1213	-	-	-	-	-109	-	-1519
TFC	-	-	2022	629	-	-	-	-	476	-	3127
INDUSTRY SECTOR	-	-	795	462	-	-	-	-	34	-	1291
Iron and Steel	-	-	-	-	-	-	-	-	-	-	-
Chemical and Petrochemical	-	-	-	-	-	-	-	-	-	-	-
Non-Metallic Minerals	-	-	-	52	-	-	-	-	-	-	52
Non-specified	-	-	795	411	-	-	-	-	34	-	1240
TRANSPORT SECTOR	-	-	939	-	-	-	-	-	-	-	939
Air	-	-	162	-	-	-	-	-	-	-	162
Road	-	-	777	-	-	-	-	-	-	-	777
Non-specified	-	-	-	-	-	-	-	-	-	-	-
OTHER SECTORS	-	-	248	167	-	-	-	-	443	-	858
Agriculture	-	-	-	-	-	-	-	-	-	-	-
Comm. and Publ. Services	-	-	-	-	-	-	-	-	173	-	173
Residential	-	-	71	167	-	-	-	-	262	-	500
Non-specified	-	-	177	-	-	-	-	-	8	-	185
NON-ENERGY USE	-	-	39	-	-	-	-	-	-	-	39
Electricity Generated - GWh	-	-	1187	5615	-	-	-	-	-	-	6802
Heat Generated - TJ	-	-	-	-	-	-	-	-	-	-	-

APPROVISIONNEMENT ET DEMANDE 1997	Charbon	Pétrole brut	Produits pétroliers	Gaz	Nucléaire	Hydro	Géotherm. solaire etc.	En. ren. combust. & déchets	Electricité	Chaleur	Total
Indigenous Production	-	46921	-	4699	-	-	-	-	-	-	51620
Imports	-	-	340	-	-	-	-	-	-	-	340
Exports	-	-43384	-1241	-405	-	-	-	-	-	-	-45030
Intl. Marine Bunkers	-	-	-29	-	-	-	-	-	-	-	-29
Stock changes	-	-163	37	-	-	-	-	-	-	-	-126
TPES	-	3374	-893	4294	-	-	-	-	-	-	6775
Electricity and CHP Plants	-	-	-414	-1827	-	-	-	-	629	-	-1612
Petroleum Refineries	-	-3375	3242	-	-	-	-	-	-	-	-133
Other Transformation *	-	1	-212	-1236	-	-	-	-	-122	-	-1570
TFC	-	-	1723	1231	-	-	-	-	507	-	3461
INDUSTRY SECTOR	-	-	429	845	-	-	-	-	39	-	1313
Iron and Steel	-	-	-	-	-	-	-	-	-	-	-
Chemical and Petrochemical	-	-	-	-	-	-	-	-	-	-	-
Non-Metallic Minerals	-	-	-	47	-	-	-	-	-	-	47
Non-specified	-	-	429	798	-	-	-	-	39	-	1265
TRANSPORT SECTOR	-	-	1008	-	-	-	-	-	-	-	1008
Air	-	-	183	-	-	-	-	-	-	-	183
Road	-	-	825	-	-	-	-	-	-	-	825
Non-specified	-	-	-	-	-	-	-	-	-	-	-
OTHER SECTORS	-	-	247	386	-	-	-	-	468	-	1101
Agriculture	-	-	-	-	-	-	-	-	-	-	-
Comm. and Publ. Services	-	-	-	-	-	-	-	-	181	-	181
Residential	-	-	59	386	-	-	-	-	279	-	723
Non-specified	-	-	188	-	-	-	-	-	8	-	197
NON-ENERGY USE	-	-	39	-	-	-	-	-	-	-	39
Electricity Generated - GWh	-	-	1205	6113	-	-	-	-	-	-	7318
Heat Generated - TJ	-	-	-	-	-	-	-	-	-	-	-

* Includes Transfers, Statistical Differences, Own Use and Distribution Losses.

Pakistan

Total Production of Energy

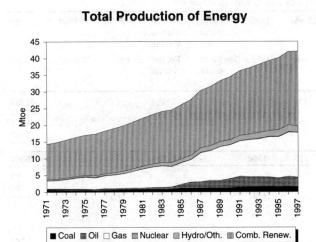

■ Coal ▥ Oil □ Gas ▧ Nuclear ▦ Hydro/Oth. ▨ Comb. Renew.

Total Primary Energy Supply*

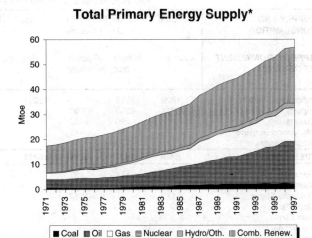

■ Coal □ Oil □ Gas ▧ Nuclear ▦ Hydro/Oth. ▨ Comb. Renew.

Electricity Generation by Fuel

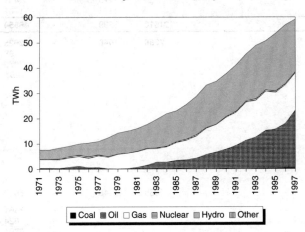

■ Coal ▥ Oil □ Gas ▧ Nuclear ▦ Hydro ▨ Other

Oil Products Consumption

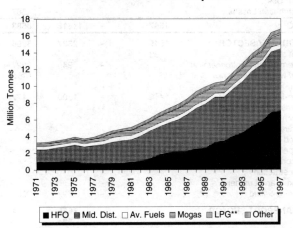

■ HFO ▥ Mid. Dist. □ Av. Fuels ▧ Mogas ▦ LPG** ▨ Other

Electricity Consumption/GDP***, TPES/GDP*** and Production/TPES

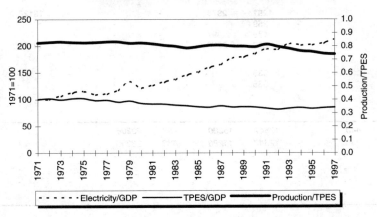

- - - - Electricity/GDP ——— TPES/GDP ——— Production/TPES

* Excluding electricity trade.
** Includes LPG, NGL, ethane and naphtha.
*** GDP in 1990 US dollars.

Pakistan : 1996

	Thousand tonnes of oil equivalent / Milliers de tonnes d'équivalent pétrole										
SUPPLY AND CONSUMPTION	Coal	Crude Oil	Petroleum Products	Gas	Nuclear	Hydro	Geotherm. Solar etc.	Combust. Renew. & Waste	Electricity	Heat	Total
APPROVISIONNEMENT ET DEMANDE	Charbon	Pétrole brut	Produits pétroliers	Gaz	Nucléaire	Hydro	Géotherm. solaire	En. ren. combust.	Electricité	Chaleur	Total
Indigenous Production	1628	3018	-	13211	126	1996	-	21955	-	-	41933
Imports	711	4333	10091	-	-	-	-	-	-	-	15134
Exports	-	-296	-139	-	-	-	-	-	-	-	-435
Intl. Marine Bunkers	-	-	-12	-	-	-	-	-	-	-	-12
Stock Changes	-	-133	-146	-	-	-	-	-	-	-	-279
TPES	2338	6921	9794	13211	126	1996	-	21955	-	-	56341
Transfers	-	-124	137	-	-	-	-	-	-	-	13
Statistical Differences	-	-	16	-	-	-	-	-	-	-	16
Electricity Plants	-179	-	-4619	-4061	-126	-1996	-	-	4898	-	-6082
CHP Plants	-	-	-	-	-	-	-	-	-	-	-
Heat Plants	-	-	-	-	-	-	-	-	-	-	-
Gas Works	-	-	-	-	-	-	-	-	-	-	-
Petroleum Refineries	-	-6797	6516	-	-	-	-	-	-	-	-281
Coal Transformation	-612	-	-	-	-	-	-	-	-	-	-612
Liquefaction	-	-	-	-	-	-	-	-	-	-	-
Other Transformation	-	-	-	-	-	-	-	-439	-	-	-439
Own Use	-	-	-226	-	-	-	-	-	-175	-	-401
Distribution Losses	-	-	-	-567	-	-	-	-	-1133	-	-1701
TFC	1547	-	11618	8583	-	-	-	21516	3589	-	46854
INDUSTRY SECTOR	1546	-	2337	5606	-	-	-	2388	1048	-	12925
Iron and Steel	98	-	-	-	-	-	-	-	-	-	98
Chemical and Petrochemical	-	-	245	2827	-	-	-	-	-	-	3073
of which: Feedstocks	-	-	-	1861	-	-	-	-	-	-	1861
Non-Ferrous Metals	-	-	-	-	-	-	-	-	-	-	-
Non-Metallic Minerals	1448	-	1500	177	-	-	-	-	-	-	3124
Transport Equipment	-	-	-	-	-	-	-	-	-	-	-
Machinery	-	-	-	-	-	-	-	-	-	-	-
Mining and Quarrying	-	-	-	-	-	-	-	-	-	-	-
Food and Tobacco	-	-	-	-	-	-	-	-	-	-	-
Paper, Pulp and Printing	-	-	-	-	-	-	-	-	-	-	-
Wood and Wood Products	-	-	-	-	-	-	-	-	-	-	-
Construction	-	-	-	-	-	-	-	-	-	-	-
Textile and Leather	-	-	-	-	-	-	-	-	-	-	-
Non-specified	-	-	592	2602	-	-	-	2388	1048	-	6630
TRANSPORT SECTOR	-	-	7655	4	-	-	-	-	2	-	7661
Air	-	-	624	-	-	-	-	-	-	-	624
Road	-	-	6791	-	-	-	-	-	-	-	6791
Rail	-	-	241	-	-	-	-	-	2	-	242
Pipeline Transport	-	-	-	4	-	-	-	-	-	-	4
Internal Navigation	-	-	-	-	-	-	-	-	-	-	-
Non-specified	-	-	-	-	-	-	-	-	-	-	-
OTHER SECTORS	1	-	1291	2973	-	-	-	19128	2540	-	25933
Agriculture	-	-	259	-	-	-	-	-	576	-	835
Comm. and Publ. Services	-	-	175	397	-	-	-	-	492	-	1064
Residential	1	-	857	2576	-	-	-	19128	1472	-	24034
Non-specified	-	-	-	-	-	-	-	-	-	-	-
NON-ENERGY USE	-	-	335	-	-	-	-	-	-	-	335
in Industry/Transf./Energy	-	-	335	-	-	-	-	-	-	-	335
in Transport	-	-	-	-	-	-	-	-	-	-	-
in Other Sectors	-	-	-	-	-	-	-	-	-	-	-
Electr. Generated - GWh	440	-	17547	15280	483	23206	-	-	-	-	56956
Electricity Plants	440	-	17547	15280	483	23206	-	-	-	-	56956
CHP Plants	-	-	-	-	-	-	-	-	-	-	-
Heat Generated - TJ	-	-	-	-	-	-	-	-	-	-	-
CHP Plants	-	-	-	-	-	-	-	-	-	-	-
Heat Plants	-	-	-	-	-	-	-	-	-	-	-

Pakistan : 1997

Thousand tonnes of oil equivalent / *Milliers de tonnes d'équivalent pétrole*

SUPPLY AND CONSUMPTION / *APPROVISIONNEMENT ET DEMANDE*	Coal / *Charbon*	Crude Oil / *Pétrole brut*	Petroleum Products / *Produits pétroliers*	Gas / *Gaz*	Nuclear / *Nucléaire*	Hydro / *Hydro*	Geotherm. Solar etc. / *Géotherm. solaire*	Combust. Renew. & Waste / *En. ren. combust.*	Electricity / *Electricité*	Heat / *Chaleur*	Total / *Total*
Indigenous Production	1413	2918	-	13434	90	1794	-	22398	-	-	42048
Imports	632	4591	9888	-	-	-	-	-	-	-	15110
Exports	-	-142	-99	-	-	-	-	-	-	-	-241
Intl. Marine Bunkers	-	-	-15	-	-	-	-	-	-	-	-15
Stock Changes	-	-131	46	-	-	-	-	-	-	-	-85
TPES	2045	7236	9821	13434	90	1794	-	22398	-	-	56818
Transfers	-	-124	137	-	-	-	-	-	-	-	13
Statistical Differences	-	-	-84	-	-	-	-	-	-	-	-84
Electricity Plants	-155	-	-5398	-3920	-90	-1794	-	-	5085	-	-6273
CHP Plants	-	-	-	-	-	-	-	-	-	-	-
Heat Plants	-	-	-	-	-	-	-	-	-	-	-
Gas Works	-	-	-	-	-	-	-	-	-	-	-
Petroleum Refineries	-	-7112	7034	-	-	-	-	-	-	-	-78
Coal Transformation	-544	-	-	-	-	-	-	-	-	-	-544
Liquefaction	-	-	-	-	-	-	-	-	-	-	-
Other Transformation	-	-	-	-	-	-	-	-447	-	-	-447
Own Use	-	-	-221	-	-	-	-	-	-203	-	-424
Distribution Losses	-	-	-	-567	-	-	-	-	-1209	-	-1776
TFC	1345	-	11287	8947	-	-	-	21951	3673	-	47203
INDUSTRY SECTOR	1344	-	1688	5844	-	-	-	2436	1030	-	12342
Iron and Steel	87	-	-	-	-	-	-	-	-	-	87
Chemical and Petrochemical	-	-	250	2947	-	-	-	-	-	-	3198
of which: Feedstocks	-	-	-	1940	-	-	-	-	-	-	1940
Non-Ferrous Metals	-	-	-	-	-	-	-	-	-	-	-
Non-Metallic Minerals	1257	-	1030	185	-	-	-	-	-	-	2471
Transport Equipment	-	-	-	-	-	-	-	-	-	-	-
Machinery	-	-	-	-	-	-	-	-	-	-	-
Mining and Quarrying	-	-	-	-	-	-	-	-	-	-	-
Food and Tobacco	-	-	-	-	-	-	-	-	-	-	-
Paper, Pulp and Printing	-	-	-	-	-	-	-	-	-	-	-
Wood and Wood Products	-	-	-	-	-	-	-	-	-	-	-
Construction	-	-	-	-	-	-	-	-	-	-	-
Textile and Leather	-	-	-	-	-	-	-	-	-	-	-
Non-specified	-	-	407	2712	-	-	-	2436	1030	-	6586
TRANSPORT SECTOR	-	-	7822	4	-	-	-	-	2	-	7827
Air	-	-	611	-	-	-	-	-	-	-	611
Road	-	-	6976	-	-	-	-	-	-	-	6976
Rail	-	-	234	-	-	-	-	-	2	-	236
Pipeline Transport	-	-	-	4	-	-	-	-	-	-	4
Internal Navigation	-	-	-	-	-	-	-	-	-	-	-
Non-specified	-	-	-	-	-	-	-	-	-	-	-
OTHER SECTORS	1	-	1448	3099	-	-	-	19515	2641	-	26704
Agriculture	-	-	264	-	-	-	-	-	609	-	873
Comm. and Publ. Services	-	-	170	414	-	-	-	-	511	-	1095
Residential	1	-	1014	2685	-	-	-	19515	1521	-	24736
Non-specified	-	-	-	-	-	-	-	-	-	-	-
NON-ENERGY USE	-	-	330	-	-	-	-	-	-	-	330
in Industry/Transf./Energy	-	-	330	-	-	-	-	-	-	-	330
in Transport	-	-	-	-	-	-	-	-	-	-	-
in Other Sectors	-	-	-	-	-	-	-	-	-	-	-
Electr. Generated - GWh	*382*	-	*22788*	*14751*	*346*	*20858*	-	-	-	-	*59125*
Electricity Plants	*382*	-	*22788*	*14751*	*346*	*20858*	-	-	-	-	*59125*
CHP Plants	-	-	-	-	-	-	-	-	-	-	-
Heat Generated - TJ	-	-	-	-	-	-	-	-	-	-	-
CHP Plants	-	-	-	-	-	-	-	-	-	-	-
Heat Plants	-	-	-	-	-	-	-	-	-	-	-

Panama

Thousand tonnes of oil equivalent / *Milliers de tonnes d'équivalent pétrole*

SUPPLY AND CONSUMPTION 1996	Coal	Crude Oil	Petroleum Products	Gas	Nuclear	Hydro	Geotherm. Solar etc.	Combust. Renew. & Waste	Electricity	Heat	Total
Indigenous Production	-	-	-	-	-	242	-	548	-	-	791
Imports	39	2093	529	-	-	-	-	-	8	-	2669
Exports	-	-	-233	-	-	-	-	-	-13	-	-247
Intl. Marine Bunkers	-	-	-1022	-	-	-	-	-	-	-	-1022
Stock changes	-5	-9	16	-	-	-	-	-	-	-	2
TPES	34	2083	-711	-	-	242	-	548	-5	-	2192
Electricity and CHP Plants	-	-	-306	-	-	-242	-	-44	337	-	-256
Petroleum Refineries	-	-2079	2039	-	-	-	-	-	-	-	-40
Other Transformation *	-	-4	-140	-	-	-	-	-7	-69	-	-220
TFC	34	-	882	-	-	-	-	497	262	-	1676
INDUSTRY SECTOR	34	-	153	-	-	-	-	92	55	-	335
Iron and Steel	-	-	-	-	-	-	-	-	-	-	-
Chemical and Petrochemical	-	-	-	-	-	-	-	-	-	-	-
Non-Metallic Minerals	-	-	-	-	-	-	-	-	-	-	-
Non-specified	34	-	153	-	-	-	-	92	55	-	335
TRANSPORT SECTOR	-	-	594	-	-	-	-	-	7	-	601
Air	-	-	3	-	-	-	-	-	-	-	3
Road	-	-	591	-	-	-	-	-	-	-	591
Non-specified	-	-	-	-	-	-	-	-	7	-	7
OTHER SECTORS	-	-	116	-	-	-	-	405	200	-	722
Agriculture	-	-	-	-	-	-	-	-	-	-	-
Comm. and Publ. Services	-	-	38	-	-	-	-	1	-	-	39
Residential	-	-	78	-	-	-	-	404	200	-	683
Non-specified	-	-	-	-	-	-	-	-	-	-	-
NON-ENERGY USE	-	-	18	-	-	-	-	-	-	-	18
Electricity Generated - GWh	-	-	1069	-	-	2816	-	30	-	-	3915
Heat Generated - TJ	-	-	-	-	-	-	-	-	-	-	-

APPROVISIONNEMENT ET DEMANDE 1997	Charbon	Pétrole brut	Produits pétroliers	Gaz	Nucléaire	Hydro	Géotherm. solaire etc.	En. ren. combust. & déchets	Electricité	Chaleur	Total
Indigenous Production	-	-	-	-	-	238	-	570	-	-	808
Imports	35	2031	1020	-	-	-	-	-	6	-	3092
Exports	-	-	-554	-	-	-	-	-	-10	-	-564
Intl. Marine Bunkers	-	-	-1022	-	-	-	-	-	-	-	-1022
Stock changes	-	16	-2	-	-	-	-	-	-	-	14
TPES	35	2047	-558	-	-	238	-	570	-5	-	2328
Electricity and CHP Plants	-	-	-366	-	-	-238	-	-51	357	-	-299
Petroleum Refineries	-	-2047	2021	-	-	-	-	-	-	-	-26
Other Transformation *	-	-	-115	-	-	-	-	-7	-83	-	-205
TFC	35	-	982	-	-	-	-	511	269	-	1798
INDUSTRY SECTOR	35	-	187	-	-	-	-	103	43	-	368
Iron and Steel	-	-	-	-	-	-	-	-	-	-	-
Chemical and Petrochemical	-	-	-	-	-	-	-	-	-	-	-
Non-Metallic Minerals	-	-	-	-	-	-	-	-	-	-	-
Non-specified	35	-	187	-	-	-	-	103	43	-	368
TRANSPORT SECTOR	-	-	627	-	-	-	-	-	8	-	635
Air	-	-	3	-	-	-	-	-	-	-	3
Road	-	-	624	-	-	-	-	-	-	-	624
Non-specified	-	-	-	-	-	-	-	-	8	-	8
OTHER SECTORS	-	-	123	-	-	-	-	409	219	-	750
Agriculture	-	-	-	-	-	-	-	-	-	-	-
Comm. and Publ. Services	-	-	42	-	-	-	-	1	-	-	43
Residential	-	-	81	-	-	-	-	408	219	-	707
Non-specified	-	-	-	-	-	-	-	-	-	-	-
NON-ENERGY USE	-	-	45	-	-	-	-	-	-	-	45
Electricity Generated - GWh	-	-	1355	-	-	2766	-	30	-	-	4151
Heat Generated - TJ	-	-	-	-	-	-	-	-	-	-	-

* Includes Transfers, Statistical Differences, Own Use and Distribution Losses.

Paraguay

Thousand tonnes of oil equivalent / *Milliers de tonnes d'équivalent pétrole*

SUPPLY AND CONSUMPTION 1996	Coal	Crude Oil	Petroleum Products	Gas	Nuclear	Hydro	Geotherm. Solar etc.	Combust. Renew. & Waste	Electricity	Heat	Total
Indigenous Production	-	-	-	-	-	4131	-	2547	-	-	6678
Imports	-	161	789	-	-	-	-	-	-	-	950
Exports	-	-	-17	-	-	-	-	-2	-3759	-	-3778
Intl. Marine Bunkers	-	-	-	-	-	-	-	-	-	-	-
Stock changes	-	-1	140	-	-	-	-	-	-	-	139
TPES	-	160	912	-	-	4131	-	2545	-3759	-	3989
Electricity and CHP Plants	-	-	-35	-	-	-4131	-	-17	4145	-	-38
Petroleum Refineries	-	-160	158	-	-	-	-	-	-	-	-2
Other Transformation *	-	-	-	-	-	-	-	-71	-74	-	-145
TFC	-	-	1034	-	-	-	-	2457	312	-	3804
INDUSTRY SECTOR	-	-	33	-	-	-	-	1285	81	-	1399
Iron and Steel	-	-	-	-	-	-	-	-	-	-	-
Chemical and Petrochemical	-	-	-	-	-	-	-	-	-	-	-
Non-Metallic Minerals	-	-	-	-	-	-	-	-	-	-	-
Non-specified	-	-	33	-	-	-	-	1285	81	-	1399
TRANSPORT SECTOR	-	-	918	-	-	-	-	19	-	-	938
Air	-	-	3	-	-	-	-	-	-	-	3
Road	-	-	905	-	-	-	-	11	-	-	916
Non-specified	-	-	10	-	-	-	-	8	-	-	18
OTHER SECTORS	-	-	72	-	-	-	-	1153	231	-	1456
Agriculture	-	-	-	-	-	-	-	-	-	-	-
Comm. and Publ. Services	-	-	-	-	-	-	-	4	41	-	45
Residential	-	-	72	-	-	-	-	1149	191	-	1411
Non-specified	-	-	-	-	-	-	-	-	-	-	-
NON-ENERGY USE	-	-	12	-	-	-	-	-	-	-	12
Electricity Generated - GWh	-	-	*61*	-	-	*48035*	-	*104*	-	-	*48200*
Heat Generated - TJ	-	-	-	-	-	-	-	-	-	-	-

APPROVISIONNEMENT ET DEMANDE 1997	Charbon	Pétrole brut	Produits pétroliers	Gaz	Nucléaire	Hydro	Géotherm. solaire etc.	En. ren. combust. & déchets	Electricité	Chaleur	Total
Indigenous Production	-	-	-	-	-	4339	-	2621	-	-	6960
Imports	-	157	960	-	-	-	-	-	-	-	1118
Exports	-	-	-18	-	-	-	-	-2	-3928	-	-3948
Intl. Marine Bunkers	-	-	-	-	-	-	-	-	-	-	-
Stock changes	-	3	59	-	-	-	-	-	-	-	62
TPES	-	161	1001	-	-	4339	-	2619	-3928	-	4191
Electricity and CHP Plants	-	-	-35	-	-	-4339	-	-17	4353	-	-38
Petroleum Refineries	-	-161	147	-	-	-	-	-	-	-	-14
Other Transformation *	-	-	1	-	-	-	-	-68	-93	-	-160
TFC	-	-	1114	-	-	-	-	2534	332	-	3980
INDUSTRY SECTOR	-	-	39	-	-	-	-	1355	119	-	1512
Iron and Steel	-	-	-	-	-	-	-	-	-	-	-
Chemical and Petrochemical	-	-	-	-	-	-	-	-	-	-	-
Non-Metallic Minerals	-	-	-	-	-	-	-	-	-	-	-
Non-specified	-	-	39	-	-	-	-	1355	119	-	1512
TRANSPORT SECTOR	-	-	993	-	-	-	-	20	-	-	1013
Air	-	-	3	-	-	-	-	-	-	-	3
Road	-	-	979	-	-	-	-	11	-	-	990
Non-specified	-	-	11	-	-	-	-	8	-	-	20
OTHER SECTORS	-	-	71	-	-	-	-	1159	213	-	1443
Agriculture	-	-	-	-	-	-	-	-	-	-	-
Comm. and Publ. Services	-	-	-	-	-	-	-	4	65	-	69
Residential	-	-	71	-	-	-	-	1155	148	-	1374
Non-specified	-	-	-	-	-	-	-	-	-	-	-
NON-ENERGY USE	-	-	12	-	-	-	-	-	-	-	12
Electricity Generated - GWh	-	-	*61*	-	-	*50452*	-	*106*	-	-	*50619*
Heat Generated - TJ	-	-	-	-	-	-	-	-	-	-	-

* Includes Transfers, Statistical Differences, Own Use and Distribution Losses.

Peru / Pérou

Thousand tonnes of oil equivalent / *Milliers de tonnes d'équivalent pétrole*

SUPPLY AND CONSUMPTION 1996	Coal	Crude Oil	Petroleum Products	Gas	Nuclear	Hydro	Geotherm. Solar etc.	Combust. Renew. & Waste	Electricity	Heat	Total
Indigenous Production	41	6126	-	731	-	1146	35	4221	-	-	12300
Imports	329	2833	1375	-	-	-	-	-	-	-	4537
Exports	-	-1880	-937	-	-	-	-	-	-	-	-2817
Intl. Marine Bunkers	-	-	-1	-	-	-	-	-	-	-	-1
Stock changes	-63	51	2	-	-	-	-	-	-	-	-10
TPES	306	7131	439	731	-	1146	35	4221	-	-	14009
Electricity and CHP Plants	-	-	-1027	-65	-	-1146	-35	-39	1486	-	-826
Petroleum Refineries	-	-7641	7261	-	-	-	-	-	-	-	-380
Other Transformation *	-92	511	-72	-612	-	-	-	-199	-255	-	-720
TFC	215	-	6601	54	-	-	-	3983	1231	-	12083
INDUSTRY SECTOR	209	-	1253	23	-	-	-	482	513	-	2480
Iron and Steel	42	-	164	-	-	-	-	-	-	-	206
Chemical and Petrochemical	-	-	-	-	-	-	-	-	-	-	-
Non-Metallic Minerals	-	-	2	-	-	-	-	-	-	-	2
Non-specified	167	-	1087	23	-	-	-	482	513	-	2272
TRANSPORT SECTOR	-	-	3341	-	-	-	-	-	-	-	3341
Air	-	-	480	-	-	-	-	-	-	-	480
Road	-	-	2811	-	-	-	-	-	-	-	2811
Non-specified	-	-	50	-	-	-	-	-	-	-	50
OTHER SECTORS	6	-	1955	31	-	-	-	3500	718	-	6210
Agriculture	-	-	667	-	-	-	-	106	104	-	878
Comm. and Publ. Services	-	-	349	-	-	-	-	-	234	-	583
Residential	6	-	939	31	-	-	-	3394	380	-	4749
Non-specified	-	-	-	-	-	-	-	-	-	-	-
NON-ENERGY USE	-	-	52	-	-	-	-	-	-	-	52
Electricity Generated - GWh	-	-	*3204*	*230*	-	*13324*	*408*	*114*	-	-	*17280*
Heat Generated - TJ	-	-	-	-	-	-	-	-	-	-	-

APPROVISIONNEMENT ET DEMANDE 1997	Charbon	Pétrole brut	Produits pétroliers	Gaz	Nucléaire	Hydro	Géotherm. solaire etc.	En. ren. combust. & déchets	Electricité	Chaleur	Total
Indigenous Production	29	5946	-	768	-	1136	48	4297	-	-	12225
Imports	295	4021	1406	-	-	-	-	-	-	-	5721
Exports	-	-2293	-1159	-	-	-	-	-	-	-	-3453
Intl. Marine Bunkers	-	-	-22	-	-	-	-	-	-	-	-22
Stock changes	-18	615	59	-	-	-	-	-	-	-	655
TPES	305	8288	283	768	-	1136	48	4297	-	-	15127
Electricity and CHP Plants	-	-	-1124	-94	-	-1136	-48	-44	1544	-	-902
Petroleum Refineries	-	-8288	7768	-	-	-	-	-	-	-	-521
Other Transformation *	-121	-	-445	-488	-	-	-	-202	-272	-	-1528
TFC	184	-	6481	187	-	-	-	4051	1272	-	12176
INDUSTRY SECTOR	180	-	1218	114	-	-	-	489	734	-	2735
Iron and Steel	49	-	169	-	-	-	-	-	323	-	540
Chemical and Petrochemical	-	-	-	-	-	-	-	-	-	-	-
Non-Metallic Minerals	-	-	3	-	-	-	-	-	-	-	3
Non-specified	132	-	1046	114	-	-	-	489	411	-	2191
TRANSPORT SECTOR	-	-	3313	-	-	-	-	-	-	-	3313
Air	-	-	469	-	-	-	-	-	-	-	469
Road	-	-	2799	-	-	-	-	-	-	-	2799
Non-specified	-	-	46	-	-	-	-	-	-	-	46
OTHER SECTORS	4	-	1804	73	-	-	-	3562	538	-	5982
Agriculture	-	-	575	-	-	-	-	125	50	-	751
Comm. and Publ. Services	-	-	305	-	-	-	-	-	62	-	367
Residential	4	-	924	73	-	-	-	3437	426	-	4865
Non-specified	-	-	-	-	-	-	-	-	-	-	-
NON-ENERGY USE	-	-	146	-	-	-	-	-	-	-	146
Electricity Generated - GWh	-	-	*3724*	*331*	-	*13215*	*556*	*128*	-	-	*17954*
Heat Generated - TJ	-	-	-	-	-	-	-	-	-	-	-

* Includes Transfers, Statistical Differences, Own Use and Distribution Losses.

INTERNATIONAL ENERGY AGENCY

Philippines

Total Production of Energy

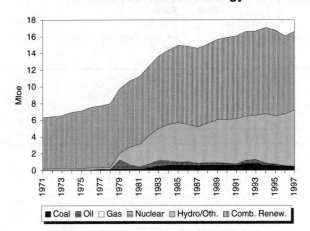

■ Coal 🏛 Oil ☐ Gas ▤ Nuclear ▨ Hydro/Oth. ▥ Comb. Renew.

Total Primary Energy Supply*

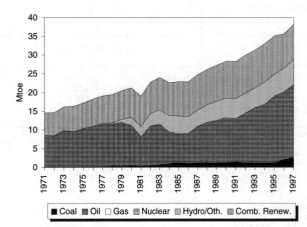

■ Coal 🏛 Oil ☐ Gas ▤ Nuclear ▨ Hydro/Oth. ▥ Comb. Renew.

Electricity Generation by Fuel

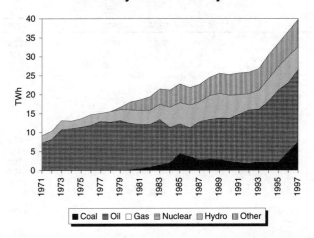

■ Coal 🏛 Oil ☐ Gas ▤ Nuclear ▨ Hydro ▥ Other

Oil Products Consumption

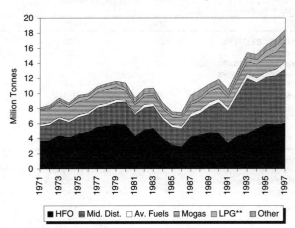

■ HFO 🏛 Mid. Dist. ☐ Av. Fuels ▤ Mogas ▨ LPG** ▥ Other

Electricity Consumption/GDP***, TPES/GDP*** and Production/TPES

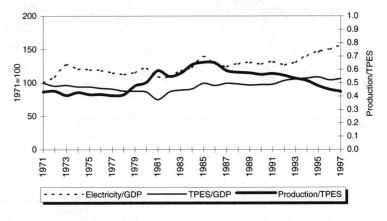

- - - - Electricity/GDP ——— TPES/GDP ▬▬▬ Production/TPES

* Excluding electricity trade.
** Includes LPG, NGL, ethane and naphtha.
*** GDP in 1990 US dollars.

Philippines : 1996

Thousand tonnes of oil equivalent / *Milliers de tonnes d'équivalent pétrole*

SUPPLY AND CONSUMPTION *APPROVISIONNEMENT ET DEMANDE*	Coal *Charbon*	Crude Oil *Pétrole brut*	Petroleum Products *Produits pétroliers*	Gas *Gaz*	Nuclear *Nucléaire*	Hydro *Hydro*	Geotherm. Solar etc. *Géotherm. solaire*	Combust. Renew. & Waste *En. ren. combust.*	Electricity *Electricité*	Heat *Chaleur*	Total *Total*
Indigenous Production	468	46	-	9	-	608	5619	9310	-	-	16061
Imports	1582	17860	1308	-	-	-	-	-	-	-	20750
Exports	-	-	-1403	-	-	-	-	-	-	-	-1403
Intl. Marine Bunkers	-	-	-72	-	-	-	-	-	-	-	-72
Stock Changes	-	22	338	-	-	-	-	-	-	-	360
TPES	**2051**	**17928**	**171**	**9**	**-**	**608**	**5619**	**9310**	**-**	**-**	**35696**
Transfers	-	-	-	-	-	-	-	-	-	-	-
Statistical Differences	-29	-	-420	-	-	-	-	-	-	-	-450
Electricity Plants	-793	-	-4749	-9	-	-608	-5619	-	3153	-	-8625
CHP Plants	-	-	-	-	-	-	-	-	-	-	-
Heat Plants	-	-	-	-	-	-	-	-	-	-	-
Gas Works	-	-	-8	4	-	-	-	-	-	-	-3
Petroleum Refineries	-	-17928	17762	-	-	-	-	-	-	-	-166
Coal Transformation	-96	-	-	-	-	-	-	-	-	-	-96
Liquefaction	-	-	-	-	-	-	-	-	-	-	-
Other Transformation	-	-	-	-	-	-	-	-916	-	-	-916
Own Use	-307	-	-1321	-	-	-	-	-	-115	-	-1743
Distribution Losses	-	-	-	-1	-	-	-	-	-527	-	-528
TFC	**826**	**-**	**11435**	**3**	**-**	**-**	**-**	**8394**	**2511**	**-**	**23169**
INDUSTRY SECTOR	**826**	**-**	**3332**	**-**	**-**	**-**	**-**	**3608**	**1019**	**-**	**8785**
Iron and Steel	37	-	391	-	-	-	-	-	76	-	504
Chemical and Petrochemical	23	-	64	-	-	-	-	-	118	-	205
of which: Feedstocks	-	-	-	-	-	-	-	-	-	-	-
Non-Ferrous Metals	-	-	46	-	-	-	-	-	-	-	46
Non-Metallic Minerals	720	-	371	-	-	-	-	-	-	-	1091
Transport Equipment	-	-	-	-	-	-	-	-	-	-	-
Machinery	-	-	-	-	-	-	-	-	-	-	-
Mining and Quarrying	-	-	-	-	-	-	-	-	-	-	-
Food and Tobacco	45	-	420	-	-	-	-	-	-	-	465
Paper, Pulp and Printing	-	-	272	-	-	-	-	-	-	-	272
Wood and Wood Products	-	-	161	-	-	-	-	-	-	-	161
Construction	-	-	134	-	-	-	-	-	-	-	134
Textile and Leather	-	-	130	-	-	-	-	-	-	-	130
Non-specified	-	-	1344	-	-	-	-	3608	825	-	5777
TRANSPORT SECTOR	**-**	**-**	**4139**	**-**	**-**	**-**	**-**	**-**	**-**	**-**	**4139**
Air	-	-	845	-	-	-	-	-	-	-	845
Road	-	-	2818	-	-	-	-	-	-	-	2818
Rail	-	-	-	-	-	-	-	-	-	-	-
Pipeline Transport	-	-	-	-	-	-	-	-	-	-	-
Internal Navigation	-	-	477	-	-	-	-	-	-	-	477
Non-specified	-	-	-	-	-	-	-	-	-	-	-
OTHER SECTORS	**-**	**-**	**3619**	**3**	**-**	**-**	**-**	**4786**	**1492**	**-**	**9900**
Agriculture	-	-	1854	-	-	-	-	-	-	-	1854
Comm. and Publ. Services	-	-	202	-	-	-	-	-	608	-	810
Residential	-	-	1561	1	-	-	-	4447	783	-	6791
Non-specified	-	-	3	2	-	-	-	339	100	-	445
NON-ENERGY USE	**-**	**-**	**345**	**-**	**-**	**-**	**-**	**-**	**-**	**-**	**345**
in Industry/Transf./Energy	-	-	345	-	-	-	-	-	-	-	345
in Transport	-	-	-	-	-	-	-	-	-	-	-
in Other Sectors	-	-	-	-	-	-	-	-	-	-	-
Electr. Generated - GWh	*4855*	*-*	*18180*	*20*	*-*	*7074*	*6534*	*-*	*-*	*-*	*36663*
Electricity Plants	*4855*	*-*	*18180*	*20*	*-*	*7074*	*6534*	*-*	*-*	*-*	*36663*
CHP Plants	*-*	*-*	*-*	*-*	*-*	*-*	*-*	*-*	*-*	*-*	*-*
Heat Generated - TJ	*-*	*-*	*-*	*-*	*-*	*-*	*-*	*-*	*-*	*-*	*-*
CHP Plants	*-*	*-*	*-*	*-*	*-*	*-*	*-*	*-*	*-*	*-*	*-*
Heat Plants	*-*	*-*	*-*	*-*	*-*	*-*	*-*	*-*	*-*	*-*	*-*

Philippines : 1997

Thousand tonnes of oil equivalent / Milliers de tonnes d'équivalent pétrole											
SUPPLY AND CONSUMPTION APPROVISIONNEMENT ET DEMANDE	Coal Charbon	Crude Oil Pétrole brut	Petroleum Products Produits pétroliers	Gas Gaz	Nuclear Nucléaire	Hydro Hydro	Geotherm. Solar etc. Géotherm. solaire	Combust. Renew. & Waste En. ren. combust.	Electricity Electricité	Heat Chaleur	Total Total
Indigenous Production	446	16	-	5	-	522	6224	9403	-	-	16616
Imports	2202	17988	2079	-	-	-	-	-	-	-	22269
Exports	-	-1	-1343	-	-	-	-	-	-	-	-1344
Intl. Marine Bunkers	-	-	-93	-	-	-	-	-	-	-	-93
Stock Changes	-	258	545	-	-	-	-	-	-	-	803
TPES	2648	18261	1188	5	-	522	6224	9403	-	-	38251
Transfers	-	-	-	-	-	-	-	-	-	-	-
Statistical Differences	-42	-	-401	-	-	-	-	-	-	-	-443
Electricity Plants	-1037	-	-5071	-5	-	-522	-6224	-	3424	-	-9435
CHP Plants	-	-	-	-	-	-	-	-	-	-	-
Heat Plants	-	-	-	-	-	-	-	-	-	-	-
Gas Works	-	-	-8	4	-	-	-	-	-	-	-3
Petroleum Refineries	-	-18261	18012	-	-	-	-	-	-	-	-249
Coal Transformation	-96	-	-	-	-	-	-	-	-	-	-96
Liquefaction	-	-	-	-	-	-	-	-	-	-	-
Other Transformation	-	-	-	-	-	-	-	-2099	-	-	-2099
Own Use	-401	-	-1336	-	-	-	-	-	-124	-	-1860
Distribution Losses	-	-	-	-1	-	-	-	-	-567	-	-568
TFC	1072	-	12385	3	-	-	-	7304	2734	-	23498
INDUSTRY SECTOR	1072	-	3507	-	-	-	-	3566	1081	-	9226
Iron and Steel	41	-	412	-	-	-	-	-	81	-	534
Chemical and Petrochemical	30	-	68	-	-	-	-	-	125	-	223
of which: Feedstocks	-	-	-	-	-	-	-	-	-	-	-
Non-Ferrous Metals	-	-	51	-	-	-	-	-	-	-	51
Non-Metallic Minerals	941	-	389	-	-	-	-	-	-	-	1330
Transport Equipment	-	-	-	-	-	-	-	-	-	-	-
Machinery	-	-	-	-	-	-	-	-	-	-	-
Mining and Quarrying	-	-	-	-	-	-	-	-	-	-	-
Food and Tobacco	59	-	451	-	-	-	-	-	-	-	510
Paper, Pulp and Printing	-	-	284	-	-	-	-	-	-	-	284
Wood and Wood Products	-	-	176	-	-	-	-	-	-	-	176
Construction	-	-	148	-	-	-	-	-	-	-	148
Textile and Leather	-	-	136	-	-	-	-	-	-	-	136
Non-specified	-	-	1393	-	-	-	-	3566	875	-	5834
TRANSPORT SECTOR	-	-	4640	-	-	-	-	-	-	-	4640
Air	-	-	1063	-	-	-	-	-	-	-	1063
Road	-	-	3056	-	-	-	-	-	-	-	3056
Rail	-	-	-	-	-	-	-	-	-	-	-
Pipeline Transport	-	-	-	-	-	-	-	-	-	-	-
Internal Navigation	-	-	521	-	-	-	-	-	-	-	521
Non-specified	-	-	-	-	-	-	-	-	-	-	-
OTHER SECTORS	-	-	3899	3	-	-	-	3738	1653	-	9293
Agriculture	-	-	2014	-	-	-	-	-	-	-	2014
Comm. and Publ. Services	-	-	209	-	-	-	-	-	664	-	873
Residential	-	-	1673	1	-	-	-	2996	870	-	5540
Non-specified	-	-	3	2	-	-	-	742	118	-	865
NON-ENERGY USE	-	-	339	-	-	-	-	-	-	-	339
in Industry/Transf./Energy	-	-	339	-	-	-	-	-	-	-	339
in Transport	-	-	-	-	-	-	-	-	-	-	-
in Other Sectors	-	-	-	-	-	-	-	-	-	-	-
Electr. Generated - GWh	7363	-	19129	13	-	6074	7237	-	-	-	39816
Electricity Plants	7363	-	19129	13	-	6074	7237	-	-	-	39816
CHP Plants	-	-	-	-	-	-	-	-	-	-	-
Heat Generated - TJ	-	-	-	-	-	-	-	-	-	-	-
CHP Plants	-	-	-	-	-	-	-	-	-	-	-
Heat Plants	-	-	-	-	-	-	-	-	-	-	-

Qatar

Thousand tonnes of oil equivalent / *Milliers de tonnes d'équivalent pétrole*

SUPPLY AND CONSUMPTION 1996	Coal	Crude Oil	Petroleum Products	Gas	Nuclear	Hydro	Geotherm. Solar etc.	Combust. Renew. & Waste	Electricity	Heat	Total
Indigenous Production	-	21856	-	11099	-	-	-	-	-	-	32956
Imports	-	-	-	-	-	-	-	2	-	-	2
Exports	-	-19328	-3344	-	-	-	-	-	-	-	-22672
Intl. Marine Bunkers	-	-	-	-	-	-	-	-	-	-	-
Stock changes	-	1872	-39	-	-	-	-	-	-	-	1833
TPES	-	4400	-3383	11099	-	-	-	2	-	-	12118
Electricity and CHP Plants	-	-	-	-1695	-	-	-	-	565	-	-1130
Petroleum Refineries	-	-3137	2995	-	-	-	-	-	-	-	-141
Other Transformation *	-	-1263	1328	-4648	-	-	-	-	-34	-	-4617
TFC	-	-	941	4756	-	-	-	2	532	-	6230
INDUSTRY SECTOR	-	-	29	4756	-	-	-	-	122	-	4908
Iron and Steel	-	-	-	356	-	-	-	-	122	-	478
Chemical and Petrochemical	-	-	29	3996	-	-	-	-	-	-	4025
Non-Metallic Minerals	-	-	-	126	-	-	-	-	-	-	126
Non-specified	-	-	-	279	-	-	-	-	-	-	279
TRANSPORT SECTOR	-	-	889	-	-	-	-	-	-	-	889
Air	-	-	185	-	-	-	-	-	-	-	185
Road	-	-	704	-	-	-	-	-	-	-	704
Non-specified	-	-	-	-	-	-	-	-	-	-	-
OTHER SECTORS	-	-	22	-	-	-	-	2	409	-	433
Agriculture	-	-	-	-	-	-	-	-	-	-	-
Comm. and Publ. Services	-	-	-	-	-	-	-	-	53	-	53
Residential	-	-	22	-	-	-	-	2	-	-	23
Non-specified	-	-	-	-	-	-	-	-	356	-	356
NON-ENERGY USE	-	-	-	-	-	-	-	-	-	-	-
Electricity Generated - GWh	-	-	-	*6575*	-	-	-	-	-	-	*6575*
Heat Generated - TJ	-	-	-	-	-	-	-	-	-	-	-

APPROVISIONNEMENT ET DEMANDE 1997	Charbon	Pétrole brut	Produits pétroliers	Gaz	Nucléaire	Hydro	Géotherm. solaire etc.	En. ren. combust. & déchets	Electricité	Chaleur	Total
Indigenous Production	-	29465	-	14502	-	-	-	-	-	-	43967
Imports	-	-	-	-	-	-	-	2	-	-	2
Exports	-	-23947	-3008	-2317	-	-	-	-	-	-	-29272
Intl. Marine Bunkers	-	-	-	-	-	-	-	-	-	-	-
Stock changes	-	-1121	-	-	-	-	-	-	-	-	-1121
TPES	-	4397	-3008	12185	-	-	-	2	-	-	13575
Electricity and CHP Plants	-	-	-	-1771	-	-	-	-	591	-	-1180
Petroleum Refineries	-	-3137	2732	-	-	-	-	-	-	-	-404
Other Transformation *	-	-1260	1326	-5147	-	-	-	-	-35	-	-5117
TFC	-	-	1050	5267	-	-	-	2	555	-	6874
INDUSTRY SECTOR	-	-	33	5267	-	-	-	-	128	-	5427
Iron and Steel	-	-	-	394	-	-	-	-	128	-	522
Chemical and Petrochemical	-	-	33	4425	-	-	-	-	-	-	4458
Non-Metallic Minerals	-	-	-	139	-	-	-	-	-	-	139
Non-specified	-	-	-	308	-	-	-	-	-	-	308
TRANSPORT SECTOR	-	-	995	-	-	-	-	-	-	-	995
Air	-	-	210	-	-	-	-	-	-	-	210
Road	-	-	786	-	-	-	-	-	-	-	786
Non-specified	-	-	-	-	-	-	-	-	-	-	-
OTHER SECTORS	-	-	22	-	-	-	-	2	428	-	451
Agriculture	-	-	-	-	-	-	-	-	-	-	-
Comm. and Publ. Services	-	-	-	-	-	-	-	-	56	-	56
Residential	-	-	22	-	-	-	-	2	-	-	23
Non-specified	-	-	-	-	-	-	-	-	372	-	372
NON-ENERGY USE	-	-	-	-	-	-	-	-	-	-	-
Electricity Generated - GWh	-	-	-	*6868*	-	-	-	-	-	-	*6868*
Heat Generated - TJ	-	-	-	-	-	-	-	-	-	-	-

* Includes Transfers, Statistical Differences, Own Use and Distribution Losses.

INTERNATIONAL ENERGY AGENCY

Romania / Roumanie

Total Production of Energy

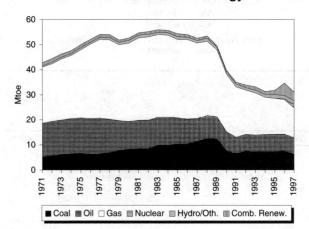

■ Coal ▦ Oil ☐ Gas ▤ Nuclear ▥ Hydro/Oth. ▧ Comb. Renew.

Total Primary Energy Supply*

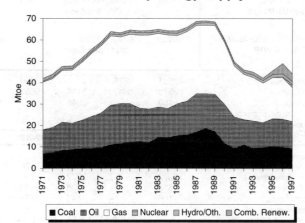

■ Coal ▦ Oil ☐ Gas ▤ Nuclear ▥ Hydro/Oth. ▧ Comb. Renew.

Electricity Generation by Fuel

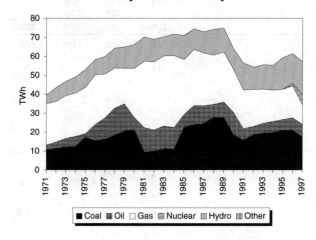

■ Coal ▦ Oil ☐ Gas ▤ Nuclear ▥ Hydro ▧ Other

Oil Products Consumption

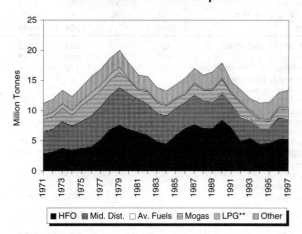

■ HFO ▦ Mid. Dist. ☐ Av. Fuels ▤ Mogas ▥ LPG** ▧ Other

Electricity Consumption/GDP***, TPES/GDP*** and Production/TPES

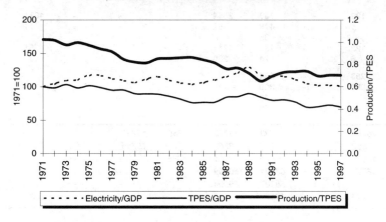

· · · · Electricity/GDP ——— TPES/GDP ▬▬ Production/TPES

* Excluding electricity trade.
** Includes LPG, NGL, ethane and naphtha.
*** GDP in 1990 US dollars.

Romania / Roumanie : 1996

Thousand tonnes of oil equivalent / *Milliers de tonnes d'équivalent pétrole*

SUPPLY AND CONSUMPTION *APPROVISIONNEMENT ET DEMANDE*	Coal *Charbon*	Crude Oil *Pétrole brut*	Petroleum Products *Produits pétroliers*	Gas *Gaz*	Nuclear *Nucléaire*	Hydro *Hydro*	Geotherm. Solar etc. *Géotherm. solaire*	Combust. Renew. & Waste *En. ren. combust.*	Electricity *Electricité*	Heat *Chaleur*	Total *Total*
Indigenous Production	7531	6676	-	13760	361	1355	-	4922	-	-	34605
Imports	2579	6947	3114	5653	-	-	-	-	193	-	18486
Exports	-109	-	-3846	-	-	-	-	-8	-123	-	-4087
Intl. Marine Bunkers	-	-	-	-	-	-	-	-	-	-	-
Stock Changes	-78	-77	268	-	-	-	-	-3	-	-	110
TPES	**9923**	**13547**	**-465**	**19413**	**361**	**1355**	**-**	**4910**	**69**	**-**	**49114**
Transfers	-	-	-	-	-	-	-	-	-	-	-
Statistical Differences	-339	-75	-270	-76	-	-	-	64	-	510	-186
Electricity Plants	-2072	-	-277	-991	-	-1355	-	-	2384	-	-2311
CHP Plants	-4616	-	-3038	-5184	-361	-	-	-30	2892	5126	-5211
Heat Plants	-106	-	-448	-1656	-	-	-	-70	-	1948	-333
Gas Works	-	-	-	-	-	-	-	-	-	-	-
Petroleum Refineries	-	-13176	13529								352
Coal Transformation	-991	-	-	-	-	-	-	-	-	-	-991
Liquefaction	-	-	-	-	-	-	-	-	-	-	-
Other Transformation	-	-	-	-	-	-	-	41	-	-	41
Own Use	-114	-99	-1653	-1054	-	-	-	-7	-1311	-945	-5184
Distribution Losses	-71	-196	-172	-477	-	-	-	-1	-618	-1027	-2562
TFC	**1615**	**-**	**7206**	**9974**	**-**	**-**	**-**	**4905**	**3417**	**5613**	**32729**
INDUSTRY SECTOR	**1197**	**-**	**2095**	**7842**	**-**	**-**	**-**	**228**	**2108**	**1277**	**14747**
Iron and Steel	991	-	229	1353	-	-	-	5	632	119	3328
Chemical and Petrochemical	134	-	342	3595	-	-	-	13	444	553	5081
of which: Feedstocks	-	-	-	*970*	-	-	-	-	-	-	*970*
Non-Ferrous Metals	-	-	-	-	-	-	-	-	-	-	-
Non-Metallic Minerals	23	-	405	731	-	-	-	1	150	27	1339
Transport Equipment	-	-	-	-	-	-	-	-	-	-	-
Machinery	11	-	436	669	-	-	-	1	330	175	1623
Mining and Quarrying	6	-	20	44	-	-	-	-	86	2	158
Food and Tobacco	20	-	146	856	-	-	-	19	112	163	1315
Paper, Pulp and Printing	-	-	62	151	-	-	-	56	65	35	369
Wood and Wood Products	-	-	15	45	-	-	-	82	24	29	195
Construction	10	-	273	47	-	-	-	4	55	38	426
Textile and Leather	-	-	37	211	-	-	-	2	65	82	397
Non-specified	1	-	130	141	-	-	-	44	145	55	516
TRANSPORT SECTOR	**1**	**-**	**3934**	**88**	**-**	**-**	**-**	**6**	**200**	**-**	**4229**
Air	-	-	94	-	-	-	-	-	-	-	94
Road	-	-	3388	85	-	-	-	-	-	-	3473
Rail	1	-	297	-	-	-	-	5	139	-	442
Pipeline Transport	-	-	5	-	-	-	-	-	5	-	11
Internal Navigation	-	-	149	-	-	-	-	-	-	-	149
Non-specified	-	-	-	4	-	-	-	1	56	-	60
OTHER SECTORS	**129**	**-**	**783**	**2044**	**-**	**-**	**-**	**4671**	**1108**	**4336**	**13071**
Agriculture	1	-	478	52	-	-	-	11	115	195	852
Comm. and Publ. Services	-	-	35	-	-	-	-	-	-	-	35
Residential	122	-	196	1711	-	-	-	4551	698	3202	10480
Non-specified	6	-	74	280	-	-	-	109	295	940	1704
NON-ENERGY USE	**288**	**-**	**394**	**-**	**-**	**-**	**-**	**-**	**-**	**-**	**682**
in Industry/Transf./Energy	288	-	309	-	-	-	-	-	-	-	597
in Transport	-	-	48	-	-	-	-	-	-	-	48
in Other Sectors	-	-	36	-	-	-	-	-	-	-	36
Electr. Generated - GWh	*20793*	*-*	*6703*	*16713*	*1386*	*15755*	*-*	*-*	*-*	*-*	*61350*
Electricity Plants	*7267*	*-*	*833*	*3862*	*-*	*15755*	*-*	*-*	*-*	*-*	*27717*
CHP Plants	*13526*	*-*	*5870*	*12851*	*1386*	*-*	*-*	*-*	*-*	*-*	*33633*
Heat Generated - TJ	*58752*	*-*	*83204*	*150809*	*-*	*-*	*-*	*3499*	*-*	*-*	*296264*
CHP Plants	*55306*	*-*	*68587*	*89904*	*-*	*-*	*-*	*879*	*-*	*-*	*214676*
Heat Plants	*3446*	*-*	*14617*	*60905*	*-*	*-*	*-*	*2620*	*-*	*-*	*81588*

Romania / Roumanie : 1997

	Coal	Crude Oil	Petroleum Products	Gas	Nuclear	Hydro	Geotherm. Solar etc.	Combust. Renew. & Waste	Electricity	Heat	Total
SUPPLY AND CONSUMPTION *APPROVISIONNEMENT ET DEMANDE*	*Charbon*	*Pétrole brut*	*Produits pétroliers*	*Gaz*	*Nucléaire*	*Hydro*	*Géotherm. solaire*	*En. ren. combust.*	*Electricité*	*Chaleur*	*Total*
Indigenous Production	6229	6579	-	11905	1407	1506	-	3388	-	-	31013
Imports	3263	6063	3881	4029	-	-	-	-	89	-	17325
Exports	-73	-	-2962	-	-	-	-	-8	-70	-	-3113
Intl. Marine Bunkers	-	-	-	-	-	-	-	-	-	-	-
Stock Changes	-250	13	-847	-	-	-	-	-5	-	-	-1089
TPES	**9169**	**12655**	**72**	**15934**	**1407**	**1506**	**-**	**3374**	**19**	**-**	**44135**
Transfers	-	-	-15	-	-	-	-	-	-	-	-15
Statistical Differences	-464	-158	484	715	-	-	-	-27	-25	-40	484
Electricity Plants	-1724	-	-126	-569	-	-1506	-	-3	2203	-	-1725
CHP Plants	-3883	-	-3235	-3663	-1407	-	-	-21	2711	4963	-4535
Heat Plants	-77	-	-461	-1581	-	-	-	-25	-	1834	-310
Gas Works	-	-	-	-	-	-	-	-	-	-	-
Petroleum Refineries	-	-12205	12636							-	431
Coal Transformation	-1086	-	-	-	-	-	-	-	-	-	-1086
Liquefaction	-	-	-	-	-	-	-	-	-	-	-
Other Transformation	-	-	-	-	-	-	-	41	-	-	41
Own Use	-260	-184	-1484	-1809	-	-	-	-5	-1038	-949	-5729
Distribution Losses	-92	-106	-139	-132	-	-	-	-	-566	-922	-1957
TFC	**1583**	**2**	**7731**	**8895**	**-**	**-**	**-**	**3335**	**3305**	**4884**	**29735**
INDUSTRY SECTOR	**1118**	**2**	**2009**	**6493**	**-**	**-**	**-**	**298**	**2161**	**990**	**13071**
Iron and Steel	906	-	246	1138	-	-	-	2	680	79	3051
Chemical and Petrochemical	36	-	590	2611	-	-	-	19	414	348	4018
of which: Feedstocks	-	-	*2*	*1192*	-	-	-	-	-	-	*1194*
Non-Ferrous Metals	-	-	-	-	-	-	-	-	-	-	-
Non-Metallic Minerals	96	-	309	768	-	-	-	1	148	21	1343
Transport Equipment	-	-	-	-	-	-	-	-	-	-	-
Machinery	63	-	118	828	-	-	-	2	317	192	1519
Mining and Quarrying	2	-	32	60	-	-	-	1	92	-	187
Food and Tobacco	13	-	289	567	-	-	-	22	125	130	1145
Paper, Pulp and Printing	-	-	47	140	-	-	-	61	67	39	353
Wood and Wood Products	-	-	32	30	-	-	-	115	30	38	245
Construction	-	2	218	33	-	-	-	6	58	26	343
Textile and Leather	1	-	95	237	-	-	-	3	78	76	490
Non-specified	1	-	33	81	-	-	-	67	152	42	376
TRANSPORT SECTOR	**2**	**-**	**4038**	**30**	**-**	**-**	**-**	**10**	**192**	**-**	**4272**
Air	-	-	136	-	-	-	-	-	-	-	136
Road	-	-	3264	28	-	-	-	-	-	-	3292
Rail	2	-	297	-	-	-	-	5	130	-	434
Pipeline Transport	-	-	1	-	-	-	-	-	5	-	6
Internal Navigation	-	-	340	-	-	-	-	-	-	-	340
Non-specified	-	-	-	2	-	-	-	5	56	-	64
OTHER SECTORS	**111**	**-**	**1322**	**2372**	**-**	**-**	**-**	**3027**	**952**	**3894**	**11678**
Agriculture	1	-	537	71	-	-	-	31	154	126	920
Comm. and Publ. Services	-	-	-	-	-	-	-	-	-	-	-
Residential	109	-	422	2057	-	-	-	2910	683	3458	9638
Non-specified	2	-	363	244	-	-	-	87	115	311	1120
NON-ENERGY USE	**352**	**-**	**361**	**-**	**-**	**-**	**-**	**-**	**-**	**-**	**713**
in Industry/Transf./Energy	352	-	239	-	-	-	-	-	-	-	591
in Transport	-	-	68	-	-	-	-	-	-	-	68
in Other Sectors	-	-	54	-	-	-	-	-	-	-	54
Electr. Generated - GWh	*17281*	*-*	*6863*	*10084*	*5400*	*17509*	*-*	*11*	*-*	*-*	*57148*
Electricity Plants	*5395*	*-*	*446*	*2260*	*-*	*17509*	*-*	*11*	*-*	*-*	*25621*
CHP Plants	*11886*	*-*	*6417*	*7824*	*5400*	*-*	*-*	*-*	*-*	*-*	*31527*
Heat Generated - TJ	*55305*	*-*	*89831*	*137704*	*-*	*-*	*-*	*1772*	*-*	*-*	*284612*
CHP Plants	*52910*	*-*	*75159*	*78906*	*-*	*-*	*-*	*849*	*-*	*-*	*207824*
Heat Plants	*2395*	*-*	*14672*	*58798*	*-*	*-*	*-*	*923*	*-*	*-*	*76788*

Thousand tonnes of oil equivalent / *Milliers de tonnes d'équivalent pétrole*

Russia / Russie : 1996

Thousand tonnes of oil equivalent / *Milliers de tonnes d'équivalent pétrole*

SUPPLY AND CONSUMPTION *APPROVISIONNEMENT ET DEMANDE*	Coal *Charbon*	Crude Oil *Pétrole brut*	Petroleum Products *Produits pétroliers*	Gas *Gaz*	Nuclear *Nucléaire*	Hydro *Hydro*	Geotherm. Solar etc. *Géotherm. solaire*	Combust. Renew. & Waste *En. ren. combust.*	Electricity *Electricité*	Heat *Chaleur*	Total *Total*
Indigenous Production	106283	300995	-	480075	28781	13271	24	17283	-	-	946711
Imports	8984	6771	2692	2794	-	-	-	1	1063	-	22304
Exports	-12272	-127083	-56684	-158649	-	-	-	-287	-2739	-	-357713
Intl. Marine Bunkers	-	-	-	-	-	-	-	-	-	-	-
Stock Changes	7279	1334	2088	-11299	-	-	-	27	-	-	-572
TPES	**110275**	**182017**	**-51904**	**312921**	**28781**	**13271**	**24**	**17023**	**-1676**	**-**	**610730**
Transfers	-	-	-	-	-	-	-	-	-	-	-
Statistical Differences	-8403	-1545	-	-2517	-	-	-	-90	-	-	-12554
Electricity Plants	-2	-	-	-	-28413	-13271	-24	-	22779	-	-18930
CHP Plants	-46759	-599	-4026	-147006	-368	-	-	-2408	49994	72893	-78278
Heat Plants	-18294	-	-34348	-41562	-	-	-	-4084	-	87325	-10964
Gas Works	-38	-	-	-	-	-	-	-	-	-	-38
Petroleum Refineries	-	-176754	175046		-	-	-	-	-	-	-1708
Coal Transformation	-1012	-	-	-	-	-	-	-	-	-	-1012
Liquefaction	-	-	-	-	-	-	-	-	-	-	-
Other Transformation	-	-	-	-	-	-	-	-16	-	-	-16
Own Use	-660	-543	-4140	-24150	-	-	-	-43	-12149	-12451	-54135
Distribution Losses	-4990	-	-	-6681	-	-	-	-143	-7263	-5122	-24199
TFC	**30116**	**2576**	**80628**	**91005**	**-**	**-**	**-**	**10240**	**51684**	**142645**	**408895**
INDUSTRY SECTOR	**12641**	**94**	**13587**	**34480**	**-**	**-**	**-**	**1510**	**22933**	**56111**	**141356**
Iron and Steel	8731	-	577	10934	-	-	-	11	4336	6276	30864
Chemical and Petrochemical	272	-	4943	13362	-	-	-	869	3201	13354	36001
of which: Feedstocks	-	-	*3247*	*10953*	-	-	-	-	-	-	*14200*
Non-Ferrous Metals	1099	5	732	1607	-	-	-	-	7160	4489	15092
Non-Metallic Minerals	735	-	-	-	-	-	-	-	-	-	735
Transport Equipment	-	-	-	-	-	-	-	-	-	-	-
Machinery	922	5	839	1956	-	-	-	2	3466	12580	19770
Mining and Quarrying	-	-	-	-	-	-	-	30	-	-	30
Food and Tobacco	189	5	780	427	-	-	-	87	889	5383	7760
Paper, Pulp and Printing	-	1	-	-	-	-	-	380	1093	5356	6830
Wood and Wood Products	72	-	613	1089	-	-	-	2	252	-	2028
Construction	475	56	4259	4959	-	-	-	49	1871	5292	16962
Textile and Leather	58	-	226	29	-	-	-	-	333	1617	2263
Non-specified	86	22	618	117	-	-	-	80	334	1765	3022
TRANSPORT SECTOR	**376**	**9**	**33389**	**14384**	**-**	**-**	**-**	**8**	**5584**	**-**	**53751**
Air	-	-	7716	-	-	-	-	-	-	-	7716
Road	-	-	18936	72	-	-	-	-	-	-	19007
Rail	311	-	3716	-	-	-	-	-	2625	-	6653
Pipeline Transport	-	8	181	14212	-	-	-	-	1625	-	16026
Internal Navigation	10	-	1937	-	-	-	-	-	-	-	1947
Non-specified	55	1	903	100	-	-	-	8	1335	-	2402
OTHER SECTORS	**15142**	**933**	**25830**	**42142**	**-**	**-**	**-**	**8722**	**23167**	**86534**	**202469**
Agriculture	2296	68	16077	3737	-	-	-	673	4189	6049	33088
Comm. and Publ. Services	-	-	-	-	-	-	-	4325	-	3694	8019
Residential	12692	119	5492	36956	-	-	-	3312	12455	66071	137097
Non-specified	154	746	4260	1449	-	-	-	412	6523	10720	24264
NON-ENERGY USE	**1957**	**1540**	**7822**		**-**	**-**	**-**	**-**	**-**	**-**	**11318**
in Industry/Transf./Energy	1957	1540	7822		-	-	-	-	-	-	11318
in Transport	-	-	-	-	-	-	-	-	-	-	-
in Other Sectors	-	-	-	-	-	-	-	-	-	-	-
Electr. Generated - GWh	*157758*	*-*	*44705*	*377840*	*109026*	*154309*	*28*	*2528*	*-*	*-*	*846194*
Electricity Plants	*-*	*-*	*-*	*-*	*109026*	*154309*	*28*	*1507*	*-*	*-*	*264870*
CHP Plants	*157758*	*-*	*44705*	*377840*	*-*	*-*	*-*	*1021*	*-*	*-*	*581324*
Heat Generated - TJ	*1402115*	*-*	*1103883*	*4048487*	*15400*	*-*	*-*	*97415*	*42000*	*-*	*6709300*
CHP Plants	*571425*	*-*	*165758*	*2299907*	*15400*	*-*	*-*	*-*	*-*	*-*	*3052490*
Heat Plants	*830690*	*-*	*938125*	*1748580*	*-*	*-*	*-*	*97415*	*42000*	*-*	*3656810*

Russia / Russie : 1997

	Coal	Crude Oil	Petroleum Products	Gas	Nuclear	Hydro	Geotherm. Solar etc.	Combust. Renew. & Waste	Electricity	Heat	Total
SUPPLY AND CONSUMPTION / APPROVISIONNEMENT ET DEMANDE	Charbon	Pétrole brut	Produits pétroliers	Gaz	Nucléaire	Hydro	Géotherm. solaire	En. ren. combust.	Electricité	Chaleur	Total
Indigenous Production	101118	305520	-	461124	28774	13499	24	17283	-	-	927341
Imports	6889	7538	3254	1872	-	-	-	1	611	-	20163
Exports	-10790	-126630	-60241	-153277	-	-	-	-287	-2305	-	-353530
Intl. Marine Bunkers	-	-	-	-	-	-	-	-	-	-	-
Stock Changes	-	-2553	534	-	-	-	-	27	-	-	-1992
TPES	97216	183875	-56453	309719	28774	13499	24	17023	-1694	-	591982
Transfers	-	-	-	-	-	-	-	-	-	-	-
Statistical Differences	-5409	-2512	-	-	-	-	-	-90	-	-	-8011
Electricity Plants	-2	-	-	-	-28406	-13499	-24	-	23341	-	-18590
CHP Plants	-41548	-601	-4307	-146681	-368	-	-	-2408	48303	66040	-81569
Heat Plants	-16651	-	-32301	-41470	-	-	-	-4084	-	78157	-16348
Gas Works	-39	-	-	-	-	-	-	-	-	-	-39
Petroleum Refineries	-	-177634	175996	-	-	-	-	-	-	-	-1638
Coal Transformation	-1109	-	-	-	-	-	-	-	-	-	-1109
Liquefaction	-	-	-	-	-	-	-	-	-	-	-
Other Transformation	-	-	-	-	-	-	-	-16	-	-	-16
Own Use	-608	-544	-4188	-24097	-	-	-	-43	-12518	-11103	-53100
Distribution Losses	-4386	-	-	-6666	-	-	-	-143	-7004	-4326	-22524
TFC	27465	2584	78748	90804	-	-	-	10240	50428	128768	389037
INDUSTRY SECTOR	12134	94	13426	34403	-	-	-	1510	22744	51507	135819
Iron and Steel	-	-	-	-	-	-	-	-	-	-	-
Chemical and Petrochemical	-	-	-	-	-	-	-	-	-	-	-
of which: Feedstocks	-	-	-	-	-	-	-	-	-	-	-
Non-Ferrous Metals	-	-	-	-	-	-	-	-	-	-	-
Non-Metallic Minerals	-	-	-	-	-	-	-	-	-	-	-
Transport Equipment	-	-	-	-	-	-	-	-	-	-	-
Machinery	-	-	-	-	-	-	-	-	-	-	-
Mining and Quarrying	-	-	-	-	-	-	-	-	-	-	-
Food and Tobacco	-	-	-	-	-	-	-	-	-	-	-
Paper, Pulp and Printing	-	-	-	-	-	-	-	-	-	-	-
Wood and Wood Products	-	-	-	-	-	-	-	-	-	-	-
Construction	-	-	-	-	-	-	-	-	-	-	-
Textile and Leather	-	-	-	-	-	-	-	-	-	-	-
Non-specified	-	-	-	-	-	-	-	-	-	-	-
TRANSPORT SECTOR	327	9	32304	14352	-	-	-	8	5461	-	52462
Air	-	-	7271	-	-	-	-	-	-	-	7271
Road	-	-	18358	71	-	-	-	-	-	-	18430
Rail	-	-	-	-	-	-	-	-	-	-	-
Pipeline Transport	-	-	-	-	-	-	-	-	-	-	-
Internal Navigation	-	-	-	-	-	-	-	-	-	-	-
Non-specified	-	-	-	-	-	-	-	-	-	-	-
OTHER SECTORS	13036	936	25401	42049	-	-	-	8722	22222	77261	189627
Agriculture	-	-	-	-	-	-	-	-	-	-	-
Comm. and Publ. Services	-	-	-	-	-	-	-	-	-	-	-
Residential	-	-	-	-	-	-	-	-	-	-	-
Non-specified	-	-	-	-	-	-	-	-	-	-	-
NON-ENERGY USE	1968	1545	7616	-	-	-	-	-	-	-	11129
in Industry/Transf./Energy	-	-	-	-	-	-	-	-	-	-	-
in Transport	-	-	-	-	-	-	-	-	-	-	-
in Other Sectors	-	-	-	-	-	-	-	-	-	-	-
Electr. Generated - GWh	139629	-	44013	377000	109000	156965	28	6430	-	-	833065
Electricity Plants	-	-	-	-	109000	156965	28	5409	-	-	271402
CHP Plants	139629	-	44013	377000	-	-	-	1021	-	-	561663
Heat Generated - TJ	1242000	-	993500	3803480	15400	-	-	70200	40000	-	6038400
CHP Plants	501420	-	189900	2058780	15400	-	-	-	-	-	2765500
Heat Plants	740580	-	803600	1744700	-	-	-	70200	40000	-	3272900

Thousand tonnes of oil equivalent / *Milliers de tonnes d'équivalent pétrole*

Saudi Arabia / Arabie saoudite

Total Production of Energy

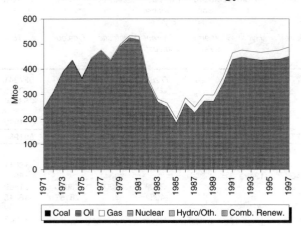

Total Primary Energy Supply*

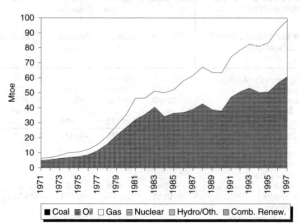

Electricity Generation by Fuel

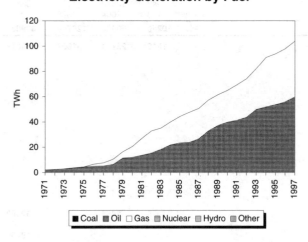

Oil Products Consumption

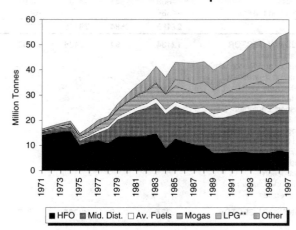

Electricity Consumption/GDP***, TPES/GDP*** and Production/TPES

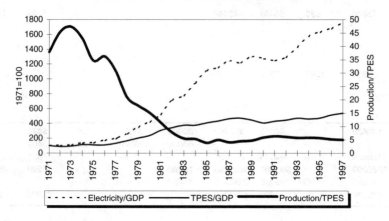

* Excluding electricity trade.
** Includes LPG, NGL, ethane and naphtha.
*** GDP in 1990 US dollars.

INTERNATIONAL ENERGY AGENCY

Saudi Arabia / Arabie saoudite

Thousand tonnes of oil equivalent / *Milliers de tonnes d'équivalent pétrole*

SUPPLY AND CONSUMPTION 1996	Coal	Crude Oil	Petroleum Products	Gas	Nuclear	Hydro	Geotherm. Solar etc.	Combust. Renew. & Waste	Electricity	Heat	Total
Indigenous Production	-	439197	-	35800	-	-	-	-	-	-	474997
Imports	-	-	-	-	-	-	-	4	-	-	4
Exports	-	-332308	-48592	-	-	-	-	-	-	-	-380900
Intl. Marine Bunkers	-	-	-1859	-	-	-	-	-	-	-	-1859
Stock changes	-	-	-	-	-	-	-	-	-	-	-
TPES	-	106889	-50450	35800	-	-	-	4	-	-	92243
Electricity and CHP Plants	-	-8855	-5725	-6462	-	-	-	-	8412	-	-12630
Petroleum Refineries	-	-86850	83720	-	-	-	-	-	-	-	-3130
Other Transformation *	-	-6354	1799	-17759	-	-	-	-	-1770	-	-24083
TFC	-	4829	29344	11580	-	-	-	4	6643	-	52400
INDUSTRY SECTOR	-	4829	3947	-	-	-	-	-	871	-	9647
Iron and Steel	-	-	-	-	-	-	-	-	-	-	-
Chemical and Petrochemical	-	4829	-	-	-	-	-	-	-	-	4829
Non-Metallic Minerals	-	-	-	-	-	-	-	-	-	-	-
Non-specified	-	-	3947	-	-	-	-	-	871	-	4818
TRANSPORT SECTOR	-	-	12844	-	-	-	-	-	-	-	12844
Air	-	-	2616	-	-	-	-	-	-	-	2616
Road	-	-	10228	-	-	-	-	-	-	-	10228
Non-specified	-	-	-	-	-	-	-	-	-	-	-
OTHER SECTORS	-	-	11609	11580	-	-	-	4	5771	-	28965
Agriculture	-	-	-	-	-	-	-	-	138	-	138
Comm. and Publ. Services	-	-	-	-	-	-	-	-	2009	-	2009
Residential	-	-	1152	-	-	-	-	4	3624	-	4781
Non-specified	-	-	10457	11580	-	-	-	-	-	-	22036
NON-ENERGY USE	-	-	945	-	-	-	-	-	-	-	945
Electricity Generated - GWh	-	-	55807	42012	-	-	-	-	-	-	97819
Heat Generated - TJ	-	-	-	-	-	-	-	-	-	-	-

APPROVISIONNEMENT ET DEMANDE 1997	Charbon	Pétrole brut	Produits pétroliers	Gaz	Nucléaire	Hydro	Géotherm. solaire etc.	En. ren. combust. & déchets	Electricité	Chaleur	Total
Indigenous Production	-	449511	-	37584	-	-	-	-	-	-	487095
Imports	-	-	-	-	-	-	-	4	-	-	4
Exports	-	-344260	-42532	-	-	-	-	-	-	-	-386792
Intl. Marine Bunkers	-	-	-1859	-	-	-	-	-	-	-	-1859
Stock changes	-	-	-	-	-	-	-	-	-	-	-
TPES	-	105252	-44391	37584	-	-	-	4	-	-	98449
Electricity and CHP Plants	-	-9442	-5875	-6784	-	-	-	-	8927	-	-13174
Petroleum Refineries	-	-83855	78979	-	-	-	-	-	-	-	-4875
Other Transformation *	-	-5898	759	-18643	-	-	-	-	-1878	-	-25661
TFC	-	6057	29473	12157	-	-	-	4	7049	-	54740
INDUSTRY SECTOR	-	6057	3420	-	-	-	-	-	924	-	10402
Iron and Steel	-	-	-	-	-	-	-	-	-	-	-
Chemical and Petrochemical	-	6057	-	-	-	-	-	-	-	-	6057
Non-Metallic Minerals	-	-	-	-	-	-	-	-	-	-	-
Non-specified	-	-	3420	-	-	-	-	-	924	-	4345
TRANSPORT SECTOR	-	-	12925	-	-	-	-	-	-	-	12925
Air	-	-	2607	-	-	-	-	-	-	-	2607
Road	-	-	10318	-	-	-	-	-	-	-	10318
Non-specified	-	-	-	-	-	-	-	-	-	-	-
OTHER SECTORS	-	-	12138	12157	-	-	-	4	6124	-	30424
Agriculture	-	-	-	-	-	-	-	-	146	-	146
Comm. and Publ. Services	-	-	-	-	-	-	-	-	2132	-	2132
Residential	-	-	1194	-	-	-	-	4	3846	-	5044
Non-specified	-	-	10944	12157	-	-	-	-	-	-	23101
NON-ENERGY USE	-	-	989	-	-	-	-	-	-	-	989
Electricity Generated - GWh	-	-	59696	44105	-	-	-	-	-	-	103801
Heat Generated - TJ	-	-	-	-	-	-	-	-	-	-	-

* Includes Transfers, Statistical Differences, Own Use and Distribution Losses.

INTERNATIONAL ENERGY AGENCY

Senegal / Sénégal

Thousand tonnes of oil equivalent / *Milliers de tonnes d'équivalent pétrole*

SUPPLY AND CONSUMPTION 1996	Coal	Crude Oil	Petroleum Products	Gas	Nuclear	Hydro	Geotherm. Solar etc.	Combust. Renew. & Waste	Electricity	Heat	Total
Indigenous Production	-	1	-	36	-	-	-	1580	-	-	1618
Imports	-	655	428	-	-	-	-	-	-	-	1082
Exports	-	-	-57	-	-	-	-	-	-	-	-57
Intl. Marine Bunkers	-	-	-60	-	-	-	-	-	-	-	-60
Stock changes	-	33	33	-	-	-	-	-	-	-	65
TPES	-	688	343	36	-	-	-	1580	-	-	2648
Electricity and CHP Plants	-	-	-291	-	-	-	-	-	101	-	-190
Petroleum Refineries	-	-688	675	-	-	-	-	-	-	-	-13
Other Transformation *	-	-	11	-	-	-	-	-375	-25	-	-390
TFC	-	-	737	36	-	-	-	1205	76	-	2054
INDUSTRY SECTOR	-	-	107	36	-	-	-	161	44	-	348
Iron and Steel	-	-	-	-	-	-	-	-	-	-	-
Chemical and Petrochemical	-	-	-	-	-	-	-	-	-	-	-
Non-Metallic Minerals	-	-	-	-	-	-	-	-	-	-	-
Non-specified	-	-	107	36	-	-	-	161	44	-	348
TRANSPORT SECTOR	-	-	491	-	-	-	-	-	-	-	491
Air	-	-	212	-	-	-	-	-	-	-	212
Road	-	-	253	-	-	-	-	-	-	-	253
Non-specified	-	-	26	-	-	-	-	-	-	-	26
OTHER SECTORS	-	-	121	-	-	-	-	1044	32	-	1198
Agriculture	-	-	35	-	-	-	-	-	4	-	39
Comm. and Publ. Services	-	-	-	-	-	-	-	-	12	-	12
Residential	-	-	86	-	-	-	-	1044	17	-	1148
Non-specified	-	-	-	-	-	-	-	-	-	-	-
NON-ENERGY USE	-	-	18	-	-	-	-	-	-	-	18
Electricity Generated - GWh	-	-	1034	140	-	-	-	-	-	-	1174
Heat Generated - TJ	-	-	-	-	-	-	-	-	-	-	-

APPROVISIONNEMENT ET DEMANDE 1997	Charbon	Pétrole brut	Produits pétroliers	Gaz	Nucléaire	Hydro	Géotherm. solaire etc.	En. ren. combust. & déchets	Electricité	Chaleur	Total
Indigenous Production	-	-	-	20	-	-	-	1634	-	-	1654
Imports	-	785	407	-	-	-	-	-	-	-	1192
Exports	-	-	-65	-	-	-	-	-	-	-	-65
Intl. Marine Bunkers	-	-	-60	-	-	-	-	-	-	-	-60
Stock changes	-	-12	62	-	-	-	-	-	-	-	49
TPES	-	773	344	20	-	-	-	1634	-	-	2770
Electricity and CHP Plants	-	-	-337	-	-	-	-	-	108	-	-229
Petroleum Refineries	-	-773	771	-	-	-	-	-	-	-	-1
Other Transformation *	-	-	35	-	-	-	-	-388	-27	-	-380
TFC	-	-	813	20	-	-	-	1246	81	-	2160
INDUSTRY SECTOR	-	-	133	20	-	-	-	166	47	-	366
Iron and Steel	-	-	-	-	-	-	-	-	-	-	-
Chemical and Petrochemical	-	-	-	-	-	-	-	-	-	-	-
Non-Metallic Minerals	-	-	-	-	-	-	-	-	-	-	-
Non-specified	-	-	133	20	-	-	-	166	47	-	366
TRANSPORT SECTOR	-	-	519	-	-	-	-	-	-	-	519
Air	-	-	216	-	-	-	-	-	-	-	216
Road	-	-	273	-	-	-	-	-	-	-	273
Non-specified	-	-	30	-	-	-	-	-	-	-	30
OTHER SECTORS	-	-	142	-	-	-	-	1080	34	-	1256
Agriculture	-	-	45	-	-	-	-	-	4	-	49
Comm. and Publ. Services	-	-	-	-	-	-	-	-	12	-	12
Residential	-	-	97	-	-	-	-	1080	18	-	1195
Non-specified	-	-	-	-	-	-	-	-	-	-	-
NON-ENERGY USE	-	-	18	-	-	-	-	-	-	-	18
Electricity Generated - GWh	-	-	1184	77	-	-	-	-	-	-	1261
Heat Generated - TJ	-	-	-	-	-	-	-	-	-	-	-

* Includes Transfers, Statistical Differences, Own Use and Distribution Losses.

Singapore / Singapour

Thousand tonnes of oil equivalent / *Milliers de tonnes d'équivalent pétrole*

SUPPLY AND CONSUMPTION 1996	Coal	Crude Oil	Petroleum Products	Gas	Nuclear	Hydro	Geotherm. Solar etc.	Combust. Renew. & Waste	Electricity	Heat	Total
Indigenous Production	-	-	-	-	-	-	55	-	-	-	55
Imports	8	56296	29545	1294	-	-	-	-	-	-	87143
Exports	-9	-	-49480	-	-	-	-	-	-	-	-49489
Intl. Marine Bunkers	-	-	-13859	-	-	-	-	-	-	-	-13859
Stock changes	-	-	-	-	-	-	-	-	-	-	-
TPES	-1	56296	-33793	1294	-	-	55	-	-	-	23851
Electricity and CHP Plants	-	-	-3733	-1294	-	-	-55	-	2073	-	-3010
Petroleum Refineries	-	-57102	51873	-	-	-	-	-	-	-	-5229
Other Transformation *	1	806	-6878	88	-	-	-	-	-189	-	-6171
TFC	-	-	7469	88	-	-	-	-	1884	-	9441
INDUSTRY SECTOR	-	-	2344	45	-	-	-	-	842	-	3230
Iron and Steel	-	-	-	-	-	-	-	-	74	-	74
Chemical and Petrochemical	-	-	2299	-	-	-	-	-	44	-	2343
Non-Metallic Minerals	-	-	-	-	-	-	-	-	-	-	-
Non-specified	-	-	45	45	-	-	-	-	724	-	814
TRANSPORT SECTOR	-	-	4590	-	-	-	-	-	20	-	4610
Air	-	-	2727	-	-	-	-	-	-	-	2727
Road	-	-	1863	-	-	-	-	-	-	-	1863
Non-specified	-	-	-	-	-	-	-	-	20	-	20
OTHER SECTORS	-	-	-	43	-	-	-	-	1022	-	1065
Agriculture	-	-	-	-	-	-	-	-	2	-	2
Comm. and Publ. Services	-	-	-	-	-	-	-	-	621	-	621
Residential	-	-	-	43	-	-	-	-	399	-	442
Non-specified	-	-	-	-	-	-	-	-	-	-	-
NON-ENERGY USE	-	-	536	-	-	-	-	-	-	-	536
Electricity Generated - GWh	-	-	18964	4494	-	-	642	-	-	-	24100
Heat Generated - TJ	-	-	-	-	-	-	-	-	-	-	-

APPROVISIONNEMENT ET DEMANDE 1997	Charbon	Pétrole brut	Produits pétroliers	Gaz	Nucléaire	Hydro	Géotherm. solaire etc.	En. ren. combust. & déchets	Electricité	Chaleur	Total
Indigenous Production	-	-	-	-	-	-	61	-	-	-	61
Imports	6	56617	32403	1275	-	-	-	-	-	-	90301
Exports	-6	-14	-47705	-	-	-	-	-	-	-	-47725
Intl. Marine Bunkers	-	-	-15760	-	-	-	-	-	-	-	-15760
Stock changes	-	-	-	-	-	-	-	-	-	-	-
TPES	-	56603	-31062	1275	-	-	61	-	-	-	26878
Electricity and CHP Plants	-	-	-4083	-1275	-	-	-61	-	2313	-	-3107
Petroleum Refineries	-	-56603	49454	-	-	-	-	-	-	-	-7149
Other Transformation *	-	-	-5923	89	-	-	-	-	-193	-	-6027
TFC	-	-	8385	89	-	-	-	-	2120	-	10595
INDUSTRY SECTOR	-	-	2886	45	-	-	-	-	953	-	3884
Iron and Steel	-	-	-	-	-	-	-	-	84	-	84
Chemical and Petrochemical	-	-	2841	-	-	-	-	-	50	-	2891
Non-Metallic Minerals	-	-	-	-	-	-	-	-	-	-	-
Non-specified	-	-	45	45	-	-	-	-	819	-	909
TRANSPORT SECTOR	-	-	4964	-	-	-	-	-	21	-	4985
Air	-	-	2988	-	-	-	-	-	-	-	2988
Road	-	-	1975	-	-	-	-	-	-	-	1975
Non-specified	-	-	-	-	-	-	-	-	21	-	21
OTHER SECTORS	-	-	-	44	-	-	-	-	1146	-	1191
Agriculture	-	-	-	-	-	-	-	-	2	-	2
Comm. and Publ. Services	-	-	-	-	-	-	-	-	697	-	697
Residential	-	-	-	44	-	-	-	-	447	-	492
Non-specified	-	-	-	-	-	-	-	-	-	-	-
NON-ENERGY USE	-	-	536	-	-	-	-	-	-	-	536
Electricity Generated - GWh	-	-	21758	4430	-	-	710	-	-	-	26898
Heat Generated - TJ	-	-	-	-	-	-	-	-	-	-	-

* Includes Transfers, Statistical Differences, Own Use and Distribution Losses.

Slovak Republic / République slovaque : 1996

Thousand tonnes of oil equivalent / Milliers de tonnes d'équivalent pétrole											
SUPPLY AND CONSUMPTION APPROVISIONNEMENT ET DEMANDE	Coal Charbon	Crude Oil Pétrole brut	Petroleum Products Produits pétroliers	Gas Gaz	Nuclear Nucléaire	Hydro Hydro	Geotherm. Solar etc. Géotherm. solaire	Combust. Renew. & Waste En. ren. combust.	Electricity Electricité	Heat Chaleur	Total Total
Indigenous Production	1121	72	-	244	2935	370	-	76	-	-	4818
Imports	4237	5330	36	5112	-	-	-	-	511	-	15227
Exports	-93	-	-1912	-	-	-	-	-	-202	-	-2207
Intl. Marine Bunkers	-	-	-	-	-	-	-	-	-	-	-
Stock Changes	-239	-160	17	196	-	-	-	-1	-	-	-186
TPES	**5027**	**5243**	**-1859**	**5552**	**2935**	**370**	**-**	**75**	**309**	**-**	**17651**
Transfers	-	-1	1	-	-	-	-	-	-	-	-
Statistical Differences	-32	-38	-	-3	-	-	-	-	-	-	-73
Electricity Plants	-1808	-	-	-	-2935	-370	-	-	1849	-	-3263
CHP Plants	-55	-	-233	-671	-	-	-	-	306	984	331
Heat Plants	-24	-	-41	-203	-	-	-	-	-86	-	-354
Gas Works	-	-	-	-	-	-	-	-	-	-	-
Petroleum Refineries	-	-5204	5333	-	-	-	-	-	-	-	130
Coal Transformation	-552	-	-	-	-	-	-	-	-	-	-552
Liquefaction	-	-	-	-	-	-	-	-	-	-	-
Other Transformation	-	-	-	-	-	-	-	-70	-	-	-70
Own Use	-114	-	-	-	-	-	-	-	-185	-133	-431
Distribution Losses	-27	-	-	-138	-	-	-	-	-174	-50	-390
TFC	**2414**	**-**	**3201**	**4537**	**-**	**-**	**-**	**6**	**2019**	**801**	**12979**
INDUSTRY SECTOR	**1680**	**-**	**1139**	**2480**	**-**	**-**	**-**	**-**	**903**	**24**	**6226**
Iron and Steel	822	-	43	251	-	-	-	-	301	-	1418
Chemical and Petrochemical	170	-	738	1193	-	-	-	-	134	-	2236
of which: Feedstocks	-	-	720	597	-	-	-	-	-	-	1317
Non-Ferrous Metals	-	-	-	-	-	-	-	-	-	-	-
Non-Metallic Minerals	140	-	26	273	-	-	-	-	94	-	533
Transport Equipment	36	-	8	51	-	-	-	-	23	6	124
Machinery	67	-	12	147	-	-	-	-	78	5	310
Mining and Quarrying	5	-	1	71	-	-	-	-	12	-	89
Food and Tobacco	46	-	36	224	-	-	-	-	105	2	413
Paper, Pulp and Printing	86	-	85	61	-	-	-	-	63	-	296
Wood and Wood Products	12	-	4	23	-	-	-	-	19	4	63
Construction	8	-	6	1	-	-	-	-	14	1	30
Textile and Leather	42	-	66	68	-	-	-	-	30	2	208
Non-specified	245	-	112	117	-	-	-	-	29	4	507
TRANSPORT SECTOR	**-**	**-**	**1171**	**-**	**-**	**-**	**-**	**-**	**85**	**-**	**1256**
Air	-	-	40	-	-	-	-	-	-	-	40
Road	-	-	1130	-	-	-	-	-	-	-	1130
Rail	-	-	-	-	-	-	-	-	85	-	85
Pipeline Transport	-	-	-	-	-	-	-	-	-	-	-
Internal Navigation	-	-	-	-	-	-	-	-	-	-	-
Non-specified	-	-	-	-	-	-	-	-	-	-	-
OTHER SECTORS	**734**	**-**	**266**	**2057**	**-**	**-**	**-**	**5**	**1031**	**777**	**4872**
Agriculture	27	-	206	49	-	-	-	4	73	22	380
Comm. and Publ. Services	424	-	32	771	-	-	-	1	489	383	2100
Residential	284	-	28	1237	-	-	-	1	469	372	2391
Non-specified	-	-	-	-	-	-	-	-	-	-	-
NON-ENERGY USE	**-**	**-**	**625**	**-**	**-**	**-**	**-**	**-**	**-**	**-**	**625**
in Industry/Transf./Energy	-	-	625	-	-	-	-	-	-	-	625
in Transport	-	-	-	-	-	-	-	-	-	-	-
in Other Sectors	-	-	-	-	-	-	-	-	-	-	-
Electr. Generated - GWh	*5936*	*-*	*1219*	*2341*	*11261*	*4303*	*-*	*-*	*-*	*-*	*25060*
Electricity Plants	*5936*	*-*	*-*	*-*	*11261*	*4303*	*-*	*-*	*-*	*-*	*21500*
CHP Plants	*-*	*-*	*1219*	*2341*	*-*	*-*	*-*	*-*	*-*	*-*	*3560*
Heat Generated - TJ	*-*	*-*	*-*	*21474*	*-*	*-*	*-*	*-*	*-*	*-*	*41198*
CHP Plants	*-*	*-*	*-*	*21474*	*-*	*-*	*-*	*-*	*-*	*-*	*41198*
Heat Plants	*-*	*-*	*-*	*-*	*-*	*-*	*-*	*-*	*-*	*-*	*-*

Slovak Republic / République slovaque : 1997

Thousand tonnes of oil equivalent / *Milliers de tonnes d'équivalent pétrole*

SUPPLY AND CONSUMPTION / *APPROVISIONNEMENT ET DEMANDE*	Coal / *Charbon*	Crude Oil / *Pétrole brut*	Petroleum Products / *Produits pétroliers*	Gas / *Gaz*	Nuclear / *Nucléaire*	Hydro / *Hydro*	Geotherm. Solar etc. / *Géotherm. solaire*	Combust. Renew. & Waste / *En. ren. combust.*	Electricity / *Electricité*	Heat / *Chaleur*	Total / *Total*
Indigenous Production	1146	65	-	225	2814	356	-	82	-	-	4688
Imports	3728	5259	214	5213	-	-	-	-	587	-	15001
Exports	-17	-	-2204	-	-	-	-	-	-236	-	-2457
Intl. Marine Bunkers	-	-	-	-	-	-	-	-	-	-	-
Stock Changes	-163	-8	-42	196	-	-	-	1	-	-	-16
TPES	**4694**	**5316**	**-2033**	**5634**	**2814**	**356**	**-**	**83**	**351**	**-**	**17216**
Transfers	-	-1	1	-	-	-	-	-	-	-	-
Statistical Differences	-29	73	-	-3	-	-	-	-	-	-	40
Electricity Plants	-1795	-	-	-	-2814	-356	-	-	1787	-	-3178
CHP Plants	-60	-	-146	-723	-	-	-	-	305	868	245
Heat Plants	-21	-	-19	-253	-	-	-	-	-145	-	-439
Gas Works	-	-	-	-	-	-	-	-	-	-	-
Petroleum Refineries	-	-5388	5506	-	-	-	-	-	-	-	118
Coal Transformation	-561	-	-	-	-	-	-	-	-	-	-561
Liquefaction	-	-	-	-	-	-	-	-	-	-	-
Other Transformation	-	-	-	-	-	-	-	-78	-	-	-78
Own Use	-121	-	-43	-	-	-	-	-	-154	-119	-438
Distribution Losses	-12	-	-	-89	-	-	-	-	-179	-48	-328
TFC	**2095**	**-**	**3266**	**4566**	**-**	**-**	**-**	**5**	**1964**	**701**	**12596**
INDUSTRY SECTOR	**1638**	**-**	**1097**	**2442**	**-**	**-**	**-**	**-**	**865**	**30**	**6071**
Iron and Steel	821	-	36	248	-	-	-	-	288	-	1393
Chemical and Petrochemical	148	-	737	1173	-	-	-	-	128	-	2186
of which: Feedstocks	-	-	*720*	*585*	-	-	-	-	-	-	*1305*
Non-Ferrous Metals	-	-	-	-	-	-	-	-	-	-	-
Non-Metallic Minerals	149	-	21	269	-	-	-	-	90	-	529
Transport Equipment	30	-	7	50	-	-	-	-	22	7	116
Machinery	58	-	11	145	-	-	-	-	74	7	295
Mining and Quarrying	5	-	1	70	-	-	-	-	12	-	87
Food and Tobacco	42	-	32	220	-	-	-	-	101	2	398
Paper, Pulp and Printing	74	-	74	60	-	-	-	-	61	-	268
Wood and Wood Products	10	-	3	23	-	-	-	-	18	5	59
Construction	7	-	5	1	-	-	-	-	16	1	29
Textile and Leather	34	-	58	67	-	-	-	-	29	2	190
Non-specified	261	-	114	115	-	-	-	-	26	5	521
TRANSPORT SECTOR	**-**	**-**	**1183**	**-**	**-**	**-**	**-**	**-**	**87**	**-**	**1270**
Air	-	-	37	-	-	-	-	-	-	-	37
Road	-	-	1145	-	-	-	-	-	-	-	1145
Rail	-	-	-	-	-	-	-	-	87	-	87
Pipeline Transport	-	-	-	-	-	-	-	-	-	-	-
Internal Navigation	-	-	-	-	-	-	-	-	-	-	-
Non-specified	-	-	-	-	-	-	-	-	-	-	-
OTHER SECTORS	**457**	**-**	**294**	**2124**	**-**	**-**	**-**	**5**	**1013**	**671**	**4563**
Agriculture	25	-	202	51	-	-	-	3	98	22	401
Comm. and Publ. Services	284	-	66	700	-	-	-	1	442	271	1764
Residential	147	-	26	1373	-	-	-	-	474	378	2398
Non-specified	-	-	-	-	-	-	-	-	-	-	-
NON-ENERGY USE	**-**	**-**	**692**	**-**	**-**	**-**	**-**	**-**	**-**	**-**	**692**
in Industry/Transf./Energy	-	-	692	-	-	-	-	-	-	-	692
in Transport	-	-	-	-	-	-	-	-	-	-	-
in Other Sectors	-	-	-	-	-	-	-	-	-	-	-
Electr. Generated - GWh	***5841***	**-**	***1216***	***2335***	***10797***	***4137***	**-**	**-**	**-**	**-**	***24326***
Electricity Plants	*5841*	-	-	-	*10797*	*4137*	-	-	-	-	*20775*
CHP Plants	-	-	*1216*	*2335*	-	-	-	-	-	-	*3551*
Heat Generated - TJ	**-**	**-**	**-**	***26811***	**-**	**-**	**-**	**-**	**-**	**-**	***36339***
CHP Plants	-	-	-	*26811*	-	-	-	-	-	-	*36339*
Heat Plants	-	-	-	-	-	-	-	-	-	-	-

Slovenia / Slovénie : 1996

	Coal	Crude Oil	Petroleum Products	Gas	Nuclear	Hydro	Geotherm. Solar etc.	Combust. Renew. & Waste	Electricity	Heat	Total
					Thousand tonnes of oil equivalent / Milliers de tonnes d'équivalent pétrole						
SUPPLY AND CONSUMPTION	Coal	Crude Oil	Petroleum Products	Gas	Nuclear	Hydro	Geotherm. Solar etc.	Combust. Renew. & Waste	Electricity	Heat	Total
APPROVISIONNEMENT ET DEMANDE	*Charbon*	*Pétrole brut*	*Produits pétroliers*	*Gaz*	*Nucléaire*	*Hydro*	*Géotherm. solaire*	*En. ren. combust.*	*Electricité*	*Chaleur*	*Total*
Indigenous Production	1013	1	-	10	1189	316	-	234	-	-	2762
Imports	220	535	2232	646	-	-	-	30	74	-	3735
Exports	-1	-	-57	-	-	-	-	-	-216	-	-274
Intl. Marine Bunkers	-	-	-	-	-	-	-	-	-	-	-
Stock Changes	-70	-	15	-	-	-	-	-1	-	-	-56
TPES	**1162**	**536**	**2190**	**656**	**1189**	**316**	**-**	**263**	**-143**	**-**	**6168**
Transfers	-	-	-	-	-	-	-	-	-	-	-
Statistical Differences	-92	28	-	20	-	-	-	-	-	-	-44
Electricity Plants	-104	-	-3	-2	-1189	-316	-	-	762	-	-852
CHP Plants	-834	-	-124	-63	-	-	-	-	336	143	-541
Heat Plants	-	-	-15	-64	-	-	-	-2	-	89	7
Gas Works	-	-	-	-	-	-	-	-	-	-	-
Petroleum Refineries	-	-563	537	-	-	-	-	-	-	-	-26
Coal Transformation	-16	-	-	-	-	-	-	-	-	-	-16
Liquefaction	-	-	-	-	-	-	-	-	-	-	-
Other Transformation	-	-	-	-	-	-	-	-	-	-	-
Own Use	-2	-	-19	-1	-	-	-	-	-77	-27	-126
Distribution Losses	-	-	-	-	-	-	-	-	-62	-	-62
TFC	**114**	**-**	**2567**	**546**	**-**	**-**	**-**	**260**	**816**	**205**	**4508**
INDUSTRY SECTOR	**66**	**-**	**218**	**487**	**-**	**-**	**-**	**61**	**411**	**31**	**1275**
Iron and Steel	4	-	3	78	-	-	-	-	55	3	143
Chemical and Petrochemical	-	-	94	110	-	-	-	-	35	3	243
of which: Feedstocks	-	-	*85*	*59*	-	-	-	-	-	-	*144*
Non-Ferrous Metals	1	-	3	3	-	-	-	-	102	-	109
Non-Metallic Minerals	35	-	40	93	-	-	-	-	20	1	190
Transport Equipment	-	-	7	11	-	-	-	-	-	-	19
Machinery	-	-	14	20	-	-	-	-	59	-	94
Mining and Quarrying	-	-	3	-	-	-	-	-	-	-	3
Food and Tobacco	1	-	22	29	-	-	-	-	19	-	72
Paper, Pulp and Printing	21	-	4	30	-	-	-	-	48	-	104
Wood and Wood Products	-	-	5	8	-	-	-	-	21	-	34
Construction	-	-	-	-	-	-	-	-	17	-	17
Textile and Leather	2	-	16	36	-	-	-	-	25	6	84
Non-specified	1	-	6	68	-	-	-	61	8	18	163
TRANSPORT SECTOR	**-**	**-**	**1511**	**-**	**-**	**-**	**-**	**-**	**14**	**-**	**1525**
Air	-	-	19	-	-	-	-	-	-	-	19
Road	-	-	1479	-	-	-	-	-	-	-	1479
Rail	-	-	12	-	-	-	-	-	14	-	26
Pipeline Transport	-	-	-	-	-	-	-	-	-	-	-
Internal Navigation	-	-	-	-	-	-	-	-	-	-	-
Non-specified	-	-	-	-	-	-	-	-	-	-	-
OTHER SECTORS	**47**	**-**	**833**	**58**	**-**	**-**	**-**	**199**	**391**	**174**	**1704**
Agriculture	-	-	-	-	-	-	-	-	-	-	-
Comm. and Publ. Services	17	-	320	18	-	-	-	51	165	53	624
Residential	30	-	513	41	-	-	-	148	226	121	1080
Non-specified	-	-	-	-	-	-	-	-	-	-	-
NON-ENERGY USE	**-**	**-**	**4**	**-**	**-**	**-**	**-**	**-**	**-**	**-**	**4**
in Industry/Transf./Energy	-	-	4	-	-	-	-	-	-	-	4
in Transport	-	-	-	-	-	-	-	-	-	-	-
in Other Sectors	-	-	-	-	-	-	-	-	-	-	-
Electr. Generated - GWh	*4189*	*-*	*346*	*-*	*4562*	*3673*	*-*	*-*	*-*	*-*	*12770*
Electricity Plants	*532*	*-*	*94*	*-*	*4562*	*3673*	*-*	*-*	*-*	*-*	*8861*
CHP Plants	*3657*	*-*	*252*	*-*	*-*	*-*	*-*	*-*	*-*	*-*	*3909*
Heat Generated - TJ	*5831*	*-*	*786*	*2989*	*-*	*-*	*-*	*96*	*-*	*-*	*9702*
CHP Plants	*5821*	*-*	*175*	*-*	*-*	*-*	*-*	*-*	*-*	*-*	*5996*
Heat Plants	*10*	*-*	*611*	*2989*	*-*	*-*	*-*	*96*	*-*	*-*	*3706*

Slovenia / Slovénie : 1997

	Coal	Crude Oil	Petroleum Products	Gas	Nuclear	Hydro	Geotherm. Solar etc.	Combust. Renew. & Waste	Electricity	Heat	Total
			Thousand tonnes of oil equivalent / *Milliers de tonnes d'équivalent pétrole*								
SUPPLY AND CONSUMPTION / ***APPROVISIONNEMENT ET DEMANDE***	Charbon	Pétrole brut	Produits pétroliers	Gaz	Nucléaire	Hydro	Géotherm. solaire	En. ren. combust.	Electricité	Chaleur	Total
Indigenous Production	1052	1	-	9	1308	266	-	234	-	-	2870
Imports	184	635	2225	706	-	-	-	30	71	-	3850
Exports	-1	-	-75	-	-	-	-	-	-217	-	-292
Intl. Marine Bunkers	-	-	-	-	-	-	-	-	-	-	-
Stock Changes	33	-16	-64	-	-	-	-	-1	-	-	-48
TPES	**1269**	**619**	**2086**	**715**	**1308**	**266**	**-**	**263**	**-146**		**6380**
Transfers	-	-	-	-	-	-	-	-	-	-	-
Statistical Differences	-125	-	1	3	-	-	-	-	-	-	-121
Electricity Plants	-123	-	-3	-2	-1308	-266	-	-	756	-	-946
CHP Plants	-913	-	-50	-13	-	-	-	-	376	137	-463
Heat Plants	-	-	-8	-63	-	-	-	-2	-	79	6
Gas Works	-	-	-	-	-	-	-	-	-	-	-
Petroleum Refineries	-	-619	553	-	-	-	-	-	-	-	-67
Coal Transformation	-13	-	-	-	-	-	-	-	-	-	-13
Liquefaction	-	-	-	-	-	-	-	-	-	-	-
Other Transformation	-	-	-	-	-	-	-	-	-	-	-
Own Use	-1	-	-22	-1	-	-	-	-	-81	-22	-127
Distribution Losses	-	-	-	-	-	-	-	-	-59	-	-59
TFC	**94**	**-**	**2556**	**639**	**-**	**-**	**-**	**260**	**846**	**194**	**4590**
INDUSTRY SECTOR	**66**	**-**	**163**	**582**	**-**	**-**	**-**	**61**	**406**	**27**	**1305**
Iron and Steel	3	-	2	111	-	-	-	-	64	3	183
Chemical and Petrochemical	-	-	28	180	-	-	-	-	35	4	247
of which: Feedstocks	-	-	15	96	-	-	-	-	-	-	110
Non-Ferrous Metals	1	-	5	16	-	-	-	-	104	-	126
Non-Metallic Minerals	38	-	38	87	-	-	-	-	37	1	201
Transport Equipment	1	-	4	18	-	-	-	-	8	-	31
Machinery	-	-	16	15	-	-	-	-	43	-	75
Mining and Quarrying	-	-	2	1	-	-	-	-	2	-	5
Food and Tobacco	1	-	35	30	-	-	-	-	19	-	84
Paper, Pulp and Printing	19	-	1	62	-	-	-	-	32	-	114
Wood and Wood Products	-	-	8	7	-	-	-	-	14	-	29
Construction	-	-	-	-	-	-	-	-	-	-	-
Textile and Leather	1	-	16	31	-	-	-	-	25	5	77
Non-specified	1	-	9	25	-	-	-	61	22	14	133
TRANSPORT SECTOR	**-**	**-**	**1533**	**-**	**-**	**-**	**-**	**-**	**14**	**-**	**1548**
Air	-	-	20	-	-	-	-	-	-	-	20
Road	-	-	1501	-	-	-	-	-	-	-	1501
Rail	-	-	12	-	-	-	-	-	14	-	27
Pipeline Transport	-	-	-	-	-	-	-	-	-	-	-
Internal Navigation	-	-	-	-	-	-	-	-	-	-	-
Non-specified	-	-	-	-	-	-	-	-	-	-	-
OTHER SECTORS	**27**	**-**	**859**	**58**	**-**	**-**	**-**	**199**	**426**	**168**	**1736**
Agriculture	-	-	-	-	-	-	-	-	-	-	-
Comm. and Publ. Services	10	-	359	17	-	-	-	51	195	51	682
Residential	17	-	500	40	-	-	-	148	231	117	1054
Non-specified	-	-	-	-	-	-	-	-	-	-	-
NON-ENERGY USE	**-**	**-**	**1**	**-**	**-**	**-**	**-**	**-**	**-**	**-**	**1**
in Industry/Transf./Energy	-	-	1	-	-	-	-	-	-	-	1
in Transport	-	-	-	-	-	-	-	-	-	-	-
in Other Sectors	-	-	-	-	-	-	-	-	-	-	-
Electr. Generated - GWh	*4710*	*-*	*345*	*-*	*5019*	*3092*	*-*	*-*	*-*	*-*	*13166*
Electricity Plants	*672*	*-*	*13*	*-*	*5019*	*3092*	*-*	*-*	*-*	*-*	*8796*
CHP Plants	*4038*	*-*	*332*	*-*	*-*	*-*	*-*	*-*	*-*	*-*	*4370*
Heat Generated - TJ	*5618*	*-*	*433*	*2921*	*-*	*-*	*-*	*94*	*-*	*-*	*9066*
CHP Plants	*5618*	*-*	*132*	*-*	*-*	*-*	*-*	*-*	*-*	*-*	*5750*
Heat Plants	*-*	*-*	*301*	*2921*	*-*	*-*	*-*	*94*	*-*	*-*	*3316*

South Africa / Afrique du Sud

Total Production of Energy

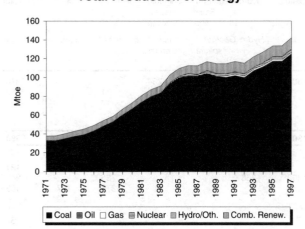

Total Primary Energy Supply*

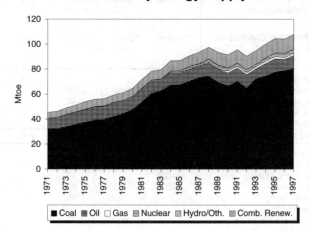

Electricity Generation by Fuel

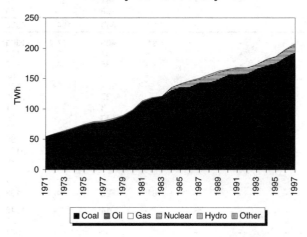

Oil Products Consumption

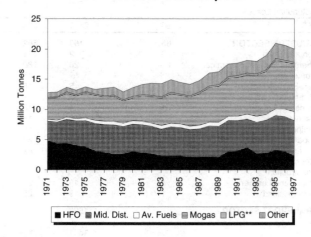

Electricity Consumption/GDP***, TPES/GDP*** and Production/TPES

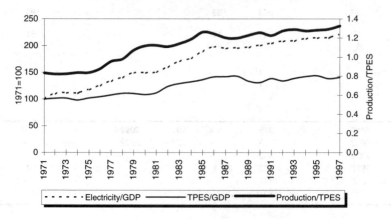

* Excluding electricity trade.
** Includes LPG, NGL, ethane and naphtha.
*** GDP in 1990 US dollars.

INTERNATIONAL ENERGY AGENCY

South Africa / Afrique du Sud : 1996

	Coal	Crude Oil	Petroleum Products	Gas	Nuclear	Hydro	Geotherm. Solar etc.	Combust. Renew. & Waste	Electricity	Heat	Total
SUPPLY AND CONSUMPTION											
APPROVISIONNEMENT ET DEMANDE	Charbon	Pétrole brut	Produits pétroliers	Gaz	Nucléaire	Hydro	Géotherm. solaire	En. ren. combust.	Electricité	Chaleur	Total
Indigenous Production	116930	401	-	1543	3069	113	-	11852	-	-	133908
Imports	315	13522	996	-	-	-	-	-	2	-	14835
Exports	-40266	-	-3249	-	-	-	-	-219	-480	-	-44213
Intl. Marine Bunkers	-	-	-2764	-	-	-	-	-	-	-	-2764
Stock Changes	1829	-	-	-	-	-	-	-	-	-	1829
TPES	**78808**	**13923**	**-5018**	**1543**	**3069**	**113**	**-**	**11633**	**-477**	**-**	**103595**
Transfers	-	-	-	-	-	-	-	-	-	-	-
Statistical Differences	-374	1581	-18	-	-	-	-	-	55	-	1244
Electricity Plants	-44235	-	-100	-	-3069	-113	-	-	17032	-	-30485
CHP Plants	-	-	-	-	-	-	-	-	26	-	26
Heat Plants	-	-	-	-	-	-	-	-	-	-	-
Gas Works	-3212	-	-	644	-	-	-	-	-	-	-2568
Petroleum Refineries	-	-20941	21795	-	-	-	-	-	-	-	854
Coal Transformation	-1309	-	-	-	-	-	-	-	-	-	-1309
Liquefaction	-14172	7018	-	-1543	-	-	-	-	-	-	-8697
Other Transformation	-	-	-	-	-	-	-	-3036	-	-	-3036
Own Use	-	-1581	-7	-1	-	-	-	-	-2540	-	-4130
Distribution Losses	-	-	-	-	-	-	-	-	-1315	-	-1315
TFC	**15507**	**-**	**16652**	**643**	**-**	**-**	**-**	**8597**	**12780**	**-**	**54179**
INDUSTRY SECTOR	**8161**	**-**	**1425**	**614**	**-**	**-**	**-**	**1554**	**7732**	**-**	**19487**
Iron and Steel	3192	-	167	231	-	-	-	-	1344	-	4934
Chemical and Petrochemical	-	-	96	62	-	-	-	-	217	-	374
of which: Feedstocks	-	-	56	-	-	-	-	-	-	-	56
Non-Ferrous Metals	1074	-	-	21	-	-	-	-	1122	-	2216
Non-Metallic Minerals	778	-	73	123	-	-	-	-	98	-	1072
Transport Equipment	-	-	-	4	-	-	-	-	1	-	5
Machinery	-	-	11	102	-	-	-	-	10	-	123
Mining and Quarrying	358	-	423	7	-	-	-	-	2995	-	3783
Food and Tobacco	-	-	27	20	-	-	-	-	43	-	91
Paper, Pulp and Printing	-	-	-	8	-	-	-	-	83	-	92
Wood and Wood Products	-	-	3	6	-	-	-	-	51	-	60
Construction	-	-	300	-	-	-	-	-	1	-	301
Textile and Leather	-	-	7	2	-	-	-	-	42	-	51
Non-specified	2760	-	319	28	-	-	-	1554	1723	-	6384
TRANSPORT SECTOR	**15**	**-**	**12328**	**-**	**-**	**-**	**-**	**-**	**368**	**-**	**12711**
Air	-	-	1265	-	-	-	-	-	1	-	1267
Road	-	-	10881	-	-	-	-	-	1	-	10881
Rail	15	-	164	-	-	-	-	-	296	-	475
Pipeline Transport	-	-	-	-	-	-	-	-	5	-	5
Internal Navigation	-	-	-	-	-	-	-	-	-	-	-
Non-specified	-	-	18	-	-	-	-	-	64	-	82
OTHER SECTORS	**2414**	**-**	**2175**	**28**	**-**	**-**	**-**	**7043**	**4680**	**-**	**16341**
Agriculture	156	-	1377	-	-	-	-	-	439	-	1972
Comm. and Publ. Services	840	-	55	18	-	-	-	-	1700	-	2612
Residential	1419	-	646	10	-	-	-	7043	2541	-	11659
Non-specified	-	-	98	-	-	-	-	-	-	-	98
NON-ENERGY USE	**4917**	**-**	**724**	**-**	**-**	**-**	**-**	**-**	**-**	**-**	**5641**
in Industry/Transf./Energy	4917	-	724	-	-	-	-	-	-	-	5641
in Transport	-	-	-	-	-	-	-	-	-	-	-
in Other Sectors	-	-	-	-	-	-	-	-	-	-	-
Electr. Generated - GWh	**185252**	**-**	**-**	**-**	**11775**	**1319**	**-**	**-**	**-**	**-**	**198346**
Electricity Plants	184952	-	-	-	11775	1319	-	-	-	-	198046
CHP Plants	300	-	-	-	-	-	-	-	-	-	300
Heat Generated - TJ	**-**	**-**	**-**	**-**	**-**	**-**	**-**	**-**	**-**	**-**	**-**
CHP Plants	-	-	-	-	-	-	-	-	-	-	-
Heat Plants	-	-	-	-	-	-	-	-	-	-	-

South Africa / Afrique du Sud : 1997

Thousand tonnes of oil equivalent / Milliers de tonnes d'équivalent pétrole

SUPPLY AND CONSUMPTION / APPROVISIONNEMENT ET DEMANDE	Coal / Charbon	Crude Oil / Pétrole brut	Petroleum Products / Produits pétroliers	Gas / Gaz	Nuclear / Nucléaire	Hydro / Hydro	Geotherm. Solar etc. / Géotherm. solaire	Combust. Renew. & Waste / En. ren. combust.	Electricity / Electricité	Heat / Chaleur	Total / Total
Indigenous Production	124678	401	-	1543	3296	180	-	12041	-	-	142139
Imports	315	12210	-	-	-	-	-	-	-	-	12525
Exports	-42924	-	-	-	-	-	-	-222	-569	-	-43715
Intl. Marine Bunkers	-	-	-2152	-	-	-	-	-	-	-	-2152
Stock Changes	-1577	-	-	-	-	-	-	-	-	-	-1577
TPES	**80492**	**12611**	**-2152**	**1543**	**3296**	**180**	**-**	**11819**	**-569**	**-**	**107220**
Transfers	-	-	-	-	-	-	-	-	-	-	-
Statistical Differences	1177	1378	-65	-	-	-	-	-	-92	-	2398
Electricity Plants	-46536	-	-88	-	-3296	-180	-	-	17840	-	-32259
CHP Plants	-	-	-	-	-	-	-	-	26	-	26
Heat Plants	-	-	-	-	-	-	-	-	-	-	-
Gas Works	-3458	-	-	643	-	-	-	-	-	-	-2815
Petroleum Refineries	-	-19628	19214	-	-	-	-	-	-	-	-415
Coal Transformation	-1493	-	-	-	-	-	-	-	-	-	-1493
Liquefaction	-13794	7018	-	-1543	-	-	-	-	-	-	-8320
Other Transformation	-	-	-	-	-	-	-	-3084	-	-	-3084
Own Use	-	-1378	-	-	-	-	-	-	-2529	-	-3908
Distribution Losses	-	-	-	-	-	-	-	-	-1384	-	-1384
TFC	**16387**	**-**	**16909**	**642**	**-**	**-**	**-**	**8735**	**13292**	**-**	**55965**
INDUSTRY SECTOR	**8994**	**-**	**1455**	**612**	**-**	**-**	**-**	**1579**	**7866**	**-**	**20506**
Iron and Steel	2479	-	147	180	-	-	-	-	1537	-	4343
Chemical and Petrochemical	-	-	86	63	-	-	-	-	209	-	358
of which: Feedstocks	-	-	51	-	-	-	-	-	-	-	51
Non-Ferrous Metals	874	-	-	26	-	-	-	-	1254	-	2155
Non-Metallic Minerals	691	-	64	127	-	-	-	-	102	-	985
Transport Equipment	-	-	-	4	-	-	-	-	1	-	5
Machinery	-	-	10	116	-	-	-	-	11	-	137
Mining and Quarrying	805	-	447	12	-	-	-	-	2614	-	3877
Food and Tobacco	-	-	24	25	-	-	-	-	46	-	96
Paper, Pulp and Printing	-	-	-	8	-	-	-	-	88	-	97
Wood and Wood Products	-	-	3	19	-	-	-	-	51	-	73
Construction	-	-	304	-	-	-	-	-	1	-	305
Textile and Leather	-	-	6	-	-	-	-	-	44	-	51
Non-specified	4145	-	366	30	-	-	-	1579	1906	-	8025
TRANSPORT SECTOR	**1**	**-**	**12746**	**-**	**-**	**-**	**-**	**-**	**392**	**-**	**13140**
Air	-	-	1404	-	-	-	-	-	1	-	1405
Road	-	-	10761	-	-	-	-	-	1	-	10761
Rail	1	-	164	-	-	-	-	-	291	-	456
Pipeline Transport	-	-	-	-	-	-	-	-	6	-	6
Internal Navigation	-	-	-	-	-	-	-	-	-	-	-
Non-specified	-	-	417	-	-	-	-	-	94	-	511
OTHER SECTORS	**2481**	**-**	**2016**	**30**	**-**	**-**	**-**	**7156**	**5034**	**-**	**16716**
Agriculture	155	-	1262	-	-	-	-	-	485	-	1903
Comm. and Publ. Services	874	-	88	19	-	-	-	-	1907	-	2888
Residential	1451	-	665	11	-	-	-	7156	2642	-	11924
Non-specified	-	-	1	-	-	-	-	-	-	-	1
NON-ENERGY USE	**4911**	**-**	**692**	**-**	**-**	**-**	**-**	**-**	**-**	**-**	**5603**
in Industry/Transf./Energy	4911	-	692	-	-	-	-	-	-	-	5603
in Transport	-	-	-	-	-	-	-	-	-	-	-
in Other Sectors	-	-	-	-	-	-	-	-	-	-	-
Electr. Generated - GWh	*193005*	*-*	*-*	*-*	*12647*	*2092*	*-*	*-*	*-*	*-*	*207744*
Electricity Plants	*192705*	*-*	*-*	*-*	*12647*	*2092*	*-*	*-*	*-*	*-*	*207444*
CHP Plants	*300*	*-*	*-*	*-*	*-*	*-*	*-*	*-*	*-*	*-*	*300*
Heat Generated - TJ	*-*	*-*	*-*	*-*	*-*	*-*	*-*	*-*	*-*	*-*	*-*
CHP Plants	*-*	*-*	*-*	*-*	*-*	*-*	*-*	*-*	*-*	*-*	*-*
Heat Plants	*-*	*-*	*-*	*-*	*-*	*-*	*-*	*-*	*-*	*-*	*-*

Sri Lanka

Thousand tonnes of oil equivalent / *Milliers de tonnes d'équivalent pétrole*

SUPPLY AND CONSUMPTION 1996	Coal	Crude Oil	Petroleum Products	Gas	Nuclear	Hydro	Geotherm. Solar etc.	Combust. Renew. & Waste	Electricity	Heat	Total
Indigenous Production	-	-	-	-	-	280	-	3924	-	-	4204
Imports	1	2054	881	-	-	-	-	-	-	-	2936
Exports	-	-	-76	-	-	-	-	-	-	-	-76
Intl. Marine Bunkers	-	-	-373	-	-	-	-	-	-	-	-373
Stock changes	-	4	98	-	-	-	-	-	-	-	102
TPES	1	2058	529	-	-	280	-	3924	-	-	6792
Electricity and CHP Plants	-	-	-297	-	-	-280	-	-	390	-	-188
Petroleum Refineries	-	-2058	1993	-	-	-	-	-	-	-	-65
Other Transformation *	-	-	-125	-	-	-	-	-	-70	-	-195
TFC	1	-	2100	-	-	-	-	3924	320	-	6344
INDUSTRY SECTOR	-	-	434	-	-	-	-	641	142	-	1217
Iron and Steel	-	-	-	-	-	-	-	-	-	-	-
Chemical and Petrochemical	-	-	-	-	-	-	-	-	-	-	-
Non-Metallic Minerals	-	-	-	-	-	-	-	-	-	-	-
Non-specified	-	-	434	-	-	-	-	641	142	-	1217
TRANSPORT SECTOR	1	-	1291	-	-	-	-	-	-	-	1291
Air	-	-	208	-	-	-	-	-	-	-	208
Road	-	-	1037	-	-	-	-	-	-	-	1037
Non-specified	1	-	46	-	-	-	-	-	-	-	47
OTHER SECTORS	-	-	322	-	-	-	-	3284	178	-	3783
Agriculture	-	-	14	-	-	-	-	-	-	-	14
Comm. and Publ. Services	-	-	15	-	-	-	-	129	67	-	211
Residential	-	-	294	-	-	-	-	3155	111	-	3559
Non-specified	-	-	-	-	-	-	-	-	-	-	-
NON-ENERGY USE	-	-	53	-	-	-	-	-	-	-	53
Electricity Generated - GWh	-	-	*1278*	-	-	*3252*	-	-	-	-	*4530*
Heat Generated - TJ	-	-	-	-	-	-	-	-	-	-	-

APPROVISIONNEMENT ET DEMANDE 1997	Charbon	Pétrole brut	Produits pétroliers	Gaz	Nucléaire	Hydro	Géotherm. solaire etc.	En. ren. combust. & déchets	Electricité	Chaleur	Total
Indigenous Production	-	-	-	-	-	296	-	4049	-	-	4345
Imports	1	2097	1191	-	-	-	-	-	-	-	3288
Exports	-	-	-203	-	-	-	-	-	-	-	-203
Intl. Marine Bunkers	-	-	-250	-	-	-	-	-	-	-	-250
Stock changes	-	-	-22	-	-	-	-	-	-	-	-22
TPES	1	2097	716	-	-	296	-	4049	-	-	7159
Electricity and CHP Plants	-	-	-437	-	-	-296	-	-	442	-	-291
Petroleum Refineries	-	-2097	1816	-	-	-	-	-	-	-	-281
Other Transformation *	-	-	-50	-	-	-	-	-	-81	-	-131
TFC	1	-	2044	-	-	-	-	4049	362	-	6456
INDUSTRY SECTOR	-	-	403	-	-	-	-	733	156	-	1292
Iron and Steel	-	-	-	-	-	-	-	-	-	-	-
Chemical and Petrochemical	-	-	-	-	-	-	-	-	-	-	-
Non-Metallic Minerals	-	-	-	-	-	-	-	-	-	-	-
Non-specified	-	-	403	-	-	-	-	733	156	-	1292
TRANSPORT SECTOR	1	-	1328	-	-	-	-	-	-	-	1329
Air	-	-	293	-	-	-	-	-	-	-	293
Road	-	-	994	-	-	-	-	-	-	-	994
Non-specified	1	-	41	-	-	-	-	-	-	-	42
OTHER SECTORS	-	-	275	-	-	-	-	3316	205	-	3796
Agriculture	-	-	14	-	-	-	-	-	-	-	14
Comm. and Publ. Services	-	-	19	-	-	-	-	130	66	-	216
Residential	-	-	242	-	-	-	-	3186	139	-	3567
Non-specified	-	-	-	-	-	-	-	-	-	-	-
NON-ENERGY USE	-	-	38	-	-	-	-	-	-	-	38
Electricity Generated - GWh	-	-	*1698*	-	-	*3447*	-	-	-	-	*5145*
Heat Generated - TJ	-	-	-	-	-	-	-	-	-	-	-

* Includes Transfers, Statistical Differences, Own Use and Distribution Losses.

Sudan / Soudan

Thousand tonnes of oil equivalent / *Milliers de tonnes d'équivalent pétrole*

SUPPLY AND CONSUMPTION 1996	Coal	Crude Oil	Petroleum Products	Gas	Nuclear	Hydro	Geotherm. Solar etc.	Combust. Renew. & Waste	Electricity	Heat	Total
Indigenous Production	-	104	-	-	-	92	-	9410	-	-	9606
Imports	-	725	799	-	-	-	-	-	-	-	1523
Exports	-	-	-	-	-	-	-	-	-	-	-
Intl. Marine Bunkers	-	-	-8	-	-	-	-	-	-	-	-8
Stock changes	-	-90	29	-	-	-	-	-	-	-	-60
TPES	-	739	820	-	-	92	-	9410	-	-	11061
Electricity and CHP Plants	-	-	-313	-	-	-92	-	-	177	-	-228
Petroleum Refineries	-	-739	698	-	-	-	-	-	-	-	-41
Other Transformation *	-	-	-8	-	-	-	-	-5261	-57	-	-5326
TFC	-	-	1197	-	-	-	-	4149	120	-	5466
INDUSTRY SECTOR	-	-	77	-	-	-	-	169	33	-	280
Iron and Steel	-	-	-	-	-	-	-	-	-	-	-
Chemical and Petrochemical	-	-	26	-	-	-	-	-	-	-	26
Non-Metallic Minerals	-	-	-	-	-	-	-	-	-	-	-
Non-specified	-	-	51	-	-	-	-	169	33	-	254
TRANSPORT SECTOR	-	-	924	-	-	-	-	-	-	-	924
Air	-	-	55	-	-	-	-	-	-	-	55
Road	-	-	868	-	-	-	-	-	-	-	868
Non-specified	-	-	-	-	-	-	-	-	-	-	-
OTHER SECTORS	-	-	96	-	-	-	-	3979	87	-	4162
Agriculture	-	-	2	-	-	-	-	-	3	-	5
Comm. and Publ. Services	-	-	-	-	-	-	-	97	17	-	114
Residential	-	-	94	-	-	-	-	3883	67	-	4043
Non-specified	-	-	-	-	-	-	-	-	-	-	-
NON-ENERGY USE	-	-	100	-	-	-	-	-	-	-	100
Electricity Generated - GWh	-	-	989	-	-	1074	-	-	-	-	2063
Heat Generated - TJ	-	-	-	-	-	-	-	-	-	-	-

APPROVISIONNEMENT ET DEMANDE 1997	Charbon	Pétrole brut	Produits pétroliers	Gaz	Nucléaire	Hydro	Géotherm. solaire etc.	En. ren. combust. & déchets	Electricité	Chaleur	Total
Indigenous Production	-	259	-	-	-	90	-	9532	-	-	9881
Imports	-	700	901	-	-	-	-	-	-	-	1601
Exports	-	-	-	-	-	-	-	-	-	-	-
Intl. Marine Bunkers	-	-	-8	-	-	-	-	-	-	-	-8
Stock changes	-	2	4	-	-	-	-	-	-	-	6
TPES	-	961	897	-	-	90	-	9532	-	-	11480
Electricity and CHP Plants	-	-	-346	-	-	-90	-	-	169	-	-266
Petroleum Refineries	-	-961	899	-	-	-	-	-	-	-	-62
Other Transformation *	-	-	-8	-	-	-	-	-5330	-53	-	-5391
TFC	-	-	1442	-	-	-	-	4203	116	-	5760
INDUSTRY SECTOR	-	-	220	-	-	-	-	172	37	-	429
Iron and Steel	-	-	-	-	-	-	-	-	-	-	-
Chemical and Petrochemical	-	-	26	-	-	-	-	-	-	-	26
Non-Metallic Minerals	-	-	-	-	-	-	-	-	-	-	-
Non-specified	-	-	194	-	-	-	-	172	37	-	403
TRANSPORT SECTOR	-	-	1026	-	-	-	-	-	-	-	1026
Air	-	-	55	-	-	-	-	-	-	-	55
Road	-	-	971	-	-	-	-	-	-	-	971
Non-specified	-	-	-	-	-	-	-	-	-	-	-
OTHER SECTORS	-	-	96	-	-	-	-	4031	78	-	4206
Agriculture	-	-	3	-	-	-	-	-	3	-	6
Comm. and Publ. Services	-	-	-	-	-	-	-	98	18	-	116
Residential	-	-	93	-	-	-	-	3933	57	-	4083
Non-specified	-	-	-	-	-	-	-	-	-	-	-
NON-ENERGY USE	-	-	100	-	-	-	-	-	-	-	100
Electricity Generated - GWh	-	-	924	-	-	1042	-	-	-	-	1966
Heat Generated - TJ	-	-	-	-	-	-	-	-	-	-	-

* Includes Transfers, Statistical Differences, Own Use and Distribution Losses.

Syria / Syrie

Thousand tonnes of oil equivalent / *Milliers de tonnes d'équivalent pétrole*

SUPPLY AND CONSUMPTION 1996	Coal	Crude Oil	Petroleum Products	Gas	Nuclear	Hydro	Geotherm. Solar etc.	Combust. Renew. & Waste	Electricity	Heat	Total
Indigenous Production	-	30868	-	2235	-	597	-	5	-	-	33705
Imports	1	-	564	-	-	-	-	-	-	-	566
Exports	-	-17722	-1632	-	-	-	-	-	-	-	-19353
Intl. Marine Bunkers	-	-	-	-	-	-	-	-	-	-	-
Stock changes	-	-376	-	-	-	-	-	-	-	-	-376
TPES	1	12770	-1068	2235	-	597	-	5	-	-	14541
Electricity and CHP Plants	-	-	-2546	-1225	-	-597	-	-	1452	-	-2916
Petroleum Refineries	-	-12770	12216	-	-	-	-	-	-	-	-554
Other Transformation *	-1	-	545	-85	-	-	-	-	-517	-	-58
TFC	-	-	9148	925	-	-	-	4	936	-	11013
INDUSTRY SECTOR	-	-	1053	823	-	-	-	-	433	-	2308
Iron and Steel	-	-	-	-	-	-	-	-	-	-	-
Chemical and Petrochemical	-	-	113	-	-	-	-	-	-	-	113
Non-Metallic Minerals	-	-	-	-	-	-	-	-	-	-	-
Non-specified	-	-	940	823	-	-	-	-	433	-	2196
TRANSPORT SECTOR	-	-	1351	-	-	-	-	-	-	-	1351
Air	-	-	173	-	-	-	-	-	-	-	173
Road	-	-	1178	-	-	-	-	-	-	-	1178
Non-specified	-	-	-	-	-	-	-	-	-	-	-
OTHER SECTORS	-	-	6371	103	-	-	-	4	502	-	6981
Agriculture	-	-	-	-	-	-	-	-	-	-	-
Comm. and Publ. Services	-	-	-	-	-	-	-	-	-	-	-
Residential	-	-	674	-	-	-	-	-	502	-	1176
Non-specified	-	-	5697	103	-	-	-	4	-	-	5804
NON-ENERGY USE	-	-	373	-	-	-	-	-	-	-	373
Electricity Generated - GWh	-	-	4834	5107	-	6944	-	-	-	-	16885
Heat Generated - TJ	-	-	-	-	-	-	-	-	-	-	-

APPROVISIONNEMENT ET DEMANDE 1997	Charbon	Pétrole brut	Produits pétroliers	Gaz	Nucléaire	Hydro	Géotherm. solaire etc.	En. ren. combust. & déchets	Electricité	Chaleur	Total
Indigenous Production	-	29860	-	2066	-	863	-	5	-	-	32794
Imports	1	-	736	-	-	-	-	-	-	-	737
Exports	-	-17090	-1799	-	-	-	-	-	-	-	-18890
Intl. Marine Bunkers	-	-	-	-	-	-	-	-	-	-	-
Stock changes	-	-	-	-	-	-	-	-	-	-	-
TPES	1	12770	-1064	2066	-	863	-	5	-	-	14642
Electricity and CHP Plants	-	-	-2446	-1132	-	-863	-	-	1544	-	-2897
Petroleum Refineries	-	-12770	12281	-	-	-	-	-	-	-	-489
Other Transformation *	-1	-	878	-79	-	-	-	-	-549	-	248
TFC	-	-	9649	855	-	-	-	4	995	-	11503
INDUSTRY SECTOR	-	-	1008	760	-	-	-	-	461	-	2229
Iron and Steel	-	-	-	-	-	-	-	-	-	-	-
Chemical and Petrochemical	-	-	116	-	-	-	-	-	-	-	116
Non-Metallic Minerals	-	-	-	-	-	-	-	-	-	-	-
Non-specified	-	-	892	760	-	-	-	-	461	-	2113
TRANSPORT SECTOR	-	-	1454	-	-	-	-	-	-	-	1454
Air	-	-	239	-	-	-	-	-	-	-	239
Road	-	-	1216	-	-	-	-	-	-	-	1216
Non-specified	-	-	-	-	-	-	-	-	-	-	-
OTHER SECTORS	-	-	6814	95	-	-	-	4	534	-	7447
Agriculture	-	-	-	-	-	-	-	-	-	-	-
Comm. and Publ. Services	-	-	-	-	-	-	-	-	-	-	-
Residential	-	-	707	-	-	-	-	-	534	-	1241
Non-specified	-	-	6107	95	-	-	-	4	-	-	6206
NON-ENERGY USE	-	-	373	-	-	-	-	-	-	-	373
Electricity Generated - GWh	-	-	4728	3184	-	10038	-	-	-	-	17950
Heat Generated - TJ	-	-	-	-	-	-	-	-	-	-	-

* Includes Transfers, Statistical Differences, Own Use and Distribution Losses.

INTERNATIONAL ENERGY AGENCY

Tajikistan / Tadjikistan

Thousand tonnes of oil equivalent / *Milliers de tonnes d'équivalent pétrole*

SUPPLY AND CONSUMPTION 1996	Coal	Crude Oil	Petroleum Products	Gas	Nuclear	Hydro	Geotherm. Solar etc.	Combust. Renew. & Waste	Electricity	Heat	Total
Indigenous Production	7	21	-	40	-	1275	-	-	-	-	1343
Imports	44	-	1230	898	-	-	-	-	424	-	2596
Exports	-	-	-23	-	-	-	-	-	-396	-	-419
Intl. Marine Bunkers	-	-	-	-	-	-	-	-	-	-	-
Stock changes	-	-	-	-	-	-	-	-	-	-	-
TPES	51	21	1207	938	-	1275	-	-	28	-	3520
Electricity and CHP Plants	-	-	-	-442	-	-1275	-	-	1290	-	-427
Petroleum Refineries	-	-21	23	-	-	-	-	-	-	-	1
Other Transformation *	-	-	-	-	-	-	-	-	-151	-	-151
TFC	51	-	1230	497	-	-	-	-	1166	-	2944
INDUSTRY SECTOR	-	-	-	-	-	-	-	-	581	-	581
Iron and Steel	-	-	-	-	-	-	-	-	-	-	-
Chemical and Petrochemical	-	-	-	-	-	-	-	-	19	-	19
Non-Metallic Minerals	-	-	-	-	-	-	-	-	-	-	-
Non-specified	-	-	-	-	-	-	-	-	561	-	561
TRANSPORT SECTOR	-	-	1071	-	-	-	-	-	7	-	1078
Air	-	-	5	-	-	-	-	-	-	-	5
Road	-	-	1066	-	-	-	-	-	-	-	1066
Non-specified	-	-	-	-	-	-	-	-	7	-	7
OTHER SECTORS	51	-	159	497	-	-	-	-	579	-	1285
Agriculture	-	-	-	-	-	-	-	-	385	-	385
Comm. and Publ. Services	-	-	-	-	-	-	-	-	-	-	-
Residential	-	-	-	-	-	-	-	-	170	-	170
Non-specified	51	-	159	497	-	-	-	-	23	-	730
NON-ENERGY USE	-	-	-	-	-	-	-	-	-	-	-
Electricity Generated - GWh	-	-	-	*175*	-	*14825*	-	-	-	-	*15000*
Heat Generated - TJ	-	-	-	-	-	-	-	-	-	-	-

APPROVISIONNEMENT ET DEMANDE 1997	Charbon	Pétrole brut	Produits pétroliers	Gaz	Nucléaire	Hydro	Géotherm. solaire etc.	En. ren. combust. & déchets	Electricité	Chaleur	Total
Indigenous Production	4	26	-	34	-	1189	-	-	-	-	1253
Imports	44	-	1230	817	-	-	-	-	421	-	2512
Exports	-	-	-28	-	-	-	-	-	-353	-	-381
Intl. Marine Bunkers	-	-	-	-	-	-	-	-	-	-	-
Stock changes	-	-	-	-	-	-	-	-	-	-	-
TPES	48	26	1202	851	-	1189	-	-	69	-	3384
Electricity and CHP Plants	-	-	-	-400	-	-1189	-	-	1204	-	-385
Petroleum Refineries	-	-26	28	-	-	-	-	-	-	-	2
Other Transformation *	-	-	-	-	-	-	-	-	-146	-	-146
TFC	48	-	1230	451	-	-	-	-	1127	-	2855
INDUSTRY SECTOR	-	-	-	-	-	-	-	-	561	-	561
Iron and Steel	-	-	-	-	-	-	-	-	-	-	-
Chemical and Petrochemical	-	-	-	-	-	-	-	-	18	-	18
Non-Metallic Minerals	-	-	-	-	-	-	-	-	-	-	-
Non-specified	-	-	-	-	-	-	-	-	542	-	542
TRANSPORT SECTOR	-	-	1071	-	-	-	-	-	7	-	1078
Air	-	-	5	-	-	-	-	-	-	-	5
Road	-	-	1066	-	-	-	-	-	-	-	1066
Non-specified	-	-	-	-	-	-	-	-	7	-	7
OTHER SECTORS	48	-	159	451	-	-	-	-	559	-	1216
Agriculture	-	-	-	-	-	-	-	-	372	-	372
Comm. and Publ. Services	-	-	-	-	-	-	-	-	-	-	-
Residential	-	-	-	-	-	-	-	-	164	-	164
Non-specified	48	-	159	451	-	-	-	-	22	-	680
NON-ENERGY USE	-	-	-	-	-	-	-	-	-	-	-
Electricity Generated - GWh	-	-	-	*175*	-	*13825*	-	-	-	-	*14000*
Heat Generated - TJ	-	-	-	-	-	-	-	-	-	-	-

* Includes Transfers, Statistical Differences, Own Use and Distribution Losses.

Tanzania / Tanzanie

Thousand tonnes of oil equivalent / *Milliers de tonnes d'équivalent pétrole*

SUPPLY AND CONSUMPTION 1996	Coal	Crude Oil	Petroleum Products	Gas	Nuclear	Hydro	Geotherm. Solar etc.	Combust. Renew. & Waste	Electricity	Heat	Total
Indigenous Production	3	-	-	-	-	150	-	13152	-	-	13305
Imports	-	600	177	-	-	-	-	-	3	-	780
Exports	-	-	-29	-	-	-	-	-	-	-	-29
Intl. Marine Bunkers	-	-	-22	-	-	-	-	-	-	-	-22
Stock changes	-	-	-	-	-	-	-	-	-	-	-
TPES	3	600	126	-	-	150	-	13152	3	-	14033
Electricity and CHP Plants	-	-	-83	-	-	-150	-	-	171	-	-62
Petroleum Refineries	-	-599	609	-	-	-	-	-	-	-	10
Other Transformation *	-1	-1	-185	-	-	-	-	-1070	-20	-	-1276
TFC	2	-	467	-	-	-	-	12082	154	-	12705
INDUSTRY SECTOR	2	-	101	-	-	-	-	1337	42	-	1482
Iron and Steel	-	-	-	-	-	-	-	-	-	-	-
Chemical and Petrochemical	-	-	-	-	-	-	-	-	-	-	-
Non-Metallic Minerals	1	-	-	-	-	-	-	-	-	-	1
Non-specified	2	-	101	-	-	-	-	1337	42	-	1481
TRANSPORT SECTOR	-	-	268	-	-	-	-	-	-	-	268
Air	-	-	38	-	-	-	-	-	-	-	38
Road	-	-	230	-	-	-	-	-	-	-	230
Non-specified	-	-	-	-	-	-	-	-	-	-	-
OTHER SECTORS	-	-	82	-	-	-	-	10745	112	-	10939
Agriculture	-	-	-	-	-	-	-	401	7	-	408
Comm. and Publ. Services	-	-	-	-	-	-	-	-	40	-	40
Residential	-	-	82	-	-	-	-	9858	59	-	9999
Non-specified	-	-	-	-	-	-	-	486	6	-	492
NON-ENERGY USE	-	-	15	-	-	-	-	-	-	-	15
Electricity Generated - GWh	-	-	*243*	-	-	*1748*	-	-	-	-	*1991*
Heat Generated - TJ	-	-	-	-	-	-	-	-	-	-	-

APPROVISIONNEMENT ET DEMANDE 1997	Charbon	Pétrole brut	Produits pétroliers	Gaz	Nucléaire	Hydro	Géotherm. solaire etc.	En. ren. combust. & déchets	Electricité	Chaleur	Total
Indigenous Production	3	-	-	-	-	125	-	13402	-	-	13529
Imports	-	600	177	-	-	-	-	-	4	-	781
Exports	-	-	-29	-	-	-	-	-	-	-	-29
Intl. Marine Bunkers	-	-	-22	-	-	-	-	-	-	-	-22
Stock changes	-	-	-	-	-	-	-	-	-	-	-
TPES	3	600	126	-	-	125	-	13402	4	-	14258
Electricity and CHP Plants	-	-	-166	-	-	-125	-	-	166	-	-124
Petroleum Refineries	-	-599	609	-	-	-	-	-	-	-	10
Other Transformation *	-1	-1	-148	-	-	-	-	-1090	-24	-	-1263
TFC	2	-	421	-	-	-	-	12312	146	-	12881
INDUSTRY SECTOR	2	-	101	-	-	-	-	1362	40	-	1505
Iron and Steel	-	-	-	-	-	-	-	-	-	-	-
Chemical and Petrochemical	-	-	-	-	-	-	-	-	-	-	-
Non-Metallic Minerals	1	-	-	-	-	-	-	-	-	-	1
Non-specified	2	-	101	-	-	-	-	1362	40	-	1505
TRANSPORT SECTOR	-	-	223	-	-	-	-	-	-	-	223
Air	-	-	38	-	-	-	-	-	-	-	38
Road	-	-	185	-	-	-	-	-	-	-	185
Non-specified	-	-	-	-	-	-	-	-	-	-	-
OTHER SECTORS	-	-	82	-	-	-	-	10949	107	-	11138
Agriculture	-	-	-	-	-	-	-	409	7	-	415
Comm. and Publ. Services	-	-	-	-	-	-	-	-	38	-	38
Residential	-	-	82	-	-	-	-	10045	56	-	10183
Non-specified	-	-	-	-	-	-	-	495	6	-	501
NON-ENERGY USE	-	-	15	-	-	-	-	-	-	-	15
Electricity Generated - GWh	-	-	*485*	-	-	*1449*	-	-	-	-	*1934*
Heat Generated - TJ	-	-	-	-	-	-	-	-	-	-	-

* Includes Transfers, Statistical Differences, Own Use and Distribution Losses.

INTERNATIONAL ENERGY AGENCY

Thailand / Thailande

Total Production of Energy

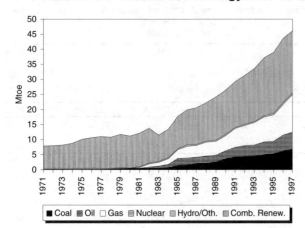

■ Coal ▦ Oil □ Gas ▤ Nuclear ▒ Hydro/Oth. ▥ Comb. Renew.

Total Primary Energy Supply*

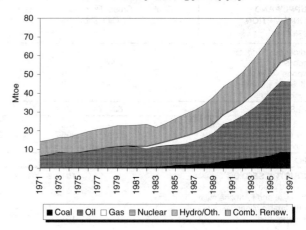

■ Coal ▦ Oil □ Gas ▤ Nuclear ▒ Hydro/Oth. ▥ Comb. Renew.

Electricity Generation by Fuel

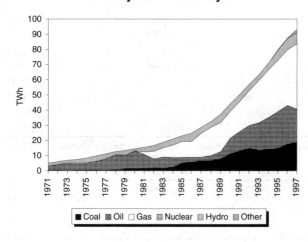

■ Coal ▦ Oil □ Gas ▤ Nuclear ▒ Hydro ▥ Other

Oil Products Consumption

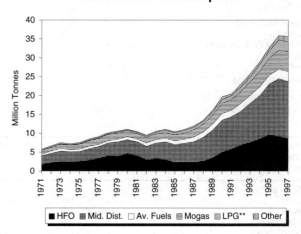

■ HFO ▦ Mid. Dist. □ Av. Fuels ▤ Mogas ▒ LPG** ▥ Other

Electricity Consumption/GDP***,
TPES/GDP*** and Production/TPES

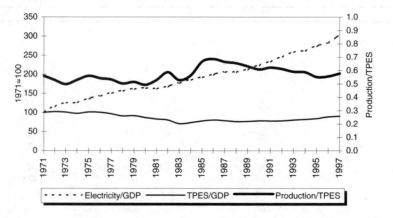

- - - - Electricity/GDP ——— TPES/GDP ━━━ Production/TPES

* Excluding electricity trade.
** Includes LPG, NGL, ethane and naphtha.
*** GDP in 1990 US dollars.

Thailand / Thaïlande : 1996

	Coal	Crude Oil	Petroleum Products	Gas	Nuclear	Hydro	Geotherm. Solar etc.	Combust. Renew. & Waste	Electricity	Heat	Total
SUPPLY AND CONSUMPTION											
APPROVISIONNEMENT ET DEMANDE	*Charbon*	*Pétrole brut*	*Produits pétroliers*	*Gaz*	*Nucléaire*	*Hydro*	*Géotherm. solaire*	*En. ren. combust.*	*Electricité*	*Chaleur*	*Total*
Indigenous Production	6229	5261	-	10186	-	631	-	21371	-	-	43678
Imports	2494	30937	7609	-	-	-	-	17	69	-	41126
Exports	-	-1117	-3237	-	-	-	-	-6	-8	-	-4368
Intl. Marine Bunkers	-	-	-738	-	-	-	-	-	-	-	-738
Stock Changes	-141	-178	-781	-	-	-	-	-	-	-	-1100
TPES	**8582**	**34903**	**2854**	**10186**	**-**	**631**	**-**	**21382**	**62**	**-**	**78599**
Transfers	-	-2187	2208	-	-	-	-	-	-	-	21
Statistical Differences	137	-	-85	-	-	-	-	-	-	-	52
Electricity Plants	-4260	-	-5758	-7552	-	-631	-	-213	7522	-	-10892
CHP Plants	-	-	-	-	-	-	-	-	-	-	-
Heat Plants	-	-	-	-	-	-	-	-	-	-	-
Gas Works	-	-	-	-	-	-	-	-	-	-	-
Petroleum Refineries	-	-32339	30921	-	-	-	-	-	-	-	-1419
Coal Transformation	-65	-	-	-	-	-	-	-	-	-	-65
Liquefaction	-	-	-	-	-	-	-	-	-	-	-
Other Transformation	-	-	-	-	-	-	-	-8713	-	-	-8713
Own Use	-	-	-366	-1788	-	-	-	-	-277	-	-2432
Distribution Losses	-	-104	-15	-	-	-	-	-	-654	-	-773
TFC	**4395**	**273**	**29759**	**845**	**-**	**-**	**-**	**12456**	**6652**	**-**	**54380**
INDUSTRY SECTOR	**4393**	**273**	**6054**	**840**	**-**	**-**	**-**	**4434**	**2979**	**-**	**18973**
Iron and Steel	16	-	351	-	-	-	-	-	287	-	654
Chemical and Petrochemical	-	273	1645	145	-	-	-	169	487	-	2719
of which: Feedstocks	-	*273*	*294*	*145*	-	-	-	-	-	-	*712*
Non-Ferrous Metals	-	-	-	-	-	-	-	-	-	-	-
Non-Metallic Minerals	3313	-	826	11	-	-	-	189	496	-	4835
Transport Equipment	-	-	-	-	-	-	-	-	-	-	-
Machinery	-	-	181	-	-	-	-	-	501	-	682
Mining and Quarrying	-	-	45	-	-	-	-	-	-	-	45
Food and Tobacco	215	-	749	-	-	-	-	4066	450	-	5480
Paper, Pulp and Printing	-	-	201	-	-	-	-	-	104	-	305
Wood and Wood Products	-	-	34	-	-	-	-	10	57	-	101
Construction	-	-	318	-	-	-	-	-	-	-	318
Textile and Leather	-	-	577	-	-	-	-	-	497	-	1074
Non-specified	848	-	1128	684	-	-	-	-	101	-	2761
TRANSPORT SECTOR	**-**	**-**	**19891**	**5**	**-**	**-**	**-**	**-**	**-**	**-**	**19896**
Air	-	-	2654	-	-	-	-	-	-	-	2654
Road	-	-	16973	5	-	-	-	-	-	-	16977
Rail	-	-	137	-	-	-	-	-	-	-	137
Pipeline Transport	-	-	-	-	-	-	-	-	-	-	-
Internal Navigation	-	-	128	-	-	-	-	-	-	-	128
Non-specified	-	-	-	-	-	-	-	-	-	-	-
OTHER SECTORS	**-**	**-**	**3266**	**-**	**-**	**-**	**-**	**8022**	**3673**	**-**	**14961**
Agriculture	-	-	1830	-	-	-	-	-	11	-	1841
Comm. and Publ. Services	-	-	-	-	-	-	-	-	2217	-	2217
Residential	-	-	1436	-	-	-	-	8022	1380	-	10837
Non-specified	-	-	-	-	-	-	-	-	65	-	65
NON-ENERGY USE	**2**	**-**	**548**	**-**	**-**	**-**	**-**	**-**	**-**	**-**	**550**
in Industry/Transf./Energy	2	-	548	-	-	-	-	-	-	-	550
in Transport	-	-	-	-	-	-	-	-	-	-	-
in Other Sectors	-	-	-	-	-	-	-	-	-	-	-
Electr. Generated - GWh	*17507*	*-*	*25604*	*36749*	*-*	*7340*	*2*	*265*	*-*	*-*	*87467*
Electricity Plants	*17507*	*-*	*25604*	*36749*	*-*	*7340*	*2*	*265*	*-*	*-*	*87467*
CHP Plants	*-*	*-*	*-*	*-*	*-*	*-*	*-*	*-*	*-*	*-*	*-*
Heat Generated - TJ	*-*	*-*	*-*	*-*	*-*	*-*	*-*	*-*	*-*	*-*	*-*
CHP Plants	*-*	*-*	*-*	*-*	*-*	*-*	*-*	*-*	*-*	*-*	*-*
Heat Plants	*-*	*-*	*-*	*-*	*-*	*-*	*-*	*-*	*-*	*-*	*-*

Thousand tonnes of oil equivalent / *Milliers de tonnes d'équivalent pétrole*

Thailand / Thailande : 1997

	Coal	Crude Oil	Petroleum Products	Gas	Nuclear	Hydro	Geotherm. Solar etc.	Combust. Renew. & Waste	Electricity	Heat	Total
SUPPLY AND CONSUMPTION											
APPROVISIONNEMENT ET DEMANDE	*Charbon*	*Pétrole brut*	*Produits pétroliers*	*Gaz*	*Nucléaire*	*Hydro*	*Géotherm. solaire*	*En. ren. combust.*	*Electricité*	*Chaleur*	*Total*
Indigenous Production	6785	5707	-	12406	-	619	-	20649	-	-	46166
Imports	2070	35560	2842	-	-	-	-	10	64	-	40546
Exports	-	-1156	-4662	-	-	-	-	-3	-9	-	-5830
Intl. Marine Bunkers	-	-	-784	-	-	-	-	-	-	-	-784
Stock Changes	-223	-145	234	-	-	-	-	-	-	-	-134
TPES	**8631**	**39966**	**-2371**	**12406**	**-**	**619**	**-**	**20656**	**55**	**-**	**79963**
Transfers	-	-2120	2141	-	-	-	-	-	-	-	21
Statistical Differences	71	-	-90	-	-	-	-	33	-	-	14
Electricity Plants	-4716	-	-4891	-9746	-	-619	-	-383	8020	-	-12335
CHP Plants	-	-	-	-	-	-	-	-	-	-	-
Heat Plants	-	-	-	-	-	-	-	-	-	-	-
Gas Works	-	-	-	-	-	-	-	-	-	-	-
Petroleum Refineries	-	-37382	35836	-	-	-	-	-	-	-	-1546
Coal Transformation	-43	-	-	-	-	-	-	-	-	-	-43
Liquefaction	-	-	-	-	-	-	-	-	-	-	-
Other Transformation	-	-	-	-	-	-	-	-8159	-	-	-8159
Own Use	-	-	-422	-1806	-	-	-	-	-291	-	-2518
Distribution Losses	-	-84	-96	-	-	-	-	-	-696	-	-875
TFC	**3943**	**379**	**30108**	**855**	**-**	**-**	**-**	**12146**	**7089**	**-**	**54521**
INDUSTRY SECTOR	**3943**	**379**	**5260**	**850**	**-**	**-**	**-**	**4663**	**2971**	**-**	**18066**
Iron and Steel	11	-	337	-	-	-	-	-	293	-	641
Chemical and Petrochemical	-	379	1264	129	-	-	-	134	467	-	2374
of which: Feedstocks	-	*379*	*160*	*129*	-	-	-	-	-	-	*669*
Non-Ferrous Metals	-	-	-	-	-	-	-	-	-	-	-
Non-Metallic Minerals	2925	-	712	7	-	-	-	231	489	-	4364
Transport Equipment	-	-	-	-	-	-	-	-	-	-	-
Machinery	-	-	141	-	-	-	-	-	493	-	635
Mining and Quarrying	-	-	45	-	-	-	-	-	-	-	45
Food and Tobacco	191	-	724	-	-	-	-	4285	468	-	5668
Paper, Pulp and Printing	-	-	199	-	-	-	-	-	104	-	303
Wood and Wood Products	-	-	29	-	-	-	-	13	54	-	96
Construction	-	-	372	-	-	-	-	-	-	-	372
Textile and Leather	-	-	529	-	-	-	-	-	500	-	1028
Non-specified	816	-	907	714	-	-	-	-	102	-	2539
TRANSPORT SECTOR	**-**	**-**	**21194**	**5**	**-**	**-**	**-**	**-**	**-**	**-**	**21199**
Air	-	-	2840	-	-	-	-	-	-	-	2840
Road	-	-	18131	5	-	-	-	-	-	-	18136
Rail	-	-	118	-	-	-	-	-	-	-	118
Pipeline Transport	-	-	-	-	-	-	-	-	-	-	-
Internal Navigation	-	-	105	-	-	-	-	-	-	-	105
Non-specified	-	-	-	-	-	-	-	-	-	-	-
OTHER SECTORS	**-**	**-**	**2965**	**-**	**-**	**-**	**-**	**7483**	**4118**	**-**	**14566**
Agriculture	-	-	1527	-	-	-	-	-	14	-	1541
Comm. and Publ. Services	-	-	-	-	-	-	-	-	2512	-	2512
Residential	-	-	1438	-	-	-	-	7483	1519	-	10440
Non-specified	-	-	-	-	-	-	-	-	73	-	73
NON-ENERGY USE	**1**	**-**	**690**	**-**	**-**	**-**	**-**	**-**	**-**	**-**	**691**
in Industry/Transf./Energy	1	-	690	-	-	-	-	-	-	-	691
in Transport	-	-	-	-	-	-	-	-	-	-	-
in Other Sectors	-	-	-	-	-	-	-	-	-	-	-
Electr. Generated - GWh	*18925*	*-*	*21790*	*43179*	*-*	*7200*	*1*	*2158*	*-*	*-*	*93253*
Electricity Plants	*18925*	*-*	*21790*	*43179*	*-*	*7200*	*1*	*2158*	*-*	*-*	*93253*
CHP Plants	*-*	*-*	*-*	*-*	*-*	*-*	*-*	*-*	*-*	*-*	*-*
Heat Generated - TJ	*-*	*-*	*-*	*-*	*-*	*-*	*-*	*-*	*-*	*-*	*-*
CHP Plants	*-*	*-*	*-*	*-*	*-*	*-*	*-*	*-*	*-*	*-*	*-*
Heat Plants	*-*	*-*	*-*	*-*	*-*	*-*	*-*	*-*	*-*	*-*	*-*

Trinidad-&-Tobago / Trinité-et-Tobago

Thousand tonnes of oil equivalent / *Milliers de tonnes d'équivalent pétrole*

SUPPLY AND CONSUMPTION 1996	Coal	Crude Oil	Petroleum Products	Gas	Nuclear	Hydro	Geotherm. Solar etc.	Combust. Renew. & Waste	Electricity	Heat	Total
Indigenous Production	-	7133	-	6413	-	-	-	29	-	-	13575
Imports	-	2172	250	-	-	-	-	-	-	-	2423
Exports	-	-3480	-3880	-	-	-	-	-	-	-	-7361
Intl. Marine Bunkers	-	-	-51	-	-	-	-	-	-	-	-51
Stock changes	-	-109	-590	-	-	-	-	-	-	-	-699
TPES	-	5716	-4271	6413	-	-	-	29	-	-	7887
Electricity and CHP Plants	-	-	-	-1467	-	-	-	-29	391	-	-1106
Petroleum Refineries	-	-5715	5188	-	-	-	-	-	-	-	-527
Other Transformation *	-	-1	-216	-840	-	-	-	-	-51	-	-1109
TFC	-	-	700	4106	-	-	-	-	339	-	5145
INDUSTRY SECTOR	-	-	89	4106	-	-	-	-	221	-	4415
Iron and Steel	-	-	-	384	-	-	-	-	-	-	384
Chemical and Petrochemical	-	-	-	2910	-	-	-	-	-	-	2910
Non-Metallic Minerals	-	-	-	127	-	-	-	-	-	-	127
Non-specified	-	-	89	684	-	-	-	-	221	-	994
TRANSPORT SECTOR	-	-	532	-	-	-	-	-	-	-	532
Air	-	-	65	-	-	-	-	-	-	-	65
Road	-	-	466	-	-	-	-	-	-	-	466
Non-specified	-	-	1	-	-	-	-	-	-	-	1
OTHER SECTORS	-	-	59	-	-	-	-	-	119	-	177
Agriculture	-	-	-	-	-	-	-	-	-	-	-
Comm. and Publ. Services	-	-	29	-	-	-	-	-	31	-	60
Residential	-	-	30	-	-	-	-	-	88	-	117
Non-specified	-	-	-	-	-	-	-	-	-	-	-
NON-ENERGY USE	-	-	20	-	-	-	-	-	-	-	20
Electricity Generated - GWh	-	-	-	4508	-	-	-	33	-	-	4541
Heat Generated - TJ	-	-	-	-	-	-	-	-	-	-	-

APPROVISIONNEMENT ET DEMANDE 1997	Charbon	Pétrole brut	Produits pétroliers	Gaz	Nucléaire	Hydro	Géotherm. solaire etc.	En. ren. combust. & déchets	Electricité	Chaleur	Total
Indigenous Production	-	6719	-	6828	-	-	-	32	-	-	13579
Imports	-	2057	82	-	-	-	-	-	-	-	2140
Exports	-	-3177	-3827	-	-	-	-	-	-	-	-7003
Intl. Marine Bunkers	-	-	-51	-	-	-	-	-	-	-	-51
Stock changes	-	-	-468	-	-	-	-	-	-	-	-468
TPES	-	5600	-4263	6828	-	-	-	32	-	-	8196
Electricity and CHP Plants	-	-	-	-1612	-	-	-	-32	429	-	-1215
Petroleum Refineries	-	-5600	5357	-	-	-	-	-	-	-	-243
Other Transformation *	-	-	-305	-804	-	-	-	-	-59	-	-1168
TFC	-	-	788	4412	-	-	-	-	370	-	5570
INDUSTRY SECTOR	-	-	107	4412	-	-	-	-	248	-	4767
Iron and Steel	-	-	-	413	-	-	-	-	-	-	413
Chemical and Petrochemical	-	-	-	3127	-	-	-	-	-	-	3127
Non-Metallic Minerals	-	-	-	136	-	-	-	-	-	-	136
Non-specified	-	-	107	735	-	-	-	-	248	-	1091
TRANSPORT SECTOR	-	-	605	-	-	-	-	-	-	-	605
Air	-	-	71	-	-	-	-	-	-	-	71
Road	-	-	532	-	-	-	-	-	-	-	532
Non-specified	-	-	1	-	-	-	-	-	-	-	1
OTHER SECTORS	-	-	60	-	-	-	-	-	122	-	182
Agriculture	-	-	-	-	-	-	-	-	-	-	-
Comm. and Publ. Services	-	-	28	-	-	-	-	-	36	-	64
Residential	-	-	32	-	-	-	-	-	87	-	118
Non-specified	-	-	-	-	-	-	-	-	-	-	-
NON-ENERGY USE	-	-	16	-	-	-	-	-	-	-	16
Electricity Generated - GWh	-	-	-	4952	-	-	-	36	-	-	4988
Heat Generated - TJ	-	-	-	-	-	-	-	-	-	-	-

* Includes Transfers, Statistical Differences, Own Use and Distribution Losses.

Tunisia / Tunisie

Total Production of Energy

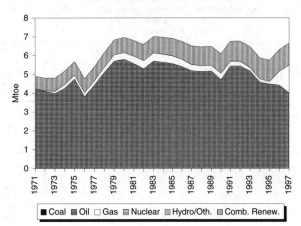

■ Coal ▦ Oil □ Gas ▤ Nuclear ▨ Hydro/Oth. ▥ Comb. Renew.

Total Primary Energy Supply*

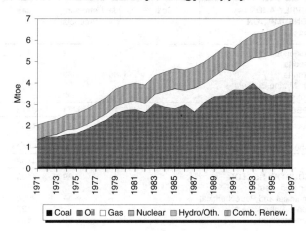

■ Coal ▦ Oil □ Gas ▤ Nuclear ▨ Hydro/Oth. ▥ Comb. Renew.

Electricity Generation by Fuel

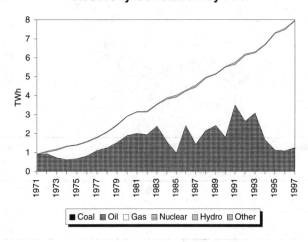

■ Coal ▦ Oil □ Gas ▤ Nuclear ▨ Hydro ▥ Other

Oil Products Consumption

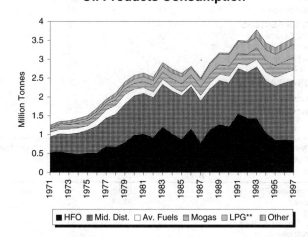

■ HFO ▦ Mid. Dist. □ Av. Fuels ▤ Mogas ▨ LPG** ▥ Other

Electricity Consumption/GDP***, TPES/GDP*** and Production/TPES

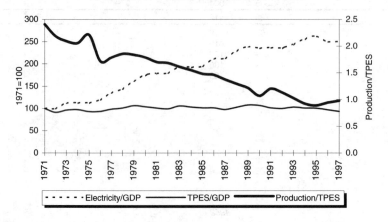

- - - Electricity/GDP   ——— TPES/GDP   ━━━ Production/TPES

* Excluding electricity trade.
** Includes LPG, NGL, ethane and naphtha.
*** GDP in 1990 US dollars.

Tunisia / Tunisie : 1996

Thousand tonnes of oil equivalent / *Milliers de tonnes d'équivalent pétrole*

SUPPLY AND CONSUMPTION / *APPROVISIONNEMENT ET DEMANDE*	Coal / *Charbon*	Crude Oil / *Pétrole brut*	Petroleum Products / *Produits pétroliers*	Gas / *Gaz*	Nuclear / *Nucléaire*	Hydro / *Hydro*	Geotherm. Solar etc. / *Géotherm. solaire*	Combust. Renew. & Waste / *En. ren. combust.*	Electricity / *Electricité*	Heat / *Chaleur*	Total / *Total*
Indigenous Production	-	4420	-	726	-	6	-	1155	-	-	6306
Imports	69	950	2344	1222	-	-	-	-	3	-	4588
Exports	-	-3397	-705	-	-	-	-	-	-	-	-4102
Intl. Marine Bunkers	-	-	-18	-	-	-	-	-	-	-	-18
Stock Changes	-	-14	-66	-	-	-	-	-	-	-	-81
TPES	**69**	**1958**	**1555**	**1948**	**-**	**6**	**-**	**1155**	**3**	**-**	**6693**
Transfers	-	-125	137	-	-	-	-	-	-	-	12
Statistical Differences	-	36	-8	57	-	-	-	-	12	-	98
Electricity Plants	-	-	-304	-1534	-	-6	-	-	648	-	-1196
CHP Plants	-	-	-	-	-	-	-	-	-	-	-
Heat Plants	-	-	-	-	-	-	-	-	-	-	-
Gas Works	-	-	-	-	-	-	-	-	-	-	-
Petroleum Refineries	-	-1869	1848	-	-	-	-	-	-	-	-22
Coal Transformation	-55	-	-	-	-	-	-	-	-	-	-55
Liquefaction	-	-	-	-	-	-	-	-	-	-	-
Other Transformation	-	-	-	-	-	-	-	-170	-	-	-170
Own Use	-	-	-72	-	-	-	-	-	-68	-	-139
Distribution Losses	-	-	-	-	-	-	-	-	-69	-	-69
TFC	**14**	**-**	**3156**	**471**	**-**	**-**	**-**	**985**	**526**	**-**	**5152**
INDUSTRY SECTOR	**14**	**-**	**602**	**361**	**-**	**-**	**-**	**-**	**241**	**-**	**1218**
Iron and Steel	14	-	-	-	-	-	-	-	15	-	29
Chemical and Petrochemical	-	-	-	-	-	-	-	-	12	-	12
of which: Feedstocks	-	-	-	-	-	-	-	-	-	-	-
Non-Ferrous Metals	-	-	-	-	-	-	-	-	-	-	-
Non-Metallic Minerals	-	-	-	-	-	-	-	-	76	-	76
Transport Equipment	-	-	-	-	-	-	-	-	-	-	-
Machinery	-	-	-	-	-	-	-	-	-	-	-
Mining and Quarrying	-	-	-	-	-	-	-	-	22	-	22
Food and Tobacco	-	-	-	-	-	-	-	-	26	-	26
Paper, Pulp and Printing	-	-	-	-	-	-	-	-	9	-	9
Wood and Wood Products	-	-	-	-	-	-	-	-	-	-	-
Construction	-	-	-	-	-	-	-	-	-	-	-
Textile and Leather	-	-	-	-	-	-	-	-	28	-	28
Non-specified	-	-	602	361	-	-	-	-	53	-	1016
TRANSPORT SECTOR	**-**	**-**	**1370**	**-**	**-**	**-**	**-**	**-**	**13**	**-**	**1383**
Air	-	-	265	-	-	-	-	-	-	-	265
Road	-	-	1088	-	-	-	-	-	-	-	1088
Rail	-	-	-	-	-	-	-	-	-	-	-
Pipeline Transport	-	-	-	-	-	-	-	-	-	-	-
Internal Navigation	-	-	-	-	-	-	-	-	-	-	-
Non-specified	-	-	17	-	-	-	-	-	13	-	30
OTHER SECTORS	**-**	**-**	**1039**	**110**	**-**	**-**	**-**	**985**	**273**	**-**	**2406**
Agriculture	-	-	304	-	-	-	-	-	32	-	336
Comm. and Publ. Services	-	-	254	32	-	-	-	-	97	-	383
Residential	-	-	481	78	-	-	-	985	144	-	1687
Non-specified	-	-	-	-	-	-	-	-	-	-	-
NON-ENERGY USE	**-**	**-**	**145**	**-**	**-**	**-**	**-**	**-**	**-**	**-**	**145**
in Industry/Transf./Energy	-	-	145	-	-	-	-	-	-	-	145
in Transport	-	-	-	-	-	-	-	-	-	-	-
in Other Sectors	-	-	-	-	-	-	-	-	-	-	-
Electr. Generated - GWh	**-**	**-**	***1066***	***6400***	**-**	***67***	**-**	**-**	**-**	**-**	***7533***
Electricity Plants	-	-	*1066*	*6400*	-	*67*	-	-	-	-	*7533*
CHP Plants	-	-	-	-	-	-	-	-	-	-	-
Heat Generated - TJ	**-**	**-**	**-**	**-**	**-**	**-**	**-**	**-**	**-**	**-**	**-**
CHP Plants	-	-	-	-	-	-	-	-	-	-	-
Heat Plants	-	-	-	-	-	-	-	-	-	-	-

Tunisia / Tunisie : 1997

	\multicolumn{11}{c	}{Thousand tonnes of oil equivalent / *Milliers de tonnes d'équivalent pétrole*}

SUPPLY AND CONSUMPTION	Coal	Crude Oil	Petroleum Products	Gas	Nuclear	Hydro	Geotherm. Solar etc.	Combust. Renew. & Waste	Electricity	Heat	Total
APPROVISIONNEMENT ET DEMANDE	*Charbon*	*Pétrole brut*	*Produits pétroliers*	*Gaz*	*Nucléaire*	*Hydro*	*Géotherm. solaire*	*En. ren. combust.*	*Electricité*	*Chaleur*	*Total*
Indigenous Production	-	4013	-	1466	-	4	-	1172	-	-	6655
Imports	69	942	2601	621	-	-	-	-	3	-	4236
Exports	-	-2863	-821	-	-	-	-	-	-	-	-3685
Intl. Marine Bunkers	-	-	-17	-	-	-	-	-	-	-	-17
Stock Changes	-	-119	-264	-	-	-	-	-	-	-	-384
TPES	**69**	**1972**	**1498**	**2087**	**-**	**4**	**-**	**1172**	**3**	**-**	**6805**
Transfers	-	-110	121	-	-	-	-	-	-	-	11
Statistical Differences	-	207	-1	-3	-	-	-	-	26	-	230
Electricity Plants	-	-	-357	-1601	-	-4	-	-	686	-	-1276
CHP Plants	-	-	-	-	-	-	-	-	-	-	-
Heat Plants	-	-	-	-	-	-	-	-	-	-	-
Gas Works	-	-	-	-	-	-	-	-	-	-	-
Petroleum Refineries	-	-2069	2055	-	-	-	-	-	-	-	-14
Coal Transformation	-55	-	-	-	-	-	-	-	-	-	-55
Liquefaction	-	-	-	-	-	-	-	-	-	-	-
Other Transformation	-	-	-	-	-	-	-	-173	-	-	-173
Own Use	-	-	-72	-	-	-	-	-	-78	-	-149
Distribution Losses	-	-	-	-	-	-	-	-	-76	-	-76
TFC	**14**	**-**	**3244**	**483**	**-**	**-**	**-**	**1000**	**562**	**-**	**5302**
INDUSTRY SECTOR	**14**	**-**	**574**	**368**	**-**	**-**	**-**	**-**	**250**	**-**	**1205**
Iron and Steel	14	-	-	-	-	-	-	-	16	-	29
Chemical and Petrochemical	-	-	-	-	-	-	-	-	12	-	12
of which: Feedstocks	-	-	-	-	-	-	-	-	-	-	-
Non-Ferrous Metals	-	-	-	-	-	-	-	-	-	-	-
Non-Metallic Minerals	-	-	-	-	-	-	-	-	79	-	79
Transport Equipment	-	-	-	-	-	-	-	-	-	-	-
Machinery	-	-	-	-	-	-	-	-	-	-	-
Mining and Quarrying	-	-	-	-	-	-	-	-	22	-	22
Food and Tobacco	-	-	-	-	-	-	-	-	27	-	27
Paper, Pulp and Printing	-	-	-	-	-	-	-	-	9	-	9
Wood and Wood Products	-	-	-	-	-	-	-	-	-	-	-
Construction	-	-	-	-	-	-	-	-	-	-	-
Textile and Leather	-	-	-	-	-	-	-	-	29	-	29
Non-specified	-	-	574	368	-	-	-	-	55	-	997
TRANSPORT SECTOR	**-**	**-**	**1471**	**-**	**-**	**-**	**-**	**-**	**13**	**-**	**1483**
Air	-	-	301	-	-	-	-	-	-	-	301
Road	-	-	1163	-	-	-	-	-	-	-	1163
Rail	-	-	-	-	-	-	-	-	-	-	-
Pipeline Transport	-	-	-	-	-	-	-	-	-	-	-
Internal Navigation	-	-	-	-	-	-	-	-	-	-	-
Non-specified	-	-	6	-	-	-	-	-	13	-	18
OTHER SECTORS	**-**	**-**	**1063**	**115**	**-**	**-**	**-**	**1000**	**299**	**-**	**2477**
Agriculture	-	-	287	-	-	-	-	-	41	-	328
Comm. and Publ. Services	-	-	268	34	-	-	-	-	104	-	406
Residential	-	-	508	81	-	-	-	1000	154	-	1743
Non-specified	-	-	-	-	-	-	-	-	-	-	-
NON-ENERGY USE	**-**	**-**	**137**	**-**	**-**	**-**	**-**	**-**	**-**	**-**	**137**
in Industry/Transf./Energy	-	-	137	-	-	-	-	-	-	-	137
in Transport	-	-	-	-	-	-	-	-	-	-	-
in Other Sectors	-	-	-	-	-	-	-	-	-	-	-
Electr. Generated - GWh	**-**	**-**	***1253***	***6680***	**-**	***44***	**-**	**-**	**-**	**-**	***7977***
Electricity Plants	-	-	*1253*	*6680*	-	*44*	-	-	-	-	*7977*
CHP Plants	-	-	-	-	-	-	-	-	-	-	-
Heat Generated - TJ	**-**	**-**	**-**	**-**	**-**	**-**	**-**	**-**	**-**	**-**	**-**
CHP Plants	-	-	-	-	-	-	-	-	-	-	-
Heat Plants	-	-	-	-	-	-	-	-	-	-	-

Turkmenistan / Turkménistan

Thousand tonnes of oil equivalent / *Milliers de tonnes d'équivalent pétrole*

SUPPLY AND CONSUMPTION 1996	Coal	Crude Oil	Petroleum Products	Gas	Nuclear	Hydro	Geotherm. Solar etc.	Combust. Renew. & Waste	Electricity	Heat	Total
Indigenous Production	-	4352	-	28504	-	-	-	-	-	-	32856
Imports	44	101	81	-	-	-	-	-	82	-	308
Exports	-	-201	-1014	-19444	-	-	-	-	-322	-	-20981
Intl. Marine Bunkers	-	-	-	-	-	-	-	-	-	-	-
Stock changes	-	-	-	-	-	-	-	-	-	-	-
TPES	44	4252	-932	9059	-	-	-	-	-241	-	12183
Electricity and CHP Plants	-	-	-	-3156	-	-	-	-	869	-	-2288
Petroleum Refineries	-	-3945	3492	-	-	-	-	-	-	-	-452
Other Transformation *	-	-307	-222	-624	-	-	-	-	-224	-	-1378
TFC	44	-	2338	5279	-	-	-	-	403	-	8064
INDUSTRY SECTOR	-	-	-	-	-	-	-	-	146	-	146
Iron and Steel	-	-	-	-	-	-	-	-	-	-	-
Chemical and Petrochemical	-	-	-	-	-	-	-	-	47	-	47
Non-Metallic Minerals	-	-	-	-	-	-	-	-	-	-	-
Non-specified	-	-	-	-	-	-	-	-	98	-	98
TRANSPORT SECTOR	-	-	517	-	-	-	-	-	10	-	527
Air	-	-	-	-	-	-	-	-	-	-	-
Road	-	-	517	-	-	-	-	-	-	-	517
Non-specified	-	-	-	-	-	-	-	-	10	-	10
OTHER SECTORS	44	-	1821	5279	-	-	-	-	247	-	7391
Agriculture	-	-	-	-	-	-	-	-	128	-	128
Comm. and Publ. Services	-	-	-	-	-	-	-	-	-	-	-
Residential	-	-	-	-	-	-	-	-	85	-	85
Non-specified	44	-	1821	5279	-	-	-	-	34	-	7179
NON-ENERGY USE	-	-	-	-	-	-	-	-	-	-	-
Electricity Generated - GWh	-	-	-	*10095*	-	*5*	-	-	-	-	*10100*
Heat Generated - TJ	-	-	-	-	-	-	-	-	-	-	-

APPROVISIONNEMENT ET DEMANDE 1997	Charbon	Pétrole brut	Produits pétroliers	Gaz	Nucléaire	Hydro	Géotherm. solaire etc.	En. ren. combust. & déchets	Electricité	Chaleur	Total
Indigenous Production	-	4723	-	14016	-	-	-	-	-	-	18739
Imports	-	704	60	-	-	-	-	-	82	-	845
Exports	-	-201	-1626	-5266	-	-	-	-	-310	-	-7403
Intl. Marine Bunkers	-	-	-	-	-	-	-	-	-	-	-
Stock changes	-	-	-	-	-	-	-	-	-	-	-
TPES	-	5225	-1567	8750	-	-	-	-	-228	-	12181
Electricity and CHP Plants	-	-	-	-3046	-	-	-	-	808	-	-2238
Petroleum Refineries	-	-5006	4216	-	-	-	-	-	-	-	-790
Other Transformation *	-	-219	-287	-665	-	-	-	-	-206	-	-1378
TFC	-	-	2362	5039	-	-	-	-	374	-	7775
INDUSTRY SECTOR	-	-	-	-	-	-	-	-	135	-	135
Iron and Steel	-	-	-	-	-	-	-	-	-	-	-
Chemical and Petrochemical	-	-	-	-	-	-	-	-	44	-	44
Non-Metallic Minerals	-	-	-	-	-	-	-	-	-	-	-
Non-specified	-	-	-	-	-	-	-	-	91	-	91
TRANSPORT SECTOR	-	-	474	-	-	-	-	-	9	-	483
Air	-	-	-	-	-	-	-	-	-	-	-
Road	-	-	474	-	-	-	-	-	-	-	474
Non-specified	-	-	-	-	-	-	-	-	9	-	9
OTHER SECTORS	-	-	1888	5039	-	-	-	-	230	-	7156
Agriculture	-	-	-	-	-	-	-	-	119	-	119
Comm. and Publ. Services	-	-	-	-	-	-	-	-	-	-	-
Residential	-	-	-	-	-	-	-	-	79	-	79
Non-specified	-	-	1888	5039	-	-	-	-	32	-	6959
NON-ENERGY USE	-	-	-	-	-	-	-	-	-	-	-
Electricity Generated - GWh	-	-	-	*9395*	-	*5*	-	-	-	-	*9400*
Heat Generated - TJ	-	-	-	-	-	-	-	-	-	-	-

* Includes Transfers, Statistical Differences, Own Use and Distribution Losses.

INTERNATIONAL ENERGY AGENCY

Ukraine : 1996

Thousand tonnes of oil equivalent / Milliers de tonnes d'équivalent pétrole										
SUPPLY AND CONSUMPTION	Coal	Crude Oil	Petroleum Products	Gas	Nuclear	Hydro	Geotherm. Solar etc.	Combust. Renew. & Waste	Electricity	Heat	Total
APPROVISIONNEMENT ET DEMANDE	*Charbon*	*Pétrole brut*	*Produits pétroliers*	*Gaz*	*Nucléaire*	*Hydro*	*Géotherm. solaire*	*En. ren. combust.*	*Electricité*	*Chaleur*	*Total*
Indigenous Production	38983	4503	-	15436	20738	742	-	252	-	-	80655
Imports	7650	9361	8897	59704	-	-	-	-	358	-	85970
Exports	-1412	-	-2532	-668	-	-	-	-	-532	-	-5144
Intl. Marine Bunkers	-	-	-	-	-	-	-	-	-	-	-
Stock Changes	-	-	-	-	-	-	-	-	-	-	-
TPES	**45222**	**13864**	**6365**	**74472**	**20738**	**742**	**-**	**252**	**-173**	**-**	**161481**
Transfers	-	-	-	-	-	-	-	-	-	-	-
Statistical Differences	-1020	-	-	-	-	-	-	-	-	-	-1020
Electricity Plants	-328	-	-2118	-9414	-20738	-742	-	-	7586	-	-25755
CHP Plants	-14004	-	-	-	-	-	-	-	8134	3714	-2156
Heat Plants	-1589	-	-867	-15902	-	-	-	-	-	11236	-7122
Gas Works	-	-	-	-	-	-	-	-	-	-	-
Petroleum Refineries	-	-13864	13270	-69	-	-	-	-	-	-	-663
Coal Transformation	-16433	-	-	-	-	-	-	-	-	-	-16433
Liquefaction	-	-	-	-	-	-	-	-	-	-	-
Other Transformation	-	-	-	-	-	-	-	-	-	-	-
Own Use	-107	-	-428	-1375	-	-	-	-	-2286	-119	-4315
Distribution Losses	-	-	-	-1375	-	-	-	-	-2150	-4478	-8003
TFC	**11741**	**-**	**16222**	**46336**	**-**	**-**	**-**	**252**	**11111**	**10352**	**96015**
INDUSTRY SECTOR	**6468**	**-**	**4108**	**17463**	**-**	**-**	**-**	**-**	**5666**	**6685**	**40391**
Iron and Steel	1825	-	-	-	-	-	-	-	1735	-	3560
Chemical and Petrochemical	-	-	-	-	-	-	-	-	724	-	724
of which: Feedstocks	-	-	-	-	-	-	-	-	-	-	-
Non-Ferrous Metals	-	-	-	-	-	-	-	-	293	-	293
Non-Metallic Minerals	-	-	-	-	-	-	-	-	248	-	248
Transport Equipment	-	-	-	-	-	-	-	-	83	-	83
Machinery	-	-	-	-	-	-	-	-	569	-	569
Mining and Quarrying	-	-	-	-	-	-	-	-	688	-	688
Food and Tobacco	-	-	-	-	-	-	-	-	394	-	394
Paper, Pulp and Printing	-	-	-	-	-	-	-	-	49	-	49
Wood and Wood Products	-	-	-	-	-	-	-	-	28	-	28
Construction	-	-	-	-	-	-	-	-	129	-	129
Textile and Leather	-	-	-	-	-	-	-	-	55	-	55
Non-specified	4643	-	4108	17463	-	-	-	-	671	6685	33571
TRANSPORT SECTOR	**-**	**-**	**6525**	**-**	**-**	**-**	**-**	**-**	**839**	**-**	**7364**
Air	-	-	836	-	-	-	-	-	-	-	836
Road	-	-	3836	-	-	-	-	-	-	-	3836
Rail	-	-	1854	-	-	-	-	-	404	-	2258
Pipeline Transport	-	-	-	-	-	-	-	-	143	-	143
Internal Navigation	-	-	-	-	-	-	-	-	-	-	-
Non-specified	-	-	-	-	-	-	-	-	292	-	292
OTHER SECTORS	**5273**	**-**	**4246**	**28873**	**-**	**-**	**-**	**252**	**4606**	**3667**	**46916**
Agriculture	-	-	2323	361	-	-	-	-	945	421	4050
Comm. and Publ. Services	-	-	-	9510	-	-	-	-	831	-	10341
Residential	5273	-	1923	19001	-	-	-	-	-	3247	29444
Non-specified	-	-	-	-	-	-	-	252	2829	-	3082
NON-ENERGY USE	**-**	**-**	**1344**	**-**	**-**	**-**	**-**	**-**	**-**	**-**	**1344**
in Industry/Transf./Energy	-	-	1344	-	-	-	-	-	-	-	1344
in Transport	-	-	-	-	-	-	-	-	-	-	-
in Other Sectors	-	-	-	-	-	-	-	-	-	-	-
Electr. Generated - GWh	***52501***	***-***	***8097***	***33978***	***79577***	***8632***	***-***	***-***	***-***	***-***	***182785***
Electricity Plants	-	-	-	-	79577	8632	-	-	-	-	88209
CHP Plants	52501	-	8097	33978	-	-	-	-	-	-	94576
Heat Generated - TJ	***-***	***-***	***-***	***-***	***-***	***-***	***-***	***-***	***-***	***-***	***626043***
CHP Plants	-	-	-	-	-	-	-	-	-	-	155523
Heat Plants	-	-	-	-	-	-	-	-	-	-	470520

Ukraine : 1997

	Thousand tonnes of oil equivalent / Milliers de tonnes d'équivalent pétrole										
SUPPLY AND CONSUMPTION *APPROVISIONNEMENT ET DEMANDE*	Coal *Charbon*	Crude Oil *Pétrole brut*	Petroleum Products *Produits pétroliers*	Gas *Gaz*	Nuclear *Nucléaire*	Hydro *Hydro*	Geotherm. Solar etc. *Géotherm. solaire*	Combust. Renew. & Waste *En. ren. combust.*	Electricity *Electricité*	Heat *Chaleur*	Total *Total*
Indigenous Production	39590	4588	-	15204	20701	848	-	246	-	-	81175
Imports	5383	9007	6188	52288	-	-	-	-	836	-	73701
Exports	-1440	-5	-1347	-1177	-	-	-	-	-849	-	-4818
Intl. Marine Bunkers	-	-	-	-	-	-	-	-	-	-	-
Stock Changes	-	-	-	-	-	-	-	-	-	-	-
TPES	**43532**	**13590**	**4842**	**66314**	**20701**	**848**	**-**	**246**	**-13**	**-**	**150059**
Transfers	-	-	-	-	-	-	-	-	-	-	-
Statistical Differences	-790	-	-	-	-	-	-	-	-	-	-790
Electricity Plants	-320	-	-1908	-7891	-20701	-848	-	-	7679	-	-23989
CHP Plants	-13426	-	-	-	-	-	-	-	7614	3613	-2199
Heat Plants	-1524	-	-785	-13329	-	-	-	-	-	10930	-4708
Gas Works	-	-	-	-	-	-	-	-	-	-	-
Petroleum Refineries	-	-13590	13051	-74	-	-	-	-	-	-	-612
Coal Transformation	-16514	-	-	-	-	-	-	-	-	-	-16514
Liquefaction	-	-	-	-	-	-	-	-	-	-	-
Other Transformation	-	-	-	-	-	-	-	-	-	-	-
Own Use	-102	-	-403	-1460	-	-	-	-	-2160	-119	-4244
Distribution Losses	-	-	-	-1461	-	-	-	-	-2443	-4353	-8257
TFC	**10857**	**-**	**14797**	**42099**	**-**	**-**	**-**	**246**	**10677**	**10071**	**88746**
INDUSTRY SECTOR	**6230**	**-**	**3761**	**15516**	**-**	**-**	**-**	**-**	**5575**	**6503**	**37586**
Iron and Steel	1778	-	-	-	-	-	-	-	1922	-	3700
Chemical and Petrochemical	-	-	-	-	-	-	-	-	651	-	651
of which: Feedstocks	-	-	-	-	-	-	-	-	-	-	-
Non-Ferrous Metals	-	-	-	-	-	-	-	-	292	-	292
Non-Metallic Minerals	-	-	-	-	-	-	-	-	234	-	234
Transport Equipment	-	-	-	-	-	-	-	-	76	-	76
Machinery	-	-	-	-	-	-	-	-	520	-	520
Mining and Quarrying	-	-	-	-	-	-	-	-	750	-	750
Food and Tobacco	-	-	-	-	-	-	-	-	335	-	335
Paper, Pulp and Printing	-	-	-	-	-	-	-	-	46	-	46
Wood and Wood Products	-	-	-	-	-	-	-	-	27	-	27
Construction	-	-	-	-	-	-	-	-	125	-	125
Textile and Leather	-	-	-	-	-	-	-	-	49	-	49
Non-specified	4452	-	3761	15516	-	-	-	-	548	6503	30781
TRANSPORT SECTOR	**-**	**-**	**5900**	**-**	**-**	**-**	**-**	**-**	**821**	**-**	**6721**
Air	-	-	718	-	-	-	-	-	-	-	718
Road	-	-	3424	-	-	-	-	-	-	-	3424
Rail	-	-	1758	-	-	-	-	-	387	-	2145
Pipeline Transport	-	-	-	-	-	-	-	-	150	-	150
Internal Navigation	-	-	-	-	-	-	-	-	-	-	-
Non-specified	-	-	-	-	-	-	-	-	284	-	284
OTHER SECTORS	**4627**	**-**	**3993**	**26583**	**-**	**-**	**-**	**246**	**4281**	**3567**	**43298**
Agriculture	-	-	2204	322	-	-	-	-	790	409	3725
Comm. and Publ. Services	-	-	-	8754	-	-	-	-	798	-	9551
Residential	4627	-	1790	17507	-	-	-	-	-	3158	27082
Non-specified	-	-	-	-	-	-	-	246	2694	-	2939
NON-ENERGY USE	**-**	**-**	**1141**	**-**	**-**	**-**	**-**	**-**	**-**	**-**	**1141**
in Industry/Transf./Energy	-	-	1141	-	-	-	-	-	-	-	1141
in Transport	-	-	-	-	-	-	-	-	-	-	-
in Other Sectors	-	-	-	-	-	-	-	-	-	-	-
Electr. Generated - GWh	*49149*	*-*	*7580*	*31808*	*79433*	*9856*	*-*	*-*	*-*	*-*	*177826*
Electricity Plants	*-*	*-*	*-*	*-*	*79433*	*9856*	*-*	*-*	*-*	*-*	*89289*
CHP Plants	*49149*	*-*	*7580*	*31808*	*-*	*-*	*-*	*-*	*-*	*-*	*88537*
Heat Generated - TJ	*-*	*-*	*-*	*-*	*-*	*-*	*-*	*-*	*-*	*-*	*609015*
CHP Plants	*-*	*-*	*-*	*-*	*-*	*-*	*-*	*-*	*-*	*-*	*151293*
Heat Plants	*-*	*-*	*-*	*-*	*-*	*-*	*-*	*-*	*-*	*-*	*457722*

United Arab Emirates / Emirats arabes unis

Thousand tonnes of oil equivalent / *Milliers de tonnes d'équivalent pétrole*

SUPPLY AND CONSUMPTION 1996	Coal	Crude Oil	Petroleum Products	Gas	Nuclear	Hydro	Geotherm. Solar etc.	Combust. Renew. & Waste	Electricity	Heat	Total
Indigenous Production	-	120907	-	27911	-	-	-	-	-	-	148818
Imports	-	-	12358	200	-	-	-	17	-	-	12575
Exports	-	-102061	-11036	-5617	-	-	-	-	-	-	-118714
Intl. Marine Bunkers	-	-	-10343	-	-	-	-	-	-	-	-10343
Stock changes	-	-	-	-	-	-	-	-	-	-	-
TPES	-	18846	-9021	22494	-	-	-	17	-	-	32336
Electricity and CHP Plants	-	-	-1564	-4471	-	-	-	-	1697	-	-4337
Petroleum Refineries	-	-12145	10922	-	-	-	-	-	-	-	-1223
Other Transformation *	-	-6701	4211	-16273	-	-	-	-	-	-	-18763
TFC	-	-	4548	1750	-	-	-	17	1697	-	8013
INDUSTRY SECTOR	-	-	404	1750	-	-	-	-	-	-	2154
Iron and Steel	-	-	-	-	-	-	-	-	-	-	-
Chemical and Petrochemical	-	-	-	1750	-	-	-	-	-	-	1750
Non-Metallic Minerals	-	-	-	-	-	-	-	-	-	-	-
Non-specified	-	-	404	-	-	-	-	-	-	-	404
TRANSPORT SECTOR	-	-	3694	-	-	-	-	-	-	-	3694
Air	-	-	223	-	-	-	-	-	-	-	223
Road	-	-	3472	-	-	-	-	-	-	-	3472
Non-specified	-	-	-	-	-	-	-	-	-	-	-
OTHER SECTORS	-	-	449	-	-	-	-	17	1697	-	2164
Agriculture	-	-	-	-	-	-	-	-	-	-	-
Comm. and Publ. Services	-	-	-	-	-	-	-	-	-	-	-
Residential	-	-	449	-	-	-	-	-	-	-	449
Non-specified	-	-	-	-	-	-	-	17	1697	-	1715
NON-ENERGY USE	-	-	-	-	-	-	-	-	-	-	-
Electricity Generated - GWh	-	-	*3378*	*16359*	-	-	-	-	-	-	*19737*
Heat Generated - TJ	-	-	-	-	-	-	-	-	-	-	-

APPROVISIONNEMENT ET DEMANDE 1997	Charbon	Pétrole brut	Produits pétroliers	Gaz	Nucléaire	Hydro	Géotherm. solaire etc.	En. ren. combust. & déchets	Electricité	Chaleur	Total
Indigenous Production	-	123003	-	30552	-	-	-	-	-	-	153555
Imports	-	-	12358	200	-	-	-	17	-	-	12575
Exports	-	-104382	-13487	-7044	-	-	-	-	-	-	-124913
Intl. Marine Bunkers	-	-	-10343	-	-	-	-	-	-	-	-10343
Stock changes	-	-	-	-	-	-	-	-	-	-	-
TPES	-	18621	-11472	23708	-	-	-	17	-	-	30874
Electricity and CHP Plants	-	-	-1545	-4709	-	-	-	-	1769	-	-4484
Petroleum Refineries	-	-13478	12296	-	-	-	-	-	-	-	-1183
Other Transformation *	-	-5143	4818	-17156	-	-	-	-	-	-	-17481
TFC	-	-	4097	1843	-	-	-	17	1769	-	7726
INDUSTRY SECTOR	-	-	407	1843	-	-	-	-	-	-	2250
Iron and Steel	-	-	-	-	-	-	-	-	-	-	-
Chemical and Petrochemical	-	-	-	1843	-	-	-	-	-	-	1843
Non-Metallic Minerals	-	-	-	-	-	-	-	-	-	-	-
Non-specified	-	-	407	-	-	-	-	-	-	-	407
TRANSPORT SECTOR	-	-	3223	-	-	-	-	-	-	-	3223
Air	-	-	204	-	-	-	-	-	-	-	204
Road	-	-	3019	-	-	-	-	-	-	-	3019
Non-specified	-	-	-	-	-	-	-	-	-	-	-
OTHER SECTORS	-	-	465	-	-	-	-	17	1769	-	2252
Agriculture	-	-	-	-	-	-	-	-	-	-	-
Comm. and Publ. Services	-	-	-	-	-	-	-	-	-	-	-
Residential	-	-	465	-	-	-	-	-	-	-	465
Non-specified	-	-	-	-	-	-	-	17	1769	-	1787
NON-ENERGY USE	-	-	1	-	-	-	-	-	-	-	1
Electricity Generated - GWh	-	-	*3340*	*17231*	-	-	-	-	-	-	*20571*
Heat Generated - TJ	-	-	-	-	-	-	-	-	-	-	-

* Includes Transfers, Statistical Differences, Own Use and Distribution Losses.

Uruguay

Thousand tonnes of oil equivalent / *Milliers de tonnes d'équivalent pétrole*

SUPPLY AND CONSUMPTION 1996	Coal	Crude Oil	Petroleum Products	Gas	Nuclear	Hydro	Geotherm. Solar etc.	Combust. Renew. & Waste	Electricity	Heat	Total
Indigenous Production	-	-	-	-	-	496	-	525	-	-	1021
Imports	-	1527	678	-	-	-	-	1	26	-	2232
Exports	-	-	-65	-	-	-	-	-	-29	-	-95
Intl. Marine Bunkers	-	-	-378	-	-	-	-	-	-	-	-378
Stock changes	-	138	-88	-	-	-	-	-	-	-	50
TPES	-	1665	147	-	-	496	-	526	-3	-	2831
Electricity and CHP Plants	-	-	-205	-	-	-496	-	-12	574	-	-140
Petroleum Refineries	-	-1664	1610	-	-	-	-	-	-	-	-53
Other Transformation *	-	-1	-142	-	-	-	-	-1	-123	-	-268
TFC	-	-	1410	-	-	-	-	512	448	-	2369
INDUSTRY SECTOR	-	-	165	-	-	-	-	209	111	-	485
Iron and Steel	-	-	-	-	-	-	-	-	-	-	-
Chemical and Petrochemical	-	-	-	-	-	-	-	-	-	-	-
Non-Metallic Minerals	-	-	-	-	-	-	-	-	-	-	-
Non-specified	-	-	165	-	-	-	-	209	111	-	485
TRANSPORT SECTOR	-	-	765	-	-	-	-	-	-	-	765
Air	-	-	9	-	-	-	-	-	-	-	9
Road	-	-	757	-	-	-	-	-	-	-	757
Non-specified	-	-	-	-	-	-	-	-	-	-	-
OTHER SECTORS	-	-	402	-	-	-	-	303	337	-	1042
Agriculture	-	-	196	-	-	-	-	-	6	-	202
Comm. and Publ. Services	-	-	36	-	-	-	-	3	125	-	164
Residential	-	-	163	-	-	-	-	300	206	-	670
Non-specified	-	-	7	-	-	-	-	-	-	-	7
NON-ENERGY USE	-	-	77	-	-	-	-	-	-	-	77
Electricity Generated - GWh	-	-	*863*	-	-	*5768*	-	*39*	-	-	*6670*
Heat Generated - TJ	-	-	-	-	-	-	-	-	-	-	-

APPROVISIONNEMENT ET DEMANDE 1997	Charbon	Pétrole brut	Produits pétroliers	Gaz	Nucléaire	Hydro	Géotherm. solaire etc.	En. ren. combust. & déchets	Electricité	Chaleur	Total
Indigenous Production	-	-	-	-	-	558	-	528	-	-	1086
Imports	1	1494	670	-	-	-	-	1	23	-	2189
Exports	-	-	-57	-	-	-	-	-	-36	-	-93
Intl. Marine Bunkers	-	-	-304	-	-	-	-	-	-	-	-304
Stock changes	-	-72	78	-	-	-	-	-	-	-	6
TPES	1	1422	387	-	-	558	-	529	-12	-	2883
Electricity and CHP Plants	-	-	-151	-	-	-558	-	-13	615	-	-107
Petroleum Refineries	-	-1422	1389	-	-	-	-	-	-	-	-32
Other Transformation *	-1	-	-124	-	-	-	-	-2	-122	-	-249
TFC	-	-	1501	-	-	-	-	514	480	-	2495
INDUSTRY SECTOR	-	-	202	-	-	-	-	211	112	-	524
Iron and Steel	-	-	-	-	-	-	-	-	-	-	-
Chemical and Petrochemical	-	-	-	-	-	-	-	-	-	-	-
Non-Metallic Minerals	-	-	-	-	-	-	-	-	-	-	-
Non-specified	-	-	202	-	-	-	-	211	112	-	524
TRANSPORT SECTOR	-	-	831	-	-	-	-	-	-	-	831
Air	-	-	7	-	-	-	-	-	-	-	7
Road	-	-	823	-	-	-	-	-	-	-	823
Non-specified	-	-	-	-	-	-	-	-	-	-	-
OTHER SECTORS	-	-	388	-	-	-	-	303	368	-	1059
Agriculture	-	-	185	-	-	-	-	-	16	-	201
Comm. and Publ. Services	-	-	44	-	-	-	-	3	140	-	187
Residential	-	-	157	-	-	-	-	300	212	-	669
Non-specified	-	-	2	-	-	-	-	-	-	-	2
NON-ENERGY USE	-	-	81	-	-	-	-	-	-	-	81
Electricity Generated - GWh	-	-	*620*	-	-	*6486*	-	*42*	-	-	*7148*
Heat Generated - TJ	-	-	-	-	-	-	-	-	-	-	-

* Includes Transfers, Statistical Differences, Own Use and Distribution Losses.

Uzbekistan / Ouzbékistan : 1996

Thousand tonnes of oil equivalent / *Milliers de tonnes d'équivalent pétrole*

SUPPLY AND CONSUMPTION / *APPROVISIONNEMENT ET DEMANDE*	Coal / *Charbon*	Crude Oil / *Pétrole brut*	Petroleum Products / *Produits pétroliers*	Gas / *Gaz*	Nuclear / *Nucléaire*	Hydro / *Hydro*	Geotherm. Solar etc. / *Géotherm. solaire*	Combust. Renew. & Waste / *En. ren. combust.*	Electricity / *Electricité*	Heat / *Chaleur*	Total / *Total*
Indigenous Production	1000	7523	-	38321	-	561	-	-	-	-	47406
Imports	18	4	44	3420	-	-	-	-	1197	-	4683
Exports	-3	-785	-356	-7390	-	-	-	-	-1103	-	-9637
Intl. Marine Bunkers	-	-	-	-	-	-	-	-	-	-	-
Stock Changes	181	62	110	-420	-	-	-	-	-	-	-66
TPES	**1197**	**6805**	**-202**	**33931**	**-**	**561**	**-**	**-**	**94**	**-**	**42386**
Transfers	-	-27	31	-	-	-	-	-	-	-	3
Statistical Differences	-	-32	-	-	-	-	-	-	-	-	-32
Electricity Plants	-430	-	-610	-4162	-	-561	-	-	2325	-	-3438
CHP Plants	-373	-	-938	-4592	-	-	-	-	1581	1317	-3005
Heat Plants	-2	-	-366	-1363	-	-	-	-	-	1259	-472
Gas Works	-	-	-	-	-	-	-	-	-	-	-
Petroleum Refineries	-	-6502	6651	-	-	-	-	-	-	-	149
Coal Transformation	12	-	-	-	-	-	-	-	-	-	12
Liquefaction	-	-	-	-	-	-	-	-	-	-	-
Other Transformation	-	-	-	-	-	-	-	-	-	-	-
Own Use	-13	-16	-256	-1666	-	-	-	-	-341	-	-2292
Distribution Losses	-51	-82	-	-1745	-	-	-	-	-349	-	-2228
TFC	**340**	**145**	**4309**	**20403**	**-**	**-**	**-**	**-**	**3310**	**2576**	**31083**
INDUSTRY SECTOR	**73**	**145**	**280**	**5242**	**-**	**-**	**-**	**-**	**1210**	**-**	**6949**
Iron and Steel	-	-	-	-	-	-	-	-	-	-	-
Chemical and Petrochemical	-	145	-	1288	-	-	-	-	-	-	1433
of which: Feedstocks	-	*145*	-	*1288*	-	-	-	-	-	-	*1433*
Non-Ferrous Metals	-	-	-	-	-	-	-	-	-	-	-
Non-Metallic Minerals	-	-	6	-	-	-	-	-	-	-	6
Transport Equipment	-	-	-	-	-	-	-	-	-	-	-
Machinery	-	-	2	-	-	-	-	-	-	-	2
Mining and Quarrying	-	-	-	-	-	-	-	-	-	-	-
Food and Tobacco	-	-	-	-	-	-	-	-	-	-	-
Paper, Pulp and Printing	-	-	-	-	-	-	-	-	-	-	-
Wood and Wood Products	-	-	-	-	-	-	-	-	-	-	-
Construction	-	-	254	-	-	-	-	-	-	-	254
Textile and Leather	-	-	18	-	-	-	-	-	-	-	18
Non-specified	73	-	1	3954	-	-	-	-	1210	-	5237
TRANSPORT SECTOR	**-**	**-**	**2134**	**1340**	**-**	**-**	**-**	**-**	**120**	**-**	**3594**
Air	-	-	332	-	-	-	-	-	-	-	332
Road	-	-	1666	61	-	-	-	-	-	-	1727
Rail	-	-	134	-	-	-	-	-	12	-	145
Pipeline Transport	-	-	-	1279	-	-	-	-	77	-	1356
Internal Navigation	-	-	-	-	-	-	-	-	-	-	-
Non-specified	-	-	2	-	-	-	-	-	32	-	34
OTHER SECTORS	**267**	**-**	**1398**	**13821**	**-**	**-**	**-**	**-**	**1980**	**2576**	**20042**
Agriculture	7	-	1101	182	-	-	-	-	1096	-	2386
Comm. and Publ. Services	-	-	-	1260	-	-	-	-	272	-	1532
Residential	20	-	16	12378	-	-	-	-	613	-	13026
Non-specified	241	-	282	-	-	-	-	-	-	2576	3099
NON-ENERGY USE	**-**	**-**	**497**	**-**	**-**	**-**	**-**	**-**	**-**	**-**	**497**
in Industry/Transf./Energy	-	-	350	-	-	-	-	-	-	-	350
in Transport	-	-	-	-	-	-	-	-	-	-	-
in Other Sectors	-	-	147	-	-	-	-	-	-	-	147
Electr. Generated - GWh	*1827*	*-*	*5069*	*31998*	*-*	*6525*	*-*	*-*	*-*	*-*	*45419*
Electricity Plants	*1238*	*-*	*2345*	*16929*	*-*	*6525*	*-*	*-*	*-*	*-*	*27037*
CHP Plants	*589*	*-*	*2724*	*15069*	*-*	*-*	*-*	*-*	*-*	*-*	*18382*
Heat Generated - TJ	*3939*	*-*	*12429*	*91506*	*-*	*-*	*-*	*-*	*-*	*-*	*107874*
CHP Plants	*3874*	*-*	*10594*	*40677*	*-*	*-*	*-*	*-*	*-*	*-*	*55145*
Heat Plants	*65*	*-*	*1835*	*50829*	*-*	*-*	*-*	*-*	*-*	*-*	*52729*

Uzbekistan / Ouzbékistan : 1997

Thousand tonnes of oil equivalent / *Milliers de tonnes d'équivalent pétrole*

SUPPLY AND CONSUMPTION / *APPROVISIONNEMENT ET DEMANDE*	Coal / *Charbon*	Crude Oil / *Pétrole brut*	Petroleum Products / *Produits pétroliers*	Gas / *Gaz*	Nuclear / *Nucléaire*	Hydro / *Hydro*	Geotherm. Solar etc. / *Géotherm. solaire*	Combust. Renew. & Waste / *En. ren. combust.*	Electricity / *Electricité*	Heat / *Chaleur*	Total / *Total*
Indigenous Production	1037	7798	-	39722	-	497	-	-	-	-	49054
Imports	9	-	36	2247	-	-	-	-	1068	-	3361
Exports	-10	-915	-373	-8058	-	-	-	-	-988	-	-10346
Intl. Marine Bunkers	-	-	-	-	-	-	-	-	-	-	-
Stock Changes	-52	41	366	129	-	-	-	-	-	-	484
TPES	**984**	**6924**	**29**	**34039**	**-**	**497**	**-**	**-**	**80**	**-**	**42553**
Transfers	-	-32	36	-	-	-	-	-	-	-	4
Statistical Differences	-	-23	-	-	-	-	-	-	-	-	-23
Electricity Plants	-417	-	-714	-4456	-	-497	-	-	2363	-	-3721
CHP Plants	-360	-	-916	-4555	-	-	-	-	1598	1219	-3014
Heat Plants	-1	-	-323	-1272	-	-	-	-	-	1171	-426
Gas Works	-	-	-	-	-	-	-	-	-	-	-
Petroleum Refineries	-	-6646	6865	-	-	-	-	-	-	-	219
Coal Transformation	13	-	-	-	-	-	-	-	-	-	13
Liquefaction	-	-	-	-	-	-	-	-	-	-	-
Other Transformation	-	-	-	-	-	-	-	-	-	-	-
Own Use	-2	-15	-280	-1618	-	-	-	-	-347	-	-2262
Distribution Losses	-13	-88	-	-1415	-	-	-	-	-345	-	-1862
TFC	**204**	**120**	**4697**	**20723**	**-**	**-**	**-**	**-**	**3349**	**2390**	**31482**
INDUSTRY SECTOR	**53**	**120**	**290**	**4085**	**-**	**-**	**-**	**-**	**1283**	**-**	**5832**
Iron and Steel	-	-	-	-	-	-	-	-	-	-	-
Chemical and Petrochemical	-	120	-	1289	-	-	-	-	-	-	1408
of which: Feedstocks	-	*120*	-	*1289*	-	-	-	-	-	-	*1408*
Non-Ferrous Metals	-	-	-	-	-	-	-	-	-	-	-
Non-Metallic Minerals	-	-	5	-	-	-	-	-	-	-	5
Transport Equipment	-	-	-	-	-	-	-	-	-	-	-
Machinery	-	-	4	-	-	-	-	-	-	-	4
Mining and Quarrying	-	-	-	-	-	-	-	-	-	-	-
Food and Tobacco	-	-	-	-	-	-	-	-	-	-	-
Paper, Pulp and Printing	-	-	-	-	-	-	-	-	-	-	-
Wood and Wood Products	-	-	-	-	-	-	-	-	-	-	-
Construction	-	-	265	-	-	-	-	-	-	-	265
Textile and Leather	-	-	15	-	-	-	-	-	-	-	15
Non-specified	53	-	1	2797	-	-	-	-	1283	-	4134
TRANSPORT SECTOR	**-**	**-**	**2269**	**1235**	**-**	**-**	**-**	**-**	**110**	**-**	**3615**
Air	-	-	295	-	-	-	-	-	-	-	295
Road	-	-	1849	55	-	-	-	-	-	-	1904
Rail	-	-	125	-	-	-	-	-	13	-	139
Pipeline Transport	-	-	-	1181	-	-	-	-	71	-	1251
Internal Navigation	-	-	-	-	-	-	-	-	-	-	-
Non-specified	-	-	-	-	-	-	-	-	26	-	26
OTHER SECTORS	**150**	**-**	**1453**	**15402**	**-**	**-**	**-**	**-**	**1955**	**2390**	**21351**
Agriculture	3	-	1144	136	-	-	-	-	1088	-	2370
Comm. and Publ. Services	-	-	-	2544	-	-	-	-	260	-	2804
Residential	12	-	25	12722	-	-	-	-	608	-	13367
Non-specified	135	-	284	-	-	-	-	-	-	2390	2810
NON-ENERGY USE	**-**	**-**	**684**	**-**	**-**	**-**	**-**	**-**	**-**	**-**	**684**
in Industry/Transf./Energy	-	-	516	-	-	-	-	-	-	-	516
in Transport	-	-	-	-	-	-	-	-	-	-	-
in Other Sectors	-	-	168	-	-	-	-	-	-	-	168
Electr. Generated - GWh	*1881*	*-*	*5483*	*32913*	*-*	*5777*	*-*	*-*	*-*	*-*	*46054*
Electricity Plants	*1234*	*-*	*2661*	*17799*	*-*	*5777*	*-*	*-*	*-*	*-*	*27471*
CHP Plants	*647*	*-*	*2822*	*15114*	*-*	*-*	*-*	*-*	*-*	*-*	*18583*
Heat Generated - TJ	*3907*	*-*	*10922*	*85271*	*-*	*-*	*-*	*-*	*-*	*-*	*100100*
CHP Plants	*3865*	*-*	*8931*	*38264*	*-*	*-*	*-*	*-*	*-*	*-*	*51060*
Heat Plants	*42*	*-*	*1991*	*47007*	*-*	*-*	*-*	*-*	*-*	*-*	*49040*

INTERNATIONAL ENERGY AGENCY

Venezuela / Vénézuela

Total Production of Energy

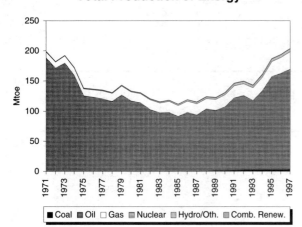

■ Coal ▦ Oil □ Gas ▤ Nuclear ▨ Hydro/Oth. ▥ Comb. Renew.

Total Primary Energy Supply*

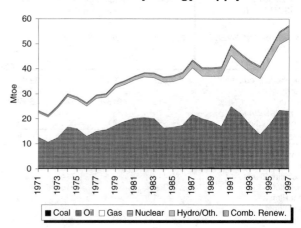

■ Coal ▦ Oil □ Gas ▤ Nuclear ▨ Hydro/Oth. ▥ Comb. Renew.

Electricity Generation by Fuel

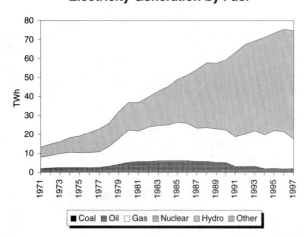

■ Coal ▦ Oil □ Gas ▤ Nuclear ▨ Hydro ▥ Other

Oil Products Consumption

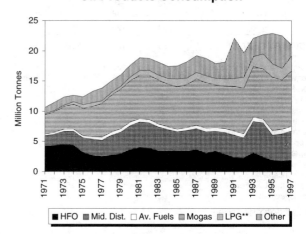

■ HFO ▦ Mid. Dist. □ Av. Fuels ▤ Mogas ▨ LPG** ▥ Other

Electricity Consumption/GDP***, TPES/GDP*** and Production/TPES

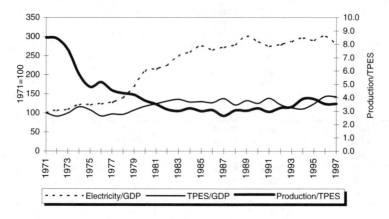

- - - - Electricity/GDP ——— TPES/GDP ▬▬ Production/TPES

* Excluding electricity trade.
** Includes LPG, NGL, ethane and naphtha.
*** GDP in 1990 US dollars.

Venezuela / Vénézuela : 1996

	Thousand tonnes of oil equivalent / *Milliers de tonnes d'équivalent pétrole*										
SUPPLY AND CONSUMPTION ***APPROVISIONNEMENT ET DEMANDE***	Coal *Charbon*	Crude Oil *Pétrole brut*	Petroleum Products *Produits pétroliers*	Gas *Gaz*	Nuclear *Nucléaire*	Hydro *Hydro*	Geotherm. Solar etc. *Géotherm. solaire*	Combust. Renew. & Waste *En. ren. combust.*	Electricity *Electricité*	Heat *Chaleur*	Total *Total*
Indigenous Production	2144	160408	-	26270	-	4631	-	541	-	-	193993
Imports	181	-	-	-	-	-	-	-	-	-	181
Exports	-2123	-98692	-36740	-	-	-	-	-	-13	-	-137569
Intl. Marine Bunkers	-	-	-644	-	-	-	-	-	-	-	-644
Stock Changes	-	-930	-70	-	-	-	-	-	-	-	-1000
TPES	**202**	**60786**	**-37454**	**26270**	**-**	**4631**	**-**	**541**	**-13**	**-**	**54962**
Transfers	-	-4876	5303	-	-	-	-	-	-	-	427
Statistical Differences	-	-795	959	-691	-	-	-	-	-78	-	-606
Electricity Plants	-	-	-786	-5687	-	-4631	-	-	6482	-	-4622
CHP Plants	-	-	-	-	-	-	-	-	-	-	-
Heat Plants	-	-	-	-	-	-	-	-	-	-	-
Gas Works	-	-	-	-	-	-	-	-	-	-	-
Petroleum Refineries	-	-51575	50474	-	-	-	-	-	-	-	-1101
Coal Transformation	-	-	-	-	-	-	-	-	-	-	-
Liquefaction	-	-	-	-	-	-	-	-	-	-	-
Other Transformation	-	-	-	-	-	-	-	-14	-	-	-14
Own Use	-	-	-892	-9554	-	-	-	-	-278	-	-10724
Distribution Losses	-	-3540	-	-	-	-	-	-	-1320	-	-4860
TFC	**202**	**-**	**17604**	**10337**	**-**	**-**	**-**	**526**	**4793**	**-**	**33463**
INDUSTRY SECTOR	**202**	**-**	**2537**	**9566**	**-**	**-**	**-**	**331**	**2373**	**-**	**15008**
Iron and Steel	-	-	67	2753	-	-	-	-	548	-	3368
Chemical and Petrochemical	-	-	1237	3082	-	-	-	-	96	-	4416
of which: Feedstocks	-	-	-	-	-	-	-	-	-	-	-
Non-Ferrous Metals	-	-	4	514	-	-	-	-	948	-	1466
Non-Metallic Minerals	202	-	-	893	-	-	-	-	30	-	1125
Transport Equipment	-	-	-	-	-	-	-	-	-	-	-
Machinery	-	-	-	-	-	-	-	-	-	-	-
Mining and Quarrying	-	-	-	-	-	-	-	-	-	-	-
Food and Tobacco	-	-	-	-	-	-	-	331	-	-	331
Paper, Pulp and Printing	-	-	-	-	-	-	-	-	-	-	-
Wood and Wood Products	-	-	-	-	-	-	-	-	-	-	-
Construction	-	-	-	-	-	-	-	-	-	-	-
Textile and Leather	-	-	-	-	-	-	-	-	-	-	-
Non-specified	-	-	1228	2323	-	-	-	-	751	-	4302
TRANSPORT SECTOR	**-**	**-**	**11243**	**-**	**-**	**-**	**-**	**-**	**7**	**-**	**11250**
Air	-	-	811	-	-	-	-	-	-	-	811
Road	-	-	10427	-	-	-	-	-	-	-	10427
Rail	-	-	6	-	-	-	-	-	7	-	13
Pipeline Transport	-	-	-	-	-	-	-	-	-	-	-
Internal Navigation	-	-	-	-	-	-	-	-	-	-	-
Non-specified	-	-	-	-	-	-	-	-	-	-	-
OTHER SECTORS	**-**	**-**	**2388**	**771**	**-**	**-**	**-**	**196**	**2413**	**-**	**5768**
Agriculture	-	-	560	-	-	-	-	-	-	-	560
Comm. and Publ. Services	-	-	734	543	-	-	-	-	1313	-	2590
Residential	-	-	1029	229	-	-	-	196	1100	-	2554
Non-specified	-	-	65	-	-	-	-	-	-	-	65
NON-ENERGY USE	**-**	**-**	**1436**	**-**	**-**	**-**	**-**	**-**	**-**	**-**	**1436**
in Industry/Transf./Energy	-	-	1436	-	-	-	-	-	-	-	1436
in Transport	-	-	-	-	-	-	-	-	-	-	-
in Other Sectors	-	-	-	-	-	-	-	-	-	-	-
Electr. Generated - GWh	*-*	*-*	*1862*	*19666*	*-*	*53844*	*-*	*-*	*-*	*-*	*75372*
Electricity Plants	*-*	*-*	*1862*	*19666*	*-*	*53844*	*-*	*-*	*-*	*-*	*75372*
CHP Plants	*-*	*-*	*-*	*-*	*-*	*-*	*-*	*-*	*-*	*-*	*-*
Heat Generated - TJ	*-*	*-*	*-*	*-*	*-*	*-*	*-*	*-*	*-*	*-*	*-*
CHP Plants	*-*	*-*	*-*	*-*	*-*	*-*	*-*	*-*	*-*	*-*	*-*
Heat Plants	*-*	*-*	*-*	*-*	*-*	*-*	*-*	*-*	*-*	*-*	*-*

Venezuela / Vénézuela : 1997

Thousand tonnes of oil equivalent / *Milliers de tonnes d'équivalent pétrole*

SUPPLY AND CONSUMPTION / *APPROVISIONNEMENT ET DEMANDE*	Coal / *Charbon*	Crude Oil / *Pétrole brut*	Petroleum Products / *Produits pétroliers*	Gas / *Gaz*	Nuclear / *Nucléaire*	Hydro / *Hydro*	Geotherm. Solar etc. / *Géotherm. solaire*	Combust. Renew. & Waste / *En. ren. combust.*	Electricity / *Electricité*	Heat / *Chaleur*	Total / *Total*
Indigenous Production	3414	166146	-	28962	-	4915	-	541	-	-	203979
Imports	196	-	-	-	-	-	-	-	-	-	196
Exports	-3352	-106038	-38725	-	-	-	-	-	-	-	-148115
Intl. Marine Bunkers	-	-	-644	-	-	-	-	-	-	-	-644
Stock Changes	-	2192	-77	-	-	-	-	-	-	-	2115
TPES	**258**	**62301**	**-39446**	**28962**	**-**	**4915**	**-**	**541**	**-**	**-**	**57530**
Transfers	-	-5043	5501	-	-	-	-	-	-	-	458
Statistical Differences	-	-1212	549	-644	-	-	-	-	92	-	-1216
Electricity Plants	-	-	-851	-4540	-	-4915	-	-	6438	-	-3868
CHP Plants	-	-	-	-	-	-	-	-	-	-	-
Heat Plants	-	-	-	-	-	-	-	-	-	-	-
Gas Works	-	-	-	-	-	-	-	-	-	-	-
Petroleum Refineries	-	-56045	54877		-	-	-	-	-	-	-1168
Coal Transformation	-	-	-	-	-	-	-	-	-	-	-
Liquefaction	-	-	-	-	-	-	-	-	-	-	-
Other Transformation	-	-	-	-	-	-	-	-14	-	-	-14
Own Use	-	-	-957	-12672	-	-	-	-	-288	-	-13918
Distribution Losses	-	-	-	-	-	-	-	-	-1369	-	-1369
TFC	**258**	**-**	**19673**	**11106**	**-**	**-**	**-**	**526**	**4873**	**-**	**36435**
INDUSTRY SECTOR	**258**	**-**	**2970**	**10277**	**-**	**-**	**-**	**331**	**2271**	**-**	**16107**
Iron and Steel	-	-	72	2958	-	-	-	-	525	-	3555
Chemical and Petrochemical	-	-	1535	3312	-	-	-	-	92	-	4939
of which: Feedstocks	-	-	-	-	-	-	-	-	-	-	-
Non-Ferrous Metals	-	-	5	552	-	-	-	-	907	-	1465
Non-Metallic Minerals	258	-	-	960	-	-	-	-	29	-	1247
Transport Equipment	-	-	-	-	-	-	-	-	-	-	-
Machinery	-	-	-	-	-	-	-	-	-	-	-
Mining and Quarrying	-	-	-	-	-	-	-	-	-	-	-
Food and Tobacco	-	-	-	-	-	-	-	331	-	-	331
Paper, Pulp and Printing	-	-	-	-	-	-	-	-	-	-	-
Wood and Wood Products	-	-	-	-	-	-	-	-	-	-	-
Construction	-	-	-	-	-	-	-	-	-	-	-
Textile and Leather	-	-	-	-	-	-	-	-	-	-	-
Non-specified	-	-	1358	2496	-	-	-	-	718	-	4572
TRANSPORT SECTOR	**-**	**-**	**12566**	**-**	**-**	**-**	**-**	**-**	**72**	**-**	**12638**
Air	-	-	829	-	-	-	-	-	-	-	829
Road	-	-	11731	-	-	-	-	-	65	-	11795
Rail	-	-	7	-	-	-	-	-	7	-	14
Pipeline Transport	-	-	-	-	-	-	-	-	-	-	-
Internal Navigation	-	-	-	-	-	-	-	-	-	-	-
Non-specified	-	-	-	-	-	-	-	-	-	-	-
OTHER SECTORS	**-**	**-**	**2700**	**829**	**-**	**-**	**-**	**196**	**2530**	**-**	**6255**
Agriculture	-	-	606	-	-	-	-	-	-	-	606
Comm. and Publ. Services	-	-	794	583	-	-	-	-	1377	-	2754
Residential	-	-	1230	246	-	-	-	196	1153	-	2825
Non-specified	-	-	71	-	-	-	-	-	-	-	71
NON-ENERGY USE	**-**	**-**	**1436**	**-**	**-**	**-**	**-**	**-**	**-**	**-**	**1436**
in Industry/Transf./Energy	-	-	1436	-	-	-	-	-	-	-	1436
in Transport	-	-	-	-	-	-	-	-	-	-	-
in Other Sectors	-	-	-	-	-	-	-	-	-	-	-
Electr. Generated - GWh	*-*	*-*	*2013*	*15700*	*-*	*57152*	*-*	*-*	*-*	*-*	*74865*
Electricity Plants	*-*	*-*	*2013*	*15700*	*-*	*57152*	*-*	*-*	*-*	*-*	*74865*
CHP Plants	*-*	*-*	*-*	*-*	*-*	*-*	*-*	*-*	*-*	*-*	*-*
Heat Generated - TJ	*-*	*-*	*-*	*-*	*-*	*-*	*-*	*-*	*-*	*-*	*-*
CHP Plants	*-*	*-*	*-*	*-*	*-*	*-*	*-*	*-*	*-*	*-*	*-*
Heat Plants	*-*	*-*	*-*	*-*	*-*	*-*	*-*	*-*	*-*	*-*	*-*

Vietnam / Viêt-Nam

Thousand tonnes of oil equivalent / *Milliers de tonnes d'équivalent pétrole*

SUPPLY AND CONSUMPTION 1996	Coal	Crude Oil	Petroleum Products	Gas	Nuclear	Hydro	Geotherm. Solar etc.	Combust. Renew. & Waste	Electricity	Heat	Total
Indigenous Production	5598	8754	-	3118	-	1257	-	21955	-	-	40682
Imports	-	-	6220	-	-	-	-	-	-	-	6220
Exports	-2194	-8713	-	-	-	-	-	-	-	-	-10908
Intl. Marine Bunkers	-	-	-	-	-	-	-	-	-	-	-
Stock changes	-	-	-	-	-	-	-	-	-	-	-
TPES	3403	41	6220	3118	-	1257	-	21955	-	-	35995
Electricity and CHP Plants	-826	-	-415	-206	-	-1257	-	-	1457	-	-1247
Petroleum Refineries	-	-41	42	-	-	-	-	-	-	-	2
Other Transformation *	-	-	-2	-	-	-	-	-700	-307	-	-1010
TFC	2577	-	5845	2912	-	-	-	21255	1150	-	33740
INDUSTRY SECTOR	1911	-	742	2912	-	-	-	-	473	-	6039
Iron and Steel	-	-	-	-	-	-	-	-	-	-	-
Chemical and Petrochemical	-	-	-	-	-	-	-	-	-	-	-
Non-Metallic Minerals	-	-	-	-	-	-	-	-	-	-	-
Non-specified	1911	-	742	2912	-	-	-	-	473	-	6039
TRANSPORT SECTOR	5	-	3892	-	-	-	-	-	10	-	3907
Air	-	-	271	-	-	-	-	-	-	-	271
Road	-	-	3509	-	-	-	-	-	-	-	3509
Non-specified	5	-	112	-	-	-	-	-	10	-	127
OTHER SECTORS	662	-	1018	-	-	-	-	21255	667	-	23601
Agriculture	35	-	252	-	-	-	-	-	62	-	349
Comm. and Publ. Services	-	-	-	-	-	-	-	-	-	-	-
Residential	504	-	449	-	-	-	-	21255	521	-	22728
Non-specified	122	-	317	-	-	-	-	-	84	-	524
NON-ENERGY USE	-	-	193	-	-	-	-	-	-	-	193
Electricity Generated - GWh	-	-	1460	865	-	14620	-	-	-	-	16945
Heat Generated - TJ	-	-	-	-	-	-	-	-	-	-	-

APPROVISIONNEMENT ET DEMANDE 1997	Charbon	Pétrole brut	Produits pétroliers	Gaz	Nucléaire	Hydro	Géotherm. solaire etc.	En. ren. combust. & déchets	Electricité	Chaleur	Total
Indigenous Production	6550	9875	-	3311	-	1392	-	22398	-	-	43525
Imports	-	-	7730	-	-	-	-	-	-	-	7730
Exports	-2115	-9834	-	-	-	-	-	-	-	-	-11949
Intl. Marine Bunkers	-	-	-	-	-	-	-	-	-	-	-
Stock changes	-	-	-	-	-	-	-	-	-	-	-
TPES	4435	41	7730	3311	-	1392	-	22398	-	-	39306
Electricity and CHP Plants	-1387	-	-587	-219	-	-1392	-	-	1647	-	-1937
Petroleum Refineries	-	-41	41	-	-	-	-	-	-	-	-
Other Transformation *	1	-	-	-	-	-	-	-714	-331	-	-1045
TFC	3049	-	7183	3092	-	-	-	21684	1316	-	36324
INDUSTRY SECTOR	2329	-	936	3092	-	-	-	-	530	-	6887
Iron and Steel	-	-	-	-	-	-	-	-	-	-	-
Chemical and Petrochemical	-	-	-	-	-	-	-	-	-	-	-
Non-Metallic Minerals	-	-	-	-	-	-	-	-	-	-	-
Non-specified	2329	-	936	3092	-	-	-	-	530	-	6887
TRANSPORT SECTOR	5	-	4915	-	-	-	-	-	11	-	4930
Air	-	-	297	-	-	-	-	-	-	-	297
Road	-	-	4477	-	-	-	-	-	-	-	4477
Non-specified	5	-	141	-	-	-	-	-	11	-	157
OTHER SECTORS	716	-	1219	-	-	-	-	21684	775	-	24394
Agriculture	45	-	320	-	-	-	-	-	59	-	424
Comm. and Publ. Services	-	-	-	-	-	-	-	-	-	-	-
Residential	535	-	732	-	-	-	-	21684	621	-	23573
Non-specified	135	-	167	-	-	-	-	-	95	-	397
NON-ENERGY USE	-	-	113	-	-	-	-	-	-	-	113
Electricity Generated - GWh	-	-	2049	919	-	16183	-	-	-	-	19151
Heat Generated - TJ	-	-	-	-	-	-	-	-	-	-	-

* Includes Transfers, Statistical Differences, Own Use and Distribution Losses.

Yemen / Yémen

Thousand tonnes of oil equivalent / *Milliers de tonnes d'équivalent pétrole*

SUPPLY AND CONSUMPTION 1996	Coal	Crude Oil	Petroleum Products	Gas	Nuclear	Hydro	Geotherm. Solar etc.	Combust. Renew. & Waste	Electricity	Heat	Total
Indigenous Production	-	18175	-	-	-	-	-	77	-	-	18252
Imports	-	-	1	-	-	-	-	-	-	-	1
Exports	-	-14607	-1760	-	-	-	-	-	-	-	-16367
Intl. Marine Bunkers	-	-	-99	-	-	-	-	-	-	-	-99
Stock changes	-	1479	-	-	-	-	-	-	-	-	1479
TPES	-	5047	-1858	-	-	-	-	77	-	-	3266
Electricity and CHP Plants	-	-	-435	-	-	-	-	-	201	-	-234
Petroleum Refineries	-	-4673	4425	-	-	-	-	-	-	-	-248
Other Transformation *	-	-374	292	-	-	-	-	-40	-73	-	-195
TFC	-	-	2424	-	-	-	-	37	128	-	2589
INDUSTRY SECTOR	-	-	156	-	-	-	-	-	-	-	156
Iron and Steel	-	-	-	-	-	-	-	-	-	-	-
Chemical and Petrochemical	-	-	-	-	-	-	-	-	-	-	-
Non-Metallic Minerals	-	-	-	-	-	-	-	-	-	-	-
Non-specified	-	-	156	-	-	-	-	-	-	-	156
TRANSPORT SECTOR	-	-	1654	-	-	-	-	-	-	-	1654
Air	-	-	94	-	-	-	-	-	-	-	94
Road	-	-	1560	-	-	-	-	-	-	-	1560
Non-specified	-	-	-	-	-	-	-	-	-	-	-
OTHER SECTORS	-	-	556	-	-	-	-	37	128	-	721
Agriculture	-	-	-	-	-	-	-	-	-	-	-
Comm. and Publ. Services	-	-	-	-	-	-	-	-	-	-	-
Residential	-	-	556	-	-	-	-	-	-	-	556
Non-specified	-	-	-	-	-	-	-	37	128	-	165
NON-ENERGY USE	-	-	58	-	-	-	-	-	-	-	58
Electricity Generated - GWh	-	-	*2334*	-	-	-	-	-	-	-	*2334*
Heat Generated - TJ	-	-	-	-	-	-	-	-	-	-	-

APPROVISIONNEMENT ET DEMANDE 1997	Charbon	Pétrole brut	Produits pétroliers	Gaz	Nucléaire	Hydro	Géotherm. solaire etc.	En. ren. combust. & déchets	Electricité	Chaleur	Total
Indigenous Production	-	19028	-	-	-	-	-	77	-	-	19105
Imports	-	-	1	-	-	-	-	-	-	-	1
Exports	-	-12096	-1689	-	-	-	-	-	-	-	-13785
Intl. Marine Bunkers	-	-	-99	-	-	-	-	-	-	-	-99
Stock changes	-	-1867	-	-	-	-	-	-	-	-	-1867
TPES	-	5065	-1787	-	-	-	-	77	-	-	3355
Electricity and CHP Plants	-	-	-452	-	-	-	-	-	203	-	-249
Petroleum Refineries	-	-4650	4406	-	-	-	-	-	-	-	-244
Other Transformation *	-	-415	329	-	-	-	-	-40	-74	-	-199
TFC	-	-	2497	-	-	-	-	37	129	-	2663
INDUSTRY SECTOR	-	-	162	-	-	-	-	-	-	-	162
Iron and Steel	-	-	-	-	-	-	-	-	-	-	-
Chemical and Petrochemical	-	-	-	-	-	-	-	-	-	-	-
Non-Metallic Minerals	-	-	-	-	-	-	-	-	-	-	-
Non-specified	-	-	162	-	-	-	-	-	-	-	162
TRANSPORT SECTOR	-	-	1690	-	-	-	-	-	-	-	1690
Air	-	-	99	-	-	-	-	-	-	-	99
Road	-	-	1590	-	-	-	-	-	-	-	1590
Non-specified	-	-	-	-	-	-	-	-	-	-	-
OTHER SECTORS	-	-	603	-	-	-	-	37	129	-	769
Agriculture	-	-	-	-	-	-	-	-	-	-	-
Comm. and Publ. Services	-	-	-	-	-	-	-	-	-	-	-
Residential	-	-	603	-	-	-	-	-	-	-	603
Non-specified	-	-	-	-	-	-	-	37	129	-	166
NON-ENERGY USE	-	-	42	-	-	-	-	-	-	-	42
Electricity Generated - GWh	-	-	*2358*	-	-	-	-	-	-	-	*2358*
Heat Generated - TJ	-	-	-	-	-	-	-	-	-	-	-

* Includes Transfers, Statistical Differences, Own Use and Distribution Losses.

Federal Republic of Yugoslavia / République fédérative de Yougoslavie

Thousand tonnes of oil equivalent / *Milliers de tonnes d'équivalent pétrole*

SUPPLY AND CONSUMPTION 1996	Coal	Crude Oil	Petroleum Products	Gas	Nuclear	Hydro	Geotherm. Solar etc.	Combust. Renew. & Waste	Electricity	Heat	Total
Indigenous Production	8193	1052	-	544	-	989	-	210	-	-	10987
Imports	44	1349	352	1701	-	-	-	-	-	-	3446
Exports	-	-	-	-	-	-	-	-	-13	-	-13
Intl. Marine Bunkers	-	-	-	-	-	-	-	-	-	-	-
Stock changes	-	-	-	-	-	-	-	-	-	-	-
TPES	8237	2400	352	2245	-	989	-	210	-13	-	14420
Electricity and CHP Plants	-6888	-	-289	-259	-	-989	-	-	3276	454	-4695
Petroleum Refineries	-	-2400	1882	-	-	-	-	-	-	-	-518
Other Transformation *	-1	-	-	-22	-	-	-	-	-560	-	-583
TFC	1349	-	1945	1963	-	-	-	210	2703	454	8624
INDUSTRY SECTOR	938	-	723	647	-	-	-	-	536	-	2844
Iron and Steel	-	-	-	-	-	-	-	-	29	-	29
Chemical and Petrochemical	-	-	243	-	-	-	-	-	35	-	278
Non-Metallic Minerals	-	-	-	-	-	-	-	-	-	-	-
Non-specified	938	-	480	647	-	-	-	-	472	-	2537
TRANSPORT SECTOR	-	-	764	-	-	-	-	-	23	-	787
Air	-	-	59	-	-	-	-	-	-	-	59
Road	-	-	706	-	-	-	-	-	5	-	710
Non-specified	-	-	-	-	-	-	-	-	19	-	19
OTHER SECTORS	411	-	218	1316	-	-	-	210	2144	454	4753
Agriculture	-	-	-	-	-	-	-	-	16	-	16
Comm. and Publ. Services	-	-	-	-	-	-	-	-	25	-	25
Residential	-	-	39	-	-	-	-	-	1414	-	1453
Non-specified	411	-	180	1316	-	-	-	210	689	454	3260
NON-ENERGY USE	-	-	240	-	-	-	-	-	-	-	240
Electricity Generated - GWh	24266	-	910	1420	-	11497	-	-	-	-	38093
Heat Generated - TJ	-	-	-	-	-	-	-	-	-	-	19000

APPROVISIONNEMENT ET DEMANDE 1997	Charbon	Pétrole brut	Produits pétroliers	Gaz	Nucléaire	Hydro	Géotherm. solaire etc.	En. ren. combust. & déchets	Electricité	Chaleur	Total
Indigenous Production	8668	1000	-	557	-	1046	-	210	-	-	11481
Imports	38	2340	310	1672	-	-	-	-	-	-	4360
Exports	-	-	-	-	-	-	-	-	-	-	-
Intl. Marine Bunkers	-	-	-	-	-	-	-	-	-	-	-
Stock changes	-	-	-	-	-	-	-	-	-	-	-
TPES	8706	3340	310	2230	-	1046	-	210	-	-	15842
Electricity and CHP Plants	-7282	-	-336	-258	-	-1046	-	-	3467	454	-5002
Petroleum Refineries	-	-3340	2723	-	-	-	-	-	-	-	-617
Other Transformation *	-1	-	-	-23	-	-	-	-	-595	-	-619
TFC	1422	-	2697	1949	-	-	-	210	2872	454	9604
INDUSTRY SECTOR	979	-	1058	642	-	-	-	-	569	-	3248
Iron and Steel	-	-	-	-	-	-	-	-	31	-	31
Chemical and Petrochemical	-	-	414	-	-	-	-	-	37	-	451
Non-Metallic Minerals	-	-	-	-	-	-	-	-	-	-	-
Non-specified	979	-	644	642	-	-	-	-	501	-	2766
TRANSPORT SECTOR	-	-	1149	-	-	-	-	-	25	-	1174
Air	-	-	101	-	-	-	-	-	-	-	101
Road	-	-	1048	-	-	-	-	-	5	-	1053
Non-specified	-	-	-	-	-	-	-	-	20	-	20
OTHER SECTORS	443	-	323	1308	-	-	-	210	2278	454	5016
Agriculture	-	-	-	-	-	-	-	-	17	-	17
Comm. and Publ. Services	-	-	-	-	-	-	-	-	27	-	27
Residential	-	-	56	-	-	-	-	-	1503	-	1559
Non-specified	443	-	267	1308	-	-	-	210	732	454	3414
NON-ENERGY USE	-	-	166	-	-	-	-	-	-	-	166
Electricity Generated - GWh	25679	-	963	1503	-	12167	-	-	-	-	40312
Heat Generated - TJ	-	-	-	-	-	-	-	-	-	-	19000

* Includes Transfers, Statistical Differences, Own Use and Distribution Losses.

Former Yugoslavia / ex-Yougoslavie

Thousand tonnes of oil equivalent / *Milliers de tonnes d'équivalent pétrole*

SUPPLY AND CONSUMPTION 1996	Coal	Crude Oil	Petroleum Products	Gas	Nuclear	Hydro	Geotherm. Solar etc.	Combust. Renew. & Waste	Electricity	Heat	Total
Indigenous Production	11108	2798	-	2012	1189	2121	13	1115	-	-	20355
Imports	455	6222	3440	3276	-	-	-	-	62	-	13455
Exports	-1	-	-1250	-	-	-	-	-	-	-	-1250
Intl. Marine Bunkers	-	-	-29	-	-	-	-	-	-	-	-29
Stock changes	-58	-136	353	-9	-	-	-	32	-	-	183
TPES	11504	8883	2515	5279	1189	2121	13	1147	62	-	32713
Electricity and CHP Plants	-9506	-	-1221	-720	-1189	-2121	-	-2	6029	889	-7842
Petroleum Refineries	-	-8596	8024	-	-	-	-	-	-	-	-573
Other Transformation *	-212	-287	-1061	-618	-	-	-13	-2	-1099	178	-3115
TFC	1787	-	8256	3941	-	-	-	1142	4991	1066	21183
INDUSTRY SECTOR	1213	-	1814	2059	-	-	-	61	1348	186	6680
Iron and Steel	73	-	68	118	-	-	-	-	199	6	464
Chemical and Petrochemical	5	-	695	650	-	-	-	-	129	36	1514
Non-Metallic Minerals	83	-	138	251	-	-	-	-	59	4	535
Non-specified	1051	-	912	1040	-	-	-	61	962	141	4166
TRANSPORT SECTOR	91	-	4390	-	-	-	-	-	59	-	4540
Air	-	-	227	-	-	-	-	-	-	-	227
Road	-	-	4056	-	-	-	-	-	5	-	4060
Non-specified	91	-	107	-	-	-	-	-	54	-	252
OTHER SECTORS	484	-	1685	1882	-	-	-	1081	3584	880	9596
Agriculture	-	-	251	21	-	-	-	-	24	11	306
Comm. and Publ. Services	28	-	416	119	-	-	-	51	443	91	1148
Residential	45	-	839	426	-	-	-	817	2279	301	4706
Non-specified	411	-	180	1316	-	-	-	214	839	477	3436
NON-ENERGY USE	-	-	368	-	-	-	-	-	-	-	368
Electricity Generated - GWh	*34959*	-	*3166*	*2738*	*4562*	*24668*	-	*10*	-	-	*70103*
Heat Generated - TJ	*7466*	-	*13661*	*8862*	-	-	*528*	*96*	-	-	*49613*

APPROVISIONNEMENT ET DEMANDE 1997	Charbon	Pétrole brut	Produits pétroliers	Gaz	Nucléaire	Hydro	Géotherm. solaire etc.	En. ren. combust. & déchets	Electricité	Chaleur	Total
Indigenous Production	11519	2800	-	1969	1308	1968	13	1109	-	-	20685
Imports	462	7136	3273	3443	-	-	-	-	211	-	14526
Exports	-3	-17	-610	-	-	-	-	-	-	-	-630
Intl. Marine Bunkers	-	-	-24	-	-	-	-	-	-	-	-24
Stock changes	103	-60	-157	-10	-	-	-	32	-	-	-92
TPES	12081	9859	2483	5402	1308	1968	13	1142	211	-	34465
Electricity and CHP Plants	-10006	-	-1134	-687	-1308	-1967	-	-	6199	875	-8029
Petroleum Refineries	-	-9587	8840	-	-	-	-	-	-	-	-747
Other Transformation *	-249	-272	-1013	-600	-	-	-13	-2	-1141	164	-3126
TFC	1826	-	9176	4115	-	-	-	1139	5269	1039	22564
INDUSTRY SECTOR	1241	-	1992	2236	-	-	-	61	1414	170	7113
Iron and Steel	75	-	58	150	-	-	-	-	216	6	505
Chemical and Petrochemical	2	-	666	769	-	-	-	-	130	41	1610
Non-Metallic Minerals	78	-	164	251	-	-	-	-	78	4	575
Non-specified	1086	-	1105	1064	-	-	-	61	989	119	4424
TRANSPORT SECTOR	91	-	4892	-	-	-	-	-	61	-	5044
Air	-	-	279	-	-	-	-	-	-	-	279
Road	-	-	4465	-	-	-	-	-	5	-	4470
Non-specified	91	-	148	-	-	-	-	-	56	-	295
OTHER SECTORS	494	-	1892	1879	-	-	-	1078	3794	869	10006
Agriculture	-	-	247	18	-	-	-	-	24	11	300
Comm. and Publ. Services	20	-	483	111	-	-	-	51	483	91	1238
Residential	31	-	895	443	-	-	-	813	2405	291	4879
Non-specified	443	-	267	1308	-	-	-	214	882	477	3590
NON-ENERGY USE	-	-	399	-	-	-	-	-	-	-	399
Electricity Generated - GWh	*37453*	-	*4058*	*2675*	*5019*	*22878*	-	*2*	-	-	*72085*
Heat Generated - TJ	*7253*	-	*11416*	*10276*	-	-	*528*	*94*	-	-	*48567*

* Includes Transfers, Statistical Differences, Own Use and Distribution Losses.

Zambia / Zambie

Thousand tonnes of oil equivalent / *Milliers de tonnes d'équivalent pétrole*

SUPPLY AND CONSUMPTION 1996	Coal	Crude Oil	Petroleum Products	Gas	Nuclear	Hydro	Geotherm. Solar etc.	Combust. Renew. & Waste	Electricity	Heat	Total
Indigenous Production	106	-	-	-	-	667	-	4663	-	-	5435
Imports	-	592	14	-	-	-	-	-	2	-	607
Exports	-4	-	-42	-	-	-	-	-	-129	-	-175
Intl. Marine Bunkers	-	-	-	-	-	-	-	-	-	-	-
Stock changes	-	-	-	-	-	-	-	-	-	-	-
TPES	102	592	-28	-	-	667	-	4663	-127	-	5868
Electricity and CHP Plants	-4	-	-1	-	-	-667	-	-	670	-	-1
Petroleum Refineries	-	-592	581	-	-	-	-	-	-	-	-12
Other Transformation *	-24	-	-45	-	-	-	-	-1257	-99	-	-1425
TFC	74	-	507	-	-	-	-	3406	444	-	4430
INDUSTRY SECTOR	74	-	120	-	-	-	-	510	344	-	1047
Iron and Steel	4	-	-	-	-	-	-	-	-	-	4
Chemical and Petrochemical	-	-	-	-	-	-	-	-	-	-	-
Non-Metallic Minerals	-	-	-	-	-	-	-	-	-	-	-
Non-specified	70	-	120	-	-	-	-	510	344	-	1043
TRANSPORT SECTOR	-	-	231	-	-	-	-	-	-	-	231
Air	-	-	40	-	-	-	-	-	-	-	40
Road	-	-	190	-	-	-	-	-	-	-	190
Non-specified	-	-	-	-	-	-	-	-	-	-	-
OTHER SECTORS	-	-	126	-	-	-	-	2896	100	-	3122
Agriculture	-	-	16	-	-	-	-	-	17	-	33
Comm. and Publ. Services	-	-	58	-	-	-	-	-	34	-	92
Residential	-	-	-	-	-	-	-	2896	49	-	2945
Non-specified	-	-	52	-	-	-	-	-	-	-	52
NON-ENERGY USE	-	-	31	-	-	-	-	-	-	-	31
Electricity Generated - GWh	*40*	-	*2*	-	-	*7754*	-	-	-	-	*7796*
Heat Generated - TJ	-	-	-	-	-	-	-	-	-	-	-

APPROVISIONNEMENT ET DEMANDE 1997	Charbon	Pétrole brut	Produits pétroliers	Gaz	Nucléaire	Hydro	Géotherm. solaire etc.	En. ren. combust. & déchets	Electricité	Chaleur	Total
Indigenous Production	106	-	-	-	-	685	-	4765	-	-	5556
Imports	-	592	14	-	-	-	-	-	-	-	606
Exports	-4	-	-42	-	-	-	-	-	-129	-	-175
Intl. Marine Bunkers	-	-	-	-	-	-	-	-	-	-	-
Stock changes	-	-	-	-	-	-	-	-	-	-	-
TPES	102	592	-28	-	-	685	-	4765	-129	-	5987
Electricity and CHP Plants	-4	-	-1	-	-	-685	-	-	689	-	-1
Petroleum Refineries	-	-592	581	-	-	-	-	-	-	-	-12
Other Transformation *	-24	-	-45	-	-	-	-	-1284	-102	-	-1456
TFC	74	-	507	-	-	-	-	3481	457	-	4518
INDUSTRY SECTOR	74	-	120	-	-	-	-	521	354	-	1069
Iron and Steel	4	-	-	-	-	-	-	-	-	-	4
Chemical and Petrochemical	-	-	-	-	-	-	-	-	-	-	-
Non-Metallic Minerals	-	-	-	-	-	-	-	-	-	-	-
Non-specified	70	-	120	-	-	-	-	521	354	-	1065
TRANSPORT SECTOR	-	-	231	-	-	-	-	-	-	-	231
Air	-	-	40	-	-	-	-	-	-	-	40
Road	-	-	190	-	-	-	-	-	-	-	190
Non-specified	-	-	-	-	-	-	-	-	-	-	-
OTHER SECTORS	-	-	126	-	-	-	-	2960	103	-	3188
Agriculture	-	-	16	-	-	-	-	-	18	-	33
Comm. and Publ. Services	-	-	58	-	-	-	-	-	35	-	93
Residential	-	-	-	-	-	-	-	2960	51	-	3011
Non-specified	-	-	52	-	-	-	-	-	-	-	52
NON-ENERGY USE	-	-	31	-	-	-	-	-	-	-	31
Electricity Generated - GWh	*40*	-	*2*	-	-	*7964*	-	-	-	-	*8006*
Heat Generated - TJ	-	-	-	-	-	-	-	-	-	-	-

* Includes Transfers, Statistical Differences, Own Use and Distribution Losses.

INTERNATIONAL ENERGY AGENCY

Zimbabwe

Thousand tonnes of oil equivalent / *Milliers de tonnes d'équivalent pétrole*

SUPPLY AND CONSUMPTION 1996	Coal	Crude Oil	Petroleum Products	Gas	Nuclear	Hydro	Geotherm. Solar etc.	Combust. Renew. & Waste	Electricity	Heat	Total
Indigenous Production	2875	-	-	-	-	179	-	5220	-	-	8274
Imports	13	-	1535	-	-	-	-	-	273	-	1821
Exports	-112	-	-	-	-	-	-	-	-	-	-112
Intl. Marine Bunkers	-	-	-	-	-	-	-	-	-	-	-
Stock changes	-	-	-31	-	-	-	-	-	-	-	-31
TPES	2775	-	1504	-	-	179	-	5220	273	-	9951
Electricity and CHP Plants	-1566	-	-	-	-	-179	-	-	630	-	-1116
Petroleum Refineries	-	-	-	-	-	-	-	-	-	-	-
Other Transformation *	-361	-	-194	-	-	-	-	-	-39	-	-593
TFC	848	-	1310	-	-	-	-	5220	864	-	8242
INDUSTRY SECTOR	343	-	105	-	-	-	-	97	490	-	1034
Iron and Steel	191	-	-	-	-	-	-	-	-	-	191
Chemical and Petrochemical	-	-	-	-	-	-	-	-	-	-	-
Non-Metallic Minerals	-	-	-	-	-	-	-	-	-	-	-
Non-specified	151	-	105	-	-	-	-	97	490	-	843
TRANSPORT SECTOR	19	-	813	-	-	-	-	-	-	-	832
Air	-	-	126	-	-	-	-	-	-	-	126
Road	-	-	687	-	-	-	-	-	-	-	687
Non-specified	19	-	-	-	-	-	-	-	-	-	19
OTHER SECTORS	486	-	367	-	-	-	-	5123	374	-	6350
Agriculture	337	-	101	-	-	-	-	266	85	-	789
Comm. and Publ. Services	142	-	-	-	-	-	-	-	131	-	273
Residential	7	-	28	-	-	-	-	4857	158	-	5050
Non-specified	-	-	237	-	-	-	-	-	-	-	237
NON-ENERGY USE	-	-	26	-	-	-	-	-	-	-	26
Electricity Generated - GWh	*5236*	-	-	-	-	*2087*	-	-	-	-	*7323*
Heat Generated - TJ	-	-	-	-	-	-	-	-	-	-	-

APPROVISIONNEMENT ET DEMANDE 1997	Charbon	Pétrole brut	Produits pétroliers	Gaz	Nucléaire	Hydro	Géotherm. solaire etc.	En. ren. combust. & déchets	Electricité	Chaleur	Total
Indigenous Production	2712	-	-	-	-	153	-	5288	-	-	8152
Imports	13	-	1535	-	-	-	-	-	345	-	1893
Exports	-88	-	-	-	-	-	-	-	-	-	-88
Intl. Marine Bunkers	-	-	-	-	-	-	-	-	-	-	-
Stock changes	-	-	-31	-	-	-	-	-	-	-	-31
TPES	2636	-	1504	-	-	153	-	5288	345	-	9926
Electricity and CHP Plants	-1396	-	-	-	-	-153	-	-	628	-	-921
Petroleum Refineries	-	-	-	-	-	-	-	-	-	-	-
Other Transformation *	-448	-	-194	-	-	-	-	-	-67	-	-708
TFC	792	-	1310	-	-	-	-	5288	906	-	8296
INDUSTRY SECTOR	320	-	105	-	-	-	-	98	488	-	1010
Iron and Steel	166	-	-	-	-	-	-	-	-	-	166
Chemical and Petrochemical	-	-	-	-	-	-	-	-	-	-	-
Non-Metallic Minerals	-	-	-	-	-	-	-	-	-	-	-
Non-specified	154	-	105	-	-	-	-	98	488	-	844
TRANSPORT SECTOR	20	-	813	-	-	-	-	-	-	-	832
Air	-	-	126	-	-	-	-	-	-	-	126
Road	-	-	687	-	-	-	-	-	-	-	687
Non-specified	20	-	-	-	-	-	-	-	-	-	20
OTHER SECTORS	453	-	367	-	-	-	-	5190	419	-	6428
Agriculture	301	-	101	-	-	-	-	269	91	-	762
Comm. and Publ. Services	145	-	-	-	-	-	-	-	157	-	302
Residential	7	-	28	-	-	-	-	4921	171	-	5127
Non-specified	-	-	237	-	-	-	-	-	-	-	237
NON-ENERGY USE	-	-	26	-	-	-	-	-	-	-	26
Electricity Generated - GWh	*5520*	-	-	-	-	*1777*	-	-	-	-	*7297*
Heat Generated - TJ	-	-	-	-	-	-	-	-	-	-	-

* Includes Transfers, Statistical Differences, Own Use and Distribution Losses.

SUMMARY TABLES AND ENERGY INDICATORS

TABLEAUX RECAPITULATIFS ET INDICATEURS ENERGETIQUES

INTERNATIONAL ENERGY AGENCY

Production of Coal (Mtoe)
Production de charbon (Mtep)
Erzeugung von Kohle (Mtoe)
Produzione di carbone (Mtep)
石炭の生産量（石油換算百万トン）
Producción de carbón (Mtep)
Производство угля (мил.тон нефтяного эквивалента (Мтнэ))

	1971	1973	1978	1983	1984	1985	1986	1987	1988
Algérie	0.24	0.20	-	0.01	0.01	0.01	0.01	0.01	0.00
RD du Congo	0.07	0.08	0.06	0.07	0.07	0.07	0.07	0.07	0.07
Maroc	0.29	0.35	0.44	0.42	0.47	0.43	0.43	0.36	0.36
Mozambique	0.20	0.24	0.09	0.04	0.02	0.02	0.00	0.03	0.03
Nigéria	0.12	0.20	0.13	0.03	0.05	0.09	0.09	0.09	0.05
Afrique du Sud	33.06	35.14	52.64	83.69	93.72	99.75	101.73	101.45	104.21
Tanzanie	-	-	0.00	0.01	0.01	0.01	0.00	0.00	0.00
Zambie	0.48	0.55	0.36	0.27	0.30	0.30	0.33	0.27	0.37
Zimbabwe	1.76	1.73	1.55	1.54	1.67	1.91	2.18	2.99	2.81
Autre Afrique	0.09	0.10	0.30	0.39	0.41	0.47	0.49	0.56	0.57
Afrique	36.31	38.59	55.58	86.46	96.74	103.07	105.34	105.83	108.48
Argentine	0.37	0.27	0.26	0.29	0.30	0.24	0.22	0.22	0.30
Brésil	1.15	1.10	2.06	3.03	3.46	3.47	3.25	3.03	3.22
Chili	1.13	0.96	0.78	0.69	0.82	0.89	1.12	1.07	1.32
Colombie	1.78	1.84	2.75	3.37	4.31	5.83	6.98	8.75	9.93
Pérou	0.02	0.02	0.02	0.06	0.07	0.09	0.10	0.07	0.09
Vénézuela	0.03	0.03	0.05	0.02	0.03	0.03	0.04	0.04	0.67
Amérique latine	4.49	4.22	5.92	7.47	9.00	10.55	11.71	13.18	15.53
Inde	37.08	39.86	51.74	67.53	72.02	75.47	81.33	88.48	95.64
Indonésie	0.12	0.09	0.16	0.30	0.67	1.23	1.57	1.86	2.84
RPD de Corée	23.70	21.39	25.31	27.99	27.99	29.03	29.54	29.54	29.85
Malaisie	-	-	-	-	-	-	-	-	0.01
Myanmar	0.01	0.01	0.02	0.03	0.04	0.04	0.04	0.03	0.03
Pakistan	0.60	0.51	0.56	0.72	0.84	1.00	0.99	1.01	1.23
Philippines	0.02	0.02	0.12	0.49	0.58	0.61	0.60	0.56	0.64
Taipei chinois	2.58	2.10	1.82	1.39	1.25	1.15	1.07	0.93	0.76
Thaïlande	0.10	0.08	0.18	0.58	0.67	1.50	1.59	2.00	2.11
Viêt-Nam	1.50	1.50	3.00	3.15	2.50	2.80	3.06	3.17	3.03
Autre Asie	0.08	0.07	0.13	0.09	0.09	0.09	0.10	0.10	0.08
Asie	65.79	65.63	83.04	102.26	106.65	112.91	119.88	127.69	136.22
Rép. populaire de Chine	192.08	204.33	302.82	350.12	386.72	427.42	438.08	454.70	480.14
Chine	192.08	204.33	302.82	350.12	386.72	427.42	438.08	454.70	480.14
Albanie	0.24	0.28	0.42	0.65	0.70	0.75	0.76	0.75	0.76
Bulgarie	4.70	4.65	4.45	5.54	5.75	5.28	6.00	6.27	5.82
Roumanie	5.20	6.05	6.90	9.73	9.83	10.29	10.47	11.28	12.59
République slovaque	1.70	1.70	1.70	1.70	1.69	1.68	1.57	1.64	1.65
Bosnie-Herzégovine	-	-	-	-	-	-	-	-	-
Croatie	-	-	-	-	-	-	-	-	-
Ex-RYM	-	-	-	-	-	-	-	-	-
Slovénie	-	-	-	-	-	-	-	-	-
RF de Yougoslavie	-	-	-	-	-	-	-	-	-
Ex-Yougoslavie	7.33	7.65	9.23	13.93	14.91	15.56	15.73	15.89	15.18
Europe non-OCDE	19.17	20.33	22.70	31.55	32.88	33.57	34.53	35.82	36.00

Production of Coal (Mtoe)
Production de charbon (Mtep)
Erzeugung von Kohle (Mtoe)
Produzione di carbone (Mtep)
石炭の生産量（石油換算百万トン）
Producción de carbón (Mtep)
Производство угля (мил.тон нефтяного эквивалента (Мтнэ))

1989	1990	1991	1992	1993	1994	1995	1996	1997	
0.00	0.00	0.03	0.03	-	-	-	-	-	Algeria
0.08	0.08	0.08	0.08	0.06	0.06	0.06	0.06	0.06	DR of Congo
0.28	0.29	0.31	0.32	0.34	0.36	0.36	0.28	0.21	Morocco
0.03	0.02	0.03	0.02	0.02	0.02	0.02	-	-	Mozambique
0.05	0.06	0.08	0.06	0.07	0.08	0.09	0.09	0.09	Nigeria
101.09	100.16	102.06	100.03	106.91	111.11	116.90	116.93	124.68	South Africa
0.00	0.00	0.00	0.00	0.00	0.00	0.00	0.00	0.00	Tanzania
0.23	0.26	0.21	0.24	0.18	0.11	0.11	0.11	0.11	Zambia
2.82	3.05	3.44	3.41	3.25	3.03	2.88	2.87	2.71	Zimbabwe
0.60	0.69	0.68	0.72	0.69	0.77	0.72	0.66	0.66	Other Africa
105.20	104.62	106.92	104.93	111.52	115.55	121.14	121.00	128.51	**Africa**
0.30	0.16	0.17	0.12	0.10	0.21	0.18	0.18	0.15	Argentina
2.84	1.88	2.06	1.84	1.79	1.99	2.00	1.85	2.16	Brazil
1.34	1.50	1.50	1.10	0.92	0.80	0.70	0.68	0.71	Chile
12.29	13.30	15.34	14.24	13.76	14.73	16.73	19.54	19.93	Colombia
0.07	0.07	0.04	0.06	0.07	0.05	0.10	0.04	0.03	Peru
1.31	1.35	1.55	1.89	2.39	2.92	2.85	2.14	3.41	Venezuela
18.15	18.26	20.67	19.25	19.03	20.69	22.57	24.45	26.40	**Latin America**
98.79	104.29	113.21	117.72	122.93	127.55	135.69	141.61	147.23	India
5.42	6.45	8.43	13.75	18.04	19.85	25.30	30.85	33.89	Indonesia
30.37	28.52	26.66	24.81	22.96	21.13	19.56	18.11	17.56	DPR of Korea
0.08	0.07	0.11	0.12	0.18	0.10	0.08	0.15	0.15	Malaysia
0.02	0.03	0.03	0.03	0.03	0.03	0.01	0.01	0.01	Myanmar
1.13	1.23	1.37	1.39	1.46	1.58	1.36	1.63	1.41	Pakistan
0.64	0.60	0.61	0.80	0.80	0.59	0.54	0.47	0.45	Philippines
0.49	0.29	0.25	0.21	0.20	0.18	0.15	0.09	0.06	Chinese Taipei
2.58	3.60	4.26	4.46	4.51	4.96	5.34	6.23	6.78	Thailand
2.56	2.56	2.60	2.61	2.78	3.07	3.27	5.60	6.55	Vietnam
0.08	0.07	0.06	0.01	0.01	0.00	0.00	0.00	0.00	Other Asia
142.16	147.71	157.59	165.90	173.91	179.05	191.31	204.74	214.11	**Asia**
516.53	529.14	532.83	547.03	570.73	622.43	680.37	698.35	686.41	People's Rep.of China
516.53	529.14	532.83	547.03	570.73	622.43	680.37	698.35	686.41	**China**
0.77	0.49	0.26	0.19	0.14	0.04	0.04	0.02	0.02	Albania
5.84	5.38	4.70	5.06	4.85	4.77	5.12	5.17	4.90	Bulgaria
12.41	7.57	6.44	7.53	7.15	7.31	7.36	7.53	6.23	Romania
1.54	1.40	1.21	1.02	1.04	1.06	1.10	1.12	1.15	Slovak Republic
-	-	-	3.19	2.76	0.30	0.35	0.35	0.35	Bosnia-Herzegovina
-	-	-	0.07	0.06	0.06	0.05	0.04	0.03	Croatia
-	-	-	1.48	1.47	1.46	1.54	1.52	1.42	FYROM
-	-	-	1.18	1.09	1.03	1.04	1.01	1.05	Slovenia
-	-	-	8.55	7.98	8.17	8.52	8.19	8.67	FR of Yugoslavia
15.22	16.21	15.24	14.47	13.36	11.02	11.49	11.11	11.52	Former Yugoslavia
35.78	31.05	27.84	28.27	26.53	24.20	25.11	24.95	23.81	**Non-OECD Europe**

Production of Coal (Mtoe)
Production de charbon (Mtep)
Erzeugung von Kohle (Mtoe)
Produzione di carbone (Mtep)
石炭の生産量（石油換算百万トン）
Producción de carbón (Mtep)
Производство угля (мил.тон нефтяного эквивалента (Мтнэ))

	1971	1973	1978	1983	1984	1985	1986	1987	1988
Bélarus	-	-	-	-	-	-	-	-	-
Estonie	-	-	-	-	-	-	-	-	-
Géorgie	-	-	-	-	-	-	-	-	-
Kazakhstan	-	-	-	-	-	-	-	-	-
Kirghizistan	-	-	-	-	-	-	-	-	-
Lettonie	-	-	-	-	-	-	-	-	-
Lituanie	-	-	-	-	-	-	-	-	-
Russie	-	-	-	-	-	-	-	-	-
Tadjikistan	-	-	-	-	-	-	-	-	-
Ukraine	-	-	-	-	-	-	-	-	-
Ouzbékistan	-	-	-	-	-	-	-	-	-
Ex-URSS	315.28	329.33	347.46	317.83	310.46	312.48	323.88	325.24	331.82
Iran	0.37	0.65	0.55	0.49	0.76	0.77	0.78	0.76	0.77
Israël	-	-	-	-	-	-	-	-	-
Moyen-Orient	0.37	0.65	0.55	0.49	0.76	0.77	0.78	0.76	0.77
Total non-OCDE	633.48	663.08	818.08	896.19	943.21	1 000.76	1 034.19	1 063.22	1 108.96
OCDE Amérique du N.	324.12	346.56	387.62	443.39	511.22	502.47	502.37	523.98	537.12
OCDE Pacifique	63.51	66.09	69.78	89.42	93.78	102.46	111.04	119.07	110.41
OCDE Europe	423.21	404.02	411.20	420.72	384.54	413.57	417.84	410.12	403.91
Total OCDE	810.83	816.68	868.61	953.53	989.54	1 018.51	1 031.25	1 053.17	1 051.44
Monde	1 444.32	1 479.76	1 686.69	1 849.71	1 932.75	2 019.26	2 065.44	2 116.39	2 160.41

Production of Coal (Mtoe)
Production de charbon (Mtep)
Erzeugung von Kohle (Mtoe)
Produzione di carbone (Mtep)
石炭の生産量（石油換算百万トン）
Producción de carbón (Mtep)
Производство угля (мил.тон нефтяного эквивалента (Мтнэ))

1989	1990	1991	1992	1993	1994	1995	1996	1997	
-	-	-	1.25	0.58	0.70	0.63	0.57	0.55	Belarus
-	-	-	3.99	2.91	2.88	2.60	2.91	2.80	Estonia
-	-	-	0.09	0.04	0.02	0.02	0.01	0.00	Georgia
-	-	-	55.77	49.21	45.98	36.64	33.76	32.01	Kazakhstan
-	-	-	0.85	0.67	0.52	0.30	0.26	0.32	Kyrgyzstan
-	-	-	0.40	0.07	0.13	0.07	0.08	0.08	Latvia
-	-	-	0.02	0.01	0.02	0.01	0.02	0.02	Lithuania
-	-	-	138.89	126.72	112.38	109.49	106.28	101.12	Russia
-	-	-	0.07	0.07	0.04	0.01	0.01	0.00	Tajikistan
-	-	-	68.55	59.57	48.58	44.19	38.98	39.59	Ukraine
-	-	-	1.66	1.35	1.35	1.08	1.00	1.04	Uzbekistan
318.30	300.51	276.67	264.47	234.74	207.55	190.41	180.09	173.58	**Former USSR**
0.74	0.80	0.86	0.92	0.94	0.55	0.64	0.57	0.57	Iran
-	0.03	0.03	0.04	0.04	0.04	0.05	0.04	0.04	Israel
0.74	0.83	0.89	0.96	0.98	0.59	0.69	0.61	0.61	**Middle East**
1 136.85	1 132.12	1 123.42	1 137.87	1 143.90	1 175.11	1 236.22	1 257.98	1 257.36	**Non-OECD Total**
555.92	580.16	560.89	556.16	528.54	579.17	576.67	593.58	609.93	OECD North America
117.54	120.18	123.82	128.80	129.05	127.62	135.94	137.85	145.32	OECD Pacific
391.77	355.63	324.39	313.29	291.11	270.61	273.91	268.70	261.59	OECD Europe
1 065.23	1 055.97	1 009.10	998.25	948.70	977.40	986.52	1 000.14	1 016.85	**OECD Total**
2 202.08	2 188.09	2 132.51	2 136.12	2 092.60	2 152.51	2 222.74	2 258.11	2 274.21	**World**

Production of Crude Oil and NGL (Mtoe)
Production de pétrole brut et LGN (Mtep)
Erzeugung von Rohöl und Kondensaten (Mtoe)
Produzione di petrolio grezzo e LGN (Mtep)
原油及びNGLの生産量（石油換算百万トン）
Producción de petróleo crudo y líquidos de gas natural (Mtep)
Производство сырой нефти и газовых конденсатов (Мтнэ)

	1971	1973	1978	1983	1984	1985	1986	1987	1988
Algérie	39.03	52.50	59.60	47.76	50.62	52.03	53.51	54.21	54.85
Angola	5.84	8.33	6.60	8.97	10.29	11.69	14.22	17.68	22.74
Bénin	-	-	-	0.02	0.01	0.33	0.37	0.32	0.24
Cameroun			0.51	5.61	7.12	8.54	8.39	8.36	8.30
Congo	0.01	2.14	2.63	5.50	6.18	6.09	6.10	6.47	7.21
RD du Congo	-	-	1.15	1.28	1.38	1.68	1.63	1.57	1.51
Egypte	15.20	8.64	25.60	36.80	42.28	45.54	41.57	46.74	44.70
Gabon	5.89	7.73	10.54	7.85	7.95	8.67	8.32	7.80	7.93
Ghana	-	-	-	0.06	0.03	0.01	-	-	-
Côte d'Ivoire	-	-	-	1.04	1.02	1.03	0.98	0.81	0.53
Libye	136.11	109.04	99.68	54.98	54.94	52.85	52.20	50.97	54.08
Maroc	0.02	0.04	0.02	0.02	0.01	0.02	0.02	0.02	0.02
Nigéria	77.11	103.54	95.51	62.31	70.18	75.59	73.96	67.68	74.77
Sénégal	-	-	-	-	-	-	-	-	0.00
Afrique du Sud	-	-	-	-	-	-	-	:	-
Soudan	-	-	-	-	-	-	-	-	-
Tunisie	4.23	3.99	5.09	5.70	5.64	5.57	5.41	5.19	5.15
Autre Afrique	-	-	-	-	-	-	-	-	-
Afrique	283.45	295.96	306.95	237.90	257.66	269.65	266.68	267.83	282.04
Argentine	22.14	22.16	23.79	26.56	26.11	25.06	23.73	23.43	23.66
Bolivie	1.79	2.29	1.61	1.41	1.29	1.23	1.20	1.23	1.18
Brésil	8.93	8.92	8.50	17.23	24.13	28.58	30.96	29.31	30.79
Chili	1.80	1.85	1.09	2.18	2.14	1.98	1.87	1.80	1.42
Colombie	11.48	9.84	6.93	8.10	8.87	9.38	15.75	20.28	19.76
Cuba	0.13	0.15	0.32	0.83	0.76	0.86	0.93	0.89	0.71
Equateur	0.20	10.78	10.64	12.44	13.55	14.60	15.03	9.03	15.52
Guatemala	-	-	0.00	0.35	0.24	0.15	0.25	0.19	0.19
Pérou	3.21	3.64	7.80	8.78	9.63	9.75	9.19	8.33	7.22
Trinité-et-Tobago	6.53	8.37	11.55	8.04	8.83	9.17	8.79	8.03	7.87
Vénézuela	188.08	179.48	116.05	97.30	97.67	90.97	97.71	93.44	103.07
Autre Amérique latine	-	0.00	0.07	0.12	0.22	0.25	0.28	0.29	0.35
Amérique latine	244.28	247.49	188.34	183.35	193.43	191.96	205.69	196.22	211.73
Bangladesh		0.01	0.01	-		0.08	0.09	0.11	0.12
Brunei	6.54	11.61	12.04	10.02	9.71	8.80	8.63	7.63	7.38
Inde	7.34	7.36	11.52	25.91	28.82	30.97	32.50	31.60	33.40
Indonésie	44.95	67.43	81.41	67.05	75.56	67.27	72.36	71.42	68.98
Malaisie	3.34	4.43	10.94	19.08	22.27	22.39	25.33	25.12	26.85
Myanmar	0.88	0.99	1.37	1.45	1.60	1.47	1.44	0.99	0.74
Pakistan	0.42	0.43	0.49	0.66	0.68	1.31	1.98	2.07	2.20
Philippines	-	-	-	0.68	0.55	0.37	0.44	0.27	0.29
Taipei chinois	0.11	0.15	0.22	0.24	0.24	0.22	0.19	0.22	0.25
Thaïlande	0.01	0.01	0.01	0.64	1.13	2.25	2.31	2.16	2.44
Viêt-Nam	-	-	-	-	-	-	-	-	0.69
Autre Asie	0.00	0.00	0.00	0.01	0.01	0.01	0.01	0.01	0.00
Asie	63.61	92.41	118.00	125.74	140.56	135.15	145.29	141.60	143.35
Rép. populaire de Chine	40.12	54.57	105.92	107.98	116.68	127.14	133.04	136.55	139.51
Chine	40.12	54.57	105.92	107.98	116.68	127.14	133.04	136.55	139.51

Production of Crude Oil and NGL (Mtoe)
Production de pétrole brut et LGN (Mtep)
Erzeugung von Rohöl und Kondensaten (Mtoe)
Produzione di petrolio grezzo e LGN (Mtep)
原油及びNGLの生産量（石油換算百万トン）
Producción de petróleo crudo y líquidos de gas natural (Mtep)
Производство сырой нефти и газовых конденсатов (Мтнэ)

1989	1990	1991	1992	1993	1994	1995	1996	1997	
55.94	58.76	58.04	57.97	58.02	57.25	57.61	60.72	61.67	Algeria
23.12	24.05	25.25	24.21	24.70	24.75	25.97	35.42	35.70	Angola
0.19	0.21	0.20	0.14	0.16	0.13	0.10	0.07	0.07	Benin
8.38	8.55	7.74	7.14	6.95	6.58	6.05	6.07	6.36	Cameroon
8.17	8.23	8.26	8.87	9.64	9.80	9.51	10.63	12.64	Congo
1.51	1.42	1.38	1.25	1.29	1.35	1.38	1.38	1.32	DR of Congo
45.00	46.23	46.14	46.60	48.29	47.24	47.49	45.80	44.64	Egypt
10.32	13.59	14.91	14.75	15.76	15.00	18.46	18.61	18.79	Gabon
-	-	-	-	0.01	0.31	0.31	0.36	0.36	Ghana
0.11	0.10	0.06	0.30	0.33	0.34	0.35	0.90	0.80	Ivory Coast
55.44	67.99	76.21	72.33	68.78	69.61	72.28	72.00	73.07	Libya
0.01	0.01	0.01	0.01	0.01	0.01	0.00	0.33	0.24	Morocco
88.48	90.18	94.08	97.58	98.54	98.24	99.59	110.30	117.25	Nigeria
0.00	0.00	0.00	0.00	0.00	0.00	0.00	0.00	-	Senegal
-	-	-	-	0.40	0.40	0.40	0.40	0.40	South Africa
-	-	-	-	-	-	-	0.10	0.26	Sudan
5.17	4.71	5.43	5.44	5.17	4.58	4.48	4.42	4.01	Tunisia
-	-	-	-	0.22	0.24	0.34	0.86	2.18	Other Africa
301.83	324.02	337.71	336.59	338.27	335.84	344.32	368.37	379.76	**Africa**
24.59	25.83	26.09	30.18	32.33	35.94	39.01	42.38	44.80	Argentina
1.25	1.36	1.42	1.31	1.36	1.56	1.75	1.80	1.88	Bolivia
32.92	35.15	34.83	35.07	36.49	37.91	37.87	42.95	45.99	Brazil
1.30	1.14	1.05	0.94	0.92	0.92	0.78	0.61	0.52	Chile
20.77	22.96	22.29	23.09	23.98	23.94	30.70	32.93	34.16	Colombia
0.71	0.66	0.52	0.87	1.09	1.28	1.45	1.45	1.68	Cuba
14.39	14.90	15.68	16.79	18.11	19.98	20.16	20.10	21.07	Ecuador
0.18	0.20	0.19	0.28	0.34	0.37	0.47	0.81	1.08	Guatemala
6.62	6.53	5.84	5.89	6.43	6.48	6.20	6.13	5.95	Peru
7.77	7.87	7.45	7.46	6.78	7.16	7.20	7.13	6.72	Trinidad-and-Tobago
99.86	105.75	120.04	124.02	115.13	130.74	154.15	160.41	166.15	Venezuela
0.37	0.38	0.43	0.43	0.43	0.43	0.44	0.40	0.40	Other Latin America
210.73	222.72	235.82	246.34	243.38	266.70	300.19	317.09	330.39	**Latin America**
0.12	0.11	0.12	0.12	0.13	0.11	0.01	0.01	0.01	Bangladesh
7.76	7.87	8.44	9.45	9.15	9.44	9.27	8.77	8.72	Brunei
35.74	35.56	33.86	30.87	29.60	33.96	38.01	34.91	37.59	India
73.51	74.79	81.85	77.62	77.58	77.72	76.96	78.51	77.29	Indonesia
29.25	31.02	32.14	32.57	34.70	36.72	37.95	36.35	38.04	Malaysia
0.74	0.73	0.77	0.73	0.68	0.69	0.48	0.41	0.40	Myanmar
2.31	2.72	3.33	3.17	3.09	2.88	2.82	3.02	2.92	Pakistan
0.27	0.24	0.16	0.41	0.48	0.27	0.15	0.05	0.02	Philippines
0.13	0.17	0.10	0.07	0.06	0.06	0.06	0.06	0.05	Chinese Taipei
2.24	2.73	3.17	3.56	3.66	4.27	4.13	5.26	5.71	Thailand
1.52	2.75	3.99	5.60	6.42	7.02	8.65	8.75	9.87	Vietnam
0.00	4.54	5.04	5.34	5.44	5.54	5.04	4.03	4.03	Other Asia
153.59	163.23	172.97	169.51	170.99	178.67	183.51	180.12	184.64	**Asia**
140.13	140.80	143.53	144.30	145.17	146.08	150.04	157.33	160.74	People's Rep.of China
140.13	140.80	143.53	144.30	145.17	146.08	150.04	157.33	160.74	**China**

Production of Crude Oil and NGL (Mtoe)
Production de pétrole brut et LGN (Mtep)
Erzeugung von Rohöl und Kondensaten (Mtoe)
Produzione di petrolio grezzo e LGN (Mtep)
原油及びNGLの生産量（石油換算百万トン）
Producción de petróleo crudo y líquidos de gas natural (Mtep)
Производство сырой нефти и газовых конденсатов (Мтнэ)

	1971	1973	1978	1983	1984	1985	1986	1987	1988
Albanie	1.64	2.09	2.18	1.19	1.37	1.18	1.19	1.17	1.16
Bulgarie	0.31	0.19	0.25	0.15	0.15	0.20	0.10	0.09	0.08
Roumanie	13.39	13.87	13.32	11.25	11.12	10.41	9.83	9.23	9.11
République slovaque	0.16	0.13	0.06	0.04	0.03	0.07	0.09	0.10	0.10
Croatie	-	-	-	-	-	-	-	-	-
Slovénie									
RF de Yougoslavie	-	-	-	-	-	-	-	-	-
Ex-Yougoslavie	3.02	3.40	4.16	4.28	4.14	4.24	4.24	3.95	3.76
Europe non-OCDE	18.53	19.68	19.98	16.91	16.82	16.09	15.45	14.53	14.21
Azerbaïdjan	-	-	-	-	-	-	-	-	-
Bélarus	-	-	-	-	-	-	-	-	-
Estonie	-	-	-	-	-	-	-	-	-
Géorgie	-	-	-	-	-	-	-	-	-
Kazakhstan	-	-	-	-	-	-	-	-	-
Kirghizistan	-	-	-	-	-	-	-	-	-
Lituanie	-	-	-	-	-	-	-	-	-
Russie	-	-	-	-	-	-	-	-	-
Tadjikistan	-	-	-	-	-	-	-	-	-
Turkménistan	-	-	-	-	-	-	-	-	-
Ukraine	-	-	-	-	-	-	-	-	-
Ouzbékistan	-	-	-	-	-	-	-	-	-
Ex-URSS	378.96	431.21	570.30	619.33	615.71	598.22	617.76	627.23	627.37
Bahrein	3.81	3.48	2.81	2.42	2.42	2.40	2.44	2.43	2.48
Iran	231.25	298.72	268.74	125.39	111.17	113.87	93.24	118.31	114.37
Irak	85.18	101.83	130.00	49.16	60.61	72.28	85.91	106.08	132.33
Israël	5.83	6.10	0.02	0.01	0.01	0.01	0.01	0.01	0.02
Jordanie	-	-	-	-	-	-	0.02	0.02	0.02
Koweit	166.01	154.95	111.42	55.79	60.09	55.37	75.95	71.71	76.55
Oman	14.89	14.89	16.34	20.18	21.60	25.84	29.15	30.33	32.33
Qatar	21.20	28.24	24.01	15.37	21.02	16.00	16.79	15.28	16.72
Arabie saoudite	243.00	387.02	432.19	269.77	248.64	185.04	264.79	225.98	272.83
Syrie	5.31	5.57	9.96	9.40	9.45	9.28	9.34	12.41	14.22
Emirats arabes unis	52.02	78.18	93.75	60.27	63.40	62.84	71.73	77.62	83.53
Yémen	-	-	-	-	-	-	-	0.97	8.31
Moyen-Orient	828.51	1 078.99	1 089.26	607.75	598.42	542.93	649.38	661.17	753.71
Total non-OCDE	1 857.46	2 220.32	2 398.75	1 898.96	1 939.28	1 881.15	2 033.29	2 045.12	2 171.93
OCDE Amérique du N.	646.71	657.65	650.34	736.10	755.75	757.11	732.40	730.51	724.88
OCDE Pacifique	15.67	20.82	23.93	21.12	25.36	29.72	30.68	30.39	30.60
OCDE Europe	22.65	23.30	92.83	176.19	192.55	201.01	206.40	211.57	210.29
Total OCDE	685.03	701.77	767.11	933.41	973.66	987.84	969.47	972.47	965.77
Monde	2 542.49	2 922.09	3 165.86	2 832.38	2 912.95	2 868.99	3 002.76	3 017.59	3 137.70
Pour mémoire: OPEP	1 283.95	1 560.96	1 512.38	905.15	913.91	844.10	958.15	952.71	1 052.08

Production of Crude Oil and NGL (Mtoe)
Production de pétrole brut et LGN (Mtep)
Erzeugung von Rohöl und Kondensaten (Mtoe)
Produzione di petrolio grezzo e LGN (Mtep)
原油及びNGLの生産量（石油換算百万トン）
Producción de petróleo crudo y líquidos de gas natural (Mtep)
Производство сырой нефти и газовых конденсатов (Мтнэ)

1989	1990	1991	1992	1993	1994	1995	1996	1997	
1.12	1.06	0.84	0.58	0.56	0.53	0.52	0.48	0.36	Albania
0.08	0.06	0.06	0.05	0.04	0.04	0.04	0.03	0.03	Bulgaria
8.91	7.70	6.59	6.65	6.75	6.80	6.77	6.68	6.58	Romania
0.10	0.08	0.12	0.09	0.07	0.07	0.08	0.07	0.06	Slovak Republic
-	-	-	2.17	2.20	1.88	1.81	1.75	1.80	*Croatia*
-	-	-	0.00	0.00	0.00	0.00	0.00	0.00	*Slovenia*
-	-	-	1.19	1.17	1.10	1.09	1.05	1.00	*FR of Yugoslavia*
3.47	3.21	3.15	3.36	3.37	2.98	2.90	2.80	2.80	Former Yugoslavia
13.67	12.10	10.76	10.74	10.80	10.41	10.31	10.06	9.83	**Non-OECD Europe**
-	-	-	11.62	10.67	9.61	9.21	9.14	9.07	Azerbaijan
-	-	-	2.01	2.02	2.01	1.94	1.87	1.83	Belarus
-	-	-	0.28	0.18	0.31	0.32	0.35	0.37	Estonia
-	-	-	0.10	0.10	0.07	0.05	0.13	0.13	Georgia
-	-	-	27.90	23.11	20.37	20.54	23.08	25.57	Kazakhstan
-	-	-	0.11	0.09	0.09	0.09	0.10	0.09	Kyrgyzstan
-	-	-	0.06	0.07	0.09	0.13	0.16	0.21	Lithuania
-	-	-	401.33	353.25	317.35	306.63	301.00	305.52	Russia
-	-	-	0.06	0.05	0.03	0.03	0.02	0.03	Tajikistan
-	-	-	4.97	4.17	4.08	4.45	4.35	4.72	Turkmenistan
-	-	-	4.50	4.27	4.22	4.08	4.50	4.59	Ukraine
-	-	-	3.31	3.96	5.53	7.48	7.52	7.80	Uzbekistan
610.22	573.53	514.82	456.00	401.78	363.47	354.65	351.90	359.59	**Former USSR**
2.27	2.45	2.47	2.54	2.49	2.47	2.42	2.38	2.37	Bahrain
145.82	158.86	170.80	177.23	186.34	184.38	186.49	187.30	184.55	Iran
142.61	101.44	14.93	21.76	24.19	26.60	27.81	29.20	58.31	Iraq
0.02	0.01	0.01	0.01	0.03	0.03	0.03	0.03	0.03	Israel
-	-	-	0.00	0.00	0.00	0.00	0.00	-	Jordan
93.42	62.56	9.79	55.67	98.09	106.85	108.27	108.47	111.63	Kuwait
33.35	35.62	36.80	38.53	40.49	42.23	44.56	46.21	46.92	Oman
19.79	21.01	20.59	22.73	21.92	21.25	21.36	21.86	29.47	Qatar
271.71	343.39	437.68	447.05	440.78	435.63	437.79	439.20	449.51	Saudi Arabia
17.57	20.71	24.70	25.85	26.87	29.12	31.45	30.87	29.86	Syria
98.07	107.57	122.17	116.27	110.44	118.62	117.56	120.91	123.00	United Arab Emirates
9.38	9.71	10.19	9.24	11.34	17.60	17.94	18.17	19.03	Yemen
834.01	863.34	850.12	916.88	962.99	984.77	995.69	1 004.59	1 054.68	**Middle East**
2 264.17	2 299.75	2 265.73	2 280.62	2 273.55	2 286.23	2 339.02	2 389.78	2 479.96	**Non-OECD Total**
692.97	680.08	685.85	683.88	674.38	672.97	669.44	676.90	688.93	OECD North America
28.09	31.16	31.51	31.03	30.83	29.08	30.64	30.74	31.83	OECD Pacific
205.49	212.84	225.19	242.06	255.43	299.55	312.77	330.97	329.75	OECD Europe
926.56	924.08	942.55	956.97	960.64	1 001.60	1 012.85	1 038.61	1 050.51	**OECD Total**
3 190.73	3 223.82	3 208.28	3 237.59	3 234.19	3 287.83	3 351.87	3 428.39	3 530.47	**World**
1 144.64	1 192.29	1 206.16	1 270.21	1 299.80	1 326.89	1 359.87	1 388.87	1 451.90	*Memo: OPEC*

Production of Petroleum Products (Mtoe)
Production de produits pétroliers (Mtep)
Erzeugung von Ölprodukten (Mtoe)

Produzione di prodotti petroliferi (Mtep)

石油製品の生産量（石油換算百万トン）

Producción de productos petrolíferos (Mtep)

Производство нефтепродуктов (Мтнэ)

	1971	1973	1978	1983	1984	1985	1986	1987	1988
Algérie	2.50	6.30	5.66	19.98	21.65	20.95	22.47	21.55	22.77
Angola	0.66	0.73	0.96	1.11	1.15	1.45	1.34	1.34	1.36
Cameroun	-	-	-	1.25	1.28	1.17	1.17	1.18	1.24
Congo	-	-	-	0.62	0.51	0.46	0.51	0.50	0.50
RD du Congo	0.68	0.73	0.20	0.47	0.43	0.33	0.19	0.31	0.36
Egypte	5.02	6.90	11.83	18.20	19.64	20.18	20.96	22.11	22.33
Ethiopie	0.67	0.62	0.59	0.75	0.72	0.72	0.75	0.77	0.66
Gabon	1.02	1.07	1.69	1.00	0.93	0.76	0.75	0.60	0.64
Ghana	0.89	1.01	1.19	0.64	0.78	0.94	0.91	0.84	0.92
Côte d'Ivoire	0.82	1.17	1.64	1.87	1.87	1.97	1.88	1.86	1.89
Kenya	2.56	2.67	2.62	2.03	2.09	2.03	1.97	2.17	2.08
Libye	0.44	1.63	5.29	5.69	6.64	10.49	11.07	12.46	13.89
Maroc	1.52	2.26	2.99	4.54	4.68	4.80	4.54	4.59	5.13
Mozambique	0.79	0.74	0.60	0.29	0.12	-	-	-	-
Nigéria	1.98	2.82	2.69	7.91	8.69	8.41	7.43	9.38	10.99
Sénégal	0.56	0.68	0.77	0.29	0.36	0.30	0.49	0.54	0.72
Afrique du Sud	12.85	13.37	13.54	14.63	15.32	14.86	14.56	15.23	16.53
Soudan	0.69	1.15	1.00	0.72	0.59	0.61	0.79	0.47	0.79
Tanzanie	0.76	0.79	0.55	0.54	0.54	0.58	0.57	0.61	0.61
Tunisie	1.14	1.05	1.20	1.52	1.62	1.59	1.60	1.69	1.73
Zambie	-	0.42	0.80	0.72	0.64	0.60	0.56	0.59	0.60
Autre Afrique	0.88	0.85	0.62	0.76	0.50	0.73	0.65	1.78	1.80
Afrique	36.43	46.96	56.44	85.49	90.76	93.92	95.17	100.57	107.54
Argentine	23.18	23.63	24.36	24.10	23.58	22.89	21.44	21.14	21.28
Bolivie	0.67	0.76	1.17	1.05	1.03	1.05	0.98	1.04	1.08
Brésil	25.57	36.46	50.83	48.76	51.80	52.19	56.13	57.39	58.10
Chili	4.77	4.75	4.81	3.93	3.90	3.87	4.16	4.37	4.94
Colombie	7.84	8.30	8.14	9.59	9.45	9.46	10.97	11.15	12.10
Costa Rica	0.41	0.39	0.41	0.32	0.40	0.40	0.62	0.61	0.60
Cuba	4.41	5.47	6.62	7.06	6.67	6.86	6.79	7.10	7.89
République dominicaine	-	1.18	1.51	1.68	1.73	1.47	1.74	1.62	1.58
Equateur	1.38	1.61	4.12	3.75	4.43	4.36	4.91	4.19	5.92
El Salvador	0.45	0.60	0.71	0.58	0.61	0.65	0.65	0.71	0.69
Guatemala	0.80	0.94	0.82	0.59	0.69	0.68	0.50	0.63	0.63
Honduras	0.49	0.53	0.33	0.28	0.33	0.30	0.16	0.25	0.34
Jamaïque	1.62	1.76	0.99	0.99	0.57	0.89	0.71	0.60	0.71
Antilles néerlandaises	40.34	42.90	26.55	21.90	19.26	8.34	8.05	9.60	9.88
Nicaragua	0.48	0.63	0.62	0.53	0.41	0.48	0.50	0.52	0.52
Panama	3.68	3.31	2.33	1.61	1.44	1.22	0.92	1.36	0.95
Paraguay	0.18	0.23	0.30	0.21	0.15	0.20	0.21	0.24	0.28
Pérou	4.29	4.78	5.99	7.32	8.37	8.33	8.23	8.37	8.18
Trinité-et-Tobago	20.21	19.50	11.76	4.26	4.06	4.28	4.19	4.93	4.40
Uruguay	1.68	1.66	1.84	1.23	1.20	1.04	1.01	1.23	1.15
Vénézuela	63.72	66.41	51.06	45.10	44.11	45.92	45.31	42.63	47.98
Autre Amérique latine	9.74	12.74	8.48	8.25	6.57	1.58	0.76	0.79	0.79
Amérique latine	215.92	238.56	213.76	193.10	190.79	176.47	178.94	180.46	189.99

Production of Petroleum Products (Mtoe)
Production de produits pétroliers (Mtep)
Erzeugung von Ölprodukten (Mtoe)
Produzione di prodotti petroliferi (Mtep)
石油製品の生産量（石油換算百万トン）
Producción de productos petrolíferos (Mtep)
Производство нефтепродуктов (Мтнэ)

1989	1990	1991	1992	1993	1994	1995	1996	1997	
21.89	22.08	22.17	22.22	22.38	20.31	20.95	19.89	21.26	Algeria
1.38	1.40	1.40	1.39	1.38	1.38	1.36	1.54	1.92	Angola
1.25	1.24	0.92	0.86	0.85	0.89	0.89	0.88	0.91	Cameroon
0.62	0.63	0.65	0.61	0.65	0.63	0.61	0.63	0.63	Congo
0.34	0.35	0.36	0.33	0.37	0.37	0.35	0.35	0.35	DR of Congo
22.61	23.76	24.10	23.72	25.03	26.09	26.86	27.80	28.12	Egypt
0.67	0.66	0.69	0.76	0.76	0.76	0.76	0.77	0.43	Ethiopia
0.82	0.43	0.38	0.63	0.66	0.67	0.69	0.73	0.73	Gabon
1.01	0.84	1.07	1.12	0.90	1.18	1.19	1.20	1.20	Ghana
1.83	1.88	1.86	1.66	1.95	2.04	2.06	2.19	2.37	Ivory Coast
2.20	2.25	2.10	2.27	2.14	2.09	1.86	1.78	1.67	Kenya
13.55	13.61	13.76	13.83	13.36	14.61	14.91	15.12	15.26	Libya
5.51	5.66	5.51	6.08	6.12	6.50	6.12	5.51	5.83	Morocco
-	0.01	0.01	0.00	0.00	0.00	0.00	0.00	0.00	Mozambique
11.85	13.30	14.34	13.96	13.70	13.42	14.38	17.22	17.49	Nigeria
0.60	0.68	0.54	0.61	0.56	0.17	0.69	0.67	0.77	Senegal
16.77	18.05	18.20	18.33	19.40	20.66	22.70	21.79	19.21	South Africa
0.55	0.82	0.97	0.74	0.32	0.62	0.72	0.70	0.90	Sudan
0.58	0.59	0.60	0.60	0.60	0.60	0.61	0.61	0.61	Tanzania
1.74	1.84	1.81	1.73	1.75	1.70	1.83	1.85	2.06	Tunisia
0.59	0.55	0.56	0.53	0.57	0.56	0.57	0.58	0.58	Zambia
1.53	1.17	1.22	1.31	1.33	1.34	1.35	1.36	1.39	Other Africa
107.87	111.80	113.20	113.29	114.75	116.61	121.48	123.18	123.68	**Africa**
22.10	22.75	23.40	24.81	23.41	22.97	21.81	23.99	24.85	Argentina
1.15	1.21	1.30	1.29	1.30	1.42	1.52	1.61	1.66	Bolivia
58.66	58.37	56.17	58.20	58.61	62.28	61.10	65.20	70.30	Brazil
5.96	6.18	6.21	6.47	7.00	7.34	7.97	8.32	8.83	Chile
11.43	11.33	11.92	11.99	13.16	12.14	13.28	14.38	13.90	Colombia
0.65	0.44	0.35	0.53	0.54	0.57	0.74	0.63	0.61	Costa Rica
8.33	6.85	4.76	1.73	1.96	1.71	1.77	2.13	2.21	Cuba
1.81	1.52	1.82	1.96	1.90	1.98	2.05	2.12	2.22	Dominican Republic
5.73	6.11	6.39	6.06	6.06	6.48	6.52	7.13	7.24	Ecuador
0.64	0.67	0.76	0.82	0.92	0.85	0.76	0.76	0.91	El Salvador
0.62	0.46	0.57	0.78	0.77	0.82	0.84	0.79	0.83	Guatemala
0.37	0.33	0.32	0.28	-	-	-	-	-	Honduras
0.85	1.00	0.86	1.19	0.70	0.75	1.00	1.05	1.10	Jamaica
9.66	10.01	9.82	11.57	12.24	12.33	12.36	12.38	12.38	Netherlands Antilles
0.58	0.63	0.63	0.68	0.66	0.68	0.59	0.62	0.77	Nicaragua
0.93	1.15	1.09	1.71	1.71	1.06	1.21	2.04	2.02	Panama
0.28	0.31	0.28	0.31	0.25	0.26	0.20	0.16	0.15	Paraguay
7.42	7.35	7.52	7.57	7.62	7.26	6.86	7.26	7.77	Peru
4.01	4.30	5.23	5.20	4.82	5.16	4.83	5.19	5.36	Trinidad-and-Tobago
1.13	1.17	1.23	1.14	0.38	-	1.30	1.61	1.39	Uruguay
47.12	49.93	51.47	49.17	52.93	53.94	50.77	50.47	54.88	Venezuela
0.86	0.97	0.98	0.96	1.00	0.99	1.02	1.01	1.03	Other Latin America
190.29	193.03	193.08	194.43	197.95	200.99	198.52	208.85	220.41	**Latin America**

Production of Petroleum Products (Mtoe)
Production de produits pétroliers (Mtep)
Erzeugung von Ölprodukten (Mtoe)
Produzione di prodotti petroliferi (Mtep)
石油製品の生産量（石油換算百万トン）
Producción de productos petrolíferos (Mtep)
Производство нефтепродуктов (Мтнэ)

	1971	1973	1978	1983	1984	1985	1986	1987	1988
Bangladesh	0.74	0.61	1.05	0.95	1.15	1.05	1.00	1.07	1.08
Brunei	-	-	-	0.04	0.27	0.25	0.25	0.26	0.27
Inde	19.64	20.60	26.19	34.87	35.70	41.53	45.65	48.28	48.06
Indonésie	9.00	10.18	14.06	12.45	21.91	23.56	26.57	31.27	32.92
RPD de Corée	-	-	1.27	2.30	2.47	2.64	2.81	2.81	2.81
Malaisie	3.79	3.91	5.80	6.65	7.45	7.22	7.48	7.56	8.12
Myanmar	1.29	1.00	1.23	1.18	1.23	1.19	1.23	0.81	0.73
Pakistan	3.34	3.41	4.04	4.81	4.91	5.30	5.56	5.76	6.04
Philippines	8.50	8.76	9.51	8.99	8.02	6.59	7.34	8.42	9.50
Singapour	16.50	22.91	27.49	38.17	34.66	30.74	33.44	34.48	32.80
Sri Lanka	1.54	1.76	1.48	1.54	1.79	1.63	1.68	1.71	1.77
Taipei chinois	5.48	8.98	17.18	18.91	20.11	20.32	20.27	22.10	23.14
Thaïlande	4.98	7.07	8.04	8.02	7.80	7.92	8.49	8.57	8.98
Viêt-Nam	-	-	-	-	-	-	-	0.00	0.04
Autre Asie	-	-	-	-	-	-	-	-	-
Asie	74.80	89.19	117.33	138.90	147.46	149.94	161.76	173.08	176.27
Rép. populaire de Chine	32.40	41.78	67.46	80.98	83.13	86.90	93.64	98.99	103.27
Chine	32.40	41.78	67.46	80.98	83.13	86.90	93.64	98.99	103.27
Albanie	1.44	1.61	2.10	1.18	1.26	1.11	1.13	1.13	1.08
Bulgarie	7.77	9.37	11.88	11.72	11.31	11.63	11.48	11.67	12.12
Chypre	-	0.67	0.44	0.56	0.56	0.45	0.55	0.63	0.73
Roumanie	16.49	18.30	26.34	23.11	23.88	24.04	25.92	29.03	29.20
République slovaque	4.45	5.04	5.66	6.78	6.65	6.63	6.68	6.86	6.62
Croatie	-	-	-	-	-	-	-	-	-
Ex-RYM	-	-	-	-	-	-	-	-	-
Slovénie	-	-	-	-	-	-	-	-	-
RF de Yougoslavie	-	-	-	-	-	-	-	-	-
Ex-Yougoslavie	8.09	9.14	14.50	13.83	13.12	12.93	14.91	15.50	17.09
Europe non-OCDE	38.23	44.13	60.92	57.18	56.79	56.79	60.68	64.81	66.85
Azerbaïdjan	-	-	-	-	-	-	-	-	-
Bélarus	-	-	-	-	-	-	-	-	-
Estonie	-	-	-	-	-	-	-	-	-
Géorgie	-	-	-	-	-	-	-	-	-
Kazakhstan	-	-	-	-	-	-	-	-	-
Kirghizistan	-	-	-	-	-	-	-	-	-
Lituanie	-	-	-	-	-	-	-	-	-
Russie	-	-	-	-	-	-	-	-	-
Tadjikistan	-	-	-	-	-	-	-	-	-
Turkménistan	-	-	-	-	-	-	-	-	-
Ukraine	-	-	-	-	-	-	-	-	-
Ouzbékistan	-	-	-	-	-	-	-	-	-
Ex-URSS	293.18	336.54	424.60	460.98	455.88	453.58	450.85	466.67	467.06

Production of Petroleum Products (Mtoe)
Production de produits pétroliers (Mtep)
Erzeugung von Ölprodukten (Mtoe)
Produzione di prodotti petroliferi (Mtep)
石油製品の生産量（石油換算百万トン）
Producción de productos petrolíferos (Mtep)
Производство нефтепродуктов (Мтнэ)

1989	1990	1991	1992	1993	1994	1995	1996	1997	
1.16	1.11	1.16	1.07	1.40	1.31	1.20	1.03	1.04	Bangladesh
0.29	0.30	0.33	0.40	0.41	0.43	0.49	0.48	0.51	Brunei
52.21	51.60	50.85	54.30	53.89	56.83	59.59	64.36	67.56	India
33.69	37.37	38.21	40.29	40.52	39.99	40.48	44.19	44.96	Indonesia
2.97	2.97	2.46	1.95	1.43	1.14	1.14	1.15	1.11	DPR of Korea
8.85	9.93	10.32	10.13	10.94	14.00	16.88	17.90	16.35	Malaysia
0.72	0.72	0.68	0.76	0.79	0.87	0.96	0.78	1.02	Myanmar
5.80	5.81	6.52	6.44	6.34	6.43	6.05	6.52	7.03	Pakistan
9.77	11.07	10.63	11.91	11.65	12.02	15.66	17.76	18.01	Philippines
34.54	41.21	43.55	45.12	48.85	48.87	47.92	51.87	49.45	Singapore
1.18	1.75	1.59	1.57	1.69	1.96	1.90	1.99	1.82	Sri Lanka
25.03	23.86	24.11	24.21	26.78	27.87	31.87	33.56	34.73	Chinese Taipei
10.96	11.56	11.96	14.80	17.08	19.78	22.70	30.92	35.84	Thailand
0.04	0.04	0.04	0.04	0.00	0.00	-	0.04	0.04	Vietnam
-	-	0.05	0.05	0.05	0.05	0.05	0.05	0.05	Other Asia
187.20	199.30	202.47	213.04	221.84	231.55	246.88	272.62	279.52	**Asia**
107.01	109.27	115.82	122.25	131.55	131.88	143.42	153.82	166.43	People's Rep.of China
107.01	109.27	115.82	122.25	131.55	131.88	143.42	153.82	166.43	**China**
1.07	1.06	1.01	0.75	0.64	0.56	0.45	0.43	0.36	Albania
12.19	8.15	4.32	2.48	5.50	6.75	7.46	6.90	6.00	Bulgaria
0.67	0.64	0.76	0.74	0.79	0.92	0.84	0.77	1.05	Cyprus
28.80	22.96	14.96	13.19	13.48	15.10	14.91	13.53	12.64	Romania
8.23	6.46	5.24	4.58	4.45	5.07	5.09	5.33	5.51	Slovak Republic
-	-	-	4.03	5.07	5.03	5.43	4.93	5.18	*Croatia*
-	-	-	0.89	0.91	0.17	0.11	0.68	0.39	*FYROM*
-	-	-	0.61	0.58	0.38	0.61	0.54	0.55	*Slovenia*
-	-	-	1.52	1.27	1.20	1.34	1.88	2.72	*FR of Yugoslavia*
16.22	15.03	12.45	7.06	7.83	6.79	7.50	8.02	8.84	Former Yugoslavia
67.17	54.29	38.74	28.80	32.69	35.19	36.25	34.99	34.38	**Non-OECD Europe**
-	-	-	11.29	10.44	9.24	8.94	7.95	7.52	Azerbaijan
-	-	-	20.28	13.86	12.01	12.37	11.32	11.20	Belarus
-	-	-	0.26	0.17	0.29	0.30	0.34	0.37	Estonia
-	-	-	0.72	0.30	0.20	0.04	0.01	0.02	Georgia
-	-	-	17.70	15.10	11.98	10.94	11.11	9.14	Kazakhstan
-	-	-	-	-	-	0.07	0.09	0.11	Kyrgyzstan
-	-	-	4.15	5.29	4.19	3.55	4.25	5.68	Lithuania
-	-	-	251.96	225.73	188.54	181.33	175.05	176.00	Russia
-	-	-	0.06	0.04	0.03	0.03	0.02	0.03	Tajikistan
-	-	-	5.72	3.88	3.58	3.27	3.49	4.22	Turkmenistan
-	-	-	36.68	22.43	19.50	16.67	13.27	13.05	Ukraine
-	-	-	6.55	6.32	5.92	6.69	6.65	6.86	Uzbekistan
461.49	435.63	423.05	355.37	303.57	255.48	244.19	233.55	234.20	**Former USSR**

Production of Petroleum Products (Mtoe)
Production de produits pétroliers (Mtep)
Erzeugung von Ölprodukten (Mtoe)
Produzione di prodotti petroliferi (Mtep)
石 油 製 品 の 生 産 量 （ 石 油 換 算 百 万 ト ン ）
Producción de productos petrolíferos (Mtep)
Производство нефтепродуктов (Мтнэ)

	1971	1973	1978	1983	1984	1985	1986	1987	1988
Bahrein	12.80	12.21	12.78	9.14	10.38	9.42	12.57	12.41	12.53
Iran	27.64	29.41	37.19	28.06	33.60	35.77	35.83	35.10	36.10
Irak	3.60	4.07	9.21	15.53	17.73	18.86	19.74	20.40	21.66
Israël	5.21	6.13	7.42	7.34	6.81	6.68	6.93	7.59	7.17
Jordanie	0.57	0.69	1.41	2.46	2.49	2.46	2.28	2.43	2.37
Koweit	21.46	20.04	20.75	25.27	25.05	28.31	30.99	34.65	40.07
Liban	2.05	2.44	1.86	0.50	0.74	0.63	1.06	1.18	0.55
Oman	-	-	-	2.27	2.48	2.89	3.36	2.79	2.87
Qatar	0.02	0.02	0.32	0.45	0.76	1.27	1.61	1.79	2.14
Arabie saoudite	27.26	28.19	31.09	43.35	43.10	51.16	56.09	67.45	71.46
Syrie	2.13	1.99	4.12	9.19	10.19	10.07	10.12	11.14	11.54
Emirats arabes unis	-	-	0.59	5.94	7.11	7.08	7.43	7.54	7.94
Yémen	3.52	2.86	1.85	4.15	4.39	3.73	3.34	4.98	5.00
Moyen-Orient	106.26	108.04	128.58	153.67	164.82	178.34	191.34	209.45	221.40
Total non-OCDE	797.23	905.19	1 069.09	1 170.28	1 189.63	1 195.94	1 232.38	1 294.03	1 332.38
OCDE Amérique du N.	717.52	803.73	945.10	785.99	815.13	810.86	844.20	859.64	882.34
OCDE Pacifique	216.82	268.27	275.37	225.62	232.56	218.96	215.39	211.99	226.90
OCDE Europe	675.58	786.16	735.00	613.92	616.26	602.29	641.81	638.93	664.25
Total OCDE	1 609.92	1 858.16	1 955.47	1 625.53	1 663.95	1 632.11	1 701.40	1 710.56	1 773.49
Monde	2 407.14	2 763.35	3 024.56	2 795.82	2 853.58	2 828.05	2 933.78	3 004.59	3 105.87

Production of Petroleum Products (Mtoe)
Production de produits pétroliers (Mtep)
Erzeugung von Ölprodukten (Mtoe)
Produzione di prodotti petroliferi (Mtep)
石油製品の生産量（石油換算百万トン）
Producción de productos petrolíferos (Mtep)
Производство нефтепродуктов (Мтнэ)

1989	1990	1991	1992	1993	1994	1995	1996	1997	
12.56	12.79	12.89	13.27	12.60	12.41	12.72	11.45	12.02	Bahrain
40.08	43.56	45.33	46.75	53.17	55.20	56.75	56.84	56.07	Iran
22.60	20.15	13.56	17.83	20.52	22.56	22.10	21.63	22.16	Iraq
7.44	8.29	8.55	10.30	11.62	11.81	11.17	9.86	11.28	Israel
2.38	2.72	2.43	2.98	2.95	3.07	3.23	3.32	3.54	Jordan
47.04	24.64	2.83	16.79	20.79	34.44	38.48	36.65	41.18	Kuwait
-	0.10	0.53	0.43	-	-	-	-	-	Lebanon
2.98	3.39	3.19	2.94	2.41	3.63	3.64	3.65	3.24	Oman
2.50	2.97	2.63	2.74	2.48	2.89	2.71	3.00	2.73	Qatar
70.36	79.87	70.94	75.72	75.76	76.81	73.76	83.72	78.98	Saudi Arabia
11.24	11.36	11.71	11.66	11.49	11.71	11.79	12.22	12.28	Syria
8.18	8.53	8.79	8.56	8.64	9.37	10.39	10.92	12.30	United Arab Emirates
3.69	4.73	4.62	5.28	3.60	3.61	3.61	4.43	4.41	Yemen
231.04	223.09	187.98	215.24	226.03	247.53	250.34	257.68	260.19	**Middle East**
1 352.08	1 326.42	1 274.35	1 242.41	1 228.39	1 219.22	1 241.08	1 284.69	1 318.80	**Non-OECD Total**
896.44	908.69	899.50	908.27	930.03	939.12	940.77	958.86	980.77	OECD North America
241.00	259.03	286.49	311.24	323.63	337.79	349.91	361.20	389.93	OECD Pacific
667.19	674.45	682.75	702.79	715.43	724.48	722.45	743.31	757.26	OECD Europe
1 804.63	1 842.17	1 868.74	1 922.30	1 969.09	2 001.39	2 013.13	2 063.38	2 127.97	**OECD Total**
3 156.71	3 168.59	3 143.08	3 164.71	3 197.48	3 220.62	3 254.21	3 348.07	3 446.77	**World**

Production of Natural Gas (Mtoe)
Production de gaz naturel (Mtep)
Erzeugung von Erdgas (Mtoe)

Produzione di gas naturale (Mtep)

天然ガスの生産量（石油換算百万トン）

Producción de gas natural (Mtep)

Производство природного газа (Мтнэ)

	1971	1973	1978	1983	1984	1985	1986	1987	1988
Algérie	2.40	4.04	10.75	28.45	28.32	30.84	32.83	37.05	38.73
Angola	0.04	0.05	0.06	0.09	0.10	0.10	0.11	0.13	0.13
Congo	0.06	0.01	0.00	0.00	0.00	0.00	0.00	0.00	0.00
Egypte	0.07	0.07	0.57	2.33	2.99	3.73	4.30	4.78	5.17
Gabon	0.08	0.40	0.03	0.08	0.03	0.04	0.08	0.10	0.15
Libye	1.30	3.42	4.29	3.11	3.86	4.36	3.97	4.07	4.59
Maroc	0.04	0.06	0.07	0.07	0.07	0.08	0.08	0.06	0.06
Mozambique	-	-	-	-	-	-	-	-	-
Nigéria	0.17	0.35	0.78	2.60	2.51	2.96	2.67	3.00	2.97
Sénégal	-	-	-	-	-	-	-	0.00	0.01
Afrique du Sud	-	-	-	-	-	-	-	-	-
Tunisie	0.00	0.11	0.28	0.43	0.42	0.40	0.38	0.32	0.30
Autre Afrique	0.00	0.00	0.00	0.00	0.00	0.00	0.00	0.00	0.00
Afrique	4.17	8.52	16.84	37.16	38.30	42.52	44.41	49.50	52.11
Argentine	5.65	5.75	6.37	10.62	11.52	11.95	12.74	13.13	15.72
Bolivie	0.06	1.69	1.83	2.55	2.49	2.54	2.52	2.50	2.61
Brésil	0.12	0.13	0.78	1.52	1.86	2.24	2.69	2.98	3.08
Chili	0.03	0.39	1.05	0.65	0.66	0.66	0.60	0.59	0.82
Colombie	1.12	1.41	2.18	3.41	3.47	3.47	3.45	3.55	3.51
Cuba	0.00	0.01	0.01	0.01	0.00	0.01	0.00	0.02	0.02
Pérou	0.44	0.42	0.60	0.47	0.64	0.99	1.05	1.05	0.99
Trinité-et-Tobago	1.60	1.59	2.09	3.38	3.26	3.22	3.37	3.52	4.16
Vénézuela	9.88	11.94	13.17	16.25	18.44	18.30	18.72	18.64	16.98
Autre Amérique latine	0.00	0.00	0.01	0.01	0.02	0.02	0.02	0.02	0.02
Amérique latine	18.91	23.34	28.07	38.86	42.35	43.40	45.17	46.00	47.91
Bangladesh	0.38	0.52	0.73	1.70	1.97	2.19	2.49	2.95	3.29
Brunei	0.04	1.33	6.06	7.79	7.32	7.19	6.93	7.31	7.20
Inde	0.61	0.63	1.41	2.60	3.15	3.83	5.32	6.13	7.30
Indonésie	0.24	0.33	8.90	16.89	24.37	29.01	30.98	33.04	35.12
Malaisie	0.07	0.10	0.83	3.76	6.36	9.63	12.95	14.00	14.45
Myanmar	0.06	0.09	0.23	0.43	0.58	0.79	0.88	0.91	0.92
Pakistan	2.36	2.86	4.06	6.42	6.08	6.46	6.68	8.63	8.89
Philippines	-	-	-	-	-	-	-	-	-
Taipei chinois	0.91	1.22	1.59	1.19	1.20	1.07	0.98	0.99	1.13
Thaïlande	-	-	-	1.23	1.86	2.87	3.88	3.88	4.60
Viêt-Nam	-	-	-	0.06	0.05	0.03	0.03	0.03	0.02
Autre Asie	2.08	2.19	1.94	2.14	2.28	2.39	2.39	1.52	1.27
Asie	6.74	9.27	25.75	44.22	55.21	65.46	73.51	79.40	84.18
Rép. populaire de Chine	3.13	5.01	11.48	10.23	10.58	10.83	11.53	11.64	11.95
Chine	3.13	5.01	11.48	10.23	10.58	10.83	11.53	11.64	11.95
Albanie	0.11	0.16	0.30	0.32	0.32	0.32	0.32	0.32	0.34
Bulgarie	0.25	0.17	0.02	0.05	0.04	0.02	0.01	0.01	0.01
Roumanie	22.44	24.30	31.69	33.02	32.78	31.34	31.71	30.14	29.65
République slovaque	0.50	0.39	0.48	0.20	0.32	0.31	0.34	0.38	0.52
Croatie	-	-	-	-	-	-	-	-	-
Slovénie	-	-	-	-	-	-	-	-	-
RF de Yougoslavie	-	-	-	-	-	-	-	-	-
Ex-Yougoslavie	1.00	1.33	1.76	1.67	1.57	2.01	2.01	2.14	2.47
Europe non-OCDE	24.29	26.35	34.25	35.27	35.03	33.99	34.40	32.99	32.98

Production of Natural Gas (Mtoe)
Production de gaz naturel (Mtep)

Erzeugung von Erdgas (Mtoe)
Produzione di gas naturale (Mtep)
天然ガスの生産量（石油換算百万トン）
Producción de gas natural (Mtep)
Производство природного газа (Мтнэ)

1989	1990	1991	1992	1993	1994	1995	1996	1997	
41.76	44.30	47.82	49.49	49.49	45.50	51.50	54.98	63.38	Algeria
0.14	0.44	0.47	0.47	0.46	0.42	0.46	0.46	0.47	Angola
0.00	0.00	0.00	0.00	0.00	0.00	0.00	0.00	0.00	Congo
6.49	6.73	7.41	7.99	9.43	9.93	10.30	10.82	11.07	Egypt
0.07	0.07	0.07	0.07	0.07	0.06	0.07	0.07	0.07	Gabon
5.69	5.16	5.65	5.93	5.57	5.59	5.55	5.62	5.75	Libya
0.05	0.04	0.03	0.02	0.02	0.02	0.02	0.02	0.02	Morocco
-	-	-	0.00	0.00	0.00	0.00	0.00	0.00	Mozambique
3.47	3.27	3.98	4.19	4.58	4.49	4.40	4.46	4.43	Nigeria
0.01	0.00	0.00	0.00	0.01	0.02	0.04	0.04	0.02	Senegal
-	1.50	1.53	1.54	1.71	1.71	1.71	1.54	1.54	South Africa
0.30	0.33	0.26	0.24	0.18	0.17	0.13	0.73	1.47	Tunisia
0.00	0.00	0.00	0.00	0.00	0.00	0.00	0.00	0.00	Other Africa
57.98	61.85	67.23	69.93	71.51	67.92	74.18	78.73	88.22	**Africa**
16.80	15.91	17.12	19.39	20.33	21.37	23.20	25.69	28.06	Argentina
2.61	2.67	2.67	2.74	2.77	2.93	2.65	2.49	3.00	Bolivia
3.29	3.34	3.26	3.61	3.83	3.97	4.26	4.78	5.29	Brazil
1.29	1.38	1.19	1.41	1.36	1.42	1.38	1.39	1.62	Chile
3.54	3.67	3.71	3.64	3.72	3.84	4.03	4.34	5.34	Colombia
0.03	0.03	0.02	0.02	0.02	0.02	0.01	0.02	0.02	Cuba
0.84	0.80	0.75	0.68	0.71	0.75	0.72	0.73	0.77	Peru
4.16	4.34	4.60	4.72	4.68	5.72	5.79	6.41	6.83	Trinidad-and-Tobago
18.11	20.17	20.32	19.35	21.08	22.51	25.08	26.27	28.96	Venezuela
0.02	0.02	0.02	0.02	0.02	0.02	0.02	0.02	0.02	Other Latin America
50.69	52.33	53.67	55.58	58.53	62.53	67.16	72.14	79.91	**Latin America**
3.68	3.71	3.82	4.43	4.74	5.10	5.69	5.90	5.79	Bangladesh
7.32	7.58	7.78	7.89	7.93	8.15	8.93	8.86	8.86	Brunei
8.89	10.13	11.31	13.08	12.96	13.92	15.59	17.30	17.81	India
38.03	43.96	46.78	50.14	51.52	58.49	59.34	63.00	63.04	Indonesia
15.64	15.48	18.39	19.64	20.55	20.93	23.82	30.25	33.10	Malaysia
0.89	0.76	0.72	0.71	0.87	1.09	1.22	1.32	1.43	Myanmar
10.11	10.10	10.51	11.03	11.26	12.01	12.33	13.21	13.43	Pakistan
-	-	-	-	-	0.01	0.01	0.01	0.00	Philippines
1.14	1.06	0.79	0.68	0.66	0.73	0.75	0.72	0.69	Chinese Taipei
4.59	5.00	6.20	6.63	7.43	8.23	8.70	10.19	12.41	Thailand
0.01	0.00	0.00	-	-	-	0.42	3.12	3.31	Vietnam
0.27	0.24	0.23	0.22	0.22	0.21	0.21	0.20	0.20	Other Asia
90.56	98.03	106.53	114.45	118.14	128.88	137.00	154.08	160.08	**Asia**
12.60	12.80	13.44	13.62	15.61	16.35	16.71	18.72	21.13	People's Rep.of China
12.60	12.80	13.44	13.62	15.61	16.35	16.71	18.72	21.13	**China**
0.28	0.20	0.12	0.09	0.08	0.04	0.02	0.02	0.02	Albania
0.01	0.01	0.01	0.03	0.05	0.05	0.04	0.03	0.03	Bulgaria
26.63	22.90	20.05	17.61	16.76	14.82	14.44	13.76	11.90	Romania
0.47	0.34	0.24	0.32	0.20	0.23	0.27	0.24	0.22	Slovak Republic
-	-	-	1.47	1.67	1.46	1.61	1.46	1.40	*Croatia*
-	-	-	0.01	0.01	0.01	0.01	0.01	0.01	*Slovenia*
-	-	-	0.69	0.70	0.67	0.72	0.54	0.56	*FR of Yugoslavia*
2.37	2.17	2.13	2.18	2.39	2.15	2.34	2.01	1.97	Former Yugoslavia
29.76	25.63	22.55	20.22	19.47	17.28	17.11	16.07	14.14	**Non-OECD Europe**

Production of Natural Gas (Mtoe)
Production de gaz naturel (Mtep)
Erzeugung von Erdgas (Mtoe)
Produzione di gas naturale (Mtep)
天然ガスの生産量（石油換算百万トン）
Producción de gas natural (Mtep)
Производство природного газа (Мтнэ)

	1971	1973	1978	1983	1984	1985	1986	1987	1988
Azerbaïdjan	-	-	-	-	-	-	-	-	-
Bélarus	-	-	-	-	-	-	-	-	-
Géorgie	-	-	-	-	-	-	-	-	-
Kazakhstan	-	-	-	-	-	-	-	-	-
Kirghizistan	-	-	-	-	-	-	-	-	-
Russie	-	-	-	-	-	-	-	-	-
Tadjikistan	-	-	-	-	-	-	-	-	-
Turkménistan	-	-	-	-	-	-	-	-	-
Ukraine	-	-	-	-	-	-	-	-	-
Ouzbékistan	-	-	-	-	-	-	-	-	-
Ex-URSS	175.70	195.40	307.80	436.18	475.96	520.05	554.84	587.93	622.72
Bahrein	0.76	1.34	2.17	2.95	3.08	3.78	4.52	4.30	4.64
Iran	12.41	15.60	15.23	8.98	11.02	11.92	12.41	13.07	16.94
Irak	0.76	0.99	1.39	2.22	2.46	2.61	3.72	6.25	8.08
Israël	0.10	0.05	0.05	0.05	0.04	0.04	0.03	0.03	0.03
Jordanie	-	-	-	-	-	-	-	-	-
Koweit	2.81	3.14	4.63	3.13	3.44	3.52	4.80	4.44	3.94
Oman	-	-	0.13	0.57	0.80	0.91	1.12	1.23	1.29
Qatar	0.82	1.29	1.21	3.88	4.84	4.49	4.72	4.73	5.10
Arabie saoudite	1.43	1.94	5.07	10.66	15.75	15.75	21.12	22.47	24.38
Syrie	-	-	0.03	0.06	0.11	0.13	0.33	0.32	0.76
Emirats arabes unis	0.88	1.05	4.65	6.86	8.98	10.78	12.41	13.80	14.21
Moyen-Orient	19.97	25.39	34.56	39.36	50.54	53.93	65.18	70.65	79.36
Total non-OCDE	252.91	293.28	458.76	641.27	707.97	770.19	829.04	878.11	931.21
OCDE Amérique du N.	565.23	574.52	517.30	461.29	497.29	479.98	462.68	481.00	503.60
OCDE Pacifique	4.04	5.95	9.72	13.76	14.77	16.21	17.75	18.06	18.76
OCDE Europe	87.67	124.79	163.43	157.46	160.79	166.46	163.20	169.06	159.50
Total OCDE	656.94	705.27	690.45	632.51	672.85	662.65	643.62	668.12	681.86
Monde	909.85	998.54	1 149.21	1 273.78	1 380.82	1 432.84	1 472.66	1 546.23	1 613.07

Production of Natural Gas (Mtoe)
Production de gaz naturel (Mtep)
Erzeugung von Erdgas (Mtoe)
Produzione di gas naturale (Mtep)
天然ガスの生産量（石油換算百万トン）
Producción de gas natural (Mtep)
Производство природного газа (Мтнэ)

1989	1990	1991	1992	1993	1994	1995	1996	1997	
-	-	-	6.38	5.51	5.17	5.38	5.11	4.83	Azerbaijan
-	-	-	0.24	0.24	0.24	0.22	0.21	0.20	Belarus
-	-	-	0.03	0.04	0.01	0.01	0.00	0.00	Georgia
-	-	-	6.54	5.43	3.64	4.79	5.29	6.57	Kazakhstan
-	-	-	0.06	0.03	0.03	0.03	0.02	0.03	Kyrgyzstan
-	-	-	518.51	497.68	480.85	476.07	480.07	461.12	Russia
-	-	-	0.05	0.04	0.03	0.03	0.04	0.03	Tajikistan
-	-	-	43.85	53.24	29.24	29.07	28.50	14.02	Turkmenistan
-	-	-	16.92	15.56	14.83	15.23	15.44	15.20	Ukraine
-	-	-	34.95	36.78	37.51	38.13	38.32	39.72	Uzbekistan
643.82	656.31	659.35	627.53	614.55	571.54	568.97	573.00	541.74	**Former USSR**
4.60	4.75	4.54	4.56	4.86	4.76	5.05	5.27	5.11	Bahrain
18.13	20.47	27.23	29.69	31.21	34.17	32.65	34.18	38.40	Iran
9.34	5.03	1.40	2.34	2.62	3.26	3.24	3.31	3.70	Iraq
0.03	0.03	0.02	0.02	0.02	0.02	0.02	0.01	0.01	Israel
0.05	0.10	0.10	0.11	0.13	0.18	0.19	0.18	0.19	Jordan
4.69	4.30	1.61	3.01	3.75	4.13	4.13	4.43	4.45	Kuwait
1.33	2.44	2.50	3.06	3.71	3.87	3.84	4.00	4.70	Oman
4.83	5.10	6.18	10.22	10.94	10.94	10.94	11.10	14.50	Qatar
24.93	25.36	26.82	27.77	29.32	30.78	32.94	35.80	37.58	Saudi Arabia
1.24	1.37	1.60	1.65	1.64	1.77	2.24	2.24	2.07	Syria
16.66	16.41	19.44	18.13	18.70	20.66	24.12	27.91	30.55	United Arab Emirates
85.83	85.36	91.44	100.55	106.89	114.54	119.34	128.42	141.27	**Middle East**
971.23	992.30	1 014.22	1 001.88	1 004.69	979.03	1 000.47	1 041.17	1 046.50	**Non-OECD Total**
513.47	530.09	531.87	543.84	560.55	589.77	589.98	602.60	607.77	OECD North America
19.15	22.81	23.91	25.77	27.15	28.54	30.76	32.01	32.26	OECD Pacific
164.55	163.42	176.38	179.70	190.85	194.38	202.50	233.27	230.23	OECD Europe
697.17	716.32	732.16	749.30	778.54	812.69	823.24	867.88	870.26	**OECD Total**
1 668.41	1 708.62	1 746.38	1 751.17	1 783.24	1 791.72	1 823.71	1 909.04	1 916.76	**World**

Production of Nuclear Energy (Mtoe)
Production d'énergie nucléaire (Mtep)
Erzeugung von Kernenergie (Mtoe)
Produzione di energia nucleare (Mtep)
原子力の生産量（石油換算百万トン）
Producción de energía nuclear (Mtep)
Производство атомной энергии (Мтнэ)

	1971	1973	1978	1983	1984	1985	1986	1987	1988
Afrique du Sud	-	-	-	-	1.02	1.39	2.29	1.61	2.73
Afrique	-	-	-	-	1.02	1.39	2.29	1.61	2.73
Argentine	-	-	0.75	0.89	1.21	1.50	1.49	1.68	1.51
Brésil	-	-	-	-	0.43	0.88	0.04	0.25	0.16
Amérique latine	-	-	0.75	0.89	1.64	2.38	1.53	1.94	1.67
Inde	0.31	0.62	0.72	0.92	1.06	1.30	1.31	1.31	1.52
Pakistan	0.03	0.08	0.06	0.06	0.08	0.09	0.11	0.13	0.07
Taipei chinois	-	-	0.70	4.93	6.41	7.49	7.02	8.63	7.99
Asie	0.34	0.70	1.48	5.91	7.55	8.87	8.44	10.08	9.57
Rép. populaire de Chine	-	-	-	-	-	-	-	-	-
Chine	-	-	-	-	-	-	-	-	-
Bulgarie	-	-	1.54	3.21	3.32	3.42	3.15	3.24	4.18
Roumanie	-	-	-	-	-	-	-	-	-
République slovaque	-	0.06	0.00	1.60	1.89	2.45	3.05	3.00	2.99
Slovénie	-	-	-	-	-	-	-	-	-
Ex-Yougoslavie	-	-	-	1.02	1.15	1.06	1.05	1.17	1.08
Europe non-OCDE	-	0.06	1.55	5.83	6.36	6.92	7.25	7.41	8.25
Arménie	-	-	-	-	-	-	-	-	-
Lituanie	-	-	-	-	-	-	-	-	-
Russie	-	-	-	-	-	-	-	-	-
Ukraine	-	-	-	-	-	-	-	-	-
Ex-URSS	1.59	3.13	13.03	28.67	37.01	43.52	41.96	49.25	56.21
Total non-OCDE	1.93	3.89	16.81	41.30	53.58	63.09	61.47	70.29	78.43
OCDE Amérique du N.	11.68	27.31	84.89	94.68	104.97	122.33	133.62	146.56	167.48
OCDE Pacifique	2.08	2.53	16.06	32.12	38.06	45.95	51.24	59.18	57.01
OCDE Europe	13.26	19.32	45.78	102.61	131.49	158.15	171.77	177.66	190.40
Total OCDE	27.02	49.16	146.74	229.41	274.53	326.43	356.63	383.39	414.89
Monde	28.95	53.05	163.54	270.71	328.11	389.52	418.10	453.68	493.32

Le Kazakhstan a produit une quantité inconnue d' électricité d' origine nucléaire de 1992 à 1994.

Production of Nuclear Energy (Mtoe)
Production d'énergie nucléaire (Mtep)
Erzeugung von Kernenergie (Mtoe)
Produzione di energia nucleare (Mtep)
原子力の生産量（石油換算百万トン）
Producción de energía nuclear (Mtep)
Производство атомной энергии (Мтнэ)

1989	1990	1991	1992	1993	1994	1995	1996	1997	
2.89	2.20	2.38	2.42	1.89	2.53	2.95	3.07	3.30	South Africa
2.89	2.20	2.38	2.42	1.89	2.53	2.95	3.07	3.30	**Africa**
1.31	1.90	2.02	1.85	2.02	2.15	1.84	1.94	2.06	Argentina
0.48	0.58	0.38	0.46	0.12	0.01	0.66	0.63	0.83	Brazil
1.79	2.48	2.40	2.30	2.13	2.16	2.50	2.58	2.88	**Latin America**
1.21	1.60	1.44	1.75	1.41	1.47	1.98	2.19	2.63	India
0.01	0.08	0.10	0.11	0.15	0.13	0.13	0.13	0.09	Pakistan
7.37	8.57	9.20	8.82	8.95	9.09	9.20	9.85	9.45	Chinese Taipei
8.58	10.24	10.74	10.68	10.51	10.69	11.32	12.16	12.17	**Asia**
-	-	-	-	0.42	3.62	3.34	3.74	3.76	People's Rep.of China
-	-	-	-	0.42	3.62	3.34	3.74	3.76	**China**
3.80	3.82	3.44	3.01	3.64	4.00	4.50	4.72	4.63	Bulgaria
-	-	-	-	-	-	-	0.36	1.41	Romania
3.17	3.14	3.05	2.88	2.87	3.16	2.98	2.93	2.81	Slovak Republic
-	-	-	1.03	1.03	1.20	1.25	1.19	1.31	*Slovenia*
1.22	1.20	1.29	1.03	1.03	1.20	1.25	1.19	1.31	Former Yugoslavia
8.19	8.16	7.77	6.93	7.55	8.36	8.73	9.20	10.16	**Non-OECD Europe**
-	-	-	-	-	-	0.08	0.61	0.42	Armenia
-	-	-	3.91	3.28	2.07	3.14	3.70	3.20	Lithuania
-	-	-	31.18	31.39	25.94	26.32	28.78	28.77	Russia
-	-	-	19.22	19.61	17.94	18.38	20.74	20.70	Ukraine
55.40	55.12	55.42	54.30	54.28	45.95	47.92	53.82	53.09	**Former USSR**
76.86	78.21	78.71	76.63	76.78	73.31	76.75	84.57	85.36	**Non-OECD Total**
167.67	179.53	192.97	193.27	194.77	206.23	213.78	212.68	197.92	OECD North America
60.00	66.50	70.30	72.91	80.11	85.42	93.37	98.02	103.24	OECD Pacific
201.46	200.83	207.46	211.01	219.95	219.69	224.27	235.04	237.65	OECD Europe
429.13	446.85	470.73	477.20	494.83	511.34	531.43	545.74	538.80	**OECD Total**
505.99	525.05	549.44	553.82	571.60	584.65	608.18	630.31	624.16	**World**

Kazakhstan produced an unknown quantity of nuclear electricity in 1992-1994.

Production of Hydro Energy (Mtoe)
Production d'énergie hydraulique (Mtep)
Erzeugung von Wasserkraft (Mtoe)
Produzione di energia idroelettrica (Mtep)
水力の生産量（石油換算百万トン）
Producción de energía hidráulica (Mtep)
Производство гидроэнергии (Мтнэ)

	1971	1973	1978	1983	1984	1985	1986	1987	1988
Algérie	0.03	0.06	0.02	0.02	0.04	0.06	0.02	0.04	0.02
Angola	0.05	0.07	0.05	0.05	0.05	0.06	0.06	0.06	0.06
Cameroun	0.09	0.09	0.11	0.18	0.19	0.21	0.21	0.21	0.22
Congo	0.00	0.00	0.00	0.02	0.02	0.03	0.02	0.02	0.02
RD du Congo	0.30	0.32	0.34	0.39	0.40	0.43	0.45	0.45	0.45
Egypte	0.43	0.44	0.80	0.84	0.83	0.75	0.80	0.74	0.77
Ethiopie	0.03	0.03	0.03	0.05	0.06	0.06	0.06	0.06	0.06
Gabon	-	0.00	0.03	0.05	0.05	0.06	0.06	0.06	0.06
Ghana	0.25	0.33	0.32	0.22	0.15	0.26	0.38	0.40	0.41
Côte d'Ivoire	0.01	0.01	0.01	0.08	0.03	0.12	0.12	0.08	0.11
Kenya	0.03	0.04	0.09	0.15	0.15	0.17	0.14	0.16	0.20
Maroc	0.13	0.10	0.12	0.04	0.03	0.04	0.06	0.07	0.08
Mozambique	0.02	0.02	0.07	0.02	0.02	0.03	0.01	0.02	0.01
Nigéria	0.14	0.16	0.20	0.17	0.23	0.27	0.31	0.28	0.34
Afrique du Sud	0.01	0.08	0.16	0.05	0.05	0.05	0.14	0.14	0.27
Soudan	0.01	0.03	0.04	0.06	0.06	0.06	0.07	0.08	0.08
Tanzanie	0.03	0.03	0.04	0.06	0.07	0.08	0.09	0.10	0.11
Tunisie	0.00	0.01	0.00	0.00	0.01	0.01	0.00	0.01	0.00
Zambie	0.08	0.27	0.68	0.88	0.86	0.88	0.86	0.74	0.72
Zimbabwe	0.21	0.30	0.48	0.30	0.26	0.25	0.20	0.10	0.24
Autre Afrique	0.07	0.09	0.12	0.23	0.25	0.27	0.28	0.30	0.28
Afrique	1.92	2.50	3.73	3.87	3.81	4.13	4.36	4.12	4.53
Argentine	0.13	0.26	0.67	1.58	1.71	1.78	1.81	1.88	1.36
Bolivie	0.08	0.08	0.11	0.10	0.10	0.10	0.10	0.09	0.10
Brésil	3.72	4.98	8.84	13.03	14.33	15.34	15.69	15.96	17.12
Chili	0.38	0.46	0.59	0.77	0.80	0.89	0.97	1.06	0.99
Colombie	0.57	0.68	1.04	1.31	1.46	1.57	1.86	2.00	2.10
Costa Rica	0.09	0.10	0.13	0.24	0.26	0.24	0.25	0.26	0.26
Cuba	0.01	0.01	0.01	0.01	0.01	0.00	0.01	0.00	0.01
République dominicaine	0.05	0.05	0.07	0.09	0.11	0.11	0.13	0.15	0.12
Equateur	0.04	0.04	0.07	0.15	0.28	0.28	0.35	0.39	0.42
El Salvador	0.04	0.04	0.06	0.07	0.07	0.08	0.09	0.08	0.09
Guatemala	0.02	0.02	0.02	0.11	0.11	0.11	0.15	0.15	0.17
Haiti	0.00	0.01	0.02	0.02	0.02	0.02	0.03	0.03	0.03
Honduras	0.02	0.03	0.06	0.08	0.09	0.11	0.12	0.15	0.16
Jamaïque	0.01	0.01	0.01	0.01	0.01	0.01	0.01	0.01	0.01
Nicaragua	0.02	0.03	0.03	0.03	0.03	0.03	0.03	0.04	0.04
Panama	0.01	0.01	0.06	0.07	0.13	0.17	0.18	0.17	0.19
Paraguay	0.01	0.02	0.02	0.06	0.09	0.35	1.02	1.59	1.71
Pérou	0.37	0.41	0.53	0.70	0.75	0.81	0.86	0.94	0.90
Uruguay	0.13	0.13	0.14	0.62	0.61	0.55	0.63	0.62	0.47
Vénézuela	0.46	0.54	1.05	1.55	1.74	1.95	2.16	2.65	2.94
Autre Amérique latine	0.09	0.09	0.08	0.07	0.08	0.09	0.07	0.03	0.05
Amérique latine	6.25	7.99	13.58	20.66	22.77	24.59	26.51	28.28	29.24

Ne comprend pas la production d' électricité des centrales à accumulation par pompage.

Production of Hydro Energy (Mtoe)
Production d'énergie hydraulique (Mtep)
Erzeugung von Wasserkraft (Mtoe)

Produzione di energia idroelettrica (Mtep)

水力の生産量（石油換算百万トン）

Producción de energía hidráulica (Mtep)

Производство гидроэнергии (Mтнэ)

1989	1990	1991	1992	1993	1994	1995	1996	1997	
0.02	0.01	0.03	0.02	0.03	0.01	0.02	0.01	0.01	Algeria
0.06	0.06	0.07	0.07	0.08	0.08	0.08	0.08	0.09	Angola
0.23	0.23	0.23	0.23	0.24	0.23	0.24	0.25	0.27	Cameroon
0.03	0.04	0.04	0.04	0.04	0.04	0.04	0.04	0.04	Congo
0.59	0.48	0.45	0.52	0.50	0.47	0.51	0.53	0.51	DR of Congo
0.86	0.85	0.85	0.83	0.83	0.86	0.93	0.99	1.03	Egypt
0.06	0.09	0.09	0.09	0.10	0.10	0.10	0.10	0.10	Ethiopia
0.06	0.06	0.06	0.06	0.06	0.06	0.06	0.06	0.06	Gabon
0.45	0.49	0.53	0.57	0.54	0.52	0.53	0.51	0.53	Ghana
0.14	0.13	0.11	0.01	0.09	0.08	0.15	0.16	0.17	Ivory Coast
0.21	0.21	0.24	0.24	0.26	0.26	0.27	0.28	0.30	Kenya
0.10	0.10	0.11	0.08	0.04	0.07	0.05	0.17	0.18	Morocco
0.02	0.02	0.03	0.03	0.03	0.03	0.03	0.04	0.07	Mozambique
0.36	0.38	0.51	0.52	0.48	0.48	0.47	0.47	0.48	Nigeria
0.24	0.09	0.17	0.06	0.01	0.09	0.05	0.11	0.18	South Africa
0.08	0.08	0.09	0.09	0.09	0.10	0.10	0.09	0.09	Sudan
0.12	0.13	0.15	0.14	0.15	0.13	0.13	0.15	0.12	Tanzania
0.00	0.00	0.01	0.01	0.01	0.00	0.00	0.01	0.00	Tunisia
0.58	0.54	0.67	0.67	0.67	0.67	0.67	0.67	0.68	Zambia
0.29	0.33	0.31	0.26	0.17	0.17	0.20	0.18	0.15	Zimbabwe
0.30	0.31	0.31	0.32	0.32	0.32	0.34	0.34	0.34	Other Africa
4.79	4.66	5.04	4.86	4.73	4.78	4.94	5.24	5.40	**Africa**
1.43	1.52	1.38	1.65	2.03	2.31	2.29	1.98	2.42	Argentina
0.10	0.10	0.11	0.10	0.13	0.14	0.15	0.18	0.20	Bolivia
17.60	17.78	18.73	19.21	20.22	20.87	21.84	22.86	24.00	Brazil
0.83	0.77	1.13	1.44	1.48	1.46	1.58	1.45	1.63	Chile
2.29	2.35	2.37	1.90	2.38	2.74	2.75	3.03	2.82	Colombia
0.29	0.29	0.31	0.31	0.34	0.34	0.31	0.33	0.41	Costa Rica
0.01	0.01	0.01	0.01	0.01	0.00	0.01	0.01	0.01	Cuba
0.04	0.03	0.05	0.17	0.13	0.14	0.15	0.11	0.12	Dominican Republic
0.43	0.43	0.44	0.43	0.50	0.57	0.45	0.54	0.58	Ecuador
0.10	0.12	0.07	0.08	0.10	0.10	0.11	0.14	0.10	El Salvador
0.18	0.18	0.20	0.17	0.19	0.19	0.21	0.27	0.32	Guatemala
0.03	0.04	0.03	0.03	0.02	0.02	0.02	0.02	0.02	Haiti
0.17	0.20	0.20	0.19	0.20	0.17	0.21	0.26	0.28	Honduras
0.01	0.01	0.01	0.01	0.01	0.01	0.01	0.01	0.01	Jamaica
0.05	0.03	0.03	0.02	0.04	0.03	0.04	0.04	0.04	Nicaragua
0.19	0.19	0.16	0.20	0.20	0.18	0.20	0.24	0.24	Panama
2.09	2.34	2.52	2.33	2.70	3.13	3.62	4.13	4.34	Paraguay
0.90	0.87	0.97	0.83	1.00	1.10	1.11	1.15	1.14	Peru
0.34	0.60	0.53	0.68	0.63	0.64	0.50	0.50	0.56	Uruguay
2.98	3.18	3.83	4.06	4.08	4.41	4.42	4.63	4.92	Venezuela
0.12	0.11	0.11	0.11	0.11	0.11	0.11	0.12	0.12	Other Latin America
30.19	31.16	33.18	33.93	36.49	38.68	40.09	41.99	44.27	**Latin America**

Does not include electricity output from pumped storage plants.

Production of Hydro Energy (Mtoe)
Production d'énergie hydraulique (Mtep)
Erzeugung von Wasserkraft (Mtoe)
Produzione di energia idroelettrica (Mtep)
水力の生産量（石油換算百万トン）
Producción de energía hidráulica (Mtep)
Производство гидроэнергии (Мтнэ)

	1971	1973	1978	1983	1984	1985	1986	1987	1988
Bangladesh	0.02	0.03	0.04	0.06	0.08	0.06	0.04	0.04	0.06
Inde	2.41	2.49	4.06	4.30	4.64	4.39	4.63	4.08	4.98
Indonésie	0.12	0.14	0.11	0.16	0.18	0.26	0.42	0.45	0.51
RPD de Corée	1.00	1.08	1.68	2.24	2.32	2.41	2.49	2.51	2.49
Malaisie	0.09	0.10	0.08	0.15	0.29	0.32	0.35	0.42	0.49
Myanmar	0.04	0.05	0.06	0.09	0.09	0.09	0.09	0.09	0.08
Népal	0.01	0.01	0.01	0.02	0.03	0.03	0.04	0.05	0.05
Pakistan	0.32	0.37	0.64	0.98	1.10	1.05	1.19	1.31	1.44
Philippines	0.17	0.20	0.24	0.34	0.45	0.48	0.52	0.45	0.54
Sri Lanka	0.07	0.06	0.12	0.10	0.18	0.21	0.23	0.19	0.22
Taipei chinois	0.27	0.29	0.43	0.43	0.38	0.60	0.64	0.61	0.57
Thaïlande	0.18	0.16	0.18	0.31	0.35	0.32	0.48	0.35	0.32
Viêt-Nam	0.05	0.04	0.07	0.11	0.14	0.13	0.12	0.12	0.15
Autre Asie	0.08	0.07	0.11	0.13	0.16	0.17	0.20	0.26	0.32
Asie	4.81	5.09	7.83	9.41	10.40	10.51	11.43	10.93	12.22
Rép. populaire de Chine	2.58	3.27	3.84	7.43	7.46	7.94	8.13	8.60	9.39
Chine	2.58	3.27	3.84	7.43	7.46	7.94	8.13	8.60	9.39
Albanie	0.06	0.10	0.21	0.28	0.28	0.25	0.29	0.29	0.31
Bulgarie	0.19	0.22	0.25	0.29	0.28	0.19	0.20	0.22	0.22
Roumanie	0.39	0.65	0.91	0.86	0.97	1.09	0.93	0.96	1.17
République slovaque	0.13	0.11	0.19	0.16	0.13	0.18	0.14	0.17	0.15
Bosnie-Herzégovine	-	-	-	-	-	-	-	-	-
Croatie	-	-	-	-	-	-	-	-	-
Ex-RYM	-	-	-	-	-	-	-	-	-
Slovénie	-	-	-	-	-	-	-	-	-
RF de Yougoslavie	-	-	-	-	-	-	-	-	-
Ex-Yougoslavie	1.35	1.41	2.17	1.88	2.23	2.09	2.37	2.26	2.22
Europe non-OCDE	2.11	2.49	3.73	3.46	3.90	3.81	3.93	3.89	4.08
Arménie	-	-	-	-	-	-	-	-	-
Azerbaïdjan	-	-	-	-	-	-	-	-	-
Bélarus	-	-	-	-	-	-	-	-	-
Estonie	-	-	-	-	-	-	-	-	-
Géorgie	-	-	-	-	-	-	-	-	-
Kazakhstan	-	-	-	-	-	-	-	-	-
Kirghizistan	-	-	-	-	-	-	-	-	-
Lettonie	-	-	-	-	-	-	-	-	-
Lituanie	-	-	-	-	-	-	-	-	-
Moldova	-	-	-	-	-	-	-	-	-
Russie	-	-	-	-	-	-	-	-	-
Tadjikistan	-	-	-	-	-	-	-	-	-
Turkménistan	-	-	-	-	-	-	-	-	-
Ukraine	-	-	-	-	-	-	-	-	-
Ouzbékistan	-	-	-	-	-	-	-	-	-
Ex-URSS	10.84	10.52	14.59	15.48	17.46	18.44	18.55	18.90	19.94

Ne comprend pas la production d' électricité des centrales à accumulation par pompage.

Production of Hydro Energy (Mtoe)
Production d'énergie hydraulique (Mtep)
Erzeugung von Wasserkraft (Mtoe)
Produzione di energia idroelettrica (Mtep)
水力の生産量（石油換算百万トン）
Producción de energía hidráulica (Mtep)
Производство гидроэнергии (Мтнэ)

1989	1990	1991	1992	1993	1994	1995	1996	1997	
0.08	0.08	0.07	0.07	0.05	0.07	0.03	0.06	0.06	Bangladesh
5.34	6.16	6.26	6.01	6.06	7.11	6.25	5.94	6.43	India
0.64	0.56	0.64	0.85	0.76	0.59	0.75	0.77	0.52	Indonesia
2.49	2.49	2.73	2.06	2.06	2.02	1.98	1.94	1.88	DPR of Korea
0.48	0.34	0.38	0.37	0.42	0.56	0.54	0.45	0.28	Malaysia
0.11	0.10	0.11	0.13	0.15	0.14	0.14	0.14	0.14	Myanmar
0.05	0.06	0.07	0.07	0.07	0.07	0.07	0.09	0.10	Nepal
1.46	1.46	1.57	1.60	1.82	1.67	1.97	2.00	1.79	Pakistan
0.56	0.52	0.44	0.37	0.43	0.51	0.53	0.61	0.52	Philippines
0.24	0.27	0.27	0.25	0.33	0.35	0.38	0.28	0.30	Sri Lanka
0.57	0.70	0.47	0.72	0.58	0.76	0.76	0.78	0.82	Chinese Taipei
0.48	0.43	0.39	0.36	0.32	0.39	0.58	0.63	0.62	Thailand
0.33	0.46	0.54	0.70	0.75	0.88	1.05	1.26	1.39	Vietnam
0.32	0.32	0.30	0.29	0.29	0.30	0.31	0.32	0.33	Other Asia
13.15	13.96	14.26	13.86	14.09	15.44	15.34	15.26	15.18	**Asia**
10.18	10.90	10.76	11.29	13.14	14.47	16.39	16.17	16.85	People's Rep.of China
10.18	10.90	10.76	11.29	13.14	14.47	16.39	16.17	16.85	**China**
0.31	0.24	0.30	0.28	0.29	0.32	0.36	0.49	0.46	Albania
0.23	0.16	0.21	0.18	0.10	0.07	0.11	0.14	0.15	Bulgaria
1.09	0.94	1.22	1.01	1.10	1.12	1.44	1.35	1.51	Romania
0.17	0.16	0.12	0.17	0.30	0.37	0.43	0.37	0.36	Slovak Republic
-	-	-	0.26	0.17	0.11	0.12	0.12	0.12	*Bosnia-Herzegovina*
-	-	-	0.37	0.37	0.42	0.45	0.62	0.46	*Croatia*
-	-	-	0.07	0.04	0.06	0.07	0.07	0.08	*FYROM*
-	-	-	0.29	0.26	0.29	0.28	0.32	0.27	*Slovenia*
-	-	-	0.98	0.86	0.96	0.96	0.99	1.05	*FR of Yugoslavia*
2.02	1.70	2.20	1.97	1.71	1.84	1.89	2.12	1.97	Former Yugoslavia
3.82	3.22	4.06	3.60	3.49	3.73	4.22	4.48	4.44	**Non-OECD Europe**
-	-	-	0.26	0.37	0.30	0.17	0.14	0.12	Armenia
-	-	-	0.15	0.21	0.16	0.13	0.13	0.13	Azerbaijan
-	-	-	0.00	0.00	0.00	0.00	0.00	0.00	Belarus
-	-	-	0.00	0.00	0.00	0.00	0.00	0.00	Estonia
-	-	-	0.56	0.60	0.41	0.46	0.52	0.52	Georgia
-	-	-	0.59	0.66	0.79	0.72	0.63	0.56	Kazakhstan
-	-	-	0.79	0.78	1.01	0.96	1.05	0.97	Kyrgyzstan
-	-	-	0.22	0.25	0.28	0.25	0.16	0.25	Latvia
-	-	-	0.03	0.03	0.04	0.03	0.03	0.03	Lithuania
-	-	-	0.02	0.03	0.02	0.03	0.03	0.03	Moldova
-	-	-	14.84	14.94	15.13	15.16	13.27	13.50	Russia
-	-	-	1.37	1.47	1.44	1.26	1.27	1.19	Tajikistan
-	-	-	0.00	0.00	0.00	0.00	0.00	0.00	Turkmenistan
-	-	-	0.69	0.95	1.04	0.86	0.74	0.85	Ukraine
-	-	-	0.54	0.63	0.62	0.53	0.56	0.50	Uzbekistan
19.26	20.04	20.18	20.07	20.93	21.23	20.54	18.54	18.64	**Former USSR**

Does not include electricity output from pumped storage plants.

Production of Hydro Energy (Mtoe)
Production d'énergie hydraulique (Mtep)
Erzeugung von Wasserkraft (Mtoe)
Produzione di energia idroelettrica (Mtep)
水力の生産量（石油換算百万トン）
Producción de energía hidráulica (Mtep)
Производство гидроэнергии (Мтнэ)

	1971	1973	1978	1983	1984	1985	1986	1987	1988
Iran	0.23	0.26	0.54	0.53	0.49	0.48	0.65	0.72	0.63
Irak	0.02	0.02	0.06	0.05	0.05	0.05	0.05	0.22	0.22
Israël	-	-	-	-	-	-	-	-	-
Jordanie	-	-	-	-	-	-	0.00	0.00	0.00
Liban	0.07	0.04	0.07	0.05	0.05	0.05	0.05	0.05	0.05
Syrie	0.00	0.00	0.18	0.24	0.25	0.25	0.14	0.18	0.41
Moyen-Orient	0.32	0.32	0.85	0.87	0.84	0.83	0.89	1.18	1.32
Total non-OCDE	28.83	32.18	48.15	61.18	66.65	70.24	73.81	75.91	80.71
OCDE Amérique du N.	37.87	40.96	45.99	53.50	54.53	52.79	53.70	49.14	46.16
OCDE Pacifique	9.48	8.07	8.77	10.26	9.13	10.31	10.42	9.96	11.29
OCDE Europe	27.70	29.33	35.41	37.71	37.94	37.65	36.01	39.60	42.37
Total OCDE	75.05	78.35	90.16	101.47	101.60	100.75	100.12	98.70	99.82
Monde	103.88	110.53	138.31	162.65	168.25	170.98	173.93	174.61	180.53

Ne comprend pas la production d' électricité des centrales à accumulation par pompage.

Production of Hydro Energy (Mtoe)
Production d'énergie hydraulique (Mtep)
Erzeugung von Wasserkraft (Mtoe)
Produzione di energia idroelettrica (Mtep)
水力の生産量（石油換算百万トン）
Producción de energía hidráulica (Mtep)
Производство гидроэнергии (Мтнэ)

1989	1990	1991	1992	1993	1994	1995	1996	1997	
0.65	0.52	0.61	0.80	0.84	0.64	0.63	0.63	0.63	Iran
0.22	0.22	0.08	0.06	0.05	0.05	0.05	0.05	0.05	Iraq
-	-	0.00	0.01	0.01	0.00	0.01	0.01	0.01	Israel
0.00	0.00	0.00	0.00	0.00	0.00	0.00	0.00	0.00	Jordan
0.04	0.04	0.05	0.06	0.06	0.07	0.06	0.07	0.08	Lebanon
0.41	0.49	0.54	0.64	0.58	0.58	0.60	0.60	0.86	Syria
1.32	1.28	1.27	1.57	1.54	1.35	1.34	1.36	1.63	**Middle East**
82.70	85.20	88.75	89.18	94.41	99.68	102.86	103.02	106.42	**Non-OECD Total**
49.18	51.04	53.18	51.33	54.45	52.68	58.23	63.50	60.86	OECD North America
11.43	11.45	12.03	10.46	12.03	9.61	11.01	10.71	11.40	OECD Pacific
36.61	37.95	37.75	40.23	41.71	41.76	42.21	40.57	41.97	OECD Europe
97.22	100.44	102.97	102.02	108.18	104.05	111.45	114.77	114.24	**OECD Total**
179.92	185.64	191.72	191.20	202.59	203.73	214.31	217.80	220.66	**World**

Does not include electricity output from pumped storage plants.

Production of Geothermal Energy (Mtoe)
Production d'énergie géothermique (Mtep)
Erzeugung von geothermischer Energie (Mtoe)

Produzione di energia geotermica (Mtep)

地熱エネルギーの生産量（石油換算百万トン）

Producción de energía geotérmica (Mtep)

Производство геотермальной энергии (Мтнэ)

	1971	1973	1978	1983	1984	1985	1986	1987	1988
Ethiopie	-	-	-	0.05	0.05	0.05	0.05	0.05	0.05
Kenya	-	-	-	-	-	-	0.32	0.31	0.28
Afrique	-	-	-	0.05	0.05	0.05	0.37	0.36	0.33
Costa Rica	-	-	-	-	-	-	-	-	-
El Salvador	-	-	0.52	0.58	0.58	0.56	0.50	0.56	0.55
Nicaragua	-	-	-	0.00	0.23	0.26	0.26	0.20	0.16
Amérique latine	-	-	0.52	0.58	0.82	0.82	0.75	0.76	0.72
Indonésie	-	-	-	0.18	0.19	0.19	0.20	0.62	0.87
Philippines	-	-	0.00	3.51	3.90	4.24	3.94	3.90	4.17
Asie	-	-	0.00	3.69	4.08	4.43	4.13	4.52	5.04
Ex-RYM	-	-	-	-	-	-	-	-	-
Ex-Yougoslavie	-	-	-	-	-	-	-	-	-
Europe non-OCDE	-	-	-	-	-	-	-	-	-
Russie	-	-	-	-	-	-	-	-	-
Ex-URSS	-	-	-	-	-	-	-	-	-
Total non-OCDE	-	-	0.52	4.31	4.95	5.30	5.26	5.64	6.08
OCDE Amérique du N.	0.50	2.25	3.22	6.44	7.97	9.91	12.32	13.62	13.37
OCDE Pacifique	1.08	1.30	1.56	2.25	2.23	2.26	2.20	2.23	2.23
OCDE Europe	2.19	2.11	2.24	2.62	2.59	2.50	2.59	2.78	2.88
Total OCDE	3.77	5.66	7.02	11.32	12.79	14.67	17.11	18.63	18.49
Monde	3.77	5.66	7.53	15.63	17.74	19.97	22.36	24.27	24.57

Production of Geothermal Energy (Mtoe)
Production d'énergie géothermique (Mtep)
Erzeugung von geothermischer Energie (Mtoe)
Produzione di energia geotermica (Mtep)
地熱エネルギーの生産量（石油換算百万トン）
Producción de energía geotérmica (Mtep)
Производство геотермальной энергии (Мтнэ)

1989	1990	1991	1992	1993	1994	1995	1996	1997	
0.06	0.06	0.06	0.05	0.06	0.05	0.05	0.04	0.09	Ethiopia
0.28	0.28	0.26	0.23	0.23	0.22	0.25	0.34	0.34	Kenya
0.33	0.34	0.31	0.29	0.30	0.28	0.30	0.38	0.43	**Africa**
-	-	-	-	-	-	0.40	0.44	0.47	Costa Rica
0.58	0.54	0.73	0.71	0.63	0.56	0.58	0.58	0.62	El Salvador
0.33	0.33	0.39	0.40	0.35	0.31	0.27	0.24	0.18	Nicaragua
0.90	0.87	1.12	1.11	0.98	0.87	1.25	1.26	1.27	**Latin America**
0.87	0.97	0.90	0.93	0.94	1.66	1.87	1.99	2.22	Indonesia
4.57	4.70	4.95	4.90	4.87	5.42	5.28	5.62	6.22	Philippines
5.44	5.67	5.85	5.83	5.81	7.08	7.15	7.61	8.44	**Asia**
-	-	-	-	-	-	0.01	0.01	0.01	*FYROM*
-	-	-	-	-	-	0.01	0.01	0.01	Former Yugoslavia
-	-	-	-	-	-	0.01	0.01	0.01	**Non-OECD Europe**
-	-	-	0.02	0.02	0.02	0.03	0.02	0.02	Russia
-	-	-	0.02	0.02	0.02	0.03	0.02	0.02	**Former USSR**
6.68	6.88	7.29	7.26	7.11	8.25	8.73	9.28	10.17	**Non-OECD Total**
16.88	18.18	18.66	19.76	20.34	19.85	17.72	18.47	17.52	OECD North America
2.83	3.67	3.70	3.81	3.71	3.67	4.59	5.13	5.13	OECD Pacific
3.29	3.23	3.20	3.36	3.48	3.38	3.47	3.70	3.83	OECD Europe
22.99	25.08	25.56	26.92	27.53	26.89	25.78	27.30	26.49	**OECD Total**
29.67	31.95	32.85	34.18	34.64	35.15	34.52	36.58	36.66	**World**

Production of Energy from Solar, Wind, Tide, etc. (Mtoe)
Production d'énergie d'origine solaire, éolienne, marémotrice, etc. (Mtep)
Erzeugung von Sonnenenergie, Windenergie, Gezeitenenergie usw. (Mtoe)
Produzione di energia solare, eolica, dalle maree, etc. (Mtep)
太陽光、風力、潮力、その他エネルギーの生産量（石油換算百万トン）
Producción de energía solar, eólica, maremotriz, etc. (Mtep)
Производство солнечной энергии, энергии ветра, приливов, и т.д. (Мтнэ)

	1971	1973	1978	1983	1984	1985	1986	1987	1988
Costa Rica	-	-	-	-	-	-	-	-	-
Pérou	-	-	-	-	-	-	-	-	-
Amérique latine	-	-	-	-	-	-	-	-	-
Inde	-	-	-	-	-	-	0.000	0.000	0.001
Singapour	-	-	-	-	-	-	-	-	-
Thaïlande	-	-	-	-	-	-	-	-	-
Asie	-	-	-	-	-	-	0.000	0.000	0.001
Lettonie	-	-	-	-	-	-	-	-	-
Ex-URSS	-	-	-	-	-	-	-	-	-
Israël	-	-	-	-	-	0.231	0.295	0.311	0.326
Jordanie	-	-	-	-	-	-	-	-	-
Liban	-	-	-	-	-	-	-	-	-
Moyen-Orient	-	-	-	-	-	0.231	0.295	0.311	0.326
Total non-OCDE	-	-	-	-	-	0.231	0.295	0.311	0.327
OCDE Amérique du N.	-	-	-	0.000	0.002	0.004	0.005	0.004	0.003
OCDE Pacifique	-	-	0.009	0.037	0.043	0.050	0.083	0.081	0.081
OCDE Europe	0.043	0.048	0.041	0.056	0.057	0.058	0.068	0.077	0.093
Total OCDE	0.043	0.048	0.050	0.093	0.101	0.112	0.156	0.163	0.177
Monde	0.043	0.048	0.050	0.093	0.101	0.343	0.451	0.474	0.504

Production of Energy from Solar, Wind, Tide, etc. (Mtoe)
Production d'énergie d'origine solaire, éolienne, marémotrice, etc. (Mtep)
Erzeugung von Sonnenenergie, Windenergie, Gezeitenenergie usw. (Mtoe)
Produzione di energia solare, eolica, dalle maree, etc. (Mtep)
太陽光、風力、潮力、その他エネルギーの生産量（石油換算百万トン）
Producción de energía solar, eólica, maremotriz, etc. (Mtep)
Производство солнечной энергии, энергии ветра, приливов, и т.д. (Мтнэ)

1989	1990	1991	1992	1993	1994	1995	1996	1997	
-	-	-	-	-	-	-	0.002	0.007	Costa Rica
-	-	-	-	-	-	-	0.035	0.048	Peru
-	-	-	-	-	-	-	0.037	0.054	**Latin America**
0.001	0.003	0.003	0.004	0.005	0.005	0.005	0.005	0.005	India
-	-	-	-	-	-	-	0.055	0.061	Singapore
-	0.000	0.000	0.000	0.000	0.000	0.000	0.000	0.000	Thailand
0.001	0.003	0.003	0.005	0.005	0.005	0.005	0.060	0.066	**Asia**
-	-	-	-	-	-	-	0.000	0.000	Latvia
-	-	-	-	-	-	-	0.000	0.000	**Former USSR**
0.341	0.358	0.379	0.398	0.419	0.440	0.462	0.485	0.511	Israel
-	0.000	0.000	0.000	0.000	0.000	0.000	0.000	0.000	Jordan
-	-	-	0.006	0.006	0.006	0.006	0.006	0.006	Lebanon
0.341	0.358	0.379	0.404	0.425	0.446	0.468	0.491	0.517	**Middle East**
0.342	0.361	0.383	0.409	0.430	0.451	0.473	0.588	0.637	**Non-OECD Total**
0.234	0.257	0.298	0.319	0.346	0.379	0.355	0.380	0.377	OECD North America
0.081	0.084	0.085	0.087	0.087	0.090	0.095	0.102	0.108	OECD Pacific
0.206	0.243	0.284	0.363	0.451	0.562	0.683	0.818	1.112	OECD Europe
0.522	0.584	0.667	0.768	0.884	1.031	1.133	1.299	1.597	**OECD Total**
0.863	0.945	1.050	1.177	1.314	1.482	1.605	1.888	2.235	**World**

Production of Combustible Renewables and Waste (Mtoe)
Production d'énergies renouvelables combustibles et de déchets (Mtep)
Erzeugung von erneuerbaren Brennstoffen und Abfällen (Mtoe)
Produzione di energia da combustibili rinnovabili e da rifiuti (Mtep)
可燃性再生可能エネルギー及び廃棄物の生産量（石油換算百万トン）
Producción de combustibles renovables y desechos (Mtep)
Производство возобновляемых видов топлива и отходов (Мтнэ)

	1971	1973	1978	1983	1984	1985	1986	1987	1988
Algérie	0.26	0.28	0.31	0.37	0.38	0.39	0.40	0.41	0.42
Angola	3.19	3.23	3.47	3.79	3.85	3.92	3.98	4.06	4.13
Bénin	1.00	1.04	1.16	1.30	1.34	1.37	1.41	1.45	1.49
Cameroun	2.36	2.46	2.81	3.20	3.28	3.36	3.44	3.53	3.62
Congo	0.48	0.51	0.56	0.62	0.63	0.65	0.66	0.68	0.70
RD du Congo	5.57	5.88	6.79	7.93	8.20	8.47	8.76	9.05	9.36
Egypte	0.65	0.68	0.73	0.83	0.87	0.94	0.98	1.01	1.04
Ethiopie	8.45	8.86	10.01	11.38	11.69	12.00	12.34	12.72	13.13
Gabon	0.49	0.51	0.57	0.63	0.64	0.66	0.67	0.68	0.70
Ghana	2.09	2.29	2.71	3.07	3.18	3.30	3.42	3.55	3.68
Côte d'Ivoire	1.63	1.74	2.08	2.48	2.57	2.66	2.75	2.85	2.95
Kenya	6.34	6.61	7.42	8.43	8.64	8.85	9.05	9.24	9.42
Libye	0.10	0.11	0.13	0.13	0.13	0.13	0.13	0.13	0.13
Maroc	0.13	0.13	0.21	0.29	0.29	0.30	0.30	0.31	0.31
Mozambique	7.28	7.19	7.20	7.24	7.22	7.18	7.10	7.02	6.93
Nigéria	33.96	35.67	40.64	46.83	48.19	49.55	50.91	52.30	53.70
Sénégal	0.83	0.87	0.99	1.13	1.15	1.18	1.22	1.25	1.28
Afrique du Sud	4.70	5.13	5.85	7.40	8.21	8.33	8.48	9.12	9.65
Soudan	5.65	5.87	6.67	7.76	7.79	7.97	8.13	8.28	8.42
Tanzanie	8.93	8.98	9.25	10.05	10.27	10.49	10.71	10.94	11.17
Tunisie	0.67	0.69	0.77	0.88	0.90	0.93	0.96	0.98	1.00
Zambie	2.47	2.57	2.87	3.42	3.54	3.65	3.72	3.79	3.86
Zimbabwe	3.05	3.17	3.50	3.97	4.08	4.19	4.30	4.41	4.52
Autre Afrique	24.26	25.13	27.97	31.41	32.15	32.92	33.72	34.56	35.42
Afrique	124.52	129.62	144.65	164.53	169.19	173.38	177.53	182.29	187.02
Argentine	2.26	2.10	2.09	2.06	2.08	2.03	1.77	1.90	1.82
Bolivie	0.19	0.22	0.29	0.91	0.90	0.92	0.93	0.66	0.73
Brésil	35.52	36.07	35.56	40.47	44.20	44.72	43.94	45.93	44.81
Chili	1.40	1.32	1.66	2.05	2.19	2.23	2.32	2.42	2.51
Colombie	4.42	3.40	4.70	4.93	4.94	5.17	5.30	5.26	5.43
Costa Rica	0.57	0.59	0.59	0.59	0.74	0.77	0.74	0.75	0.75
Cuba	3.28	3.03	3.87	3.88	3.99	3.72	3.92	3.98	4.31
République dominicaine	1.17	1.16	1.29	1.68	1.82	1.25	1.26	1.28	1.03
Equateur	1.11	1.05	0.98	1.05	1.10	1.06	1.01	1.03	1.11
El Salvador	1.21	1.29	1.52	1.51	1.54	1.56	1.24	1.34	1.20
Guatemala	1.91	2.02	2.64	2.42	2.47	2.53	2.59	2.68	2.74
Haiti	1.38	1.46	1.72	1.59	1.60	1.61	1.20	1.22	1.24
Honduras	0.99	1.03	1.15	1.32	1.33	1.36	1.36	1.41	1.44
Jamaïque	0.27	0.24	0.24	0.22	0.26	0.24	0.24	0.25	0.25
Nicaragua	0.72	0.76	0.86	0.90	0.96	1.02	0.97	1.00	1.06
Panama	0.34	0.34	0.42	0.46	0.46	0.44	0.42	0.41	0.42
Paraguay	1.16	1.25	1.40	1.67	1.65	1.69	1.76	2.07	2.16
Pérou	3.57	3.63	3.59	3.69	3.73	3.81	3.83	3.85	3.76
Trinité-et-Tobago	0.02	0.02	0.02	0.01	0.01	0.03	0.04	0.05	0.04
Uruguay	0.39	0.40	0.45	0.52	0.54	0.58	0.64	0.63	0.58
Vénézuela	0.41	0.40	0.37	0.37	0.38	0.45	0.51	0.54	0.48
Autre Amérique latine	0.47	0.48	0.58	0.69	0.69	0.68	0.69	0.62	0.56
Amérique latine	62.76	62.28	65.99	73.01	77.60	77.86	76.69	79.28	78.44

Production of Combustible Renewables and Waste (Mtoe)
Production d'énergies renouvelables combustibles et de déchets (Mtep)
Erzeugung von erneuerbaren Brennstoffen und Abfällen (Mtoe)
Produzione di energia da combustibili rinnovabili e da rifiuti (Mtep)
可燃性再生可能エネルギー及び廃棄物の生産量（石油換算百万トン）
Producción de combustibles renovables y desechos (Mtep)
Производство возобновляемых видов топлива и отходов (Мтнэ)

1989	1990	1991	1992	1993	1994	1995	1996	1997	
0.43	0.44	0.45	0.46	0.47	0.49	0.50	0.51	0.52	Algeria
4.22	4.32	4.42	4.53	4.63	4.73	4.85	5.03	5.18	Angola
1.52	1.56	1.60	1.64	1.68	1.72	1.76	1.79	1.83	Benin
3.72	3.82	3.92	4.03	4.14	4.25	4.37	4.50	4.63	Cameroon
0.71	0.73	0.75	0.77	0.79	0.81	0.82	0.84	0.86	Congo
9.68	10.00	10.33	10.66	11.01	11.37	11.73	12.10	12.49	DR of Congo
1.01	1.06	1.12	1.14	1.16	1.24	1.21	1.23	1.26	Egypt
13.58	14.07	14.53	15.01	14.57	14.97	15.38	15.76	16.13	Ethiopia
0.71	0.74	0.77	0.79	0.81	0.81	0.84	0.85	0.86	Gabon
3.82	3.90	4.11	4.27	4.43	4.60	4.73	4.84	4.95	Ghana
3.06	3.18	3.29	3.40	3.50	3.61	3.70	3.82	3.93	Ivory Coast
9.60	9.78	9.95	10.12	10.30	10.46	10.63	10.81	11.01	Kenya
0.13	0.13	0.13	0.13	0.13	0.13	0.13	0.13	0.13	Libya
0.32	0.32	0.38	0.38	0.39	0.39	0.40	0.41	0.41	Morocco
6.86	6.80	6.79	6.82	6.90	7.02	6.93	6.93	6.93	Mozambique
55.12	56.58	58.08	59.64	61.27	62.95	64.71	66.72	68.79	Nigeria
1.32	1.36	1.39	1.43	1.46	1.49	1.53	1.58	1.63	Senegal
10.36	10.58	10.79	10.98	11.20	11.41	11.65	11.85	12.04	South Africa
8.56	8.69	8.81	8.94	9.06	9.17	9.29	9.41	9.53	Sudan
11.40	11.65	11.89	12.14	12.40	12.66	12.92	13.15	13.40	Tanzania
1.01	1.04	1.06	1.08	1.10	1.12	1.14	1.15	1.17	Tunisia
3.94	4.02	4.12	4.23	4.35	4.47	4.58	4.66	4.77	Zambia
4.62	4.73	4.82	4.91	5.00	5.09	5.16	5.22	5.29	Zimbabwe
36.32	37.25	37.97	38.96	39.53	40.60	41.67	42.75	43.87	Other Africa
192.03	196.73	201.49	206.47	210.28	215.54	220.60	226.05	231.60	**Africa**
1.56	1.72	1.80	1.86	1.97	2.31	2.66	2.69	2.65	Argentina
0.70	0.75	0.79	0.77	0.78	0.79	0.82	0.77	0.88	Bolivia
44.58	40.36	39.62	38.97	38.61	40.57	39.79	39.30	41.97	Brazil
2.53	2.68	2.99	3.32	3.08	3.23	3.46	3.71	3.69	Chile
5.53	5.82	5.69	5.90	6.04	6.27	6.98	6.89	5.27	Colombia
0.71	0.74	0.74	0.69	0.35	0.36	0.37	0.37	0.27	Costa Rica
4.87	5.43	5.41	5.42	5.43	5.45	5.46	5.47	5.54	Cuba
0.99	1.02	1.02	1.30	1.30	1.30	1.30	1.30	1.31	Dominican Republic
1.08	1.07	1.10	1.12	1.19	1.18	1.22	1.22	1.14	Ecuador
1.17	1.22	1.30	1.33	1.38	1.41	1.83	1.86	1.93	El Salvador
2.86	2.92	2.98	3.02	2.94	2.89	2.93	2.98	3.03	Guatemala
1.25	1.21	1.24	1.32	1.32	1.35	1.37	1.57	1.28	Haiti
1.46	1.50	1.50	1.51	1.51	1.51	1.52	1.52	1.72	Honduras
0.30	0.45	0.36	0.37	0.41	0.53	0.49	0.54	0.59	Jamaica
1.06	1.13	1.12	1.10	1.12	1.14	1.19	1.22	1.31	Nicaragua
0.41	0.40	0.41	0.41	0.47	0.49	0.52	0.55	0.57	Panama
2.22	2.24	2.34	2.19	2.12	2.25	2.41	2.55	2.62	Paraguay
3.80	3.95	3.99	4.05	4.06	4.12	4.17	4.22	4.30	Peru
0.04	0.05	0.04	0.04	0.04	0.04	0.03	0.03	0.03	Trinidad-and-Tobago
0.59	0.57	0.59	0.60	0.59	0.56	0.53	0.52	0.53	Uruguay
0.51	0.53	0.55	0.56	0.55	0.56	0.54	0.54	0.54	Venezuela
0.56	0.58	0.58	0.58	0.58	0.59	0.62	0.63	0.61	Other Latin America
78.75	76.34	76.16	76.42	75.82	78.93	80.21	80.46	81.77	**Latin America**

Production of Combustible Renewables and Waste (Mtoe)
Production d'énergies renouvelables combustibles et de déchets (Mtep)
Erzeugung von erneuerbaren Brennstoffen und Abfällen (Mtoe)
Produzione di energia da combustibili rinnovabili e da rifiuti (Mtep)
可燃性再生可能エネルギー及び廃棄物の生産量（石油換算百万トン）
Producción de combustibles renovables y desechos (Mtep)
Производство возобновляемых видов топлива и отходов (Мтнэ)

	1971	1973	1978	1983	1984	1985	1986	1987	1988
Bangladesh	9.55	10.26	11.56	12.82	13.11	13.41	13.78	14.10	14.42
Brunei	0.02	0.02	0.02	0.02	0.02	0.02	0.02	0.02	0.02
Inde	120.74	126.34	141.58	156.49	159.57	162.33	165.45	168.58	170.82
Indonésie	27.50	28.87	32.31	35.88	36.62	37.32	38.05	38.76	39.43
RPD de Corée	0.68	0.72	0.83	0.90	0.91	0.92	0.92	0.93	0.93
Malaisie	1.27	1.35	1.54	1.72	1.78	1.80	1.84	1.93	1.99
Myanmar	6.36	6.68	7.33	7.99	8.13	8.29	8.49	8.71	9.00
Népal	2.50	2.61	2.95	4.81	4.90	4.97	5.09	5.20	5.33
Pakistan	10.63	11.33	13.21	15.28	15.74	16.19	16.68	17.18	17.68
Philippines	6.12	6.35	7.61	8.66	8.89	9.25	9.37	9.42	9.46
Sri Lanka	2.74	2.79	3.02	3.31	3.55	3.57	3.69	3.80	3.88
Taipei chinois	0.08	0.10	0.02	0.02	0.02	0.01	0.01	0.01	0.01
Thaïlande	7.60	7.91	10.42	8.73	9.51	10.75	11.59	12.23	12.78
Viêt-Nam	12.48	13.08	14.68	16.17	16.46	16.80	17.19	17.62	18.05
Autre Asie	1.89	1.99	2.30	2.40	2.37	2.36	2.42	2.44	2.48
Asie	210.15	220.39	249.37	275.17	281.56	288.00	294.59	300.92	306.29
Rép. populaire de Chine	154.24	161.72	175.34	185.96	187.13	188.53	190.65	192.77	194.94
Hong-Kong, Chine	0.03	0.03	0.04	0.04	0.04	0.04	0.04	0.04	0.04
Chine	154.27	161.76	175.37	186.00	187.17	188.57	190.69	192.81	194.99
Albanie	0.38	0.38	0.38	0.38	0.38	0.38	0.38	0.38	0.38
Bulgarie	0.27	0.24	0.21	0.40	0.40	0.41	0.41	0.21	0.19
Chypre	0.01	0.01	0.01	0.01	0.01	0.01	0.01	0.01	0.01
Roumanie	1.39	1.37	1.20	1.06	0.97	1.19	1.08	0.95	0.90
République slovaque	0.20	0.18	0.19	0.12	0.12	0.13	0.13	0.14	0.14
Bosnie-Herzégovine	-	-	-	-	-	-	-	-	-
Croatie	-	-	-	-	-	-	-	-	-
Ex-RYM	-	-	-	-	-	-	-	-	-
Slovénie	-	-	-	-	-	-	-	-	-
RF de Yougoslavie	-	-	-	-	-	-	-	-	-
Ex-Yougoslavie	1.67	0.89	0.78	0.93	0.99	0.98	1.01	0.88	0.90
Europe non-OCDE	3.91	3.07	2.76	2.91	2.86	3.09	3.01	2.57	2.51
Arménie	-	-	-	-	-	-	-	-	-
Azerbaïdjan	-	-	-	-	-	-	-	-	-
Bélarus	-	-	-	-	-	-	-	-	-
Estonie	-	-	-	-	-	-	-	-	-
Géorgie	-	-	-	-	-	-	-	-	-
Kazakhstan	-	-	-	-	-	-	-	-	-
Kirghizistan	-	-	-	-	-	-	-	-	-
Lettonie	-	-	-	-	-	-	-	-	-
Lituanie	-	-	-	-	-	-	-	-	-
Moldova	-	-	-	-	-	-	-	-	-
Russie	-	-	-	-	-	-	-	-	-
Ukraine	-	-	-	-	-	-	-	-	-
Ouzbékistan	-	-	-	-	-	-	-	-	-
Ex-URSS	20.15	19.48	18.15	18.73	19.87	20.24	18.82	20.13	18.91

Production of Combustible Renewables and Waste (Mtoe)
Production d'énergies renouvelables combustibles et de déchets (Mtep)
Erzeugung von erneuerbaren Brennstoffen und Abfällen (Mtoe)
Produzione di energia da combustibili rinnovabili e da rifiuti (Mtep)
可燃性再生可能エネルギー及び廃棄物の生産量（石油換算百万トン）
Producción de combustibles renovables y desechos (Mtep)
Производство возобновляемых видов топлива и отходов (Мтнэ)

1989	1990	1991	1992	1993	1994	1995	1996	1997	
14.72	14.88	15.11	15.29	15.45	15.62	15.76	15.90	16.03	Bangladesh
0.02	0.02	0.02	0.02	0.02	0.02	0.02	0.02	0.02	Brunei
173.22	175.82	179.74	182.08	184.77	186.74	188.65	190.47	192.80	India
40.11	40.78	41.46	42.10	42.75	43.39	44.07	44.60	44.60	Indonesia
0.94	0.95	0.96	0.98	0.99	1.00	1.01	1.02	1.02	DPR of Korea
2.04	2.10	2.15	2.20	2.25	2.31	2.36	2.41	2.41	Malaysia
9.16	9.31	9.50	9.60	9.72	9.84	9.96	10.11	10.27	Myanmar
5.42	5.57	5.74	5.88	6.11	6.26	6.28	6.46	6.46	Nepal
18.21	18.77	19.31	19.80	20.40	20.97	21.52	21.95	22.40	Pakistan
9.61	9.84	9.93	10.14	10.08	10.29	10.27	9.31	9.40	Philippines
3.94	3.92	3.93	4.08	3.93	3.65	3.64	3.92	4.05	Sri Lanka
0.01	0.01	0.01	0.01	0.01	0.01	0.01	0.01	0.01	Chinese Taipei
14.34	14.78	15.28	16.28	17.73	19.37	20.23	21.37	20.65	Thailand
18.48	18.90	19.32	19.73	20.15	20.56	20.97	21.95	22.40	Vietnam
2.43	2.47	2.52	2.58	2.65	2.74	2.83	2.92	2.92	Other Asia
312.66	318.13	324.98	330.76	337.01	342.76	347.56	352.44	355.44	**Asia**
197.71	200.41	202.35	203.49	204.54	205.49	206.08	207.15	208.26	People's Rep.of China
0.04	0.04	0.04	0.04	0.05	0.05	0.05	0.05	0.05	Hong Kong, China
197.75	200.45	202.39	203.53	204.59	205.53	206.13	207.20	208.31	**China**
0.38	0.36	0.36	0.36	0.13	0.06	0.06	0.06	0.06	Albania
0.36	0.36	0.14	0.17	0.16	0.17	0.22	0.25	0.25	Bulgaria
0.01	0.01	0.01	0.02	0.02	0.02	0.02	0.01	0.01	Cyprus
0.65	0.60	0.72	1.06	1.39	1.19	1.73	4.92	3.39	Romania
0.13	0.17	0.14	0.12	0.12	0.17	0.08	0.08	0.08	Slovak Republic
-	-	-	0.16	0.16	0.16	0.16	0.16	0.16	Bosnia-Herzegovina
-	-	-	0.26	0.24	0.26	0.27	0.33	0.32	Croatia
-	-	-	0.19	0.19	0.19	0.19	0.19	0.19	FYROM
-	-	-	0.24	0.23	0.23	0.23	0.23	0.23	Slovenia
-	-	-	0.21	0.21	0.21	0.21	0.21	0.21	FR of Yugoslavia
0.94	0.77	0.76	1.06	1.05	1.05	1.05	1.11	1.11	Former Yugoslavia
2.47	2.27	2.13	2.79	2.86	2.66	3.15	6.44	4.90	**Non-OECD Europe**
-	-	-	0.00	0.00	0.00	0.00	0.00	0.00	Armenia
-	-	-	0.00	0.00	0.00	0.00	0.00	0.00	Azerbaijan
-	-	-	0.60	0.57	0.55	0.53	0.52	0.69	Belarus
-	-	-	0.18	0.18	0.31	0.35	0.59	0.63	Estonia
-	-	-	0.05	0.05	0.03	0.03	0.03	0.03	Georgia
-	-	-	0.11	0.08	0.08	0.08	0.07	0.07	Kazakhstan
-	-	-	0.00	0.00	0.00	0.00	0.00	0.00	Kyrgyzstan
-	-	-	0.48	0.48	0.59	0.40	0.76	1.30	Latvia
-	-	-	0.19	0.25	0.23	0.24	0.27	0.52	Lithuania
-	-	-	0.04	0.03	0.03	0.03	0.06	0.07	Moldova
-	-	-	17.28	17.28	17.28	17.28	17.28	17.28	Russia
-	-	-	0.30	0.27	0.27	0.26	0.25	0.25	Ukraine
-	-	-	0.00	0.00	0.00	0.00	0.00	0.00	Uzbekistan
17.39	19.10	19.10	19.24	19.20	19.37	19.20	19.85	20.84	**Former USSR**

Production of Combustible Renewables and Waste (Mtoe)
Production d'énergies renouvelables combustibles et de déchets (Mtep)
Erzeugung von erneuerbaren Brennstoffen und Abfällen (Mtoe)
Produzione di energia da combustibili rinnovabili e da rifiuti (Mtep)
可燃性再生可能エネルギー及び廃棄物の生産量（石油換算百万トン）
Producción de combustibles renovables y desechos (Mtep)
Производство возобновляемых видов топлива и отходов (Мтнэ)

	1971	1973	1978	1983	1984	1985	1986	1987	1988
Iran	0.33	0.35	0.58	0.63	0.65	0.68	0.69	0.65	0.71
Irak	0.02	0.02	0.02	0.02	0.02	0.02	0.02	0.02	0.02
Israël	0.00	0.00	0.00	0.00	0.00	0.00	0.00	0.00	0.00
Jordanie	0.00	0.00	0.00	0.00	0.00	0.00	0.00	0.00	0.00
Koweit	0.00	0.00	0.00	0.00	0.00	0.00	0.00	0.00	0.00
Liban	0.10	0.10	0.11	0.10	0.10	0.10	0.10	0.10	0.10
Syrie	0.00	0.00	0.00	0.01	0.00	0.00	0.00	0.00	0.00
Yémen	0.05	0.05	0.06	0.06	0.07	0.07	0.07	0.07	0.07
Moyen-Orient	0.51	0.53	0.77	0.82	0.84	0.88	0.89	0.86	0.92
Total non-OCDE	576.27	597.12	657.07	721.18	739.10	752.03	762.23	778.86	789.07
OCDE Amérique du N.	50.86	53.51	65.43	74.37	81.67	80.94	83.03	89.83	88.88
OCDE Pacifique	3.56	3.53	4.18	9.26	9.69	10.02	9.76	10.08	10.63
OCDE Europe	21.53	22.73	26.72	33.49	35.00	35.59	35.92	36.02	38.26
Total OCDE	75.94	79.76	96.33	117.12	126.36	126.55	128.72	135.93	137.77
Monde	652.21	676.88	753.39	838.30	865.46	878.57	890.94	914.79	926.84

Avant 1978 les données pour les pays de l' OCDE sont incomplètes.

Production of Combustible Renewables and Waste (Mtoe)
Production d'énergies renouvelables combustibles et de déchets (Mtep)
Erzeugung von erneuerbaren Brennstoffen und Abfällen (Mtoe)
Produzione di energia da combustibili rinnovabili e da rifiuti (Mtep)
可燃性再生可能エネルギー及び廃棄物の生産量（石油換算百万トン）
Producción de combustibles renovables y desechos (Mtep)
Производство возобновляемых видов топлива и отходов (Мтнэ)

1989	1990	1991	1992	1993	1994	1995	1996	1997	
0.66	0.68	0.69	0.72	0.73	0.73	0.72	0.79	0.79	Iran
0.02	0.02	0.02	0.02	0.02	0.03	0.03	0.03	0.03	Iraq
0.00	0.00	0.00	0.00	0.00	0.00	0.00	0.00	0.00	Israel
0.00	0.00	0.00	0.00	0.00	0.00	0.00	0.00	0.00	Jordan
0.00	0.00	0.00	-	-	-	-	-	-	Kuwait
0.11	0.10	0.10	0.11	0.11	0.11	0.12	0.12	0.12	Lebanon
0.00	0.00	0.01	0.01	0.00	0.00	0.00	0.00	0.00	Syria
0.08	0.08	0.08	0.08	0.08	0.08	0.08	0.08	0.08	Yemen
0.87	0.89	0.91	0.94	0.95	0.96	0.95	1.02	1.02	**Middle East**
801.92	813.91	827.16	840.14	850.71	865.75	877.81	893.46	903.89	**Non-OECD Total**
85.12	78.18	80.39	89.70	83.43	85.47	88.79	89.56	86.34	OECD North America
11.13	11.33	11.45	11.00	11.60	12.27	12.95	13.87	15.04	OECD Pacific
47.54	46.56	47.48	47.89	51.40	52.81	54.19	56.18	58.41	OECD Europe
143.79	136.07	139.32	148.59	146.43	150.55	155.93	159.61	159.80	**OECD Total**
945.71	949.98	966.48	988.73	997.14	1 016.30	1 033.74	1 053.07	1 063.68	**World**

Prior to 1978 data for OECD countries are incomplete.

Total Production of Energy (Mtoe)
Production totale d'énergie (Mtep)
Gesamte Energieerzeugung (Mtoe)
Produzione totale di energia (Mtep)
エネルギー総生産量（石油換算百万トン）
Producción total de energía (Mtep)
Общее производство топлива и энергии (Мтнэ)

	1971	1973	1978	1983	1984	1985	1986	1987	1988
Algérie	41.96	57.09	70.69	76.62	79.37	83.33	86.76	91.73	94.02
Angola	9.12	11.68	10.18	12.90	14.28	15.77	18.37	21.92	27.07
Bénin	1.00	1.04	1.16	1.33	1.34	1.70	1.78	1.76	1.73
Cameroun	2.46	2.55	3.43	8.99	10.59	12.11	12.04	12.10	12.14
Congo	0.56	2.67	3.20	6.14	6.83	6.76	6.79	7.18	7.94
RD du Congo	5.93	6.28	8.35	9.67	10.06	10.66	10.91	11.15	11.39
Egypte	16.36	9.84	27.70	40.81	46.96	50.96	47.65	53.28	51.69
Ethiopie	8.47	8.89	10.05	11.49	11.79	12.10	12.45	12.83	13.25
Gabon	6.45	8.64	11.17	8.61	8.68	9.43	9.13	8.64	8.84
Ghana	2.34	2.63	3.03	3.35	3.37	3.57	3.79	3.95	4.09
Côte d'Ivoire	1.64	1.76	2.09	3.60	3.62	3.80	3.86	3.74	3.59
Kenya	6.37	6.65	7.51	8.58	8.79	9.02	9.51	9.71	9.90
Libye	137.51	112.56	104.10	58.22	58.92	57.33	56.30	55.16	58.79
Maroc	0.61	0.68	0.86	0.84	0.88	0.87	0.89	0.81	0.83
Mozambique	7.50	7.45	7.36	7.29	7.27	7.24	7.12	7.06	6.97
Nigéria	111.49	139.93	137.27	111.94	121.16	128.45	127.94	123.34	131.84
Sénégal	0.83	0.87	0.99	1.13	1.15	1.18	1.22	1.25	1.29
Afrique du Sud	37.77	40.36	58.66	91.14	103.01	109.53	112.64	112.31	116.87
Soudan	5.66	5.90	6.71	7.82	7.85	8.03	8.20	8.35	8.50
Tanzanie	8.96	9.01	9.30	10.12	10.34	10.58	10.80	11.04	11.28
Tunisie	4.91	4.81	6.15	7.01	6.97	6.91	6.74	6.50	6.45
Zambie	3.03	3.39	3.92	4.57	4.70	4.83	4.91	4.81	4.96
Zimbabwe	5.02	5.20	5.52	5.81	6.01	6.34	6.68	7.49	7.56
Autre Afrique	24.42	25.33	28.38	32.03	32.82	33.66	34.50	35.41	36.28
Afrique	450.37	475.19	527.75	529.97	566.77	594.18	600.98	611.53	637.25
Argentine	30.56	30.53	33.92	41.99	42.93	42.56	41.75	42.25	44.37
Bolivie	2.12	4.29	3.83	4.96	4.78	4.79	4.75	4.48	4.63
Brésil	49.45	51.20	55.73	75.28	88.40	95.23	96.57	97.46	99.16
Chili	4.73	4.98	5.16	6.33	6.61	6.65	6.88	6.94	7.06
Colombie	19.37	17.19	17.59	21.13	23.05	25.42	33.34	39.83	40.72
Costa Rica	0.66	0.69	0.72	0.83	0.99	1.00	0.99	1.01	1.01
Cuba	3.43	3.21	4.21	4.73	4.76	4.59	4.86	4.89	5.04
République dominicaine	1.22	1.21	1.36	1.77	1.93	1.36	1.39	1.43	1.16
Equateur	1.34	11.87	11.69	13.64	14.92	15.95	16.38	10.45	17.05
El Salvador	1.25	1.33	2.09	2.15	2.19	2.20	1.83	1.98	1.84
Guatemala	1.93	2.04	2.66	2.89	2.82	2.79	2.99	3.02	3.09
Haiti	1.38	1.47	1.74	1.61	1.62	1.63	1.22	1.24	1.27
Honduras	1.01	1.06	1.21	1.40	1.42	1.47	1.48	1.56	1.60
Jamaïque	0.28	0.25	0.25	0.24	0.27	0.25	0.26	0.26	0.26
Nicaragua	0.74	0.80	0.89	0.93	1.23	1.31	1.27	1.24	1.27
Panama	0.34	0.35	0.48	0.53	0.59	0.60	0.60	0.59	0.61
Paraguay	1.17	1.27	1.42	1.73	1.74	2.03	2.78	3.67	3.88
Pérou	7.61	8.12	12.54	13.70	14.81	15.45	15.04	14.24	12.96
Trinité-et-Tobago	8.15	9.97	13.65	11.43	12.10	12.41	12.20	11.60	12.07
Uruguay	0.52	0.54	0.59	1.14	1.15	1.14	1.26	1.26	1.05
Vénézuela	198.86	192.39	130.70	115.50	118.26	111.69	119.15	115.31	124.14
Autre Amérique latine	0.57	0.57	0.74	0.89	1.01	1.04	1.06	0.96	0.98
Amérique latine	336.69	345.32	303.17	324.81	347.61	351.56	368.05	365.66	385.23

Total Production of Energy (Mtoe)
Production totale d'énergie (Mtep)
Gesamte Energieerzeugung (Mtoe)
Produzione totale di energia (Mtep)
エネルギー総生産量（石油換算百万トン）
Producción total de energía (Mtep)
Общее производство топлива и энергии (Мтнэ)

1989	1990	1991	1992	1993	1994	1995	1996	1997	
98.16	103.52	106.36	107.97	108.01	103.25	109.63	116.22	125.58	Algeria
27.54	28.87	30.21	29.28	29.86	29.98	31.35	40.99	41.43	Angola
1.72	1.77	1.80	1.78	1.84	1.85	1.85	1.87	1.90	Benin
12.33	12.60	11.90	11.41	11.33	11.06	10.65	10.81	11.25	Cameroon
8.92	9.01	9.05	9.68	10.47	10.65	10.37	11.51	13.54	Congo
11.85	11.98	12.24	12.51	12.85	13.24	13.67	14.06	14.36	DR of Congo
53.35	54.87	55.53	56.56	59.71	59.26	59.92	58.85	58.00	Egypt
13.70	14.22	14.68	15.15	14.74	15.12	15.52	15.90	16.32	Ethiopia
11.16	14.46	15.81	15.67	16.70	15.93	19.42	19.59	19.79	Gabon
4.27	4.39	4.64	4.84	4.99	5.44	5.56	5.72	5.84	Ghana
3.31	3.40	3.46	3.71	3.92	4.03	4.20	4.88	4.91	Ivory Coast
10.09	10.27	10.44	10.60	10.79	10.95	11.14	11.43	11.65	Kenya
61.26	73.27	81.99	78.38	74.48	75.33	77.95	77.74	78.94	Libya
0.76	0.77	0.84	0.82	0.79	0.85	0.84	1.20	1.07	Morocco
6.91	6.85	6.84	6.87	6.96	7.08	6.98	6.96	6.99	Mozambique
147.48	150.45	156.73	162.00	164.93	166.23	169.26	182.04	191.03	Nigeria
1.33	1.36	1.40	1.43	1.47	1.51	1.57	1.62	1.65	Senegal
114.59	114.53	116.93	115.04	122.13	127.26	133.66	133.91	142.14	South Africa
8.64	8.77	8.90	9.03	9.15	9.27	9.37	9.61	9.88	Sudan
11.53	11.78	12.04	12.29	12.55	12.79	13.05	13.31	13.53	Tanzania
6.48	6.08	6.76	6.76	6.46	5.88	5.76	6.31	6.65	Tunisia
4.75	4.82	5.01	5.14	5.19	5.24	5.36	5.44	5.56	Zambia
7.73	8.10	8.57	8.58	8.42	8.29	8.24	8.27	8.15	Zimbabwe
37.21	38.25	38.95	40.00	40.76	41.94	43.06	44.61	47.05	Other Africa
665.05	694.41	721.08	725.49	738.49	742.42	768.43	802.83	837.21	**Africa**
46.00	47.04	48.58	55.03	58.77	64.28	69.18	74.86	80.13	Argentina
4.66	4.88	4.99	4.91	5.04	5.43	5.36	5.24	5.95	Bolivia
101.71	99.09	98.88	99.15	101.05	105.32	106.42	112.37	120.24	Brazil
7.27	7.47	7.87	8.22	7.76	7.83	7.91	7.84	8.17	Chile
44.41	48.10	49.40	48.76	49.87	51.52	61.19	66.74	67.52	Colombia
0.99	1.03	1.06	0.99	0.69	0.70	1.07	1.14	1.16	Costa Rica
5.61	6.12	5.96	6.31	6.55	6.74	6.92	6.95	7.25	Cuba
1.04	1.05	1.06	1.46	1.43	1.44	1.45	1.41	1.42	Dominican Republic
15.90	16.40	17.22	18.35	19.80	21.73	21.84	21.85	22.79	Ecuador
1.85	1.88	2.10	2.12	2.11	2.07	2.52	2.59	2.65	El Salvador
3.22	3.31	3.37	3.48	3.47	3.45	3.62	4.06	4.43	Guatemala
1.28	1.25	1.27	1.35	1.34	1.37	1.39	1.60	1.30	Haiti
1.63	1.69	1.70	1.70	1.71	1.68	1.73	1.78	2.00	Honduras
0.31	0.46	0.37	0.38	0.42	0.54	0.50	0.55	0.60	Jamaica
1.44	1.49	1.54	1.52	1.51	1.49	1.49	1.49	1.53	Nicaragua
0.60	0.60	0.57	0.61	0.66	0.67	0.72	0.79	0.81	Panama
4.31	4.58	4.86	4.52	4.82	5.38	6.03	6.68	6.96	Paraguay
12.23	12.22	11.59	11.52	12.27	12.50	12.31	12.30	12.22	Peru
11.97	12.26	12.09	12.22	11.49	12.92	13.02	13.57	13.58	Trinidad-and-Tobago
0.92	1.17	1.11	1.29	1.22	1.21	1.04	1.02	1.09	Uruguay
122.77	130.98	146.29	149.88	143.23	161.13	187.05	193.99	203.98	Venezuela
1.08	1.09	1.14	1.14	1.14	1.15	1.20	1.17	1.16	Other Latin America
391.19	404.16	423.03	434.92	436.35	470.57	513.97	540.00	566.94	**Latin America**

Total Production of Energy (Mtoe)
Production totale d'énergie (Mtep)
Gesamte Energieerzeugung (Mtoe)

Produzione totale di energia (Mtep)

エネルギー総生産量（石油換算百万トン）

Producción total de energía (Mtep)

Общее производство топлива и энергии (Мтнэ)

	1971	1973	1978	1983	1984	1985	1986	1987	1988
Bangladesh	9.94	10.81	12.34	14.57	15.15	15.74	16.40	17.21	17.89
Brunei	6.59	12.95	18.12	17.83	17.05	16.01	15.58	14.96	14.59
Inde	168.49	177.30	211.03	257.76	269.26	278.28	290.54	300.19	313.66
Indonésie	72.94	96.87	122.89	120.46	137.59	135.28	143.59	146.14	147.75
RPD de Corée	25.38	23.19	27.81	31.12	31.22	32.35	32.95	32.98	33.28
Malaisie	4.77	5.97	13.38	24.70	30.71	34.14	40.47	41.47	43.79
Myanmar	7.34	7.81	9.00	9.98	10.43	10.67	10.95	10.73	10.76
Népal	2.51	2.62	2.97	4.83	4.92	5.00	5.12	5.25	5.38
Pakistan	14.36	15.59	19.03	24.11	24.51	26.11	27.62	30.33	31.51
Philippines	6.31	6.57	7.98	13.68	14.37	14.95	14.86	14.60	15.10
Singapour	-	-	-	-	-	-	-	-	-
Sri Lanka	2.81	2.85	3.14	3.42	3.73	3.78	3.92	3.98	4.10
Taipei chinois	3.95	3.86	4.77	8.19	9.49	10.54	9.92	11.40	10.72
Thaïlande	7.89	8.16	10.80	11.50	13.52	17.70	19.85	20.62	22.25
Viêt-Nam	14.03	14.61	17.75	19.48	19.15	19.75	20.41	20.94	21.95
Autre Asie	4.13	4.33	4.49	4.76	4.91	5.03	5.12	4.33	4.15
Asie	351.43	393.49	485.48	566.40	606.02	625.34	657.28	675.14	696.87
Rép. populaire de Chine	392.15	428.91	599.39	661.72	708.57	761.86	781.43	804.26	835.93
Hong-Kong, Chine	0.03	0.03	0.04	0.04	0.04	0.04	0.04	0.04	0.04
Chine	392.18	428.94	599.43	661.76	708.61	761.91	781.47	804.30	835.97
Albanie	2.42	3.00	3.48	2.81	3.06	2.88	2.94	2.90	2.95
Bulgarie	5.71	5.47	6.73	9.65	9.94	9.53	9.88	10.04	10.50
Chypre	0.01	0.01	0.01	0.01	0.01	0.01	0.01	0.01	0.01
Roumanie	42.81	46.24	54.03	55.93	55.68	54.32	54.02	52.56	53.41
République slovaque	2.69	2.57	2.63	3.83	4.19	4.82	5.33	5.42	5.55
Bosnie-Herzégovine	-	-	-	-	-	-	-	-	-
Croatie	-	-	-	-	-	-	-	-	-
Ex-RYM	-	-	-	-	-	-	-	-	-
Slovénie	-	-	-	-	-	-	-	-	-
RF de Yougoslavie	-	-	-	-	-	-	-	-	-
Ex-Yougoslavie	14.37	14.68	18.10	23.72	24.98	25.92	26.40	26.28	25.61
Europe non-OCDE	68.01	71.98	84.97	95.94	97.86	97.47	98.57	97.21	98.03
Arménie	-	-	-	-	-	-	-	-	-
Azerbaïdjan	-	-	-	-	-	-	-	-	-
Bélarus	-	-	-	-	-	-	-	-	-
Estonie	-	-	-	-	-	-	-	-	-
Géorgie	-	-	-	-	-	-	-	-	-
Kazakhstan	-	-	-	-	-	-	-	-	-
Kirghizistan	-	-	-	-	-	-	-	-	-
Lettonie	-	-	-	-	-	-	-	-	-
Lituanie	-	-	-	-	-	-	-	-	-
Moldova	-	-	-	-	-	-	-	-	-
Russie	-	-	-	-	-	-	-	-	-
Tadjikistan	-	-	-	-	-	-	-	-	-
Turkménistan	-	-	-	-	-	-	-	-	-
Ukraine	-	-	-	-	-	-	-	-	-
Ouzbékistan	-	-	-	-	-	-	-	-	-
Ex-URSS	902.52	989.06	1 271.33	1 436.21	1 476.47	1 512.95	1 575.82	1 628.69	1 676.98

Total Production of Energy (Mtoe)
Production totale d'énergie (Mtep)
Gesamte Energieerzeugung (Mtoe)

Produzione totale di energia (Mtep)

エネルギー総生産量（石油換算百万トン）

Producción total de energía (Mtep)

Общее производство топлива и энергии (Мтнэ)

1989	1990	1991	1992	1993	1994	1995	1996	1997	
18.60	18.78	19.12	19.91	20.37	20.90	21.49	21.87	21.89	Bangladesh
15.10	15.48	16.24	17.36	17.10	17.61	18.21	17.65	17.59	Brunei
323.18	333.57	345.82	351.52	357.73	370.76	386.18	392.42	404.50	India
158.57	167.50	180.06	185.38	191.59	201.71	208.28	219.72	221.55	Indonesia
33.81	31.96	30.36	27.85	26.01	24.14	22.55	21.06	20.46	DPR of Korea
47.49	49.02	53.16	54.91	58.10	60.63	64.74	69.61	73.98	Malaysia
10.91	10.94	11.13	11.19	11.45	11.78	11.80	11.99	12.25	Myanmar
5.47	5.64	5.82	5.95	6.18	6.34	6.35	6.56	6.56	Nepal
33.23	34.34	36.18	37.10	38.17	39.24	40.13	41.93	42.05	Pakistan
15.64	15.90	16.09	16.61	16.67	17.08	16.77	16.06	16.62	Philippines
-	-	-	-	-	-	-	0.06	0.06	Singapore
4.18	4.19	4.20	4.33	4.26	4.00	4.02	4.20	4.35	Sri Lanka
9.71	10.81	10.83	10.51	10.47	10.83	10.94	11.51	11.09	Chinese Taipei
24.24	26.54	29.31	31.30	33.65	37.21	38.97	43.68	46.17	Thailand
22.90	24.68	26.45	28.64	30.11	31.54	34.36	40.68	43.53	Vietnam
3.11	7.63	8.16	8.44	8.61	8.80	8.39	7.47	7.49	Other Asia
726.14	756.97	792.93	811.00	830.47	862.57	893.20	926.47	950.13	**Asia**
877.16	894.04	902.91	919.73	949.62	1 008.44	1 072.93	1 101.46	1 097.16	People's Rep.of China
0.04	0.04	0.04	0.04	0.05	0.05	0.05	0.05	0.05	Hong Kong, China
877.20	894.09	902.95	919.77	949.66	1 008.48	1 072.98	1 101.51	1 097.21	**China**
2.85	2.35	1.88	1.49	1.19	1.00	1.00	1.08	0.91	Albania
10.32	9.80	8.55	8.49	8.84	9.09	10.03	10.35	9.98	Bulgaria
0.01	0.01	0.01	0.02	0.02	0.02	0.02	0.01	0.01	Cyprus
49.69	39.72	35.02	33.85	33.15	31.24	31.74	34.61	31.01	Romania
5.58	5.28	4.88	4.60	4.59	5.06	4.93	4.82	4.69	Slovak Republic
-	-	-	3.61	3.10	0.57	0.63	0.63	0.63	*Bosnia-Herzegovina*
-	-	-	4.35	4.55	4.08	4.18	4.19	4.01	*Croatia*
-	-	-	1.74	1.71	1.70	1.81	1.79	1.70	*FYROM*
-	-	-	2.76	2.62	2.77	2.81	2.76	2.87	*Slovenia*
-	-	-	11.62	10.93	11.11	11.50	10.99	11.48	*FR of Yugoslavia*
25.24	25.27	24.77	24.08	22.91	20.23	20.93	20.35	20.69	Former Yugoslavia
93.68	82.43	75.12	72.54	70.69	66.64	68.64	71.22	67.29	**Non-OECD Europe**
-	-	-	0.26	0.37	0.30	0.25	0.74	0.54	Armenia
-	-	-	18.15	16.39	14.94	14.73	14.39	14.03	Azerbaijan
-	-	-	4.10	3.41	3.50	3.32	3.16	3.27	Belarus
-	-	-	4.45	3.27	3.49	3.26	3.85	3.79	Estonia
-	-	-	0.83	0.83	0.54	0.57	0.70	0.69	Georgia
-	-	-	90.92	78.48	70.86	62.77	62.83	64.78	Kazakhstan
-	-	-	1.82	1.58	1.66	1.38	1.44	1.41	Kyrgyzstan
-	-	-	1.10	0.79	1.00	0.72	1.00	1.64	Latvia
-	-	-	4.21	3.64	2.45	3.55	4.17	3.97	Lithuania
-	-	-	0.06	0.06	0.05	0.05	0.09	0.10	Moldova
-	-	-	1 122.06	1 041.30	968.95	950.98	946.71	927.34	Russia
-	-	-	1.55	1.63	1.53	1.32	1.34	1.25	Tajikistan
-	-	-	48.82	57.42	33.32	33.52	32.86	18.74	Turkmenistan
-	-	-	110.17	100.23	86.88	82.99	80.65	81.18	Ukraine
-	-	-	40.46	42.72	45.00	47.22	47.41	49.05	Uzbekistan
1 664.38	1 624.61	1 545.54	1 441.62	1 345.50	1 229.14	1 201.71	1 197.22	1 167.51	**Former USSR**

Total Production of Energy (Mtoe)
Production totale d'énergie (Mtep)
Gesamte Energieerzeugung (Mtoe)

Produzione totale di energia (Mtep)

エネルギー総生産量（石油換算百万トン）

Producción total de energía (Mtep)

Общее производство топлива и энергии (Мтнэ)

	1971	1973	1978	1983	1984	1985	1986	1987	1988
Bahrein	4.57	4.82	4.99	5.37	5.51	6.18	6.96	6.73	7.12
Iran	244.60	315.57	285.64	136.02	124.10	127.72	107.76	133.51	133.43
Irak	85.98	102.87	131.48	51.45	63.14	74.96	89.70	112.58	140.66
Israël	5.94	6.15	0.07	0.07	0.06	0.28	0.34	0.36	0.38
Jordanie	0.00	0.00	0.00	0.00	0.00	0.00	0.02	0.02	0.02
Koweit	168.82	158.09	116.05	58.92	63.54	58.89	80.76	76.15	80.48
Liban	0.17	0.14	0.18	0.15	0.15	0.15	0.15	0.16	0.16
Oman	14.89	14.89	16.47	20.74	22.40	26.75	30.27	31.56	33.62
Qatar	22.03	29.53	25.22	19.25	25.86	20.49	21.51	20.01	21.82
Arabie saoudite	244.43	388.96	437.26	280.43	264.39	200.79	285.91	248.45	297.20
Syrie	5.32	5.57	10.17	9.71	9.81	9.66	9.81	12.92	15.40
Emirats arabes unis	52.90	79.23	98.41	67.13	72.38	73.62	84.14	91.42	97.74
Yémen	0.05	0.05	0.06	0.06	0.07	0.07	0.07	1.05	8.39
Moyen-Orient	849.68	1 105.88	1 126.00	649.31	651.41	599.57	717.40	734.92	836.41
Total non-OCDE	3 350.87	3 809.87	4 398.13	4 264.39	4 454.75	4 542.98	4 799.57	4 917.46	5 166.74
OCDE Amérique du N.	1 636.97	1 702.76	1 754.79	1 869.77	2 013.41	2 005.54	1 980.12	2 034.65	2 081.50
OCDE Pacifique	99.43	108.29	134.02	178.24	193.07	216.98	233.16	249.04	241.01
OCDE Europe	598.24	625.65	777.65	930.85	944.95	1 014.98	1 033.79	1 046.89	1 047.70
Total OCDE	2 334.64	2 436.69	2 666.46	2 978.86	3 151.43	3 237.50	3 247.08	3 330.58	3 370.21
Monde	5 685.51	6 246.55	7 064.59	7 243.25	7 606.17	7 780.48	8 046.64	8 248.04	8 536.95

Total Production of Energy (Mtoe)
Production totale d'énergie (Mtep)
Gesamte Energieerzeugung (Mtoe)
Produzione totale di energia (Mtep)
エネルギー総生産量（石油換算百万トン）
Producción total de energía (Mtep)
Общее производство топлива и энергии (Мтнэ)

1989	1990	1991	1992	1993	1994	1995	1996	1997	
6.87	7.20	7.01	7.10	7.35	7.23	7.47	7.64	7.49	Bahrain
166.00	181.33	200.19	209.36	220.06	220.48	221.12	223.46	224.93	Iran
152.19	106.71	16.42	24.18	26.89	29.94	31.13	32.58	62.09	Iraq
0.39	0.43	0.45	0.47	0.52	0.53	0.57	0.57	0.60	Israel
0.05	0.10	0.11	0.11	0.13	0.18	0.19	0.18	0.19	Jordan
98.11	66.86	11.40	58.67	101.84	110.98	112.39	112.90	116.09	Kuwait
0.15	0.14	0.15	0.17	0.18	0.19	0.19	0.20	0.21	Lebanon
34.68	38.06	39.30	41.59	44.21	46.10	48.40	50.21	51.62	Oman
24.62	26.11	26.77	32.95	32.86	32.19	32.30	32.96	43.97	Qatar
296.64	368.75	464.50	474.81	470.09	466.41	470.73	475.00	487.10	Saudi Arabia
19.22	22.57	26.84	28.15	29.09	31.48	34.29	33.71	32.79	Syria
114.73	123.99	141.61	134.40	129.14	139.28	141.67	148.82	153.55	United Arab Emirates
9.46	9.79	10.27	9.32	11.42	17.68	18.02	18.25	19.11	Yemen
923.11	952.05	945.01	1 021.29	1 073.77	1 102.66	1 118.48	1 136.48	1 199.73	**Middle East**
5 340.75	5 408.72	5 405.66	5 433.97	5 451.57	5 487.82	5 642.32	5 779.85	5 890.29	**Non-OECD Total**
2 081.44	2 117.51	2 124.11	2 138.25	2 116.80	2 206.52	2 214.97	2 257.66	2 269.64	OECD North America
250.27	267.18	276.81	283.87	294.56	296.29	319.37	328.44	344.34	OECD Pacific
1 050.91	1 020.69	1 022.15	1 038.22	1 054.71	1 083.05	1 114.30	1 169.57	1 164.87	OECD Europe
3 382.62	3 405.39	3 423.06	3 460.33	3 466.07	3 585.86	3 648.64	3 755.66	3 778.85	**OECD Total**
8 723.37	8 814.10	8 828.72	8 894.31	8 917.64	9 073.68	9 290.97	9 535.51	9 669.15	**World**

Net Imports of Coal (Mtoe)
Importations nettes de charbon (Mtep)
Nettoimporte von Kohle (Mtoe)
Importazioni nette di carbone (Mtep)
石炭の純輸入量（石油換算百万トン）
Importaciones netas de carbón (Mtep)
Чистый импорт угля (Мтнэ)

	1971	1973	1978	1983	1984	1985	1986	1987	1988
Algérie	0.23	0.27	0.17	0.73	0.81	0.85	0.88	1.05	0.99
RD du Congo	0.14	0.08	0.09	0.13	0.13	0.13	0.14	0.14	0.14
Egypte	0.25	0.24	0.53	0.60	0.67	0.73	0.74	0.77	0.86
Ghana	0.01	0.01	0.00	0.00	0.00	0.00	0.00	0.00	0.00
Kenya	0.05	0.04	0.03	0.02	0.05	0.06	0.05	0.06	0.07
Maroc	0.01	0.00	-0.01	0.18	0.12	0.28	0.53	0.69	0.72
Mozambique	0.18	0.12	0.06	0.08	0.05	0.04	0.04	0.01	0.01
Nigéria	0.01	-0.02	-0.00	0.00	-0.00	-0.00	-0.02	-0.02	-0.02
Afrique du Sud	-0.97	-1.30	-10.48	-21.11	-26.32	-32.21	-31.30	-28.22	-29.89
Tanzanie	-	-	0.00	0.00	0.00	0.00	0.00	0.00	0.00
Tunisie	0.07	0.07	0.08	0.05	0.07	0.07	0.06	0.08	0.08
Zambie	0.09	0.04	0.06	-	-	-	-0.01	-	-
Zimbabwe	-0.20	-0.08	-0.18	-0.12	-0.12	-0.05	0.03	0.03	-0.01
Autre Afrique	0.03	0.04	0.04	0.06	0.05	0.04	0.07	0.05	0.05
Afrique	-0.11	-0.49	-9.61	-19.37	-24.48	-30.06	-28.80	-25.38	-27.00
Argentine	0.50	0.56	0.61	0.35	0.39	0.55	0.83	0.90	1.00
Bolivie	-	-	-	0.17	0.10	0.06	0.04	-	-
Brésil	1.30	1.39	2.91	4.14	5.57	5.94	6.18	7.54	7.40
Chili	0.18	0.20	0.18	0.23	0.28	0.27	0.34	0.32	0.34
Colombie	-0.02	-0.05	-0.16	-0.33	-0.65	-2.31	-3.76	-6.30	-7.06
Costa Rica	0.00	0.00	0.00	0.00	0.00	0.00	0.00	0.00	0.00
Cuba	0.08	0.07	0.09	0.09	0.09	0.12	0.11	0.11	0.10
République dominicaine	-	-	-	0.00	0.01	0.14	0.02	0.14	0.06
El Salvador	-	-	0.00	-	-	-	-	-	-
Haiti	-	-	-	0.04	0.04	0.04	0.01	0.01	0.02
Honduras	-	-	-	-	-	-	0.00	0.00	0.00
Jamaïque	-	-	-	-	-	-	-	-	-
Panama	-	-	-	0.00	0.01	0.04	0.02	0.02	0.01
Pérou	0.08	0.14	0.13	0.06	0.03	0.07	0.11	0.10	0.05
Uruguay	0.02	0.02	0.00	0.00	0.00	0.00	0.00	0.00	-
Vénézuela	0.10	0.22	0.02	0.09	0.09	0.14	0.15	0.20	-0.50
Autre Amérique latine	0.02	0.03	0.02	0.01	0.01	-	-	-	-
Amérique latine	2.27	2.57	3.80	4.84	5.97	5.04	4.05	3.05	1.43
Bangladesh	0.09	0.12	0.16	0.09	0.03	0.05	0.07	0.12	0.12
Inde	-0.13	-0.24	0.00	0.29	0.16	1.15	1.22	1.75	2.18
Indonésie	0.02	0.00	-0.01	-0.24	-0.52	-0.64	0.22	0.45	0.22
RPD de Corée	0.42	0.22	0.43	1.39	1.52	1.70	1.70	1.70	1.70
Malaisie	0.04	0.03	0.02	0.22	0.24	0.32	0.24	0.29	0.44
Myanmar	0.13	0.04	0.13	0.12	0.11	0.12	0.12	0.03	0.03
Népal	0.01	0.05	0.01	0.03	0.06	0.05	0.03	0.05	0.05
Pakistan	0.02	0.02	0.01	0.34	0.32	0.47	0.56	0.60	0.56
Philippines	0.01	0.01	0.17	0.23	0.39	0.66	0.52	0.37	0.73
Singapour	0.00	0.00	0.00	0.00	0.00	0.01	0.01	0.01	0.01
Sri Lanka	0.00	0.00	0.00	0.00	0.00	0.00	0.00	0.00	-
Taipei chinois	-0.00	0.10	0.91	4.25	4.95	6.61	7.01	9.19	11.45
Thaïlande	0.01	0.01	0.03	0.11	0.17	0.16	0.14	0.18	0.24
Viêt-Nam	-0.22	-0.11	-0.70	-0.19	-0.23	-0.29	-0.30	-0.09	-0.15
Autre Asie	0.23	0.06	0.04	0.07	0.08	0.12	0.12	0.11	0.12
Asie	0.63	0.32	1.22	6.70	7.29	10.50	11.65	14.75	17.72

Un chiffre négatif correspond à des exportations nettes.

INTERNATIONAL ENERGY AGENCY

Net Imports of Coal (Mtoe)
Importations nettes de charbon (Mtep)

Nettoimporte von Kohle (Mtoe)

Importazioni nette di carbone (Mtep)

石炭の純輸入量（石油換算百万トン）

Importaciones netas de carbón (Mtep)

Чистый импорт угля (Мтнэ)

1989	1990	1991	1992	1993	1994	1995	1996	1997	
0.84	0.65	0.65	0.78	0.50	0.46	0.46	0.46	0.38	Algeria
0.14	0.15	0.15	0.15	0.15	0.15	0.18	0.18	0.18	DR of Congo
0.89	0.76	0.72	0.70	0.91	0.95	0.64	0.64	0.64	Egypt
0.00	0.00	0.00	0.00	0.00	0.00	0.00	0.00	0.00	Ghana
0.08	0.09	0.09	0.10	0.08	0.07	0.06	0.06	0.06	Kenya
0.88	0.83	0.94	0.80	0.86	0.99	1.27	1.80	1.81	Morocco
0.01	0.01	0.01	0.01	0.01	0.01	0.01	0.01	0.01	Mozambique
-0.02	-0.02	-0.02	-0.02	-0.01	-	-	-	-	Nigeria
-31.68	-33.62	-31.92	-35.17	-34.57	-36.66	-39.63	-39.95	-42.61	South Africa
0.00	0.00	0.00	0.00	0.00	0.00	0.00	-	-	Tanzania
0.09	0.08	0.07	0.08	0.08	0.07	0.07	0.07	0.07	Tunisia
-	-0.00	-0.01	-0.01	-0.01	-0.01	-0.00	-0.00	-0.00	Zambia
0.02	-0.01	-0.05	-0.01	-0.01	-0.14	-0.13	-0.10	-0.08	Zimbabwe
0.09	0.09	0.07	0.07	0.07	0.06	0.07	0.07	0.08	Other Africa
-28.66	-31.01	-29.29	-32.53	-31.94	-34.06	-36.99	-36.76	-39.46	**Africa**
0.74	0.74	0.53	0.73	0.64	0.91	0.49	0.76	0.63	Argentina
-	-	-	-	-	-	-	-	-	Bolivia
7.91	7.86	8.67	8.34	8.89	9.36	9.94	10.67	10.23	Brazil
1.09	1.16	1.01	0.60	0.75	1.33	1.52	2.49	3.17	Chile
-8.75	-9.52	-10.64	-10.49	-11.52	-11.98	-11.88	-16.11	-17.22	Colombia
0.00	0.00	-	-	-	-	-	0.01	-	Costa Rica
0.18	0.14	0.07	0.04	0.05	0.07	0.06	0.01	0.05	Cuba
0.01	0.01	0.04	0.09	0.07	0.06	0.07	0.08	0.09	Dominican Republic
-	-	-	-	0.00	0.00	0.00	0.00	0.00	El Salvador
0.00	0.01	0.02	0.02	-	-	-	-	-	Haiti
0.00	0.00	0.00	0.00	0.00	0.00	0.00	0.00	0.00	Honduras
0.03	0.04	0.02	0.02	0.03	0.03	0.03	0.04	0.04	Jamaica
0.01	0.02	0.03	0.03	0.05	0.04	0.04	0.04	0.04	Panama
0.09	0.08	0.23	0.23	0.35	0.38	0.35	0.33	0.29	Peru
-	-	0.00	0.00	0.00	-	-	-	0.00	Uruguay
-0.85	-1.13	-1.30	-1.61	-2.21	-2.70	-2.64	-1.94	-3.16	Venezuela
0.00	0.00	0.00	0.00	0.00	0.00	0.00	0.00	0.00	Other Latin America
0.47	-0.60	-1.31	-2.00	-2.90	-2.50	-2.01	-3.63	-5.84	**Latin America**
0.12	0.28	0.09	0.08	0.03	-	-	-	-	Bangladesh
2.64	3.46	3.59	3.79	4.32	5.03	5.41	5.75	5.78	India
-0.87	-2.50	-4.24	-9.58	-10.99	-12.88	-18.87	-22.16	-25.27	Indonesia
1.70	1.70	1.44	1.46	1.43	1.43	1.43	1.44	1.40	DPR of Korea
1.12	1.45	1.25	1.29	1.21	1.56	1.54	1.92	1.45	Malaysia
0.03	0.03	0.03	0.01	0.01	0.01	-	-	-	Myanmar
0.05	0.01	0.04	0.06	0.06	0.07	0.07	0.07	0.07	Nepal
0.59	0.59	0.64	0.65	0.65	0.72	0.72	0.71	0.63	Pakistan
0.59	0.79	0.78	0.43	0.46	0.63	0.63	1.58	2.20	Philippines
0.01	0.02	0.01	0.02	0.02	0.02	0.02	-0.00	-	Singapore
0.00	0.00	0.00	0.00	0.00	0.00	0.00	0.00	0.00	Sri Lanka
10.99	12.10	12.06	14.52	16.61	17.55	18.83	20.40	23.75	Chinese Taipei
0.30	0.20	0.33	0.34	0.66	0.96	1.53	2.49	2.07	Thailand
-0.25	-0.36	-0.46	-0.64	-0.86	-1.02	-0.89	-2.19	-2.11	Vietnam
0.13	0.12	0.13	0.13	0.13	0.13	0.13	0.13	0.14	Other Asia
17.15	17.91	15.69	12.55	13.74	14.21	10.54	10.14	10.10	**Asia**

A negative number shows net exports.

Net Imports of Coal (Mtoe)
Importations nettes de charbon (Mtep)
Nettoimporte von Kohle (Mtoe)
Importazioni nette di carbone (Mtep)
石炭の純輸入量（石油換算百万トン）
Importaciones netas de carbón (Mtep)
Чистый импорт угля (Мтнэ)

	1971	1973	1978	1983	1984	1985	1986	1987	1988
Rép. populaire de Chine	-1.67	-1.69	-0.54	-2.40	-2.44	-2.91	-2.71	-4.45	-7.54
Hong-Kong, Chine	0.02	0.01	0.01	2.10	2.75	3.40	3.93	4.93	5.70
Chine	-1.65	-1.68	-0.53	-0.30	0.31	0.49	1.23	0.48	-1.84
Albanie	0.06	0.07	0.12	0.14	0.14	0.15	0.16	0.16	0.16
Bulgarie	3.79	3.70	3.85	4.38	4.43	5.18	4.42	4.39	3.89
Chypre	-	-	-	-	0.03	0.05	0.03	0.09	0.06
Malte	-	-	-	0.03	0.06	0.13	0.09	0.11	0.15
Roumanie	1.61	2.25	4.10	4.58	4.34	4.93	5.15	5.61	6.09
République slovaque	6.29	6.21	6.52	6.28	6.23	6.50	6.89	6.97	7.06
Croatie	-	-	-	-	-	-	-	-	-
Ex-RYM	-	-	-	-	-	-	-	-	-
Slovénie	-	-	-	-	-	-	-	-	-
RF de Yougoslavie	-	-	-	-	-	-	-	-	-
Ex-Yougoslavie	1.66	1.66	0.94	2.91	3.44	3.19	3.37	2.66	2.88
Europe non-OCDE	13.41	13.88	15.52	18.30	18.67	20.12	20.12	19.98	20.28
Ex-URSS	-10.25	-9.46	-9.19	-6.16	-6.46	-8.80	-10.17	-11.88	-14.55
Iran	-	0.04	0.06	0.04	0.03	0.06	0.09	0.12	0.12
Israël	0.01	0.00	0.00	1.35	1.68	1.93	1.96	2.24	2.25
Liban	0.00	0.01	0.00	-	-	-	-	-	-
Syrie	0.00	0.00	0.00	0.01	0.00	0.00	0.00	0.00	0.00
Moyen-Orient	0.01	0.06	0.07	1.39	1.72	2.00	2.05	2.36	2.37
Total non-OCDE	4.32	5.20	1.28	5.41	3.01	-0.71	0.13	3.37	-1.59
OCDE Amérique du N.	-26.78	-27.21	-20.08	-48.87	-54.54	-64.94	-60.15	-56.00	-65.71
OCDE Pacifique	18.98	23.55	13.02	22.19	24.36	21.22	15.85	11.32	15.65
OCDE Europe	14.63	8.07	4.71	26.07	34.06	49.88	49.81	50.95	50.88
Total OCDE	6.83	4.42	-2.35	-0.61	3.88	6.16	5.50	6.26	0.82
Monde	11.14	9.62	-1.07	4.80	6.89	5.45	5.63	9.63	-0.76

Un chiffre négatif correspond à des exportations nettes.
La ligne Monde montre la divergence entre le total des exportations et importations mondiales.

INTERNATIONAL ENERGY AGENCY

Net Imports of Coal (Mtoe)
Importations nettes de charbon (Mtep)
Nettoimporte von Kohle (Mtoe)
Importazioni nette di carbone (Mtep)
石炭の純輸入量（石油換算百万トン）
Importaciones netas de carbón (Mtep)
Чистый импорт угля (Мтнэ)

1989	1990	1991	1992	1993	1994	1995	1996	1997	
-7.52	-8.36	-9.87	-9.95	-11.56	-14.91	-20.40	-23.02	-22.33	People's Rep.of China
6.11	5.49	5.93	6.29	7.28	5.20	5.60	4.16	3.51	Hong Kong, China
-1.42	-2.87	-3.94	-3.66	-4.29	-9.72	-14.80	-18.86	-18.82	**China**
0.16	0.14	0.11	0.01	-0.00	-	-	-	-	Albania
3.68	3.37	2.87	2.25	2.62	2.12	2.20	2.38	2.39	Bulgaria
0.06	0.06	0.06	0.02	0.02	0.02	0.02	0.01	0.02	Cyprus
0.19	0.18	0.15	0.13	0.18	0.13	0.02	-	-	Malta
4.64	4.30	2.70	3.82	1.97	2.47	2.71	2.47	3.19	Romania
6.73	5.68	5.24	4.70	4.66	4.16	3.89	4.14	3.71	Slovak Republic
-	-	-	0.39	0.27	0.17	0.16	0.10	0.16	*Croatia*
-	-	-	0.11	0.09	0.06	0.08	0.08	0.07	*FYROM*
-	-	-	0.07	0.16	0.14	0.19	0.22	0.18	*Slovenia*
-	-	-	0.73	0.04	0.04	0.04	0.04	0.04	*FR of Yugoslavia*
2.55	2.36	1.71	1.36	0.61	0.44	0.48	0.45	0.46	Former Yugoslavia
18.01	16.10	12.84	12.28	10.07	9.34	9.33	9.46	9.77	**Non-OECD Europe**
-12.29	-11.72	-9.80	-9.09	-10.67	-8.67	-9.21	-7.79	-7.50	**Former USSR**
0.18	0.25	0.25	0.31	-	0.25	0.31	0.36	0.36	Iran
2.35	2.54	2.46	3.40	3.37	3.83	4.44	4.48	5.06	Israel
-	-	-	-	0.07	0.07	0.12	0.13	0.13	Lebanon
-	-	0.00	0.00	0.00	0.00	0.00	0.00	0.00	Syria
2.53	2.79	2.70	3.71	3.44	4.15	4.87	4.97	5.56	**Middle East**
-4.21	-9.40	-13.11	-18.73	-22.54	-27.24	-38.27	-42.47	-46.19	**Non-OECD Total**
-73.56	-76.33	-81.38	-71.23	-54.73	-54.23	-65.28	-65.94	-60.87	OECD North America
17.86	16.55	16.90	10.03	11.66	16.09	17.03	19.56	18.49	OECD Pacific
56.34	68.02	83.67	82.48	68.40	67.61	73.14	79.99	85.17	OECD Europe
0.64	8.25	19.20	21.28	25.33	29.47	24.90	33.61	42.80	**OECD Total**
-3.57	-1.15	6.09	2.55	2.79	2.22	-13.37	-8.86	-3.39	**World**

A negative number shows net exports.
The row World shows the discrepancy between total world exports and imports.

INTERNATIONAL ENERGY AGENCY

Net Imports of Oil (Mtoe)
Importations nettes de pétrole (Mtep)
Netto-Ölimporte (Mtoe)

Importazioni nette di petrolio (Mtep)

石油の純輸入量（石油換算百万トン）

Importaciones netas de petróleo (Mtep)

Чистый импорт нефти и нефтепродуктов (Мтнэ)

	1971	1973	1978	1983	1984	1985	1986	1987	1988
Algérie	-35.35	-49.04	-51.61	-40.39	-42.50	-43.34	-45.12	-45.73	-46.53
Angola	-4.66	-7.27	-5.54	-7.66	-8.95	-10.25	-12.82	-16.05	-21.29
Bénin	0.11	0.14	0.14	0.12	0.14	-0.15	-0.22	-0.16	-0.10
Cameroun	0.30	0.33	0.04	-4.70	-6.14	-7.58	-7.46	-7.44	-7.33
Congo	0.14	-1.33	-1.83	-5.23	-5.77	-5.79	-5.87	-5.79	-6.15
RD du Congo	0.83	0.90	-0.29	-0.15	-0.29	-0.67	-0.73	-0.55	-0.38
Egypte	-8.86	-1.87	-14.18	-17.99	-22.35	-25.47	-22.19	-25.37	-22.55
Ethiopie	0.59	0.57	0.51	0.62	0.63	0.58	0.80	0.90	0.94
Gabon	-5.47	-7.13	-9.56	-6.91	-7.05	-8.27	-7.59	-6.93	-7.30
Ghana	0.75	0.90	0.92	0.38	0.74	0.73	0.73	0.89	0.87
Côte d'Ivoire	0.82	1.08	1.33	-0.06	0.03	0.20	0.42	0.53	1.14
Kenya	1.67	1.75	1.92	1.73	1.61	1.73	1.79	1.98	2.14
Libye	-135.74	-109.39	-95.92	-48.14	-48.07	-45.18	-45.50	-42.59	-46.97
Maroc	2.02	2.22	3.44	3.98	4.19	4.41	4.21	4.40	4.58
Mozambique	0.84	0.78	0.51	0.56	0.52	0.46	0.49	0.50	0.50
Nigéria	-74.97	-100.33	-87.96	-52.57	-58.09	-70.59	-64.82	-58.10	-63.91
Sénégal	1.44	1.55	1.27	0.92	1.01	0.80	1.09	0.92	0.89
Afrique du Sud	12.18	13.01	14.18	11.41	12.27	12.57	12.11	11.66	10.26
Soudan	1.47	1.60	1.14	1.35	1.28	1.52	1.47	1.15	1.64
Tanzanie	0.87	1.00	0.79	0.71	0.71	0.76	0.74	0.72	0.74
Tunisie	-2.63	-2.32	-2.48	-2.24	-2.76	-2.79	-2.45	-2.92	-2.17
Zambie	0.50	0.95	0.81	0.80	0.63	0.61	0.57	0.57	0.61
Zimbabwe	0.56	0.71	0.70	0.77	0.76	0.79	0.82	0.76	0.94
Autre Afrique	3.26	3.27	4.26	5.02	4.87	4.80	5.37	6.12	6.38
Afrique	-239.33	-247.94	-237.40	-157.66	-172.59	-190.15	-184.13	-180.52	-193.06
Argentine	3.27	3.84	2.19	-1.78	-1.33	-3.74	-1.20	1.03	-0.07
Bolivie	-1.12	-1.54	-0.39	-0.19	-0.06	-0.02	-0.06	-0.01	-0.01
Brésil	21.32	34.46	45.56	34.95	27.53	23.69	28.39	28.61	32.91
Chili	3.84	3.50	3.79	2.45	2.57	2.50	3.31	2.99	4.41
Colombie	-4.94	-2.84	0.64	0.54	-0.39	-0.69	-6.32	-10.31	-10.34
Costa Rica	0.46	0.59	0.89	0.63	0.63	0.67	0.73	0.77	0.84
Cuba	7.36	8.74	10.26	10.93	9.56	10.04	9.69	10.29	10.84
République dominicaine	1.16	1.65	2.00	2.29	2.42	2.19	2.45	2.77	2.91
Equateur	1.15	-9.15	-7.34	-8.25	-9.08	-9.84	-10.66	-4.45	-10.87
El Salvador	0.50	0.69	0.80	0.61	0.58	0.63	0.68	0.76	0.74
Guatemala	0.92	1.05	1.40	0.78	1.05	1.17	0.77	0.93	1.03
Haiti	0.13	0.13	0.23	0.21	0.22	0.23	0.26	0.28	0.29
Honduras	0.38	0.41	0.53	0.59	0.66	0.57	0.57	0.66	0.73
Jamaïque	1.92	2.93	2.29	1.92	1.85	1.62	1.64	1.70	1.76
Antilles néerlandaises	6.82	8.01	6.34	5.86	4.80	3.49	3.25	2.64	3.05
Nicaragua	0.53	0.58	0.82	0.70	0.61	0.67	0.72	0.78	0.73
Panama	7.05	6.87	3.28	1.88	1.67	1.59	1.61	1.83	1.49
Paraguay	0.21	0.28	0.48	0.42	0.54	0.46	0.56	0.65	0.63
Pérou	1.91	1.89	-1.61	-2.61	-3.23	-3.91	-2.93	-1.76	-0.34
Trinité-et-Tobago	-3.62	-5.05	-9.43	-6.82	-7.19	-7.43	-7.30	-7.04	-6.97
Uruguay	1.89	1.88	2.26	1.27	1.29	1.12	1.49	1.26	1.59
Vénézuela	-173.06	-167.49	-101.77	-77.16	-80.24	-71.71	-79.58	-71.08	-83.11
Autre Amérique latine	7.34	8.21	6.43	3.38	3.49	3.44	3.33	3.99	3.93
Amérique latine	-114.58	-100.38	-30.35	-27.40	-42.05	-43.24	-48.60	-32.69	-43.84

Un chiffre négatif correspond à des exportations nettes.

Net Imports of Oil (Mtoe)
Importations nettes de pétrole (Mtep)
Netto-Ölimporte (Mtoe)
Importazioni nette di petrolio (Mtep)
石油の純輸入量（石油換算百万トン）
Importaciones netas de petróleo (Mtep)
Чистый импорт нефти и нефтепродуктов (Мтнэ)

1989	1990	1991	1992	1993	1994	1995	1996	1997	
-47.22	-50.27	-48.96	-48.72	-49.19	-48.95	-49.52	-52.88	-52.43	Algeria
-21.64	-22.58	-23.86	-22.72	-23.17	-23.07	-24.44	-33.82	-33.93	Angola
-0.08	-0.11	-0.12	-0.06	-0.07	-0.04	-0.00	0.23	0.25	Benin
-7.33	-7.52	-6.80	-6.27	-6.09	-5.68	-5.15	-5.17	-5.44	Cameroon
-7.15	-7.91	-7.80	-8.40	-9.53	-9.13	-8.28	-9.51	-12.30	Congo
-0.39	-0.22	-0.17	-0.13	-0.03	-0.08	-0.03	-0.03	0.03	DR of Congo
-22.02	-21.54	-21.51	-23.56	-24.19	-14.61	-13.72	-11.03	-9.79	Egypt
0.95	1.00	1.04	1.12	1.15	1.21	1.31	1.41	0.88	Ethiopia
-8.97	-12.20	-14.04	-13.81	-15.03	-15.80	-17.21	-17.80	-17.90	Gabon
0.93	0.96	0.86	0.85	0.88	1.10	1.12	1.11	1.11	Ghana
1.65	1.34	1.36	1.47	1.23	1.25	1.23	1.40	0.85	Ivory Coast
2.20	2.14	2.03	2.06	2.28	2.41	2.10	2.17	2.62	Kenya
-54.88	-63.56	-68.35	-66.49	-61.97	-62.76	-61.24	-61.54	-62.56	Libya
5.07	5.21	5.10	6.04	6.29	6.75	6.41	5.89	6.57	Morocco
0.48	0.47	0.42	0.44	0.47	0.47	0.48	0.52	0.70	Mozambique
-77.76	-80.08	-83.02	-85.38	-92.71	-81.05	-85.86	-97.28	-102.30	Nigeria
0.88	0.85	0.82	0.98	0.88	1.05	0.93	1.02	1.13	Senegal
9.14	11.31	9.90	10.80	8.95	10.57	12.21	11.27	12.21	South Africa
1.48	1.93	1.76	1.66	1.19	1.72	1.63	1.52	1.60	Sudan
0.74	0.77	0.77	0.69	0.73	0.74	0.74	0.75	0.75	Tanzania
-1.78	-1.41	-1.53	-1.84	-1.04	-0.91	-1.32	-0.81	-0.14	Tunisia
0.59	0.53	0.53	0.54	0.53	0.54	0.55	0.56	0.56	Zambia
0.87	0.81	0.98	1.07	1.15	1.35	1.58	1.53	1.53	Zimbabwe
6.27	5.95	6.13	6.28	6.26	6.35	6.40	5.92	4.81	Other Africa
-217.99	-234.14	-244.46	-243.36	-251.04	-226.57	-230.09	-254.54	-261.15	**Africa**
-3.06	-4.82	-4.55	-6.76	-7.03	-11.86	-14.38	-18.05	-20.06	Argentina
-0.00	-0.01	-0.02	0.03	0.06	0.03	0.20	0.14	0.24	Bolivia
30.79	30.87	30.42	31.23	36.78	36.92	35.89	41.18	42.64	Brazil
5.00	5.72	6.14	7.26	7.60	8.87	9.62	10.69	11.16	Chile
-10.79	-11.51	-11.36	-10.49	-11.56	-10.64	-16.99	-18.51	-19.37	Colombia
0.86	0.94	1.00	1.40	1.44	1.53	1.55	1.46	1.51	Costa Rica
10.78	10.07	8.06	5.98	5.51	5.69	6.12	6.25	6.93	Cuba
2.94	2.94	2.91	3.31	3.30	3.45	3.60	3.68	3.94	Dominican Republic
-10.14	-9.92	-10.07	-11.38	-12.59	-13.55	-13.45	-12.66	-13.97	Ecuador
0.77	0.79	1.02	1.12	1.32	1.40	1.65	1.46	1.38	El Salvador
1.03	1.15	1.34	1.44	1.43	1.68	1.72	1.46	1.31	Guatemala
0.33	0.32	0.29	0.30	0.22	0.06	0.32	0.37	0.48	Haiti
0.79	0.74	0.79	0.90	0.81	0.95	1.14	1.19	1.15	Honduras
1.98	2.53	2.44	2.59	2.63	2.66	2.92	3.05	3.24	Jamaica
2.86	3.70	4.04	3.84	4.43	4.51	4.48	4.44	4.44	Netherlands Antilles
0.61	0.64	0.63	0.75	0.72	0.65	0.87	0.93	0.96	Nicaragua
1.59	1.96	2.12	2.39	2.20	2.40	2.40	2.39	2.50	Panama
0.74	0.68	0.66	0.77	0.81	1.07	1.11	0.93	1.10	Paraguay
-0.69	-0.67	-0.16	-0.44	-0.63	-0.17	1.29	1.39	1.97	Peru
-6.74	-6.38	-5.88	-5.64	-5.33	-6.31	-6.40	-4.94	-4.86	Trinidad-and-Tobago
1.55	1.35	1.67	1.93	1.55	1.65	1.99	2.14	2.11	Uruguay
-79.62	-86.16	-95.30	-93.88	-97.89	-121.22	-134.40	-135.43	-144.76	Venezuela
4.00	4.26	4.17	3.97	4.04	4.02	4.07	4.10	4.19	Other Latin America
-44.44	-50.79	-59.63	-59.38	-60.15	-86.22	-104.68	-102.34	-111.76	**Latin America**

A negative number shows net exports.

Net Imports of Oil (Mtoe)
Importations nettes de pétrole (Mtep)
Netto-Ölimporte (Mtoe)

Importazioni nette di petrolio (Mtep)

石油の純輸入量（石油換算百万トン）

Importaciones netas de petróleo (Mtep)

Чистый импорт нефти и нефтепродуктов (Мтнэ)

	1971	1973	1978	1983	1984	1985	1986	1987	1988
Bangladesh	0.96	0.90	1.37	1.42	1.48	1.54	1.71	1.66	1.93
Brunei	-5.95	-11.77	-12.11	-9.84	-9.44	-8.57	-8.25	-7.42	-7.22
Inde	14.75	17.28	19.17	14.01	13.26	15.66	15.60	18.50	21.35
Indonésie	-33.88	-50.82	-67.16	-44.31	-53.45	-43.41	-45.77	-43.80	-41.47
RPD de Corée	0.71	0.80	1.85	2.78	2.94	3.11	3.28	3.36	3.36
Malaisie	1.45	0.28	-3.84	-9.04	-12.37	-12.65	-15.53	-14.58	-16.15
Myanmar	0.28	0.05	-0.01	-0.01	-0.01	-0.01	-0.01	-0.00	-0.00
Népal	0.06	0.07	0.09	0.13	0.15	0.16	0.18	0.18	0.22
Pakistan	2.91	3.18	4.03	5.91	6.49	6.34	6.23	6.87	7.55
Philippines	8.73	9.25	11.19	10.17	10.31	7.23	7.47	9.45	10.17
Singapour	6.61	12.23	11.83	12.87	11.82	10.52	14.84	20.58	22.04
Sri Lanka	1.32	1.64	1.46	1.64	1.69	1.59	1.58	1.85	1.77
Taipei chinois	7.29	10.40	18.65	17.99	18.30	16.50	19.14	19.50	23.23
Thaïlande	6.37	8.31	10.66	10.76	10.54	8.86	8.94	10.82	11.38
Viêt-Nam	5.90	5.86	0.95	1.92	1.88	1.92	2.14	2.49	1.88
Autre Asie	1.64	2.07	1.96	2.19	2.11	2.34	2.24	2.60	2.48
Asie	19.15	9.73	0.07	18.57	5.71	11.13	13.82	32.05	42.50
Rép. populaire de Chine	-0.25	-1.83	-12.31	-19.91	-28.01	-36.30	-31.61	-29.56	-27.17
Hong-Kong, Chine	4.02	4.83	6.14	5.80	4.98	5.10	5.55	4.96	5.65
Chine	3.77	2.99	-6.17	-14.11	-23.03	-31.20	-26.06	-24.60	-21.51
Albanie	-0.78	-1.32	-1.01	-0.06	-0.16	-0.20	-0.20	-0.17	-0.11
Bulgarie	9.78	11.28	14.27	12.20	11.89	11.50	12.64	11.77	12.02
Chypre	0.64	0.85	0.85	0.96	0.99	0.92	1.02	1.26	1.27
Gibraltar	0.21	0.23	0.17	0.16	0.15	0.32	0.40	0.44	0.56
Malte	0.33	0.35	0.39	0.35	0.46	0.30	0.60	0.60	0.62
Roumanie	-2.22	-0.61	5.21	3.03	2.74	4.56	6.00	8.82	7.10
République slovaque	4.17	5.25	6.97	6.12	5.89	5.88	5.84	5.71	5.58
Bosnie-Herzégovine	-	-	-	-	-	-	-	-	-
Croatie	-	-	-	-	-	-	-	-	-
Ex-RYM	-	-	-	-	-	-	-	-	-
Slovénie	-	-	-	-	-	-	-	-	-
RF de Yougoslavie	-	-	-	-	-	-	-	-	-
Ex-Yougoslavie	5.34	9.26	11.46	9.62	10.20	9.61	10.95	12.16	13.56
Europe non-OCDE	17.48	25.31	38.30	32.36	32.17	32.89	37.26	40.58	40.61

Un chiffre négatif correspond à des exportations nettes.

Net Imports of Oil (Mtoe)
Importations nettes de pétrole (Mtep)
Netto-Ölimporte (Mtoe)
Importazioni nette di petrolio (Mtep)
石油の純輸入量（石油換算百万トン）
Importaciones netas de petróleo (Mtep)
Чистый импорт нефти и нефтепродуктов (Мтнэ)

1989	1990	1991	1992	1993	1994	1995	1996	1997	
2.06	1.93	1.64	1.79	1.92	2.22	2.28	2.43	2.44	Bangladesh
-7.55	-7.73	-8.39	-9.13	-8.90	-9.18	-8.87	-8.46	-8.06	Brunei
23.38	26.15	29.68	37.19	39.27	39.52	44.83	52.61	50.87	India
-43.76	-41.47	-43.91	-39.71	-34.13	-39.70	-35.04	-33.26	-26.46	Indonesia
3.56	3.51	3.02	2.62	2.14	1.85	1.64	1.51	1.47	DPR of Korea
-17.49	-17.56	-17.70	-17.72	-15.00	-15.22	-18.42	-14.76	-12.83	Malaysia
-0.00	-0.02	0.06	0.17	0.28	0.39	0.66	0.68	0.95	Myanmar
0.20	0.20	0.25	0.31	0.30	0.44	0.52	0.52	0.52	Nepal
7.83	8.42	7.90	8.96	10.10	11.76	12.09	13.99	14.24	Pakistan
11.00	11.74	11.67	13.63	13.47	14.56	16.34	17.76	18.72	Philippines
21.17	24.50	26.58	28.91	32.54	34.44	31.02	36.36	41.30	Singapore
1.54	1.66	1.64	1.71	2.09	2.08	2.50	2.86	3.08	Sri Lanka
25.94	28.94	27.85	30.19	31.64	33.96	37.42	38.05	37.95	Chinese Taipei
14.55	17.55	17.84	20.61	23.88	25.61	30.55	34.19	32.58	Thailand
0.86	0.14	-1.40	-2.31	-2.41	-2.59	-4.47	-2.49	-2.10	Vietnam
2.56	-1.87	-2.37	-2.92	-3.05	-3.25	-2.75	-1.74	-1.66	Other Asia
45.85	56.09	54.34	74.30	94.14	96.89	110.30	140.25	153.00	**Asia**
-21.51	-23.82	-18.05	-7.54	11.31	5.26	10.76	16.22	38.01	People's Rep.of China
6.49	6.37	6.49	8.31	8.83	9.18	9.57	9.79	9.77	Hong Kong, China
-15.01	-17.44	-11.56	0.77	20.14	14.44	20.32	26.01	47.78	**China**
-0.12	0.01	-0.07	-0.05	0.02	0.07	0.10	0.11	0.12	Albania
12.34	8.57	6.09	5.36	6.76	6.00	6.56	5.89	5.02	Bulgaria
1.53	1.59	1.70	1.90	2.03	2.15	2.04	2.19	2.14	Cyprus
0.48	0.50	0.91	0.92	0.95	0.96	0.96	0.96	0.96	Gibraltar
0.62	0.62	0.71	0.76	0.77	0.75	0.89	0.97	1.02	Malta
7.58	10.59	7.26	5.72	6.21	4.37	6.46	6.21	6.98	Romania
5.45	4.72	3.75	3.46	3.17	3.06	3.65	3.45	3.27	Slovak Republic
-	-	-	0.55	0.75	0.90	0.90	0.90	0.90	*Bosnia-Herzegovina*
-	-	-	1.30	1.33	1.89	2.24	2.22	2.39	*Croatia*
-	-	-	1.10	1.18	0.82	0.98	1.24	1.06	*FYROM*
-	-	-	1.62	1.96	2.15	2.29	2.71	2.78	*Slovenia*
-	-	-	0.86	0.36	0.36	0.47	1.70	2.65	*FR of Yugoslavia*
13.23	12.80	10.63	5.48	5.56	6.00	6.55	8.41	9.78	Former Yugoslavia
41.11	39.40	30.99	23.56	25.48	23.35	27.20	28.20	29.29	**Non-OECD Europe**

A negative number shows net exports.

Net Imports of Oil (Mtoe)
Importations nettes de pétrole (Mtep)
Netto-Ölimporte (Mtoe)

Importazioni nette di petrolio (Mtep)

石油の純輸入量（石油換算百万トン）

Importaciones netas de petróleo (Mtep)

Чистый импорт нефти и нефтепродуктов (Мтнэ)

	1971	1973	1978	1983	1984	1985	1986	1987	1988
Arménie	-	-	-	-	-	-	-	-	-
Azerbaïdjan	-	-	-	-	-	-	-	-	-
Bélarus	-	-	-	-	-	-	-	-	-
Estonie	-	-	-	-	-	-	-	-	-
Géorgie	-	-	-	-	-	-	-	-	-
Kazakhstan	-	-	-	-	-	-	-	-	-
Kirghizistan	-	-	-	-	-	-	-	-	-
Lettonie	-	-	-	-	-	-	-	-	-
Lituanie	-	-	-	-	-	-	-	-	-
Moldova	-	-	-	-	-	-	-	-	-
Russie	-	-	-	-	-	-	-	-	-
Tadjikistan	-	-	-	-	-	-	-	-	-
Turkménistan	-	-	-	-	-	-	-	-	-
Ukraine	-	-	-	-	-	-	-	-	-
Ouzbékistan	-	-	-	-	-	-	-	-	-
Ex-URSS	-99.19	-104.70	-158.13	-174.50	-171.93	-153.10	-173.56	-181.33	-184.74
Bahrein	-1.98	-1.58	-2.30	-1.82	-1.27	-1.01	-1.56	-1.55	-1.43
Iran	-212.88	-275.63	-197.50	-90.45	-70.18	-72.08	-49.53	-73.06	-68.93
Irak	-81.47	-97.13	-121.62	-35.58	-46.27	-56.76	-70.73	-89.76	-115.25
Israël	0.40	2.44	7.53	6.91	6.52	6.22	7.25	7.01	10.13
Jordanie	0.62	0.68	1.34	2.54	2.80	2.92	2.93	3.25	3.06
Koweit	-158.20	-149.60	-104.71	-47.42	-52.18	-44.85	-68.45	-65.18	-69.97
Liban	2.04	2.41	2.06	2.33	2.25	2.60	2.74	2.80	2.12
Oman	-13.46	-13.61	-15.19	-19.57	-20.65	-24.89	-28.20	-29.15	-31.59
Qatar	-21.04	-28.11	-23.96	-13.92	-18.94	-15.28	-17.30	-14.62	-16.83
Arabie saoudite	-225.49	-367.48	-409.41	-227.36	-210.00	-139.82	-221.18	-180.70	-224.89
Syrie	-2.54	-3.11	-5.41	-0.85	-0.14	-1.80	-0.15	-3.08	-4.85
Emirats arabes unis	-51.79	-77.73	-91.04	-50.86	-55.05	-52.45	-59.62	-65.56	-70.08
Yémen	1.07	1.27	1.72	2.11	2.28	2.72	2.67	2.03	-5.22
Moyen-Orient	-764.71	-1007.18	-958.48	-473.94	-460.84	-394.47	-501.11	-507.58	-593.73
Total non-OCDE	-1177.41	-1422.16	-1352.15	-796.69	-832.56	-768.14	-882.38	-854.09	-953.78
OCDE Amérique du N.	193.25	296.85	420.35	135.47	155.90	134.83	202.13	224.63	251.84
OCDE Pacifique	245.93	300.76	300.83	255.42	259.60	242.51	250.34	255.01	273.59
OCDE Europe	690.54	784.94	679.34	440.03	435.71	420.77	448.70	442.06	448.13
Total OCDE	1 129.72	1 382.55	1 400.53	830.91	851.21	798.12	901.17	921.70	973.56
Monde	-47.69	-39.61	48.38	34.22	18.65	29.98	18.79	67.61	19.79
Pour mémoire: OPEP	-1203.86	-1472.75	-1352.66	-728.18	-734.96	-655.46	-767.59	-750.18	-847.94

Un chiffre négatif correspond à des exportations nettes.

Ex-URSS: la somme des républiques peut être différente du total.

La ligne Monde montre la divergence entre le total des exportations et importations mondiales.

INTERNATIONAL ENERGY AGENCY

Net Imports of Oil (Mtoe)
Importations nettes de pétrole (Mtep)
Netto-Ölimporte (Mtoe)
Importazioni nette di petrolio (Mtep)
石油の純輸入量（石油換算百万トン）
Importaciones netas de petróleo (Mtep)
Чистый импорт нефти и нефтепродуктов (Мтнэ)

1989	1990	1991	1992	1993	1994	1995	1996	1997	
-	-	-	2.43	1.22	0.39	0.28	0.15	0.15	Armenia
-	-	-	-4.87	-2.99	-0.88	-2.20	-2.21	-2.11	Azerbaijan
-	-	-	19.01	12.22	9.59	9.10	7.33	7.75	Belarus
-	-	-	1.33	1.53	1.34	1.04	1.06	0.94	Estonia
-	-	-	1.29	0.66	0.30	0.10	0.81	0.88	Georgia
-	-	-	-3.87	-7.05	-7.30	-8.98	-12.60	-16.45	Kazakhstan
-	-	-	1.82	1.09	0.33	0.53	0.59	0.41	Kyrgyzstan
-	-	-	2.75	2.51	2.46	2.07	2.32	1.56	Latvia
-	-	-	3.93	3.69	3.19	3.64	2.98	3.14	Lithuania
-	-	-	2.87	1.86	1.14	1.09	0.84	0.97	Moldova
-	-	-	-159.32	-158.49	-165.58	-157.92	-174.30	-176.08	Russia
-	-	-	5.85	3.66	1.19	1.20	1.21	1.20	Tajikistan
-	-	-	0.08	-1.32	-0.50	-0.91	-1.03	-1.06	Turkmenistan
-	-	-	36.59	24.94	19.58	21.18	15.73	13.84	Ukraine
-	-	-	5.02	5.42	2.55	-0.48	-1.09	-1.25	Uzbekistan
-171.06	-158.37	-104.53	-79.55	-105.14	-134.66	-139.44	-167.66	-175.27	**Former USSR**
-1.43	-1.63	-1.83	-1.68	-1.80	-1.81	-2.06	-0.67	0.92	Bahrain
-94.64	-105.42	-114.56	-121.74	-128.64	-120.23	-120.71	-117.16	-116.59	Iran
-124.68	-84.12	-2.67	-5.50	-5.49	-6.06	-6.04	-7.56	-35.00	Iraq
9.68	8.96	8.73	10.46	10.73	9.43	11.51	10.92	11.82	Israel
3.11	3.52	3.14	3.85	3.89	3.96	4.11	4.31	4.60	Jordan
-84.55	-55.08	-7.82	-49.92	-92.41	-96.63	-96.51	-98.56	-99.74	Kuwait
1.72	2.15	2.80	2.76	3.38	3.72	4.27	4.36	4.85	Lebanon
-32.18	-33.97	-33.51	-35.98	-37.68	-39.88	-41.61	-43.35	-44.28	Oman
-19.16	-19.85	-19.33	-22.56	-21.14	-20.62	-20.54	-22.67	-26.96	Qatar
-231.05	-303.70	-388.14	-393.74	-385.06	-383.41	-385.22	-380.90	-386.79	Saudi Arabia
-8.88	-11.14	-14.23	-14.73	-15.68	-18.34	-19.20	-18.79	-18.15	Syria
-83.70	-93.75	-104.60	-98.74	-92.79	-98.05	-97.85	-100.74	-105.51	United Arab Emirates
-6.11	-6.74	-6.94	-5.46	-8.55	-14.06	-15.28	-16.37	-13.78	Yemen
-671.88	-700.76	-678.94	-732.97	-771.23	-781.99	-785.13	-787.17	-824.61	**Middle East**
-1033.42	-1066.02	-1013.80	-1016.65	-1047.81	-1094.76	-1101.52	-1117.25	-1142.72	**Non-OECD Total**
294.03	290.71	257.25	276.25	308.86	334.57	315.29	339.75	367.41	OECD North America
300.36	318.09	324.22	345.80	356.55	382.86	386.21	402.13	407.35	OECD Pacific
459.50	457.83	455.12	452.28	432.82	394.64	387.64	393.60	398.35	OECD Europe
1 053.89	1 066.63	1 036.59	1 074.33	1 098.23	1 112.07	1 089.14	1 135.47	1 173.10	**OECD Total**
20.47	0.61	22.79	57.68	50.42	17.31	-12.38	18.22	30.38	**World**
-941.03	-983.46	-976.65	-1026.37	-1061.41	-1078.69	-1092.93	-1107.97	-1159.09	*Memo: OPEC*

A negative number shows net exports.
Former USSR: data for individual republics may not add to the total.
The row World shows the discrepancy between total world exports and imports.

INTERNATIONAL ENERGY AGENCY

Net Imports of Gas (Mtoe)
Importations nettes de gaz naturel (Mtep)
Nettoimporte von Erdgas (Mtoe)
Importazioni nette di gas naturale (Mtep)
ガスの純輸入量（石油換算百万トン）
Importaciones netas de gas natural (Mtep)
Чистый импорт природного газа (Мтнэ)

	1971	1973	1978	1983	1984	1985	1986	1987	1988
Algérie	-1.26	-2.32	-6.20	-17.02	-17.88	-20.38	-20.05	-24.52	-25.03
Libye	-0.40	-2.57	-3.30	-0.66	-0.95	-0.95	-0.84	-0.72	-0.89
Tunisie	-	-	-	-	0.30	0.52	0.29	0.78	0.60
Afrique	-1.66	-4.89	-9.49	-17.68	-18.53	-20.81	-20.60	-24.46	-25.31
Argentine	-	1.45	2.02	2.00	1.99	1.97	1.95	1.90	1.99
Bolivie	-	-1.61	-1.62	-2.04	-2.02	-2.02	-2.02	-1.94	-2.04
Chili	-	-	-0.56	-	-	-	-	-	-
Uruguay	0.00	0.00	0.00	0.00	0.00	0.00	0.00	0.00	-
Amérique latine	0.00	-0.16	-0.17	-0.04	-0.03	-0.05	-0.07	-0.04	-0.04
Brunei	-	-1.27	-5.97	-6.12	-6.24	-6.01	-5.99	-6.12	-6.28
Indonésie	-	-	-4.28	-11.31	-16.74	-17.59	-17.68	-19.54	-21.57
Malaisie	-	-	-0.01	-2.37	-4.51	-5.08	-6.79	-7.85	-8.18
Singapour	-	-	-	-	-	-	-	-	-
Taipei chinois	-	-	-	-	-	-	-	-	-
Autre Asie	-1.88	-2.05	-1.81	-1.87	-1.87	-1.87	-1.87	-1.35	-1.07
Asie	-1.88	-3.32	-12.08	-21.67	-29.35	-30.55	-32.32	-34.86	-37.11
Rép. populaire de Chine	-	-	-	-	-	-	-	-	-
Hong-Kong, Chine	-	-	-	-	-	-	-	-	-
Chine									
Bulgarie	-	-	2.29	3.97	4.19	4.58	4.77	5.10	5.08
Roumanie	-0.16	-0.16	0.65	1.43	1.54	1.52	2.37	2.57	3.14
République slovaque	0.82	1.18	2.59	4.10	3.97	3.92	3.93	3.94	3.87
Bosnie-Herzégovine	-	-	-	-	-	-	-	-	-
Croatie	-	-	-	-	-	-	-	-	-
Ex-RYM	-	-	-	-	-	-	-	-	-
Slovénie	-	-	-	-	-	-	-	-	-
RF de Yougoslavie	-	-	-	-	-	-	-	-	-
Ex-Yougoslavie	-	-	-	2.25	2.81	2.95	3.14	3.53	3.31
Europe non-OCDE	0.66	1.02	5.53	11.75	12.50	12.97	14.21	15.13	15.39
Arménie	-	-	-	-	-	-	-	-	-
Azerbaïdjan	-	-	-	-	-	-	-	-	-
Bélarus	-	-	-	-	-	-	-	-	-
Estonie	-	-	-	-	-	-	-	-	-
Géorgie	-	-	-	-	-	-	-	-	-
Kazakhstan	-	-	-	-	-	-	-	-	-
Kirghizistan	-	-	-	-	-	-	-	-	-
Lettonie	-	-	-	-	-	-	-	-	-
Lituanie	-	-	-	-	-	-	-	-	-
Moldova	-	-	-	-	-	-	-	-	-
Russie	-	-	-	-	-	-	-	-	-
Tadjikistan	-	-	-	-	-	-	-	-	-
Turkménistan	-	-	-	-	-	-	-	-	-
Ukraine	-	-	-	-	-	-	-	-	-
Ouzbékistan	-	-	-	-	-	-	-	-	-
Ex-URSS	2.90	3.80	-22.00	-47.60	-53.30	-54.80	-63.70	-68.40	-70.28

Un chiffre négatif correspond à des exportations nettes.
Ex-URSS: la somme des républiques peut être différente du total.

INTERNATIONAL ENERGY AGENCY

Net Imports of Gas (Mtoe)
Importations nettes de gaz naturel (Mtep)
Nettoimporte von Erdgas (Mtoe)
Importazioni nette di gas naturale (Mtep)
ガスの純輸入量（石油換算百万トン）
Importaciones netas de gas natural (Mtep)
Чистый импорт природного газа (Мтнэ)

1989	1990	1991	1992	1993	1994	1995	1996	1997	
-28.25	-29.64	-32.32	-33.54	-33.68	-30.36	-35.80	-39.23	-46.46	Algeria
-1.03	-0.90	-1.59	-1.85	-1.61	-1.49	-1.50	-1.21	-1.21	Libya
0.64	0.90	0.60	0.98	1.00	1.49	1.79	1.22	0.62	Tunisia
-28.63	-29.64	-33.31	-34.40	-34.29	-30.36	-35.51	-39.21	-47.05	**Africa**
1.99	1.82	1.81	1.76	1.74	1.87	1.76	1.76	0.80	Argentina
-2.06	-2.05	-2.03	-1.98	-1.95	-2.02	-1.91	-1.77	-1.92	Bolivia
-	-	-	-	-	-	-	-	0.54	Chile
-	-	-	-	0.00	-	-	-	0.00	Uruguay
-0.07	-0.23	-0.22	-0.21	-0.21	-0.15	-0.15	-0.01	-0.58	**Latin America**
-6.26	-6.26	-6.24	-6.30	-6.48	-6.76	-7.32	-7.28	-7.23	Brunei
-21.77	-24.10	-26.33	-27.85	-28.47	-30.83	-29.18	-31.17	-31.54	Indonesia
-8.46	-8.69	-8.28	-8.26	-9.91	-10.53	-12.26	-15.30	-13.91	Malaysia
-	-	-	0.43	1.17	1.39	1.45	1.29	1.28	Singapore
-	0.76	1.84	1.93	2.07	2.59	2.95	3.06	3.75	Chinese Taipei
-0.06	-	-	-	-	-	-	-	-	Other Asia
-36.56	-38.29	-39.00	-40.06	-41.63	-44.13	-44.36	-49.39	-47.65	**Asia**
-	-	-	-	-	-	-0.02	-1.48	-2.31	People's Rep.of China
-	-	-	-	-	-	0.02	1.48	2.31	Hong Kong, China
-	-	-	-	-	-	-	-	-	**China**
4.92	4.89	4.50	4.06	3.78	3.74	4.56	4.73	3.85	Bulgaria
5.97	5.93	3.74	3.58	3.61	3.73	4.79	5.65	4.03	Romania
4.19	5.60	4.81	4.76	4.29	4.32	4.57	5.11	5.21	Slovak Republic
-	-	-	0.34	0.12	0.21	0.21	0.21	0.21	*Bosnia-Herzegovina*
-	-	-	0.59	0.65	0.61	0.22	0.72	0.85	*Croatia*
-	-	-	0.21	0.23	-	-	-	-	*FYROM*
-	-	-	0.54	0.54	0.57	0.68	0.65	0.71	*Slovenia*
-	-	-	0.96	0.79	-	0.16	1.70	1.67	*FR of Yugoslavia*
3.40	3.87	3.96	2.65	2.31	1.39	1.27	3.28	3.44	Former Yugoslavia
18.47	20.29	17.01	15.06	13.99	13.18	15.20	18.77	16.54	**Non-OECD Europe**
-	-	-	1.52	0.66	0.71	1.14	0.89	1.11	Armenia
-	-	-	3.50	2.04	2.12	0.43	0.02	-	Azerbaijan
-	-	-	14.57	13.53	11.87	11.23	11.91	13.48	Belarus
-	-	-	0.72	0.36	0.51	0.58	0.64	0.62	Estonia
-	-	-	3.93	2.97	1.99	0.73	0.63	0.76	Georgia
-	-	-	8.35	5.12	4.49	5.31	2.55	-0.37	Kazakhstan
-	-	-	1.47	1.10	0.69	0.71	0.86	0.48	Kyrgyzstan
-	-	-	2.13	0.76	0.81	1.00	0.87	1.06	Latvia
-	-	-	2.88	1.56	1.73	2.03	2.17	2.00	Lithuania
-	-	-	2.93	2.71	2.54	2.56	2.94	3.13	Moldova
-	-	-	-149.16	-134.96	-148.81	-151.77	-155.86	-151.41	Russia
-	-	-	1.35	1.13	0.62	0.70	0.90	0.82	Tajikistan
-	-	-	-38.34	-45.41	-18.60	-17.87	-19.44	-5.27	Turkmenistan
-	-	-	72.31	61.91	53.61	53.24	59.04	51.11	Ukraine
-	-	-	-0.81	-1.15	-3.37	-3.43	-3.97	-5.81	Uzbekistan
-81.62	-86.87	-82.23	-77.49	-78.96	-83.11	-94.88	-100.06	-94.55	**Former USSR**

A negative number shows net exports.
Former USSR: data for individual republics may not add to the total.

Net Imports of Gas (Mtoe)
Importations nettes de gaz naturel (Mtep)
Nettoimporte von Erdgas (Mtoe)
Importazioni nette di gas naturale (Mtep)
ガスの純輸入量（石油換算百万トン）
Importaciones netas de gas natural (Mtep)
Чистый импорт природного газа (Мтнэ)

	1971	1973	1978	1983	1984	1985	1986	1987	1988
Iran	-4.21	-6.81	-5.92	-	-	-	-	-	-
Irak	-	-	-	-	-	-	-0.53	-2.25	-2.53
Koweit	-	-	-	-	-	-	1.07	2.30	2.36
Oman	-	-	-	-	-	-	-	-	-
Qatar	-	-	-	-	-	-	-	-	-
Emirats arabes unis	-	-	-1.39	-1.96	-2.29	-2.53	-2.37	-2.37	-2.61
Moyen-Orient	-4.21	-6.81	-7.31	-1.96	-2.29	-2.53	-1.82	-2.31	-2.78
Total non-OCDE	-4.20	-10.37	-45.52	-77.19	-91.01	-95.78	-104.32	-114.94	-120.13
OCDE Amérique du N.	-0.68	-0.72	1.09	1.73	-0.77	-1.00	-1.33	-1.42	-1.19
OCDE Pacifique	1.17	2.78	13.44	22.44	30.80	32.97	33.96	36.08	38.39
OCDE Europe	4.52	8.93	30.50	46.06	53.24	58.01	66.75	73.87	76.44
Total OCDE	5.01	11.00	45.03	70.23	83.27	89.98	99.38	108.54	113.65
Monde	0.81	0.63	-0.49	-6.96	-7.73	-5.79	-4.93	-6.41	-6.48

Un chiffre négatif correspond à des exportations nettes.
La ligne Monde montre la divergence entre le total des exportations et importations mondiales.

Net Imports of Gas (Mtoe)
Importations nettes de gaz naturel (Mtep)
Nettoimporte von Erdgas (Mtoe)
Importazioni nette di gas naturale (Mtep)
ガスの純輸入量（石油換算百万トン）
Importaciones netas de gas natural (Mtep)
Чистый импорт природного газа (Мтнэ)

1989	1990	1991	1992	1993	1994	1995	1996	1997	
-	-1.22	-2.46	-	-0.41	-0.65	-0.08	-0.08	-0.08	Iran
-2.94	-1.63	-	-	-	-	-	-	-	Iraq
3.01	1.63	-	-	-	-	-	-	-	Kuwait
-	-	-	-	-	-	-0.41	-0.41	-0.41	Oman
-	-	-	-	-	-	-	-	-2.32	Qatar
-2.53	-2.61	-2.86	-2.78	-2.69	-3.34	-5.25	-5.42	-6.84	United Arab Emirates
-2.46	-3.84	-5.32	-2.78	-3.10	-3.99	-5.74	-5.90	-9.65	**Middle East**
-130.88	-138.58	-143.06	-139.89	-144.19	-148.55	-165.45	-175.80	-182.94	**Non-OECD Total**
-0.79	1.04	0.66	-0.08	1.11	-0.33	-0.67	0.41	0.50	OECD North America
41.19	42.00	43.79	44.60	45.14	49.35	50.01	56.65	57.98	OECD Pacific
84.77	91.26	93.35	93.70	94.66	96.30	110.00	114.89	118.24	OECD Europe
125.16	134.29	137.81	138.22	140.91	145.32	159.34	171.95	176.72	**OECD Total**
-5.71	-4.28	-5.25	-1.67	-3.28	-3.23	-6.11	-3.85	-6.21	**World**

A negative number shows net exports.
The row World shows the discrepancy between total world exports and imports.

Net Imports of Electricity (Mtoe)
Importations nettes d'électricité (Mtep)
Nettoimporte von Elektrizität (Mtoe)
Importazioni nette di energia elettrica (Mtep)
電力の純輸入量（石油換算百万トン）
Importaciones netas de electricidad (Mtep)
Чистый импорт электроэнергии (Мтнэ)

	1971	1973	1978	1983	1984	1985	1986	1987	1988
Algérie	0.000	-0.000	0.000	-0.002	0.006	0.003	0.007	-0.004	-0.004
Bénin	0.003	0.004	0.014	0.014	0.009	0.013	0.012	0.013	0.017
Congo	-	-	0.006	0.005	0.005	0.005	0.011	0.013	0.013
RD du Congo	-0.002	-0.002	-0.005	-0.009	-0.011	-0.013	-0.016	-0.016	-0.009
Ghana	-	-0.009	-0.019	-0.042	-0.053	-0.053	-0.057	-0.042	-0.018
Côte d'Ivoire	-	-	-	0.015	0.033	0.019	-	-	-
Kenya	0.025	0.026	0.019	0.015	0.019	0.018	0.019	0.015	0.009
Maroc	-	-	-	-	-	-	-	-	-
Mozambique	0.000	0.013	0.007	0.009	0.003	-0.001	0.019	0.023	0.016
Nigéria	-	-	-0.004	-0.011	-0.012	-0.012	-	-	-
Afrique du Sud	-0.001	-0.017	0.568	0.322	-0.035	-0.027	-0.105	-0.102	-0.086
Tanzanie	-	-	-	-	-	-	-	-	-
Tunisie	-	-	-	0.001	-0.006	0.002	0.001	-0.000	-0.002
Zambie	0.295	0.174	-0.149	-0.260	-0.249	-0.250	-0.265	-0.129	-0.129
Zimbabwe	0.015	0.008	0.005	0.305	0.295	0.344	0.284	0.218	0.074
Autre Afrique	0.002	0.009	0.019	0.017	0.009	0.016	0.017	0.025	0.031
Afrique	0.337	0.207	0.462	0.380	0.012	0.065	-0.072	0.013	-0.090
Argentine	-0.001	0.004	0.006	-0.001	-0.000	-0.001	0.271	0.263	0.188
Bolivie	-	-	-	-	-	-	-	-	-
Brésil	-0.002	-0.001	-0.011	-0.021	-0.007	0.165	0.885	1.445	1.543
Chili	0.000	0.000	0.000	-	-	-	-	-	-
Colombie	-	-	0.002	0.003	0.003	0.000	0.000	-	-
Costa Rica	-	-	-	-0.042	-0.037	-0.005	0.007	0.015	0.016
El Salvador	-	-	-	-	-	-	0.008	0.001	0.003
Guatemala	-	-	-	-	-	-	-0.008	-0.001	-
Nicaragua	-	-	-0.000	0.028	0.023	0.016	0.006	0.007	0.007
Panama	-0.004	-0.004	-0.001	0.002	0.005	0.002	-0.002	0.010	0.002
Paraguay	-	-0.007	0.005	0.021	0.005	-0.243	-0.900	-1.446	-1.545
Pérou	-	-	-	-	-	-	-	-	-
Uruguay	0.003	0.002	0.002	-0.309	-0.287	-0.230	-0.271	-0.254	-0.180
Vénézuela	-	-	-0.002	-	-	-	-	-	-
Autre Amérique latine	-	-	-	-	-	-	-	-	-
Amérique latine	-0.003	-0.006	0.003	-0.318	-0.296	-0.296	-0.003	0.040	0.034
Inde	-	-0.001	-0.004	-0.007	-0.009	-0.008	0.009	0.074	0.103
Malaisie	-	-	0.001	0.003	0.007	0.005	-	-0.002	-0.006
Népal	0.000	-	0.002	0.005	0.005	0.006	0.003	0.001	0.004
Singapour	-	-	-0.004	-0.005	-0.006	-0.004	-	-	-
Thaïlande	-0.004	0.013	0.019	0.058	0.059	0.060	0.064	0.034	0.035
Autre Asie	-	-	0.000	0.000	0.001	0.003	-0.025	-0.094	-0.120
Asie	-0.003	0.012	0.014	0.054	0.057	0.062	0.050	0.013	0.017
Rép. populaire de Chine	-	-	-	0.037	0.066	0.092	0.100	0.108	0.126
Hong-Kong, Chine	-	-	-	-0.032	-0.064	-0.090	-0.104	-0.116	-0.123
Chine	-	-	-	0.005	0.003	0.002	-0.003	-0.009	0.003

Un chiffre négatif correspond à des exportations nettes.

Net Imports of Electricity (Mtoe)
Importations nettes d'électricité (Mtep)
Nettoimporte von Elektrizität (Mtoe)
Importazioni nette di energia elettrica (Mtep)
電力の純輸入量（石油換算百万トン）
Importaciones netas de electricidad (Mtep)
Чистый импорт электроэнергии (Мтнэ)

1989	1990	1991	1992	1993	1994	1995	1996	1997	
-0.001	-0.005	-0.057	-0.080	-0.107	-0.097	-0.023	-0.012	-0.000	Algeria
0.016	0.017	0.018	0.018	0.020	0.018	0.022	0.023	0.020	Benin
0.005	0.005	0.006	0.009	0.009	0.010	0.010	0.010	0.010	Congo
-0.046	-0.000	-0.004	-0.012	-0.012	-0.012	-0.012	-0.012	-0.012	DR of Congo
-0.045	-0.065	-0.069	-0.077	-0.066	-0.033	-0.034	-0.034	-0.034	Ghana
-	0.025	0.032	0.042	0.008	0.003	0.003	0.003	0.003	Ivory Coast
0.010	0.016	0.012	0.021	0.023	0.023	0.015	0.013	0.012	Kenya
-	0.009	0.055	0.080	0.088	0.068	0.021	0.011	0.011	Morocco
0.010	0.014	0.026	0.034	0.036	0.040	0.043	0.052	0.017	Mozambique
-	-	-0.008	-0.008	-0.009	-0.012	-	-	-	Nigeria
-0.096	-0.109	-0.140	-0.126	-0.214	-0.226	-0.245	-0.477	-0.569	South Africa
-	-	-	-	-	0.001	0.003	0.003	0.004	Tanzania
0.002	-0.001	0.002	-0.000	0.019	0.028	0.007	0.003	0.003	Tunisia
-0.129	-0.127	-0.127	-0.127	-0.127	-0.127	-0.127	-0.127	-0.129	Zambia
0.072	0.029	0.098	0.174	0.104	0.173	0.199	0.273	0.345	Zimbabwe
0.038	0.041	0.049	0.048	0.052	0.062	0.066	0.097	0.102	Other Africa
-0.164	-0.152	-0.108	-0.004	-0.174	-0.081	-0.053	-0.176	-0.216	**Africa**
0.111	0.224	0.161	0.293	0.194	0.151	0.182	0.289	0.446	Argentina
0.001	0.001	0.001	0.001	0.001	0.001	0.001	-0.000	0.000	Bolivia
1.901	2.282	2.329	2.065	2.369	2.732	3.040	3.144	3.480	Brazil
-	-	-	-	-	-	-	-	-	Chile
0.019	0.017	0.018	0.030	0.026	0.024	0.032	0.014	0.017	Colombia
0.012	0.014	0.002	-0.006	-0.000	-0.001	0.002	0.011	-0.012	Costa Rica
0.000	0.000	0.000	0.005	0.007	0.001	0.003	0.002	0.008	El Salvador
-	-	-	-	-	-	-	-	-	Guatemala
0.001	0.006	0.008	0.003	-0.004	-0.002	-	0.001	0.014	Nicaragua
0.006	0.010	0.012	0.012	0.007	0.007	0.007	-0.005	-0.005	Panama
-1.765	-2.147	-2.320	-2.117	-2.418	-2.804	-3.253	-3.759	-3.928	Paraguay
-	-	-	-	-	0.000	-	-	0.000	Peru
-0.099	-0.218	-0.154	-0.291	-0.193	-0.143	-0.004	-0.003	-0.012	Uruguay
-	-	-0.017	-0.031	-	-0.011	-0.017	-0.013	-	Venezuela
-	-	0.000	0.001	0.001	0.001	0.001	0.001	0.001	Other Latin America
0.188	0.188	0.040	-0.035	-0.010	-0.042	-0.004	-0.318	0.011	**Latin America**
0.112	0.118	0.125	0.104	0.107	0.122	0.133	0.133	0.133	India
-0.014	-0.005	-0.002	-0.002	0.002	0.004	-0.002	-0.001	0.001	Malaysia
0.009	0.003	-0.004	-0.000	0.003	0.004	0.006	-0.002	-0.002	Nepal
-	-	-	-	-0.004	-0.008	-	-	-	Singapore
0.053	0.053	0.048	0.038	0.051	0.071	0.053	0.062	0.055	Thailand
-0.120	-0.120	-0.120	-0.113	-0.113	-0.114	-0.116	-0.117	-0.123	Other Asia
0.040	0.050	0.046	0.026	0.046	0.080	0.074	0.074	0.064	**Asia**
0.148	0.158	0.245	0.245	0.377	-0.176	-0.463	-0.309	-0.612	People's Rep.of China
-0.152	-0.155	-0.263	-0.427	-0.349	0.559	0.521	0.623	0.629	Hong Kong, China
-0.004	0.004	-0.018	-0.182	0.029	0.383	0.058	0.315	0.017	**China**

A negative number shows net exports.

Net Imports of Electricity (Mtoe)
Importations nettes d'électricité (Mtep)
Nettoimporte von Elektrizität (Mtoe)
Importazioni nette di energia elettrica (Mtep)
電力の純輸入量（石油換算百万トン）
Importaciones netas de electricidad (Mtep)
Чистый импорт электроэнергии (Мтнэ)

	1971	1973	1978	1983	1984	1985	1986	1987	1988
Albanie	-	-0.023	-0.011	-0.049	-0.052	-0.054	-0.056	-0.056	-0.056
Bulgarie	0.019	0.280	0.328	0.203	0.206	0.370	0.342	0.374	0.357
Gibraltar	-	-	-	-	-	-	-	-	-
Roumanie	-0.271	-0.305	-0.156	0.058	0.168	0.280	0.381	0.443	0.619
République slovaque	0.205	0.244	0.245	0.436	0.435	0.360	0.280	0.407	0.502
Bosnie-Herzégovine	-	-	-	-	-	-	-	-	-
Croatie	-	-	-	-	-	-	-	-	-
Ex-RYM	-	-	-	-	-	-	-	-	-
Slovénie	-	-	-	-	-	-	-	-	-
RF de Yougoslavie	-	-	-	-	-	-	-	-	-
Ex-Yougoslavie	-0.014	-0.004	-0.056	0.091	-0.097	0.054	0.040	0.032	-0.121
Europe non-OCDE	-0.062	0.192	0.350	0.738	0.660	1.010	0.987	1.200	1.300
Arménie	-	-	-	-	-	-	-	-	-
Azerbaïdjan	-	-	-	-	-	-	-	-	-
Bélarus	-	-	-	-	-	-	-	-	-
Estonie	-	-	-	-	-	-	-	-	-
Géorgie	-	-	-	-	-	-	-	-	-
Kazakhstan	-	-	-	-	-	-	-	-	-
Kirghizistan	-	-	-	-	-	-	-	-	-
Lettonie	-	-	-	-	-	-	-	-	-
Lituanie	-	-	-	-	-	-	-	-	-
Moldova	-	-	-	-	-	-	-	-	-
Russie	-	-	-	-	-	-	-	-	-
Tadjikistan	-	-	-	-	-	-	-	-	-
Turkménistan	-	-	-	-	-	-	-	-	-
Ukraine	-	-	-	-	-	-	-	-	-
Ouzbékistan	-	-	-	-	-	-	-	-	-
Ex-URSS	-0.576	-0.834	-1.049	-2.055	-2.124	-2.485	-2.494	-2.976	-3.345
Israël	-0.003	-0.005	-0.014	-0.021	-0.022	-0.026	-0.032	-0.031	-0.046
Jordanie	0.003	0.005	-	-	-	-0.002	-0.018	-0.029	-
Liban	-	-	0.004	0.003	0.003	0.003	0.003	0.003	-
Syrie	-0.006	-	-	-0.012	-0.012	-0.012	-0.011	-0.011	-
Moyen-Orient	-0.006	-	-0.010	-0.029	-0.031	-0.036	-0.059	-0.067	-0.046
Total non-OCDE	-0.314	-0.428	-0.230	-1.225	-1.720	-1.678	-1.595	-1.786	-2.127
OCDE Amérique du N.	0.009	0.048	0.006	-0.072	0.042	0.043	0.048	0.039	0.180
OCDE Europe	0.430	0.250	0.205	1.166	1.339	1.423	1.497	1.729	1.949
Total OCDE	0.440	0.298	0.211	1.095	1.380	1.467	1.544	1.768	2.129
Monde	0.126	-0.130	-0.019	-0.130	-0.340	-0.211	-0.050	-0.018	0.003

Un chiffre négatif correspond à des exportations nettes.
Ex-URSS: la somme des républiques peut être différente du total.
La ligne Monde montre la divergence entre le total des exportations et importations mondiales.

Net Imports of Electricity (Mtoe)
Importations nettes d'électricité (Mtep)
Nettoimporte von Elektrizität (Mtoe)
Importazioni nette di energia elettrica (Mtep)
電力の純輸入量（石油換算百万トン）
Importaciones netas de electricidad (Mtep)
Чистый импорт электроэнергии (Мтнэ)

1989	1990	1991	1992	1993	1994	1995	1996	1997	
-0.052	0.018	-0.101	-0.044	-0.012	-0.016	-0.006	0.017	0.017	Albania
0.377	0.326	0.183	0.233	0.009	-0.006	-0.014	-0.039	-0.305	Bulgaria
-0.000	-0.000	-0.000	-	-0.000	-	-	-	-	Gibraltar
0.672	0.815	0.606	0.361	0.161	0.062	0.026	0.069	0.019	Romania
0.480	0.447	0.373	0.298	0.096	0.038	0.119	0.309	0.351	Slovak Republic
-	-	-	-	-	0.014	0.018	0.018	0.018	*Bosnia-Herzegovina*
-	-	-	0.240	0.200	0.307	0.301	0.200	0.340	*Croatia*
-	-	-	0.024	0.052	0.014	-	-	-	*FYROM*
-	-	-	-0.156	-0.122	-0.166	-0.142	-0.143	-0.146	*Slovenia*
-	-	-	-0.034	-	-	-	-0.013	-	*FR of Yugoslavia*
-0.035	-0.031	-0.026	0.073	0.130	0.168	0.176	0.062	0.211	Former Yugoslavia
1.442	1.574	1.035	0.921	0.384	0.246	0.301	0.419	0.293	**Non-OECD Europe**
-	-	-	0.024	0.010	0.001	0.001	-	-	Armenia
-	-	-	-0.044	0.009	0.022	0.034	0.038	0.069	Azerbaijan
-	-	-	0.559	0.516	0.329	0.616	0.735	0.655	Belarus
-	-	-	-0.278	-0.137	-0.102	-0.065	-0.074	-0.084	Estonia
-	-	-	0.087	0.061	0.069	0.058	0.007	0.016	Georgia
-	-	-	1.219	0.516	1.121	0.636	0.589	0.576	Kazakhstan
-	-	-	-0.180	-0.088	-0.215	-0.118	-0.179	-0.146	Kyrgyzstan
-	-	-	0.351	0.215	0.156	0.194	0.278	0.157	Latvia
-	-	-	-0.456	-0.235	0.098	-0.230	-0.444	-0.303	Lithuania
-	-	-	-0.066	0.007	0.053	0.161	0.138	0.168	Moldova
-	-	-	-1.397	-1.611	-1.763	-1.686	-1.676	-1.694	Russia
-	-	-	0.072	-0.101	-0.043	0.057	0.028	0.069	Tajikistan
-	-	-	-0.369	-0.274	-0.228	-0.174	-0.241	-0.228	Turkmenistan
-	-	-	-0.437	-0.133	-0.090	-0.254	-0.173	-0.013	Ukraine
-	-	-	-0.042	-0.035	-0.034	-0.111	0.094	0.080	Uzbekistan
-3.380	-3.010	-1.860	-0.958	-1.279	-0.626	-0.882	-0.880	-0.678	**Former USSR**
-0.036	-0.039	-0.064	-0.053	-0.054	-0.063	-0.079	-0.084	-0.091	Israel
-	-	-	-0.006	-0.004	-	-	-0.000	-	Jordan
-	-	-	-	-	-	0.025	0.059	0.052	Lebanon
-	-	-	-	-	-	-	-	-	Syria
-0.036	-0.039	-0.064	-0.059	-0.058	-0.063	-0.053	-0.026	-0.039	**Middle East**
-1.914	-1.385	-0.929	-0.289	-1.063	-0.104	-0.559	-0.593	-0.548	**Non-OECD Total**
0.025	0.013	0.210	0.193	-0.008	-0.009	0.070	0.039	0.363	OECD North America
1.744	1.527	0.964	0.176	0.385	0.285	0.240	0.468	0.502	OECD Europe
1.770	1.540	1.174	0.369	0.377	0.276	0.310	0.507	0.865	**OECD Total**
-0.144	0.155	0.245	0.080	-0.686	0.172	-0.249	-0.086	0.317	**World**

A negative number shows net exports.
Former USSR: data for individual republics may not add to the total.
The row World shows the discrepancy between total world exports and imports.

Total Net Imports of Energy (Mtoe)
Importations nettes totales d'énergie (Mtep)
Gesamte Nettoimporte von Energie (Mtoe)
Importazioni nette totali di energia (Mtep)
エネルギー純輸入量（石油換算百万トン）
Importaciones netas totales de energía (Mtep)
Общий чистый импорт топлива и энергии (Мтнэ)

	1971	1973	1978	1983	1984	1985	1986	1987	1988
Algérie	-36.38	-51.10	-57.64	-56.68	-59.56	-62.88	-64.29	-69.21	-70.57
Angola	-4.66	-7.27	-5.54	-7.66	-8.95	-10.25	-12.82	-16.05	-21.29
Bénin	0.11	0.14	0.15	0.13	0.15	-0.14	-0.20	-0.14	-0.09
Cameroun	0.30	0.33	0.04	-4.70	-6.14	-7.58	-7.46	-7.44	-7.33
Congo	0.14	-1.33	-1.82	-5.22	-5.76	-5.79	-5.86	-5.77	-6.13
RD du Congo	0.96	0.97	-0.20	-0.02	-0.17	-0.55	-0.61	-0.43	-0.25
Egypte	-8.61	-1.63	-13.64	-17.39	-21.67	-24.73	-21.45	-24.61	-21.70
Ethiopie	0.59	0.57	0.51	0.62	0.63	0.58	0.80	0.90	0.94
Gabon	-5.47	-7.13	-9.56	-6.91	-7.05	-8.27	-7.59	-6.93	-7.30
Ghana	0.76	0.90	0.91	0.34	0.69	0.67	0.68	0.85	0.86
Côte d'Ivoire	0.82	1.08	1.33	-0.05	0.06	0.21	0.42	0.53	1.14
Kenya	1.74	1.82	1.97	1.77	1.68	1.80	1.86	2.06	2.22
Libye	-136.15	-111.97	-99.22	-48.80	-49.02	-46.13	-46.34	-43.31	-47.86
Maroc	2.03	2.22	3.43	4.16	4.31	4.69	4.74	5.08	5.30
Mozambique	1.02	0.92	0.57	0.64	0.57	0.51	0.55	0.53	0.53
Nigéria	-74.97	-100.35	-87.97	-52.58	-58.10	-70.60	-64.84	-58.12	-63.93
Sénégal	1.44	1.55	1.27	0.92	1.01	0.80	1.09	0.92	0.89
Afrique du Sud	11.21	11.69	4.27	-9.40	-14.14	-19.72	-19.34	-16.76	-19.84
Soudan	1.47	1.60	1.14	1.35	1.28	1.52	1.47	1.15	1.64
Tanzanie	0.87	1.00	0.79	0.71	0.71	0.76	0.74	0.72	0.74
Tunisie	-2.56	-2.25	-2.40	-2.18	-2.40	-2.20	-2.10	-2.06	-1.49
Zambie	0.89	1.16	0.73	0.54	0.38	0.36	0.30	0.44	0.48
Zimbabwe	0.37	0.64	0.52	0.96	0.94	1.08	1.13	1.02	1.00
Autre Afrique	3.29	3.31	4.32	5.09	4.93	4.86	5.46	6.19	6.46
Afrique	-240.77	-253.11	-256.04	-194.36	-215.65	-241.00	-233.65	-230.44	-245.58
Argentine	3.76	5.85	4.82	0.57	1.05	-1.21	1.86	4.10	3.11
Bolivie	-1.12	-3.14	-2.01	-2.05	-1.99	-1.98	-2.04	-1.95	-2.04
Brésil	22.62	35.81	48.45	38.85	32.60	29.58	35.30	37.58	41.81
Chili	4.02	3.70	3.41	2.68	2.86	2.76	3.65	3.32	4.74
Colombie	-4.96	-2.90	0.49	0.21	-1.03	-3.00	-10.08	-16.61	-17.40
Costa Rica	0.46	0.59	0.84	0.59	0.59	0.67	0.74	0.79	0.85
Cuba	7.44	8.80	10.35	11.02	9.65	10.16	9.80	10.40	10.95
République dominicaine	1.16	1.65	2.00	2.30	2.43	2.33	2.46	2.92	2.97
Equateur	1.15	-9.15	-7.34	-8.25	-9.08	-9.84	-10.66	-4.45	-10.87
El Salvador	0.50	0.69	0.80	0.61	0.58	0.63	0.68	0.77	0.74
Guatemala	0.92	1.05	1.40	0.78	1.05	1.17	0.76	0.93	1.03
Haiti	0.13	0.13	0.23	0.24	0.26	0.26	0.27	0.29	0.32
Honduras	0.38	0.41	0.53	0.59	0.66	0.57	0.57	0.66	0.73
Jamaïque	1.92	2.93	2.29	1.92	1.85	1.62	1.64	1.70	1.76
Antilles néerlandaises	6.82	8.01	6.34	5.86	4.80	3.49	3.25	2.64	3.05
Nicaragua	0.53	0.58	0.82	0.73	0.63	0.69	0.73	0.78	0.74
Panama	7.04	6.86	3.28	1.88	1.68	1.63	1.63	1.86	1.51
Paraguay	0.21	0.27	0.49	0.44	0.55	0.22	-0.34	-0.79	-0.92
Pérou	1.99	2.02	-1.48	-2.55	-3.20	-3.83	-2.83	-1.66	-0.29
Trinité-et-Tobago	-3.62	-5.05	-9.43	-6.82	-7.19	-7.43	-7.30	-7.04	-6.97
Uruguay	1.92	1.90	2.27	0.96	1.00	0.89	1.22	1.01	1.41
Vénézuela	-172.96	-167.27	-101.75	-77.07	-80.15	-71.57	-79.44	-70.88	-83.61
Autre Amérique latine	7.37	8.24	6.45	3.39	3.50	3.44	3.33	3.99	3.93
Amérique latine	-112.32	-98.01	-26.72	-23.13	-36.90	-38.77	-44.79	-29.65	-42.47

Un chiffre négatif correspond à des exportations nettes.

Total Net Imports of Energy (Mtoe)
Importations nettes totales d'énergie (Mtep)
Gesamte Nettoimporte von Energie (Mtoe)
Importazioni nette totali di energia (Mtep)
エネルギー純輸入量（石油換算百万トン）
Importaciones netas totales de energía (Mtep)
Общий чистый импорт топлива и энергии (Мтнэ)

1989	1990	1991	1992	1993	1994	1995	1996	1997	
-74.63	-79.26	-80.68	-81.56	-82.48	-78.95	-84.88	-91.65	-98.51	Algeria
-21.64	-22.58	-23.86	-22.72	-23.17	-23.07	-24.44	-33.82	-33.93	Angola
-0.07	-0.09	-0.10	-0.04	-0.05	-0.02	0.02	0.26	0.27	Benin
-7.33	-7.52	-6.80	-6.27	-6.09	-5.68	-5.15	-5.17	-5.44	Cameroon
-7.14	-7.90	-7.80	-8.39	-9.52	-9.12	-8.27	-9.50	-12.29	Congo
-0.30	-0.08	-0.02	0.01	0.06	0.11	0.14	0.14	0.21	DR of Congo
-21.13	-20.79	-20.79	-22.86	-23.28	-13.66	-13.09	-10.40	-9.16	Egypt
0.95	1.00	1.04	1.12	1.15	1.21	1.31	1.41	0.88	Ethiopia
-8.97	-12.20	-14.04	-13.81	-15.03	-15.80	-17.21	-17.80	-17.90	Gabon
0.88	0.89	0.80	0.77	0.82	1.07	1.09	1.08	1.08	Ghana
1.65	1.36	1.39	1.51	1.24	1.26	1.23	1.40	0.85	Ivory Coast
2.29	2.25	2.14	2.17	2.38	2.50	2.17	2.24	2.69	Kenya
-55.91	-64.46	-69.94	-68.34	-63.58	-64.25	-62.74	-62.74	-63.77	Libya
5.94	6.04	6.09	6.92	7.24	7.80	7.70	7.70	8.39	Morocco
0.50	0.49	0.46	0.49	0.52	0.52	0.53	0.58	0.73	Mozambique
-77.78	-80.10	-83.04	-85.41	-92.73	-81.07	-85.86	-97.28	-102.30	Nigeria
0.88	0.85	0.82	0.98	0.88	1.05	0.93	1.02	1.13	Senegal
-22.80	-22.59	-22.33	-24.69	-26.05	-26.54	-27.89	-29.38	-31.19	South Africa
1.48	1.93	1.76	1.66	1.19	1.72	1.63	1.52	1.60	Sudan
0.74	0.77	0.77	0.69	0.73	0.74	0.75	0.75	0.75	Tanzania
-1.05	-0.43	-0.87	-0.77	0.06	0.68	0.55	0.49	0.55	Tunisia
0.46	0.40	0.39	0.41	0.40	0.41	0.42	0.43	0.43	Zambia
0.96	0.83	1.03	1.23	1.25	1.37	1.65	1.71	1.80	Zimbabwe
6.39	6.08	6.25	6.40	6.38	6.47	6.54	6.09	4.99	Other Africa
-275.62	-295.11	-307.34	-310.49	-317.66	-291.28	-302.87	-330.92	-348.11	**Africa**
-0.23	-2.04	-2.05	-3.98	-4.46	-8.93	-11.94	-15.25	-18.18	Argentina
-2.06	-2.06	-2.04	-1.95	-1.88	-1.99	-1.71	-1.63	-1.67	Bolivia
40.61	41.59	41.96	41.92	48.67	49.80	49.89	55.55	56.11	Brazil
6.08	6.88	7.16	7.86	8.35	10.20	11.14	13.18	14.88	Chile
-19.52	-21.01	-21.98	-20.95	-23.06	-22.60	-28.84	-34.60	-36.57	Colombia
0.87	0.96	1.00	1.39	1.44	1.53	1.56	1.48	1.50	Costa Rica
10.95	10.21	8.13	6.02	5.56	5.75	6.18	6.26	6.98	Cuba
2.94	2.95	2.95	3.39	3.38	3.52	3.66	3.76	4.02	Dominican Republic
-10.14	-9.92	-10.07	-11.38	-12.59	-13.55	-13.45	-12.66	-13.97	Ecuador
0.77	0.79	1.02	1.13	1.33	1.40	1.65	1.46	1.38	El Salvador
1.03	1.15	1.34	1.44	1.43	1.68	1.72	1.46	1.31	Guatemala
0.33	0.33	0.31	0.31	0.22	0.06	0.32	0.37	0.48	Haiti
0.79	0.74	0.79	0.91	0.81	0.95	1.14	1.19	1.15	Honduras
2.01	2.57	2.47	2.61	2.66	2.69	2.95	3.09	3.29	Jamaica
2.86	3.70	4.04	3.84	4.43	4.51	4.48	4.44	4.44	Netherlands Antilles
0.62	0.65	0.63	0.75	0.72	0.65	0.87	0.93	0.97	Nicaragua
1.61	1.99	2.16	2.43	2.25	2.45	2.44	2.42	2.53	Panama
-1.02	-1.47	-1.66	-1.35	-1.61	-1.73	-2.14	-2.83	-2.83	Paraguay
-0.60	-0.59	0.07	-0.21	-0.28	0.22	1.64	1.72	2.27	Peru
-6.74	-6.38	-5.88	-5.64	-5.33	-6.31	-6.40	-4.94	-4.86	Trinidad-and-Tobago
1.45	1.14	1.51	1.64	1.35	1.51	1.99	2.14	2.10	Uruguay
-80.47	-87.29	-96.62	-95.53	-100.10	-123.93	-137.06	-137.39	-147.92	Venezuela
4.00	4.26	4.18	3.98	4.05	4.02	4.07	4.10	4.19	Other Latin America
-43.85	-50.85	-60.58	-61.36	-62.65	-88.12	-105.83	-105.74	-118.41	**Latin America**

A negative number shows net exports.

Total Net Imports of Energy (Mtoe)
Importations nettes totales d'énergie (Mtep)
Gesamte Nettoimporte von Energie (Mtoe)
Importazioni nette totali di energia (Mtep)
エネルギー純輸入量（石油換算百万トン）
Importaciones netas totales de energía (Mtep)
Общий чистый импорт топлива и энергии (Мтнэ)

	1971	1973	1978	1983	1984	1985	1986	1987	1988
Bangladesh	1.05	1.02	1.52	1.51	1.52	1.59	1.79	1.77	2.05
Brunei	-5.95	-13.04	-18.09	-15.96	-15.67	-14.58	-14.23	-13.54	-13.50
Inde	14.62	17.04	19.17	14.29	13.41	16.80	16.83	20.32	23.64
Indonésie	-33.87	-50.84	-71.48	-55.89	-70.73	-61.67	-63.26	-62.90	-62.85
RPD de Corée	1.13	1.02	2.28	4.17	4.46	4.81	4.98	5.06	5.06
Malaisie	1.46	0.31	-3.80	-11.15	-16.61	-17.38	-22.05	-22.13	-23.87
Myanmar	0.41	0.09	0.12	0.10	0.10	0.10	0.11	0.03	0.03
Népal	0.06	0.12	0.10	0.16	0.21	0.21	0.22	0.24	0.28
Pakistan	2.93	3.20	4.04	6.25	6.81	6.81	6.79	7.47	8.11
Philippines	8.67	9.25	11.36	10.40	10.70	7.90	8.00	9.82	10.90
Singapour	6.62	12.24	11.83	12.87	11.83	10.53	14.85	20.59	22.05
Sri Lanka	1.32	1.65	1.46	1.62	1.67	1.58	1.56	1.84	1.77
Taipei chinois	7.29	10.50	19.56	22.24	23.26	23.12	26.15	28.69	34.68
Thaïlande	6.38	8.33	10.71	10.88	10.73	9.07	9.13	11.01	11.65
Viêt-Nam	5.69	5.75	0.25	1.73	1.66	1.63	1.84	2.39	1.73
Autre Asie	-0.01	0.08	0.20	0.39	0.31	0.60	0.46	1.26	1.41
Asie	17.80	6.72	-10.77	3.61	-16.34	-8.89	-6.84	11.90	23.11
Rép. populaire de Chine	-1.92	-3.52	-12.84	-22.28	-30.38	-39.13	-34.22	-33.90	-34.58
Hong-Kong, Chine	4.06	4.85	6.16	7.89	7.67	8.42	9.39	9.78	11.24
Chine	2.14	1.33	-6.68	-14.39	-22.71	-30.70	-24.83	-24.12	-23.34
Albanie	-0.72	-1.27	-0.91	0.02	-0.07	-0.11	-0.10	-0.07	-0.01
Bulgarie	13.59	15.27	20.73	20.75	20.72	21.63	22.17	21.62	21.34
Chypre	0.64	0.85	0.85	0.96	1.02	0.96	1.06	1.35	1.33
Gibraltar	0.21	0.23	0.17	0.16	0.15	0.32	0.40	0.44	0.56
Malte	0.33	0.35	0.39	0.38	0.52	0.43	0.69	0.71	0.77
Roumanie	-1.06	1.16	9.81	9.08	8.78	11.28	13.90	17.43	16.94
République slovaque	11.49	12.88	16.32	16.94	16.52	16.66	16.94	17.02	17.01
Bosnie-Herzégovine	-	-	-	-	-	-	-	-	-
Croatie	-	-	-	-	-	-	-	-	-
Ex-RYM	-	-	-	-	-	-	-	-	-
Slovénie	-	-	-	-	-	-	-	-	-
RF de Yougoslavie	-	-	-	-	-	-	-	-	-
Ex-Yougoslavie	6.95	10.85	12.35	14.84	16.35	15.81	17.49	18.36	19.62
Europe non-OCDE	31.43	40.32	59.71	63.13	63.99	66.99	72.55	76.87	77.57

Un chiffre négatif correspond à des exportations nettes.

Total Net Imports of Energy (Mtoe)
Importations nettes totales d'énergie (Mtep)
Gesamte Nettoimporte von Energie (Mtoe)
Importazioni nette totali di energia (Mtep)
エネルギー純輸入量（石油換算百万トン）
Importaciones netas totales de energía (Mtep)
Общий чистый импорт топлива и энергии (Мтнэ)

1989	1990	1991	1992	1993	1994	1995	1996	1997	
2.19	2.21	1.73	1.87	1.95	2.22	2.28	2.43	2.44	Bangladesh
-13.80	-13.99	-14.62	-15.43	-15.39	-15.93	-16.20	-15.74	-15.30	Brunei
26.13	29.73	33.39	41.08	43.70	44.67	50.38	58.49	56.79	India
-66.42	-68.12	-74.57	-77.23	-73.70	-83.52	-83.20	-86.69	-83.37	Indonesia
5.26	5.21	4.46	4.07	3.57	3.27	3.06	2.95	2.86	DPR of Korea
-24.82	-24.79	-24.74	-24.70	-23.71	-24.19	-29.17	-28.16	-25.31	Malaysia
0.03	0.01	0.09	0.18	0.28	0.40	0.66	0.68	0.95	Myanmar
0.26	0.21	0.28	0.36	0.36	0.51	0.60	0.59	0.59	Nepal
8.42	9.01	8.54	9.61	10.75	12.48	12.81	14.70	14.87	Pakistan
11.59	12.53	12.46	14.06	13.93	15.19	16.97	19.35	20.92	Philippines
21.18	24.52	26.59	29.36	33.73	35.85	32.49	37.65	42.58	Singapore
1.52	1.66	1.64	1.70	2.09	2.07	2.50	2.86	3.09	Sri Lanka
36.93	41.80	41.76	46.64	50.33	54.10	59.20	61.51	65.44	Chinese Taipei
14.84	17.81	18.21	20.99	24.59	26.65	32.14	36.76	34.72	Thailand
0.60	-0.23	-1.87	-2.95	-3.26	-3.61	-5.36	-4.69	-4.22	Vietnam
2.51	-1.86	-2.36	-2.91	-3.04	-3.24	-2.74	-1.72	-1.65	Other Asia
26.41	35.72	30.99	46.71	66.18	66.93	76.44	100.96	115.40	**Asia**
-28.88	-32.02	-27.67	-17.25	0.13	-9.83	-10.13	-8.59	12.76	People's Rep.of China
12.46	11.72	12.17	14.19	15.77	14.95	15.72	16.06	16.23	Hong Kong, China
-16.42	-20.30	-15.50	-3.06	15.90	5.12	5.59	7.47	28.99	**China**
-0.01	0.17	-0.07	-0.09	0.00	0.05	0.09	0.13	0.14	Albania
21.31	17.15	13.64	11.90	13.17	11.85	13.30	12.95	10.94	Bulgaria
1.60	1.65	1.76	1.92	2.06	2.17	2.05	2.20	2.16	Cyprus
0.48	0.50	0.91	0.92	0.95	0.96	0.96	0.96	0.96	Gibraltar
0.81	0.80	0.86	0.89	0.95	0.88	0.91	0.97	1.02	Malta
18.86	21.63	14.30	13.48	11.95	10.63	13.99	14.40	14.21	Romania
16.85	16.46	14.18	13.22	12.22	11.58	12.23	13.02	12.54	Slovak Republic
-	-	-	0.90	0.86	1.12	1.12	1.12	1.12	*Bosnia-Herzegovina*
-	-	-	2.52	2.44	2.97	2.92	3.24	3.74	*Croatia*
-	-	-	1.45	1.55	0.89	1.06	1.33	1.14	*FYROM*
-	-	-	2.10	2.56	2.73	3.04	3.46	3.56	*Slovenia*
-	-	-	2.52	1.19	0.40	0.67	3.43	4.36	*FR of Yugoslavia*
19.11	18.95	16.23	9.57	8.61	7.99	8.48	12.20	13.90	Former Yugoslavia
79.00	77.31	61.82	51.81	49.92	46.11	52.01	56.83	55.87	**Non-OECD Europe**

A negative number shows net exports.

Total Net Imports of Energy (Mtoe)
Importations nettes totales d'énergie (Mtep)
Gesamte Nettoimporte von Energie (Mtoe)
Importazioni nette totali di energia (Mtep)
エネルギー純輸入量（石油換算百万トン）
Importaciones netas totales de energía (Mtep)
Общий чистый импорт топлива и энергии (Мтнэ)

	1971	1973	1978	1983	1984	1985	1986	1987	1988
Arménie	-	-	-	-	-	-	-	-	-
Azerbaïdjan	-	-	-	-	-	-	-	-	-
Bélarus	-	-	-	-	-	-	-	-	-
Estonie	-	-	-	-	-	-	-	-	-
Géorgie	-	-	-	-	-	-	-	-	-
Kazakhstan	-	-	-	-	-	-	-	-	-
Kirghizistan	-	-	-	-	-	-	-	-	-
Lettonie	-	-	-	-	-	-	-	-	-
Lituanie	-	-	-	-	-	-	-	-	-
Moldova	-	-	-	-	-	-	-	-	-
Russie	-	-	-	-	-	-	-	-	-
Tadjikistan	-	-	-	-	-	-	-	-	-
Turkménistan	-	-	-	-	-	-	-	-	-
Ukraine	-	-	-	-	-	-	-	-	-
Ouzbékistan	-	-	-	-	-	-	-	-	-
Ex-URSS	-107.12	-111.20	-190.37	-230.32	-233.82	-219.18	-249.93	-264.59	-272.91
Bahrein	-1.98	-1.58	-2.30	-1.82	-1.27	-1.01	-1.56	-1.55	-1.43
Iran	-217.09	-282.40	-203.36	-90.41	-70.15	-72.02	-49.44	-72.94	-68.81
Irak	-81.46	-97.13	-121.62	-35.58	-46.27	-56.76	-71.26	-92.00	-117.78
Israël	0.40	2.44	7.51	8.24	8.18	8.12	9.18	9.22	12.33
Jordanie	0.62	0.69	1.34	2.55	2.80	2.92	2.91	3.22	3.06
Koweit	-158.19	-149.58	-104.71	-47.42	-52.18	-44.84	-67.37	-62.87	-67.60
Liban	2.05	2.42	2.07	2.34	2.25	2.61	2.75	2.80	2.12
Oman	-13.46	-13.61	-15.19	-19.57	-20.65	-24.89	-28.20	-29.15	-31.59
Qatar	-21.04	-28.10	-23.96	-13.92	-18.94	-15.27	-17.29	-14.62	-16.83
Arabie saoudite	-225.49	-367.48	-409.40	-227.35	-209.99	-139.81	-221.17	-180.69	-224.88
Syrie	-2.54	-3.11	-5.41	-0.86	-0.15	-1.82	-0.16	-3.10	-4.85
Emirats arabes unis	-51.79	-77.73	-92.42	-52.82	-57.33	-54.98	-61.99	-67.93	-72.69
Yémen	1.07	1.27	1.72	2.11	2.28	2.72	2.67	2.03	-5.22
Moyen-Orient	-768.90	-1013.90	-965.72	-474.53	-461.41	-395.03	-500.92	-507.57	-594.17
Total non-OCDE	-1177.73	-1427.85	-1396.60	-869.99	-922.83	-866.58	-988.39	-967.60	-1077.78
OCDE Amérique du N.	165.80	268.97	401.37	88.26	100.63	68.94	140.69	167.25	185.13
OCDE Pacifique	266.08	327.09	327.29	300.04	314.76	296.71	300.14	302.41	327.63
OCDE Europe	710.13	802.22	714.88	513.41	524.46	530.21	567.07	568.91	577.60
Total OCDE	1 142.01	1 398.28	1 443.54	901.71	939.85	895.86	1 007.90	1 038.56	1 090.36
Monde	-35.72	-29.57	46.95	31.72	17.02	29.28	19.51	70.96	12.58

Un chiffre négatif correspond à des exportations nettes.
Ex-URSS: la somme des républiques peut être différente du total.
La ligne Monde montre la divergence entre le total des exportations et importations mondiales.

INTERNATIONAL ENERGY AGENCY

Total Net Imports of Energy (Mtoe)
Importations nettes totales d'énergie (Mtep)
Gesamte Nettoimporte von Energie (Mtoe)
Importazioni nette totali di energia (Mtep)
エネルギー純輸入量（石油換算百万トン）
Importaciones netas totales de energía (Mtep)
Общий чистый импорт топлива и энергии (Мтнэ)

1989	1990	1991	1992	1993	1994	1995	1996	1997	
-	-	-	4.04	1.89	1.12	1.43	1.05	1.27	Armenia
-	-	-	-1.40	-0.94	1.26	-1.73	-2.15	-2.04	Azerbaijan
-	-	-	35.12	27.27	22.38	21.69	20.55	22.33	Belarus
-	-	-	2.32	2.14	2.13	1.89	1.81	1.77	Estonia
-	-	-	5.40	3.79	2.41	0.90	1.47	1.66	Georgia
-	-	-	-11.25	-15.33	-12.66	-7.76	-18.15	-26.37	Kazakhstan
-	-	-	3.26	2.32	1.34	1.27	1.60	1.39	Kyrgyzstan
-	-	-	5.50	3.72	3.52	3.25	3.41	2.64	Latvia
-	-	-	6.77	5.15	5.23	5.56	4.89	4.94	Lithuania
-	-	-	6.75	5.48	4.70	4.41	4.40	4.44	Moldova
-	-	-	-314.17	-294.99	-315.43	-314.54	-335.41	-333.37	Russia
-	-	-	7.46	4.76	1.76	1.96	2.18	2.13	Tajikistan
-	-	-	-38.36	-46.95	-19.33	-18.95	-20.67	-6.56	Turkmenistan
-	-	-	111.52	90.27	75.30	82.50	80.83	68.88	Ukraine
-	-	-	4.72	4.47	-0.59	-4.03	-4.95	-6.98	Uzbekistan
-268.35	-259.97	-198.42	-167.38	-196.33	-227.45	-244.83	-276.83	-278.00	**Former USSR**
-1.43	-1.63	-1.83	-1.68	-1.80	-1.81	-2.06	-0.67	0.92	Bahrain
-94.46	-106.39	-116.77	-121.43	-129.05	-120.64	-120.49	-116.88	-116.31	Iran
-127.62	-85.75	-2.67	-5.50	-5.49	-6.06	-6.04	-7.56	-35.00	Iraq
11.99	11.46	11.12	13.81	14.04	13.20	15.88	15.32	16.80	Israel
3.11	3.52	3.15	3.85	3.88	3.96	4.11	4.31	4.60	Jordan
-81.54	-53.44	-7.81	-49.92	-92.40	-96.63	-96.50	-98.56	-99.74	Kuwait
1.72	2.15	2.80	2.76	3.45	3.79	4.42	4.55	5.04	Lebanon
-32.18	-33.97	-33.51	-35.98	-37.68	-39.88	-42.02	-43.75	-44.69	Oman
-19.16	-19.85	-19.32	-22.56	-21.14	-20.62	-20.54	-22.67	-29.27	Qatar
-231.04	-303.69	-388.13	-393.73	-385.05	-383.40	-385.21	-380.90	-386.79	Saudi Arabia
-8.88	-11.14	-14.23	-14.73	-15.68	-18.34	-19.20	-18.79	-18.15	Syria
-86.23	-96.36	-107.45	-101.48	-95.46	-101.37	-103.08	-106.14	-112.34	United Arab Emirates
-6.11	-6.74	-6.94	-5.46	-8.55	-14.06	-15.28	-16.37	-13.78	Yemen
-671.81	-701.83	-681.59	-732.05	-770.91	-781.85	-786.01	-788.09	-828.71	**Middle East**
-1170.64	-1215.02	-1170.64	-1175.81	-1215.55	-1270.55	-1305.49	-1336.32	-1372.97	**Non-OECD Total**
219.70	215.43	176.75	205.13	255.25	280.03	249.44	274.28	307.41	OECD North America
359.40	376.64	384.92	400.43	413.35	448.30	453.25	478.34	483.82	OECD Pacific
602.55	618.80	633.40	628.92	596.74	559.20	571.43	589.30	602.71	OECD Europe
1 181.66	1 210.87	1 195.07	1 234.48	1 265.34	1 287.54	1 274.12	1 341.91	1 393.94	**OECD Total**
11.01	-4.15	24.43	58.67	49.79	16.99	-31.37	5.59	20.97	**World**

A negative number shows net exports.
Former USSR: data for individual republics may not add to the total.
The row World shows the discrepancy between total world exports and imports.

Primary Supply of Coal (Mtoe)
Approvisionnement primaire en charbon (Mtep)
Primäraufkommen von Kohle (Mtoe)
Disponibilità primaria di carbone (Mtep)
石炭の一次供給量（石油換算百万トン）
Suministro primario de carbón (Mtep)
Первичная поставка угля (Мтнэ)

	1971	1973	1978	1983	1984	1985	1986	1987	1988
Algérie	0.198	0.228	0.192	0.731	0.754	0.754	0.918	0.993	1.002
RD du Congo	0.207	0.157	0.159	0.200	0.204	0.204	0.207	0.210	0.212
Egypte	0.352	0.270	0.604	0.682	0.692	0.730	0.736	0.766	0.858
Ghana	0.007	0.011	0.001	0.001	0.001	0.001	0.001	0.002	0.002
Kenya	0.050	0.044	0.032	0.020	0.052	0.056	0.053	0.057	0.070
Maroc	0.295	0.364	0.475	0.586	0.573	0.684	0.915	1.072	1.048
Mozambique	0.385	0.361	0.152	0.111	0.077	0.065	0.041	0.039	0.040
Nigéria	0.128	0.179	0.107	0.032	0.044	0.059	0.069	0.067	0.031
Afrique du Sud	32.099	33.842	42.160	62.576	67.395	67.539	70.429	73.224	74.322
Tanzanie	-	-	0.001	0.005	0.007	0.010	0.003	0.003	0.003
Tunisie	0.065	0.072	0.078	0.052	0.069	0.066	0.060	0.079	0.080
Zambie	0.570	0.594	0.427	0.296	0.275	0.278	0.303	0.336	0.368
Zimbabwe	1.554	1.644	1.370	1.422	1.555	1.860	2.214	3.024	2.798
Autre Afrique	0.125	0.137	0.334	0.448	0.463	0.509	0.557	0.609	0.624
Afrique	36.036	37.902	46.094	67.163	72.162	72.816	76.507	80.481	81.457
Argentine	0.813	0.713	1.136	0.715	0.605	0.857	1.116	1.125	1.213
Bolivie	-	-	-	0.168	0.098	0.059	0.036	-	-
Brésil	2.368	2.528	4.866	6.858	8.552	9.852	9.860	10.619	10.667
Chili	1.321	1.199	1.010	1.003	1.302	1.254	1.306	1.268	1.774
Colombie	1.744	1.846	2.629	3.071	3.034	3.088	3.130	2.945	3.167
Costa Rica	0.001	0.001	0.001	0.001	0.001	0.001	0.001	0.001	0.001
Cuba	0.081	0.066	0.088	0.092	0.092	0.119	0.113	0.107	0.101
République dominicaine	-	-	-	0.005	0.008	0.138	0.017	0.143	0.060
El Salvador	-	-	0.001	-	-	-	-	-	-
Haiti	-	-	-	0.035	0.037	0.038	0.011	0.011	0.019
Honduras	-	-	-	-	-	-	0.001	0.001	0.001
Jamaïque	-	-	-	-	-	-	-	-	-
Panama	-	-	-	0.002	0.006	0.023	0.023	0.023	0.011
Pérou	0.139	0.149	0.127	0.139	0.109	0.151	0.185	0.175	0.143
Uruguay	0.027	0.021	0.003	0.001	0.001	0.001	0.001	0.001	-
Vénézuela	0.129	0.250	0.069	0.112	0.122	0.164	0.181	0.239	0.171
Autre Amérique latine	0.023	0.028	0.020	0.011	0.007	-	-	-	-
Amérique latine	6.646	6.801	9.949	12.212	13.974	15.743	15.980	16.658	17.326
Bangladesh	0.093	0.121	0.159	0.087	0.035	0.049	0.074	0.116	0.120
Inde	37.815	39.405	50.463	67.939	69.183	76.338	82.173	88.583	95.073
Indonésie	0.130	0.083	0.129	0.180	0.216	0.594	1.796	2.317	2.704
RPD de Corée	24.114	21.612	25.736	29.384	29.507	30.727	31.244	31.244	31.552
Malaisie	0.040	0.033	0.024	0.221	0.239	0.319	0.237	0.289	0.456
Myanmar	0.139	0.047	0.148	0.144	0.150	0.152	0.158	0.062	0.057
Népal	0.007	0.047	0.010	0.025	0.060	0.048	0.032	0.051	0.053
Pakistan	0.655	0.575	0.577	1.062	1.159	1.472	1.546	1.616	1.791
Philippines	0.029	0.025	0.287	0.638	0.941	1.229	0.984	1.098	1.298
Singapour	0.004	0.004	0.004	0.003	0.005	0.010	0.007	0.011	0.013
Sri Lanka	0.001	0.002	0.002	0.001	0.001	0.002	0.001	0.001	-
Taipei chinois	2.541	2.277	2.535	5.697	6.730	7.142	8.761	9.592	11.023
Thaïlande	0.114	0.096	0.217	0.690	0.839	1.653	1.715	2.157	2.329
Viêt-Nam	1.280	1.386	2.303	2.517	2.519	2.505	2.761	3.076	2.879
Autre Asie	0.311	0.135	0.176	0.161	0.168	0.212	0.216	0.211	0.205
Asie	67.274	65.848	82.770	108.750	111.753	122.453	131.703	140.424	149.552

Primary Supply of Coal (Mtoe)
Approvisionnement primaire en charbon (Mtep)
Primäraufkommen von Kohle (Mtoe)
Disponibilità primaria di carbone (Mtep)
石炭の一次供給量（石油換算百万トン）
Suministro primario de carbón (Mtep)
Первичная поставка угля (Мтнэ)

1989	1990	1991	1992	1993	1994	1995	1996	1997	
0.904	0.757	0.546	0.557	0.491	0.517	0.517	0.517	0.272	Algeria
0.218	0.220	0.228	0.228	0.206	0.210	0.241	0.241	0.241	DR of Congo
0.889	0.759	0.716	0.735	0.886	0.949	0.644	0.644	0.644	Egypt
0.002	0.002	0.002	0.002	0.002	0.002	0.002	0.002	0.002	Ghana
0.081	0.093	0.094	0.098	0.081	0.068	0.058	0.055	0.057	Kenya
1.143	1.151	1.242	1.124	1.197	1.354	1.638	2.079	2.019	Morocco
0.044	0.036	0.038	0.037	0.037	0.037	0.034	0.011	0.012	Mozambique
0.033	0.039	0.069	0.040	0.062	0.080	0.086	0.086	0.086	Nigeria
69.418	66.538	70.141	64.862	72.334	74.448	77.470	78.808	80.492	South Africa
0.003	0.003	0.003	0.003	0.003	0.003	0.004	0.003	0.003	Tanzania
0.089	0.079	0.065	0.079	0.079	0.066	0.069	0.069	0.069	Tunisia
0.231	0.255	0.201	0.229	0.171	0.100	0.104	0.102	0.102	Zambia
2.841	3.038	3.395	3.399	3.147	2.882	2.755	2.775	2.636	Zimbabwe
0.685	0.782	0.745	0.795	0.758	0.830	0.788	0.731	0.734	Other Africa
76.580	73.752	77.485	72.189	79.453	81.546	84.412	86.123	87.370	**Africa**
1.065	0.943	0.773	0.983	0.588	0.825	0.653	0.888	0.856	Argentina
-	-	-	-	-	-	-	-	-	Bolivia
10.676	9.400	10.718	10.618	10.954	11.250	11.731	12.167	12.315	Brazil
2.363	2.569	2.037	1.808	1.829	2.178	2.396	3.288	4.240	Chile
3.426	3.663	4.214	2.861	2.917	3.591	3.646	3.132	2.704	Colombia
0.001	0.001	-	-	-	-	-	0.008	-	Costa Rica
0.146	0.136	0.071	0.042	0.050	0.067	0.061	0.011	0.047	Cuba
0.006	0.006	0.035	0.085	0.074	0.064	0.068	0.079	0.085	Dominican Republic
-	-	-	-	0.001	0.001	0.001	0.001	0.001	El Salvador
0.007	0.007	0.015	0.016	-	-	-	-	-	Haiti
0.001	0.001	0.001	0.001	0.001	0.001	0.001	0.001	0.001	Honduras
0.034	0.038	0.006	0.040	0.044	0.033	0.034	0.039	0.041	Jamaica
0.012	0.020	0.028	0.036	0.039	0.032	0.033	0.034	0.035	Panama
0.175	0.161	0.292	0.289	0.438	0.405	0.385	0.306	0.305	Peru
-	-	0.001	0.001	0.001	-	-	-	0.001	Uruguay
0.452	0.218	0.247	0.277	0.179	0.218	0.213	0.202	0.258	Venezuela
0.001	0.001	0.001	0.001	0.001	0.001	0.001	0.001	0.001	Other Latin America
18.366	17.163	18.439	17.057	17.115	18.666	19.223	20.157	20.890	**Latin America**
0.125	0.281	0.090	0.084	0.031	-	-	-	-	Bangladesh
100.659	105.911	113.906	120.143	127.555	134.137	141.982	149.085	153.279	India
3.402	3.952	4.088	4.385	4.910	6.232	6.358	8.941	9.545	Indonesia
32.069	30.217	28.103	26.266	24.383	22.551	20.985	19.546	18.959	DPR of Korea
1.196	1.379	1.356	1.410	1.510	1.851	1.612	2.046	1.581	Malaysia
0.056	0.063	0.066	0.033	0.034	0.037	0.008	0.009	0.009	Myanmar
0.049	0.007	0.040	0.055	0.060	0.067	0.073	0.071	0.071	Nepal
1.762	2.089	2.005	2.072	2.116	2.301	2.083	2.338	2.045	Pakistan
1.117	1.297	1.349	1.218	1.247	1.219	1.164	2.051	2.648	Philippines
0.012	0.021	0.012	0.018	0.020	0.024	0.024	-0.001	-	Singapore
0.001	0.005	0.001	0.001	0.001	0.001	0.003	0.001	0.001	Sri Lanka
11.616	11.232	12.241	14.329	15.845	16.734	17.199	19.554	22.223	Chinese Taipei
2.788	3.817	4.547	4.855	5.244	5.920	6.896	8.582	8.631	Thailand
2.311	2.200	2.136	1.970	1.928	2.056	2.382	3.403	4.435	Vietnam
0.214	0.189	0.189	0.133	0.133	0.130	0.131	0.134	0.140	Other Asia
157.377	162.658	170.129	176.972	185.015	193.259	200.898	215.762	223.568	**Asia**

INTERNATIONAL ENERGY AGENCY

Primary Supply of Coal (Mtoe)
Approvisionnement primaire en charbon (Mtep)
Primäraufkommen von Kohle (Mtoe)
Disponibilità primaria di carbone (Mtep)
石炭の一次供給量（石油換算百万トン）
Suministro primario de carbón (Mtep)
Первичная поставка угля (Мтнэ)

	1971	1973	1978	1983	1984	1985	1986	1987	1988
Rép. populaire de Chine	190.409	202.642	302.284	341.364	375.311	404.841	429.034	457.721	486.268
Hong-Kong, Chine	0.024	0.010	0.009	2.105	2.746	3.399	3.933	4.927	5.701
Chine	190.433	202.652	302.293	343.469	378.057	408.240	432.967	462.648	491.969
Albanie	0.299	0.353	0.535	0.784	0.847	0.902	0.914	0.904	0.928
Bulgarie	8.426	8.330	8.326	9.920	10.181	10.465	10.422	10.740	9.706
Chypre	-	-	-	-	0.032	0.046	0.034	0.093	0.056
Malte	-	-	-	0.032	0.058	0.118	0.084	0.112	0.141
Roumanie	6.811	8.288	11.000	14.305	14.171	15.214	15.617	16.884	18.673
République slovaque	7.990	7.907	8.217	7.797	8.022	8.315	8.416	8.547	8.507
Bosnie-Herzégovine	-	-	-	-	-	-	-	-	-
Croatie	-	-	-	-	-	-	-	-	-
Ex-RYM	-	-	-	-	-	-	-	-	-
Slovénie	-	-	-	-	-	-	-	-	-
RF de Yougoslavie	-	-	-	-	-	-	-	-	-
Ex-Yougoslavie	8.997	9.307	10.173	16.836	18.233	18.868	19.100	18.550	18.059
Europe non-OCDE	32.522	34.186	38.252	49.674	51.542	53.929	54.587	55.830	56.070
Arménie	-	-	-	-	-	-	-	-	-
Azerbaïdjan	-	-	-	-	-	-	-	-	-
Bélarus	-	-	-	-	-	-	-	-	-
Estonie	-	-	-	-	-	-	-	-	-
Géorgie	-	-	-	-	-	-	-	-	-
Kazakhstan	-	-	-	-	-	-	-	-	-
Kirghizistan	-	-	-	-	-	-	-	-	-
Lettonie	-	-	-	-	-	-	-	-	-
Lituanie	-	-	-	-	-	-	-	-	-
Moldova	-	-	-	-	-	-	-	-	-
Russie	-	-	-	-	-	-	-	-	-
Tadjikistan	-	-	-	-	-	-	-	-	-
Turkménistan	-	-	-	-	-	-	-	-	-
Ukraine	-	-	-	-	-	-	-	-	-
Ouzbékistan	-	-	-	-	-	-	-	-	-
Ex-URSS	304.113	318.958	337.366	310.802	302.269	301.117	309.429	314.216	317.268
Iran	0.369	0.688	0.615	0.529	0.794	0.831	0.868	0.886	0.898
Israël	0.005	0.001	0.001	1.060	1.690	1.848	2.047	2.180	2.137
Liban	0.003	0.010	0.004	-	-	-	-	-	-
Syrie	0.004	0.003	0.005	0.005	0.003	0.001	0.001	0.001	0.001
Moyen-Orient	0.381	0.702	0.624	1.594	2.488	2.680	2.916	3.066	3.036
Total non-OCDE	637.404	667.048	817.347	893.664	932.244	976.979	1 024.090	1 073.322	1 116.678
OCDE Amérique du N.	296.753	328.138	377.391	415.241	439.198	454.537	439.931	465.925	484.570
OCDE Pacifique	84.016	89.266	81.509	107.602	119.677	126.318	122.775	123.158	131.830
OCDE Europe	428.208	417.523	419.617	443.298	440.602	467.054	458.986	465.980	454.700
Total OCDE	808.977	834.926	878.517	966.140	999.477	1 047.908	1 021.692	1 055.063	1 071.100
Monde	1 446.381	1 501.975	1 695.864	1 859.804	1 931.721	2 024.887	2 045.782	2 128.385	2 187.779

Primary Supply of Coal (Mtoe)
Approvisionnement primaire en charbon (Mtep)
Primäraufkommen von Kohle (Mtoe)
Disponibilità primaria di carbone (Mtep)
石炭の一次供給量（石油換算百万トン）
Suministro primario de carbón (Mtep)
Первичная поставка угля (Мтнэ)

1989	1990	1991	1992	1993	1994	1995	1996	1997	
505.054	515.414	532.131	541.355	569.114	614.745	658.142	678.978	658.083	People's Rep.of China
6.106	5.492	5.926	6.291	7.277	5.199	5.603	4.163	3.512	Hong Kong, China
511.160	520.906	538.057	547.646	576.390	619.943	663.745	683.141	661.595	**China**
0.931	0.630	0.375	0.197	0.139	0.040	0.038	0.024	0.016	Albania
9.383	8.782	7.589	7.259	7.777	7.117	7.221	7.283	7.492	Bulgaria
0.063	0.060	0.060	0.016	0.019	0.017	0.012	0.010	0.013	Cyprus
0.186	0.185	0.149	0.126	0.185	0.125	0.032	-	-	Malta
17.197	11.683	9.353	10.814	9.360	9.760	10.100	9.923	9.169	Romania
8.210	7.395	6.598	5.833	5.798	5.093	5.232	5.027	4.694	Slovak Republic
-	-	-	3.186	2.761	0.297	0.348	0.348	0.348	*Bosnia-Herzegovina*
-	-	-	0.415	0.346	0.214	0.181	0.145	0.258	*Croatia*
-	-	-	1.546	1.524	1.622	1.654	1.606	1.496	*FYROM*
-	-	-	1.321	1.288	1.213	1.224	1.162	1.269	*Slovenia*
-	-	-	9.287	8.020	8.211	8.567	8.237	8.706	*FR of Yugoslavia*
17.765	18.673	17.107	15.812	13.988	11.589	11.985	11.504	12.081	Former Yugoslavia
53.736	47.407	41.231	40.057	37.265	33.741	34.620	33.771	33.465	**Non-OECD Europe**
-	-	-	0.063	0.001	0.016	0.001	0.002	0.002	Armenia
-	-	-	0.012	0.003	0.004	0.003	0.003	0.003	Azerbaijan
-	-	-	2.328	1.678	1.416	1.300	1.225	1.011	Belarus
-	-	-	4.440	3.295	3.259	3.092	3.209	3.168	Estonia
-	-	-	0.180	0.128	0.067	0.032	0.032	0.002	Georgia
-	-	-	38.818	35.295	35.012	31.907	25.065	21.888	Kazakhstan
-	-	-	0.997	0.892	1.050	0.449	0.592	0.959	Kyrgyzstan
-	-	-	0.673	0.345	0.293	0.215	0.185	0.160	Latvia
-	-	-	0.311	0.309	0.229	0.184	0.166	0.139	Lithuania
-	-	-	1.069	1.000	0.969	0.592	0.511	0.151	Moldova
-	-	-	126.906	125.012	113.989	109.094	110.275	97.216	Russia
-	-	-	0.253	0.148	0.037	0.013	0.051	0.048	Tajikistan
-	-	-	0.266	0.060	-	-	0.044	-	Turkmenistan
-	-	-	70.592	64.595	50.782	52.526	45.222	43.532	Ukraine
-	-	-	2.181	1.585	1.610	1.067	1.197	0.984	Uzbekistan
306.005	288.795	257.687	240.714	226.399	202.803	193.531	182.308	164.127	**Former USSR**
0.923	1.046	1.107	1.230	0.896	0.800	0.950	0.927	0.927	Iran
2.309	2.396	2.576	3.169	3.582	3.795	4.160	4.950	5.478	Israel
-	-	-	-	0.073	0.074	0.119	0.132	0.132	Lebanon
-	-	0.001	0.001	0.001	0.001	0.001	0.001	0.001	Syria
3.232	3.442	3.684	4.400	4.552	4.670	5.230	6.011	6.538	**Middle East**
1 126.456	1 114.122	1 106.712	1 107.353	1 134.087	1 160.526	1 208.591	1 232.736	1 202.685	**Non-OECD Total**
491.435	484.317	478.546	482.630	503.300	503.016	506.138	529.769	547.266	OECD North America
133.896	134.410	138.700	137.181	141.558	143.647	148.171	157.095	163.034	OECD Pacific
449.616	427.288	408.032	387.115	365.747	356.869	353.792	356.094	341.984	OECD Europe
1 074.946	1 046.015	1 025.278	1 006.926	1 010.605	1 003.532	1 008.100	1 042.958	1 052.284	**OECD Total**
2 201.402	2 160.138	2 131.990	2 114.279	2 144.693	2 164.058	2 216.692	2 275.694	2 254.969	**World**

Primary Supply of Oil (Mtoe)
Approvisionnement primaire en pétrole (Mtep)
Primäraufkommen von Öl (Mtoe)

Disponibilità primaria di petrolio (Mtep)

石油の一次供給量（石油換算百万トン）

Suministro primario de petróleo (Mtep)

Первичная поставка нефти и нефтепродуктов (Мтнэ)

	1971	1973	1978	1983	1984	1985	1986	1987	1988
Algérie	2.28	3.08	4.54	6.35	7.02	7.72	7.73	7.55	7.59
Angola	0.65	0.86	0.90	0.78	0.65	0.85	0.78	0.77	0.81
Bénin	0.11	0.14	0.14	0.14	0.14	0.18	0.16	0.14	0.13
Cameroun	0.30	0.33	0.55	0.90	0.97	0.96	0.92	0.91	0.96
Congo	0.16	0.17	0.22	0.36	0.33	0.31	0.22	0.69	1.07
RD du Congo	0.70	0.78	0.84	1.10	1.06	0.99	0.87	0.99	1.10
Egypte	6.33	6.61	10.26	16.79	18.43	19.34	19.44	20.86	21.28
Ethiopie	0.53	0.51	0.49	0.61	0.60	0.62	0.70	0.88	0.90
Gabon	0.36	0.55	1.04	1.02	0.94	0.74	0.73	0.64	0.68
Ghana	0.69	0.75	0.89	0.36	0.65	0.83	0.74	0.89	0.76
Côte d'Ivoire	0.86	0.99	1.47	0.98	1.01	1.02	1.18	1.10	1.47
Kenya	1.26	1.42	1.73	1.50	1.68	1.76	2.00	2.11	2.08
Libye	0.66	1.77	3.75	6.36	6.48	6.96	6.41	7.84	6.98
Maroc	1.82	2.29	3.50	4.16	4.34	4.32	4.26	4.23	4.52
Mozambique	0.59	0.58	0.44	0.45	0.42	0.43	0.46	0.47	0.47
Nigéria	1.76	2.68	6.81	10.04	8.96	9.06	8.50	9.31	10.06
Sénégal	0.52	0.60	0.71	0.79	0.85	0.79	0.85	0.93	0.84
Afrique du Sud	8.52	10.02	11.13	9.49	10.02	9.51	9.38	9.48	10.72
Soudan	1.28	1.48	1.16	1.33	1.26	1.47	1.43	1.15	1.58
Tanzanie	0.73	0.81	0.76	0.69	0.68	0.68	0.72	0.70	0.71
Tunisie	1.30	1.41	2.18	2.99	2.81	2.74	2.93	2.58	3.01
Zambie	0.50	0.69	0.70	0.72	0.61	0.60	0.57	0.57	0.61
Zimbabwe	0.56	0.71	0.70	0.77	0.75	0.79	0.83	0.77	0.94
Autre Afrique	2.32	2.53	3.33	4.47	4.28	4.22	4.75	5.50	5.71
Afrique	34.79	41.75	58.24	73.15	74.95	76.89	76.58	81.06	84.97
Argentine	24.79	25.33	25.87	23.91	24.22	21.26	22.00	23.62	22.76
Bolivie	0.66	0.77	1.14	1.26	1.20	1.16	1.19	1.23	1.25
Brésil	28.91	39.21	53.11	46.67	46.19	49.21	53.90	55.01	60.58
Chili	5.12	5.16	4.97	4.67	4.69	4.56	4.89	5.07	5.62
Colombie	6.24	6.88	7.65	8.64	8.48	8.84	9.26	9.82	9.42
Costa Rica	0.47	0.57	0.83	0.60	0.64	0.67	0.71	0.77	0.79
Cuba	7.26	8.46	10.45	11.81	10.28	10.37	10.32	10.98	11.23
République dominicaine	1.16	1.66	1.99	2.30	2.40	2.17	2.45	2.80	2.90
Equateur	1.20	1.42	3.54	4.20	4.44	4.56	4.52	4.51	4.76
El Salvador	0.51	0.66	0.78	0.59	0.60	0.65	0.65	0.77	0.77
Guatemala	0.86	0.98	1.28	1.01	1.17	1.19	0.91	0.99	1.08
Haiti	0.13	0.13	0.23	0.22	0.23	0.23	0.25	0.28	0.29
Honduras	0.38	0.43	0.55	0.59	0.59	0.57	0.54	0.65	0.73
Jamaïque	1.87	2.82	2.31	1.88	1.84	1.60	1.62	1.68	1.71
Antilles néerlandaises	5.52	6.03	3.76	4.41	3.06	1.83	1.77	1.54	1.54
Nicaragua	0.55	0.60	0.84	0.69	0.65	0.66	0.77	0.79	0.71
Panama	1.39	1.80	1.24	1.21	0.97	0.89	0.93	0.99	0.79
Paraguay	0.21	0.28	0.47	0.46	0.50	0.54	0.56	0.60	0.63
Pérou	4.97	5.29	6.19	6.02	6.19	5.77	6.16	6.61	6.76
Trinité-et-Tobago	1.08	1.07	1.43	0.89	1.28	1.49	1.64	1.06	0.96
Uruguay	1.87	1.82	1.94	1.29	1.17	1.11	1.10	1.20	1.51
Vénézuela	12.50	12.22	15.46	20.09	16.15	16.43	17.20	21.52	19.94
Autre Amérique latine	4.50	5.36	5.17	3.19	3.43	3.26	3.48	3.91	3.95
Amérique latine	112.15	128.96	151.19	146.60	140.37	139.01	146.82	156.41	160.67

Primary Supply of Oil (Mtoe)
Approvisionnement primaire en pétrole (Mtep)
Primäraufkommen von Öl (Mtoe)
Disponibilità primaria di petrolio (Mtep)
石油の一次供給量（石油換算百万トン）
Suministro primario de petróleo (Mtep)
Первичная поставка нефти и нефтепродуктов (Мтнэ)

1989	1990	1991	1992	1993	1994	1995	1996	1997	
7.69	8.09	8.43	8.09	8.35	8.08	7.59	7.38	8.78	Algeria
0.80	0.79	0.75	0.74	0.74	0.74	0.74	0.95	1.12	Angola
0.11	0.10	0.08	0.08	0.09	0.09	0.09	0.34	0.33	Benin
1.04	1.01	0.92	0.85	0.84	0.88	0.87	0.85	0.86	Cameroon
0.43	0.33	0.36	0.36	0.34	0.36	0.32	0.33	0.33	Congo
1.08	1.16	1.18	1.09	1.22	1.24	1.32	1.32	1.32	DR of Congo
21.29	22.49	22.37	22.59	23.25	20.59	22.26	23.68	25.60	Egypt
0.95	0.99	0.99	0.96	0.97	1.01	1.15	1.27	0.82	Ethiopia
0.54	0.40	0.44	0.49	0.54	0.53	0.53	0.54	0.64	Gabon
0.91	0.90	0.83	0.78	0.97	1.34	1.48	1.45	1.45	Ghana
1.51	1.27	1.19	1.62	1.29	1.26	1.28	1.44	1.49	Ivory Coast
2.18	2.10	1.99	2.08	2.18	2.29	1.97	1.97	2.42	Kenya
6.32	7.37	9.62	8.73	9.71	9.10	11.73	10.38	10.42	Libya
4.95	5.12	5.24	5.85	6.17	6.63	6.15	6.48	6.63	Morocco
0.45	0.45	0.41	0.45	0.48	0.45	0.46	0.50	0.64	Mozambique
10.45	10.65	11.61	12.48	5.64	16.56	13.51	12.94	14.87	Nigeria
0.84	0.85	0.83	0.95	0.88	0.95	0.93	1.03	1.12	Senegal
10.55	10.59	10.69	10.68	7.98	10.12	10.83	8.91	10.46	South Africa
1.45	1.86	1.69	1.59	1.18	1.66	1.58	1.56	1.86	Sudan
0.72	0.74	0.74	0.68	0.71	0.71	0.72	0.73	0.73	Tanzania
3.28	3.33	3.61	3.59	3.91	3.49	3.34	3.51	3.47	Tunisia
0.59	0.53	0.53	0.54	0.52	0.54	0.55	0.56	0.56	Zambia
0.87	0.81	0.98	1.07	1.15	1.29	1.51	1.50	1.50	Zimbabwe
5.54	5.34	5.49	5.67	5.91	5.90	6.04	6.12	6.32	Other Africa
84.56	87.28	90.96	92.03	85.01	95.81	96.96	95.71	103.73	**Africa**
20.14	19.28	20.40	21.95	24.01	23.77	23.49	23.68	24.43	Argentina
1.33	1.42	1.54	1.44	1.47	1.61	1.93	1.96	2.10	Bolivia
61.70	62.28	63.35	65.15	68.59	74.30	74.36	80.13	85.67	Brazil
6.19	6.48	6.86	7.72	8.33	9.10	9.86	10.62	11.29	Chile
9.95	11.24	10.88	12.47	12.41	13.01	13.37	13.98	14.33	Colombia
0.94	0.98	1.04	1.34	1.46	1.55	1.55	1.44	1.52	Costa Rica
11.52	10.86	8.58	6.75	6.37	6.88	7.18	7.74	8.66	Cuba
2.94	2.92	2.94	3.34	3.31	3.46	3.60	3.69	3.94	Dominican Republic
4.40	5.06	5.16	5.45	5.12	5.66	6.00	6.81	6.79	Ecuador
0.81	0.81	1.02	1.12	1.29	1.45	1.61	1.47	1.44	El Salvador
1.11	1.27	1.34	1.53	1.57	1.85	2.03	2.15	2.28	Guatemala
0.33	0.33	0.30	0.27	0.24	0.06	0.32	0.37	0.48	Haiti
0.79	0.75	0.75	0.90	0.81	0.96	1.17	1.16	1.18	Honduras
2.11	2.54	2.50	2.50	2.60	2.67	2.96	3.13	3.33	Jamaica
1.89	2.06	2.62	1.62	2.62	2.74	2.71	2.66	2.66	Netherlands Antilles
0.62	0.67	0.66	0.73	0.71	0.81	0.81	0.92	1.03	Nicaragua
0.78	0.91	1.03	1.14	1.15	1.36	1.28	1.37	1.49	Panama
0.68	0.68	0.65	0.81	0.89	1.04	1.19	1.07	1.16	Paraguay
6.08	5.77	5.63	5.39	5.90	6.33	7.41	7.57	8.57	Peru
0.78	1.30	1.25	1.77	1.54	1.21	1.13	1.45	1.34	Trinidad-and-Tobago
1.61	1.28	1.46	1.69	1.49	1.35	1.53	1.81	1.81	Uruguay
18.45	16.75	24.73	21.90	17.32	13.40	17.63	23.33	22.85	Venezuela
4.12	4.30	4.36	4.19	4.24	4.24	4.28	4.32	4.41	Other Latin America
159.29	159.92	169.05	171.17	173.46	178.81	187.39	202.85	212.76	**Latin America**

Primary Supply of Oil (Mtoe)
Approvisionnement primaire en pétrole (Mtep)
Primäraufkommen von Öl (Mtoe)
Disponibilità primaria di petrolio (Mtep)
石油の一次供給量（石油換算百万トン）
Suministro primario de petróleo (Mtep)
Первичная поставка нефти и нефтепродуктов (Мтнэ)

	1971	1973	1978	1983	1984	1985	1986	1987	1988
Bangladesh	0.83	0.94	1.28	1.42	1.51	1.62	1.80	1.76	1.83
Brunei	0.07	0.07	0.10	0.16	0.30	0.26	0.34	0.62	1.08
Inde	21.88	24.24	29.60	38.70	40.96	44.10	47.48	49.64	53.80
Indonésie	8.33	10.81	17.44	22.82	22.70	23.49	26.36	25.55	27.16
RPD de Corée	0.71	0.80	1.85	2.78	2.94	3.11	3.28	3.36	3.36
Malaisie	4.76	4.68	6.82	9.92	9.76	10.02	9.85	10.27	10.69
Myanmar	1.32	1.06	1.23	1.24	1.28	1.25	1.33	0.80	0.73
Népal	0.06	0.07	0.09	0.13	0.15	0.16	0.18	0.20	0.23
Pakistan	3.44	3.50	4.30	6.37	7.05	7.57	8.07	8.70	9.51
Philippines	8.36	9.60	11.28	10.83	8.46	7.69	8.06	9.87	10.65
Singapour	2.94	4.07	6.04	6.95	7.77	7.81	8.33	8.87	9.78
Sri Lanka	0.95	1.24	1.15	1.48	1.32	1.20	1.23	1.36	1.30
Taipei chinois	6.82	10.31	18.69	17.84	17.76	16.84	19.45	18.40	21.19
Thaïlande	6.25	8.23	10.66	10.83	11.41	10.93	11.29	12.46	13.96
Viêt-Nam	5.90	5.86	0.95	1.92	1.88	1.92	2.14	2.49	2.57
Autre Asie	1.47	1.90	1.86	2.03	2.04	2.27	2.14	2.50	2.41
Asie	74.10	87.38	113.33	135.43	137.28	140.23	151.34	156.83	170.26
Rép. populaire de Chine	39.87	52.74	93.61	86.07	88.29	93.29	99.16	105.74	112.83
Hong-Kong, Chine	3.40	3.67	5.65	4.97	4.70	4.25	4.53	4.08	4.81
Chine	43.27	56.41	99.27	91.04	92.99	97.54	103.69	109.82	117.64
Albanie	0.86	0.77	1.17	1.12	1.21	0.98	1.00	1.00	1.06
Bulgarie	10.09	11.48	14.52	12.35	12.04	11.49	12.14	11.04	12.10
Chypre	0.64	0.85	0.81	1.00	1.03	1.02	1.16	1.27	1.33
Gibraltar	0.04	0.04	0.03	0.04	0.04	0.04	0.05	0.05	0.06
Malte	0.27	0.33	0.36	0.33	0.44	0.28	0.57	0.57	0.59
Roumanie	11.17	13.26	18.53	14.28	13.84	14.95	15.86	18.03	16.22
République slovaque	4.35	5.37	7.02	6.11	5.98	5.99	5.89	5.81	5.72
Bosnie-Herzégovine	-	-	-	-	-	-	-	-	-
Croatie	-	-	-	-	-	-	-	-	-
Ex-RYM	-	-	-	-	-	-	-	-	-
Slovénie	-	-	-	-	-	-	-	-	-
RF de Yougoslavie	-	-	-	-	-	-	-	-	-
Ex-Yougoslavie	9.18	10.62	15.62	13.63	13.90	13.46	14.87	15.63	16.99
Europe non-OCDE	36.60	42.71	58.06	48.88	48.49	48.21	51.54	53.41	54.07

Primary Supply of Oil (Mtoe)
Approvisionnement primaire en pétrole (Mtep)
Primäraufkommen von Öl (Mtoe)

Disponibilità primaria di petrolio (Mtep)

石油の一次供給量（石油換算百万トン）

Suministro primario de petróleo (Mtep)

Первичная поставка нефти и нефтепродуктов (Мтнэ)

1989	1990	1991	1992	1993	1994	1995	1996	1997	
1.96	1.99	1.65	1.79	2.06	2.23	2.33	2.44	2.44	Bangladesh
0.21	0.13	0.13	0.26	0.22	0.27	0.35	0.39	0.46	Brunei
57.54	60.11	60.77	63.91	64.99	70.68	82.11	86.98	87.94	India
28.90	32.78	37.63	37.66	43.22	37.82	41.83	44.92	50.50	Indonesia
3.56	3.51	3.02	2.62	2.14	1.85	1.64	1.51	1.47	DPR of Korea
11.54	13.35	14.15	14.87	19.47	20.74	19.29	21.43	25.03	Malaysia
0.73	0.73	0.76	0.74	0.93	1.05	1.09	1.13	1.31	Myanmar
0.20	0.19	0.25	0.31	0.33	0.44	0.53	0.53	0.53	Nepal
10.07	10.75	10.99	12.01	13.03	14.26	14.94	16.72	17.06	Pakistan
11.39	11.93	11.60	13.32	14.67	15.61	17.88	18.10	19.45	Philippines
9.83	13.34	14.45	15.92	18.99	23.33	19.97	22.50	25.54	Singapore
1.26	1.28	1.37	1.39	1.76	1.81	2.08	2.59	2.81	Sri Lanka
25.10	26.02	27.97	28.17	30.28	31.27	34.42	34.84	35.76	Chinese Taipei
16.20	19.63	20.65	23.11	26.19	29.44	34.33	37.76	37.60	Thailand
2.37	2.89	2.59	3.29	4.02	4.43	4.18	6.26	7.77	Vietnam
2.48	2.62	2.60	2.33	2.30	2.19	2.20	2.20	2.28	Other Asia
183.36	201.24	210.57	221.71	244.62	257.44	279.15	300.29	317.94	**Asia**
117.65	116.53	124.19	132.54	146.65	146.14	158.44	173.42	193.76	People's Rep.of China
4.94	5.06	5.25	6.79	7.08	7.54	7.61	7.34	7.61	Hong Kong, China
122.59	121.60	129.44	139.34	153.74	153.67	166.04	180.76	201.37	**China**
0.99	1.11	0.77	0.53	0.58	0.68	0.61	0.60	0.48	Albania
11.77	8.82	5.97	5.61	6.39	5.86	6.29	5.57	4.71	Bulgaria
1.42	1.47	1.67	1.82	1.85	2.13	1.95	2.10	2.04	Cyprus
0.06	0.07	0.08	0.09	0.10	0.14	0.13	0.13	0.13	Gibraltar
0.59	0.59	0.61	0.63	0.74	0.72	0.81	0.89	0.94	Malta
17.47	18.25	14.66	11.94	12.74	11.61	13.16	13.08	12.73	Romania
5.45	4.71	4.07	3.58	2.94	3.18	3.43	3.38	3.28	Slovak Republic
-	-	-	0.55	1.52	0.90	0.90	0.90	0.90	*Bosnia-Herzegovina*
-	-	-	3.29	3.44	3.64	3.98	3.78	4.03	*Croatia*
-	-	-	1.10	1.17	0.76	0.84	1.24	1.06	*FYROM*
-	-	-	1.63	1.97	2.12	2.34	2.73	2.71	*Slovenia*
-	-	-	2.05	1.54	1.46	1.55	2.75	3.65	*FR of Yugoslavia*
16.54	16.43	13.70	8.61	9.63	8.88	9.61	11.40	12.34	Former Yugoslavia
54.28	51.46	41.51	32.81	34.97	33.19	36.00	37.16	36.66	**Non-OECD Europe**

Primary Supply of Oil (Mtoe)
Approvisionnement primaire en pétrole (Mtep)
Primäraufkommen von Öl (Mtoe)

Disponibilità primaria di petrolio (Mtep)

石油の一次供給量（石油換算百万トン）

Suministro primario de petróleo (Mtep)

Первичная поставка нефти и нефтепродуктов (Мтнэ)

	1971	1973	1978	1983	1984	1985	1986	1987	1988
Arménie	-	-	-	-	-	-	-	-	-
Azerbaïdjan	-	-	-	-	-	-	-	-	-
Bélarus	-	-	-	-	-	-	-	-	-
Estonie	-	-	-	-	-	-	-	-	-
Géorgie	-	-	-	-	-	-	-	-	-
Kazakhstan	-	-	-	-	-	-	-	-	-
Kirghizistan	-	-	-	-	-	-	-	-	-
Lettonie	-	-	-	-	-	-	-	-	-
Lituanie	-	-	-	-	-	-	-	-	-
Moldova	-	-	-	-	-	-	-	-	-
Russie	-	-	-	-	-	-	-	-	-
Tadjikistan	-	-	-	-	-	-	-	-	-
Turkménistan	-	-	-	-	-	-	-	-	-
Ukraine	-	-	-	-	-	-	-	-	-
Ouzbékistan	-	-	-	-	-	-	-	-	-
Ex-URSS	275.65	322.00	407.46	434.11	432.36	430.71	427.66	438.74	438.16
Bahrein	0.79	0.91	0.79	0.82	1.00	0.88	0.85	1.25	0.91
Iran	13.24	16.60	28.98	33.28	39.37	40.18	42.15	43.94	43.98
Irak	3.71	4.25	7.59	13.49	14.25	15.40	15.09	16.08	16.90
Israël	6.23	8.54	7.55	6.92	6.53	6.11	7.11	7.87	9.33
Jordanie	0.53	0.65	1.34	2.58	2.80	2.82	2.99	3.17	3.12
Koweit	3.18	2.57	3.42	7.61	7.20	9.78	6.77	5.79	5.84
Liban	1.81	2.42	2.05	2.35	2.27	2.62	2.76	2.82	2.12
Oman	0.09	0.10	0.54	1.09	1.36	1.56	0.72	0.62	1.04
Qatar	0.10	0.16	0.50	0.55	0.53	0.64	0.46	0.64	0.65
Arabie saoudite	5.02	6.28	15.83	40.60	34.26	36.46	36.93	39.06	42.82
Syrie	2.71	2.50	4.84	8.05	9.15	8.54	8.58	9.51	9.38
Emirats arabes unis	0.22	0.44	2.52	8.26	7.10	8.11	7.88	7.66	7.77
Yémen	0.72	0.96	1.08	1.85	1.93	2.23	2.35	2.51	2.62
Moyen-Orient	38.33	46.40	77.02	127.46	127.76	135.36	134.65	140.92	146.48
Total non-OCDE	614.89	725.61	964.57	1 056.66	1 054.20	1 067.96	1 092.26	1 137.19	1 172.25
OCDE Amérique du N.	826.98	938.13	1 054.89	854.39	879.20	880.47	908.94	932.12	958.65
OCDE Pacifique	240.04	297.15	316.20	270.42	276.24	261.61	271.05	273.66	298.36
OCDE Europe	660.69	755.77	745.84	604.03	607.48	601.43	616.63	621.08	627.52
Total OCDE	1 727.70	1 991.05	2 116.94	1 728.85	1 762.91	1 743.52	1 796.63	1 826.86	1 884.53
Monde	2 342.59	2 716.67	3 081.51	2 785.51	2 817.11	2 811.48	2 888.89	2 964.04	3 056.78

Le total Monde ne comprend pas les soutages maritimes internationaux.

INTERNATIONAL ENERGY AGENCY

Primary Supply of Oil (Mtoe)
Approvisionnement primaire en pétrole (Mtep)
Primäraufkommen von Öl (Mtoe)
Disponibilità primaria di petrolio (Mtep)
石油の一次供給量（石油換算百万トン）
Suministro primario de petróleo (Mtep)
Первичная поставка нефти и нефтепродуктов (Мтнэ)

1989	1990	1991	1992	1993	1994	1995	1996	1997	
-	-	-	2.43	1.22	0.39	0.28	0.15	0.15	Armenia
-	-	-	6.76	7.68	8.73	7.01	6.94	6.95	Azerbaijan
-	-	-	21.40	14.33	12.07	10.75	10.50	9.01	Belarus
-	-	-	1.51	1.57	1.54	1.23	1.29	1.26	Estonia
-	-	-	1.41	0.77	0.38	0.15	0.89	0.95	Georgia
-	-	-	24.04	16.03	13.07	11.56	10.48	9.12	Kazakhstan
-	-	-	1.93	1.17	0.42	0.62	0.69	0.50	Kyrgyzstan
-	-	-	2.87	2.48	2.52	2.03	2.06	1.74	Latvia
-	-	-	4.35	3.82	3.46	3.38	3.05	3.23	Lithuania
-	-	-	2.90	2.09	1.14	1.05	0.97	0.90	Moldova
-	-	-	237.98	200.12	150.57	146.98	130.11	127.42	Russia
-	-	-	5.92	3.71	1.22	1.23	1.23	1.23	Tajikistan
-	-	-	5.05	2.85	3.57	3.54	3.32	3.66	Turkmenistan
-	-	-	41.90	29.21	23.80	25.26	20.23	18.43	Ukraine
-	-	-	8.76	9.38	8.08	6.88	6.60	6.95	Uzbekistan
434.55	408.31	405.87	368.94	296.27	230.63	221.62	198.17	191.14	**Former USSR**
0.76	0.77	0.71	0.80	0.81	0.96	0.82	2.43	3.37	Bahrain
49.06	50.85	52.94	55.97	56.88	62.74	63.08	65.91	67.63	Iran
17.80	17.20	14.05	17.15	18.71	20.56	21.77	21.64	23.31	Iraq
8.94	9.18	9.03	10.11	10.43	10.78	12.52	10.81	11.67	Israel
3.14	3.34	3.38	3.83	3.85	4.13	4.13	4.31	4.60	Jordan
8.29	7.19	1.95	5.66	5.53	10.03	11.53	9.73	11.71	Kuwait
1.72	2.15	2.71	2.71	3.38	3.72	4.27	4.36	4.85	Lebanon
1.10	1.91	3.45	3.04	2.71	2.43	2.84	2.78	2.48	Oman
1.42	0.64	0.86	0.81	0.86	0.70	0.83	1.02	1.39	Qatar
38.82	37.90	47.13	50.77	53.30	50.22	50.71	56.44	60.86	Saudi Arabia
9.02	10.07	10.90	10.76	10.68	11.03	11.29	11.70	11.71	Syria
8.65	7.70	7.93	8.47	9.09	9.95	8.57	9.82	7.15	United Arab Emirates
2.86	2.59	2.98	3.39	2.68	2.84	3.08	3.19	3.28	Yemen
151.58	151.50	158.00	173.49	178.88	190.08	195.44	204.13	214.01	**Middle East**
1 190.20	1 181.30	1 205.40	1 199.76	1 167.12	1 139.95	1 182.92	1 219.41	1 277.98	**Non-OECD Total**
968.94	931.29	914.28	933.30	949.96	974.53	968.26	995.10	1 023.55	OECD North America
315.12	339.42	344.76	366.53	372.98	394.30	404.08	418.54	421.74	OECD Pacific
630.14	632.14	647.97	653.74	652.64	654.76	667.32	684.09	685.98	OECD Europe
1 914.21	1 902.85	1 907.02	1 953.57	1 975.58	2 023.59	2 039.66	2 097.72	2 131.26	**OECD Total**
3 104.41	3 084.15	3 112.43	3 153.32	3 142.70	3 163.55	3 222.58	3 317.14	3 409.24	**World**

World Total excludes international marine bunkers.

Primary Supply of Gas (Mtoe)
Approvisionnement primaire en gaz (Mtep)
Primäraufkommen von Gas (Mtoe)

Disponibilità primaria di gas (Mtep)

ガスの一次供給量（石油換算百万トン）

Suministro primario de gas natural (Mtep)

Первичная поставка газа (Мтнэ)

	1971	1973	1978	1983	1984	1985	1986	1987	1988
Algérie	1.14	1.72	4.56	11.44	10.44	10.46	12.78	12.53	13.70
Angola	0.04	0.05	0.06	0.09	0.10	0.10	0.11	0.13	0.13
Congo	0.06	0.01	0.00	0.00	0.00	0.00	0.00	0.00	0.00
Egypte	0.07	0.07	0.57	2.33	2.97	3.69	4.30	4.78	5.17
Gabon	0.08	0.40	0.03	0.08	0.03	0.04	0.08	0.10	0.15
Libye	0.90	0.84	1.00	2.45	2.90	3.41	3.14	3.34	3.70
Maroc	0.04	0.06	0.07	0.07	0.07	0.08	0.08	0.06	0.06
Mozambique	-	-	-	-	-	-	-	-	-
Nigéria	0.17	0.35	0.78	2.60	2.51	2.96	2.67	3.00	2.97
Sénégal	-	-	-	-	-	-	-	0.00	0.01
Afrique du Sud	-	-	-	-	-	-	-	-	-
Tunisie	0.00	0.11	0.28	0.43	0.72	0.92	0.66	1.10	0.90
Autre Afrique	0.00	0.00	0.00	0.00	0.00	0.00	0.00	0.00	0.00
Afrique	2.51	3.63	7.35	19.48	19.75	21.67	23.80	25.04	26.79
Argentine	5.65	7.20	8.38	12.61	13.51	13.92	14.69	15.03	17.71
Bolivie	0.06	0.09	0.21	0.51	0.47	0.52	0.50	0.57	0.58
Brésil	0.12	0.13	0.78	1.52	1.86	2.24	2.69	2.98	3.08
Chili	0.03	0.39	0.49	0.65	0.66	0.66	0.60	0.59	0.82
Colombie	1.12	1.41	2.18	3.41	3.47	3.47	3.45	3.55	3.51
Cuba	0.00	0.01	0.01	0.01	0.00	0.01	0.00	0.02	0.02
Pérou	0.44	0.42	0.60	0.47	0.64	0.99	1.05	1.05	0.99
Trinité-et-Tobago	1.60	1.59	2.09	3.38	3.26	3.22	3.37	3.52	4.16
Uruguay	0.00	0.00	0.00	0.00	0.00	0.00	0.00	0.00	-
Vénézuela	9.88	11.94	13.17	16.25	18.44	18.30	18.72	18.64	16.98
Autre Amérique latine	0.00	0.00	0.01	0.01	0.02	0.02	0.02	0.02	0.02
Amérique latine	18.91	23.18	27.90	38.82	42.32	43.35	45.10	45.96	47.87
Bangladesh	0.31	0.50	0.73	1.70	1.97	2.19	2.49	2.95	3.29
Brunei	0.04	0.06	0.09	1.67	1.08	1.18	0.94	1.19	0.92
Inde	0.61	0.63	1.41	2.60	3.15	3.83	5.32	6.13	7.30
Indonésie	0.23	0.33	4.61	5.59	7.62	11.42	13.29	13.50	13.55
Malaisie	0.07	0.10	0.82	1.39	1.85	4.55	6.07	5.63	6.13
Myanmar	0.06	0.09	0.23	0.43	0.58	0.79	0.88	0.91	0.92
Pakistan	2.36	2.86	4.06	6.42	6.08	6.46	6.68	8.63	8.89
Philippines	-	-	-	-	-	-	-	-	-
Singapour	-	-	-	-	-	-	-	-	-
Taipei chinois	0.91	1.22	1.59	1.19	1.20	1.07	0.98	0.99	1.13
Thaïlande	-	-	-	1.23	1.86	2.87	3.88	3.88	4.60
Viêt-Nam	-	-	-	0.06	0.05	0.03	0.03	0.03	0.02
Autre Asie	0.19	0.14	0.13	0.27	0.41	0.52	0.53	0.17	0.19
Asie	4.78	5.93	13.68	22.55	25.85	34.92	41.10	44.02	46.93
Rép. populaire de Chine	3.13	5.01	11.48	10.23	10.58	10.83	11.53	11.64	11.95
Hong-Kong, Chine	-	-	-	-	-	-	-	-	-
Chine	3.13	5.01	11.48	10.23	10.58	10.83	11.53	11.64	11.95

Primary Supply of Gas (Mtoe)
Approvisionnement primaire en gaz (Mtep)
Primäraufkommen von Gas (Mtoe)
Disponibilità primaria di gas (Mtep)
ガスの一次供給量（石油換算百万トン）
Suministro primario de gas natural (Mtep)
Первичная поставка газа (Мтнэ)

1989	1990	1991	1992	1993	1994	1995	1996	1997	
13.52	14.66	15.50	15.95	15.81	15.13	15.70	15.76	16.92	Algeria
0.14	0.44	0.47	0.47	0.46	0.42	0.46	0.46	0.47	Angola
0.00	0.00	0.00	0.00	0.00	0.00	0.00	0.00	0.00	Congo
6.49	6.73	7.41	7.99	9.43	9.93	10.30	10.82	11.07	Egypt
0.07	0.07	0.07	0.07	0.07	0.06	0.07	0.07	0.07	Gabon
4.66	4.26	4.06	4.08	3.95	4.10	4.05	4.41	4.54	Libya
0.05	0.04	0.03	0.02	0.02	0.02	0.02	0.02	0.02	Morocco
-	-	-	0.00	0.00	0.00	0.00	0.00	0.00	Mozambique
3.47	3.27	3.98	4.19	4.58	4.49	4.40	4.46	4.43	Nigeria
0.01	0.00	0.00	0.00	0.01	0.02	0.04	0.04	0.02	Senegal
-	1.50	1.53	1.54	1.71	1.71	1.71	1.54	1.54	South Africa
0.94	1.23	0.86	1.22	1.18	1.66	1.92	1.95	2.09	Tunisia
0.00	0.00	0.00	0.00	0.00	0.00	0.00	0.00	0.00	Other Africa
29.34	32.21	33.92	35.52	37.22	37.55	38.67	39.52	41.17	**Africa**
18.79	17.73	18.93	21.15	22.06	23.24	24.97	27.45	28.86	Argentina
0.56	0.62	0.64	0.76	0.83	0.92	0.74	0.72	1.08	Bolivia
3.29	3.34	3.26	3.61	3.83	3.97	4.26	4.78	5.29	Brazil
1.29	1.38	1.19	1.41	1.36	1.42	1.38	1.39	2.16	Chile
3.54	3.67	3.71	3.64	3.72	3.84	4.03	4.34	5.34	Colombia
0.03	0.03	0.02	0.02	0.02	0.02	0.01	0.02	0.02	Cuba
0.84	0.80	0.75	0.68	0.71	0.75	0.72	0.73	0.77	Peru
4.16	4.34	4.60	4.72	4.68	5.72	5.79	6.41	6.83	Trinidad-and-Tobago
-	-	-	-	0.00	-	-	-	0.00	Uruguay
18.11	20.17	20.32	19.35	21.08	22.51	25.08	26.27	28.96	Venezuela
0.02	0.02	0.02	0.02	0.02	0.02	0.02	0.02	0.02	Other Latin America
50.62	52.10	53.45	55.36	58.31	62.39	67.01	72.14	79.33	**Latin America**
3.68	3.71	3.82	4.43	4.74	5.10	5.69	5.90	5.79	Bangladesh
1.06	1.33	1.54	1.58	1.44	1.40	1.60	1.58	1.62	Brunei
8.89	10.13	11.31	13.08	12.96	13.92	15.59	17.30	17.81	India
16.25	19.86	20.45	22.28	23.05	27.66	30.16	31.83	31.51	Indonesia
7.18	6.79	10.11	11.39	10.64	10.26	11.10	14.95	19.19	Malaysia
0.89	0.57	0.60	0.57	0.72	0.91	1.04	1.16	1.28	Myanmar
10.11	10.10	10.51	11.03	11.26	12.01	12.33	13.21	13.43	Pakistan
-	-	-	-	-	0.01	0.01	0.01	0.00	Philippines
-	-	-	0.43	1.17	1.39	1.45	1.29	1.28	Singapore
1.14	1.74	2.50	2.68	2.59	3.47	3.64	3.70	4.27	Chinese Taipei
4.59	5.00	6.20	6.63	7.43	8.23	8.70	10.19	12.41	Thailand
0.01	0.00	0.00	-	-	-	0.42	3.12	3.31	Vietnam
0.21	0.24	0.23	0.22	0.22	0.21	0.21	0.20	0.20	Other Asia
54.01	59.47	67.27	74.33	76.22	84.57	91.94	104.45	112.12	**Asia**
12.60	12.80	13.44	13.62	15.61	16.35	16.68	17.25	18.82	People's Rep.of China
-	-	-	-	-	-	0.02	1.48	2.31	Hong Kong, China
12.60	12.80	13.44	13.62	15.61	16.35	16.71	18.72	21.13	**China**

Primary Supply of Gas (Mtoe)
Approvisionnement primaire en gaz (Mtep)
Primäraufkommen von Gas (Mtoe)
Disponibilità primaria di gas (Mtep)
ガスの一次供給量（石油換算百万トン）
Suministro primario de gas natural (Mtep)
Первичная поставка газа (Мтнэ)

	1971	1973	1978	1983	1984	1985	1986	1987	1988
Albanie	0.11	0.16	0.30	0.32	0.32	0.32	0.32	0.32	0.34
Bulgarie	0.25	0.17	2.31	4.02	4.23	4.61	4.58	5.11	4.96
Roumanie	22.28	24.14	32.34	34.45	34.32	31.91	32.48	31.90	32.00
République slovaque	1.31	1.56	3.06	4.31	4.29	4.23	4.27	4.32	4.39
Bosnie-Herzégovine	-	-	-	-	-	-	-	-	-
Croatie	-	-	-	-	-	-	-	-	-
Ex-RYM	-	-	-	-	-	-	-	-	-
Slovénie	-	-	-	-	-	-	-	-	-
RF de Yougoslavie	-	-	-	-	-	-	-	-	-
Ex-Yougoslavie	1.00	1.33	1.76	3.92	4.37	4.95	5.15	5.67	5.77
Europe non-OCDE	24.95	27.36	39.78	47.02	47.53	46.03	46.81	47.32	47.47
Arménie	-	-	-	-	-	-	-	-	-
Azerbaïdjan	-	-	-	-	-	-	-	-	-
Bélarus	-	-	-	-	-	-	-	-	-
Estonie	-	-	-	-	-	-	-	-	-
Géorgie	-	-	-	-	-	-	-	-	-
Kazakhstan	-	-	-	-	-	-	-	-	-
Kirghizistan	-	-	-	-	-	-	-	-	-
Lettonie	-	-	-	-	-	-	-	-	-
Lituanie	-	-	-	-	-	-	-	-	-
Moldova	-	-	-	-	-	-	-	-	-
Russie	-	-	-	-	-	-	-	-	-
Tadjikistan	-	-	-	-	-	-	-	-	-
Turkménistan	-	-	-	-	-	-	-	-	-
Ukraine	-	-	-	-	-	-	-	-	-
Ouzbékistan	-	-	-	-	-	-	-	-	-
Ex-URSS	176.60	196.20	281.80	386.58	424.56	460.85	482.14	512.33	541.94
Bahrein	0.76	1.34	2.17	2.95	3.08	3.78	4.52	4.30	4.64
Iran	8.20	8.78	9.31	8.98	11.02	11.92	12.41	13.07	16.94
Irak	0.76	0.99	1.39	2.22	2.46	2.61	3.19	4.01	5.55
Israël	0.10	0.05	0.05	0.05	0.04	0.04	0.03	0.03	0.03
Jordanie	-	-	-	-	-	-	-	-	-
Koweit	2.81	3.14	4.63	3.13	3.44	3.52	5.88	6.75	6.30
Oman	-	-	0.13	0.57	0.80	0.91	1.12	1.23	1.29
Qatar	0.82	1.29	1.21	3.88	4.84	4.49	4.72	4.73	5.10
Arabie saoudite	1.43	1.94	5.07	10.66	15.75	15.75	21.12	22.47	24.38
Syrie	-	-	0.03	0.06	0.11	0.13	0.33	0.32	0.76
Emirats arabes unis	0.88	1.05	3.27	4.90	6.70	8.25	10.04	11.43	11.60
Moyen-Orient	15.76	18.58	27.25	37.40	48.25	51.40	63.35	68.34	76.58
Total non-OCDE	246.63	279.89	409.24	562.08	618.84	669.04	713.83	754.65	799.53
OCDE Amérique du N.	556.87	562.28	514.98	474.20	491.66	485.45	457.03	479.37	500.49
OCDE Pacifique	5.21	8.74	23.06	35.86	45.41	49.24	51.45	54.18	56.91
OCDE Europe	91.51	133.27	191.57	201.60	212.44	221.95	227.72	240.45	237.07
Total OCDE	653.59	704.28	729.61	711.66	749.51	756.64	736.21	774.00	794.47
Monde	900.22	984.17	1 138.85	1 273.74	1 368.34	1 425.68	1 450.04	1 528.65	1 594.00

Primary Supply of Gas (Mtoe)
Approvisionnement primaire en gaz (Mtep)
Primäraufkommen von Gas (Mtoe)
Disponibilità primaria di gas (Mtep)
ガスの一次供給量（石油換算百万トン）
Suministro primario de gas natural (Mtep)
Первичная поставка газа (Мтнэ)

1989	1990	1991	1992	1993	1994	1995	1996	1997	
0.28	0.20	0.12	0.09	0.08	0.04	0.02	0.02	0.02	Albania
5.11	4.85	4.62	4.07	3.80	3.82	4.58	4.68	3.70	Bulgaria
32.10	28.83	23.79	21.19	20.37	18.55	19.23	19.41	15.93	Romania
4.66	5.34	5.07	5.10	4.98	4.74	5.08	5.55	5.63	Slovak Republic
-	-	-	0.34	0.12	0.21	0.21	0.21	0.21	*Bosnia-Herzegovina*
-	-	-	2.11	2.22	2.09	1.93	2.17	2.25	*Croatia*
-	-	-	0.21	0.23	-	-	-	-	*FYROM*
-	-	-	0.55	0.55	0.57	0.67	0.66	0.72	*Slovenia*
-	-	-	1.65	1.49	0.67	0.88	2.25	2.23	*FR of Yugoslavia*
5.76	6.06	5.99	4.87	4.61	3.54	3.70	5.28	5.40	Former Yugoslavia
47.92	45.29	39.59	35.32	33.83	30.69	32.62	34.94	30.68	**Non-OECD Europe**
-	-	-	1.52	0.66	0.71	1.14	0.89	1.11	Armenia
-	-	-	9.87	7.55	7.28	5.81	5.13	4.83	Azerbaijan
-	-	-	14.82	14.09	12.17	11.49	12.11	13.78	Belarus
-	-	-	0.72	0.36	0.51	0.58	0.64	0.62	Estonia
-	-	-	3.96	3.01	2.00	0.74	0.63	0.77	Georgia
-	-	-	14.89	10.55	8.12	10.11	7.84	6.20	Kazakhstan
-	-	-	1.53	1.14	0.73	0.74	0.88	0.51	Kyrgyzstan
-	-	-	1.72	1.14	0.83	1.01	0.87	1.06	Latvia
-	-	-	2.88	1.56	1.73	2.03	2.17	2.00	Lithuania
-	-	-	2.93	2.69	2.58	2.56	2.93	3.13	Moldova
-	-	-	368.46	352.49	316.88	311.50	312.92	309.72	Russia
-	-	-	1.46	1.16	0.65	0.73	0.94	0.85	Tajikistan
-	-	-	5.51	7.83	10.63	11.20	9.06	8.75	Turkmenistan
-	-	-	86.12	80.41	71.50	68.46	74.47	66.31	Ukraine
-	-	-	33.57	35.63	34.14	32.73	33.93	34.04	Uzbekistan
548.90	559.44	584.84	549.96	520.26	470.46	460.84	465.41	453.68	**Former USSR**
4.60	4.75	4.54	4.56	4.86	4.76	5.05	5.27	5.11	Bahrain
18.13	19.24	24.78	29.69	30.80	33.52	32.56	34.10	38.32	Iran
6.40	3.40	1.40	2.34	2.62	3.26	3.24	3.31	3.70	Iraq
0.03	0.03	0.02	0.02	0.02	0.02	0.02	0.01	0.01	Israel
0.05	0.10	0.10	0.11	0.13	0.18	0.19	0.18	0.19	Jordan
7.70	5.93	1.61	3.01	3.75	4.13	4.13	4.43	4.45	Kuwait
1.33	2.44	2.50	3.06	3.71	3.87	3.43	3.59	4.29	Oman
4.83	5.10	6.18	10.22	10.94	10.94	10.94	11.10	12.19	Qatar
24.93	25.36	26.82	27.77	29.32	30.78	32.94	35.80	37.58	Saudi Arabia
1.24	1.37	1.60	1.65	1.64	1.77	2.24	2.24	2.07	Syria
14.13	13.80	16.58	15.35	16.01	17.32	18.86	22.49	23.71	United Arab Emirates
83.37	81.52	86.12	97.77	103.79	110.55	113.60	122.52	131.62	**Middle East**
826.75	842.83	878.64	861.89	845.24	812.56	821.39	857.69	869.74	**Non-OECD Total**
521.06	516.78	538.74	551.49	567.74	581.74	600.95	602.38	607.54	OECD North America
60.36	64.65	67.71	70.13	72.22	77.67	80.90	88.17	89.89	OECD Pacific
246.68	252.87	269.30	268.69	284.29	286.55	311.02	345.42	344.01	OECD Europe
828.10	834.29	875.75	890.31	924.24	945.96	992.88	1 035.97	1 041.43	**OECD Total**
1 654.85	1 677.13	1 754.39	1 752.20	1 769.48	1 758.52	1 814.27	1 893.66	1 911.17	**World**

Primary Supply of Combustible Renewables and Waste (Mtoe)
Approvisionnement primaire en énergies renouv. combustibles et en déchets (Mtep)

Primäraufkommen an erneuerbaren Brennstoffen und Abfällen (Mtoe)

Totale disponibilita' primaria da Fonti rinnovabili e combustibili derivati dai rifiuti (Mtep)

可燃性再生可能エネルギー及び廃棄物の一次供給量（石油換算百万トン）

Suministro primario de combustibles renovables y desechos (Mtep)

Первичная поставка возобновляемых видов топлива и отходов (Мтнэ)

	1971	1973	1978	1983	1984	1985	1986	1987	1988
Algérie	0.26	0.28	0.31	0.37	0.38	0.39	0.40	0.41	0.42
Angola	3.19	3.23	3.47	3.79	3.85	3.92	3.98	4.06	4.13
Bénin	1.00	1.04	1.16	1.30	1.34	1.37	1.41	1.45	1.49
Cameroun	2.36	2.46	2.81	3.20	3.28	3.36	3.44	3.53	3.62
Congo	0.48	0.51	0.56	0.62	0.63	0.65	0.66	0.68	0.70
RD du Congo	5.57	5.88	6.79	7.93	8.20	8.47	8.76	9.05	9.36
Egypte	0.65	0.68	0.73	0.83	0.87	0.94	0.98	1.01	1.04
Ethiopie	8.45	8.86	10.01	11.38	11.69	12.00	12.34	12.72	13.13
Gabon	0.49	0.51	0.57	0.63	0.64	0.66	0.67	0.68	0.70
Ghana	2.09	2.29	2.71	3.07	3.18	3.30	3.42	3.55	3.68
Côte d'Ivoire	1.63	1.74	2.08	2.48	2.57	2.66	2.75	2.85	2.95
Kenya	6.34	6.61	7.42	8.43	8.64	8.85	9.05	9.24	9.42
Libye	0.10	0.11	0.13	0.13	0.13	0.13	0.13	0.13	0.13
Maroc	0.13	0.13	0.21	0.29	0.29	0.30	0.30	0.31	0.31
Mozambique	7.28	7.19	7.20	7.24	7.22	7.18	7.10	7.02	6.93
Nigéria	33.96	35.67	40.64	46.83	48.19	49.55	50.91	52.30	53.70
Sénégal	0.83	0.87	0.99	1.13	1.15	1.18	1.22	1.25	1.28
Afrique du Sud	4.70	5.13	5.85	7.38	8.17	8.29	8.43	9.03	9.53
Soudan	5.65	5.87	6.67	7.76	7.79	7.97	8.13	8.28	8.42
Tanzanie	8.93	8.98	9.25	10.05	10.27	10.49	10.71	10.94	11.17
Tunisie	0.67	0.69	0.77	0.88	0.90	0.93	0.96	0.98	1.00
Zambie	2.47	2.57	2.87	3.42	3.54	3.65	3.72	3.79	3.86
Zimbabwe	3.05	3.17	3.50	3.97	4.08	4.19	4.30	4.41	4.52
Autre Afrique	24.26	25.13	27.97	31.41	32.15	32.92	33.72	34.56	35.42
Afrique	124.52	129.62	144.65	164.51	169.14	173.33	177.49	182.19	186.90
Argentine	2.26	2.10	2.09	2.06	2.08	2.03	1.77	1.90	1.82
Bolivie	0.19	0.22	0.29	0.91	0.90	0.92	0.93	0.66	0.73
Brésil	35.30	35.80	35.02	39.16	42.97	43.12	44.41	45.31	44.90
Chili	1.40	1.32	1.66	2.05	2.19	2.23	2.32	2.42	2.51
Colombie	4.42	3.40	4.70	4.93	4.94	5.17	5.30	5.26	5.43
Costa Rica	0.57	0.59	0.59	0.59	0.73	0.77	0.74	0.75	0.75
Cuba	3.27	3.02	3.86	3.89	4.01	3.73	3.90	3.98	4.29
République dominicaine	1.17	1.16	1.29	1.68	1.82	1.25	1.26	1.28	1.03
Equateur	1.11	1.05	0.98	1.05	1.10	1.06	1.01	1.03	1.11
El Salvador	1.21	1.29	1.52	1.51	1.54	1.56	1.24	1.34	1.20
Guatemala	1.91	2.02	2.64	2.42	2.47	2.53	2.59	2.68	2.74
Haiti	1.38	1.46	1.72	1.59	1.60	1.61	1.20	1.22	1.24
Honduras	0.99	1.03	1.15	1.32	1.33	1.36	1.36	1.41	1.44
Jamaïque	0.27	0.24	0.24	0.22	0.26	0.24	0.24	0.25	0.25
Nicaragua	0.72	0.76	0.86	0.90	0.96	1.02	0.97	1.00	1.06
Panama	0.34	0.34	0.42	0.46	0.46	0.44	0.42	0.41	0.42
Paraguay	1.16	1.25	1.40	1.67	1.65	1.69	1.76	2.06	2.15
Pérou	3.57	3.63	3.59	3.69	3.73	3.81	3.83	3.85	3.76
Trinité-et-Tobago	0.02	0.02	0.02	0.01	0.01	0.03	0.04	0.05	0.04
Uruguay	0.39	0.40	0.45	0.52	0.54	0.58	0.64	0.63	0.58
Vénézuela	0.41	0.40	0.37	0.37	0.38	0.45	0.51	0.54	0.48
Autre Amérique latine	0.47	0.48	0.58	0.69	0.69	0.68	0.69	0.62	0.57
Amérique latine	62.53	61.99	65.46	71.70	76.38	76.27	77.15	78.65	78.50

Primary Supply of Combustible Renewables and Waste (Mtoe)
Approvisionnement primaire en énergies renouv. combustibles et en déchets (Mtep)
Primäraufkommen an erneuerbaren Brennstoffen und Abfällen (Mtoe)
Totale disponibilita' primaria da Fonti rinnovabili e combustibili derivati dai rifiuti (Mtep)
可燃性再生可能エネルギー及び廃棄物の一次供給量（石油換算百万トン）
Suministro primario de combustibles renovables y desechos (Mtep)
Первичная поставка возобновляемых видов топлива и отходов (Мтнэ)

1989	1990	1991	1992	1993	1994	1995	1996	1997	
0.43	0.44	0.45	0.46	0.47	0.49	0.50	0.51	0.52	Algeria
4.22	4.32	4.42	4.53	4.63	4.73	4.85	5.03	5.18	Angola
1.52	1.56	1.60	1.64	1.68	1.72	1.76	1.79	1.83	Benin
3.72	3.82	3.92	4.03	4.14	4.25	4.37	4.50	4.63	Cameroon
0.71	0.73	0.75	0.77	0.79	0.81	0.82	0.84	0.86	Congo
9.68	10.00	10.33	10.66	11.01	11.37	11.73	12.10	12.49	DR of Congo
1.01	1.06	1.12	1.14	1.17	1.24	1.19	1.21	1.24	Egypt
13.58	14.07	14.53	15.01	14.57	14.97	15.38	15.76	16.13	Ethiopia
0.71	0.74	0.77	0.79	0.81	0.81	0.84	0.85	0.86	Gabon
3.82	3.90	4.11	4.27	4.43	4.60	4.73	4.84	4.95	Ghana
3.06	3.18	3.29	3.40	3.50	3.61	3.70	3.82	3.93	Ivory Coast
9.60	9.78	9.95	10.12	10.30	10.46	10.63	10.81	11.01	Kenya
0.13	0.13	0.13	0.13	0.13	0.13	0.13	0.13	0.13	Libya
0.32	0.32	0.38	0.38	0.39	0.39	0.40	0.41	0.41	Morocco
6.86	6.80	6.79	6.82	6.90	7.02	6.93	6.93	6.93	Mozambique
55.12	56.58	58.08	59.64	61.27	62.95	64.71	66.72	68.79	Nigeria
1.32	1.36	1.39	1.43	1.46	1.49	1.53	1.58	1.63	Senegal
10.20	10.41	10.63	10.79	10.98	11.20	11.44	11.63	11.82	South Africa
8.56	8.69	8.81	8.94	9.06	9.17	9.29	9.41	9.53	Sudan
11.40	11.65	11.89	12.14	12.40	12.66	12.92	13.15	13.40	Tanzania
1.01	1.04	1.06	1.08	1.10	1.12	1.14	1.15	1.17	Tunisia
3.94	4.02	4.12	4.23	4.35	4.47	4.58	4.66	4.77	Zambia
4.62	4.73	4.82	4.91	5.00	5.09	5.16	5.22	5.29	Zimbabwe
36.32	37.25	37.97	38.96	39.53	40.60	41.67	42.75	43.87	Other Africa
191.86	196.56	201.32	206.28	210.07	215.33	220.37	225.81	231.36	**Africa**
1.56	1.72	1.80	1.86	1.97	2.31	2.66	2.69	2.65	Argentina
0.70	0.75	0.79	0.77	0.78	0.79	0.82	0.77	0.88	Bolivia
45.00	40.48	39.21	38.99	39.18	41.08	40.49	39.63	40.45	Brazil
2.53	2.68	2.99	3.32	3.08	3.23	3.46	3.71	3.69	Chile
5.53	5.82	5.69	5.90	6.04	6.27	6.98	6.89	5.27	Colombia
0.71	0.74	0.74	0.69	0.35	0.36	0.37	0.37	0.27	Costa Rica
4.89	5.43	5.41	5.42	5.43	5.45	5.46	5.47	5.54	Cuba
0.99	1.02	1.02	1.30	1.30	1.30	1.30	1.30	1.31	Dominican Republic
1.08	1.07	1.10	1.12	1.19	1.18	1.22	1.22	1.14	Ecuador
1.17	1.22	1.30	1.33	1.38	1.41	1.83	1.86	1.93	El Salvador
2.86	2.92	2.98	3.02	2.94	2.89	2.93	2.98	3.03	Guatemala
1.25	1.21	1.24	1.32	1.32	1.35	1.37	1.57	1.28	Haiti
1.46	1.50	1.50	1.51	1.51	1.51	1.52	1.52	1.72	Honduras
0.30	0.45	0.36	0.37	0.41	0.53	0.49	0.54	0.59	Jamaica
1.06	1.13	1.12	1.10	1.12	1.14	1.19	1.22	1.31	Nicaragua
0.41	0.40	0.41	0.41	0.47	0.49	0.52	0.55	0.57	Panama
2.22	2.23	2.34	2.19	2.13	2.24	2.40	2.55	2.62	Paraguay
3.80	3.95	3.99	4.05	4.06	4.12	4.17	4.22	4.30	Peru
0.04	0.05	0.04	0.04	0.04	0.04	0.03	0.03	0.03	Trinidad-and-Tobago
0.59	0.57	0.59	0.60	0.59	0.56	0.53	0.53	0.53	Uruguay
0.51	0.53	0.55	0.56	0.55	0.56	0.54	0.54	0.54	Venezuela
0.56	0.58	0.58	0.58	0.58	0.59	0.63	0.63	0.62	Other Latin America
79.20	76.45	75.75	76.44	76.40	79.43	80.91	80.79	80.25	**Latin America**

Primary Supply of Combustible Renewables and Waste (Mtoe)
Approvisionnement primaire en énergies renouv. combustibles et en déchets (Mtep)
Primäraufkommen an erneuerbaren Brennstoffen und Abfällen (Mtoe)
Totale disponibilita' primaria da Fonti rinnovabili e combustibili derivati dai rifiuti (Mtep)
可燃性再生可能エネルギー及び廃棄物の一次供給量（石油換算百万トン）
Suministro primario de combustibles renovables y desechos (Mtep)
Первичная поставка возобновляемых видов топлива и отходов (Мтнэ)

	1971	1973	1978	1983	1984	1985	1986	1987	1988
Bangladesh	9.55	10.26	11.56	12.82	13.11	13.41	13.78	14.10	14.42
Brunei	0.02	0.02	0.02	0.02	0.02	0.02	0.02	0.02	0.02
Inde	120.74	126.34	141.58	156.49	159.57	162.33	165.45	168.58	170.82
Indonésie	27.48	28.85	32.28	35.86	36.60	37.30	38.03	38.73	39.41
RPD de Corée	0.68	0.72	0.83	0.90	0.91	0.92	0.92	0.93	0.93
Malaisie	1.24	1.34	1.56	1.75	1.81	1.83	1.87	1.96	2.01
Myanmar	6.36	6.68	7.33	7.99	8.13	8.29	8.49	8.71	9.00
Népal	2.50	2.61	2.95	4.81	4.90	4.97	5.09	5.20	5.33
Pakistan	10.63	11.33	13.21	15.28	15.74	16.19	16.68	17.18	17.68
Philippines	6.05	6.34	7.61	8.66	8.89	9.25	9.37	9.42	9.46
Singapour	0.01	0.00	0.01	0.01	0.01	-	-	-	-
Sri Lanka	2.74	2.79	3.02	3.29	3.53	3.55	3.67	3.77	3.87
Taipei chinois	0.08	0.10	0.02	0.02	0.02	0.01	0.01	0.01	0.01
Thaïlande	7.60	7.91	10.42	8.68	9.48	10.72	11.57	12.20	12.77
Viêt-Nam	12.48	13.08	14.68	16.17	16.46	16.80	17.19	17.62	18.05
Autre Asie	1.89	1.99	2.30	2.40	2.37	2.36	2.42	2.44	2.48
Asie	210.05	220.36	249.38	275.12	281.52	287.97	294.55	300.87	306.27
Rép. populaire de Chine	154.24	161.72	175.34	185.96	187.13	188.53	190.65	192.77	194.94
Hong-Kong, Chine	0.05	0.05	0.05	0.06	0.06	0.06	0.06	0.06	0.06
Chine	154.29	161.77	175.38	186.02	187.18	188.58	190.71	192.82	195.00
Albanie	0.38	0.38	0.38	0.38	0.38	0.38	0.38	0.38	0.38
Bulgarie	0.27	0.24	0.21	0.40	0.40	0.41	0.41	0.21	0.19
Chypre	0.01	0.01	0.01	0.01	0.01	0.01	0.01	0.01	0.01
Roumanie	1.37	1.36	1.20	1.06	0.97	1.19	1.08	0.95	0.89
République slovaque	0.20	0.18	0.19	0.12	0.12	0.13	0.13	0.14	0.14
Bosnie-Herzégovine	-	-	-	-	-	-	-	-	-
Croatie	-	-	-	-	-	-	-	-	-
Ex-RYM	-	-	-	-	-	-	-	-	-
Slovénie	-	-	-	-	-	-	-	-	-
RF de Yougoslavie	-	-	-	-	-	-	-	-	-
Ex-Yougoslavie	1.63	0.82	0.79	0.91	0.99	0.98	0.99	0.86	0.89
Europe non-OCDE	3.85	2.99	2.77	2.88	2.86	3.09	2.99	2.54	2.49
Arménie	-	-	-	-	-	-	-	-	-
Azerbaïdjan	-	-	-	-	-	-	-	-	-
Bélarus	-	-	-	-	-	-	-	-	-
Estonie	-	-	-	-	-	-	-	-	-
Géorgie	-	-	-	-	-	-	-	-	-
Kazakhstan	-	-	-	-	-	-	-	-	-
Kirghizistan	-	-	-	-	-	-	-	-	-
Lettonie	-	-	-	-	-	-	-	-	-
Lituanie	-	-	-	-	-	-	-	-	-
Moldova	-	-	-	-	-	-	-	-	-
Russie	-	-	-	-	-	-	-	-	-
Ukraine	-	-	-	-	-	-	-	-	-
Ouzbékistan	-	-	-	-	-	-	-	-	-
Ex-URSS	20.15	19.48	18.15	18.73	19.87	20.24	18.82	20.13	18.91

Primary Supply of Combustible Renewables and Waste (Mtoe)
Approvisionnement primaire en énergies renouv. combustibles et en déchets (Mtep)
Primäraufkommen an erneuerbaren Brennstoffen und Abfällen (Mtoe)
Totale disponibilita' primaria da Fonti rinnovabili e combustibili derivati dai rifiuti (Mtep)
可燃性再生可能エネルギー及び廃棄物の一次供給量（石油換算百万トン）
Suministro primario de combustibles renovables y desechos (Mtep)
Первичная поставка возобновляемых видов топлива и отходов (Мтнэ)

1989	1990	1991	1992	1993	1994	1995	1996	1997	
14.72	14.88	15.11	15.29	15.45	15.62	15.76	15.90	16.03	Bangladesh
0.02	0.02	0.02	0.02	0.02	0.02	0.02	0.02	0.02	Brunei
173.22	175.82	179.74	182.08	184.77	186.74	188.65	190.47	192.80	India
40.08	40.73	41.37	42.01	42.64	43.28	43.96	44.49	44.49	Indonesia
0.94	0.95	0.96	0.98	0.99	1.00	1.01	1.02	1.02	DPR of Korea
2.07	2.12	2.14	2.19	2.25	2.29	2.34	2.39	2.39	Malaysia
9.16	9.31	9.50	9.60	9.72	9.84	9.96	10.11	10.27	Myanmar
5.42	5.57	5.74	5.88	6.11	6.26	6.28	6.46	6.46	Nepal
18.21	18.77	19.31	19.80	20.40	20.97	21.52	21.95	22.40	Pakistan
9.61	9.84	9.93	10.14	10.08	10.29	10.27	9.31	9.40	Philippines
-	-	-	-	-	-	-	-	-	Singapore
3.93	3.92	3.92	4.07	3.93	3.66	3.64	3.92	4.05	Sri Lanka
0.01	0.01	0.01	0.01	0.01	0.01	0.01	0.01	0.01	Chinese Taipei
14.28	14.78	15.28	16.28	17.73	19.37	20.24	21.38	20.66	Thailand
18.48	18.90	19.32	19.73	20.15	20.56	20.97	21.95	22.40	Vietnam
2.43	2.47	2.52	2.58	2.65	2.74	2.83	2.92	2.92	Other Asia
312.59	318.09	324.88	330.65	336.89	342.65	347.45	352.34	355.33	**Asia**
197.71	200.41	202.35	203.49	204.54	205.49	206.08	207.15	208.26	People's Rep.of China
0.06	0.05	0.06	0.05	0.06	0.06	0.06	0.06	0.06	Hong Kong, China
197.76	200.46	202.40	203.55	204.60	205.54	206.14	207.21	208.32	**China**
0.38	0.36	0.36	0.36	0.13	0.06	0.06	0.06	0.06	Albania
0.36	0.36	0.14	0.16	0.15	0.17	0.21	0.24	0.24	Bulgaria
0.01	0.01	0.01	0.02	0.02	0.02	0.02	0.02	0.02	Cyprus
0.65	0.60	0.71	1.05	1.39	1.18	1.72	4.91	3.37	Romania
0.13	0.17	0.14	0.12	0.12	0.17	0.08	0.08	0.08	Slovak Republic
-	-	-	0.16	0.16	0.16	0.16	0.16	0.16	*Bosnia-Herzegovina*
-	-	-	0.26	0.24	0.26	0.27	0.33	0.32	*Croatia*
-	-	-	0.19	0.20	0.19	0.19	0.19	0.19	*FYROM*
-	-	-	0.27	0.26	0.26	0.26	0.26	0.26	*Slovenia*
-	-	-	0.21	0.21	0.21	0.21	0.21	0.21	*FR of Yugoslavia*
0.92	0.72	0.72	1.09	1.08	1.09	1.08	1.15	1.14	Former Yugoslavia
2.44	2.21	2.08	2.80	2.88	2.68	3.17	6.45	4.92	**Non-OECD Europe**
-	-	-	0.00	0.00	0.00	0.00	0.00	0.00	Armenia
-	-	-	0.00	0.00	0.00	0.00	0.00	0.00	Azerbaijan
-	-	-	0.60	0.57	0.55	0.53	0.52	0.69	Belarus
-	-	-	0.18	0.18	0.29	0.33	0.58	0.59	Estonia
-	-	-	0.05	0.05	0.03	0.03	0.03	0.03	Georgia
-	-	-	0.11	0.08	0.08	0.08	0.07	0.07	Kazakhstan
-	-	-	0.00	0.00	0.00	0.00	0.00	0.00	Kyrgyzstan
-	-	-	0.48	0.49	0.52	0.28	0.59	1.08	Latvia
-	-	-	0.19	0.25	0.24	0.24	0.27	0.52	Lithuania
-	-	-	0.04	0.03	0.03	0.03	0.06	0.06	Moldova
-	-	-	17.02	17.02	17.02	17.02	17.02	17.02	Russia
-	-	-	0.30	0.27	0.27	0.26	0.25	0.25	Ukraine
-	-	-	0.00	0.00	0.00	0.00	0.00	0.00	Uzbekistan
17.39	19.10	19.10	18.98	18.96	19.05	18.80	19.41	20.32	**Former USSR**

Primary Supply of Combustible Renewables and Waste (Mtoe)
Approvisionnement primaire en énergies renouv. combustibles et en déchets (Mtep)
Primäraufkommen an erneuerbaren Brennstoffen und Abfällen (Mtoe)
Totale disponibilita' primaria da Fonti rinnovabili e combustibili derivati dai rifiuti (Mtep)
可燃性再生可能エネルギー及び廃棄物の一次供給量（石油換算百万トン）
Suministro primario de combustibles renovables y desechos (Mtep)
Первичная поставка возобновляемых видов топлива и отходов (Мтнэ)

	1971	1973	1978	1983	1984	1985	1986	1987	1988
Iran	0.33	0.35	0.58	0.63	0.65	0.68	0.69	0.65	0.71
Irak	0.03	0.03	0.03	0.02	0.02	0.02	0.02	0.02	0.02
Israël	0.00	0.00	0.00	0.00	0.00	0.00	0.00	0.00	0.00
Jordanie	0.00	0.00	0.00	0.00	0.00	0.00	0.00	0.00	0.00
Koweit	0.01	0.02	0.00	0.00	0.01	0.01	0.01	0.01	0.01
Liban	0.10	0.10	0.11	0.11	0.11	0.11	0.11	0.11	0.11
Qatar	0.00	0.00	0.00	0.00	0.00	0.00	0.00	0.00	0.00
Arabie saoudite	0.00	0.00	0.00	0.01	0.01	0.00	0.01	0.01	0.01
Syrie	0.00	0.00	0.00	0.00	0.00	0.00	0.00	0.00	0.00
Emirats arabes unis	-	-	-	-	-	-	-	-	-
Yémen	0.05	0.05	0.06	0.06	0.07	0.07	0.07	0.07	0.07
Moyen-Orient	0.52	0.56	0.79	0.84	0.86	0.90	0.91	0.88	0.94
Total non-OCDE	575.91	596.77	656.57	719.81	737.83	750.38	762.62	778.09	789.02
OCDE Amérique du N.	50.86	53.51	65.43	74.37	81.67	80.94	83.03	89.83	88.88
OCDE Pacifique	3.56	3.53	4.18	9.26	9.69	10.02	9.76	10.08	10.63
OCDE Europe	21.52	22.75	26.87	33.68	35.30	35.85	36.36	36.51	38.43
Total OCDE	75.93	79.79	96.47	117.31	126.66	126.81	129.16	136.42	137.94
Monde	651.84	676.56	753.04	837.11	864.49	877.20	891.78	914.51	926.96

Avant 1978 les données pour les pays de l' OCDE sont incomplètes.

Primary Supply of Combustible Renewables and Waste (Mtoe)
Approvisionnement primaire en énergies renouv. combustibles et en déchets (Mtep)
Primäraufkommen an erneuerbaren Brennstoffen und Abfällen (Mtoe)
Totale disponibilita' primaria da Fonti rinnovabili e combustibili derivati dai rifiuti (Mtep)
可燃性再生可能エネルギー及び廃棄物の一次供給量（石油換算百万トン）
Suministro primario de combustibles renovables y desechos (Mtep)
Первичная поставка возобновляемых видов топлива и отходов (Мтнэ)

1989	1990	1991	1992	1993	1994	1995	1996	1997	
0.66	0.68	0.69	0.72	0.73	0.73	0.72	0.79	0.79	Iran
0.02	0.02	0.02	0.02	0.02	0.03	0.03	0.03	0.03	Iraq
0.00	0.00	0.00	0.00	0.00	0.00	0.01	0.01	0.01	Israel
0.00	0.00	0.00	0.00	0.00	0.00	0.00	0.00	0.00	Jordan
0.01	0.01	0.01	0.01	0.00	0.00	0.01	0.00	0.00	Kuwait
0.11	0.10	0.11	0.11	0.11	0.11	0.12	0.12	0.12	Lebanon
0.00	0.00	0.00	0.00	0.00	0.00	0.00	0.00	0.00	Qatar
0.01	0.01	0.01	0.01	0.01	0.01	0.01	0.00	0.00	Saudi Arabia
0.00	0.00	0.00	0.01	0.00	0.00	0.00	0.00	0.00	Syria
-	-	-	0.03	0.03	0.02	0.02	0.02	0.02	United Arab Emirates
0.08	0.08	0.08	0.08	0.08	0.08	0.08	0.08	0.08	Yemen
0.90	0.92	0.93	0.99	0.99	1.00	1.00	1.05	1.05	**Middle East**
802.14	813.80	826.47	839.69	850.78	865.68	877.83	893.06	901.54	**Non-OECD Total**
85.12	78.18	80.39	89.70	83.48	85.48	88.87	89.57	86.30	OECD North America
11.13	11.33	11.45	11.00	11.60	12.27	12.95	13.87	15.02	OECD Pacific
47.89	46.94	48.03	48.26	51.92	53.19	54.72	56.53	58.91	OECD Europe
144.15	136.45	139.87	148.96	147.00	150.94	156.54	159.98	160.23	**OECD Total**
946.29	950.25	966.34	988.65	997.78	1 016.62	1 034.37	1 053.04	1 061.78	**World**

Prior to 1978 data for OECD countries are incomplete.

Total Primary Energy Supply (TPES) (Mtoe)
Approvisionnements totaux en énergie primaire (ATEP) (Mtep)
Gesamtaufkommen von Primärenergie (TPES) (Mtoe)
Disponibilità totale di energia primaria (DTEP) (Mtep)
一次エネルギー総供給量（石油換算百万トン）
Suministro total de energía primaria (TPES) (Mtep)
Общая первичная поставка топлива и энергии (ОППТЭ) (Мтнэ)

	1971	1973	1978	1983	1984	1985	1986	1987	1988
Algérie	3.91	5.37	9.62	18.90	18.63	19.38	21.85	21.52	22.73
Angola	3.93	4.21	4.48	4.71	4.65	4.92	4.93	5.02	5.13
Bénin	1.11	1.19	1.31	1.46	1.49	1.56	1.58	1.60	1.64
Cameroun	2.76	2.88	3.47	4.28	4.44	4.52	4.57	4.65	4.80
Congo	0.71	0.69	0.79	1.01	0.98	1.00	0.92	1.41	1.80
RD du Congo	6.77	7.14	8.13	9.61	9.86	10.08	10.27	10.69	11.11
Egypte	7.83	8.08	12.97	21.47	23.79	25.45	26.26	28.17	29.12
Ethiopie	9.00	9.40	10.54	12.09	12.39	12.72	13.16	13.71	14.14
Gabon	0.92	1.45	1.66	1.78	1.67	1.50	1.54	1.48	1.59
Ghana	3.03	3.38	3.90	3.61	3.93	4.33	4.48	4.80	4.84
Côte d'Ivoire	2.50	2.75	3.56	3.56	3.65	3.81	4.05	4.03	4.53
Kenya	7.70	8.14	9.29	10.11	10.54	10.86	11.58	11.89	12.06
Libye	1.66	2.72	4.87	8.94	9.51	10.50	9.67	11.31	10.80
Maroc	2.42	2.95	4.38	5.15	5.31	5.42	5.61	5.75	6.02
Mozambique	8.28	8.16	7.86	7.83	7.74	7.71	7.64	7.56	7.47
Nigéria	36.15	39.04	48.53	59.66	59.92	61.88	62.47	64.95	67.10
Sénégal	1.34	1.47	1.70	1.91	2.00	1.97	2.07	2.18	2.14
Afrique du Sud	45.33	49.06	59.87	79.82	86.62	86.74	90.56	93.37	97.49
Soudan	6.94	7.39	7.86	9.14	9.12	9.50	9.63	9.50	10.08
Tanzanie	9.69	9.82	10.06	10.81	11.03	11.26	11.52	11.74	11.98
Tunisie	2.04	2.30	3.31	4.36	4.50	4.67	4.61	4.74	4.99
Zambie	3.91	4.29	4.53	5.06	5.04	5.17	5.19	5.31	5.43
Zimbabwe	5.39	5.83	6.05	6.76	6.94	7.43	7.83	8.51	8.57
Autre Afrique	26.78	27.90	31.77	36.57	37.16	37.94	39.33	40.99	42.07
Afrique	200.10	215.60	260.51	328.60	340.90	350.34	361.33	374.88	387.63
Argentine	33.65	35.60	38.91	41.77	43.33	41.35	43.14	45.51	46.56
Bolivie	0.99	1.16	1.74	2.94	2.77	2.76	2.76	2.54	2.66
Brésil	70.42	82.64	102.61	107.21	114.32	120.81	127.48	131.58	138.05
Chili	8.24	8.53	8.71	9.14	9.64	9.59	10.09	10.41	11.72
Colombie	14.10	14.23	18.19	21.37	21.39	22.14	23.00	23.58	23.62
Costa Rica	1.13	1.26	1.55	1.39	1.59	1.67	1.70	1.80	1.81
Cuba	10.62	11.56	14.42	15.81	14.39	14.22	14.34	15.09	15.65
République dominicaine	2.37	2.87	3.36	4.07	4.35	3.67	3.86	4.36	4.12
Equateur	2.35	2.50	4.59	5.40	5.82	5.90	5.88	5.93	6.29
El Salvador	1.76	2.00	2.87	2.74	2.80	2.85	2.48	2.75	2.61
Guatemala	2.80	3.03	3.94	3.54	3.76	3.83	3.64	3.83	3.99
Haiti	1.51	1.60	1.96	1.86	1.88	1.90	1.48	1.53	1.59
Honduras	1.39	1.49	1.76	1.99	2.01	2.03	2.02	2.21	2.33
Jamaïque	2.15	3.07	2.56	2.12	2.11	1.86	1.88	1.94	1.97
Antilles néerlandaises	5.52	6.03	3.76	4.41	3.06	1.83	1.77	1.54	1.54
Nicaragua	1.29	1.39	1.73	1.64	1.90	1.98	2.04	2.04	1.99
Panama	1.73	2.14	1.72	1.75	1.57	1.52	1.55	1.61	1.41
Paraguay	1.38	1.54	1.90	2.22	2.25	2.33	2.44	2.81	2.95
Pérou	9.49	9.90	11.03	11.01	11.41	11.53	12.09	12.62	12.55
Trinité-et-Tobago	2.70	2.68	3.54	4.28	4.55	4.74	5.05	4.63	5.16
Uruguay	2.41	2.38	2.54	2.12	2.03	2.02	2.09	2.21	2.37
Vénézuela	23.38	25.35	30.12	38.38	36.84	37.30	38.77	43.59	40.52
Autre Amérique latine	5.09	5.96	5.86	3.97	4.22	4.05	4.26	4.58	4.58
Amérique latine	206.48	228.92	269.36	291.14	297.99	301.87	313.83	328.69	336.03

Total Primary Energy Supply (TPES) (Mtoe)
Approvisionnements totaux en énergie primaire (ATEP) (Mtep)
Gesamtaufkommen von Primärenergie (TPES) (Mtoe)
Disponibilità totale di energia primaria (DTEP) (Mtep)
一次エネルギー総供給量（石油換算百万トン）
Suministro total de energía primaria (TPES) (Mtep)
Общая первичная поставка топлива и энергии (ОПППТЭ) (Мтнэ)

1989	1990	1991	1992	1993	1994	1995	1996	1997	
22.56	23.96	24.89	25.00	25.05	24.13	24.30	24.16	26.50	Algeria
5.22	5.62	5.71	5.80	5.90	5.97	6.12	6.52	6.85	Angola
1.65	1.68	1.70	1.74	1.79	1.83	1.87	2.15	2.18	Benin
4.99	5.06	5.07	5.12	5.23	5.36	5.48	5.59	5.76	Cameroon
1.19	1.12	1.16	1.18	1.17	1.22	1.20	1.22	1.24	Congo
11.52	11.86	12.18	12.49	12.93	13.27	13.78	14.17	14.54	DR of Congo
30.54	31.89	32.47	33.29	35.56	33.57	35.32	37.35	39.58	Egypt
14.64	15.21	15.67	16.11	15.71	16.13	16.67	17.17	17.13	Ethiopia
1.38	1.27	1.34	1.42	1.48	1.46	1.50	1.53	1.63	Gabon
5.14	5.23	5.40	5.54	5.89	6.44	6.70	6.77	6.90	Ghana
4.71	4.60	4.62	5.07	4.89	4.96	5.13	5.42	5.60	Ivory Coast
12.36	12.48	12.54	12.80	13.07	13.33	13.18	13.46	14.14	Kenya
11.11	11.75	13.80	12.94	13.79	13.33	15.91	14.91	15.09	Libya
6.56	6.74	7.05	7.54	7.90	8.53	8.28	9.15	9.27	Morocco
7.39	7.32	7.29	7.37	7.48	7.58	7.49	7.52	7.66	Mozambique
69.43	70.91	74.24	76.87	72.01	84.54	83.18	84.68	88.65	Nigeria
2.16	2.21	2.22	2.38	2.35	2.46	2.50	2.65	2.77	Senegal
93.20	91.23	95.39	90.23	94.70	99.88	104.20	103.59	107.22	South Africa
10.09	10.63	10.59	10.62	10.33	10.93	10.96	11.06	11.48	Sudan
12.25	12.53	12.78	12.97	13.25	13.50	13.78	14.03	14.26	Tanzania
5.33	5.68	5.61	5.97	6.29	6.37	6.48	6.69	6.80	Tunisia
5.20	5.22	5.40	5.54	5.58	5.65	5.78	5.87	5.99	Zambia
8.70	8.93	9.60	9.82	9.57	9.60	9.83	9.95	9.93	Zimbabwe
42.88	43.72	44.56	45.79	46.57	47.72	48.91	50.04	51.36	Other Africa
390.20	396.84	411.31	413.60	418.50	437.74	448.54	455.67	472.53	**Africa**
44.41	43.31	45.46	49.73	52.87	54.76	56.09	58.92	61.71	Argentina
2.69	2.90	3.07	3.07	3.21	3.46	3.63	3.63	4.25	Bolivia
140.65	136.13	137.97	140.11	145.25	154.22	156.37	163.33	172.03	Brazil
13.19	13.88	14.21	15.70	16.08	17.39	18.68	20.46	23.01	Chile
24.76	26.76	26.89	26.80	27.49	29.47	30.80	31.39	30.48	Colombia
1.94	2.03	2.10	2.33	2.16	2.25	2.63	2.60	2.66	Costa Rica
16.60	16.46	14.10	12.24	11.88	12.41	12.71	13.25	14.27	Cuba
3.98	3.97	4.04	4.89	4.81	4.96	5.12	5.18	5.45	Dominican Republic
5.91	6.56	6.71	7.00	6.81	7.42	7.67	8.57	8.51	Ecuador
2.66	2.70	3.12	3.25	3.41	3.53	4.14	4.06	4.09	El Salvador
4.15	4.38	4.52	4.73	4.70	4.93	5.17	5.40	5.63	Guatemala
1.62	1.59	1.58	1.63	1.58	1.44	1.72	1.97	1.78	Haiti
2.42	2.44	2.46	2.60	2.52	2.64	2.89	2.95	3.18	Honduras
2.45	3.04	2.87	2.92	3.06	3.24	3.49	3.72	3.96	Jamaica
1.89	2.06	2.62	1.62	2.62	2.74	2.71	2.66	2.66	Netherlands Antilles
2.06	2.17	2.20	2.25	2.22	2.29	2.31	2.42	2.57	Nicaragua
1.40	1.54	1.64	1.80	1.86	2.07	2.04	2.19	2.33	Panama
3.22	3.10	3.18	3.21	3.30	3.61	3.96	3.99	4.19	Paraguay
11.79	11.55	11.63	11.24	12.11	12.70	13.80	14.01	15.13	Peru
4.98	5.69	5.89	6.53	6.25	6.97	6.95	7.89	8.20	Trinidad-and-Tobago
2.43	2.23	2.41	2.68	2.52	2.41	2.57	2.83	2.88	Uruguay
40.51	40.85	49.66	46.12	43.21	41.08	47.87	54.96	57.53	Venezuela
4.82	5.01	5.08	4.90	4.96	4.96	5.04	5.10	5.18	Other Latin America
340.55	340.33	353.43	357.34	364.88	380.96	398.36	421.48	441.71	**Latin America**

Total Primary Energy Supply (TPES) (Mtoe)
Approvisionnements totaux en énergie primaire (ATEP) (Mtep)
Gesamtaufkommen von Primärenergie (TPES) (Mtoe)

Disponibilità totale di energia primaria (DTEP) (Mtep)

一次エネルギー総供給量（石油換算百万トン）

Suministro total de energía primaria (TPES) (Mtep)

Общая первичная поставка топлива и энергии (ОППТЭ) (Мтнэ)

	1971	1973	1978	1983	1984	1985	1986	1987	1988
Bangladesh	10.80	11.84	13.78	16.08	16.70	17.33	18.18	18.97	19.72
Brunei	0.12	0.15	0.21	1.85	1.40	1.45	1.30	1.83	2.01
Inde	183.77	193.72	227.83	270.95	278.55	292.28	306.38	318.40	333.59
Indonésie	36.30	40.21	54.57	64.78	67.50	73.25	80.10	81.16	84.20
RPD de Corée	26.50	24.21	30.09	35.29	35.68	37.17	37.94	38.04	38.34
Malaisie	6.20	6.26	9.30	13.43	13.96	17.05	18.38	18.57	19.77
Myanmar	7.91	7.93	9.00	9.88	10.23	10.57	10.95	10.58	10.79
Népal	2.57	2.74	3.07	4.99	5.14	5.21	5.34	5.50	5.66
Pakistan	17.43	18.72	22.85	30.16	31.20	32.84	34.28	37.57	39.37
Philippines	14.61	16.17	19.42	23.97	22.64	22.88	22.87	24.73	26.12
Singapour	2.96	4.08	6.05	6.96	7.77	7.82	8.33	8.89	9.80
Sri Lanka	3.76	4.09	4.29	4.88	5.03	4.96	5.12	5.32	5.40
Taipei chinois	10.62	14.20	23.96	30.10	32.49	33.15	36.86	38.24	41.92
Thaïlande	14.14	16.41	21.50	21.80	24.00	26.56	29.00	31.08	34.02
Viêt-Nam	19.71	20.36	18.00	20.77	21.05	21.38	22.25	23.33	23.68
Autre Asie	3.95	4.24	4.57	4.99	5.15	5.54	5.47	5.48	5.48
Asie	361.35	385.33	468.48	560.90	578.50	609.44	642.75	667.69	699.86
Rép. populaire de Chine	390.23	425.38	586.55	631.10	668.84	705.53	738.60	776.57	815.50
Hong-Kong, Chine	3.47	3.73	5.71	7.10	7.43	7.62	8.42	8.95	10.44
Chine	393.71	429.11	592.26	638.19	676.27	713.14	747.02	785.52	825.94
Albanie	1.70	1.73	2.57	2.83	2.99	2.77	2.84	2.83	2.96
Bulgarie	19.24	20.71	27.49	30.40	30.66	30.96	31.24	30.93	31.72
Chypre	0.65	0.86	0.82	1.01	1.07	1.08	1.20	1.37	1.39
Gibraltar	0.04	0.04	0.03	0.04	0.04	0.04	0.05	0.05	0.06
Malte	0.27	0.33	0.36	0.36	0.50	0.40	0.65	0.68	0.73
Roumanie	41.75	47.40	63.83	65.02	64.43	64.64	66.35	69.17	69.57
République slovaque	14.19	15.44	18.93	20.54	20.86	21.65	22.18	22.39	22.40
Bosnie-Herzégovine	-	-	-	-	-	-	-	-	-
Croatie	-	-	-	-	-	-	-	-	-
Ex-RYM	-	-	-	-	-	-	-	-	-
Slovénie	-	-	-	-	-	-	-	-	-
RF de Yougoslavie	-	-	-	-	-	-	-	-	-
Ex-Yougoslavie	22.13	23.48	30.45	38.29	40.78	41.46	43.57	44.18	44.89
Europe non-OCDE	99.97	109.99	144.49	158.49	161.33	163.00	168.09	171.61	173.72

Total Primary Energy Supply (TPES) (Mtoe)
Approvisionnements totaux en énergie primaire (ATEP) (Mtep)
Gesamtaufkommen von Primärenergie (TPES) (Mtoe)
Disponibilità totale di energia primaria (DTEP) (Mtep)
一次エネルギー総供給量（石油換算百万トン）
Suministro total de energía primaria (TPES) (Mtep)
Общая первичная поставка топлива и энергии (ОППТЭ) (Мтнэ)

1989	1990	1991	1992	1993	1994	1995	1996	1997	
20.57	20.94	20.74	21.66	22.34	23.03	23.81	24.30	24.33	Bangladesh
1.29	1.48	1.69	1.86	1.69	1.69	1.97	1.99	2.11	Brunei
346.96	359.85	373.55	387.09	397.85	414.19	436.70	452.11	461.03	India
90.15	98.85	105.08	108.11	115.52	117.25	124.92	132.95	138.78	Indonesia
39.06	37.17	34.82	31.92	29.57	27.42	25.61	24.01	23.32	DPR of Korea
22.45	23.97	28.14	30.23	34.29	35.71	34.87	41.26	48.47	Malaysia
10.94	10.79	11.03	11.08	11.55	11.96	12.24	12.56	13.01	Myanmar
5.73	5.83	6.10	6.32	6.57	6.85	6.97	7.16	7.16	Nepal
41.62	43.24	44.49	46.62	48.77	51.34	52.97	56.34	56.82	Pakistan
27.24	28.29	28.27	29.94	31.31	33.06	35.13	35.70	38.25	Philippines
9.84	13.36	14.46	16.37	20.18	24.74	21.44	23.85	26.88	Singapore
5.43	5.48	5.56	5.72	6.02	5.82	6.11	6.79	7.16	Sri Lanka
45.82	48.27	52.39	54.73	58.26	61.34	65.24	68.73	72.54	Chinese Taipei
38.39	43.71	47.12	51.28	56.96	63.43	70.79	78.60	79.96	Thailand
23.51	24.45	24.59	25.69	26.84	27.93	29.00	35.99	39.31	Vietnam
5.54	5.72	5.73	5.44	5.49	5.47	5.56	5.66	5.75	Other Asia
734.53	771.39	803.76	834.07	873.22	911.22	953.33	1 008.00	1 044.88	**Asia**
843.34	856.21	883.11	902.55	949.85	1 000.63	1 058.62	1 096.39	1 098.93	People's Rep.of China
10.95	10.45	10.97	12.71	14.07	13.35	13.81	13.66	14.12	Hong Kong, China
854.29	866.67	894.08	915.26	963.92	1 013.98	1 072.43	1 110.06	1 113.05	**China**
2.83	2.57	1.83	1.41	1.19	1.13	1.09	1.20	1.05	Albania
31.02	27.13	22.14	20.52	21.87	21.02	22.90	22.60	20.62	Bulgaria
1.48	1.54	1.73	1.85	1.89	2.16	1.98	2.12	2.07	Cyprus
0.06	0.07	0.08	0.09	0.10	0.14	0.13	0.13	0.13	Gibraltar
0.78	0.77	0.76	0.75	0.92	0.84	0.84	0.89	0.94	Malta
69.18	61.12	50.34	46.37	45.12	42.29	45.67	49.11	44.14	Romania
22.27	21.36	19.42	17.98	17.10	16.75	17.35	17.65	17.22	Slovak Republic
-	-	-	4.50	4.73	1.69	1.75	1.75	1.75	*Bosnia-Herzegovina*
-	-	-	6.68	6.83	6.93	7.12	7.24	7.65	*Croatia*
-	-	-	3.14	3.21	2.65	2.76	3.13	2.84	*FYROM*
-	-	-	4.94	5.24	5.49	5.88	6.17	6.38	*Slovenia*
-	-	-	14.15	12.12	11.51	12.18	14.42	15.84	*FR of Yugoslavia*
44.20	44.76	40.98	33.47	32.18	28.31	29.70	32.71	34.47	Former Yugoslavia
171.82	159.32	137.28	122.43	120.37	112.65	119.66	126.43	120.63	**Non-OECD Europe**

Total Primary Energy Supply (TPES) (Mtoe)
Approvisionnements totaux en énergie primaire (ATEP) (Mtep)
Gesamtaufkommen von Primärenergie (TPES) (Mtoe)
Disponibilità totale di energia primaria (DTEP) (Mtep)
一次エネルギー総供給量（石油換算百万トン）
Suministro total de energía primaria (TPES) (Mtep)
Общая первичная поставка топлива и энергии (ОППТЭ) (Мтнэ)

	1971	1973	1978	1983	1984	1985	1986	1987	1988
Arménie	-	-	-	-	-	-	-	-	-
Azerbaïdjan	-	-	-	-	-	-	-	-	-
Bélarus	-	-	-	-	-	-	-	-	-
Estonie	-	-	-	-	-	-	-	-	-
Géorgie	-	-	-	-	-	-	-	-	-
Kazakhstan	-	-	-	-	-	-	-	-	-
Kirghizistan	-	-	-	-	-	-	-	-	-
Lettonie	-	-	-	-	-	-	-	-	-
Lituanie	-	-	-	-	-	-	-	-	-
Moldova	-	-	-	-	-	-	-	-	-
Russie	-	-	-	-	-	-	-	-	-
Tadjikistan	-	-	-	-	-	-	-	-	-
Turkménistan	-	-	-	-	-	-	-	-	-
Ukraine	-	-	-	-	-	-	-	-	-
Ouzbékistan	-	-	-	-	-	-	-	-	-
Ex-URSS	788.36	869.45	1 071.35	1 192.31	1 231.41	1 272.40	1 296.07	1 350.60	1 389.09
Bahrein	1.55	2.25	2.96	3.76	4.08	4.66	5.37	5.55	5.55
Iran	22.37	26.68	40.03	43.95	52.33	54.10	56.77	59.26	63.16
Irak	4.51	5.29	9.07	15.77	16.78	18.08	18.35	20.34	22.69
Israël	6.34	8.59	7.59	8.02	8.24	8.21	9.45	10.37	11.79
Jordanie	0.53	0.66	1.34	2.59	2.80	2.82	2.97	3.15	3.12
Koweit	5.99	5.74	8.05	10.74	10.65	13.31	12.65	12.54	12.15
Liban	1.98	2.57	2.24	2.51	2.43	2.78	2.92	2.99	2.28
Oman	0.09	0.10	0.67	1.66	2.16	2.47	1.84	1.85	2.34
Qatar	0.93	1.46	1.71	4.44	5.38	5.13	5.19	5.38	5.75
Arabie saoudite	6.45	8.22	20.90	51.27	50.03	52.21	58.06	61.53	67.21
Syrie	2.71	2.50	5.05	8.35	9.50	8.91	9.04	10.01	10.56
Emirats arabes unis	1.10	1.49	5.78	13.16	13.80	16.36	17.92	19.09	19.37
Yémen	0.76	1.01	1.14	1.92	1.99	2.30	2.42	2.58	2.69
Moyen-Orient	55.31	66.57	106.53	168.14	180.18	191.36	202.95	214.63	228.64
Total non-OCDE	2 105.28	2 304.96	2 912.97	3 337.77	3 466.57	3 601.54	3 732.03	3 893.62	4 040.91
OCDE Amérique du N.	1 781.51	1 952.61	2 146.79	1 972.76	2 059.24	2 086.49	2 088.62	2 176.60	2 259.78
OCDE Pacifique	345.47	410.58	451.34	467.81	500.48	505.76	518.98	532.52	568.35
OCDE Europe	1 245.54	1 379.80	1 467.57	1 426.77	1 469.22	1 526.05	1 551.64	1 585.86	1 595.41
Total OCDE	3 372.52	3 742.99	4 065.71	3 867.35	4 028.95	4 118.30	4 159.24	4 294.99	4 423.54
Monde	5 477.81	6 047.95	6 978.68	7 205.11	7 495.52	7 719.85	7 891.28	8 188.60	8 464.45

Le total Monde ne comprend pas les soutages maritimes internationaux.

Total Primary Energy Supply (TPES) (Mtoe)
Approvisionnements totaux en énergie primaire (ATEP) (Mtep)
Gesamtaufkommen von Primärenergie (TPES) (Mtoe)
Disponibilità totale di energia primaria (DTEP) (Mtep)
一次エネルギー総供給量（石油換算百万トン）
Suministro total de energía primaria (TPES) (Mtep)
Общая первичная поставка топлива и энергии (ОППТЭ) (Мтнэ)

1989	1990	1991	1992	1993	1994	1995	1996	1997	
-	-	-	4.30	2.26	1.42	1.67	1.79	1.80	Armenia
-	-	-	16.75	15.45	16.20	13.00	12.24	11.99	Azerbaijan
-	-	-	39.70	31.19	26.53	24.68	25.09	25.14	Belarus
-	-	-	6.58	5.26	5.50	5.17	5.65	5.56	Estonia
-	-	-	6.24	4.63	2.96	1.47	2.11	2.30	Georgia
-	-	-	79.67	63.13	58.19	55.01	44.68	38.42	Kazakhstan
-	-	-	5.07	3.90	2.99	2.65	3.04	2.79	Kyrgyzstan
-	-	-	6.32	4.91	4.60	3.98	4.15	4.46	Latvia
-	-	-	11.21	9.01	7.86	8.77	8.94	8.81	Lithuania
-	-	-	6.88	5.85	4.79	4.41	4.63	4.44	Moldova
-	-	-	795.02	739.40	637.79	624.42	610.73	591.98	Russia
-	-	-	9.08	6.39	3.30	3.28	3.52	3.38	Tajikistan
-	-	-	10.46	10.47	13.98	14.57	12.18	12.18	Turkmenistan
-	-	-	218.38	194.91	165.24	165.49	161.48	150.06	Ukraine
-	-	-	45.01	47.19	44.41	41.10	42.39	42.55	Uzbekistan
1 378.11	1 347.80	1 341.25	1 252.03	1 135.83	989.51	962.40	936.80	900.34	**Former USSR**
5.36	5.52	5.25	5.36	5.67	5.72	5.87	7.70	8.49	Bahrain
69.42	72.34	80.13	88.41	90.14	98.43	97.94	102.35	108.29	Iran
24.44	20.84	15.55	19.57	21.41	23.89	25.09	25.03	27.09	Iraq
11.58	11.92	11.95	13.66	14.40	14.98	17.10	16.18	17.59	Israel
3.20	3.45	3.48	3.94	3.97	4.31	4.32	4.49	4.79	Jordan
15.99	13.13	3.57	8.68	9.28	14.16	15.67	14.16	16.16	Kuwait
1.87	2.30	2.86	2.89	3.63	3.98	4.60	4.75	5.24	Lebanon
2.43	4.34	5.95	6.10	6.42	6.30	6.27	6.37	6.78	Oman
6.26	5.75	7.04	11.04	11.80	11.63	11.77	12.12	13.58	Qatar
63.76	63.28	73.96	78.55	82.62	81.01	83.66	92.24	98.45	Saudi Arabia
10.68	11.93	13.04	13.05	12.90	13.39	14.13	14.54	14.64	Syria
22.78	21.50	24.50	23.86	25.12	27.30	27.46	32.34	30.87	United Arab Emirates
2.94	2.67	3.05	3.46	2.76	2.92	3.16	3.27	3.36	Yemen
240.70	238.97	250.32	278.56	290.12	308.02	317.02	335.53	355.33	**Middle East**
4 110.20	4 121.31	4 191.43	4 181.87	4 174.89	4 160.31	4 278.99	4 399.77	4 453.99	**Non-OECD Total**
2 300.54	2 259.58	2 277.28	2 321.99	2 374.37	2 423.90	2 454.38	2 511.89	2 541.69	OECD North America
594.86	631.51	648.75	672.10	694.29	726.69	755.18	791.63	809.55	OECD Pacific
1 617.64	1 603.01	1 623.00	1 613.26	1 620.89	1 617.34	1 658.04	1 723.05	1 716.27	OECD Europe
4 513.04	4 494.10	4 549.02	4 607.36	4 689.55	4 767.93	4 867.59	5 026.57	5 067.52	**OECD Total**
8 623.25	8 615.41	8 740.45	8 789.23	8 864.45	8 928.24	9 146.58	9 426.34	9 521.51	**World**

World Total excludes international marine bunkers.

Total Primary Energy Supply excluding Combustible Renewables and Waste (Mtoe)
ATEP non compris les énergies renouvelables combustibles et les déchets (Mtep)
Gesamtaufkommen von Primärenergie, ohne erneuerbare Brennstoffe und Abfälle (Mtoe)
Totale disponibilita' primaria a netto di Fonti rinnovabili e combustibili derivati dai rifiuti (Mtep)
一次エネルギー総供給量（可燃性再生可能エネルギー及び廃棄物除く、石油換算百万トン）
TPES excluyendo combustibles renovables y desechos (Mtep)
ОПҐГЭ исключая возобновляемые виды топлива и отходов (Мтнэ)

	1971	1973	1978	1983	1984	1985	1986	1987	1988
Algérie	3.65	5.09	9.31	18.54	18.25	18.99	21.46	21.11	22.31
Angola	0.74	0.98	1.01	0.92	0.80	1.01	0.95	0.96	1.00
Bénin	0.11	0.14	0.15	0.15	0.15	0.19	0.17	0.15	0.15
Cameroun	0.39	0.42	0.66	1.08	1.16	1.16	1.13	1.12	1.18
Congo	0.23	0.19	0.23	0.39	0.35	0.35	0.26	0.73	1.11
RD du Congo	1.20	1.26	1.34	1.68	1.66	1.61	1.52	1.64	1.75
Egypte	7.18	7.40	12.24	20.64	22.92	24.51	25.28	27.16	28.08
Ethiopie	0.55	0.54	0.53	0.71	0.71	0.72	0.82	1.00	1.01
Gabon	0.43	0.95	1.09	1.15	1.03	0.84	0.87	0.80	0.89
Ghana	0.94	1.09	1.20	0.53	0.75	1.03	1.06	1.25	1.16
Côte d'Ivoire	0.87	1.00	1.49	1.08	1.08	1.15	1.30	1.18	1.58
Kenya	1.36	1.52	1.87	1.68	1.90	2.01	2.53	2.65	2.64
Libye	1.56	2.61	4.75	8.82	9.38	10.37	9.55	11.18	10.68
Maroc	2.29	2.82	4.17	4.86	5.01	5.12	5.30	5.44	5.71
Mozambique	1.00	0.97	0.67	0.59	0.52	0.53	0.54	0.54	0.54
Nigéria	2.20	3.37	7.89	12.83	11.73	12.33	11.55	12.66	13.40
Sénégal	0.52	0.60	0.71	0.79	0.85	0.79	0.85	0.93	0.85
Afrique du Sud	40.63	43.93	54.02	72.44	78.45	78.46	82.13	84.35	87.96
Soudan	1.30	1.51	1.19	1.38	1.32	1.53	1.51	1.22	1.66
Tanzanie	0.76	0.84	0.80	0.76	0.76	0.77	0.81	0.80	0.82
Tunisie	1.37	1.60	2.54	3.47	3.60	3.74	3.65	3.76	3.99
Zambie	1.44	1.72	1.66	1.64	1.50	1.51	1.47	1.52	1.57
Zimbabwe	2.34	2.66	2.55	2.80	2.86	3.24	3.53	4.10	4.05
Autre Afrique	2.52	2.77	3.80	5.16	5.00	5.02	5.60	6.43	6.65
Afrique	75.59	85.98	115.87	164.08	171.75	177.00	183.84	192.68	200.72
Argentine	31.39	33.51	36.82	39.71	41.25	39.31	41.37	43.61	44.74
Bolivie	0.79	0.94	1.46	2.04	1.87	1.84	1.82	1.89	1.93
Brésil	35.12	46.84	67.58	68.06	71.35	77.69	83.07	86.27	93.15
Chili	6.84	7.20	7.05	7.09	7.45	7.36	7.77	7.99	9.21
Colombie	9.67	10.83	13.49	16.44	16.45	16.97	17.70	18.32	18.19
Costa Rica	0.56	0.67	0.96	0.80	0.86	0.90	0.96	1.04	1.07
Cuba	7.35	8.54	10.55	11.91	10.38	10.49	10.44	11.11	11.36
République dominicaine	1.21	1.71	2.06	2.39	2.52	2.42	2.60	3.08	3.08
Equateur	1.24	1.45	3.61	4.34	4.72	4.84	4.87	4.90	5.17
El Salvador	0.55	0.71	1.35	1.23	1.26	1.30	1.24	1.41	1.42
Guatemala	0.89	1.00	1.31	1.12	1.28	1.30	1.05	1.14	1.25
Haiti	0.14	0.14	0.25	0.27	0.28	0.28	0.29	0.32	0.35
Honduras	0.40	0.46	0.60	0.67	0.68	0.68	0.67	0.80	0.90
Jamaïque	1.89	2.83	2.32	1.90	1.85	1.62	1.63	1.69	1.72
Antilles néerlandaises	5.52	6.03	3.76	4.41	3.06	1.83	1.77	1.54	1.54
Nicaragua	0.57	0.63	0.86	0.75	0.93	0.97	1.07	1.04	0.92
Panama	1.39	1.81	1.31	1.29	1.11	1.09	1.13	1.20	0.99
Paraguay	0.22	0.29	0.50	0.54	0.59	0.64	0.68	0.75	0.80
Pérou	5.92	6.26	7.44	7.32	7.69	7.72	8.26	8.77	8.78
Trinité-et-Tobago	2.68	2.66	3.52	4.27	4.54	4.71	5.01	4.58	5.12
Uruguay	2.02	1.98	2.09	1.60	1.49	1.44	1.45	1.58	1.79
Vénézuela	22.97	24.95	29.75	38.01	36.46	36.84	38.26	43.05	40.04
Autre Amérique latine	4.62	5.47	5.28	3.27	3.53	3.37	3.57	3.96	4.02
Amérique latine	143.96	166.92	203.90	219.43	221.60	225.60	236.68	250.04	257.52

Total Primary Energy Supply excluding Combustible Renewables and Waste (Mtoe)
ATEP non compris les énergies renouvelables combustibles et les déchets (Mtep)
Gesamtaufkommen von Primärenergie, ohne erneuerbare Brennstoffe und Abfälle (Mtoe)
Totale disponibilita' primaria a netto di Fonti rinnovabili e combustibili derivati dai rifiuti (Mtep)
一次エネルギー総供給量（可燃性再生可能エネルギー及び廃棄物除く、石油換算百万トン）
TPES excluyendo combustibles renovables y desechos (Mtep)
ОПTrue исключая возобновляемые виды топлива и отходов (Мтнэ)

1989	1990	1991	1992	1993	1994	1995	1996	1997	
22.13	23.52	24.44	24.54	24.57	23.64	23.80	23.65	25.98	Algeria
1.00	1.29	1.29	1.27	1.27	1.24	1.28	1.49	1.67	Angola
0.13	0.11	0.10	0.10	0.11	0.11	0.11	0.36	0.35	Benin
1.27	1.24	1.15	1.09	1.08	1.11	1.11	1.10	1.13	Cameroon
0.48	0.38	0.41	0.41	0.39	0.41	0.37	0.38	0.38	Congo
1.85	1.86	1.85	1.83	1.92	1.90	2.06	2.07	2.05	DR of Congo
29.53	30.84	31.35	32.15	34.39	32.32	34.13	36.13	38.34	Egypt
1.06	1.14	1.14	1.10	1.14	1.16	1.30	1.40	1.01	Ethiopia
0.67	0.53	0.57	0.62	0.67	0.65	0.66	0.68	0.77	Gabon
1.32	1.33	1.29	1.28	1.45	1.83	1.97	1.93	1.94	Ghana
1.65	1.42	1.33	1.67	1.39	1.35	1.43	1.60	1.67	Ivory Coast
2.76	2.70	2.59	2.68	2.77	2.87	2.56	2.65	3.13	Kenya
10.99	11.62	13.68	12.82	13.66	13.20	15.78	14.79	14.96	Libya
6.24	6.43	6.67	7.16	7.51	8.14	7.88	8.75	8.86	Morocco
0.53	0.52	0.50	0.55	0.58	0.56	0.57	0.60	0.74	Mozambique
14.31	14.33	16.17	17.23	10.75	21.59	18.46	17.95	19.86	Nigeria
0.84	0.86	0.83	0.95	0.89	0.97	0.97	1.07	1.14	Senegal
83.00	80.82	84.77	79.44	83.72	88.68	92.76	91.96	95.40	South Africa
1.53	1.94	1.78	1.68	1.27	1.76	1.67	1.65	1.95	Sudan
0.84	0.88	0.89	0.83	0.86	0.85	0.86	0.88	0.86	Tanzania
4.32	4.65	4.55	4.89	5.19	5.25	5.34	5.54	5.63	Tunisia
1.26	1.20	1.28	1.31	1.23	1.18	1.20	1.20	1.22	Zambia
4.07	4.21	4.78	4.91	4.57	4.51	4.67	4.73	4.64	Zimbabwe
6.56	6.47	6.59	6.83	7.04	7.12	7.23	7.29	7.50	Other Africa
198.34	200.28	209.99	207.32	208.43	222.41	228.17	229.86	241.18	**Africa**
42.85	41.59	43.66	47.87	50.90	52.44	53.43	56.23	59.06	Argentina
1.99	2.14	2.29	2.31	2.43	2.67	2.81	2.86	3.38	Bolivia
95.65	95.65	98.76	101.11	106.07	113.14	115.88	123.70	131.58	Brazil
10.67	11.20	11.22	12.38	13.00	14.15	15.22	16.75	19.32	Chile
19.23	20.94	21.20	20.90	21.46	23.20	23.82	24.50	25.21	Colombia
1.24	1.28	1.36	1.64	1.80	1.89	2.26	2.23	2.39	Costa Rica
11.70	11.03	8.69	6.82	6.44	6.97	7.26	7.78	8.73	Cuba
2.99	2.96	3.03	3.59	3.52	3.67	3.82	3.88	4.15	Dominican Republic
4.83	5.49	5.60	5.88	5.62	6.24	6.45	7.35	7.38	Ecuador
1.49	1.47	1.83	1.92	2.02	2.12	2.30	2.19	2.17	El Salvador
1.29	1.46	1.54	1.70	1.76	2.04	2.24	2.42	2.60	Guatemala
0.37	0.37	0.34	0.31	0.26	0.08	0.35	0.39	0.50	Haiti
0.97	0.94	0.95	1.09	1.01	1.13	1.38	1.42	1.46	Honduras
2.15	2.59	2.51	2.55	2.65	2.71	3.00	3.18	3.38	Jamaica
1.89	2.06	2.62	1.62	2.62	2.74	2.71	2.66	2.66	Netherlands Antilles
1.00	1.05	1.09	1.16	1.10	1.15	1.12	1.20	1.26	Nicaragua
0.99	1.13	1.23	1.39	1.39	1.57	1.52	1.64	1.76	Panama
1.00	0.87	0.85	1.02	1.18	1.37	1.56	1.44	1.57	Paraguay
7.99	7.60	7.64	7.19	8.06	8.58	9.63	9.79	10.83	Peru
4.94	5.64	5.85	6.49	6.22	6.92	6.92	7.86	8.16	Trinidad-and-Tobago
1.85	1.67	1.83	2.08	1.93	1.85	2.03	2.31	2.35	Uruguay
40.00	40.32	49.11	45.56	42.66	40.52	47.33	54.42	56.99	Venezuela
4.26	4.43	4.49	4.32	4.38	4.37	4.42	4.47	4.57	Other Latin America
261.35	263.88	277.68	280.89	288.48	301.54	317.45	340.68	361.46	**Latin America**

Total Primary Energy Supply excluding Combustible Renewables and Waste (Mtoe)
ATEP non compris les énergies renouvelables combustibles et les déchets (Mtep)
Gesamtaufkommen von Primärenergie, ohne erneuerbare Brennstoffe und Abfälle (Mtoe)

Totale disponibilita' primaria a netto di Fonti rinnovabili e combustibili derivati dai rifiuti (Mtep)

一次エネルギー総供給量（可燃性再生可能エネルギー及び廃棄物除く、石油換算百万トン）

TPES excluyendo combustibles renovables y desechos (Mtep)

ОПTrue исключая возобновляемые виды топлива и отходов (Мтнэ)

	1971	1973	1978	1983	1984	1985	1986	1987	1988
Bangladesh	1.25	1.59	2.22	3.26	3.59	3.92	4.40	4.87	5.30
Brunei	0.11	0.13	0.19	1.83	1.38	1.44	1.28	1.81	1.99
Inde	63.02	67.38	86.25	114.46	118.98	129.95	140.93	149.82	162.77
Indonésie	8.82	11.36	22.30	28.92	30.91	35.95	42.08	42.43	44.79
RPD de Corée	25.82	23.48	29.26	34.40	34.77	36.25	37.02	37.11	37.40
Malaisie	4.96	4.91	7.74	11.68	12.15	15.22	16.51	16.61	17.76
Myanmar	1.56	1.24	1.67	1.90	2.10	2.28	2.46	1.86	1.79
Népal	0.07	0.13	0.12	0.18	0.24	0.24	0.25	0.30	0.33
Pakistan	6.80	7.39	9.64	14.89	15.47	16.65	17.59	20.39	21.69
Philippines	8.56	9.83	11.81	15.32	13.75	13.63	13.50	15.32	16.66
Singapour	2.95	4.07	6.04	6.95	7.77	7.82	8.33	8.89	9.80
Sri Lanka	1.02	1.31	1.27	1.59	1.50	1.40	1.45	1.55	1.53
Taipei chinois	10.53	14.10	23.94	30.08	32.47	33.14	36.84	38.23	41.91
Thaïlande	6.54	8.50	11.08	13.12	14.52	15.84	17.43	18.88	21.24
Viêt-Nam	7.24	7.28	3.32	4.60	4.59	4.58	5.06	5.71	5.63
Autre Asie	2.05	2.25	2.27	2.59	2.78	3.18	3.05	3.05	3.00
Asie	151.29	164.97	219.10	285.78	296.98	321.47	348.20	366.82	393.59
Rép. populaire de Chine	235.99	263.66	411.21	445.13	481.71	517.00	547.95	583.80	620.56
Hong-Kong, Chine	3.43	3.68	5.66	7.04	7.38	7.56	8.36	8.89	10.38
Chine	239.42	267.34	416.87	452.17	489.09	524.56	556.31	592.69	630.94
Albanie	1.32	1.35	2.19	2.46	2.61	2.40	2.47	2.45	2.58
Bulgarie	18.97	20.47	27.28	29.99	30.25	30.55	30.83	30.72	31.53
Chypre	0.64	0.85	0.81	1.00	1.06	1.07	1.19	1.36	1.39
Gibraltar	0.04	0.04	0.03	0.04	0.04	0.04	0.05	0.05	0.06
Malte	0.27	0.33	0.36	0.36	0.50	0.40	0.65	0.68	0.73
Roumanie	40.38	46.04	62.63	63.95	63.47	63.45	65.28	68.22	68.68
République slovaque	13.99	15.26	18.75	20.41	20.74	21.53	22.06	22.25	22.25
Bosnie-Herzégovine	-	-	-	-	-	-	-	-	-
Croatie	-	-	-	-	-	-	-	-	-
Ex-RYM	-	-	-	-	-	-	-	-	-
Slovénie	-	-	-	-	-	-	-	-	-
RF de Yougoslavie	-	-	-	-	-	-	-	-	-
Ex-Yougoslavie	20.51	22.66	29.67	37.38	39.79	40.48	42.57	43.31	44.00
Europe non-OCDE	96.11	107.00	141.72	155.60	158.48	159.91	165.10	169.06	171.23

Total Primary Energy Supply excluding Combustible Renewables and Waste (Mtoe)
ATEP non compris les énergies renouvelables combustibles et les déchets (Mtep)
Gesamtaufkommen von Primärenergie, ohne erneuerbare Brennstoffe und Abfälle (Mtoe)
Totale disponibilita' primaria a netto di Fonti rinnovabili e combustibili derivati dai rifiuti (Mtep)
一次エネルギー総供給量（可燃性再生可能エネルギー及び廃棄物除く、石油換算百万トン）
TPES excluyendo combustibles renovables y desechos (Mtep)
ОПΠТЭ исключая возобновляемые виды топлива и отходов (Мтнэ)

1989	1990	1991	1992	1993	1994	1995	1996	1997	
5.85	6.06	5.63	6.37	6.89	7.41	8.05	8.40	8.30	Bangladesh
1.27	1.46	1.67	1.84	1.67	1.67	1.96	1.97	2.09	Brunei
173.74	184.03	193.81	205.01	213.08	227.45	248.05	261.63	268.23	India
50.07	58.12	63.71	66.11	72.88	73.97	80.96	88.46	94.29	Indonesia
38.12	36.22	33.86	30.95	28.59	26.42	24.60	22.99	22.30	DPR of Korea
20.38	21.85	26.00	28.03	32.04	33.41	32.53	38.87	46.08	Malaysia
1.78	1.47	1.53	1.48	1.83	2.13	2.28	2.44	2.74	Myanmar
0.31	0.26	0.36	0.44	0.46	0.58	0.68	0.69	0.70	Nepal
23.41	24.47	25.18	26.82	28.37	30.37	31.45	34.39	34.42	Pakistan
17.63	18.45	18.34	19.81	21.22	22.77	24.86	26.39	28.85	Philippines
9.84	13.36	14.46	16.37	20.18	24.74	21.44	23.85	26.88	Singapore
1.50	1.56	1.64	1.64	2.09	2.16	2.47	2.87	3.11	Sri Lanka
45.81	48.26	52.38	54.72	58.25	61.33	65.23	68.72	72.53	Chinese Taipei
24.11	28.93	31.84	35.00	39.23	44.05	50.55	57.22	59.31	Thailand
5.02	5.55	5.27	5.96	6.70	7.37	8.04	14.04	16.91	Vietnam
3.11	3.25	3.21	2.86	2.83	2.73	2.73	2.74	2.83	Other Asia
421.95	453.29	478.87	503.42	536.32	568.56	605.88	655.66	689.55	**Asia**
645.63	655.80	680.76	699.06	745.31	795.15	852.53	889.24	890.67	People's Rep.of China
10.89	10.40	10.91	12.66	14.01	13.29	13.75	13.61	14.06	Hong Kong, China
656.53	666.20	691.68	711.71	759.32	808.44	866.29	902.85	904.73	**China**
2.45	2.20	1.46	1.04	1.06	1.07	1.03	1.14	0.99	Albania
30.66	26.77	22.00	20.35	21.72	20.86	22.69	22.36	20.38	Bulgaria
1.48	1.53	1.72	1.84	1.87	2.15	1.96	2.11	2.06	Cyprus
0.06	0.07	0.08	0.09	0.10	0.14	0.13	0.13	0.13	Gibraltar
0.78	0.77	0.76	0.75	0.92	0.84	0.84	0.89	0.94	Malta
68.53	60.52	49.63	45.31	43.73	41.11	43.95	44.20	40.76	Romania
22.14	21.20	19.28	17.86	16.98	16.58	17.27	17.58	17.13	Slovak Republic
-	-	-	4.34	4.57	1.53	1.59	1.59	1.59	*Bosnia-Herzegovina*
-	-	-	6.42	6.59	6.67	6.85	6.91	7.33	*Croatia*
-	-	-	2.95	3.01	2.46	2.57	2.94	2.65	*FYROM*
-	-	-	4.68	4.98	5.23	5.62	5.90	6.12	*Slovenia*
-	-	-	13.93	11.91	11.30	11.97	14.21	15.63	*FR of Yugoslavia*
43.28	44.04	40.26	32.38	31.09	27.22	28.61	31.57	33.32	Former Yugoslavia
169.38	157.11	135.20	119.63	117.49	109.97	116.49	119.98	115.72	**Non-OECD Europe**

Total Primary Energy Supply excluding Combustible Renewables and Waste (Mtoe)
ATEP non compris les énergies renouvelables combustibles et les déchets (Mtep)
Gesamtaufkommen von Primärenergie, ohne erneuerbare Brennstoffe und Abfälle (Mtoe)
Totale disponibilita' primaria a netto di Fonti rinnovabili e combustibili derivati dai rifiuti (Mtep)
一次エネルギー総供給量（可燃性再生可能エネルギー及び廃棄物除く、石油換算百万トン）
TPES excluyendo combustibles renovables y desechos (Mtep)
ОПППЭ исключая возобновляемые виды топлива и отходов (Мтнэ)

	1971	1973	1978	1983	1984	1985	1986	1987	1988
Arménie	-	-	-	-	-	-	-	-	-
Azerbaïdjan	-	-	-	-	-	-	-	-	-
Bélarus	-	-	-	-	-	-	-	-	-
Estonie	-	-	-	-	-	-	-	-	-
Géorgie	-	-	-	-	-	-	-	-	-
Kazakhstan	-	-	-	-	-	-	-	-	-
Kirghizistan	-	-	-	-	-	-	-	-	-
Lettonie	-	-	-	-	-	-	-	-	-
Lituanie	-	-	-	-	-	-	-	-	-
Moldova	-	-	-	-	-	-	-	-	-
Russie	-	-	-	-	-	-	-	-	-
Tadjikistan	-	-	-	-	-	-	-	-	-
Turkménistan	-	-	-	-	-	-	-	-	-
Ukraine	-	-	-	-	-	-	-	-	-
Ouzbékistan	-	-	-	-	-	-	-	-	-
Ex-URSS	768.21	849.97	1 053.20	1 173.58	1 211.54	1 252.16	1 277.25	1 330.47	1 370.18
Bahrein	1.55	2.25	2.96	3.76	4.08	4.66	5.37	5.55	5.55
Iran	22.04	26.33	39.45	43.32	51.68	53.42	56.08	58.61	62.45
Irak	4.48	5.27	9.04	15.75	16.76	18.06	18.33	20.31	22.67
Israël	6.34	8.59	7.59	8.02	8.24	8.21	9.45	10.36	11.78
Jordanie	0.53	0.66	1.34	2.58	2.80	2.82	2.97	3.15	3.12
Koweit	5.98	5.71	8.05	10.74	10.65	13.30	12.65	12.54	12.14
Liban	1.89	2.47	2.13	2.40	2.32	2.68	2.81	2.88	2.17
Oman	0.09	0.10	0.67	1.66	2.16	2.47	1.84	1.85	2.34
Qatar	0.93	1.45	1.70	4.43	5.38	5.13	5.18	5.37	5.75
Arabie saoudite	6.44	8.22	20.90	51.26	50.02	52.21	58.05	61.52	67.20
Syrie	2.71	2.50	5.05	8.35	9.50	8.91	9.04	10.01	10.56
Emirats arabes unis	1.10	1.49	5.78	13.16	13.80	16.36	17.92	19.09	19.37
Yémen	0.72	0.96	1.08	1.85	1.93	2.23	2.35	2.51	2.62
Moyen-Orient	54.79	66.00	105.74	167.30	179.31	190.46	202.04	213.75	227.70
Total non-OCDE	1 529.37	1 708.19	2 256.40	2 617.96	2 728.74	2 851.16	2 969.41	3 115.52	3 251.89
OCDE Amérique du N.	1 730.66	1 899.11	2 081.36	1 898.39	1 977.57	2 005.55	2 005.58	2 086.77	2 170.90
OCDE Pacifique	341.91	407.05	447.17	458.56	490.79	495.74	509.22	522.44	557.72
OCDE Europe	1 224.02	1 357.05	1 440.71	1 393.10	1 433.93	1 490.20	1 515.28	1 549.36	1 556.98
Total OCDE	3 296.59	3 663.20	3 969.24	3 750.04	3 902.29	3 991.49	4 030.08	4 158.57	4 285.60
Monde	4 825.96	5 371.39	6 225.64	6 368.00	6 631.03	6 842.65	6 999.50	7 274.10	7 537.49

Le total Monde ne comprend pas les soutages maritimes internationaux.

Total Primary Energy Supply excluding Combustible Renewables and Waste (Mtoe)
ATEP non compris les énergies renouvelables combustibles et les déchets (Mtep)
Gesamtaufkommen von Primärenergie, ohne erneuerbare Brennstoffe und Abfälle (Mtoe)
Totale disponibilita' primaria a netto di Fonti rinnovabili e combustibili derivati dai rifiuti (Mtep)
一次エネルギー総供給量（可燃性再生可能エネルギー及び廃棄物除く、石油換算百万トン）
TPES excluyendo combustibles renovables y desechos (Mtep)
ОПuТЭ исключая возобновляемые виды топлива и отходов (Мтнэ)

1989	1990	1991	1992	1993	1994	1995	1996	1997	
-	-	-	4.30	2.26	1.42	1.67	1.79	1.80	Armenia
-	-	-	16.75	15.45	16.19	12.99	12.24	11.98	Azerbaijan
-	-	-	39.11	30.62	25.98	24.15	24.57	24.46	Belarus
-	-	-	6.39	5.08	5.21	4.84	5.06	4.96	Estonia
-	-	-	6.19	4.58	2.92	1.43	2.08	2.26	Georgia
-	-	-	79.56	63.05	58.12	54.93	44.61	38.34	Kazakhstan
-	-	-	5.07	3.90	2.99	2.65	3.04	2.79	Kyrgyzstan
-	-	-	5.83	4.42	4.08	3.70	3.56	3.38	Latvia
-	-	-	11.02	8.76	7.62	8.54	8.67	8.29	Lithuania
-	-	-	6.85	5.82	4.76	4.38	4.57	4.38	Moldova
-	-	-	778.00	722.37	620.77	607.39	593.71	574.96	Russia
-	-	-	9.08	6.39	3.30	3.28	3.52	3.38	Tajikistan
-	-	-	10.46	10.47	13.98	14.57	12.18	12.18	Turkmenistan
-	-	-	218.08	194.64	164.97	165.23	161.23	149.81	Ukraine
-	-	-	45.01	47.19	44.41	41.10	42.39	42.55	Uzbekistan
1 360.73	1 328.70	1 322.14	1 233.05	1 116.88	970.47	943.60	917.39	880.02	**Former USSR**
5.36	5.52	5.25	5.36	5.67	5.72	5.87	7.70	8.49	Bahrain
68.76	71.66	79.43	87.69	89.42	97.69	97.22	101.56	107.50	Iran
24.42	20.82	15.52	19.54	21.38	23.86	25.06	25.00	27.06	Iraq
11.58	11.92	11.95	13.66	14.40	14.98	17.09	16.18	17.58	Israel
3.19	3.44	3.47	3.93	3.97	4.31	4.32	4.48	4.79	Jordan
15.98	13.13	3.56	8.67	9.27	14.15	15.66	14.16	16.16	Kuwait
1.76	2.19	2.76	2.78	3.52	3.87	4.49	4.62	5.12	Lebanon
2.43	4.34	5.95	6.10	6.42	6.30	6.27	6.37	6.78	Oman
6.25	5.75	7.04	11.04	11.80	11.63	11.76	12.12	13.57	Qatar
63.74	63.26	73.94	78.54	82.61	81.00	83.65	92.24	98.44	Saudi Arabia
10.67	11.93	13.03	13.05	12.89	13.39	14.12	14.54	14.64	Syria
22.78	21.50	24.50	23.82	25.09	27.28	27.43	32.32	30.86	United Arab Emirates
2.86	2.59	2.98	3.39	2.68	2.84	3.08	3.19	3.28	Yemen
239.81	238.05	249.39	277.57	289.13	307.03	316.03	334.48	354.28	**Middle East**
3 308.06	3 307.51	3 364.96	3 342.18	3 324.11	3 294.63	3 401.16	3 506.71	3 552.44	**Non-OECD Total**
2 215.42	2 181.40	2 196.89	2 232.30	2 290.89	2 338.42	2 365.50	2 422.31	2 455.40	OECD North America
583.73	620.18	637.30	661.10	682.70	714.42	742.22	777.76	794.53	OECD Pacific
1 569.75	1 556.07	1 574.97	1 565.00	1 568.97	1 564.16	1 603.32	1 666.52	1 657.36	OECD Europe
4 368.90	4 357.65	4 409.16	4 458.40	4 542.56	4 616.99	4 711.05	4 866.59	4 907.29	**OECD Total**
7 676.96	7 665.16	7 774.11	7 800.58	7 866.66	7 911.62	8 112.20	8 373.30	8 459.73	**World**

World Total excludes international marine bunkers.

Electricity Generation from Coal (% of total)
Production d'électricité à partir du charbon (% du total)
Elektrizitätserzeugung auf Kohlebasis (in %)
Produzione di energia termoelettrica da carbone (% del totale)
石炭からの発電量(%)
Generación de electricidad a partir del carbón (% del total)
Производство электроэнергии за счет потребления угля (в % к общему производству)

	1971	1973	1978	1983	1984	1985	1986	1987	1988
Maroc	13.41	27.51	27.73	20.40	18.19	15.58	27.07	34.35	29.19
Mozambique	-	-	-	23.02	20.35	19.64	22.58	24.40	28.78
Nigéria	-	-	0.95	0.11	0.14	0.17	0.13	0.11	0.09
Afrique du Sud	99.80	98.47	97.74	99.51	96.68	95.80	92.88	94.83	91.29
Zambie	20.31	6.71	1.15	0.43	0.48	0.35	0.36	0.41	0.42
Zimbabwe	32.68	32.58	11.31	16.62	26.21	40.29	54.44	82.54	66.08
Afrique	62.44	60.54	54.75	55.26	54.71	53.02	51.41	52.51	50.15
Argentine	4.03	2.87	2.69	3.15	2.14	2.39	3.51	1.82	1.58
Brésil	3.83	2.43	2.48	1.79	1.90	2.05	2.51	1.98	1.50
Chili	14.85	13.02	10.65	11.83	17.48	14.01	12.81	11.54	22.20
Colombie	7.71	10.75	9.30	12.34	11.34	10.52	8.10	8.89	7.31
République dominicaine	-	-	-	-	-	14.30	1.73	13.27	7.12
Amérique latine	3.66	2.98	2.61	2.53	2.60	2.63	2.61	2.28	2.25
Inde	48.45	48.13	47.28	58.06	58.87	62.58	65.36	69.75	69.17
Indonésie	-	-	-	-	-	2.21	26.54	31.32	29.49
RPD de Corée	31.40	37.50	39.06	36.59	40.00	41.67	42.00	41.83	45.28
Malaisie	-	-	-	-	-	-	-	-	1.45
Myanmar	3.91	2.56	1.93	2.03	1.53	1.98	2.27	1.72	1.66
Pakistan	1.22	0.79	0.35	0.21	0.16	0.20	0.16	0.09	0.08
Philippines	0.13	0.09	0.51	6.62	9.02	19.05	16.69	11.57	12.06
Taipei chinois	12.28	6.94	3.54	18.63	25.61	26.39	32.86	32.76	31.20
Thaïlande	6.10	3.50	3.87	9.57	11.02	23.03	22.43	23.38	20.95
Viêt-Nam	73.30	82.13	77.78	39.61	42.19	45.45	48.07	50.65	50.70
Asie	30.16	28.27	27.60	33.19	35.51	38.82	41.97	43.71	43.31
Rép. populaire de Chine	71.65	61.52	65.25	59.21	62.43	64.85	66.26	68.40	68.94
Hong-Kong, Chine	-	-	-	53.79	69.45	77.96	84.95	95.21	98.35
Chine	68.88	59.11	62.71	58.97	62.75	65.44	67.11	69.62	70.26
Albanie	-	-	-	-	-	-	-	-	-
Bulgarie	81.50	77.28	45.87	45.06	45.56	36.45	42.10	44.17	36.58
Malte	-	-	-	17.63	20.86	43.62	29.06	36.86	42.43
Roumanie	26.20	26.02	28.11	15.77	15.13	31.21	31.46	33.13	36.98
République slovaque	66.34	64.40	40.86	40.45	41.36	37.15	32.92	33.36	34.42
Bosnie-Herzégovine	-	-	-	-	-	-	-	-	-
Croatie	-	-	-	-	-	-	-	-	-
Ex-RYM	-	-	-	-	-	-	-	-	-
Slovénie	-	-	-	-	-	-	-	-	-
RF de Yougoslavie	-	-	-	-	-	-	-	-	-
Ex-Yougoslavie	43.31	46.03	41.01	52.84	53.37	56.92	54.72	55.81	56.57
Europe non-OCDE	46.04	44.69	35.84	35.89	36.53	41.14	40.66	42.43	42.80

Electricity Generation from Coal (% of total)
Production d'électricité à partir du charbon (% du total)

Elektrizitätserzeugung auf Kohlebasis (in %)

Produzione di energia termoelettrica da carbone (% del totale)

石炭からの発電量(%)

Generación de electricidad a partir del carbón (% del total)

Производство электроэнергии за счет потребления угля (в % к общему производству)

1989	1990	1991	1992	1993	1994	1995	1996	1997	
21.92	22.97	19.73	17.16	25.58	24.97	46.85	44.92	45.00	Morocco
20.97	13.88	16.77	11.78	4.08	-	-	-	-	Mozambique
0.10	0.15	0.07	0.13	-	-	-	-	-	Nigeria
91.46	94.28	93.39	94.02	95.73	94.08	93.67	93.40	92.91	South Africa
0.59	0.62	0.52	0.51	0.51	0.51	0.51	0.51	0.50	Zambia
61.15	59.51	60.07	63.38	73.62	73.31	69.98	71.50	75.65	Zimbabwe
49.70	51.10	49.74	49.41	50.63	50.10	50.46	51.23	51.22	**Africa**
1.14	0.93	1.42	1.14	1.60	3.26	2.51	2.18	1.05	Argentina
1.98	1.46	1.69	1.58	1.44	1.42	1.54	1.64	1.78	Brazil
31.73	36.59	22.41	16.48	17.77	26.12	24.33	35.08	33.85	Chile
8.02	8.80	9.66	14.61	9.73	7.78	10.92	6.91	7.95	Colombia
0.83	0.65	3.90	6.65	5.45	4.48	4.55	4.32	4.38	Dominican Republic
2.80	2.80	2.52	2.54	2.31	2.71	2.86	3.11	3.16	**Latin America**
69.54	65.16	67.13	68.40	70.22	68.28	72.03	73.17	73.13	India
31.55	29.45	31.44	27.37	23.79	24.36	27.77	25.86	30.74	Indonesia
45.79	45.79	40.65	36.84	36.84	36.49	36.11	35.73	35.73	DPR of Korea
10.58	13.04	12.55	11.56	8.89	9.31	7.28	6.39	5.27	Malaysia
1.48	1.61	1.87	0.17	0.15	0.14	-	-	-	Myanmar
0.07	0.10	0.07	0.07	0.09	0.09	0.07	0.77	0.65	Pakistan
10.97	9.37	7.57	6.92	7.99	7.02	6.02	13.24	18.49	Philippines
30.44	27.08	28.96	34.61	37.97	37.88	36.85	42.27	42.50	Chinese Taipei
21.06	25.02	25.98	25.95	21.30	19.85	18.46	20.02	20.29	Thailand
35.02	23.02	15.59	-	-	-	-	-	-	Vietnam
43.89	41.30	42.07	42.55	43.04	41.91	43.28	44.21	44.51	**Asia**
71.64	72.46	74.02	74.14	72.58	74.15	73.43	74.09	73.56	People's Rep.of China
97.17	98.31	96.58	95.91	95.09	98.13	97.71	98.02	98.86	Hong Kong, China
72.78	73.61	75.04	75.10	73.51	74.82	74.09	74.70	74.19	**China**
0.49	0.63	0.54	0.60	0.57	-	-	-	-	Albania
38.26	35.42	48.96	50.59	46.81	45.75	43.13	41.95	44.29	Bulgaria
56.18	55.91	41.58	37.58	40.00	23.30	5.51	-	-	Malta
36.41	28.77	27.49	34.18	34.62	35.80	35.10	33.89	30.24	Romania
33.17	32.24	26.89	26.72	25.16	20.04	23.13	23.69	24.01	Slovak Republic
-	-	-	72.13	78.63	34.93	35.54	35.54	35.54	*Bosnia-Herzegovina*
-	-	-		5.04	1.22	2.73	1.14	5.28	*Croatia*
-	-	-	85.05	89.03	86.99	86.31	86.32	85.88	*FYROM*
-	-	-	32.77	37.91	34.33	34.33	32.80	35.77	*Slovenia*
-	-	-	62.89	63.66	61.84	63.70	63.70	63.70	*FR of Yugoslavia*
59.17	63.51	56.55	53.96	56.05	49.71	51.23	49.87	51.96	Former Yugoslavia
43.71	43.02	42.28	43.04	42.65	39.01	39.08	37.98	38.49	**Non-OECD Europe**

Electricity Generation from Coal (% of total)
Production d'électricité à partir du charbon (% du total)
Elektrizitätserzeugung auf Kohlebasis (in %)
Produzione di energia termoelettrica da carbone (% del totale)
石炭からの発電量(%)
Generación de electricidad a partir del carbón (% del total)
Производство электроэнергии за счет потребления угля (в % к общему производству)

	1971	1973	1978	1983	1984	1985	1986	1987	1988
Bélarus	-	-	-	-	-	-	-	-	-
Estonie	-	-	-	-	-	-	-	-	-
Kazakhstan	-	-	-	-	-	-	-	-	-
Kirghizistan	-	-	-	-	-	-	-	-	-
Lettonie	-	-	-	-	-	-	-	-	-
Moldova	-	-	-	-	-	-	-	-	-
Russie	-	-	-	-	-	-	-	-	-
Ukraine	-	-	-	-	-	-	-	-	-
Ouzbékistan	-	-	-	-	-	-	-	-	-
Ex-URSS	43.38	42.45	34.94	28.04	26.49	25.37	26.31	26.67	26.01
Israël	-	-	-	32.36	49.16	53.48	54.67	51.90	47.04
Moyen-Orient	-	-	-	3.31	4.63	4.87	4.82	4.61	4.27
Total non-OCDE	41.52	39.41	34.27	30.82	30.92	31.46	32.51	33.62	33.51
OCDE Amérique du N.	41.14	41.47	39.36	48.46	49.36	49.90	48.69	49.97	50.33
OCDE Pacifique	18.30	15.71	15.73	20.14	22.32	23.74	23.55	23.70	23.70
OCDE Europe	44.38	40.81	41.59	43.75	39.70	40.53	40.77	39.82	38.41
Total OCDE	39.55	37.96	37.07	42.91	42.19	42.90	42.33	42.56	42.22
Monde	40.09	38.36	36.22	38.77	38.28	38.89	38.80	39.31	39.04

Electricity Generation from Coal (% of total)
Production d'électricité à partir du charbon (% du total)
Elektrizitätserzeugung auf Kohlebasis (in %)
Produzione di energia termoelettrica da carbone (% del totale)
石炭からの発電量(%)
Generación de electricidad a partir del carbón (% del total)
Производство электроэнергии за счет потребления угля (в % к общему производству)

1989	1990	1991	1992	1993	1994	1995	1996	1997	
-	-	-	12.18	10.94	-	-	-	-	Belarus
-	-	-	90.01	89.61	94.80	96.41	96.65	95.34	Estonia
-	-	-	72.31	71.43	70.94	72.03	72.03	72.03	Kazakhstan
-	-	-	6.05	5.19	6.03	6.03	6.61	6.61	Kyrgyzstan
-	-	-	-	2.01	1.76	2.46	3.07	2.31	Latvia
-	-	-	34.37	34.67	38.33	26.29	22.21	9.25	Moldova
-	-	-	16.66	16.80	18.26	18.26	18.64	16.76	Russia
-	-	-	26.24	28.23	27.16	32.90	28.72	27.64	Ukraine
-	-	-	4.89	6.48	8.31	3.45	4.02	4.08	Uzbekistan
25.71	24.96	24.27	20.28	20.54	21.02	21.78	21.06	19.31	**Former USSR**
48.00	49.92	52.83	56.94	61.60	60.00	62.07	68.92	70.70	Israel
4.33	4.35	4.78	5.32	5.55	5.38	5.76	6.48	6.78	**Middle East**
34.23	34.02	34.53	34.01	34.79	35.74	36.70	37.31	37.19	**Non-OECD Total**
47.86	47.29	46.53	46.94	46.86	46.06	45.33	46.40	47.35	OECD North America
23.45	22.93	23.10	23.52	24.66	26.27	27.06	27.92	29.19	OECD Pacific
38.68	38.65	38.27	36.19	34.41	34.18	33.53	32.96	31.04	OECD Europe
41.13	40.60	40.09	39.63	39.22	38.89	38.42	38.90	38.89	**OECD Total**
38.64	38.21	38.08	37.61	37.62	37.77	37.80	38.32	38.26	**World**

Electricity Generation from Oil (% of total)
Production d'électricité à partir du pétrole (% du total)
Elektrizitätserzeugung auf Ölbasis (in %)
Produzione di energia termoelettrica da prodotti petroliferi (% del totale)
石油からの発電量 (%)
Generación de electricidad a partir del petróleo (% del total)
Производство электроэнергии за счет потребления нефти (в % к общему производству)

	1971	1973	1978	1983	1984	1985	1986	1987	1988
Algérie	46.34	18.25	19.38	11.72	10.52	9.34	9.23	8.65	7.93
Angola	18.46	17.28	9.74	26.18	25.43	13.03	13.03	13.46	13.99
Bénin	-	100.00	100.00	100.00	100.00	100.00	100.00	100.00	100.00
Cameroun	2.57	4.56	5.83	3.47	1.67	2.17	1.90	2.27	1.94
Congo	-	-	-	0.86	0.85	0.64	0.74	1.07	1.03
RD du Congo	3.05	2.08	1.96	3.01	2.90	2.78	2.72	2.76	2.67
Egypte	36.97	36.39	38.48	35.09	37.63	40.71	40.62	43.19	43.23
Ethiopie	48.74	43.65	28.40	20.35	18.13	17.57	15.19	12.33	11.63
Gabon	100.00	96.97	32.11	24.97	23.27	22.53	21.96	22.01	22.09
Ghana	1.19	0.97	1.38	0.78	1.75	0.99	0.68	3.90	0.70
Côte d'Ivoire	76.36	78.89	88.22	39.24	77.87	26.28	32.02	57.71	46.30
Kenya	45.93	48.55	22.36	8.61	11.54	6.45	11.23	14.47	4.55
Libye	100.00	100.00	100.00	100.00	100.00	100.00	100.00	100.00	100.00
Maroc	20.22	31.03	40.00	72.41	76.52	77.81	64.64	55.34	60.40
Mozambique	66.77	70.20	8.22	20.38	18.34	11.64	21.29	22.02	21.96
Nigéria	7.16	17.11	32.92	38.06	38.68	36.64	39.56	36.64	39.73
Sénégal	100.00	100.00	100.00	100.00	100.00	100.00	100.00	100.00	97.55
Soudan	60.24	34.53	41.56	30.01	30.01	40.03	34.83	34.81	35.27
Tanzanie	37.07	49.14	25.18	12.96	15.18	12.81	9.77	9.51	9.14
Tunisie	94.38	61.07	59.46	67.30	39.95	23.99	56.55	31.37	43.39
Zambie	5.02	1.34	0.57	0.40	0.41	0.40	0.50	0.43	0.47
Autre Afrique	68.46	64.43	67.28	54.89	54.49	52.24	53.55	54.79	57.65
Afrique	11.63	10.80	13.50	15.33	15.95	16.28	16.98	17.80	18.18
Argentine	74.33	66.99	49.52	22.44	20.01	13.69	15.87	16.55	23.79
Bolivie	8.20	9.64	8.90	9.74	9.43	9.61	8.43	6.45	6.57
Brésil	9.40	5.34	4.97	2.52	1.92	2.02	5.04	3.83	3.35
Chili	30.30	23.91	20.90	13.66	9.62	7.34	6.35	5.28	5.81
Colombie	12.11	13.97	4.04	1.96	2.22	1.30	0.72	0.67	0.60
Costa Rica	10.45	16.04	23.22	2.93	2.05	2.13	1.97	4.25	4.60
Cuba	82.12	86.74	88.19	90.27	90.51	90.51	90.78	91.03	90.62
République dominicaine	44.76	70.57	68.68	66.85	62.72	54.32	61.82	49.37	52.64
Equateur	58.10	65.37	69.64	60.00	28.73	28.06	19.95	15.12	14.45
El Salvador	27.59	43.64	12.89	4.46	4.20	6.90	4.28	17.26	12.61
Guatemala	57.04	67.96	74.44	13.05	16.88	19.89	3.15	7.33	5.99
Haiti	52.44	21.31	17.55	49.15	49.76	47.62	23.70	31.98	23.74
Honduras	38.11	26.13	14.76	14.61	13.31	6.80	3.30	3.30	3.49
Jamaïque	86.34	86.10	88.53	87.46	87.48	87.78	88.39	89.43	89.62
Antilles néerlandaises	100.00	100.00	100.00	100.00	100.00	100.00	100.00	100.00	100.00
Nicaragua	66.21	46.03	69.22	61.06	34.70	32.92	38.73	41.31	41.45
Panama	90.74	91.78	53.88	59.28	33.33	21.97	19.63	24.50	14.85
Paraguay	15.68	14.51	49.91	17.42	4.86	0.44	0.18	0.10	0.14
Pérou	23.06	23.86	23.69	21.61	22.51	18.82	18.84	19.06	20.11
Trinité-et-Tobago	3.94	1.99	1.48	0.55	0.96	0.33	0.82	0.20	0.09
Uruguay	37.84	38.06	46.08	2.09	1.50	1.55	1.08	3.43	21.81
Vénézuela	14.83	15.48	12.49	14.13	13.48	12.45	12.13	10.61	9.77
Autre Amérique latine	62.43	69.87	75.75	79.38	77.05	75.96	80.35	89.59	87.31
Amérique latine	31.43	28.41	21.49	14.05	12.28	10.68	11.95	11.27	11.81

Electricity Generation from Oil (% of total)
Production d'électricité à partir du pétrole (% du total)
Elektrizitätserzeugung auf Ölbasis (in %)
Produzione di energia termoelettrica da prodotti petroliferi (% del totale)
石油からの発電量(%)
Generación de electricidad a partir del petróleo (% del total)
Производство электроэнергии за счет потребления нефти (в % к общему производству)

1989	1990	1991	1992	1993	1994	1995	1996	1997	
7.35	5.40	4.38	4.63	4.09	3.65	3.34	3.61	3.56	Algeria
13.31	9.03	17.34	11.30	6.32	6.28	6.25	10.02	9.29	Angola
100.00	100.00	100.00	100.00	100.00	100.00	100.00	100.00	100.00	Benin
1.55	1.52	1.44	1.28	1.06	1.18	1.15	1.17	1.15	Cameroon
0.76	0.59	0.62	0.70	0.70	0.69	0.68	0.69	0.70	Congo
0.46	0.44	0.42	0.31	1.78	2.13	3.80	2.08	2.16	DR of Congo
42.53	43.11	42.06	40.12	38.90	37.01	37.32	33.08	35.24	Egypt
10.69	9.47	9.53	9.94	7.93	9.45	10.62	11.25	4.93	Ethiopia
11.42	12.02	11.38	11.21	11.82	11.90	11.81	12.89	15.99	Gabon
0.55	0.17	0.11	0.02	0.35	0.46	0.03	0.12	0.11	Ghana
27.59	27.56	32.00	85.20	51.26	57.51	39.83	37.91	37.07	Ivory Coast
3.76	7.61	4.96	4.57	3.86	5.93	8.91	8.79	8.38	Kenya
100.00	100.00	100.00	100.00	100.00	100.00	100.00	100.00	100.00	Libya
65.26	64.35	66.95	72.92	69.95	67.37	48.07	39.40	39.29	Morocco
18.64	23.57	14.65	10.10	7.91	7.09	6.86	6.09	21.09	Mozambique
37.68	36.47	35.10	36.78	28.09	28.79	24.06	26.14	25.82	Nigeria
97.18	98.03	98.51	99.22	95.83	94.33	86.22	88.07	93.89	Senegal
34.43	36.77	37.51	33.35	35.29	39.67	47.85	47.94	47.00	Sudan
5.96	4.85	5.27	9.14	9.44	12.39	16.40	12.20	25.08	Tanzania
46.99	32.35	60.89	42.67	48.84	24.61	15.29	14.15	15.71	Tunisia
0.81	-	-	-	0.03	0.03	0.03	0.03	0.02	Zambia
56.73	56.58	58.41	58.18	59.40	59.48	58.68	58.20	58.12	Other Africa
17.84	17.62	18.00	17.98	17.48	16.97	16.07	14.81	15.04	**Africa**
15.69	10.97	14.62	14.22	10.21	6.34	4.83	5.45	3.62	Argentina
6.37	5.86	5.27	5.10	5.74	5.88	6.13	6.19	5.83	Bolivia
2.77	2.24	2.17	2.62	2.47	2.48	2.75	3.20	3.16	Brazil
10.07	10.06	7.77	6.90	8.60	5.04	8.38	8.42	7.67	Chile
0.70	0.74	0.91	1.88	1.07	0.94	0.55	0.54	0.53	Colombia
2.40	2.48	5.38	14.11	10.70	17.37	17.33	9.31	3.05	Costa Rica
91.22	88.34	88.44	86.02	89.41	90.54	92.17	90.49	88.72	Cuba
82.15	87.97	81.10	58.22	67.65	68.36	68.03	76.76	76.97	Dominican Republic
14.13	21.34	27.04	30.70	21.45	19.44	38.67	32.52	29.39	Ecuador
8.07	6.87	26.29	23.84	31.89	41.36	42.95	31.74	46.83	El Salvador
5.14	3.95	4.29	29.01	22.53	23.88	23.86	20.67	18.73	Guatemala
26.61	20.60	25.43	20.09	23.56	15.89	46.99	53.28	64.80	Haiti
3.01	0.69	0.39	5.39	8.95	13.91	9.12	0.29	1.46	Honduras
87.80	88.59	90.26	91.45	92.46	92.77	93.15	93.26	92.95	Jamaica
100.00	100.00	100.00	100.00	100.00	100.00	100.00	100.00	100.00	Netherlands Antilles
27.00	40.53	42.76	51.93	45.21	54.33	57.66	60.89	65.44	Nicaragua
17.35	19.92	31.38	29.78	32.17	37.85	32.85	27.31	32.64	Panama
0.02	0.03	0.02	0.03	0.03	0.01	0.14	0.13	0.12	Paraguay
20.51	19.96	16.52	23.23	18.18	11.04	21.32	18.54	20.74	Peru
0.09	0.08	0.08	0.08	0.08	-	-	-	-	Trinidad-and-Tobago
31.25	5.31	12.20	10.54	8.15	1.46	6.52	12.94	8.67	Uruguay
9.26	8.69	4.89	4.66	4.66	2.71	2.86	2.47	2.69	Venezuela
74.46	78.13	78.22	78.57	79.07	79.26	79.66	79.13	79.19	Other Latin America
10.82	9.75	9.35	9.69	9.13	8.38	9.15	9.09	8.92	**Latin America**

Electricity Generation from Oil (% of total)
Production d'électricité à partir du pétrole (% du total)
Elektrizitätserzeugung auf Ölbasis (in %)
Produzione di energia termoelettrica da prodotti petroliferi (% del totale)
石油からの発電量 (%)
Generación de electricidad a partir del petróleo (% del total)
Производство электроэнергии за счет потребления нефти (в % к общему производству)

	1971	1973	1978	1983	1984	1985	1986	1987	1988
Bangladesh	43.11	41.74	33.84	18.41	18.53	19.43	27.46	15.05	12.09
Brunei	2.67	-	1.11	0.59	0.67	0.65	0.77	0.77	0.82
Inde	7.07	8.30	6.88	4.98	5.56	5.67	4.19	4.73	2.94
Indonésie	53.20	55.77	76.53	84.88	84.20	78.75	45.89	40.77	41.81
Malaisie	72.44	76.63	87.99	84.32	72.63	57.78	53.93	49.49	44.61
Myanmar	23.15	20.71	30.33	21.61	19.84	19.63	15.72	19.27	13.48
Népal	22.09	22.12	12.07	9.49	8.93	7.73	5.38	1.26	1.05
Pakistan	4.16	3.21	0.86	13.51	12.41	14.50	13.89	14.65	17.16
Philippines	78.73	81.91	81.38	55.88	44.66	34.91	34.71	45.24	42.67
Singapour	100.00	100.00	100.00	100.00	100.00	100.00	100.00	100.00	100.00
Sri Lanka	7.33	31.33	1.37	42.43	7.52	2.80	0.26	19.58	7.22
Taipei chinois	68.17	76.66	75.16	31.90	18.81	9.43	12.01	9.06	19.84
Thaïlande	53.61	69.53	79.43	38.31	30.27	14.75	13.62	7.87	9.75
Viêt-Nam	-	-	-	26.64	21.98	23.99	25.28	25.38	21.96
Autre Asie	59.62	75.29	66.57	60.74	54.10	54.50	54.30	49.50	42.09
Asie	25.19	31.17	31.23	23.04	19.21	16.08	14.19	13.98	14.54
Rép. populaire de Chine	6.67	15.70	17.37	16.04	14.33	12.38	12.11	11.00	10.64
Hong-Kong, Chine	100.00	100.00	100.00	46.21	30.55	22.04	15.05	4.79	1.65
Chine	10.29	19.00	20.58	17.39	15.07	12.81	12.24	10.72	10.24
Albanie	42.86	33.78	22.58	15.34	11.89	6.45	33.41	23.78	9.89
Bulgarie	8.17	11.02	26.13	18.20	18.64	18.82	17.87	14.95	2.72
Chypre	100.00	100.00	100.00	100.00	100.00	100.00	100.00	100.00	100.00
Gibraltar	100.00	100.00	100.00	100.00	100.00	100.00	100.00	100.00	100.00
Malte	100.00	100.00	100.00	82.37	79.14	56.38	70.94	63.14	57.57
Roumanie	7.05	9.56	22.58	17.28	16.04	9.41	13.52	12.59	8.89
République slovaque	15.09	17.71	32.89	10.96	8.34	7.57	5.31	5.55	5.03
Bosnie-Herzégovine	-	-	-	-	-	-	-	-	-
Croatie	-	-	-	-	-	-	-	-	-
Ex-RYM	-	-	-	-	-	-	-	-	-
Slovénie	-	-	-	-	-	-	-	-	-
RF de Yougoslavie	-	-	-	-	-	-	-	-	-
Ex-Yougoslavie	2.88	5.34	7.96	5.58	3.16	2.86	2.68	4.15	5.15
Europe non-OCDE	8.27	10.72	20.38	13.76	12.14	9.47	10.96	10.37	6.89
Arménie	-	-	-	-	-	-	-	-	-
Azerbaïdjan	-	-	-	-	-	-	-	-	-
Bélarus	-	-	-	-	-	-	-	-	-
Estonie	-	-	-	-	-	-	-	-	-
Géorgie	-	-	-	-	-	-	-	-	-
Kazakhstan	-	-	-	-	-	-	-	-	-
Kirghizistan	-	-	-	-	-	-	-	-	-
Lettonie	-	-	-	-	-	-	-	-	-
Lituanie	-	-	-	-	-	-	-	-	-
Moldova	-	-	-	-	-	-	-	-	-
Russie	-	-	-	-	-	-	-	-	-
Ukraine	-	-	-	-	-	-	-	-	-
Ouzbékistan	-	-	-	-	-	-	-	-	-
Ex-URSS	18.14	21.92	23.79	22.99	20.35	18.14	16.58	14.93	13.31

Electricity Generation from Oil (% of total)
Production d'électricité à partir du pétrole (% du total)
Elektrizitätserzeugung auf Ölbasis (in %)
Produzione di energia termoelettrica da prodotti petroliferi (% del totale)
石油からの発電量 (%)
Generación de electricidad a partir del petróleo (% del total)
Производство электроэнергии за счет потребления нефти (в % к общему производству)

1989	1990	1991	1992	1993	1994	1995	1996	1997	
8.19	4.31	1.77	4.45	6.79	6.70	8.53	6.46	9.43	Bangladesh
0.88	0.94	0.95	0.85	0.90	0.84	0.76	0.75	0.79	Brunei
4.22	4.51	3.82	3.67	3.45	3.11	2.87	2.76	2.59	India
37.12	45.60	43.00	43.05	46.95	30.48	17.08	25.38	30.00	Indonesia
41.28	43.56	28.77	26.29	21.57	16.28	16.83	12.30	10.20	Malaysia
7.82	10.94	12.74	3.57	8.06	10.57	15.07	8.90	12.15	Myanmar
3.90	0.14	0.34	6.15	8.74	9.92	15.44	9.82	9.54	Nepal
19.06	20.57	22.39	24.65	25.54	29.71	29.41	30.81	38.54	Pakistan
42.88	44.96	49.93	54.61	52.19	52.75	57.16	49.59	48.04	Philippines
100.00	100.00	100.00	91.51	78.59	76.80	77.35	78.69	80.89	Singapore
1.96	0.16	7.73	18.08	4.60	6.79	7.29	28.21	33.00	Sri Lanka
27.94	26.32	27.34	23.06	24.62	23.23	26.30	18.68	21.86	Chinese Taipei
12.73	23.49	25.43	26.45	28.78	29.67	30.48	29.27	23.37	Thailand
15.18	15.17	16.27	15.95	17.75	16.39	4.26	8.62	10.70	Vietnam
42.54	42.99	45.17	45.31	45.45	45.61	47.23	45.78	44.73	Other Asia
16.03	17.54	17.20	17.14	17.28	16.22	15.94	15.16	15.94	**Asia**
7.52	6.60	7.08	8.12	8.68	5.93	6.13	6.96	7.31	People's Rep.of China
2.83	1.69	3.42	4.09	4.91	1.87	2.29	1.98	1.14	Hong Kong, China
7.31	6.38	6.91	7.94	8.53	5.82	6.02	6.83	7.15	**China**
11.96	9.53	3.58	2.83	3.85	3.41	4.76	4.08	3.73	Albania
4.13	4.67	3.47	3.41	4.42	4.34	3.54	3.22	3.05	Bulgaria
100.00	100.00	100.00	100.00	100.00	100.00	100.00	100.00	100.00	Cyprus
100.00	100.00	100.00	100.00	100.00	100.00	100.00	100.00	100.00	Gibraltar
43.82	44.09	58.42	62.42	60.00	76.70	94.49	100.00	100.00	Malta
10.92	18.38	10.80	8.20	9.80	10.49	9.78	10.93	12.01	Romania
4.24	3.44	7.37	7.43	6.77	3.91	4.76	4.86	5.00	Slovak Republic
-	-	-	3.28	4.27	-	-	-	-	*Bosnia-Herzegovina*
-	-	-	35.75	28.83	16.91	27.74	17.75	27.89	*Croatia*
-	-	-	0.97	0.89	0.40	0.59	0.59	0.73	*FYROM*
-	-	-	2.94	2.41	2.26	2.26	2.71	2.62	*Slovenia*
-	-	-	2.42	2.52	2.32	2.39	2.39	2.39	*FR of Yugoslavia*
4.43	4.16	2.20	6.44	6.12	3.99	5.48	4.52	5.63	Former Yugoslavia
7.55	9.53	6.78	8.05	8.58	7.95	8.17	8.17	8.82	**Non-OECD Europe**
-	-	-	43.31	8.67	11.24	7.64	2.04	2.11	Armenia
-	-	-	-	0.52	71.70	72.78	72.86	72.83	Azerbaijan
-	-	-	39.90	38.96	28.13	28.45	29.12	14.83	Belarus
-	-	-	4.47	6.80	1.42	1.18	1.10	1.93	Estonia
-	-	-	3.72	2.77	2.79	1.30	1.11	0.13	Georgia
-	-	-	8.83	8.70	7.18	7.29	7.29	7.29	Kazakhstan
-	-	-	4.47	3.84	3.31	3.94	4.31	4.31	Kyrgyzstan
-	-	-	7.90	10.42	16.78	10.51	20.58	5.31	Latvia
-	-	-	15.19	10.40	13.09	8.26	8.08	10.87	Lithuania
-	-	-	26.41	25.39	11.07	4.25	6.27	4.04	Moldova
-	-	-	9.24	8.68	5.36	9.23	5.28	5.28	Russia
-	-	-	10.14	7.42	4.05	4.29	4.43	4.26	Ukraine
-	-	-	6.87	3.74	3.25	11.23	11.16	11.91	Uzbekistan
13.23	14.26	13.91	10.01	8.77	6.58	9.31	6.72	6.47	**Former USSR**

Electricity Generation from Oil (% of total)
Production d'électricité à partir du pétrole (% du total)
Elektrizitätserzeugung auf Ölbasis (in %)
Produzione di energia termoelettrica da prodotti petroliferi (% del totale)
石油からの発電量 (%)
Generación de electricidad a partir del petróleo (% del total)
Производство электроэнергии за счет потребления нефти (в % к общему производству)

	1971	1973	1978	1983	1984	1985	1986	1987	1988
Iran	60.70	56.21	50.30	58.96	59.11	66.57	61.70	43.18	45.20
Irak	92.86	91.76	90.96	96.29	96.87	97.09	97.31	88.45	88.91
Israël	99.23	100.00	100.00	67.64	50.84	46.52	45.33	48.10	52.96
Jordanie	100.00	100.00	100.00	100.00	100.00	100.00	99.90	99.45	99.17
Koweit	5.67	8.14	4.06	80.42	68.21	80.36	59.26	51.91	18.62
Liban	38.98	73.31	65.22	84.44	84.61	84.85	86.57	86.74	85.00
Oman	100.00	100.00	43.38	27.12	27.13	26.06	22.20	21.17	18.95
Qatar	9.49	9.55	11.34	1.30	1.57	0.79	0.88	0.87	0.80
Arabie saoudite	100.00	100.00	60.21	51.66	54.67	52.58	49.81	52.39	57.40
Syrie	96.21	98.81	23.69	57.59	55.66	58.31	73.08	66.92	41.44
Emirats arabes unis	100.00	100.00	64.56	39.27	35.64	35.39	35.32	29.58	27.83
Yémen	100.00	100.00	100.00	100.00	100.00	100.00	100.00	100.00	100.00
Moyen-Orient	71.47	71.01	59.09	61.57	59.69	61.41	58.23	52.29	50.00
Total non-OCDE	19.18	22.69	24.11	21.93	19.69	17.85	16.98	15.67	14.79
OCDE Amérique du N.	12.93	15.85	16.17	7.22	6.00	5.27	6.59	6.06	6.86
OCDE Pacifique	54.20	63.19	51.37	34.51	29.07	23.91	22.16	21.11	22.44
OCDE Europe	22.71	25.31	19.30	10.58	11.02	8.53	7.81	7.53	7.03
Total OCDE	21.50	25.27	21.90	12.15	10.96	9.01	9.18	8.72	9.15
Monde	20.87	24.56	22.58	15.50	13.99	12.11	11.99	11.24	11.21

Electricity Generation from Oil (% of total)
Production d'électricité à partir du pétrole (% du total)
Elektrizitätserzeugung auf Ölbasis (in %)
Produzione di energia termoelettrica da prodotti petroliferi (% del totale)
石油からの発電量(%)
Generación de electricidad a partir del petróleo (% del total)
Производство электроэнергии за счет потребления нефти (в % к общему производству)

1989	1990	1991	1992	1993	1994	1995	1996	1997	
41.27	37.26	37.54	34.04	38.40	36.92	35.49	37.14	33.89	Iran
89.10	89.17	95.68	97.23	97.72	98.00	98.03	98.03	98.03	Iraq
52.00	50.08	47.07	42.75	38.14	39.80	37.72	30.89	29.11	Israel
91.32	87.77	88.53	89.33	88.20	84.48	86.40	87.59	87.15	Jordan
17.83	17.52	29.68	16.68	16.35	18.49	21.68	21.68	25.95	Kuwait
80.00	66.67	81.33	80.91	84.75	84.24	86.42	88.54	89.42	Lebanon
17.83	18.37	18.89	18.76	18.86	19.17	19.64	17.45	16.47	Oman
0.78	-	-	-	-	-	-	-	-	Qatar
60.40	61.48	59.54	58.95	60.65	57.02	57.18	57.05	57.51	Saudi Arabia
36.81	32.41	26.85	17.15	26.97	37.22	32.10	28.63	26.34	Syria
26.71	26.29	21.55	21.26	19.22	17.90	17.71	17.12	16.24	United Arab Emirates
100.00	100.00	100.00	100.00	100.00	100.00	100.00	100.00	100.00	Yemen
48.80	47.23	48.04	45.78	47.25	46.37	45.97	45.33	44.55	**Middle East**
14.43	14.89	14.59	13.44	13.37	12.05	12.85	12.12	12.40	**Non-OECD Total**
7.03	5.73	5.46	4.91	5.21	5.23	4.13	4.29	4.91	OECD North America
23.54	24.09	23.15	23.88	20.00	21.99	19.13	17.72	15.46	OECD Pacific
7.96	7.82	8.22	8.89	7.63	7.43	7.66	7.38	6.87	OECD Europe
9.74	9.24	9.13	9.21	8.32	8.69	7.76	7.53	7.34	**OECD Total**
11.43	11.29	11.10	10.74	10.14	9.89	9.59	9.19	9.19	**World**

Electricity Generation from Gas (% of total)
Production d'électricité à partir du gaz (% du total)
Elektrizitätserzeugung auf Gasbasis (in %)
Produzione di energia termoelettrica da gas (% del totale)
ガスからの発電量 (%)
Generación de electricidad a partir del gas natural (% del total)
Производство электроэнергии за счет потребления газа (в % к общему производству)

	1971	1973	1978	1983	1984	1985	1986	1987	1988
Algérie	38.85	54.95	76.06	85.98	85.44	85.40	88.85	87.43	90.76
Congo	37.50	47.92	10.91	0.86	0.85	0.64	0.74	1.07	1.03
Egypte	-	-	-	26.97	29.21	31.75	31.65	33.93	34.03
Gabon	-	-	-	-	-	-	-	-	-
Mozambique	-	-	-	-	-	0.55	0.32	-	-
Nigéria	6.17	11.62	16.14	39.00	31.64	33.55	26.59	34.10	25.79
Sénégal	-	-	-	-	-	-	-	-	2.45
Tunisie	-	32.74	39.15	31.88	58.35	73.33	42.26	66.15	55.66
Afrique	1.12	2.09	3.71	9.14	9.51	10.29	9.76	10.72	10.59
Argentine	14.49	18.38	15.48	22.74	22.79	24.92	19.90	23.57	31.34
Bolivie	2.64	4.08	14.03	19.41	19.10	20.92	24.53	29.05	28.63
Brésil	-	-	-	-	-	-	-	-	0.00
Chili	1.01	1.03	0.97	1.24	1.30	1.14	0.92	0.94	1.06
Colombie	8.76	11.03	16.00	22.06	21.07	19.00	17.59	15.59	17.77
Cuba	-	-	-	-	-	-	-	0.04	0.07
Pérou	-	-	2.43	0.64	1.96	1.82	1.94	2.00	1.70
Trinité-et-Tobago	92.53	95.38	96.66	98.90	98.60	99.11	98.39	98.91	99.14
Vénézuela	44.61	45.80	40.25	43.63	41.80	41.31	38.63	32.42	31.03
Autre Amérique latine	-	-	-	-	-	-	0.41	0.27	0.63
Amérique latine	8.38	9.19	8.35	10.76	10.29	10.18	8.99	8.67	9.50
Bangladesh	39.90	34.69	43.35	62.31	58.85	64.24	63.17	75.69	77.59
Brunei	97.33	100.00	98.89	99.41	99.33	99.35	99.23	99.23	99.18
Inde	0.46	0.47	0.50	1.52	1.27	1.20	1.19	1.55	1.49
Indonésie	-	-	-	-	-	0.02	1.02	1.43	1.73
Malaisie	-	-	1.14	2.07	2.45	16.57	20.05	21.56	23.66
Myanmar	3.91	6.58	8.53	17.07	25.13	31.05	35.59	34.87	42.86
Pakistan	44.66	40.38	36.79	27.42	27.31	30.58	30.33	30.38	31.55
Philippines	-	-	-	-	-	-	-	-	-
Singapour	-	-	-	-	-	-	-	-	-
Taipei chinois	-	-	-	-	-	-	-	-	-
Thaïlande	-	-	-	32.71	39.30	46.23	41.48	54.53	57.66
Viêt-Nam	-	-	-	3.06	2.37	1.42	1.19	1.09	0.94
Asie	3.10	2.73	2.64	4.94	5.27	6.41	6.27	7.67	8.23
Rép. populaire de Chine	-	-	-	0.18	0.24	0.28	0.60	0.49	0.40
Chine	-	-	-	0.17	0.23	0.27	0.57	0.46	0.38
Albanie	-	-	-	-	-	-	-	-	-
Bulgarie	-	-	-	-	-	7.82	5.60	6.44	19.32
Roumanie	55.36	48.29	32.79	52.66	53.02	40.44	39.27	37.52	34.51
République slovaque	4.99	5.26	11.48	7.09	6.21	2.88	5.34	2.94	1.73
Croatie	-	-	-	-	-	-	-	-	-
Slovénie	-	-	-	-	-	-	-	-	-
RF de Yougoslavie	-	-	-	-	-	-	-	-	-
Ex-Yougoslavie	0.79	1.87	1.86	2.88	1.93	2.36	2.14	1.98	2.41
Europe non-OCDE	21.94	20.07	14.25	19.70	18.90	16.10	15.43	14.39	15.91

Electricity Generation from Gas (% of total)
Production d'électricité à partir du gaz (% du total)
Elektrizitätserzeugung auf Gasbasis (in %)
Produzione di energia termoelettrica da gas (% del totale)
ガスからの発電量 (%)
Generación de electricidad a partir del gas natural (% del total)
Производство электроэнергии за счет потребления газа (в % к общему производству)

1989	1990	1991	1992	1993	1994	1995	1996	1997	
91.18	93.77	93.93	94.29	94.09	95.51	95.68	95.74	96.09	Algeria
0.76	0.59	0.62	0.70	0.70	0.69	0.68	0.69	0.70	Congo
33.52	34.05	36.45	39.23	41.52	43.73	42.97	45.70	43.97	Egypt
11.53	10.93	11.27	11.32	11.17	11.15	11.06	11.44	10.53	Gabon
0.21	-	-	-	0.26	0.25	0.25	0.21	0.20	Mozambique
29.91	28.46	22.96	22.25	33.50	35.40	37.96	37.17	37.33	Nigeria
2.82	1.97	1.49	0.78	4.17	5.67	13.78	11.93	6.11	Senegal
52.35	66.86	37.28	56.28	50.15	74.80	84.18	84.96	83.74	Tunisia
11.03	11.63	11.67	12.76	13.67	14.63	14.73	14.86	14.71	**Africa**
39.93	35.01	36.55	36.13	34.56	32.96	40.01	48.32	44.93	Argentina
33.23	36.52	37.14	43.70	37.79	35.80	36.69	29.58	25.72	Bolivia
0.10	0.17	0.17	0.16	0.15	0.18	0.20	0.33	0.36	Brazil
1.12	1.09	1.30	1.25	1.50	1.19	1.08	1.04	2.15	Chile
13.73	14.38	14.87	17.07	16.47	13.83	14.71	12.79	19.75	Colombia
0.07	0.16	0.11	0.04	0.13	0.34	0.06	0.08	0.07	Cuba
1.67	1.75	1.73	1.38	1.53	1.56	1.36	1.33	1.84	Peru
99.15	99.05	99.25	99.17	99.19	99.21	99.23	99.27	99.28	Trinidad-and-Tobago
30.57	28.97	24.78	25.25	26.91	25.28	27.09	26.09	20.97	Venezuela
0.49	0.42	0.40	0.39	0.38	0.33	0.37	0.28	0.27	Other Latin America
9.90	9.07	8.84	9.18	9.17	8.60	9.47	10.06	9.47	**Latin America**
78.88	84.26	88.10	86.60	86.61	84.64	88.03	87.10	84.51	Bangladesh
99.12	99.06	99.05	99.15	99.10	99.16	99.24	99.25	99.21	Brunei
1.41	3.44	4.23	4.89	5.02	5.68	5.86	6.25	5.95	India
2.72	3.88	3.63	3.85	7.96	28.36	35.71	31.98	27.82	Indonesia
21.32	26.07	42.07	47.27	55.51	57.72	62.21	71.23	78.83	Malaysia
40.98	39.31	39.07	45.59	41.42	44.38	44.88	49.25	48.49	Myanmar
31.74	33.63	32.05	33.34	29.93	30.84	26.88	26.83	24.95	Pakistan
-	-	-	-	-	0.04	0.04	0.05	0.03	Philippines
-	-	-	8.49	21.41	23.20	22.65	18.65	16.47	Singapore
0.03	1.23	2.78	2.73	2.14	4.14	4.18	4.43	5.74	Chinese Taipei
51.32	40.22	39.45	40.18	44.09	44.13	42.34	42.01	46.30	Thailand
0.49	0.07	0.11	-	-	-	6.47	5.10	4.80	Vietnam
7.88	8.91	10.18	11.63	12.84	15.01	15.81	16.67	17.14	**Asia**
0.59	0.54	0.44	0.34	0.32	0.29	0.25	0.22	0.59	People's Rep.of China
0.56	0.52	0.42	0.33	0.31	0.28	0.25	0.22	0.57	**China**
0.49	0.53	0.43	0.48	0.46	-	-	-	-	Albania
18.68	20.64	7.41	7.77	8.18	6.82	7.88	7.19	5.87	Bulgaria
34.63	35.10	36.03	35.93	32.44	30.05	26.95	27.24	17.65	Romania
2.26	4.93	6.90	6.70	5.13	8.93	9.14	9.34	9.60	Slovak Republic
-	-	-	15.37	19.50	22.22	10.06	12.50	12.10	*Croatia*
-	-	-	3.21	-	-	-	-	-	*Slovenia*
-	-	-	3.60	3.89	3.62	3.73	3.73	3.73	*FR of Yugoslavia*
2.63	2.87	3.90	4.05	4.38	4.91	3.40	3.91	3.71	Former Yugoslavia
15.93	15.89	13.57	13.75	13.02	12.96	11.82	11.90	8.54	**Non-OECD Europe**

Electricity Generation from Gas (% of total)
Production d'électricité à partir du gaz (% du total)
Elektrizitätserzeugung auf Gasbasis (in %)
Produzione di energia termoelettrica da gas (% del totale)
ガスからの発電量(%)
Generación de electricidad a partir del gas natural (% del total)
Производство электроэнергии за счет потребления газа (в % к общему производству)

	1971	1973	1978	1983	1984	1985	1986	1987	1988
Arménie	-	-	-	-	-	-	-	-	-
Azerbaïdjan	-	-	-	-	-	-	-	-	-
Bélarus	-	-	-	-	-	-	-	-	-
Estonie	-	-	-	-	-	-	-	-	-
Géorgie	-	-	-	-	-	-	-	-	-
Kazakhstan	-	-	-	-	-	-	-	-	-
Kirghizistan	-	-	-	-	-	-	-	-	-
Lettonie	-	-	-	-	-	-	-	-	-
Lituanie	-	-	-	-	-	-	-	-	-
Moldova	-	-	-	-	-	-	-	-	-
Russie	-	-	-	-	-	-	-	-	-
Tadjikistan	-	-	-	-	-	-	-	-	-
Turkménistan	-	-	-	-	-	-	-	-	-
Ukraine	-	-	-	-	-	-	-	-	-
Ouzbékistan	-	-	-	-	-	-	-	-	-
Ex-URSS	19.15	18.45	20.83	26.82	28.42	30.29	32.19	32.58	33.20
Bahrein	100.00	100.00	100.00	100.00	100.00	100.00	100.00	100.00	100.00
Iran	5.39	17.89	17.56	22.25	25.18	19.28	20.21	38.66	39.44
Israël	0.77	-	-	-	-	-	-	-	-
Jordanie	-	-	-	-	-	-	-	-	-
Koweit	94.33	91.86	95.94	19.58	31.79	19.64	40.74	48.09	81.38
Oman	-	-	56.62	72.88	72.87	73.94	77.80	78.83	81.05
Qatar	90.51	90.45	88.66	98.70	98.43	99.21	99.12	99.13	99.20
Arabie saoudite	-	-	39.79	48.34	45.33	47.42	50.19	47.61	42.60
Syrie	-	-	3.34	3.04	4.98	5.48	6.36	6.51	8.56
Emirats arabes unis	-	-	35.44	60.73	64.36	64.61	64.68	70.42	72.17
Moyen-Orient	14.89	18.59	26.87	27.97	29.49	28.15	31.35	36.12	38.48
Total non-OCDE	13.67	13.17	13.83	17.56	18.21	18.80	19.37	19.71	20.03
OCDE Amérique du N.	21.09	17.00	12.66	10.56	10.77	10.31	8.77	9.27	8.32
OCDE Pacifique	1.58	2.40	9.08	13.16	16.67	16.64	16.71	17.15	16.99
OCDE Europe	5.47	7.44	7.88	5.76	5.96	5.26	5.28	5.45	5.24
Total OCDE	13.02	11.69	10.47	9.18	9.84	9.34	8.58	8.99	8.45
Monde	13.20	12.10	11.50	12.05	12.74	12.65	12.46	12.88	12.69

Electricity Generation from Gas (% of total)
Production d'électricité à partir du gaz (% du total)
Elektrizitätserzeugung auf Gasbasis (in %)
Produzione di energia termoelettrica da gas (% del totale)
ガスからの発電量 (%)
Generación de electricidad a partir del gas natural (% del total)
Производство электроэнергии за счет потребления газа (в % к общему производству)

1989	1990	1991	1992	1993	1994	1995	1996	1997	
-	-	-	22.88	23.13	26.65	52.38	35.26	48.25	Armenia
-	-	-	91.12	86.91	17.89	18.09	18.14	18.13	Azerbaijan
-	-	-	47.88	50.05	71.81	71.47	70.81	85.09	Belarus
-	-	-	5.51	3.58	3.75	2.31	2.18	2.61	Estonia
-	-	-	39.73	27.93	27.93	21.74	15.29	15.60	Georgia
-	-	-	10.56	10.02	8.06	8.18	8.18	8.18	Kazakhstan
-	-	-	12.11	10.38	-	-	-	-	Kyrgyzstan
-	-	-	26.37	14.30	7.03	13.22	16.77	26.77	Latvia
-	-	-	4.90	-	3.28	1.54	4.07	3.50	Lithuania
-	-	-	36.91	36.29	47.22	64.12	65.53	79.46	Moldova
-	-	-	44.62	40.66	42.09	40.12	44.65	45.25	Russia
-	-	-	5.31	3.51	1.71	1.16	1.17	1.25	Tajikistan
-	-	-	99.97	99.96	99.96	99.96	99.95	99.95	Turkmenistan
-	-	-	31.23	26.78	28.85	21.29	18.59	17.89	Ukraine
-	-	-	75.90	74.80	73.47	72.28	70.45	71.47	Uzbekistan
34.49	33.83	33.96	41.09	37.76	38.63	36.09	38.67	39.77	**Former USSR**
100.00	100.00	100.00	100.00	100.00	100.00	100.00	100.00	100.00	Bahrain
44.46	52.45	51.46	52.32	48.68	54.00	55.95	54.74	58.41	Iran
-	-	-	-	-	-	-	-	-	Israel
8.19	11.90	11.25	10.31	11.32	15.23	13.27	12.03	12.48	Jordan
82.17	82.48	70.32	83.32	83.65	81.51	78.32	78.32	74.05	Kuwait
82.17	81.63	81.11	81.24	81.14	80.83	80.36	82.55	83.53	Oman
99.22	100.00	100.00	100.00	100.00	100.00	100.00	100.00	100.00	Qatar
39.60	38.52	40.46	41.05	39.35	42.98	42.82	42.95	42.49	Saudi Arabia
17.98	18.95	21.84	24.06	19.94	17.99	22.52	30.25	17.74	Syria
73.29	73.71	78.45	78.74	80.78	82.10	82.29	82.88	83.76	United Arab Emirates
40.03	42.17	40.94	41.97	40.97	43.27	43.50	43.63	43.49	**Middle East**
20.41	20.10	19.52	21.76	19.90	19.79	18.74	19.26	19.15	**Non-OECD Total**
10.60	10.70	11.00	11.61	11.62	12.66	13.37	11.77	12.46	OECD North America
16.98	17.15	17.31	17.31	17.14	16.95	16.77	17.46	17.80	OECD Pacific
6.10	6.25	6.18	6.09	7.33	8.44	9.28	10.56	12.37	OECD Europe
9.95	10.14	10.31	10.60	11.03	11.95	12.57	12.31	13.33	**OECD Total**
13.73	13.75	13.64	14.62	14.23	14.75	14.79	14.82	15.47	**World**

Electricity Generation from Nuclear Energy (% of total)
Production d'électricité à partir d'énergie nucléaire (% du total)
Elektrizitätserzeugung auf Kernkraftbasis (in %)
Produzione di energia nucleotermoelettrica (% del totale)
原子力からの発電量 (%)
Generación de electricidad a partir de energía nuclear (% del total)
Производство атомной электроэнергии (в % к общему производству)

	1971	1973	1978	1983	1984	1985	1986	1987	1988
Afrique du Sud	-	-	-	-	2.90	3.76	6.01	4.09	6.69
Afrique	-	-	-	-	1.61	2.03	3.21	2.15	3.48
Argentine	-	-	8.66	7.92	10.32	12.74	12.83	13.09	11.46
Brésil	-	-	-	-	0.92	1.75	0.07	0.48	0.28
Amérique latine	-	-	1.19	0.99	1.68	2.31	1.39	1.68	1.38
Inde	1.79	3.29	2.52	2.35	2.41	2.72	2.50	2.30	2.41
Pakistan	1.37	3.63	1.87	1.16	1.48	1.50	1.68	1.75	0.77
Taipei chinois	-	-	7.45	39.14	47.09	51.71	43.22	47.89	40.19
Asie	0.94	1.65	2.26	6.46	7.50	8.17	7.15	7.83	6.74
Rép. populaire de Chine	-	-	-	-	-	-	-	-	-
Chine	-	-	-	-	-	-	-	-	-
Bulgarie	-	-	18.77	28.88	28.51	31.54	28.86	28.61	35.61
Roumanie	-	-	-	-	-	-	-	-	-
République slovaque	-	1.89	0.12	31.98	36.29	42.75	49.37	49.79	51.01
Slovénie	-	-	-	-	-	-	-	-	-
Ex-Yougoslavie	-	-	-	5.88	6.05	5.42	5.16	5.56	4.94
Europe non-OCDE	-	0.19	3.55	10.95	11.34	12.33	12.28	12.46	13.57
Arménie	-	-	-	-	-	-	-	-	-
Lituanie	-	-	-	-	-	-	-	-	-
Russie	-	-	-	-	-	-	-	-	-
Ukraine	-	-	-	-	-	-	-	-	-
Ex-URSS	0.76	1.31	4.16	7.76	9.52	10.82	10.07	11.35	12.65
Total non-OCDE	0.51	0.89	2.74	5.19	6.30	7.04	6.50	7.03	7.47
OCDE Amérique du N.	2.29	4.59	11.93	12.25	12.94	14.72	15.92	16.88	18.47
OCDE Pacifique	1.73	1.72	8.84	15.50	17.47	20.21	22.09	23.96	21.83
OCDE Europe	3.66	4.62	9.15	18.58	22.75	26.15	27.79	27.72	29.01
Total OCDE	2.72	4.24	10.53	14.98	17.10	19.64	21.12	21.84	22.74
Monde	2.12	3.32	8.14	11.63	13.35	15.22	15.86	16.46	17.16

Electricity Generation from Nuclear Energy (% of total)
Production d'électricité à partir d'énergie nucléaire (% du total)
Elektrizitätserzeugung auf Kernkraftbasis (in %)
Produzione di energia nucleotermoelettrica (% del totale)
原子力からの発電量 (%)
Generación de electricidad a partir de energía nuclear (% del total)
Производство атомной электроэнергии (в % к общему производству)

1989	1990	1991	1992	1993	1994	1995	1996	1997	
6.84	5.11	5.43	5.53	4.19	5.33	6.05	5.94	6.09	South Africa
3.54	2.64	2.77	2.79	2.11	2.71	3.06	3.08	3.17	**Africa**
9.91	15.36	15.32	12.99	13.17	13.37	10.94	10.69	10.97	Argentina
0.83	1.00	0.62	0.73	0.18	0.02	0.91	0.83	1.03	Brazil
1.43	1.96	1.80	1.68	1.47	1.42	1.55	1.50	1.61	**Latin America**
1.72	2.12	1.75	2.02	1.51	1.46	1.82	1.93	2.18	India
0.09	0.78	0.94	0.92	1.19	0.98	0.95	0.85	0.59	Pakistan
33.64	36.32	35.40	31.76	29.50	27.69	26.11	27.93	23.66	Chinese Taipei
5.52	6.09	5.89	5.62	5.13	4.81	4.71	4.78	4.43	**Asia**
-	-	-	-	0.19	1.50	1.27	1.33	1.27	People's Rep.of China
-	-	-	-	0.18	1.46	1.24	1.29	1.24	**China**
32.86	34.80	33.88	32.44	37.59	40.91	42.39	43.60	42.71	Bulgaria
-	-	-	-	-	-	-	2.26	9.45	Romania
51.95	51.37	52.51	50.34	47.88	49.56	43.92	44.94	44.38	Slovak Republic
-	-	-	32.86	33.84	36.49	37.78	35.72	38.12	*Slovenia*
5.62	5.58	6.05	5.24	5.54	7.33	7.13	6.51	6.96	Former Yugoslavia
13.41	14.29	14.40	13.64	14.87	17.05	16.60	16.95	18.99	**Non-OECD Europe**
-	-	-	-	-	-	5.47	37.40	26.57	Armenia
-	-	-	78.25	86.81	79.00	87.44	85.84	83.58	Lithuania
-	-	-	11.86	12.47	11.18	11.59	12.88	13.08	Russia
-	-	-	29.20	32.76	33.96	36.39	43.54	44.67	Ukraine
12.35	12.25	12.65	13.33	14.16	13.16	14.12	16.25	16.37	**Former USSR**
7.05	7.00	6.90	6.68	6.51	6.10	6.14	6.54	6.38	**Non-OECD Total**
17.13	18.16	19.00	18.88	18.41	19.03	19.19	18.58	17.18	OECD North America
21.58	22.28	22.73	23.15	24.86	24.46	25.77	26.16	26.54	OECD Pacific
29.98	29.56	29.99	30.24	31.32	30.79	30.56	31.19	31.17	OECD Europe
22.25	22.73	23.35	23.44	23.77	23.82	24.04	24.01	23.39	**OECD Total**
16.76	17.02	17.40	17.40	17.54	17.49	17.60	17.69	17.16	**World**

Electricity Generation from Hydro Energy (% of total)
Production d'électricité à partir d'énergie hydraulique (% du total)
Elektrizitätserzeugung auf Wasserkraftbasis (in %)
Produzione di energia idroelettrica (% del totale)
水力からの発電量 (%)
Generación de electricidad a partir de energía hidráulica (% del total)
Производство гидроэлектроэнергии (в % к общему производству)

	1971	1973	1978	1983	1984	1985	1986	1987	1988
Algérie	14.80	26.80	4.56	2.30	4.04	5.26	1.93	3.92	1.31
Angola	81.54	82.72	90.26	73.82	74.57	86.97	86.97	86.54	86.01
Cameroun	97.43	95.44	94.17	96.53	98.33	97.83	98.10	97.73	98.06
Congo	62.50	52.08	89.09	98.28	98.30	98.73	98.52	97.86	97.95
RD du Congo	96.95	97.92	98.04	96.99	97.10	97.22	97.28	97.24	97.33
Egypte	63.03	63.61	61.52	37.93	33.16	27.54	27.73	22.88	22.74
Ethiopie	51.26	56.35	71.60	73.26	75.82	76.01	78.36	80.83	81.19
Gabon	-	3.03	67.89	75.03	76.73	77.47	78.04	77.99	77.91
Ghana	98.81	99.03	98.62	99.22	98.25	99.01	99.32	96.10	99.30
Côte d'Ivoire	23.64	21.11	11.78	60.76	22.13	73.72	67.98	42.29	53.70
Kenya	54.07	51.45	77.64	91.39	88.46	93.55	72.82	72.00	83.80
Maroc	66.38	41.46	32.27	7.19	5.29	6.62	8.28	10.31	10.42
Mozambique	33.23	29.80	91.78	56.59	61.31	68.18	55.81	53.57	49.26
Nigéria	86.67	71.27	50.00	22.83	29.53	29.63	33.72	29.15	34.39
Afrique du Sud	0.20	1.53	2.26	0.49	0.41	0.44	1.11	1.07	2.02
Soudan	39.76	65.47	58.44	69.99	69.99	59.97	65.17	65.19	64.73
Tanzanie	62.93	50.86	74.82	87.04	84.82	87.19	90.23	90.49	90.86
Tunisie	5.62	6.19	1.39	0.82	1.70	2.69	1.19	2.48	0.95
Zambie	74.67	91.95	98.28	99.18	99.12	99.25	99.14	99.16	99.10
Zimbabwe	67.32	67.42	88.69	83.38	73.79	59.71	45.56	17.46	33.92
Autre Afrique	31.54	35.57	32.72	45.11	45.51	47.76	46.45	45.21	42.35
Afrique	24.82	26.58	28.04	20.24	18.19	18.35	18.49	16.68	17.47
Argentine	6.54	11.23	23.19	42.83	44.21	45.61	47.20	44.34	31.22
Bolivie	87.70	84.45	75.16	68.28	69.04	67.30	64.77	61.65	62.13
Brésil	84.91	90.66	91.34	93.22	92.87	92.10	90.25	91.28	92.62
Chili	51.58	60.68	65.85	70.76	69.14	73.77	76.46	78.92	67.81
Colombie	70.14	63.03	69.93	62.98	64.40	68.67	72.85	74.35	73.57
Costa Rica	89.55	83.96	76.78	97.07	97.95	97.87	98.03	95.75	95.40
Cuba	2.19	1.09	0.98	0.54	0.57	0.44	0.45	0.32	0.50
République dominicaine	55.24	25.96	28.46	30.33	33.64	30.13	35.24	36.22	39.45
Equateur	41.90	34.63	30.36	40.00	71.27	71.94	80.05	84.88	85.55
El Salvador	70.26	53.95	44.73	51.11	52.89	54.14	60.33	47.81	54.26
Guatemala	37.74	26.66	17.73	83.17	79.50	76.63	93.54	89.28	90.57
Haiti	32.93	68.85	77.55	47.93	47.39	48.98	71.78	64.78	73.27
Honduras	61.89	73.87	85.24	85.39	86.69	93.20	96.70	96.70	96.51
Jamaïque	7.52	4.53	7.58	9.00	9.12	9.17	8.71	6.74	5.52
Nicaragua	30.90	51.74	27.94	34.68	32.85	34.72	33.53	37.54	39.43
Panama	8.74	7.83	44.33	34.78	60.50	74.11	76.33	71.93	80.23
Paraguay	65.25	73.88	43.76	77.61	90.92	98.53	99.41	99.60	99.73
Pérou	72.00	71.66	70.72	75.99	73.95	77.56	77.67	77.67	76.88
Uruguay	61.70	61.50	53.53	97.64	98.12	97.76	98.17	95.83	77.44
Vénézuela	40.56	38.72	47.26	42.24	44.72	46.24	49.24	56.98	59.20
Autre Amérique latine	34.19	27.19	20.97	17.66	20.19	21.49	16.56	7.84	10.42
Amérique latine	54.36	57.55	64.64	69.50	70.96	72.21	73.11	74.11	73.16

Electricity Generation from Hydro Energy (% of total)
Production d'électricité à partir d'énergie hydraulique (% du total)
Elektrizitätserzeugung auf Wasserkraftbasis (in %)
Produzione di energia idroelettrica (% del totale)
水力からの発電量 (%)
Generación de electricidad a partir de energía hidráulica (% del total)
Производство гидроэлектроэнергии (в % к общему производству)

1989	1990	1991	1992	1993	1994	1995	1996	1997	
1.47	0.84	1.69	1.09	1.82	0.83	0.98	0.65	0.35	Algeria
86.69	90.97	82.66	88.70	93.68	93.72	93.75	89.98	90.71	Angola
98.45	98.48	98.56	98.72	98.94	98.82	98.85	98.83	98.85	Cameroon
98.49	98.82	98.76	98.60	98.61	98.62	98.63	98.62	98.61	Congo
99.54	99.56	99.58	99.69	98.22	97.87	96.20	97.92	97.84	DR of Congo
23.95	22.84	21.48	20.65	19.58	19.25	19.71	21.22	20.79	Egypt
82.03	85.23	85.20	85.08	86.70	85.79	85.09	85.03	87.31	Ethiopia
77.05	77.05	77.35	77.48	77.01	76.96	77.13	75.67	73.49	Gabon
99.45	99.83	99.89	99.98	99.65	99.54	99.97	99.88	99.89	Ghana
72.41	72.44	68.00	14.80	48.74	42.49	60.17	62.09	62.93	Ivory Coast
85.11	81.64	85.84	86.97	88.13	86.69	83.35	81.56	82.35	Kenya
12.82	12.67	13.32	9.92	4.47	7.66	5.08	15.68	15.71	Morocco
60.17	62.56	68.58	78.13	87.76	92.66	92.89	93.70	78.71	Mozambique
32.31	34.92	41.86	40.85	38.41	35.81	37.98	36.69	36.85	Nigeria
1.70	0.61	1.18	0.45	0.08	0.59	0.28	0.66	1.01	South Africa
65.57	63.23	62.49	66.65	64.71	60.33	52.15	52.06	53.00	Sudan
94.04	95.15	94.73	90.86	90.56	87.61	83.60	87.80	74.92	Tanzania
0.66	0.80	1.83	1.05	1.01	0.60	0.53	0.89	0.55	Tunisia
98.60	99.38	99.48	99.49	99.46	99.46	99.46	99.46	99.48	Zambia
38.85	40.49	39.93	36.62	26.38	26.69	30.02	28.50	24.35	Zimbabwe
43.27	43.42	41.59	41.82	40.60	40.52	41.32	41.80	41.88	Other Africa
17.77	16.89	17.71	16.96	16.01	15.51	15.57	15.92	15.74	**Africa**
32.80	37.16	31.60	35.12	40.05	43.68	41.27	32.95	39.14	Argentina
58.06	55.32	55.21	48.80	54.99	57.01	55.89	63.03	67.17	Bolivia
92.31	92.77	92.92	92.39	93.29	93.32	92.11	91.27	90.81	Brazil
53.92	48.60	65.77	74.84	71.63	67.17	65.65	54.82	55.73	Chile
76.82	75.69	74.20	65.91	72.28	76.96	73.21	79.10	71.12	Colombia
97.60	97.52	94.62	85.89	89.30	82.63	73.05	79.57	85.87	Costa Rica
0.54	0.61	0.79	0.70	0.75	0.41	0.59	0.72	0.66	Cuba
16.01	10.68	14.33	34.65	26.42	26.69	26.96	18.48	18.24	Dominican Republic
85.87	78.66	72.96	69.30	78.55	80.56	61.33	67.48	70.61	Ecuador
58.12	64.23	36.42	40.75	41.57	37.89	36.86	47.78	32.78	El Salvador
90.79	92.27	91.78	70.66	70.59	69.31	70.04	74.24	76.78	Guatemala
70.61	76.55	70.94	75.89	71.73	77.15	49.13	43.52	32.00	Haiti
96.99	99.31	99.61	94.61	91.05	86.09	90.88	99.71	98.54	Honduras
7.04	5.46	5.84	4.96	2.37	2.32	2.06	2.12	1.93	Jamaica
42.31	28.81	22.91	16.23	28.73	22.72	22.67	22.45	21.34	Nicaragua
80.01	77.08	65.79	67.42	66.03	61.65	66.33	71.93	66.63	Panama
99.89	99.90	99.92	99.91	99.87	99.94	99.66	99.66	99.67	Paraguay
76.57	77.26	80.79	74.29	79.55	86.68	76.65	77.11	73.60	Peru
67.88	94.16	87.09	89.02	91.49	98.02	92.85	86.48	90.74	Uruguay
60.16	62.34	70.32	70.09	68.43	72.01	70.05	71.44	76.34	Venezuela
23.67	20.03	20.00	19.70	19.24	19.10	18.70	19.34	19.34	Other Latin America
73.26	74.42	75.50	74.98	76.15	77.14	75.21	74.29	74.78	**Latin America**

Electricity Generation from Hydro Energy (% of total)
Production d'électricité à partir d'énergie hydraulique (% du total)
Elektrizitätserzeugung auf Wasserkraftbasis (in %)
Produzione di energia idroelettrica (% del totale)
水力からの発電量 (%)
Generación de electricidad a partir de energía hidráulica (% del total)
Производство гидроэлектроэнергии (в % к общему производству)

	1971	1973	1978	1983	1984	1985	1986	1987	1988
Bangladesh	16.99	23.58	22.80	19.28	22.62	16.32	9.38	9.25	10.32
Inde	42.23	39.81	42.83	33.10	31.89	27.83	26.76	21.67	23.99
Indonésie	46.80	44.23	23.47	13.56	14.33	17.69	25.37	23.26	23.01
RPD de Corée	68.60	62.50	60.94	63.41	60.00	58.33	58.00	58.17	54.72
Malaisie	27.56	23.37	10.87	13.61	24.92	25.65	26.02	28.95	30.29
Myanmar	69.03	70.16	59.21	59.28	53.49	47.33	46.41	44.14	42.00
Népal	77.91	77.88	87.93	90.51	91.07	92.27	94.62	98.74	98.95
Pakistan	48.59	51.99	60.14	57.70	58.64	53.21	53.95	53.13	50.44
Philippines	21.14	18.00	18.09	18.50	24.92	24.39	27.60	23.17	25.53
Sri Lanka	92.67	68.67	98.63	57.57	92.48	97.20	99.74	80.42	92.78
Taipei chinois	19.55	16.39	13.85	10.33	8.48	12.47	11.90	10.29	8.76
Thaïlande	40.29	26.97	16.70	19.41	19.41	16.00	22.47	14.22	11.64
Viêt-Nam	26.70	17.87	22.22	30.68	33.47	29.14	25.46	22.87	26.40
Autre Asie	40.38	24.71	33.43	39.26	45.90	45.50	45.70	50.50	57.91
Asie	40.61	36.18	36.27	31.15	31.28	29.30	29.35	25.74	26.10
Rép. populaire de Chine	21.68	22.78	17.38	24.57	22.99	22.49	21.03	20.11	20.02
Chine	20.84	21.89	16.71	23.47	21.95	21.49	20.07	19.20	19.13
Albanie	57.14	66.22	77.42	84.66	88.11	93.55	66.59	76.22	90.11
Bulgarie	10.33	11.71	9.24	7.86	7.30	5.37	5.56	5.84	5.77
Roumanie	11.39	16.13	16.52	14.29	15.80	17.70	14.32	15.13	18.09
République slovaque	13.58	10.75	14.64	9.53	7.80	9.65	7.06	8.36	7.80
Bosnie-Herzégovine	-	-	-	-	-	-	-	-	-
Croatie	-	-	-	-	-	-	-	-	-
Ex-RYM	-	-	-	-	-	-	-	-	-
Slovénie	-	-	-	-	-	-	-	-	-
RF de Yougoslavie	-	-	-	-	-	-	-	-	-
Ex-Yougoslavie	53.01	46.76	49.17	32.82	35.50	32.45	35.30	32.49	30.93
Europe non-OCDE	23.75	24.33	25.98	19.70	21.08	20.55	20.19	19.83	20.34
Arménie	-	-	-	-	-	-	-	-	-
Azerbaïdjan	-	-	-	-	-	-	-	-	-
Bélarus	-	-	-	-	-	-	-	-	-
Estonie	-	-	-	-	-	-	-	-	-
Géorgie	-	-	-	-	-	-	-	-	-
Kazakhstan	-	-	-	-	-	-	-	-	-
Kirghizistan	-	-	-	-	-	-	-	-	-
Lettonie	-	-	-	-	-	-	-	-	-
Lituanie	-	-	-	-	-	-	-	-	-
Moldova	-	-	-	-	-	-	-	-	-
Russie	-	-	-	-	-	-	-	-	-
Tadjikistan	-	-	-	-	-	-	-	-	-
Turkménistan	-	-	-	-	-	-	-	-	-
Ukraine	-	-	-	-	-	-	-	-	-
Ouzbékistan	-	-	-	-	-	-	-	-	-
Ex-URSS	15.74	13.37	14.12	12.69	13.61	13.89	13.49	13.20	13.60

Electricity Generation from Hydro Energy (% of total)
Production d'électricité à partir d'énergie hydraulique (% du total)
Elektrizitätserzeugung auf Wasserkraftbasis (in %)
Produzione di energia idroelettrica (% del totale)
水力からの発電量 (%)
Generación de electricidad a partir de energía hidráulica (% del total)
Производство гидроэлектроэнергии (в % к общему производству)

1989	1990	1991	1992	1993	1994	1995	1996	1997	
12.93	11.43	10.13	8.95	6.60	8.66	3.44	6.44	6.06	Bangladesh
23.10	24.76	23.06	21.00	19.78	21.46	17.40	15.88	16.14	India
25.21	17.97	19.21	23.17	18.96	13.10	15.54	13.33	8.00	Indonesia
54.21	54.21	59.35	63.16	63.16	63.51	63.89	64.27	64.27	DPR of Korea
26.83	17.33	16.61	14.87	14.03	16.69	13.69	10.09	5.69	Malaysia
49.72	48.14	46.32	50.67	50.37	44.91	40.05	41.85	39.36	Myanmar
96.10	99.86	99.66	93.85	91.26	90.08	84.56	90.18	90.46	Nepal
49.04	44.93	44.56	41.01	43.25	38.38	42.68	40.74	35.28	Pakistan
25.36	24.01	20.06	16.44	18.65	19.52	18.49	19.29	15.26	Philippines
98.04	99.84	92.27	81.92	95.40	93.21	92.71	71.79	67.00	Sri Lanka
7.95	9.05	5.53	7.84	5.77	7.06	6.56	6.69	6.24	Chinese Taipei
14.89	11.26	9.14	7.42	5.84	6.34	8.38	8.39	7.72	Thailand
49.31	61.74	68.03	84.05	82.25	83.61	89.27	86.28	84.50	Vietnam
57.46	57.01	54.83	54.69	54.55	54.39	52.77	54.22	55.27	Other Asia
25.62	25.13	23.69	22.12	20.84	21.07	19.33	18.18	16.76	**Asia**
20.25	20.40	18.46	17.40	18.23	18.13	18.91	17.40	17.28	People's Rep.of China
19.34	19.49	17.63	16.63	17.48	17.62	18.40	16.96	16.85	**China**
87.07	89.31	95.44	96.10	95.12	96.59	95.24	95.92	96.27	Albania
6.07	4.46	6.27	5.79	3.00	2.18	3.06	4.04	4.07	Bulgaria
16.65	17.08	25.06	21.59	23.02	23.66	28.17	25.68	30.64	Romania
8.38	8.02	6.33	8.82	15.06	17.56	19.05	17.17	17.01	Slovak Republic
-	-	-	24.59	17.09	65.07	64.46	64.46	64.46	Bosnia-Herzegovina
-	-	-	48.81	46.43	59.58	59.40	68.52	54.71	Croatia
-	-	-	13.98	10.08	12.61	13.10	13.10	13.39	FYROM
-	-	-	28.22	25.84	26.91	25.62	28.76	23.48	Slovenia
-	-	-	31.09	29.93	32.23	30.18	30.18	30.18	FR of Yugoslavia
28.14	23.88	31.30	30.29	27.88	34.04	32.75	35.19	31.74	Former Yugoslavia
18.94	17.06	22.80	21.49	20.84	23.03	24.33	25.01	25.15	**Non-OECD Europe**
-	-	-	33.81	68.20	62.11	34.51	25.30	23.07	Armenia
-	-	-	8.88	12.57	10.41	9.13	9.00	9.05	Azerbaijan
-	-	-	0.05	0.06	0.06	0.08	0.07	0.08	Belarus
-	-	-	0.01	0.01	0.03	0.02	0.02	0.03	Estonia
-	-	-	56.55	69.30	69.28	76.96	83.60	84.27	Georgia
-	-	-	8.30	9.85	13.82	12.50	12.50	12.50	Kazakhstan
-	-	-	77.36	80.59	90.66	90.03	89.08	89.08	Kyrgyzstan
-	-	-	65.73	73.27	74.44	73.81	59.54	65.59	Latvia
-	-	-	1.66	2.78	4.63	2.76	2.01	2.05	Lithuania
-	-	-	2.30	3.65	3.38	5.34	5.98	7.24	Moldova
-	-	-	17.11	18.18	20.11	20.52	18.24	18.84	Russia
-	-	-	94.69	96.49	98.29	98.84	98.83	98.75	Tajikistan
-	-	-	0.03	0.04	0.04	0.04	0.05	0.05	Turkmenistan
-	-	-	3.20	4.81	5.98	5.14	4.72	5.54	Ukraine
-	-	-	12.34	14.97	14.97	13.04	14.37	12.54	Uzbekistan
13.00	13.49	13.96	14.96	16.67	18.63	18.51	17.10	17.56	**Former USSR**

Electricity Generation from Hydro Energy (% of total)
Production d'électricité à partir d'énergie hydraulique (% du total)
Elektrizitätserzeugung auf Wasserkraftbasis (in %)

Produzione di energia idroelettrica (% del totale)

水力からの発電量 (%)

Generación de electricidad a partir de energía hidráulica (% del total)

Производство гидроэлектроэнергии (в % к общему производству)

	1971	1973	1978	1983	1984	1985	1986	1987	1988
Iran	33.91	25.90	32.14	18.79	15.71	14.15	18.08	18.16	15.36
Irak	7.14	8.24	9.04	3.71	3.13	2.91	2.69	11.55	11.09
Israël	-	-	-	-	-	-	-	-	-
Jordanie	-	-	-	-	-	-	0.10	0.55	0.83
Liban	61.02	26.69	34.78	15.56	15.39	15.15	13.43	13.26	15.00
Syrie	3.79	1.19	72.97	39.38	39.36	36.21	20.56	26.57	50.00
Moyen-Orient	13.64	10.40	14.04	7.14	6.20	5.57	5.59	6.98	7.25
Total non-OCDE	23.34	22.30	23.76	23.31	23.74	23.77	23.64	23.01	23.28
OCDE Amérique du N.	22.51	20.95	19.69	21.17	20.54	19.34	19.48	17.22	15.45
OCDE Pacifique	23.86	16.65	14.62	15.00	12.70	13.74	13.61	12.22	13.11
OCDE Europe	23.14	21.23	21.45	20.70	19.91	18.88	17.66	18.74	19.57
Total OCDE	22.90	20.51	19.66	20.16	19.24	18.40	18.01	17.07	16.60
Monde	23.02	21.00	20.92	21.24	20.81	20.28	20.03	19.23	19.04

Electricity Generation from Hydro Energy (% of total)
Production d'électricité à partir d'énergie hydraulique (% du total)
Elektrizitätserzeugung auf Wasserkraftbasis (in %)
Produzione di energia idroelettrica (% del totale)
水力からの発電量(%)
Generación de electricidad a partir de energía hidráulica (% del total)
Производство гидроэлектроэнергии (в % к общему производству)

1989	1990	1991	1992	1993	1994	1995	1996	1997	
14.27	10.29	11.00	13.64	12.92	9.08	8.56	8.12	7.70	Iran
10.90	10.83	4.32	2.77	2.28	2.00	1.97	1.97	1.97	Iraq
-	-	0.09	0.31	0.26	0.21	0.21	0.19	0.19	Israel
0.50	0.30	0.19	0.34	0.46	0.28	0.32	0.36	0.35	Jordan
20.00	33.33	18.67	19.09	15.25	15.76	13.58	11.46	10.58	Lebanon
45.21	48.64	51.31	58.80	53.09	44.79	45.39	41.13	55.92	Syria
6.83	6.25	6.24	6.92	6.23	4.98	4.77	4.56	5.18	**Middle East**
22.99	23.11	23.58	23.59	24.37	25.33	25.09	24.25	24.22	**Non-OECD Total**
15.27	15.67	15.91	15.22	15.61	14.74	15.84	16.82	16.01	OECD North America
12.46	11.63	11.79	10.06	11.31	8.34	9.21	8.66	8.88	OECD Pacific
16.52	16.94	16.54	17.48	18.00	17.73	17.43	16.32	16.68	OECD Europe
15.30	15.50	15.49	15.19	15.75	14.69	15.28	15.31	15.03	**OECD Total**
18.08	18.26	18.42	18.22	18.86	18.49	18.81	18.54	18.39	**World**

Total Electricity Generation (GWh)
Production totale d'électricité (GWh)
Gesamte Elektrizitätserzeugung (GWh)

Produzione totale di energia elettrica (GWh)

総発電量(GWh)

Generación total de electricidad (GWh)

Общее производство электроэнергии (ГВт. ч)

	1971	1973	1978	1983	1984	1985	1986	1987	1988
Algérie	2 229	2 806	5 480	10 218	11 182	12 274	12 981	12 722	13 966
Angola	742	984	585	783	806	806	806	810	815
Bénin	-	9	5	21	61	24	26	26	20
Cameroun	1 129	1 119	1 337	2 164	2 215	2 441	2 478	2 508	2 623
Congo	88	96	55	233	235	314	270	281	292
RD du Congo	3 545	3 848	4 080	4 618	4 835	5 171	5 406	5 391	5 392
Egypte	7 998	8 106	15 150	25 879	29 049	31 458	33 464	37 845	39 580
Ethiopie	593	591	567	860	910	888	915	892	877
Gabon	114	165	436	729	795	861	888	895	910
Ghana	2 944	3 910	3 773	2 568	1 831	3 017	4 402	4 866	4 842
Côte d'Ivoire	588	796	1 418	1 608	1 690	1 842	2 086	2 074	2 309
Kenya	627	793	1 382	1 904	1 949	2 155	2 307	2 654	2 772
Libye	508	1 127	3 363	7 150	9 200	11 844	13 280	15 600	16 551
Maroc	2 290	2 875	4 388	6 686	6 921	7 345	7 764	8 001	8 984
Mozambique	674	641	876	417	398	550	310	336	337
Nigéria	1 816	2 607	4 648	8 713	9 035	10 431	10 765	11 265	11 654
Sénégal	347	407	586	714	774	772	769	831	857
Afrique du Sud	54 647	64 390	84 496	121 122	135 260	141 384	146 456	150 599	156 738
Soudan	415	530	770	933	993	1 229	1 338	1 379	1 378
Tanzanie	491	582	695	849	922	1 015	1 146	1 272	1 378
Tunisie	926	1 179	2 082	3 551	3 880	4 019	4 278	4 549	4 953
Zambie	1 216	3 368	8 092	10 336	10 061	10 324	10 101	8 685	8 485
Zimbabwe	3 602	5 172	6 252	4 164	4 120	4 785	5 204	6 345	8 121
Autre Afrique	2 590	3 101	4 095	5 866	6 432	6 595	7 012	7 594	7 676
Afrique	90 119	109 202	154 611	222 086	243 554	261 544	274 452	287 420	301 510
Argentine	23 624	26 661	33 434	43 003	44 966	45 265	44 523	49 400	50 576
Bolivie	1 024	1 151	1 675	1 674	1 686	1 697	1 720	1 721	1 872
Brésil	50 878	63 854	112 483	162 490	179 390	193 681	202 129	203 332	214 948
Chili	8 524	8 766	10 360	12 624	13 497	14 040	14 744	15 637	16 915
Colombie	9 500	12 596	17 209	24 243	26 361	26 633	29 660	31 309	33 201
Costa Rica	1 148	1 347	1 925	2 870	3 067	2 816	2 949	3 133	3 193
Cuba	5 021	5 703	8 481	11 551	12 292	12 199	13 167	13 594	14 542
République dominicaine	1 068	2 246	2 874	3 409	3 769	4 175	4 327	4 663	3 640
Equateur	1 050	1 256	2 605	4 303	4 536	4 558	5 012	5 384	5 641
El Salvador	743	912	1 489	1 571	1 643	1 738	1 704	1 918	2 007
Guatemala	689	874	1 596	1 533	1 605	1 669	1 841	1 978	2 153
Haiti	82	122	245	411	422	441	443	494	535
Honduras	391	486	759	1 150	1 210	1 411	1 487	1 816	1 976
Jamaïque	1 676	2 187	1 517	1 667	1 645	1 636	1 722	1 855	1 831
Antilles néerlandaises	710	775	825	883	818	760	637	692	700
Nicaragua	657	717	1 163	940	974	1 060	1 211	1 300	1 187
Panama	950	1 302	1 622	2 490	2 466	2 603	2 746	2 825	2 741
Paraguay	236	379	569	844	1 090	4 089	11 912	18 598	19 988
Pérou	5 949	6 655	8 765	10 674	11 770	12 115	12 949	14 010	13 560
Trinité-et-Tobago	991	1 105	1 557	2 905	3 006	3 023	3 291	3 481	3 485
Uruguay	2 381	2 530	3 047	7 366	7 244	6 600	7 430	7 579	7 000
Vénézuela	13 290	16 077	25 885	42 764	45 271	48 980	51 093	54 132	57 773
Autre Amérique latine	3 197	3 741	4 259	4 297	4 448	4 733	4 890	4 793	5 239
Amérique latine	133 779	161 442	244 344	345 662	373 176	395 922	421 587	443 644	464 703

Total Electricity Generation (GWh)
Production totale d'électricité (GWh)
Gesamte Elektrizitätserzeugung (GWh)
Produzione totale di energia elettrica (GWh)
総発電量(GWh)
Generación total de electricidad (GWh)
Общее производство электроэнергии (ГВт. ч)

1989	1990	1991	1992	1993	1994	1995	1996	1997	
15 324	16 105	17 345	18 286	19 415	19 883	19 714	20 650	21 685	Algeria
819	797	934	947	950	955	960	1 028	1 109	Angola
20	21	23	25	26	52	33	47	50	Benin
2 706	2 697	2 710	2 737	2 832	2 720	2 785	2 902	3 128	Cameroon
397	508	482	428	431	435	438	435	431	Congo
6 937	5 650	5 281	6 073	5 885	5 545	6 180	6 261	6 010	DR of Congo
41 649	43 478	46 086	46 978	49 435	51 947	54 833	54 444	57 656	Egypt
879	1 246	1 270	1 267	1 399	1 302	1 328	1 316	1 340	Ethiopia
876	915	914	919	922	933	940	970	1 007	Gabon
5 260	5 731	6 116	6 603	6 335	6 105	6 109	5 987	6 155	Ghana
2 200	2 021	1 894	784	2 150	2 285	2 870	3 015	3 213	Ivory Coast
2 901	3 034	3 227	3 215	3 396	3 539	3 747	4 040	4 238	Kenya
16 700	16 800	16 800	16 950	17 000	17 800	18 000	18 180	18 180	Libya
9 024	9 628	9 205	9 719	9 910	10 966	11 907	12 356	13 128	Morocco
472	454	471	416	392	395	408	476	1 005	Mozambique
12 812	12 564	14 167	14 834	14 505	15 531	14 483	14 991	15 179	Nigeria
888	913	937	1 021	1 008	1 041	1 103	1 174	1 261	Senegal
162 320	165 385	168 316	167 816	173 236	181 953	186 817	198 346	207 744	South Africa
1 365	1 515	1 661	1 634	1 686	1 858	1 864	2 063	1 966	Sudan
1 509	1 628	1 822	1 816	1 875	1 703	1 841	1 991	1 934	Tanzania
5 156	5 534	5 743	6 180	6 313	6 714	7 306	7 533	7 977	Tunisia
6 782	6 330	7 834	7 780	7 785	7 785	7 790	7 796	8 006	Zambia
8 569	9 361	8 924	8 237	7 467	7 535	7 806	7 323	7 297	Zimbabwe
7 990	8 304	8 540	8 812	9 178	9 258	9 454	9 430	9 500	Other Africa
313 555	320 619	330 702	333 477	343 531	358 240	368 716	382 754	399 199	**Africa**
50 837	47 412	50 619	54 521	58 858	61 589	64 601	69 759	71 947	Argentina
2 010	2 133	2 275	2 412	2 646	2 824	3 020	3 232	3 433	Bolivia
221 737	222 821	234 366	241 731	251 971	260 076	275 649	291 204	307 302	Brazil
17 811	18 372	19 961	22 362	24 004	25 276	28 027	30 790	33 994	Chile
34 602	36 157	37 191	33 445	38 219	41 359	43 616	44 605	46 115	Colombia
3 410	3 468	3 864	4 173	4 431	4 772	4 865	4 791	5 599	Costa Rica
15 240	15 025	13 247	11 538	11 004	11 964	12 459	13 236	14 087	Cuba
3 255	3 698	3 895	5 581	5 874	6 182	6 506	6 847	7 335	Dominican Republic
5 769	6 373	7 013	7 199	7 432	8 257	8 526	9 260	9 595	Ecuador
2 056	2 242	2 320	2 412	2 810	3 170	3 364	3 418	3 643	El Salvador
2 313	2 330	2 493	2 802	3 094	3 161	3 541	4 247	4 897	Guatemala
575	597	468	423	382	302	515	625	625	Haiti
2 062	2 303	2 326	2 320	2 536	2 307	2 698	3 059	3 294	Honduras
1 803	2 016	2 054	2 199	3 791	4 775	5 829	6 038	6 255	Jamaica
737	790	801	853	909	971	1 009	1 017	1 052	Netherlands Antilles
1 352	1 399	1 471	1 583	1 681	1 686	1 795	1 920	1 907	Nicaragua
2 726	2 871	2 897	3 432	3 444	3 361	3 519	3 915	4 151	Panama
24 333	27 185	29 328	27 141	31 449	36 415	42 236	48 200	50 619	Paraguay
13 737	13 163	13 901	13 044	14 678	14 786	16 880	17 280	17 954	Peru
3 428	3 577	3 720	3 976	3 817	4 069	4 307	4 541	4 988	Trinidad-and-Tobago
5 750	7 444	7 019	8 899	7 978	7 619	6 307	6 670	7 148	Uruguay
57 620	59 309	63 325	67 427	69 368	71 207	73 432	75 372	74 865	Venezuela
5 970	6 131	6 420	6 636	6 824	6 894	7 081	7 186	7 491	Other Latin America
479 133	486 816	510 974	526 109	557 200	583 022	619 782	657 212	688 296	**Latin America**

Total Electricity Generation (GWh)
Production totale d'électricité (GWh)
Gesamte Elektrizitätserzeugung (GWh)

Produzione totale di energia elettrica (GWh)

総発電量(GWh)

Generación total de electricidad (GWh)

Общее производство электроэнергии (ГВт. ч)

	1971	1973	1978	1983	1984	1985	1986	1987	1988
Bangladesh	1 030	1 404	2 219	3 433	3 966	4 528	4 800	5 587	6 541
Brunei	150	200	361	682	750	766	776	1 043	1 098
Inde	66 384	72 796	110 130	150 994	169 205	183 390	201 281	218 986	241 314
Indonésie	3 045	3 624	5 534	13 392	14 777	16 899	19 455	22 285	25 606
RPD de Corée	16 910	20 000	32 000	41 000	45 000	48 000	50 000	50 200	53 000
Malaisie	3 795	4 783	8 241	12 733	13 721	14 554	15 682	16 986	18 735
Myanmar	691	821	1 243	1 675	1 890	2 119	2 245	2 320	2 226
Népal	86	104	174	316	336	362	465	557	569
Pakistan	7 572	8 377	12 375	19 697	21 873	23 003	25 589	28 703	33 091
Philippines	9 145	13 186	15 542	21 454	21 180	22 766	21 797	22 642	24 539
Singapour	2 585	3 719	5 898	8 704	9 490	9 960	10 640	11 909	13 113
Sri Lanka	900	1 031	1 385	2 114	2 261	2 464	2 652	2 707	2 799
Taipei chinois	15 811	20 735	35 846	48 294	52 213	55 554	62 329	69 176	76 258
Thaïlande	5 083	6 971	12 637	18 857	21 024	23 074	24 717	28 652	32 464
Viêt-Nam	2 300	2 350	3 600	3 986	4 778	5 069	5 527	6 051	6 785
Autre Asie	2 256	3 500	3 751	3 836	4 163	4 466	5 044	5 978	6 375
Asie	137 743	163 601	250 936	351 167	386 627	416 974	452 999	493 782	544 513
Rép. populaire de Chine	138 400	166 800	256 552	351 440	377 390	410 690	449 530	497 270	545 210
Hong-Kong, Chine	5 574	6 799	10 370	16 467	17 918	19 230	21 406	23 746	25 501
Chine	143 974	173 599	266 922	367 907	395 308	429 920	470 936	521 016	570 711
Albanie	1 225	1 702	3 100	3 780	3 717	3 147	5 106	4 395	3 984
Bulgarie	21 016	21 956	31 492	42 648	44 672	41 632	41 820	43 473	45 021
Chypre	665	830	919	1 221	1 250	1 319	1 423	1 501	1 647
Gibraltar	47	49	52	61	61	63	65	70	74
Malte	305	365	459	675	700	784	850	944	1 030
Roumanie	39 454	46 779	64 255	70 260	71 667	71 818	75 478	74 077	75 322
République slovaque	10 865	12 299	15 432	19 232	19 947	21 945	23 729	23 122	22 492
Bosnie-Herzégovine	-	-	-	-	-	-	-	-	-
Croatie	-	-	-	-	-	-	-	-	-
Ex-RYM	-	-	-	-	-	-	-	-	-
Slovénie	-	-	-	-	-	-	-	-	-
RF de Yougoslavie	-	-	-	-	-	-	-	-	-
Ex-Yougoslavie	29 509	35 062	51 250	66 571	73 008	74 802	77 916	80 792	83 651
Europe non-OCDE	103 086	119 042	166 959	204 448	215 022	215 510	226 387	228 374	233 221

Total Electricity Generation (GWh)
Production totale d'électricité (GWh)
Gesamte Elektrizitätserzeugung (GWh)
Produzione totale di energia elettrica (GWh)
総発電量(GWh)
Generación total de electricidad (GWh)
Общее производство электроэнергии (ГВт. ч)

1989	1990	1991	1992	1993	1994	1995	1996	1997	
7 115	7 732	8 270	8 894	9 206	9 784	10 806	11 474	11 858	Bangladesh
1 131	1 172	1 269	1 408	1 547	1 663	1 966	2 123	2 407	Brunei
268 664	289 439	315 631	332 713	356 335	385 557	417 836	435 075	463 402	India
29 554	36 117	38 498	42 445	46 639	52 269	55 822	67 062	74 832	Indonesia
53 500	53 500	53 500	38 000	38 000	37 000	36 000	35 036	33 985	DPR of Korea
20 799	23 012	26 535	29 314	34 733	39 093	45 453	51 407	57 875	Malaysia
2 494	2 478	2 677	2 996	3 385	3 594	4 055	3 945	4 205	Myanmar
564	713	873	927	881	927	1 004	1 191	1 226	Nepal
34 601	37 673	41 076	45 465	48 810	50 646	53 555	56 956	59 125	Pakistan
25 573	25 245	25 654	25 869	26 777	30 465	33 531	36 663	39 816	Philippines
14 136	15 714	16 921	17 674	18 962	20 849	22 244	24 100	26 898	Singapore
2 858	3 150	3 377	3 540	3 979	4 387	4 802	4 530	5 145	Sri Lanka
84 055	90 479	99 687	106 557	116 471	125 932	135 277	135 277	153 294	Chinese Taipei
37 406	44 175	50 186	57 098	63 406	71 177	80 061	87 467	93 253	Thailand
7 784	8 722	9 300	9 691	10 659	12 270	13 677	16 945	19 151	Vietnam
6 477	6 520	6 467	6 233	6 251	6 470	6 869	6 789	6 958	Other Asia
596 711	645 841	699 921	728 824	786 041	852 083	922 958	976 040	1 053 430	**Asia**
584 810	621 200	677 550	754 195	838 256	928 083	1 007 726	1 080 017	1 134 471	People's Rep.of China
27 361	28 938	31 807	34 907	35 950	26 743	27 916	28 442	28 945	Hong Kong, China
612 171	650 138	709 357	789 102	874 206	954 826	1 035 642	1 108 459	1 163 416	**China**
4 123	3 189	3 686	3 357	3 484	3 904	4 414	5 926	5 600	Albania
44 328	42 141	38 917	35 610	37 171	37 484	40 719	41 472	41 560	Bulgaria
1 831	1 974	2 077	2 404	2 581	2 681	2 473	2 592	2 711	Cyprus
81	82	82	88	88	88	88	93	93	Gibraltar
1 100	1 100	1 419	1 490	1 500	1 541	1 632	1 658	1 686	Malta
75 851	64 309	56 803	54 195	55 476	55 136	59 266	61 350	57 148	Romania
23 400	23 432	22 260	21 952	23 019	24 486	26 041	25 060	24 326	Slovak Republic
-	-	-	12 200	11 700	1 921	2 203	2 203	2 203	*Bosnia-Herzegovina*
-	-	-	8 894	9 359	8 275	8 863	10 548	9 685	*Croatia*
-	-	-	6 065	5 180	5 511	6 114	6 489	6 719	*FYROM*
-	-	-	12 086	11 692	12 630	12 649	12 770	13 166	*Slovenia*
-	-	-	36 488	33 457	34 528	37 176	38 093	40 312	*FR of Yugoslavia*
83 468	82 905	81 885	75 733	71 388	62 865	67 005	70 103	72 085	Former Yugoslavia
234 182	219 132	207 129	194 829	194 707	188 185	201 638	208 254	205 209	**Non-OECD Europe**

Total Electricity Generation (GWh)
Production totale d'électricité (GWh)
Gesamte Elektrizitätserzeugung (GWh)
Produzione totale di energia elettrica (GWh)
総発電量(GWh)
Generación total de electricidad (GWh)
Общее производство электроэнергии (ГВт. ч)

	1971	1973	1978	1983	1984	1985	1986	1987	1988
Arménie	-	-	-	-	-	-	-	-	-
Azerbaïdjan	-	-	-	-	-	-	-	-	-
Bélarus	-	-	-	-	-	-	-	-	-
Estonie	-	-	-	-	-	-	-	-	-
Géorgie	-	-	-	-	-	-	-	-	-
Kazakhstan	-	-	-	-	-	-	-	-	-
Kirghizistan	-	-	-	-	-	-	-	-	-
Lettonie	-	-	-	-	-	-	-	-	-
Lituanie	-	-	-	-	-	-	-	-	-
Moldova	-	-	-	-	-	-	-	-	-
Russie	-	-	-	-	-	-	-	-	-
Tadjikistan	-	-	-	-	-	-	-	-	-
Turkménistan	-	-	-	-	-	-	-	-	-
Ukraine	-	-	-	-	-	-	-	-	-
Ouzbékistan	-	-	-	-	-	-	-	-	-
Ex-URSS	800 400	914 600	1 201 900	1 418 000	1 492 000	1 544 000	1 599 000	1 664 900	1 705 000
Bahrein	430	500	1 212	2 216	2 417	2 937	2 970	3 316	3 320
Iran	7 900	11 540	19 446	33 009	36 594	39 220	41 571	46 197	47 600
Irak	2 800	3 519	7 835	16 185	19 459	20 973	22 287	22 510	23 450
Israël	7 639	8 720	11 874	14 578	14 909	15 701	16 277	17 491	19 215
Jordanie	230	315	704	1 918	2 265	2 495	2 955	3 486	3 263
Koweit	3 087	4 138	7 446	12 830	14 196	15 689	17 216	18 092	19 598
Liban	1 375	1 791	2 300	3 600	3 800	3 861	4 170	4 600	4 000
Oman	13	47	468	1 615	2 016	2 498	3 180	3 392	3 773
Qatar	316	419	1 349	3 235	3 563	3 949	4 303	4 371	4 501
Arabie saoudite	2 082	2 949	10 551	35 190	40 069	44 311	47 646	50 649	57 229
Syrie	1 345	1 423	2 871	7 116	7 310	7 898	7 942	7 989	9 614
Emirats arabes unis	203	720	3 764	10 141	11 000	12 000	12 814	13 657	14 840
Yémen	209	206	360	752	856	895	1 117	1 165	1 318
Moyen-Orient	27 629	36 287	70 180	142 385	158 454	172 427	184 448	196 915	211 721
Total non-OCDE	1 436 730	1 677 773	2 355 852	3 051 655	3 264 141	3 436 297	3 629 809	3 836 051	4 031 379
OCDE Amérique du N.	1 956 252	2 272 690	2 716 339	2 938 747	3 086 713	3 173 828	3 205 338	3 317 439	3 473 744
OCDE Pacifique	461 963	563 154	697 335	795 301	836 248	872 502	889 919	947 847	1 001 981
OCDE Europe	1 391 649	1 605 896	1 919 141	2 118 305	2 215 883	2 318 972	2 370 159	2 457 732	2 517 037
Total OCDE	3 809 864	4 441 740	5 332 815	5 852 353	6 138 844	6 365 302	6 465 416	6 723 018	6 992 762
Monde	5 246 594	6 119 513	7 688 667	8 904 008	9 402 985	9 801 599	10 095 225	10 559 069	11 024 141

Total Electricity Generation (GWh)
Production totale d'électricité (GWh)
Gesamte Elektrizitätserzeugung (GWh)
Produzione totale di energia elettrica (GWh)
総発電量(GWh)
Generación total de electricidad (GWh)
Общее производство электроэнергии (ГВт. ч)

1989	1990	1991	1992	1993	1994	1995	1996	1997	
-	-	-	9 004	6 295	5 658	5 561	6 214	6 021	Armenia
-	-	-	19 673	19 100	17 571	17 044	17 088	16 800	Azerbaijan
-	-	-	37 595	33 369	31 397	24 918	23 728	26 057	Belarus
-	-	-	11 831	9 118	9 151	8 693	9 103	9 218	Estonia
-	-	-	11 520	10 150	6 803	6 900	7 226	7 172	Georgia
-	-	-	82 701	77 444	66 397	66 659	58 657	52 000	Kazakhstan
-	-	-	11 892	11 273	12 932	12 349	13 758	12 600	Kyrgyzstan
-	-	-	3 834	3 924	4 440	3 979	3 124	4 501	Latvia
-	-	-	18 707	14 122	9 755	13 520	16 241	14 387	Lithuania
-	-	-	11 248	10 265	8 228	6 068	6 122	5 273	Moldova
-	-	-	1 008 450	955 702	874 881	859 026	846 194	833 065	Russia
-	-	-	16 822	17 741	16 982	14 768	15 000	14 000	Tajikistan
-	-	-	13 183	12 637	10 496	9 800	10 100	9 400	Turkmenistan
-	-	-	252 524	229 708	202 713	193 821	182 785	177 826	Ukraine
-	-	-	50 911	49 149	47 800	47 453	45 419	46 054	Uzbekistan
1 722 000	1 727 000	1 681 083	1 559 895	1 459 997	1 325 204	1 290 559	1 260 759	1 234 374	**Former USSR**
3 400	3 490	3 495	3 896	4 245	4 550	4 750	4 771	4 924	Bahrain
52 712	59 102	64 126	68 419	76 014	82 019	84 969	90 851	95 794	Iran
23 860	24 000	20 810	25 300	26 300	28 000	29 000	29 000	29 561	Iraq
20 297	20 722	21 470	24 641	25 957	28 273	30 342	32 497	35 098	Israel
3 433	3 638	3 724	4 422	4 761	5 076	5 616	6 058	6 273	Jordan
21 179	20 610	10 780	16 786	20 178	22 798	23 726	25 475	27 091	Kuwait
2 500	1 500	3 000	3 771	4 720	5 184	5 281	6 965	8 515	Lebanon
3 927	4 501	4 628	5 117	5 832	6 197	6 460	6 802	7 318	Oman
4 624	4 818	4 653	5 153	5 525	5 814	5 976	6 575	6 868	Qatar
61 568	64 899	69 212	74 009	82 183	91 019	93 898	97 819	103 801	Saudi Arabia
10 455	11 611	12 179	12 562	12 638	15 182	15 300	16 885	17 950	Syria
15 612	17 081	17 222	17 460	17 578	18 870	19 070	19 737	20 571	United Arab Emirates
1 345	1 663	1 802	1 953	2 051	2 159	2 369	2 334	2 358	Yemen
224 912	237 635	237 101	263 489	287 982	315 141	326 757	345 769	366 122	**Middle East**
4 182 664	4 287 181	4 376 267	4 395 725	4 503 664	4 576 701	4 766 052	4 939 247	5 110 046	**Non-OECD Total**
3 745 063	3 786 207	3 887 048	3 922 491	4 055 309	4 156 157	4 273 265	4 390 801	4 420 688	OECD North America
1 066 797	1 145 037	1 186 996	1 208 593	1 236 689	1 340 034	1 390 175	1 438 049	1 492 840	OECD Pacific
2 576 703	2 605 122	2 654 179	2 676 759	2 694 155	2 737 746	2 815 839	2 890 860	2 925 162	OECD Europe
7 388 563	7 536 366	7 728 223	7 807 843	7 986 153	8 233 937	8 479 279	8 719 710	8 838 690	**OECD Total**
11 571 227	11 823 547	12 104 490	12 203 568	12 489 817	12 810 638	13 245 331	13 658 957	13 948 736	**World**

Electricity Consumption (GWh)
Consommation d'électricité (GWh)
Elekrizitätsverbrauch (GWh)
Consumi finali di Energia elettrica (GWh)
電力消費量 （GWh）
Consumo de electricidad (GWh)
Потребление электроэнергии (ГВт. ч)

	1971	1973	1978	1983	1984	1985	1986	1987	1988
Algérie	1 991	2 494	5 050	8 810	9 604	10 464	11 211	10 827	11 779
Angola	556	738	439	587	604	604	604	607	611
Bénin	32	58	168	149	135	143	137	140	167
Cameroun	1 019	1 080	1 265	2 020	1 996	2 253	2 167	2 228	2 323
Congo	87	95	125	289	287	368	398	428	439
RD du Congo	3 337	3 630	3 703	3 921	4 116	4 427	4 623	4 598	4 733
Egypte	7 216	7 282	13 309	23 526	25 024	24 487	29 867	32 255	33 914
Ethiopie	552	551	535	780	842	809	835	881	864
Gabon	112	162	432	723	787	852	878	883	897
Ghana	2 809	3 728	3 542	2 060	1 198	2 382	3 727	4 340	4 601
Côte d'Ivoire	517	697	1 226	1 666	2 029	1 853	1 752	1 742	1 940
Kenya	808	959	1 396	1 789	1 888	2 060	2 192	2 463	2 407
Libye	508	1 127	3 363	7 150	9 200	11 844	13 280	15 600	16 551
Maroc	2 047	2 594	3 928	6 029	6 226	6 627	7 060	7 297	8 149
Mozambique	476	559	671	471	372	401	473	540	441
Nigéria	1 534	2 184	4 086	6 607	5 967	6 471	7 793	7 862	8 264
Sénégal	324	374	517	642	690	673	678	711	740
Afrique du Sud	50 768	59 870	84 337	117 452	128 480	134 565	140 874	143 197	149 360
Soudan	373	491	744	697	961	1 055	1 198	1 242	1 119
Tanzanie	424	503	601	692	727	787	909	946	1 061
Tunisie	803	1 036	1 833	3 111	3 290	3 501	3 756	4 027	4 364
Zambie	4 411	5 174	5 838	6 708	6 660	6 565	6 183	6 335	6 135
Zimbabwe	3 560	4 962	5 942	7 282	7 108	8 308	8 019	8 373	8 452
Autre Afrique	2 591	3 173	4 247	5 839	6 234	6 417	6 871	7 452	7 428
Afrique	86 855	103 521	147 297	209 000	224 425	237 916	255 485	264 974	276 739
Argentine	21 214	24 111	29 589	36 846	38 682	38 945	39 755	44 307	45 274
Bolivie	919	1 035	1 512	1 465	1 477	1 488	1 499	1 488	1 628
Brésil	44 132	55 844	98 793	143 911	160 001	173 563	187 070	192 757	203 899
Chili	7 569	7 780	9 249	10 794	11 678	12 112	12 796	13 331	14 409
Colombie	8 386	10 932	14 493	20 446	22 226	22 430	24 978	26 395	25 587
Costa Rica	1 148	1 347	1 925	2 382	2 637	2 757	3 027	3 303	3 383
Cuba	4 519	5 133	7 554	10 133	10 794	10 621	11 433	11 749	12 436
République dominicaine	1 068	1 955	2 269	2 270	2 815	3 082	3 199	3 222	2 617
Equateur	903	1 080	2 221	3 524	3 641	3 744	3 884	4 303	4 408
El Salvador	644	802	1 315	1 357	1 423	1 495	1 539	1 643	1 713
Guatemala	666	851	1 446	1 349	1 400	1 456	1 511	1 679	1 753
Haiti	59	83	187	300	309	323	351	350	367
Honduras	354	426	628	957	1 015	1 235	1 257	1 516	1 621
Jamaïque	1 552	1 969	1 317	1 382	1 369	1 356	1 432	1 527	1 467
Antilles néerlandaises	603	658	701	750	695	646	525	580	588
Nicaragua	603	624	1 023	1 097	1 111	1 098	1 095	1 177	1 071
Panama	819	1 096	1 407	2 168	2 114	2 156	2 197	2 384	2 230
Paraguay	209	279	567	924	1 046	1 081	1 270	1 498	1 728
Pérou	5 399	6 018	7 768	9 664	10 593	10 702	11 396	12 154	11 636
Trinité-et-Tobago	991	1 105	1 557	2 905	3 006	3 023	3 291	3 481	3 157
Uruguay	2 033	1 969	2 604	3 107	3 132	3 224	3 367	3 561	3 874
Vénézuela	12 172	14 251	22 496	36 457	38 566	40 995	42 096	44 670	48 400
Autre Amérique latine	3 134	3 646	4 100	4 157	4 234	4 537	4 712	4 565	5 000
Amérique latine	119 096	142 994	214 721	298 345	323 964	342 069	363 680	381 640	398 246

Electricity Consumption (GWh)
Consommation d'électricité (GWh)
Elekrizitätsverbrauch (GWh)
Consumi finali di Energia elettrica (GWh)
電力消費量（GW h）
Consumo de electricidad (GWh)
Потребление электроэнергии (ГВт. ч)

1989	1990	1991	1992	1993	1994	1995	1996	1997	
13 109	13 845	14 484	14 688	15 145	15 545	16 102	16 694	18 506	Algeria
614	598	667	677	680	684	687	735	794	Angola
162	172	189	210	216	225	246	270	250	Benin
2 402	2 345	2 334	2 360	2 457	2 213	2 177	2 315	2 517	Cameroon
458	570	546	535	540	546	552	546	541	Congo
5 672	4 922	4 944	5 700	5 204	5 205	5 840	5 916	5 673	DR of Congo
36 015	38 459	40 766	41 555	43 728	45 950	48 503	48 159	51 000	Egypt
866	1 231	1 255	1 252	1 384	1 287	1 313	1 301	1 325	Ethiopia
771	810	814	814	816	847	846	873	906	Gabon
4 706	4 927	5 263	5 657	5 521	5 705	5 703	5 577	5 745	Ghana
1 848	1 945	1 899	1 065	1 882	1 948	2 439	2 561	2 727	Ivory Coast
2 558	2 729	2 854	2 945	3 103	3 234	3 321	3 550	3 675	Kenya
16 700	16 800	16 800	16 950	17 000	17 800	18 000	18 180	18 180	Libya
8 174	8 910	9 385	10 339	10 517	11 302	11 678	11 995	12 692	Morocco
507	548	742	746	728	662	714	732	893	Mozambique
9 304	7 833	9 063	9 693	8 695	11 065	9 876	10 222	10 350	Nigeria
735	781	796	873	859	931	949	988	1 050	Senegal
153 469	155 988	157 518	156 741	159 333	166 814	173 708	179 722	187 646	South Africa
1 075	1 282	1 264	1 295	1 271	1 345	1 362	1 413	1 361	Sudan
1 151	1 303	1 428	1 441	1 433	1 456	1 623	1 793	1 700	Tanzania
4 611	4 926	5 140	5 527	5 838	6 329	6 630	6 767	7 130	Tunisia
4 632	4 164	5 484	5 410	5 423	5 423	5 427	5 431	5 596	Zambia
8 898	9 133	9 436	9 814	8 242	8 998	9 524	9 743	10 333	Zimbabwe
7 879	8 225	8 554	8 813	9 268	9 436	9 710	10 002	10 102	Other Africa
286 316	292 446	301 625	305 100	309 283	324 950	336 930	345 485	360 692	**Africa**
43 992	41 784	43 908	46 533	48 625	51 753	55 075	60 361	64 568	Argentina
1 757	1 829	1 963	2 082	2 353	2 484	2 673	2 851	3 067	Bolivia
212 380	217 658	225 372	230 472	241 165	249 828	264 853	277 645	295 524	Brazil
15 785	16 428	17 732	19 990	21 124	22 506	25 100	28 102	30 772	Chile
27 304	28 652	29 612	27 306	30 288	32 501	34 425	35 023	36 131	Colombia
3 272	3 344	3 592	3 791	4 084	4 389	4 526	4 529	5 045	Costa Rica
12 955	12 848	11 195	9 711	9 067	9 737	10 010	10 660	11 689	Cuba
2 348	2 791	2 918	4 035	4 380	4 610	4 852	5 106	5 299	Dominican Republic
4 454	4 896	5 362	5 617	5 908	6 629	6 844	7 345	7 432	Ecuador
1 740	1 880	2 004	2 116	2 430	2 679	2 969	2 977	3 251	El Salvador
1 903	1 958	2 120	2 407	2 694	2 752	3 083	3 698	4 263	Guatemala
391	414	323	295	231	163	254	287	358	Haiti
1 637	1 834	1 801	1 728	1 855	1 695	2 026	2 386	2 494	Honduras
1 393	1 658	1 671	1 760	3 791	4 257	5 197	5 385	5 579	Jamaica
625	678	689	743	799	851	884	891	922	Netherlands Antilles
1 104	1 220	1 243	1 267	1 247	1 211	1 298	1 402	1 564	Nicaragua
2 146	2 282	2 324	2 443	2 649	2 749	2 940	3 129	3 184	Panama
1 837	2 131	2 200	2 392	2 854	3 269	3 743	3 816	4 058	Paraguay
11 838	10 818	11 426	10 721	12 065	12 155	13 874	14 612	15 071	Peru
3 056	3 278	3 257	3 522	3 502	3 611	3 892	4 091	4 587	Trinidad-and-Tobago
3 753	3 870	4 212	4 309	4 623	4 759	5 089	5 313	5 665	Uruguay
47 856	48 633	51 027	54 916	56 685	57 104	57 946	59 878	58 949	Venezuela
5 695	5 849	6 094	6 311	6 484	6 541	6 684	6 792	7 081	Other Latin America
409 221	416 733	432 045	444 467	468 903	488 233	518 237	546 279	576 553	**Latin America**

Electricity Consumption (GWh)
Consommation d'électricité (GWh)
Elekrizitätsverbrauch (GWh)
Consumi finali di Energia elettrica (GWh)
電力消費量 （GW h ）
Consumo de electricidad (GWh)
Потребление электроэнергии (ГВт. ч)

	1971	1973	1978	1983	1984	1985	1986	1987	1988
Bangladesh	710	1 042	1 651	2 537	2 866	3 040	3 534	3 771	4 173
Brunei	150	200	343	651	716	731	740	995	1 048
Inde	55 522	59 854	90 727	123 220	137 886	149 105	163 600	177 610	196 482
Indonésie	2 530	3 010	4 404	10 681	11 735	13 561	15 614	18 107	21 280
RPD de Corée	15 320	18 120	28 992	37 100	40 750	43 470	45 000	45 200	48 000
Malaisie	3 464	4 347	7 437	11 461	12 551	13 103	14 056	15 096	16 643
Myanmar	553	666	1 020	1 192	1 313	1 595	1 626	1 665	1 503
Népal	74	74	151	263	282	315	346	408	423
Pakistan	5 584	6 389	8 913	14 757	16 384	18 336	20 398	22 454	25 919
Philippines	8 688	12 562	14 740	18 650	18 047	19 020	18 199	18 304	20 191
Singapour	2 440	3 506	5 574	8 166	8 896	9 422	10 104	11 440	12 616
Sri Lanka	735	881	1 178	1 811	1 886	2 073	2 243	2 285	2 383
Taipei chinois	14 520	18 913	33 345	45 196	49 019	52 681	59 392	65 474	72 373
Thaïlande	4 568	6 485	11 886	17 514	19 456	21 117	23 047	26 085	29 560
Viêt-Nam	1 794	1 833	2 808	3 260	3 866	4 141	4 423	4 949	5 515
Autre Asie	2 170	3 386	3 559	3 516	3 829	4 145	4 441	4 562	4 670
Asie	101 708	121 315	184 928	259 615	284 866	308 244	337 340	368 256	409 264
Rép. populaire de Chine	138 400	166 800	256 552	324 410	348 800	381 330	417 488	462 400	508 730
Hong-Kong, Chine	5 250	6 009	9 107	14 031	15 041	15 923	17 660	21 430	22 918
Chine	143 650	172 809	265 659	338 441	363 841	397 253	435 148	483 830	531 648
Albanie	1 164	1 363	2 819	3 045	2 957	2 361	4 294	3 583	3 161
Bulgarie	19 571	23 131	32 157	40 963	42 763	41 920	41 622	43 409	44 373
Chypre	615	774	865	1 149	1 175	1 243	1 334	1 411	1 540
Gibraltar	45	47	50	59	59	61	63	62	66
Malte	305	365	459	615	635	710	770	857	935
Roumanie	33 395	39 732	58 495	67 115	69 732	71 222	75 795	75 175	78 226
République slovaque	12 290	14 046	16 962	23 020	23 708	24 990	25 666	26 559	27 043
Bosnie-Herzégovine	-	-	-	-	-	-	-	-	-
Croatie	-	-	-	-	-	-	-	-	-
Ex-RYM	-	-	-	-	-	-	-	-	-
Slovénie	-	-	-	-	-	-	-	-	-
RF de Yougoslavie	-	-	-	-	-	-	-	-	-
Ex-Yougoslavie	26 127	31 260	45 608	61 815	65 990	67 899	70 645	73 736	75 956
Europe non-OCDE	92 348	109 355	154 596	194 736	204 062	208 045	215 895	221 209	228 139

L' Albanie, la Rép.pop.dém. de Corée et le Viêt-Nam ne sont pas pris en compte dans les totaux régionaux.

Electricity Consumption (GWh)
Consommation d'électricité (GWh)
Elekrizitätsverbrauch (GWh)
Consumi finali di Energia elettrica (GWh)
電力消費量 (GW h)
Consumo de electricidad (GWh)
Потребление электроэнергии (ГВт. ч)

1989	1990	1991	1992	1993	1994	1995	1996	1997	
5 093	5 135	5 324	6 522	7 411	8 010	9 011	9 637	10 062	Bangladesh
1 079	1 118	1 192	1 320	1 483	1 615	1 897	2 080	2 378	Brunei
216 701	238 238	260 917	277 706	297 795	317 411	344 723	358 893	382 177	India
24 878	30 663	32 809	37 402	40 542	45 625	48 215	58 782	66 216	Indonesia
48 500	48 500	28 500	6 080	6 080	5 920	5 760	5 612	5 444	DPR of Korea
18 378	20 745	22 779	26 500	28 919	36 059	41 058	45 686	52 552	Malaysia
1 638	1 823	1 712	1 875	2 173	2 342	2 510	2 562	2 717	Myanmar
484	543	588	702	715	713	774	841	866	Nepal
27 652	29 865	32 922	35 381	37 979	39 099	41 192	43 777	45 072	Pakistan
21 776	22 020	22 451	21 778	22 862	25 725	28 109	30 535	33 223	Philippines
13 596	15 184	16 136	16 849	18 035	19 838	21 258	23 016	25 780	Singapore
2 367	2 624	2 765	2 946	3 272	3 613	3 936	3 756	4 245	Sri Lanka
79 243	85 124	93 931	100 639	110 556	119 495	128 198	127 621	146 398	Chinese Taipei
34 297	40 130	45 334	51 647	58 834	65 136	74 178	80 581	85 807	Thailand
6 099	6 610	6 970	7 358	8 159	9 586	10 956	13 742	15 767	Vietnam
4 764	4 820	4 735	4 592	4 609	4 809	5 261	5 161	5 251	Other Asia
451 946	498 032	543 595	585 859	635 185	689 490	750 320	792 928	862 744	**Asia**
545 200	579 580	631 650	701 582	780 735	866 430	927 888	1 002 172	1 037 699	People's Rep.of China
22 385	23 833	25 316	26 148	27 727	29 184	29 856	31 635	32 245	Hong Kong, China
567 585	603 413	656 966	727 730	808 462	895 614	957 744	1 033 807	1 069 944	**China**
3 297	2 828	2 063	2 397	2 866	1 876	2 094	3 021	2 860	Albania
44 058	41 488	35 702	33 367	33 333	33 316	36 203	36 690	33 243	Bulgaria
1 724	1 863	1 943	2 246	2 451	2 556	2 358	2 443	2 538	Cyprus
76	76	76	85	83	88	88	93	93	Gibraltar
1 000	1 000	1 309	1 370	1 380	1 400	1 480	1 440	1 465	Malta
79 027	67 856	57 986	52 873	51 721	50 796	52 827	54 974	50 788	Romania
27 784	27 436	25 261	23 752	22 663	23 198	25 974	26 857	26 544	Slovak Republic
-	-	-	12 200	11 700	1 675	1 896	1 896	1 896	*Bosnia-Herzegovina*
-	-	-	10 194	10 205	10 256	10 699	10 970	11 811	*Croatia*
-	-	-	5 900	5 186	4 999	5 351	5 679	5 880	*FYROM*
-	-	-	9 611	9 607	10 160	10 308	10 383	10 783	*Slovenia*
-	-	-	32 685	30 150	31 108	33 722	34 412	36 566	*FR of Yugoslavia*
75 711	74 794	75 542	70 590	66 848	58 198	61 976	63 340	66 936	Former Yugoslavia
229 380	214 513	197 819	184 283	178 479	169 552	180 906	185 837	181 607	**Non-OECD Europe**

Albania, the Dem. Ppl's Rep. of Korea and Vietnam are not included in the regional aggregates.

Electricity Consumption (GWh)
Consommation d'électricité (GWh)
Elekrizitätsverbrauch (GWh)
Consumi finali di Energia elettrica (GWh)
電力消費量 （GWh）
Consumo de electricidad (GWh)
Потребление электроэнергии (ГВт. ч)

	1971	1973	1978	1983	1984	1985	1986	1987	1988
Arménie	-	-	-	-	-	-	-	-	-
Azerbaïdjan	-	-	-	-	-	-	-	-	-
Bélarus	-	-	-	-	-	-	-	-	-
Estonie	-	-	-	-	-	-	-	-	-
Géorgie	-	-	-	-	-	-	-	-	-
Kazakhstan	-	-	-	-	-	-	-	-	-
Kirghizistan	-	-	-	-	-	-	-	-	-
Lettonie	-	-	-	-	-	-	-	-	-
Lituanie	-	-	-	-	-	-	-	-	-
Moldova	-	-	-	-	-	-	-	-	-
Russie	-	-	-	-	-	-	-	-	-
Tadjikistan	-	-	-	-	-	-	-	-	-
Turkménistan	-	-	-	-	-	-	-	-	-
Ukraine	-	-	-	-	-	-	-	-	-
Ouzbékistan	-	-	-	-	-	-	-	-	-
Ex-URSS	730 500	832 300	1 091 900	1 278 800	1 341 200	1 381 400	1 432 700	1 487 700	1 526 200
Bahrein	430	500	1 212	2 216	2 417	2 937	2 970	3 316	3 320
Iran	7 199	10 533	17 404	28 850	31 984	34 279	36 334	40 397	41 603
Irak	2 660	3 343	7 443	15 376	18 486	19 924	21 173	21 332	22 250
Israël	7 024	8 187	11 114	13 601	13 972	14 753	15 223	16 411	17 890
Jordanie	237	338	574	1 664	1 961	2 160	2 343	2 673	2 780
Koweit	2 810	3 807	6 740	11 654	12 930	14 364	15 763	16 592	18 098
Liban	1 237	1 612	2 115	3 276	3 447	3 511	3 780	4 176	3 600
Oman	11	40	399	1 378	1 720	2 132	2 714	2 895	3 220
Qatar	316	419	1 213	2 945	3 263	3 734	4 088	4 151	4 271
Arabie saoudite	1 927	2 716	9 381	31 177	36 986	41 904	45 866	48 906	51 531
Syrie	1 158	1 283	2 406	5 968	6 068	6 583	6 674	6 709	7 360
Emirats arabes unis	189	670	3 501	9 431	10 230	11 160	11 904	12 657	13 840
Yémen	209	206	360	693	776	823	972	1 041	1 185
Moyen-Orient	25 407	33 654	63 862	128 229	144 240	158 264	169 804	181 256	190 948
Total non-OCDE	1 299 564	1 515 948	2 122 963	2 707 166	2 886 598	3 033 191	3 210 052	3 388 865	3 561 184
OCDE Amérique du N.	1 789 307	2 079 898	2 473 291	2 691 663	2 869 343	2 943 602	2 989 260	3 145 986	3 296 548
OCDE Pacifique	432 675	528 815	662 578	758 730	801 277	833 158	851 119	906 767	960 575
OCDE Europe	1 302 702	1 501 891	1 796 939	1 994 650	2 091 138	2 186 838	2 241 103	2 324 636	2 386 674
Total OCDE	3 524 684	4 110 604	4 932 808	5 445 043	5 761 758	5 963 598	6 081 482	6 377 389	6 643 797
Monde	4 824 248	5 626 552	7 055 771	8 152 209	8 648 356	8 996 789	9 291 534	9 766 254	10 204 981

L' Albanie, la Rép.pop.dém. de Corée et le Viêt-Nam ne sont pas pris en compte dans les totaux régionaux.

Electricity Consumption (GWh)
Consommation d'électricité (GWh)
Elekrizitätsverbrauch (GWh)
Consumi finali di Energia elettrica (GWh)
電力消費量 （GW h ）
Consumo de electricidad (GWh)
Потребление электроэнергии (ГВт. ч)

1989	1990	1991	1992	1993	1994	1995	1996	1997	
-	-	-	6 763	4 019	3 433	3 379	3 865	4 770	Armenia
-	-	-	16 374	15 400	14 354	13 485	13 745	13 800	Azerbaijan
-	-	-	39 974	35 327	31 380	28 441	28 514	29 876	Belarus
-	-	-	7 564	6 052	6 433	6 160	6 533	6 734	Estonia
-	-	-	9 736	8 331	5 903	5 870	6 238	6 200	Georgia
-	-	-	88 023	74 170	69 865	63 906	56 535	50 657	Kazakhstan
-	-	-	8 616	8 466	8 190	7 524	7 127	6 623	Kyrgyzstan
-	-	-	6 898	5 250	4 974	4 963	4 895	5 000	Latvia
-	-	-	11 702	9 185	9 184	9 212	9 852	9 751	Lithuania
-	-	-	9 281	8 901	7 316	6 510	6 336	5 800	Moldova
-	-	-	908 115	850 142	769 972	756 947	743 264	732 960	Russia
-	-	-	16 454	14 373	14 326	13 636	13 559	13 100	Tajikistan
-	-	-	7 793	8 274	6 868	6 574	6 169	5 700	Turkmenistan
-	-	-	224 662	206 002	180 141	172 235	155 973	149 441	Ukraine
-	-	-	45 651	44 284	43 064	42 020	42 450	42 975	Uzbekistan
1 541 100	1 550 000	1 518 528	1 407 606	1 298 176	1 175 403	1 140 862	1 105 055	1 083 387	**Former USSR**
3 400	3 490	3 495	3 700	4 044	4 386	4 579	4 599	4 747	Bahrain
42 468	47 838	56 559	60 151	61 521	66 256	68 661	72 472	75 145	Iran
22 660	22 800	19 610	25 300	26 300	28 000	29 000	29 000	29 561	Iraq
18 948	19 285	19 659	22 901	24 242	26 352	28 167	30 213	31 047	Israel
2 930	3 330	3 387	3 969	4 291	4 640	5 088	5 462	5 658	Jordan
19 679	18 768	9 280	15 286	20 178	22 798	23 726	25 475	27 091	Kuwait
2 250	1 400	2 800	3 371	3 964	4 355	4 681	6 730	8 000	Lebanon
3 351	3 841	3 950	4 316	4 909	5 209	5 495	5 702	6 073	Oman
4 394	4 568	4 381	4 873	5 199	5 455	5 314	6 184	6 456	Qatar
55 201	58 973	63 632	67 492	74 171	82 182	85 890	89 619	95 100	Saudi Arabia
8 179	8 573	8 978	9 262	12 638	15 182	15 300	16 885	17 950	Syria
14 612	16 081	16 222	16 460	17 578	18 870	19 070	19 737	20 571	United Arab Emirates
1 212	1 471	1 586	1 738	1 826	1 664	1 760	1 734	1 752	Yemen
199 284	210 418	213 539	238 819	260 861	285 349	296 731	313 812	329 151	**Middle East**
3 684 832	3 785 555	3 864 117	3 893 864	3 959 349	4 028 591	4 181 730	4 323 203	4 464 078	**Non-OECD Total**
3 396 906	3 474 521	3 630 870	3 653 676	3 771 761	3 876 360	3 985 474	4 125 131	4 169 194	OECD North America
1 021 573	1 099 032	1 140 825	1 161 241	1 190 802	1 290 636	1 339 905	1 388 824	1 441 300	OECD Pacific
2 444 406	2 470 522	2 514 907	2 534 951	2 538 629	2 569 411	2 638 688	2 710 967	2 751 817	OECD Europe
6 862 885	7 044 075	7 286 602	7 349 868	7 501 192	7 736 407	7 964 067	8 224 922	8 362 311	**OECD Total**
10 547 717	10 829 630	11 150 719	11 243 732	11 460 541	11 764 998	12 145 797	12 548 125	12 826 389	**World**

Albania, the Dem. Ppl's Rep. of Korea and Vietnam are not included in the regional aggregates.

Final Consumption of Coal (Mtoe)
Consommation finale de charbon (Mtep)
Endverbrauch von Kohle (Mtoe)
Consumo finale di carbone (Mtep)
石炭の最終消費量（石油換算百万トン）
Consumo final de carbón (Mtep)
Конечное потребление угля (Мтнэ)

	1971	1973	1978	1983	1984	1985	1986	1987	1988
Algérie	0.04	0.04	0.04	0.11	0.11	0.11	0.14	0.15	0.15
RD du Congo	0.16	0.11	0.11	0.15	0.16	0.16	0.16	0.16	0.16
Egypte	0.09	0.06	0.14	0.16	0.16	0.17	0.17	0.13	0.15
Ghana	0.01	0.01	0.00	0.00	0.00	0.00	0.00	0.00	0.00
Kenya	0.05	0.04	0.03	0.02	0.05	0.06	0.05	0.06	0.07
Maroc	0.19	0.08	0.02	0.15	0.17	0.32	0.30	0.26	0.27
Mozambique	0.38	0.36	0.15	0.09	0.06	0.04	0.02	0.02	0.02
Nigéria	0.12	0.17	0.10	0.03	0.04	0.06	0.06	0.06	0.02
Afrique du Sud	15.56	16.63	18.13	14.78	15.41	15.73	15.05	15.33	16.02
Tanzanie	-	-	0.00	0.00	0.01	0.01	0.00	0.00	0.00
Tunisie	0.03	0.03	0.03	0.02	0.03	0.02	0.02	0.02	0.03
Zambie	0.45	0.49	0.29	0.25	0.23	0.24	0.27	0.30	0.32
Zimbabwe	1.01	1.12	0.87	1.21	1.06	1.12	1.15	1.18	1.13
Autre Afrique	0.12	0.13	0.16	0.22	0.18	0.21	0.27	0.27	0.27
Afrique	18.20	19.28	20.09	17.20	17.66	18.24	17.66	17.95	18.62
Argentine	0.34	0.34	0.46	0.13	0.21	0.28	0.20	0.30	0.37
Bolivie	-	-	-	0.03	0.02	0.01	0.01	-	-
Brésil	0.53	0.65	1.16	2.68	3.08	3.32	3.45	3.80	3.99
Chili	0.75	0.68	0.52	0.45	0.56	0.60	0.67	0.64	0.72
Colombie	0.98	1.04	1.66	1.74	1.75	1.89	1.96	2.03	2.01
Costa Rica	0.00	0.00	0.00	0.00	0.00	0.00	0.00	0.00	0.00
Cuba	0.04	0.03	0.06	0.06	0.06	0.09	0.08	0.07	0.07
République dominicaine	-	-	-	0.00	0.00	-	-	-	-
Haiti	-	-	-	0.03	0.03	0.04	0.01	0.01	0.02
Jamaïque	-	-	-	-	-	-	-	-	0.00
Panama	-	-	-	-	-	0.02	0.02	0.02	0.01
Pérou	0.08	0.09	0.07	0.10	0.10	0.13	0.13	0.12	0.12
Uruguay	0.00	0.00	0.00	0.00	0.00	-	-	0.00	-
Vénézuela	0.05	0.08	0.04	0.04	0.05	0.05	0.06	0.08	0.07
Autre Amérique latine	0.02	0.03	0.02	0.01	0.01	-	-	-	-
Amérique latine	2.80	2.93	4.00	5.28	5.87	6.43	6.59	7.07	7.38
Bangladesh	0.09	0.12	0.12	0.08	0.03	0.05	0.07	0.12	0.12
Inde	20.61	22.92	24.92	28.62	29.34	36.99	34.81	35.29	38.84
Indonésie	0.09	0.05	0.09	0.15	0.19	0.21	0.18	0.19	0.47
RPD de Corée	19.90	16.76	18.69	20.77	19.94	20.42	20.63	20.63	20.90
Malaisie	0.02	0.01	0.01	0.21	0.23	0.30	0.22	0.27	0.38
Myanmar	0.12	0.04	0.13	0.13	0.14	0.13	0.14	0.04	0.04
Népal	0.01	0.05	0.01	0.03	0.06	0.05	0.03	0.05	0.05
Pakistan	0.58	0.53	0.56	0.75	0.87	1.05	1.05	1.09	1.30
Philippines	0.00	0.00	0.11	0.35	0.42	0.47	0.37	0.47	0.57
Sri Lanka	0.00	0.00	0.00	0.00	0.00	0.00	0.00	0.00	-
Taipei chinois	2.17	1.99	1.56	2.61	2.59	2.93	3.17	3.32	4.25
Thaïlande	0.02	0.02	0.07	0.20	0.25	0.37	0.44	0.63	0.77
Viêt-Nam	0.86	0.90	1.59	1.71	1.62	1.61	1.76	1.94	1.44
Autre Asie	0.33	0.16	0.21	0.19	0.20	0.24	0.24	0.24	0.23
Asie	44.79	43.55	48.08	55.81	55.87	64.84	63.12	64.28	69.37
Rép. populaire de Chine	135.11	144.40	212.57	241.07	264.57	281.60	298.21	311.53	323.93
Hong-Kong, Chine	0.02	0.01	0.00	0.00	-	0.00	0.00	0.00	-
Chine	135.13	144.41	212.57	241.07	264.57	281.61	298.22	311.53	323.93

Final Consumption of Coal (Mtoe)
Consommation finale de charbon (Mtep)
Endverbrauch von Kohle (Mtoe)
Consumo finale di carbone (Mtep)
石炭の最終消費量（石油換算百万トン）
Consumo final de carbón (Mtep)
Конечное потребление угля (Мтнэ)

1989	1990	1991	1992	1993	1994	1995	1996	1997	
0.13	0.12	0.08	0.08	0.15	0.14	0.14	0.14	0.10	Algeria
0.17	0.17	0.17	0.17	0.15	0.15	0.16	0.16	0.16	DR of Congo
0.15	0.13	0.15	0.16	0.19	0.16	0.13	0.13	0.13	Egypt
0.00	0.00	0.00	0.00	0.00	0.00	0.00	0.00	0.00	Ghana
0.08	0.09	0.09	0.10	0.08	0.07	0.06	0.06	0.06	Kenya
0.46	0.36	0.47	0.43	0.44	0.45	0.45	0.56	0.59	Morocco
0.02	0.02	0.02	0.03	0.03	0.04	0.03	0.01	0.01	Mozambique
0.02	0.03	0.03	0.03	0.04	0.04	0.01	0.01	0.01	Nigeria
16.21	15.79	14.99	13.50	13.52	13.61	16.92	15.51	16.39	South Africa
0.00	0.00	0.00	0.00	0.00	0.00	0.00	0.00	0.00	Tanzania
0.03	0.02	0.02	0.02	0.02	0.01	0.01	0.01	0.01	Tunisia
0.19	0.18	0.14	0.16	0.12	0.07	0.08	0.07	0.07	Zambia
1.14	1.49	1.37	1.42	1.29	0.73	0.78	0.85	0.79	Zimbabwe
0.27	0.25	0.14	0.23	0.23	0.10	0.09	0.09	0.09	Other Africa
18.88	18.65	17.68	16.33	16.26	15.58	18.87	17.60	18.42	**Africa**
0.38	0.32	0.28	0.27	0.30	0.28	0.23	0.34	0.36	Argentina
-	-	-	-	-	-	-	-	-	Bolivia
3.70	3.18	3.74	3.33	3.57	3.78	3.96	4.51	4.61	Brazil
0.73	0.67	0.75	0.86	0.81	0.74	0.74	0.85	1.33	Chile
2.12	2.33	2.64	2.32	2.27	2.11	1.87	1.98	2.03	Colombia
0.00	0.00	-	-	-	-	-	0.00	-	Costa Rica
0.11	0.10	0.08	0.07	0.05	0.05	0.04	0.01	0.03	Cuba
-	-	-	-	-	-	-	-	-	Dominican Republic
0.01	0.01	0.02	0.02	-	-	-	-	-	Haiti
0.03	0.04	0.01	0.04	0.04	0.03	0.03	0.04	0.04	Jamaica
0.01	0.02	0.03	0.04	0.04	0.03	0.03	0.03	0.04	Panama
0.14	0.11	0.24	0.23	0.40	0.34	0.30	0.21	0.18	Peru
-	-	0.00	0.00	-	-	-	-	-	Uruguay
0.31	0.22	0.25	0.28	0.18	0.22	0.21	0.20	0.26	Venezuela
0.00	0.00	0.00	0.00	0.00	0.00	0.00	0.00	0.00	Other Latin America
7.54	6.99	8.02	7.44	7.66	7.57	7.44	8.17	8.88	**Latin America**
0.12	0.28	0.09	0.08	0.03	-	-	-	-	Bangladesh
38.25	41.06	42.33	42.77	41.91	45.07	42.81	41.33	42.61	India
0.94	1.12	0.87	0.96	1.08	1.37	1.15	3.15	2.74	Indonesia
21.27	19.58	17.61	17.75	16.01	14.31	13.98	13.77	13.36	DPR of Korea
0.56	0.52	0.46	0.49	0.56	0.69	0.66	0.74	0.75	Malaysia
0.04	0.05	0.05	0.03	0.03	0.03	0.01	0.01	0.01	Myanmar
0.05	0.01	0.04	0.06	0.06	0.07	0.07	0.07	0.07	Nepal
1.25	1.49	1.44	1.47	1.53	1.66	1.44	1.55	1.35	Pakistan
0.37	0.41	0.62	0.50	0.55	0.48	0.46	0.83	1.07	Philippines
0.00	0.00	0.00	0.00	0.00	0.00	0.00	0.00	0.00	Sri Lanka
4.04	4.07	4.27	4.61	4.98	5.13	5.04	5.27	5.71	Chinese Taipei
1.04	1.29	1.51	1.68	2.49	3.05	3.62	4.39	3.94	Thailand
1.32	1.42	1.65	1.61	1.62	1.60	1.68	2.58	3.05	Vietnam
0.24	0.21	0.21	0.13	0.13	0.13	0.13	0.13	0.14	Other Asia
69.49	71.51	71.14	72.15	70.97	73.60	71.04	73.83	74.81	**Asia**
326.83	318.83	316.67	307.44	326.30	330.09	359.04	362.48	333.61	People's Rep.of China
-	-	-	-	-	-	-	-	-	Hong Kong, China
326.83	318.83	316.67	307.44	326.30	330.09	359.04	362.48	333.61	**China**

Final Consumption of Coal (Mtoe)
Consommation finale de charbon (Mtep)
Endverbrauch von Kohle (Mtoe)

Consumo finale di carbone (Mtep)

石炭の最終消費量（石油換算百万トン）

Consumo final de carbón (Mtep)

Конечное потребление угля (Мтнэ)

	1971	1973	1978	1983	1984	1985	1986	1987	1988
Albanie	0.26	0.30	0.45	0.66	0.72	0.80	0.79	0.83	0.84
Bulgarie	3.57	3.78	3.24	3.77	3.76	4.55	3.96	3.99	1.86
Chypre	-	-	-	-	0.03	0.05	0.03	0.09	0.06
Roumanie	2.75	2.86	4.19	5.51	4.65	5.28	5.02	6.04	6.21
République slovaque	3.57	3.48	4.01	3.59	3.74	3.87	3.85	3.90	3.93
Bosnie-Herzégovine	-	-	-	-	-	-	-	-	-
Croatie	-	-	-	-	-	-	-	-	-
Ex-RYM	-	-	-	-	-	-	-	-	-
Slovénie	-	-	-	-	-	-	-	-	-
RF de Yougoslavie	-	-	-	-	-	-	-	-	-
Ex-Yougoslavie	4.77	4.20	3.27	4.24	3.89	3.83	3.78	3.04	2.54
Europe non-OCDE	14.93	14.62	15.16	17.79	16.80	18.37	17.43	17.89	15.44
Arménie	-	-	-	-	-	-	-	-	-
Azerbaïdjan	-	-	-	-	-	-	-	-	-
Bélarus	-	-	-	-	-	-	-	-	-
Estonie	-	-	-	-	-	-	-	-	-
Géorgie	-	-	-	-	-	-	-	-	-
Kazakhstan	-	-	-	-	-	-	-	-	-
Kirghizistan	-	-	-	-	-	-	-	-	-
Lettonie	-	-	-	-	-	-	-	-	-
Lituanie	-	-	-	-	-	-	-	-	-
Moldova	-	-	-	-	-	-	-	-	-
Russie	-	-	-	-	-	-	-	-	-
Tadjikistan	-	-	-	-	-	-	-	-	-
Turkménistan	-	-	-	-	-	-	-	-	-
Ukraine	-	-	-	-	-	-	-	-	-
Ouzbékistan	-	-	-	-	-	-	-	-	-
Ex-URSS	102.08	109.75	133.43	113.84	109.70	107.04	102.53	107.08	110.57
Iran	0.34	0.66	0.38	0.37	0.61	0.62	0.65	0.67	0.68
Israël	0.00	-	-	-	-	-	-	0.10	0.09
Liban	0.00	0.01	0.00	-	-	-	-	-	-
Syrie	0.00	0.00	0.00	0.00	0.00	-	-	-	-
Moyen-Orient	0.35	0.67	0.39	0.37	0.61	0.62	0.65	0.77	0.77
Total non-OCDE	318.28	335.22	433.71	451.36	471.08	497.13	506.20	526.58	546.07
OCDE Amérique du N.	84.44	80.34	70.83	59.83	64.46	63.52	61.29	59.73	61.96
OCDE Pacifique	30.57	33.04	31.33	37.30	40.64	42.49	41.65	40.50	43.11
OCDE Europe	187.13	170.91	146.43	145.09	150.70	160.34	152.52	155.71	151.35
Total OCDE	302.15	284.29	248.59	242.22	255.80	266.35	255.47	255.94	256.42
Monde	620.43	619.51	682.30	693.58	726.88	763.48	761.67	782.52	802.48

Ex-URSS: les séries antérieures à 1990-1992 ne sont pas comparables aux années récentes; 1991 a été estimé.

Final Consumption of Coal (Mtoe)
Consommation finale de charbon (Mtep)
Endverbrauch von Kohle (Mtoe)
Consumo finale di carbone (Mtep)
石炭の最終消費量（石油換算百万トン）
Consumo final de carbón (Mtep)
Конечное потребление угля (Мтнэ)

1989	1990	1991	1992	1993	1994	1995	1996	1997	
0.84	0.51	0.27	0.12	0.06	0.04	0.04	0.02	0.02	Albania
1.77	1.36	1.16	1.13	1.15	1.09	1.05	1.10	1.03	Bulgaria
0.06	0.06	0.06	0.02	0.02	0.02	0.01	0.01	0.01	Cyprus
4.35	2.89	2.35	1.96	1.23	1.30	1.58	1.61	1.58	Romania
4.56	4.21	3.89	3.34	3.17	2.67	2.41	2.41	2.09	Slovak Republic
-	-	-	1.75	1.69	0.11	0.09	0.09	0.09	*Bosnia-Herzegovina*
-	-	-	0.17	0.15	0.13	0.12	0.11	0.10	*Croatia*
-	-	-	0.12	0.17	0.16	0.15	0.12	0.12	*FYROM*
-	-	-	0.23	0.18	0.14	0.11	0.11	0.09	*Slovenia*
-	-	-	2.52	1.32	1.35	1.41	1.35	1.42	*FR of Yugoslavia*
3.46	2.94	2.25	4.78	3.50	1.89	1.88	1.79	1.83	Former Yugoslavia
15.05	11.97	9.98	11.35	9.13	7.00	6.97	6.95	6.57	**Non-OECD Europe**
-	-	-	0.06	0.00	0.02	0.00	0.00	0.00	Armenia
-	-	-	0.01	0.00	0.00	0.00	0.00	0.00	Azerbaijan
-	-	-	1.04	1.12	0.87	0.79	0.78	0.63	Belarus
-	-	-	0.19	0.10	0.12	0.18	0.20	0.15	Estonia
-	-	-	0.18	0.13	0.07	0.03	0.03	0.00	Georgia
-	-	-	12.40	11.18	10.62	8.92	7.15	6.09	Kazakhstan
-	-	-	0.83	0.74	0.87	0.28	0.37	0.73	Kyrgyzstan
-	-	-	0.45	0.19	0.21	0.14	0.11	0.10	Latvia
-	-	-	0.28	0.27	0.19	0.15	0.15	0.12	Lithuania
-	-	-	0.28	0.23	0.20	0.12	0.16	0.12	Moldova
-	-	-	45.02	43.72	35.41	33.97	30.12	27.47	Russia
-	-	-	0.26	0.15	0.04	0.01	0.05	0.05	Tajikistan
-	-	-	0.27	0.06	-	-	0.04	-	Turkmenistan
-	-	-	26.46	25.77	17.19	14.94	11.74	10.86	Ukraine
-	-	-	0.81	0.58	0.60	0.33	0.34	0.20	Uzbekistan
109.17	95.61	73.78	86.75	82.13	65.06	58.89	50.63	45.93	**Former USSR**
0.69	0.82	0.89	0.98	0.47	0.49	0.58	0.58	0.58	Iran
0.10	0.01	0.01	0.01	0.01	0.01	0.01	0.02	0.02	Israel
-	-	-	-	0.07	0.07	0.12	0.13	0.13	Lebanon
-	-	-	-	-	-	-	-	-	Syria
0.79	0.83	0.90	0.99	0.55	0.57	0.70	0.73	0.73	**Middle East**
547.75	524.41	498.17	502.45	513.01	499.47	522.96	520.39	488.95	**Non-OECD Total**
60.18	61.49	58.67	32.52	33.92	33.28	33.29	31.14	31.32	OECD North America
42.80	42.66	43.02	39.89	39.35	37.97	36.11	37.49	37.28	OECD Pacific
141.98	121.66	103.42	94.61	87.89	80.87	79.23	83.45	77.36	OECD Europe
244.96	225.81	205.11	167.02	161.15	152.12	148.62	152.07	145.96	**OECD Total**
792.71	750.22	703.28	669.47	674.16	651.60	671.58	672.47	634.90	**World**

Former USSR: series up to 1990-1992 are not comparable with the recent years; 1991 is estimated.

Final Consumption of Oil (Mtoe)
Consommation finale de pétrole (Mtep)
Endverbrauch von Öl (Mtoe)
Consumo finale di petrolio (Mtep)
石油の最終消費量（石油換算百万トン）
Consumo final de petróleo (Mtep)
Конечное потребление нефти и нефтепродуктов (Мтнэ)

	1971	1973	1978	1983	1984	1985	1986	1987	1988
Algérie	1.69	2.40	3.93	6.37	7.12	7.41	7.62	7.76	7.68
Angola	0.54	0.71	0.68	0.48	0.38	0.56	0.45	0.42	0.43
Bénin	0.11	0.13	0.13	0.12	0.14	0.17	0.15	0.13	0.13
Cameroun	0.29	0.32	0.53	0.82	0.89	0.89	0.86	0.86	0.90
Congo	0.16	0.17	0.22	0.27	0.25	0.26	0.25	0.23	0.21
RD du Congo	0.66	0.73	0.78	1.04	1.00	0.94	0.82	0.95	1.01
Egypte	4.99	5.25	7.68	12.02	13.28	14.05	13.66	14.49	14.75
Ethiopie	0.37	0.39	0.40	0.48	0.48	0.49	0.60	0.72	0.75
Gabon	0.13	0.17	0.57	0.60	0.56	0.57	0.52	0.45	0.49
Ghana	0.65	0.67	0.84	0.67	0.65	0.70	0.71	0.74	0.91
Côte d'Ivoire	0.67	0.71	0.97	0.98	0.86	0.85	0.93	0.97	0.97
Kenya	1.06	1.23	1.54	1.42	1.63	1.69	1.80	1.87	1.85
Libye	0.47	0.93	1.98	3.61	3.55	3.49	3.31	3.45	3.32
Maroc	1.74	2.12	3.04	3.10	3.06	3.09	3.07	3.05	3.21
Mozambique	0.36	0.37	0.34	0.32	0.29	0.32	0.34	0.35	0.35
Nigéria	1.55	2.42	6.48	8.93	8.00	7.98	7.42	7.57	7.97
Sénégal	0.38	0.45	0.50	0.56	0.59	0.59	0.56	0.53	0.51
Afrique du Sud	8.98	10.60	12.13	12.67	13.46	13.04	12.83	13.51	14.64
Soudan	1.10	1.55	0.98	1.24	1.17	1.32	1.29	1.00	1.42
Tanzanie	0.45	0.52	0.51	0.42	0.42	0.44	0.46	0.46	0.50
Tunisie	0.91	1.09	1.67	2.14	2.21	2.26	2.17	2.03	2.28
Zambie	0.48	0.66	0.70	0.65	0.59	0.55	0.49	0.50	0.52
Zimbabwe	0.56	0.71	0.69	0.78	0.74	0.79	0.84	0.84	0.88
Autre Afrique	2.31	2.59	2.71	2.78	2.74	2.86	2.92	4.25	4.32
Afrique	30.63	36.88	50.02	62.49	64.06	65.29	64.07	67.15	69.99
Argentine	16.52	17.06	18.00	18.23	18.05	16.68	16.72	18.09	16.52
Bolivie	0.57	0.68	1.02	0.97	0.91	0.88	0.91	0.98	0.98
Brésil	23.98	32.42	46.13	40.67	40.24	42.93	46.40	47.71	48.85
Chili	3.95	3.97	3.96	3.98	4.02	3.92	4.18	4.36	4.88
Colombie	5.61	6.06	6.99	7.86	7.79	7.95	8.02	8.76	8.86
Costa Rica	0.37	0.45	0.68	0.54	0.58	0.63	0.65	0.69	0.72
Cuba	4.27	5.68	6.23	6.45	6.53	7.02	6.87	7.09	7.31
République dominicaine	0.78	0.91	1.11	1.17	1.23	1.04	1.18	1.40	1.46
Equateur	1.04	1.18	2.46	3.06	3.38	3.73	3.92	3.82	4.22
El Salvador	0.36	0.44	0.63	0.49	0.50	0.52	0.52	0.56	0.64
Guatemala	0.66	0.73	0.92	0.75	0.80	0.82	0.83	0.91	0.98
Haiti	0.11	0.12	0.21	0.17	0.17	0.18	0.19	0.20	0.21
Honduras	0.31	0.36	0.47	0.52	0.55	0.54	0.56	0.62	0.69
Jamaïque	1.43	2.06	1.93	1.44	1.35	1.21	1.15	1.18	1.20
Antilles néerlandaises	2.13	2.21	1.25	1.03	0.95	0.87	0.78	0.76	0.70
Nicaragua	0.41	0.51	0.57	0.49	0.49	0.50	0.54	0.58	0.51
Panama	0.50	0.55	0.64	0.53	0.59	0.61	0.65	0.64	0.56
Paraguay	0.18	0.23	0.38	0.44	0.49	0.51	0.53	0.56	0.58
Pérou	4.17	4.40	4.98	4.76	4.83	4.79	5.17	5.72	5.62
Trinité-et-Tobago	0.43	0.54	0.78	0.86	0.91	0.86	0.80	0.74	0.72
Uruguay	1.35	1.35	1.38	1.11	1.00	0.93	0.95	1.02	1.04
Vénézuela	6.13	7.27	11.41	13.11	12.72	13.50	14.02	15.13	14.98
Autre Amérique latine	1.77	2.22	2.00	1.73	1.91	1.81	1.96	2.32	2.25
Amérique latine	77.00	91.43	114.15	110.37	110.00	112.44	117.48	123.85	124.48

Final Consumption of Oil (Mtoe)
Consommation finale de pétrole (Mtep)
Endverbrauch von Öl (Mtoe)
Consumo finale di petrolio (Mtep)
石油の最終消費量（石油換算百万トン）
Consumo final de petróleo (Mtep)
Конечное потребление нефти и нефтепродуктов (Мтнэ)

1989	1990	1991	1992	1993	1994	1995	1996	1997	
7.98	7.80	8.44	8.33	8.59	7.93	7.70	7.53	8.97	Algeria
0.41	0.42	0.42	0.42	0.42	0.42	0.43	0.51	0.77	Angola
0.11	0.09	0.07	0.07	0.08	0.08	0.09	0.33	0.32	Benin
0.99	0.96	0.89	0.83	0.81	0.85	0.84	0.83	0.84	Cameroon
0.28	0.27	0.28	0.26	0.27	0.26	0.29	0.29	0.29	Congo
1.00	1.08	1.11	1.03	1.15	1.17	1.19	1.19	1.19	DR of Congo
14.92	15.30	14.43	14.81	13.05	12.79	14.53	15.48	16.53	Egypt
0.79	0.82	0.85	0.86	0.89	0.94	1.01	1.16	0.71	Ethiopia
0.38	0.30	0.35	0.36	0.40	0.39	0.42	0.46	0.54	Gabon
0.97	0.92	0.87	0.94	1.00	1.07	1.11	1.11	1.11	Ghana
0.95	0.80	0.79	0.75	0.78	0.80	0.84	0.90	0.92	Ivory Coast
1.94	1.89	1.79	1.86	1.89	1.98	1.93	2.10	2.01	Kenya
3.78	3.79	3.97	3.95	4.45	5.15	5.66	5.72	5.93	Libya
3.55	3.70	3.86	4.39	4.38	4.78	4.93	4.99	5.08	Morocco
0.33	0.38	0.33	0.40	0.45	0.41	0.39	0.40	0.52	Mozambique
8.22	6.80	8.29	10.85	8.66	7.86	7.75	8.53	8.64	Nigeria
0.56	0.58	0.57	0.60	0.53	0.59	0.65	0.74	0.81	Senegal
15.03	15.46	15.43	15.28	14.57	15.17	16.57	16.65	16.91	South Africa
1.32	1.71	1.50	1.43	0.90	1.45	1.29	1.20	1.44	Sudan
0.51	0.54	0.53	0.52	0.48	0.48	0.48	0.47	0.42	Tanzania
2.53	2.58	2.50	2.65	2.93	2.93	2.99	3.16	3.24	Tunisia
0.51	0.49	0.50	0.47	0.48	0.50	0.50	0.51	0.51	Zambia
0.82	0.96	1.03	1.01	1.05	1.16	1.32	1.31	1.31	Zimbabwe
4.56	4.38	4.58	4.74	4.92	4.91	5.00	5.05	5.25	Other Africa
72.44	72.02	73.37	76.80	73.14	74.07	77.89	80.60	84.27	**Africa**
15.68	15.58	15.96	16.61	18.00	18.60	18.81	19.84	19.89	Argentina
1.04	1.08	1.10	1.11	1.15	1.21	1.34	1.45	1.49	Bolivia
50.43	51.42	52.20	53.66	55.62	58.68	63.00	67.49	72.47	Brazil
5.33	5.54	5.85	6.41	7.02	7.48	8.17	8.76	9.58	Chile
9.21	9.36	9.33	10.09	10.96	11.62	12.50	12.88	13.10	Colombia
0.85	0.84	0.87	1.16	1.21	1.33	1.33	1.33	1.40	Costa Rica
7.69	6.68	4.86	3.77	3.62	3.88	3.95	4.34	4.41	Cuba
1.68	1.66	1.62	2.02	1.87	1.99	2.07	2.15	2.31	Dominican Republic
4.14	4.17	4.22	4.01	3.97	4.23	4.47	4.87	5.21	Ecuador
0.65	0.70	0.74	0.91	0.97	1.03	1.20	1.16	1.21	El Salvador
1.02	1.10	1.14	1.23	1.37	1.44	1.73	1.73	1.83	Guatemala
0.24	0.25	0.23	0.23	0.21	0.05	0.27	0.29	0.34	Haiti
0.77	0.71	0.72	0.83	0.81	0.96	1.16	1.16	1.18	Honduras
1.57	1.77	1.83	1.85	0.98	1.35	1.30	1.39	1.48	Jamaica
0.64	0.61	0.65	0.69	0.73	0.82	0.82	0.82	0.82	Netherlands Antilles
0.49	0.47	0.45	0.48	0.48	0.54	0.54	0.57	0.65	Nicaragua
0.58	0.61	0.65	0.70	0.71	0.79	0.85	0.88	0.98	Panama
0.65	0.65	0.64	0.78	0.88	1.04	1.14	1.03	1.11	Paraguay
5.02	4.93	4.90	4.80	5.13	5.14	6.16	6.60	6.48	Peru
0.65	0.69	0.76	0.74	0.65	0.64	0.66	0.70	0.79	Trinidad-and-Tobago
1.06	1.05	1.11	1.17	1.23	1.29	1.30	1.41	1.50	Uruguay
14.11	14.47	15.17	15.81	17.68	18.81	18.54	17.60	19.67	Venezuela
2.58	2.57	2.50	2.29	2.32	2.41	2.44	2.51	2.59	Other Latin America
126.09	126.90	127.52	131.35	137.57	145.34	153.76	160.97	170.49	**Latin America**

Final Consumption of Oil (Mtoe)
Consommation finale de pétrole (Mtep)
Endverbrauch von Öl (Mtoe)

Consumo finale di petrolio (Mtep)

石油の最終消費量（石油換算百万トン）

Consumo final de petróleo (Mtep)

Конечное потребление нефти и нефтепродуктов (Мтнэ)

	1971	1973	1978	1983	1984	1985	1986	1987	1988
Bangladesh	0.58	0.61	1.01	1.15	1.16	1.28	1.28	1.36	1.38
Brunei	0.06	0.07	0.10	0.23	0.22	0.23	0.24	0.25	0.28
Inde	18.24	20.76	25.34	33.18	35.48	38.39	40.86	43.67	47.11
Indonésie	6.42	7.75	13.95	18.78	18.00	18.61	18.67	19.37	21.19
RPD de Corée	0.71	0.80	1.79	2.68	2.84	3.00	3.16	3.19	3.19
Malaisie	3.58	3.49	4.72	6.79	6.83	6.89	7.04	7.46	7.96
Myanmar	1.20	0.97	1.08	1.01	1.06	1.02	1.05	0.65	0.63
Népal	0.05	0.06	0.08	0.12	0.14	0.14	0.17	0.20	0.23
Pakistan	3.03	3.19	4.14	5.35	5.98	6.36	6.80	7.30	7.76
Philippines	5.37	6.01	7.07	6.14	5.20	4.70	4.98	5.68	6.18
Singapour	1.06	1.42	2.37	2.98	3.27	3.97	4.03	4.47	4.50
Sri Lanka	0.82	1.04	1.06	1.11	1.17	1.12	1.12	1.13	1.17
Taipei chinois	3.74	5.82	10.65	12.69	13.62	14.45	15.77	16.91	17.98
Thaïlande	5.00	6.27	7.70	8.75	9.51	9.57	10.13	11.29	12.70
Viêt-Nam	5.90	5.86	0.95	1.62	1.58	1.57	1.77	2.07	2.13
Autre Asie	1.43	1.65	1.33	1.19	1.26	1.45	1.25	1.48	1.42
Asie	57.19	65.76	83.34	103.75	107.32	112.73	118.31	126.48	135.79
Rép. populaire de Chine	34.70	41.79	72.69	59.88	63.36	67.16	72.65	78.76	83.21
Hong-Kong, Chine	2.05	2.02	3.20	3.14	3.30	3.05	3.57	3.65	4.35
Chine	36.75	43.80	75.89	63.02	66.66	70.21	76.22	82.42	87.56
Albanie	0.58	0.47	0.87	0.86	0.85	0.64	0.66	0.68	0.72
Bulgarie	8.78	9.35	8.15	6.99	6.64	6.09	6.16	5.94	7.72
Chypre	0.44	0.57	0.53	0.65	0.67	0.66	0.74	0.80	0.87
Gibraltar	0.03	0.03	0.02	0.02	0.02	0.03	0.03	0.04	0.05
Malte	0.17	0.19	0.17	0.18	0.23	0.13	0.25	0.27	0.27
Roumanie	9.47	10.76	12.62	8.75	8.29	10.08	9.94	11.60	11.08
République slovaque	2.89	3.28	3.34	4.24	4.11	4.05	3.94	3.91	3.81
Bosnie-Herzégovine	-	-	-	-	-	-	-	-	-
Croatie	-	-	-	-	-	-	-	-	-
Ex-RYM	-	-	-	-	-	-	-	-	-
Slovénie	-	-	-	-	-	-	-	-	-
RF de Yougoslavie	-	-	-	-	-	-	-	-	-
Ex-Yougoslavie	7.80	8.89	12.20	10.73	9.43	9.93	11.12	11.09	11.70
Europe non-OCDE	30.14	33.55	37.90	32.44	30.24	31.62	32.84	34.32	36.21

Final Consumption of Oil (Mtoe)
Consommation finale de pétrole (Mtep)
Endverbrauch von Öl (Mtoe)
Consumo finale di petrolio (Mtep)
石油の最終消費量（石油換算百万トン）
Consumo final de petróleo (Mtep)
Конечное потребление нефти и нефтепродуктов (Мтнэ)

1989	1990	1991	1992	1993	1994	1995	1996	1997	
1.58	1.63	1.58	1.71	1.88	2.03	2.42	2.52	2.54	Bangladesh
0.29	0.30	0.32	0.36	0.39	0.41	0.45	0.49	0.53	Brunei
51.16	51.86	55.42	58.18	59.83	63.93	71.45	77.23	79.87	India
23.91	25.36	26.52	27.93	30.40	32.98	35.31	38.39	42.56	Indonesia
3.43	3.39	3.08	2.86	2.57	1.76	1.53	1.33	1.29	DPR of Korea
8.79	10.19	11.08	11.78	13.76	14.93	16.24	17.99	18.64	Malaysia
0.62	0.60	0.56	0.65	0.78	0.90	1.18	1.28	1.46	Myanmar
0.20	0.19	0.25	0.30	0.31	0.42	0.52	0.52	0.52	Nepal
8.07	8.16	8.13	8.47	9.59	9.99	10.49	11.62	11.29	Pakistan
7.03	7.38	7.18	8.12	9.22	9.80	10.73	11.44	12.38	Philippines
4.84	5.76	5.57	5.32	5.87	6.54	7.02	7.47	8.39	Singapore
1.14	1.18	1.19	1.39	1.48	1.65	1.82	2.10	2.04	Sri Lanka
19.03	20.04	20.38	21.71	22.70	24.80	25.50	26.71	27.36	Chinese Taipei
14.89	16.73	17.30	19.10	21.09	23.74	26.82	30.03	30.49	Thailand
2.04	2.52	2.19	2.84	3.43	3.84	4.01	5.85	7.18	Vietnam
1.61	1.73	1.73	1.50	1.48	1.41	1.51	1.49	1.53	Other Asia
148.62	157.02	162.48	172.23	184.77	199.12	216.99	236.44	248.08	**Asia**
87.03	86.75	93.94	100.72	109.37	113.26	126.87	142.64	158.62	People's Rep.of China
4.52	4.75	4.57	5.88	6.18	6.92	6.87	6.72	7.05	Hong Kong, China
91.55	91.50	98.51	106.60	115.55	120.18	133.74	149.36	165.66	**China**
0.67	0.83	0.68	0.52	0.47	0.60	0.45	0.45	0.40	Albania
7.75	6.29	3.34	3.46	3.91	3.72	3.64	3.60	3.12	Bulgaria
0.91	0.92	1.08	1.17	1.15	1.19	1.25	1.30	1.30	Cyprus
0.05	0.05	0.06	0.07	0.07	0.09	0.09	0.09	0.09	Gibraltar
0.27	0.27	0.30	0.30	0.31	0.31	0.34	0.40	0.44	Malta
9.95	9.10	8.49	7.80	5.25	6.05	6.08	7.21	7.73	Romania
5.27	4.57	3.81	3.40	2.67	3.18	3.13	3.20	3.27	Slovak Republic
-	-	-	0.46	0.44	0.33	0.33	0.33	0.33	*Bosnia-Herzegovina*
-	-	-	2.16	2.24	2.46	2.51	2.39	2.74	*Croatia*
-	-	-	0.78	0.78	0.70	0.66	1.02	0.85	*FYROM*
-	-	-	1.49	1.81	1.95	2.18	2.57	2.56	*Slovenia*
-	-	-	1.63	1.22	1.16	1.28	1.95	2.70	*FR of Yugoslavia*
10.96	10.74	9.92	6.51	6.50	6.60	6.95	8.26	9.18	Former Yugoslavia
35.82	32.75	27.69	23.25	20.33	21.73	21.92	24.50	25.52	**Non-OECD Europe**

Final Consumption of Oil (Mtoe)
Consommation finale de pétrole (Mtep)
Endverbrauch von Öl (Mtoe)

Consumo finale di petrolio (Mtep)

石油の最終消費量（石油換算百万トン）

Consumo final de petróleo (Mtep)

Конечное потребление нефти и нефтепродуктов (Мтнэ)

	1971	1973	1978	1983	1984	1985	1986	1987	1988
Arménie	-	-	-	-	-	-	-	-	-
Azerbaïdjan	-	-	-	-	-	-	-	-	-
Bélarus	-	-	-	-	-	-	-	-	-
Estonie	-	-	-	-	-	-	-	-	-
Géorgie	-	-	-	-	-	-	-	-	-
Kazakhstan	-	-	-	-	-	-	-	-	-
Kirghizistan	-	-	-	-	-	-	-	-	-
Lettonie	-	-	-	-	-	-	-	-	-
Lituanie	-	-	-	-	-	-	-	-	-
Moldova	-	-	-	-	-	-	-	-	-
Russie	-	-	-	-	-	-	-	-	-
Tadjikistan	-	-	-	-	-	-	-	-	-
Turkménistan	-	-	-	-	-	-	-	-	-
Ukraine	-	-	-	-	-	-	-	-	-
Ouzbékistan	-	-	-	-	-	-	-	-	-
Ex-URSS	203.21	230.00	285.90	306.58	306.93	304.26	309.09	312.98	320.91
Bahrein	0.22	0.31	0.68	0.74	0.77	0.77	0.77	0.76	0.79
Iran	7.87	11.21	21.67	25.62	30.46	30.26	31.83	35.26	35.17
Irak	2.03	2.35	5.28	8.90	9.57	10.11	10.72	10.77	11.67
Israël	3.23	3.73	4.07	4.42	4.31	4.27	4.77	5.12	5.23
Jordanie	0.41	0.48	1.04	1.83	1.79	1.85	1.96	1.96	1.97
Koweit	0.67	0.75	1.79	2.90	2.25	2.84	2.27	2.23	2.40
Liban	1.52	1.83	1.49	1.60	1.51	1.71	1.63	1.72	1.17
Oman	0.09	0.09	0.47	0.75	0.96	1.03	0.99	0.89	0.93
Qatar	0.08	0.14	0.37	0.55	0.53	0.57	0.53	0.57	0.55
Arabie saoudite	2.25	3.34	11.64	31.51	24.25	25.76	26.66	25.56	27.17
Syrie	1.90	1.80	4.31	6.23	6.81	6.47	6.58	7.17	6.98
Emirats arabes unis	0.21	0.42	2.36	3.97	3.64	4.19	4.21	4.27	3.93
Yémen	0.31	0.51	0.82	1.47	1.48	1.74	1.82	1.95	2.05
Moyen-Orient	20.78	26.96	56.00	90.50	88.33	91.57	94.75	98.22	100.00
Total non-OCDE	455.71	528.38	703.21	769.13	773.53	788.12	812.75	845.41	874.95
OCDE Amérique du N.	728.24	801.54	879.18	747.47	777.47	779.94	794.70	817.85	841.39
OCDE Pacifique	177.80	209.64	218.91	199.29	210.35	209.07	216.61	225.50	240.74
OCDE Europe	525.88	598.01	596.38	507.08	509.14	511.31	530.27	535.84	546.06
Total OCDE	1 431.92	1 609.19	1 694.47	1 453.84	1 496.96	1 500.32	1 541.58	1 579.18	1 628.19
Monde	1 887.63	2 137.57	2 397.67	2 222.98	2 270.49	2 288.44	2 354.33	2 424.58	2 503.14

Ex-URSS: les séries antérieures à 1990-1992 ne sont pas comparables aux années récentes; 1991 a été estimé.

Final Consumption of Oil (Mtoe)
Consommation finale de pétrole (Mtep)
Endverbrauch von Öl (Mtoe)
Consumo finale di petrolio (Mtep)

石油の最終消費量（石油換算百万トン）

Consumo final de petróleo (Mtep)

Конечное потребление нефти и нефтепродуктов (Мтнэ)

1989	1990	1991	1992	1993	1994	1995	1996	1997	
-	-	-	1.28	0.88	0.24	0.17	0.09	0.09	Armenia
-	-	-	5.06	3.58	3.59	2.86	2.08	1.79	Azerbaijan
-	-	-	11.99	8.31	6.21	5.68	5.47	5.60	Belarus
-	-	-	0.75	0.76	0.85	0.90	0.98	0.95	Estonia
-	-	-	0.97	0.56	0.21	0.10	0.78	0.85	Georgia
-	-	-	18.60	13.72	9.67	9.50	7.80	6.97	Kazakhstan
-	-	-	1.93	1.17	0.42	0.62	0.69	0.50	Kyrgyzstan
-	-	-	1.87	1.69	1.56	1.31	1.10	1.21	Latvia
-	-	-	2.83	1.82	1.58	1.70	1.75	1.83	Lithuania
-	-	-	1.26	0.90	0.71	0.71	0.65	0.66	Moldova
-	-	-	141.07	121.18	80.12	89.14	83.20	81.33	Russia
-	-	-	5.92	3.69	1.23	1.23	1.23	1.23	Tajikistan
-	-	-	4.41	2.40	2.42	2.16	2.34	2.36	Turkmenistan
-	-	-	30.79	21.31	18.63	19.97	16.22	14.80	Ukraine
-	-	-	6.31	5.73	5.38	4.82	4.45	4.82	Uzbekistan
321.65	311.77	307.45	235.03	187.71	132.82	140.88	128.83	125.00	**Former USSR**
0.81	0.89	0.77	0.78	0.78	0.85	0.84	0.84	1.01	Bahrain
36.66	39.08	42.07	43.57	45.46	46.89	48.26	50.30	50.73	Iran
12.92	12.99	10.35	11.24	13.58	14.75	14.52	14.18	14.56	Iraq
5.23	5.53	5.65	6.10	6.67	7.04	7.93	8.31	8.55	Israel
2.00	2.26	2.09	2.43	2.52	2.61	2.71	2.83	3.03	Jordan
2.55	2.23	0.60	4.06	3.40	3.55	3.96	3.92	4.45	Kuwait
1.24	1.28	1.72	1.92	2.39	2.59	3.06	2.87	3.16	Lebanon
0.99	1.58	2.72	2.46	2.11	1.70	2.11	2.02	1.72	Oman
1.38	0.65	0.78	0.67	0.71	0.79	0.85	0.94	1.05	Qatar
26.62	29.29	30.88	31.61	33.94	33.86	32.01	34.17	35.53	Saudi Arabia
7.01	7.53	8.10	8.22	8.28	8.67	8.84	9.15	9.65	Syria
4.11	3.97	4.02	4.41	4.86	5.12	4.92	4.55	4.10	United Arab Emirates
2.01	1.75	2.31	2.60	2.03	2.16	2.43	2.42	2.50	Yemen
103.55	109.04	112.06	120.08	126.71	130.57	132.44	136.50	140.04	**Middle East**
899.71	901.01	909.08	865.33	845.77	823.82	877.62	917.20	959.06	**Non-OECD Total**
841.83	830.42	808.11	831.95	842.31	869.90	873.48	899.46	917.52	OECD North America
252.19	267.04	278.10	293.51	301.13	314.64	329.62	339.49	347.69	OECD Pacific
540.17	544.64	559.19	565.82	565.92	569.51	580.09	593.85	598.33	OECD Europe
1 634.19	1 642.10	1 645.40	1 691.28	1 709.36	1 754.06	1 783.19	1 832.80	1 863.54	**OECD Total**
2 533.91	2 543.11	2 554.47	2 556.61	2 555.13	2 577.88	2 660.81	2 750.00	2 822.60	**World**

Former USSR: series up to 1990-1992 are not comparable with the recent years; 1991 is estimated.

Final Consumption of Gas (Mtoe)
Consommation finale de gaz (Mtep)
Endverbrauch von Gas (Mtoe)
Consumo finale di gas (Mtep)
ガスの最終消費量（石油換算百万トン）
Consumo final de gas natural (Mtep)
Конечное потребление газа (Мтнэ)

	1971	1973	1978	1983	1984	1985	1986	1987	1988
Algérie	0.26	0.30	0.92	1.39	1.54	1.90	2.34	2.55	2.57
Angola	0.04	0.05	0.06	0.09	0.10	0.10	0.11	0.13	0.13
Congo	0.05	-	-	-	-	-	-	-	-
Egypte	0.00	0.00	0.33	1.16	1.44	1.47	1.53	1.77	1.75
Gabon	-	-	-	0.03	0.03	0.04	0.08	0.10	0.15
Libye	-	-	0.65	1.18	1.29	1.44	1.48	1.50	1.10
Maroc	0.04	0.06	0.07	0.07	0.07	0.08	0.08	0.06	0.06
Nigéria	0.03	0.03	0.03	0.12	0.31	0.37	0.39	0.46	0.58
Sénégal	-	-	-	-	-	-	-	0.00	0.01
Afrique du Sud	0.14	0.22	0.34	0.48	0.48	0.45	0.48	0.44	0.44
Tunisie	0.01	0.01	0.07	0.11	0.12	0.14	0.18	0.29	0.15
Autre Afrique	0.00	0.00	0.00	0.00	0.00	0.00	0.00	0.00	0.00
Afrique	0.57	0.69	2.47	4.64	5.38	5.98	6.67	7.29	6.94
Argentine	2.89	3.68	4.76	6.96	7.74	7.65	8.12	9.05	9.76
Bolivie	-	0.00	0.05	0.04	0.04	0.06	0.07	0.08	0.08
Brésil	0.17	0.24	0.84	1.51	1.66	1.91	1.94	2.24	2.39
Chili	0.05	0.09	0.15	0.17	0.18	0.18	0.19	0.18	0.41
Colombie	0.11	0.21	0.55	0.70	0.78	0.81	0.89	0.88	0.95
Cuba	0.03	0.05	0.06	0.06	0.06	0.11	0.11	0.13	0.13
Panama	0.01	0.00	0.00	0.00	0.00	-	-	-	-
Pérou	0.05	0.06	0.09	0.05	0.10	0.11	0.11	0.11	0.10
Trinité-et-Tobago	0.46	0.52	0.86	1.67	1.64	1.64	1.77	1.79	2.39
Uruguay	0.00	0.00	0.00	0.00	0.00	0.00	0.00	0.00	0.00
Vénézuela	2.33	4.35	6.01	7.36	8.58	9.05	8.11	8.77	7.05
Autre Amérique latine	0.00	0.00	0.01	0.01	0.02	0.02	0.01	0.01	0.01
Amérique latine	6.10	9.21	13.37	18.55	20.81	21.53	21.31	23.25	23.28
Bangladesh	0.19	0.33	0.45	1.01	1.12	1.19	1.38	1.54	1.72
Inde	0.20	0.19	0.74	1.28	1.58	2.43	3.21	3.37	5.13
Indonésie	0.10	0.12	1.39	3.35	3.99	5.40	5.52	5.74	5.80
Malaisie	0.01	0.01	0.03	0.04	0.12	0.50	0.89	1.11	1.24
Myanmar	0.02	0.03	0.07	0.13	0.17	0.23	0.26	0.27	0.27
Pakistan	1.50	1.80	2.58	4.40	4.34	4.48	4.57	5.30	5.52
Philippines	0.01	0.01	0.01	0.00	0.00	0.00	0.00	0.00	0.00
Singapour	0.02	0.02	0.04	0.05	0.05	0.05	0.05	0.05	0.05
Taipei chinois	0.65	0.94	1.46	0.97	0.99	0.88	0.81	0.84	0.90
Thaïlande	-	-	-	0.03	0.17	0.16	0.08	0.04	0.05
Viêt-Nam	-	-	-	-	-	-	-	-	-
Autre Asie	0.19	0.14	0.13	0.27	0.41	0.52	0.53	0.17	0.19
Asie	2.88	3.59	6.89	11.53	12.94	15.84	17.28	18.41	20.88
Rép. populaire de Chine	1.41	2.25	6.31	6.36	6.80	7.93	8.32	8.84	9.28
Hong-Kong, Chine	0.02	0.03	0.05	0.13	0.15	0.17	0.19	0.23	0.26
Chine	1.43	2.28	6.37	6.49	6.95	8.10	8.51	9.06	9.54

Final Consumption of Gas (Mtoe)
Consommation finale de gaz (Mtep)
Endverbrauch von Gas (Mtoe)

Consumo finale di gas (Mtep)

ガスの最終消費量（石油換算百万トン）

Consumo final de gas natural (Mtep)

Конечное потребление газа (Мтнэ)

1989	1990	1991	1992	1993	1994	1995	1996	1997	
2.35	2.44	2.89	3.04	4.94	4.74	4.80	4.85	5.03	Algeria
0.14	0.44	0.47	0.47	0.46	0.42	0.46	0.46	0.47	Angola
-	-	-	-	-	-	-	-	-	Congo
2.59	2.42	2.46	2.58	3.10	2.87	3.10	3.15	3.32	Egypt
0.00	0.00	0.00	0.00	0.00	0.00	0.00	0.00	0.00	Gabon
1.12	1.29	1.40	1.40	1.44	1.49	1.47	1.58	1.69	Libya
0.05	0.04	0.03	0.02	0.02	0.02	0.02	0.02	0.02	Morocco
0.68	0.72	0.65	0.70	0.74	0.81	0.74	0.75	0.72	Nigeria
0.01	0.00	0.00	0.00	0.01	0.02	0.04	0.04	0.02	Senegal
0.45	0.50	0.49	0.48	0.53	0.53	0.57	0.64	0.64	South Africa
0.21	0.32	0.34	0.38	0.40	0.42	0.47	0.47	0.48	Tunisia
0.00	0.00	0.00	0.00	0.00	0.00	0.00	0.00	0.00	Other Africa
7.61	8.17	8.74	9.06	11.64	11.32	11.66	11.96	12.40	**Africa**
8.46	8.94	9.72	10.22	11.55	12.21	12.76	12.83	13.80	Argentina
0.15	0.16	0.18	0.21	0.24	0.26	0.29	0.26	0.35	Bolivia
2.49	2.49	2.60	2.77	2.81	2.91	3.07	3.45	3.83	Brazil
0.85	0.94	0.73	0.93	0.92	0.98	0.96	1.00	1.68	Chile
0.98	1.05	1.10	1.13	1.18	1.27	1.34	1.42	1.46	Colombia
0.08	0.14	0.13	0.14	0.14	0.13	0.15	0.15	0.15	Cuba
-	-	-	-	-	-	-	-	-	Panama
0.09	0.07	0.08	0.05	0.05	0.05	0.05	0.05	0.19	Peru
2.44	2.51	2.67	2.74	2.68	3.60	3.64	4.11	4.41	Trinidad-and-Tobago
0.00	-	0.00	0.00	0.00	0.00	-	-	0.00	Uruguay
7.21	8.34	9.54	9.02	9.33	9.39	10.26	10.34	11.11	Venezuela
0.01	0.02	0.01	0.01	0.01	0.01	0.01	0.02	0.02	Other Latin America
22.76	24.65	26.76	27.21	28.91	30.82	32.54	33.62	37.00	**Latin America**
1.84	1.79	1.75	2.14	2.28	2.61	2.99	3.08	2.97	Bangladesh
6.28	6.02	6.21	7.17	6.90	7.33	8.21	9.11	9.42	India
5.80	7.23	7.68	8.58	7.64	7.76	8.50	9.32	9.87	Indonesia
1.07	1.12	1.12	1.37	1.72	1.86	1.88	2.43	2.22	Malaysia
0.26	0.22	0.18	0.18	0.23	0.29	0.29	0.28	0.33	Myanmar
5.99	5.99	6.33	6.43	7.10	7.61	7.92	8.58	8.95	Pakistan
0.00	0.00	0.00	0.00	0.00	0.00	0.00	0.00	0.00	Philippines
0.06	0.06	0.07	0.07	0.08	0.08	0.08	0.09	0.09	Singapore
0.91	0.94	1.12	1.26	1.28	1.36	1.53	1.61	1.63	Chinese Taipei
0.10	0.24	0.32	0.40	0.44	0.53	0.71	0.84	0.85	Thailand
-	-	-	-	-	-	0.21	2.91	3.09	Vietnam
0.21	0.24	0.23	0.22	0.22	0.21	0.21	0.20	0.20	Other Asia
22.52	23.86	25.01	27.82	27.89	29.64	32.52	38.46	39.63	**Asia**
10.48	10.81	11.37	11.53	11.88	12.36	12.58	14.74	13.13	People's Rep.of China
0.29	0.32	0.35	0.39	0.41	0.45	0.50	1.97	2.82	Hong Kong, China
10.78	11.13	11.72	11.92	12.30	12.80	13.08	16.71	15.96	**China**

Final Consumption of Gas (Mtoe)
Consommation finale de gaz (Mtep)
Endverbrauch von Gas (Mtoe)

Consumo finale di gas (Mtep)

ガスの最終消費量（石油換算百万トン）

Consumo final de gas natural (Mtep)

Конечное потребление газа (Мтнэ)

	1971	1973	1978	1983	1984	1985	1986	1987	1988
Albanie	0.11	0.16	0.30	0.32	0.32	0.32	0.32	0.32	0.34
Bulgarie	0.25	0.17	2.31	4.02	4.23	3.69	3.81	4.03	2.15
Roumanie	13.01	13.89	25.15	27.59	27.73	21.82	22.26	22.12	22.48
République slovaque	1.18	1.41	2.76	3.88	3.87	3.82	3.85	3.89	3.95
Bosnie-Herzégovine	-	-	-	-	-	-	-	-	-
Croatie	-	-	-	-	-	-	-	-	-
Ex-RYM	-	-	-	-	-	-	-	-	-
Slovénie	-	-	-	-	-	-	-	-	-
RF de Yougoslavie	-	-	-	-	-	-	-	-	-
Ex-Yougoslavie	0.80	0.98	1.46	2.00	2.71	2.95	3.10	4.13	3.17
Europe non-OCDE	15.35	16.62	31.98	37.82	38.86	32.60	33.34	34.50	32.10
Arménie	-	-	-	-	-	-	-	-	-
Azerbaïdjan	-	-	-	-	-	-	-	-	-
Bélarus	-	-	-	-	-	-	-	-	-
Estonie	-	-	-	-	-	-	-	-	-
Géorgie	-	-	-	-	-	-	-	-	-
Kazakhstan	-	-	-	-	-	-	-	-	-
Kirghizistan	-	-	-	-	-	-	-	-	-
Lettonie	-	-	-	-	-	-	-	-	-
Lituanie	-	-	-	-	-	-	-	-	-
Moldova	-	-	-	-	-	-	-	-	-
Russie	-	-	-	-	-	-	-	-	-
Tadjikistan	-	-	-	-	-	-	-	-	-
Turkménistan	-	-	-	-	-	-	-	-	-
Ukraine	-	-	-	-	-	-	-	-	-
Ouzbékistan	-	-	-	-	-	-	-	-	-
Ex-URSS	100.90	116.20	155.30	206.57	216.56	227.86	237.76	249.06	259.96
Bahrein	0.18	0.59	0.93	1.10	1.16	1.67	2.19	1.95	2.25
Iran	8.09	8.25	8.37	7.11	8.67	9.99	10.26	8.50	10.81
Irak	0.76	0.99	1.39	2.22	2.46	2.61	3.19	4.01	5.55
Israël	0.09	0.05	0.05	0.05	0.04	0.04	0.03	0.03	0.03
Koweit	1.93	1.99	2.48	2.25	1.86	2.44	3.41	3.69	2.76
Oman	-	-	-	0.02	0.26	0.37	0.15	0.20	0.28
Qatar	0.38	0.60	0.46	1.55	1.99	1.76	1.83	1.83	1.99
Arabie saoudite	0.32	0.32	0.32	1.98	1.63	2.37	6.75	9.25	9.24
Syrie	-	-	-	-	-	-	0.18	0.17	0.50
Emirats arabes unis	-	-	-	-	0.54	0.50	0.56	0.63	0.72
Moyen-Orient	11.75	12.78	13.99	16.27	18.61	21.74	28.55	30.25	34.13
Total non-OCDE	138.98	161.37	230.36	301.86	320.10	333.66	353.42	371.82	386.84
OCDE Amérique du N.	385.14	398.30	364.69	341.54	363.55	353.02	335.19	346.62	375.46
OCDE Pacifique	7.58	9.54	13.26	17.40	19.14	20.34	21.26	21.95	23.13
OCDE Europe	76.53	106.30	153.34	163.72	172.15	180.61	183.41	194.25	191.46
Total OCDE	469.25	514.14	531.29	522.66	554.83	553.98	539.87	562.82	590.05
Monde	608.23	675.50	761.66	824.52	874.93	887.63	893.29	934.65	976.89

Ex-URSS: les séries antérieures à 1990-1992 ne sont pas comparables aux années récentes; 1991 a été estimé.

Final Consumption of Gas (Mtoe)
Consommation finale de gaz (Mtep)
Endverbrauch von Gas (Mtoe)
Consumo finale di gas (Mtep)
ガスの最終消費量（石油換算百万トン）
Consumo final de gas natural (Mtep)
Конечное потребление газа (Мтнэ)

1989	1990	1991	1992	1993	1994	1995	1996	1997	
0.27	0.19	0.11	0.08	0.07	0.04	0.02	0.02	0.02	Albania
2.23	2.29	1.95	1.61	1.59	1.79	2.10	2.12	1.76	Bulgaria
22.58	19.85	14.97	7.36	7.15	9.25	10.13	9.97	8.89	Romania
4.17	4.61	4.10	4.17	4.02	3.74	4.07	4.54	4.57	Slovak Republic
-	-	-	0.34	-	-	-	-	-	Bosnia-Herzegovina
-	-	-	1.28	1.28	1.31	1.36	1.43	1.53	Croatia
-	-	-	0.21	0.23	-	-	-	-	FYROM
-	-	-	0.40	0.39	0.47	0.50	0.55	0.64	Slovenia
-	-	-	1.32	1.35	0.61	0.80	1.96	1.95	FR of Yugoslavia
2.88	5.87	4.67	3.56	3.26	2.39	2.66	3.94	4.11	Former Yugoslavia
32.12	32.81	25.80	16.78	16.08	17.21	18.98	20.59	19.35	**Non-OECD Europe**
-	-	-	1.08	0.39	0.44	0.72	0.21	0.22	Armenia
-	-	-	6.17	5.00	4.97	3.93	3.59	3.48	Azerbaijan
-	-	-	3.75	3.57	2.64	3.13	3.06	3.03	Belarus
-	-	-	0.28	0.16	0.26	0.38	0.38	0.38	Estonia
-	-	-	2.76	2.22	1.30	0.48	0.35	0.46	Georgia
-	-	-	8.84	6.07	4.67	5.81	4.51	3.57	Kazakhstan
-	-	-	0.77	0.50	0.31	0.32	0.38	0.20	Kyrgyzstan
-	-	-	0.61	0.37	0.32	0.29	0.46	0.34	Latvia
-	-	-	1.11	0.64	0.70	0.85	0.92	0.81	Lithuania
-	-	-	0.85	1.29	1.24	1.24	1.47	1.57	Moldova
-	-	-	153.28	128.28	118.22	125.31	91.01	90.80	Russia
-	-	-	0.74	0.61	0.34	0.39	0.50	0.45	Tajikistan
-	-	-	3.30	4.55	6.47	6.97	5.28	5.04	Turkmenistan
-	-	-	48.53	47.84	40.42	42.05	46.34	42.10	Ukraine
-	-	-	19.02	20.42	20.66	20.39	20.40	20.72	Uzbekistan
265.93	277.03	262.99	251.09	221.91	202.97	212.25	178.84	173.17	**Former USSR**
2.19	2.31	2.19	2.22	2.64	2.45	2.59	2.75	2.64	Bahrain
10.91	8.11	14.05	17.43	18.21	18.72	21.95	25.52	28.71	Iran
6.40	3.40	1.40	2.34	2.62	3.26	3.24	3.31	3.70	Iraq
0.03	0.03	0.02	0.02	0.02	0.02	0.02	0.01	0.01	Israel
3.83	2.92	-	-	-	-	-	-	-	Kuwait
0.30	0.66	0.70	1.17	1.66	1.71	1.21	0.63	1.23	Oman
1.85	1.95	2.52	4.50	4.81	4.77	4.75	4.76	5.27	Qatar
8.54	8.97	8.78	8.87	9.41	9.58	10.66	11.58	12.16	Saudi Arabia
0.76	0.75	0.58	0.62	0.68	0.73	0.93	0.93	0.86	Syria
0.78	0.80	1.09	1.21	1.25	1.35	1.47	1.75	1.84	United Arab Emirates
35.59	29.90	31.34	38.38	41.31	42.59	46.82	51.23	56.41	**Middle East**
397.30	407.54	392.35	382.25	360.03	347.35	367.85	351.42	353.92	**Non-OECD Total**
370.87	360.31	359.11	363.94	380.61	384.50	394.13	410.08	403.61	OECD North America
24.13	25.50	27.46	28.77	30.82	32.68	35.25	38.45	40.54	OECD Pacific
196.24	199.57	214.75	213.26	220.27	220.55	233.79	254.09	246.03	OECD Europe
591.25	585.37	601.32	605.96	631.69	637.73	663.17	702.62	690.18	**OECD Total**
988.56	992.91	993.68	988.21	991.73	985.08	1 031.02	1 054.04	1 044.10	**World**

Former USSR: series up to 1990-1992 are not comparable with the recent years; 1991 is estimated.

Final Consumption of Electricity (Mtoe)
Consommation finale d'électricité (Mtep)
Endverbrauch von Elektrizität (Mtoe)
Consumo finale di energia elettrica (Mtep)
電力の最終消費量（石油換算百万トン）
Consumo final de electricidad (Mtep)
Конечное потребление электроэнергии (Мтнэ)

	1971	1973	1978	1983	1984	1985	1986	1987	1988
Algérie	0.14	0.17	0.35	0.59	0.66	0.72	0.78	0.75	0.81
Angola	0.04	0.06	0.04	0.05	0.05	0.05	0.05	0.05	0.05
Bénin	0.00	0.00	0.01	0.01	0.01	0.01	0.01	0.01	0.01
Cameroun	0.09	0.09	0.11	0.17	0.17	0.19	0.19	0.19	0.20
Congo	0.01	0.01	0.01	0.02	0.02	0.03	0.03	0.04	0.04
RD du Congo	0.31	0.30	0.29	0.34	0.35	0.38	0.40	0.39	0.41
Egypte	0.59	0.60	1.08	1.91	2.04	1.99	2.45	2.63	2.76
Ethiopie	0.05	0.05	0.04	0.06	0.07	0.07	0.07	0.08	0.07
Gabon	0.01	0.01	0.03	0.06	0.06	0.07	0.07	0.07	0.07
Ghana	0.24	0.31	0.29	0.17	0.10	0.18	0.29	0.33	0.36
Côte d'Ivoire	0.04	0.06	0.10	0.14	0.17	0.15	0.14	0.14	0.16
Kenya	0.07	0.08	0.12	0.15	0.16	0.17	0.18	0.20	0.20
Libye	0.04	0.10	0.29	0.61	0.79	1.02	1.14	1.34	1.42
Maroc	0.16	0.20	0.31	0.47	0.48	0.51	0.55	0.57	0.64
Mozambique	0.04	0.05	0.05	0.04	0.03	0.03	0.04	0.05	0.04
Nigéria	0.13	0.18	0.34	0.53	0.48	0.52	0.63	0.64	0.68
Sénégal	0.03	0.03	0.04	0.05	0.05	0.06	0.05	0.05	0.06
Afrique du Sud	4.01	4.74	6.59	8.12	8.87	9.27	9.70	10.00	10.48
Soudan	0.03	0.04	0.06	0.06	0.08	0.09	0.10	0.10	0.09
Tanzanie	0.04	0.04	0.05	0.06	0.06	0.07	0.08	0.08	0.09
Tunisie	0.07	0.08	0.15	0.23	0.25	0.27	0.28	0.30	0.32
Zambie	0.38	0.44	0.48	0.55	0.55	0.54	0.51	0.52	0.51
Zimbabwe	0.30	0.42	0.51	0.60	0.62	0.66	0.69	0.71	0.73
Autre Afrique	0.22	0.27	0.35	0.44	0.47	0.52	0.56	0.61	0.61
Afrique	7.02	8.33	11.68	15.44	16.60	17.56	19.00	19.87	20.81
Argentine	1.70	1.96	2.40	3.07	3.20	3.16	3.22	3.66	3.74
Bolivie	0.07	0.08	0.11	0.13	0.13	0.13	0.13	0.13	0.14
Brésil	3.65	4.66	8.21	11.94	13.27	14.39	15.54	16.03	16.97
Chili	0.61	0.63	0.75	0.88	0.96	0.99	1.05	1.09	1.19
Colombie	0.70	0.77	1.18	1.56	1.69	1.71	1.92	2.03	2.09
Costa Rica	0.09	0.10	0.15	0.19	0.21	0.22	0.24	0.25	0.26
Cuba	0.34	0.39	0.57	0.79	0.84	0.85	0.91	0.93	0.98
République dominicaine	0.09	0.16	0.18	0.18	0.23	0.26	0.25	0.29	0.22
Equateur	0.07	0.08	0.19	0.30	0.31	0.30	0.33	0.36	0.37
El Salvador	0.05	0.07	0.11	0.11	0.12	0.12	0.13	0.14	0.14
Guatemala	0.06	0.07	0.12	0.11	0.12	0.12	0.13	0.14	0.15
Haiti	0.00	0.01	0.02	0.02	0.03	0.03	0.03	0.03	0.03
Honduras	0.03	0.04	0.05	0.08	0.08	0.10	0.10	0.12	0.13
Jamaïque	0.13	0.16	0.17	0.10	0.10	0.11	0.11	0.12	0.12
Antilles néerlandaises	0.05	0.05	0.05	0.06	0.05	0.05	0.04	0.04	0.04
Nicaragua	0.05	0.05	0.08	0.09	0.09	0.09	0.09	0.10	0.10
Panama	0.07	0.09	0.12	0.18	0.17	0.17	0.17	0.18	0.17
Paraguay	0.02	0.02	0.05	0.08	0.09	0.09	0.10	0.12	0.14
Pérou	0.46	0.52	0.67	0.83	0.91	0.92	0.98	1.03	0.98
Trinité-et-Tobago	0.05	0.07	0.11	0.23	0.23	0.22	0.25	0.26	0.26
Uruguay	0.17	0.16	0.21	0.26	0.26	0.27	0.28	0.30	0.32
Vénézuela	0.96	1.11	1.80	2.95	3.13	3.33	3.44	3.49	3.82
Autre Amérique latine	0.27	0.31	0.35	0.35	0.36	0.38	0.39	0.38	0.41
Amérique latine	9.67	11.53	17.65	24.49	26.57	28.01	29.82	31.24	32.77

Final Consumption of Electricity (Mtoe)
Consommation finale d'électricité (Mtep)
Endverbrauch von Elektrizität (Mtoe)
Consumo finale di energia elettrica (Mtep)
電力の最終消費量（石油換算百万トン）
Consumo final de electricidad (Mtep)
Конечное потребление электроэнергии (Мтнэ)

1989	1990	1991	1992	1993	1994	1995	1996	1997	
0.92	0.97	1.03	1.05	1.17	1.20	1.25	1.29	1.43	Algeria
0.05	0.05	0.05	0.05	0.05	0.06	0.06	0.06	0.06	Angola
0.01	0.01	0.02	0.02	0.02	0.02	0.02	0.02	0.02	Benin
0.21	0.20	0.20	0.20	0.21	0.19	0.19	0.20	0.22	Cameroon
0.04	0.05	0.05	0.05	0.05	0.05	0.05	0.05	0.05	Congo
0.41	0.39	0.44	0.45	0.52	0.53	0.50	0.50	0.48	DR of Congo
2.94	3.14	3.33	3.40	3.57	3.76	3.97	3.94	4.17	Egypt
0.07	0.09	0.10	0.10	0.09	0.11	0.11	0.11	0.11	Ethiopia
0.06	0.07	0.07	0.07	0.07	0.07	0.07	0.07	0.07	Gabon
0.37	0.38	0.41	0.44	0.46	0.44	0.42	0.41	0.43	Ghana
0.15	0.16	0.15	0.09	0.15	0.16	0.20	0.21	0.22	Ivory Coast
0.21	0.23	0.24	0.25	0.26	0.28	0.28	0.30	0.31	Kenya
1.44	1.44	1.44	1.46	1.46	1.53	1.55	1.56	1.56	Libya
0.64	0.70	0.75	0.82	0.83	0.89	0.92	0.94	0.99	Morocco
0.04	0.04	0.06	0.06	0.06	0.05	0.06	0.06	0.07	Mozambique
0.76	0.64	0.69	0.75	0.86	0.83	0.81	0.84	0.85	Nigeria
0.06	0.06	0.06	0.06	0.07	0.07	0.07	0.08	0.08	Senegal
10.92	11.13	11.18	11.20	11.54	11.62	12.31	12.78	13.29	South Africa
0.09	0.11	0.11	0.11	0.11	0.11	0.12	0.12	0.12	Sudan
0.10	0.11	0.12	0.12	0.12	0.13	0.14	0.15	0.15	Tanzania
0.34	0.37	0.39	0.42	0.45	0.49	0.51	0.53	0.56	Tunisia
0.38	0.34	0.45	0.44	0.44	0.44	0.44	0.44	0.46	Zambia
0.77	0.78	0.71	0.80	0.71	0.76	0.80	0.86	0.91	Zimbabwe
0.64	0.67	0.70	0.72	0.69	0.70	0.75	0.77	0.77	Other Africa
21.64	22.15	22.76	23.11	23.98	24.48	25.58	26.30	27.38	**Africa**
3.49	3.14	3.32	3.56	3.83	4.13	4.42	4.67	5.01	Argentina
0.15	0.15	0.17	0.18	0.20	0.21	0.23	0.24	0.26	Bolivia
17.69	18.13	18.71	19.14	20.06	20.82	22.06	23.11	24.54	Brazil
1.28	1.33	1.46	1.66	1.75	1.86	2.07	2.31	2.53	Chile
2.21	2.26	2.38	2.49	2.72	2.87	3.00	3.11	3.05	Colombia
0.27	0.29	0.30	0.32	0.34	0.36	0.37	0.38	0.40	Costa Rica
1.02	1.03	0.89	0.77	0.72	0.77	0.79	0.84	0.99	Cuba
0.21	0.27	0.29	0.33	0.36	0.38	0.40	0.42	0.43	Dominican Republic
0.38	0.41	0.45	0.47	0.49	0.55	0.56	0.60	0.63	Ecuador
0.14	0.16	0.17	0.18	0.20	0.22	0.24	0.25	0.27	El Salvador
0.16	0.18	0.18	0.20	0.23	0.24	0.26	0.32	0.37	Guatemala
0.03	0.03	0.03	0.02	0.02	0.01	0.02	0.02	0.03	Haiti
0.13	0.15	0.15	0.14	0.15	0.14	0.17	0.20	0.21	Honduras
0.12	0.14	0.14	0.15	0.33	0.36	0.44	0.46	0.48	Jamaica
0.05	0.05	0.05	0.06	0.06	0.07	0.07	0.07	0.07	Netherlands Antilles
0.09	0.09	0.09	0.09	0.09	0.09	0.09	0.10	0.11	Nicaragua
0.17	0.18	0.19	0.20	0.22	0.23	0.25	0.26	0.27	Panama
0.15	0.17	0.18	0.19	0.24	0.27	0.31	0.31	0.33	Paraguay
1.00	0.91	0.96	0.90	1.02	1.02	1.17	1.23	1.27	Peru
0.25	0.27	0.27	0.29	0.29	0.30	0.31	0.34	0.37	Trinidad-and-Tobago
0.31	0.33	0.36	0.36	0.39	0.41	0.43	0.45	0.48	Uruguay
3.82	3.87	4.12	4.42	4.59	4.59	4.73	4.79	4.87	Venezuela
0.47	0.48	0.50	0.52	0.53	0.55	0.56	0.56	0.59	Other Latin America
33.59	34.02	35.37	36.67	38.83	40.43	42.95	45.05	47.57	**Latin America**

Final Consumption of Electricity (Mtoe)
Consommation finale d'électricité (Mtep)
Endverbrauch von Elektrizität (Mtoe)
Consumo finale di energia elettrica (Mtep)
電力の最終消費量（石油換算百万トン）
Consumo final de electricidad (Mtep)
Конечное потребление электроэнергии (Мтнэ)

	1971	1973	1978	1983	1984	1985	1986	1987	1988
Bangladesh	0.06	0.09	0.13	0.21	0.23	0.24	0.28	0.30	0.32
Brunei	0.01	0.02	0.03	0.05	0.06	0.06	0.06	0.08	0.08
Inde	4.42	4.73	7.09	9.52	10.54	11.41	12.70	13.71	15.21
Indonésie	0.17	0.21	0.37	0.88	0.95	1.09	1.27	1.47	1.72
RPD de Corée	1.23	1.46	2.33	3.02	3.29	3.51	3.63	3.65	3.87
Malaisie	0.28	0.35	0.60	0.93	1.02	1.06	1.14	1.24	1.37
Myanmar	0.04	0.05	0.08	0.10	0.11	0.13	0.13	0.14	0.12
Népal	0.01	0.01	0.01	0.02	0.02	0.03	0.03	0.03	0.04
Pakistan	0.46	0.53	0.74	1.22	1.35	1.51	1.69	1.87	2.16
Philippines	0.71	1.03	1.20	1.47	1.47	1.55	1.14	1.48	1.65
Singapour	0.18	0.25	0.40	0.58	0.64	0.68	0.73	0.83	0.92
Sri Lanka	0.06	0.08	0.10	0.15	0.16	0.18	0.19	0.20	0.20
Taipei chinois	1.15	1.50	2.67	3.61	3.89	4.07	4.56	5.02	5.60
Thaïlande	0.37	0.53	0.97	1.43	1.58	1.71	1.88	2.12	2.43
Viêt-Nam	0.15	0.16	0.24	0.27	0.31	0.33	0.36	0.40	0.43
Autre Asie	0.19	0.29	0.30	0.30	0.32	0.35	0.37	0.38	0.39
Asie	9.49	11.26	17.26	23.75	25.94	27.91	30.16	32.91	36.51
Rép. populaire de Chine	9.82	11.83	18.20	25.13	26.99	29.93	32.73	34.95	38.74
Hong-Kong, Chine	0.42	0.52	0.78	1.21	1.29	1.37	1.52	1.69	1.80
Chine	10.24	12.35	18.99	26.34	28.28	31.30	34.25	36.64	40.55
Albanie	0.09	0.11	0.23	0.25	0.25	0.19	0.36	0.29	0.26
Bulgarie	1.50	1.78	2.49	3.14	3.28	3.02	3.18	3.21	3.28
Chypre	0.05	0.06	0.07	0.09	0.09	0.10	0.11	0.12	0.13
Gibraltar	0.00	0.00	0.00	0.00	0.00	0.00	0.00	0.00	0.00
Malte	0.02	0.03	0.04	0.05	0.05	0.06	0.06	0.07	0.07
Roumanie	2.43	2.89	4.26	5.34	5.54	5.11	5.43	5.37	5.52
République slovaque	0.93	1.06	1.28	1.71	1.76	1.84	1.89	1.96	2.01
Bosnie-Herzégovine	-	-	-	-	-	-	-	-	-
Croatie	-	-	-	-	-	-	-	-	-
Ex-RYM	-	-	-	-	-	-	-	-	-
Slovénie	-	-	-	-	-	-	-	-	-
RF de Yougoslavie	-	-	-	-	-	-	-	-	-
Ex-Yougoslavie	2.05	2.46	3.62	4.70	5.03	5.25	5.46	5.62	5.37
Europe non-OCDE	7.08	8.40	11.99	15.29	16.01	15.57	16.50	16.64	16.64

Final Consumption of Electricity (Mtoe)
Consommation finale d'électricité (Mtep)
Endverbrauch von Elektrizität (Mtoe)
Consumo finale di energia elettrica (Mtep)
電力の最終消費量（石油換算百万トン）
Consumo final de electricidad (Mtep)
Конечное потребление электроэнергии (Мтнэ)

1989	1990	1991	1992	1993	1994	1995	1996	1997	
0.40	0.40	0.42	0.52	0.59	0.64	0.72	0.77	0.81	Bangladesh
0.08	0.09	0.09	0.10	0.12	0.13	0.15	0.17	0.20	Brunei
16.78	18.53	20.32	21.67	23.24	25.24	27.08	28.19	30.02	India
2.02	2.39	2.71	3.01	3.57	3.84	4.39	5.02	5.67	Indonesia
3.91	3.91	2.21	0.52	0.52	0.51	0.50	0.48	0.47	DPR of Korea
1.51	1.72	1.84	2.22	2.45	2.93	3.38	3.78	4.38	Malaysia
0.14	0.15	0.14	0.16	0.17	0.19	0.20	0.20	0.22	Myanmar
0.04	0.05	0.05	0.06	0.06	0.06	0.07	0.07	0.07	Nepal
2.30	2.48	2.71	2.91	3.14	3.21	3.39	3.59	3.67	Pakistan
1.78	1.81	1.84	1.78	1.87	2.11	2.31	2.51	2.73	Philippines
1.00	1.11	1.19	1.24	1.33	1.46	1.56	1.88	2.12	Singapore
0.20	0.22	0.24	0.25	0.28	0.31	0.32	0.32	0.36	Sri Lanka
6.10	6.54	7.21	7.69	8.38	9.00	9.70	10.32	11.06	Chinese Taipei
2.82	3.30	3.73	4.24	4.84	5.38	6.13	6.65	7.09	Thailand
0.49	0.54	0.56	0.60	0.67	0.79	0.91	1.15	1.32	Vietnam
0.40	0.41	0.40	0.39	0.39	0.41	0.44	0.44	0.44	Other Asia
39.98	43.63	45.66	47.35	51.63	56.22	61.25	65.56	70.64	**Asia**
41.36	43.95	48.25	53.42	55.29	61.85	65.92	72.35	72.53	People's Rep.of China
1.93	2.05	2.18	2.25	2.38	2.51	2.57	2.72	2.77	Hong Kong, China
43.29	46.00	50.42	55.67	57.68	64.36	68.49	75.07	75.31	**China**
0.27	0.23	0.16	0.19	0.23	0.16	0.17	0.26	0.24	Albania
3.26	3.03	2.60	2.25	2.25	2.28	2.47	2.57	2.29	Bulgaria
0.14	0.15	0.16	0.18	0.20	0.21	0.19	0.20	0.20	Cyprus
0.01	0.01	0.01	0.01	0.01	0.01	0.01	0.01	0.01	Gibraltar
0.08	0.08	0.10	0.11	0.11	0.10	0.11	0.12	0.12	Malta
5.58	4.66	3.92	3.56	3.14	2.94	3.13	3.42	3.30	Romania
2.08	2.01	1.88	1.75	1.74	1.78	1.87	2.02	1.96	Slovak Republic
-	-	-	1.05	1.01	0.13	0.15	0.15	0.15	*Bosnia-Herzegovina*
-	-	-	0.81	0.80	0.82	0.85	0.88	0.95	*Croatia*
-	-	-	0.45	0.43	0.41	0.41	0.44	0.45	*FYROM*
-	-	-	0.75	0.75	0.80	0.81	0.82	0.85	*Slovenia*
-	-	-	2.55	2.34	2.42	2.65	2.70	2.87	*FR of Yugoslavia*
5.37	5.55	6.02	5.61	5.34	4.59	4.87	4.99	5.27	Former Yugoslavia
16.79	15.73	14.85	13.66	13.02	12.06	12.81	13.57	13.40	**Non-OECD Europe**

Final Consumption of Electricity (Mtoe)
Consommation finale d'électricité (Mtep)
Endverbrauch von Elektrizität (Mtoe)

Consumo finale di energia elettrica (Mtep)

電力の最終消費量（石油換算百万トン）

Consumo final de electricidad (Mtep)

Конечное потребление электроэнергии (Мтнэ)

	1971	1973	1978	1983	1984	1985	1986	1987	1988
Arménie	-	-	-	-	-	-	-	-	-
Azerbaïdjan	-	-	-	-	-	-	-	-	-
Bélarus	-	-	-	-	-	-	-	-	-
Estonie	-	-	-	-	-	-	-	-	-
Géorgie	-	-	-	-	-	-	-	-	-
Kazakhstan	-	-	-	-	-	-	-	-	-
Kirghizistan	-	-	-	-	-	-	-	-	-
Lettonie	-	-	-	-	-	-	-	-	-
Lituanie	-	-	-	-	-	-	-	-	-
Moldova	-	-	-	-	-	-	-	-	-
Russie	-	-	-	-	-	-	-	-	-
Tadjikistan	-	-	-	-	-	-	-	-	-
Turkménistan	-	-	-	-	-	-	-	-	-
Ukraine	-	-	-	-	-	-	-	-	-
Ouzbékistan	-	-	-	-	-	-	-	-	-
Ex-URSS	54.67	62.41	77.73	90.35	95.04	97.51	100.75	104.09	106.96
Bahrein	0.04	0.04	0.10	0.19	0.21	0.25	0.26	0.29	0.29
Iran	0.60	0.87	1.43	2.34	2.60	2.79	2.95	3.29	3.38
Irak	0.23	0.29	0.64	1.32	1.59	1.71	1.82	1.83	1.91
Israël	0.56	0.65	0.89	1.09	1.12	1.18	1.17	1.31	1.45
Jordanie	0.01	0.01	0.05	0.14	0.17	0.18	0.20	0.23	0.24
Koweit	0.18	0.25	0.44	0.76	0.84	0.94	1.05	1.10	1.21
Liban	0.10	0.13	0.17	0.26	0.28	0.28	0.30	0.34	0.29
Oman	0.00	0.00	0.03	0.11	0.14	0.18	0.23	0.24	0.27
Qatar	0.03	0.04	0.10	0.25	0.28	0.32	0.35	0.36	0.37
Arabie saoudite	0.17	0.23	0.51	2.08	2.63	3.13	3.36	3.62	3.86
Syrie	0.09	0.11	0.20	0.49	0.50	0.54	0.55	0.56	0.61
Emirats arabes unis	0.02	0.06	0.30	0.81	0.88	0.96	1.02	1.09	1.19
Yémen	0.02	0.02	0.03	0.05	0.06	0.06	0.07	0.08	0.09
Moyen-Orient	2.03	2.69	4.90	9.93	11.30	12.54	13.33	14.33	15.14
Total non-OCDE	100.21	116.98	160.19	205.59	219.73	230.40	243.81	255.72	269.38
OCDE Amérique du N.	142.64	165.03	196.49	212.21	226.12	231.95	235.49	245.29	257.29
OCDE Pacifique	35.07	42.69	52.88	60.04	63.38	65.71	67.04	71.29	75.77
OCDE Europe	98.78	114.27	137.61	151.18	158.34	165.06	169.52	176.37	181.26
Total OCDE	276.49	321.99	386.98	423.43	447.84	462.72	472.06	492.96	514.31
Monde	376.70	438.97	547.17	629.02	667.57	693.13	715.87	748.68	783.69

Ex-URSS: les séries antérieures à 1990-1992 ne sont pas comparables aux années récentes; 1991 a été estimé.

Final Consumption of Electricity (Mtoe)
Consommation finale d'électricité (Mtep)
Endverbrauch von Elektrizität (Mtoe)
Consumo finale di energia elettrica (Mtep)
電力の最終消費量（石油換算百万トン）
Consumo final de electricidad (Mtep)
Конечное потребление электроэнергии (Мтнэ)

1989	1990	1991	1992	1993	1994	1995	1996	1997	
-	-	-	0.54	0.33	0.28	0.26	0.29	0.37	Armenia
-	-	-	1.04	1.14	1.23	1.17	1.18	1.10	Azerbaijan
-	-	-	2.92	2.71	2.40	2.18	2.19	2.30	Belarus
-	-	-	0.49	0.36	0.41	0.39	0.42	0.43	Estonia
-	-	-	0.84	0.69	0.46	0.49	0.54	0.53	Georgia
-	-	-	7.57	6.38	4.96	4.39	3.92	3.52	Kazakhstan
-	-	-	0.70	0.69	0.70	0.62	0.58	0.55	Kyrgyzstan
-	-	-	0.57	0.40	0.39	0.39	0.38	0.37	Latvia
-	-	-	0.84	0.60	0.57	0.55	0.56	0.58	Lithuania
-	-	-	0.71	0.69	0.56	0.50	0.49	0.45	Moldova
-	-	-	65.78	60.70	54.63	53.18	51.68	50.43	Russia
-	-	-	1.32	1.14	1.19	1.19	1.17	1.13	Tajikistan
-	-	-	0.47	0.52	0.43	0.43	0.40	0.37	Turkmenistan
-	-	-	16.05	14.90	12.92	12.34	11.11	10.68	Ukraine
-	-	-	3.46	3.38	3.29	3.26	3.31	3.35	Uzbekistan
107.24	107.30	107.38	103.31	94.63	84.42	81.32	78.23	76.16	**Former USSR**
0.29	0.30	0.30	0.30	0.32	0.35	0.37	0.37	0.38	Bahrain
3.44	3.88	4.61	4.90	5.00	5.39	5.58	5.89	6.09	Iran
1.95	1.96	1.69	2.18	2.26	2.41	2.49	2.49	2.54	Iraq
1.54	1.56	1.60	1.85	1.95	2.13	2.31	2.49	2.54	Israel
0.25	0.26	0.27	0.32	0.34	0.37	0.41	0.44	0.46	Jordan
1.33	1.26	0.55	0.99	1.48	1.68	1.75	1.87	2.00	Kuwait
0.18	0.12	0.23	0.29	0.34	0.37	0.40	0.58	0.69	Lebanon
0.28	0.32	0.33	0.36	0.41	0.43	0.46	0.48	0.51	Oman
0.38	0.39	0.38	0.42	0.45	0.47	0.46	0.53	0.56	Qatar
4.03	4.32	4.71	4.95	5.49	6.09	6.37	6.64	7.05	Saudi Arabia
0.68	0.71	0.73	0.75	0.76	0.84	0.85	0.94	0.99	Syria
1.26	1.38	1.40	1.42	1.51	1.62	1.64	1.70	1.77	United Arab Emirates
0.09	0.11	0.12	0.13	0.14	0.12	0.13	0.13	0.13	Yemen
15.68	16.58	16.92	18.84	20.45	22.29	23.22	24.54	25.71	**Middle East**
278.21	285.41	293.36	298.60	300.21	304.25	315.63	328.33	336.18	**Non-OECD Total**
264.84	270.89	283.47	284.17	293.32	301.55	310.19	320.43	324.90	OECD North America
80.64	86.70	89.88	91.50	93.62	101.79	105.57	109.54	113.65	OECD Pacific
186.18	188.47	191.70	193.21	194.54	197.29	202.62	208.35	211.96	OECD Europe
531.66	546.05	565.05	568.88	581.47	600.63	618.39	638.32	650.50	**OECD Total**
809.87	831.46	858.41	867.49	881.68	904.88	934.02	966.65	986.68	**World**

Former USSR: series up to 1990-1992 are not comparable with the recent years; 1991 is estimated.

Total Final Consumption of Energy* (Mtoe)
Consommation finale totale d'énergie (Mtep)
Gesamter Endverbrauch von Energie (Mtoe)
Consumo finale totale di energia (Mtep)
最終エネルギー総消費量（石油換算百万トン）
Consumo final total de energía (Mtep)
Общее конечное потребление топлива и энергии (Мтнэ)

	1971	1973	1978	1983	1984	1985	1986	1987	1988
Algérie	2.14	2.92	5.23	8.46	9.44	10.14	10.89	11.21	11.21
Angola	0.62	0.82	0.77	0.61	0.53	0.70	0.60	0.60	0.61
Bénin	0.11	0.13	0.15	0.13	0.15	0.18	0.16	0.14	0.14
Cameroun	0.38	0.41	0.64	1.00	1.06	1.08	1.05	1.05	1.10
Congo	0.22	0.18	0.23	0.30	0.27	0.29	0.29	0.27	0.25
RD du Congo	1.13	1.14	1.18	1.53	1.51	1.47	1.37	1.51	1.58
Egypte	5.67	5.91	9.23	15.25	16.92	17.67	17.81	19.02	19.41
Ethiopie	0.42	0.44	0.45	0.54	0.55	0.56	0.67	0.80	0.82
Gabon	0.14	0.18	0.60	0.69	0.65	0.69	0.67	0.62	0.71
Ghana	0.89	1.00	1.14	0.84	0.75	0.88	1.00	1.08	1.27
Côte d'Ivoire	0.72	0.77	1.07	1.12	1.03	1.00	1.07	1.11	1.13
Kenya	1.18	1.35	1.69	1.59	1.84	1.92	2.03	2.13	2.12
Libye	0.51	1.03	2.92	5.40	5.63	5.95	5.94	6.30	5.84
Maroc	2.14	2.46	3.44	3.79	3.78	3.99	3.99	3.95	4.18
Mozambique	0.79	0.78	0.55	0.45	0.38	0.39	0.41	0.41	0.40
Nigéria	1.83	2.80	6.95	9.62	8.83	8.92	8.51	8.73	9.25
Sénégal	0.41	0.48	0.54	0.61	0.64	0.64	0.61	0.59	0.57
Afrique du Sud	28.69	32.19	37.19	36.05	38.22	38.50	38.06	39.27	41.59
Soudan	1.13	1.59	1.04	1.29	1.25	1.40	1.39	1.10	1.51
Tanzanie	0.48	0.56	0.56	0.49	0.49	0.52	0.54	0.55	0.60
Tunisie	1.01	1.22	1.92	2.51	2.60	2.68	2.65	2.65	2.78
Zambie	1.30	1.58	1.48	1.45	1.37	1.34	1.27	1.33	1.35
Zimbabwe	1.88	2.25	2.07	2.59	2.42	2.57	2.67	2.74	2.74
Autre Afrique	2.65	2.99	3.22	3.45	3.40	3.58	3.74	5.13	5.20
Afrique	**56.42**	**65.18**	**84.25**	**99.76**	**103.70**	**107.08**	**107.40**	**112.26**	**116.36**
Argentine	21.46	23.04	25.62	28.39	29.21	27.76	28.24	31.11	30.40
Bolivie	0.64	0.76	1.18	1.17	1.10	1.09	1.11	1.18	1.20
Brésil	28.32	37.97	56.33	56.81	58.25	62.54	67.33	69.77	72.20
Chili	5.36	5.37	5.39	5.49	5.72	5.69	6.08	6.28	7.20
Colombie	7.40	8.08	10.38	11.87	12.01	12.35	12.78	13.71	13.90
Costa Rica	0.45	0.55	0.83	0.73	0.79	0.85	0.89	0.94	0.98
Cuba	4.68	6.15	6.93	7.36	7.49	8.07	7.97	8.22	8.50
République dominicaine	0.87	1.06	1.30	1.35	1.46	1.31	1.43	1.69	1.68
Equateur	1.11	1.26	2.65	3.36	3.69	4.03	4.24	4.17	4.59
El Salvador	0.41	0.51	0.74	0.60	0.62	0.65	0.65	0.70	0.78
Guatemala	0.72	0.80	1.04	0.86	0.92	0.94	0.96	1.06	1.13
Haiti	0.11	0.13	0.22	0.22	0.23	0.24	0.23	0.24	0.26
Honduras	0.34	0.40	0.53	0.60	0.63	0.64	0.66	0.74	0.82
Jamaïque	1.56	2.22	2.10	1.54	1.45	1.32	1.26	1.30	1.33
Antilles néerlandaises	2.18	2.26	1.30	1.08	1.00	0.92	0.82	0.80	0.75
Nicaragua	0.46	0.56	0.65	0.58	0.59	0.59	0.63	0.67	0.60
Panama	0.58	0.64	0.76	0.71	0.77	0.80	0.84	0.85	0.74
Paraguay	0.19	0.25	0.43	0.52	0.57	0.60	0.63	0.68	0.72
Pérou	4.77	5.06	5.81	5.74	5.94	5.94	6.39	6.99	6.82
Trinité-et-Tobago	0.94	1.13	1.74	2.76	2.77	2.72	2.82	2.80	3.37
Uruguay	1.52	1.51	1.59	1.37	1.26	1.20	1.23	1.32	1.36
Vénézuela	9.45	12.80	19.27	23.47	24.48	25.94	25.63	27.47	25.92
Autre Amérique latine	2.07	2.56	2.38	2.10	2.29	2.21	2.36	2.71	2.67
Amérique latine	**95.57**	**115.11**	**149.18**	**158.68**	**163.24**	**168.40**	**175.20**	**185.40**	**187.92**

* Note: à cause de données manquantes avant 1994, les combustibles renouvelables et les déchets ne sont pas inclus.

Total Final Consumption of Energy* (Mtoe)
Consommation finale totale d'énergie (Mtep)
Gesamter Endverbrauch von Energie (Mtoe)
Consumo finale totale di energia (Mtep)
最終エネルギー総消費量（石油換算百万トン）
Consumo final total de energía (Mtep)
Общее конечное потребление топлива и энергии (Мтнэ)

1989	1990	1991	1992	1993	1994	1995	1996	1997	
11.39	11.32	12.45	12.50	14.85	14.01	13.89	13.82	15.53	Algeria
0.60	0.91	0.95	0.94	0.94	0.90	0.94	1.03	1.30	Angola
0.12	0.11	0.09	0.09	0.10	0.10	0.11	0.35	0.35	Benin
1.19	1.16	1.09	1.03	1.03	1.04	1.03	1.03	1.06	Cameroon
0.32	0.31	0.32	0.31	0.31	0.31	0.33	0.34	0.34	Congo
1.58	1.64	1.72	1.65	1.82	1.85	1.85	1.85	1.83	DR of Congo
20.60	20.99	20.37	20.95	19.91	19.57	21.73	22.70	24.15	Egypt
0.87	0.92	0.94	0.95	0.98	1.05	1.12	1.27	0.82	Ethiopia
0.44	0.37	0.41	0.43	0.46	0.46	0.49	0.53	0.62	Gabon
1.34	1.31	1.28	1.38	1.47	1.51	1.53	1.52	1.53	Ghana
1.10	0.95	0.94	0.84	0.94	0.96	1.04	1.10	1.14	Ivory Coast
2.23	2.21	2.12	2.21	2.23	2.32	2.27	2.46	2.38	Kenya
6.34	6.53	6.81	6.81	7.36	8.17	8.67	8.87	9.18	Libya
4.70	4.81	5.10	5.66	5.66	6.13	6.31	6.51	6.69	Morocco
0.40	0.44	0.41	0.48	0.54	0.50	0.48	0.47	0.60	Mozambique
9.69	8.18	9.67	12.33	10.30	9.54	9.30	10.12	10.22	Nigeria
0.62	0.64	0.64	0.67	0.61	0.67	0.77	0.85	0.91	Senegal
42.61	42.88	42.09	40.45	40.16	40.94	46.38	45.58	47.23	South Africa
1.41	1.81	1.61	1.54	1.00	1.57	1.41	1.32	1.56	Sudan
0.61	0.65	0.65	0.65	0.61	0.61	0.62	0.62	0.57	Tanzania
3.12	3.29	3.25	3.47	3.80	3.86	3.98	4.17	4.30	Tunisia
1.07	1.00	1.09	1.07	1.05	1.01	1.02	1.02	1.04	Zambia
2.73	3.23	3.12	3.22	3.05	2.65	2.90	3.02	3.01	Zimbabwe
5.47	5.31	5.42	5.69	5.83	5.71	5.83	5.91	6.11	Other Africa
120.56	121.00	122.56	125.30	125.02	125.45	134.00	136.46	142.47	**Africa**
28.01	27.97	29.28	30.65	33.68	35.22	36.22	37.68	39.06	Argentina
1.34	1.39	1.45	1.50	1.59	1.68	1.85	1.94	2.10	Bolivia
74.31	75.21	77.25	78.89	82.07	86.18	92.09	98.56	105.45	Brazil
8.20	8.48	8.78	9.87	10.49	11.06	11.95	12.92	15.12	Chile
14.52	14.99	15.45	16.03	17.14	17.86	18.70	19.39	19.63	Colombia
1.12	1.13	1.16	1.47	1.55	1.69	1.70	1.71	1.81	Costa Rica
8.91	7.95	5.96	4.75	4.53	4.82	4.93	5.34	5.58	Cuba
1.89	1.93	1.91	2.34	2.22	2.37	2.46	2.56	2.74	Dominican Republic
4.52	4.58	4.67	4.48	4.46	4.78	5.03	5.48	5.84	Ecuador
0.80	0.85	0.91	1.08	1.17	1.25	1.44	1.41	1.49	El Salvador
1.18	1.28	1.32	1.43	1.60	1.68	2.00	2.05	2.20	Guatemala
0.28	0.29	0.27	0.26	0.23	0.07	0.29	0.31	0.36	Haiti
0.89	0.87	0.87	0.97	0.96	1.10	1.33	1.36	1.39	Honduras
1.72	1.95	1.98	2.04	1.35	1.75	1.78	1.89	2.00	Jamaica
0.68	0.66	0.71	0.75	0.79	0.89	0.89	0.88	0.89	Netherlands Antilles
0.58	0.56	0.54	0.58	0.58	0.63	0.64	0.67	0.76	Nicaragua
0.77	0.81	0.87	0.94	0.98	1.06	1.13	1.18	1.29	Panama
0.80	0.82	0.82	0.98	1.11	1.31	1.45	1.35	1.45	Paraguay
6.24	6.03	6.18	5.99	6.60	6.56	7.68	8.10	8.12	Peru
3.34	3.47	3.70	3.77	3.61	4.53	4.61	5.15	5.57	Trinidad-and-Tobago
1.37	1.37	1.47	1.53	1.62	1.70	1.73	1.86	1.98	Uruguay
25.44	26.89	29.08	29.53	31.78	33.01	33.74	32.94	35.91	Venezuela
3.06	3.06	3.02	2.83	2.87	2.97	3.02	3.09	3.20	Other Latin America
189.97	192.57	197.67	202.67	212.97	224.16	236.68	247.82	263.94	**Latin America**

* Note: due to missing data prior to 1994 combustible renewables and waste are not included.

INTERNATIONAL ENERGY AGENCY

Total Final Consumption of Energy* (Mtoe)
Consommation finale totale d'énergie (Mtep)
Gesamter Endverbrauch von Energie (Mtoe)
Consumo finale totale di energia (Mtep)
最終エネルギー総消費量（石油換算百万トン）
Consumo final total de energía (Mtep)
Общее конечное потребление топлива и энергии (Мтнэ)

	1971	1973	1978	1983	1984	1985	1986	1987	1988
Bangladesh	0.93	1.15	1.72	2.44	2.54	2.76	3.02	3.31	3.55
Brunei	0.08	0.09	0.13	0.28	0.27	0.29	0.30	0.33	0.36
Inde	43.46	48.59	58.09	72.61	76.94	89.22	91.57	96.03	106.29
Indonésie	6.77	8.12	15.80	23.15	23.14	25.31	25.64	26.77	29.19
RPD de Corée	21.84	19.02	22.81	26.46	26.07	26.93	27.41	27.47	27.96
Malaisie	3.88	3.87	5.36	7.98	8.19	8.76	9.29	10.08	10.94
Myanmar	1.38	1.08	1.36	1.37	1.47	1.51	1.58	1.10	1.06
Népal	0.06	0.12	0.10	0.16	0.22	0.22	0.23	0.28	0.32
Pakistan	5.56	6.05	8.01	11.72	12.53	13.40	14.11	15.56	16.73
Philippines	6.09	7.05	8.38	7.96	7.09	6.73	6.48	7.64	8.40
Singapour	1.26	1.69	2.81	3.60	3.95	4.69	4.81	5.34	5.47
Sri Lanka	0.89	1.12	1.17	1.27	1.34	1.30	1.32	1.32	1.37
Taipei chinois	7.71	10.24	16.33	19.89	21.09	22.33	24.31	26.09	28.73
Thaïlande	5.39	6.81	8.73	10.41	11.51	11.81	12.53	14.07	15.95
Viêt-Nam	6.92	6.92	2.78	3.60	3.52	3.51	3.89	4.40	4.00
Autre Asie	2.13	2.24	1.97	1.94	2.19	2.56	2.39	2.28	2.24
Asie	114.36	124.16	155.57	194.84	202.08	221.32	228.87	242.08	262.56
Rép. populaire de Chine	181.05	200.28	309.78	340.42	370.17	395.75	421.78	445.16	466.44
Hong-Kong, Chine	2.51	2.57	4.04	4.47	4.74	4.59	5.29	5.57	6.41
Chine	183.56	202.85	313.82	344.90	374.91	400.34	427.06	450.73	472.85
Albanie	1.04	1.04	1.85	2.10	2.14	1.95	2.13	2.13	2.16
Bulgarie	14.73	15.84	18.23	19.64	19.52	18.78	18.62	19.16	16.47
Chypre	0.49	0.63	0.60	0.74	0.80	0.80	0.89	1.01	1.05
Gibraltar	0.03	0.03	0.02	0.03	0.03	0.03	0.04	0.04	0.05
Malte	0.19	0.22	0.21	0.23	0.28	0.18	0.31	0.34	0.34
Roumanie	31.25	33.99	50.28	51.97	50.99	47.40	48.10	50.82	51.57
République slovaque	9.08	9.75	11.96	13.91	13.98	14.12	14.09	14.26	14.30
Bosnie-Herzégovine	-	-	-	-	-	-	-	-	-
Croatie	-	-	-	-	-	-	-	-	-
Ex-RYM	-	-	-	-	-	-	-	-	-
Slovénie	-	-	-	-	-	-	-	-	-
RF de Yougoslavie	-	-	-	-	-	-	-	-	-
Ex-Yougoslavie	15.42	16.54	20.55	23.64	22.87	23.51	25.07	25.42	24.21
Europe non-OCDE	72.23	78.06	103.70	112.26	110.59	106.78	109.24	113.17	110.15

* Note: à cause de données manquantes avant 1994, les combustibles renouvelables et les déchets ne sont pas inclus.

Total Final Consumption of Energy* (Mtoe)
Consommation finale totale d'énergie (Mtep)
Gesamter Endverbrauch von Energie (Mtoe)
Consumo finale totale di energia (Mtep)
最終エネルギー総消費量（石油換算百万トン）
Consumo final total de energía (Mtep)
Общее конечное потребление топлива и энергии (Мтнэ)

1989	1990	1991	1992	1993	1994	1995	1996	1997	
3.95	4.11	3.84	4.45	4.79	5.28	6.13	6.37	6.32	Bangladesh
0.37	0.38	0.40	0.46	0.51	0.54	0.60	0.66	0.73	Brunei
112.46	117.47	124.27	129.79	131.87	141.56	149.55	155.85	161.93	India
32.67	36.09	37.78	40.48	42.69	45.94	49.35	55.88	60.84	Indonesia
28.61	26.88	22.90	21.14	19.10	16.59	16.00	15.59	15.12	DPR of Korea
11.94	13.55	14.50	15.86	18.49	20.42	22.15	24.94	26.00	Malaysia
1.06	1.02	0.93	1.01	1.21	1.41	1.67	1.78	2.02	Myanmar
0.29	0.24	0.34	0.42	0.43	0.54	0.66	0.67	0.67	Nepal
17.60	18.11	18.62	19.29	21.36	22.47	23.24	25.34	25.25	Pakistan
9.18	9.60	9.64	10.40	11.64	12.40	13.50	14.77	16.19	Philippines
5.89	6.94	6.83	6.63	7.28	8.08	8.66	9.44	10.59	Singapore
1.35	1.41	1.42	1.64	1.76	1.95	2.14	2.42	2.41	Sri Lanka
30.07	31.58	32.97	35.25	37.34	40.29	41.77	43.91	45.76	Chinese Taipei
18.86	21.56	22.87	25.42	28.86	32.70	37.27	41.92	42.37	Thailand
3.85	4.47	4.40	5.05	5.72	6.23	6.81	12.49	14.64	Vietnam
2.46	2.59	2.57	2.24	2.22	2.16	2.29	2.26	2.32	Other Asia
280.60	296.02	304.29	319.54	335.26	358.58	381.81	414.29	433.17	**Asia**
478.25	475.13	486.53	490.23	521.30	538.97	583.91	613.49	600.19	People's Rep.of China
6.74	7.12	7.10	8.52	8.97	9.87	9.93	11.41	12.64	Hong Kong, China
484.99	482.25	493.63	498.75	530.27	548.84	593.84	624.90	612.84	**China**
2.05	1.88	1.34	1.02	0.94	0.85	0.69	0.75	0.67	Albania
19.06	17.51	12.66	11.45	11.66	11.48	12.05	12.32	10.84	Bulgaria
1.11	1.13	1.31	1.37	1.37	1.41	1.45	1.51	1.51	Cyprus
0.05	0.05	0.07	0.08	0.08	0.09	0.09	0.09	0.09	Gibraltar
0.34	0.34	0.40	0.41	0.42	0.42	0.44	0.52	0.56	Malta
48.91	42.66	34.56	29.58	25.39	24.90	25.59	27.82	26.40	Romania
16.69	16.05	14.34	13.32	12.19	11.94	12.13	12.97	12.59	Slovak Republic
-	-	-	3.84	3.37	0.58	0.59	0.59	0.59	*Bosnia-Herzegovina*
-	-	-	4.68	4.71	4.95	5.08	5.08	5.56	*Croatia*
-	-	-	1.64	1.78	1.40	1.35	1.71	1.55	*FYROM*
-	-	-	3.06	3.33	3.54	3.79	4.25	4.33	*Slovenia*
-	-	-	8.49	6.72	6.00	6.59	8.41	9.39	*FR of Yugoslavia*
24.02	26.46	24.26	21.71	19.91	16.46	17.40	20.04	21.42	Former Yugoslavia
112.24	106.09	88.92	78.93	71.96	67.55	69.86	76.03	74.10	**Non-OECD Europe**

* Note: due to missing data prior to 1994 combustible renewables and waste are not included.

Total Final Consumption of Energy* (Mtoe)
Consommation finale totale d'énergie (Mtep)
Gesamter Endverbrauch von Energie (Mtoe)
Consumo finale totale di energia (Mtep)
最終エネルギー総消費量（石油換算百万トン）
Consumo final total de energía (Mtep)
Общее конечное потребление топлива и энергии (Мтнэ)

	1971	1973	1978	1983	1984	1985	1986	1987	1988
Arménie	-	-	-	-	-	-	-	-	-
Azerbaïdjan	-	-	-	-	-	-	-	-	-
Bélarus	-	-	-	-	-	-	-	-	-
Estonie	-	-	-	-	-	-	-	-	-
Géorgie	-	-	-	-	-	-	-	-	-
Kazakhstan	-	-	-	-	-	-	-	-	-
Kirghizistan	-	-	-	-	-	-	-	-	-
Lettonie	-	-	-	-	-	-	-	-	-
Lituanie	-	-	-	-	-	-	-	-	-
Moldova	-	-	-	-	-	-	-	-	-
Russie	-	-	-	-	-	-	-	-	-
Tadjikistan	-	-	-	-	-	-	-	-	-
Turkménistan	-	-	-	-	-	-	-	-	-
Ukraine	-	-	-	-	-	-	-	-	-
Ouzbékistan	-	-	-	-	-	-	-	-	-
Ex-URSS	507.03	569.93	718.72	806.22	822.82	838.55	868.09	880.99	909.48
Bahrein	0.44	0.94	1.71	2.03	2.14	2.68	3.21	3.00	3.32
Iran	16.89	20.98	31.84	35.44	42.34	43.66	45.70	47.72	50.04
Irak	3.01	3.62	7.31	12.44	13.62	14.43	15.73	16.61	19.13
Israël	3.88	4.43	5.01	5.57	5.48	5.72	6.26	6.88	7.12
Jordanie	0.42	0.49	1.09	1.96	1.96	2.04	2.16	2.18	2.20
Koweit	2.78	3.00	4.71	5.91	4.95	6.22	6.73	7.03	6.36
Liban	1.62	1.97	1.67	1.86	1.78	1.99	1.94	2.06	1.45
Oman	0.09	0.09	0.50	0.88	1.36	1.58	1.37	1.32	1.49
Qatar	0.49	0.78	0.93	2.35	2.80	2.65	2.71	2.75	2.91
Arabie saoudite	2.73	3.89	12.47	35.57	28.51	31.27	36.77	38.42	40.28
Syrie	2.00	1.91	4.52	6.73	7.31	7.02	7.32	7.90	8.09
Emirats arabes unis	0.23	0.48	2.67	4.78	5.06	5.65	5.79	5.99	5.84
Yémen	0.33	0.53	0.85	1.53	1.54	1.81	1.89	2.03	2.14
Moyen-Orient	34.91	43.11	75.28	117.07	118.84	126.71	137.58	143.88	150.37
Total non-OCDE	1 064.08	1 198.39	1 600.52	1 833.74	1 896.19	1 969.17	2 053.45	2 128.51	2 209.69
OCDE Amérique du N.	1 340.46	1 445.30	1 511.60	1 363.84	1 434.42	1 430.62	1 429.03	1 471.92	1 538.08
OCDE Pacifique	251.02	294.94	316.48	314.19	333.67	337.80	346.79	359.47	382.98
OCDE Europe	905.88	1 009.78	1 063.76	1 001.44	1 026.09	1 056.04	1 074.24	1 101.65	1 107.52
Total OCDE	2 497.36	2 750.01	2 891.83	2 679.48	2 794.18	2 824.47	2 850.06	2 933.04	3 028.58
Monde	3 561.44	3 948.40	4 492.35	4 513.22	4 690.37	4 793.63	4 903.51	5 061.55	5 238.26

* Note: à cause de données manquantes avant 1994, les combustibles renouvelables et les déchets ne sont pas inclus.
Ex-URSS: les séries antérieures à 1990-1992 ne sont pas comparables aux années récentes; 1991 a été estimé.

Total Final Consumption of Energy* (Mtoe)
Consommation finale totale d'énergie (Mtep)
Gesamter Endverbrauch von Energie (Mtoe)
Consumo finale totale di energia (Mtep)
最終エネルギー総消費量（石油換算百万トン）
Consumo final total de energía (Mtep)
Общее конечное потребление топлива и энергии (Мтнэ)

1989	1990	1991	1992	1993	1994	1995	1996	1997	
-	-	-	3.16	1.67	1.03	1.21	0.67	0.76	Armenia
-	-	-	12.29	10.09	10.99	8.93	7.60	7.11	Azerbaijan
-	-	-	28.94	24.80	19.62	18.78	18.80	18.97	Belarus
-	-	-	3.04	2.41	2.59	2.43	2.61	2.55	Estonia
-	-	-	5.77	4.30	2.51	1.58	2.17	2.33	Georgia
-	-	-	47.42	37.35	29.92	28.63	23.38	20.15	Kazakhstan
-	-	-	4.67	3.25	2.41	2.27	2.48	2.43	Kyrgyzstan
-	-	-	4.71	3.81	3.46	3.01	3.16	2.94	Latvia
-	-	-	7.87	5.46	4.76	4.80	4.77	4.67	Lithuania
-	-	-	3.64	3.54	3.04	2.89	3.10	3.10	Moldova
-	-	-	612.91	558.34	473.48	473.49	398.65	378.80	Russia
-	-	-	8.24	5.60	2.80	2.82	2.94	2.85	Tajikistan
-	-	-	8.45	7.54	9.33	9.56	8.06	7.77	Turkmenistan
-	-	-	138.74	124.07	102.34	100.28	95.76	88.50	Ukraine
-	-	-	32.38	32.78	32.60	31.41	31.08	31.48	Uzbekistan
916.07	902.49	866.48	920.43	822.89	699.55	691.11	604.61	573.83	**Former USSR**
3.30	3.50	3.26	3.30	3.75	3.64	3.81	3.96	4.03	Bahrain
51.70	51.90	61.61	66.88	69.14	71.49	76.38	82.28	86.11	Iran
21.27	18.35	13.44	15.75	18.46	20.41	20.25	19.98	20.80	Iraq
7.25	7.49	7.66	8.38	9.07	9.64	10.72	11.31	11.63	Israel
2.25	2.53	2.36	2.75	2.86	2.99	3.12	3.27	3.49	Jordan
7.71	6.41	1.15	5.05	4.88	5.23	5.71	5.79	6.45	Kuwait
1.42	1.40	1.95	2.21	2.81	3.04	3.59	3.59	3.99	Lebanon
1.57	2.55	3.75	3.99	4.18	3.84	3.78	3.13	3.46	Oman
3.61	2.99	3.67	5.59	5.97	6.03	6.06	6.23	6.87	Qatar
39.20	42.58	44.37	45.43	48.84	49.53	49.04	52.40	54.74	Saudi Arabia
8.44	8.99	9.42	9.59	9.71	10.25	10.61	11.01	11.50	Syria
6.14	6.14	6.51	7.04	7.62	8.09	8.03	8.00	7.71	United Arab Emirates
2.10	1.86	2.43	2.73	2.16	2.29	2.56	2.55	2.63	Yemen
155.96	156.70	161.59	178.69	189.45	196.46	203.65	213.49	223.41	**Middle East**
2 260.40	2 257.12	2 235.13	2 324.33	2 287.81	2 220.58	2 310.95	2 317.59	2 323.74	**Non-OECD Total**
1 540.20	1 525.28	1 516.03	1 520.10	1 557.64	1 596.71	1 618.91	1 669.22	1 685.70	OECD North America
400.02	422.45	439.06	454.33	465.61	488.72	508.51	527.25	541.69	OECD Pacific
1 101.95	1 091.57	1 106.80	1 104.01	1 106.35	1 105.68	1 131.86	1 178.58	1 171.41	OECD Europe
3 042.16	3 039.30	3 061.89	3 078.45	3 129.60	3 191.11	3 259.29	3 375.05	3 398.81	**OECD Total**
5 302.56	5 296.42	5 297.02	5 402.77	5 417.41	5 411.69	5 570.24	5 692.64	5 722.56	**World**

* Note: due to missing data prior to 1994 combustible renewables and waste are not included.
Former USSR: series up to 1990-1992 are not comparable with the recent years; 1991 is estimated.

INTERNATIONAL ENERGY AGENCY

Industry Consumption of Coal (Mtoe)
Consommation industrielle de charbon (Mtep)
Industrieverbrauch von Kohle (Mtoe)
Consumo di carbone nell'industria (Mtep)
石炭の産業用消費量（石油換算百万トン）
Consumo industrial de carbón (Mtep)
Потребление угля промышленным сектором (Мтнэ)

	1971	1973	1978	1983	1984	1985	1986	1987	1988
Algérie	0.04	0.04	0.04	0.11	0.11	0.11	0.14	0.15	0.15
RD du Congo	0.16	0.11	0.11	0.10	0.11	0.11	0.11	0.11	0.11
Egypte	0.06	0.04	0.10	0.11	0.11	0.12	0.12	0.13	0.15
Kenya	0.01	0.01	0.01	0.02	0.05	0.06	0.05	0.05	0.07
Maroc	0.17	0.07	0.02	0.15	0.17	0.32	0.30	0.26	0.27
Mozambique	0.38	0.36	0.15	0.09	0.06	0.04	0.02	0.02	0.02
Nigéria	0.03	0.07	0.08	0.02	0.03	0.05	0.05	0.05	0.02
Afrique du Sud	8.30	9.92	12.57	10.79	10.58	11.10	10.83	11.26	11.68
Tanzanie	-	-	0.00	0.00	0.01	0.01	0.00	0.00	0.00
Tunisie	0.03	0.03	0.03	0.02	0.03	0.02	0.02	0.02	0.03
Zambie	0.43	0.47	0.28	0.24	0.22	0.23	0.25	0.30	0.32
Zimbabwe	0.56	0.64	0.51	0.80	0.67	0.66	0.63	0.67	0.63
Autre Afrique	0.02	0.03	0.14	0.15	0.16	0.18	0.16	0.16	0.18
Afrique	10.19	11.81	14.03	12.62	12.31	13.02	12.68	13.19	13.63
Argentine	0.32	0.34	0.45	0.13	0.21	0.28	0.20	0.30	0.37
Bolivie	-	-	-	0.03	0.02	0.01	0.01	-	-
Brésil	0.45	0.58	1.06	2.53	2.89	3.14	3.27	3.69	3.93
Chili	0.53	0.49	0.42	0.44	0.55	0.59	0.66	0.63	0.70
Colombie	0.82	0.88	1.51	1.55	1.56	1.70	1.78	1.85	1.83
Costa Rica	0.00	0.00	0.00	0.00	0.00	0.00	0.00	0.00	0.00
Cuba	0.04	0.03	0.06	0.06	0.05	0.08	0.08	0.06	0.06
République dominicaine	-	-	-	0.00	0.00	-	-	-	-
Haiti	-	-	-	0.03	0.03	0.04	0.01	0.01	0.02
Jamaïque	-	-	-	-	-	-	-	-	0.00
Panama	-	-	-	-	-	0.02	0.02	0.02	0.01
Pérou	0.08	0.09	0.07	0.10	0.10	0.13	0.13	0.12	0.12
Uruguay	0.00	0.00	0.00	0.00	0.00	-	-	0.00	-
Vénézuela	0.05	0.08	0.04	0.04	0.05	0.05	0.06	0.08	0.07
Amérique latine	2.29	2.48	3.63	4.91	5.46	6.04	6.21	6.76	7.11
Bangladesh	0.09	0.12	0.12	0.08	0.03	0.05	0.07	0.12	0.12
Inde	10.86	13.62	17.34	22.66	23.65	31.28	30.13	30.68	34.67
Indonésie	0.06	0.03	0.07	0.14	0.18	0.18	0.18	0.19	0.47
RPD de Corée	19.90	16.76	18.69	20.77	19.94	20.42	20.63	20.63	20.90
Malaisie	0.02	0.01	0.01	0.21	0.23	0.30	0.22	0.27	0.38
Myanmar	0.12	0.04	0.13	0.13	0.14	0.13	0.14	0.04	0.04
Népal	0.01	0.05	0.01	0.02	0.06	0.05	0.03	0.03	0.04
Pakistan	0.48	0.50	0.53	0.74	0.86	1.04	1.04	1.08	1.29
Philippines	0.00	0.00	0.11	0.35	0.42	0.47	0.37	0.47	0.57
Sri Lanka	0.00	0.00	0.00	0.00	0.00	0.00	0.00	0.00	-
Taipei chinois	1.87	1.73	1.35	2.51	2.49	2.86	3.11	3.28	4.23
Thaïlande	0.02	0.02	0.07	0.20	0.25	0.37	0.44	0.63	0.77
Viêt-Nam	0.00	-	0.00	1.03	1.12	1.29	1.43	1.57	1.16
Autre Asie	0.10	-	-	0.16	0.17	0.21	0.22	0.21	0.20
Asie	33.53	32.88	38.45	49.01	49.53	58.66	58.01	59.20	64.85
Rép. populaire de Chine	2.94	3.20	4.41	145.54	161.70	170.93	185.22	195.03	200.10
Hong-Kong, Chine	0.02	0.01	0.00	0.00	-	0.00	0.00	0.00	-
Chine	2.96	3.21	4.41	145.54	161.70	170.93	185.22	195.03	200.10

Rép.populaire de Chine: jusqu'en 1978 la ventilation de la consommation finale par secteur est incomplète.

Industry Consumption of Coal (Mtoe)
Consommation industrielle de charbon (Mtep)
Industrieverbrauch von Kohle (Mtoe)
Consumo di carbone nell'industria (Mtep)
石炭の産業用消費量（石油換算百万トン）
Consumo industrial de carbón (Mtep)
Потребление угля промышленным сектором (Мтнэ)

1989	1990	1991	1992	1993	1994	1995	1996	1997	
0.13	0.12	0.08	0.08	0.15	0.14	0.14	0.14	0.10	Algeria
0.11	0.11	0.12	0.12	0.09	0.10	0.10	0.10	0.10	DR of Congo
0.15	0.13	0.15	0.16	0.19	0.16	0.13	0.13	0.13	Egypt
0.08	0.09	0.09	0.10	0.08	0.07	0.06	0.05	0.06	Kenya
0.46	0.36	0.47	0.43	0.44	0.45	0.45	0.56	0.59	Morocco
0.02	0.02	0.02	0.03	0.03	0.04	0.03	0.01	0.01	Mozambique
0.02	0.03	0.03	0.03	0.04	0.04	0.01	0.01	0.01	Nigeria
10.91	10.54	9.30	8.16	6.73	7.01	7.49	8.16	8.99	South Africa
0.00	0.00	0.00	0.00	0.00	0.00	0.00	0.00	0.00	Tanzania
0.03	0.02	0.02	0.02	0.02	0.01	0.01	0.01	0.01	Tunisia
0.19	0.18	0.14	0.16	0.12	0.07	0.08	0.07	0.07	Zambia
0.64	0.95	0.81	0.89	0.79	0.31	0.35	0.34	0.32	Zimbabwe
0.16	0.14	0.03	0.13	0.13	0.01	0.01	0.01	0.01	Other Africa
12.92	12.70	11.26	10.31	8.81	8.43	8.87	9.61	10.41	**Africa**
0.38	0.32	0.28	0.27	0.30	0.28	0.23	0.34	0.36	Argentina
-	-	-	-	-	-	-	-	-	Bolivia
3.63	3.08	3.64	3.29	3.53	3.71	3.90	4.37	4.42	Brazil
0.71	0.62	0.65	0.76	0.74	0.67	0.71	0.76	1.22	Chile
1.93	2.14	2.46	2.13	2.09	1.95	1.74	1.84	1.89	Colombia
0.00	0.00	-	-	-	-	-	0.00	-	Costa Rica
0.11	0.10	0.08	0.07	0.05	0.05	0.04	0.01	0.03	Cuba
-	-	-	-	-	-	-	-	-	Dominican Republic
0.01	0.01	0.02	0.02	-	-	-	-	-	Haiti
0.03	0.04	0.01	0.04	0.04	0.03	0.03	0.04	0.04	Jamaica
0.01	0.02	0.03	0.04	0.04	0.03	0.03	0.03	0.04	Panama
0.11	0.09	0.22	0.21	0.39	0.34	0.30	0.21	0.18	Peru
-	-	0.00	0.00	-	-	-	-	-	Uruguay
0.31	0.22	0.25	0.28	0.18	0.22	0.21	0.20	0.26	Venezuela
7.23	6.63	7.62	7.10	7.36	7.28	7.20	7.80	8.44	**Latin America**
0.12	0.28	0.09	0.08	0.03	-	-	-	-	Bangladesh
34.11	37.30	39.44	40.39	40.68	44.59	42.57	41.21	42.54	India
0.87	1.12	0.87	0.96	1.08	1.37	1.15	3.15	2.74	Indonesia
21.27	19.58	17.61	17.75	16.01	14.31	13.98	13.77	13.36	DPR of Korea
0.56	0.52	0.46	0.49	0.56	0.69	0.66	0.74	0.75	Malaysia
0.04	0.05	0.05	0.03	0.03	0.03	0.01	0.01	0.01	Myanmar
0.03	0.01	0.04	0.05	0.06	0.06	0.06	0.06	0.06	Nepal
1.24	1.49	1.44	1.46	1.53	1.66	1.44	1.55	1.34	Pakistan
0.37	0.41	0.62	0.50	0.55	0.48	0.46	0.83	1.07	Philippines
-	-	-	-	-	-	-	-	-	Sri Lanka
4.04	4.07	4.27	4.61	4.98	5.13	5.04	5.27	5.71	Chinese Taipei
1.04	1.29	1.51	1.68	2.48	3.05	3.62	4.39	3.94	Thailand
1.08	1.18	1.42	1.40	1.25	1.21	1.31	1.91	2.33	Vietnam
0.21	0.19	0.19	0.13	0.13	0.13	0.13	0.13	0.14	Other Asia
64.99	67.48	68.00	69.54	69.36	72.73	70.42	73.03	74.01	**Asia**
205.73	199.75	199.27	199.91	217.10	227.50	243.56	243.38	226.53	People's Rep.of China
-	-	-	-	-	-	-	-	-	Hong Kong, China
205.73	199.75	199.27	199.91	217.10	227.50	243.56	243.38	226.53	**China**

People's Rep. of China: up to 1978, the breakdown of final consumption by sector is incomplete.

Industry Consumption of Coal (Mtoe)
Consommation industrielle de charbon (Mtep)
Industrieverbrauch von Kohle (Mtoe)
Consumo di carbone nell'industria (Mtep)
石炭の産業用消費量（石油換算百万トン）
Consumo industrial de carbón (Mtep)
Потребление угля промышленным сектором (Мтнэ)

	1971	1973	1978	1983	1984	1985	1986	1987	1988
Albanie	0.16	0.19	0.31	0.44	0.47	0.52	0.49	0.53	0.52
Bulgarie	2.32	2.30	2.20	2.68	2.75	3.49	2.82	2.81	1.00
Chypre	-	-	-	-	0.03	0.05	0.03	0.09	0.06
Roumanie	1.47	1.62	2.33	2.55	2.65	3.23	3.36	3.40	3.60
République slovaque	2.43	2.32	2.85	2.44	2.58	2.67	2.64	2.69	2.66
Bosnie-Herzégovine	-	-	-	-	-	-	-	-	-
Croatie	-	-	-	-	-	-	-	-	-
Ex-RYM	-	-	-	-	-	-	-	-	-
Slovénie	-	-	-	-	-	-	-	-	-
RF de Yougoslavie	-	-	-	-	-	-	-	-	-
Ex-Yougoslavie	3.28	3.05	1.82	2.05	2.13	2.03	2.35	1.62	1.55
Europe non-OCDE	9.66	9.48	9.51	10.16	10.61	11.98	11.70	11.14	9.38
Azerbaïdjan	-	-	-	-	-	-	-	-	-
Bélarus	-	-	-	-	-	-	-	-	-
Estonie	-	-	-	-	-	-	-	-	-
Kazakhstan	-	-	-	-	-	-	-	-	-
Kirghizistan	-	-	-	-	-	-	-	-	-
Lettonie	-	-	-	-	-	-	-	-	-
Lituanie	-	-	-	-	-	-	-	-	-
Moldova	-	-	-	-	-	-	-	-	-
Russie	-	-	-	-	-	-	-	-	-
Ukraine	-	-	-	-	-	-	-	-	-
Ouzbékistan	-	-	-	-	-	-	-	-	-
Ex-URSS	35.97	39.99	56.57	44.23	40.20	36.78	33.92	39.55	41.82
Iran	0.34	0.66	0.38	0.37	0.61	0.62	0.65	0.67	0.68
Israël	0.00	-	-	-	-	-	-	0.10	0.09
Liban	0.00	0.01	0.00	-	-	-	-	-	-
Syrie	0.00	0.00	0.00	0.00	0.00	-	-	-	-
Moyen-Orient	0.35	0.67	0.39	0.37	0.61	0.62	0.65	0.77	0.77
Total non-OCDE	94.95	100.52	126.99	266.85	280.42	298.03	308.39	325.64	337.65
OCDE Amérique du N.	66.68	66.22	58.80	49.15	54.08	54.91	52.94	51.70	53.52
OCDE Pacifique	21.95	24.12	22.01	26.53	28.48	29.76	28.61	27.97	30.61
OCDE Europe	89.51	85.49	74.47	77.64	83.05	87.19	79.94	81.09	81.81
Total OCDE	178.14	175.83	155.27	153.32	165.61	171.86	161.50	160.76	165.94
Monde	273.09	276.36	282.26	420.16	446.03	469.89	469.89	486.40	503.59

Ex-URSS: les séries antérieures à 1990-1992 ne sont pas comparables aux années récentes; 1991 a été estimé.

Industry Consumption of Coal (Mtoe)
Consommation industrielle de charbon (Mtep)
Industrieverbrauch von Kohle (Mtoe)
Consumo di carbone nell'industria (Mtep)
石炭の産業用消費量（石油換算百万トン）
Consumo industrial de carbón (Mtep)
Потребление угля промышленным сектором (Мтнэ)

1989	1990	1991	1992	1993	1994	1995	1996	1997	
0.52	0.38	0.20	0.04	-	-	-	-	-	Albania
0.93	0.55	0.44	0.37	0.36	0.51	0.58	0.53	0.54	Bulgaria
0.06	0.06	0.06	0.02	0.02	0.02	0.01	0.01	0.01	Cyprus
3.49	2.07	1.47	0.86	0.73	0.96	1.18	1.20	1.12	Romania
2.57	2.40	2.40	2.19	2.01	1.85	1.74	1.68	1.64	Slovak Republic
-	-	-	0.19	0.13	-	-	-	-	*Bosnia-Herzegovina*
-	-	-	0.15	0.12	0.12	0.10	0.10	0.09	*Croatia*
-	-	-	0.10	0.16	0.15	0.14	0.11	0.11	*FYROM*
-	-	-	0.09	0.08	0.06	0.06	0.07	0.07	*Slovenia*
-	-	-	0.16	0.92	0.94	0.98	0.94	0.98	*FR of Yugoslavia*
2.11	1.62	0.95	0.69	1.40	1.27	1.29	1.21	1.24	Former Yugoslavia
9.69	7.08	5.52	4.15	4.53	4.61	4.80	4.63	4.55	**Non-OECD Europe**
-	-	-	-	0.00	-	-	-	-	Azerbaijan
-	-	-	0.05	0.05	0.03	0.02	0.02	0.02	Belarus
-	-	-	0.11	0.06	0.08	0.14	0.13	0.08	Estonia
-	-	-	12.40	11.18	10.58	8.92	7.15	6.09	Kazakhstan
-	-	-	0.83	0.74	0.87	0.28	0.37	0.73	Kyrgyzstan
-	-	-	0.02	0.02	0.02	0.02	0.01	0.01	Latvia
-	-	-	0.03	0.03	0.03	0.01	0.01	0.01	Lithuania
-	-	-	0.03	0.02	0.01	0.01	0.01	0.01	Moldova
-	-	-	14.25	17.50	14.25	14.57	12.64	12.13	Russia
-	-	-	14.81	14.94	8.71	8.13	6.47	6.23	Ukraine
-	-	-	-	-	-	0.08	0.07	0.05	Uzbekistan
40.93	38.88	30.22	41.47	43.03	33.68	31.37	26.30	24.79	**Former USSR**
0.69	0.82	0.89	0.98	0.47	0.49	0.58	0.58	0.58	Iran
0.10	0.01	0.01	0.01	0.01	0.01	0.01	0.02	0.02	Israel
-	-	-	-	0.07	0.07	0.12	0.13	0.13	Lebanon
-	-	-	-	-	-	-	-	-	Syria
0.79	0.83	0.90	0.99	0.55	0.57	0.70	0.73	0.73	**Middle East**
342.27	333.34	322.78	333.46	350.74	354.78	366.91	365.47	349.46	**Non-OECD Total**
52.07	52.23	49.73	30.10	31.87	31.19	31.35	29.20	29.11	OECD North America
31.74	32.11	34.11	33.03	33.96	34.12	32.94	34.92	35.17	OECD Pacific
79.96	72.74	61.68	57.74	53.71	52.47	53.25	58.36	54.47	OECD Europe
163.77	157.08	145.51	120.87	119.54	117.79	117.55	122.48	118.76	**OECD Total**
506.05	490.42	468.29	454.33	470.27	472.57	484.46	487.95	468.22	**World**

Former USSR: series up to 1990-1992 are not comparable with the recent years; 1991 is estimated.

Industry Consumption of Oil (Mtoe)
Consommation industrielle de pétrole (Mtep)
Industrieverbrauch von Öl (Mtoe)
Consumo di petrolio nell'industria (Mtep)
石油の産業用消費量（石油換算百万トン）
Consumo industrial de petróleo (Mtep)
Потребление нефти и нефтепродуктов промышленным сектором (Мтнэ)

	1971	1973	1978	1983	1984	1985	1986	1987	1988
Algérie	0.11	0.32	0.34	0.52	0.50	0.53	0.43	0.44	0.42
Angola	0.05	0.07	0.10	0.08	0.07	0.06	0.03	0.01	0.01
Bénin	0.00	0.00	0.00	0.01	0.01	0.01	0.01	0.01	0.01
Cameroun	0.02	0.02	0.05	0.05	0.06	0.07	0.07	0.05	0.05
Congo	0.03	0.03	0.01	0.02	0.02	0.02	0.02	-	-
RD du Congo	0.00	0.00	0.00	0.00	0.00	0.00	0.04	0.05	0.05
Egypte	2.08	2.09	3.15	4.73	5.12	5.22	4.68	6.28	6.54
Ethiopie	0.09	0.09	0.07	0.07	0.07	0.07	0.10	0.10	0.13
Gabon	0.12	0.13	0.34	0.33	0.32	0.30	0.20	0.19	0.20
Ghana	0.11	0.11	0.13	0.08	0.09	0.07	0.09	0.08	0.08
Côte d'Ivoire	0.22	0.23	0.20	0.25	0.07	0.13	0.16	0.16	0.17
Kenya	0.19	0.19	0.28	0.27	0.30	0.30	0.30	0.37	0.34
Libye	-	0.06	0.16	0.60	0.64	0.71	0.68	0.58	0.50
Maroc	0.57	0.74	1.12	1.03	1.00	0.95	0.86	0.74	0.45
Mozambique	-	-	-	-	-	-	-	-	-
Nigéria	0.21	0.37	0.82	1.32	0.93	1.01	0.72	0.52	0.77
Sénégal	0.10	0.11	0.11	0.12	0.14	0.14	0.13	0.14	0.12
Afrique du Sud	1.86	2.20	2.26	3.09	3.18	3.05	2.92	2.92	3.14
Soudan	0.18	0.50	0.18	0.24	0.20	0.21	0.24	0.14	0.17
Tanzanie	0.07	0.10	0.09	0.08	0.08	0.10	0.10	0.09	0.09
Tunisie	0.22	0.25	0.35	0.56	0.57	0.60	0.59	0.39	0.57
Zambie	0.15	0.26	0.31	0.26	0.20	0.15	0.13	0.13	0.12
Zimbabwe	0.05	0.06	0.05	0.06	0.09	0.09	0.11	0.09	0.10
Autre Afrique	0.02	0.02	0.10	0.17	0.20	0.20	0.21	0.22	0.26
Afrique	6.46	7.96	10.23	13.95	13.84	13.98	12.82	13.70	14.29
Argentine	3.72	3.05	3.11	2.68	3.03	2.72	1.89	2.40	2.24
Bolivie	0.13	0.14	0.13	0.10	0.07	0.05	0.03	0.02	0.04
Brésil	6.64	9.77	15.26	10.37	10.19	10.67	11.14	12.32	12.68
Chili	1.15	1.20	1.28	1.05	1.08	1.05	1.09	1.14	1.28
Colombie	1.44	1.35	1.13	1.03	0.97	0.88	0.93	1.10	1.17
Costa Rica	0.11	0.13	0.17	0.13	0.15	0.17	0.16	0.16	0.13
Cuba	1.76	2.16	2.53	2.48	2.45	2.24	2.11	2.27	2.32
République dominicaine	0.21	0.28	0.38	0.38	0.35	0.15	0.23	0.25	0.39
Equateur	0.15	0.13	0.34	0.65	0.66	0.79	0.78	0.80	0.71
El Salvador	0.11	0.14	0.17	0.11	0.12	0.11	0.12	0.13	0.16
Guatemala	0.27	0.22	0.27	0.14	0.14	0.14	0.16	0.18	0.20
Haiti	0.05	0.05	0.08	0.04	0.04	0.04	0.04	0.05	0.04
Honduras	0.09	0.11	0.16	0.17	0.20	0.17	0.16	0.19	0.22
Jamaïque	0.60	1.15	1.09	0.81	0.78	0.60	0.51	0.51	0.48
Antilles néerlandaises	1.43	1.57	0.51	0.37	0.34	0.33	0.31	0.30	0.25
Nicaragua	0.08	0.11	0.09	0.11	0.11	0.11	0.12	0.12	0.11
Panama	0.03	0.04	0.07	0.11	0.13	0.18	0.20	0.20	0.13
Paraguay	0.01	0.04	0.01	0.02	0.01	0.02	0.02	0.03	0.04
Pérou	0.67	0.92	1.24	1.05	1.06	1.14	1.18	0.97	0.83
Trinité-et-Tobago	0.03	0.02	0.04	0.05	0.02	0.01	0.00	0.01	0.08
Uruguay	0.41	0.40	0.41	0.22	0.19	0.15	0.17	0.18	0.18
Vénézuela	0.71	0.75	0.57	1.39	1.26	1.25	1.62	2.44	1.83
Autre Amérique latine	0.02	0.02	0.02	0.03	0.03	0.03	0.05	0.17	0.17
Amérique latine	19.79	23.75	29.07	23.50	23.40	23.00	23.03	25.93	25.68

Ne comprend pas l' usage non énergétique, sauf les produits d'alimentation.

Industry Consumption of Oil (Mtoe)
Consommation industrielle de pétrole (Mtep)
Industrieverbrauch von Öl (Mtoe)
Consumo di petrolio nell'industria (Mtep)
石油の産業用消費量（石油換算百万トン）
Consumo industrial de petróleo (Mtep)
Потребление нефти и нефтепродуктов промышленным сектором (Мтнэ)

1989	1990	1991	1992	1993	1994	1995	1996	1997	
0.32	0.00	-	-	0.06	0.06	0.04	0.05	0.05	Algeria
0.01	0.01	0.01	0.01	0.01	0.01	0.01	0.01	0.02	Angola
0.01	0.01	0.01	0.01	0.01	0.01	0.01	0.02	0.02	Benin
0.06	0.05	0.05	0.04	0.04	0.04	0.05	0.05	0.05	Cameroon
0.00	0.00	0.00	0.00	0.00	0.00	0.00	0.00	0.00	Congo
0.05	0.06	0.06	0.05	0.06	0.06	0.06	0.06	0.06	DR of Congo
6.73	6.99	6.33	6.77	5.24	4.81	6.04	6.44	7.01	Egypt
0.15	0.17	0.18	0.18	0.15	0.17	0.20	0.21	0.10	Ethiopia
0.09	0.05	0.05	0.09	0.10	0.09	0.10	0.12	0.13	Gabon
0.11	0.09	0.09	0.09	0.09	0.10	0.11	0.10	0.10	Ghana
0.17	0.18	0.13	0.14	0.15	0.15	0.16	0.14	0.14	Ivory Coast
0.31	0.33	0.32	0.34	0.39	0.38	0.35	0.41	0.38	Kenya
0.82	0.90	1.03	0.93	1.01	1.11	1.29	1.30	1.46	Libya
0.66	0.60	0.57	0.69	0.38	0.51	0.53	0.54	0.55	Morocco
-	0.01	0.01	0.01	0.01	0.01	0.01	0.01	0.01	Mozambique
0.78	0.67	0.63	0.57	0.58	0.63	0.56	0.70	0.72	Nigeria
0.06	0.07	0.08	0.09	0.07	0.10	0.11	0.11	0.13	Senegal
3.15	3.32	3.28	3.00	1.31	1.40	1.31	1.42	1.46	South Africa
0.19	0.23	0.29	0.22	0.03	0.26	0.16	0.08	0.22	Sudan
0.10	0.10	0.10	0.10	0.10	0.10	0.10	0.10	0.10	Tanzania
0.73	0.77	0.69	0.75	0.63	0.57	0.55	0.60	0.57	Tunisia
0.12	0.11	0.12	0.11	0.11	0.12	0.12	0.12	0.12	Zambia
0.11	0.10	0.22	0.07	0.08	0.08	0.10	0.10	0.10	Zimbabwe
0.27	0.28	0.27	0.26	0.25	0.26	0.23	0.24	0.24	Other Africa
15.01	15.11	14.49	14.53	10.88	11.05	12.21	12.91	13.75	**Africa**
1.95	2.05	1.94	1.86	1.71	1.72	1.41	1.59	1.45	Argentina
0.05	0.07	0.07	0.07	0.07	0.07	0.08	0.09	0.08	Bolivia
13.17	12.98	12.82	13.73	14.25	15.24	15.92	16.92	18.89	Brazil
1.45	1.43	1.43	1.58	1.78	1.79	2.08	2.26	2.71	Chile
1.18	1.36	1.09	1.38	1.17	1.38	1.55	1.60	1.70	Colombia
0.20	0.21	0.21	0.18	0.18	0.21	0.21	0.19	0.22	Costa Rica
2.55	2.82	1.93	1.62	1.73	2.02	2.02	2.22	2.22	Cuba
0.29	0.25	0.24	0.27	0.31	0.33	0.34	0.35	0.38	Dominican Republic
0.73	0.91	0.91	0.67	0.60	0.63	0.69	0.68	0.77	Ecuador
0.16	0.16	0.17	0.23	0.23	0.25	0.30	0.28	0.29	El Salvador
0.22	0.23	0.24	0.27	0.33	0.34	0.40	0.38	0.40	Guatemala
0.05	0.05	0.05	0.05	0.04	0.01	0.04	0.05	0.08	Haiti
0.26	0.24	0.25	0.28	0.28	0.29	0.37	0.48	0.49	Honduras
0.87	0.15	0.16	0.16	0.13	0.22	0.19	0.19	0.19	Jamaica
0.20	0.17	0.17	0.17	0.13	0.13	0.13	0.13	0.13	Netherlands Antilles
0.09	0.10	0.08	0.08	0.07	0.10	0.09	0.09	0.07	Nicaragua
0.14	0.13	0.17	0.18	0.17	0.20	0.22	0.15	0.19	Panama
0.04	0.05	0.04	0.08	0.06	0.10	0.08	0.03	0.04	Paraguay
0.67	0.67	0.62	0.51	0.52	0.59	0.93	1.25	1.22	Peru
0.09	0.10	0.10	0.09	0.07	0.07	0.08	0.09	0.11	Trinidad-and-Tobago
0.17	0.16	0.17	0.17	0.15	0.14	0.14	0.17	0.20	Uruguay
1.71	1.70	1.85	1.58	2.47	2.97	2.72	2.54	2.97	Venezuela
0.17	0.15	0.14	0.13	0.13	0.13	0.08	0.08	0.09	Other Latin America
26.41	26.12	24.85	25.33	26.59	28.91	30.08	31.82	34.87	**Latin America**

Excluding non-energy use, except feedstocks.

Industry Consumption of Oil (Mtoe)
Consommation industrielle de pétrole (Mtep)
Industrieverbrauch von Öl (Mtoe)
Consumo di petrolio nell'industria (Mtep)
石油の産業用消費量（石油換算百万トン）
Consumo industrial de petróleo (Mtep)
Потребление нефти и нефтепродуктов промышленным сектором (Мтнэ)

	1971	1973	1978	1983	1984	1985	1986	1987	1988
Bangladesh	0.17	0.23	0.26	0.23	0.22	0.15	0.09	0.06	0.04
Brunei	0.02	0.02	0.02	0.04	0.03	0.04	0.03	0.03	0.04
Inde	4.55	5.82	6.90	8.04	8.80	9.32	9.53	9.47	10.17
Indonésie	1.45	1.70	3.71	5.40	4.78	5.23	4.73	4.65	5.06
RPD de Corée	0.09	0.09	0.31	0.46	0.51	0.55	0.59	0.60	0.60
Malaisie	1.91	1.51	1.93	2.62	2.48	2.39	2.24	2.44	2.54
Myanmar	0.35	0.25	0.37	0.36	0.34	0.34	0.30	0.19	0.17
Népal	-	0.00	0.00	0.00	0.00	0.01	0.02	0.01	0.01
Pakistan	0.15	0.24	0.22	0.38	0.66	0.79	0.91	1.18	1.18
Philippines	1.54	1.97	2.43	2.07	1.57	1.39	1.43	1.70	1.87
Singapour	0.07	0.10	0.23	0.25	0.63	1.26	1.19	1.49	1.49
Sri Lanka	0.09	0.17	0.17	0.20	0.19	0.10	0.09	0.08	0.10
Taipei chinois	1.12	1.93	5.36	5.95	6.53	7.02	7.86	8.30	8.58
Thaïlande	1.30	1.54	1.82	1.60	1.60	1.60	1.61	1.83	1.88
Viêt-Nam				0.56	0.65	0.38	0.40	0.41	0.37
Autre Asie	0.04	0.05	0.06	0.13	0.12	0.12	0.11	0.11	0.10
Asie	12.87	15.62	23.79	28.30	29.12	30.66	31.14	32.55	34.20
Rép. populaire de Chine	6.82	8.93	23.24	20.79	21.18	24.24	24.62	27.63	29.35
Hong-Kong, Chine	0.78	0.67	1.47	0.71	0.86	0.75	0.77	0.87	1.01
Chine	7.59	9.61	24.71	21.50	22.04	24.99	25.39	28.50	30.36
Albanie	0.20	0.05	-	0.20	0.18	0.11	0.15	0.18	0.11
Bulgarie	-	-	-	-	-	-	-	-	2.31
Chypre	0.12	0.16	0.21	0.20	0.18	0.15	0.21	0.21	0.26
Malte	-	-	-	-	-	-	-	-	-
Roumanie	0.80	0.83	1.16	-	-	1.12	1.27	2.12	2.21
République slovaque	0.78	0.86	0.97	2.32	2.20	2.19	2.08	2.02	1.89
Bosnie-Herzégovine	-	-	-	-	-	-	-	-	-
Croatie	-	-	-	-	-	-	-	-	-
Ex-RYM	-	-	-	-	-	-	-	-	-
Slovénie	-	-	-	-	-	-	-	-	-
RF de Yougoslavie	-	-	-	-	-	-	-	-	-
Ex-Yougoslavie	2.28	3.18	4.14	3.26	2.96	2.82	3.07	3.06	3.31
Europe non-OCDE	4.19	5.08	6.48	5.98	5.52	6.40	6.78	7.58	10.10
Arménie	-	-	-	-	-	-	-	-	-
Azerbaïdjan	-	-	-	-	-	-	-	-	-
Bélarus	-	-	-	-	-	-	-	-	-
Estonie	-	-	-	-	-	-	-	-	-
Géorgie	-	-	-	-	-	-	-	-	-
Lettonie	-	-	-	-	-	-	-	-	-
Lituanie	-	-	-	-	-	-	-	-	-
Moldova	-	-	-	-	-	-	-	-	-
Russie	-	-	-	-	-	-	-	-	-
Ukraine	-	-	-	-	-	-	-	-	-
Ouzbékistan	-	-	-	-	-	-	-	-	-
Ex-URSS	56.06	66.10	81.50	80.41	79.95	75.90	76.01	75.43	76.10

Ne comprend pas l' usage non énergétique, sauf les produits d'alimentation.
Rép.populaire de Chine: jusqu'en 1978 la ventilation de la consommation finale par secteur est incomplète.
Ex-URSS: les séries antérieures à 1990-1992 ne sont pas comparables aux années récentes; 1991 a été estimé.

INTERNATIONAL ENERGY AGENCY

Industry Consumption of Oil (Mtoe)
Consommation industrielle de pétrole (Mtep)
Industrieverbrauch von Öl (Mtoe)
Consumo di petrolio nell'industria (Mtep)
石油の産業用消費量（石油換算百万トン）
Consumo industrial de petróleo (Mtep)
Потребление нефти и нефтепродуктов промышленным сектором (Мтнэ)

1989	1990	1991	1992	1993	1994	1995	1996	1997	
0.03	0.03	0.06	0.07	0.13	0.20	0.21	0.22	0.21	Bangladesh
0.04	0.05	0.05	0.06	0.06	0.06	0.07	0.08	0.09	Brunei
11.79	12.36	12.48	12.80	12.67	13.58	14.77	16.48	18.00	India
4.97	5.38	5.42	5.72	6.48	6.98	7.37	7.78	10.65	Indonesia
0.67	0.67	0.58	0.48	0.35	0.29	0.23	0.19	0.18	DPR of Korea
2.93	3.47	3.75	3.84	5.00	4.77	4.95	5.44	5.96	Malaysia
0.16	0.12	0.13	0.22	0.26	0.28	0.38	0.43	0.48	Myanmar
0.01	0.00	0.00	0.01	0.01	0.03	0.11	0.11	0.11	Nepal
1.24	1.25	1.11	1.12	1.43	1.60	1.81	2.34	1.69	Pakistan
2.08	1.99	1.93	2.06	2.30	2.59	3.30	3.33	3.51	Philippines
1.58	1.84	1.69	1.49	1.63	1.64	2.03	2.34	2.89	Singapore
0.13	0.12	0.11	0.19	0.20	0.19	0.17	0.43	0.40	Sri Lanka
8.37	8.67	8.59	8.79	9.07	10.31	10.59	10.89	11.52	Chinese Taipei
2.27	2.68	2.89	3.49	3.83	4.79	5.19	6.33	5.64	Thailand
0.37	0.57	0.48	1.19	1.19	1.33	1.39	0.74	0.94	Vietnam
0.10	0.11	0.11	0.11	0.10	0.10	0.02	0.02	0.02	Other Asia
36.75	39.30	39.37	41.62	44.70	48.75	52.57	57.14	62.29	**Asia**
31.25	30.17	31.81	33.91	33.66	37.32	38.34	41.25	41.53	People's Rep.of China
1.01	0.65	0.28	0.69	0.35	0.35	0.28	0.22	0.18	Hong Kong, China
32.26	30.82	32.10	34.59	34.01	37.67	38.62	41.47	41.71	**China**
0.15	0.30	0.19	0.11	0.14	0.20	0.05	0.05	0.02	Albania
2.25	2.05	0.76	1.23	1.24	1.12	1.25	1.42	1.39	Bulgaria
0.23	0.14	0.30	0.31	0.32	0.33	0.34	0.38	0.35	Cyprus
-	-	-	-	-	-	0.02			Malta
2.80	2.06	2.44	1.48	0.93	1.87	1.90	2.09	2.01	Romania
3.00	2.13	1.86	1.85	1.22	1.43	1.19	1.14	1.10	Slovak Republic
-	-	-	0.08	0.19	0.06	0.06	0.06	0.06	*Bosnia-Herzegovina*
-	-	-	0.58	0.60	0.65	0.64	0.55	0.53	*Croatia*
-	-	-	0.33	0.21	0.14	0.13	0.26	0.19	*FYROM*
-	-	-	0.18	0.18	0.20	0.22	0.22	0.16	*Slovenia*
-	-	-	0.67	0.52	0.49	0.54	0.72	1.06	*FR of Yugoslavia*
3.08	2.92	2.49	1.85	1.70	1.54	1.59	1.81	1.99	Former Yugoslavia
11.50	9.59	8.05	6.82	5.54	6.50	6.34	6.90	6.87	**Non-OECD Europe**
-	-	-	0.19	0.23	-	-	-	-	Armenia
-	-	-	-	0.29	0.19	0.16	0.15	0.14	Azerbaijan
-	-	-	1.51	2.01	1.69	1.57	1.43	1.52	Belarus
-	-	-	0.17	0.12	0.16	0.26	0.25	0.20	Estonia
-	-	-	-	-	-	-	0.06	0.07	Georgia
-	-	-	0.18	0.20	0.12	0.14	0.06	0.28	Latvia
-	-	-	0.95	0.21	0.20	0.19	0.18	0.18	Lithuania
-	-	-	0.04	0.09	0.03	0.03	0.03	0.02	Moldova
-	-	-	26.18	29.61	18.19	17.38	13.68	13.52	Russia
-	-	-	8.82	6.07	5.04	5.51	4.11	3.76	Ukraine
-	-	-	0.29	-	-	0.14	0.42	0.41	Uzbekistan
73.97	72.08	72.93	38.34	38.83	25.62	25.37	20.38	20.11	**Former USSR**

Excluding non-energy use, except feedstocks.
People's Rep. of China: up to 1978, the breakdown of final consumption by sector is incomplete.
Former USSR: series up to 1990-1992 are not comparable with the recent years; 1991 is estimated.

Industry Consumption of Oil (Mtoe)
Consommation industrielle de pétrole (Mtep)
Industrieverbrauch von Öl (Mtoe)
Consumo di petrolio nell'industria (Mtep)
石油の産業用消費量（石油換算百万トン）
Consumo industrial de petróleo (Mtep)
Потребление нефти и нефтепродуктов промышленным сектором (Мтнэ)

	1971	1973	1978	1983	1984	1985	1986	1987	1988
Iran	1.27	1.72	3.36	1.99	4.97	4.84	6.96	8.00	7.97
Irak	0.07	0.03	0.10	0.03	0.30	0.23	0.36	-	0.10
Israël	0.81	0.92	1.20	1.26	1.30	1.29	1.49	1.51	1.56
Jordanie	0.05	0.07	0.11	0.35	0.32	0.37	0.42	0.41	0.37
Koweit	-	-	0.15	0.64	0.02	0.56	0.01	0.01	0.20
Liban	0.45	0.47	0.41	0.04	0.03	0.05	0.05	0.09	0.06
Oman	0.01	0.01	0.01	0.01	0.01	0.01	0.01	0.01	0.01
Qatar	-	-	0.01	0.01	-	0.06	-	0.03	-
Arabie saoudite	0.61	1.13	3.47	14.81	5.82	7.02	7.60	7.43	6.58
Syrie	0.51	0.24	1.59	1.95	2.04	2.11	2.04	2.07	1.76
Emirats arabes unis	-	-	-	0.44	0.09	0.46	0.33	0.34	0.29
Yémen	-	0.00	0.10	0.21	0.22	0.36	0.35	0.30	0.29
Moyen-Orient	3.77	4.61	10.52	21.74	15.11	17.37	19.62	20.19	19.18
Total non-OCDE	110.73	132.73	186.31	195.37	188.98	192.30	194.79	203.87	209.91
OCDE Amérique du N.	117.34	136.99	148.25	120.08	129.12	123.91	126.31	128.30	127.22
OCDE Pacifique	85.78	98.55	87.91	65.63	69.68	67.91	68.67	71.73	75.15
OCDE Europe	172.51	191.22	170.38	126.25	119.46	115.36	116.73	117.11	117.12
Total OCDE	375.63	426.77	406.54	311.96	318.26	307.19	311.71	317.14	319.48
Monde	486.36	559.49	592.85	507.34	507.24	499.48	506.50	521.01	529.39

Ne comprend pas l' usage non énergétique, sauf les produits d'alimentation.

Industry Consumption of Oil (Mtoe)
Consommation industrielle de pétrole (Mtep)
Industrieverbrauch von Öl (Mtoe)
Consumo di petrolio nell'industria (Mtep)
石油の産業用消費量（石油換算百万トン）
Consumo industrial de petróleo (Mtep)
Потребление нефти и нефтепродуктов промышленным сектором (Мтнэ)

1989	1990	1991	1992	1993	1994	1995	1996	1997	
7.44	0.14	0.15	0.15	6.60	6.70	7.06	7.65	7.46	Iran
0.85	0.93	2.06	1.33	1.88	2.37	2.27	2.18	2.24	Iraq
1.44	1.43	1.48	1.35	1.41	1.54	1.83	1.94	2.36	Israel
0.37	0.40	0.35	0.40	0.44	0.46	0.48	0.48	0.51	Jordan
0.21	0.21	0.05	2.09	1.34	1.30	1.58	1.46	1.75	Kuwait
0.07	0.10	0.10	0.07	0.42	0.55	0.73	0.63	0.83	Lebanon
0.01	0.49	1.60	1.38	0.93	0.50	0.92	0.80	0.43	Oman
0.84	0.02	0.18	0.02	0.03	0.03	0.03	0.03	0.03	Qatar
7.03	7.39	7.52	7.63	8.00	7.61	7.97	8.78	9.48	Saudi Arabia
0.65	0.63	0.73	0.78	0.80	0.92	1.02	1.05	1.01	Syria
0.34	0.34	0.34	0.48	0.62	0.60	0.43	0.40	0.41	United Arab Emirates
0.30	0.10	0.31	0.39	0.13	0.14	0.14	0.16	0.16	Yemen
19.54	12.18	14.87	16.06	22.61	22.72	24.45	25.56	26.67	**Middle East**
215.43	205.20	206.66	177.31	183.15	181.22	189.64	196.17	206.28	**Non-OECD Total**
124.86	122.19	113.11	119.17	117.83	122.66	119.56	124.81	126.35	OECD North America
78.96	83.30	86.47	91.76	94.61	98.16	101.84	103.45	108.68	OECD Pacific
112.83	107.09	108.23	106.42	102.45	104.98	110.01	108.01	109.94	OECD Europe
316.66	312.58	307.80	317.35	314.89	325.80	331.41	336.28	344.96	**OECD Total**
532.09	517.78	514.46	494.66	498.04	507.02	521.05	532.45	551.24	**World**

Excluding non-energy use, except feedstocks.

Industry Consumption of Gas (Mtoe)
Consommation industrielle de gaz (Mtep)
Industrieverbrauch von Gas (Mtoe)

Consumo di gas nell'industria (Mtep)

ガスの産業用消費量（石油換算百万トン）

Consumo industrial de gas natural (Mtep)

Потребление газа промышленным сектором (Мтнэ)

	1971	1973	1978	1983	1984	1985	1986	1987	1988
Algérie	0.18	0.21	0.62	0.84	0.91	1.18	1.39	1.51	1.48
Angola	0.04	0.05	0.06	0.09	0.10	0.10	0.11	0.13	0.13
Congo	0.05	-	-	-	-	-	-	-	-
Egypte	0.00	0.00	0.33	1.14	1.41	1.44	1.49	1.72	1.69
Gabon	-	-	-	0.03	0.03	0.04	0.05	0.05	0.07
Libye	-	-	0.65	1.18	1.29	1.44	1.48	1.50	1.10
Maroc	0.04	0.06	0.07	0.07	0.07	0.08	0.08	0.06	0.06
Nigéria	0.03	0.03	0.03	0.12	0.31	0.37	0.39	0.46	0.58
Sénégal	-	-	-	-	-	-	-	0.00	0.01
Afrique du Sud	0.13	0.22	0.33	0.46	0.46	0.44	0.46	0.42	0.43
Tunisie	0.00	0.01	0.06	0.10	0.11	0.13	0.16	0.27	0.13
Afrique	0.48	0.57	2.16	4.04	4.70	5.21	5.61	6.12	5.67
Argentine	1.71	2.31	2.79	3.65	3.75	3.76	3.99	4.43	4.46
Bolivie	-	0.00	0.05	0.04	0.04	0.06	0.07	0.08	0.07
Brésil	0.04	0.11	0.67	1.32	1.47	1.70	1.73	2.02	2.17
Chili	0.00	0.00	0.01	0.01	0.01	0.01	0.01	0.01	0.23
Colombie	0.11	0.21	0.55	0.69	0.75	0.78	0.84	0.82	0.86
Cuba	0.01	0.02	0.02	0.01	0.01	0.02	0.02	0.03	0.03
Pérou	0.01	0.01	0.05	0.03	0.04	0.05	0.07	0.07	0.06
Trinité-et-Tobago	0.46	0.52	0.86	1.67	1.64	1.64	1.77	1.79	2.39
Uruguay	0.00	0.00	0.00	0.00	0.00	0.00	0.00	0.00	0.00
Vénézuela	1.86	3.79	5.33	6.38	7.65	8.15	7.61	8.13	6.48
Autre Amérique latine	-	-	0.00	0.00	0.00	0.01	0.00	0.00	0.00
Amérique latine	4.19	6.97	10.33	13.80	15.36	16.17	16.10	17.38	16.77
Bangladesh	0.19	0.33	0.42	0.84	0.94	0.99	1.18	1.32	1.47
Inde	0.18	0.16	0.69	1.21	1.51	2.33	3.10	3.24	5.01
Indonésie	0.10	0.11	1.12	2.88	3.92	2.04	2.09	2.21	2.05
Malaisie	-	-	0.00	0.00	0.08	0.46	0.84	1.04	1.18
Myanmar	0.02	0.03	0.07	0.13	0.17	0.23	0.26	0.27	0.27
Pakistan	1.41	1.69	2.24	3.62	3.46	3.48	3.47	3.95	4.05
Philippines	0.00	0.00	0.00	-	-	-	-	-	-
Singapour	0.00	0.01	0.02	0.02	0.02	0.02	0.02	0.02	0.03
Taipei chinois	0.62	0.87	1.23	0.63	0.58	0.46	0.36	0.37	0.40
Thaïlande	-	-	-	0.03	0.17	0.16	0.08	0.04	0.05
Viêt-Nam	-	-	-	-	-	-	-	-	-
Asie	2.52	3.19	5.79	9.36	10.86	10.18	11.38	12.45	14.50
Rép. populaire de Chine	1.41	2.25	6.31	5.67	5.78	7.20	7.33	7.64	7.64
Hong-Kong, Chine	0.00	0.00	0.00	0.01	0.01	0.01	0.01	0.01	0.01
Chine	1.41	2.26	6.32	5.68	5.79	7.21	7.34	7.65	7.65
Albanie	-	-	-	-	-	-	-	-	-
Bulgarie	-	-	-	-	-	3.69	3.81	4.03	2.15
Roumanie	13.01	11.45	25.15	27.59	27.73	18.76	19.29	19.07	19.77
République slovaque	0.70	0.83	1.62	2.29	2.27	2.25	2.27	2.29	2.33
Croatie	-	-	-	-	-	-	-	-	-
Ex-RYM	-	-	-	-	-	-	-	-	-
Slovénie	-	-	-	-	-	-	-	-	-
RF de Yougoslavie	-	-	-	-	-	-	-	-	-
Ex-Yougoslavie	0.70	0.86	1.22	1.84	2.40	2.57	2.72	3.02	2.25
Europe non-OCDE	14.42	13.14	27.99	31.72	32.41	27.27	28.09	28.42	26.50

Rép.populaire de Chine: jusqu'en 1978 la ventilation de la consommation finale par secteur est incomplète.

Industry Consumption of Gas (Mtoe)
Consommation industrielle de gaz (Mtep)
Industrieverbrauch von Gas (Mtoe)
Consumo di gas nell'industria (Mtep)
ガスの産業用消費量（石油換算百万トン）
Consumo industrial de gas natural (Mtep)
Потребление газа промышленным сектором (Мтнэ)

1989	1990	1991	1992	1993	1994	1995	1996	1997	
1.19	1.26	1.38	1.44	2.75	2.45	2.40	2.30	2.50	Algeria
0.14	0.44	0.47	0.47	0.46	0.42	0.46	0.46	0.47	Angola
-	-	-	-	-	-	-	-	-	Congo
2.52	2.35	2.38	2.49	3.01	2.18	2.28	2.28	2.40	Egypt
0.00	0.00	0.00	0.00	0.00	0.00	0.00	0.00	0.00	Gabon
1.12	1.29	1.40	1.40	1.44	1.49	1.47	1.58	1.69	Libya
0.05	0.04	0.03	0.02	0.02	0.02	0.02	0.02	0.02	Morocco
0.68	0.72	0.65	0.70	0.74	0.81	0.74	0.75	0.72	Nigeria
0.01	0.00	0.00	0.00	0.01	0.02	0.04	0.04	0.02	Senegal
0.44	0.48	0.47	0.46	0.47	0.50	0.54	0.61	0.61	South Africa
0.19	0.26	0.27	0.30	0.31	0.33	0.36	0.36	0.37	Tunisia
6.33	6.84	7.05	7.27	9.21	8.21	8.31	8.39	8.80	**Africa**
3.76	3.91	4.40	4.64	4.70	5.36	5.88	5.90	6.24	Argentina
0.14	0.16	0.18	0.21	0.24	0.26	0.29	0.25	0.34	Bolivia
2.27	2.28	2.38	2.57	2.58	2.70	2.84	3.20	3.55	Brazil
0.66	0.72	0.52	0.71	0.69	0.75	0.72	0.75	1.42	Chile
0.88	0.91	0.92	0.90	0.89	0.97	1.01	1.05	1.01	Colombia
0.03	0.04	0.03	0.02	0.03	0.02	0.02	0.02	0.03	Cuba
0.04	0.03	0.04	0.02	0.02	0.02	0.02	0.02	0.11	Peru
2.44	2.51	2.67	2.74	2.68	3.60	3.64	4.11	4.41	Trinidad-and-Tobago
0.00		0.00	0.00	0.00	0.00	-	-	0.00	Uruguay
6.71	7.73	8.86	7.77	8.63	8.65	9.49	9.57	10.28	Venezuela
0.00	0.00	0.00	0.00	0.00	0.00	0.00	0.00	0.00	Other Latin America
16.94	18.30	20.00	19.58	20.44	22.33	23.91	24.87	27.39	**Latin America**
1.56	1.50	1.44	1.80	1.92	2.16	2.50	2.53	2.39	Bangladesh
6.17	5.89	6.04	6.90	6.61	7.03	7.87	8.74	9.02	India
1.52	6.65	6.81	7.58	6.58	6.52	6.95	7.57	7.87	Indonesia
1.03	1.09	1.09	1.34	1.69	1.65	1.58	2.05	1.87	Malaysia
0.26	0.22	0.18	0.18	0.23	0.28	0.28	0.28	0.33	Myanmar
4.32	4.32	4.47	4.47	4.99	5.32	5.27	5.61	5.84	Pakistan
-	-	-	-	-	0.00	0.00	0.00	0.00	Philippines
0.03	0.03	0.03	0.04	0.04	0.04	0.04	0.05	0.05	Singapore
0.36	0.39	0.53	0.63	0.66	0.69	0.82	0.86	0.87	Chinese Taipei
0.10	0.24	0.32	0.40	0.44	0.52	0.70	0.84	0.85	Thailand
-	-	-	-	-	-	0.21	2.91	3.09	Vietnam
15.35	20.32	20.93	23.33	23.16	24.21	26.23	31.42	32.18	**Asia**
8.55	8.57	9.06	9.14	8.82	9.35	9.80	12.02	10.38	People's Rep.of China
0.01	0.01	0.02	0.02	0.02	0.02	0.04	1.50	2.33	Hong Kong, China
8.56	8.59	9.08	9.15	8.84	9.37	9.84	13.52	12.71	**China**
0.21	0.16	0.09	0.06	0.05	0.03	0.02	0.02	0.01	Albania
2.23	2.27	1.95	1.61	1.58	1.78	2.08	2.08	1.75	Bulgaria
19.55	16.76	11.43	5.43	5.21	7.24	7.89	7.84	6.49	Romania
2.45	2.88	2.28	2.26	2.14	1.63	2.17	2.48	2.44	Slovak Republic
-	-	-	0.98	0.91	0.95	0.92	0.92	1.01	*Croatia*
-	-	-	0.13	0.13	-	-	-	-	*FYROM*
-	-	-	0.36	0.36	0.44	0.45	0.49	0.58	*Slovenia*
-	-	-	0.52	0.46	0.21	0.28	0.65	0.64	*FR of Yugoslavia*
2.27	0.66	2.02	1.99	1.86	1.60	1.65	2.06	2.24	Former Yugoslavia
26.72	22.74	17.77	11.34	10.85	12.28	13.79	14.48	12.93	**Non-OECD Europe**

People's Rep. of China: up to 1978, the breakdown of final consumption by sector is incomplete.

Industry Consumption of Gas (Mtoe)
Consommation industrielle de gaz (Mtep)
Industrieverbrauch von Gas (Mtoe)
Consumo di gas nell'industria (Mtep)
ガスの産業用消費量（石油換算百万トン）
Consumo industrial de gas natural (Mtep)
Потребление газа промышленным сектором (Мтнэ)

	1971	1973	1978	1983	1984	1985	1986	1987	1988
Arménie	-	-	-	-	-	-	-	-	-
Azerbaïdjan	-	-	-	-	-	-	-	-	-
Bélarus	-	-	-	-	-	-	-	-	-
Estonie	-	-	-	-	-	-	-	-	-
Géorgie	-	-	-	-	-	-	-	-	-
Lettonie	-	-	-	-	-	-	-	-	-
Lituanie	-	-	-	-	-	-	-	-	-
Moldova	-	-	-	-	-	-	-	-	-
Russie	-	-	-	-	-	-	-	-	-
Ukraine	-	-	-	-	-	-	-	-	-
Ouzbékistan	-	-	-	-	-	-	-	-	-
Ex-URSS	77.20	88.80	117.00	139.78	145.38	151.28	156.78	162.58	168.48
Bahrein	0.18	0.59	0.93	1.10	1.16	1.67	2.19	1.95	2.25
Iran	8.09	8.25	8.37	7.11	8.67	9.99	10.26	8.50	6.16
Irak	0.76	0.99	1.39	2.22	2.46	2.61	3.19	4.01	5.55
Israël	0.09	0.05	0.05	0.05	0.04	0.04	0.03	0.03	0.03
Koweit	1.93	1.99	2.48	2.25	1.86	2.44	3.41	3.69	2.76
Oman	-	-	-	0.01	0.21	0.32	0.15	0.19	0.21
Qatar	0.38	0.60	0.46	1.55	1.99	1.76	1.83	1.83	1.99
Arabie saoudite	0.32	0.32	0.32	0.74	1.63	2.37	2.79	-	-
Syrie	-	-	-	-	-	-	-	-	-
Emirats arabes unis	-	-	-	-	0.54	0.50	0.56	0.63	0.72
Moyen-Orient	11.75	12.78	13.99	15.02	18.56	21.69	24.41	20.83	19.67
Total non-OCDE	111.96	127.71	183.57	219.39	233.06	239.01	249.71	255.42	259.24
OCDE Amérique du N.	181.97	196.04	162.71	150.97	164.58	157.71	147.15	156.82	168.12
OCDE Pacifique	3.14	3.87	5.73	7.65	8.53	9.60	9.85	10.29	10.62
OCDE Europe	41.44	57.40	76.53	74.91	79.07	79.41	78.26	84.10	85.34
Total OCDE	226.55	257.32	244.97	233.53	252.17	246.72	235.25	251.21	264.08
Monde	338.51	385.03	428.54	452.92	485.23	485.73	484.97	506.63	523.32

Ex-URSS: les séries antérieures à 1990-1992 ne sont pas comparables aux années récentes; 1991 a été estimé.

Industry Consumption of Gas (Mtoe)
Consommation industrielle de gaz (Mtep)
Industrieverbrauch von Gas (Mtoe)
Consumo di gas nell'industria (Mtep)
ガスの産業用消費量（石油換算百万トン）
Consumo industrial de gas natural (Mtep)
Потребление газа промышленным сектором (Мтнэ)

1989	1990	1991	1992	1993	1994	1995	1996	1997	
-	-	-	0.40	0.15	0.14	0.23	0.11	0.11	Armenia
-	-	-	-	3.49	3.41	2.69	1.50	1.51	Azerbaijan
-	-	-	2.12	2.01	1.17	1.70	1.63	1.57	Belarus
-	-	-	0.19	0.10	0.19	0.33	0.34	0.33	Estonia
-	-	-	-	1.61	0.63	0.39	0.25	0.35	Georgia
-	-	-	0.33	0.16	0.24	0.17	0.22	0.21	Latvia
-	-	-	0.62	0.32	0.45	0.62	0.69	0.62	Lithuania
-	-	-	0.42	0.38	0.36	0.36	0.43	0.45	Moldova
-	-	-	82.99	48.21	39.23	40.76	34.48	34.40	Russia
-	-	-	27.53	26.15	16.37	16.39	17.46	15.52	Ukraine
-	-	-	-	-	-	6.72	5.24	4.09	Uzbekistan
171.46	179.05	169.98	114.58	82.58	62.20	70.35	62.35	59.17	**Former USSR**
2.19	2.31	2.19	2.22	2.64	2.45	2.59	2.75	2.64	Bahrain
6.22	5.69	8.65	11.63	12.01	11.53	12.81	15.19	17.08	Iran
6.40	3.40	1.40	2.34	2.62	3.26	3.24	3.31	3.70	Iraq
0.03	0.03	0.02	0.02	0.02	0.02	0.02	0.01	0.01	Israel
3.83	2.92	-	-	-	-	-	-	-	Kuwait
0.23	0.59	0.63	1.09	1.57	1.61	1.12	0.46	0.85	Oman
1.85	1.95	2.52	4.50	4.81	4.77	4.75	4.76	5.27	Qatar
-	-	-	-	-	-	-	-	-	Saudi Arabia
-	-	0.37	0.58	0.60	0.65	0.82	0.82	0.76	Syria
0.78	0.80	1.09	1.21	1.25	1.35	1.47	1.75	1.84	United Arab Emirates
21.52	17.69	16.87	23.59	25.54	25.64	26.82	29.05	32.15	**Middle East**
266.87	273.53	261.68	208.85	180.61	164.25	179.26	184.09	185.31	**Non-OECD Total**
156.72	157.18	150.06	148.57	155.69	159.34	164.77	166.40	164.92	OECD North America
11.26	11.89	12.83	13.11	14.05	15.18	16.14	17.31	18.50	OECD Pacific
89.65	89.31	87.73	88.49	89.16	90.79	95.48	98.92	100.96	OECD Europe
257.63	258.38	250.62	250.17	258.90	265.32	276.39	282.63	284.39	**OECD Total**
524.50	531.90	512.31	459.02	439.51	429.56	455.65	466.72	469.70	**World**

Former USSR: series up to 1990-1992 are not comparable with the recent years; 1991 is estimated.

Industry Consumption of Electricity (Mtoe)
Consommation industrielle d'électricité (Mtep)
Industrieverbrauch von Elektrizität (Mtoe)
Consumo di energia elettrica nell'industria (Mtep)
電力の産業用消費量（石油換算百万トン）
Consumo industrial de electricidad (Mtep)
Потребление электроэнергии промышленным сектором (Мтнэ)

	1971	1973	1978	1983	1984	1985	1986	1987	1988
Algérie	0.068	0.081	0.170	0.328	0.339	0.369	0.440	0.372	0.400
Angola	0.011	0.015	0.009	0.012	0.012	0.012	0.012	0.012	0.012
Bénin	0.001	0.002	0.003	0.006	0.006	0.006	0.006	0.005	0.007
Cameroun	0.076	0.072	0.071	0.107	0.102	0.111	0.106	0.103	0.111
Congo	0.004	0.004	0.005	0.012	0.012	0.012	0.016	0.018	0.018
RD du Congo	-	-	0.211	0.228	0.231	0.221	0.216	0.217	0.217
Egypte	0.354	0.361	0.700	0.982	1.007	1.093	1.187	1.265	1.324
Ethiopie	0.031	0.031	0.027	0.037	0.040	0.034	0.036	0.045	0.040
Gabon	-	0.003	0.015	0.021	0.030	0.032	0.030	0.033	0.033
Ghana	0.210	0.278	0.243	0.111	0.047	0.125	0.236	0.271	0.297
Côte d'Ivoire	0.011	0.016	0.033	0.028	0.077	0.050	0.073	0.072	0.081
Kenya	0.031	0.037	0.061	0.098	0.104	0.116	0.127	0.131	0.135
Maroc	0.084	0.105	0.159	0.233	0.245	0.251	0.275	0.272	0.326
Mozambique	0.017	0.020	0.024	0.017	0.013	0.014	0.016	0.022	0.014
Nigéria	0.051	0.073	0.113	0.176	0.149	0.164	0.211	0.222	0.204
Sénégal	0.016	0.019	0.027	0.030	0.034	0.037	0.035	0.035	0.037
Afrique du Sud	2.797	3.058	4.412	5.278	5.859	6.137	6.382	6.580	6.837
Soudan	0.018	0.020	0.031	0.014	0.016	0.016	0.015	0.020	0.021
Tanzanie	0.013	0.015	0.018	0.021	0.022	0.024	0.028	0.029	0.030
Tunisie	0.039	0.050	0.083	0.123	0.131	0.139	0.147	0.160	0.169
Zambie	0.311	0.361	0.401	0.460	0.457	0.449	0.422	0.441	0.410
Zimbabwe	0.190	0.293	0.343	0.393	0.407	0.439	0.457	0.463	0.480
Autre Afrique	0.022	0.027	0.087	0.117	0.126	0.167	0.157	0.172	0.185
Afrique	**4.355**	**4.943**	**7.245**	**8.834**	**9.466**	**10.020**	**10.630**	**10.958**	**11.389**
Argentine	0.892	1.059	1.244	1.666	1.717	1.656	1.753	1.840	2.004
Bolivie	0.013	0.015	0.029	0.075	0.072	0.069	0.061	0.051	0.055
Brésil	1.918	2.538	4.685	6.476	7.498	8.276	8.975	9.022	9.585
Chili	0.422	0.409	0.495	0.567	0.624	0.644	0.674	0.700	0.777
Colombie	0.257	0.297	0.406	0.505	0.574	0.530	0.574	0.641	0.724
Costa Rica	0.027	0.031	0.049	0.053	0.064	0.066	0.071	0.073	0.073
Cuba	0.173	0.196	0.290	0.418	0.446	0.432	0.472	0.488	0.514
République dominicaine	0.026	0.058	0.071	0.076	0.112	0.119	0.108	0.131	0.064
Equateur	0.025	0.028	0.054	0.114	0.117	0.120	0.121	0.117	0.119
El Salvador	0.024	0.030	0.051	0.040	0.040	0.041	0.041	0.044	0.045
Guatemala	0.030	0.031	0.050	0.038	0.038	0.042	0.044	0.043	0.055
Haiti	0.001	0.002	0.007	0.014	0.014	0.015	0.015	0.014	0.014
Honduras	0.016	0.020	0.026	0.038	0.040	0.050	0.049	0.064	0.067
Jamaïque	0.104	0.130	0.130	0.058	0.057	0.066	0.063	0.075	0.074
Antilles néerlandaises	0.027	0.029	0.031	0.032	0.029	0.027	0.021	0.024	0.024
Nicaragua	0.023	0.026	0.034	0.030	0.030	0.027	0.026	0.029	0.026
Panama	0.020	0.022	0.023	0.041	0.040	0.028	0.029	0.031	0.024
Paraguay	0.009	0.011	0.024	0.030	0.032	0.019	0.021	0.039	0.049
Pérou	0.315	0.357	0.452	0.482	0.558	0.550	0.585	0.613	0.583
Trinité-et-Tobago	0.029	0.044	0.060	0.141	0.131	0.131	0.150	0.162	0.153
Uruguay	0.057	0.056	0.084	0.092	0.095	0.098	0.106	0.109	0.117
Vénézuela	0.354	0.390	0.654	1.279	1.429	1.596	1.628	1.876	2.039
Autre Amérique latine	0.010	0.012	0.018	0.023	0.025	0.026	0.028	0.030	0.032
Amérique latine	**4.772**	**5.791**	**8.968**	**12.288**	**13.784**	**14.629**	**15.615**	**16.216**	**17.215**

Industry Consumption of Electricity (Mtoe)
Consommation industrielle d'électricité (Mtep)
Industrieverbrauch von Elektrizität (Mtoe)
Consumo di energia elettrica nell'industria (Mtep)
電力の産業用消費量（石油換算百万トン）
Consumo industrial de electricidad (Mtep)
Потребление электроэнергии промышленным сектором (Мтнэ)

1989	1990	1991	1992	1993	1994	1995	1996	1997	
0.453	0.469	0.484	0.480	0.535	0.524	0.535	0.538	0.661	Algeria
0.012	0.011	0.016	0.017	0.017	0.017	0.017	0.019	0.020	Angola
0.007	0.007	0.009	0.010	0.010	0.009	0.010	0.011	0.010	Benin
0.117	0.117	0.115	0.112	0.118	0.111	0.110	0.113	0.123	Cameroon
0.020	0.025	0.025	0.024	0.025	0.025	0.025	0.025	0.025	Congo
0.260	0.261	0.284	0.292	0.364	0.372	0.323	0.327	0.314	DR of Congo
1.400	1.482	1.571	1.601	1.685	1.771	1.869	1.856	1.965	Egypt
0.040	0.058	0.059	0.059	0.052	0.070	0.070	0.070	0.071	Ethiopia
0.033	0.034	0.034	0.034	0.034	0.036	0.036	0.037	0.039	Gabon
0.303	0.306	0.319	0.332	0.341	0.309	0.292	0.291	0.300	Ghana
0.079	0.079	0.077	0.043	0.076	0.079	0.099	0.104	0.110	Ivory Coast
0.140	0.151	0.158	0.158	0.165	0.168	0.172	0.184	0.195	Kenya
0.312	0.345	0.359	0.388	0.379	0.407	0.414	0.427	0.452	Morocco
0.016	0.015	0.027	0.025	0.023	0.021	0.022	0.021	0.025	Mozambique
0.230	0.164	0.192	0.188	0.178	0.176	0.175	0.181	0.184	Nigeria
0.039	0.038	0.039	0.040	0.044	0.040	0.043	0.044	0.047	Senegal
6.996	7.081	7.089	6.833	6.511	6.509	6.937	7.732	7.866	South Africa
0.016	0.018	0.019	0.020	0.020	0.036	0.033	0.033	0.037	Sudan
0.033	0.034	0.040	0.034	0.033	0.034	0.038	0.042	0.040	Tanzania
0.179	0.192	0.199	0.206	0.216	0.228	0.238	0.241	0.250	Tunisia
0.298	0.240	0.346	0.343	0.343	0.343	0.344	0.344	0.354	Zambia
0.496	0.487	0.426	0.458	0.388	0.426	0.462	0.490	0.488	Zimbabwe
0.181	0.188	0.175	0.170	0.135	0.138	0.108	0.079	0.079	Other Africa
11.661	11.801	12.062	11.869	11.692	11.848	12.372	13.210	13.653	**Africa**
1.896	1.506	1.513	1.564	1.684	1.732	1.850	1.916	2.073	Argentina
0.060	0.063	0.071	0.073	0.096	0.096	0.102	0.098	0.106	Bolivia
9.851	9.661	9.894	10.026	10.532	10.851	10.937	11.159	11.670	Brazil
0.849	0.868	0.975	1.135	1.182	1.242	1.418	1.578	1.744	Chile
0.733	0.684	0.723	0.727	0.837	0.920	0.951	0.975	0.970	Colombia
0.061	0.064	0.065	0.070	0.086	0.095	0.086	0.088	0.094	Costa Rica
0.523	0.488	0.389	0.337	0.286	0.304	0.305	0.342	0.452	Cuba
0.062	0.053	0.058	0.071	0.092	0.096	0.102	0.107	0.120	Dominican Republic
0.121	0.132	0.145	0.141	0.156	0.177	0.179	0.195	0.203	Ecuador
0.044	0.049	0.051	0.054	0.063	0.067	0.071	0.072	0.078	El Salvador
0.051	0.068	0.068	0.071	0.078	0.080	0.089	0.107	0.124	Guatemala
0.015	0.015	0.011	0.009	0.007	0.005	0.009	0.010	0.011	Haiti
0.064	0.073	0.066	0.051	0.055	0.049	0.050	0.067	0.063	Honduras
0.064	0.018	0.025	0.027	0.244	0.233	0.284	0.295	0.305	Jamaica
0.026	0.028	0.029	0.031	0.034	0.036	0.038	0.038	0.039	Netherlands Antilles
0.028	0.031	0.029	0.025	0.021	0.022	0.026	0.030	0.033	Nicaragua
0.028	0.032	0.036	0.039	0.045	0.048	0.052	0.055	0.043	Panama
0.048	0.054	0.045	0.055	0.062	0.067	0.080	0.081	0.119	Paraguay
0.605	0.390	0.412	0.386	0.388	0.429	0.470	0.513	0.734	Peru
0.149	0.165	0.166	0.180	0.177	0.190	0.195	0.221	0.248	Trinidad-and-Tobago
0.123	0.127	0.136	0.132	0.133	0.133	0.113	0.111	0.112	Uruguay
2.016	2.116	2.199	2.313	2.245	2.226	2.334	2.373	2.271	Venezuela
0.031	0.033	0.034	0.035	0.036	0.036	0.037	0.037	0.039	Other Latin America
17.448	16.718	17.138	17.551	18.541	19.134	19.775	20.466	21.652	**Latin America**

Industry Consumption of Electricity (Mtoe)
Consommation industrielle d'électricité (Mtep)
Industrieverbrauch von Elektrizität (Mtoe)
Consumo di energia elettrica nell'industria (Mtep)
電力の産業用消費量（石油換算百万トン）
Consumo industrial de electricidad (Mtep)
Потребление электроэнергии промышленным сектором (Мтнэ)

	1971	1973	1978	1983	1984	1985	1986	1987	1988
Bangladesh	0.041	0.067	0.106	0.135	0.145	0.144	0.159	0.164	0.175
Brunei	0.006	0.009	0.013	0.012	0.013	0.013	0.014	0.018	0.019
Inde	2.985	3.200	4.380	5.312	5.817	6.216	6.564	6.512	7.163
Indonésie	0.044	0.055	0.099	0.296	0.345	0.419	0.532	0.637	0.781
RPD de Corée	0.615	0.728	1.164	1.509	1.645	1.753	1.815	1.827	1.935
Malaisie	0.156	0.205	0.322	0.422	0.454	0.444	0.472	0.525	0.603
Myanmar	0.026	0.031	0.047	0.050	0.055	0.081	0.079	0.078	0.063
Népal	0.001	0.001	0.003	0.006	0.013	0.009	0.012	0.015	0.014
Pakistan	0.246	0.315	0.395	0.479	0.504	0.537	0.627	0.689	0.772
Philippines	0.292	0.408	0.595	0.750	0.736	0.775	0.503	0.667	0.737
Singapour	0.079	0.116	0.224	0.244	0.281	0.292	0.302	0.346	0.395
Sri Lanka	0.033	0.043	0.046	0.065	0.074	0.075	0.080	0.075	0.078
Taipei chinois	0.776	1.005	1.770	2.188	2.375	2.428	2.772	2.993	3.263
Thaïlande	0.258	0.339	0.438	0.670	0.734	0.783	0.856	0.955	1.114
Viêt-Nam	-	-	-	0.149	0.174	0.182	0.189	0.205	0.223
Autre Asie	-	-	0.163	0.164	0.185	0.203	0.200	0.198	0.209
Asie	5.558	6.520	9.765	12.452	13.551	14.355	15.175	15.901	17.544
Rép. populaire de Chine	-	-	-	19.324	20.714	23.372	25.810	27.032	29.802
Hong-Kong, Chine	0.175	0.208	0.305	0.402	0.443	0.449	0.511	0.550	0.572
Chine	0.175	0.208	0.305	19.726	21.157	23.822	26.321	27.582	30.373
Albanie	-	-	-	-	-	-	0.105	0.091	0.076
Bulgarie	1.066	1.207	1.499	1.799	1.862	1.625	1.838	1.762	1.810
Chypre	0.015	0.020	0.020	0.023	0.025	0.018	0.021	0.023	0.025
Malte	-	-	-	-	-	-	-	-	-
Roumanie	1.930	2.297	3.467	4.433	4.707	3.815	4.083	4.043	4.194
République slovaque	0.626	0.716	0.864	1.143	1.182	1.246	1.277	1.315	1.342
Croatie	-	-	-	-	-	-	-	-	-
Ex-RYM	-	-	-	-	-	-	-	-	-
Slovénie	-	-	-	-	-	-	-	-	-
RF de Yougoslavie	-	-	-	-	-	-	-	-	-
Ex-Yougoslavie	1.164	1.354	2.063	2.599	3.040	3.157	3.263	3.192	2.987
Europe non-OCDE	4.802	5.594	7.914	9.997	10.816	9.861	10.588	10.426	10.434
Arménie	-	-	-	-	-	-	-	-	-
Azerbaïdjan	-	-	-	-	-	-	-	-	-
Bélarus	-	-	-	-	-	-	-	-	-
Estonie	-	-	-	-	-	-	-	-	-
Géorgie	-	-	-	-	-	-	-	-	-
Kazakhstan	-	-	-	-	-	-	-	-	-
Kirghizistan	-	-	-	-	-	-	-	-	-
Lettonie	-	-	-	-	-	-	-	-	-
Lituanie	-	-	-	-	-	-	-	-	-
Moldova	-	-	-	-	-	-	-	-	-
Russie	-	-	-	-	-	-	-	-	-
Tadjikistan	-	-	-	-	-	-	-	-	-
Turkménistan	-	-	-	-	-	-	-	-	-
Ukraine	-	-	-	-	-	-	-	-	-
Ouzbékistan	-	-	-	-	-	-	-	-	-
Ex-URSS	38.820	43.765	54.963	59.968	62.780	63.821	65.068	67.364	69.239

Rép.populaire de Chine: jusqu'en 1978 la ventilation de la consommation finale par secteur est incomplète.
Ex-URSS: les séries antérieures à 1990-1992 ne sont pas comparables aux années récentes; 1991 a été estimé.

Industry Consumption of Electricity (Mtoe)
Consommation industrielle d'électricité (Mtep)
Industrieverbrauch von Elektrizität (Mtoe)
Consumo di energia elettrica nell'industria (Mtep)
電力の産業用消費量（石油換算百万トン）
Consumo industrial de electricidad (Mtep)
Потребление электроэнергии промышленным сектором (Мтнэ)

1989	1990	1991	1992	1993	1994	1995	1996	1997	
0.221	0.234	0.225	0.361	0.442	0.484	0.554	0.600	0.644	Bangladesh
0.020	0.020	0.021	0.022	0.024	0.024	0.025	0.024	0.023	Brunei
8.632	9.401	9.967	10.451	10.848	11.518	12.358	12.866	13.701	India
0.982	1.218	1.378	1.527	1.882	1.932	2.161	2.443	2.689	Indonesia
1.978	1.978	-	-	-	-	-	-	-	DPR of Korea
0.690	0.830	0.920	1.137	1.303	1.567	1.826	2.029	2.422	Malaysia
0.069	0.074	0.061	0.066	0.071	0.073	0.079	0.077	0.079	Myanmar
0.016	0.015	0.018	0.021	0.024	0.026	0.028	0.031	0.031	Nepal
0.813	0.889	0.969	1.083	1.122	1.087	1.077	1.048	1.030	Pakistan
0.840	0.772	0.803	0.744	0.811	0.919	0.966	1.019	1.081	Philippines
0.439	0.485	0.622	0.640	0.682	0.750	0.804	0.842	0.953	Singapore
0.078	0.078	0.090	0.091	0.105	0.134	0.144	0.142	0.156	Sri Lanka
3.531	3.679	3.979	4.227	4.455	4.744	5.086	5.332	5.841	Chinese Taipei
1.327	1.542	1.704	1.755	1.924	2.487	2.826	2.979	2.971	Thailand
0.226	0.245	0.267	0.275	0.299	0.330	0.388	0.473	0.530	Vietnam
0.217	0.214	0.205	0.204	0.204	0.207	0.124	0.117	0.117	Other Asia
20.078	21.676	21.230	22.604	24.196	26.281	28.446	30.023	32.268	**Asia**
31.442	32.840	35.818	39.501	38.697	42.622	45.976	50.257	48.076	People's Rep.of China
0.601	0.596	0.598	0.578	0.555	0.512	0.483	0.476	0.453	Hong Kong, China
32.044	33.436	36.417	40.079	39.252	43.134	46.460	50.733	48.529	**China**
0.129	0.118	0.091	0.107	0.129	0.045	0.044	0.069	0.065	Albania
1.678	1.595	1.204	1.028	0.939	0.986	1.046	1.054	1.009	Bulgaria
0.027	0.029	0.029	0.031	0.033	0.034	0.034	0.035	0.034	Cyprus
-	-	0.020	0.022	0.022	0.042	0.042	0.044	0.039	Malta
4.218	3.316	2.656	2.192	2.026	1.882	2.008	2.108	2.161	Romania
1.376	1.291	0.953	0.966	0.856	0.871	0.787	0.903	0.865	Slovak Republic
-	-	-	0.294	0.264	0.263	0.236	0.228	0.259	*Croatia*
-	-	-	0.218	0.195	0.183	0.163	0.173	0.179	*FYROM*
-	-	-	0.401	0.392	0.426	0.424	0.411	0.406	*Slovenia*
-	-	-	0.940	0.502	0.521	0.525	0.536	0.569	*FR of Yugoslavia*
3.186	3.324	1.898	1.853	1.352	1.393	1.348	1.348	1.414	Former Yugoslavia
10.615	9.673	6.851	6.200	5.357	5.252	5.309	5.560	5.587	**Non-OECD Europe**
-	-	-	0.155	0.071	0.057	0.084	0.070	0.062	Armenia
-	-	-	0.384	0.378	0.490	0.417	0.412	0.383	Azerbaijan
-	-	-	1.716	1.288	1.026	0.913	0.928	1.081	Belarus
-	-	-	0.192	0.119	0.147	0.151	0.164	0.182	Estonia
-	-	-	0.468	0.355	0.238	0.253	0.075	0.071	Georgia
-	-	-	4.950	3.864	2.078	2.007	1.793	1.607	Kazakhstan
-	-	-	0.298	0.216	0.168	0.127	0.137	0.128	Kyrgyzstan
-	-	-	0.225	0.124	0.128	0.123	0.146	0.128	Latvia
-	-	-	0.389	0.268	0.240	0.241	0.217	0.239	Lithuania
-	-	-	0.256	0.258	0.222	0.159	0.149	0.155	Moldova
-	-	-	40.403	32.985	28.016	25.220	22.933	22.744	Russia
-	-	-	0.808	0.608	0.585	0.591	0.581	0.561	Tajikistan
-	-	-	0.166	0.188	0.157	0.155	0.146	0.135	Turkmenistan
-	-	-	9.343	8.177	6.592	6.124	5.666	5.575	Ukraine
-	-	-	1.644	1.562	1.519	1.238	1.210	1.283	Uzbekistan
68.757	68.172	68.220	61.396	50.460	41.663	37.804	34.628	34.334	**Former USSR**

People's Rep. of China: up to 1978, the breakdown of final consumption by sector is incomplete.
Former USSR: series up to 1990-1992 are not comparable with the recent years; 1991 is estimated.

Industry Consumption of Electricity (Mtoe)
Consommation industrielle d'électricité (Mtep)
Industrieverbrauch von Elektrizität (Mtoe)
Consumo di energia elettrica nell'industria (Mtep)
電力の産業用消費量（石油換算百万トン）
Consumo industrial de electricidad (Mtep)
Потребление электроэнергии промышленным сектором (Мтнэ)

	1971	1973	1978	1983	1984	1985	1986	1987	1988
Bahrein	0.015	0.015	0.019	0.020	0.020	0.026	0.028	0.028	0.030
Iran	0.164	0.324	0.506	0.724	0.993	1.152	1.380	0.989	0.913
Irak	0.078	0.098	0.282	0.469	0.564	0.608	0.646	0.683	0.778
Israël	0.171	0.198	0.278	0.342	0.355	0.361	0.378	0.416	0.416
Jordanie	0.007	0.009	0.018	0.061	0.073	0.078	0.078	0.091	0.089
Koweit	0.026	0.031	0.051	0.058	0.060	0.061	0.064	0.066	0.090
Liban	-	-	-	-	-	-	-	-	-
Oman	-	0.000	0.003	0.010	0.012	0.015	0.019	0.020	0.023
Qatar	-	-	-	0.039	0.041	0.043	0.041	0.040	0.042
Arabie saoudite	0.110	0.150	0.127	0.154	0.251	0.533	0.453	0.480	0.510
Syrie	0.067	0.075	0.124	0.293	0.276	0.304	0.319	0.321	0.326
Yémen	0.002	0.003	0.010	0.003	0.002	0.003	0.003	0.006	0.006
Moyen-Orient	0.638	0.904	1.418	2.173	2.648	3.184	3.408	3.140	3.223
Total non-OCDE	59.120	67.726	90.578	125.437	134.202	139.690	146.805	151.587	159.417
OCDE Amérique du N.	58.829	66.199	77.176	75.008	82.772	83.429	81.319	84.823	89.640
OCDE Pacifique	23.449	28.298	32.363	30.882	32.880	33.866	34.061	36.039	38.562
OCDE Europe	54.500	62.857	70.817	71.711	75.483	76.945	78.000	80.411	83.432
Total OCDE	136.778	157.353	180.357	177.601	191.134	194.240	193.380	201.273	211.634
Monde	195.898	225.079	270.935	303.038	325.336	333.930	340.185	352.860	371.051

Industry Consumption of Electricity (Mtoe)
Consommation industrielle d'électricité (Mtep)
Industrieverbrauch von Elektrizität (Mtoe)
Consumo di energia elettrica nell'industria (Mtep)
電力の産業用消費量（石油換算百万トン）
Consumo industrial de electricidad (Mtep)
Потребление электроэнергии промышленным сектором (Мтнэ)

1989	1990	1991	1992	1993	1994	1995	1996	1997	
0.034	0.036	0.034	0.035	0.053	0.057	0.060	0.060	0.062	Bahrain
0.728	0.879	1.295	1.539	1.339	1.677	1.759	1.868	1.933	Iran
0.774	-	-	-	-	-	-	-	-	Iraq
0.436	0.455	0.460	0.489	0.515	0.568	0.614	0.673	0.688	Israel
0.094	0.102	0.102	0.115	0.125	0.131	0.144	0.152	0.158	Jordan
0.097	0.090	-	-	-	-	-	-	-	Kuwait
-	-	-	-	0.102	0.116	0.105	0.151	0.181	Lebanon
0.024	0.027	0.029	0.031	0.035	0.035	0.032	0.034	0.039	Oman
0.043	0.052	0.060	0.069	0.065	0.069	0.065	0.122	0.128	Qatar
0.618	0.685	0.703	0.722	0.752	0.824	0.828	0.871	0.924	Saudi Arabia
0.335	0.361	0.364	0.330	0.329	0.389	0.393	0.433	0.461	Syria
-	-	-	-	-	-	-	-	-	Yemen
3.183	2.687	3.047	3.331	3.315	3.865	3.999	4.365	4.574	**Middle East**
163.786	164.163	164.964	163.030	152.814	151.179	154.164	158.984	160.596	**Non-OECD Total**
91.604	93.470	101.400	103.641	104.955	109.368	111.212	115.120	118.025	OECD North America
40.881	43.456	44.577	44.574	44.895	48.832	50.370	52.241	54.135	OECD Pacific
86.008	85.219	83.927	83.479	82.679	83.588	86.127	86.790	89.799	OECD Europe
218.493	222.144	229.905	231.694	232.528	241.788	247.710	254.151	261.959	**OECD Total**
382.279	386.307	394.869	394.724	385.342	392.967	401.874	413.135	422.555	**World**

Total Industry Consumption of Energy* (Mtoe)
Consommation industrielle totale d'énergie (Mtep)
Gesamtindustrieverbrauch von Energie (Mtoe)
Consumo totale di energia nell'industria (Mtep)
産業用エネルギー総消費量（石油換算百万トン）
Consumo total industrial de energía (Mtep)
Общее потребление топлива и энергии промышленным сектором (Мтнэ)

	1971	1973	1978	1983	1984	1985	1986	1987	1988
Algérie	0.39	0.65	1.17	1.80	1.86	2.19	2.40	2.47	2.44
Angola	0.10	0.14	0.17	0.18	0.18	0.17	0.15	0.15	0.15
Bénin	0.01	0.00	0.01	0.02	0.02	0.02	0.02	0.01	0.01
Cameroun	0.10	0.09	0.12	0.16	0.16	0.18	0.17	0.16	0.16
Congo	0.08	0.03	0.01	0.03	0.03	0.03	0.03	0.02	0.02
RD du Congo	0.16	0.12	0.32	0.33	0.34	0.33	0.37	0.37	0.38
Egypte	2.49	2.49	4.28	6.97	7.66	7.87	7.48	9.39	9.70
Ethiopie	0.12	0.12	0.10	0.10	0.11	0.11	0.14	0.15	0.17
Gabon	0.12	0.13	0.35	0.39	0.38	0.37	0.28	0.27	0.30
Ghana	0.32	0.39	0.37	0.19	0.13	0.19	0.32	0.35	0.38
Côte d'Ivoire	0.23	0.24	0.24	0.28	0.14	0.18	0.23	0.23	0.25
Kenya	0.22	0.24	0.35	0.39	0.45	0.47	0.48	0.56	0.54
Libye	-	0.06	0.82	1.78	1.93	2.15	2.16	2.09	1.59
Maroc	0.87	0.97	1.37	1.49	1.49	1.60	1.51	1.34	1.11
Mozambique	0.40	0.38	0.18	0.11	0.07	0.05	0.04	0.04	0.03
Nigéria	0.32	0.54	1.04	1.64	1.42	1.59	1.37	1.25	1.57
Sénégal	0.12	0.13	0.14	0.15	0.17	0.17	0.16	0.17	0.17
Afrique du Sud	13.09	15.39	19.57	19.62	20.08	20.73	20.59	21.18	22.09
Soudan	0.20	0.52	0.21	0.26	0.22	0.22	0.25	0.16	0.19
Tanzanie	0.09	0.12	0.11	0.11	0.11	0.13	0.13	0.12	0.13
Tunisie	0.29	0.34	0.52	0.81	0.83	0.89	0.91	0.84	0.91
Zambie	0.89	1.10	0.99	0.96	0.87	0.83	0.81	0.87	0.85
Zimbabwe	0.80	1.00	0.90	1.26	1.16	1.20	1.20	1.23	1.21
Autre Afrique	0.06	0.08	0.33	0.44	0.49	0.54	0.52	0.54	0.62
Afrique	21.47	25.28	33.66	39.44	40.31	42.22	41.73	43.97	44.97
Argentine	6.64	6.75	7.60	8.12	8.71	8.41	7.83	8.97	9.08
Bolivie	0.14	0.16	0.21	0.25	0.21	0.19	0.17	0.15	0.17
Brésil	9.05	12.99	21.68	20.70	22.04	23.78	25.12	27.06	28.37
Chili	2.10	2.10	2.20	2.06	2.26	2.29	2.43	2.48	2.98
Colombie	2.62	2.74	3.60	3.77	3.85	3.90	4.12	4.41	4.58
Costa Rica	0.13	0.16	0.22	0.19	0.21	0.24	0.23	0.24	0.21
Cuba	1.97	2.41	2.90	2.98	2.96	2.77	2.68	2.85	2.93
République dominicaine	0.23	0.34	0.45	0.46	0.46	0.27	0.34	0.38	0.45
Equateur	0.17	0.16	0.40	0.77	0.78	0.91	0.90	0.91	0.83
El Salvador	0.13	0.17	0.22	0.15	0.16	0.15	0.16	0.17	0.20
Guatemala	0.30	0.25	0.32	0.18	0.18	0.18	0.21	0.23	0.26
Haiti	0.05	0.05	0.09	0.08	0.09	0.09	0.07	0.07	0.08
Honduras	0.11	0.13	0.19	0.20	0.24	0.22	0.21	0.25	0.29
Jamaïque	0.71	1.28	1.22	0.87	0.84	0.67	0.57	0.59	0.56
Antilles néerlandaises	1.46	1.60	0.54	0.40	0.37	0.36	0.34	0.32	0.27
Nicaragua	0.10	0.13	0.13	0.14	0.14	0.14	0.14	0.14	0.14
Panama	0.05	0.06	0.10	0.15	0.17	0.24	0.25	0.25	0.17
Paraguay	0.02	0.05	0.03	0.05	0.04	0.04	0.04	0.06	0.09
Pérou	1.07	1.38	1.81	1.66	1.75	1.87	1.96	1.78	1.58
Trinité-et-Tobago	0.52	0.59	0.95	1.86	1.79	1.77	1.92	1.97	2.63
Uruguay	0.47	0.46	0.50	0.31	0.29	0.25	0.28	0.29	0.29
Vénézuela	2.98	5.01	6.60	9.09	10.39	11.04	10.92	12.52	10.42
Autre Amérique latine	0.03	0.03	0.04	0.06	0.06	0.07	0.08	0.20	0.20
Amérique latine	31.05	39.00	52.00	54.49	58.00	59.84	60.96	66.29	66.77

* Note: à cause de données manquantes avant 1994, les combustibles renouvelables et les déchets ne sont pas inclus.

Total Industry Consumption of Energy* (Mtoe)
Consommation industrielle totale d'énergie (Mtep)
Gesamtindustrieverbrauch von Energie (Mtoe)
Consumo totale di energia nell'industria (Mtep)
産業用エネルギー総消費量（石油換算百万トン）
Consumo total industrial de energía (Mtep)
Общее потребление топлива и энергии промышленным сектором (Мтнэ)

1989	1990	1991	1992	1993	1994	1995	1996	1997	
2.10	1.85	1.95	2.01	3.49	3.17	3.13	3.03	3.31	Algeria
0.16	0.46	0.50	0.49	0.48	0.45	0.48	0.48	0.50	Angola
0.02	0.02	0.02	0.02	0.02	0.02	0.02	0.03	0.03	Benin
0.17	0.17	0.16	0.16	0.16	0.15	0.16	0.16	0.17	Cameroon
0.02	0.03	0.03	0.03	0.03	0.03	0.03	0.03	0.03	Congo
0.43	0.43	0.46	0.46	0.52	0.53	0.48	0.49	0.47	DR of Congo
10.80	10.95	10.43	11.02	10.12	8.92	10.32	10.70	11.51	Egypt
0.19	0.23	0.24	0.24	0.21	0.24	0.27	0.28	0.17	Ethiopia
0.12	0.08	0.09	0.13	0.14	0.13	0.14	0.15	0.17	Gabon
0.41	0.39	0.41	0.42	0.43	0.41	0.40	0.39	0.40	Ghana
0.25	0.25	0.21	0.18	0.22	0.23	0.26	0.24	0.25	Ivory Coast
0.53	0.57	0.57	0.60	0.64	0.62	0.59	0.65	0.63	Kenya
1.94	2.19	2.42	2.32	2.45	2.60	2.76	2.88	3.15	Libya
1.48	1.35	1.43	1.54	1.22	1.39	1.41	1.54	1.62	Morocco
0.04	0.04	0.05	0.06	0.07	0.07	0.07	0.04	0.05	Mozambique
1.71	1.58	1.50	1.49	1.54	1.66	1.48	1.64	1.64	Nigeria
0.11	0.12	0.12	0.13	0.12	0.15	0.19	0.19	0.20	Senegal
21.49	21.43	20.13	18.44	15.03	15.43	16.27	17.93	18.93	South Africa
0.21	0.25	0.30	0.24	0.05	0.30	0.19	0.11	0.26	Sudan
0.13	0.14	0.15	0.14	0.14	0.14	0.14	0.14	0.14	Tanzania
1.13	1.24	1.17	1.28	1.18	1.14	1.16	1.22	1.21	Tunisia
0.61	0.53	0.61	0.61	0.58	0.53	0.54	0.54	0.55	Zambia
1.24	1.53	1.46	1.42	1.25	0.82	0.92	0.94	0.91	Zimbabwe
0.61	0.61	0.47	0.56	0.51	0.41	0.35	0.33	0.33	Other Africa
45.92	46.45	44.87	43.99	40.59	39.54	41.75	44.13	46.62	**Africa**
7.99	7.78	8.14	8.33	8.39	9.09	9.36	9.74	10.12	Argentina
0.25	0.29	0.32	0.35	0.40	0.43	0.47	0.44	0.53	Bolivia
28.93	28.00	28.73	29.61	30.89	32.51	33.60	35.65	38.53	Brazil
3.68	3.64	3.57	4.19	4.39	4.45	4.92	5.35	7.09	Chile
4.72	5.10	5.19	5.14	4.98	5.21	5.25	5.46	5.57	Colombia
0.26	0.28	0.27	0.25	0.27	0.30	0.30	0.28	0.31	Costa Rica
3.21	3.45	2.43	2.05	2.10	2.39	2.39	2.60	2.73	Cuba
0.35	0.31	0.30	0.34	0.40	0.42	0.44	0.46	0.50	Dominican Republic
0.85	1.04	1.05	0.81	0.75	0.81	0.87	0.88	0.97	Ecuador
0.20	0.21	0.23	0.28	0.29	0.31	0.37	0.35	0.36	El Salvador
0.27	0.30	0.31	0.34	0.41	0.42	0.49	0.49	0.52	Guatemala
0.08	0.07	0.08	0.08	0.05	0.01	0.05	0.06	0.09	Haiti
0.32	0.32	0.31	0.33	0.33	0.34	0.42	0.54	0.55	Honduras
0.97	0.20	0.19	0.22	0.42	0.48	0.50	0.52	0.54	Jamaica
0.23	0.20	0.20	0.20	0.17	0.17	0.17	0.17	0.17	Netherlands Antilles
0.12	0.13	0.11	0.10	0.10	0.12	0.11	0.12	0.10	Nicaragua
0.18	0.18	0.23	0.25	0.26	0.28	0.30	0.24	0.27	Panama
0.09	0.10	0.09	0.13	0.13	0.17	0.16	0.11	0.16	Paraguay
1.43	1.18	1.29	1.13	1.32	1.38	1.72	2.00	2.25	Peru
2.68	2.77	2.94	3.01	2.92	3.86	3.91	4.42	4.77	Trinidad-and-Tobago
0.29	0.29	0.30	0.30	0.29	0.27	0.26	0.28	0.31	Uruguay
10.74	11.77	13.15	11.94	13.52	14.06	14.76	14.68	15.78	Venezuela
0.20	0.18	0.18	0.17	0.17	0.17	0.12	0.12	0.13	Other Latin America
68.03	67.77	69.60	69.56	72.93	77.66	80.96	84.95	92.35	**Latin America**

* Note: due to missing data prior to 1994 combustible renewables and waste are not included.

Total Industry Consumption of Energy* (Mtoe)
Consommation industrielle totale d'énergie (Mtep)
Gesamtindustrieverbrauch von Energie (Mtoe)
Consumo totale di energia nell'industria (Mtep)
産業用エネルギー総消費量（石油換算百万トン）
Consumo total industrial de energía (Mtep)
Общее потребление топлива и энергии промышленным сектором (Мтнэ)

	1971	1973	1978	1983	1984	1985	1986	1987	1988
Bangladesh	0.50	0.74	0.91	1.29	1.33	1.33	1.50	1.65	1.80
Brunei	0.03	0.03	0.03	0.06	0.05	0.05	0.05	0.05	0.06
Inde	18.57	22.80	29.31	37.23	39.77	49.15	49.32	49.90	57.02
Indonésie	1.65	1.89	5.00	8.72	9.23	7.87	7.54	7.68	8.36
RPD de Corée	20.60	17.58	20.17	22.74	22.10	22.73	23.04	23.05	23.43
Malaisie	2.08	1.73	2.27	3.25	3.24	3.60	3.78	4.27	4.70
Myanmar	0.51	0.34	0.62	0.66	0.70	0.78	0.77	0.58	0.54
Népal	0.01	0.05	0.01	0.04	0.07	0.06	0.06	0.06	0.06
Pakistan	2.28	2.74	3.39	5.22	5.49	5.84	6.04	6.90	7.29
Philippines	1.84	2.38	3.13	3.17	2.73	2.63	2.30	2.84	3.18
Singapour	0.16	0.22	0.47	0.51	0.93	1.57	1.52	1.86	1.91
Sri Lanka	0.13	0.21	0.22	0.27	0.26	0.17	0.17	0.16	0.18
Taipei chinois	4.39	5.54	9.72	11.27	11.98	12.76	14.10	14.94	16.48
Thaïlande	1.57	1.89	2.33	2.50	2.75	2.92	2.99	3.45	3.81
Viêt-Nam	0.00	-	0.00	1.75	1.94	1.85	2.01	2.19	1.75
Autre Asie	0.14	0.05	0.22	0.45	0.48	0.54	0.53	0.52	0.52
Asie	54.48	58.21	77.79	99.11	103.06	113.86	115.71	120.10	131.10
Rép. populaire de Chine	11.17	14.39	33.96	197.62	216.05	233.32	251.03	266.22	276.00
Hong-Kong, Chine	0.97	0.89	1.78	1.12	1.31	1.20	1.29	1.43	1.59
Chine	12.14	15.28	35.74	198.74	217.36	234.52	252.33	267.64	277.60
Albanie	0.37	0.24	0.31	0.64	0.65	0.63	0.75	0.80	0.70
Bulgarie	3.89	4.10	4.95	5.87	5.70	9.68	9.61	9.99	8.18
Chypre	0.14	0.18	0.23	0.23	0.23	0.21	0.26	0.32	0.34
Malte	-	-	-	-	-	-	-	-	-
Roumanie	17.21	16.20	32.11	34.58	35.09	26.93	28.01	28.63	29.77
République slovaque	4.55	4.74	6.33	8.20	8.25	8.36	8.27	8.33	8.24
Bosnie-Herzégovine	-	-	-	-	-	-	-	-	-
Croatie	-	-	-	-	-	-	-	-	-
Ex-RYM	-	-	-	-	-	-	-	-	-
Slovénie	-	-	-	-	-	-	-	-	-
RF de Yougoslavie	-	-	-	-	-	-	-	-	-
Ex-Yougoslavie	7.43	8.44	9.25	10.73	11.44	11.35	12.21	11.66	10.82
Europe non-OCDE	33.59	33.90	53.17	60.25	61.37	57.17	59.13	59.73	58.05
Arménie	-	-	-	-	-	-	-	-	-
Azerbaïdjan	-	-	-	-	-	-	-	-	-
Bélarus	-	-	-	-	-	-	-	-	-
Estonie	-	-	-	-	-	-	-	-	-
Géorgie	-	-	-	-	-	-	-	-	-
Kazakhstan	-	-	-	-	-	-	-	-	-
Kirghizistan	-	-	-	-	-	-	-	-	-
Lettonie	-	-	-	-	-	-	-	-	-
Lituanie	-	-	-	-	-	-	-	-	-
Moldova	-	-	-	-	-	-	-	-	-
Russie	-	-	-	-	-	-	-	-	-
Tadjikistan	-	-	-	-	-	-	-	-	-
Turkménistan	-	-	-	-	-	-	-	-	-
Ukraine	-	-	-	-	-	-	-	-	-
Ouzbékistan	-	-	-	-	-	-	-	-	-
Ex-URSS	244.14	279.14	359.61	363.28	367.91	368.28	386.76	387.71	399.74

* Note: à cause de données manquantes avant 1994, les combustibles renouvelables et les déchets ne sont pas inclus.
Rép.populaire de Chine: jusqu'en 1978 la ventilation de la consommation finale par secteur est incomplète.
Ex-URSS: les séries antérieures à 1990-1992 ne sont pas comparables aux années récentes; 1991 a été estimé.

Total Industry Consumption of Energy* (Mtoe)
Consommation industrielle totale d'énergie (Mtep)
Gesamtindustrieverbrauch von Energie (Mtoe)

Consumo totale di energia nell'industria (Mtep)

産業用エネルギー総消費量（石油換算百万トン）

Consumo total industrial de energía (Mtep)

Общее потребление топлива и энергии промышленным сектором (Мтнэ)

1989	1990	1991	1992	1993	1994	1995	1996	1997	
1.94	2.04	1.82	2.32	2.52	2.84	3.26	3.35	3.24	Bangladesh
0.06	0.07	0.07	0.08	0.08	0.09	0.09	0.10	0.11	Brunei
60.69	64.95	67.92	70.54	70.80	76.72	77.57	79.29	83.26	India
8.35	14.36	14.48	15.79	16.02	16.80	17.63	20.94	23.95	Indonesia
23.92	22.23	18.19	18.23	16.36	14.60	14.21	13.96	13.55	DPR of Korea
5.21	5.91	6.21	6.80	8.56	8.68	9.01	10.26	11.01	Malaysia
0.53	0.47	0.41	0.49	0.58	0.66	0.75	0.79	0.90	Myanmar
0.05	0.03	0.06	0.08	0.09	0.11	0.20	0.20	0.20	Nepal
7.61	7.94	7.99	8.14	9.07	9.67	9.60	10.54	9.91	Pakistan
3.29	3.17	3.35	3.31	3.66	4.00	4.72	5.18	5.66	Philippines
2.05	2.36	2.35	2.16	2.35	2.43	2.87	3.23	3.88	Singapore
0.20	0.20	0.20	0.28	0.31	0.32	0.31	0.58	0.56	Sri Lanka
16.30	16.81	17.37	18.25	19.16	20.88	21.53	22.36	23.95	Chinese Taipei
4.74	5.75	6.43	7.32	8.68	10.86	12.33	14.54	13.40	Thailand
1.68	1.99	2.17	2.87	2.74	2.88	3.30	6.04	6.89	Vietnam
0.53	0.51	0.50	0.44	0.44	0.44	0.28	0.27	0.28	Other Asia
137.16	148.78	149.53	157.09	161.42	171.97	177.66	191.62	200.74	**Asia**
287.22	283.63	289.54	296.27	313.44	334.01	353.42	363.51	343.89	People's Rep.of China
1.62	1.26	0.90	1.28	0.92	0.88	0.80	2.20	2.96	Hong Kong, China
288.84	284.89	290.43	297.55	314.36	334.89	354.22	365.71	346.85	**China**
1.01	1.07	0.68	0.41	0.42	0.28	0.11	0.13	0.10	Albania
7.08	10.03	7.02	6.31	5.98	6.18	6.91	7.06	6.49	Bulgaria
0.32	0.23	0.39	0.36	0.37	0.38	0.39	0.43	0.40	Cyprus
-	-	0.02	0.02	0.02	0.04	0.07	0.04	0.04	Malta
30.07	24.21	19.79	15.19	13.59	13.71	14.23	14.52	12.77	Romania
9.41	8.76	7.50	7.26	6.25	5.81	5.91	6.23	6.07	Slovak Republic
-	-	-	0.28	0.32	0.06	0.06	0.06	0.06	Bosnia-Herzegovina
-	-	-	2.09	1.99	2.07	1.99	1.89	1.96	Croatia
-	-	-	0.80	0.83	0.55	0.50	0.61	0.54	FYROM
-	-	-	1.08	1.04	1.16	1.18	1.21	1.24	Slovenia
-	-	-	2.29	2.41	2.16	2.32	2.84	3.25	FR of Yugoslavia
11.31	9.23	7.76	6.53	6.59	6.00	6.05	6.62	7.05	Former Yugoslavia
59.21	53.53	43.15	36.08	33.22	32.41	33.67	35.03	32.93	**Non-OECD Europe**
-	-	-	0.80	0.49	0.24	0.35	0.21	0.21	Armenia
-	-	-	0.38	4.53	5.23	4.20	2.81	2.78	Azerbaijan
-	-	-	9.99	9.96	7.42	7.10	7.01	7.19	Belarus
-	-	-	1.25	0.96	0.97	0.91	0.93	0.86	Estonia
-	-	-	0.47	1.96	0.87	0.64	0.39	0.49	Georgia
-	-	-	17.35	15.04	12.65	10.93	8.94	7.70	Kazakhstan
-	-	-	1.13	0.96	1.04	0.40	0.51	0.86	Kyrgyzstan
-	-	-	1.31	0.66	0.61	0.55	0.77	0.90	Latvia
-	-	-	3.36	1.67	1.43	1.54	1.54	1.46	Lithuania
-	-	-	0.89	0.81	0.65	0.58	0.62	0.64	Moldova
-	-	-	163.82	226.65	179.08	171.23	139.85	134.31	Russia
-	-	-	0.81	0.61	0.58	0.59	0.58	0.56	Tajikistan
-	-	-	0.17	0.19	0.16	0.16	0.15	0.14	Turkmenistan
-	-	-	72.01	65.02	45.42	43.24	40.39	37.59	Ukraine
-	-	-	1.94	1.56	1.52	8.18	6.95	5.83	Uzbekistan
399.20	400.98	345.09	274.63	329.56	256.96	249.80	211.05	200.92	**Former USSR**

* Note: due to missing data prior to 1994 combustible renewables and waste are not included.

People's Rep. of China: up to 1978, the breakdown of final consumption by sector is incomplete.

Former USSR: series up to 1990-1992 are not comparable with the recent years; 1991 is estimated.

Total Industry Consumption of Energy* (Mtoe)
Consommation industrielle totale d'énergie (Mtep)
Gesamtindustrieverbrauch von Energie (Mtoe)
Consumo totale di energia nell'industria (Mtep)
産業用エネルギー総消費量（石油換算百万トン）
Consumo total industrial de energía (Mtep)
Общее потребление топлива и энергии промышленным сектором (Мтнэ)

	1971	1973	1978	1983	1984	1985	1986	1987	1988
Bahrein	0.19	0.60	0.95	1.12	1.18	1.69	2.21	1.98	2.28
Iran	9.86	10.94	12.61	10.19	15.24	16.60	19.25	18.15	15.72
Irak	0.91	1.12	1.77	2.72	3.32	3.45	4.20	4.69	6.42
Israël	1.08	1.17	1.53	1.65	1.70	1.69	1.90	2.06	2.09
Jordanie	0.05	0.08	0.13	0.41	0.39	0.45	0.50	0.50	0.46
Koweit	1.96	2.02	2.68	2.95	1.94	3.06	3.48	3.76	3.04
Liban	0.45	0.48	0.42	0.04	0.03	0.05	0.05	0.09	0.06
Oman	0.01	0.01	0.02	0.03	0.24	0.34	0.18	0.22	0.24
Qatar	0.38	0.60	0.46	1.59	2.03	1.86	1.87	1.89	2.04
Arabie saoudite	1.03	1.60	3.92	15.70	7.70	9.93	10.85	7.91	7.09
Syrie	0.58	0.32	1.72	2.25	2.32	2.41	2.36	2.39	2.09
Emirats arabes unis	-	-	-	0.44	0.63	0.96	0.89	0.97	1.01
Yémen	0.00	0.01	0.11	0.21	0.22	0.37	0.35	0.31	0.30
Moyen-Orient	16.50	18.96	26.31	39.30	36.94	42.87	48.09	44.93	42.84
Total non-OCDE	413.36	469.78	638.29	854.62	884.94	918.75	964.70	990.37	1 021.06
OCDE Amérique du N.	424.82	465.55	447.34	396.40	431.65	420.76	408.47	422.40	438.88
OCDE Pacifique	134.32	154.84	148.00	130.69	139.56	141.13	141.20	146.03	154.93
OCDE Europe	369.03	409.55	408.77	366.13	373.69	376.37	370.54	380.60	382.02
Total OCDE	928.17	1 029.94	1 004.12	893.22	944.91	938.26	920.21	949.02	975.84
Monde	1 341.53	1 499.71	1 642.41	1 747.84	1 829.84	1 857.01	1 884.91	1 939.39	1 996.90

* Note: à cause de données manquantes avant 1994, les combustibles renouvelables et les déchets ne sont pas inclus.

Total Industry Consumption of Energy* (Mtoe)
Consommation industrielle totale d'énergie (Mtep)
Gesamtindustrieverbrauch von Energie (Mtoe)
Consumo totale di energia nell'industria (Mtep)
産業用エネルギー総消費量（石油換算百万トン）
Consumo total industrial de energía (Mtep)
Общее потребление топлива и энергии промышленным сектором (Мтнэ)

1989	1990	1991	1992	1993	1994	1995	1996	1997	
2.23	2.35	2.23	2.26	2.69	2.50	2.65	2.81	2.70	Bahrain
15.08	7.53	10.99	14.30	20.42	20.39	22.20	25.28	27.05	Iran
8.02	4.33	3.46	3.66	4.50	5.63	5.51	5.50	5.94	Iraq
2.01	1.92	1.97	1.86	1.95	2.13	2.47	2.64	3.08	Israel
0.47	0.50	0.45	0.51	0.57	0.59	0.62	0.64	0.67	Jordan
4.13	3.22	0.05	2.09	1.34	1.30	1.58	1.46	1.75	Kuwait
0.07	0.10	0.10	0.07	0.59	0.74	0.95	0.91	1.14	Lebanon
0.26	1.10	2.26	2.50	2.54	2.15	2.07	1.29	1.31	Oman
2.73	2.03	2.76	4.59	4.90	4.87	4.84	4.91	5.43	Qatar
7.64	8.07	8.23	8.36	8.76	8.43	8.80	9.65	10.40	Saudi Arabia
0.99	0.99	1.46	1.69	1.74	1.96	2.23	2.31	2.23	Syria
1.12	1.13	1.43	1.69	1.87	1.95	1.90	2.15	2.25	United Arab Emirates
0.30	0.10	0.31	0.39	0.13	0.14	0.14	0.16	0.16	Yemen
45.04	33.39	35.68	43.98	52.01	52.80	55.98	59.70	64.12	**Middle East**
1 043.38	1 035.77	978.36	922.87	1 004.09	966.22	994.05	992.19	984.54	**Non-OECD Total**
425.83	425.57	419.33	407.48	416.25	427.93	432.57	441.37	444.53	OECD North America
162.84	170.97	178.20	182.73	187.78	196.59	201.63	208.28	216.85	OECD Pacific
382.08	366.89	353.17	346.42	338.26	341.25	352.60	360.95	364.04	OECD Europe
970.75	963.43	950.70	936.63	942.29	965.78	986.80	1 010.61	1 025.42	**OECD Total**
2 014.14	1 999.20	1 929.07	1 859.50	1 946.38	1 932.00	1 980.85	2 002.79	2 009.96	**World**

* Note: due to missing data prior to 1994 combustible renewables and waste are not included.

Consumption of Oil in Transport (Mtoe)
Consommation de pétrole dans les transports (Mtep)
Ölverbrauch im Verkehrssektor (Mtoe)
Consumo di petrolio nel settore dei trasporti (Mtep)
運輸部門の石油消費量（石油換算百万トン）
Consumo de petróleo en el transporte (Mtep)
Потребление нефти и нефтепродуктов в транспорте (Мтнэ)

	1971	1973	1978	1983	1984	1985	1986	1987	1988
Algérie	0.90	1.13	2.00	3.34	3.79	4.17	4.46	4.52	4.51
Angola	0.41	0.55	0.53	0.33	0.23	0.41	0.34	0.32	0.33
Bénin	0.09	0.11	0.10	0.10	0.11	0.14	0.13	0.11	0.11
Cameroun	0.24	0.28	0.41	0.55	0.58	0.63	0.60	0.61	0.67
Congo	0.12	0.13	0.18	0.23	0.21	0.20	0.20	0.22	0.20
RD du Congo	0.54	0.58	0.64	0.72	0.71	0.66	0.56	0.64	0.66
Egypte	1.38	1.54	2.21	3.63	4.06	4.39	4.33	3.98	4.03
Ethiopie	0.23	0.24	0.25	0.30	0.31	0.33	0.37	0.45	0.47
Gabon	0.01	0.02	0.20	0.17	0.11	0.14	0.20	0.15	0.17
Ghana	0.39	0.40	0.49	0.35	0.38	0.46	0.44	0.42	0.57
Côte d'Ivoire	0.34	0.36	0.55	0.51	0.49	0.49	0.49	0.50	0.54
Kenya	0.64	0.77	0.91	0.87	1.00	1.05	1.14	1.12	1.10
Libye	0.32	0.67	1.44	2.37	2.39	2.32	2.23	2.36	2.31
Maroc	0.70	0.83	1.06	1.10	1.07	1.10	1.13	1.18	0.62
Mozambique	0.15	0.14	0.12	0.10	0.10	0.07	0.08	0.08	0.08
Nigéria	0.97	1.40	4.20	5.71	5.29	5.04	4.72	4.64	5.06
Sénégal	0.25	0.29	0.34	0.40	0.40	0.39	0.38	0.33	0.32
Afrique du Sud	5.48	6.74	8.08	8.02	8.62	8.36	8.29	8.91	9.63
Soudan	0.78	0.93	0.74	0.92	0.82	0.97	0.92	0.73	1.11
Tanzanie	0.27	0.30	0.28	0.25	0.25	0.26	0.25	0.26	0.29
Tunisie	0.42	0.50	0.76	0.81	0.83	0.84	0.82	0.87	0.92
Zambie	0.22	0.27	0.27	0.24	0.25	0.25	0.24	0.24	0.26
Zimbabwe	0.36	0.47	0.47	0.58	0.55	0.59	0.61	0.56	0.64
Autre Afrique	0.67	0.82	1.21	1.57	1.58	1.62	1.77	2.31	2.38
Afrique	15.87	19.47	27.42	33.16	34.11	34.88	34.71	35.51	36.98
Argentine	8.01	8.76	9.31	10.55	10.03	9.05	9.97	10.43	9.18
Bolivie	0.31	0.38	0.66	0.57	0.56	0.56	0.62	0.69	0.69
Brésil	13.42	17.92	23.09	21.18	20.89	22.05	24.09	23.71	23.92
Chili	1.98	1.82	1.90	2.26	2.26	2.23	2.33	2.49	2.81
Colombie	2.57	2.94	4.27	5.08	5.17	5.32	5.37	5.56	6.04
Costa Rica	0.23	0.29	0.45	0.37	0.40	0.41	0.45	0.49	0.55
Cuba	1.51	1.91	2.19	2.10	2.15	2.86	2.84	2.94	3.06
République dominicaine	0.51	0.56	0.64	0.67	0.75	0.73	0.80	0.99	0.89
Equateur	0.71	0.83	1.51	1.73	1.93	2.05	2.13	2.06	2.48
El Salvador	0.20	0.24	0.38	0.30	0.32	0.35	0.34	0.37	0.41
Guatemala	0.28	0.39	0.48	0.42	0.45	0.48	0.46	0.54	0.58
Haiti	0.05	0.07	0.11	0.11	0.11	0.12	0.13	0.13	0.15
Honduras	0.15	0.18	0.21	0.26	0.26	0.28	0.30	0.32	0.35
Jamaïque	0.68	0.70	0.41	0.42	0.38	0.41	0.40	0.45	0.49
Antilles néerlandaises	0.54	0.51	0.60	0.49	0.47	0.41	0.34	0.35	0.37
Nicaragua	0.25	0.29	0.37	0.30	0.30	0.29	0.31	0.34	0.30
Panama	0.36	0.39	0.47	0.34	0.38	0.35	0.37	0.36	0.35
Paraguay	0.14	0.16	0.34	0.37	0.42	0.43	0.45	0.47	0.49
Pérou	1.94	2.29	2.26	2.43	2.44	2.28	2.42	2.73	2.75
Trinité-et-Tobago	0.32	0.42	0.47	0.62	0.64	0.64	0.67	0.63	0.51
Uruguay	0.57	0.59	0.58	0.49	0.45	0.44	0.44	0.46	0.47
Vénézuela	4.37	5.31	8.71	9.54	9.30	9.51	9.59	9.89	10.24
Autre Amérique latine	0.43	0.31	0.52	0.65	0.68	0.67	0.62	0.80	0.86
Amérique latine	39.55	47.26	59.94	61.25	60.76	61.91	65.46	67.23	67.93

Consumption of Oil in Transport (Mtoe)
Consommation de pétrole dans les transports (Mtep)
Ölverbrauch im Verkehrssektor (Mtoe)
Consumo di petrolio nel settore dei trasporti (Mtep)
運輸部門の石油消費量（石油換算百万トン）
Consumo de petróleo en el transporte (Mtep)
Потребление нефти и нефтепродуктов в транспорте (Мтнэ)

1989	1990	1991	1992	1993	1994	1995	1996	1997	
4.64	4.71	4.82	2.75	2.87	2.73	2.58	2.57	2.30	Algeria
0.31	0.31	0.31	0.32	0.31	0.31	0.31	0.39	0.63	Angola
0.09	0.07	0.06	0.05	0.06	0.06	0.06	0.27	0.27	Benin
0.66	0.63	0.57	0.57	0.60	0.65	0.62	0.63	0.64	Cameroon
0.23	0.21	0.22	0.20	0.20	0.20	0.21	0.21	0.21	Congo
0.67	0.68	0.69	0.63	0.71	0.73	0.74	0.74	0.74	DR of Congo
3.84	3.97	3.94	3.93	3.97	4.19	4.53	4.96	5.22	Egypt
0.49	0.48	0.49	0.50	0.55	0.57	0.61	0.73	0.47	Ethiopia
0.23	0.19	0.19	0.18	0.19	0.19	0.20	0.22	0.29	Gabon
0.58	0.57	0.51	0.57	0.63	0.67	0.70	0.70	0.70	Ghana
0.52	0.46	0.47	0.45	0.48	0.49	0.51	0.57	0.58	Ivory Coast
1.18	1.18	1.16	1.18	1.16	1.24	1.22	1.32	1.29	Kenya
2.38	2.28	2.31	2.35	2.73	3.26	3.49	3.52	3.54	Libya
0.63	0.67	0.61	0.65	0.82	0.80	0.80	0.81	0.82	Morocco
0.08	0.11	0.10	0.09	0.10	0.11	0.08	0.09	0.10	Mozambique
5.10	4.29	5.56	6.70	5.81	4.94	4.80	5.07	5.15	Nigeria
0.39	0.39	0.38	0.40	0.36	0.38	0.41	0.49	0.52	Senegal
9.98	10.28	10.33	10.57	10.61	11.02	12.39	12.33	12.75	South Africa
0.99	1.33	1.07	1.07	0.72	1.03	0.97	0.92	1.03	Sudan
0.29	0.31	0.29	0.30	0.29	0.29	0.28	0.27	0.22	Tanzania
0.96	1.01	0.99	1.07	1.20	1.27	1.32	1.37	1.47	Tunisia
0.25	0.23	0.24	0.21	0.22	0.23	0.23	0.23	0.23	Zambia
0.63	0.61	0.51	0.66	0.59	0.66	0.81	0.81	0.81	Zimbabwe
2.51	2.51	2.39	2.48	2.72	2.65	2.47	2.48	2.58	Other Africa
37.60	37.50	38.20	37.87	37.88	38.65	40.36	41.70	42.56	**Africa**
8.71	8.65	9.64	10.10	11.12	11.77	12.05	12.64	12.90	Argentina
0.73	0.76	0.78	0.77	0.80	0.83	0.92	1.01	1.02	Bolivia
24.91	25.71	26.99	27.19	28.26	29.55	32.68	35.66	38.33	Brazil
3.02	3.14	3.36	3.60	3.94	4.34	4.74	5.17	5.51	Chile
6.14	6.22	6.46	6.86	6.93	6.60	7.78	7.94	8.10	Colombia
0.59	0.57	0.59	0.88	0.92	0.98	0.99	0.98	1.02	Costa Rica
3.19	1.85	1.35	0.95	0.78	0.73	0.78	0.93	0.99	Cuba
1.09	0.94	0.93	1.10	0.99	1.03	1.06	1.10	1.15	Dominican Republic
2.40	2.28	2.26	2.26	2.26	2.34	2.45	2.81	2.96	Ecuador
0.42	0.46	0.48	0.57	0.63	0.66	0.75	0.73	0.78	El Salvador
0.58	0.63	0.65	0.69	0.76	0.81	1.00	1.01	1.08	Guatemala
0.16	0.17	0.15	0.15	0.14	0.04	0.19	0.21	0.21	Haiti
0.38	0.35	0.36	0.40	0.39	0.53	0.63	0.53	0.54	Honduras
0.49	0.50	0.49	0.48	0.54	0.64	0.73	0.80	0.86	Jamaica
0.34	0.33	0.38	0.41	0.34	0.43	0.44	0.44	0.44	Netherlands Antilles
0.26	0.26	0.29	0.30	0.30	0.35	0.37	0.39	0.44	Nicaragua
0.36	0.40	0.39	0.42	0.44	0.48	0.51	0.59	0.63	Panama
0.55	0.54	0.53	0.63	0.74	0.86	0.98	0.92	0.99	Paraguay
2.43	2.62	2.67	2.75	2.94	2.74	3.23	3.34	3.31	Peru
0.49	0.52	0.55	0.54	0.47	0.48	0.51	0.53	0.60	Trinidad-and-Tobago
0.50	0.50	0.53	0.57	0.64	0.73	0.72	0.77	0.83	Uruguay
9.70	10.00	10.41	10.46	10.91	11.32	11.93	11.24	12.57	Venezuela
0.88	0.88	0.88	0.85	0.84	0.92	0.93	0.93	0.94	Other Latin America
68.31	68.28	71.14	72.95	76.07	79.16	86.39	90.67	96.20	**Latin America**

Consumption of Oil in Transport (Mtoe)
Consommation de pétrole dans les transports (Mtep)
Ölverbrauch im Verkehrssektor (Mtoe)
Consumo di petrolio nel settore dei trasporti (Mtep)
運輸部門の石油消費量（石油換算百万トン）
Consumo de petróleo en el transporte (Mtep)
Потребление нефти и нефтепродуктов в транспорте (Мтнэ)

	1971	1973	1978	1983	1984	1985	1986	1987	1988
Bangladesh	0.10	0.10	0.25	0.41	0.42	0.51	0.51	0.54	0.58
Brunei	0.03	0.04	0.08	0.16	0.16	0.17	0.18	0.19	0.21
Inde	6.84	7.62	10.72	14.89	15.77	17.27	18.74	19.59	20.65
Indonésie	2.72	3.13	5.14	6.90	7.03	7.22	7.67	8.43	9.14
RPD de Corée	0.59	0.67	1.38	2.00	2.10	2.21	2.31	2.33	2.33
Malaisie	1.50	1.74	1.97	3.29	3.40	3.58	3.82	4.05	4.35
Myanmar	0.52	0.44	0.56	0.62	0.68	0.64	0.70	0.42	0.42
Népal	0.02	0.03	0.04	0.06	0.07	0.08	0.08	0.11	0.13
Pakistan	1.42	1.43	2.15	3.24	3.51	3.68	3.89	4.38	4.68
Philippines	2.40	2.49	2.39	1.90	1.82	1.74	1.84	2.08	2.28
Singapour	0.74	0.93	1.68	2.19	2.26	2.20	2.36	2.48	2.54
Sri Lanka	0.45	0.54	0.60	0.68	0.77	0.80	0.80	0.82	0.83
Taipei chinois	1.20	2.05	2.97	4.02	4.26	4.49	4.85	5.39	6.02
Thaïlande	2.39	3.23	3.89	4.95	5.68	5.88	6.27	7.20	8.36
Viêt-Nam	3.38	3.12	0.20	0.55	0.53	0.91	1.04	1.19	1.31
Autre Asie	0.59	0.58	0.73	0.90	1.00	1.01	0.91	1.04	0.99
Asie	24.89	28.14	34.75	46.77	49.47	52.39	55.98	60.25	64.81
Rép. populaire de Chine	4.78	6.54	10.23	19.53	21.39	24.19	27.11	29.38	31.89
Hong-Kong, Chine	0.90	1.06	1.30	2.17	2.17	2.06	2.54	2.52	3.10
Chine	5.68	7.60	11.53	21.71	23.56	26.25	29.65	31.91	35.00
Albanie	0.24	0.26	0.38	0.31	0.45	0.36	0.34	0.35	0.39
Bulgarie	1.66	1.63	1.73	1.71	1.70	1.69	1.68	1.70	2.79
Chypre	0.26	0.33	0.26	0.35	0.40	0.42	0.44	0.49	0.51
Gibraltar	0.02	0.02	0.01	0.01	0.01	0.02	0.02	0.03	0.04
Malte	0.13	0.15	0.13	0.15	0.19	0.10	0.22	0.23	0.23
Roumanie	2.17	2.45	2.78	1.85	1.77	1.50	1.40	1.95	1.82
République slovaque	1.43	1.65	1.56	0.91	0.92	0.91	0.93	0.96	1.01
Bosnie-Herzégovine	-	-	-	-	-	-	-	-	-
Croatie	-	-	-	-	-	-	-	-	-
Ex-RYM	-	-	-	-	-	-	-	-	-
Slovénie	-	-	-	-	-	-	-	-	-
RF de Yougoslavie	-	-	-	-	-	-	-	-	-
Ex-Yougoslavie	3.74	3.82	5.05	4.37	4.19	4.73	5.83	5.22	5.75
Europe non-OCDE	9.65	10.31	11.90	9.68	9.64	9.73	10.87	10.93	12.53

Rép.populaire de Chine: jusqu'en 1978 la ventilation de la consommation finale par secteur est incomplète.

Consumption of Oil in Transport (Mtoe)
Consommation de pétrole dans les transports (Mtep)
Ölverbrauch im Verkehrssektor (Mtoe)
Consumo di petrolio nel settore dei trasporti (Mtep)
運輸部門の石油消費量（石油換算百万トン）
Consumo de petróleo en el transporte (Mtep)
Потребление нефти и нефтепродуктов в транспорте (Мтнэ)

1989	1990	1991	1992	1993	1994	1995	1996	1997	
0.62	0.64	0.68	0.77	0.82	0.88	1.07	1.12	1.13	Bangladesh
0.21	0.23	0.24	0.27	0.29	0.31	0.33	0.36	0.39	Brunei
22.87	23.59	26.81	28.37	30.08	32.34	36.66	39.42	40.91	India
9.95	11.38	12.21	12.95	13.82	15.50	16.97	18.87	19.87	Indonesia
2.49	2.44	2.24	2.17	2.04	1.32	1.16	1.02	0.99	DPR of Korea
4.67	5.41	5.93	6.27	7.20	7.97	7.82	8.79	9.16	Malaysia
0.41	0.44	0.40	0.35	0.42	0.50	0.63	0.68	0.78	Myanmar
0.10	0.11	0.14	0.17	0.18	0.18	0.20	0.20	0.20	Nepal
4.84	4.92	5.13	5.75	6.52	6.84	7.09	7.66	7.82	Pakistan
2.57	2.66	2.44	2.81	3.16	3.34	3.66	4.14	4.64	Philippines
2.76	3.24	3.22	3.29	3.77	4.39	4.50	4.59	4.96	Singapore
0.79	0.82	0.82	0.90	0.96	1.13	1.20	1.29	1.33	Sri Lanka
6.87	7.50	7.94	9.22	9.87	10.65	11.28	11.71	11.98	Chinese Taipei
9.91	11.11	11.39	12.39	14.20	15.83	18.32	19.89	21.19	Thailand
1.23	1.46	1.23	1.20	1.64	1.76	1.85	3.89	4.91	Vietnam
0.96	0.99	0.95	0.89	0.88	0.79	0.69	0.70	0.70	Other Asia
71.25	76.91	81.79	87.76	95.83	103.73	113.43	124.33	130.97	**Asia**
33.14	33.89	38.66	41.46	49.54	45.14	52.51	56.31	63.45	People's Rep.of China
3.26	3.86	4.09	4.97	5.62	6.31	6.36	6.26	6.51	Hong Kong, China
36.40	37.75	42.75	46.43	55.15	51.46	58.87	62.57	69.96	**China**
0.32	0.28	0.24	0.18	0.22	0.25	0.29	0.30	0.30	Albania
2.96	2.52	1.43	1.45	1.70	1.57	1.62	1.50	1.09	Bulgaria
0.57	0.64	0.69	0.73	0.69	0.72	0.77	0.77	0.79	Cyprus
0.03	0.04	0.05	0.06	0.06	0.07	0.07	0.07	0.07	Gibraltar
0.23	0.23	0.26	0.26	0.28	0.28	0.30	0.35	0.36	Malta
3.80	4.12	3.60	3.46	3.02	3.15	2.93	3.93	4.04	Romania
1.00	0.98	0.82	0.86	0.89	1.10	1.25	1.17	1.18	Slovak Republic
-	-	-	0.37	0.25	0.27	0.27	0.27	0.27	*Bosnia-Herzegovina*
-	-	-	0.97	1.05	1.14	1.20	1.25	1.41	*Croatia*
-	-	-	0.27	0.50	0.34	0.34	0.59	0.53	*FYROM*
-	-	-	0.89	1.08	1.20	1.34	1.51	1.53	*Slovenia*
-	-	-	0.61	0.39	0.37	0.41	0.76	1.15	*FR of Yugoslavia*
5.42	5.44	4.87	3.11	3.26	3.32	3.54	4.39	4.89	Former Yugoslavia
14.34	14.23	11.97	10.12	10.12	10.47	10.77	12.49	12.72	**Non-OECD Europe**

People's Rep. of China: up to 1978, the breakdown of final consumption by sector is incomplete.

Consumption of Oil in Transport (Mtoe)
Consommation de pétrole dans les transports (Mtep)
Ölverbrauch im Verkehrssektor (Mtoe)
Consumo di petrolio nel settore dei trasporti (Mtep)
運輸部門の石油消費量（石油換算百万トン）
Consumo de petróleo en el transporte (Mtep)
Потребление нефти и нефтепродуктов в транспорте (Мтнэ)

	1971	1973	1978	1983	1984	1985	1986	1987	1988
Arménie	-	-	-	-	-	-	-	-	-
Azerbaïdjan	-	-	-	-	-	-	-	-	-
Bélarus	-	-	-	-	-	-	-	-	-
Estonie	-	-	-	-	-	-	-	-	-
Géorgie	-	-	-	-	-	-	-	-	-
Kazakhstan	-	-	-	-	-	-	-	-	-
Kirghizistan	-	-	-	-	-	-	-	-	-
Lettonie	-	-	-	-	-	-	-	-	-
Lituanie	-	-	-	-	-	-	-	-	-
Moldova	-	-	-	-	-	-	-	-	-
Russie	-	-	-	-	-	-	-	-	-
Tadjikistan	-	-	-	-	-	-	-	-	-
Turkménistan	-	-	-	-	-	-	-	-	-
Ukraine	-	-	-	-	-	-	-	-	-
Ouzbékistan	-	-	-	-	-	-	-	-	-
Ex-URSS	76.74	85.34	105.44	117.03	118.59	119.64	124.03	126.46	128.07
Bahrein	0.20	0.29	0.63	0.68	0.70	0.70	0.69	0.70	0.73
Iran	1.72	2.88	5.30	5.09	5.68	6.16	11.01	11.90	11.66
Irak	1.09	1.35	3.39	6.35	6.77	7.23	7.60	7.83	8.06
Israël	1.67	1.99	2.13	2.34	2.24	2.36	2.31	2.53	2.55
Jordanie	0.21	0.27	0.64	1.01	0.99	1.00	1.11	1.13	1.15
Koweit	0.61	0.67	1.43	2.03	1.97	2.00	2.02	1.96	1.98
Liban	0.98	1.25	0.98	1.42	1.35	1.49	1.40	1.46	0.97
Oman	0.04	0.04	0.29	0.47	0.57	0.62	0.65	0.57	0.58
Qatar	0.08	0.13	0.36	0.54	0.52	0.50	0.51	0.52	0.53
Arabie saoudite	0.77	1.09	3.86	11.63	11.76	12.55	12.63	12.40	8.66
Syrie	0.81	1.02	2.01	2.49	2.70	2.72	2.77	3.02	3.00
Emirats arabes unis	0.20	0.41	2.34	3.47	3.49	3.66	3.81	3.86	3.52
Yémen	0.25	0.45	0.61	1.05	1.04	1.11	1.20	1.29	1.37
Moyen-Orient	8.64	11.85	23.99	38.57	39.78	42.09	47.72	49.16	44.76
Total non-OCDE	181.02	209.97	274.97	328.16	335.90	346.89	368.42	381.45	390.07
OCDE Amérique du N.	402.39	451.21	510.21	481.82	496.58	501.15	513.11	530.07	552.52
OCDE Pacifique	51.06	59.12	74.84	79.62	82.81	84.88	88.63	93.05	99.04
OCDE Europe	157.24	180.39	213.35	222.79	229.69	233.76	245.76	256.09	270.36
Total OCDE	610.68	690.72	798.40	784.23	809.08	819.79	847.50	879.21	921.93
Monde	791.70	900.70	1 073.37	1 112.40	1 144.97	1 166.67	1 215.92	1 260.66	1 312.00

Ex-URSS: les séries antérieures à 1990-1992 ne sont pas comparables aux années récentes; 1991 a été estimé.

Consumption of Oil in Transport (Mtoe)
Consommation de pétrole dans les transports (Mtep)
Ölverbrauch im Verkehrssektor (Mtoe)
Consumo di petrolio nel settore dei trasporti (Mtep)
運輸部門の石油消費量（石油換算百万トン）
Consumo de petróleo en el transporte (Mtep)
Потребление нефти и нефтепродуктов в транспорте (Мтнэ)

1989	1990	1991	1992	1993	1994	1995	1996	1997	
-	-	-	0.80	0.49	0.11	0.08	0.04	0.04	Armenia
-	-	-	1.57	1.66	1.81	1.56	1.36	1.02	Azerbaijan
-	-	-	4.83	3.60	2.56	1.99	1.83	1.89	Belarus
-	-	-	0.37	0.44	0.52	0.49	0.54	0.56	Estonia
-	-	-	0.43	0.26	0.08	0.05	0.54	0.51	Georgia
-	-	-	5.15	4.00	3.09	2.83	2.77	2.36	Kazakhstan
-	-	-	0.50	0.31	0.12	0.31	0.34	0.19	Kyrgyzstan
-	-	-	1.00	0.89	0.80	0.82	0.82	0.80	Latvia
-	-	-	1.11	1.11	1.00	1.18	1.25	1.25	Lithuania
-	-	-	0.78	0.49	0.39	0.40	0.38	0.40	Moldova
-	-	-	72.65	49.81	31.77	35.15	33.40	32.31	Russia
-	-	-	5.11	3.22	1.07	1.07	1.07	1.07	Tajikistan
-	-	-	0.76	0.56	0.53	0.48	0.52	0.47	Turkmenistan
-	-	-	10.79	8.86	7.70	7.70	6.52	5.90	Ukraine
-	-	-	2.34	2.13	2.12	1.65	2.13	2.27	Uzbekistan
130.50	122.21	148.31	108.20	77.82	53.68	55.76	53.51	51.03	**Former USSR**
0.75	0.82	0.70	0.70	0.71	0.77	0.77	0.76	0.93	Bahrain
12.62	7.10	7.90	8.19	18.13	18.78	19.30	19.44	20.14	Iran
8.47	8.84	6.86	7.98	8.78	9.22	9.16	9.01	9.25	Iraq
2.62	2.76	2.82	2.90	3.12	3.31	3.61	3.78	3.67	Israel
1.20	1.16	1.02	1.12	1.16	1.19	1.26	1.35	1.44	Jordan
2.13	1.81	0.46	1.85	1.93	2.04	2.16	2.26	2.47	Kuwait
1.04	0.71	0.88	1.23	1.43	1.37	1.66	1.61	1.54	Lebanon
0.62	0.89	0.89	0.84	0.92	0.90	0.91	0.94	1.01	Oman
0.53	0.61	0.59	0.64	0.67	0.74	0.80	0.89	1.00	Qatar
7.97	9.88	10.98	10.33	11.34	12.30	12.09	12.84	12.93	Saudi Arabia
1.53	1.72	1.69	1.49	1.42	1.42	1.30	1.35	1.45	Syria
3.62	3.47	3.54	3.60	3.81	4.03	4.01	3.69	3.22	United Arab Emirates
1.32	1.34	1.63	1.77	1.40	1.47	1.71	1.65	1.69	Yemen
44.43	41.09	39.97	42.65	54.80	57.55	58.74	59.58	60.74	**Middle East**
402.83	397.97	434.12	405.97	407.68	394.70	424.32	444.84	464.18	**Non-OECD Total**
560.01	558.72	551.52	564.26	573.65	593.62	605.01	618.97	631.65	OECD North America
106.39	112.47	116.74	122.57	126.01	133.39	140.53	147.29	150.87	OECD Pacific
280.80	289.01	292.54	301.73	307.78	311.12	315.95	325.66	330.55	OECD Europe
947.20	960.19	960.80	988.56	1 007.44	1 038.14	1 061.49	1 091.93	1 113.07	**OECD Total**
1 350.02	1 358.16	1 394.92	1 394.53	1 415.12	1 432.84	1 485.81	1 536.77	1 577.25	**World**

Former USSR: series up to 1990-1992 are not comparable with the recent years; 1991 is estimated.

Consumption of Electricity in Transport (Mtoe)
Consommation d'électricité dans les transports (Mtep)
Elektrizitätsverbrauch im Verkehrssektor (Mtoe)
Consumo di energia elettrica nel settore dei trasporti (Mtep)
運輸部門の電力消費量（石油換算百万トン）
Consumo de electricidad en el transporte (Mtep)
Потребление электроэнергии в транспорте (Мтнэ)

	1971	1973	1978	1983	1984	1985	1986	1987	1988
Algérie	0.002	0.002	0.003	0.003	0.003	0.003	0.015	0.017	0.014
Maroc	0.007	0.007	0.009	0.010	0.010	0.013	0.013	0.014	0.016
Afrique du Sud	0.282	0.249	0.302	0.373	0.395	0.394	0.387	0.348	0.354
Tunisie	-	-	-	0.006	0.006	0.007	0.007	0.007	0.007
Afrique	0.290	0.259	0.314	0.392	0.414	0.418	0.423	0.387	0.392
Argentine	0.025	0.025	0.022	0.024	0.023	0.023	0.023	0.029	0.029
Brésil	0.053	0.052	0.058	0.091	0.096	0.099	0.100	0.102	0.103
Chili	0.019	0.017	0.018	0.019	0.019	0.019	0.019	0.020	0.019
Colombie	-	-	-	0.000	0.000	0.000	0.000	0.000	0.000
Costa Rica	0.001	0.001	0.001	0.001	0.001	0.001	0.001	0.001	0.001
Cuba	-	-	-	-	-	0.001	0.000	0.001	0.000
Panama	-	-	-	0.007	0.008	0.006	-	-	-
Paraguay	-	-	-	-	-	0.000	0.000	0.000	0.000
Vénézuela	-	-	-	-	-	-	-	0.015	0.016
Amérique latine	0.098	0.094	0.099	0.143	0.147	0.149	0.143	0.167	0.168
Inde	0.141	0.132	0.198	0.248	0.264	0.285	0.305	0.343	0.350
Népal	-	-	-	-	-	-	-	-	-
Pakistan	0.002	0.002	0.003	0.004	0.003	0.003	0.003	0.003	0.003
Philippines	-	-	-	-	-	-	-	0.002	0.002
Singapour	-	-	-	-	-	-	-	0.002	0.009
Taipei chinois	0.000	0.001	0.005	0.022	0.022	0.022	0.024	0.023	0.022
Viêt-Nam	-	-	-	-	-	-	-	-	0.003
Asie	0.144	0.135	0.206	0.273	0.289	0.310	0.332	0.373	0.390
Rép. populaire de Chine	-	-	-	0.308	0.356	0.545	0.575	0.660	0.770
Hong-Kong, Chine	-	-	-	-	-	-	-	0.052	0.051
Chine	-	-	-	0.308	0.356	0.545	0.575	0.711	0.821
Bulgarie	-	-	-	-	0.116	0.113	0.115	0.124	0.128
Chypre	0.003	0.003	0.000	0.001	0.001	0.002	0.003	0.003	0.004
Roumanie	-	-	-	-	-	0.209	0.222	0.222	0.243
République slovaque	0.048	0.054	0.066	0.091	0.094	0.093	0.094	0.095	0.100
Croatie	-	-	-	-	-	-	-	-	-
Ex-RYM	-	-	-	-	-	-	-	-	-
Slovénie	-	-	-	-	-	-	-	-	-
RF de Yougoslavie	-	-	-	-	-	-	-	-	-
Ex-Yougoslavie	0.046	0.062	0.071	0.103	0.105	0.106	0.128	0.106	0.112
Europe non-OCDE	0.096	0.119	0.137	0.195	0.316	0.524	0.562	0.550	0.586

Rép.populaire de Chine: jusqu'en 1978 la ventilation de la consommation finale par secteur est incomplète.

Consumption of Electricity in Transport (Mtoe)
Consommation d'électricité dans les transports (Mtep)
Elektrizitätsverbrauch im Verkehrssektor (Mtoe)
Consumo di energia elettrica nel settore dei trasporti (Mtep)
運輸部門の電力消費量（石油換算百万トン）
Consumo de electricidad en el transporte (Mtep)
Потребление электроэнергии в транспорте (Мтнэ)

1989	1990	1991	1992	1993	1994	1995	1996	1997	
0.018	0.016	0.017	0.017	0.027	0.026	0.028	0.029	0.030	Algeria
0.015	0.017	0.019	0.023	0.016	0.016	0.017	0.018	0.018	Morocco
0.364	0.340	0.317	0.307	0.345	0.377	0.369	0.368	0.392	South Africa
0.008	0.009	0.009	0.010	0.011	0.011	0.012	0.013	0.013	Tunisia
0.405	0.382	0.362	0.357	0.399	0.431	0.426	0.427	0.454	**Africa**
0.019	0.027	0.021	0.024	0.024	0.026	0.026	0.036	0.041	Argentina
0.111	0.103	0.093	0.103	0.103	0.101	0.104	0.099	0.100	Brazil
0.017	0.018	0.019	0.019	0.022	0.019	0.017	0.017	0.018	Chile
0.000	0.000	-	-	-	-	-	-	0.004	Colombia
0.001	0.001	0.001	0.000	0.000	0.001	-	-	-	Costa Rica
0.000	0.008	0.007	0.005	0.005	0.004	0.006	0.006	0.008	Cuba
0.005	0.005	0.005	0.002	0.001	0.006	0.006	0.007	0.008	Panama
0.000	0.000	0.000	0.000	0.000	0.000	0.000	-	-	Paraguay
0.019	0.024	0.017	0.017	0.012	0.011	0.009	0.007	0.072	Venezuela
0.174	0.187	0.163	0.170	0.167	0.167	0.168	0.172	0.250	**Latin America**
0.352	0.354	0.389	0.436	0.483	0.506	0.543	0.565	0.602	India
-	0.000	0.000	0.000	0.000	0.000	0.000	0.000	0.000	Nepal
0.003	0.003	0.003	0.002	0.002	0.002	0.002	0.002	0.002	Pakistan
0.002	0.002	0.002	0.002	0.002	-	-	-	-	Philippines
0.011	0.016	0.016	0.016	0.016	0.017	0.017	0.020	0.021	Singapore
0.033	0.037	0.037	0.038	0.039	0.040	0.040	0.044	0.051	Chinese Taipei
0.004	0.004	0.003	0.003	0.004	0.007	0.008	0.010	0.011	Vietnam
0.405	0.417	0.450	0.498	0.547	0.573	0.610	0.641	0.687	**Asia**
0.849	0.911	0.799	1.019	1.258	1.411	0.939	1.019	1.317	People's Rep.of China
-	-	-	-	-	-	-	-	-	Hong Kong, China
0.849	0.911	0.799	1.019	1.258	1.411	0.939	1.019	1.317	**China**
0.113	0.112	0.104	0.089	0.062	0.055	0.069	0.070	0.057	Bulgaria
0.004	0.004	0.005	0.002	0.002	0.003	0.003	0.003	0.002	Cyprus
0.251	0.225	0.154	0.243	0.190	0.162	0.187	0.200	0.192	Romania
0.102	0.100	0.124	0.082	0.097	0.126	0.119	0.085	0.087	Slovak Republic
-	-	-	0.019	0.020	0.020	0.021	0.021	0.021	*Croatia*
-	-	-	0.002	0.002	-	0.001	0.001	0.001	*FYROM*
-	-	-	0.013	0.013	0.013	0.015	0.014	0.014	*Slovenia*
-	-	-	0.029	0.022	0.023	0.023	0.023	0.025	*FR of Yugoslavia*
0.121	0.091	0.064	0.063	0.056	0.056	0.059	0.059	0.061	Former Yugoslavia
0.592	0.533	0.449	0.480	0.407	0.401	0.437	0.417	0.398	**Non-OECD Europe**

People's Rep. of China: up to 1978, the breakdown of final consumption by sector is incomplete.

Consumption of Electricity in Transport (Mtoe)
Consommation d'électricité dans les transports (Mtep)
Elektrizitätsverbrauch im Verkehrssektor (Mtoe)
Consumo di energia elettrica nel settore dei trasporti (Mtep)
運輸部門の電力消費量（石油換算百万トン）
Consumo de electricidad en el transporte (Mtep)
Потребление электроэнергии в транспорте (Мтнэ)

	1971	1973	1978	1983	1984	1985	1986	1987	1988
Arménie	-	-	-	-	-	-	-	-	-
Azerbaïdjan	-	-	-	-	-	-	-	-	-
Bélarus	-	-	-	-	-	-	-	-	-
Estonie	-	-	-	-	-	-	-	-	-
Géorgie	-	-	-	-	-	-	-	-	-
Kazakhstan	-	-	-	-	-	-	-	-	-
Kirghizistan	-	-	-	-	-	-	-	-	-
Lettonie	-	-	-	-	-	-	-	-	-
Lituanie	-	-	-	-	-	-	-	-	-
Moldova	-	-	-	-	-	-	-	-	-
Russie	-	-	-	-	-	-	-	-	-
Tadjikistan	-	-	-	-	-	-	-	-	-
Turkménistan	-	-	-	-	-	-	-	-	-
Ukraine	-	-	-	-	-	-	-	-	-
Ouzbékistan	-	-	-	-	-	-	-	-	-
Ex-URSS	4.197	4.627	6.029	6.846	6.949	7.052	7.155	7.258	7.362
Israël	-	-	-	-	-	-	-	-	0.041
Moyen-Orient	-	-	-	-	-	-	-	-	0.041
Total non-OCDE	4.825	5.234	6.785	8.156	8.471	8.998	9.191	9.447	9.760
OCDE Amérique du N.	0.577	0.678	0.428	0.472	0.553	0.632	0.646	0.664	0.701
OCDE Pacifique	1.058	1.209	1.390	1.513	1.512	1.568	1.614	1.660	1.767
OCDE Europe	3.202	3.353	3.796	4.212	4.373	4.560	4.705	4.803	4.886
Total OCDE	4.837	5.240	5.614	6.197	6.438	6.759	6.965	7.127	7.354
Monde	9.662	10.474	12.399	14.353	14.909	15.757	16.155	16.574	17.114

Ex-URSS: les séries antérieures à 1990-1992 ne sont pas comparables aux années récentes; 1991 a été estimé.

Consumption of Electricity in Transport (Mtoe)
Consommation d'électricité dans les transports (Mtep)
Elektrizitätsverbrauch im Verkehrssektor (Mtoe)
Consumo di energia elettrica nel settore dei trasporti (Mtep)
運輸部門の電力消費量（石油換算百万トン）
Consumo de electricidad en el transporte (Mtep)
Потребление электроэнергии в транспорте (Мтнэ)

1989	1990	1991	1992	1993	1994	1995	1996	1997	
-	-	-	0.027	0.016	0.015	0.016	0.015	0.013	Armenia
-	-	-	0.073	0.060	0.047	0.041	0.039	0.036	Azerbaijan
-	-	-	0.083	0.174	0.187	0.158	0.157	0.147	Belarus
-	-	-	0.028	0.013	0.010	0.010	0.009	0.009	Estonia
-	-	-	0.053	0.052	0.035	0.037	0.022	0.020	Georgia
-	-	-	0.518	0.427	0.376	0.330	0.295	0.264	Kazakhstan
-	-	-	0.008	0.012	0.011	0.011	0.011	0.010	Kyrgyzstan
-	-	-	0.025	0.019	0.020	0.017	0.015	0.015	Latvia
-	-	-	0.012	0.008	0.009	0.008	0.009	0.009	Lithuania
-	-	-	0.018	0.008	0.007	0.012	0.011	0.010	Moldova
-	-	-	7.463	6.598	5.882	5.604	5.584	5.461	Russia
-	-	-	-	-	0.008	0.007	0.007	0.007	Tajikistan
-	-	-	0.013	0.013	0.011	0.011	0.010	0.009	Turkmenistan
-	-	-	1.092	1.027	0.935	0.927	0.839	0.821	Ukraine
-	-	-	0.108	0.094	0.096	0.117	0.120	0.110	Uzbekistan
7.439	7.482	7.487	9.522	8.522	7.650	7.307	7.143	6.943	**Former USSR**
0.045	0.046	0.048	0.052	0.058	0.063	0.069	0.075	0.077	Israel
0.045	0.046	0.048	0.052	0.058	0.063	0.069	0.075	0.077	**Middle East**
9.908	9.957	9.758	12.097	11.358	10.696	9.954	9.894	10.126	**Non-OECD Total**
0.699	0.706	0.683	0.713	0.721	0.739	0.749	0.753	0.790	OECD North America
1.866	1.962	2.025	2.054	2.077	2.129	2.165	2.170	2.207	OECD Pacific
4.972	5.241	5.312	5.501	5.636	5.745	5.817	5.996	5.994	OECD Europe
7.537	7.909	8.019	8.268	8.434	8.613	8.731	8.919	8.991	**OECD Total**
17.445	17.866	17.777	20.365	19.792	19.309	18.686	18.812	19.118	**World**

Former USSR: series up to 1990-1992 are not comparable with the recent years; 1991 is estimated.

Total Consumption of Energy in Transport (Mtoe)
Consommation totale d'énergie dans les transports (Mtep)
Gesamtenergieverbrauch im Verkehrssektor (Mtoe)
Consumo totale di energia nel settore dei trasporti (Mtep)
運輸部門のエネルギー総消費量（石油換算百万トン）
Consumo total de energía en el transporte (Mtep)
Общее потребление топлива и энергии в транспорте (Мтнэ)

	1971	1973	1978	1983	1984	1985	1986	1987	1988
Algérie	0.90	1.14	2.00	3.34	3.79	4.17	4.48	4.54	4.52
Angola	0.41	0.55	0.53	0.33	0.23	0.41	0.34	0.32	0.33
Bénin	0.09	0.11	0.10	0.10	0.11	0.14	0.13	0.11	0.11
Cameroun	0.24	0.28	0.41	0.55	0.58	0.63	0.60	0.61	0.67
Congo	0.12	0.13	0.18	0.23	0.21	0.20	0.20	0.22	0.20
RD du Congo	0.54	0.58	0.64	0.72	0.71	0.66	0.56	0.64	0.66
Egypte	1.38	1.54	2.21	3.63	4.06	4.39	4.33	3.98	4.03
Ethiopie	0.23	0.24	0.25	0.30	0.31	0.33	0.37	0.45	0.47
Gabon	0.01	0.02	0.20	0.17	0.11	0.14	0.20	0.15	0.17
Ghana	0.40	0.41	0.49	0.35	0.39	0.46	0.44	0.42	0.57
Côte d'Ivoire	0.34	0.36	0.55	0.51	0.49	0.49	0.49	0.50	0.54
Kenya	0.68	0.81	0.94	0.87	1.00	1.05	1.15	1.12	1.10
Libye	0.32	0.67	1.44	2.37	2.39	2.32	2.23	2.36	2.31
Maroc	0.71	0.84	1.06	1.11	1.08	1.11	1.15	1.19	0.64
Mozambique	0.15	0.14	0.12	0.10	0.10	0.07	0.08	0.08	0.08
Nigéria	1.05	1.49	4.21	5.71	5.29	5.04	4.72	4.64	5.06
Sénégal	0.25	0.29	0.34	0.40	0.40	0.39	0.38	0.33	0.32
Afrique du Sud	9.05	9.73	9.90	9.23	9.78	9.39	9.27	9.75	10.27
Soudan	0.78	0.93	0.74	0.92	0.82	0.97	0.92	0.73	1.11
Tanzanie	0.27	0.30	0.28	0.25	0.25	0.26	0.25	0.26	0.29
Tunisie	0.42	0.50	0.76	0.82	0.84	0.85	0.82	0.88	0.93
Zambie	0.23	0.27	0.27	0.24	0.25	0.25	0.24	0.24	0.26
Zimbabwe	0.62	0.71	0.60	0.71	0.66	0.70	0.77	0.73	0.77
Autre Afrique	0.67	0.82	1.21	1.57	1.58	1.62	1.78	2.32	2.39
Afrique	19.86	22.85	29.42	34.52	35.41	36.06	35.90	36.58	37.80
Argentine	8.10	8.82	9.33	10.57	10.06	9.08	10.01	10.50	9.28
Bolivie	0.31	0.38	0.66	0.57	0.56	0.56	0.62	0.69	0.69
Brésil	13.66	18.17	23.95	23.90	24.36	26.28	29.64	29.36	29.90
Chili	2.16	1.97	1.98	2.28	2.28	2.25	2.35	2.51	2.83
Colombie	2.58	2.95	4.27	5.09	5.17	5.32	5.37	5.57	6.05
Costa Rica	0.23	0.29	0.45	0.37	0.40	0.41	0.46	0.49	0.55
Cuba	1.59	1.97	2.28	2.20	2.26	2.96	2.95	3.06	3.17
République dominicaine	0.51	0.56	0.64	0.67	0.75	0.73	0.80	0.99	0.89
Equateur	0.71	0.83	1.51	1.73	1.93	2.05	2.13	2.06	2.48
El Salvador	0.20	0.24	0.38	0.30	0.32	0.35	0.34	0.37	0.41
Guatemala	0.28	0.39	0.48	0.42	0.45	0.48	0.46	0.54	0.58
Haiti	0.05	0.07	0.11	0.11	0.11	0.12	0.13	0.13	0.15
Honduras	0.15	0.18	0.21	0.26	0.26	0.28	0.30	0.32	0.35
Jamaïque	0.68	0.70	0.41	0.42	0.38	0.41	0.40	0.45	0.49
Antilles néerlandaises	0.54	0.51	0.60	0.49	0.47	0.41	0.34	0.35	0.37
Nicaragua	0.25	0.29	0.37	0.30	0.30	0.29	0.31	0.34	0.30
Panama	0.36	0.39	0.47	0.35	0.39	0.36	0.37	0.36	0.35
Paraguay	0.15	0.17	0.35	0.39	0.45	0.44	0.47	0.50	0.52
Pérou	1.94	2.29	2.26	2.43	2.44	2.28	2.42	2.73	2.75
Trinité-et-Tobago	0.32	0.42	0.47	0.62	0.64	0.64	0.67	0.63	0.51
Uruguay	0.57	0.59	0.58	0.49	0.45	0.44	0.44	0.46	0.47
Vénézuela	4.37	5.31	8.71	9.54	9.30	9.51	9.59	9.90	10.25
Autre Amérique latine	0.43	0.31	0.52	0.65	0.68	0.67	0.62	0.80	0.86
Amérique latine	40.14	47.79	61.02	64.15	64.42	66.32	71.20	73.13	74.20

Total Consumption of Energy in Transport (Mtoe)
Consommation totale d'énergie dans les transports (Mtep)
Gesamtenergieverbrauch im Verkehrssektor (Mtoe)
Consumo totale di energia nel settore dei trasporti (Mtep)
運輸部門のエネルギー総消費量（石油換算百万トン）
Consumo total de energía en el transporte (Mtep)
Общее потребление топлива и энергии в транспорте (Мтнэ)

1989	1990	1991	1992	1993	1994	1995	1996	1997	
4.66	4.72	4.84	2.77	3.36	3.22	3.12	3.16	3.01	Algeria
0.31	0.31	0.31	0.32	0.31	0.31	0.31	0.39	0.63	Angola
0.09	0.07	0.06	0.05	0.06	0.06	0.06	0.27	0.27	Benin
0.66	0.63	0.57	0.57	0.60	0.65	0.62	0.63	0.64	Cameroon
0.23	0.21	0.22	0.20	0.20	0.20	0.21	0.21	0.21	Congo
0.67	0.68	0.69	0.63	0.71	0.73	0.74	0.74	0.74	DR of Congo
3.84	3.97	3.94	3.93	3.97	4.19	4.53	4.96	5.22	Egypt
0.49	0.48	0.49	0.50	0.55	0.57	0.61	0.73	0.47	Ethiopia
0.23	0.19	0.19	0.18	0.19	0.19	0.20	0.22	0.29	Gabon
0.58	0.57	0.51	0.57	0.63	0.67	0.70	0.70	0.70	Ghana
0.52	0.46	0.47	0.45	0.48	0.49	0.51	0.57	0.58	Ivory Coast
1.18	1.18	1.16	1.18	1.16	1.24	1.22	1.32	1.29	Kenya
2.38	2.28	2.31	2.35	2.73	3.26	3.49	3.52	3.54	Libya
0.65	0.69	0.63	0.68	0.83	0.82	0.82	0.83	0.84	Morocco
0.08	0.11	0.10	0.09	0.10	0.11	0.08	0.09	0.10	Mozambique
5.10	4.29	5.56	6.70	5.81	4.94	4.80	5.07	5.15	Nigeria
0.39	0.39	0.38	0.40	0.36	0.38	0.41	0.49	0.52	Senegal
10.45	10.67	10.67	10.88	11.01	11.42	12.79	12.71	13.14	South Africa
0.99	1.33	1.07	1.07	0.72	1.03	0.97	0.92	1.03	Sudan
0.29	0.31	0.29	0.30	0.29	0.29	0.28	0.27	0.22	Tanzania
0.97	1.02	0.99	1.08	1.21	1.28	1.33	1.38	1.48	Tunisia
0.25	0.23	0.24	0.21	0.22	0.23	0.23	0.23	0.23	Zambia
0.75	0.72	0.62	0.77	0.69	0.68	0.84	0.83	0.83	Zimbabwe
2.52	2.52	2.40	2.48	2.72	2.65	2.47	2.48	2.58	Other Africa
38.25	38.05	38.71	38.35	38.91	39.60	41.36	42.73	43.71	**Africa**
8.84	8.80	9.98	10.59	11.77	12.57	12.91	13.58	14.00	Argentina
0.73	0.76	0.78	0.77	0.80	0.83	0.92	1.01	1.02	Bolivia
31.37	31.52	33.03	33.11	34.44	36.15	39.52	42.75	45.18	Brazil
3.04	3.16	3.38	3.62	3.96	4.37	4.76	5.20	5.53	Chile
6.15	6.24	6.49	6.89	6.97	6.64	7.83	7.99	8.15	Colombia
0.59	0.57	0.59	0.88	0.92	0.98	0.99	0.98	1.02	Costa Rica
3.30	1.95	1.44	1.04	0.87	0.82	0.87	1.02	1.08	Cuba
1.09	0.94	0.93	1.10	0.99	1.03	1.06	1.10	1.15	Dominican Republic
2.40	2.28	2.26	2.26	2.26	2.34	2.45	2.81	2.96	Ecuador
0.42	0.46	0.48	0.57	0.63	0.66	0.75	0.73	0.78	El Salvador
0.58	0.63	0.71	0.72	0.76	0.81	1.00	1.01	1.08	Guatemala
0.16	0.17	0.15	0.15	0.14	0.04	0.19	0.21	0.21	Haiti
0.38	0.35	0.36	0.40	0.39	0.53	0.63	0.53	0.54	Honduras
0.49	0.50	0.49	0.48	0.54	0.64	0.73	0.80	0.86	Jamaica
0.34	0.33	0.38	0.41	0.34	0.43	0.44	0.44	0.44	Netherlands Antilles
0.26	0.26	0.29	0.30	0.30	0.35	0.37	0.39	0.44	Nicaragua
0.37	0.40	0.40	0.42	0.44	0.48	0.52	0.60	0.63	Panama
0.58	0.56	0.55	0.66	0.77	0.88	1.00	0.94	1.01	Paraguay
2.43	2.62	2.67	2.75	2.94	2.74	3.23	3.34	3.31	Peru
0.49	0.52	0.55	0.54	0.47	0.48	0.51	0.53	0.60	Trinidad-and-Tobago
0.50	0.50	0.53	0.57	0.64	0.73	0.72	0.77	0.83	Uruguay
9.72	10.02	10.43	10.48	10.93	11.33	11.93	11.25	12.64	Venezuela
0.88	0.88	0.88	0.85	0.84	0.92	0.93	0.93	0.94	Other Latin America
75.09	74.43	77.75	79.57	83.10	86.76	94.26	98.89	104.40	**Latin America**

Total Consumption of Energy in Transport (Mtoe)
Consommation totale d'énergie dans les transports (Mtep)
Gesamtenergieverbrauch im Verkehrssektor (Mtoe)
Consumo totale di energia nel settore dei trasporti (Mtep)
運輸部門のエネルギー総消費量（石油換算百万トン）
Consumo total de energía en el transporte (Mtep)
Общее потребление топлива и энергии в транспорте (Мтнэ)

	1971	1973	1978	1983	1984	1985	1986	1987	1988
Bangladesh	0.10	0.10	0.25	0.41	0.42	0.51	0.51	0.54	0.58
Brunei	0.03	0.04	0.08	0.16	0.16	0.17	0.18	0.19	0.21
Inde	14.94	14.71	16.98	20.10	20.59	22.14	22.92	23.54	24.22
Indonésie	2.75	3.15	5.16	6.91	7.04	7.24	7.67	8.43	9.14
RPD de Corée	0.59	0.67	1.38	2.00	2.10	2.21	2.31	2.33	2.33
Malaisie	1.50	1.74	1.97	3.29	3.40	3.58	3.82	4.05	4.35
Myanmar	0.52	0.44	0.56	0.62	0.68	0.64	0.71	0.42	0.42
Népal	0.02	0.03	0.04	0.06	0.07	0.08	0.08	0.11	0.13
Pakistan	1.42	1.43	2.16	3.24	3.52	3.68	3.89	4.38	4.68
Philippines	2.40	2.49	2.39	1.90	1.82	1.74	1.84	2.09	2.28
Singapour	0.74	0.93	1.68	2.19	2.26	2.20	2.36	2.48	2.55
Sri Lanka	0.45	0.54	0.60	0.68	0.77	0.80	0.80	0.82	0.83
Taipei chinois	1.29	2.15	3.03	4.04	4.28	4.51	4.88	5.41	6.04
Thaïlande	2.39	3.23	3.89	4.95	5.68	5.88	6.27	7.20	8.36
Viêt-Nam	3.38	3.12	0.20	0.62	0.60	0.96	1.08	1.25	1.35
Autre Asie	0.59	0.58	0.73	0.90	1.00	1.01	0.91	1.04	0.99
Asie	33.11	35.35	41.10	52.08	54.38	57.36	60.24	64.29	68.45
Rép. populaire de Chine	4.78	6.54	10.23	30.62	32.96	36.09	39.06	41.10	43.85
Hong-Kong, Chine	0.90	1.06	1.30	2.17	2.17	2.06	2.54	2.57	3.16
Chine	5.68	7.60	11.53	32.79	35.13	38.15	41.60	43.68	47.01
Albanie	0.24	0.26	0.38	0.31	0.45	0.36	0.34	0.35	0.39
Bulgarie	1.92	1.82	1.81	1.71	1.82	1.80	1.80	1.82	2.92
Chypre	0.26	0.33	0.26	0.35	0.40	0.43	0.45	0.49	0.51
Gibraltar	0.02	0.02	0.01	0.01	0.01	0.02	0.02	0.03	0.04
Malte	0.13	0.15	0.13	0.15	0.19	0.10	0.22	0.23	0.23
Roumanie	2.72	2.91	2.78	1.85	1.77	1.71	1.62	2.17	2.06
République slovaque	1.48	1.70	1.63	1.00	1.01	1.00	1.03	1.06	1.11
Bosnie-Herzégovine	-	-	-	-	-	-	-	-	-
Croatie	-	-	-	-	-	-	-	-	-
Ex-RYM	-	-	-	-	-	-	-	-	-
Slovénie	-	-	-	-	-	-	-	-	-
RF de Yougoslavie	-	-	-	-	-	-	-	-	-
Ex-Yougoslavie	4.22	4.13	5.20	4.50	4.32	4.85	5.97	5.33	5.87
Europe non-OCDE	10.99	11.33	12.20	9.89	9.97	10.27	11.45	11.48	13.12

Rép.populaire de Chine: jusqu'en 1978 la ventilation de la consommation finale par secteur est incomplète.

Total Consumption of Energy in Transport (Mtoe)
Consommation totale d'énergie dans les transports (Mtep)
Gesamtenergieverbrauch im Verkehrssektor (Mtoe)
Consumo totale di energia nel settore dei trasporti (Mtep)
運輸部門のエネルギー総消費量（石油換算百万トン）
Consumo total de energía en el transporte (Mtep)
Общее потребление топлива и энергии в транспорте (Мтнэ)

1989	1990	1991	1992	1993	1994	1995	1996	1997	
0.62	0.64	0.68	0.77	0.82	0.88	1.07	1.12	1.13	Bangladesh
0.21	0.23	0.24	0.27	0.29	0.31	0.33	0.36	0.39	Brunei
25.99	26.44	29.62	30.88	31.52	33.17	37.33	40.05	41.54	India
9.95	11.38	12.21	12.95	13.82	15.50	16.97	18.87	19.87	Indonesia
2.49	2.44	2.24	2.17	2.04	1.32	1.16	1.02	0.99	DPR of Korea
4.67	5.41	5.93	6.27	7.20	7.98	7.83	8.79	9.17	Malaysia
0.41	0.44	0.40	0.35	0.42	0.50	0.63	0.68	0.78	Myanmar
0.10	0.11	0.14	0.17	0.18	0.18	0.20	0.20	0.20	Nepal
4.84	4.92	5.14	5.76	6.53	6.84	7.09	7.66	7.83	Pakistan
2.57	2.66	2.44	2.82	3.16	3.34	3.66	4.14	4.64	Philippines
2.77	3.26	3.23	3.30	3.78	4.41	4.52	4.61	4.98	Singapore
0.79	0.82	0.82	0.90	0.96	1.13	1.20	1.29	1.33	Sri Lanka
6.91	7.54	7.97	9.25	9.91	10.69	11.32	11.75	12.03	Chinese Taipei
9.91	11.11	11.39	12.39	14.20	15.83	18.33	19.90	21.20	Thailand
1.24	1.47	1.25	1.21	1.65	1.78	1.87	3.91	4.93	Vietnam
0.96	0.99	0.95	0.89	0.88	0.79	0.69	0.70	0.70	Other Asia
74.44	79.85	84.68	90.34	97.35	104.65	114.20	125.06	131.71	**Asia**
45.27	45.57	49.57	51.86	59.66	54.24	59.10	62.51	70.96	People's Rep.of China
3.26	3.86	4.09	4.97	5.62	6.31	6.36	6.26	6.51	Hong Kong, China
48.53	49.43	53.66	56.83	65.27	60.55	65.46	68.77	77.46	**China**
0.32	0.28	0.24	0.18	0.22	0.25	0.29	0.30	0.30	Albania
3.07	2.63	1.54	1.55	1.77	1.63	1.69	1.57	1.15	Bulgaria
0.58	0.65	0.69	0.73	0.69	0.72	0.77	0.78	0.79	Cyprus
0.03	0.04	0.05	0.06	0.06	0.07	0.07	0.07	0.07	Gibraltar
0.23	0.23	0.26	0.26	0.28	0.28	0.30	0.35	0.36	Malta
4.09	4.37	3.78	3.72	3.24	3.34	3.32	4.23	4.27	Romania
1.10	1.08	0.95	0.95	0.99	1.23	1.37	1.26	1.27	Slovak Republic
-	-	-	0.37	0.25	0.38	0.36	0.36	0.36	*Bosnia-Herzegovina*
-	-	-	0.98	1.07	1.16	1.22	1.27	1.43	*Croatia*
-	-	-	0.27	0.50	0.34	0.34	0.59	0.53	*FYROM*
-	-	-	0.90	1.09	1.21	1.35	1.52	1.55	*Slovenia*
-	-	-	0.64	0.41	0.39	0.43	0.79	1.17	*FR of Yugoslavia*
5.55	5.54	4.94	3.18	3.32	3.48	3.69	4.54	5.04	Former Yugoslavia
14.98	14.81	12.44	10.63	10.57	11.01	11.51	13.09	13.25	**Non-OECD Europe**

People's Rep. of China: up to 1978, the breakdown of final consumption by sector is incomplete.

Total Consumption of Energy in Transport (Mtoe)
Consommation totale d'énergie dans les transports (Mtep)
Gesamtenergieverbrauch im Verkehrssektor (Mtoe)
Consumo totale di energia nel settore dei trasporti (Mtep)
運輸部門のエネルギー総消費量（石油換算百万トン）
Consumo total de energía en el transporte (Mtep)
Общее потребление топлива и энергии в транспорте (Мтнэ)

	1971	1973	1978	1983	1984	1985	1986	1987	1988
Arménie	-	-	-	-	-	-	-	-	-
Azerbaïdjan	-	-	-	-	-	-	-	-	-
Bélarus	-	-	-	-	-	-	-	-	-
Estonie	-	-	-	-	-	-	-	-	-
Géorgie	-	-	-	-	-	-	-	-	-
Kazakhstan	-	-	-	-	-	-	-	-	-
Kirghizistan	-	-	-	-	-	-	-	-	-
Lettonie	-	-	-	-	-	-	-	-	-
Lituanie	-	-	-	-	-	-	-	-	-
Moldova	-	-	-	-	-	-	-	-	-
Russie	-	-	-	-	-	-	-	-	-
Tadjikistan	-	-	-	-	-	-	-	-	-
Turkménistan	-	-	-	-	-	-	-	-	-
Ukraine	-	-	-	-	-	-	-	-	-
Ouzbékistan	-	-	-	-	-	-	-	-	-
Ex-URSS	89.66	98.61	119.18	130.47	132.34	134.06	139.30	142.66	144.63
Bahrein	0.20	0.29	0.63	0.68	0.70	0.70	0.69	0.70	0.73
Iran	1.72	2.88	5.30	5.09	5.68	6.16	11.01	11.90	11.66
Irak	1.09	1.35	3.39	6.35	6.77	7.23	7.60	7.83	8.06
Israël	1.67	1.99	2.13	2.34	2.24	2.36	2.31	2.53	2.59
Jordanie	0.21	0.27	0.64	1.01	0.99	1.00	1.11	1.13	1.15
Koweit	0.61	0.67	1.43	2.03	1.97	2.00	2.02	1.96	1.98
Liban	0.98	1.25	0.98	1.42	1.35	1.49	1.40	1.46	0.97
Oman	0.04	0.04	0.29	0.47	0.57	0.62	0.65	0.57	0.58
Qatar	0.08	0.13	0.36	0.54	0.52	0.50	0.51	0.52	0.53
Arabie saoudite	0.77	1.09	3.86	11.63	11.76	12.55	12.63	12.40	8.66
Syrie	0.81	1.02	2.01	2.49	2.70	2.72	2.77	3.02	3.00
Emirats arabes unis	0.20	0.41	2.34	3.47	3.49	3.66	3.81	3.86	3.52
Yémen	0.25	0.45	0.61	1.05	1.04	1.11	1.20	1.29	1.37
Moyen-Orient	8.64	11.85	23.99	38.57	39.78	42.09	47.72	49.16	44.80
Total non-OCDE	208.08	235.38	298.44	362.48	371.42	384.31	407.40	420.98	430.00
OCDE Amérique du N.	420.45	468.85	522.99	494.86	510.98	515.50	526.76	544.93	570.22
OCDE Pacifique	52.84	60.57	76.24	81.21	84.48	86.65	90.44	94.91	100.98
OCDE Europe	165.30	187.61	219.10	228.38	235.40	239.55	251.61	262.03	276.28
Total OCDE	638.59	717.03	818.33	804.45	830.87	841.70	868.81	901.86	947.48
Monde	846.67	952.41	1 116.77	1 166.93	1 202.29	1 226.01	1 276.21	1 322.84	1 377.48

Ex-URSS: les séries antérieures à 1990-1992 ne sont pas comparables aux années récentes; 1991 a été estimé.

Total Consumption of Energy in Transport (Mtoe)
Consommation totale d'énergie dans les transports (Mtep)
Gesamtenergieverbrauch im Verkehrssektor (Mtoe)
Consumo totale di energia nel settore dei trasporti (Mtep)
運輸部門のエネルギー総消費量（石油換算百万トン）
Consumo total de energía en el transporte (Mtep)
Общее потребление топлива и энергии в транспорте (Мтнэ)

1989	1990	1991	1992	1993	1994	1995	1996	1997	
-	-	-	0.83	0.50	0.12	0.09	0.06	0.05	Armenia
-	-	-	1.64	1.81	1.91	1.65	1.44	1.09	Azerbaijan
-	-	-	5.06	3.93	2.84	2.23	2.08	2.11	Belarus
-	-	-	0.41	0.46	0.53	0.50	0.55	0.57	Estonia
-	-	-	0.48	0.31	0.11	0.09	0.56	0.53	Georgia
-	-	-	5.66	4.42	3.47	3.16	3.07	2.62	Kazakhstan
-	-	-	0.51	0.32	0.13	0.32	0.35	0.20	Kyrgyzstan
-	-	-	1.05	0.93	0.82	0.84	0.84	0.81	Latvia
-	-	-	1.16	1.13	1.02	1.19	1.26	1.26	Lithuania
-	-	-	0.80	0.51	0.48	0.48	0.47	0.50	Moldova
-	-	-	123.38	70.67	51.30	53.67	53.75	52.46	Russia
-	-	-	5.11	3.22	1.08	1.08	1.08	1.08	Tajikistan
-	-	-	0.78	0.58	0.54	0.49	0.53	0.48	Turkmenistan
-	-	-	11.88	9.89	8.63	8.62	7.36	6.72	Ukraine
-	-	-	2.45	2.23	2.21	2.85	3.59	3.61	Uzbekistan
147.61	139.87	164.70	161.20	100.89	75.20	77.26	76.97	74.10	**Former USSR**
0.75	0.82	0.70	0.70	0.71	0.77	0.77	0.76	0.93	Bahrain
12.62	7.10	7.90	8.19	18.13	18.78	19.30	19.44	20.14	Iran
8.47	8.84	6.86	7.98	8.78	9.22	9.16	9.01	9.25	Iraq
2.67	2.80	2.86	2.95	3.18	3.37	3.68	3.86	3.75	Israel
1.20	1.16	1.02	1.12	1.16	1.19	1.26	1.35	1.44	Jordan
2.13	1.81	0.46	1.85	1.93	2.04	2.16	2.26	2.47	Kuwait
1.04	0.71	0.88	1.23	1.43	1.37	1.66	1.61	1.54	Lebanon
0.62	0.89	0.89	0.84	0.92	0.90	0.91	0.94	1.01	Oman
0.53	0.61	0.59	0.64	0.67	0.74	0.80	0.89	1.00	Qatar
7.97	9.88	10.98	10.33	11.34	12.30	12.09	12.84	12.93	Saudi Arabia
1.53	1.72	1.69	1.49	1.42	1.42	1.30	1.35	1.45	Syria
3.62	3.47	3.54	3.60	3.81	4.03	4.01	3.69	3.22	United Arab Emirates
1.32	1.34	1.63	1.77	1.40	1.47	1.71	1.65	1.69	Yemen
44.47	41.13	40.02	42.70	54.86	57.61	58.81	59.65	60.81	**Middle East**
443.38	437.58	471.96	479.62	450.96	435.39	462.85	485.16	505.46	**Non-OECD Total**
578.25	577.75	569.50	582.91	595.32	617.27	629.27	642.74	656.87	OECD North America
108.42	114.57	119.00	124.88	128.37	135.81	143.00	149.77	153.40	OECD Pacific
286.60	295.00	298.50	307.95	314.02	317.23	322.14	332.08	336.99	OECD Europe
973.27	987.32	987.01	1 015.74	1 037.71	1 070.32	1 094.41	1 124.59	1 147.26	**OECD Total**
1 416.65	1 424.90	1 458.96	1 495.36	1 488.67	1 505.70	1 557.27	1 609.75	1 652.71	**World**

Former USSR: series up to 1990-1992 are not comparable with the recent years; 1991 is estimated.

Other Sectors' Consumption of Coal (Mtoe)
Consommation de charbon des autres secteurs (Mtep)
Kohleverbrauch in sonstigen Sektoren (Mtoe)
Consumo di carbone siderurgico negli altri settori (Mtep)
他の部門の石炭消費量（石油換算百万トン）
Consumo de carbón de otros sectores (Mtep)
Потребление угля другими секторами (Мтнэ)

	1971	1973	1978	1983	1984	1985	1986	1987	1988
RD du Congo	-	-	-	0.051	0.050	0.051	0.051	0.050	0.050
Egypte	0.027	0.019	0.042	0.047	0.047	0.047	0.047	-	-
Ghana	-	-	-	-	-	-	-	-	0.002
Kenya	-	-	-	-	-	-	-	-	-
Maroc	0.023	0.012	0.005	-	-	-	-	-	-
Nigéria	0.006	0.006	0.017	0.005	0.006	0.002	0.014	0.008	-
Afrique du Sud	3.217	3.209	3.370	1.614	2.120	2.137	1.691	1.707	1.880
Zambie	0.012	0.012	0.006	0.006	0.006	0.006	0.008	-	-
Zimbabwe	0.189	0.228	0.242	0.282	0.279	0.341	0.353	0.338	0.371
Autre Afrique	0.097	0.102	0.013	0.066	0.019	0.024	0.102	0.103	0.088
Afrique	3.570	3.588	3.695	2.071	2.526	2.609	2.266	2.207	2.391
Chili	0.066	0.058	0.038	0.012	0.013	0.012	0.007	0.016	0.020
Colombie	0.153	0.151	0.143	0.189	0.193	0.183	0.179	0.177	0.179
Cuba	-	-	-	-	0.003	0.001	-	-	-
Pérou	-	-	-	-	-	-	-	-	-
Autre Amérique latine	0.023	0.028	0.020	0.011	0.007	-	-	-	-
Amérique latine	0.242	0.237	0.201	0.211	0.216	0.195	0.186	0.194	0.200
Bangladesh	0.002	0.002	-	-	-	-	-	-	-
Inde	1.786	2.337	1.514	1.007	1.138	1.129	0.802	1.008	0.951
Indonésie	-	-	-	-	0.004	0.010	-	-	-
Népal	-	-	-	0.001	0.003	0.001	0.001	0.017	0.017
Pakistan	0.098	0.033	0.024	0.013	0.015	0.014	0.013	0.009	0.011
Taipei chinois	0.212	0.165	0.153	0.105	0.104	0.070	0.054	0.043	0.020
Viêt-Nam	0.854	0.903	1.586	0.610	0.428	0.264	0.290	0.309	0.241
Autre Asie	0.234	0.158	0.210	0.029	0.028	0.028	0.029	0.029	0.025
Asie	3.186	3.598	3.487	1.765	1.721	1.517	1.189	1.415	1.266
Rép. populaire de Chine	132.174	141.199	208.164	84.762	91.663	99.333	101.707	105.484	112.727
Chine	132.174	141.199	208.164	84.762	91.663	99.333	101.707	105.484	112.727
Albanie	0.095	0.108	0.145	0.227	0.250	0.274	0.301	0.301	0.316
Bulgarie	0.988	1.285	0.955	1.089	1.015	1.059	1.142	1.182	0.861
Roumanie	0.740	0.784	1.859	2.962	1.999	2.046	1.653	2.646	2.618
République slovaque	1.144	1.162	1.160	1.154	1.156	1.202	1.212	1.206	1.267
Bosnie-Herzégovine	-	-	-	-	-	-	-	-	-
Croatie	-	-	-	-	-	-	-	-	-
Ex-RYM	-	-	-	-	-	-	-	-	-
Slovénie	-	-	-	-	-	-	-	-	-
RF de Yougoslavie	-	-	-	-	-	-	-	-	-
Ex-Yougoslavie	1.064	0.908	1.371	2.171	1.743	1.790	1.418	1.415	0.985
Europe non-OCDE	4.031	4.247	5.489	7.602	6.163	6.372	5.727	6.750	6.047

Comprend agriculture, services privés et publics, secteur résidentiel et autres secteurs non spécifiés.
Rép.populaire de Chine: jusqu'en 1978 la ventilation de la consommation finale par secteur est incomplète.

Other Sectors' Consumption of Coal (Mtoe)
Consommation de charbon des autres secteurs (Mtep)
Kohleverbrauch in sonstigen Sektoren (Mtoe)
Consumo di carbone siderurgico negli altri settori (Mtep)
他の部門の石炭消費量（石油換算百万トン）
Consumo de carbón de otros sectores (Mtep)
Потребление угля другими секторами (Мтнэ)

1989	1990	1991	1992	1993	1994	1995	1996	1997	
0.053	0.055	0.057	0.057	0.057	0.059	0.060	0.060	0.060	DR of Congo
-	-	-	-	-	-	-	-	-	Egypt
0.002	0.002	0.002	0.002	0.002	0.002	0.002	0.002	0.002	Ghana
-	-	0.001	0.001	0.001	0.001	0.001	0.001	0.001	Kenya
-	-	-	-	-	-	-	-	-	Morocco
-	-	-	-	-	-	-	-	-	Nigeria
2.196	2.419	2.543	2.600	3.393	3.212	4.655	2.414	2.481	South Africa
-	-	-	-	-	-	-	-	-	Zambia
0.370	0.424	0.437	0.411	0.405	0.392	0.410	0.486	0.453	Zimbabwe
0.099	0.100	0.101	0.095	0.095	0.083	0.078	0.078	0.079	Other Africa
2.719	3.000	3.140	3.166	3.953	3.749	5.205	3.040	3.074	**Africa**
0.021	0.058	0.100	0.098	0.067	0.062	0.039	0.087	0.107	Chile
0.189	0.182	0.184	0.185	0.183	0.157	0.133	0.140	0.137	Colombia
-	-	-	-	-	-	-	-	-	Cuba
0.025	0.026	0.016	0.020	0.009	0.009	0.008	0.006	0.004	Peru
0.001	0.001	0.001	0.001	0.001	0.001	0.001	0.001	0.001	Other Latin America
0.235	0.267	0.302	0.305	0.260	0.230	0.182	0.234	0.249	**Latin America**
-	-	-	-	-	-	-	-	-	Bangladesh
1.375	1.259	0.475	0.312	0.280	0.163	0.111	0.046	0.046	India
0.068	-	-	-	-	-	-	-	-	Indonesia
0.017	-	0.004	-	0.003	0.007	0.009	0.008	0.008	Nepal
0.007	0.003	0.002	0.003	0.001	0.001	0.001	0.001	0.001	Pakistan
-	-	-	-	-	-	-	-	-	Chinese Taipei
0.223	0.224	0.209	0.207	0.354	0.370	0.357	0.662	0.716	Vietnam
0.026	0.026	0.025	-	-	-	-	-	-	Other Asia
1.716	1.512	0.714	0.522	0.638	0.541	0.478	0.717	0.771	**Asia**
109.879	108.469	107.448	98.305	100.408	94.985	95.853	99.427	87.074	People's Rep.of China
109.879	108.469	107.448	98.305	100.408	94.985	95.853	99.427	87.074	**China**
0.316	0.130	0.073	0.082	0.057	0.040	0.038	0.024	0.016	Albania
0.841	0.808	0.715	0.759	0.788	0.574	0.462	0.574	0.490	Bulgaria
0.832	0.802	0.499	0.810	0.207	0.066	0.058	0.129	0.111	Romania
1.991	1.812	1.485	1.155	1.157	0.816	0.675	0.734	0.457	Slovak Republic
-	-	-	1.559	1.559	-	-	-	-	*Bosnia-Herzegovina*
-	-	-	0.020	0.029	0.015	0.013	0.013	0.011	*Croatia*
-	-	-	0.012	0.010	0.015	0.014	0.014	0.014	*FYROM*
-	-	-	0.140	0.100	0.075	0.052	0.047	0.027	*Slovenia*
-	-	-	2.360	0.405	0.414	0.429	0.411	0.443	*FR of Yugoslavia*
1.348	1.313	1.298	4.089	2.099	0.517	0.506	0.484	0.494	Former Yugoslavia
5.329	4.865	4.070	6.895	4.308	2.013	1.739	1.945	1.568	**Non-OECD Europe**

Includes agriculture, commercial and public services, residential and non-specified other sectors.
People's Rep. of China: up to 1978, the breakdown of final consumption by sector is incomplete.

Other Sectors' Consumption of Coal (Mtoe)
Consommation de charbon des autres secteurs (Mtep)
Kohleverbrauch in sonstigen Sektoren (Mtoe)
Consumo di carbone siderurgico negli altri settori (Mtep)
他 の 部 門 の 石 炭 消 費 量（石 油 換 算 百 万 ト ン）
Consumo de carbón de otros sectores (Mtep)
Потребление угля другими секторами (Мтнэ)

	1971	1973	1978	1983	1984	1985	1986	1987	1988
Arménie	-	-	-	-	-	-	-	-	-
Azerbaïdjan	-	-	-	-	-	-	-	-	-
Bélarus	-	-	-	-	-	-	-	-	-
Estonie	-	-	-	-	-	-	-	-	-
Géorgie	-	-	-	-	-	-	-	-	-
Kazakhstan	-	-	-	-	-	-	-	-	-
Lettonie	-	-	-	-	-	-	-	-	-
Lituanie	-	-	-	-	-	-	-	-	-
Moldova	-	-	-	-	-	-	-	-	-
Russie	-	-	-	-	-	-	-	-	-
Tadjikistan	-	-	-	-	-	-	-	-	-
Turkménistan	-	-	-	-	-	-	-	-	-
Ukraine	-	-	-	-	-	-	-	-	-
Ouzbékistan	-	-	-	-	-	-	-	-	-
Ex-URSS	56.060	59.713	65.392	59.105	59.708	61.184	60.214	59.751	61.190
Total non-OCDE	199.264	212.581	286.427	155.516	161.997	171.210	171.289	175.801	183.821
OCDE Amérique du N.	17.621	13.994	11.886	10.552	10.215	8.320	8.050	7.682	8.038
OCDE Pacifique	7.898	8.690	9.317	10.747	12.081	12.663	12.970	12.447	12.428
OCDE Europe	90.379	79.317	68.548	64.815	64.928	70.580	70.175	72.355	66.932
Total OCDE	115.897	102.001	89.750	86.114	87.225	91.563	91.194	92.484	87.398
Monde	315.162	314.581	376.177	241.631	249.222	262.773	262.483	268.285	271.220

Comprend agriculture, services privés et publics, secteur résidentiel et autres secteurs non spécifiés.
Ex-URSS: les séries antérieures à 1990-1992 ne sont pas comparables aux années récentes; 1991 a été estimé.

Other Sectors' Consumption of Coal (Mtoe)
Consommation de charbon des autres secteurs (Mtep)
Kohleverbrauch in sonstigen Sektoren (Mtoe)
Consumo di carbone siderurgico negli altri settori (Mtep)
他の部門の石炭消費量（石油換算百万トン）
Consumo de carbón de otros sectores (Mtep)
Потребление угля другими секторами (Мтнэ)

1989	1990	1991	1992	1993	1994	1995	1996	1997	
-	-	-	0.063	0.001	0.016	0.001	0.002	0.002	Armenia
-	-	-	0.012	-	-	-	-	-	Azerbaijan
-	-	-	0.983	1.061	0.828	0.764	0.745	0.609	Belarus
-	-	-	0.077	0.036	0.019	0.019	0.046	0.041	Estonia
-	-	-	0.180	0.128	0.067	0.032	0.032	0.002	Georgia
-	-	-	-	-	0.040	-	-	-	Kazakhstan
-	-	-	0.425	0.177	0.191	0.110	0.096	0.090	Latvia
-	-	-	0.252	0.232	0.164	0.136	0.134	0.110	Lithuania
-	-	-	0.258	0.209	0.187	0.104	0.152	0.105	Moldova
-	-	-	29.867	24.675	19.898	17.064	15.142	13.036	Russia
-	-	-	0.260	0.151	0.027	0.007	0.051	0.048	Tajikistan
-	-	-	0.266	0.060	-	-	0.044	-	Turkmenistan
-	-	-	11.651	10.836	8.478	6.810	5.273	4.627	Ukraine
-	-	-	0.814	0.584	0.596	0.251	0.267	0.150	Uzbekistan
60.715	52.548	40.351	44.373	37.550	30.065	25.129	21.943	18.798	**Former USSR**
180.594	170.662	156.024	153.565	147.117	131.583	128.586	127.306	111.533	**Non-OECD Total**
7.706	8.895	8.693	2.279	1.900	1.813	1.666	1.673	1.958	OECD North America
10.979	10.481	8.840	6.781	5.301	3.764	3.075	2.478	2.017	OECD Pacific
59.546	47.068	40.573	35.718	33.224	27.657	25.234	24.306	22.104	OECD Europe
78.232	66.444	58.106	44.778	40.424	33.234	29.974	28.457	26.079	**OECD Total**
258.826	237.106	214.130	198.343	187.541	164.816	158.559	155.763	137.612	**World**

Includes agriculture, commercial and public services, residential and non-specified other sectors.
Former USSR: series up to 1990-1992 are not comparable with the recent years; 1991 is estimated.

Other Sectors' Consumption of Oil (Mtoe)
Consommation de pétrole des autres secteurs (Mtep)
Ölverbrauch in sonstigen Sektoren (Mtoe)
Consumo di petrolio negli altri settori (Mtep)
他の部門の石油消費量（石油換算百万トン）
Consumo de petróleo de otros sectores (Mtep)
Потребление нефти и нефтепродуктов другими секторами (Мтнэ)

	1971	1973	1978	1983	1984	1985	1986	1987	1988
Algérie	0.560	0.793	1.327	2.055	2.266	2.174	2.198	2.267	2.223
Angola	0.042	0.050	0.037	0.049	0.062	0.072	0.062	0.067	0.071
Bénin	0.014	0.020	0.028	0.014	0.014	0.019	0.009	0.012	0.013
Cameroun	0.019	0.015	0.058	0.096	0.106	0.129	0.133	0.132	0.141
Congo	0.011	0.013	0.032	0.025	0.025	0.024	0.026	0.008	0.007
RD du Congo	0.110	0.149	0.140	0.305	0.277	0.252	0.195	0.235	0.277
Egypte	1.271	1.411	1.984	3.030	3.286	3.485	3.557	3.212	3.153
Ethiopie	0.028	0.035	0.044	0.085	0.067	0.057	0.089	0.131	0.115
Gabon	-	0.014	0.020	0.086	0.105	0.108	0.090	0.077	0.087
Ghana	0.126	0.145	0.193	0.188	0.130	0.144	0.154	0.172	0.190
Côte d'Ivoire	0.093	0.091	0.151	0.150	0.150	0.153	0.146	0.139	0.141
Kenya	0.201	0.231	0.282	0.230	0.265	0.274	0.285	0.295	0.339
Libye	0.126	0.128	0.254	0.166	0.189	0.223	0.202	0.338	0.336
Maroc	0.365	0.433	0.723	0.832	0.861	0.913	0.934	0.987	1.993
Mozambique	0.184	0.189	0.222	0.212	0.189	0.245	0.264	0.264	0.265
Nigéria	0.291	0.377	0.992	1.550	1.523	1.660	1.682	2.049	1.898
Sénégal	0.023	0.024	0.031	0.038	0.040	0.045	0.045	0.048	0.056
Afrique du Sud	1.455	1.486	1.567	1.255	1.365	1.342	1.329	1.418	1.537
Soudan	0.105	0.083	0.031	0.051	0.048	0.043	0.050	0.045	0.048
Tanzanie	0.064	0.081	0.101	0.060	0.063	0.064	0.101	0.096	0.108
Tunisie	0.233	0.296	0.497	0.689	0.733	0.759	0.757	0.768	0.776
Zambie	0.072	0.091	0.087	0.093	0.094	0.094	0.083	0.092	0.100
Zimbabwe	0.110	0.133	0.128	0.104	0.090	0.095	0.102	0.176	0.118
Autre Afrique	1.536	1.612	1.228	0.709	0.660	0.768	0.652	1.306	1.256
Afrique	7.038	7.898	10.156	12.071	12.605	13.142	13.148	14.332	15.246
Argentine	3.414	3.891	4.493	4.095	3.732	3.658	3.484	3.823	3.679
Bolivie	0.123	0.149	0.215	0.280	0.269	0.266	0.246	0.257	0.239
Brésil	2.766	3.515	4.851	6.529	6.329	6.771	7.682	8.365	8.848
Chili	0.792	0.938	0.773	0.661	0.664	0.640	0.742	0.706	0.769
Colombie	1.508	1.678	1.451	1.564	1.470	1.577	1.544	1.929	1.378
Costa Rica	0.029	0.031	0.038	0.022	0.025	0.023	0.024	0.024	0.023
Cuba	0.876	1.432	1.224	1.512	1.663	1.541	1.537	1.489	1.535
République dominicaine	0.067	0.071	0.089	0.123	0.123	0.163	0.145	0.157	0.171
Equateur	0.146	0.177	0.529	0.618	0.708	0.790	0.864	0.824	0.853
El Salvador	0.051	0.052	0.057	0.052	0.045	0.051	0.057	0.055	0.053
Guatemala	0.074	0.094	0.126	0.147	0.155	0.154	0.156	0.151	0.149
Haiti	0.007	0.006	0.014	0.018	0.019	0.020	0.022	0.020	0.020
Honduras	0.065	0.076	0.107	0.099	0.090	0.093	0.099	0.106	0.119
Jamaïque	0.102	0.169	0.399	0.176	0.165	0.182	0.238	0.201	0.210
Antilles néerlandaises	0.136	0.112	0.114	0.117	0.103	0.099	0.099	0.074	0.058
Nicaragua	0.062	0.071	0.077	0.079	0.074	0.078	0.089	0.084	0.074
Panama	0.098	0.102	0.087	0.055	0.057	0.061	0.065	0.075	0.073
Paraguay	0.019	0.023	0.026	0.043	0.044	0.055	0.050	0.051	0.041
Pérou	1.465	1.089	1.351	1.196	1.229	1.232	1.413	1.857	1.951
Trinité-et-Tobago	0.048	0.065	0.064	0.066	0.079	0.065	0.069	0.051	0.088
Uruguay	0.314	0.303	0.308	0.336	0.313	0.301	0.287	0.310	0.323
Vénézuela	0.717	0.793	0.988	1.172	1.182	1.592	1.342	1.169	1.182
Autre Amérique latine	1.317	1.880	1.442	1.013	1.168	1.062	1.237	1.312	1.189
Amérique latine	14.197	16.716	18.822	19.973	19.704	20.475	21.489	23.090	23.027

Comprend agriculture, services privés et publics, secteur résidentiel et autres secteurs non spécifiés.

Other Sectors' Consumption of Oil (Mtoe)
Consommation de pétrole des autres secteurs (Mtep)
Ölverbrauch in sonstigen Sektoren (Mtoe)
Consumo di petrolio negli altri settori (Mtep)
他の部門の石油消費量（石油換算百万トン）
Consumo de petróleo de otros sectores (Mtep)
Потребление нефти и нефтепродуктов другими секторами (Мтнэ)

1989	1990	1991	1992	1993	1994	1995	1996	1997	
2.479	2.634	3.203	5.170	4.989	4.572	4.507	4.379	6.124	Algeria
0.075	0.079	0.078	0.074	0.076	0.074	0.077	0.089	0.098	Angola
0.012	0.012	0.008	0.013	0.010	0.013	0.014	0.042	0.037	Benin
0.227	0.229	0.226	0.174	0.124	0.123	0.130	0.125	0.128	Cameroon
0.040	0.040	0.041	0.041	0.040	0.040	0.055	0.054	0.054	Congo
0.234	0.294	0.316	0.296	0.323	0.325	0.330	0.330	0.330	DR of Congo
3.282	3.203	3.046	3.003	2.691	2.624	2.709	2.831	2.975	Egypt
0.115	0.132	0.134	0.137	0.147	0.149	0.157	0.176	0.091	Ethiopia
0.047	0.049	0.054	0.053	0.055	0.051	0.056	0.059	0.065	Gabon
0.211	0.190	0.198	0.207	0.223	0.243	0.245	0.246	0.246	Ghana
0.139	0.114	0.127	0.114	0.109	0.113	0.118	0.137	0.140	Ivory Coast
0.394	0.325	0.246	0.282	0.282	0.310	0.303	0.323	0.303	Kenya
0.392	0.471	0.516	0.540	0.583	0.649	0.706	0.732	0.760	Libya
2.089	2.222	2.448	2.810	2.988	3.236	3.369	3.444	3.492	Morocco
0.251	0.251	0.209	0.295	0.332	0.284	0.289	0.292	0.402	Mozambique
1.914	1.402	1.373	1.562	1.288	0.822	0.795	0.897	0.906	Nigeria
0.096	0.100	0.099	0.107	0.096	0.095	0.110	0.121	0.142	Senegal
1.583	1.562	1.588	1.465	2.041	2.066	2.221	2.175	2.016	South Africa
0.047	0.052	0.059	0.057	0.053	0.063	0.068	0.096	0.096	Sudan
0.111	0.115	0.117	0.106	0.080	0.080	0.082	0.082	0.082	Tanzania
0.838	0.796	0.822	0.821	0.939	0.946	0.983	1.039	1.063	Tunisia
0.101	0.110	0.113	0.113	0.119	0.123	0.126	0.126	0.126	Zambia
0.071	0.229	0.278	0.253	0.359	0.406	0.371	0.367	0.367	Zimbabwe
1.371	1.228	1.526	1.594	1.581	1.617	1.922	1.962	2.029	Other Africa
16.115	15.838	16.826	19.287	19.529	19.024	19.742	20.122	22.070	**Africa**
3.854	3.920	3.338	3.596	3.477	3.778	3.909	4.142	4.061	Argentina
0.242	0.233	0.242	0.259	0.274	0.294	0.322	0.343	0.385	Bolivia
9.135	9.267	9.183	9.641	9.998	10.591	11.168	11.415	11.617	Brazil
0.831	0.946	1.037	1.205	1.274	1.352	1.353	1.326	1.363	Chile
1.596	1.467	1.433	1.455	1.534	2.191	1.403	1.582	1.449	Colombia
0.044	0.051	0.070	0.082	0.096	0.103	0.113	0.135	0.146	Costa Rica
1.510	1.693	1.397	1.083	0.842	0.851	0.847	0.870	0.865	Cuba
0.307	0.311	0.349	0.467	0.478	0.542	0.574	0.606	0.693	Dominican Republic
0.862	0.875	0.944	0.997	1.012	1.113	1.208	1.259	1.367	Ecuador
0.057	0.061	0.064	0.074	0.086	0.093	0.111	0.103	0.111	El Salvador
0.181	0.191	0.200	0.215	0.222	0.232	0.264	0.275	0.291	Guatemala
0.026	0.027	0.023	0.017	0.021	0.011	0.033	0.033	0.035	Haiti
0.125	0.117	0.121	0.149	0.146	0.138	0.162	0.150	0.157	Honduras
0.195	1.109	1.163	1.195	0.298	0.482	0.372	0.394	0.414	Jamaica
0.065	0.077	0.077	0.082	0.164	0.170	0.172	0.178	0.178	Netherlands Antilles
0.066	0.069	0.060	0.059	0.060	0.082	0.060	0.067	0.112	Nicaragua
0.073	0.082	0.087	0.093	0.099	0.106	0.112	0.116	0.123	Panama
0.054	0.054	0.056	0.063	0.066	0.069	0.069	0.072	0.071	Paraguay
1.851	1.578	1.534	1.475	1.575	1.718	1.915	1.955	1.804	Peru
0.053	0.053	0.055	0.085	0.083	0.065	0.055	0.059	0.060	Trinidad-and-Tobago
0.330	0.334	0.345	0.368	0.374	0.374	0.384	0.402	0.388	Uruguay
1.328	1.388	1.390	1.919	2.702	2.468	2.463	2.388	2.700	Venezuela
1.488	1.503	1.439	1.277	1.317	1.324	1.397	1.466	1.530	Other Latin America
24.273	25.407	24.605	25.854	26.199	28.148	28.466	29.335	29.919	**Latin America**

Includes agriculture, commercial and public services, residential and non-specified other sectors.

Other Sectors' Consumption of Oil (Mtoe)
Consommation de pétrole des autres secteurs (Mtep)
Ölverbrauch in sonstigen Sektoren (Mtoe)
Consumo di petrolio negli altri settori (Mtep)
他の部門の石油消費量（石油換算百万トン）
Consumo de petróleo de otros sectores (Mtep)
Потребление нефти и нефтепродуктов другими секторами (Мтнэ)

	1971	1973	1978	1983	1984	1985	1986	1987	1988
Bangladesh	0.287	0.241	0.459	0.470	0.450	0.530	0.595	0.648	0.621
Brunei	0.007	0.007	0.007	0.013	0.013	0.014	0.013	0.013	0.015
Inde	4.878	5.021	5.601	7.870	8.707	9.307	9.906	11.828	13.183
Indonésie	2.170	2.820	4.908	6.270	6.100	5.945	6.025	6.038	6.383
RPD de Corée	0.031	0.037	0.106	0.219	0.230	0.240	0.251	0.261	0.261
Malaisie	0.139	0.181	0.476	0.514	0.523	0.503	0.542	0.546	0.652
Myanmar	0.306	0.260	0.123	0.023	0.022	0.012	0.014	0.010	0.010
Népal	0.023	0.037	0.038	0.041	0.050	0.055	0.067	0.076	0.086
Pakistan	1.320	1.384	1.478	1.436	1.515	1.565	1.638	1.393	1.535
Philippines	1.195	1.248	1.960	1.867	1.625	1.458	1.558	1.727	1.840
Singapour	0.093	0.068	0.062	0.069	0.070	0.073	0.088	0.079	0.079
Sri Lanka	0.282	0.287	0.254	0.203	0.177	0.187	0.182	0.181	0.193
Taipei chinois	1.017	1.381	1.934	2.325	2.405	2.467	2.576	2.725	2.800
Thaïlande	1.045	1.259	1.765	1.992	1.983	1.926	2.063	2.088	2.306
Viêt-Nam	2.497	2.731	0.704	0.474	0.373	0.282	0.277	0.393	0.378
Autre Asie	0.777	0.988	0.494	0.119	0.104	0.275	0.182	0.299	0.294
Asie	16.067	17.951	20.366	23.906	24.347	24.842	25.978	28.305	30.635
Rép. populaire de Chine	20.687	23.589	34.882	11.931	12.325	11.620	12.228	13.009	13.995
Hong-Kong, Chine	0.304	0.200	0.355	0.162	0.157	0.135	0.146	0.148	0.131
Chine	20.991	23.789	35.237	12.092	12.482	11.755	12.374	13.157	14.126
Albanie	0.108	0.149	0.284	0.099	0.084	0.057	0.057	0.059	0.063
Bulgarie	6.850	7.371	5.780	4.602	4.201	3.704	3.720	3.517	1.872
Chypre	0.042	0.046	0.042	0.054	0.055	0.055	0.056	0.061	0.062
Malte	0.035	0.042	0.045	0.025	0.035	0.019	0.027	0.033	0.033
Roumanie	5.276	6.168	7.235	5.660	5.335	6.298	5.941	6.333	5.973
République slovaque	0.439	0.485	0.521	0.785	0.747	0.746	0.727	0.729	0.712
Croatie	-	-	-	-	-	-	-	-	-
Ex-RYM	-	-	-	-	-	-	-	-	-
Slovénie	-	-	-	-	-	-	-	-	-
RF de Yougoslavie	-	-	-	-	-	-	-	-	-
Ex-Yougoslavie	1.158	1.277	1.883	1.456	1.356	1.149	1.148	1.329	1.243
Europe non-OCDE	13.908	15.539	15.791	12.680	11.812	12.029	11.676	12.061	9.959
Arménie	-	-	-	-	-	-	-	-	-
Azerbaïdjan	-	-	-	-	-	-	-	-	-
Bélarus	-	-	-	-	-	-	-	-	-
Estonie	-	-	-	-	-	-	-	-	-
Géorgie	-	-	-	-	-	-	-	-	-
Kazakhstan	-	-	-	-	-	-	-	-	-
Kirghizistan	-	-	-	-	-	-	-	-	-
Lettonie	-	-	-	-	-	-	-	-	-
Lituanie	-	-	-	-	-	-	-	-	-
Moldova	-	-	-	-	-	-	-	-	-
Russie	-	-	-	-	-	-	-	-	-
Tadjikistan	-	-	-	-	-	-	-	-	-
Turkménistan	-	-	-	-	-	-	-	-	-
Ukraine	-	-	-	-	-	-	-	-	-
Ouzbékistan	-	-	-	-	-	-	-	-	-
Ex-URSS	49.233	54.576	67.528	74.321	75.695	75.357	75.673	75.883	75.778

Comprend agriculture, services privés et publics, secteur résidentiel et autres secteurs non spécifiés.
Rép.populaire de Chine: jusqu'en 1978 la ventilation de la consommation finale par secteur est incomplète.
Ex-URSS: les séries antérieures à 1990-1992 ne sont pas comparables aux années récentes; 1991 a été estimé.

Other Sectors' Consumption of Oil (Mtoe)
Consommation de pétrole des autres secteurs (Mtep)
Ölverbrauch in sonstigen Sektoren (Mtoe)

Consumo di petrolio negli altri settori (Mtep)

他の部門の石油消費量（石油換算百万トン）

Consumo de petróleo de otros sectores (Mtep)

Потребление нефти и нефтепродуктов другими секторами (Мтнэ)

1989	1990	1991	1992	1993	1994	1995	1996	1997	
0.786	0.758	0.682	0.693	0.729	0.704	0.896	0.934	0.955	Bangladesh
0.017	0.015	0.017	0.019	0.022	0.023	0.021	0.022	0.024	Brunei
13.121	12.460	12.690	13.354	13.692	14.577	15.599	16.434	16.909	India
7.508	8.010	8.158	8.706	9.267	9.594	10.015	10.647	10.954	Indonesia
0.272	0.272	0.256	0.218	0.180	0.158	0.139	0.122	0.118	DPR of Korea
0.667	0.739	0.761	0.992	0.932	1.542	1.445	1.581	1.743	Malaysia
0.015	0.012	0.011	0.067	0.086	0.109	0.151	0.164	0.183	Myanmar
0.080	0.075	0.095	0.119	0.123	0.202	0.207	0.207	0.207	Nepal
1.624	1.791	1.588	1.275	1.275	1.226	1.248	1.291	1.448	Pakistan
2.143	2.477	2.523	2.972	3.464	3.568	3.465	3.619	3.899	Philippines
0.090	0.043	0.068	-	-	-	-	-	-	Singapore
0.202	0.213	0.211	0.268	0.281	0.290	0.302	0.322	0.275	Sri Lanka
2.922	2.962	2.860	2.665	2.643	2.747	2.596	2.871	2.602	Chinese Taipei
2.543	2.731	2.755	2.907	2.721	2.767	2.890	3.266	2.965	Thailand
0.383	0.466	0.452	0.450	0.606	0.593	0.608	1.018	1.219	Vietnam
0.509	0.598	0.636	0.476	0.469	0.492	0.771	0.740	0.774	Other Asia
32.883	33.623	33.762	35.180	36.491	38.592	40.354	43.239	44.276	**Asia**
14.969	15.435	16.613	18.156	20.889	21.540	26.269	30.822	34.435	People's Rep.of China
0.104	0.094	0.083	0.087	0.073	0.068	0.066	0.059	0.057	Hong Kong, China
15.073	15.529	16.696	18.242	20.962	21.608	26.335	30.880	34.492	**China**
0.064	0.059	0.064	0.043	0.030	0.030	0.077	0.072	0.060	Albania
1.820	1.216	0.941	0.699	0.875	0.949	0.718	0.668	0.630	Bulgaria
0.061	0.068	0.068	0.080	0.074	0.074	0.075	0.076	0.080	Cyprus
0.033	0.034	0.035	0.037	0.020	0.020	0.012	0.053	0.088	Malta
1.942	1.861	1.557	2.286	1.097	0.771	0.969	0.783	1.322	Romania
0.682	0.642	0.441	0.395	0.371	0.314	0.276	0.266	0.294	Slovak Republic
-	-	-	0.506	0.505	0.563	0.577	0.501	0.615	*Croatia*
-	-	-	0.128	0.047	0.179	0.148	0.132	0.094	*FYROM*
-	-	-	0.409	0.555	0.547	0.622	0.833	0.859	*Slovenia*
-	-	-	0.155	0.136	0.132	0.147	0.218	0.323	*FR of Yugoslavia*
0.967	0.775	1.171	1.197	1.242	1.420	1.493	1.685	1.892	Former Yugoslavia
5.568	4.655	4.277	4.737	3.710	3.580	3.621	3.604	4.365	**Non-OECD Europe**
-	-	-	0.239	0.113	0.095	0.068	0.038	0.038	Armenia
-	-	-	2.280	1.480	1.377	1.098	0.500	0.566	Azerbaijan
-	-	-	2.130	1.961	1.377	1.517	1.732	1.650	Belarus
-	-	-	0.167	0.159	0.119	0.106	0.149	0.146	Estonia
-	-	-	0.540	0.297	0.135	0.050	0.183	0.282	Georgia
-	-	-	11.016	8.006	6.522	6.556	4.921	4.534	Kazakhstan
-	-	-	1.427	0.865	0.293	0.292	0.332	0.294	Kyrgyzstan
-	-	-	0.676	0.575	0.643	0.351	0.216	0.142	Latvia
-	-	-	0.745	0.468	0.327	0.294	0.264	0.226	Lithuania
-	-	-	0.443	0.314	0.268	0.271	0.233	0.224	Moldova
-	-	-	29.887	30.291	21.041	26.676	26.763	26.337	Russia
-	-	-	0.805	0.480	0.159	0.159	0.159	0.159	Tajikistan
-	-	-	3.644	1.835	1.897	1.673	1.821	1.888	Turkmenistan
-	-	-	7.908	4.763	4.532	5.691	4.246	3.993	Ukraine
-	-	-	3.682	3.600	3.268	2.742	1.398	1.453	Uzbekistan
75.778	75.364	52.679	65.590	55.205	42.054	47.544	42.954	41.931	**Former USSR**

Includes agriculture, commercial and public services, residential and non-specified other sectors.

People's Rep. of China: up to 1978, the breakdown of final consumption by sector is incomplete.

Former USSR: series up to 1990-1992 are not comparable with the recent years; 1991 is estimated.

Other Sectors' Consumption of Oil (Mtoe)
Consommation de pétrole des autres secteurs (Mtep)
Ölverbrauch in sonstigen Sektoren (Mtoe)
Consumo di petrolio negli altri settori (Mtep)
他の部門の石油消費量（石油換算百万トン）
Consumo de petróleo de otros sectores (Mtep)
Потребление нефти и нефтепродуктов другими секторами (Мтнэ)

	1971	1973	1978	1983	1984	1985	1986	1987	1988
Bahrein	0.016	0.013	0.020	0.026	0.030	0.031	0.033	0.037	0.039
Iran	4.261	5.877	11.759	16.618	18.464	17.843	11.872	13.169	13.163
Irak	0.726	0.728	1.296	1.459	1.433	1.505	1.620	1.798	2.018
Israël	0.546	0.618	0.582	0.636	0.627	0.517	0.840	0.919	0.948
Jordanie	0.122	0.115	0.191	0.346	0.361	0.350	0.294	0.273	0.308
Koweit	0.031	0.031	0.093	0.103	0.115	0.118	0.116	0.124	0.116
Liban	0.102	0.107	0.099	0.132	0.127	0.175	0.186	0.170	0.137
Oman	0.042	0.033	0.154	0.250	0.300	0.346	0.328	0.307	0.343
Qatar	0.004	0.006	0.005	0.006	0.010	0.016	0.016	0.016	0.016
Arabie saoudite	0.692	0.871	3.951	3.313	5.032	4.610	4.932	4.824	10.926
Syrie	0.369	0.351	0.495	1.350	1.607	1.112	1.410	1.748	1.859
Emirats arabes unis	0.008	0.010	0.022	0.060	0.065	0.069	0.070	0.082	0.127
Yémen	0.058	0.057	0.108	0.213	0.228	0.276	0.223	0.258	0.292
Moyen-Orient	6.978	8.818	18.776	24.513	28.398	26.966	21.940	23.724	30.293
Total non-OCDE	128.411	145.287	186.675	179.556	185.044	184.565	182.278	190.552	199.064
OCDE Amérique du N.	165.060	162.614	152.929	94.430	96.949	97.392	95.747	97.108	99.157
OCDE Pacifique	31.938	40.750	45.413	42.628	45.861	44.447	46.577	47.601	52.571
OCDE Europe	164.361	192.551	181.576	129.752	130.701	133.205	137.602	131.158	125.101
Total OCDE	361.359	395.915	379.918	266.809	273.511	275.044	279.926	275.868	276.829
Monde	489.769	541.202	566.593	446.365	458.554	459.609	462.204	466.420	475.893

Comprend agriculture, services privés et publics, secteur résidentiel et autres secteurs non spécifiés.

Other Sectors' Consumption of Oil (Mtoe)
Consommation de pétrole des autres secteurs (Mtep)
Ölverbrauch in sonstigen Sektoren (Mtoe)
Consumo di petrolio negli altri settori (Mtep)
他の部門の石油消費量（石油換算百万トン）
Consumo de petróleo de otros sectores (Mtep)
Потребление нефти и нефтепродуктов другими секторами (Мтнэ)

1989	1990	1991	1992	1993	1994	1995	1996	1997	
0.042	0.045	0.049	0.052	0.052	0.055	0.048	0.051	0.051	Bahrain
14.313	28.962	30.845	32.013	17.220	17.899	19.057	20.265	20.190	Iran
2.167	1.780	0.924	1.291	2.260	2.398	2.340	2.269	2.326	Iraq
0.996	1.067	1.079	1.526	1.620	1.719	1.970	2.085	2.112	Israel
0.322	0.582	0.595	0.770	0.746	0.806	0.814	0.827	0.902	Jordan
0.133	0.165	0.090	0.095	0.109	0.125	0.130	0.123	0.140	Kuwait
0.137	0.476	0.739	0.599	0.501	0.596	0.605	0.531	0.710	Lebanon
0.348	0.174	0.208	0.214	0.220	0.255	0.243	0.248	0.247	Oman
0.016	0.010	0.004	0.010	0.014	0.021	0.021	0.022	0.022	Qatar
10.999	11.137	11.496	12.585	13.439	12.813	11.061	11.609	12.138	Saudi Arabia
4.438	4.795	5.398	5.611	5.722	5.965	6.156	6.371	6.814	Syria
0.150	0.161	0.145	0.330	0.434	0.479	0.483	0.449	0.465	United Arab Emirates
0.323	0.249	0.323	0.401	0.444	0.494	0.524	0.556	0.603	Yemen
34.382	49.603	51.896	55.497	42.782	43.625	43.451	45.407	46.721	**Middle East**
204.072	220.019	200.740	224.387	204.879	196.631	209.513	215.541	223.774	**Non-OECD Total**
96.600	83.349	79.184	84.357	85.551	84.490	83.674	86.555	85.506	OECD North America
52.858	56.998	60.723	64.640	65.829	67.960	72.144	73.396	71.453	OECD Pacific
113.686	116.289	124.785	123.252	122.416	117.640	119.062	126.187	120.365	OECD Europe
263.145	256.636	264.693	272.249	273.796	270.090	274.881	286.138	277.323	**OECD Total**
467.217	476.655	465.434	496.636	478.675	466.721	484.394	501.680	501.098	**World**

Includes agriculture, commercial and public services, residential and non-specified other sectors.

Other Sectors' Consumption of Gas (Mtoe)
Consommation de gaz des autres secteurs (Mtep)
Gasverbrauch in sonstigen Sektoren (Mtoe)

Consumo di gas negli altri settori (Mtep)

他の部門のガス消費量（石油換算百万トン）

Consumo de gas natural de otros sectores (Mtep)

Потребление газа другими секторами (Мтнэ)

	1971	1973	1978	1983	1984	1985	1986	1987	1988
Algérie	0.082	0.099	0.292	0.557	0.632	0.719	0.952	1.038	1.098
Egypte	0.002	0.001	0.002	0.016	0.025	0.032	0.041	0.047	0.058
Gabon	-	-	-	-	-	0.000	0.027	0.049	0.079
Afrique du Sud	0.005	0.008	0.009	0.017	0.017	0.016	0.016	0.016	0.017
Tunisie	0.006	0.007	0.007	0.010	0.009	0.010	0.020	0.022	0.021
Autre Afrique	0.001	0.001	0.001	0.001	0.000	0.000	0.000	0.000	0.000
Afrique	0.096	0.116	0.311	0.600	0.683	0.777	1.056	1.172	1.274
Argentine	1.182	1.371	1.969	3.311	3.995	3.882	4.110	4.586	5.229
Bolivie	-	-	-	-	-	-	-	0.002	0.007
Brésil	0.127	0.133	0.162	0.192	0.193	0.204	0.204	0.214	0.217
Chili	0.052	0.092	0.145	0.169	0.175	0.171	0.181	0.174	0.185
Colombie	-	-	0.002	0.015	0.028	0.031	0.046	0.059	0.081
Cuba	0.027	0.030	0.044	0.048	0.052	0.094	0.095	0.097	0.101
Panama	0.007	0.005	0.001	0.000	0.000	-	-	-	-
Pérou	0.042	0.042	0.037	0.028	0.063	0.055	0.045	0.046	0.040
Vénézuela	0.460	0.557	0.684	0.984	0.929	0.903	0.504	0.641	0.567
Autre Amérique latine	0.004	0.004	0.005	0.006	0.012	0.013	0.014	0.012	0.009
Amérique latine	1.903	2.235	3.047	4.754	5.448	5.351	5.198	5.831	6.435
Bangladesh	0.002	0.003	0.038	0.163	0.179	0.196	0.202	0.220	0.249
Inde	0.018	0.020	0.052	0.066	0.074	0.091	0.109	0.123	0.119
Indonésie	-	0.014	0.263	0.475	0.069	3.355	3.428	3.530	3.754
Malaisie	0.006	0.009	0.029	0.043	0.042	0.046	0.053	0.073	0.060
Myanmar	0.000	0.000	0.000	0.001	0.001	0.001	0.001	0.001	0.001
Pakistan	0.089	0.117	0.336	0.785	0.871	0.994	1.104	1.350	1.470
Philippines	0.008	0.008	0.006	0.004	0.003	0.002	0.002	0.003	0.003
Singapour	0.015	0.017	0.024	0.027	0.028	0.026	0.025	0.023	0.024
Taipei chinois	0.030	0.066	0.225	0.344	0.402	0.428	0.448	0.470	0.505
Autre Asie	0.193	0.145	0.131	0.268	0.407	0.521	0.525	0.171	0.192
Asie	0.361	0.401	1.104	2.176	2.075	5.661	5.898	5.963	6.377
Rép. populaire de Chine	-	-	-	0.680	1.005	0.718	0.899	1.155	1.545
Hong-Kong, Chine	0.020	0.025	0.050	0.122	0.143	0.165	0.187	0.219	0.254
Chine	0.020	0.025	0.050	0.802	1.148	0.883	1.086	1.373	1.798
Albanie	0.106	0.159	0.297	0.322	0.322	0.322	0.322	0.322	0.344
Bulgarie	0.247	0.168	2.312	4.022	4.229	-	-	-	-
Roumanie	-	2.449	-	-	-	3.061	2.971	3.047	2.714
République slovaque	0.486	0.579	1.137	1.598	1.591	1.569	1.584	1.603	1.628
Bosnie-Herzégovine	-	-	-	-	-	-	-	-	-
Croatie	-	-	-	-	-	-	-	-	-
Ex-RYM	-	-	-	-	-	-	-	-	-
Slovénie	-	-	-	-	-	-	-	-	-
RF de Yougoslavie	-	-	-	-	-	-	-	-	-
Ex-Yougoslavie	0.094	0.124	0.237	0.154	0.308	0.380	0.375	1.109	0.916
Europe non-OCDE	0.932	3.480	3.983	6.097	6.451	5.333	5.252	6.081	5.601

Comprend agriculture, services privés et publics, secteur résidentiel et autres secteurs non spécifiés.
Rép.populaire de Chine: jusqu'en 1978 la ventilation de la consommation finale par secteur est incomplète.

INTERNATIONAL ENERGY AGENCY

Other Sectors' Consumption of Gas (Mtoe)
Consommation de gaz des autres secteurs (Mtep)
Gasverbrauch in sonstigen Sektoren (Mtoe)
Consumo di gas negli altri settori (Mtep)
他の部門のガス消費量（石油換算百万トン）
Consumo de gas natural de otros sectores (Mtep)
Потребление газа другими секторами (Мтнэ)

1989	1990	1991	1992	1993	1994	1995	1996	1997	
1.167	1.182	1.511	1.597	1.724	1.819	1.878	1.989	1.857	Algeria
0.069	0.071	0.081	0.086	0.093	0.689	0.821	0.876	0.924	Egypt
-	-	-	-	-	-	-	-	-	Gabon
0.016	0.017	0.025	0.024	0.054	0.034	0.028	0.028	0.030	South Africa
0.023	0.060	0.071	0.081	0.089	0.092	0.106	0.110	0.115	Tunisia
0.000	0.000	0.000	0.000	0.000	0.000	0.000	0.000	0.000	Other Africa
1.275	1.330	1.687	1.788	1.960	2.634	2.833	3.003	2.926	**Africa**
4.578	4.914	5.001	5.109	6.217	6.074	6.046	6.028	6.517	Argentina
0.007	-	0.001	0.001	0.001	0.002	0.002	0.002	0.002	Bolivia
0.223	0.206	0.216	0.199	0.215	0.174	0.190	0.215	0.246	Brazil
0.190	0.207	0.204	0.220	0.225	0.229	0.238	0.245	0.257	Chile
0.092	0.119	0.160	0.202	0.255	0.258	0.285	0.327	0.405	Colombia
0.058	0.104	0.100	0.113	0.110	0.106	0.131	0.122	0.122	Cuba
-	-	-	-	-	-	-	-	-	Panama
0.044	0.041	0.041	0.029	0.031	0.031	0.031	0.031	0.073	Peru
0.500	0.605	0.685	1.244	0.707	0.740	0.766	0.771	0.829	Venezuela
0.013	0.015	0.010	0.009	0.014	0.011	0.014	0.017	0.017	Other Latin America
5.706	6.212	6.417	7.126	7.775	7.626	7.703	7.759	8.468	**Latin America**
0.282	0.292	0.302	0.332	0.364	0.451	0.491	0.544	0.579	Bangladesh
0.110	0.128	0.166	0.269	0.286	0.299	0.334	0.371	0.405	India
4.284	0.584	0.866	0.999	1.058	1.241	1.551	1.759	2.003	Indonesia
0.045	0.032	0.034	0.032	0.023	0.211	0.289	0.374	0.341	Malaysia
0.001	0.001	0.001	0.002	0.002	0.002	0.002	0.002	0.002	Myanmar
1.668	1.668	1.851	1.961	2.108	2.286	2.646	2.973	3.099	Pakistan
0.003	0.003	0.003	0.003	0.003	0.003	0.003	0.003	0.003	Philippines
0.024	0.030	0.032	0.034	0.037	0.040	0.040	0.043	0.044	Singapore
0.545	0.553	0.586	0.630	0.628	0.673	0.715	0.749	0.757	Chinese Taipei
0.210	0.240	0.235	0.223	0.220	0.213	0.210	0.204	0.204	Other Asia
7.171	3.531	4.077	4.485	4.729	5.417	6.282	7.023	7.437	**Asia**
1.873	2.074	2.150	2.229	2.989	2.936	2.728	2.606	2.737	People's Rep.of China
0.283	0.311	0.334	0.374	0.393	0.426	0.451	0.474	0.494	Hong Kong, China
2.156	2.385	2.484	2.602	3.382	3.361	3.179	3.081	3.231	**China**
0.056	0.031	0.017	0.017	0.015	0.008	0.004	0.003	0.003	Albania
-	0.016	0.005	0.007	0.007	0.007	0.020	0.035	0.012	Bulgaria
3.020	3.087	3.535	1.915	1.909	1.986	2.045	2.044	2.372	Romania
1.716	1.732	1.817	1.908	1.888	2.110	1.903	2.057	2.124	Slovak Republic
-	-	-	0.344	-	-	-	-	-	*Bosnia-Herzegovina*
-	-	-	0.305	0.374	0.357	0.437	0.507	0.514	*Croatia*
-	-	-	0.086	0.091	-	-	-	-	*FYROM*
-	-	-	0.036	0.037	0.033	0.049	0.058	0.058	*Slovenia*
-	-	-	0.802	0.889	0.402	0.526	1.316	1.308	*FR of Yugoslavia*
0.614	5.203	2.647	1.573	1.391	0.792	1.013	1.882	1.879	Former Yugoslavia
5.407	10.069	8.020	5.420	5.210	4.903	4.984	6.022	6.390	**Non-OECD Europe**

Includes agriculture, commercial and public services, residential and non-specified other sectors.
People's Rep. of China: up to 1978, the breakdown of final consumption by sector is incomplete.

Other Sectors' Consumption of Gas (Mtoe)
Consommation de gaz des autres secteurs (Mtep)
Gasverbrauch in sonstigen Sektoren (Mtoe)
Consumo di gas negli altri settori (Mtep)
他の部門のガス消費量（石油換算百万トン）
Consumo de gas natural de otros sectores (Mtep)
Потребление газа другими секторами (Мтнэ)

	1971	1973	1978	1983	1984	1985	1986	1987	1988
Arménie	-	-	-	-	-	-	-	-	-
Azerbaïdjan	-	-	-	-	-	-	-	-	-
Bélarus	-	-	-	-	-	-	-	-	-
Estonie	-	-	-	-	-	-	-	-	-
Géorgie	-	-	-	-	-	-	-	-	-
Kazakhstan	-	-	-	-	-	-	-	-	-
Kirghizistan	-	-	-	-	-	-	-	-	-
Lettonie	-	-	-	-	-	-	-	-	-
Lituanie	-	-	-	-	-	-	-	-	-
Moldova	-	-	-	-	-	-	-	-	-
Russie	-	-	-	-	-	-	-	-	-
Tadjikistan	-	-	-	-	-	-	-	-	-
Turkménistan	-	-	-	-	-	-	-	-	-
Ukraine	-	-	-	-	-	-	-	-	-
Ouzbékistan	-	-	-	-	-	-	-	-	-
Ex-URSS	23.100	26.700	37.300	65.483	69.482	74.081	77.480	81.979	86.478
Iran	-	-	-	-	-	-	-	-	4.651
Israël	-	-	0.001	0.000	0.000	0.000	0.000	0.000	0.000
Oman	-	-	-	0.008	0.044	0.050	0.000	0.004	0.072
Arabie saoudite	-	-	-	1.246	-	-	3.955	9.252	9.242
Syrie	-	-	-	-	-	-	0.182	0.172	0.501
Moyen-Orient	-	-	0.001	1.255	0.044	0.050	4.138	9.428	14.466
Total non-OCDE	26.412	32.956	45.796	81.166	85.331	92.136	100.107	111.827	122.429
OCDE Amérique du N.	185.829	185.414	189.631	178.000	185.127	181.600	175.040	175.604	190.337
OCDE Pacifique	4.435	5.666	7.536	9.701	10.519	10.606	11.283	11.544	12.411
OCDE Europe	34.983	48.758	76.510	88.526	92.802	100.943	104.891	109.884	105.877
Total OCDE	225.247	239.838	273.676	276.227	288.447	293.148	291.214	297.032	308.624
Monde	251.659	272.794	319.472	357.394	373.778	385.284	391.321	408.858	431.053

Comprend agriculture, services privés et publics, secteur résidentiel et autres secteurs non spécifiés.
Ex-URSS: les séries antérieures à 1990-1992 ne sont pas comparables aux années récentes; 1991 a été estimé.

Other Sectors' Consumption of Gas (Mtoe)
Consommation de gaz des autres secteurs (Mtep)
Gasverbrauch in sonstigen Sektoren (Mtoe)
Consumo di gas negli altri settori (Mtep)
他の部門のガス消費量（石油換算百万トン）
Consumo de gas natural de otros sectores (Mtep)
Потребление газа другими секторами (Мтнэ)

1989	1990	1991	1992	1993	1994	1995	1996	1997	
-	-	-	0.686	0.242	0.301	0.487	0.100	0.107	Armenia
-	-	-	6.174	1.416	1.509	1.193	2.050	1.932	Azerbaijan
-	-	-	1.482	1.409	1.391	1.360	1.355	1.394	Belarus
-	-	-	0.090	0.063	0.069	0.048	0.046	0.045	Estonia
-	-	-	2.764	0.606	0.664	0.092	0.099	0.113	Georgia
-	-	-	8.841	6.070	4.671	5.813	4.509	3.566	Kazakhstan
-	-	-	0.768	0.495	0.309	0.315	0.376	0.201	Kyrgyzstan
-	-	-	0.262	0.189	0.077	0.126	0.247	0.125	Latvia
-	-	-	0.459	0.321	0.247	0.225	0.226	0.191	Lithuania
-	-	-	0.424	0.899	0.805	0.816	0.966	1.032	Moldova
-	-	-	27.246	66.264	65.807	72.020	42.142	42.049	Russia
-	-	-	0.742	0.614	0.343	0.387	0.497	0.451	Tajikistan
-	-	-	3.301	4.553	6.472	6.972	5.279	5.039	Turkmenistan
-	-	-	21.001	21.690	24.047	25.661	28.873	26.583	Ukraine
-	-	-	19.017	20.416	20.664	12.589	13.821	15.402	Uzbekistan
88.975	91.974	87.314	93.257	125.246	127.376	128.104	100.584	98.229	**Former USSR**
4.695	2.421	5.395	5.802	6.204	7.189	9.145	10.332	11.623	Iran
0.000	0.000	0.000	0.000	0.000	0.000	0.000	0.000	0.000	Israel
0.069	0.070	0.070	0.077	0.086	0.097	0.094	0.167	0.386	Oman
8.542	8.968	8.782	8.867	9.406	9.580	10.655	11.580	12.157	Saudi Arabia
0.758	0.750	0.219	0.043	0.075	0.082	0.103	0.103	0.095	Syria
14.064	12.209	14.466	14.789	15.772	16.948	19.998	22.182	24.261	**Middle East**
124.753	127.710	124.465	129.466	164.074	168.265	173.083	149.652	150.943	**Non-OECD Total**
196.608	184.806	191.751	197.433	205.931	204.472	207.802	221.687	215.749	OECD North America
12.796	13.539	14.477	15.478	16.575	17.291	18.896	20.915	21.803	OECD Pacific
106.360	110.027	126.781	124.429	130.779	129.454	137.978	154.798	144.675	OECD Europe
315.765	308.372	333.009	337.340	353.285	351.217	364.677	397.401	382.228	**OECD Total**
440.518	436.081	457.474	466.807	517.359	519.482	537.759	547.053	533.171	**World**

Includes agriculture, commercial and public services, residential and non-specified other sectors.
Former USSR: series up to 1990-1992 are not comparable with the recent years; 1991 is estimated.

Total Other Sectors' Consumption of Energy* (Mtoe)
Consommation totale d'énergie des autres secteurs (Mtep)
Gesamtenergieverbrauch in sonstigen Sektoren (Mtoe)

Consumo totale di energia negli altri settori (Mtep)

他の部門のエネルギー総消費量（石油換算百万トン）

Consumo total de energía de otros sectores (Mtep)

Общее потребление топлива и энергии другими секторами (Мтнэ)

	1971	1973	1978	1983	1984	1985	1986	1987	1988
Algérie	0.71	0.98	1.79	2.87	3.22	3.24	3.48	3.67	3.72
Angola	0.08	0.09	0.06	0.08	0.10	0.11	0.10	0.10	0.11
Bénin	0.01	0.02	0.04	0.02	0.02	0.02	0.02	0.02	0.02
Cameroun	0.03	0.04	0.10	0.16	0.18	0.21	0.21	0.22	0.23
Congo	0.01	0.02	0.04	0.04	0.04	0.04	0.04	0.03	0.03
RD du Congo	0.42	0.45	0.22	0.46	0.45	0.46	0.43	0.46	0.52
Egypte	1.54	1.67	2.41	4.02	4.39	4.46	4.91	4.62	4.65
Ethiopie	0.04	0.05	0.06	0.11	0.10	0.09	0.12	0.16	0.15
Gabon	0.01	0.02	0.03	0.12	0.14	0.15	0.16	0.17	0.21
Ghana	0.15	0.18	0.24	0.24	0.18	0.20	0.21	0.23	0.25
Côte d'Ivoire	0.12	0.13	0.22	0.26	0.24	0.26	0.22	0.21	0.22
Kenya	0.24	0.27	0.34	0.28	0.32	0.33	0.34	0.37	0.40
Libye	0.17	0.22	0.54	0.78	0.98	1.24	1.34	1.68	1.76
Maroc	0.46	0.53	0.87	1.06	1.08	1.16	1.20	1.27	2.29
Mozambique	0.21	0.21	0.25	0.23	0.21	0.26	0.29	0.29	0.29
Nigéria	0.37	0.49	1.23	1.91	1.86	2.02	2.12	2.48	2.37
Sénégal	0.03	0.03	0.04	0.05	0.06	0.06	0.06	0.07	0.08
Afrique du Sud	5.61	6.14	6.82	5.36	6.12	6.23	5.97	6.21	6.72
Soudan	0.12	0.10	0.06	0.09	0.11	0.11	0.14	0.13	0.12
Tanzanie	0.09	0.11	0.13	0.10	0.10	0.11	0.15	0.15	0.17
Tunisie	0.27	0.34	0.57	0.80	0.85	0.89	0.90	0.93	0.94
Zambie	0.15	0.18	0.18	0.19	0.19	0.19	0.18	0.18	0.20
Zimbabwe	0.41	0.49	0.54	0.59	0.58	0.66	0.69	0.76	0.74
Autre Afrique	1.83	1.95	1.50	1.10	1.02	1.14	1.15	1.85	1.77
Afrique	13.08	14.73	18.29	20.95	22.53	23.65	24.42	26.24	27.94
Argentine	5.38	6.14	7.59	8.78	9.19	9.02	9.03	10.20	10.61
Bolivie	0.18	0.21	0.30	0.33	0.33	0.33	0.32	0.34	0.33
Brésil	4.57	5.72	8.48	12.10	12.20	12.99	14.35	15.48	16.35
Chili	1.08	1.29	1.20	1.14	1.16	1.15	1.29	1.27	1.37
Colombie	2.10	2.30	2.37	2.83	2.81	2.97	3.12	3.56	3.00
Costa Rica	0.09	0.10	0.13	0.16	0.17	0.18	0.19	0.20	0.21
Cuba	1.07	1.65	1.55	1.93	2.12	2.05	2.07	2.03	2.11
République dominicaine	0.13	0.17	0.20	0.23	0.24	0.31	0.29	0.32	0.33
Equateur	0.19	0.23	0.66	0.81	0.90	0.97	1.07	1.07	1.11
El Salvador	0.08	0.09	0.12	0.12	0.12	0.13	0.14	0.15	0.15
Guatemala	0.10	0.13	0.20	0.22	0.23	0.23	0.24	0.25	0.24
Haiti	0.01	0.01	0.02	0.03	0.03	0.03	0.03	0.03	0.03
Honduras	0.08	0.09	0.13	0.14	0.13	0.14	0.15	0.16	0.18
Jamaïque	0.13	0.20	0.44	0.22	0.21	0.22	0.28	0.25	0.26
Antilles néerlandaises	0.15	0.13	0.14	0.14	0.13	0.12	0.12	0.09	0.08
Nicaragua	0.09	0.10	0.13	0.14	0.14	0.14	0.15	0.15	0.14
Panama	0.16	0.17	0.18	0.18	0.18	0.20	0.21	0.23	0.22
Paraguay	0.03	0.03	0.05	0.09	0.10	0.13	0.13	0.13	0.13
Pérou	1.65	1.29	1.60	1.57	1.64	1.65	1.85	2.32	2.39
Trinité-et-Tobago	0.07	0.09	0.11	0.15	0.17	0.16	0.17	0.15	0.19
Uruguay	0.42	0.41	0.44	0.50	0.48	0.47	0.46	0.50	0.53
Vénézuela	1.78	2.07	2.82	3.83	3.81	4.23	3.65	3.41	3.52
Autre Amérique latine	1.60	2.21	1.80	1.36	1.52	1.43	1.62	1.67	1.58
Amérique latine	21.15	24.84	30.65	37.00	38.00	39.25	40.93	43.97	45.05

* Note: à cause de données manquantes avant 1994, les combustibles renouvelables et les déchets ne sont pas inclus.
Comprend agriculture, services privés et publics, secteur résidentiel et autres secteurs non spécifiés.

Total Other Sectors' Consumption of Energy* (Mtoe)
Consommation totale d'énergie des autres secteurs (Mtep)
Gesamtenergieverbrauch in sonstigen Sektoren (Mtoe)
Consumo totale di energia negli altri settori (Mtep)
他の部門のエネルギー総消費量（石油換算百万トン）
Consumo total de energía de otros sectores (Mtep)
Общее потребление топлива и энергии другими секторами (Мтнэ)

1989	1990	1991	1992	1993	1994	1995	1996	1997	
4.10	4.30	5.24	7.32	7.33	7.05	7.07	7.09	8.72	Algeria
0.11	0.12	0.12	0.11	0.11	0.11	0.12	0.13	0.14	Angola
0.02	0.02	0.02	0.02	0.02	0.02	0.02	0.05	0.05	Benin
0.32	0.31	0.31	0.26	0.22	0.20	0.21	0.21	0.22	Cameroon
0.06	0.06	0.06	0.06	0.06	0.06	0.08	0.08	0.07	Congo
0.43	0.48	0.53	0.51	0.54	0.54	0.56	0.57	0.56	DR of Congo
4.89	4.94	4.89	4.89	4.67	5.30	5.63	5.79	6.10	Egypt
0.15	0.17	0.17	0.17	0.18	0.19	0.19	0.21	0.13	Ethiopia
0.08	0.08	0.09	0.09	0.09	0.08	0.09	0.09	0.10	Gabon
0.28	0.27	0.29	0.31	0.34	0.37	0.37	0.37	0.37	Ghana
0.21	0.19	0.20	0.16	0.19	0.19	0.22	0.24	0.25	Ivory Coast
0.47	0.41	0.33	0.38	0.38	0.42	0.41	0.44	0.42	Kenya
1.83	1.92	1.96	2.00	2.04	2.18	2.25	2.30	2.32	Libya
2.40	2.56	2.82	3.22	3.42	3.70	3.85	3.94	4.02	Morocco
0.28	0.28	0.24	0.33	0.37	0.32	0.33	0.33	0.44	Mozambique
2.44	1.88	1.88	2.12	1.97	1.47	1.43	1.56	1.57	Nigeria
0.12	0.12	0.12	0.13	0.12	0.12	0.14	0.15	0.18	Senegal
7.36	7.70	7.93	8.14	10.17	10.05	11.91	9.30	9.56	South Africa
0.12	0.14	0.15	0.15	0.14	0.14	0.15	0.18	0.17	Sudan
0.18	0.19	0.20	0.20	0.17	0.17	0.18	0.19	0.19	Tanzania
1.02	1.03	1.07	1.10	1.25	1.29	1.35	1.42	1.48	Tunisia
0.18	0.21	0.22	0.21	0.22	0.22	0.23	0.23	0.23	Zambia
0.71	0.95	1.00	1.00	1.09	1.14	1.12	1.23	1.24	Zimbabwe
1.93	1.81	2.15	2.24	2.23	2.26	2.64	2.73	2.80	Other Africa
29.68	30.13	31.99	35.12	37.33	37.60	40.57	38.83	41.34	**Africa**
10.01	10.44	10.12	10.67	11.82	12.22	12.50	12.89	13.48	Argentina
0.34	0.32	0.34	0.36	0.38	0.41	0.45	0.49	0.54	Bolivia
17.09	17.84	18.13	18.85	19.64	20.63	22.38	23.48	24.63	Brazil
1.46	1.65	1.80	2.03	2.11	2.24	2.27	2.37	2.49	Chile
3.36	3.34	3.43	3.61	3.86	4.56	3.87	4.19	4.06	Colombia
0.26	0.27	0.30	0.33	0.35	0.37	0.40	0.43	0.46	Costa Rica
2.07	2.33	1.99	1.63	1.38	1.42	1.45	1.49	1.52	Cuba
0.45	0.53	0.58	0.72	0.74	0.82	0.87	0.92	1.01	Dominican Republic
1.12	1.16	1.25	1.33	1.34	1.48	1.59	1.67	1.79	Ecuador
0.16	0.17	0.18	0.20	0.23	0.25	0.28	0.28	0.31	El Salvador
0.29	0.31	0.31	0.35	0.37	0.39	0.44	0.48	0.53	Guatemala
0.04	0.05	0.04	0.03	0.03	0.02	0.04	0.04	0.05	Haiti
0.19	0.20	0.21	0.24	0.24	0.23	0.28	0.28	0.31	Honduras
0.25	1.23	1.28	1.32	0.38	0.61	0.53	0.56	0.59	Jamaica
0.09	0.10	0.10	0.11	0.19	0.20	0.20	0.21	0.21	Netherlands Antilles
0.13	0.13	0.12	0.13	0.13	0.15	0.13	0.14	0.19	Nicaragua
0.21	0.23	0.24	0.26	0.27	0.28	0.30	0.32	0.34	Panama
0.15	0.17	0.19	0.20	0.24	0.27	0.30	0.30	0.28	Paraguay
2.31	2.17	2.14	2.04	2.24	2.35	2.65	2.71	2.42	Peru
0.15	0.16	0.16	0.19	0.19	0.17	0.17	0.18	0.18	Trinidad-and-Tobago
0.52	0.53	0.57	0.60	0.63	0.65	0.70	0.74	0.76	Uruguay
3.61	3.72	3.98	5.26	5.74	5.56	5.61	5.57	6.06	Venezuela
1.94	1.97	1.92	1.78	1.83	1.85	1.94	2.01	2.10	Other Latin America
46.18	49.00	49.39	52.23	54.35	57.13	59.36	61.74	64.30	**Latin America**

* Note: due to missing data prior to 1994 combustible renewables and waste are not included.
Includes agriculture, commercial and public services, residential and non-specified other sectors.

Total Other Sectors' Consumption of Energy* (Mtoe)
Consommation totale d'énergie des autres secteurs (Mtep)
Gesamtenergieverbrauch in sonstigen Sektoren (Mtoe)
Consumo totale di energia negli altri settori (Mtep)
他の部門のエネルギー総消費量（石油換算百万トン）
Consumo total de energía de otros sectores (Mtep)
Общее потребление топлива и энергии другими секторами (Мтнэ)

	1971	1973	1978	1983	1984	1985	1986	1987	1988
Bangladesh	0.31	0.26	0.53	0.70	0.72	0.83	0.92	1.00	1.02
Brunei	0.01	0.01	0.02	0.05	0.06	0.06	0.06	0.07	0.08
Inde	7.97	8.77	9.67	12.90	14.38	15.43	16.65	19.81	21.95
Indonésie	2.29	2.99	5.45	7.32	6.78	9.98	10.19	10.40	11.08
RPD de Corée	0.65	0.76	1.27	1.73	1.87	1.99	2.07	2.09	2.20
Malaisie	0.27	0.34	0.79	1.07	1.13	1.17	1.26	1.33	1.48
Myanmar	0.32	0.28	0.16	0.07	0.08	0.06	0.07	0.07	0.07
Népal	0.03	0.04	0.05	0.06	0.06	0.07	0.09	0.11	0.12
Pakistan	1.72	1.75	2.18	2.97	3.24	3.55	3.82	3.93	4.40
Philippines	1.62	1.88	2.57	2.59	2.36	2.24	2.19	2.55	2.75
Singapour	0.21	0.22	0.26	0.43	0.45	0.49	0.54	0.58	0.62
Sri Lanka	0.31	0.32	0.31	0.29	0.26	0.29	0.29	0.30	0.32
Taipei chinois	1.64	2.10	3.21	4.18	4.41	4.58	4.84	5.24	5.64
Thaïlande	1.16	1.45	2.29	2.75	2.83	2.85	3.08	3.26	3.62
Viêt-Nam	3.51	3.79	2.53	1.20	0.94	0.70	0.73	0.89	0.83
Autre Asie	1.39	1.58	0.97	0.55	0.68	0.97	0.91	0.68	0.70
Asie	23.41	26.55	32.24	38.87	40.25	45.27	47.72	52.32	56.86
Rép. populaire de Chine	162.68	176.62	261.25	104.56	112.70	119.24	122.99	129.10	138.61
Hong-Kong, Chine	0.57	0.53	0.88	1.09	1.15	1.22	1.34	1.45	1.57
Chine	163.25	177.16	262.13	105.64	113.85	120.46	124.33	130.55	140.17
Albanie	0.40	0.53	0.96	0.90	0.90	0.84	0.93	0.89	0.90
Bulgarie	8.66	9.57	10.82	11.38	11.27	6.59	6.45	6.62	4.63
Chypre	0.07	0.09	0.09	0.12	0.12	0.14	0.14	0.15	0.16
Gibraltar	0.00	0.00	0.00	0.00	0.00	0.00	0.00	0.00	0.00
Malte	0.06	0.07	0.08	0.07	0.08	0.08	0.09	0.10	0.11
Roumanie	10.10	13.58	13.94	14.31	12.95	17.60	17.14	18.82	18.67
République slovaque	2.81	3.02	3.73	4.48	4.46	4.55	4.59	4.67	4.75
Bosnie-Herzégovine	-	-	-	-	-	-	-	-	-
Croatie	-	-	-	-	-	-	-	-	-
Ex-RYM	-	-	-	-	-	-	-	-	-
Slovénie	-	-	-	-	-	-	-	-	-
RF de Yougoslavie	-	-	-	-	-	-	-	-	-
Ex-Yougoslavie	3.15	3.35	4.98	6.76	6.19	6.07	5.82	6.95	6.13
Europe non-OCDE	25.26	30.21	34.61	38.02	35.98	35.88	35.17	38.20	35.35

* Note: à cause de données manquantes avant 1994, les combustibles renouvelables et les déchets ne sont pas inclus.
Comprend agriculture, services privés et publics, secteur résidentiel et autres secteurs non spécifiés.
Rép.populaire de Chine: jusqu'en 1978 la ventilation de la consommation finale par secteur est incomplète.

Total Other Sectors' Consumption of Energy* (Mtoe)
Consommation totale d'énergie des autres secteurs (Mtep)
Gesamtenergieverbrauch in sonstigen Sektoren (Mtoe)
Consumo totale di energia negli altri settori (Mtep)
他の部門のエネルギー総消費量（石油換算百万トン）
Consumo total de energía de otros sectores (Mtep)
Общее потребление топлива и энергии другими секторами (Мтнэ)

1989	1990	1991	1992	1993	1994	1995	1996	1997	
1.25	1.22	1.18	1.18	1.25	1.31	1.55	1.65	1.70	Bangladesh
0.08	0.08	0.08	0.10	0.11	0.13	0.15	0.17	0.20	Brunei
22.40	22.62	23.29	24.72	26.16	28.25	30.22	31.61	33.07	India
12.89	9.76	10.35	11.18	12.02	12.74	13.79	14.98	15.94	Indonesia
2.21	2.21	2.47	0.74	0.70	0.67	0.63	0.60	0.59	DPR of Korea
1.53	1.66	1.72	2.10	2.10	3.12	3.28	3.70	4.05	Malaysia
0.09	0.09	0.09	0.16	0.19	0.23	0.28	0.29	0.32	Myanmar
0.12	0.10	0.13	0.15	0.16	0.24	0.26	0.26	0.26	Nepal
4.78	5.04	5.18	5.07	5.40	5.64	6.21	6.80	7.19	Pakistan
3.08	3.51	3.56	4.01	4.53	4.77	4.81	5.11	5.55	Philippines
0.66	0.69	0.65	0.62	0.67	0.73	0.78	1.07	1.19	Singapore
0.33	0.36	0.36	0.43	0.45	0.46	0.48	0.50	0.48	Sri Lanka
6.00	6.33	6.64	6.72	7.16	7.64	7.89	8.57	8.53	Chinese Taipei
4.04	4.49	4.78	5.39	5.64	5.66	6.19	6.94	7.08	Thailand
0.87	0.98	0.95	0.98	1.33	1.42	1.48	2.35	2.71	Vietnam
0.93	1.06	1.09	0.88	0.87	0.91	1.30	1.26	1.31	Other Asia
61.27	60.20	62.54	64.43	68.74	73.91	79.30	85.87	90.17	**Asia**
138.09	138.66	140.58	134.91	142.92	141.47	147.61	158.60	152.31	People's Rep.of China
1.71	1.86	2.00	2.13	2.30	2.49	2.60	2.78	2.87	Hong Kong, China
139.80	140.52	142.57	137.04	145.22	143.96	150.21	161.38	155.18	**China**
0.58	0.34	0.23	0.23	0.21	0.20	0.25	0.29	0.26	Albania
8.19	4.34	3.89	3.52	3.82	3.60	3.39	3.67	3.19	Bulgaria
0.17	0.19	0.20	0.23	0.24	0.25	0.23	0.24	0.25	Cyprus
0.01	0.01	0.01	0.01	0.01	0.01	0.01	0.01	0.01	Gibraltar
0.11	0.11	0.12	0.12	0.11	0.08	0.08	0.12	0.17	Malta
13.35	13.04	9.74	9.80	8.07	7.32	7.43	8.40	8.65	Romania
5.59	5.39	5.20	4.82	4.77	4.56	4.44	4.87	4.56	Slovak Republic
-	-	-	3.19	2.80	0.14	0.17	0.17	0.17	*Bosnia-Herzegovina*
-	-	-	1.49	1.56	1.61	1.78	1.83	1.98	*Croatia*
-	-	-	0.52	0.42	0.47	0.47	0.47	0.44	*FYROM*
-	-	-	1.07	1.20	1.16	1.25	1.50	1.54	*Slovenia*
-	-	-	5.37	3.73	3.28	3.66	4.54	4.81	*FR of Yugoslavia*
5.67	10.07	10.17	11.64	9.71	6.67	7.33	8.52	8.93	Former Yugoslavia
33.66	33.48	29.55	30.37	26.92	22.69	23.16	26.11	26.01	**Non-OECD Europe**

* Note: due to missing data prior to 1994 combustible renewables and waste are not included.
Includes agriculture, commercial and public services, residential and non-specified other sectors.
People's Rep. of China: up to 1978, the breakdown of final consumption by sector is incomplete.

Total Other Sectors' Consumption of Energy* (Mtoe)
Consommation totale d'énergie des autres secteurs (Mtep)
Gesamtenergieverbrauch in sonstigen Sektoren (Mtoe)

Consumo totale di energia negli altri settori (Mtep)

他の部門のエネルギー総消費量（石油換算百万トン）

Consumo total de energía de otros sectores (Mtep)

Общее потребление топлива и энергии другими секторами (Мтнэ)

	1971	1973	1978	1983	1984	1985	1986	1987	1988
Arménie	-	-	-	-	-	-	-	-	-
Azerbaïdjan	-	-	-	-	-	-	-	-	-
Bélarus	-	-	-	-	-	-	-	-	-
Estonie	-	-	-	-	-	-	-	-	-
Géorgie	-	-	-	-	-	-	-	-	-
Kazakhstan	-	-	-	-	-	-	-	-	-
Kirghizistan	-	-	-	-	-	-	-	-	-
Lettonie	-	-	-	-	-	-	-	-	-
Lituanie	-	-	-	-	-	-	-	-	-
Moldova	-	-	-	-	-	-	-	-	-
Russie	-	-	-	-	-	-	-	-	-
Tadjikistan	-	-	-	-	-	-	-	-	-
Turkménistan	-	-	-	-	-	-	-	-	-
Ukraine	-	-	-	-	-	-	-	-	-
Ouzbékistan	-	-	-	-	-	-	-	-	-
Ex-URSS	150.14	166.10	203.75	272.44	285.18	298.64	304.88	312.07	320.79
Bahrein	0.04	0.04	0.10	0.20	0.22	0.26	0.26	0.29	0.29
Iran	4.70	6.42	12.68	18.24	20.07	19.48	13.44	15.47	20.28
Irak	0.88	0.92	1.65	2.31	2.46	2.61	2.79	2.95	3.15
Israël	0.93	1.07	1.19	1.39	1.40	1.57	1.92	2.12	2.27
Jordanie	0.12	0.12	0.22	0.42	0.46	0.46	0.42	0.41	0.46
Koweit	0.19	0.25	0.48	0.81	0.89	1.00	1.10	1.16	1.23
Liban	0.20	0.24	0.27	0.40	0.40	0.46	0.49	0.51	0.43
Oman	0.04	0.04	0.18	0.36	0.47	0.56	0.53	0.53	0.66
Qatar	0.03	0.04	0.11	0.22	0.25	0.29	0.33	0.33	0.34
Arabie saoudite	0.75	0.95	4.34	6.49	7.41	7.21	11.80	17.21	23.52
Syrie	0.40	0.38	0.57	1.55	1.83	1.35	1.82	2.15	2.64
Emirats arabes unis	0.02	0.07	0.32	0.87	0.94	1.03	1.09	1.17	1.32
Yémen	0.07	0.07	0.13	0.26	0.29	0.34	0.29	0.33	0.38
Moyen-Orient	8.37	10.61	22.26	33.53	37.09	36.60	36.30	44.65	56.96
Total non-OCDE	404.66	450.19	603.92	546.45	572.89	599.76	613.75	647.99	683.13
OCDE Amérique du N.	451.74	460.18	473.33	421.32	436.81	436.60	433.96	441.87	466.08
OCDE Pacifique	54.84	68.31	81.48	90.88	97.62	98.18	102.42	105.42	113.08
OCDE Europe	337.28	376.40	403.06	377.11	386.04	409.53	420.38	426.15	413.92
Total OCDE	843.85	904.89	957.88	889.30	920.47	944.32	956.76	973.44	993.08
Monde	1 248.51	1 355.09	1 561.80	1 435.75	1 493.37	1 544.07	1 570.51	1 621.44	1 676.21

* Note: à cause de données manquantes avant 1994, les combustibles renouvelables et les déchets ne sont pas inclus.
Comprend agriculture, services privés et publics, secteur résidentiel et autres secteurs non spécifiés.
Ex-URSS: les séries antérieures à 1990-1992 ne sont pas comparables aux années récentes; 1991 a été estimé.

INTERNATIONAL ENERGY AGENCY

Total Other Sectors' Consumption of Energy* (Mtoe)
Consommation totale d'énergie des autres secteurs (Mtep)
Gesamtenergieverbrauch in sonstigen Sektoren (Mtoe)
Consumo totale di energia negli altri settori (Mtep)
他の部門のエネルギー総消費量（石油換算百万トン）
Consumo total de energía de otros sectores (Mtep)
Общее потребление топлива и энергии другими секторами (Мтнэ)

1989	1990	1991	1992	1993	1994	1995	1996	1997	
-	-	-	1.48	0.63	0.64	0.74	0.39	0.49	Armenia
-	-	-	9.05	3.60	3.64	3.04	3.28	3.18	Azerbaijan
-	-	-	10.37	10.17	8.78	8.85	9.24	9.13	Belarus
-	-	-	1.34	0.94	1.02	0.95	1.06	1.06	Estonia
-	-	-	4.82	2.02	1.53	0.85	1.23	1.32	Georgia
-	-	-	21.97	16.17	13.74	14.43	11.27	9.75	Kazakhstan
-	-	-	3.03	1.98	1.23	1.53	1.60	1.36	Kyrgyzstan
-	-	-	2.34	2.19	2.03	1.60	1.55	1.23	Latvia
-	-	-	3.32	2.63	2.27	2.03	1.91	1.77	Lithuania
-	-	-	1.95	2.20	1.89	1.82	1.99	1.94	Moldova
-	-	-	312.67	248.47	233.19	236.71	193.75	180.90	Russia
-	-	-	2.32	1.78	1.13	1.14	1.29	1.22	Tajikistan
-	-	-	7.51	6.77	8.64	8.91	7.39	7.16	Turkmenistan
-	-	-	51.58	47.54	46.93	47.34	46.66	43.05	Ukraine
-	-	-	27.99	28.99	28.87	20.08	20.04	21.35	Uzbekistan
324.50	319.52	323.15	461.02	375.49	355.08	349.86	302.61	284.88	**Former USSR**
0.30	0.31	0.32	0.31	0.32	0.35	0.36	0.36	0.37	Bahrain
21.72	34.38	39.55	41.17	27.08	28.80	32.03	34.62	35.97	Iran
3.34	3.74	2.61	3.47	4.52	4.81	4.83	4.76	4.87	Iraq
2.40	2.49	2.56	3.23	3.42	3.66	4.06	4.31	4.40	Israel
0.47	0.74	0.76	0.97	0.96	1.05	1.08	1.12	1.20	Jordan
1.36	1.33	0.64	1.08	1.58	1.80	1.88	1.99	2.14	Kuwait
0.32	0.59	0.97	0.90	0.75	0.86	0.91	0.96	1.22	Lebanon
0.67	0.54	0.58	0.62	0.68	0.75	0.76	0.86	1.10	Oman
0.35	0.35	0.32	0.36	0.40	0.42	0.41	0.43	0.45	Qatar
22.95	23.74	24.29	25.68	27.58	27.66	27.26	28.96	30.42	Saudi Arabia
5.54	5.90	5.98	6.08	6.23	6.50	6.71	6.98	7.44	Syria
1.41	1.54	1.54	1.75	1.95	2.10	2.12	2.15	2.23	United Arab Emirates
0.42	0.36	0.44	0.53	0.58	0.62	0.65	0.68	0.73	Yemen
61.24	76.01	80.56	86.15	76.06	79.38	83.07	88.18	92.56	**Middle East**
696.33	708.88	719.76	866.36	784.12	769.75	785.52	764.73	754.45	**Non-OECD Total**
475.36	455.44	462.66	465.41	482.61	484.33	493.51	516.73	511.52	OECD North America
114.78	122.63	127.69	132.18	134.78	141.18	148.78	153.85	154.75	OECD Pacific
398.54	396.10	420.74	414.47	420.12	410.75	421.36	450.82	432.18	OECD Europe
988.67	974.18	1 011.09	1 012.05	1 037.51	1 036.26	1 063.65	1 121.40	1 098.46	**OECD Total**
1 685.01	1 683.06	1 730.84	1 878.41	1 821.63	1 806.01	1 849.17	1 886.13	1 852.91	**World**

* Note: due to missing data prior to 1994 combustible renewables and waste are not included.
Includes agriculture, commercial and public services, residential and non-specified other sectors.
Former USSR: series up to 1990-1992 are not comparable with the recent years; 1991 is estimated.

GDP using Exchange Rates (billion US$90)
PIB basé sur les taux de change (milliards US$90)
BIP auf Wechselkursbasis (Milliarden US$90)
PIL utilizzando i tassi di cambio (miliardi de US$ 1990)
為替換算による国内総生産（十億米ドル，1990年価格）
PIB basado en los tipos de cambio (millardos US$90)
ВВП по валютному курсу (млрд.долл.США в ценах 1990 г.)

	1971	1973	1978	1983	1984	1985	1986	1987	1988
Algérie	23.7	31.3	44.1	55.2	58.2	61.5	61.4	60.9	59.8
Angola	14.0	15.1	8.5	8.3	8.4	8.7	9.0	9.7	10.2
Bénin	1.0	1.1	1.2	1.5	1.6	1.7	1.8	1.7	1.8
Cameroun	4.9	5.3	8.5	11.9	12.8	13.9	14.8	14.5	13.4
Congo	1.0	1.2	1.4	2.7	2.9	2.8	2.6	2.7	2.7
RD du Congo	8.7	9.4	8.3	8.8	9.3	9.4	9.8	10.1	10.1
Egypte	19.9	20.4	31.2	44.6	47.3	50.4	51.8	53.1	55.9
Ethiopie	4.4	4.6	4.9	7.8	7.4	6.7	7.3	8.3	8.3
Gabon	2.6	3.2	4.8	5.4	5.8	5.6	5.6	4.6	5.2
Ghana	4.8	4.8	4.9	4.1	4.4	4.7	4.9	5.1	5.4
Côte d'Ivoire	6.6	7.3	11.0	10.0	9.8	10.2	10.5	10.5	10.6
Kenya	3.2	4.0	5.0	6.1	6.2	6.5	7.0	7.4	7.8
Libye	33.8	29.7	37.7	33.4	30.7	28.0	25.6	24.8	24.9
Maroc	11.0	11.5	15.5	18.8	19.6	20.8	22.5	22.0	24.3
Mozambique	1.4	1.6	1.2	1.0	0.9	1.0	1.0	1.2	1.3
Nigéria	21.0	22.8	26.2	23.9	22.8	25.0	25.6	25.4	28.0
Sénégal	3.5	3.6	4.1	4.9	4.7	4.9	5.1	5.3	5.6
Afrique du Sud	69.4	73.8	82.7	94.7	100.3	97.6	98.0	100.8	104.5
Soudan	5.9	5.3	8.9	9.4	9.0	8.4	8.7	8.9	8.5
Tanzanie	2.2	2.4	2.9	3.0	3.1	3.1	3.2	3.3	3.4
Tunisie	4.7	5.5	7.6	9.5	10.1	10.7	10.5	11.2	11.2
Zambie	2.9	3.2	3.4	3.4	3.4	3.5	3.5	3.6	3.8
Zimbabwe	4.6	5.1	4.8	6.7	6.6	7.0	7.2	7.2	7.8
Autre Afrique	27.2	28.0	33.5	36.6	36.9	38.3	39.6	40.5	42.7
Afrique	282.6	300.3	362.4	411.7	422.1	430.3	436.9	442.8	457.3
Argentine	126.5	132.9	140.6	151.2	154.2	143.5	153.7	157.7	154.7
Bolivie	3.4	3.8	4.9	4.4	4.4	4.4	4.3	4.4	4.5
Brésil	197.1	251.7	342.4	370.4	390.0	421.0	454.6	470.9	470.5
Chili	17.2	16.2	17.8	18.9	20.4	21.9	23.1	24.6	26.4
Colombie	17.9	20.6	26.3	30.3	31.3	32.3	34.1	36.0	37.4
Costa Rica	2.8	3.2	4.3	4.2	4.5	4.6	4.8	5.0	5.2
Cuba	6.3	7.4	10.1	12.9	13.8	14.5	14.6	14.1	14.4
République dominicaine	3.1	3.9	5.0	6.1	6.1	6.2	6.4	7.0	7.2
Equateur	3.9	5.6	7.9	8.9	9.3	9.7	10.0	9.4	10.3
El Salvador	4.6	5.1	6.6	4.7	4.8	4.8	4.8	4.9	5.0
Guatemala	4.3	4.9	6.5	6.6	6.7	6.6	6.6	6.9	7.1
Haiti	2.2	2.3	2.7	2.9	3.0	3.0	3.0	2.9	3.0
Honduras	1.5	1.7	2.3	2.4	2.5	2.6	2.6	2.8	2.9
Jamaïque	3.7	4.1	3.6	3.5	3.5	3.3	3.4	3.7	3.8
Antilles néerlandaises	1.2	1.3	1.4	1.5	1.5	1.4	1.4	1.5	1.5
Nicaragua	1.3	1.4	1.7	1.4	1.4	1.3	1.3	1.1	1.1
Panama	3.4	3.7	4.4	5.1	5.2	5.5	5.7	5.6	4.8
Paraguay	1.8	2.1	3.1	4.1	4.2	4.4	4.4	4.5	4.8
Pérou	29.2	31.7	36.8	37.4	38.9	39.8	43.5	47.2	43.2
Trinité-et-Tobago	2.7	3.0	4.5	5.6	5.9	5.7	5.5	5.2	5.0
Uruguay	6.3	6.2	7.5	6.9	6.9	7.0	7.6	8.2	8.2
Vénézuela	34.8	37.8	46.5	42.1	42.7	42.8	45.6	47.2	49.9
Autre Amérique latine	10.4	11.1	12.4	13.7	14.3	14.5	15.2	15.5	16.5
Amérique latine	485.6	561.7	699.1	745.3	775.4	800.4	856.2	886.5	887.7

GDP using Exchange Rates (billion US$90)
PIB basé sur les taux de change (milliards US$90)
BIP auf Wechselkursbasis (Milliarden US$90)
PIL utilizzando i tassi di cambio (miliardi de US$ 1990)
為替換算による国内総生産（十億米ドル，1990年価格）
PIB basado en los tipos de cambio (millardos US$90)
ВВП по валютному курсу (млрд.долл.США в ценах 1990 г.)

1989	1990	1991	1992	1993	1994	1995	1996	1997	
62.7	61.9	61.2	62.1	60.8	60.1	62.3	64.7	65.6	Algeria
10.3	10.3	10.3	9.8	7.5	7.6	8.4	9.4	10.1	Angola
1.9	1.8	1.9	2.0	2.1	2.2	2.3	2.4	2.5	Benin
13.1	12.3	11.8	11.5	11.1	10.8	11.2	11.8	12.4	Cameroon
2.8	2.8	2.9	2.9	2.9	2.8	2.9	3.0	3.0	Congo
10.0	9.4	8.6	7.7	6.6	6.4	6.4	6.4	6.0	DR of Congo
58.7	62.0	62.7	65.5	67.4	70.0	73.3	77.0	81.2	Egypt
8.4	8.6	8.1	7.8	8.7	8.8	9.4	10.4	10.9	Ethiopia
5.7	6.0	6.3	6.1	6.3	6.5	6.9	7.2	7.5	Gabon
5.7	5.9	6.2	6.4	6.8	7.0	7.3	7.6	7.9	Ghana
10.9	10.8	10.8	10.8	10.8	11.0	11.8	12.5	13.3	Ivory Coast
8.2	8.5	8.7	8.6	8.6	8.8	9.2	9.6	9.8	Kenya
25.1	27.4	28.4	27.7	26.4	25.3	25.8	26.2	26.3	Libya
24.9	25.8	27.6	26.5	26.2	29.0	27.1	30.3	29.7	Morocco
1.4	1.4	1.5	1.6	1.9	2.0	2.0	2.1	2.4	Mozambique
30.0	32.4	34.0	35.0	35.7	35.5	36.4	38.8	40.3	Nigeria
5.5	5.7	5.7	5.8	5.7	5.8	6.1	6.5	6.8	Senegal
107.0	106.7	105.6	103.3	104.6	107.5	111.1	114.7	116.6	South Africa
9.2	9.0	9.1	10.1	10.9	11.5	12.0	12.5	13.0	Sudan
3.6	3.8	3.9	3.6	4.0	4.1	4.2	4.3	4.5	Tanzania
11.4	12.3	12.8	13.8	14.1	14.6	14.9	16.0	16.8	Tunisia
3.8	3.7	3.7	3.7	3.9	3.8	3.7	3.9	4.1	Zambia
8.2	8.8	9.3	8.4	8.5	9.1	9.1	9.7	10.0	Zimbabwe
44.4	45.7	46.8	47.4	47.7	47.7	49.9	52.5	55.3	Other Africa
472.6	483.1	487.8	488.0	489.1	497.6	513.6	539.4	556.0	**Africa**
144.0	141.4	156.3	171.3	181.1	195.6	187.8	196.7	213.7	Argentina
4.7	4.9	5.1	5.2	5.4	5.7	6.0	6.2	6.5	Bolivia
485.9	465.0	471.0	468.7	491.7	520.7	542.5	557.7	575.6	Brazil
29.2	30.3	32.7	36.8	39.3	41.6	46.0	49.4	52.9	Chile
38.7	40.3	41.2	42.8	45.2	47.8	50.6	51.6	53.2	Colombia
5.5	5.7	5.8	6.3	6.7	7.0	7.2	7.1	7.3	Costa Rica
14.5	14.1	12.5	11.1	9.5	9.6	9.8	10.5	10.7	Cuba
7.5	7.1	7.1	7.7	7.9	8.3	8.7	9.3	10.1	Dominican Republic
10.4	10.7	11.2	11.6	11.9	12.4	12.7	12.9	13.3	Ecuador
5.1	5.3	5.5	5.9	6.4	6.8	7.2	7.3	7.6	El Salvador
7.4	7.7	7.9	8.3	8.6	9.0	9.4	9.7	10.1	Guatemala
3.0	3.0	3.1	2.7	2.6	2.4	2.5	2.6	2.6	Haiti
3.0	3.0	3.1	3.3	3.5	3.5	3.6	3.7	3.9	Honduras
4.0	4.2	4.3	4.3	4.4	4.5	4.5	4.4	4.3	Jamaica
1.6	1.7	1.7	1.7	1.7	1.8	1.8	1.9	1.9	Netherlands Antilles
1.1	1.1	1.1	1.1	1.1	1.1	1.2	1.5	1.6	Nicaragua
4.9	5.3	5.8	6.3	6.6	6.8	6.9	7.1	7.4	Panama
5.1	5.3	5.4	5.5	5.7	5.9	6.2	6.3	6.5	Paraguay
38.2	36.1	38.7	38.0	40.4	45.7	49.1	50.4	54.0	Peru
5.0	5.1	5.8	5.1	5.0	5.2	5.4	5.6	5.8	Trinidad-and-Tobago
8.3	8.4	8.6	9.3	9.6	10.2	10.0	10.6	11.1	Uruguay
45.6	48.6	53.3	56.6	56.7	55.4	57.4	57.1	60.1	Venezuela
17.4	18.0	18.8	18.9	19.0	19.2	19.5	20.2	21.0	Other Latin America
890.2	872.1	906.4	928.7	970.2	1 026.0	1 055.9	1 089.8	1 141.1	**Latin America**

GDP using Exchange Rates (billion US$90)
PIB basé sur les taux de change (milliards US$90)
BIP auf Wechselkursbasis (Milliarden US$90)
PIL utilizzando i tassi di cambio (miliardi de US$ 1990)
為替換算による国内総生産（十億米ドル，1990年価格）
PIB basado en los tipos de cambio (millardos US$90)
ВВП по валютному курсу (млрд.долл.США в ценах 1990 г.)

	1971	1973	1978	1983	1984	1985	1986	1987	1988
Bangladesh	11.6	10.3	13.1	16.7	17.5	18.1	18.9	19.7	20.3
Brunei	2.3	2.5	3.7	3.6	3.6	3.5	3.4	3.5	3.5
Inde	131.4	134.5	171.4	205.7	213.3	224.9	236.0	247.2	271.7
Indonésie	30.9	36.6	53.0	73.1	78.4	81.1	85.9	90.5	96.2
Malaisie	11.9	14.6	20.4	28.9	31.1	30.8	31.1	32.8	35.8
Myanmar	13.9	14.1	18.6	24.7	26.0	26.7	26.4	25.4	22.5
Népal	1.8	1.8	2.2	2.4	2.7	2.8	2.9	3.0	3.2
Pakistan	13.6	14.7	18.7	26.3	27.7	29.8	31.4	33.4	36.0
Philippines	22.2	25.5	33.8	40.9	37.9	35.2	36.4	37.9	40.5
Singapour	9.0	11.3	15.5	23.6	25.5	25.1	25.6	28.1	31.3
Sri Lanka	3.5	3.7	4.7	6.2	6.5	6.8	7.1	7.2	7.4
Taipei chinois	34.3	43.9	66.3	91.8	101.5	106.6	119.0	134.1	144.6
Thaïlande	21.7	24.9	36.2	47.2	49.9	52.3	55.2	60.4	68.4
Viêt-Nam	-	-	-	-	4.9	5.1	5.3	5.5	5.7
Autre Asie	10.7	11.9	14.7	17.3	17.7	17.9	18.7	18.1	17.9
Asie	319.0	350.4	472.3	608.4	639.3	661.6	698.0	741.3	799.3
Rép. populaire de Chine	102.0	113.4	140.1	203.1	234.0	265.5	288.9	322.4	358.9
Hong-Kong, Chine	17.5	21.9	32.1	47.3	51.9	52.0	57.8	65.3	70.5
Chine	119.5	135.4	172.3	250.4	285.9	317.6	346.7	387.7	429.4
Albanie	-	-	-	2.0	2.0	2.1	2.2	2.2	2.1
Bulgarie	12.8	13.9	16.2	18.1	18.7	19.2	20.0	21.3	23.6
Chypre	2.2	2.5	2.6	3.5	3.8	4.0	4.1	4.4	4.8
Gibraltar	0.2	0.2	0.3	0.3	0.3	0.3	0.3	0.3	0.3
Malte	0.6	0.7	1.3	1.7	1.7	1.7	1.8	1.9	2.0
Roumanie	22.0	24.2	35.4	42.1	44.6	44.5	45.6	43.3	43.1
République slovaque	10.5	11.3	12.9	13.7	14.0	14.4	15.1	15.4	15.7
Bosnie-Herzégovine	-	-	-	-	-	-	-	-	-
Croatie	-	-	-	-	-	-	-	-	-
Ex-RYM	-	-	-	-	-	-	-	-	-
Slovénie	-	-	-	-	-	-	-	-	-
RF de Yougoslavie	-	-	-	-	-	-	-	-	-
Ex-Yougoslavie	56.5	60.0	86.4	94.0	95.6	95.0	98.4	97.5	94.9
Europe non-OCDE	104.8	112.7	155.2	173.2	178.6	179.2	185.2	184.0	184.4
Arménie	-	-	-	-	-	-	-	-	-
Azerbaïdjan	-	-	-	-	-	-	-	-	-
Bélarus	-	-	-	-	-	-	-	-	-
Estonie	-	-	-	-	-	-	-	-	-
Géorgie	-	-	-	-	-	-	-	-	-
Kazakhstan	-	-	-	-	-	-	-	-	-
Kirghizistan	-	-	-	-	-	-	-	-	-
Lettonie	-	-	-	-	-	-	-	-	-
Lituanie	-	-	-	-	-	-	-	-	-
Moldova	-	-	-	-	-	-	-	-	-
Russie	-	-	-	-	-	-	-	-	-
Tadjikistan	-	-	-	-	-	-	-	-	-
Turkménistan	-	-	-	-	-	-	-	-	-
Ukraine	-	-	-	-	-	-	-	-	-
Ouzbékistan	-	-	-	-	-	-	-	-	-
Ex-URSS	592.8	648.1	765.9	846.1	858.0	867.4	903.8	915.6	929.3

Faute de séries complètes, l'Albanie, la Rép.pop.dém. de Corée et le Viêt-Nam ne sont pas pris en compte dans les totaux régionaux.
Ex-URSS: la somme des républiques peut être différente du total.

GDP using Exchange Rates (billion US$90)
PIB basé sur les taux de change (milliards US$90)
BIP auf Wechselkursbasis (Milliarden US$90)
PIL utilizzando i tassi di cambio (miliardi de US$ 1990)
為替換算による国内総生産（十億米ドル, 1990年価格）
PIB basado en los tipos de cambio (millardos US$90)
ВВП по валютному курсу (млрд.долл.США в ценах 1990 г.)

1989	1990	1991	1992	1993	1994	1995	1996	1997	
20.8	22.1	22.9	23.9	24.9	25.9	27.1	28.5	30.1	Bangladesh
3.5	3.6	3.7	3.7	3.5	3.6	3.7	3.8	3.9	Brunei
289.6	305.9	307.3	323.9	339.9	367.6	394.7	423.8	445.9	India
105.0	114.4	124.6	133.6	143.3	154.1	166.8	179.9	188.7	Indonesia
39.0	42.8	46.5	50.1	54.3	59.3	64.9	70.4	76.0	Malaysia
23.3	24.0	23.8	26.1	27.7	29.8	31.8	33.8	35.2	Myanmar
3.4	3.5	3.7	3.9	4.0	4.4	4.5	4.8	5.0	Nepal
37.7	39.4	41.6	44.8	45.7	47.5	49.9	52.2	52.1	Pakistan
43.0	44.3	44.1	44.2	45.1	47.1	49.3	52.2	54.9	Philippines
34.4	37.5	40.2	42.7	47.1	52.1	56.6	60.5	65.1	Singapore
7.6	8.0	8.4	8.8	9.4	9.9	10.4	10.8	11.5	Sri Lanka
156.6	165.0	177.5	189.5	201.4	214.6	227.5	240.4	256.8	Chinese Taipei
76.8	85.3	92.6	100.1	108.5	118.2	128.7	135.8	135.2	Thailand
6.2	6.5	6.9	7.5	8.1	8.8	9.6	10.5	11.4	Vietnam
17.6	17.6	18.3	18.9	19.4	19.6	21.2	22.3	23.3	Other Asia
858.2	913.6	955.2	1 014.1	1 074.4	1 153.6	1 237.1	1 319.2	1 383.7	**Asia**
373.6	387.8	423.4	483.6	548.9	618.0	682.9	748.5	814.3	People's Rep.of China
72.3	74.8	78.6	83.5	88.6	93.4	97.0	101.9	107.1	Hong Kong, China
445.9	462.6	502.0	567.1	637.5	711.4	779.9	850.4	921.5	**China**
2.3	2.1	1.5	1.4	1.5	1.7	1.8	2.0	1.9	Albania
22.8	20.7	19.0	17.6	17.3	17.7	18.2	16.3	15.2	Bulgaria
5.2	5.6	5.6	6.1	6.2	6.5	6.9	7.1	7.3	Cyprus
0.3	0.3	0.3	0.3	0.3	0.3	0.3	0.3	0.4	Gibraltar
2.2	2.3	2.5	2.6	2.7	2.8	3.0	3.1	3.2	Malta
40.6	38.2	33.3	30.4	30.8	32.1	34.3	35.7	33.3	Romania
15.9	15.5	13.2	12.3	11.9	12.5	13.3	14.2	15.1	Slovak Republic
-	-	-	1.7	1.1	0.7	1.0	1.4	1.9	*Bosnia-Herzegovina*
-	-	-	8.0	7.8	8.3	8.5	8.9	9.5	*Croatia*
-	-	-	2.0	1.8	1.8	1.8	1.8	1.8	*FYROM*
-	-	-	20.6	21.2	22.3	23.2	23.9	24.8	*Slovenia*
-	-	-	42.5	38.9	37.4	38.8	40.5	40.9	*FR of Yugoslavia*
95.4	89.6	90.7	74.8	70.8	70.5	73.2	76.5	79.0	Former Yugoslavia
182.4	172.3	164.6	144.1	140.1	142.3	149.3	153.2	153.4	**Non-OECD Europe**
-	-	-	1.8	1.5	1.6	1.7	1.8	1.9	Armenia
-	-	-	6.3	4.9	4.1	3.6	3.6	3.7	Azerbaijan
-	-	-	31.2	28.8	25.2	22.6	23.2	25.6	Belarus
-	-	-	4.9	4.5	4.4	4.6	4.8	5.3	Estonia
-	-	-	5.8	3.5	3.1	3.2	3.5	3.9	Georgia
-	-	-	32.7	29.5	25.8	23.7	23.8	24.2	Kazakhstan
-	-	-	2.3	1.9	1.5	1.4	1.5	1.7	Kyrgyzstan
-	-	-	7.3	6.2	6.2	6.2	6.4	6.8	Latvia
-	-	-	9.8	8.2	7.4	7.7	8.1	8.5	Lithuania
-	-	-	6.7	6.6	4.5	4.4	3.9	4.0	Moldova
-	-	-	469.9	429.2	375.2	359.7	347.1	349.9	Russia
-	-	-	3.1	2.8	2.2	1.9	1.8	1.8	Tajikistan
-	-	-	5.7	5.2	4.2	3.8	3.7	4.7	Turkmenistan
-	-	-	129.1	110.8	85.3	74.9	67.4	65.2	Ukraine
-	-	-	20.5	20.0	19.2	18.9	19.3	20.3	Uzbekistan
956.3	949.4	892.6	766.7	687.0	589.9	557.4	534.1	542.1	**Former USSR**

Due to lack of complete series Albania, the Dem. Ppl's Rep. of Korea and Vietnam are not included in the regional aggregates.
Former USSR: data for individual republics may not add to the total.

GDP using Exchange Rates (billion US$90)
PIB basé sur les taux de change (milliards US$90)

BIP auf Wechselkursbasis (Milliarden US$90)

PIL utilizzando i tassi di cambio (miliardi de US$ 1990)

為替換算による国内総生産（十億米ドル, 1990年価格）

PIB basado en los tipos de cambio (millardos US$90)

ВВП по валютному курсу (млрд.долл.США в ценах 1990 г.)

	1971	1973	1978	1983	1984	1985	1986	1987	1988
Bahrein	1.9	2.3	3.6	3.5	3.7	3.5	3.6	3.5	3.9
Iran	62.8	82.9	120.2	119.9	121.9	124.9	113.1	111.4	105.2
Irak	74.6	82.2	126.8	106.4	106.5	91.2	73.7	72.6	65.3
Israël	23.5	27.6	32.4	40.7	41.1	42.5	44.5	47.7	48.7
Jordanie	1.5	1.5	2.5	3.8	4.1	4.3	4.6	4.7	4.6
Koweit	20.7	21.4	20.5	13.9	14.5	14.1	15.0	16.1	14.9
Liban	5.3	6.3	3.5	3.5	3.4	3.4	3.7	3.6	3.9
Oman	2.6	2.5	4.1	6.8	8.0	9.1	9.3	9.0	9.5
Qatar	3.1	4.1	5.0	6.1	6.5	5.7	6.4	6.7	6.9
Arabie saoudite	43.5	60.0	85.0	92.0	89.5	85.9	90.7	89.4	96.2
Syrie	8.2	9.4	16.6	21.8	21.0	22.2	21.1	21.5	24.4
Emirats arabes unis	9.1	13.1	21.3	30.5	31.9	29.8	24.3	25.8	25.2
Yémen	2.6	3.1	5.7	7.6	8.0	8.2	8.6	9.0	10.5
Moyen-Orient	259.4	316.4	447.2	456.8	460.0	444.7	418.5	421.0	419.1
Total non-OCDE	2 163.7	2 425.1	3 074.2	3 491.9	3 619.2	3 701.2	3 845.5	3 978.9	4 106.5
OCDE Amérique du N.	3 770.9	4 184.3	4 810.9	5 098.2	5 399.6	5 585.7	5 730.9	5 890.2	6 112.0
OCDE Pacifique	1 608.7	1 870.3	2 190.4	2 584.1	2 700.0	2 821.3	2 914.5	3 047.7	3 237.2
OCDE Europe	4 511.1	4 976.0	5 566.7	6 012.4	6 168.4	6 330.1	6 513.4	6 701.7	6 962.0
Total OCDE	9 890.7	11 030.6	12 568.0	13 694.7	14 268.1	14 737.0	15 158.8	15 639.6	16 311.2
Monde	12 054.3	13 455.7	15 642.2	17 186.6	17 887.3	18 438.3	19 004.2	19 618.5	20 417.8

Faute de séries complètes, l'Albanie, la Rép.pop.dém. de Corée et le Viêt-Nam ne sont pas pris en compte dans les totaux régionaux.

GDP using Exchange Rates (billion US$90)
PIB basé sur les taux de change (milliards US$90)
BIP auf Wechselkursbasis (Milliarden US$90)
PIL utilizzando i tassi di cambio (miliardi de US$ 1990)
為替換算による国内総生産（十億米ドル, 1990年価格）
PIB basado en los tipos de cambio (millardos US$90)
ВВП по валютному курсу (млрд.долл.США в ценах 1990 г.)

1989	1990	1991	1992	1993	1994	1995	1996	1997	
4.0	4.0	4.2	4.5	4.9	5.0	5.1	5.3	5.4	Bahrain
109.2	120.4	132.4	141.4	144.9	146.7	150.6	158.1	163.7	Iran
70.2	48.7	24.0	24.8	25.1	23.9	23.6	23.6	29.5	Iraq
49.1	52.5	56.5	59.7	63.0	67.4	72.1	75.4	77.1	Israel
4.0	4.0	4.1	4.8	5.0	5.4	5.8	5.8	5.9	Jordan
18.0	18.4	9.3	13.9	19.5	19.8	20.4	21.5	22.6	Kuwait
2.2	2.8	3.9	4.1	4.4	4.7	5.0	5.2	5.5	Lebanon
9.8	10.5	11.5	12.3	13.1	13.6	14.0	14.8	15.5	Oman
7.1	7.4	7.6	8.1	8.3	8.3	8.1	8.2	8.9	Qatar
96.3	104.7	115.6	118.9	118.1	118.7	119.2	120.9	123.2	Saudi Arabia
22.2	23.9	25.6	28.3	30.0	32.0	34.2	34.9	36.3	Syria
28.5	33.7	36.2	36.5	36.9	38.0	38.8	40.1	41.3	United Arab Emirates
11.1	11.4	11.4	12.0	12.3	12.3	13.3	13.8	14.6	Yemen
431.9	442.3	442.4	469.2	485.6	495.7	510.2	527.9	549.6	**Middle East**
4 237.4	4 295.4	4 351.0	4 377.8	4 484.0	4 616.6	4 803.4	5 013.9	5 247.3	**Non-OECD Total**
6 310.4	6 389.5	6 334.0	6 503.8	6 660.8	6 910.5	7 058.9	7 301.1	7 595.9	OECD North America
3 393.6	3 562.5	3 694.0	3 748.0	3 789.2	3 854.8	3 946.0	4 110.9	4 171.4	OECD Pacific
7 190.8	7 365.0	7 427.7	7 504.5	7 482.2	7 688.8	7 882.2	8 030.5	8 255.2	OECD Europe
16 894.7	17 316.9	17 455.8	17 756.3	17 932.2	18 454.1	18 887.1	19 442.6	20 022.4	**OECD Total**
21 132.2	21 612.3	21 806.8	22 134.1	22 416.2	23 070.7	23 690.6	24 456.5	25 269.8	**World**

Due to lack of complete series Albania, the Dem. Ppl's Rep. of Korea and Vietnam are not included in the regional aggregates.

GDP using Purchasing Power Parities (billion US$90)
PIB basé sur les parités de pouvoir d'achat (milliards US$90)
BIP auf Kaufkraftparitätenbasis (Milliarden US$90)

PIL utilizzando i PPA (miliardi di US$ 1990)

購買力平価換算による国内総生産（十億米ドル，1990年価格）

PIB basado en la paridad de poder adquisitivo (millardos US$90)

ВВП по ППС (млрд.долл.США в ценах 1990 г.)

	1971	1973	1978	1983	1984	1985	1986	1987	1988
Algérie	31.4	41.6	58.5	73.3	77.4	81.7	81.5	81.0	79.4
Angola	12.6	13.7	7.7	7.5	7.6	7.9	8.1	8.8	9.3
Bénin	3.4	3.8	4.0	4.9	5.3	5.7	5.8	5.7	5.9
Cameroun	9.9	10.7	17.2	24.1	25.9	28.0	29.9	29.3	27.0
Congo	1.8	2.2	2.5	4.9	5.2	5.2	4.8	4.8	4.9
RD du Congo	42.6	46.2	40.9	43.3	45.7	45.9	48.1	49.4	49.6
Egypte	66.8	68.6	104.8	149.7	158.9	169.4	173.8	178.2	187.7
Ethiopie	13.1	13.7	14.7	23.3	22.0	19.9	21.8	24.8	24.8
Gabon	2.9	3.5	5.3	5.8	6.3	6.1	6.1	5.0	5.7
Ghana	20.1	20.1	20.2	16.9	18.4	19.3	20.3	21.3	22.5
Côte d'Ivoire	11.2	12.4	18.8	17.1	16.7	17.4	18.0	17.9	18.1
Kenya	12.1	15.0	18.9	22.9	23.3	24.3	26.0	27.6	29.3
Libye	31.7	27.8	35.3	31.2	28.8	26.2	23.9	23.2	23.3
Maroc	35.1	36.8	49.6	60.0	62.6	66.6	72.1	70.3	77.6
Mozambique	2.4	2.6	1.9	1.7	1.6	1.7	1.7	2.0	2.2
Nigéria	74.4	81.1	93.1	85.0	80.9	88.7	91.0	90.3	99.3
Sénégal	7.2	7.2	8.3	10.0	9.6	9.9	10.4	10.8	11.3
Afrique du Sud	108.4	115.2	129.1	147.8	156.6	152.3	152.9	157.4	163.2
Soudan	5.1	4.6	7.7	8.1	7.7	7.3	7.5	7.7	7.4
Tanzanie	8.1	8.9	10.9	11.1	11.4	11.6	11.9	12.2	12.8
Tunisie	13.3	15.6	21.4	26.9	28.5	30.1	29.6	31.6	31.6
Zambie	6.1	6.5	6.9	7.0	7.0	7.1	7.2	7.3	7.8
Zimbabwe	12.2	13.5	12.9	17.8	17.5	18.7	19.1	19.3	20.8
Autre Afrique	68.9	70.7	83.6	89.6	90.5	93.6	96.8	99.6	105.0
Afrique	600.8	642.0	774.1	890.1	915.1	944.6	968.4	985.5	1 026.4
Argentine	163.9	172.3	182.3	196.0	199.9	186.0	199.3	204.5	200.6
Bolivie	10.3	11.5	14.7	13.4	13.4	13.2	12.8	13.2	13.5
Brésil	310.1	396.1	538.8	582.9	613.7	662.4	715.3	741.1	740.3
Chili	61.2	57.7	63.3	67.3	72.6	77.8	82.2	87.6	94.0
Colombie	66.1	75.9	97.2	111.8	115.6	119.2	126.1	132.9	138.3
Costa Rica	6.9	8.0	10.5	10.4	11.2	11.3	11.9	12.5	12.9
Cuba	14.2	16.7	22.8	29.1	31.2	32.6	33.0	31.7	32.4
République dominicaine	9.3	11.6	14.8	18.2	18.2	18.4	19.1	21.0	21.4
Equateur	14.5	20.8	29.0	32.7	34.1	35.6	36.7	34.5	38.1
El Salvador	11.3	12.6	16.1	11.6	11.7	11.8	11.8	12.1	12.4
Guatemala	13.8	15.8	20.8	21.4	21.5	21.4	21.4	22.2	23.0
Haiti	3.7	3.8	4.5	5.0	5.0	5.0	5.0	4.9	5.0
Honduras	2.0	2.3	3.1	3.2	3.4	3.5	3.5	3.7	3.9
Jamaïque	8.1	8.9	7.8	7.7	7.6	7.3	7.4	8.0	8.2
Antilles néerlandaises	1.6	1.7	1.9	2.0	2.0	1.8	1.9	2.0	2.0
Nicaragua	5.0	5.3	6.5	5.3	5.3	5.1	5.1	5.0	4.4
Panama	5.8	6.3	7.4	8.6	8.9	9.3	9.6	9.5	8.2
Paraguay	4.4	5.0	7.7	9.9	10.2	10.6	10.6	11.1	11.8
Pérou	58.4	63.3	73.4	74.7	77.8	79.5	86.9	94.3	86.4
Trinité-et-Tobago	3.4	3.9	5.7	7.2	7.6	7.3	7.1	6.7	6.5
Uruguay	15.5	15.3	18.6	17.2	17.0	17.2	18.7	20.2	20.2
Vénézuela	100.0	108.5	133.5	120.7	122.4	122.7	130.7	135.3	143.2
Autre Amérique latine	12.5	13.3	14.8	16.3	16.9	17.2	18.0	18.5	19.6
Amérique latine	901.9	1 036.6	1 295.3	1 372.6	1 427.2	1 476.2	1 574.2	1 632.3	1 646.4

GDP using Purchasing Power Parities (billion US$90)
PIB basé sur les parités de pouvoir d'achat (milliards US$90)
BIP auf Kaufkraftparitätenbasis (Milliarden US$90)
PIL utilizzando i PPA (miliardi di US$ 1990)
購買力平価換算による国内総生産（十億米ドル, 1990年価格）
PIB basado en la paridad de poder adquisitivo (millardos US$90)
ВВП по ППС (млрд.долл.США в ценах 1990 г.)

1989	1990	1991	1992	1993	1994	1995	1996	1997	
83.3	82.2	81.3	82.6	80.7	79.8	82.8	86.0	87.1	Algeria
9.3	9.3	9.3	8.9	6.7	6.8	7.6	8.5	9.1	Angola
6.2	6.1	6.4	6.7	6.9	7.2	7.5	7.9	8.4	Benin
26.5	24.9	23.9	23.2	22.4	21.9	22.6	23.7	24.9	Cameroon
5.0	5.1	5.2	5.4	5.3	5.0	5.2	5.5	5.4	Congo
49.0	45.8	41.9	37.5	32.5	31.2	31.4	31.1	29.4	DR of Congo
197.0	208.2	210.5	219.8	226.1	235.1	246.0	258.4	272.5	Egypt
24.8	25.7	24.0	23.1	25.9	26.3	27.9	30.8	32.6	Ethiopia
6.2	6.5	6.9	6.7	6.8	7.1	7.5	7.8	8.2	Gabon
23.7	24.4	25.7	26.7	28.1	29.0	30.1	31.5	32.9	Ghana
18.6	18.4	18.4	18.4	18.4	18.7	20.2	21.4	22.7	Ivory Coast
30.7	32.0	32.4	32.2	32.3	33.1	34.6	36.0	36.8	Kenya
23.5	25.7	26.6	26.0	24.8	23.7	24.1	24.5	24.6	Libya
79.5	82.6	88.3	84.8	83.9	92.6	86.5	97.0	95.0	Morocco
2.3	2.4	2.6	2.6	3.1	3.3	3.3	3.5	3.9	Mozambique
106.4	115.1	120.6	124.1	126.9	126.1	129.3	137.6	143.0	Nigeria
11.2	11.6	11.6	11.8	11.6	11.9	12.5	13.2	13.8	Senegal
167.1	166.6	164.9	161.2	163.3	167.8	173.5	179.1	182.1	South Africa
7.9	7.8	7.8	8.7	9.4	9.9	10.4	10.8	11.2	Sudan
13.3	14.0	14.6	13.3	14.9	15.1	15.5	16.2	16.8	Tanzania
32.2	34.8	36.1	38.9	39.8	41.1	42.1	45.0	47.5	Tunisia
7.7	7.7	7.7	7.6	8.1	7.8	7.6	8.1	8.4	Zambia
21.9	23.4	24.7	22.4	22.7	24.3	24.1	25.9	26.7	Zimbabwe
109.3	112.9	114.7	115.9	117.8	118.2	124.4	131.5	138.1	Other Africa
1 062.5	1 093.0	1 106.2	1 108.4	1 118.4	1 142.9	1 176.9	1 241.2	1 281.3	**Africa**
186.7	183.3	202.6	222.1	234.8	253.6	243.4	255.1	277.0	Argentina
14.1	14.7	15.5	15.7	16.4	17.2	18.0	18.7	19.5	Bolivia
764.6	731.7	741.2	737.5	773.7	819.3	853.7	877.6	905.7	Brazil
103.9	107.7	116.3	130.6	139.7	147.7	163.4	175.4	187.8	Chile
143.0	148.7	152.3	158.2	166.8	176.5	186.8	190.6	196.4	Colombia
13.6	14.1	14.4	15.5	16.5	17.3	17.7	17.6	18.1	Costa Rica
32.8	31.7	28.3	25.1	21.4	21.6	22.1	23.6	24.2	Cuba
22.4	21.1	21.3	23.1	23.7	24.8	26.0	27.9	30.1	Dominican Republic
38.2	39.4	41.3	42.8	43.7	45.6	46.6	47.6	49.2	Ecuador
12.5	13.1	13.5	14.6	15.6	16.6	17.6	18.0	18.7	El Salvador
24.0	24.7	25.6	26.8	27.9	29.0	30.5	31.4	32.7	Guatemala
5.0	5.0	5.2	4.5	4.4	4.1	4.2	4.4	4.4	Haiti
4.1	4.1	4.2	4.5	4.7	4.7	4.8	5.0	5.2	Honduras
8.8	9.2	9.3	9.5	9.6	9.7	9.7	9.6	9.3	Jamaica
2.1	2.2	2.2	2.3	2.3	2.4	2.4	2.4	2.5	Netherlands Antilles
4.3	4.3	4.3	4.3	4.3	4.4	4.6	5.8	6.1	Nicaragua
8.3	9.0	9.9	10.7	11.2	11.6	11.8	12.1	12.6	Panama
12.5	12.9	13.2	13.4	14.0	14.4	15.1	15.3	15.8	Paraguay
76.3	72.2	77.2	75.9	80.7	91.3	98.2	100.6	107.9	Peru
6.4	6.5	7.5	6.6	6.5	6.7	7.0	7.2	7.5	Trinidad-and-Tobago
20.5	20.7	21.3	23.0	23.7	25.2	24.8	26.1	27.4	Uruguay
130.9	139.4	153.0	162.3	162.7	158.8	164.7	163.9	172.3	Venezuela
20.7	21.6	22.7	22.9	23.1	23.5	23.9	24.7	25.8	Other Latin America
1 655.7	1 637.4	1 702.5	1 751.9	1 827.6	1 925.8	1 996.9	2 060.4	2 156.3	**Latin America**

GDP using Purchasing Power Parities (billion US$90)
PIB basé sur les parités de pouvoir d'achat (milliards US$90)

BIP auf Kaufkraftparitätenbasis (Milliarden US$90)

PIL utilizzando i PPA (miliardi di US$ 1990)

購買力平価換算による国内総生産（十億米ドル，1990年価格）

PIB basado en la paridad de poder adquisitivo (millardos US$90)

ВВП по ППС (млрд.долл.США в ценах 1990 г.)

	1971	1973	1978	1983	1984	1985	1986	1987	1988
Bangladesh	53.1	47.1	59.7	76.2	79.8	83.0	86.6	90.1	92.7
Brunei	1.6	1.8	2.7	2.6	2.6	2.5	2.4	2.5	2.5
Inde	391.5	400.6	510.6	612.6	635.4	669.9	702.9	736.3	809.1
Indonésie	125.9	149.1	215.5	297.6	318.9	330.0	349.7	368.2	391.6
Malaisie	30.3	37.0	51.8	73.2	78.9	78.0	78.9	83.2	90.6
Myanmar	48.8	49.5	65.1	86.6	90.9	93.5	92.5	88.8	78.7
Népal	7.9	8.2	9.9	10.8	11.8	12.5	13.1	13.3	14.4
Pakistan	68.5	73.9	94.2	132.2	138.9	149.4	157.7	167.8	180.6
Philippines	68.2	78.4	103.5	125.6	116.4	107.9	111.6	116.4	124.2
Singapour	12.5	15.7	21.5	32.7	35.4	34.8	35.5	39.0	43.5
Sri Lanka	16.9	18.0	22.8	29.7	31.2	32.7	34.2	34.7	35.6
Taipei chinois	50.7	64.9	98.2	135.8	150.2	157.7	176.0	198.5	214.0
Thaïlande	64.2	73.8	107.2	140.0	148.1	154.9	163.5	179.1	202.9
Viêt-Nam	-	-	-	-	50.5	52.4	53.9	55.8	58.7
Autre Asie	28.4	31.4	38.2	44.5	45.5	45.9	47.8	45.4	44.0
Asie	968.5	1 049.3	1 400.8	1 800.0	1 883.9	1 952.8	2 052.4	2 163.2	2 324.5
Rép. populaire de Chine	526.2	585.1	723.0	1 047.8	1 207.1	1 370.0	1 490.6	1 663.5	1 851.4
Hong-Kong, Chine	23.9	29.9	43.7	64.4	70.7	70.9	78.7	88.9	96.0
Chine	550.0	615.0	766.7	1 112.2	1 277.8	1 440.9	1 569.3	1 752.4	1 947.5
Albanie	-	-	-	7.9	7.8	8.0	8.4	8.3	8.2
Bulgarie	33.8	36.8	42.8	48.0	49.6	50.9	53.1	56.3	62.4
Chypre	3.1	3.3	3.6	4.7	5.2	5.4	5.6	6.0	6.5
Gibraltar	0.2	0.2	0.3	0.3	0.3	0.3	0.3	0.3	0.3
Malte	0.9	1.0	1.9	2.4	2.4	2.5	2.6	2.7	2.9
Roumanie	46.1	50.6	74.2	88.1	93.3	93.2	95.4	90.6	90.2
République slovaque	25.7	27.6	31.7	33.4	34.1	35.3	36.8	37.7	38.4
Ex-Yougoslavie	56.5	60.0	86.4	94.0	95.6	95.0	98.4	97.5	94.9
Europe non-OCDE	166.2	179.6	240.8	270.8	280.5	282.6	292.1	291.0	295.7
Arménie	-	-	-	-	-	-	-	-	-
Azerbaïdjan	-	-	-	-	-	-	-	-	-
Bélarus	-	-	-	-	-	-	-	-	-
Estonie	-	-	-	-	-	-	-	-	-
Géorgie	-	-	-	-	-	-	-	-	-
Kazakhstan	-	-	-	-	-	-	-	-	-
Kirghizistan	-	-	-	-	-	-	-	-	-
Lettonie	-	-	-	-	-	-	-	-	-
Lituanie	-	-	-	-	-	-	-	-	-
Moldova	-	-	-	-	-	-	-	-	-
Russie	-	-	-	-	-	-	-	-	-
Tadjikistan	-	-	-	-	-	-	-	-	-
Turkménistan	-	-	-	-	-	-	-	-	-
Ukraine	-	-	-	-	-	-	-	-	-
Ouzbékistan	-	-	-	-	-	-	-	-	-
Ex-URSS	1 181.3	1 291.6	1 526.2	1 686.1	1 709.7	1 728.5	1 801.1	1 824.5	1 851.9

Faute de séries complètes, l'Albanie, la Rép.pop.dém. de Corée et le Viêt-Nam ne sont pas pris en compte dans les totaux régionaux.
Ex-URSS: la somme des républiques peut être différente du total.

GDP using Purchasing Power Parities (billion US$90)
PIB basé sur les parités de pouvoir d'achat (milliards US$90)
BIP auf Kaufkraftparitätenbasis (Milliarden US$90)
PIL utilizzando i PPA (miliardi di US$ 1990)
購買力平価換算による国内総生産（十億米ドル, 1990年価格）
PIB basado en la paridad de poder adquisitivo (millardos US$90)
ВВП по ППС (млрд.долл.США в ценах 1990 г.)

1989	1990	1991	1992	1993	1994	1995	1996	1997	
95.0	101.3	104.7	109.1	113.9	118.6	123.8	130.4	137.8	Bangladesh
2.5	2.6	2.7	2.6	2.5	2.6	2.6	2.7	2.8	Brunei
862.6	911.1	915.3	964.6	1 012.4	1 094.7	1 175.5	1 262.0	1 328.0	India
427.2	465.6	507.2	543.8	583.3	627.3	678.8	731.9	767.8	Indonesia
98.9	108.4	117.7	126.9	137.5	150.2	164.4	178.5	192.5	Malaysia
81.6	83.9	83.4	91.4	97.0	104.2	111.4	118.5	123.3	Myanmar
15.0	15.7	16.7	17.4	18.0	19.5	20.2	21.3	22.1	Nepal
189.6	198.0	208.8	225.2	229.5	238.4	250.7	262.4	261.5	Pakistan
131.9	135.9	135.2	135.6	138.5	144.6	151.3	160.2	168.4	Philippines
47.7	52.0	55.8	59.3	65.4	72.3	78.6	83.9	90.5	Singapore
36.4	38.7	40.5	42.2	45.1	47.6	50.3	52.2	55.5	Sri Lanka
231.6	244.1	262.6	280.3	298.0	317.5	336.7	355.7	380.0	Chinese Taipei
227.6	253.0	274.7	296.9	321.8	350.5	381.5	402.6	400.8	Thailand
63.0	66.2	70.2	76.2	82.4	89.7	98.2	107.4	116.9	Vietnam
42.9	42.5	43.8	45.6	47.2	47.6	52.8	55.3	57.7	Other Asia
2 490.6	2 652.9	2 768.9	2 941.0	3 110.1	3 335.6	3 578.4	3 817.4	3 988.7	**Asia**
1 927.3	2 000.6	2 184.6	2 494.9	2 831.7	3 188.5	3 523.2	3 861.5	4 201.3	People's Rep.of China
98.5	101.8	107.0	113.7	120.7	127.2	132.1	138.7	145.9	Hong Kong, China
2 025.8	2 102.4	2 291.6	2 608.5	2 952.3	3 315.6	3 655.4	4 000.3	4 347.2	**China**
9.0	8.1	5.9	5.5	6.0	6.5	7.1	7.8	7.2	Albania
60.4	54.9	50.2	46.6	45.9	46.7	48.1	43.2	40.2	Bulgaria
7.1	7.6	7.6	8.3	8.5	8.9	9.4	9.6	9.9	Cyprus
0.3	0.3	0.3	0.3	0.3	0.3	0.3	0.3	0.4	Gibraltar
3.1	3.3	3.5	3.7	3.9	4.0	4.3	4.5	4.6	Malta
84.9	80.1	69.7	63.6	64.6	67.1	71.9	74.7	69.8	Romania
38.9	37.9	32.3	30.2	29.0	30.5	32.6	34.7	37.0	Slovak Republic
95.4	89.6	90.7	74.8	70.8	70.5	73.2	76.5	79.0	Former Yugoslavia
290.1	273.7	254.5	227.5	223.1	228.1	239.9	243.6	240.7	**Non-OECD Europe**
-	-	-	3.2	2.7	2.9	3.1	3.2	3.4	Armenia
-	-	-	13.4	10.3	8.7	7.5	7.6	7.9	Azerbaijan
-	-	-	60.5	55.9	48.8	43.8	45.0	49.7	Belarus
-	-	-	9.6	8.8	8.7	9.0	9.4	10.5	Estonia
-	-	-	10.2	6.2	5.5	5.6	6.2	6.9	Georgia
-	-	-	71.7	64.8	56.6	52.0	52.2	53.1	Kazakhstan
-	-	-	4.2	3.5	2.8	2.7	2.8	3.1	Kyrgyzstan
-	-	-	11.6	9.9	10.0	9.9	10.2	10.9	Latvia
-	-	-	15.2	12.7	11.5	11.8	12.4	13.1	Lithuania
-	-	-	12.2	12.1	8.3	8.1	7.3	7.4	Moldova
-	-	-	934.4	853.4	746.1	715.2	690.2	695.7	Russia
-	-	-	6.8	6.1	4.8	4.2	4.0	4.0	Tajikistan
-	-	-	14.0	12.7	10.3	9.4	9.1	11.5	Turkmenistan
-	-	-	262.1	224.9	173.1	152.0	136.8	132.4	Ukraine
-	-	-	42.5	41.6	39.8	39.3	40.1	42.2	Uzbekistan
1 905.6	1 891.9	1 778.8	1 527.8	1 369.0	1 175.5	1 110.7	1 064.3	1 080.2	**Former USSR**

Due to lack of complete series Albania, the Dem. Ppl's Rep. of Korea and Vietnam are not included in the regional aggregates.
Former USSR: data for individual republics may not add to the total.

GDP using Purchasing Power Parities (billion US$90)
PIB basé sur les parités de pouvoir d'achat (milliards US$90)
BIP auf Kaufkraftparitätenbasis (Milliarden US$90)

PIL utilizzando i PPA (miliardi di US$ 1990)

購買力平価換算による国内総生産（十億米ドル，1990年価格）

PIB basado en la paridad de poder adquisitivo (millardos US$90)

ВВП по ППС (млрд.долл.США в ценах 1990 г.)

	1971	1973	1978	1983	1984	1985	1986	1987	1988
Bahrein	2.1	2.5	3.9	3.8	4.0	3.8	3.9	3.9	4.3
Iran	31.1	41.1	59.6	59.5	60.5	61.9	56.1	55.3	52.2
Irak	59.6	65.7	101.4	85.1	85.2	73.0	58.9	58.0	52.2
Israël	23.7	27.9	32.7	41.1	41.5	42.9	45.0	48.2	49.2
Jordanie	5.4	5.3	9.0	13.5	14.6	15.1	16.2	16.6	16.3
Koweit	30.6	31.6	30.3	20.5	21.5	20.8	22.2	23.7	22.0
Liban	20.3	23.9	13.4	13.4	13.0	12.9	14.2	13.5	14.8
Oman	5.7	5.4	8.8	14.8	17.3	19.7	20.1	19.4	20.6
Qatar	3.4	4.5	5.5	6.7	7.0	6.3	7.0	7.3	7.5
Arabie saoudite	64.9	89.4	126.7	137.1	133.4	128.0	135.1	133.2	143.3
Syrie	19.8	22.7	40.1	52.8	50.6	53.7	51.1	52.1	59.0
Emirats arabes unis	6.5	9.4	15.2	21.8	22.7	21.3	17.3	18.4	17.9
Yémen	12.4	14.9	27.8	37.1	38.7	39.6	41.8	43.7	51.0
Moyen-Orient	285.6	344.3	474.3	507.2	510.0	499.0	488.8	493.4	510.4
Total non-OCDE	4 654.3	5 158.4	6 478.2	7 639.1	8 004.2	8 324.6	8 746.3	9 142.3	9 602.8
OCDE Amérique du N.	3 898.3	4 335.4	5 011.5	5 348.2	5 657.7	5 849.9	5 982.5	6 145.6	6 368.9
OCDE Pacifique	1 327.6	1 541.9	1 818.3	2 151.6	2 252.8	2 355.4	2 440.5	2 559.0	2 722.2
OCDE Europe	4 084.8	4 520.4	5 097.9	5 473.7	5 621.4	5 772.8	5 949.7	6 135.9	6 373.6
Total OCDE	9 310.7	10 397.7	11 927.7	12 973.5	13 531.9	13 978.1	14 372.7	14 840.4	15 464.7
Monde	13 965.0	15 556.1	18 405.9	20 612.5	21 536.1	22 302.8	23 119.0	23 982.7	25 067.5

Faute de séries complètes, l'Albanie, la Rép.pop.dém. de Corée et le Viêt-Nam ne sont pas pris en compte dans les totaux régionaux.

GDP using Purchasing Power Parities (billion US$90)
PIB basé sur les parités de pouvoir d'achat (milliards US$90)
BIP auf Kaufkraftparitätenbasis (Milliarden US$90)
PIL utilizzando i PPA (miliardi di US$ 1990)
購買力平価換算による国内総生産（十億米ドル, 1990年価格）
PIB basado en la paridad de poder adquisitivo (millardos US$90)
ВВП по ППС (млрд.долл.США в ценах 1990 г.)

1989	1990	1991	1992	1993	1994	1995	1996	1997	
4.3	4.4	4.6	5.0	5.4	5.5	5.6	5.8	6.0	Bahrain
54.2	59.7	65.7	70.1	71.9	72.8	74.7	78.4	81.2	Iran
56.2	38.9	19.2	19.8	20.1	19.1	18.9	18.9	23.6	Iraq
49.6	53.0	57.1	60.3	63.6	68.0	72.8	76.2	77.9	Israel
14.1	14.3	14.5	16.9	17.8	19.3	20.5	20.6	21.0	Jordan
26.6	27.2	13.7	20.6	28.8	29.2	30.1	31.8	33.3	Kuwait
8.5	10.8	14.9	15.6	16.7	18.0	19.2	20.0	20.8	Lebanon
21.3	22.9	25.0	26.6	28.5	29.5	30.5	32.0	33.6	Oman
7.7	8.0	8.3	8.8	9.0	9.0	8.8	8.9	9.7	Qatar
143.6	156.0	172.3	177.1	176.0	176.8	177.7	180.2	183.6	Saudi Arabia
53.7	57.8	61.9	68.5	72.5	77.4	82.6	84.4	87.8	Syria
20.4	24.0	25.8	26.1	26.3	27.1	27.7	28.6	29.5	United Arab Emirates
53.8	55.1	55.3	58.0	59.7	59.4	64.3	67.2	70.8	Yemen
514.0	532.0	538.4	573.3	596.4	611.3	633.3	653.1	678.7	**Middle East**
9 944.3	10 183.3	10 440.9	10 738.4	11 196.9	11 734.8	12 391.4	13 080.2	13 773.1	**Non-OECD Total**
6 578.9	6 673.9	6 633.5	6 815.4	6 978.2	7 242.1	7 365.9	7 626.0	7 945.3	OECD North America
2 855.0	3 000.9	3 117.1	3 169.3	3 214.6	3 285.5	3 377.7	3 524.8	3 587.1	OECD Pacific
6 572.7	6 717.5	6 750.1	6 828.2	6 834.5	7 007.7	7 204.5	7 362.1	7 588.6	OECD Europe
16 006.6	16 392.3	16 500.7	16 812.8	17 027.3	17 535.2	17 948.1	18 512.9	19 121.0	**OECD Total**
25 951.0	26 575.7	26 941.6	27 551.2	28 224.1	29 270.1	30 339.5	31 593.1	32 894.1	**World**

Due to lack of complete series Albania, the Dem. Ppl's Rep. of Korea and Vietnam are not included in the regional aggregates.

Population (millions)
Population (millions)
Bevölkerung (Millionen)
Popolazione (milioni)
人口（100 万人）
Población (millones)
Численность населения (млн. человек)

	1971	1973	1978	1983	1984	1985	1986	1987	1988
Algérie	14.2	15.1	17.6	20.5	21.2	21.9	22.5	23.2	23.8
Angola	5.7	5.9	6.6	7.6	7.8	8.0	8.2	8.4	8.7
Bénin	2.7	2.9	3.3	3.8	3.9	4.0	4.2	4.3	4.4
Cameroun	6.8	7.1	8.2	9.4	9.7	10.0	10.3	10.5	10.8
Congo	1.3	1.4	1.6	1.8	1.9	1.9	2.0	2.0	2.1
RD du Congo	20.8	22.0	25.4	29.7	30.7	31.7	32.7	33.8	35.0
Egypte	33.6	34.9	38.9	44.2	45.3	46.5	47.7	48.9	50.1
Ethiopie	29.7	31.3	35.7	41.0	42.2	43.4	44.7	46.1	47.6
Gabon	0.5	0.6	0.7	0.8	0.8	0.8	0.8	0.9	0.9
Ghana	8.9	9.4	10.3	11.7	12.2	12.6	13.1	13.5	14.0
Côte d'Ivoire	5.7	6.2	7.6	9.2	9.5	9.9	10.2	10.6	10.9
Kenya	11.9	12.8	15.4	18.5	19.2	19.9	20.6	21.3	22.0
Libye	2.1	2.2	2.8	3.5	3.6	3.8	3.9	4.1	4.2
Maroc	15.7	16.5	18.5	20.7	21.2	21.6	22.1	22.6	23.1
Mozambique	9.6	10.0	11.4	13.1	13.3	13.5	13.7	13.8	13.9
Nigéria	54.7	57.8	66.9	78.2	80.7	83.2	85.7	88.3	90.9
Sénégal	4.3	4.5	5.2	6.0	6.2	6.4	6.6	6.7	6.9
Afrique du Sud	22.6	23.7	26.4	29.7	30.5	31.3	32.1	32.9	33.7
Soudan	14.2	15.1	17.6	20.8	20.9	21.5	22.0	22.5	23.0
Tanzanie	14.1	15.0	17.4	20.4	21.1	21.8	22.5	23.2	23.9
Tunisie	5.2	5.4	6.0	6.9	7.0	7.3	7.5	7.7	7.9
Zambie	4.3	4.6	5.4	6.3	6.5	6.7	6.9	7.1	7.3
Zimbabwe	5.4	5.7	6.6	7.8	8.0	8.3	8.6	8.9	9.2
Autre Afrique	70.8	73.8	83.7	96.1	98.8	101.7	104.8	108.2	111.6
Afrique	364.9	383.7	439.1	507.7	522.2	537.6	553.4	569.6	586.0
Argentine	24.4	25.2	27.3	29.4	29.9	30.3	30.8	31.2	31.7
Bolivie	4.3	4.5	5.1	5.7	5.8	5.9	6.0	6.1	6.3
Brésil	98.4	103.2	116.2	129.9	132.6	135.2	137.9	140.4	143.0
Chili	9.7	10.0	10.8	11.7	11.9	12.0	12.2	12.4	12.7
Colombie	23.1	24.2	27.2	30.4	31.0	31.7	32.3	33.0	33.6
Costa Rica	1.8	1.9	2.2	2.5	2.6	2.6	2.7	2.8	2.9
Cuba	8.7	9.0	9.6	9.9	10.0	10.1	10.2	10.3	10.4
République dominicaine	4.5	4.8	5.4	6.1	6.2	6.4	6.5	6.7	6.8
Equateur	6.1	6.5	7.5	8.6	8.9	9.1	9.3	9.6	9.8
El Salvador	3.7	3.9	4.4	4.7	4.7	4.8	4.8	4.9	4.9
Guatemala	5.4	5.7	6.5	7.4	7.5	7.7	7.9	8.1	8.3
Haiti	4.6	4.8	5.2	5.6	5.8	5.9	6.0	6.1	6.2
Honduras	2.7	2.8	3.3	3.9	4.1	4.2	4.3	4.5	4.6
Jamaïque	1.9	2.0	2.1	2.2	2.3	2.3	2.3	2.4	2.4
Antilles néerlandaises	0.2	0.2	0.2	0.2	0.2	0.2	0.2	0.2	0.2
Nicaragua	2.2	2.3	2.7	3.2	3.3	3.4	3.5	3.6	3.7
Panama	1.5	1.6	1.9	2.1	2.1	2.2	2.2	2.3	2.3
Paraguay	2.4	2.5	2.9	3.4	3.5	3.6	3.7	3.8	4.0
Pérou	13.6	14.3	16.4	18.6	19.1	19.5	19.9	20.3	20.8
Trinité-et-Tobago	1.0	1.0	1.0	1.1	1.2	1.2	1.2	1.2	1.2
Uruguay	2.8	2.8	2.9	3.0	3.0	3.0	3.0	3.0	3.1
Vénézuela	11.1	11.9	14.2	16.3	16.7	17.1	17.5	18.0	18.4
Autre Amérique latine	2.9	2.9	3.0	3.1	3.1	3.1	3.2	3.2	3.2
Amérique latine	236.9	248.2	278.1	309.0	315.3	321.5	327.8	334.1	340.4

Population (millions)
Population (millions)
Bevölkerung (Millionen)
Popolazione (milioni)
人 口 (100 万人)
Población (millones)
Численность населения (млн. человек)

1989	1990	1991	1992	1993	1994	1995	1996	1997	
24.4	25.0	25.6	26.3	26.9	27.5	28.1	28.7	29.3	Algeria
8.9	9.2	9.6	9.9	10.3	10.6	11.0	11.3	11.7	Angola
4.6	4.7	4.9	5.0	5.2	5.3	5.5	5.6	5.8	Benin
11.2	11.5	11.8	12.1	12.5	12.8	13.2	13.6	13.9	Cameroon
2.2	2.2	2.3	2.4	2.4	2.5	2.6	2.6	2.7	Congo
36.2	37.4	38.6	39.9	41.2	42.5	43.8	45.3	46.7	DR of Congo
51.3	52.4	53.6	54.8	55.9	57.1	58.2	59.3	60.3	Egypt
49.3	51.2	53.0	54.8	53.3	54.9	56.5	58.2	59.8	Ethiopia
0.9	1.0	1.0	1.0	1.0	1.1	1.1	1.1	1.2	Gabon
14.4	14.9	15.3	15.8	16.2	16.6	17.1	17.5	18.0	Ghana
11.3	11.6	12.0	12.4	12.7	13.1	13.5	13.9	14.2	Ivory Coast
22.8	23.6	24.3	25.0	25.8	26.5	27.2	27.9	28.6	Kenya
4.3	4.4	4.5	4.6	4.8	4.9	5.0	5.1	5.2	Libya
23.6	24.0	24.5	25.0	25.5	25.9	26.4	26.8	27.3	Morocco
14.0	14.2	14.4	14.7	15.0	15.4	15.8	16.2	16.6	Mozambique
93.5	96.2	99.0	101.9	104.9	108.0	111.3	114.6	117.9	Nigeria
7.1	7.3	7.5	7.7	7.9	8.1	8.3	8.6	8.8	Senegal
34.5	35.2	35.9	36.7	37.5	38.3	39.1	39.9	40.6	South Africa
23.6	24.1	24.6	25.1	25.6	26.1	26.6	27.2	27.7	Sudan
24.7	25.5	26.3	27.1	27.9	28.8	29.6	30.5	31.3	Tanzania
8.0	8.2	8.3	8.5	8.7	8.8	8.9	9.1	9.2	Tunisia
7.5	7.8	8.0	8.3	8.5	8.7	9.0	9.2	9.4	Zambia
9.5	9.7	10.0	10.3	10.5	10.8	11.0	11.2	11.5	Zimbabwe
114.8	117.5	120.4	123.3	126.4	128.0	129.6	133.0	137.9	Other Africa
602.5	618.8	635.5	652.5	666.5	682.4	698.4	716.4	735.7	**Africa**
32.1	32.5	33.0	33.4	33.8	34.3	34.8	35.2	35.7	Argentina
6.4	6.6	6.7	6.9	7.1	7.2	7.4	7.6	7.8	Bolivia
145.5	147.9	150.3	152.7	155.0	157.2	159.3	161.5	163.7	Brazil
12.9	13.1	13.3	13.5	13.8	14.0	14.2	14.4	14.6	Chile
34.3	35.0	35.7	36.4	37.1	37.8	38.5	39.3	40.0	Colombia
3.0	3.0	3.1	3.2	3.2	3.3	3.4	3.4	3.5	Costa Rica
10.5	10.6	10.7	10.8	10.9	10.9	11.0	11.0	11.1	Cuba
7.0	7.1	7.3	7.4	7.5	7.7	7.8	8.0	8.1	Dominican Republic
10.0	10.3	10.5	10.7	11.0	11.2	11.5	11.7	11.9	Ecuador
5.0	5.1	5.2	5.3	5.4	5.5	5.7	5.8	5.9	El Salvador
8.5	8.7	9.0	9.2	9.5	9.7	10.0	10.2	10.5	Guatemala
6.3	6.5	6.6	6.7	6.9	7.0	7.2	7.3	7.5	Haiti
4.7	4.9	5.0	5.2	5.3	5.5	5.7	5.8	6.0	Honduras
2.4	2.4	2.4	2.4	2.5	2.5	2.5	2.5	2.6	Jamaica
0.2	0.2	0.2	0.2	0.2	0.2	0.2	0.2	0.2	Netherlands Antilles
3.7	3.8	3.9	4.1	4.2	4.3	4.4	4.6	4.7	Nicaragua
2.4	2.4	2.4	2.5	2.5	2.6	2.6	2.7	2.7	Panama
4.1	4.2	4.3	4.5	4.6	4.7	4.8	5.0	5.1	Paraguay
21.2	21.6	22.0	22.4	22.8	23.2	23.5	23.9	24.4	Peru
1.2	1.2	1.2	1.3	1.3	1.3	1.3	1.3	1.3	Trinidad-and-Tobago
3.1	3.1	3.1	3.1	3.2	3.2	3.2	3.2	3.3	Uruguay
18.9	19.5	20.0	20.4	20.9	21.4	21.8	22.3	22.8	Venezuela
3.3	3.3	3.3	3.3	3.4	3.4	3.4	3.4	3.5	Other Latin America
346.7	353.1	359.4	365.6	371.9	378.1	384.3	390.5	396.7	**Latin America**

Population (millions)
Population (millions)
Bevölkerung (Millionen)

Popolazione (milioni)

人口（100万人）

Población (millones)

Численность населения (млн. человек)

	1971	1973	1978	1983	1984	1985	1986	1987	1988
Bangladesh	68.5	72.4	82.6	93.2	95.6	98.0	101.0	103.6	106.0
Brunei	0.1	0.1	0.2	0.2	0.2	0.2	0.2	0.2	0.2
Inde	560.3	586.2	656.9	734.1	749.7	765.1	781.9	798.7	815.6
Indonésie	120.4	126.4	142.2	157.2	160.1	163.0	166.0	169.0	172.0
RPD de Corée	14.7	15.6	17.2	18.4	18.7	18.9	19.2	19.5	19.8
Malaisie	11.1	11.7	13.1	14.8	15.3	15.7	16.1	16.6	17.1
Myanmar	27.8	29.1	32.4	36.1	36.8	37.5	38.3	38.9	39.5
Népal	11.6	12.2	13.8	15.7	16.1	16.5	16.9	17.4	17.8
Pakistan	62.5	66.7	78.0	89.8	92.3	94.8	97.4	100.0	102.6
Philippines	38.7	40.9	46.3	51.9	53.2	54.7	56.2	57.7	59.4
Singapour	1.9	2.0	2.2	2.4	2.4	2.5	2.5	2.6	2.6
Sri Lanka	12.6	13.1	14.2	15.4	15.6	15.8	16.1	16.4	16.6
Taipei chinois	15.0	15.6	17.1	18.7	19.0	19.3	19.5	19.7	19.9
Thaïlande	36.8	39.1	44.7	49.5	50.4	51.1	52.0	52.8	53.7
Viêt-Nam	43.7	45.8	51.4	56.7	57.7	58.9	60.2	61.8	63.3
Autre Asie	17.5	18.4	20.9	20.1	20.3	20.6	20.8	21.1	21.5
Asie	984.9	1 033.9	1 164.5	1 299.0	1 327.0	1 354.9	1 384.8	1 414.6	1 444.6
Rép. populaire de Chine	841.1	881.9	956.2	1 023.3	1 036.8	1 051.0	1 066.8	1 084.0	1 101.6
Hong-Kong, Chine	4.0	4.2	4.6	5.3	5.4	5.5	5.5	5.6	5.6
Chine	845.1	886.1	960.8	1 028.6	1 042.2	1 056.5	1 072.3	1 089.6	1 107.2
Albanie	2.2	2.3	2.6	2.8	2.9	3.0	3.0	3.1	3.1
Bulgarie	8.5	8.6	8.8	8.9	8.9	8.9	9.0	9.0	9.0
Chypre	0.6	0.6	0.6	0.6	0.6	0.6	0.7	0.7	0.7
Gibraltar	0.0	0.0	0.0	0.0	0.0	0.0	0.0	0.0	0.0
Malte	0.3	0.3	0.4	0.4	0.4	0.3	0.3	0.3	0.3
Roumanie	20.5	20.8	21.9	22.6	22.6	22.7	22.8	22.9	23.1
République slovaque	4.6	4.7	4.9	5.1	5.2	5.2	5.2	5.3	5.3
Bosnie-Herzégovine	-	-	-	-	-	-	-	-	-
Croatie	-	-	-	-	-	-	-	-	-
Ex-RYM	-	-	-	-	-	-	-	-	-
Slovénie	-	-	-	-	-	-	-	-	-
RF de Yougoslavie	-	-	-	-	-	-	-	-	-
Ex-Yougoslavie	20.6	21.0	22.0	22.7	22.8	23.0	23.1	23.2	23.3
Europe non-OCDE	55.1	56.0	58.6	60.3	60.6	60.8	61.1	61.4	61.6

Faute de séries complètes, l'Albanie, la Rép.pop.dém. de Corée et le Viêt-Nam ne sont pas pris en compte dans les totaux régionaux.
Ex-Yougoslavie: la somme des républiques peut être différente du total.

Population (millions)
Population (millions)
Bevölkerung (Millionen)
Popolazione (milioni)
人口（100 万人）
Población (millones)
Численность населения (млн. человек)

1989	1990	1991	1992	1993	1994	1995	1996	1997	
108.2	110.4	112.4	114.3	116.1	117.9	119.8	121.7	123.6	Bangladesh
0.2	0.3	0.3	0.3	0.3	0.3	0.3	0.3	0.3	Brunei
832.5	849.5	866.5	882.3	898.2	913.6	929.4	945.6	962.4	India
175.1	178.2	181.4	184.6	187.7	190.8	194.0	197.2	200.4	Indonesia
20.1	20.5	20.8	21.1	21.5	21.9	22.2	22.6	22.9	DPR of Korea
17.7	18.2	18.7	19.1	19.6	20.1	20.6	21.1	21.7	Malaysia
40.1	40.5	41.0	41.4	41.9	42.4	42.9	43.4	43.9	Myanmar
18.3	18.8	19.3	19.7	20.2	20.8	21.3	21.8	22.3	Nepal
105.3	108.0	110.8	113.6	116.4	119.4	122.4	125.4	128.5	Pakistan
61.0	62.6	64.1	65.6	67.1	68.7	70.3	71.9	73.5	Philippines
2.6	2.7	2.8	2.8	2.9	2.9	3.0	3.0	3.1	Singapore
16.8	17.0	17.2	17.4	17.6	17.9	18.1	18.3	18.6	Sri Lanka
20.1	20.4	20.6	20.8	20.9	21.1	21.3	21.5	21.7	Chinese Taipei
54.6	55.6	56.5	57.3	58.1	58.7	59.4	60.0	60.6	Thailand
64.8	66.2	67.8	69.4	71.0	72.5	74.0	75.4	76.7	Vietnam
22.0	22.2	22.6	24.5	26.9	28.6	30.0	31.2	32.5	Other Asia
1 474.5	1 504.3	1 534.0	1 563.6	1 594.0	1 623.1	1 652.6	1 682.4	1 713.0	**Asia**
1 118.6	1 135.2	1 150.8	1 165.0	1 178.4	1 190.9	1 203.3	1 215.4	1 227.2	People's Rep.of China
5.7	5.7	5.8	5.8	5.9	6.0	6.2	6.3	6.5	Hong Kong, China
1 124.3	1 140.9	1 156.5	1 170.8	1 184.3	1 196.9	1 209.5	1 221.7	1 233.7	**China**
3.2	3.3	3.3	3.2	3.2	3.2	3.2	3.3	3.3	Albania
8.9	8.7	8.6	8.5	8.5	8.4	8.4	8.4	8.3	Bulgaria
0.7	0.7	0.7	0.7	0.7	0.7	0.7	0.7	0.7	Cyprus
0.0	0.0	0.0	0.0	0.0	0.0	0.0	0.0	0.0	Gibraltar
0.4	0.4	0.4	0.4	0.4	0.4	0.4	0.4	0.4	Malta
23.2	23.2	23.2	22.8	22.8	22.7	22.7	22.6	22.6	Romania
5.3	5.3	5.3	5.3	5.3	5.3	5.3	5.3	5.4	Slovak Republic
-	-	-	3.8	3.4	2.9	2.6	2.3	2.3	*Bosnia-Herzegovina*
-	-	-	4.8	4.8	4.8	4.8	4.8	4.8	*Croatia*
-	-	-	1.9	1.9	1.9	2.0	2.0	2.0	*FYROM*
-	-	-	2.0	2.0	2.0	2.0	2.0	2.0	*Slovenia*
-	-	-	10.5	10.5	10.5	10.6	10.6	10.6	*FR of Yugoslavia*
23.4	23.4	23.6	23.6	23.5	23.2	23.0	23.0	23.1	Former Yugoslavia
61.7	61.7	61.8	61.4	61.2	60.8	60.6	60.4	60.5	**Non-OECD Europe**

Due to lack of complete series Albania, the Dem. Ppl's Rep. of Korea and Vietnam are not included in the regional aggregates.
Former Yugoslavia: data for individual republics may not add to the total.

Population (millions)
Population (millions)
Bevölkerung (Millionen)
Popolazione (milioni)
人口 (100 万人)
Población (millones)
Численность населения (млн. человек)

	1971	1973	1978	1983	1984	1985	1986	1987	1988
Arménie	-	-	-	-	-	-	-	-	-
Azerbaïdjan	-	-	-	-	-	-	-	-	-
Bélarus	-	-	-	-	-	-	-	-	-
Estonie	-	-	-	-	-	-	-	-	-
Géorgie	-	-	-	-	-	-	-	-	-
Kazakhstan	-	-	-	-	-	-	-	-	-
Kirghizistan	-	-	-	-	-	-	-	-	-
Lettonie	-	-	-	-	-	-	-	-	-
Lituanie	-	-	-	-	-	-	-	-	-
Moldova	-	-	-	-	-	-	-	-	-
Russie	-	-	-	-	-	-	-	-	-
Tadjikistan	-	-	-	-	-	-	-	-	-
Turkménistan	-	-	-	-	-	-	-	-	-
Ukraine	-	-	-	-	-	-	-	-	-
Ouzbékistan	-	-	-	-	-	-	-	-	-
Ex-URSS	244.9	249.7	261.5	273.0	275.6	278.1	280.6	283.1	285.5
Bahrein	0.2	0.2	0.3	0.4	0.4	0.4	0.4	0.5	0.5
Iran	29.4	31.2	36.6	43.6	45.3	47.1	48.8	50.4	51.9
Irak	9.7	10.3	12.2	14.3	14.8	15.3	15.8	16.4	16.9
Israël	3.1	3.3	3.7	4.1	4.2	4.2	4.3	4.4	4.4
Jordanie	1.6	1.7	2.0	2.5	2.6	2.6	2.7	2.8	2.9
Koweit	0.8	0.9	1.2	1.6	1.7	1.7	1.8	1.9	2.0
Liban	2.7	2.8	2.9	3.1	3.2	3.3	3.3	3.4	3.5
Oman	0.7	0.8	1.0	1.3	1.3	1.4	1.4	1.5	1.5
Qatar	0.1	0.1	0.2	0.3	0.3	0.4	0.4	0.4	0.4
Arabie saoudite	6.0	6.6	8.4	11.1	11.7	12.4	13.0	13.7	14.4
Syrie	6.5	7.0	8.2	9.7	10.0	10.4	10.7	11.0	11.3
Emirats arabes unis	0.3	0.4	0.8	1.3	1.3	1.4	1.4	1.5	1.6
Yémen	6.4	6.6	7.9	9.5	9.8	10.1	10.4	10.7	11.1
Moyen-Orient	67.4	71.9	85.4	102.8	106.6	110.7	114.6	118.5	122.5
Total non-OCDE	2 799.2	2 929.5	3 248.0	3 580.4	3 649.5	3 720.2	3 794.6	3 870.9	3 947.8
OCDE Amérique du N.	280.4	288.6	309.6	331.3	335.2	339.1	343.1	347.2	351.3
OCDE Pacifique	153.6	159.2	169.4	177.8	179.3	180.6	182.0	183.3	184.5
OCDE Europe	442.6	449.2	461.9	474.1	476.1	478.2	480.4	482.5	485.0
Total OCDE	876.6	897.1	940.8	983.2	990.6	998.0	1 005.5	1 013.0	1 020.8
Monde	3 675.7	3 826.6	4 188.8	4 563.5	4 640.0	4 718.2	4 800.1	4 883.9	4 968.6

Faute de séries complètes, l'Albanie, la Rép.pop.dém. de Corée et le Viêt-Nam ne sont pas pris en compte dans les totaux régionaux.

Population (millions)
Population (millions)
Bevölkerung (Millionen)
Popolazione (milioni)
人口 *(100 万人)*
Población (millones)
Численность населения (млн. человек)

1989	1990	1991	1992	1993	1994	1995	1996	1997	
-	-	-	3.7	3.7	3.7	3.8	3.8	3.8	Armenia
-	-	-	7.3	7.4	7.5	7.5	7.6	7.6	Azerbaijan
-	-	-	10.3	10.4	10.4	10.3	10.3	10.3	Belarus
-	-	-	1.5	1.5	1.5	1.5	1.5	1.5	Estonia
-	-	-	5.5	5.4	5.4	5.4	5.4	5.4	Georgia
-	-	-	16.5	16.5	16.3	16.1	15.9	15.8	Kazakhstan
-	-	-	4.5	4.5	4.5	4.5	4.6	4.6	Kyrgyzstan
-	-	-	2.6	2.6	2.5	2.5	2.5	2.5	Latvia
-	-	-	3.7	3.7	3.7	3.7	3.7	3.7	Lithuania
-	-	-	4.4	4.3	4.3	4.3	4.3	4.3	Moldova
-	-	-	148.7	148.5	148.3	148.1	147.7	147.3	Russia
-	-	-	5.6	5.6	5.7	5.8	5.9	6.0	Tajikistan
-	-	-	4.0	4.3	4.4	4.5	4.6	4.7	Turkmenistan
-	-	-	52.2	52.2	51.9	51.5	51.1	50.7	Ukraine
-	-	-	21.5	21.9	22.4	22.8	23.2	23.7	Uzbekistan
287.0	289.0	290.6	291.8	292.3	292.4	292.3	292.0	291.5	**Former USSR**
0.5	0.5	0.5	0.5	0.5	0.6	0.6	0.6	0.6	Bahrain
53.2	54.4	55.3	56.2	57.1	58.0	59.0	59.9	60.9	Iran
17.5	18.1	18.6	19.2	19.7	20.3	20.8	21.3	21.8	Iraq
4.5	4.7	4.9	5.1	5.3	5.4	5.5	5.7	5.8	Israel
3.1	3.2	3.5	3.7	3.9	4.1	4.2	4.3	4.4	Jordan
2.0	2.1	1.4	1.4	1.5	1.5	1.6	1.7	1.8	Kuwait
3.6	3.6	3.7	3.8	3.9	3.9	4.0	4.1	4.1	Lebanon
1.6	1.6	1.8	1.9	2.0	2.1	2.1	2.2	2.3	Oman
0.4	0.5	0.5	0.5	0.6	0.6	0.7	0.7	0.7	Qatar
15.1	15.8	16.3	16.8	17.4	18.3	19.0	19.4	20.1	Saudi Arabia
11.7	12.1	12.5	12.9	13.3	13.7	14.1	14.5	14.9	Syria
1.7	1.8	2.0	2.1	2.2	2.3	2.3	2.5	2.6	United Arab Emirates
11.4	11.9	13.4	13.9	14.3	14.8	15.3	15.7	16.1	Yemen
126.4	130.3	134.5	138.1	141.7	145.6	149.1	152.5	156.2	**Middle East**
4 023.1	4 098.1	4 172.2	4 243.7	4 311.9	4 379.3	4 446.7	4 515.9	4 587.3	**Non-OECD Total**
355.7	360.3	364.8	369.5	374.1	378.4	383.3	387.5	390.6	OECD North America
185.7	186.8	187.9	189.0	189.9	191.1	192.4	193.4	194.5	OECD Pacific
487.7	490.9	493.8	496.9	499.9	502.3	504.6	506.7	508.8	OECD Europe
1 029.1	1 038.0	1 046.6	1 055.4	1 063.9	1 071.8	1 080.3	1 087.7	1 093.9	**OECD Total**
5 052.3	5 136.1	5 218.8	5 299.1	5 375.7	5 451.1	5 527.0	5 603.6	5 681.2	**World**

Due to lack of complete series Albania, the Dem. Ppl's Rep. of Korea and Vietnam are not included in the regional aggregates.

Energy Production/TPES (Self Sufficiency)
Production d'énergie/ATEP (indépendance énergétique)
Energieerzeugung/TPES (Eigenversorgung)
Produzione di energia/ATEP (indice di autosufficienza energetica)
エネルギー生産量／一次エネルギー総供給量（自給率）
Producción energética/TPES (autosuficiencia energética)
Производство топлива и энергии/ОПППТЭ (самостоятельность)

	1971	1973	1978	1983	1984	1985	1986	1987	1988
Algérie	10.72	10.64	7.35	4.05	4.26	4.30	3.97	4.26	4.14
Angola	2.32	2.77	2.27	2.74	3.07	3.20	3.73	4.37	5.27
Bénin	0.90	0.88	0.88	0.91	0.90	1.09	1.13	1.10	1.06
Cameroun	0.89	0.89	0.99	2.10	2.39	2.68	2.63	2.60	2.53
Congo	0.79	3.85	4.03	6.10	6.94	6.79	7.36	5.10	4.40
RD du Congo	0.88	0.88	1.03	1.01	1.02	1.06	1.06	1.04	1.02
Egypte	2.09	1.22	2.14	1.90	1.97	2.00	1.81	1.89	1.78
Ethiopie	0.94	0.95	0.95	0.95	0.95	0.95	0.95	0.94	0.94
Gabon	7.02	5.94	6.73	4.84	5.19	6.27	5.92	5.82	5.57
Ghana	0.77	0.78	0.78	0.93	0.86	0.82	0.85	0.82	0.85
Côte d'Ivoire	0.66	0.64	0.59	1.01	0.99	1.00	0.95	0.93	0.79
Kenya	0.83	0.82	0.81	0.85	0.83	0.83	0.82	0.82	0.82
Libye	82.86	41.44	21.37	6.51	6.20	5.46	5.82	4.88	5.44
Maroc	0.25	0.23	0.20	0.16	0.17	0.16	0.16	0.14	0.14
Mozambique	0.91	0.91	0.94	0.93	0.94	0.94	0.93	0.93	0.93
Nigéria	3.08	3.58	2.83	1.88	2.02	2.08	2.05	1.90	1.96
Sénégal	0.61	0.59	0.58	0.59	0.58	0.60	0.59	0.57	0.61
Afrique du Sud	0.83	0.82	0.98	1.14	1.19	1.26	1.24	1.20	1.20
Soudan	0.82	0.80	0.85	0.85	0.86	0.85	0.85	0.88	0.84
Tanzanie	0.92	0.92	0.92	0.94	0.94	0.94	0.94	0.94	0.94
Tunisie	2.41	2.09	1.86	1.61	1.55	1.48	1.46	1.37	1.29
Zambie	0.77	0.79	0.86	0.90	0.93	0.94	0.95	0.90	0.91
Zimbabwe	0.93	0.89	0.91	0.86	0.87	0.85	0.85	0.88	0.88
Autre Afrique	0.91	0.91	0.89	0.88	0.88	0.89	0.88	0.86	0.86
Afrique	2.25	2.20	2.03	1.61	1.66	1.70	1.66	1.63	1.64
Argentine	0.91	0.86	0.87	1.01	0.99	1.03	0.97	0.93	0.95
Bolivie	2.14	3.69	2.20	1.69	1.73	1.74	1.72	1.76	1.74
Brésil	0.70	0.62	0.54	0.70	0.77	0.79	0.76	0.74	0.72
Chili	0.57	0.58	0.59	0.69	0.69	0.69	0.68	0.67	0.60
Colombie	1.37	1.21	0.97	0.99	1.08	1.15	1.45	1.69	1.72
Costa Rica	0.58	0.54	0.46	0.60	0.62	0.60	0.58	0.56	0.56
Cuba	0.32	0.28	0.29	0.30	0.33	0.32	0.34	0.32	0.32
République dominicaine	0.51	0.42	0.41	0.43	0.44	0.37	0.36	0.33	0.28
Equateur	0.57	4.74	2.55	2.53	2.57	2.70	2.79	1.76	2.71
El Salvador	0.71	0.67	0.73	0.79	0.78	0.77	0.74	0.72	0.71
Guatemala	0.69	0.68	0.68	0.81	0.75	0.73	0.82	0.79	0.78
Haiti	0.91	0.92	0.88	0.87	0.86	0.86	0.82	0.81	0.80
Honduras	0.73	0.71	0.69	0.70	0.71	0.72	0.73	0.71	0.69
Jamaïque	0.13	0.08	0.10	0.11	0.13	0.14	0.14	0.13	0.13
Nicaragua	0.57	0.57	0.52	0.56	0.65	0.66	0.62	0.61	0.64
Panama	0.20	0.16	0.28	0.31	0.38	0.40	0.39	0.36	0.43
Paraguay	0.85	0.83	0.75	0.78	0.77	0.87	1.14	1.31	1.31
Pérou	0.80	0.82	1.14	1.24	1.30	1.34	1.24	1.13	1.03
Trinité-et-Tobago	3.02	3.72	3.86	2.67	2.66	2.62	2.42	2.51	2.34
Uruguay	0.21	0.23	0.23	0.54	0.57	0.56	0.60	0.57	0.44
Vénézuela	8.50	7.59	4.34	3.01	3.21	2.99	3.07	2.65	3.06
Autre Amérique latine	0.11	0.10	0.13	0.22	0.24	0.26	0.25	0.21	0.21
Amérique latine	1.63	1.51	1.13	1.12	1.17	1.16	1.17	1.11	1.15

Energy Production/TPES (Self Sufficiency)
Production d'énergie/ATEP (indépendance énergétique)
Energieerzeugung/TPES (Eigenversorgung)
Produzione di energia/ATEP (indice di autosufficienza energetica)
エネルギー生産量／一次エネルギー総供給量（自給率）
Producción energética/TPES (autosuficiencia energética)
Производство топлива и энергии/ОППТЭ (самостоятельность)

1989	1990	1991	1992	1993	1994	1995	1996	1997	
4.35	4.32	4.27	4.32	4.31	4.28	4.51	4.81	4.74	Algeria
5.27	5.14	5.29	5.05	5.06	5.02	5.12	6.29	6.05	Angola
1.04	1.06	1.06	1.02	1.03	1.01	0.99	0.87	0.87	Benin
2.47	2.49	2.34	2.23	2.17	2.06	1.94	1.93	1.95	Cameroon
7.50	8.06	7.81	8.21	8.92	8.73	8.67	9.41	10.90	Congo
1.03	1.01	1.00	1.00	0.99	1.00	0.99	0.99	0.99	DR of Congo
1.75	1.72	1.71	1.70	1.68	1.77	1.70	1.58	1.47	Egypt
0.94	0.93	0.94	0.94	0.94	0.94	0.93	0.93	0.95	Ethiopia
8.08	11.41	11.83	11.07	11.26	10.89	12.97	12.84	12.10	Gabon
0.83	0.84	0.86	0.87	0.85	0.84	0.83	0.84	0.85	Ghana
0.70	0.74	0.75	0.73	0.80	0.81	0.82	0.90	0.88	Ivory Coast
0.82	0.82	0.83	0.83	0.83	0.82	0.85	0.85	0.82	Kenya
5.51	6.24	5.94	6.06	5.40	5.65	4.90	5.21	5.23	Libya
0.12	0.11	0.12	0.11	0.10	0.10	0.10	0.13	0.12	Morocco
0.94	0.94	0.94	0.93	0.93	0.93	0.93	0.93	0.91	Mozambique
2.12	2.12	2.11	2.11	2.29	1.97	2.03	2.15	2.15	Nigeria
0.61	0.62	0.63	0.60	0.63	0.61	0.63	0.61	0.60	Senegal
1.23	1.26	1.23	1.27	1.29	1.27	1.28	1.29	1.33	South Africa
0.86	0.83	0.84	0.85	0.89	0.85	0.86	0.87	0.86	Sudan
0.94	0.94	0.94	0.95	0.95	0.95	0.95	0.95	0.95	Tanzania
1.22	1.07	1.21	1.13	1.03	0.92	0.89	0.94	0.98	Tunisia
0.91	0.92	0.93	0.93	0.93	0.93	0.93	0.93	0.93	Zambia
0.89	0.91	0.89	0.87	0.88	0.86	0.84	0.83	0.82	Zimbabwe
0.87	0.88	0.87	0.87	0.88	0.88	0.88	0.89	0.92	Other Africa
1.70	1.75	1.75	1.75	1.76	1.70	1.71	1.76	1.77	**Africa**
1.04	1.09	1.07	1.11	1.11	1.17	1.23	1.27	1.30	Argentina
1.73	1.69	1.62	1.60	1.57	1.57	1.48	1.44	1.40	Bolivia
0.72	0.73	0.72	0.71	0.70	0.68	0.68	0.69	0.70	Brazil
0.55	0.54	0.55	0.52	0.48	0.45	0.42	0.38	0.35	Chile
1.79	1.80	1.84	1.82	1.81	1.75	1.99	2.13	2.22	Colombia
0.51	0.51	0.50	0.43	0.32	0.31	0.41	0.44	0.43	Costa Rica
0.34	0.37	0.42	0.52	0.55	0.54	0.54	0.52	0.51	Cuba
0.26	0.26	0.26	0.30	0.30	0.29	0.28	0.27	0.26	Dominican Republic
2.69	2.50	2.57	2.62	2.91	2.93	2.85	2.55	2.68	Ecuador
0.70	0.70	0.67	0.65	0.62	0.59	0.61	0.64	0.65	El Salvador
0.78	0.76	0.74	0.74	0.74	0.70	0.70	0.75	0.79	Guatemala
0.79	0.79	0.80	0.83	0.85	0.96	0.81	0.81	0.73	Haiti
0.67	0.69	0.69	0.65	0.68	0.64	0.60	0.61	0.63	Honduras
0.13	0.15	0.13	0.13	0.14	0.17	0.14	0.15	0.15	Jamaica
0.70	0.69	0.70	0.68	0.68	0.65	0.65	0.62	0.59	Nicaragua
0.43	0.39	0.35	0.34	0.36	0.32	0.35	0.36	0.35	Panama
1.34	1.48	1.53	1.41	1.46	1.49	1.52	1.67	1.66	Paraguay
1.04	1.06	1.00	1.03	1.01	0.98	0.89	0.88	0.81	Peru
2.40	2.16	2.05	1.87	1.84	1.85	1.87	1.72	1.66	Trinidad-and-Tobago
0.38	0.52	0.46	0.48	0.48	0.50	0.40	0.36	0.38	Uruguay
3.03	3.21	2.95	3.25	3.31	3.92	3.91	3.53	3.55	Venezuela
0.22	0.22	0.22	0.23	0.23	0.23	0.24	0.23	0.22	Other Latin America
1.15	1.19	1.20	1.22	1.20	1.24	1.29	1.28	1.28	**Latin America**

Energy Production/TPES (Self Sufficiency)
Production d'énergie/ATEP (indépendance énergétique)
Energieerzeugung/TPES (Eigenversorgung)
Produzione di energia/ATEP (indice di autosufficienza energetica)
エネルギー生産量／一次エネルギー総供給量（自給率）
Producción energética/TPES (autosuficiencia energética)
Производство топлива и энергии/ОПЛТЭ (самостоятельность)

	1971	1973	1978	1983	1984	1985	1986	1987	1988
Bangladesh	0.92	0.91	0.90	0.91	0.91	0.91	0.90	0.91	0.91
Brunei	53.79	88.61	86.45	9.63	12.16	11.00	11.96	8.17	7.26
Inde	0.92	0.92	0.93	0.95	0.97	0.95	0.95	0.94	0.94
Indonésie	2.01	2.41	2.25	1.86	2.04	1.85	1.79	1.80	1.75
RPD de Corée	0.96	0.96	0.92	0.88	0.88	0.87	0.87	0.87	0.87
Malaisie	0.77	0.95	1.44	1.84	2.20	2.00	2.20	2.23	2.22
Myanmar	0.93	0.99	1.00	1.01	1.02	1.01	1.00	1.01	1.00
Népal	0.98	0.96	0.97	0.97	0.96	0.96	0.96	0.95	0.95
Pakistan	0.82	0.83	0.83	0.80	0.79	0.80	0.81	0.81	0.80
Philippines	0.43	0.41	0.41	0.57	0.63	0.65	0.65	0.59	0.58
Singapour	-	-	-	-	-	-	-	-	-
Sri Lanka	0.75	0.70	0.73	0.70	0.74	0.76	0.76	0.75	0.76
Taipei chinois	0.37	0.27	0.20	0.27	0.29	0.32	0.27	0.30	0.26
Thaïlande	0.56	0.50	0.50	0.53	0.56	0.67	0.68	0.66	0.65
Viêt-Nam	0.71	0.72	0.99	0.94	0.91	0.92	0.92	0.90	0.93
Autre Asie	1.05	1.02	0.98	0.95	0.95	0.91	0.93	0.79	0.76
Asie	0.99	1.04	1.05	1.02	1.06	1.04	1.04	1.02	1.01
Rép. populaire de Chine	1.00	1.01	1.02	1.05	1.06	1.08	1.06	1.04	1.03
Hong-Kong, Chine	0.01	0.01	0.01	0.01	0.01	0.01	0.01	0.00	0.00
Chine	1.00	1.00	1.01	1.04	1.05	1.07	1.05	1.02	1.01
Albanie	1.42	1.73	1.35	0.99	1.02	1.04	1.03	1.03	1.00
Bulgarie	0.30	0.26	0.24	0.32	0.32	0.31	0.32	0.32	0.33
Chypre	0.01	0.01	0.01	0.01	0.01	0.01	0.01	0.00	0.00
Roumanie	1.03	0.98	0.85	0.86	0.86	0.84	0.81	0.76	0.77
République slovaque	0.19	0.17	0.14	0.19	0.20	0.22	0.24	0.24	0.25
Bosnie-Herzégovine	-	-	-	-	-	-	-	-	-
Croatie	-	-	-	-	-	-	-	-	-
Ex-RYM	-	-	-	-	-	-	-	-	-
Slovénie	-	-	-	-	-	-	-	-	-
RF de Yougoslavie	-	-	-	-	-	-	-	-	-
Ex-Yougoslavie	0.65	0.63	0.59	0.62	0.61	0.63	0.61	0.59	0.57
Europe non-OCDE	0.67	0.64	0.57	0.60	0.60	0.59	0.58	0.56	0.56
Arménie	-	-	-	-	-	-	-	-	-
Azerbaïdjan	-	-	-	-	-	-	-	-	-
Bélarus	-	-	-	-	-	-	-	-	-
Estonie	-	-	-	-	-	-	-	-	-
Géorgie	-	-	-	-	-	-	-	-	-
Kazakhstan	-	-	-	-	-	-	-	-	-
Kirghizistan	-	-	-	-	-	-	-	-	-
Lettonie	-	-	-	-	-	-	-	-	-
Lituanie	-	-	-	-	-	-	-	-	-
Moldova	-	-	-	-	-	-	-	-	-
Russie	-	-	-	-	-	-	-	-	-
Tadjikistan	-	-	-	-	-	-	-	-	-
Turkménistan	-	-	-	-	-	-	-	-	-
Ukraine	-	-	-	-	-	-	-	-	-
Ouzbékistan	-	-	-	-	-	-	-	-	-
Ex-URSS	1.14	1.14	1.19	1.20	1.20	1.19	1.22	1.21	1.21

Energy Production/TPES (Self Sufficiency)
Production d'énergie/ATEP (indépendance énergétique)
Energieerzeugung/TPES (Eigenversorgung)
Produzione di energia/ATEP (indice di autosufficienza energetica)
エネルギー生産量／一次エネルギー総供給量（自給率）
Producción energética/TPES (autosuficiencia energética)
Производство топлива и энергии/ОПТЭ (самостоятельность)

1989	1990	1991	1992	1993	1994	1995	1996	1997	
0.90	0.90	0.92	0.92	0.91	0.91	0.90	0.90	0.90	Bangladesh
11.72	10.49	9.61	9.32	10.14	10.42	9.22	8.87	8.35	Brunei
0.93	0.93	0.93	0.91	0.90	0.90	0.88	0.87	0.88	India
1.76	1.69	1.71	1.71	1.66	1.72	1.67	1.65	1.60	Indonesia
0.87	0.86	0.87	0.87	0.88	0.88	0.88	0.88	0.88	DPR of Korea
2.12	2.04	1.89	1.82	1.69	1.70	1.86	1.69	1.53	Malaysia
1.00	1.01	1.01	1.01	0.99	0.98	0.96	0.96	0.94	Myanmar
0.95	0.97	0.95	0.94	0.94	0.93	0.91	0.92	0.92	Nepal
0.80	0.79	0.81	0.80	0.78	0.76	0.76	0.74	0.74	Pakistan
0.57	0.56	0.57	0.55	0.53	0.52	0.48	0.45	0.43	Philippines
-	-	-	-	-	-	-	0.00	0.00	Singapore
0.77	0.77	0.76	0.76	0.71	0.69	0.66	0.62	0.61	Sri Lanka
0.21	0.22	0.21	0.19	0.18	0.18	0.17	0.17	0.15	Chinese Taipei
0.63	0.61	0.62	0.61	0.59	0.59	0.55	0.56	0.58	Thailand
0.97	1.01	1.08	1.11	1.12	1.13	1.18	1.13	1.11	Vietnam
0.56	1.33	1.42	1.55	1.57	1.61	1.51	1.32	1.30	Other Asia
1.00	0.99	0.99	0.97	0.95	0.94	0.93	0.91	0.90	**Asia**
1.04	1.04	1.02	1.02	1.00	1.01	1.01	1.00	1.00	People's Rep.of China
0.00	0.00	0.00	0.00	0.00	0.00	0.00	0.00	0.00	Hong Kong, China
1.03	1.03	1.01	1.00	0.99	0.99	1.00	0.99	0.99	**China**
1.01	0.92	1.03	1.06	1.00	0.88	0.92	0.89	0.87	Albania
0.33	0.36	0.39	0.41	0.40	0.43	0.44	0.46	0.48	Bulgaria
0.00	0.00	0.00	0.01	0.01	0.01	0.01	0.01	0.01	Cyprus
0.72	0.65	0.70	0.73	0.73	0.74	0.70	0.70	0.70	Romania
0.25	0.25	0.25	0.26	0.27	0.30	0.28	0.27	0.27	Slovak Republic
-	-	-	0.80	0.65	0.34	0.36	0.36	0.36	*Bosnia-Herzegovina*
-	-	-	0.65	0.67	0.59	0.59	0.58	0.52	*Croatia*
-	-	-	0.55	0.53	0.64	0.65	0.57	0.60	*FYROM*
-	-	-	0.56	0.50	0.50	0.48	0.45	0.45	*Slovenia*
-	-	-	0.82	0.90	0.97	0.94	0.76	0.72	*FR of Yugoslavia*
0.57	0.56	0.60	0.72	0.71	0.71	0.70	0.62	0.60	Former Yugoslavia
0.54	0.51	0.54	0.59	0.58	0.59	0.57	0.56	0.56	**Non-OECD Europe**
-	-	-	0.06	0.16	0.21	0.15	0.41	0.30	Armenia
-	-	-	1.08	1.06	0.92	1.13	1.18	1.17	Azerbaijan
-	-	-	0.10	0.11	0.13	0.13	0.13	0.13	Belarus
-	-	-	0.68	0.62	0.63	0.63	0.68	0.68	Estonia
-	-	-	0.13	0.18	0.18	0.39	0.33	0.30	Georgia
-	-	-	1.14	1.24	1.22	1.14	1.41	1.69	Kazakhstan
-	-	-	0.36	0.40	0.55	0.52	0.47	0.50	Kyrgyzstan
-	-	-	0.17	0.16	0.22	0.18	0.24	0.37	Latvia
-	-	-	0.38	0.40	0.31	0.41	0.47	0.45	Lithuania
-	-	-	0.01	0.01	0.01	0.01	0.02	0.02	Moldova
-	-	-	1.41	1.41	1.52	1.52	1.55	1.57	Russia
-	-	-	0.17	0.26	0.46	0.40	0.38	0.37	Tajikistan
-	-	-	4.67	5.49	2.38	2.30	2.70	1.54	Turkmenistan
-	-	-	0.50	0.51	0.53	0.50	0.50	0.54	Ukraine
-	-	-	0.90	0.91	1.01	1.15	1.12	1.15	Uzbekistan
1.21	1.21	1.15	1.15	1.18	1.24	1.25	1.28	1.30	**Former USSR**

Energy Production/TPES (Self Sufficiency)
Production d'énergie/ATEP (indépendance énergétique)
Energieerzeugung/TPES (Eigenversorgung)
Produzione di energia/ATEP (indice di autosufficienza energetica)
エネルギー生産量／一次エネルギー総供給量（自給率）
Producción energética/TPES (autosuficiencia energética)
Производство топлива и энергии/ОПТЭ (самостоятельность)

	1971	1973	1978	1983	1984	1985	1986	1987	1988
Bahrein	2.94	2.14	1.68	1.43	1.35	1.33	1.30	1.21	1.28
Iran	10.94	11.83	7.14	3.10	2.37	2.36	1.90	2.25	2.11
Irak	19.07	19.43	14.49	3.26	3.76	4.15	4.89	5.54	6.20
Israël	0.94	0.72	0.01	0.01	0.01	0.03	0.04	0.03	0.03
Jordanie	0.00	0.00	0.00	0.00	0.00	0.00	0.01	0.01	0.01
Koweit	28.17	27.56	14.41	5.49	5.97	4.42	6.38	6.07	6.63
Liban	0.09	0.06	0.08	0.06	0.06	0.06	0.05	0.05	0.07
Oman	166.91	147.62	24.63	12.49	10.38	10.81	16.46	17.02	14.39
Qatar	23.72	20.28	14.79	4.34	4.81	3.99	4.15	3.72	3.79
Arabie saoudite	37.92	47.29	20.92	5.47	5.29	3.85	4.92	4.04	4.42
Syrie	1.96	2.22	2.01	1.16	1.03	1.08	1.09	1.29	1.46
Emirats arabes unis	48.17	53.06	17.02	5.10	5.25	4.50	4.69	4.79	5.05
Yémen	0.06	0.05	0.05	0.03	0.03	0.03	0.03	0.41	3.11
Moyen-Orient	15.36	16.61	10.57	3.86	3.62	3.13	3.53	3.42	3.66
Total non-OCDE	1.61	1.67	1.52	1.28	1.29	1.27	1.29	1.27	1.28
OCDE Amérique du N.	0.92	0.87	0.82	0.95	0.98	0.96	0.95	0.93	0.92
OCDE Pacifique	0.29	0.26	0.30	0.38	0.39	0.43	0.45	0.47	0.42
OCDE Europe	0.48	0.45	0.53	0.65	0.64	0.67	0.67	0.66	0.66
Total OCDE	0.69	0.65	0.66	0.77	0.78	0.79	0.78	0.78	0.76
Monde	1.04	1.03	1.01	1.01	1.02	1.01	1.02	1.01	1.01

Le total Monde ne comprend pas les soutages maritimes internationaux.

Energy Production/TPES (Self Sufficiency)
Production d'énergie/ATEP (indépendance énergétique)
Energieerzeugung/TPES (Eigenversorgung)
Produzione di energia/ATEP (indice di autosufficienza energetica)
エネルギー生産量／一次エネルギー総供給量（自給率）
Producción energética/TPES (autosuficiencia energética)
Производство топлива и энергии/ОППТЭ (самостоятельность)

1989	1990	1991	1992	1993	1994	1995	1996	1997	
1.28	1.30	1.34	1.33	1.30	1.26	1.27	0.99	0.88	Bahrain
2.39	2.51	2.50	2.37	2.44	2.24	2.26	2.18	2.08	Iran
6.23	5.12	1.06	1.24	1.26	1.25	1.24	1.30	2.29	Iraq
0.03	0.04	0.04	0.03	0.04	0.04	0.03	0.04	0.03	Israel
0.02	0.03	0.03	0.03	0.04	0.04	0.04	0.04	0.04	Jordan
6.14	5.09	3.20	6.76	10.98	7.84	7.17	7.97	7.18	Kuwait
0.08	0.06	0.05	0.06	0.05	0.05	0.04	0.04	0.04	Lebanon
14.25	8.76	6.60	6.82	6.88	7.32	7.72	7.88	7.62	Oman
3.93	4.54	3.80	2.99	2.78	2.77	2.75	2.72	3.24	Qatar
4.65	5.83	6.28	6.05	5.69	5.76	5.63	5.15	4.95	Saudi Arabia
1.80	1.89	2.06	2.16	2.26	2.35	2.43	2.32	2.24	Syria
5.04	5.77	5.78	5.63	5.14	5.10	5.16	4.60	4.97	United Arab Emirates
3.22	3.67	3.36	2.69	4.14	6.06	5.71	5.59	5.69	Yemen
3.84	3.98	3.78	3.67	3.70	3.58	3.53	3.39	3.38	**Middle East**
1.31	1.32	1.29	1.30	1.31	1.32	1.32	1.32	1.33	**Non-OECD Total**
0.90	0.94	0.93	0.92	0.89	0.91	0.90	0.90	0.89	OECD North America
0.42	0.42	0.43	0.42	0.42	0.41	0.42	0.41	0.43	OECD Pacific
0.65	0.64	0.63	0.64	0.65	0.67	0.67	0.68	0.68	OECD Europe
0.75	0.76	0.75	0.75	0.74	0.75	0.75	0.75	0.75	**OECD Total**
1.01	1.02	1.01	1.01	1.01	1.02	1.02	1.01	1.02	**World**

World Total excludes international marine bunkers.

Net Oil Imports/GDP (toe per thousand 90 US$)
Importations nettes de pétrole/PIB (tep par millier de $US 90)

Netto-Ölimporte/BIP (toe pro tausend 1990er US$)

Importanzione nette di petrolio/PIL (tep per "1,000" US$ 90)

石油純輸入量／GDP （石油換算トン／千米ドル、1990年価格）

Importaciones netas de petóleo/PIB (tep por miles de 90 US$)

Чистый импорт нефти и нефтепродуктов / ВВП (тнэ на тыс.долл.США 1990 г.)

	1971	1973	1978	1983	1984	1985	1986	1987	1988
Algérie	-1.49	-1.57	-1.17	-0.73	-0.73	-0.70	-0.74	-0.75	-0.78
Angola	-0.33	-0.48	-0.65	-0.93	-1.06	-1.17	-1.43	-1.66	-2.08
Bénin	0.10	0.12	0.11	0.08	0.09	-0.09	-0.12	-0.09	-0.06
Cameroun	0.06	0.06	0.00	-0.39	-0.48	-0.55	-0.50	-0.51	-0.55
Congo	0.14	-1.11	-1.35	-1.94	-2.00	-2.04	-2.21	-2.18	-2.28
RD du Congo	0.10	0.10	-0.03	-0.02	-0.03	-0.07	-0.07	-0.05	-0.04
Egypte	-0.45	-0.09	-0.45	-0.40	-0.47	-0.50	-0.43	-0.48	-0.40
Ethiopie	0.13	0.12	0.10	0.08	0.08	0.09	0.11	0.11	0.11
Gabon	-2.08	-2.21	-1.98	-1.29	-1.23	-1.47	-1.36	-1.50	-1.40
Ghana	0.16	0.19	0.19	0.09	0.17	0.16	0.15	0.17	0.16
Côte d'Ivoire	0.13	0.15	0.12	-0.01	0.00	0.02	0.04	0.05	0.11
Kenya	0.51	0.44	0.38	0.28	0.26	0.27	0.26	0.27	0.27
Libye	-4.01	-3.68	-2.54	-1.44	-1.56	-1.61	-1.78	-1.72	-1.88
Maroc	0.18	0.19	0.22	0.21	0.21	0.21	0.19	0.20	0.19
Mozambique	0.58	0.50	0.43	0.55	0.55	0.45	0.49	0.42	0.38
Nigéria	-3.58	-4.39	-3.36	-2.20	-2.55	-2.82	-2.53	-2.28	-2.29
Sénégal	0.41	0.44	0.31	0.19	0.16	0.16	0.21	0.17	0.16
Afrique du Sud	0.18	0.18	0.17	0.12	0.12	0.13	0.12	0.12	0.10
Soudan	0.25	0.30	0.13	0.14	0.14	0.18	0.17	0.13	0.19
Tanzanie	0.40	0.42	0.27	0.24	0.23	0.24	0.23	0.22	0.22
Tunisie	-0.56	-0.42	-0.33	-0.23	-0.27	-0.26	-0.23	-0.26	-0.19
Zambie	0.17	0.30	0.24	0.23	0.19	0.18	0.16	0.16	0.16
Zimbabwe	0.12	0.14	0.14	0.12	0.12	0.11	0.11	0.11	0.12
Autre Afrique	0.12	0.12	0.13	0.14	0.13	0.13	0.14	0.15	0.15
Afrique	-0.85	-0.83	-0.66	-0.38	-0.41	-0.44	-0.42	-0.41	-0.42
Argentine	0.03	0.03	0.02	-0.01	-0.01	-0.03	-0.01	0.01	-0.00
Bolivie	-0.33	-0.40	-0.08	-0.04	-0.01	-0.00	-0.01	-0.00	-0.00
Brésil	0.11	0.14	0.13	0.09	0.07	0.06	0.06	0.06	0.07
Chili	0.22	0.22	0.21	0.13	0.13	0.11	0.14	0.12	0.17
Colombie	-0.28	-0.14	0.02	0.02	-0.01	-0.02	-0.19	-0.29	-0.28
Costa Rica	0.17	0.18	0.21	0.15	0.14	0.15	0.15	0.15	0.16
Cuba	1.17	1.18	1.01	0.85	0.69	0.69	0.66	0.73	0.75
République dominicaine	0.37	0.42	0.40	0.38	0.40	0.36	0.38	0.40	0.41
Equateur	0.29	-1.62	-0.93	-0.93	-0.98	-1.02	-1.07	-0.48	-1.05
El Salvador	0.11	0.14	0.12	0.13	0.12	0.13	0.14	0.15	0.15
Guatemala	0.22	0.21	0.22	0.12	0.16	0.18	0.12	0.13	0.14
Haiti	0.06	0.06	0.09	0.07	0.08	0.08	0.09	0.09	0.10
Honduras	0.25	0.24	0.23	0.24	0.26	0.22	0.22	0.24	0.25
Jamaïque	0.52	0.72	0.64	0.54	0.53	0.48	0.48	0.46	0.47
Antilles néerlandaises	5.78	6.18	4.43	3.83	3.23	2.56	2.25	1.78	1.98
Nicaragua	0.41	0.42	0.48	0.50	0.44	0.51	0.54	0.60	0.64
Panama	2.08	1.84	0.75	0.37	0.32	0.29	0.28	0.33	0.31
Paraguay	0.12	0.13	0.15	0.10	0.13	0.11	0.13	0.14	0.13
Pérou	0.07	0.06	-0.04	-0.07	-0.08	-0.10	-0.07	-0.04	-0.01
Trinité-et-Tobago	-1.35	-1.66	-2.11	-1.22	-1.21	-1.31	-1.33	-1.34	-1.39
Uruguay	0.30	0.30	0.30	0.18	0.19	0.16	0.20	0.15	0.19
Vénézuela	-4.97	-4.43	-2.19	-1.83	-1.88	-1.68	-1.75	-1.51	-1.66
Autre Amérique latine	0.70	0.74	0.52	0.25	0.24	0.24	0.22	0.26	0.24
Amérique latine	-0.24	-0.18	-0.04	-0.04	-0.05	-0.05	-0.06	-0.04	-0.05

Faute de séries complètes, l'Albanie, la Rép.pop.dém. de Corée et le Viêt-Nam ne sont pas pris en compte dans les totaux régionaux.

Net Oil Imports/GDP (toe per thousand 90 US$)
Importations nettes de pétrole/PIB (tep par millier de $US 90)
Netto-Ölimporte/BIP (toe pro tausend 1990er US$)

Importanzione nette di petrolio/PIL (tep per "1,000" US$ 90)

石油純輸入量／GDP（石油換算トン／千米ドル、1990年価格）

Importaciones netas de petóleo/PIB (tep por miles de 90 US$)

Чистый импорт нефти и нефтепродуктов / ВВП (тнэ на тыс.долл.США 1990 г.)

1989	1990	1991	1992	1993	1994	1995	1996	1997	
-0.75	-0.81	-0.80	-0.78	-0.81	-0.82	-0.79	-0.82	-0.80	Algeria
-2.10	-2.20	-2.31	-2.32	-3.10	-3.05	-2.90	-3.59	-3.35	Angola
-0.04	-0.06	-0.06	-0.03	-0.03	-0.02	-0.00	0.10	0.10	Benin
-0.56	-0.61	-0.57	-0.55	-0.55	-0.52	-0.46	-0.44	-0.44	Cameroon
-2.58	-2.82	-2.72	-2.86	-3.28	-3.32	-2.90	-3.12	-4.12	Congo
-0.04	-0.02	-0.02	-0.02	-0.01	-0.01	-0.00	-0.00	0.01	DR of Congo
-0.38	-0.35	-0.34	-0.36	-0.36	-0.21	-0.19	-0.14	-0.12	Egypt
0.11	0.12	0.13	0.14	0.13	0.14	0.14	0.14	0.08	Ethiopia
-1.59	-2.05	-2.22	-2.26	-2.40	-2.44	-2.49	-2.48	-2.39	Gabon
0.16	0.16	0.14	0.13	0.13	0.16	0.15	0.15	0.14	Ghana
0.15	0.12	0.13	0.14	0.11	0.11	0.10	0.11	0.06	Ivory Coast
0.27	0.25	0.24	0.24	0.26	0.27	0.23	0.23	0.27	Kenya
-2.19	-2.32	-2.40	-2.40	-2.34	-2.48	-2.38	-2.35	-2.38	Libya
0.20	0.20	0.18	0.23	0.24	0.23	0.24	0.19	0.22	Morocco
0.34	0.32	0.27	0.28	0.25	0.23	0.24	0.24	0.30	Mozambique
-2.59	-2.47	-2.44	-2.44	-2.60	-2.28	-2.36	-2.51	-2.54	Nigeria
0.16	0.15	0.14	0.17	0.16	0.18	0.15	0.16	0.17	Senegal
0.09	0.11	0.09	0.10	0.09	0.10	0.11	0.10	0.10	South Africa
0.16	0.21	0.19	0.16	0.11	0.15	0.14	0.12	0.12	Sudan
0.21	0.21	0.20	0.19	0.18	0.18	0.18	0.17	0.17	Tanzania
-0.16	-0.11	-0.12	-0.13	-0.07	-0.06	-0.09	-0.05	-0.01	Tunisia
0.16	0.14	0.14	0.15	0.14	0.14	0.15	0.14	0.14	Zambia
0.11	0.09	0.11	0.13	0.14	0.15	0.17	0.16	0.15	Zimbabwe
0.14	0.13	0.13	0.13	0.13	0.13	0.13	0.11	0.09	Other Africa
-0.46	-0.48	-0.50	-0.50	-0.51	-0.46	-0.45	-0.47	-0.47	**Africa**
-0.02	-0.03	-0.03	-0.04	-0.04	-0.06	-0.08	-0.09	-0.09	Argentina
-0.00	-0.00	-0.00	0.01	0.01	0.01	0.03	0.02	0.04	Bolivia
0.06	0.07	0.06	0.07	0.07	0.07	0.07	0.07	0.07	Brazil
0.17	0.19	0.19	0.20	0.19	0.21	0.21	0.22	0.21	Chile
-0.28	-0.29	-0.28	-0.24	-0.26	-0.22	-0.34	-0.36	-0.36	Colombia
0.16	0.17	0.17	0.22	0.21	0.22	0.22	0.21	0.21	Costa Rica
0.74	0.72	0.64	0.54	0.58	0.60	0.63	0.60	0.65	Cuba
0.39	0.42	0.41	0.43	0.42	0.42	0.41	0.40	0.39	Dominican Republic
-0.98	-0.93	-0.90	-0.98	-1.06	-1.09	-1.06	-0.98	-1.05	Ecuador
0.15	0.15	0.19	0.19	0.21	0.21	0.23	0.20	0.18	El Salvador
0.14	0.15	0.17	0.17	0.17	0.19	0.18	0.15	0.13	Guatemala
0.11	0.11	0.09	0.11	0.09	0.03	0.13	0.14	0.18	Haiti
0.26	0.24	0.25	0.27	0.23	0.27	0.32	0.32	0.30	Honduras
0.49	0.60	0.57	0.60	0.60	0.60	0.65	0.69	0.76	Jamaica
1.79	2.21	2.41	2.20	2.54	2.52	2.51	2.40	2.32	Netherlands Antilles
0.55	0.58	0.56	0.67	0.65	0.57	0.72	0.61	0.60	Nicaragua
0.32	0.37	0.36	0.38	0.33	0.35	0.35	0.34	0.34	Panama
0.15	0.13	0.12	0.14	0.14	0.18	0.18	0.15	0.17	Paraguay
-0.02	-0.02	-0.00	-0.01	-0.02	-0.00	0.03	0.03	0.04	Peru
-1.35	-1.26	-1.01	-1.10	-1.06	-1.21	-1.18	-0.88	-0.84	Trinidad-and-Tobago
0.19	0.16	0.19	0.21	0.16	0.16	0.20	0.20	0.19	Uruguay
-1.74	-1.77	-1.79	-1.66	-1.73	-2.19	-2.34	-2.37	-2.41	Venezuela
0.23	0.24	0.22	0.21	0.21	0.21	0.21	0.20	0.20	Other Latin America
-0.05	-0.06	-0.07	-0.06	-0.06	-0.08	-0.10	-0.09	-0.10	**Latin America**

Due to lack of complete series Albania, the Dem. Ppl's Rep. of Korea and Vietnam are not included in the regional aggregates.

Net Oil Imports/GDP (toe per thousand 90 US$)
Importations nettes de pétrole/PIB (tep par millier de $US 90)
Netto-Ölimporte/BIP (toe pro tausend 1990er US$)

Importanzione nette di petrolio/PIL (tep per "1,000" US$ 90)

石油純輸入量／GDP（石油換算トン／千米ドル、1990年価格）

Importaciones netas de petóleo/PIB (tep por miles de 90 US$)

Чистый импорт нефти и нефтепродуктов / ВВП (тнэ на тыс.долл.США 1990 г.)

	1971	1973	1978	1983	1984	1985	1986	1987	1988
Bangladesh	0.08	0.09	0.10	0.09	0.08	0.08	0.09	0.08	0.10
Brunei	-2.64	-4.73	-3.25	-2.75	-2.64	-2.43	-2.41	-2.12	-2.04
Inde	0.11	0.13	0.11	0.07	0.06	0.07	0.07	0.07	0.08
Indonésie	-1.10	-1.39	-1.27	-0.61	-0.68	-0.54	-0.53	-0.48	-0.43
Malaisie	0.12	0.02	-0.19	-0.31	-0.40	-0.41	-0.50	-0.44	-0.45
Myanmar	0.02	0.00	-0.00	-0.00	-0.00	-0.00	-0.00	-0.00	-0.00
Népal	0.03	0.04	0.04	0.05	0.06	0.06	0.06	0.06	0.07
Pakistan	0.21	0.22	0.21	0.22	0.23	0.21	0.20	0.21	0.21
Philippines	0.39	0.36	0.33	0.25	0.27	0.21	0.21	0.25	0.25
Singapour	0.74	1.08	0.76	0.55	0.46	0.42	0.58	0.73	0.70
Sri Lanka	0.38	0.44	0.31	0.27	0.26	0.23	0.22	0.26	0.24
Taipei chinois	0.21	0.24	0.28	0.20	0.18	0.15	0.16	0.15	0.16
Thaïlande	0.29	0.33	0.29	0.23	0.21	0.17	0.16	0.18	0.17
Viêt-Nam	-	-	-	-	0.38	0.37	0.41	0.46	0.33
Autre Asie	0.15	0.17	0.13	0.13	0.12	0.13	0.12	0.14	0.14
Asie	0.04	0.01	-0.01	0.02	0.00	0.01	0.01	0.04	0.05
Rép. populaire de Chine	-0.00	-0.02	-0.09	-0.10	-0.12	-0.14	-0.11	-0.09	-0.08
Hong-Kong, Chine	0.23	0.22	0.19	0.12	0.10	0.10	0.10	0.08	0.08
Chine	0.03	0.02	-0.04	-0.06	-0.08	-0.10	-0.08	-0.06	-0.05
Albanie	-	-	-	-0.03	-0.08	-0.10	-0.09	-0.08	-0.05
Bulgarie	0.77	0.81	0.88	0.67	0.64	0.60	0.63	0.55	0.51
Chypre	0.28	0.35	0.32	0.27	0.26	0.23	0.25	0.28	0.27
Gibraltar	1.17	1.06	0.66	0.62	0.57	1.19	1.43	1.50	1.87
Malte	0.56	0.50	0.29	0.21	0.28	0.17	0.34	0.32	0.31
Roumanie	-0.10	-0.03	0.15	0.07	0.06	0.10	0.13	0.20	0.16
République slovaque	0.40	0.47	0.54	0.45	0.42	0.41	0.39	0.37	0.36
Bosnie-Herzégovine	-	-	-	-	-	-	-	-	-
Croatie	-	-	-	-	-	-	-	-	-
Ex-RYM	-	-	-	-	-	-	-	-	-
Slovénie	-	-	-	-	-	-	-	-	-
RF de Yougoslavie	-	-	-	-	-	-	-	-	-
Ex-Yougoslavie	0.09	0.15	0.13	0.10	0.11	0.10	0.11	0.12	0.14
Europe non-OCDE	0.17	0.24	0.25	0.19	0.18	0.18	0.20	0.22	0.22
Arménie	-	-	-	-	-	-	-	-	-
Azerbaïdjan	-	-	-	-	-	-	-	-	-
Bélarus	-	-	-	-	-	-	-	-	-
Estonie	-	-	-	-	-	-	-	-	-
Géorgie	-	-	-	-	-	-	-	-	-
Kazakhstan	-	-	-	-	-	-	-	-	-
Kirghizistan	-	-	-	-	-	-	-	-	-
Lettonie	-	-	-	-	-	-	-	-	-
Lituanie	-	-	-	-	-	-	-	-	-
Moldova	-	-	-	-	-	-	-	-	-
Russie	-	-	-	-	-	-	-	-	-
Tadjikistan	-	-	-	-	-	-	-	-	-
Turkménistan	-	-	-	-	-	-	-	-	-
Ukraine	-	-	-	-	-	-	-	-	-
Ouzbékistan	-	-	-	-	-	-	-	-	-
Ex-URSS	-0.17	-0.16	-0.21	-0.21	-0.20	-0.18	-0.19	-0.20	-0.20

Faute de séries complètes, l'Albanie, la Rép.pop.dém. de Corée et le Viêt-Nam ne sont pas pris en compte dans les totaux régionaux.
Ex-URSS: la somme des républiques peut être différente du total.

Net Oil Imports/GDP (toe per thousand 90 US$)
Importations nettes de pétrole/PIB (tep par millier de $US 90)

Netto-Ölimporte/BIP (toe pro tausend 1990er US$)

Importanzione nette di petrolio/PIL (tep per "1,000" US$ 90)

石油純輸入量／GDP （石油換算トン／千米ドル、1990年価格）

Importaciones netas de petóleo/PIB (tep por miles de 90 US$)

Чистый импорт нефти и нефтепродуктов / ВВП (тнэ на тыс.долл.США 1990 г.)

1989	1990	1991	1992	1993	1994	1995	1996	1997	
0.10	0.09	0.07	0.08	0.08	0.09	0.08	0.09	0.08	Bangladesh
-2.16	-2.15	-2.25	-2.48	-2.52	-2.55	-2.42	-2.25	-2.06	Brunei
0.08	0.09	0.10	0.11	0.12	0.11	0.11	0.12	0.11	India
-0.42	-0.36	-0.35	-0.30	-0.24	-0.26	-0.21	-0.18	-0.14	Indonesia
-0.45	-0.41	-0.38	-0.35	-0.28	-0.26	-0.28	-0.21	-0.17	Malaysia
-0.00	-0.00	0.00	0.01	0.01	0.01	0.02	0.02	0.03	Myanmar
0.06	0.06	0.07	0.08	0.07	0.10	0.11	0.11	0.10	Nepal
0.21	0.21	0.19	0.20	0.22	0.25	0.24	0.27	0.27	Pakistan
0.26	0.27	0.26	0.31	0.30	0.31	0.33	0.34	0.34	Philippines
0.62	0.65	0.66	0.68	0.69	0.66	0.55	0.60	0.63	Singapore
0.20	0.21	0.20	0.19	0.22	0.21	0.24	0.26	0.27	Sri Lanka
0.17	0.18	0.16	0.16	0.16	0.16	0.16	0.16	0.15	Chinese Taipei
0.19	0.21	0.19	0.21	0.22	0.22	0.24	0.25	0.24	Thailand
0.14	0.02	-0.20	-0.31	-0.30	-0.30	-0.47	-0.24	-0.18	Vietnam
0.15	-0.11	-0.13	-0.15	-0.16	-0.17	-0.13	-0.08	-0.07	Other Asia
0.05	0.06	0.06	0.07	0.09	0.08	0.09	0.11	0.11	**Asia**
-0.06	-0.06	-0.04	-0.02	0.02	0.01	0.02	0.02	0.05	People's Rep.of China
0.09	0.09	0.08	0.10	0.10	0.10	0.10	0.10	0.09	Hong Kong, China
-0.03	-0.04	-0.02	0.00	0.03	0.02	0.03	0.03	0.05	**China**
-0.05	0.01	-0.05	-0.04	0.01	0.04	0.05	0.06	0.06	Albania
0.54	0.41	0.32	0.30	0.39	0.34	0.36	0.36	0.33	Bulgaria
0.30	0.29	0.30	0.31	0.33	0.33	0.29	0.31	0.30	Cyprus
1.54	1.56	2.74	2.77	2.87	2.89	2.89	2.84	2.75	Gibraltar
0.28	0.27	0.29	0.30	0.29	0.27	0.30	0.31	0.32	Malta
0.19	0.28	0.22	0.19	0.20	0.14	0.19	0.17	0.21	Romania
0.34	0.31	0.28	0.28	0.27	0.25	0.27	0.24	0.22	Slovak Republic
-	-	-	0.33	0.68	1.24	0.94	0.64	0.46	*Bosnia-Herzegovina*
-	-	-	0.16	0.17	0.23	0.26	0.25	0.25	*Croatia*
-	-	-	0.55	0.64	0.45	0.55	0.71	0.60	*FYROM*
-	-	-	0.08	0.09	0.10	0.10	0.11	0.11	*Slovenia*
-	-	-	0.02	0.01	0.01	0.01	0.04	0.06	*FR of Yugoslavia*
0.14	0.14	0.12	0.07	0.08	0.09	0.09	0.11	0.12	Former Yugoslavia
0.23	0.23	0.19	0.16	0.18	0.16	0.18	0.18	0.19	**Non-OECD Europe**
-	-	-	1.35	0.80	0.24	0.16	0.08	0.08	Armenia
-	-	-	-0.77	-0.61	-0.22	-0.62	-0.61	-0.57	Azerbaijan
-	-	-	0.61	0.42	0.38	0.40	0.32	0.30	Belarus
-	-	-	0.27	0.34	0.30	0.23	0.22	0.18	Estonia
-	-	-	0.22	0.19	0.10	0.03	0.23	0.23	Georgia
-	-	-	-0.12	-0.24	-0.28	-0.38	-0.53	-0.68	Kazakhstan
-	-	-	0.80	0.57	0.22	0.37	0.38	0.25	Kyrgyzstan
-	-	-	0.38	0.41	0.39	0.33	0.36	0.23	Latvia
-	-	-	0.40	0.45	0.43	0.47	0.37	0.37	Lithuania
-	-	-	0.43	0.28	0.25	0.25	0.21	0.24	Moldova
-	-	-	-0.34	-0.37	-0.44	-0.44	-0.50	-0.50	Russia
-	-	-	1.89	1.32	0.55	0.63	0.67	0.65	Tajikistan
-	-	-	0.01	-0.26	-0.12	-0.24	-0.28	-0.23	Turkmenistan
-	-	-	0.28	0.23	0.23	0.28	0.23	0.21	Ukraine
-	-	-	0.25	0.27	0.13	-0.03	-0.06	-0.06	Uzbekistan
-0.18	-0.17	-0.12	-0.10	-0.15	-0.23	-0.25	-0.31	-0.32	**Former USSR**

Due to lack of complete series Albania, the Dem. Ppl's Rep. of Korea and Vietnam are not included in the regional aggregates.
Former USSR: data for individual republics may not add to the total.

Net Oil Imports/GDP (toe per thousand 90 US$)
Importations nettes de pétrole/PIB (tep par millier de $US 90)
Netto-Ölimporte/BIP (toe pro tausend 1990er US$)
Importanzione nette di petrolio/PIL (tep per "1,000" US$ 90)
石油純輸入量／GDP （石油換算トン／千米ドル、1990年価格）
Importaciones netas de petóleo/PIB (tep por miles de 90 US$)
Чистый импорт нефти и нефтепродуктов / ВВП （тнэ на тыс.долл.США 1990 г.）

	1971	1973	1978	1983	1984	1985	1986	1987	1988
Bahrein	-1.03	-0.69	-0.65	-0.52	-0.35	-0.29	-0.44	-0.44	-0.37
Iran	-3.39	-3.32	-1.64	-0.75	-0.58	-0.58	-0.44	-0.66	-0.66
Irak	-1.09	-1.18	-0.96	-0.33	-0.43	-0.62	-0.96	-1.24	-1.76
Israël	0.02	0.09	0.23	0.17	0.16	0.15	0.16	0.15	0.21
Jordanie	0.41	0.46	0.53	0.67	0.68	0.69	0.64	0.69	0.67
Koweit	-7.64	-6.98	-5.11	-3.41	-3.59	-3.18	-4.56	-4.05	-4.70
Liban	0.38	0.38	0.59	0.66	0.66	0.77	0.73	0.79	0.54
Oman	-5.09	-5.46	-3.75	-2.87	-2.59	-2.74	-3.04	-3.26	-3.33
Qatar	-6.78	-6.78	-4.78	-2.28	-2.94	-2.66	-2.71	-2.18	-2.44
Arabie saoudite	-5.18	-6.13	-4.82	-2.47	-2.35	-1.63	-2.44	-2.02	-2.34
Syrie	-0.31	-0.33	-0.33	-0.04	-0.01	-0.08	-0.01	-0.14	-0.20
Emirats arabes unis	-5.71	-5.91	-4.27	-1.67	-1.73	-1.76	-2.46	-2.54	-2.79
Yémen	0.42	0.41	0.30	0.28	0.29	0.33	0.31	0.23	-0.50
Moyen-Orient	-2.95	-3.18	-2.14	-1.04	-1.00	-0.89	-1.20	-1.21	-1.42
Total non-OCDE	-0.55	-0.59	-0.44	-0.23	-0.23	-0.21	-0.23	-0.22	-0.23
OCDE Amérique du N.	0.05	0.07	0.09	0.03	0.03	0.02	0.04	0.04	0.04
OCDE Pacifique	0.15	0.16	0.14	0.10	0.10	0.09	0.09	0.08	0.08
OCDE Europe	0.15	0.16	0.12	0.07	0.07	0.07	0.07	0.07	0.06
Total OCDE	0.11	0.13	0.11	0.06	0.06	0.05	0.06	0.06	0.06
Pour mémoire: OPEP	-3.36	-3.49	-2.31	-1.22	-1.22	-1.11	-1.35	-1.31	-1.48

Faute de séries complètes, l'Albanie, la Rép.pop.dém. de Corée et le Viêt-Nam ne sont pas pris en compte dans les totaux régionaux.

Net Oil Imports/GDP (toe per thousand 90 US$)
Importations nettes de pétrole/PIB (tep par millier de $US 90)
Netto-Ölimporte/BIP (toe pro tausend 1990er US$)
Importanzione nette di petrolio/PIL (tep per "1,000" US$ 90)
石油純輸入量／GDP （石油換算トン／千米ドル、1990年価格）
Importaciones netas de petóleo/PIB (tep por miles de 90 US$)
Чистый импорт нефти и нефтепродуктов / ВВП (тнэ на тыс.долл.США 1990 г.)

1989	1990	1991	1992	1993	1994	1995	1996	1997	
-0.36	-0.41	-0.44	-0.37	-0.37	-0.36	-0.40	-0.13	0.17	Bahrain
-0.87	-0.88	-0.87	-0.86	-0.89	-0.82	-0.80	-0.74	-0.71	Iran
-1.78	-1.73	-0.11	-0.22	-0.22	-0.25	-0.26	-0.32	-1.18	Iraq
0.20	0.17	0.15	0.18	0.17	0.14	0.16	0.14	0.15	Israel
0.78	0.87	0.77	0.81	0.77	0.73	0.71	0.74	0.78	Jordan
-4.69	-2.99	-0.84	-3.58	-4.73	-4.88	-4.74	-4.58	-4.42	Kuwait
0.77	0.76	0.71	0.67	0.77	0.79	0.85	0.83	0.89	Lebanon
-3.28	-3.22	-2.91	-2.93	-2.87	-2.93	-2.96	-2.93	-2.86	Oman
-2.69	-2.70	-2.54	-2.80	-2.55	-2.49	-2.55	-2.77	-3.03	Qatar
-2.40	-2.90	-3.36	-3.31	-3.26	-3.23	-3.23	-3.15	-3.14	Saudi Arabia
-0.40	-0.47	-0.56	-0.52	-0.52	-0.57	-0.56	-0.54	-0.50	Syria
-2.93	-2.79	-2.89	-2.70	-2.51	-2.58	-2.52	-2.51	-2.55	United Arab Emirates
-0.55	-0.59	-0.61	-0.46	-0.69	-1.15	-1.15	-1.18	-0.94	Yemen
-1.56	-1.58	-1.53	-1.56	-1.59	-1.58	-1.54	-1.49	-1.50	**Middle East**
-0.24	-0.25	-0.23	-0.23	-0.23	-0.24	-0.23	-0.22	-0.22	**Non-OECD Total**
0.05	0.05	0.04	0.04	0.05	0.05	0.04	0.05	0.05	OECD North America
0.09	0.09	0.09	0.09	0.09	0.10	0.10	0.10	0.10	OECD Pacific
0.06	0.06	0.06	0.06	0.06	0.05	0.05	0.05	0.05	OECD Europe
0.06	0.06	0.06	0.06	0.06	0.06	0.06	0.06	0.06	**OECD Total**
-1.57	-1.59	-1.56	-1.56	-1.57	-1.57	-1.54	-1.50	-1.51	*Memo: OPEC*

Due to lack of complete series Albania, the Dem. Ppl's Rep. of Korea and Vietnam are not included in the regional aggregates.

TPES/GDP (toe per thousand 90 US$)
ATEP/PIB (tep par milliers de $US 90)
TPES/BIP (in toe pro tausend 1990er US$)

DTEP/PIL (tep per migliaia di $US 90)

一次エネルギー供給／GDP（石油換算トン／千米ドル、1990 年価格）

TPES/PIB (tep por miles de 90 US$)

ОППТЭ / ВВП (тнэ на тыс.долл.США в ценах и по валютному курсу 1990 г.)

	1971	1973	1978	1983	1984	1985	1986	1987	1988
Algérie	0.17	0.17	0.22	0.34	0.32	0.32	0.36	0.35	0.38
Angola	0.28	0.28	0.53	0.57	0.55	0.56	0.55	0.52	0.50
Bénin	1.07	1.04	1.08	0.99	0.94	0.91	0.90	0.93	0.92
Cameroun	0.56	0.54	0.41	0.36	0.35	0.33	0.31	0.32	0.36
Congo	0.70	0.58	0.59	0.37	0.34	0.35	0.35	0.53	0.67
RD du Congo	0.78	0.76	0.97	1.09	1.06	1.07	1.05	1.06	1.10
Egypte	0.39	0.40	0.42	0.48	0.50	0.50	0.51	0.53	0.52
Ethiopie	2.04	2.04	2.14	1.54	1.68	1.90	1.80	1.65	1.70
Gabon	0.35	0.45	0.34	0.33	0.29	0.27	0.28	0.32	0.30
Ghana	0.63	0.70	0.80	0.88	0.89	0.93	0.92	0.93	0.89
Côte d'Ivoire	0.38	0.38	0.32	0.35	0.37	0.37	0.39	0.38	0.43
Kenya	2.38	2.03	1.84	1.65	1.69	1.67	1.67	1.61	1.54
Libye	0.05	0.09	0.13	0.27	0.31	0.37	0.38	0.46	0.43
Maroc	0.22	0.26	0.28	0.27	0.27	0.26	0.25	0.26	0.25
Mozambique	5.74	5.19	6.67	7.76	8.20	7.39	7.63	6.32	5.69
Nigéria	1.72	1.71	1.85	2.49	2.63	2.48	2.44	2.55	2.40
Sénégal	0.38	0.41	0.42	0.39	0.43	0.40	0.41	0.41	0.38
Afrique du Sud	0.65	0.67	0.72	0.84	0.86	0.89	0.92	0.93	0.93
Soudan	1.17	1.39	0.89	0.97	1.02	1.13	1.11	1.07	1.18
Tanzanie	4.45	4.10	3.44	3.63	3.61	3.63	3.62	3.58	3.50
Tunisie	0.43	0.42	0.44	0.46	0.45	0.44	0.44	0.42	0.44
Zambie	1.33	1.35	1.34	1.48	1.48	1.49	1.49	1.49	1.43
Zimbabwe	1.18	1.15	1.25	1.01	1.06	1.06	1.09	1.18	1.10
Autre Afrique	0.98	1.00	0.95	1.00	1.01	0.99	0.99	1.01	0.99
Afrique	0.71	0.72	0.72	0.80	0.81	0.81	0.83	0.85	0.85
Argentine	0.27	0.27	0.28	0.28	0.28	0.29	0.28	0.29	0.30
Bolivie	0.29	0.31	0.36	0.67	0.62	0.63	0.65	0.58	0.59
Brésil	0.36	0.33	0.30	0.29	0.29	0.29	0.28	0.28	0.29
Chili	0.48	0.53	0.49	0.48	0.47	0.44	0.44	0.42	0.44
Colombie	0.79	0.69	0.69	0.71	0.68	0.69	0.67	0.66	0.63
Costa Rica	0.41	0.39	0.36	0.33	0.35	0.37	0.35	0.36	0.35
Cuba	1.69	1.56	1.43	1.23	1.04	0.98	0.98	1.07	1.09
République dominicaine	0.76	0.74	0.68	0.67	0.71	0.60	0.61	0.62	0.57
Equateur	0.60	0.44	0.58	0.61	0.63	0.61	0.59	0.63	0.61
El Salvador	0.38	0.39	0.44	0.58	0.59	0.59	0.51	0.56	0.52
Guatemala	0.66	0.62	0.61	0.53	0.56	0.58	0.55	0.56	0.56
Haiti	0.68	0.70	0.73	0.63	0.64	0.64	0.50	0.52	0.54
Honduras	0.95	0.88	0.77	0.83	0.80	0.78	0.77	0.79	0.80
Jamaïque	0.58	0.75	0.71	0.60	0.60	0.55	0.55	0.53	0.52
Antilles néerlandaises	4.68	4.65	2.63	2.88	2.06	1.34	1.23	1.04	1.00
Nicaragua	1.00	1.00	1.02	1.18	1.37	1.50	1.54	1.58	1.74
Panama	0.51	0.57	0.39	0.34	0.30	0.28	0.27	0.29	0.29
Paraguay	0.77	0.75	0.61	0.55	0.54	0.54	0.56	0.62	0.61
Pérou	0.32	0.31	0.30	0.29	0.29	0.29	0.28	0.27	0.29
Trinité-et-Tobago	1.01	0.88	0.79	0.77	0.77	0.83	0.92	0.88	1.03
Uruguay	0.39	0.38	0.34	0.30	0.30	0.29	0.28	0.27	0.29
Vénézuela	0.67	0.67	0.65	0.91	0.86	0.87	0.85	0.92	0.81
Autre Amérique latine	0.49	0.54	0.47	0.29	0.30	0.28	0.28	0.30	0.28
Amérique latine	0.43	0.41	0.39	0.39	0.38	0.38	0.37	0.37	0.38

TPES/GDP (toe per thousand 90 US$)
ATEP/PIB (tep par milliers de $US 90)
TPES/BIP (in toe pro tausend 1990er US$)

DTEP/PIL (tep per migliaia di $US 90)

一次エネルギー供給／GDP（石油換算トン／千米ドル、1990 年価格）

TPES/PIB (tep por miles de 90 US$)

ОПΠΤЭ / ВВП (тнэ на тыс.долл.США в ценах и по валютному курсу 1990 г.)

1989	1990	1991	1992	1993	1994	1995	1996	1997	
0.36	0.39	0.41	0.40	0.41	0.40	0.39	0.37	0.40	Algeria
0.51	0.55	0.55	0.59	0.79	0.79	0.73	0.69	0.68	Angola
0.88	0.91	0.88	0.87	0.86	0.84	0.82	0.90	0.86	Benin
0.38	0.41	0.43	0.45	0.47	0.49	0.49	0.48	0.47	Cameroon
0.43	0.40	0.40	0.40	0.40	0.44	0.42	0.40	0.42	Congo
1.15	1.27	1.42	1.63	1.95	2.08	2.15	2.23	2.42	DR of Congo
0.52	0.51	0.52	0.51	0.53	0.48	0.48	0.49	0.49	Egypt
1.75	1.76	1.95	2.08	1.81	1.83	1.78	1.66	1.56	Ethiopia
0.24	0.21	0.21	0.23	0.24	0.23	0.22	0.21	0.22	Gabon
0.90	0.89	0.87	0.86	0.87	0.92	0.92	0.89	0.87	Ghana
0.43	0.43	0.43	0.47	0.46	0.45	0.43	0.43	0.42	Ivory Coast
1.51	1.46	1.45	1.49	1.52	1.51	1.43	1.40	1.44	Kenya
0.44	0.43	0.49	0.47	0.52	0.53	0.62	0.57	0.57	Libya
0.26	0.26	0.26	0.28	0.30	0.29	0.31	0.30	0.31	Morocco
5.27	5.07	4.72	4.72	3.93	3.76	3.75	3.55	3.21	Mozambique
2.32	2.19	2.19	2.20	2.02	2.38	2.28	2.18	2.20	Nigeria
0.39	0.39	0.39	0.41	0.41	0.42	0.41	0.41	0.41	Senegal
0.87	0.86	0.90	0.87	0.91	0.93	0.94	0.90	0.92	South Africa
1.10	1.18	1.17	1.05	0.95	0.95	0.92	0.88	0.88	Sudan
3.44	3.34	3.26	3.63	3.31	3.32	3.30	3.23	3.16	Tanzania
0.47	0.46	0.44	0.43	0.45	0.44	0.43	0.42	0.40	Tunisia
1.38	1.39	1.44	1.51	1.42	1.49	1.56	1.49	1.47	Zambia
1.06	1.02	1.04	1.17	1.12	1.05	1.09	1.02	0.99	Zimbabwe
0.97	0.96	0.95	0.97	0.98	1.00	0.98	0.95	0.93	Other Africa
0.83	0.82	0.84	0.85	0.86	0.88	0.87	0.84	0.85	**Africa**
0.31	0.31	0.29	0.29	0.29	0.28	0.30	0.30	0.29	Argentina
0.58	0.59	0.60	0.59	0.59	0.61	0.61	0.59	0.66	Bolivia
0.29	0.29	0.29	0.30	0.30	0.30	0.29	0.29	0.30	Brazil
0.45	0.46	0.43	0.43	0.41	0.42	0.41	0.41	0.44	Chile
0.64	0.66	0.65	0.63	0.61	0.62	0.61	0.61	0.57	Colombia
0.35	0.35	0.36	0.37	0.32	0.32	0.37	0.37	0.36	Costa Rica
1.14	1.17	1.12	1.10	1.25	1.30	1.30	1.27	1.33	Cuba
0.53	0.56	0.57	0.63	0.61	0.60	0.59	0.56	0.54	Dominican Republic
0.57	0.61	0.60	0.60	0.57	0.60	0.61	0.66	0.64	Ecuador
0.52	0.51	0.57	0.55	0.53	0.52	0.58	0.55	0.54	El Salvador
0.56	0.57	0.57	0.57	0.54	0.55	0.55	0.56	0.56	Guatemala
0.54	0.53	0.51	0.60	0.60	0.60	0.68	0.76	0.68	Haiti
0.80	0.80	0.78	0.78	0.71	0.76	0.80	0.79	0.82	Honduras
0.61	0.71	0.67	0.67	0.69	0.73	0.78	0.85	0.92	Jamaica
1.19	1.23	1.56	0.93	1.50	1.53	1.51	1.44	1.39	Netherlands Antilles
1.86	1.96	1.99	2.03	2.00	2.01	1.92	1.60	1.61	Nicaragua
0.28	0.29	0.28	0.29	0.28	0.30	0.29	0.31	0.31	Panama
0.63	0.59	0.59	0.58	0.58	0.61	0.64	0.64	0.65	Paraguay
0.31	0.32	0.30	0.30	0.30	0.28	0.28	0.28	0.28	Peru
1.00	1.12	1.02	1.28	1.24	1.33	1.28	1.40	1.41	Trinidad-and-Tobago
0.29	0.27	0.28	0.29	0.26	0.24	0.26	0.27	0.26	Uruguay
0.89	0.84	0.93	0.82	0.76	0.74	0.83	0.96	0.96	Venezuela
0.28	0.28	0.27	0.26	0.26	0.26	0.26	0.25	0.25	Other Latin America
0.38	0.39	0.39	0.38	0.38	0.37	0.38	0.39	0.39	**Latin America**

TPES/GDP (toe per thousand 90 US$)
ATEP/PIB (tep par milliers de $US 90)
TPES/BIP (in toe pro tausend 1990er US$)
DTEP/PIL (tep per migliaia di $US 90)
一次エネルギー供給／GDP（石油換算トン／千米ドル、1990 年価格）
TPES/PIB (tep por miles de 90 US$)
ОПТЭ / ВВП (тнэ на тыс.долл.США в ценах и по валютному курсу 1990 г.)

	1971	1973	1978	1983	1984	1985	1986	1987	1988
Bangladesh	0.93	1.15	1.06	0.97	0.96	0.96	0.96	0.96	0.97
Brunei	0.05	0.06	0.06	0.52	0.39	0.41	0.38	0.52	0.57
Inde	1.40	1.44	1.33	1.32	1.31	1.30	1.30	1.29	1.23
Indonésie	1.17	1.10	1.03	0.89	0.86	0.90	0.93	0.90	0.87
Malaisie	0.52	0.43	0.46	0.46	0.45	0.55	0.59	0.57	0.55
Myanmar	0.57	0.56	0.48	0.40	0.39	0.40	0.41	0.42	0.48
Népal	1.44	1.49	1.38	2.06	1.94	1.85	1.81	1.84	1.75
Pakistan	1.28	1.27	1.22	1.15	1.13	1.10	1.09	1.12	1.09
Philippines	0.66	0.63	0.58	0.59	0.60	0.65	0.63	0.65	0.65
Singapour	0.33	0.36	0.39	0.30	0.30	0.31	0.33	0.32	0.31
Sri Lanka	1.07	1.09	0.90	0.79	0.78	0.73	0.72	0.74	0.73
Taipei chinois	0.31	0.32	0.36	0.33	0.32	0.31	0.31	0.29	0.29
Thaïlande	0.65	0.66	0.59	0.46	0.48	0.51	0.53	0.51	0.50
Viêt-Nam	-	-	-	-	4.26	4.17	4.23	4.28	4.13
Autre Asie	0.37	0.36	0.31	0.29	0.29	0.31	0.29	0.30	0.31
Asie	0.99	0.97	0.89	0.83	0.82	0.83	0.83	0.82	0.80
Rép. populaire de Chine	3.83	3.75	4.19	3.11	2.86	2.66	2.56	2.41	2.27
Hong-Kong, Chine	0.20	0.17	0.18	0.15	0.14	0.15	0.15	0.14	0.15
Chine	3.29	3.17	3.44	2.55	2.37	2.25	2.15	2.03	1.92
Albanie	-	-	-	1.38	1.48	1.35	1.31	1.31	1.39
Bulgarie	1.51	1.49	1.70	1.68	1.64	1.61	1.56	1.46	1.35
Chypre	0.29	0.35	0.31	0.29	0.28	0.27	0.29	0.31	0.29
Gibraltar	0.21	0.17	0.12	0.16	0.14	0.16	0.18	0.18	0.21
Malte	0.46	0.47	0.27	0.22	0.30	0.23	0.36	0.37	0.36
Roumanie	1.90	1.96	1.80	1.55	1.45	1.45	1.46	1.60	1.62
République slovaque	1.35	1.37	1.46	1.50	1.49	1.50	1.47	1.45	1.42
Bosnie-Herzégovine	-	-	-	-	-	-	-	-	-
Croatie	-	-	-	-	-	-	-	-	-
Ex-RYM	-	-	-	-	-	-	-	-	-
Slovénie	-	-	-	-	-	-	-	-	-
RF de Yougoslavie	-	-	-	-	-	-	-	-	-
Ex-Yougoslavie	0.39	0.39	0.35	0.41	0.43	0.44	0.44	0.45	0.47
Europe non-OCDE	0.94	0.96	0.91	0.90	0.89	0.89	0.89	0.92	0.93
Arménie	-	-	-	-	-	-	-	-	-
Azerbaïdjan	-	-	-	-	-	-	-	-	-
Bélarus	-	-	-	-	-	-	-	-	-
Estonie	-	-	-	-	-	-	-	-	-
Géorgie	-	-	-	-	-	-	-	-	-
Kazakhstan	-	-	-	-	-	-	-	-	-
Kirghizistan	-	-	-	-	-	-	-	-	-
Lettonie	-	-	-	-	-	-	-	-	-
Lituanie	-	-	-	-	-	-	-	-	-
Moldova	-	-	-	-	-	-	-	-	-
Russie	-	-	-	-	-	-	-	-	-
Tadjikistan	-	-	-	-	-	-	-	-	-
Turkménistan	-	-	-	-	-	-	-	-	-
Ukraine	-	-	-	-	-	-	-	-	-
Ouzbékistan	-	-	-	-	-	-	-	-	-
Ex-URSS	1.33	1.34	1.40	1.41	1.44	1.47	1.43	1.48	1.49

Faute de séries complètes, l'Albanie, la Rép.pop.dém. de Corée et le Viêt-Nam ne sont pas pris en compte dans les totaux régionaux.

TPES/GDP (toe per thousand 90 US$)
ATEP/PIB (tep par milliers de $US 90)
TPES/BIP (in toe pro tausend 1990er US$)

DTEP/PIL (tep per migliaia di $US 90)

一次エネルギー供給／GDP（石油換算トン／千米ドル、1990年価格）

TPES/PIB (tep por miles de 90 US$)

ОПptэ / ВВП (тнэ на тыс.долл.США в ценах и по валютному курсу 1990 г.)

1989	1990	1991	1992	1993	1994	1995	1996	1997	
0.99	0.95	0.91	0.91	0.90	0.89	0.88	0.85	0.81	Bangladesh
0.37	0.41	0.45	0.51	0.48	0.47	0.54	0.53	0.54	Brunei
1.20	1.18	1.22	1.20	1.17	1.13	1.11	1.07	1.03	India
0.86	0.86	0.84	0.81	0.81	0.76	0.75	0.74	0.74	Indonesia
0.57	0.56	0.61	0.60	0.63	0.60	0.54	0.59	0.64	Malaysia
0.47	0.45	0.46	0.42	0.42	0.40	0.38	0.37	0.37	Myanmar
1.70	1.66	1.63	1.62	1.62	1.56	1.54	1.50	1.44	Nepal
1.10	1.10	1.07	1.04	1.07	1.08	1.06	1.08	1.09	Pakistan
0.63	0.64	0.64	0.68	0.69	0.70	0.71	0.68	0.70	Philippines
0.29	0.36	0.36	0.38	0.43	0.48	0.38	0.39	0.41	Singapore
0.72	0.68	0.66	0.65	0.64	0.59	0.59	0.63	0.62	Sri Lanka
0.29	0.29	0.30	0.29	0.29	0.29	0.29	0.29	0.28	Chinese Taipei
0.50	0.51	0.51	0.51	0.52	0.54	0.55	0.58	0.59	Thailand
3.82	3.78	3.58	3.45	3.33	3.19	3.02	3.43	3.44	Vietnam
0.31	0.32	0.31	0.29	0.28	0.28	0.26	0.25	0.25	Other Asia
0.78	0.78	0.78	0.77	0.76	0.74	0.73	0.72	0.71	**Asia**
2.26	2.21	2.09	1.87	1.73	1.62	1.55	1.46	1.35	People's Rep.of China
0.15	0.14	0.14	0.15	0.16	0.14	0.14	0.13	0.13	Hong Kong, China
1.92	1.87	1.78	1.61	1.51	1.43	1.38	1.31	1.21	**China**
1.21	1.22	1.20	1.00	0.77	0.67	0.59	0.60	0.56	Albania
1.36	1.31	1.17	1.17	1.26	1.19	1.26	1.39	1.36	Bulgaria
0.29	0.28	0.31	0.30	0.30	0.33	0.29	0.30	0.29	Cyprus
0.20	0.22	0.24	0.27	0.31	0.41	0.40	0.39	0.38	Gibraltar
0.36	0.33	0.31	0.29	0.34	0.30	0.28	0.29	0.30	Malta
1.71	1.60	1.51	1.53	1.46	1.32	1.33	1.38	1.32	Romania
1.40	1.38	1.47	1.46	1.44	1.34	1.30	1.24	1.14	Slovak Republic
-	-	-	2.72	4.33	2.34	1.84	1.26	0.91	*Bosnia-Herzegovina*
-	-	-	0.84	0.88	0.84	0.84	0.81	0.81	*Croatia*
-	-	-	1.56	1.75	1.48	1.56	1.77	1.59	*FYROM*
-	-	-	0.24	0.25	0.25	0.25	0.26	0.26	*Slovenia*
-	-	-	0.33	0.31	0.31	0.31	0.36	0.39	*FR of Yugoslavia*
0.46	0.50	0.45	0.45	0.45	0.40	0.41	0.43	0.44	Former Yugoslavia
0.93	0.91	0.82	0.84	0.85	0.78	0.79	0.82	0.78	**Non-OECD Europe**
-	-	-	2.40	1.48	0.88	0.97	0.98	0.96	Armenia
-	-	-	2.65	3.17	3.95	3.65	3.40	3.23	Azerbaijan
-	-	-	1.27	1.08	1.05	1.09	1.08	0.98	Belarus
-	-	-	1.34	1.17	1.25	1.12	1.18	1.04	Estonia
-	-	-	1.08	1.32	0.95	0.46	0.60	0.59	Georgia
-	-	-	2.44	2.14	2.25	2.32	1.88	1.59	Kazakhstan
-	-	-	2.24	2.05	1.97	1.85	1.98	1.65	Kyrgyzstan
-	-	-	0.87	0.79	0.74	0.64	0.65	0.65	Latvia
-	-	-	1.14	1.09	1.06	1.14	1.11	1.03	Lithuania
-	-	-	1.03	0.89	1.06	1.00	1.17	1.11	Moldova
-	-	-	1.69	1.72	1.70	1.74	1.76	1.69	Russia
-	-	-	2.93	2.31	1.52	1.73	1.94	1.84	Tajikistan
-	-	-	1.83	2.02	3.34	3.81	3.27	2.60	Turkmenistan
-	-	-	1.69	1.76	1.94	2.21	2.40	2.30	Ukraine
-	-	-	2.20	2.36	2.32	2.17	2.20	2.09	Uzbekistan
1.44	1.42	1.50	1.63	1.65	1.68	1.73	1.75	1.66	**Former USSR**

Due to lack of complete series Albania, the Dem. Ppl's Rep. of Korea and Vietnam are not included in the regional aggregates.

TPES/GDP (toe per thousand 90 US$)
ATEP/PIB (tep par milliers de $US 90)
TPES/BIP (in toe pro tausend 1990er US$)
DTEP/PIL (tep per migliaia di $US 90)
一次エネルギー供給／GDP（石油換算トン／千米ドル、1990 年価格）
TPES/PIB (tep por miles de 90 US$)
ОПП ТЭ / ВВП (тнэ на тыс.долл.США в ценах и по валютному курсу 1990 г.)

	1971	1973	1978	1983	1984	1985	1986	1987	1988
Bahrein	0.81	0.98	0.83	1.08	1.12	1.34	1.51	1.58	1.42
Iran	0.36	0.32	0.33	0.37	0.43	0.43	0.50	0.53	0.60
Irak	0.06	0.06	0.07	0.15	0.16	0.20	0.25	0.28	0.35
Israël	0.27	0.31	0.23	0.20	0.20	0.19	0.21	0.22	0.24
Jordanie	0.35	0.44	0.53	0.68	0.68	0.66	0.65	0.67	0.68
Koweit	0.29	0.27	0.39	0.77	0.73	0.95	0.84	0.78	0.82
Liban	0.37	0.41	0.64	0.71	0.71	0.82	0.78	0.84	0.58
Oman	0.03	0.04	0.17	0.24	0.27	0.27	0.20	0.21	0.25
Qatar	0.30	0.35	0.34	0.73	0.83	0.89	0.81	0.80	0.83
Arabie saoudite	0.15	0.14	0.25	0.56	0.56	0.61	0.64	0.69	0.70
Syrie	0.33	0.27	0.30	0.38	0.45	0.40	0.43	0.46	0.43
Emirats arabes unis	0.12	0.11	0.27	0.43	0.43	0.55	0.74	0.74	0.77
Yémen	0.30	0.33	0.20	0.25	0.25	0.28	0.28	0.29	0.26
Moyen-Orient	0.21	0.21	0.24	0.37	0.39	0.43	0.48	0.51	0.55
Total non-OCDE	0.95	0.93	0.93	0.94	0.94	0.96	0.95	0.96	0.97
OCDE Amérique du N.	0.47	0.47	0.45	0.39	0.38	0.37	0.36	0.37	0.37
OCDE Pacifique	0.21	0.22	0.21	0.18	0.19	0.18	0.18	0.17	0.18
OCDE Europe	0.28	0.28	0.26	0.24	0.24	0.24	0.24	0.24	0.23
Total OCDE	0.34	0.34	0.32	0.28	0.28	0.28	0.27	0.27	0.27
Monde	0.45	0.45	0.44	0.42	0.42	0.42	0.41	0.41	0.41

Faute de séries complètes, l'Albanie, la Rép.pop.dém. de Corée et le Viêt-Nam ne sont pas pris en compte dans les totaux régionaux.
Le total Monde ne comprend pas les soutages maritimes internationaux.

INTERNATIONAL ENERGY AGENCY

TPES/GDP (toe per thousand 90 US$)
ATEP/PIB (tep par milliers de $US 90)
TPES/BIP (in toe pro tausend 1990er US$)

DTEP/PIL (tep per migliaia di $US 90)

一次エネルギー供給／ＧＤＰ（石油換算トン／千米ドル、1990 年価格）

TPES/PIB (tep por miles de 90 US$)

ОПЛТЭ / ВВП (тнэ на тыс.долл.США в ценах и по валютному курсу 1990 г.)

1989	1990	1991	1992	1993	1994	1995	1996	1997	
1.36	1.38	1.25	1.19	1.16	1.14	1.15	1.46	1.56	Bahrain
0.64	0.60	0.61	0.63	0.62	0.67	0.65	0.65	0.66	Iran
0.35	0.43	0.65	0.79	0.85	1.00	1.06	1.06	0.92	Iraq
0.24	0.23	0.21	0.23	0.23	0.22	0.24	0.21	0.23	Israel
0.80	0.86	0.85	0.83	0.79	0.79	0.75	0.77	0.81	Jordan
0.89	0.71	0.38	0.62	0.48	0.72	0.77	0.66	0.72	Kuwait
0.84	0.81	0.73	0.70	0.83	0.84	0.91	0.90	0.96	Lebanon
0.25	0.41	0.52	0.50	0.49	0.46	0.45	0.43	0.44	Oman
0.88	0.78	0.92	1.37	1.42	1.40	1.46	1.48	1.53	Qatar
0.66	0.60	0.64	0.66	0.70	0.68	0.70	0.76	0.80	Saudi Arabia
0.48	0.50	0.51	0.46	0.43	0.42	0.41	0.42	0.40	Syria
0.80	0.64	0.68	0.65	0.68	0.72	0.71	0.81	0.75	United Arab Emirates
0.26	0.23	0.27	0.29	0.22	0.24	0.24	0.24	0.23	Yemen
0.56	0.54	0.57	0.59	0.60	0.62	0.62	0.64	0.65	**Middle East**
0.95	0.94	0.95	0.94	0.92	0.89	0.88	0.86	0.84	**Non-OECD Total**
0.36	0.35	0.36	0.36	0.36	0.35	0.35	0.34	0.33	OECD North America
0.18	0.18	0.18	0.18	0.18	0.19	0.19	0.19	0.19	OECD Pacific
0.23	0.22	0.22	0.22	0.22	0.21	0.21	0.21	0.21	OECD Europe
0.27	0.26	0.26	0.26	0.26	0.26	0.26	0.26	0.25	**OECD Total**
0.41	0.40	0.40	0.39	0.39	0.38	0.38	0.38	0.37	**World**

Due to lack of complete series Albania, the Dem. Ppl's Rep. of Korea and Vietnam are not included in the regional aggregates.
World Total excludes international marine bunkers.

TPES/GDP (toe per thousand 90 US$ PPP)
ATEP/PIB (tep par milliers de $US 90 PPA)
TPES/BIP (in toe pro tausend 1990er US$ Kaufkraftparität)
DTEP/PIL (tep per migliaia di $US 90 PPA)
一次エネルギー供給／GDP（石油換算トン／千米ドル、1990年価格、購買力平価）
TPES/PIB (tep por miles de 90 US$ PPP)
ОПЛТЭ / ВВП (тнэ на тыс.долл.США в ценах и по ППС 1990 г.)

	1971	1973	1978	1983	1984	1985	1986	1987	1988
Algérie	0.12	0.13	0.16	0.26	0.24	0.24	0.27	0.27	0.29
Angola	0.31	0.31	0.58	0.63	0.61	0.62	0.61	0.57	0.55
Bénin	0.32	0.31	0.33	0.30	0.28	0.28	0.27	0.28	0.28
Cameroun	0.28	0.27	0.20	0.18	0.17	0.16	0.15	0.16	0.18
Congo	0.39	0.32	0.32	0.21	0.19	0.19	0.19	0.29	0.37
RD du Congo	0.16	0.15	0.20	0.22	0.22	0.22	0.21	0.22	0.22
Egypte	0.12	0.12	0.12	0.14	0.15	0.15	0.15	0.16	0.16
Ethiopie	0.69	0.69	0.72	0.52	0.56	0.64	0.60	0.55	0.57
Gabon	0.32	0.41	0.32	0.30	0.27	0.25	0.25	0.29	0.28
Ghana	0.15	0.17	0.19	0.21	0.21	0.22	0.22	0.23	0.21
Côte d'Ivoire	0.22	0.22	0.19	0.21	0.22	0.22	0.23	0.23	0.25
Kenya	0.63	0.54	0.49	0.44	0.45	0.45	0.44	0.43	0.41
Libye	0.05	0.10	0.14	0.29	0.33	0.40	0.40	0.49	0.46
Maroc	0.07	0.08	0.09	0.09	0.08	0.08	0.08	0.08	0.08
Mozambique	3.47	3.14	4.04	4.69	4.96	4.48	4.62	3.83	3.44
Nigéria	0.49	0.48	0.52	0.70	0.74	0.70	0.69	0.72	0.68
Sénégal	0.19	0.20	0.21	0.19	0.21	0.20	0.20	0.20	0.19
Afrique du Sud	0.42	0.43	0.46	0.54	0.55	0.57	0.59	0.59	0.60
Soudan	1.36	1.60	1.02	1.12	1.18	1.31	1.28	1.23	1.37
Tanzanie	1.20	1.10	0.92	0.97	0.97	0.97	0.97	0.96	0.94
Tunisie	0.15	0.15	0.15	0.16	0.16	0.16	0.16	0.15	0.16
Zambie	0.65	0.66	0.65	0.72	0.72	0.73	0.72	0.72	0.70
Zimbabwe	0.44	0.43	0.47	0.38	0.40	0.40	0.41	0.44	0.41
Autre Afrique	0.39	0.39	0.38	0.41	0.41	0.41	0.41	0.41	0.40
Afrique	0.33	0.34	0.34	0.37	0.37	0.37	0.37	0.38	0.38
Argentine	0.21	0.21	0.21	0.21	0.22	0.22	0.22	0.22	0.23
Bolivie	0.10	0.10	0.12	0.22	0.21	0.21	0.21	0.19	0.20
Brésil	0.23	0.21	0.19	0.18	0.19	0.18	0.18	0.18	0.19
Chili	0.13	0.15	0.14	0.14	0.13	0.12	0.12	0.12	0.12
Colombie	0.21	0.19	0.19	0.19	0.19	0.19	0.18	0.18	0.17
Costa Rica	0.17	0.16	0.15	0.13	0.14	0.15	0.14	0.14	0.14
Cuba	0.75	0.69	0.63	0.54	0.46	0.44	0.43	0.48	0.48
République dominicaine	0.25	0.25	0.23	0.22	0.24	0.20	0.20	0.21	0.19
Equateur	0.16	0.12	0.16	0.17	0.17	0.17	0.16	0.17	0.17
El Salvador	0.16	0.16	0.18	0.24	0.24	0.24	0.21	0.23	0.21
Guatemala	0.20	0.19	0.19	0.17	0.17	0.18	0.17	0.17	0.17
Haiti	0.41	0.42	0.43	0.38	0.38	0.38	0.30	0.31	0.32
Honduras	0.71	0.66	0.58	0.62	0.60	0.58	0.57	0.59	0.60
Jamaïque	0.27	0.35	0.33	0.28	0.28	0.26	0.25	0.24	0.24
Antilles néerlandaises	3.55	3.53	1.99	2.18	1.56	1.01	0.93	0.79	0.76
Nicaragua	0.26	0.26	0.27	0.31	0.36	0.39	0.40	0.41	0.45
Panama	0.30	0.34	0.23	0.20	0.18	0.16	0.16	0.17	0.17
Paraguay	0.31	0.31	0.25	0.22	0.22	0.22	0.23	0.25	0.25
Pérou	0.16	0.16	0.15	0.15	0.15	0.14	0.14	0.13	0.15
Trinité-et-Tobago	0.78	0.68	0.62	0.60	0.60	0.65	0.71	0.69	0.80
Uruguay	0.16	0.16	0.14	0.12	0.12	0.12	0.11	0.11	0.12
Vénézuela	0.23	0.23	0.23	0.32	0.30	0.30	0.30	0.32	0.28
Autre Amérique latine	0.41	0.45	0.40	0.24	0.25	0.24	0.24	0.25	0.23
Amérique latine	0.23	0.22	0.21	0.21	0.21	0.20	0.20	0.20	0.20

TPES/GDP (toe per thousand 90 US$ PPP)
ATEP/PIB (tep par milliers de $US 90 PPA)
TPES/BIP (in toe pro tausend 1990er US$ Kaufkraftparität)

DTEP/PIL (tep per migliaia di $US 90 PPA)

一次エネルギー供給／GDP（石油換算トン／千米ドル、1990年価格、購買力平価）

TPES/PIB (tep por miles de 90 US$ PPP)

ОПРТЭ / ВВП (тнэ на тыс.долл.США в ценах и по ППС 1990 г.)

1989	1990	1991	1992	1993	1994	1995	1996	1997	
0.27	0.29	0.31	0.30	0.31	0.30	0.29	0.28	0.30	Algeria
0.56	0.61	0.61	0.66	0.87	0.87	0.80	0.77	0.75	Angola
0.27	0.27	0.27	0.26	0.26	0.25	0.25	0.27	0.26	Benin
0.19	0.20	0.21	0.22	0.23	0.25	0.24	0.24	0.23	Cameroon
0.24	0.22	0.22	0.22	0.22	0.24	0.23	0.22	0.23	Congo
0.24	0.26	0.29	0.33	0.40	0.43	0.44	0.46	0.50	DR of Congo
0.16	0.15	0.15	0.15	0.16	0.14	0.14	0.14	0.15	Egypt
0.59	0.59	0.65	0.70	0.61	0.61	0.60	0.56	0.53	Ethiopia
0.22	0.20	0.19	0.21	0.22	0.21	0.20	0.19	0.20	Gabon
0.22	0.21	0.21	0.21	0.21	0.22	0.22	0.21	0.21	Ghana
0.25	0.25	0.25	0.28	0.27	0.26	0.25	0.25	0.25	Ivory Coast
0.40	0.39	0.39	0.40	0.40	0.40	0.38	0.37	0.38	Kenya
0.47	0.46	0.52	0.50	0.56	0.56	0.66	0.61	0.61	Libya
0.08	0.08	0.08	0.09	0.09	0.09	0.10	0.09	0.10	Morocco
3.19	3.07	2.86	2.86	2.38	2.28	2.27	2.15	1.94	Mozambique
0.65	0.62	0.62	0.62	0.57	0.67	0.64	0.62	0.62	Nigeria
0.19	0.19	0.19	0.20	0.20	0.21	0.20	0.20	0.20	Senegal
0.56	0.55	0.58	0.56	0.58	0.60	0.60	0.58	0.59	South Africa
1.27	1.36	1.35	1.22	1.10	1.10	1.06	1.02	1.02	Sudan
0.92	0.90	0.88	0.97	0.89	0.89	0.89	0.87	0.85	Tanzania
0.17	0.16	0.16	0.15	0.16	0.16	0.15	0.15	0.14	Tunisia
0.67	0.68	0.70	0.73	0.69	0.72	0.76	0.72	0.71	Zambia
0.40	0.38	0.39	0.44	0.42	0.40	0.41	0.38	0.37	Zimbabwe
0.39	0.39	0.39	0.39	0.40	0.40	0.39	0.38	0.37	Other Africa
0.37	0.36	0.37	0.37	0.37	0.38	0.38	0.37	0.37	**Africa**
0.24	0.24	0.22	0.22	0.23	0.22	0.23	0.23	0.22	Argentina
0.19	0.20	0.20	0.20	0.20	0.20	0.20	0.19	0.22	Bolivia
0.18	0.19	0.19	0.19	0.19	0.19	0.18	0.19	0.19	Brazil
0.13	0.13	0.12	0.12	0.12	0.12	0.11	0.12	0.12	Chile
0.17	0.18	0.18	0.17	0.16	0.17	0.16	0.16	0.16	Colombia
0.14	0.14	0.15	0.15	0.13	0.13	0.15	0.15	0.15	Costa Rica
0.51	0.52	0.50	0.49	0.55	0.58	0.58	0.56	0.59	Cuba
0.18	0.19	0.19	0.21	0.20	0.20	0.20	0.19	0.18	Dominican Republic
0.15	0.17	0.16	0.16	0.16	0.16	0.16	0.18	0.17	Ecuador
0.21	0.21	0.23	0.22	0.22	0.21	0.23	0.23	0.22	El Salvador
0.17	0.18	0.18	0.18	0.17	0.17	0.17	0.17	0.17	Guatemala
0.32	0.32	0.30	0.36	0.36	0.35	0.40	0.45	0.40	Haiti
0.59	0.60	0.58	0.58	0.53	0.57	0.60	0.59	0.61	Honduras
0.28	0.33	0.31	0.31	0.32	0.33	0.36	0.39	0.42	Jamaica
0.90	0.93	1.18	0.70	1.14	1.16	1.15	1.09	1.05	Netherlands Antilles
0.49	0.51	0.52	0.53	0.52	0.52	0.50	0.42	0.42	Nicaragua
0.17	0.17	0.17	0.17	0.17	0.18	0.17	0.18	0.18	Panama
0.26	0.24	0.24	0.24	0.24	0.25	0.26	0.26	0.26	Paraguay
0.15	0.16	0.15	0.15	0.15	0.14	0.14	0.14	0.14	Peru
0.78	0.87	0.79	0.99	0.96	1.04	1.00	1.09	1.10	Trinidad-and-Tobago
0.12	0.11	0.11	0.12	0.11	0.10	0.10	0.11	0.11	Uruguay
0.31	0.29	0.32	0.28	0.27	0.26	0.29	0.34	0.33	Venezuela
0.23	0.23	0.22	0.21	0.21	0.21	0.21	0.21	0.20	Other Latin America
0.21	0.21	0.21	0.20	0.20	0.20	0.20	0.20	0.20	**Latin America**

TPES/GDP (toe per thousand 90 US$ PPP)
ATEP/PIB (tep par milliers de $US 90 PPA)
TPES/BIP (in toe pro tausend 1990er US$ Kaufkraftparität)
DTEP/PIL (tep per migliaia di $US 90 PPA)
一次エネルギー供給／GDP（石油換算トン／千米ドル、1990年価格、購買力平価）
TPES/PIB (tep por miles de 90 US$ PPP)
ОПΠΤЭ / ВВП (тнэ на тыс.долл.США в ценах и по ППС 1990 г.)

	1971	1973	1978	1983	1984	1985	1986	1987	1988
Bangladesh	0.20	0.25	0.23	0.21	0.21	0.21	0.21	0.21	0.21
Brunei	0.08	0.08	0.08	0.72	0.55	0.58	0.53	0.73	0.80
Inde	0.47	0.48	0.45	0.44	0.44	0.44	0.44	0.43	0.41
Indonésie	0.29	0.27	0.25	0.22	0.21	0.22	0.23	0.22	0.22
Malaisie	0.20	0.17	0.18	0.18	0.18	0.22	0.23	0.22	0.22
Myanmar	0.16	0.16	0.14	0.11	0.11	0.11	0.12	0.12	0.14
Népal	0.32	0.34	0.31	0.46	0.44	0.42	0.41	0.41	0.39
Pakistan	0.25	0.25	0.24	0.23	0.22	0.22	0.22	0.22	0.22
Philippines	0.21	0.21	0.19	0.19	0.19	0.21	0.21	0.21	0.21
Singapour	0.24	0.26	0.28	0.21	0.22	0.22	0.23	0.23	0.23
Sri Lanka	0.22	0.23	0.19	0.16	0.16	0.15	0.15	0.15	0.15
Taipei chinois	0.21	0.22	0.24	0.22	0.22	0.21	0.21	0.19	0.20
Thaïlande	0.22	0.22	0.20	0.16	0.16	0.17	0.18	0.17	0.17
Viêt-Nam	-	-	-	-	0.42	0.41	0.41	0.42	0.40
Autre Asie	0.14	0.14	0.12	0.11	0.11	0.12	0.11	0.12	0.12
Asie	0.33	0.32	0.30	0.28	0.28	0.28	0.28	0.28	0.27
Rép. populaire de Chine	0.74	0.73	0.81	0.60	0.55	0.52	0.50	0.47	0.44
Hong-Kong, Chine	0.15	0.12	0.13	0.11	0.11	0.11	0.11	0.10	0.11
Chine	0.72	0.70	0.77	0.57	0.53	0.49	0.48	0.45	0.42
Albanie	-	-	-	0.36	0.38	0.35	0.34	0.34	0.36
Bulgarie	0.57	0.56	0.64	0.63	0.62	0.61	0.59	0.55	0.51
Chypre	0.21	0.26	0.23	0.21	0.21	0.20	0.21	0.23	0.21
Gibraltar	0.20	0.17	0.12	0.15	0.14	0.16	0.17	0.18	0.21
Malte	0.32	0.32	0.19	0.15	0.21	0.16	0.25	0.26	0.25
Roumanie	0.91	0.94	0.86	0.74	0.69	0.69	0.70	0.76	0.77
République slovaque	0.55	0.56	0.60	0.61	0.61	0.61	0.60	0.59	0.58
Ex-Yougoslavie	0.39	0.39	0.35	0.41	0.43	0.44	0.44	0.45	0.47
Europe non-OCDE	0.59	0.60	0.59	0.57	0.56	0.57	0.57	0.58	0.58
Arménie	-	-	-	-	-	-	-	-	-
Azerbaïdjan	-	-	-	-	-	-	-	-	-
Bélarus	-	-	-	-	-	-	-	-	-
Estonie	-	-	-	-	-	-	-	-	-
Géorgie	-	-	-	-	-	-	-	-	-
Kazakhstan	-	-	-	-	-	-	-	-	-
Kirghizistan	-	-	-	-	-	-	-	-	-
Lettonie	-	-	-	-	-	-	-	-	-
Lituanie	-	-	-	-	-	-	-	-	-
Moldova	-	-	-	-	-	-	-	-	-
Russie	-	-	-	-	-	-	-	-	-
Tadjikistan	-	-	-	-	-	-	-	-	-
Turkménistan	-	-	-	-	-	-	-	-	-
Ukraine	-	-	-	-	-	-	-	-	-
Ouzbékistan	-	-	-	-	-	-	-	-	-
Ex-URSS	0.67	0.67	0.70	0.71	0.72	0.74	0.72	0.74	0.75

Faute de séries complètes, l'Albanie, la Rép.pop.dém. de Corée et le Viêt-Nam ne sont pas pris en compte dans les totaux régionaux.

TPES/GDP (toe per thousand 90 US$ PPP)
ATEP/PIB (tep par milliers de $US 90 PPA)
TPES/BIP (in toe pro tausend 1990er US$ Kaufkraftparität)

DTEP/PIL (tep per migliaia di $US 90 PPA)

一次エネルギー供給／GDP（石油換算トン／千米ドル、1990年価格、購買力平価）

TPES/PIB (tep por miles de 90 US$ PPP)

ОП//ТЭ / ВВП (тнэ на тыс.долл.США в ценах и по ППС 1990 г.)

1989	1990	1991	1992	1993	1994	1995	1996	1997	
0.22	0.21	0.20	0.20	0.20	0.19	0.19	0.19	0.18	Bangladesh
0.52	0.58	0.64	0.71	0.67	0.66	0.75	0.74	0.75	Brunei
0.40	0.39	0.41	0.40	0.39	0.38	0.37	0.36	0.35	India
0.21	0.21	0.21	0.20	0.20	0.19	0.18	0.18	0.18	Indonesia
0.23	0.22	0.24	0.24	0.25	0.24	0.21	0.23	0.25	Malaysia
0.13	0.13	0.13	0.12	0.12	0.11	0.11	0.11	0.11	Myanmar
0.38	0.37	0.37	0.36	0.36	0.35	0.34	0.34	0.32	Nepal
0.22	0.22	0.21	0.21	0.21	0.22	0.21	0.21	0.22	Pakistan
0.21	0.21	0.21	0.22	0.23	0.23	0.23	0.22	0.23	Philippines
0.21	0.26	0.26	0.28	0.31	0.34	0.27	0.28	0.30	Singapore
0.15	0.14	0.14	0.14	0.13	0.12	0.12	0.13	0.13	Sri Lanka
0.20	0.20	0.20	0.20	0.20	0.19	0.19	0.19	0.19	Chinese Taipei
0.17	0.17	0.17	0.17	0.18	0.18	0.19	0.20	0.20	Thailand
0.37	0.37	0.35	0.34	0.33	0.31	0.30	0.34	0.34	Vietnam
0.13	0.13	0.13	0.12	0.12	0.11	0.11	0.10	0.10	Other Asia
0.27	0.27	0.27	0.26	0.26	0.26	0.25	0.25	0.25	**Asia**
0.44	0.43	0.40	0.36	0.34	0.31	0.30	0.28	0.26	People's Rep.of China
0.11	0.10	0.10	0.11	0.12	0.11	0.10	0.10	0.10	Hong Kong, China
0.42	0.41	0.39	0.35	0.33	0.31	0.29	0.28	0.26	**China**
0.31	0.32	0.31	0.26	0.20	0.17	0.15	0.16	0.15	Albania
0.51	0.49	0.44	0.44	0.48	0.45	0.48	0.52	0.51	Bulgaria
0.21	0.20	0.23	0.22	0.22	0.24	0.21	0.22	0.21	Cyprus
0.19	0.22	0.24	0.27	0.30	0.40	0.39	0.38	0.37	Gibraltar
0.25	0.23	0.21	0.20	0.24	0.21	0.19	0.20	0.21	Malta
0.81	0.76	0.72	0.73	0.70	0.63	0.64	0.66	0.63	Romania
0.57	0.56	0.60	0.60	0.59	0.55	0.53	0.51	0.47	Slovak Republic
0.46	0.50	0.45	0.45	0.45	0.40	0.41	0.43	0.44	Former Yugoslavia
0.58	0.57	0.53	0.53	0.53	0.49	0.49	0.51	0.50	**Non-OECD Europe**
-	-	-	1.35	0.83	0.49	0.54	0.55	0.54	Armenia
-	-	-	1.25	1.50	1.86	1.72	1.60	1.52	Azerbaijan
-	-	-	0.66	0.56	0.54	0.56	0.56	0.51	Belarus
-	-	-	0.68	0.60	0.64	0.57	0.60	0.53	Estonia
-	-	-	0.61	0.75	0.54	0.26	0.34	0.33	Georgia
-	-	-	1.11	0.97	1.03	1.06	0.86	0.72	Kazakhstan
-	-	-	1.21	1.11	1.07	1.00	1.07	0.90	Kyrgyzstan
-	-	-	0.54	0.50	0.46	0.40	0.41	0.41	Latvia
-	-	-	0.74	0.71	0.69	0.74	0.72	0.67	Lithuania
-	-	-	0.56	0.48	0.58	0.55	0.64	0.60	Moldova
-	-	-	0.85	0.87	0.85	0.87	0.88	0.85	Russia
-	-	-	1.33	1.05	0.69	0.79	0.88	0.84	Tajikistan
-	-	-	0.75	0.83	1.36	1.56	1.34	1.06	Turkmenistan
-	-	-	0.83	0.87	0.95	1.09	1.18	1.13	Ukraine
-	-	-	1.06	1.13	1.12	1.05	1.06	1.01	Uzbekistan
0.72	0.71	0.75	0.82	0.83	0.84	0.87	0.88	0.83	**Former USSR**

Due to lack of complete series Albania, the Dem. Ppl's Rep. of Korea and Vietnam are not included in the regional aggregates.

TPES/GDP (toe per thousand 90 US$ PPP)
ATEP/PIB (tep par milliers de $US 90 PPA)
TPES/BIP (in toe pro tausend 1990er US$ Kaufkraftparität)
DTEP/PIL (tep per migliaia di $US 90 PPA)
一次エネルギー供給／GDP（石油換算トン／千米ドル、1990年価格、購買力平価）
TPES/PIB (tep por miles de 90 US$ PPP)
ОППТЭ / ВВП (тнэ на тыс.долл.США в ценах и по ППС 1990 г.)

	1971	1973	1978	1983	1984	1985	1986	1987	1988
Bahrein	0.74	0.89	0.76	0.98	1.02	1.22	1.38	1.43	1.29
Iran	0.72	0.65	0.67	0.74	0.87	0.87	1.01	1.07	1.21
Irak	0.08	0.08	0.09	0.19	0.20	0.25	0.31	0.35	0.43
Israël	0.27	0.31	0.23	0.20	0.20	0.19	0.21	0.22	0.24
Jordanie	0.10	0.12	0.15	0.19	0.19	0.19	0.18	0.19	0.19
Koweit	0.20	0.18	0.27	0.52	0.50	0.64	0.57	0.53	0.55
Liban	0.10	0.11	0.17	0.19	0.19	0.22	0.21	0.22	0.15
Oman	0.02	0.02	0.08	0.11	0.12	0.13	0.09	0.10	0.11
Qatar	0.28	0.32	0.31	0.67	0.77	0.82	0.75	0.74	0.77
Arabie saoudite	0.10	0.09	0.17	0.37	0.38	0.41	0.43	0.46	0.47
Syrie	0.14	0.11	0.13	0.16	0.19	0.17	0.18	0.19	0.18
Emirats arabes unis	0.17	0.16	0.38	0.61	0.61	0.77	1.04	1.04	1.08
Yémen	0.06	0.07	0.04	0.05	0.05	0.06	0.06	0.06	0.05
Moyen-Orient	0.19	0.19	0.22	0.33	0.35	0.38	0.42	0.44	0.45
Total non-OCDE	0.44	0.44	0.44	0.43	0.43	0.43	0.42	0.42	0.41
OCDE Amérique du N.	0.46	0.45	0.43	0.37	0.36	0.36	0.35	0.35	0.35
OCDE Pacifique	0.26	0.27	0.25	0.22	0.22	0.21	0.21	0.21	0.21
OCDE Europe	0.30	0.31	0.29	0.26	0.26	0.26	0.26	0.26	0.25
Total OCDE	0.36	0.36	0.34	0.30	0.30	0.29	0.29	0.29	0.29
Monde	0.39	0.39	0.38	0.35	0.35	0.34	0.34	0.34	0.34

Faute de séries complètes, l'Albanie, la Rép.pop.dém. de Corée et le Viêt-Nam ne sont pas pris en compte dans les totaux régionaux.
Le total Monde ne comprend pas les soutages maritimes internationaux.

TPES/GDP (toe per thousand 90 US$ PPP)
ATEP/PIB (tep par milliers de $US 90 PPA)
TPES/BIP (in toe pro tausend 1990er US$ Kaufkraftparität)
DTEP/PIL (tep per migliaia di $US 90 PPA)
一次エネルギー供給／GDP（石油換算トン／千米ドル、1990年価格、購買力平価）
TPES/PIB (tep por miles de 90 US$ PPP)
ОПЛТЭ / ВВП (тнэ на тыс.долл.США в ценах и по ППС 1990 г.)

1989	1990	1991	1992	1993	1994	1995	1996	1997	
1.24	1.26	1.14	1.08	1.06	1.04	1.05	1.33	1.42	Bahrain
1.28	1.21	1.22	1.26	1.25	1.35	1.31	1.30	1.33	Iran
0.44	0.54	0.81	0.99	1.07	1.25	1.33	1.32	1.15	Iraq
0.23	0.23	0.21	0.23	0.23	0.22	0.23	0.21	0.23	Israel
0.23	0.24	0.24	0.23	0.22	0.22	0.21	0.22	0.23	Jordan
0.60	0.48	0.26	0.42	0.32	0.48	0.52	0.45	0.49	Kuwait
0.22	0.21	0.19	0.18	0.22	0.22	0.24	0.24	0.25	Lebanon
0.11	0.19	0.24	0.23	0.23	0.21	0.21	0.20	0.20	Oman
0.81	0.72	0.85	1.26	1.31	1.29	1.34	1.36	1.40	Qatar
0.44	0.41	0.43	0.44	0.47	0.46	0.47	0.51	0.54	Saudi Arabia
0.20	0.21	0.21	0.19	0.18	0.17	0.17	0.17	0.17	Syria
1.12	0.90	0.95	0.92	0.95	1.01	0.99	1.13	1.05	United Arab Emirates
0.05	0.05	0.06	0.06	0.05	0.05	0.05	0.05	0.05	Yemen
0.47	0.45	0.47	0.49	0.49	0.50	0.50	0.51	0.52	**Middle East**
0.41	0.40	0.40	0.38	0.37	0.35	0.34	0.33	0.32	**Non-OECD Total**
0.35	0.34	0.34	0.34	0.34	0.33	0.33	0.33	0.32	OECD North America
0.21	0.21	0.21	0.21	0.22	0.22	0.22	0.22	0.23	OECD Pacific
0.25	0.24	0.24	0.24	0.24	0.23	0.23	0.23	0.23	OECD Europe
0.28	0.27	0.28	0.27	0.28	0.27	0.27	0.27	0.27	**OECD Total**
0.33	0.32	0.32	0.32	0.31	0.30	0.30	0.30	0.29	**World**

Due to lack of complete series Albania, the Dem. Ppl's Rep. of Korea and Vietnam are not included in the regional aggregates.
World Total excludes international marine bunkers.

TPES/Population (toe per capita)
ATEP/Population (tep par habitant)
TPES/Bevölkerung (toe pro Kopf)
DTEP/Popolazione (tep per abitante)
一人当たり一次エネルギー供給（石油換算トン／人）
TPES/ población (tep per capita)
ОПТЭ / Численность населения (тнэ на человека)

	1971	1973	1978	1983	1984	1985	1986	1987	1988
Algérie	0.28	0.36	0.55	0.92	0.88	0.89	0.97	0.93	0.96
Angola	0.69	0.72	0.67	0.62	0.60	0.62	0.60	0.59	0.59
Bénin	0.41	0.41	0.40	0.38	0.38	0.39	0.38	0.37	0.37
Cameroun	0.41	0.40	0.42	0.45	0.46	0.45	0.45	0.44	0.44
Congo	0.55	0.51	0.50	0.55	0.53	0.52	0.47	0.69	0.86
RD du Congo	0.33	0.32	0.32	0.32	0.32	0.32	0.31	0.32	0.32
Egypte	0.23	0.23	0.33	0.49	0.52	0.55	0.55	0.58	0.58
Ethiopie	0.30	0.30	0.30	0.30	0.29	0.29	0.29	0.30	0.30
Gabon	1.77	2.63	2.55	2.33	2.12	1.85	1.83	1.70	1.76
Ghana	0.34	0.36	0.38	0.31	0.32	0.34	0.34	0.35	0.35
Côte d'Ivoire	0.44	0.44	0.47	0.39	0.38	0.39	0.40	0.38	0.41
Kenya	0.65	0.64	0.61	0.55	0.55	0.55	0.56	0.56	0.55
Libye	0.80	1.21	1.75	2.58	2.62	2.77	2.47	2.79	2.59
Maroc	0.15	0.18	0.24	0.25	0.25	0.25	0.25	0.25	0.26
Mozambique	0.86	0.82	0.69	0.60	0.58	0.57	0.56	0.55	0.54
Nigéria	0.66	0.68	0.73	0.76	0.74	0.74	0.73	0.74	0.74
Sénégal	0.31	0.32	0.33	0.32	0.32	0.31	0.32	0.32	0.31
Afrique du Sud	2.01	2.07	2.27	2.69	2.84	2.77	2.82	2.84	2.89
Soudan	0.49	0.49	0.45	0.44	0.44	0.44	0.44	0.42	0.44
Tanzanie	0.69	0.66	0.58	0.53	0.52	0.52	0.51	0.51	0.50
Tunisie	0.39	0.43	0.55	0.63	0.64	0.64	0.62	0.62	0.63
Zambie	0.91	0.94	0.85	0.80	0.78	0.77	0.75	0.75	0.74
Zimbabwe	1.00	1.02	0.92	0.87	0.86	0.89	0.91	0.96	0.93
Autre Afrique	0.38	0.38	0.38	0.38	0.38	0.37	0.38	0.38	0.38
Afrique	0.55	0.56	0.59	0.65	0.65	0.65	0.65	0.66	0.66
Argentine	1.38	1.41	1.43	1.42	1.45	1.36	1.40	1.46	1.47
Bolivie	0.23	0.26	0.34	0.52	0.48	0.47	0.46	0.41	0.42
Brésil	0.72	0.80	0.88	0.83	0.86	0.89	0.92	0.94	0.97
Chili	0.85	0.85	0.81	0.78	0.81	0.80	0.82	0.84	0.93
Colombie	0.61	0.59	0.67	0.70	0.69	0.70	0.71	0.72	0.70
Costa Rica	0.64	0.67	0.72	0.56	0.62	0.63	0.62	0.64	0.63
Cuba	1.22	1.28	1.50	1.60	1.44	1.41	1.41	1.47	1.50
République dominicaine	0.52	0.60	0.62	0.67	0.70	0.58	0.59	0.65	0.60
Equateur	0.38	0.38	0.61	0.63	0.66	0.65	0.63	0.62	0.64
El Salvador	0.48	0.51	0.65	0.58	0.59	0.60	0.51	0.56	0.53
Guatemala	0.52	0.53	0.61	0.48	0.50	0.49	0.46	0.47	0.48
Haiti	0.33	0.34	0.38	0.33	0.33	0.32	0.25	0.25	0.26
Honduras	0.52	0.52	0.53	0.51	0.50	0.49	0.47	0.50	0.51
Jamaïque	1.13	1.57	1.23	0.95	0.93	0.80	0.80	0.82	0.84
Antilles néerlandaises	34.26	37.01	21.96	24.66	17.02	10.04	9.70	8.38	8.26
Nicaragua	0.59	0.59	0.63	0.51	0.57	0.58	0.59	0.57	0.54
Panama	1.12	1.31	0.93	0.84	0.74	0.70	0.70	0.71	0.61
Paraguay	0.57	0.61	0.65	0.65	0.64	0.65	0.66	0.73	0.74
Pérou	0.70	0.69	0.67	0.59	0.60	0.59	0.61	0.62	0.60
Trinité-et-Tobago	2.75	2.69	3.37	3.75	3.92	4.02	4.23	3.84	4.24
Uruguay	0.86	0.84	0.88	0.71	0.68	0.67	0.69	0.73	0.78
Vénézuela	2.11	2.14	2.13	2.35	2.20	2.18	2.21	2.43	2.20
Autre Amérique latine	1.77	2.02	1.95	1.28	1.35	1.29	1.34	1.43	1.42
Amérique latine	0.87	0.92	0.97	0.94	0.95	0.94	0.96	0.98	0.99

TPES/Population (toe per capita)
ATEP/Population (tep par habitant)
TPES/Bevölkerung (toe pro Kopf)

DTEP/Popolazione (tep per abitante)

一人当たり一次エネルギー供給（石油換算トン／人）

TPES/ población (tep per capita)

ОПТЭ / Численность населения (тнэ на человека)

1989	1990	1991	1992	1993	1994	1995	1996	1997	
0.93	0.96	0.97	0.95	0.93	0.88	0.87	0.84	0.90	Algeria
0.58	0.61	0.60	0.58	0.57	0.56	0.56	0.58	0.59	Angola
0.36	0.35	0.35	0.35	0.35	0.34	0.34	0.38	0.38	Benin
0.45	0.44	0.43	0.42	0.42	0.42	0.42	0.41	0.41	Cameroon
0.55	0.50	0.51	0.50	0.48	0.49	0.47	0.46	0.46	Congo
0.32	0.32	0.32	0.31	0.31	0.31	0.31	0.31	0.31	DR of Congo
0.60	0.61	0.61	0.61	0.64	0.59	0.61	0.63	0.66	Egypt
0.30	0.30	0.30	0.29	0.29	0.29	0.30	0.29	0.29	Ethiopia
1.49	1.32	1.35	1.39	1.42	1.37	1.36	1.36	1.42	Gabon
0.36	0.35	0.35	0.35	0.36	0.39	0.39	0.39	0.38	Ghana
0.42	0.40	0.39	0.41	0.38	0.38	0.38	0.39	0.39	Ivory Coast
0.54	0.53	0.52	0.51	0.51	0.50	0.48	0.48	0.49	Kenya
2.59	2.66	3.05	2.79	2.90	2.74	3.20	2.94	2.90	Libya
0.28	0.28	0.29	0.30	0.31	0.33	0.31	0.34	0.34	Morocco
0.53	0.52	0.51	0.50	0.50	0.49	0.47	0.46	0.46	Mozambique
0.74	0.74	0.75	0.75	0.69	0.78	0.75	0.74	0.75	Nigeria
0.30	0.30	0.30	0.31	0.30	0.30	0.30	0.31	0.32	Senegal
2.70	2.59	2.65	2.46	2.53	2.61	2.66	2.60	2.64	South Africa
0.43	0.44	0.43	0.42	0.40	0.42	0.41	0.41	0.41	Sudan
0.50	0.49	0.49	0.48	0.47	0.47	0.46	0.46	0.46	Tanzania
0.67	0.70	0.67	0.70	0.73	0.72	0.72	0.74	0.74	Tunisia
0.69	0.67	0.67	0.67	0.66	0.65	0.64	0.64	0.63	Zambia
0.92	0.92	0.96	0.96	0.91	0.89	0.89	0.89	0.87	Zimbabwe
0.37	0.37	0.37	0.37	0.37	0.37	0.38	0.38	0.37	Other Africa
0.65	0.64	0.65	0.63	0.63	0.64	0.64	0.64	0.64	**Africa**
1.38	1.33	1.38	1.49	1.56	1.60	1.61	1.67	1.73	Argentina
0.42	0.44	0.46	0.45	0.45	0.48	0.49	0.48	0.55	Bolivia
0.97	0.92	0.92	0.92	0.94	0.98	0.98	1.01	1.05	Brazil
1.02	1.06	1.07	1.16	1.17	1.24	1.31	1.42	1.57	Chile
0.72	0.77	0.75	0.74	0.74	0.78	0.80	0.80	0.76	Colombia
0.66	0.66	0.68	0.74	0.67	0.68	0.78	0.76	0.77	Costa Rica
1.58	1.55	1.31	1.13	1.09	1.14	1.16	1.20	1.29	Cuba
0.57	0.56	0.56	0.66	0.64	0.65	0.65	0.65	0.67	Dominican Republic
0.59	0.64	0.64	0.65	0.62	0.66	0.67	0.73	0.71	Ecuador
0.53	0.53	0.60	0.61	0.63	0.64	0.73	0.70	0.69	El Salvador
0.49	0.50	0.50	0.51	0.50	0.51	0.52	0.53	0.54	Guatemala
0.25	0.24	0.24	0.24	0.23	0.21	0.24	0.27	0.24	Haiti
0.51	0.50	0.49	0.50	0.47	0.48	0.51	0.51	0.53	Honduras
1.03	1.26	1.19	1.19	1.24	1.30	1.38	1.47	1.55	Jamaica
10.07	10.84	13.58	8.24	13.17	13.57	13.22	12.79	12.67	Netherlands Antilles
0.55	0.57	0.56	0.56	0.53	0.53	0.52	0.53	0.55	Nicaragua
0.59	0.64	0.67	0.72	0.73	0.80	0.78	0.82	0.86	Panama
0.79	0.73	0.73	0.72	0.72	0.77	0.82	0.81	0.82	Paraguay
0.56	0.54	0.53	0.50	0.53	0.55	0.59	0.59	0.62	Peru
4.06	4.60	4.73	5.20	4.94	5.46	5.40	6.08	6.27	Trinidad-and-Tobago
0.79	0.72	0.77	0.85	0.79	0.75	0.80	0.87	0.88	Uruguay
2.14	2.09	2.49	2.26	2.07	1.92	2.19	2.46	2.53	Venezuela
1.48	1.53	1.53	1.47	1.47	1.47	1.48	1.48	1.49	Other Latin America
0.98	0.96	0.98	0.98	0.98	1.01	1.04	1.08	1.11	**Latin America**

TPES/Population (toe per capita)
ATEP/Population (tep par habitant)
TPES/Bevölkerung (toe pro Kopf)
DTEP/Popolazione (tep per abitante)
一人当たり一次エネルギー供給（石油換算トン／人）
TPES/ población (tep per capita)
ОПТТЭ / Численность населения (тнэ на человека)

	1971	1973	1978	1983	1984	1985	1986	1987	1988
Bangladesh	0.16	0.16	0.17	0.17	0.17	0.18	0.18	0.18	0.19
Brunei	0.90	0.99	1.16	8.78	6.46	6.52	5.69	7.79	8.31
Inde	0.33	0.33	0.35	0.37	0.37	0.38	0.39	0.40	0.41
Indonésie	0.30	0.32	0.38	0.41	0.42	0.45	0.48	0.48	0.49
RPD de Corée	1.80	1.56	1.75	1.92	1.91	1.96	1.97	1.95	1.93
Malaisie	0.56	0.54	0.71	0.90	0.92	1.09	1.14	1.12	1.15
Myanmar	0.29	0.27	0.28	0.27	0.28	0.28	0.29	0.27	0.27
Népal	0.22	0.22	0.22	0.32	0.32	0.32	0.32	0.32	0.32
Pakistan	0.28	0.28	0.29	0.34	0.34	0.35	0.35	0.38	0.38
Philippines	0.38	0.40	0.42	0.46	0.43	0.42	0.41	0.43	0.44
Singapour	1.56	2.08	2.79	2.89	3.18	3.15	3.31	3.48	3.77
Sri Lanka	0.30	0.31	0.30	0.32	0.32	0.31	0.32	0.33	0.33
Taipei chinois	0.71	0.91	1.40	1.61	1.71	1.72	1.89	1.94	2.11
Thaïlande	0.38	0.42	0.48	0.44	0.48	0.52	0.56	0.59	0.63
Viêt-Nam	0.45	0.44	0.35	0.37	0.36	0.36	0.37	0.38	0.37
Autre Asie	0.23	0.23	0.22	0.25	0.25	0.27	0.26	0.26	0.26
Asie	0.32	0.33	0.36	0.39	0.39	0.41	0.42	0.43	0.44
Rép. populaire de Chine	0.46	0.48	0.61	0.62	0.65	0.67	0.69	0.72	0.74
Hong-Kong, Chine	0.87	0.89	1.24	1.34	1.38	1.40	1.52	1.60	1.86
Chine	0.47	0.48	0.62	0.62	0.65	0.68	0.70	0.72	0.75
Albanie	0.78	0.75	1.00	1.00	1.03	0.94	0.94	0.92	0.94
Bulgarie	2.25	2.40	3.12	3.41	3.43	3.46	3.49	3.45	3.53
Chypre	1.05	1.41	1.35	1.60	1.67	1.66	1.84	2.07	2.09
Gibraltar	1.36	1.31	1.07	1.38	1.30	1.48	1.69	1.78	2.13
Malte	0.85	1.01	1.02	1.02	1.43	1.15	1.88	1.98	2.10
Roumanie	2.04	2.28	2.92	2.88	2.85	2.84	2.91	3.02	3.02
République slovaque	3.09	3.32	3.83	4.01	4.04	4.17	4.24	4.26	4.25
Bosnie-Herzégovine	-	-	-	-	-	-	-	-	-
Croatie	-	-	-	-	-	-	-	-	-
Ex-RYM	-	-	-	-	-	-	-	-	-
Slovénie	-	-	-	-	-	-	-	-	-
RF de Yougoslavie	-	-	-	-	-	-	-	-	-
Ex-Yougoslavie	1.08	1.12	1.39	1.68	1.79	1.81	1.89	1.91	1.93
Europe non-OCDE	1.78	1.93	2.42	2.58	2.61	2.63	2.70	2.75	2.77

Faute de séries complètes, l'Albanie, la Rép.pop.dém. de Corée et le Viêt-Nam ne sont pas pris en compte dans les totaux régionaux.

TPES/Population (toe per capita)
ATEP/Population (tep par habitant)
TPES/Bevölkerung (toe pro Kopf)
DTEP/Popolazione (tep per abitante)
一人当たり一次エネルギー供給（石油換算トン／人）
TPES/ población (tep per capita)
ОПППТЭ / Численность населения (тнэ на человека)

1989	1990	1991	1992	1993	1994	1995	1996	1997	
0.19	0.19	0.18	0.19	0.19	0.20	0.20	0.20	0.20	Bangladesh
5.17	5.74	6.38	6.84	6.04	5.89	6.72	6.61	6.84	Brunei
0.42	0.42	0.43	0.44	0.44	0.45	0.47	0.48	0.48	India
0.52	0.55	0.58	0.59	0.62	0.61	0.64	0.67	0.69	Indonesia
1.94	1.82	1.67	1.51	1.38	1.25	1.15	1.06	1.02	DPR of Korea
1.27	1.32	1.51	1.58	1.75	1.78	1.69	1.95	2.24	Malaysia
0.27	0.27	0.27	0.27	0.28	0.28	0.29	0.29	0.30	Myanmar
0.31	0.31	0.32	0.32	0.32	0.33	0.33	0.33	0.32	Nepal
0.40	0.40	0.40	0.41	0.42	0.43	0.43	0.45	0.44	Pakistan
0.45	0.45	0.44	0.46	0.47	0.48	0.50	0.50	0.52	Philippines
3.72	4.94	5.24	5.81	7.02	8.44	7.18	7.84	8.66	Singapore
0.32	0.32	0.32	0.33	0.34	0.33	0.34	0.37	0.39	Sri Lanka
2.28	2.37	2.55	2.64	2.78	2.90	3.06	3.20	3.35	Chinese Taipei
0.70	0.79	0.83	0.89	0.98	1.08	1.19	1.31	1.32	Thailand
0.36	0.37	0.36	0.37	0.38	0.39	0.39	0.48	0.51	Vietnam
0.25	0.26	0.25	0.22	0.20	0.19	0.19	0.18	0.18	Other Asia
0.46	0.47	0.49	0.50	0.51	0.53	0.54	0.56	0.57	**Asia**
0.75	0.75	0.77	0.77	0.81	0.84	0.88	0.90	0.90	People's Rep.of China
1.93	1.83	1.91	2.19	2.38	2.23	2.24	2.17	2.17	Hong Kong, China
0.76	0.76	0.77	0.78	0.81	0.85	0.89	0.91	0.90	**China**
0.88	0.78	0.56	0.44	0.38	0.35	0.34	0.37	0.32	Albania
3.49	3.11	2.57	2.40	2.58	2.49	2.73	2.70	2.48	Bulgaria
2.21	2.26	2.49	2.62	2.63	2.98	2.70	2.87	2.78	Cyprus
1.98	2.30	2.59	3.23	3.66	4.84	4.60	4.60	4.60	Gibraltar
2.21	2.19	2.13	2.09	2.53	2.29	2.26	2.39	2.52	Malta
2.99	2.63	2.17	2.03	1.98	1.86	2.01	2.17	1.96	Romania
4.20	4.04	3.68	3.39	3.21	3.13	3.25	3.30	3.20	Slovak Republic
-	-	-	1.17	1.41	0.57	0.68	0.78	0.75	*Bosnia-Herzegovina*
-	-	-	1.40	1.43	1.45	1.49	1.52	1.60	*Croatia*
-	-	-	1.63	1.66	1.36	1.41	1.58	1.42	*FYROM*
-	-	-	2.48	2.66	2.76	2.95	3.10	3.21	*Slovenia*
-	-	-	1.35	1.16	1.09	1.15	1.36	1.49	*FR of Yugoslavia*
1.89	1.91	1.74	1.42	1.37	1.22	1.29	1.42	1.49	Former Yugoslavia
2.74	2.54	2.19	1.97	1.95	1.83	1.96	2.07	1.98	**Non-OECD Europe**

Due to lack of complete series Albania, the Dem. Ppl's Rep. of Korea and Vietnam are not included in the regional aggregates.

TPES/Population (toe per capita)
ATEP/Population (tep par habitant)
TPES/Bevölkerung (toe pro Kopf)
DTEP/Popolazione (tep per abitante)
一人当たり一次エネルギー供給 (石油換算トン／人)
TPES/ población (tep per capita)
ОПППТЭ / Численность населения (тнэ на человека)

	1971	1973	1978	1983	1984	1985	1986	1987	1988
Arménie	-	-	-	-	-	-	-	-	-
Azerbaïdjan	-	-	-	-	-	-	-	-	-
Bélarus	-	-	-	-	-	-	-	-	-
Estonie	-	-	-	-	-	-	-	-	-
Géorgie	-	-	-	-	-	-	-	-	-
Kazakhstan	-	-	-	-	-	-	-	-	-
Kirghizistan	-	-	-	-	-	-	-	-	-
Lettonie	-	-	-	-	-	-	-	-	-
Lituanie	-	-	-	-	-	-	-	-	-
Moldova	-	-	-	-	-	-	-	-	-
Russie	-	-	-	-	-	-	-	-	-
Tadjikistan	-	-	-	-	-	-	-	-	-
Turkménistan	-	-	-	-	-	-	-	-	-
Ukraine	-	-	-	-	-	-	-	-	-
Ouzbékistan	-	-	-	-	-	-	-	-	-
Ex-URSS	3.22	3.48	4.10	4.37	4.47	4.58	4.62	4.77	4.87
Bahrein	7.19	9.46	9.78	9.65	10.03	10.97	12.15	12.11	11.72
Iran	0.76	0.86	1.10	1.01	1.16	1.15	1.16	1.18	1.22
Irak	0.47	0.51	0.75	1.10	1.13	1.18	1.16	1.24	1.34
Israël	2.07	2.62	2.06	1.95	1.98	1.94	2.20	2.37	2.65
Jordanie	0.34	0.39	0.66	1.05	1.10	1.07	1.08	1.11	1.06
Koweit	7.53	6.42	6.61	6.74	6.44	7.77	7.06	6.70	6.20
Liban	0.74	0.91	0.76	0.80	0.76	0.85	0.87	0.87	0.65
Oman	0.12	0.13	0.68	1.29	1.60	1.77	1.27	1.25	1.53
Qatar	7.61	9.91	8.53	14.84	16.40	14.34	13.91	13.82	14.20
Arabie saoudite	1.07	1.25	2.49	4.62	4.26	4.22	4.45	4.48	4.66
Syrie	0.42	0.36	0.62	0.86	0.95	0.86	0.85	0.91	0.93
Emirats arabes unis	4.36	4.23	6.93	10.24	10.29	11.87	12.51	12.70	12.16
Yémen	0.12	0.15	0.14	0.20	0.20	0.23	0.23	0.24	0.24
Moyen-Orient	0.82	0.93	1.25	1.64	1.69	1.73	1.77	1.81	1.87
Total non-OCDE	0.74	0.77	0.88	0.92	0.93	0.95	0.97	0.99	1.01
OCDE Amérique du N.	6.35	6.77	6.94	5.95	6.14	6.15	6.09	6.27	6.43
OCDE Pacifique	2.25	2.58	2.66	2.63	2.79	2.80	2.85	2.91	3.08
OCDE Europe	2.81	3.07	3.18	3.01	3.09	3.19	3.23	3.29	3.29
Total OCDE	3.85	4.17	4.32	3.93	4.07	4.13	4.14	4.24	4.33
Monde	1.48	1.57	1.65	1.57	1.60	1.62	1.63	1.66	1.69

Faute de séries complètes, l'Albanie, la Rép.pop.dém. de Corée et le Viêt-Nam ne sont pas pris en compte dans les totaux régionaux.
Le total Monde ne comprend pas les soutages maritimes internationaux.

INTERNATIONAL ENERGY AGENCY

TPES/Population (toe per capita)
ATEP/Population (tep par habitant)
TPES/Bevölkerung (toe pro Kopf)

DTEP/Popolazione (tep per abitante)

一人当たり一次エネルギー供給 （石油換算トン／人）

TPES/ población (tep per capita)

ОПП ТЭ / Численность населения (тнэ на человека)

1989	1990	1991	1992	1993	1994	1995	1996	1997	
-	-	-	1.17	0.61	0.38	0.44	0.47	0.48	Armenia
-	-	-	2.28	2.09	2.17	1.73	1.62	1.58	Azerbaijan
-	-	-	3.85	3.01	2.56	2.39	2.44	2.45	Belarus
-	-	-	4.26	3.47	3.67	3.48	3.85	3.81	Estonia
-	-	-	1.14	0.85	0.55	0.27	0.39	0.42	Georgia
-	-	-	4.82	3.83	3.57	3.42	2.81	2.43	Kazakhstan
-	-	-	1.13	0.87	0.67	0.59	0.67	0.60	Kyrgyzstan
-	-	-	2.40	1.90	1.81	1.58	1.66	1.81	Latvia
-	-	-	2.99	2.41	2.11	2.36	2.41	2.38	Lithuania
-	-	-	1.58	1.35	1.10	1.02	1.07	1.03	Moldova
-	-	-	5.35	4.98	4.30	4.22	4.13	4.02	Russia
-	-	-	1.63	1.13	0.57	0.56	0.59	0.56	Tajikistan
-	-	-	2.59	2.43	3.17	3.23	2.65	2.62	Turkmenistan
-	-	-	4.19	3.74	3.18	3.21	3.16	2.96	Ukraine
-	-	-	2.10	2.15	1.98	1.80	1.83	1.80	Uzbekistan
4.80	4.66	4.62	4.29	3.89	3.38	3.29	3.21	3.09	**Former USSR**
10.97	10.98	10.33	10.32	10.55	10.27	10.17	12.85	13.69	Bahrain
1.30	1.33	1.45	1.57	1.58	1.70	1.66	1.71	1.78	Iran
1.40	1.15	0.83	1.02	1.08	1.18	1.21	1.17	1.24	Iraq
2.56	2.56	2.41	2.67	2.74	2.77	3.08	2.84	3.01	Israel
1.05	1.09	0.98	1.05	1.02	1.06	1.03	1.04	1.08	Jordan
7.81	6.18	2.62	6.11	6.35	9.42	9.89	8.35	8.94	Kuwait
0.53	0.63	0.77	0.76	0.94	1.01	1.15	1.16	1.26	Lebanon
1.54	2.67	3.38	3.23	3.18	3.04	2.94	2.93	3.00	Oman
14.83	11.83	13.84	20.71	20.35	19.07	17.96	17.49	18.83	Qatar
4.22	4.00	4.54	4.67	4.75	4.42	4.41	4.75	4.91	Saudi Arabia
0.91	0.98	1.04	1.01	0.97	0.98	1.00	1.00	0.98	Syria
13.35	11.66	12.18	11.45	11.26	11.94	11.73	13.16	11.97	United Arab Emirates
0.26	0.22	0.23	0.25	0.19	0.20	0.21	0.21	0.21	Yemen
1.90	1.83	1.86	2.02	2.05	2.12	2.13	2.20	2.27	**Middle East**
1.01	0.99	0.99	0.97	0.95	0.94	0.95	0.96	0.96	**Non-OECD Total**
6.47	6.27	6.24	6.28	6.35	6.41	6.40	6.48	6.51	OECD North America
3.20	3.38	3.45	3.56	3.66	3.80	3.93	4.09	4.16	OECD Pacific
3.32	3.27	3.29	3.25	3.24	3.22	3.29	3.40	3.37	OECD Europe
4.39	4.33	4.35	4.37	4.41	4.45	4.51	4.62	4.63	**OECD Total**
1.69	1.66	1.66	1.65	1.64	1.63	1.64	1.67	1.66	**World**

Due to lack of complete series Albania, the Dem. Ppl's Rep. of Korea and Vietnam are not included in the regional aggregates.
World Total excludes international marine bunkers.

Oil Supply/GDP (toe per thousand 90 US$)
Approvisionnement en pétrole/PIB (tep par millier de $US 90)
Ölaufkommen/BIP (toe pro tausend 1990er US$)
Totale disponibilita' primaria petrolio/PIL (tep per migliaia US$ 90)
石油供給量／GDP （石油換算トン／千米ドル、1990年価格）
Suministro de petóleo/PIB (tep por miles de 90 US$)
Поставка нефти и нефтепродуктов / ВВП （тнэ на тыс.долл.США　1990 г.）

	1971	1973	1978	1983	1984	1985	1986	1987	1988
Algérie	0.10	0.10	0.10	0.12	0.12	0.13	0.13	0.12	0.13
Angola	0.05	0.06	0.11	0.09	0.08	0.10	0.09	0.08	0.08
Bénin	0.10	0.12	0.11	0.09	0.09	0.10	0.09	0.08	0.08
Cameroun	0.06	0.06	0.06	0.08	0.08	0.07	0.06	0.06	0.07
Congo	0.16	0.14	0.16	0.13	0.11	0.11	0.08	0.26	0.40
RD du Congo	0.08	0.08	0.10	0.12	0.11	0.11	0.09	0.10	0.11
Egypte	0.32	0.32	0.33	0.38	0.39	0.38	0.38	0.39	0.38
Ethiopie	0.12	0.11	0.10	0.08	0.08	0.09	0.10	0.11	0.11
Gabon	0.14	0.17	0.21	0.19	0.16	0.13	0.13	0.14	0.13
Ghana	0.14	0.16	0.18	0.09	0.15	0.18	0.15	0.17	0.14
Côte d'Ivoire	0.13	0.14	0.13	0.10	0.10	0.10	0.11	0.11	0.14
Kenya	0.39	0.35	0.34	0.24	0.27	0.27	0.29	0.29	0.27
Libye	0.02	0.06	0.10	0.19	0.21	0.25	0.25	0.32	0.28
Maroc	0.17	0.20	0.23	0.22	0.22	0.21	0.19	0.19	0.19
Mozambique	0.41	0.37	0.37	0.45	0.44	0.42	0.46	0.39	0.36
Nigéria	0.08	0.12	0.26	0.42	0.39	0.36	0.33	0.37	0.36
Sénégal	0.15	0.17	0.17	0.16	0.18	0.16	0.17	0.18	0.15
Afrique du Sud	0.12	0.14	0.13	0.10	0.10	0.10	0.10	0.09	0.10
Soudan	0.22	0.28	0.13	0.14	0.14	0.18	0.16	0.13	0.19
Tanzanie	0.34	0.34	0.26	0.23	0.22	0.22	0.23	0.21	0.21
Tunisie	0.27	0.26	0.29	0.31	0.28	0.26	0.28	0.23	0.27
Zambie	0.17	0.22	0.21	0.21	0.18	0.18	0.16	0.16	0.16
Zimbabwe	0.12	0.14	0.14	0.12	0.11	0.11	0.12	0.11	0.12
Autre Afrique	0.09	0.09	0.10	0.12	0.12	0.11	0.12	0.14	0.13
Afrique	0.12	0.14	0.16	0.18	0.18	0.18	0.18	0.18	0.19
Argentine	0.20	0.19	0.18	0.16	0.16	0.15	0.14	0.15	0.15
Bolivie	0.19	0.20	0.23	0.28	0.27	0.27	0.28	0.28	0.28
Brésil	0.15	0.16	0.16	0.13	0.12	0.12	0.12	0.12	0.13
Chili	0.30	0.32	0.28	0.25	0.23	0.21	0.21	0.21	0.21
Colombie	0.35	0.33	0.29	0.29	0.27	0.27	0.27	0.27	0.25
Costa Rica	0.17	0.18	0.19	0.14	0.14	0.15	0.15	0.15	0.15
Cuba	1.16	1.14	1.03	0.92	0.74	0.72	0.70	0.78	0.78
République dominicaine	0.37	0.43	0.40	0.38	0.39	0.35	0.38	0.40	0.40
Equateur	0.31	0.25	0.45	0.47	0.48	0.47	0.45	0.48	0.46
El Salvador	0.11	0.13	0.12	0.12	0.13	0.14	0.13	0.15	0.15
Guatemala	0.20	0.20	0.20	0.15	0.18	0.18	0.14	0.14	0.15
Haiti	0.06	0.06	0.09	0.07	0.08	0.08	0.08	0.09	0.10
Honduras	0.26	0.25	0.24	0.24	0.24	0.22	0.21	0.23	0.25
Jamaïque	0.50	0.69	0.64	0.53	0.52	0.48	0.48	0.46	0.45
Antilles néerlandaises	4.68	4.65	2.63	2.88	2.06	1.34	1.23	1.04	1.00
Nicaragua	0.43	0.43	0.49	0.50	0.47	0.50	0.58	0.61	0.63
Panama	0.41	0.48	0.28	0.24	0.18	0.16	0.16	0.18	0.16
Paraguay	0.12	0.13	0.15	0.11	0.12	0.12	0.13	0.13	0.13
Pérou	0.17	0.17	0.17	0.16	0.16	0.14	0.14	0.14	0.16
Trinité-et-Tobago	0.40	0.35	0.32	0.16	0.22	0.26	0.30	0.20	0.19
Uruguay	0.30	0.29	0.26	0.19	0.17	0.16	0.14	0.15	0.18
Vénézuela	0.36	0.32	0.33	0.48	0.38	0.38	0.38	0.46	0.40
Autre Amérique latine	0.43	0.48	0.42	0.23	0.24	0.22	0.23	0.25	0.24
Amérique latine	0.23	0.23	0.22	0.20	0.18	0.17	0.17	0.18	0.18

Oil Supply/GDP (toe per thousand 90 US$)
Approvisionnement en pétrole/PIB (tep par millier de $US 90)
Ölaufkommen/BIP (toe pro tausend 1990er US$)
Totale disponibilita' primaria petrolio/PIL (tep per migliaia US$ 90)
石油供給量／GDP （石油換算トン／千米ドル、1990年価格）
Suministro de petóleo/PIB (tep por miles de 90 US$)
Поставка нефти и нефтепродуктов / ВВП （тнэ на тыс.долл.США 1990 г.)

1989	1990	1991	1992	1993	1994	1995	1996	1997	
0.12	0.13	0.14	0.13	0.14	0.13	0.12	0.11	0.13	Algeria
0.08	0.08	0.07	0.08	0.10	0.10	0.09	0.10	0.11	Angola
0.06	0.05	0.04	0.04	0.04	0.04	0.04	0.14	0.13	Benin
0.08	0.08	0.08	0.07	0.08	0.08	0.08	0.07	0.07	Cameroon
0.16	0.12	0.13	0.12	0.12	0.13	0.11	0.11	0.11	Congo
0.11	0.12	0.14	0.14	0.18	0.19	0.21	0.21	0.22	DR of Congo
0.36	0.36	0.36	0.35	0.35	0.29	0.30	0.31	0.32	Egypt
0.11	0.12	0.12	0.12	0.11	0.11	0.12	0.12	0.07	Ethiopia
0.10	0.07	0.07	0.08	0.09	0.08	0.08	0.08	0.09	Gabon
0.16	0.15	0.13	0.12	0.14	0.19	0.20	0.19	0.18	Ghana
0.14	0.12	0.11	0.15	0.12	0.12	0.11	0.11	0.11	Ivory Coast
0.27	0.25	0.23	0.24	0.25	0.26	0.21	0.20	0.25	Kenya
0.25	0.27	0.34	0.32	0.37	0.36	0.46	0.40	0.40	Libya
0.20	0.20	0.19	0.22	0.24	0.23	0.23	0.21	0.22	Morocco
0.32	0.31	0.27	0.29	0.25	0.22	0.23	0.23	0.27	Mozambique
0.35	0.33	0.34	0.36	0.16	0.47	0.37	0.33	0.37	Nigeria
0.15	0.15	0.15	0.16	0.16	0.16	0.15	0.16	0.16	Senegal
0.10	0.10	0.10	0.10	0.08	0.09	0.09	0.08	0.09	South Africa
0.16	0.21	0.19	0.16	0.11	0.15	0.13	0.12	0.14	Sudan
0.20	0.20	0.19	0.19	0.18	0.18	0.17	0.17	0.16	Tanzania
0.29	0.27	0.28	0.26	0.28	0.24	0.22	0.22	0.21	Tunisia
0.16	0.14	0.14	0.15	0.13	0.14	0.15	0.14	0.14	Zambia
0.11	0.09	0.11	0.13	0.13	0.14	0.17	0.15	0.15	Zimbabwe
0.12	0.12	0.12	0.12	0.12	0.12	0.12	0.12	0.11	Other Africa
0.18	0.18	0.19	0.19	0.17	0.19	0.19	0.18	0.19	**Africa**
0.14	0.14	0.13	0.13	0.13	0.12	0.13	0.12	0.11	Argentina
0.29	0.29	0.30	0.28	0.27	0.28	0.32	0.32	0.32	Bolivia
0.13	0.13	0.13	0.14	0.14	0.14	0.14	0.14	0.15	Brazil
0.21	0.21	0.21	0.21	0.21	0.22	0.21	0.22	0.21	Chile
0.26	0.28	0.26	0.29	0.27	0.27	0.26	0.27	0.27	Colombia
0.17	0.17	0.18	0.21	0.22	0.22	0.22	0.20	0.21	Costa Rica
0.79	0.77	0.68	0.61	0.67	0.72	0.73	0.74	0.81	Cuba
0.39	0.41	0.41	0.43	0.42	0.42	0.41	0.40	0.39	Dominican Republic
0.42	0.47	0.46	0.47	0.43	0.46	0.47	0.53	0.51	Ecuador
0.16	0.15	0.19	0.19	0.20	0.22	0.22	0.20	0.19	El Salvador
0.15	0.17	0.17	0.18	0.18	0.21	0.21	0.22	0.23	Guatemala
0.11	0.11	0.10	0.10	0.09	0.03	0.13	0.14	0.18	Haiti
0.26	0.24	0.24	0.27	0.23	0.28	0.32	0.31	0.30	Honduras
0.52	0.60	0.58	0.58	0.59	0.60	0.66	0.71	0.77	Jamaica
1.19	1.23	1.56	0.93	1.50	1.53	1.51	1.44	1.39	Netherlands Antilles
0.56	0.61	0.59	0.66	0.64	0.71	0.68	0.61	0.65	Nicaragua
0.16	0.17	0.18	0.18	0.17	0.20	0.18	0.19	0.20	Panama
0.13	0.13	0.12	0.15	0.16	0.18	0.19	0.17	0.18	Paraguay
0.16	0.16	0.15	0.14	0.15	0.14	0.15	0.15	0.16	Peru
0.16	0.26	0.22	0.35	0.31	0.23	0.21	0.26	0.23	Trinidad-and-Tobago
0.19	0.15	0.17	0.18	0.16	0.13	0.15	0.17	0.16	Uruguay
0.40	0.34	0.46	0.39	0.31	0.24	0.31	0.41	0.38	Venezuela
0.24	0.24	0.23	0.22	0.22	0.22	0.22	0.21	0.21	Other Latin America
0.18	0.18	0.19	0.18	0.18	0.17	0.18	0.19	0.19	**Latin America**

Oil Supply/GDP (toe per thousand 90 US$)
Approvisionnement en pétrole/PIB (tep par millier de $US 90)
Ölaufkommen/BIP (toe pro tausend 1990er US$)
Totale disponibilita' primaria petrolio/PIL (tep per migliaia US$ 90)
石油供給量／GDP （石油換算トン／千米ドル、1990年価格）
Suministro de petóleo/PIB (tep por miles de 90 US$)
Поставка нефти и нефтепродуктов / ВВП （тнэ на тыс.долл.США 1990 г.)

	1971	1973	1978	1983	1984	1985	1986	1987	1988
Bangladesh	0.07	0.09	0.10	0.09	0.09	0.09	0.09	0.09	0.09
Brunei	0.03	0.03	0.03	0.04	0.08	0.07	0.10	0.18	0.30
Inde	0.17	0.18	0.17	0.19	0.19	0.20	0.20	0.20	0.20
Indonésie	0.27	0.30	0.33	0.31	0.29	0.29	0.31	0.28	0.28
Malaisie	0.40	0.32	0.33	0.34	0.31	0.33	0.32	0.31	0.30
Myanmar	0.09	0.08	0.07	0.05	0.05	0.05	0.05	0.03	0.03
Népal	0.03	0.04	0.04	0.05	0.06	0.06	0.06	0.07	0.07
Pakistan	0.25	0.24	0.23	0.24	0.25	0.25	0.26	0.26	0.26
Philippines	0.38	0.38	0.33	0.26	0.22	0.22	0.22	0.26	0.26
Singapour	0.33	0.36	0.39	0.30	0.30	0.31	0.33	0.32	0.31
Sri Lanka	0.27	0.33	0.24	0.24	0.20	0.18	0.17	0.19	0.18
Taipei chinois	0.20	0.24	0.28	0.19	0.17	0.16	0.16	0.14	0.15
Thaïlande	0.29	0.33	0.29	0.23	0.23	0.21	0.20	0.21	0.20
Viêt-Nam	-	-	-	-	0.38	0.37	0.41	0.46	0.45
Autre Asie	0.14	0.16	0.13	0.12	0.12	0.13	0.11	0.14	0.13
Asie	0.21	0.23	0.23	0.21	0.21	0.20	0.21	0.20	0.21
Rép. populaire de Chine	0.39	0.47	0.67	0.42	0.38	0.35	0.34	0.33	0.31
Hong-Kong, Chine	0.19	0.17	0.18	0.11	0.09	0.08	0.08	0.06	0.07
Chine	0.36	0.42	0.58	0.36	0.33	0.31	0.30	0.28	0.27
Albanie	-	-	-	0.55	0.60	0.47	0.46	0.46	0.50
Bulgarie	0.79	0.83	0.90	0.68	0.64	0.60	0.61	0.52	0.51
Chypre	0.28	0.35	0.31	0.29	0.27	0.26	0.28	0.29	0.28
Gibraltar	0.21	0.17	0.12	0.16	0.14	0.16	0.18	0.18	0.21
Malte	0.46	0.47	0.27	0.20	0.27	0.16	0.32	0.31	0.29
Roumanie	0.51	0.55	0.52	0.34	0.31	0.34	0.35	0.42	0.38
République slovaque	0.41	0.48	0.54	0.45	0.43	0.41	0.39	0.38	0.36
Bosnie-Herzégovine	-	-	-	-	-	-	-	-	-
Croatie	-	-	-	-	-	-	-	-	-
Ex-RYM	-	-	-	-	-	-	-	-	-
Slovénie	-	-	-	-	-	-	-	-	-
RF de Yougoslavie	-	-	-	-	-	-	-	-	-
Ex-Yougoslavie	0.16	0.18	0.18	0.15	0.15	0.14	0.15	0.16	0.18
Europe non-OCDE	0.34	0.37	0.37	0.28	0.26	0.26	0.27	0.28	0.29
Arménie	-	-	-	-	-	-	-	-	-
Azerbaïdjan	-	-	-	-	-	-	-	-	-
Bélarus	-	-	-	-	-	-	-	-	-
Estonie	-	-	-	-	-	-	-	-	-
Géorgie	-	-	-	-	-	-	-	-	-
Kazakhstan	-	-	-	-	-	-	-	-	-
Kirghizistan	-	-	-	-	-	-	-	-	-
Lettonie	-	-	-	-	-	-	-	-	-
Lituanie	-	-	-	-	-	-	-	-	-
Moldova	-	-	-	-	-	-	-	-	-
Russie	-	-	-	-	-	-	-	-	-
Tadjikistan	-	-	-	-	-	-	-	-	-
Turkménistan	-	-	-	-	-	-	-	-	-
Ukraine	-	-	-	-	-	-	-	-	-
Ouzbékistan	-	-	-	-	-	-	-	-	-
Ex-URSS	0.47	0.50	0.53	0.51	0.50	0.50	0.47	0.48	0.47

Faute de séries complètes, l'Albanie, la Rép.pop.dém. de Corée et le Viêt-Nam ne sont pas pris en compte dans les totaux régionaux.

Oil Supply/GDP (toe per thousand 90 US$)
Approvisionnement en pétrole/PIB (tep par millier de $US 90)
Ölaufkommen/BIP (toe pro tausend 1990er US$)
Totale disponibilita' primaria petrolio/PIL (tep per migliaia US$ 90)
石油供給量／GDP （石油換算トン／千米ドル、1990年価格）
Suministro de petóleo/PIB (tep por miles de 90 US$)
Поставка нефти и нефтепродуктов / ВВП （тнэ на тыс.долл.США 1990 г.)

1989	1990	1991	1992	1993	1994	1995	1996	1997	
0.09	0.09	0.07	0.08	0.08	0.09	0.09	0.09	0.08	Bangladesh
0.06	0.04	0.04	0.07	0.06	0.08	0.10	0.10	0.12	Brunei
0.20	0.20	0.20	0.20	0.19	0.19	0.21	0.21	0.20	India
0.28	0.29	0.30	0.28	0.30	0.25	0.25	0.25	0.27	Indonesia
0.30	0.31	0.30	0.30	0.36	0.35	0.30	0.30	0.33	Malaysia
0.03	0.03	0.03	0.03	0.03	0.04	0.03	0.03	0.04	Myanmar
0.06	0.05	0.07	0.08	0.08	0.10	0.12	0.11	0.11	Nepal
0.27	0.27	0.26	0.27	0.29	0.30	0.30	0.32	0.33	Pakistan
0.26	0.27	0.26	0.30	0.33	0.33	0.36	0.35	0.35	Philippines
0.29	0.36	0.36	0.37	0.40	0.45	0.35	0.37	0.39	Singapore
0.17	0.16	0.16	0.16	0.19	0.18	0.20	0.24	0.24	Sri Lanka
0.16	0.16	0.16	0.15	0.15	0.15	0.15	0.14	0.14	Chinese Taipei
0.21	0.23	0.22	0.23	0.24	0.25	0.27	0.28	0.28	Thailand
0.39	0.45	0.38	0.44	0.50	0.51	0.44	0.60	0.68	Vietnam
0.14	0.15	0.14	0.12	0.12	0.11	0.10	0.10	0.10	Other Asia
0.21	0.21	0.21	0.21	0.22	0.22	0.22	0.22	0.22	**Asia**
0.31	0.30	0.29	0.27	0.27	0.24	0.23	0.23	0.24	People's Rep.of China
0.07	0.07	0.07	0.08	0.08	0.08	0.08	0.07	0.07	Hong Kong, China
0.27	0.26	0.26	0.25	0.24	0.22	0.21	0.21	0.22	**China**
0.42	0.53	0.50	0.37	0.37	0.40	0.33	0.30	0.25	Albania
0.52	0.43	0.31	0.32	0.37	0.33	0.35	0.34	0.31	Bulgaria
0.27	0.26	0.30	0.30	0.30	0.33	0.28	0.30	0.28	Cyprus
0.20	0.22	0.24	0.27	0.31	0.41	0.40	0.39	0.38	Gibraltar
0.27	0.26	0.25	0.24	0.28	0.26	0.27	0.29	0.30	Malta
0.43	0.48	0.44	0.39	0.41	0.36	0.38	0.37	0.38	Romania
0.34	0.30	0.31	0.29	0.25	0.26	0.26	0.24	0.22	Slovak Republic
-	-	-	0.33	1.39	1.24	0.94	0.64	0.46	*Bosnia-Herzegovina*
-	-	-	0.41	0.44	0.44	0.47	0.42	0.42	*Croatia*
-	-	-	0.54	0.64	0.43	0.47	0.71	0.60	*FYROM*
-	-	-	0.08	0.09	0.10	0.10	0.11	0.11	*Slovenia*
-	-	-	0.05	0.04	0.04	0.04	0.07	0.09	*FR of Yugoslavia*
0.17	0.18	0.15	0.12	0.14	0.13	0.13	0.15	0.16	Former Yugoslavia
0.29	0.29	0.25	0.22	0.25	0.23	0.24	0.24	0.24	**Non-OECD Europe**
-	-	-	1.35	0.80	0.24	0.16	0.08	0.08	Armenia
-	-	-	1.07	1.58	2.13	1.97	1.93	1.87	Azerbaijan
-	-	-	0.69	0.50	0.48	0.48	0.45	0.35	Belarus
-	-	-	0.31	0.35	0.35	0.27	0.27	0.24	Estonia
-	-	-	0.24	0.22	0.12	0.05	0.25	0.24	Georgia
-	-	-	0.74	0.54	0.51	0.49	0.44	0.38	Kazakhstan
-	-	-	0.85	0.62	0.28	0.43	0.45	0.30	Kyrgyzstan
-	-	-	0.39	0.40	0.40	0.33	0.32	0.26	Latvia
-	-	-	0.44	0.46	0.46	0.44	0.38	0.38	Lithuania
-	-	-	0.44	0.32	0.25	0.24	0.25	0.22	Moldova
-	-	-	0.51	0.47	0.40	0.41	0.37	0.36	Russia
-	-	-	1.91	1.34	0.56	0.65	0.68	0.67	Tajikistan
-	-	-	0.88	0.55	0.85	0.93	0.89	0.78	Turkmenistan
-	-	-	0.32	0.26	0.28	0.34	0.30	0.28	Ukraine
-	-	-	0.43	0.47	0.42	0.36	0.34	0.34	Uzbekistan
0.45	0.43	0.45	0.48	0.43	0.39	0.40	0.37	0.35	**Former USSR**

Due to lack of complete series Albania, the Dem. Ppl's Rep. of Korea and Vietnam are not included in the regional aggregates.

Oil Supply/GDP (toe per thousand 90 US$)
Approvisionnement en pétrole/PIB (tep par millier de $US 90)
Ölaufkommen/BIP (toe pro tausend 1990er US$)
Totale disponibilita' primaria petrolio/PIL (tep per migliaia US$ 90)
石油供給量／GDP（石油換算トン／千米ドル、1990年価格）
Suministro de petóleo/PIB (tep por miles de 90 US$)
Поставка нефти и нефтепродуктов / ВВП (тнэ на тыс.долл.США 1990 г.)

	1971	1973	1978	1983	1984	1985	1986	1987	1988
Bahrein	0.41	0.40	0.22	0.23	0.27	0.25	0.24	0.35	0.23
Iran	0.21	0.20	0.24	0.28	0.32	0.32	0.37	0.39	0.42
Irak	0.05	0.05	0.06	0.13	0.13	0.17	0.20	0.22	0.26
Israël	0.27	0.31	0.23	0.17	0.16	0.14	0.16	0.16	0.19
Jordanie	0.35	0.44	0.53	0.68	0.68	0.66	0.66	0.68	0.68
Koweit	0.15	0.12	0.17	0.55	0.50	0.69	0.45	0.36	0.39
Liban	0.34	0.39	0.58	0.67	0.66	0.77	0.74	0.80	0.54
Oman	0.03	0.04	0.13	0.16	0.17	0.17	0.08	0.07	0.11
Qatar	0.03	0.04	0.10	0.09	0.08	0.11	0.07	0.10	0.09
Arabie saoudite	0.12	0.10	0.19	0.44	0.38	0.42	0.41	0.44	0.45
Syrie	0.33	0.27	0.29	0.37	0.44	0.38	0.41	0.44	0.38
Emirats arabes unis	0.02	0.03	0.12	0.27	0.22	0.27	0.32	0.30	0.31
Yémen	0.28	0.31	0.19	0.24	0.24	0.27	0.27	0.28	0.25
Moyen-Orient	0.15	0.15	0.17	0.28	0.28	0.30	0.32	0.33	0.35
Total non-OCDE	0.28	0.30	0.31	0.30	0.29	0.29	0.28	0.28	0.28
OCDE Amérique du N.	0.22	0.22	0.22	0.17	0.16	0.16	0.16	0.16	0.16
OCDE Pacifique	0.15	0.16	0.14	0.10	0.10	0.09	0.09	0.09	0.09
OCDE Europe	0.15	0.15	0.13	0.10	0.10	0.10	0.09	0.09	0.09
Total OCDE	0.17	0.18	0.17	0.13	0.12	0.12	0.12	0.12	0.12
Monde	0.19	0.20	0.20	0.16	0.16	0.15	0.15	0.15	0.15

Faute de séries complètes, l'Albanie, la Rép.pop.dém. de Corée et le Viêt-Nam ne sont pas pris en compte dans les totaux régionaux.
Le total Monde ne comprend pas les soutages maritimes internationaux.

Oil Supply/GDP (toe per thousand 90 US$)
Approvisionnement en pétrole/PIB (tep par millier de $US 90)
Ölaufkommen/BIP (toe pro tausend 1990er US$)
Totale disponibilita' primaria petrolio/PIL (tep per migliaia US$ 90)
石油供給量／GDP （石油換算トン／千米ドル、1990年価格）
Suministro de petóleo/PIB (tep por miles de 90 US$)
Поставка нефти и нефтепродуктов / ВВП (тнэ на тыс.долл.США 1990 г.)

1989	1990	1991	1992	1993	1994	1995	1996	1997	
0.19	0.19	0.17	0.18	0.17	0.19	0.16	0.46	0.62	Bahrain
0.45	0.42	0.40	0.40	0.39	0.43	0.42	0.42	0.41	Iran
0.25	0.35	0.58	0.69	0.74	0.86	0.92	0.92	0.79	Iraq
0.18	0.17	0.16	0.17	0.17	0.16	0.17	0.14	0.15	Israel
0.79	0.83	0.82	0.81	0.77	0.76	0.72	0.74	0.78	Jordan
0.46	0.39	0.21	0.41	0.28	0.51	0.57	0.45	0.52	Kuwait
0.77	0.76	0.69	0.66	0.77	0.79	0.85	0.83	0.89	Lebanon
0.11	0.18	0.30	0.25	0.21	0.18	0.20	0.19	0.16	Oman
0.20	0.09	0.11	0.10	0.10	0.08	0.10	0.12	0.16	Qatar
0.40	0.36	0.41	0.43	0.45	0.42	0.43	0.47	0.49	Saudi Arabia
0.41	0.42	0.43	0.38	0.36	0.34	0.33	0.34	0.32	Syria
0.30	0.23	0.22	0.23	0.25	0.26	0.22	0.24	0.17	United Arab Emirates
0.26	0.23	0.26	0.28	0.22	0.23	0.23	0.23	0.22	Yemen
0.35	0.34	0.36	0.37	0.37	0.38	0.38	0.39	0.39	**Middle East**
0.28	0.27	0.28	0.27	0.26	0.25	0.24	0.24	0.24	**Non-OECD Total**
0.15	0.15	0.14	0.14	0.14	0.14	0.14	0.14	0.13	OECD North America
0.09	0.10	0.09	0.10	0.10	0.10	0.10	0.10	0.10	OECD Pacific
0.09	0.09	0.09	0.09	0.09	0.09	0.08	0.09	0.08	OECD Europe
0.11	0.11	0.11	0.11	0.11	0.11	0.11	0.11	0.11	**OECD Total**
0.15	0.14	0.14	0.14	0.14	0.14	0.14	0.14	0.13	**World**

Due to lack of complete series Albania, the Dem. Ppl's Rep. of Korea and Vietnam are not included in the regional aggregates.
World Total excludes international marine bunkers.

Oil Supply/Population (toe per capita)
Approvisionnement en pétrole/Population (tep par habitant)
Ölaufkommen/Bevölkerung (toe pro Kopf)
Totale disponibilita' primaria petrolio/POP (tep per capita)
一人当たり石油供給量（石油換算トン／人）
Suministro de petróleo/población (tep per capita)
Поставка нефти и нефтепродуктов / Численность населения (тнэ на человека)

	1971	1973	1978	1983	1984	1985	1986	1987	1988
Algérie	0.16	0.20	0.26	0.31	0.33	0.35	0.34	0.33	0.32
Angola	0.11	0.15	0.14	0.10	0.08	0.11	0.09	0.09	0.09
Bénin	0.04	0.05	0.04	0.04	0.04	0.04	0.04	0.03	0.03
Cameroun	0.04	0.05	0.07	0.10	0.10	0.10	0.09	0.09	0.09
Congo	0.13	0.12	0.14	0.20	0.17	0.16	0.11	0.34	0.51
RD du Congo	0.03	0.04	0.03	0.04	0.03	0.03	0.03	0.03	0.03
Egypte	0.19	0.19	0.26	0.38	0.41	0.42	0.41	0.43	0.42
Ethiopie	0.02	0.02	0.01	0.01	0.01	0.01	0.02	0.02	0.02
Gabon	0.69	0.99	1.59	1.34	1.20	0.91	0.87	0.74	0.76
Ghana	0.08	0.08	0.09	0.03	0.05	0.07	0.06	0.07	0.05
Côte d'Ivoire	0.15	0.16	0.19	0.11	0.11	0.10	0.12	0.10	0.13
Kenya	0.11	0.11	0.11	0.08	0.09	0.09	0.10	0.10	0.09
Libye	0.32	0.79	1.35	1.83	1.79	1.84	1.63	1.94	1.67
Maroc	0.12	0.14	0.19	0.20	0.20	0.20	0.19	0.19	0.20
Mozambique	0.06	0.06	0.04	0.03	0.03	0.03	0.03	0.03	0.03
Nigéria	0.03	0.05	0.10	0.13	0.11	0.11	0.10	0.11	0.11
Sénégal	0.12	0.13	0.14	0.13	0.14	0.12	0.13	0.14	0.12
Afrique du Sud	0.38	0.42	0.42	0.32	0.33	0.30	0.29	0.29	0.32
Soudan	0.09	0.10	0.07	0.06	0.06	0.07	0.07	0.05	0.07
Tanzanie	0.05	0.05	0.04	0.03	0.03	0.03	0.03	0.03	0.03
Tunisie	0.25	0.26	0.36	0.43	0.40	0.38	0.39	0.34	0.38
Zambie	0.12	0.15	0.13	0.11	0.09	0.09	0.08	0.08	0.08
Zimbabwe	0.10	0.12	0.11	0.10	0.09	0.09	0.10	0.09	0.10
Autre Afrique	0.03	0.03	0.04	0.05	0.04	0.04	0.05	0.05	0.05
Afrique	0.10	0.11	0.13	0.14	0.14	0.14	0.14	0.14	0.15
Argentine	1.02	1.01	0.95	0.81	0.81	0.70	0.72	0.76	0.72
Bolivie	0.15	0.17	0.22	0.22	0.21	0.20	0.20	0.20	0.20
Brésil	0.29	0.38	0.46	0.36	0.35	0.36	0.39	0.39	0.42
Chili	0.53	0.52	0.46	0.40	0.40	0.38	0.40	0.41	0.44
Colombie	0.27	0.28	0.28	0.28	0.27	0.28	0.29	0.30	0.28
Costa Rica	0.26	0.31	0.39	0.24	0.25	0.25	0.26	0.28	0.27
Cuba	0.84	0.94	1.08	1.19	1.03	1.02	1.01	1.07	1.08
République dominicaine	0.25	0.35	0.37	0.38	0.39	0.34	0.38	0.42	0.43
Equateur	0.20	0.22	0.47	0.49	0.50	0.50	0.48	0.47	0.49
El Salvador	0.14	0.17	0.18	0.12	0.13	0.14	0.13	0.16	0.16
Guatemala	0.16	0.17	0.20	0.14	0.16	0.15	0.11	0.12	0.13
Haiti	0.03	0.03	0.04	0.04	0.04	0.04	0.04	0.05	0.05
Honduras	0.14	0.15	0.16	0.15	0.15	0.14	0.13	0.15	0.16
Jamaïque	0.99	1.44	1.11	0.84	0.81	0.69	0.69	0.72	0.73
Antilles néerlandaises	34.26	37.01	21.96	24.66	17.02	10.04	9.70	8.38	8.26
Nicaragua	0.25	0.25	0.30	0.21	0.20	0.19	0.22	0.22	0.20
Panama	0.90	1.10	0.67	0.58	0.46	0.41	0.42	0.44	0.34
Paraguay	0.09	0.11	0.16	0.14	0.14	0.15	0.15	0.16	0.16
Pérou	0.37	0.37	0.38	0.32	0.32	0.30	0.31	0.33	0.33
Trinité-et-Tobago	1.10	1.08	1.37	0.78	1.10	1.27	1.38	0.88	0.79
Uruguay	0.66	0.65	0.67	0.43	0.39	0.37	0.36	0.40	0.49
Vénézuela	1.13	1.03	1.09	1.23	0.96	0.96	0.98	1.20	1.08
Autre Amérique latine	1.57	1.82	1.72	1.03	1.10	1.04	1.10	1.22	1.23
Amérique latine	0.47	0.52	0.54	0.47	0.45	0.43	0.45	0.47	0.47

Oil Supply/Population (toe per capita)
Approvisionnement en pétrole/Population (tep par habitant)
Ölaufkommen/Bevölkerung (toe pro Kopf)
Totale disponibilita' primaria petrolio/POP (tep per capita)
一人当たり石油供給量（石油換算トン／人）
Suministro de petróleo/población (tep per capita)
Поставка нефти и нефтепродуктов / Численность населения (тнэ на человека)

1989	1990	1991	1992	1993	1994	1995	1996	1997	
0.32	0.32	0.33	0.31	0.31	0.29	0.27	0.26	0.30	Algeria
0.09	0.09	0.08	0.07	0.07	0.07	0.07	0.08	0.10	Angola
0.02	0.02	0.02	0.02	0.02	0.02	0.02	0.06	0.06	Benin
0.09	0.09	0.08	0.07	0.07	0.07	0.07	0.06	0.06	Cameroon
0.20	0.15	0.16	0.15	0.14	0.14	0.13	0.13	0.12	Congo
0.03	0.03	0.03	0.03	0.03	0.03	0.03	0.03	0.03	DR of Congo
0.42	0.43	0.42	0.41	0.42	0.36	0.38	0.40	0.42	Egypt
0.02	0.02	0.02	0.02	0.02	0.02	0.02	0.02	0.01	Ethiopia
0.58	0.41	0.44	0.49	0.52	0.49	0.49	0.48	0.56	Gabon
0.06	0.06	0.05	0.05	0.06	0.08	0.09	0.08	0.08	Ghana
0.13	0.11	0.10	0.13	0.10	0.10	0.09	0.10	0.10	Ivory Coast
0.10	0.09	0.08	0.08	0.08	0.09	0.07	0.07	0.08	Kenya
1.47	1.67	2.12	1.88	2.04	1.87	2.36	2.04	2.00	Libya
0.21	0.21	0.21	0.23	0.24	0.26	0.23	0.24	0.24	Morocco
0.03	0.03	0.03	0.03	0.03	0.03	0.03	0.03	0.04	Mozambique
0.11	0.11	0.12	0.12	0.05	0.15	0.12	0.11	0.13	Nigeria
0.12	0.12	0.11	0.12	0.11	0.12	0.11	0.12	0.13	Senegal
0.31	0.30	0.30	0.29	0.21	0.26	0.28	0.22	0.26	South Africa
0.06	0.08	0.07	0.06	0.05	0.06	0.06	0.06	0.07	Sudan
0.03	0.03	0.03	0.03	0.03	0.02	0.02	0.02	0.02	Tanzania
0.41	0.41	0.43	0.42	0.45	0.40	0.37	0.39	0.38	Tunisia
0.08	0.07	0.07	0.07	0.06	0.06	0.06	0.06	0.06	Zambia
0.09	0.08	0.10	0.10	0.11	0.12	0.14	0.13	0.13	Zimbabwe
0.05	0.05	0.05	0.05	0.05	0.05	0.05	0.05	0.05	Other Africa
0.14	0.14	0.14	0.14	0.13	0.14	0.14	0.13	0.14	**Africa**
0.63	0.59	0.62	0.66	0.71	0.69	0.68	0.67	0.68	Argentina
0.21	0.22	0.23	0.21	0.21	0.22	0.26	0.26	0.27	Bolivia
0.42	0.42	0.42	0.43	0.44	0.47	0.47	0.50	0.52	Brazil
0.48	0.49	0.52	0.57	0.61	0.65	0.69	0.74	0.77	Chile
0.29	0.32	0.31	0.34	0.33	0.34	0.35	0.36	0.36	Colombia
0.32	0.32	0.34	0.42	0.45	0.47	0.46	0.42	0.44	Costa Rica
1.10	1.02	0.80	0.62	0.59	0.63	0.65	0.70	0.78	Cuba
0.42	0.41	0.41	0.45	0.44	0.45	0.46	0.46	0.49	Dominican Republic
0.44	0.49	0.49	0.51	0.47	0.50	0.52	0.58	0.57	Ecuador
0.16	0.16	0.20	0.21	0.24	0.26	0.28	0.25	0.24	El Salvador
0.13	0.15	0.15	0.17	0.17	0.19	0.20	0.21	0.22	Guatemala
0.05	0.05	0.05	0.04	0.03	0.01	0.05	0.05	0.06	Haiti
0.17	0.15	0.15	0.17	0.15	0.17	0.21	0.20	0.20	Honduras
0.89	1.06	1.03	1.02	1.05	1.07	1.17	1.23	1.30	Jamaica
10.07	10.84	13.58	8.24	13.17	13.57	13.22	12.79	12.67	Netherlands Antilles
0.17	0.18	0.17	0.18	0.17	0.19	0.18	0.20	0.22	Nicaragua
0.33	0.38	0.42	0.46	0.45	0.53	0.49	0.51	0.55	Panama
0.17	0.16	0.15	0.18	0.19	0.22	0.25	0.22	0.23	Paraguay
0.29	0.27	0.26	0.24	0.26	0.27	0.31	0.32	0.35	Peru
0.64	1.05	1.01	1.41	1.22	0.95	0.88	1.11	1.02	Trinidad-and-Tobago
0.52	0.41	0.47	0.54	0.47	0.42	0.48	0.56	0.55	Uruguay
0.97	0.86	1.24	1.07	0.83	0.63	0.81	1.05	1.00	Venezuela
1.27	1.31	1.32	1.26	1.26	1.25	1.25	1.26	1.27	Other Latin America
0.46	0.45	0.47	0.47	0.47	0.47	0.49	0.52	0.54	**Latin America**

Oil Supply/Population (toe per capita)
Approvisionnement en pétrole/Population (tep par habitant)
Ölaufkommen/Bevölkerung (toe pro Kopf)
Totale disponibilita' primaria petrolio/POP (tep per capita)
一人当たり石油供給量（石油換算トン／人）
Suministro de petróleo/población (tep per capita)
Поставка нефти и нефтепродуктов / Численность населения (тнэ на человека)

	1971	1973	1978	1983	1984	1985	1986	1987	1988
Bangladesh	0.01	0.01	0.02	0.02	0.02	0.02	0.02	0.02	0.02
Brunei	0.48	0.48	0.58	0.76	1.38	1.15	1.49	2.64	4.44
Inde	0.04	0.04	0.05	0.05	0.05	0.06	0.06	0.06	0.07
Indonésie	0.07	0.09	0.12	0.15	0.14	0.14	0.16	0.15	0.16
RPD de Corée	0.05	0.05	0.11	0.15	0.16	0.16	0.17	0.17	0.17
Malaisie	0.43	0.40	0.52	0.67	0.64	0.64	0.61	0.62	0.62
Myanmar	0.05	0.04	0.04	0.03	0.03	0.03	0.03	0.02	0.02
Népal	0.00	0.01	0.01	0.01	0.01	0.01	0.01	0.01	0.01
Pakistan	0.06	0.05	0.06	0.07	0.08	0.08	0.08	0.09	0.09
Philippines	0.22	0.23	0.24	0.21	0.16	0.14	0.14	0.17	0.18
Singapour	1.55	2.07	2.78	2.89	3.18	3.15	3.31	3.47	3.76
Sri Lanka	0.08	0.10	0.08	0.10	0.08	0.08	0.08	0.08	0.08
Taipei chinois	0.45	0.66	1.09	0.95	0.93	0.87	1.00	0.94	1.06
Thaïlande	0.17	0.21	0.24	0.22	0.23	0.21	0.22	0.24	0.26
Viêt-Nam	0.14	0.13	0.02	0.03	0.03	0.03	0.04	0.04	0.04
Autre Asie	0.08	0.10	0.09	0.10	0.10	0.11	0.10	0.12	0.11
Asie	0.07	0.08	0.09	0.10	0.10	0.10	0.11	0.11	0.11
Rép. populaire de Chine	0.05	0.06	0.10	0.08	0.09	0.09	0.09	0.10	0.10
Hong-Kong, Chine	0.85	0.88	1.23	0.94	0.87	0.78	0.82	0.73	0.85
Chine	0.05	0.06	0.10	0.09	0.09	0.09	0.10	0.10	0.11
Albanie	0.39	0.33	0.45	0.40	0.42	0.33	0.33	0.32	0.34
Bulgarie	1.18	1.33	1.65	1.39	1.35	1.29	1.36	1.23	1.35
Chypre	1.04	1.39	1.34	1.59	1.61	1.58	1.77	1.92	2.00
Gibraltar	1.36	1.31	1.07	1.38	1.30	1.48	1.69	1.78	2.13
Malte	0.85	1.01	1.02	0.93	1.27	0.81	1.64	1.66	1.70
Roumanie	0.55	0.64	0.85	0.63	0.61	0.66	0.70	0.79	0.70
République slovaque	0.95	1.15	1.42	1.19	1.16	1.15	1.13	1.11	1.08
Bosnie-Herzégovine	-	-	-	-	-	-	-	-	-
Croatie	-	-	-	-	-	-	-	-	-
Ex-RYM	-	-	-	-	-	-	-	-	-
Slovénie	-	-	-	-	-	-	-	-	-
RF de Yougoslavie	-	-	-	-	-	-	-	-	-
Ex-Yougoslavie	0.45	0.51	0.71	0.60	0.61	0.59	0.64	0.67	0.73
Europe non-OCDE	0.65	0.75	0.97	0.79	0.78	0.78	0.83	0.85	0.86

Faute de séries complètes, l'Albanie, la Rép.pop.dém. de Corée et le Viêt-Nam ne sont pas pris en compte dans les totaux régionaux.

Oil Supply/Population (toe per capita)
Approvisionnement en pétrole/Population (tep par habitant)
Ölaufkommen/Bevölkerung (toe pro Kopf)
Totale disponibilita' primaria petrolio/POP (tep per capita)
一人当たり石油供給量（石油換算トン／人）
Suministro de petróleo/población (tep per capita)
Поставка нефти и нефтепродуктов / Численность населения (тнэ на человека)

1989	1990	1991	1992	1993	1994	1995	1996	1997	
0.02	0.02	0.01	0.02	0.02	0.02	0.02	0.02	0.02	Bangladesh
0.85	0.51	0.49	0.95	0.80	0.96	1.20	1.30	1.51	Brunei
0.07	0.07	0.07	0.07	0.07	0.08	0.09	0.09	0.09	India
0.17	0.18	0.21	0.20	0.23	0.20	0.22	0.23	0.25	Indonesia
0.18	0.17	0.15	0.12	0.10	0.08	0.07	0.07	0.06	DPR of Korea
0.65	0.73	0.76	0.78	0.99	1.03	0.94	1.01	1.16	Malaysia
0.02	0.02	0.02	0.02	0.02	0.02	0.03	0.03	0.03	Myanmar
0.01	0.01	0.01	0.02	0.02	0.02	0.03	0.02	0.02	Nepal
0.10	0.10	0.10	0.11	0.11	0.12	0.12	0.13	0.13	Pakistan
0.19	0.19	0.18	0.20	0.22	0.23	0.25	0.25	0.26	Philippines
3.71	4.93	5.23	5.65	6.61	7.96	6.68	7.39	8.23	Singapore
0.07	0.08	0.08	0.08	0.10	0.10	0.12	0.14	0.15	Sri Lanka
1.25	1.28	1.36	1.36	1.45	1.48	1.62	1.62	1.65	Chinese Taipei
0.30	0.35	0.37	0.40	0.45	0.50	0.58	0.63	0.62	Thailand
0.04	0.04	0.04	0.05	0.06	0.06	0.06	0.08	0.10	Vietnam
0.11	0.12	0.11	0.10	0.09	0.08	0.07	0.07	0.07	Other Asia
0.12	0.13	0.13	0.14	0.15	0.15	0.17	0.17	0.18	**Asia**
0.11	0.10	0.11	0.11	0.12	0.12	0.13	0.14	0.16	People's Rep.of China
0.87	0.89	0.91	1.17	1.20	1.26	1.24	1.16	1.17	Hong Kong, China
0.11	0.11	0.11	0.12	0.13	0.13	0.14	0.15	0.16	**China**
0.31	0.34	0.23	0.17	0.18	0.21	0.19	0.18	0.14	Albania
1.33	1.01	0.69	0.66	0.75	0.69	0.75	0.67	0.57	Bulgaria
2.11	2.16	2.40	2.58	2.58	2.93	2.66	2.83	2.74	Cyprus
1.99	2.30	2.59	3.23	3.67	4.84	4.60	4.60	4.60	Gibraltar
1.68	1.67	1.71	1.74	2.03	1.95	2.17	2.39	2.52	Malta
0.75	0.79	0.63	0.52	0.56	0.51	0.58	0.58	0.56	Romania
1.03	0.89	0.77	0.67	0.55	0.59	0.64	0.63	0.61	Slovak Republic
-	-	-	0.14	0.45	0.30	0.35	0.40	0.38	*Bosnia-Herzegovina*
-	-	-	0.69	0.72	0.76	0.83	0.79	0.84	*Croatia*
-	-	-	0.57	0.60	0.39	0.43	0.63	0.53	*FYROM*
-	-	-	0.82	1.00	1.07	1.18	1.37	1.36	*Slovenia*
-	-	-	0.20	0.15	0.14	0.15	0.26	0.34	*FR of Yugoslavia*
0.71	0.70	0.58	0.36	0.41	0.38	0.42	0.50	0.54	Former Yugoslavia
0.86	0.82	0.66	0.53	0.56	0.53	0.58	0.61	0.60	**Non-OECD Europe**

Due to lack of complete series Albania, the Dem. Ppl's Rep. of Korea and Vietnam are not included in the regional aggregates.

Oil Supply/Population (toe per capita)
Approvisionnement en pétrole/Population (tep par habitant)
Ölaufkommen/Bevölkerung (toe pro Kopf)
Totale disponibilita' primaria petrolio/POP (tep per capita)
一人当たり石油供給量（石油換算トン／人）
Suministro de petróleo/población (tep per capita)
Поставка нефти и нефтепродуктов / Численность населения (тнэ на человека)

	1971	1973	1978	1983	1984	1985	1986	1987	1988
Arménie	-	-	-	-	-	-	-	-	-
Azerbaïdjan	-	-	-	-	-	-	-	-	-
Bélarus	-	-	-	-	-	-	-	-	-
Estonie	-	-	-	-	-	-	-	-	-
Géorgie	-	-	-	-	-	-	-	-	-
Kazakhstan	-	-	-	-	-	-	-	-	-
Kirghizistan	-	-	-	-	-	-	-	-	-
Lettonie	-	-	-	-	-	-	-	-	-
Lituanie	-	-	-	-	-	-	-	-	-
Moldova	-	-	-	-	-	-	-	-	-
Russie	-	-	-	-	-	-	-	-	-
Tadjikistan	-	-	-	-	-	-	-	-	-
Turkménistan	-	-	-	-	-	-	-	-	-
Ukraine	-	-	-	-	-	-	-	-	-
Ouzbékistan	-	-	-	-	-	-	-	-	-
Ex-URSS	1.13	1.29	1.56	1.59	1.57	1.55	1.52	1.55	1.53
Bahrein	3.68	3.82	2.61	2.09	2.45	2.08	1.93	2.73	1.92
Iran	0.45	0.53	0.79	0.76	0.87	0.85	0.86	0.87	0.85
Irak	0.38	0.41	0.62	0.94	0.96	1.01	0.95	0.98	1.00
Israël	2.03	2.61	2.05	1.69	1.57	1.44	1.65	1.80	2.10
Jordanie	0.34	0.39	0.66	1.05	1.10	1.07	1.09	1.11	1.06
Koweit	3.99	2.88	2.81	4.77	4.35	5.72	3.78	3.09	2.98
Liban	0.67	0.86	0.70	0.75	0.71	0.80	0.83	0.83	0.61
Oman	0.12	0.13	0.55	0.85	1.01	1.12	0.50	0.42	0.68
Qatar	0.83	1.12	2.48	1.86	1.63	1.78	1.24	1.65	1.61
Arabie saoudite	0.84	0.96	1.89	3.66	2.92	2.95	2.83	2.85	2.97
Syrie	0.42	0.36	0.59	0.83	0.91	0.82	0.81	0.87	0.83
Emirats arabes unis	0.88	1.26	3.02	6.43	5.30	5.88	5.50	5.09	4.88
Yémen	0.11	0.14	0.14	0.20	0.20	0.22	0.23	0.23	0.24
Moyen-Orient	0.57	0.65	0.90	1.24	1.20	1.22	1.17	1.19	1.20
Total non-OCDE	0.22	0.25	0.30	0.29	0.29	0.29	0.29	0.29	0.30
OCDE Amérique du N.	2.95	3.25	3.41	2.58	2.62	2.60	2.65	2.68	2.73
OCDE Pacifique	1.56	1.87	1.87	1.52	1.54	1.45	1.49	1.49	1.62
OCDE Europe	1.49	1.68	1.61	1.27	1.28	1.26	1.28	1.29	1.29
Total OCDE	1.97	2.22	2.25	1.76	1.78	1.75	1.79	1.80	1.85
Monde	0.64	0.71	0.73	0.61	0.61	0.59	0.60	0.61	0.61

Faute de séries complètes, l'Albanie, la Rép.pop.dém. de Corée et le Viêt-Nam ne sont pas pris en compte dans les totaux régionaux.
Le total Monde ne comprend pas les soutages maritimes internationaux.

Oil Supply/Population (toe per capita)
Approvisionnement en pétrole/Population (tep par habitant)
Ölaufkommen/Bevölkerung (toe pro Kopf)
Totale disponibilita' primaria petrolio/POP (tep per capita)
一人当たり石油供給量（石油換算トン／人）
Suministro de petróleo/población (tep per capita)
Поставка нефти и нефтепродуктов / Численность населения (тнэ на человека)

1989	1990	1991	1992	1993	1994	1995	1996	1997	
-	-	-	0.66	0.33	0.10	0.07	0.04	0.04	Armenia
-	-	-	0.92	1.04	1.17	0.93	0.92	0.91	Azerbaijan
-	-	-	2.07	1.38	1.17	1.04	1.02	0.88	Belarus
-	-	-	0.98	1.04	1.03	0.83	0.88	0.87	Estonia
-	-	-	0.26	0.14	0.07	0.03	0.16	0.18	Georgia
-	-	-	1.46	0.97	0.80	0.72	0.66	0.58	Kazakhstan
-	-	-	0.43	0.26	0.09	0.14	0.15	0.11	Kyrgyzstan
-	-	-	1.09	0.96	0.99	0.81	0.83	0.71	Latvia
-	-	-	1.16	1.02	0.93	0.91	0.82	0.87	Lithuania
-	-	-	0.67	0.48	0.26	0.24	0.22	0.21	Moldova
-	-	-	1.60	1.35	1.02	0.99	0.88	0.87	Russia
-	-	-	1.06	0.66	0.21	0.21	0.21	0.20	Tajikistan
-	-	-	1.25	0.66	0.81	0.78	0.72	0.79	Turkmenistan
-	-	-	0.80	0.56	0.46	0.49	0.40	0.36	Ukraine
-	-	-	0.41	0.43	0.36	0.30	0.28	0.29	Uzbekistan
1.51	1.41	1.40	1.26	1.01	0.79	0.76	0.68	0.66	**Former USSR**
1.56	1.54	1.39	1.54	1.50	1.72	1.42	4.05	5.44	Bahrain
0.92	0.93	0.96	1.00	1.00	1.08	1.07	1.10	1.11	Iran
1.02	0.95	0.75	0.89	0.95	1.01	1.05	1.02	1.07	Iraq
1.98	1.97	1.83	1.97	1.98	2.00	2.26	1.90	2.00	Israel
1.03	1.05	0.95	1.03	0.98	1.02	0.98	1.00	1.04	Jordan
4.05	3.39	1.43	3.98	3.78	6.67	7.28	5.73	6.47	Kuwait
0.48	0.59	0.73	0.72	0.88	0.95	1.07	1.07	1.17	Lebanon
0.70	1.17	1.96	1.61	1.34	1.17	1.33	1.28	1.10	Oman
3.36	1.32	1.68	1.53	1.49	1.14	1.26	1.47	1.93	Qatar
2.57	2.40	2.89	3.02	3.07	2.74	2.67	2.91	3.03	Saudi Arabia
0.77	0.83	0.87	0.83	0.80	0.80	0.80	0.81	0.79	Syria
5.07	4.18	3.94	4.07	4.08	4.35	3.66	4.00	2.77	United Arab Emirates
0.25	0.22	0.22	0.24	0.19	0.19	0.20	0.20	0.20	Yemen
1.20	1.16	1.17	1.26	1.26	1.31	1.31	1.34	1.37	**Middle East**
0.29	0.29	0.29	0.28	0.27	0.26	0.26	0.27	0.28	**Non-OECD Total**
2.72	2.58	2.51	2.53	2.54	2.58	2.53	2.57	2.62	OECD North America
1.70	1.82	1.83	1.94	1.96	2.06	2.10	2.16	2.17	OECD Pacific
1.29	1.29	1.31	1.32	1.31	1.30	1.32	1.35	1.35	OECD Europe
1.86	1.83	1.82	1.85	1.86	1.89	1.89	1.93	1.95	**OECD Total**
0.61	0.60	0.60	0.59	0.58	0.58	0.58	0.59	0.60	**World**

Due to lack of complete series Albania, the Dem. Ppl's Rep. of Korea and Vietnam are not included in the regional aggregates.
World Total excludes international marine bunkers.

Electricity Consumption/GDP (kWh per 90 US$)
Consommation d'électricité/PIB (kWh par $US 90)
Elekrizitätsverbrauch/BIP (kWh pro 90 US$)
Totale consumi finali energia elettrica/PIL (kWh per 90 US$)
電力消費量／GDP （kWh／米ドル、1990年価格）
Consumo de electricidad/PIB (kWh por 90 US$)
Потребление электроэнергии / ВВП (КВт.ч на тыс.долл.США 1990 г.)

	1971	1973	1978	1983	1984	1985	1986	1987	1988
Algérie	0.08	0.08	0.11	0.16	0.16	0.17	0.18	0.18	0.20
Angola	0.04	0.05	0.05	0.07	0.07	0.07	0.07	0.06	0.06
Bénin	0.03	0.05	0.14	0.10	0.08	0.08	0.08	0.08	0.09
Cameroun	0.21	0.20	0.15	0.17	0.16	0.16	0.15	0.15	0.17
Congo	0.09	0.08	0.09	0.11	0.10	0.13	0.15	0.16	0.16
RD du Congo	0.38	0.39	0.44	0.44	0.44	0.47	0.47	0.46	0.47
Egypte	0.36	0.36	0.43	0.53	0.53	0.49	0.58	0.61	0.61
Ethiopie	0.13	0.12	0.11	0.10	0.11	0.12	0.11	0.11	0.10
Gabon	0.04	0.05	0.09	0.14	0.14	0.15	0.16	0.19	0.17
Ghana	0.58	0.77	0.73	0.51	0.27	0.51	0.76	0.85	0.85
Côte d'Ivoire	0.08	0.10	0.11	0.17	0.21	0.18	0.17	0.17	0.18
Kenya	0.25	0.24	0.28	0.29	0.30	0.32	0.32	0.33	0.31
Libye	0.02	0.04	0.09	0.21	0.30	0.42	0.52	0.63	0.66
Maroc	0.19	0.23	0.25	0.32	0.32	0.32	0.31	0.33	0.34
Mozambique	0.33	0.36	0.57	0.47	0.39	0.38	0.47	0.45	0.34
Nigéria	0.07	0.10	0.16	0.28	0.26	0.26	0.30	0.31	0.30
Sénégal	0.09	0.11	0.13	0.13	0.15	0.14	0.13	0.13	0.13
Afrique du Sud	0.73	0.81	1.02	1.24	1.28	1.38	1.44	1.42	1.43
Soudan	0.06	0.09	0.08	0.07	0.11	0.13	0.14	0.14	0.13
Tanzanie	0.19	0.21	0.21	0.23	0.24	0.25	0.29	0.29	0.31
Tunisie	0.17	0.19	0.24	0.33	0.33	0.33	0.36	0.36	0.39
Zambie	1.50	1.62	1.73	1.97	1.96	1.90	1.78	1.77	1.61
Zimbabwe	0.78	0.98	1.23	1.09	1.08	1.18	1.12	1.16	1.09
Autre Afrique	0.10	0.11	0.13	0.16	0.17	0.17	0.17	0.18	0.17
Afrique	0.31	0.34	0.41	0.51	0.53	0.55	0.58	0.60	0.61
Argentine	0.17	0.18	0.21	0.24	0.25	0.27	0.26	0.28	0.29
Bolivie	0.27	0.27	0.31	0.33	0.33	0.34	0.35	0.34	0.36
Brésil	0.22	0.22	0.29	0.39	0.41	0.41	0.41	0.41	0.43
Chili	0.44	0.48	0.52	0.57	0.57	0.55	0.55	0.54	0.54
Colombie	0.47	0.53	0.55	0.68	0.71	0.70	0.73	0.73	0.68
Costa Rica	0.41	0.42	0.45	0.57	0.58	0.60	0.63	0.65	0.65
Cuba	0.72	0.69	0.75	0.79	0.78	0.73	0.78	0.84	0.87
République dominicaine	0.34	0.50	0.46	0.37	0.46	0.50	0.50	0.46	0.37
Equateur	0.23	0.19	0.28	0.40	0.39	0.39	0.39	0.46	0.43
El Salvador	0.14	0.16	0.20	0.29	0.30	0.31	0.32	0.33	0.34
Guatemala	0.16	0.17	0.22	0.20	0.21	0.22	0.23	0.24	0.25
Haiti	0.03	0.04	0.07	0.10	0.10	0.11	0.12	0.12	0.12
Honduras	0.24	0.25	0.28	0.40	0.40	0.47	0.48	0.54	0.56
Jamaïque	0.42	0.48	0.37	0.39	0.39	0.41	0.42	0.42	0.39
Antilles néerlandaises	0.51	0.51	0.49	0.49	0.47	0.47	0.36	0.39	0.38
Nicaragua	0.47	0.45	0.60	0.79	0.80	0.83	0.83	0.91	0.94
Panama	0.24	0.29	0.32	0.43	0.40	0.39	0.39	0.43	0.46
Paraguay	0.12	0.14	0.18	0.23	0.25	0.25	0.29	0.33	0.36
Pérou	0.18	0.19	0.21	0.26	0.27	0.27	0.26	0.26	0.27
Trinité-et-Tobago	0.37	0.36	0.35	0.52	0.51	0.53	0.60	0.66	0.63
Uruguay	0.32	0.32	0.35	0.45	0.46	0.46	0.44	0.44	0.47
Vénézuela	0.35	0.38	0.48	0.87	0.90	0.96	0.92	0.95	0.97
Autre Amérique latine	0.30	0.33	0.33	0.30	0.30	0.31	0.31	0.29	0.30
Amérique latine	0.25	0.25	0.31	0.40	0.42	0.43	0.42	0.43	0.45

Electricity Consumption/GDP (kWh per 90 US$)
Consommation d'électricité/PIB (kWh par $US 90)
Elekrizitätsverbrauch/BIP (kWh pro 90 US$)
Totale consumi finali energia elettrica/PIL (kWh per 90 US$)
電力消費量／GDP （kWh／米ドル、1990年価格）
Consumo de electricidad/PIB (kWh por 90 US$)
Потребление электроэнергии / ВВП (КВт.ч на тыс.долл.США 1990 г.)

1989	1990	1991	1992	1993	1994	1995	1996	1997	
0.21	0.22	0.24	0.24	0.25	0.26	0.26	0.26	0.28	Algeria
0.06	0.06	0.06	0.07	0.09	0.09	0.08	0.08	0.08	Angola
0.09	0.09	0.10	0.10	0.10	0.10	0.11	0.11	0.10	Benin
0.18	0.19	0.20	0.21	0.22	0.20	0.19	0.20	0.20	Cameroon
0.17	0.20	0.19	0.18	0.19	0.20	0.19	0.18	0.18	Congo
0.57	0.53	0.58	0.74	0.78	0.82	0.91	0.93	0.95	DR of Congo
0.61	0.62	0.65	0.63	0.65	0.66	0.66	0.63	0.63	Egypt
0.10	0.14	0.16	0.16	0.16	0.15	0.14	0.13	0.12	Ethiopia
0.14	0.14	0.13	0.13	0.13	0.13	0.12	0.12	0.12	Gabon
0.83	0.84	0.85	0.88	0.82	0.82	0.79	0.73	0.73	Ghana
0.17	0.18	0.18	0.10	0.18	0.18	0.21	0.20	0.21	Ivory Coast
0.31	0.32	0.33	0.34	0.36	0.37	0.36	0.37	0.37	Kenya
0.67	0.61	0.59	0.61	0.64	0.70	0.70	0.70	0.69	Libya
0.33	0.35	0.34	0.39	0.40	0.39	0.43	0.40	0.43	Morocco
0.36	0.38	0.48	0.48	0.38	0.33	0.36	0.35	0.37	Mozambique
0.31	0.24	0.27	0.28	0.24	0.31	0.27	0.26	0.26	Nigeria
0.13	0.14	0.14	0.15	0.15	0.16	0.16	0.15	0.15	Senegal
1.43	1.46	1.49	1.52	1.52	1.55	1.56	1.57	1.61	South Africa
0.12	0.14	0.14	0.13	0.12	0.12	0.11	0.11	0.10	Sudan
0.32	0.35	0.36	0.40	0.36	0.36	0.39	0.41	0.38	Tanzania
0.40	0.40	0.40	0.40	0.41	0.43	0.45	0.42	0.42	Tunisia
1.23	1.11	1.47	1.47	1.38	1.43	1.46	1.38	1.37	Zambia
1.09	1.04	1.02	1.17	0.97	0.99	1.05	1.00	1.03	Zimbabwe
0.18	0.18	0.18	0.19	0.19	0.20	0.19	0.19	0.18	Other Africa
0.61	0.61	0.62	0.63	0.63	0.65	0.66	0.64	0.65	**Africa**
0.31	0.30	0.28	0.27	0.27	0.26	0.29	0.31	0.30	Argentina
0.38	0.38	0.38	0.40	0.43	0.44	0.45	0.46	0.48	Bolivia
0.44	0.47	0.48	0.49	0.49	0.48	0.49	0.50	0.51	Brazil
0.54	0.54	0.54	0.54	0.54	0.54	0.55	0.57	0.58	Chile
0.71	0.71	0.72	0.64	0.67	0.68	0.68	0.68	0.68	Colombia
0.59	0.59	0.62	0.60	0.61	0.63	0.63	0.64	0.69	Costa Rica
0.89	0.91	0.89	0.87	0.96	1.02	1.02	1.02	1.09	Cuba
0.31	0.39	0.41	0.52	0.55	0.56	0.56	0.55	0.53	Dominican Republic
0.43	0.46	0.48	0.48	0.50	0.54	0.54	0.57	0.56	Ecuador
0.34	0.35	0.36	0.36	0.38	0.40	0.41	0.41	0.43	El Salvador
0.26	0.26	0.27	0.29	0.31	0.31	0.33	0.38	0.42	Guatemala
0.13	0.14	0.10	0.11	0.09	0.07	0.10	0.11	0.14	Haiti
0.54	0.60	0.57	0.52	0.53	0.49	0.56	0.64	0.64	Honduras
0.35	0.39	0.39	0.41	0.86	0.96	1.16	1.22	1.30	Jamaica
0.39	0.41	0.41	0.43	0.46	0.48	0.49	0.48	0.48	Netherlands Antilles
0.99	1.10	1.12	1.14	1.12	1.06	1.08	0.93	0.98	Nicaragua
0.44	0.43	0.40	0.39	0.40	0.40	0.42	0.44	0.43	Panama
0.36	0.40	0.41	0.44	0.50	0.55	0.61	0.61	0.63	Paraguay
0.31	0.30	0.30	0.28	0.30	0.27	0.28	0.29	0.28	Peru
0.61	0.65	0.56	0.69	0.69	0.69	0.72	0.73	0.79	Trinidad-and-Tobago
0.45	0.46	0.49	0.46	0.48	0.47	0.51	0.50	0.51	Uruguay
1.05	1.00	0.96	0.97	1.00	1.03	1.01	1.05	0.98	Venezuela
0.33	0.32	0.32	0.33	0.34	0.34	0.34	0.34	0.34	Other Latin America
0.46	0.48	0.48	0.48	0.48	0.48	0.49	0.50	0.51	**Latin America**

Electricity Consumption/GDP (kWh per 90 US$)
Consommation d'électricité/PIB (kWh par $US 90)
Elekrizitätsverbrauch/BIP (kWh pro 90 US$)
Totale consumi finali energia elettrica/PIL (kWh per 90 US$)
電力消費量／GDP （kWh／米ドル、1990年価格）
Consumo de electricidad/PIB (kWh por 90 US$)
Потребление электроэнергии / ВВП (КВт.ч на тыс.долл.США 1990 г.)

	1971	1973	1978	1983	1984	1985	1986	1987	1988
Bangladesh	0.06	0.10	0.13	0.15	0.16	0.17	0.19	0.19	0.21
Brunei	0.07	0.08	0.09	0.18	0.20	0.21	0.22	0.28	0.30
Inde	0.42	0.45	0.53	0.60	0.65	0.66	0.69	0.72	0.72
Indonésie	0.08	0.08	0.08	0.15	0.15	0.17	0.18	0.20	0.22
Malaisie	0.29	0.30	0.36	0.40	0.40	0.43	0.45	0.46	0.47
Myanmar	0.04	0.05	0.05	0.05	0.05	0.06	0.06	0.07	0.07
Népal	0.04	0.04	0.07	0.11	0.11	0.11	0.12	0.14	0.13
Pakistan	0.41	0.43	0.48	0.56	0.59	0.62	0.65	0.67	0.72
Philippines	0.39	0.49	0.44	0.46	0.48	0.54	0.50	0.48	0.50
Singapour	0.27	0.31	0.36	0.35	0.35	0.38	0.39	0.41	0.40
Sri Lanka	0.21	0.24	0.25	0.29	0.29	0.31	0.32	0.32	0.32
Taipei chinois	0.42	0.43	0.50	0.49	0.48	0.49	0.50	0.49	0.50
Thaïlande	0.21	0.26	0.33	0.37	0.39	0.40	0.42	0.43	0.43
Viêt-Nam	-	-	-	-	0.78	0.81	0.84	0.91	0.96
Autre Asie	0.20	0.28	0.24	0.20	0.22	0.23	0.24	0.25	0.26
Asie	0.32	0.35	0.39	0.43	0.45	0.47	0.48	0.50	0.51
Rép. populaire de Chine	1.36	1.47	1.83	1.60	1.49	1.44	1.45	1.43	1.42
Hong-Kong, Chine	0.30	0.27	0.28	0.30	0.29	0.31	0.31	0.33	0.33
Chine	1.20	1.28	1.54	1.35	1.27	1.25	1.26	1.25	1.24
Albanie	-	-	-	1.49	1.46	1.15	1.98	1.66	1.49
Bulgarie	1.53	1.66	1.99	2.26	2.28	2.18	2.08	2.04	1.88
Chypre	0.27	0.32	0.33	0.33	0.31	0.31	0.32	0.32	0.32
Gibraltar	0.25	0.22	0.19	0.23	0.23	0.23	0.23	0.21	0.22
Malte	0.51	0.53	0.34	0.37	0.38	0.41	0.43	0.46	0.46
Roumanie	1.52	1.64	1.65	1.60	1.56	1.60	1.66	1.74	1.82
République slovaque	1.17	1.25	1.31	1.68	1.70	1.73	1.71	1.72	1.72
Bosnie-Herzégovine	-	-	-	-	-	-	-	-	-
Croatie	-	-	-	-	-	-	-	-	-
Ex-RYM	-	-	-	-	-	-	-	-	-
Slovénie	-	-	-	-	-	-	-	-	-
RF de Yougoslavie	-	-	-	-	-	-	-	-	-
Ex-Yougoslavie	0.46	0.52	0.53	0.66	0.69	0.71	0.72	0.76	0.80
Europe non-OCDE	0.88	0.97	1.00	1.12	1.14	1.16	1.17	1.20	1.24
Arménie	-	-	-	-	-	-	-	-	-
Azerbaïdjan	-	-	-	-	-	-	-	-	-
Bélarus	-	-	-	-	-	-	-	-	-
Estonie	-	-	-	-	-	-	-	-	-
Géorgie	-	-	-	-	-	-	-	-	-
Kazakhstan	-	-	-	-	-	-	-	-	-
Kirghizistan	-	-	-	-	-	-	-	-	-
Lettonie	-	-	-	-	-	-	-	-	-
Lituanie	-	-	-	-	-	-	-	-	-
Moldova	-	-	-	-	-	-	-	-	-
Russie	-	-	-	-	-	-	-	-	-
Tadjikistan	-	-	-	-	-	-	-	-	-
Turkménistan	-	-	-	-	-	-	-	-	-
Ukraine	-	-	-	-	-	-	-	-	-
Ouzbékistan	-	-	-	-	-	-	-	-	-
Ex-URSS	1.23	1.28	1.43	1.51	1.56	1.59	1.59	1.62	1.64

Faute de séries complètes, l'Albanie, la Rép.pop.dém. de Corée et le Viêt-Nam ne sont pas pris en compte dans les totaux régionaux.

Electricity Consumption/GDP (kWh per 90 US$)
Consommation d'électricité/PIB (kWh par $US 90)
Elekrizitätsverbrauch/BIP (kWh pro 90 US$)
Totale consumi finali energia elettrica/PIL (kWh per 90 US$)
電力消費量／GDP （kWh／米ドル、1990年価格）
Consumo de electricidad/PIB (kWh por 90 US$)
Потребление электроэнергии / ВВП (КВт.ч на тыс.долл.США 1990 г.)

1989	1990	1991	1992	1993	1994	1995	1996	1997	
0.25	0.23	0.23	0.27	0.30	0.31	0.33	0.34	0.33	Bangladesh
0.31	0.31	0.32	0.36	0.42	0.45	0.52	0.55	0.61	Brunei
0.75	0.78	0.85	0.86	0.88	0.86	0.87	0.85	0.86	India
0.24	0.27	0.26	0.28	0.28	0.30	0.29	0.33	0.35	Indonesia
0.47	0.49	0.49	0.53	0.53	0.61	0.63	0.65	0.69	Malaysia
0.07	0.08	0.07	0.07	0.08	0.08	0.08	0.08	0.08	Myanmar
0.14	0.15	0.16	0.18	0.18	0.16	0.17	0.18	0.17	Nepal
0.73	0.76	0.79	0.79	0.83	0.82	0.83	0.84	0.87	Pakistan
0.51	0.50	0.51	0.49	0.51	0.55	0.57	0.58	0.61	Philippines
0.40	0.41	0.40	0.39	0.38	0.38	0.38	0.38	0.40	Singapore
0.31	0.33	0.33	0.34	0.35	0.37	0.38	0.35	0.37	Sri Lanka
0.51	0.52	0.53	0.53	0.55	0.56	0.56	0.53	0.57	Chinese Taipei
0.45	0.47	0.49	0.52	0.54	0.55	0.58	0.59	0.63	Thailand
0.99	1.02	1.02	0.99	1.01	1.09	1.14	1.31	1.38	Vietnam
0.27	0.27	0.26	0.24	0.24	0.25	0.25	0.23	0.23	Other Asia
0.53	0.55	0.57	0.58	0.59	0.60	0.61	0.60	0.62	**Asia**
1.46	1.49	1.49	1.45	1.42	1.40	1.36	1.34	1.27	People's Rep.of China
0.31	0.32	0.32	0.31	0.31	0.31	0.31	0.31	0.30	Hong Kong, China
1.27	1.30	1.31	1.28	1.27	1.26	1.23	1.22	1.16	**China**
1.41	1.35	1.36	1.70	1.86	1.11	1.14	1.50	1.53	Albania
1.93	2.00	1.88	1.90	1.92	1.89	1.99	2.25	2.19	Bulgaria
0.33	0.34	0.35	0.37	0.39	0.39	0.34	0.35	0.35	Cyprus
0.24	0.24	0.23	0.26	0.25	0.27	0.27	0.27	0.27	Gibraltar
0.46	0.43	0.53	0.53	0.51	0.50	0.50	0.47	0.46	Malta
1.95	1.77	1.74	1.74	1.68	1.58	1.54	1.54	1.52	Romania
1.75	1.77	1.91	1.93	1.91	1.86	1.95	1.89	1.75	Slovak Republic
-	-	-	7.36	10.70	2.32	1.99	1.36	0.98	Bosnia-Herzegovina
-	-	-	1.28	1.31	1.24	1.26	1.23	1.25	Croatia
-	-	-	2.93	2.83	2.78	3.01	3.22	3.29	FYROM
-	-	-	0.47	0.45	0.46	0.44	0.43	0.43	Slovenia
-	-	-	0.77	0.77	0.83	0.87	0.85	0.89	FR of Yugoslavia
0.79	0.83	0.83	0.94	0.94	0.83	0.85	0.83	0.85	Former Yugoslavia
1.26	1.25	1.20	1.28	1.27	1.19	1.21	1.21	1.18	**Non-OECD Europe**
-	-	-	3.77	2.62	2.13	1.96	2.12	2.53	Armenia
-	-	-	2.59	3.16	3.50	3.79	3.82	3.72	Azerbaijan
-	-	-	1.28	1.23	1.25	1.26	1.23	1.17	Belarus
-	-	-	1.54	1.35	1.46	1.34	1.37	1.27	Estonia
-	-	-	1.68	2.37	1.89	1.84	1.77	1.58	Georgia
-	-	-	2.69	2.51	2.71	2.70	2.37	2.09	Kazakhstan
-	-	-	3.80	4.46	5.40	5.24	4.63	3.92	Kyrgyzstan
-	-	-	0.95	0.85	0.80	0.80	0.76	0.73	Latvia
-	-	-	1.19	1.11	1.23	1.20	1.22	1.15	Lithuania
-	-	-	1.39	1.35	1.62	1.48	1.60	1.45	Moldova
-	-	-	1.93	1.98	2.05	2.10	2.14	2.09	Russia
-	-	-	5.30	5.21	6.61	7.19	7.48	7.11	Tajikistan
-	-	-	1.36	1.60	1.64	1.72	1.66	1.21	Turkmenistan
-	-	-	1.74	1.86	2.11	2.30	2.31	2.29	Ukraine
-	-	-	2.23	2.21	2.25	2.22	2.20	2.11	Uzbekistan
1.61	1.63	1.70	1.84	1.89	1.99	2.05	2.07	2.00	**Former USSR**

Due to lack of complete series Albania, the Dem. Ppl's Rep. of Korea and Vietnam are not included in the regional aggregates.

Electricity Consumption/GDP (kWh per 90 US$)
Consommation d'électricité/PIB (kWh par $US 90)
Elekrizitätsverbrauch/BIP (kWh pro 90 US$)
Totale consumi finali energia elettrica/PIL (kWh per 90 US$)
電力消費量／GDP （kWh／米ドル、1990年価格）
Consumo de electricidad/PIB (kWh por 90 US$)
Потребление электроэнергии / ВВП (КВт.ч на тыс.долл.США 1990 г.)

	1971	1973	1978	1983	1984	1985	1986	1987	1988
Bahrein	0.22	0.22	0.34	0.64	0.66	0.84	0.84	0.94	0.85
Iran	0.11	0.13	0.14	0.24	0.26	0.27	0.32	0.36	0.40
Irak	0.04	0.04	0.06	0.14	0.17	0.22	0.29	0.29	0.34
Israël	0.30	0.30	0.34	0.33	0.34	0.35	0.34	0.34	0.37
Jordanie	0.16	0.23	0.23	0.44	0.48	0.51	0.51	0.57	0.60
Koweit	0.14	0.18	0.33	0.84	0.89	1.02	1.05	1.03	1.22
Liban	0.23	0.26	0.60	0.93	1.01	1.03	1.01	1.18	0.92
Oman	0.00	0.02	0.10	0.20	0.22	0.23	0.29	0.32	0.34
Qatar	0.10	0.10	0.24	0.48	0.51	0.65	0.64	0.62	0.62
Arabie saoudite	0.04	0.05	0.11	0.34	0.41	0.49	0.51	0.55	0.54
Syrie	0.14	0.14	0.15	0.27	0.29	0.30	0.32	0.31	0.30
Emirats arabes unis	0.02	0.05	0.16	0.31	0.32	0.37	0.49	0.49	0.55
Yémen	0.08	0.07	0.06	0.09	0.10	0.10	0.11	0.12	0.11
Moyen-Orient	0.10	0.11	0.14	0.28	0.31	0.36	0.41	0.43	0.46
Total non-OCDE	0.60	0.63	0.69	0.78	0.80	0.82	0.83	0.85	0.87
OCDE Amérique du N.	0.47	0.50	0.51	0.53	0.53	0.53	0.52	0.53	0.54
OCDE Pacifique	0.27	0.28	0.30	0.29	0.30	0.30	0.29	0.30	0.30
OCDE Europe	0.29	0.30	0.32	0.33	0.34	0.35	0.34	0.35	0.34
Total OCDE	0.36	0.37	0.39	0.40	0.40	0.40	0.40	0.41	0.41
Monde	0.40	0.42	0.45	0.47	0.48	0.49	0.49	0.50	0.50

Faute de séries complètes, l'Albanie, la Rép.pop.dém. de Corée et le Viêt-Nam ne sont pas pris en compte dans les totaux régionaux.

Electricity Consumption/GDP (kWh per 90 US$)
Consommation d'électricité/PIB (kWh par $US 90)
Elekrizitätsverbrauch/BIP (kWh pro 90 US$)
Totale consumi finali energia elettrica/PIL (kWh per 90 US$)
電力消費量／GDP （ｋWh／米ドル、1990年価格）
Consumo de electricidad/PIB (kWh por 90 US$)
Потребление электроэнергии / ВВП (КВт.ч на тыс.долл.США 1990 г.)

1989	1990	1991	1992	1993	1994	1995	1996	1997	
0.86	0.87	0.83	0.82	0.83	0.88	0.90	0.87	0.87	Bahrain
0.39	0.40	0.43	0.43	0.42	0.45	0.46	0.46	0.46	Iran
0.32	0.47	0.82	1.02	1.05	1.17	1.23	1.23	1.00	Iraq
0.39	0.37	0.35	0.38	0.38	0.39	0.39	0.40	0.40	Israel
0.74	0.83	0.83	0.84	0.86	0.85	0.88	0.94	0.96	Jordan
1.09	1.02	1.00	1.10	1.03	1.15	1.17	1.18	1.20	Kuwait
1.00	0.49	0.71	0.82	0.90	0.92	0.93	1.28	1.47	Lebanon
0.34	0.36	0.34	0.35	0.37	0.38	0.39	0.39	0.39	Oman
0.62	0.62	0.58	0.60	0.63	0.66	0.66	0.75	0.73	Qatar
0.57	0.56	0.55	0.57	0.63	0.69	0.72	0.74	0.77	Saudi Arabia
0.37	0.36	0.35	0.33	0.42	0.47	0.45	0.48	0.49	Syria
0.51	0.48	0.45	0.45	0.48	0.50	0.49	0.49	0.50	United Arab Emirates
0.11	0.13	0.14	0.15	0.15	0.14	0.13	0.13	0.12	Yemen
0.46	0.48	0.48	0.51	0.54	0.58	0.58	0.59	0.60	**Middle East**
0.87	0.88	0.89	0.89	0.88	0.87	0.87	0.86	0.85	**Non-OECD Total**
0.54	0.54	0.57	0.56	0.57	0.56	0.56	0.57	0.55	OECD North America
0.30	0.31	0.31	0.31	0.31	0.33	0.34	0.34	0.35	OECD Pacific
0.34	0.34	0.34	0.34	0.34	0.33	0.33	0.34	0.33	OECD Europe
0.41	0.41	0.42	0.41	0.42	0.42	0.42	0.42	0.42	**OECD Total**
0.50	0.50	0.51	0.51	0.51	0.51	0.51	0.51	0.51	**World**

Due to lack of complete series Albania, the Dem. Ppl's Rep. of Korea and Vietnam are not included in the regional aggregates.

Electricity Consumption/Population (kWh per capita)
Consommation d'électricité/Population (kWh par habitant)
Elekrizitätsverbrauch/Bevölkerung (kWh pro Kopf)
Totale consumi finali energia elettrica/POP (kWh per capita)
一人当たり電力消費量（kWh／人）
Consumo de electricidad/población (kWh per capita)
Потребление электроэнергии / Численность населения (КВт.ч на человека)

	1971	1973	1978	1983	1984	1985	1986	1987	1988
Algérie	141	166	288	429	453	478	498	468	495
Angola	98	126	66	77	77	75	73	72	70
Bénin	12	20	51	39	34	35	33	32	38
Cameroun	150	151	155	214	206	226	211	211	214
Congo	67	69	79	159	154	191	201	210	210
RD du Congo	160	165	146	132	134	140	141	136	135
Egypte	214	209	342	533	552	526	626	660	677
Ethiopie	19	18	15	19	20	19	19	19	18
Gabon	216	292	665	949	999	1 045	1 042	1 014	996
Ghana	317	397	343	175	98	189	285	321	329
Côte d'Ivoire	90	112	162	181	213	188	171	165	178
Kenya	68	75	91	97	98	104	107	116	109
Libye	246	502	1 207	2 061	2 538	3 128	3 387	3 852	3 964
Maroc	130	157	212	291	294	306	319	323	353
Mozambique	50	56	59	36	28	30	35	39	32
Nigéria	28	38	61	84	74	78	91	89	91
Sénégal	76	82	99	107	111	106	103	105	107
Afrique du Sud	2 246	2 531	3 200	3 951	4 212	4 298	4 386	4 348	4 428
Soudan	26	33	42	34	46	49	54	55	49
Tanzanie	30	34	34	34	34	36	40	41	44
Tunisie	154	192	303	450	467	482	501	524	555
Zambie	1 025	1 137	1 089	1 063	1 024	980	896	891	838
Zimbabwe	657	865	902	938	884	999	932	942	921
Autre Afrique	37	43	51	61	63	63	66	69	67
Afrique	238	270	335	412	430	443	462	465	472
Argentine	870	956	1 085	1 253	1 296	1 285	1 292	1 420	1 430
Bolivie	213	228	295	258	256	252	249	242	259
Brésil	449	541	850	1 108	1 207	1 284	1 357	1 372	1 426
Chili	783	777	855	925	985	1 005	1 045	1 071	1 138
Colombie	363	451	533	673	717	708	773	801	761
Costa Rica	646	721	895	956	1 028	1 044	1 112	1 178	1 172
Cuba	520	568	784	1 024	1 080	1 050	1 121	1 142	1 195
République dominicaine	235	408	418	372	451	483	491	483	384
Equateur	147	166	295	408	411	411	416	450	450
El Salvador	174	205	297	289	301	313	318	333	347
Guatemala	123	149	223	183	186	188	190	206	210
Haiti	13	17	36	53	54	55	59	57	59
Honduras	133	150	188	243	250	295	291	340	353
Jamaïque	819	1 007	633	617	600	587	613	650	623
Antilles néerlandaises	3 745	4 037	4 099	4 190	3 861	3 549	2 869	3 152	3 161
Nicaragua	275	266	373	341	335	323	314	329	293
Panama	529	670	756	1 042	996	995	993	1 056	968
Paraguay	87	111	194	272	299	300	341	390	436
Pérou	398	420	472	519	556	549	572	598	561
Trinité-et-Tobago	1 010	1 109	1 484	2 546	2 589	2 566	2 759	2 886	2 592
Uruguay	721	698	904	1 046	1 048	1 071	1 113	1 171	1 266
Vénézuela	1 098	1 200	1 588	2 230	2 304	2 392	2 400	2 486	2 627
Autre Amérique latine	1 091	1 238	1 367	1 346	1 358	1 444	1 485	1 427	1 552
Amérique latine	503	576	772	965	1 028	1 064	1 109	1 142	1 170

Electricity Consumption/Population (kWh per capita)
Consommation d'électricité/Population (kWh par habitant)
Flekrizitätsverbrauch/Bevölkerung (kWh pro Kopf)
Totale consumi finali energia elettrica/POP (kWh per capita)
一人当たり電力消費量（kWh／人）
Consumo de electricidad/población (kWh per capita)
Потребление электроэнергии / Численность населения (КВт.ч на человека)

1989	1990	1991	1992	1993	1994	1995	1996	1997	
537	554	565	559	564	566	574	582	631	Algeria
69	65	70	68	66	64	63	65	68	Angola
35	36	39	42	42	42	45	48	43	Benin
215	204	198	195	197	173	165	171	181	Cameroon
212	257	239	228	223	219	216	207	200	Congo
157	132	128	143	126	123	133	131	121	DR of Congo
703	733	760	759	782	805	834	813	845	Egypt
18	24	24	23	26	23	23	22	22	Ethiopia
829	844	822	799	781	791	770	776	786	Gabon
326	331	344	359	341	343	334	318	319	Ghana
164	167	158	86	148	148	180	184	192	Ivory Coast
112	116	117	118	120	122	122	127	128	Kenya
3 886	3 804	3 707	3 650	3 576	3 662	3 624	3 579	3 495	Libya
347	371	383	414	413	436	443	447	465	Morocco
36	39	51	51	49	43	45	45	54	Mozambique
100	81	92	95	83	102	89	89	88	Nigeria
103	107	106	113	109	115	114	115	119	Senegal
4 450	4 431	4 384	4 272	4 252	4 357	4 440	4 503	4 621	South Africa
46	53	51	52	50	52	51	52	49	Sudan
47	51	54	53	51	51	55	59	54	Tanzania
579	604	616	649	673	718	741	746	774	Tunisia
614	535	684	655	638	620	604	589	593	Zambia
940	937	942	955	783	835	865	867	901	Zimbabwe
69	70	71	71	73	74	75	75	73	Other Africa
475	473	475	468	464	476	482	482	490	**Africa**
1 370	1 285	1 332	1 393	1 437	1 509	1 584	1 714	1 810	Argentina
274	278	292	302	333	343	361	376	395	Bolivia
1 460	1 471	1 499	1 510	1 556	1 589	1 662	1 719	1 805	Brazil
1 226	1 254	1 331	1 476	1 534	1 608	1 766	1 949	2 105	Chile
796	819	830	751	817	860	893	892	902	Colombia
1 102	1 097	1 163	1 200	1 263	1 328	1 341	1 324	1 456	Costa Rica
1 231	1 209	1 043	897	834	892	913	967	1 057	Cuba
337	393	402	545	581	600	620	641	654	Dominican Republic
444	477	511	523	538	591	597	628	623	Ecuador
347	368	385	398	448	483	524	513	548	El Salvador
223	224	236	261	285	283	309	361	405	Guatemala
62	64	49	44	34	23	35	39	48	Haiti
346	376	358	334	348	309	358	410	417	Honduras
587	690	689	719	1 534	1 706	2 061	2 122	2 184	Jamaica
3 324	3 568	3 570	3 791	4 015	4 213	4 312	4 284	4 390	Netherlands Antilles
295	319	315	312	298	281	293	308	334	Nicaragua
913	952	951	980	1 043	1 063	1 117	1 170	1 171	Panama
449	505	506	535	621	694	775	770	798	Paraguay
559	502	520	479	530	525	590	610	618	Peru
2 491	2 652	2 616	2 806	2 768	2 830	3 024	3 154	3 510	Trinidad-and-Tobago
1 220	1 246	1 347	1 369	1 458	1 490	1 581	1 639	1 735	Uruguay
2 528	2 494	2 555	2 687	2 711	2 671	2 653	2 684	2 588	Venezuela
1 752	1 779	1 841	1 892	1 928	1 931	1 960	1 976	2 041	Other Latin America
1 180	1 180	1 202	1 216	1 261	1 291	1 349	1 399	1 453	**Latin America**

Electricity Consumption/Population (kWh per capita)
Consommation d'électricité/Population (kWh par habitant)
Elekrizitätsverbrauch/Bevölkerung (kWh pro Kopf)
Totale consumi finali energia elettrica/POP (kWh per capita)
一人当たり電力消費量（kWh／人）
Consumo de electricidad/población (kWh per capita)
Потребление электроэнергии / Численность населения (КВт.ч на человека)

	1971	1973	1978	1983	1984	1985	1986	1987	1988
Bangladesh	10	14	20	27	30	31	35	36	39
Brunei	1 103	1 351	1 906	3 085	3 300	3 278	3 231	4 234	4 331
Inde	99	102	138	168	184	195	209	222	241
Indonésie	21	24	31	68	73	83	94	107	124
RPD de Corée	1 041	1 164	1 688	2 014	2 182	2 295	2 341	2 316	2 422
Malaisie	311	372	566	772	823	836	871	908	971
Myanmar	20	23	31	33	36	42	43	43	38
Népal	6	6	11	17	18	19	20	23	24
Pakistan	89	96	114	164	178	193	210	225	253
Philippines	225	307	318	360	339	348	324	317	340
Singapour	1 284	1 784	2 570	3 394	3 640	3 795	4 011	4 479	4 854
Sri Lanka	58	67	83	117	121	131	139	140	144
Taipei chinois	968	1 215	1 946	2 413	2 578	2 736	3 053	3 328	3 636
Thaïlande	124	166	266	354	386	413	444	494	551
Viêt-Nam	41	40	55	58	67	70	73	80	87
Autre Asie	124	184	170	175	188	202	214	217	217
Asie	103	117	159	200	215	228	244	260	283
Rép. populaire de Chine	165	189	268	317	336	363	391	427	462
Hong-Kong, Chine	1 309	1 440	1 981	2 655	2 801	2 918	3 197	3 840	4 073
Chine	170	195	277	329	349	376	406	444	480
Albanie	532	593	1 100	1 073	1 021	798	1 424	1 165	1 007
Bulgarie	2 293	2 683	3 648	4 598	4 791	4 689	4 647	4 839	4 941
Chypre	1 002	1 267	1 418	1 818	1 836	1 918	2 040	2 138	2 316
Gibraltar	1 607	1 621	1 724	2 034	2 034	2 103	2 172	2 067	2 200
Malte	941	1 134	1 300	1 723	1 809	2 064	2 232	2 477	2 695
Roumanie	1 631	1 908	2 677	2 976	3 082	3 134	3 321	3 277	3 393
République slovaque	2 679	3 018	3 430	4 489	4 593	4 812	4 904	5 058	5 126
Bosnie-Herzégovine	-	-	-	-	-	-	-	-	-
Croatie	-	-	-	-	-	-	-	-	-
Ex-RYM	-	-	-	-	-	-	-	-	-
Slovénie	-	-	-	-	-	-	-	-	-
RF de Yougoslavie	-	-	-	-	-	-	-	-	-
Ex-Yougoslavie	1 270	1 491	2 076	2 720	2 889	2 958	3 063	3 183	3 264
Europe non-OCDE	1 675	1 952	2 639	3 228	3 369	3 420	3 533	3 605	3 702

Faute de séries complètes, l'Albanie, la Rép.pop.dém. de Corée et le Viêt-Nam ne sont pas pris en compte dans les totaux régionaux.

Electricity Consumption/Population (kWh per capita)
Consommation d'électricité/Population (kWh par habitant)
Elekrizitätsverbrauch/Bevölkerung (kWh pro Kopf)
Totale consumi finali energia elettrica/POP (kWh per capita)
一人当たり電力消費量（kWh／人）
Consumo de electricidad/población (kWh per capita)
Потребление электроэнергии / Численность населения (КВт.ч на человека)

1989	1990	1991	1992	1993	1994	1995	1996	1997	
47	47	47	57	64	68	75	79	81	Bangladesh
4 333	4 350	4 498	4 853	5 315	5 627	6 452	6 910	7 721	Brunei
260	280	301	315	332	347	371	380	397	India
142	172	181	203	216	239	249	298	330	Indonesia
2 409	2 370	1 370	288	283	271	259	249	238	DPR of Korea
1 040	1 140	1 221	1 385	1 475	1 794	1 992	2 162	2 425	Malaysia
41	45	42	45	52	55	59	59	62	Myanmar
26	29	31	36	35	34	36	39	39	Nepal
263	277	297	312	326	327	337	349	351	Pakistan
357	352	350	332	341	375	400	425	452	Philippines
5 134	5 613	5 840	5 979	6 275	6 771	7 117	7 561	8 305	Singapore
141	154	160	169	186	202	217	205	229	Sri Lanka
3 941	4 182	4 569	4 850	5 279	5 656	6 018	5 944	6 752	Chinese Taipei
628	722	802	901	1 013	1 109	1 249	1 343	1 416	Thailand
94	100	103	106	115	132	148	182	206	Vietnam
217	217	209	188	171	168	176	165	162	Other Asia
307	331	354	375	398	425	454	471	504	**Asia**
487	511	549	602	663	728	771	825	846	People's Rep.of China
3 937	4 178	4 399	4 507	4 699	4 866	4 850	5 013	4 959	Hong Kong, China
505	529	568	622	683	748	792	846	867	**China**
1 021	862	633	751	905	586	645	919	860	Albania
4 963	4 759	4 136	3 907	3 934	3 950	4 310	4 391	3 999	Bulgaria
2 569	2 736	2 800	3 177	3 414	3 521	3 217	3 301	3 398	Cyprus
2 452	2 452	2 452	3 036	2 964	3 143	3 034	3 207	3 207	Gibraltar
2 849	2 825	3 656	3 795	3 781	3 804	3 989	3 861	3 907	Malta
3 413	2 924	2 501	2 320	2 273	2 235	2 329	2 432	2 252	Romania
5 245	5 193	4 782	4 476	4 256	4 339	4 871	5 027	4 931	Slovak Republic
-	-	-	3 179	3 483	570	737	841	808	*Bosnia-Herzegovina*
-	-	-	2 132	2 135	2 147	2 240	2 299	2 477	*Croatia*
-	-	-	3 065	2 683	2 568	2 722	2 868	2 944	*FYROM*
-	-	-	4 813	4 884	5 108	5 180	5 215	5 430	*Slovenia*
-	-	-	3 127	2 875	2 957	3 196	3 254	3 445	*FR of Yugoslavia*
3 241	3 192	3 204	2 988	2 841	2 509	2 691	2 758	2 901	Former Yugoslavia
3 715	3 476	3 203	3 004	2 917	2 787	2 987	3 076	3 003	**Non-OECD Europe**

Due to lack of complete series Albania, the Dem. Ppl's Rep. of Korea and Vietnam are not included in the regional aggregates.

Electricity Consumption/Population (kWh per capita)
Consommation d'électricité/Population (kWh par habitant)
Elekrizitätsverbrauch/Bevölkerung (kWh pro Kopf)
Totale consumi finali energia elettrica/POP (kWh per capita)
一人当たり電力消費量（kWh／人）
Consumo de electricidad/población (kWh per capita)
Потребление электроэнергии / Численность населения (КВт.ч на человека)

	1971	1973	1978	1983	1984	1985	1986	1987	1988
Arménie	-	-	-	-	-	-	-	-	-
Azerbaïdjan	-	-	-	-	-	-	-	-	-
Bélarus	-	-	-	-	-	-	-	-	-
Estonie	-	-	-	-	-	-	-	-	-
Géorgie	-	-	-	-	-	-	-	-	-
Kazakhstan	-	-	-	-	-	-	-	-	-
Kirghizistan	-	-	-	-	-	-	-	-	-
Lettonie	-	-	-	-	-	-	-	-	-
Lituanie	-	-	-	-	-	-	-	-	-
Moldova	-	-	-	-	-	-	-	-	-
Russie	-	-	-	-	-	-	-	-	-
Tadjikistan	-	-	-	-	-	-	-	-	-
Turkménistan	-	-	-	-	-	-	-	-	-
Ukraine	-	-	-	-	-	-	-	-	-
Ouzbékistan	-	-	-	-	-	-	-	-	-
Ex-URSS	2 983	3 333	4 175	4 684	4 867	4 967	5 105	5 255	5 346
Bahrein	1 991	2 101	4 000	5 682	5 939	6 911	6 719	7 240	7 019
Iran	245	338	476	662	706	728	744	801	802
Irak	275	324	611	1 072	1 248	1 301	1 337	1 302	1 314
Israël	2 289	2 498	3 012	3 313	3 359	3 485	3 541	3 756	4 027
Jordanie	151	200	284	678	769	817	854	939	943
Koweit	3 530	4 258	5 534	7 316	7 813	8 390	8 801	8 859	9 243
Liban	461	572	717	1 040	1 073	1 072	1 131	1 223	1 032
Oman	15	51	404	1 071	1 279	1 526	1 880	1 946	2 103
Qatar	2 590	2 850	6 065	9 849	9 948	10 430	10 960	10 671	10 546
Arabie saoudite	321	413	1 117	2 808	3 151	3 385	3 518	3 564	3 574
Syrie	179	184	295	617	605	633	626	612	649
Emirats arabes unis	750	1 898	4 198	7 334	7 629	8 093	8 307	8 421	8 688
Yémen	33	31	46	73	79	82	94	97	107
Moyen-Orient	377	468	748	1 248	1 353	1 430	1 482	1 529	1 559
Total non-OCDE	464	517	654	756	791	815	846	875	902
OCDE Amérique du N.	6 382	7 206	7 990	8 124	8 560	8 680	8 712	9 061	9 383
OCDE Pacifique	2 818	3 321	3 912	4 268	4 470	4 613	4 677	4 947	5 207
OCDE Europe	2 943	3 343	3 890	4 208	4 392	4 573	4 665	4 818	4 921
Total OCDE	4 021	4 582	5 243	5 538	5 817	5 976	6 048	6 296	6 508
Monde	1 312	1 470	1 684	1 786	1 864	1 907	1 936	2 000	2 054

Faute de séries complètes, l'Albanie, la Rép.pop.dém. de Corée et le Viêt-Nam ne sont pas pris en compte dans les totaux régionaux.

Electricity Consumption/Population (kWh per capita)
Consommation d'électricité/Population (kWh par habitant)
Elekrizitätsverbrauch/Bevölkerung (kWh pro Kopf)
Totale consumi finali energia elettrica/POP (kWh per capita)
一人当たり電力消費量（kWh／人）
Consumo de electricidad/población (kWh per capita)
Потребление электроэнергии / Численность населения (КВт.ч на человека)

1989	1990	1991	1992	1993	1994	1995	1996	1997	
-	-	-	1 835	1 077	916	899	1 024	1 260	Armenia
-	-	-	2 233	2 081	1 924	1 795	1 819	1 816	Azerbaijan
-	-	-	3 876	3 411	3 030	2 754	2 769	2 910	Belarus
-	-	-	4 899	3 992	4 292	4 151	4 456	4 619	Estonia
-	-	-	1 785	1 532	1 088	1 084	1 151	1 142	Georgia
-	-	-	5 329	4 501	4 287	3 978	3 551	3 206	Kazakhstan
-	-	-	1 918	1 888	1 831	1 667	1 557	1 429	Kyrgyzstan
-	-	-	2 621	2 030	1 952	1 973	1 965	2 028	Latvia
-	-	-	3 127	2 462	2 468	2 480	2 656	2 631	Lithuania
-	-	-	2 133	2 047	1 683	1 500	1 465	1 345	Moldova
-	-	-	6 107	5 724	5 191	5 110	5 031	4 976	Russia
-	-	-	2 954	2 549	2 494	2 341	2 292	2 177	Tajikistan
-	-	-	1 933	1 921	1 559	1 458	1 342	1 224	Turkmenistan
-	-	-	4 308	3 948	3 470	3 342	3 051	2 948	Ukraine
-	-	-	2 128	2 018	1 924	1 844	1 828	1 816	Uzbekistan
5 369	5 363	5 225	4 823	4 441	4 020	3 903	3 785	3 716	**Former USSR**
6 953	6 938	6 880	7 129	7 531	7 874	7 936	7 678	7 656	Bahrain
798	879	1 023	1 071	1 078	1 142	1 165	1 209	1 233	Iran
1 294	1 261	1 052	1 318	1 332	1 382	1 396	1 361	1 353	Iraq
4 194	4 138	3 972	4 470	4 608	4 881	5 080	5 308	5 320	Israel
959	1 050	955	1 063	1 099	1 143	1 213	1 266	1 275	Jordan
9 609	8 832	6 809	10 750	13 811	15 168	14 969	15 012	14 976	Kuwait
631	385	755	892	1 028	1 108	1 169	1 651	1 930	Lebanon
2 126	2 361	2 242	2 281	2 433	2 509	2 574	2 624	2 692	Oman
10 412	9 399	8 607	9 143	8 964	8 943	8 113	8 924	8 954	Qatar
3 652	3 732	3 904	4 010	4 266	4 479	4 526	4 617	4 739	Saudi Arabia
698	708	717	717	949	1 107	1 084	1 164	1 205	Syria
8 565	8 721	8 067	7 902	7 883	8 255	8 146	8 030	7 973	United Arab Emirates
106	124	118	125	128	113	115	111	109	Yemen
1 577	1 615	1 588	1 730	1 841	1 960	1 990	2 057	2 107	**Middle East**
916	924	926	918	918	920	940	957	973	**Non-OECD Total**
9 549	9 644	9 952	9 889	10 083	10 244	10 399	10 645	10 673	OECD North America
5 501	5 882	6 070	6 144	6 269	6 755	6 964	7 180	7 412	OECD Pacific
5 012	5 033	5 093	5 102	5 079	5 115	5 229	5 350	5 409	OECD Europe
6 669	6 786	6 962	6 964	7 051	7 218	7 372	7 562	7 645	**OECD Total**
2 088	2 109	2 137	2 122	2 132	2 158	2 198	2 239	2 258	**World**

Due to lack of complete series Albania, the Dem. Ppl's Rep. of Korea and Vietnam are not included in the regional aggregates.

INTERNATIONAL ENERGY AGENCY
ENERGY STATISTICS DIVISION
POSSIBLE STAFF VACANCIES

The Division is responsible for statistical support and advice to the policy and operational Divisions of the International Energy Agency. It also produces a wide range of annual and quarterly publications complemented by a data service on microcomputer diskettes. For these purposes, the Division maintains, on a central computer and an expanding network of microcomputers, extensive international databases covering most aspects of energy supply and use.

- Vacancies for statistical assistants occur from time to time. Typically their work includes:

- Gathering and vetting data from questionnaires and publications, discussions on data issues with respondents to questionnaires in national administrations and fuel companies.

- Managing energy databases on a mainframe computer and microcomputers in order to maintain accuracy and timeliness of output.

- Preparing computer procedures for the production of tables, reports and analyses.

Seasonal adjustment of data and analysis of trends and market movements.

- Preparing studies on an ad-hoc basis as required by other Divisions of the International Energy Agency.

Nationals of any OECD Member Country are eligible for appointment. Basic salaries range from 15 400 to 20 400 French francs per month, depending on qualifications. The possibilities for advancement are good for candidates with appropriate qualifications and experience. Tentative enquiries about future vacancies are welcomed from men and women with relevant qualifications and experience. Applications in French or English, specifying the reference "ENERSTAT" and enclosing a curriculum vitae, should be sent to:

Human Resources Management Division
OECD, 2, rue André-Pascal,
75775 Paris CEDEX 16,
France

AGENCE INTERNATIONALE DE L'ENERGIE
DIVISION DES STATISTIQUES DE L'ENERGIE
VACANCES D'EMPLOI EVENTUELLES

La Division est chargée de fournir une aide et des conseils dans le domaine statistique aux Divisions administratives et opérationnelles de l'Agence internationale de l'énergie. En outre, elle diffuse une large gamme de publications annuelles et trimestrielles complétées par un service de données sur disquettes pour micro-ordinateur. A cet effet, la Division tient à jour, sur un ordinateur central et un réseau de plus en plus étendu de micro-ordinateurs, de vastes bases de données internationales portant sur la plupart des aspects de l'offre et de la consommation d'énergie.

Des postes d'assistant statisticien sont susceptibles de se libérer de temps à autre. Les fonctions dévolues aux titulaires de ces postes sont notamment les suivantes :

- Rassembler et valider les données tirées de questionnaires et de publications, ainsi que d'échanges de vues sur les données avec les personnes des Administrations nationales ou des entreprises du secteur de l'énergie qui répondent aux questionnaires.

- Gérer des bases de données relatives à l'énergie sur un ordinateur central et des micro-ordinateurs en vue de s'assurer de l'exactitude et de l'actualisation des données de sortie.

- Mettre au point des procédures informatiques pour la réalisation de tableaux, rapports et analyses. Procéder à l'ajustement saisonnier des données et analyses relatives aux tendances et aux fluctuations du marché.

- Effectuer des études en fonction des besoins des autres Divisions de l'Agence internationale de l'énergie.

Ces postes sont ouverts aux ressortissants des pays Membres de l'OCDE. Les traitements de base sont compris entre 15 400 et 20 400 francs français par mois, suivant les qualifications. Les candidats possédant les qualifications et l'expérience appropriées se verront offrir des perspectives de promotion. Les demandes de renseignements sur les postes susceptibles de se libérer qui émanent de personnes dotées des qualifications et de l'expérience voulues seront les bienvenues. Les candidatures, rédigées en français ou en anglais et accompagnées d'un curriculum vitae, doivent être envoyées, sous la référence "ENERSTAT", à l'adresse suivante :

Division de la Gestion des Ressources Humaines
OCDE, 2 rue André-Pascal,
75775 Paris CEDEX 16,
France

MULTILINGUAL PULLOUT

French
German

1.	Production nationale		Produktion
2.	Importations		Importe
3.	Exportations		Exporte
4.	Soutages maritimes internationaux		Bunker
5.	Variation des stocks		Bestandsveränderungen

6.	**APPROV. TOTAUX EN ENERGIE PRIMAIRE**		**GESAMTENERGIEAUFKOMMEN**

7.	Transferts		Transfer
8.	Ecarts statistiques		Stat. Differenzen
9.	Centrales électriques		Elektrizitätswerke
10.	Centrales de cogénération		Elektrizitäts- und Heizkraftwerke
11.	Centrales calogènes		Heizkraftwerke
12.	Usines à gaz		Ortsgas
13.	Raffineries de pétrole		Raffinerien
14.	Transformation du charbon		Kohleumwandlung
15.	Liquéfaction		Verflüssigung
16.	Autres transformations		Sonst. Umwandlungsbereich
17.	Consommation propre		Eigenverbrauch
18.	Pertes de distribution et de transport		Verteilungsverluste

19.	**CONSOMMATION FINALE TOTALE**		**ENDENERGIEVERBRAUCH**

20.	**SECTEUR INDUSTRIE**		**INDUSTRIE**
21.	Sidérurgie		Eisen- und Stahlindustrie
22.	Industrie chimique et pétrochimique		Chemische Industrie
23.	*Dont : produits d'alimentation*		*Davon: Feedstocks*
24.	Métaux non ferreux		Ne-Metallerzeugung
25.	Produits minéraux non métalliques		Glas- und Keramikindustrie
26.	Matériel de transport		Fahrzeugbau
27.	Construction mécanique		Maschinenbau
28.	Industries extractives		Bergbau- und Steinbrüche
29.	Produits alimentaires, boissons et tabacs		Nahrungs- und Genußmittel
30.	Pâtes à papier, papier et imprimerie		Zellstoff, Papier, Pappeerzeugung
31.	Bois et produits dérivés		Holz und Holzprodukte
32.	Construction		Baugewerbe
33.	Textiles et cuir		Textil- und Lederindustrie
34.	Non spécifiés		Sonstige

35.	**SECTEUR TRANSPORTS**		**VERKEHRSSEKTOR**
36.	Transport aérien		Luftverkehr
37.	Transport routier		Straßenverkehr
38.	Transport ferroviaire		Schienenverkehr
39.	Transport par conduites		Rohrleitungen
40.	Navigation intérieure		Binnenschiffahrt
41.	Non spécifiés		Sonstige

42.	**AUTRES SECTEURS**		**ANDERE SEKTOREN**
43.	Agriculture		Landwirtschaft
44.	Commerce et services publics		Handel- und öffentliche Einrichtungen
45.	Résidentiel		Wohnungssektor
46.	Non spécifiés		Sonstige

47.	**UTILISATIONS NON ENERGETIQUES**		**NICHTENERGETISCHER VERBRAUCH**
48.	Dans l'industrie/transformation/énergie		In Industrie/Umwandlung/Energiesektor
49.	Dans les transports		In Verkehr
50.	Dans les autres secteurs		In anderen Sektoren

51.	**Electricité produite - GWh**		**Elektrizitätsproduktion - GWh**
52.	*Centrales électriques*		*Elektrizitätswerke*
53.	*Centrales de cogénération*		*Elektrizitäts- und Heizkraftwerke*
54.	**Chaleur produite - TJ**		**Wärmeerzeugung - TJ**
55.	*Centrales de cogénération*		*Elektrizitäts- und Heizkraftwerke*
56.	*Centrales calogènes*		*Heizkraftwerke*

MULTILINGUAL PULLOUT

Italian
Japanese

1.	Produzione	国内生産
2.	Importazione	輸入
3.	Esportazione	輸出
4.	Bunkeraggi	バンカー
5.	Variazione di stock	在庫変動
6.	**TOTALE RISORSE PRIMARIE**	国内供給計
7.	Ritorni e trasferimenti	変換
8.	Differenza statistica	統計誤差
9.	Centrali elettriche	電気事業者・自家発
10.	Impianti di cogenerazione	コージェネレーション
11.	Impianti di riscaldamento	熱供給事業者
12.	Officine del gas	ガス業者
13.	Raffinerie di petrolio	石油精製
14.	Trasformazione del carbone	石炭変換
15.	Liquefazione	液化
16.	Altri settori di trasformazione	その他転換
17.	Autoconsumo	自家消費
18.	Perdite di distribuzione	送配電ロス
19.	**CONSUMI FINALI**	最終エネルギー消費計
20.	**SETTORE INDUSTRIALE**	産業部門
21.	Siderurgico	鉄鋼業
22.	Chimico	化学工業
23.	*Di cui: prodotti intermedi*	（含原料油・半製品）
24.	Metalli non ferrosi	非鉄金属
25.	Minerali non metallici	窯業土石
26.	Equipaggiamento per trasporti	輸送用機械
27.	Meccanico	金属機械
28.	Estrattivo	鉱業
29.	Alimentare e del tabacco	食料品・たばこ
30.	Cartario e grafico	紙・パルプ
31.	Legno e prodotti del legno	木製品
32.	Costruzioni	建設業
33.	Tessile e pelli	繊維工業
34.	Non specificato	その他製造業
35.	**SETTORE DEI TRASPORTI**	運輸部門
36.	Trasporti aerei	航空輸送業
37.	Trasporti stradali	道路運送業
38.	Trasporti ferroviari	鉄道
39.	Trasporti per pipeline	パイプライン輸送
40.	Trasporti fluviali	水運業
41.	Non specificato	その他
42.	**ALTRI SETTORI**	他の部門
43.	Agricoltura	農林業
44.	Commercio e servizi pubblici	民生・業務用
45.	Domestico	民生・家庭用
46.	Non specificato	その他
47.	**USI NON ENERGETICI**	非エネルギー
48.	Dell'industria	産業・変換・エネルギー部門
49.	Dei trasporti	運輸部門
50.	Degli altri settori	他の部門
51.	**Elettricità Prodotta - GWh**	発電実績—GWh
52.	*Centrali elettriche*	電気事業者・自家発
53.	*Impianti di cogenerazione*	コージェネレーション
54.	**Calore Prodotto - TJ**	発熱実績—TJ
55.	*Impianti di cogenerazione*	コージェネレーション
56.	*Impianti di riscaldamento*	熱供給事業者

MULTILINGUAL PULLOUT

Spanish
Russian

	Spanish	Russian
1.	Producción	Собственное производство
2.	Importaciones	Импорт
3.	Exportaciones	Экспорт
4.	Bunkers	Международная морская бункеровка
5.	Cambio de stocks	Изменение остатков
6.	**SUMINISTRO AL CONSUMO**	**ОБЩАЯ ПЕРВИЧНАЯ ПОСТАВКА ТОПЛИВА И ЭНЕРГИИ**
7.	Transferencias	Передачи
8.	Diferencias estadísticas	Статистическое расхождение
9.	Central eléctrica	Электростанции
10.	Central combinada de calor y electricidad	Тепло-электроцентрали
11.	Central de calor	Теплоцентрали
12.	Gas ciudad	Газовые заводы
13.	Refinerías de petróleo	Нефтепереработка
14.	Transformación de carbón	Переработка угля
15.	Licuefacción	Ожижение
16.	Otros Sect. de transformación	Др. отрасли преобразования и переработки топлива
17.	Consumos propios	Собственное использование в ТЭК
18.	Pérdidas de distribución	Потери при распределении
19.	**CONSUMO FINAL**	**КОНЕЧНОЕ ПОТРЕБЛЕНИЕ**
20.	**SECTOR INDUSTRIA**	**СЕКТОР ПРОМЫШЛЕННОСТИ**
21.	Siderurgia	Черная металлургия
22.	Químico	Химия и нефтехимия
23.	*Incl.: prod. de aliment.*	*в т.ч. П/фабрикаты нефтепереработки*
24.	Metales no férreos	Цветная металлургия
25.	Minerales no metálicos	Неметалл. минералы
26.	Equipos de transporte	Транспортное оборудование
27.	Maquinaria	Машиностроение
28.	Extracción y minas	Горнодобывающая промышленность
29.	Alimentación y tabaco	Пищевая и табачная промышленность
30.	Papel, pasta e impresión	Бум.-целл. и полиграф. пр-сть
31.	Madera	Пр-во древесины и деревообработка
32.	Construcción	Строительство
33.	Textil y piel	Текст.-кожевенная пр-сть
34.	No especificado	Др. отрасли промышленности
35.	**SECTOR TRANSPORTE**	**СЕКТОР ТРАНСПОРТА**
36.	Transporte aéreo	Воздушный транспорт
37.	Transporte por carretera	Автодорожный транспорт
38.	Ferrocarril	Железнодорожный транспорт
39.	Oleoducto	Транспортировка по трубопроводам
40.	Navegación interior	Внутренний водный транспорт
41.	No especificado	Неспецифицированный транспорт
42.	**OTROS SECTORES**	**ДРУГИЕ СЕКТОРЫ**
43.	Agricultura	Сельское хозяйство
44.	Comercio y servicios públicos	Торговля и услуги
45.	Residencial	Бытовой сектор
46.	No especificado	Неспецифицированные другие секторы
47.	**USOS NO ENERGETICOS**	**НЕЭНЕРГЕТИЧЕСКОЕ ИСПОЛЬЗОВАНИЕ**
48.	En la industria/tranf./energia	В промышленности/преобраз.-переработке/топл.-энергетике
49.	En el transporte	В транспорте
50.	En otros sectores	В других секторах
51.	**Electr. Producida - GWh**	**Производство электроэнергии - Гвт.ч**
52.	*Central Eléctrica*	*Электростанции*
53.	*Central combinada de calor y electricidad*	*Тепло-электроцентрали*
54.	**Calor producido - TJ**	**Производство тепла - ТДж**
55.	*Central combinada de calor y electricidad*	*Тепло-электроцентрали*
56.	*Central de calor*	*Теплоцентрали*

MULTILINGUAL PULLOUT

Chinese

1. 本国生产量
2. 进口
3. 出口
4. 国际海运加油
5. 库存变化

6. 一次能源供应总量 (TPES)

7. 转换
8. 统计差额
9. 发电厂
10. 热电联产厂
11. 热力厂
12. 制气
13. 炼油
14. 煤炭转化
15. 液化
16. 其他转化
17. 自用
18. 分配损耗

19. 最终消费合计 (TFC)

20. 工业部门
21. 钢铁
22. 化工和石化
23.   *其中：用作原料*
24. 有色金属
25. 非金属矿
26. 交通设备
27. 机械
28 采矿和挖掘
29. 食品和烟草
30. 纸、纸浆和印刷
31. 木材和木材制品
32. 建筑业
33. 纺织和皮革
34. 其他

35. 交通部门
36. 航空运输
37. 公路运输
38. 铁路运输
39. 管道运输
40. 国内海运
41. 其他

42. 其他部门
43. 农业
44. 商业和公共服务
45. 住宅
46. 其他

47. 非能源使用
48. 工业／转换／能源
49. 交通
50. 其他部门

51. **发电（百万千瓦小时）**
52.   *发电厂*
53.   *热电联产厂*
54. **供热（太焦耳）**
55.   *热电联产厂*
56.   *热力厂*

INTERNATIONAL ENERGY AGENCY

NINE ANNUAL PUBLICATIONS

Coal Information 1998

Issued annually since 1983, this publication provides comprehensive information on current world coal market trends and long-term prospects. Compiled in cooperation with the Coal Industry Advisory Board, it contains thorough analysis and current country-specific statistics for OECD Member countries and selected non-OECD countries on coal prices, demand, trade, production, productive capacity, emissions standards for coal-fired boilers, coal ports, coal-fired power stations and coal data for non-OECD countries. This publication is a key reference tool for all sectors of the coal industry as well as for OECD Member country governments. *Published July 1999.*

Electricity Information 1998

This publication brings together in one volume the IEA's data on electricity and heat supply and demand in the OECD. The report presents a comprehensive picture of electricity capacity and production, consumption, trade and prices for the OECD regions and individual countries in over 20 separate tables for each OECD country. Detailed data on the fuels used for electricity and heat production are also presented.
Published July 1999.

Natural Gas Information 1998

A detailed reference work on gas supply and demand, covering not only the OECD countries but also the rest of the world. Contains essential information on LNG and pipeline trade, gas reserves, storage capacity and prices. The main part of the book, however, concentrates on OECD countries, showing a detailed gas supply and demand balance for each individual country and for the three OECD regions: North America, Europe and Asia-Pacific, as well as a breakdown of gas consumption by end-user. Import and export data are reported by source and destination. *Published July 1999.*

Oil Information 1998

A comprehensive reference book on current developments in oil supply and demand. The first part of this publication contains key data on world production, trade, prices and consumption of major oil product groups, with time series back to the early 1970s. The second part gives a more detailed and comprehensive picture of oil supply, demand, trade, production and consumption by end-user for each OECD country individually and for the OECD regions. Trade data are reported extensively by origin and destination. *Published July 1999.*

Energy Statistics of OECD Countries 1996-1997

No other publication offers such in-depth statistical coverage. It is intended for anyone involved in analytical or policy work related to energy issues. It contains data on energy supply and consumption in original units for coal, oil, natural gas, combustible renewables/wastes and products derived from these primary fuels, as well as for electricity and heat. Data are presented for the two most recent years available in detailed supply and consumption tables. Historical tables summarise data on production, trade and final consumption. Each issue includes definitions of products and flows and explanatory notes on the individual country data. *Published June 1999.*

Energy Balances of OECD Countries 1996-1997

A companion volume to *Energy Statistics of OECD Countries*, this publication presents standardised energy balances expressed in million tonnes of oil equivalent. Energy supply and consumption data are divided by main fuel: coal, oil, gas, nuclear, hydro, geothermal/solar, combustible renewables/wastes, electricity and heat. This allows for easy comparison of the contributions each fuel makes to the economy and their interrelationships through the conversion of one fuel to another. All of this is essential for estimating total energy supply, forecasting, energy conservation, and analysing the potential for interfuel substitution. Complete energy balances are presented for the two most recent years available. Historical tables summarise key energy and economic indicators as well as data on production, trade and final consumption. Each issue includes definitions of products and flows and explanatory notes on the individual country data as well as conversion factors from original units to tonnes of oil equivalent. *Published June 1999.*

Energy Statistics of non-OECD Countries 1996-1997

This new publication offers the same in-depth statistical coverage as the homonymous publication covering OECD countries. It includes data in original units for 103 individual countries and nine main regions. The consistency of OECD and non-OECD countries' detailed statistics provides an accurate picture of the global energy situation. For a description of the content, please see *Energy Statistics of OECD Countries* above. *Published August 1999.*

Energy Balances of non-OECD Countries 1996-1997

A companion volume to the new publication*Energy Statistics of Non-OECD Countries*, this publication presents energy balances in million tonnes of oil equivalent and key economic and energy indicators for 103 individual countries and nine main regions. It offers the same statistical coverage as the homonymous publication covering OECD Countries, and thus provides an accurate picture of the global energy situation. It includes most of the historical series as well as the energy indicators and energy balances previously published in *Energy Statistics and Balances of Non-OECD Countries*. For a description of the content, please see *Energy Balances of OECD Countries* above. *Published August 1999.*

CO_2 Emissions from Fuel Combustion - 1999 Edition

In order for nations to tackle the problem of climate change, they need accurate greenhouse gas emissions data. This publication provides a new basis for comparative analysis of CO_2 emissions from fossil fuel combustion, a major source of anthropogenic emissions. The data in this book are designed to assist in understanding the evolution of these emissions from 1971 to 1997 on a country, regional and worldwide basis. They should help in the preparation and the follow-up to the Fifth Conference of the Parties (COP-5) meeting under the U.N. Climate Convention in Bonn in October/November 1999. Emissions were calculated using IEA energy databases and the default methods and emissions factors from the *Revised 1996 IPCC Guidelines for National Greenhouse Gas Inventories. Published October 1999.*

TWO QUARTERLIES
Oil, Gas, Coal and Electricity, Quarterly Statistics

Oil statistics cover OECD production, trade (by origin and destination), refinery intake and output, stock changes and consumption for crude oil, NGL and nine selected oil product groups. Statistics for natural gas show OECD supply, consumption and trade (by origin and destination). Coal data cover the main OECD and world-wide producers of hard and brown coal and major exporters and importers of steam and coking coal. Trade data for the main OECD countries are reported by origin and destination. Electricity statistics cover production (by major fuel category), consumption and trade for 28 OECD countries. Quarterly data on world oil and coal production are included, as well as world steam and coking coal trade.

Energy Prices and Taxes

This publication responds to the needs of the energy industry and OECD governments for up-to-date information on prices and taxes in national and international energy markets. It contains for OECD countries and certain non-OECD countries prices at all market levels: import prices, industry prices and consumer prices. The statistics cover the main petroleum products, gas, coal and electricity, giving for imported products an average price both for importing country and country of origin. Every issue includes full notes on sources and methods and a description of price mechanisms in each country.

For more information on the IEA statistics publications, please feel free to contact Ms. Sharon Michel in the Energy Statistics Division, Tel: (33 1) 40 57 66 25; Fax: (33 1) 40 57 66 49.

To complement its publications, the Energy Statistics Division produces diskettes containing the complete databases which are used for preparing the statistics publications. State-of-the-art software allows you to access and manipulate all these data in a very user-friendly manner and includes graphic and mapping facilities.

The diskette service includes:

Annual Diskettes
- Energy Statistics of OECD Countries, 1960-1997
- Energy Balances of OECD Countries, 1960-1997
- Energy Statistics of non-OECD Countries, 1971-1997
- Energy Balances of non-OECD Countries, 1971-1997
- CO_2 Emissions from Fuel Combustion

- Natural Gas Information
- Oil Information
- Coal Information
- Electricity Information
- IEA Energy Technology R&D Statistics 74-95

Quarterly Diskettes
- Energy Prices and Taxes.

Monthly Diskettes
- The IEA Monthly Oil Data Service (see box below).

The IEA Monthly Oil Data Service

Diskettes

The IEA Monthly Oil Data Service provides the detailed database of historical and projected information which is used in preparing the IEA's monthly Oil Market Report (OMR) and includes all the information previously contained in the Monthly Oil Statistics (MOS) diskettes. The IEA Monthly Oil Data Service comprises three packages:
- Supply, Demand, Balances and Stocks;
- Trade;
- Field-by-Field Supply;

available separately or combined, either as a subscriber service on the Internet or on diskettes.

Internet

The Internet version is available two days after the official release of the Oil Market Report. Diskettes are mailed four days afterwards.

CD-ROM and INTERNET (http://www.iea.org)

Annual Oil and Gas Statistics, Energy Statistics and Balances of OECD and Non-OECD Countries are also available on **CD Rom** in the OECD Statistical Compendium (for additional information on the CD-ROM contact the OECD Publications Service, Tel: (33 1) 45 24 82 00; Fax: (33 1) 49 10 42 76. Moreover, the IEA site on **Internet** contains key energy indicators by country, graphs on the world and OECD's energy situation evolution from 1971 to the most recent year available, as well as selected databases for demonstration. The IEA site can be accessed at: **http://www.iea.org**. For more information, please feel free to contact the Energy Statistics Division of the IEA by Fax: (33 1) 40 57 66 49 or Phone: (33 1) 40 57 66 25.

INTERNATIONAL ENERGY AGENCY

IEA Publications, PO Box 2722, London, W1A 5BL, UK
Telephone:+44 171 896 2244 - **Fax:**+44 171 896 2245
e-mail: ieaorder@pearson-pro.com

▶ *Special Discount: Order three or more products to claim a 20% discount.*

I would like to order the following publications (please tick)

ANNUAL PUBLICATIONS - 1999 Edition		QTY	PRICE	TOTAL
☐ Energy Statistics of OECD Countries 1996-1997			$110.00	
☐ Energy Balances of OECD Countries 1996-1997			$110.00	
☐ Energy Statistics of Non-OECD Countries 1996-1997	Formerly, these 2 volumes were contained in *Energy Statistics and Balances of Non-OECD Countries.*		$110.00	
☐ Energy Balances of Non-OECD Countries 1996-1997			$110.00	
☐ Coal Information 1998			$200.00	
☐ Electricity Information 1998			$130.00	
☐ Natural Gas Information 1998			$150.00	
☐ Oil Information 1998			$150.00	
☐ CO_2 Emissions from Fuel Combustion 1971-1997			$150.00	

Please enter my subscription as indicated below (please tick)

QUARTERLY PUBLICATIONS	QTY	SINGLE COPY	ANNUAL	TOTAL
☐ Energy Prices and Taxes		$100.00	$300.00	
☐ Oil, Gas, Coal and Electricity Statistics		$100.00	$300.00	

MONEY-BACK GUARANTEE:
Refunds are given on one-off publications returned in resaleable condition by registered post within 7 days of receipt.

Sub Total	
Postage and packing ($15.00)	
Less discount	
Total	

Delivery details (If your billing address differs from your delivery address please advise us)

Name

Position   Organisation

Address

Country   Postcode

Telephone   Fax

As all products are delivered by express courier service, please enter your telephone number above.

Payment details

☐ I enclose a cheque payable to IEA Publications for the sum of $_____

☐ Please debit my credit card (tick choice).

   ☐ Access/Mastercard   ☐ Diners   ☐ VISA   ☐ AMEX

Card no: ⌴⌴⌴⌴⌴⌴⌴⌴⌴⌴⌴⌴⌴⌴⌴⌴

Signature: Expiry date: ⌴⌴⌴⌴

EU registered companies	Non-EU registered companies
☐ Please send me an invoice along with the publications. I enclose a copy of my company headed stationery with this order form. To avoid extra charges EU companies (except UK) must supply VAT/TVA/MOMS/MST/IVA/FPA numbers: ⌴⌴⌴⌴⌴⌴⌴⌴⌴⌴⌴⌴	☐ Please send me a pro-forma invoice. The publications will be forwarded to me on receipt of payment.

FOR FAST PROCESSING OF YOUR ORDER, FAX THIS FORM TO CUSTOMER SERVICES ON +44 171 896 2245

INTERNATIONAL ENERGY AGENCY

IEA Publications, PO Box 2722, London, W1A 5BL, UK
Telephone: +44 171 896 2244 - **Fax:** +44 171 896 2245
e-mail: ieaorder@pearson-pro.com

▶ *Special Discount: Order three or more products to claim a 20% discount.*

I would like to order the following diskettes (please tick)

ANNUAL DISKETTES* - 1999 Edition	QTY	PRICE	TOTAL
☐ Energy Statistics of OECD Countries (1960-1997)		$500.00	
☐ Energy Balances of OECD Countries (1960-1997)		$500.00	
☐ Energy Statistics of Non-OECD Countries (1971-1997)		$500.00	
☐ Energy Balances of Non-OECD Countries (1971-1997)		$500.00	
☐ *Combined subscription of the above four series*		$1 200.00	
☐ Coal Information 1998		$500.00	
☐ Electricity Information 1998		$500.00	
☐ Natural Gas Information 1998		$500.00	
☐ Oil Information 1998		$500.00	
☐ CO_2 Emissions from Fuel Combustion (1960/71-1997)		$500.00	
☐ IEA Energy Technology R&D Statistics (1974-1995)		$500.00	

Please enter my subscription as indicated below (please tick)

QUARTERLY DISKETTES*	QTY	PRICE	TOTAL
☐ Energy Prices and Taxes (four quarters)		$800.00	

*Prices are for single user licence. Please contact us for pricing information on multi-user licences.

Sub Total	
Postage and packing ($15.00)	
Less discount	
Total	

MONEY-BACK GUARANTEE:
*Refunds are given on one-off publications returned in resaleable condition
by registered post within 7 days of receipt.*

Delivery details (If your billing address differs from your delivery address please advise us)

Name

Position   Organisation

Address

Country   Postcode

Telephone   Fax

As all products are delivered by express courier service, please enter your telephone number above.

Payment details

☐ I enclose a cheque payable to IEA Publications for the sum of $_____

☐ Please debit my credit card (tick choice).

☐ Access/Mastercard ☐ Diners ☐ VISA ☐ AMEX

Card no:

Signature: Expiry date:

EU registered companies	Non-EU registered companies
☐ Please send me an invoice along with the publications. I enclose a copy of my company headed stationery with this order form. To avoid extra charges EU companies (except UK) must supply VAT/TVA/MOMS/MST/IVA/FPA numbers:	☐ Please send me a pro-forma invoice. The publications will be forwarded to me on receipt of payment.

Data Protection Act: The information you provide will be held on our database and may be used to keep you informed of our and our associated companies' products and for selected third party mailings.

FOR FAST PROCESSING OF YOUR ORDER, FAX THIS FORM TO CUSTOMER SERVICES ON +44 171 896 2245

International Energy Agency, 9 rue de la Fédération, 75739 Paris CEDEX 15
PRINTED IN FRANCE BY CHIRAT
(61 99 16 3 P) ISBN 92-64-05867-2 - 1999